Microwaves in Organic Synthesis
Volume 1

Edited by
André Loupy

Related Titles

Wasserscheid, P., Welton, T. (eds.)

Ionic Liquids in Synthesis

2006
ISBN 3-527-31239-0

Kappe, C. O., Stadler, A.

Microwaves in Organic and Medicinal Chemistry

2005
ISBN 3-527-31210-2

de Meijere, A., Diederich, F. (eds.)

Metal-Catalyzed Cross-Coupling Reactions

2 volumes

2004
ISBN 3-527-30518-1

Yamamoto, H., Oshima, K. (eds.)

Main Group Metals in Organic Synthesis

2 volumes

2004
ISBN 3-527-30508-4

Tanaka, K.

Solvent-free Organic Synthesis

2003
ISBN 3-527-30612-9

Microwaves in Organic Synthesis

Volume 1

Edited by André Loupy

Second, Completely Revised and Enlarged Edition

WILEY-VCH Verlag GmbH & Co. KGaA

The Editor

André Loupy
Laboratoire des Réactions Sélectives sur Supports
Université Paris-Sud
Batiment 410
91405 Orsay cedex
France

All books published by **Wiley-VCH** are carefully produced. Nevertheless, authors, editors, and publisher do not warrant the information contained in these books, including this book, to be free of errors. Readers are advised to keep in mind that statements, data, illustrations, procedural details or other items may inadvertently be inaccurate.

Library of Congress Card No.: applied for
British Library Cataloguing-in-Publication Data
A catalogue record for this book is available from the British Library.

Bibliographic information published by the Deutsche Nationalbibliothek
Die Deutsche Nationalbibliothek lists this publication in the Deutsche Nationalbibliografie; detailed bibliographic data are available in the Internet at ⟨http://dnb.d-nb.de⟩.

© 2006 WILEY-VCH Verlag GmbH & Co. KGaA, Weinheim

All rights reserved (including those of translation into other languages). No part of this book may be reproduced in any form – by photoprinting, microfilm, or any other means – nor transmitted or translated into a machine language without written permission from the publishers. Registered names, trademarks, etc. used in this book, even when not specifically marked as such, are not to be considered unprotected by law.

Typesetting   Asco Typesetters, Hong Kong
Printing   betz-druck GmbH, Darmstadt
Binding   Litges & Dopf GmbH, Heppenheim
Cover Design
Grafik-Design Schulz, Fußgönheim

Printed in the Federal Republic of Germany
Printed on acid-free paper

ISBN-13: 978-3-527-31452-2
ISBN-10: 3-527-31452-0

Contents

Volume 1

Preface *XVII*

About European Cooperation in COST Chemistry Programs *XIX*

List of Authors *XXIII*

1 Microwave–Material Interactions and Dielectric Properties, Key Ingredients for Mastery of Chemical Microwave Processes *1*
Didier Stuerga
1.1 Fundamentals of Microwave–Matter Interactions *1*
1.1.1 Introduction *2*
1.1.2 The Complex Dielectric Permittivity *9*
1.1.3 Dielectric Properties and Molecular Behavior *29*
1.2 Key Ingredients for Mastery of Chemical Microwave Processes *43*
1.2.1 Systemic Approach *43*
1.2.2 The Thermal Dependence of Dielectric Loss *45*
1.2.3 The Electric Field Effects *46*
1.2.4 Hydrodynamic Aspects *49*
1.2.5 Thermodynamic and Other Effects of Electric Fields *51*
1.2.6 The Athermal and Specific Effects of Electric Fields *52*
1.2.7 The Thermal Path Effect: Anisothermal Conditions *54*
1.2.8 Hot Spots and Heterogeneous Kinetics *56*
References *57*

2 Development and Design of Laboratory and Pilot Scale Reactors for Microwave-assisted Chemistry *62*
Bernd Ondruschka, Werner Bonrath, and Didier Stuerga
2.1 Introduction *62*
2.2 Basic Concepts for Reactions and Reactors in Organic Synthesis *63*
2.3 Methods for Enhancing the Rates of Organic Reactions *64*
2.4 Microwave-assisted Organic Syntheses (MAOS) *66*

Microwaves in Organic Synthesis, Second edition. Edited by A. Loupy
Copyright © 2006 WILEY-VCH Verlag GmbH & Co. KGaA, Weinheim
ISBN: 3-527-31452-0

2.4.1	Microwave Ovens and Reactors – Background	67
2.4.2	Scale-up of Microwave Cavities	72
2.4.3	Efficiency of Energy and Power	73
2.4.4	Field Homogeneity and Penetration Depth	73
2.4.5	Continuous Tube Reactors	74
2.4.6	MAOS – An Interdisciplinary Field	74
2.5	Commercial Microwave Reactors – Market Overview	75
2.5.1	Prolabo's Products	77
2.5.2	CEM's products	78
2.5.3	Milestone's Products	80
2.5.4	Biotage's Products	84
2.6	Selected Equipment and Applications	85
2.6.1	Heterogeneous Catalysis	89
2.6.2	Hyphenated Techniques in Combination with Microwaves	90
2.6.3	Combination of Microwave Irradiation with Pressure Setup	92
2.6.4	Synthesis of Laurydone®	97
2.6.5	Industrial Equipment: Batch or Continuous Flow?	98
2.7	Qualification and Validation of Reactors and Results	101
2.8	Conclusion and Future	102
	References	103

3 **Roles of Pressurized Microwave Reactors in the Development of Microwave-assisted Organic Chemistry** *108*

Thach Le Ngoc, Brett A. Roberts, and Christopher R. Strauss

3.1	Introduction	108
3.2	Toward Dedicated Microwave Reactors	109
3.2.1	The Continuous Microwave Reactor (CMR)	110
3.2.2	Microwave Batch Reactors	111
3.3	Applications of the New Reactors	112
3.3.1	With Low-boiling Solvents	112
3.3.2	Kinetics Measurements	113
3.3.3	Core Enabling Technology	114
3.4	Commercial Release of MBRs and CMRs	114
3.5	Advantages of Pressurized Microwave Reactors	116
3.6	Applications	117
3.6.1	General Synthesis	117
3.6.2	Kinetic Products	121
3.6.3	Selective Synthesis of *O*-Glycofuranosides and Pyranosides from Unprotected Sugars	121
3.6.4	Synthesis in High-temperature Water	123
3.6.5	Biomimetic Reactions	125
3.6.6	Indoles	126
3.7	Effect of the Properties of Microwave Heating on the Scale-up of Methods in Pressurized Reactors	127
3.8	Software Technology for Translation of Reaction Conditions	129

3.9	Conclusion *131*	
	Acknowledgments *132*	
	References *132*	

4	**Nonthermal Effects of Microwaves in Organic Synthesis** *134*	
	Laurence Perreux and André Loupy	
4.1	Introduction *134*	
4.2	Origin of Microwave Effects *135*	
4.3	Specific Nonthermal Microwave Effects *137*	
4.4	Effects of the Medium *141*	
4.4.1	Polar Solvents *141*	
4.4.2	Nonpolar Solvents *143*	
4.4.3	Solvent-free Reactions *146*	
4.5	Effects Depending on Reaction Mechanisms *147*	
4.5.1	Isopolar Transition-state Reactions *148*	
4.5.2	Bimolecular Reactions Between Neutral Reactants Leading to Charged Products *152*	
4.5.3	Anionic Bimolecular Reactions Involving Neutral Electrophiles *153*	
4.5.4	Unimolecular Reactions *154*	
4.6	Effects Depending on the Position of the Transition State Along the Reaction Coordinates *155*	
4.7	Effects on Selectivity *156*	
4.8	Some Illustrative Examples *157*	
4.8.1	Bimolecular Reactions Between Neutral Reactants *157*	
4.8.2	Bimolecular Reactions with One Charged Reactant *184*	
4.8.3	Unimolecular Reactions *198*	
4.8.4	Some Illustrative Examples of Effects on Selectivity *204*	
4.9	Concerning the Absence of Microwave Effects *209*	
4.10	Conclusions: Suitable Conditions for Observation of Specific MW Effects *210*	
	References *212*	

5	**Selectivity Under the Action of Microwave Irradiation** *219*	
	Antonio de la Hoz, Angel Díaz-Ortiz, and Andrés Moreno	
5.1	Introduction *219*	
5.2	Selective Heating *220*	
5.2.1	Solvents *220*	
5.2.2	Catalysts *223*	
5.2.3	Reagents; Molecular Radiators *230*	
5.2.4	Susceptors *232*	
5.3	Modification of Chemoselectivity and Regioselectivity *233*	
5.3.1	Protection and Deprotection of Alcohols *233*	
5.3.2	Electrophilic Aromatic Substitution *236*	
5.3.3	Synthesis and Reactivity of Heterocyclic Compounds *241*	
5.3.4	Cycloaddition Reactions *247*	

5.3.5	Polymerization 252
5.3.6	Miscellaneous 257
5.4	Modification of Stereo and Enantioselectivity 264
5.5	Conclusions 272
	Acknowledgments 273
	References 273

6 Microwaves and Phase-transfer Catalysis 278

André Loupy, Alain Petit, and Dariusz Bogdal

6.1	Phase-transfer Catalysis 278
6.2	Synthetic Applications of Phase-transfer Processes 281
6.2.1	O-Alkylations 281
6.2.2	N-Alkylations 293
6.2.3	C-Alkylations of Active Methylene Groups 297
6.2.4	Alkylations with Dihalogenoalkanes 300
6.2.5	Nucleophilic Addition to Carbonyl Compounds 302
6.2.6	Deprotonations 307
6.2.7	Miscellaneous Reactions 309
6.3	Conclusion 322
	References 322

7 Microwaves and Ionic Liquids 327

Nicholas E. Leadbeater and Hanna M. Torenius

7.1	Introduction 327
7.2	Ionic Liquids in Conjunction with Microwave Activation 330
7.2.1	Synthesis of Ionic Liquids Using Microwave Heating 330
7.2.2	Reactions Using Microwave Irradiation and Ionic Liquids as Solvents and Reagents 333
7.2.3	Use of Ionic Liquids and Microwaves in Multicomponent Reactions 349
7.2.4	Use of Ionic Liquids as Heating Aids 354
7.3	Conclusions 357
	Abbreviations 357
	References 358

8 Organic Synthesis Using Microwaves and Supported Reagents 362

Rajender S. Varma and Yuhong Ju

8.1	Introduction 362
8.2	Microwave-accelerated Solvent-free Organic Reactions 363
8.2.1	Protection–Deprotection Reactions 364
8.2.2	Condensation Reactions 372
8.2.3	Isomerization and Rearrangement Reactions 379
8.2.4	Oxidation Reactions – Oxidation of Alcohols and Sulfides 382
8.2.5	Reduction Reactions 388
8.2.6	Synthesis of Heterocyclic Compounds 391
8.2.7	Miscellaneous Reactions 401

8.3	Conclusions *406*	
	References *407*	
9	**Microwave-assisted Reactions on Graphite** *416*	
	Thierry Besson, Valérie Thiery, and Jacques Dubac	
9.1	Introduction *416*	
9.2	Graphite as a Sensitizer *418*	
9.2.1	Diels–Alder Reactions *418*	
9.2.2	Ene Reactions *423*	
9.2.3	Oxidation of Propan-2-ol *425*	
9.2.4	Thermolysis of Esters *425*	
9.2.5	Thermal Reactions in Heterocyclic Syntheses *427*	
9.2.6	Decomplexation of Metal Complexes *433*	
9.2.7	Redistribution Reactions between Tetraalkyl or Tetraarylgermanes, and Germanium Tetrahalides *434*	
9.2.8	Pyrolysis of Urea *436*	
9.2.9	Esterification of Stearic Acid with *n*-Butanol *437*	
9.3	Graphite as Sensitizer and Catalyst *437*	
9.3.1	Analysis of Two Synthetic Commercial Graphites *438*	
9.3.2	Acylation of Aromatic Compounds *438*	
9.3.3	Acylative Cleavage of Ethers *443*	
9.3.4	Ketodecarboxylation of Carboxylic Diacids *444*	
9.4	Notes *447*	
9.4.1	MW Apparatus, Typical Procedures, and Safety Measures *447*	
9.4.2	Temperature Measurement *448*	
9.4.3	Mechanism of Retention of Reactants on Graphite *449*	
9.4.4	Graphite or Amorphous Carbon for C/MW Coupling *449*	
9.5	Conclusion *450*	
	References *451*	
10	**Microwaves in Heterocyclic Chemistry** *456*	
	Jean Pierre Bazureau, Jack Hamelin, Florence Mongin, and Françoise Texier-Boullet	
10.1	Introduction *456*	
10.2	Microwave-assisted Reactions in Solvents *456*	
10.2.1	Three, Four, and Five-membered Systems with One Heteroatom *456*	
10.2.2	Five-membered Systems with Two Heteroatoms *463*	
10.2.3	Five-membered Systems with More than Two Heteroatoms *468*	
10.2.4	Six-membered Systems with One Heteroatom *469*	
10.2.5	Six-membered Systems with More than One Heteroatom *479*	
10.2.6	Synthesis of More Complex or Polyheterocyclic Systems *485*	
10.3	Solvent-free Synthesis *492*	
10.3.1	Three, Four and Five-membered Systems with One Heteroatom *492*	
10.3.2	Five-membered Systems with Two Heteroatoms *499*	
10.3.3	Five-membered Systems with More than Two Heteroatoms – Synthesis of Triazoles and Related Compounds *506*	

10.3.4	Six-membered Systems with One Heteroatom 506
10.3.5	Six-membered Systems with More than One Heteroatom 511
10.3.6	Synthesis of More Complex or Polyheterocyclic Systems 515
10.4	Conclusion 517
	References and Notes 517

Volume 2

11	**Microwaves in Cycloadditions** 524
	Khalid Bougrin, Mohamed Soufiaoui, and George Bashiardes
11.1	Cycloaddition Reactions 524
11.2	Reactions with Solvent 525
11.3	Reactions under Solvent-free Conditions 526
11.3.1	Reaction on Mineral Supports 527
11.3.2	Reaction with Neat Reactants 528
11.4	[4+2] Cycloadditions 532
11.4.1	Intramolecular Hetero and Diels–Alder Reactions 533
11.4.2	Intermolecular Hetero and Diels–Alder Reactions 538
11.5	[3+2] Cycloadditions 546
11.5.1	Nitrile Oxides, Nitrile Sulfides, and Nitrones 547
11.5.2	Azomethine Ylides, and Nitrile Imines 556
11.5.3	Azides 562
11.6	[2+2] Cycloadditions 567
11.6.1	Cycloadditions of Ketenes with Imines 567
11.6.2	Cycloadditions of β-Formyl Enamides with Alkynes 569
11.7	Other Cycloadditions 570
11.8	Conclusions 571
	Acknowledgments 572
	References 573

12	**Microwave-assisted Chemistry of Carbohydrates** 579
	Antonino Corsaro, Ugo Chiacchio, Venerando Pistarà, and Giovanni Romeo
12.1	Introduction 579
12.2	Protection 580
12.2.1	Acylation 580
12.2.2	Acetalation 583
12.2.3	Silylation 585
12.2.4	Methylation of Carbohydrate Carboxylic Acids 586
12.2.5	Synthesis of 1,6-Anhydro-β-D-hexopyranoses 586
12.3	Deprotection 586
12.3.1	Deacylation 587
12.3.2	Deacetalation 588
12.3.3	Desilylation 588
12.4	Glycosylation 589
12.4.1	*O*-Glycosylation 589
12.4.2	*C*-Glycosylation Reactions 594

12.4.3	N-Glycosylation	594
12.5	Hydrogenation (Catalytic Transfer Hydrogenation)	594
12.6	Oxidation	595
12.7	Halogenation	595
12.7.1	Chlorination and Bromination	595
12.7.2	Fluorination	596
12.7.3	Iodination	597
12.8	Stereospecific C–H Bond Activation for Rapid Deuterium Labeling	597
12.9	Reaction of Carbohydrates with Amino-derivatized Labels	598
12.10	Ferrier (II) Rearrangement to Carbasugars	599
12.11	Synthesis of Unsaturated Monosaccharides	599
12.12	Synthesis of Dimers and Polysaccharides, and their Derivatives	601
12.13	Synthesis of Heterocycles and Amino Acids	602
12.13.1	Spiroisoxazoli(di)ne Derivatives	602
12.13.2	Triazole-linked Glycodendrimers	603
12.13.3	5,6-Dihydro-1,2,3,4-tetrazenes	604
12.13.4	C-Nucleosides	605
12.13.5	Pterins	605
12.13.6	Epoxides	606
12.13.7	Azetidinones	607
12.13.8	Substituted Pyrazoles	608
12.13.9	3H-Pyrido[3,4-b]indole Derivatives	608
12.14	Enzymatic Reactions	608
12.15	Conclusion	611
	References	611

13	**Microwave Catalysis in Organic Synthesis**	**615**
	Milan Hájek	
13.1	Introduction	615
13.1.1	Definitions	616
13.2	Preparation of Heterogeneous Catalysts	617
13.2.1	Drying and Calcination	617
13.2.2	Catalyst Activation and Reactivation (Regeneration)	620
13.3	Microwave Activation of Catalytic Reactions	621
13.3.1	Reactions in the Liquid Phase	622
13.3.2	Reactions in the Gas Phase	628
13.3.3	Microwave Effects	634
13.3.4	Microwave Catalytic Reactors	641
13.4	Industrial Applications	645
	References	647

14	**Polymer Chemistry Under the Action of Microwave Irradiation**	**653**
	Dariusz Bogdal and Katarzyna Matras	
14.1	Introduction	653
14.2	Synthesis of Polymers Under the Action of Microwave Irradiation	653
14.2.1	Chain Polymerizations	654

14.2.2 Step-growth Polymerization 663
14.2.3 Miscellaneous Polymers 675
14.2.4 Polymer Modification 679
14.3 Conclusion 681
References 682

15 Microwave-assisted Transition Metal-catalyzed Coupling Reactions 685
Kristofer Olofsson, Peter Nilsson, and Mats Larhed
15.1 Introduction 685
15.2 Cross-coupling Reactions 686
15.2.1 The Suzuki–Miyaura Reaction 686
15.2.2 The Stille Reaction 700
15.2.3 The Negishi Reaction 703
15.2.4 The Kumada Reaction 704
15.2.5 The Hiyama Reaction 705
15.3 Arylation of C, N, O, S, P and Halogen Nucleophiles 706
15.3.1 The Sonogashira Coupling Reaction 706
15.3.2 The Nitrile Coupling 707
15.3.3 Aryl–Nitrogen Coupling 708
15.3.4 Aryl–Oxygen Bond Formation 713
15.3.5 Aryl–Phosphorus Coupling 713
15.3.6 Aryl–Sulfur Bond Formation 715
15.3.7 Aryl Halide Exchange Reactions 716
15.4 The Heck Reaction 717
15.5 Carbonylative Coupling Reactions 719
15.6 Summary 721
Acknowledgment 721
References 722

16 Microwave-assisted Combinatorial and High-throughput Synthesis 726
Alexander Stadler and C. Oliver Kappe
16.1 Solid-phase Organic Synthesis 726
16.1.1 Introduction 726
16.1.2 Microwave Chemistry and Solid-phase Organic Synthesis 727
16.1.3 Peptide Synthesis and Related Examples 728
16.1.4 Resin Functionalization 729
16.1.5 Transition-metal Catalysis 734
16.1.6 Substitution Reactions 740
16.1.7 Multicomponent Chemistry 745
16.1.8 Microwave-assisted Condensation Reactions 747
16.1.9 Rearrangements 748
16.1.10 Cleavage Reactions 749
16.1.11 Miscellaneous 752
16.2 Soluble Polymer-supported Synthesis 756
16.3 Fluorous-phase Organic Synthesis 762

16.4	Polymer-supported Reagents 769
16.5	Polymer-supported Catalysts 778
16.6	Polymer-supported Scavengers 781
16.7	Conclusion 783
	References 784

17 Multicomponent Reactions Under Microwave Irradiation Conditions 788

Tijmen de Boer, Alessia Amore, and Romano V.A. Orru

17.1	Introduction 788
17.1.1	General 788
17.1.2	Tandem Reactions and Multicomponent Reactions 789
17.1.3	Microwaves and Multicomponent Reactions 791
17.2	Nitrogen-containing Heterocycles 793
17.2.1	Dihydropyridines 793
17.2.2	Pyridones 797
17.2.3	Pyridines 798
17.2.4	Dihydropyrimidinones 800
17.2.5	Imidazoles 802
17.2.6	Pyrroles 805
17.2.7	Other N-containing Heterocycles 807
17.3	Oxygen-containing Heterocycles 809
17.4	Other Ring Systems 812
17.5	Linear Structures 814
17.6	Conclusions and Outlook 816
	References 816

18 Microwave-enhanced Radiochemistry 820

John R. Jones and Shui-Yu Lu

18.1	Introduction 820
18.1.1	Methods for Incorporating Tritium into Organic Compounds 821
18.1.2	Problems and Possible Solutions 822
18.1.3	Use of Microwaves 824
18.1.4	Instrumentation 826
18.2	Microwave-enhanced Tritiation Reactions 827
18.2.1	Hydrogen Isotope Exchange 827
18.2.2	Hydrogenation 832
18.2.3	Aromatic Dehalogenation 833
18.2.4	Borohydride Reductions 834
18.2.5	Methylation Reactions 835
18.2.6	Aromatic Decarboxylation 836
18.2.7	The Development of Parallel Procedures 838
18.2.8	Combined Methodology 839
18.3	Microwave-enhanced Detritiation Reactions 841
18.4	Microwave-enhanced PET Radiochemistry 842
18.4.1	^{11}C-labeled Compounds 843

| XIV | Contents

| 18.4.2 | ^{18}F-labeled Compounds 846
| 18.4.3 | Compounds Labeled with Other Positron Emitters 852
| 18.5 | Conclusion 854
| | Acknowledgments 854
| | References 855

19 Microwaves in Photochemistry 860
Petr Klán and Vladimír Církva
| 19.1 | Introduction 860
| 19.2 | Ultraviolet Discharge in Electrodeless Lamps 861
| 19.2.1 | Theoretical Aspects of the Discharge in EDL 862
| 19.2.2 | The Fundamentals of EDL Construction and Performance 863
| 19.2.3 | EDL Manufacture and Performance Testing 865
| 19.2.4 | Spectral Characteristics of EDL 866
| 19.3 | Photochemical Reactor and Microwaves 869
| 19.4 | Interactions of Ultraviolet and Microwave Radiation with Matter 877
| 19.5 | Photochemical Reactions in the Microwave Field 878
| 19.5.1 | Thermal Effects 878
| 19.5.2 | Microwaves and Catalyzed Photoreactions 883
| 19.5.3 | Intersystem Crossing in Radical Recombination Reactions in the Microwave Field – Nonthermal Microwave Effects 885
| 19.6 | Applications 888
| 19.6.1 | Analytical Applications 888
| 19.6.2 | Environmental Applications 888
| 19.6.3 | Other Applications 891
| 19.7 | Concluding Remarks 891
| | Acknowledgments 892
| | References 892

20 Microwave-enhanced Solid-phase Peptide Synthesis 898
Jonathan M. Collins and Michael J. Collins
| 20.1 | Introduction 898
| 20.2 | Solid-phase Peptide Synthesis 899
| 20.2.1 | Boc Chemistry 900
| 20.2.2 | Fmoc Chemistry 901
| 20.2.3 | Microwave Synthesis 905
| 20.2.4 | Tools for Microwave SPPS 906
| 20.2.5 | Microwave Enhanced *N*-Fmoc Deprotection 907
| 20.2.6 | Microwave-enhanced Coupling 912
| 20.2.7 | Longer Peptides 922
| 20.3 | Conclusion 925
| 20.4 | Future Trends 926
| | Abbreviations 927
| | References 928

21	**Application of Microwave Irradiation in Fullerene and Carbon Nanotube Chemistry** *931*
	Fernando Langa and Pilar de la Cruz
21.1	Fullerenes Under the Action of Microwave Irradiation *931*
21.1.1	Introduction *931*
21.1.2	Functionalization of Fullerenes *932*
21.2	Microwave Irradiation in Carbon Nanotube Chemistry *949*
21.2.1	Synthesis and Purification *949*
21.2.2	Functionalization of Carbon Nanotubes *950*
21.3	Conclusions *953*
	References *954*

22	**Microwave-assisted Extraction of Essential Oils** *959*
	Farid Chemat and Marie-Elisabeth Lucchesi
22.1	Introduction *959*
22.2	Essential Oils: Composition, Properties, and Applications *960*
22.3	Essential Oils: Conventional Recovery Methods *963*
22.4	Microwave Extraction Techniques *965*
22.4.1	Microwave-assisted Solvent Extraction (MASE) *965*
22.4.2	Compressed Air Microwave Distillation (CAMD) *968*
22.4.3	Vacuum Microwave Hydrodistillation (VMHD) *968*
22.4.4	Microwave Headspace (MHS) *969*
22.4.5	Microwave Hydrodistillation (MWHD) *969*
22.4.6	Solvent-free Microwave Hydrodistillation (SFME) *969*
22.5	Importance of the Extraction Step *970*
22.6	Solvent-free Microwave Extraction: Concept, Application, and Future *970*
22.6.1	Concept and Design *971*
22.6.2	Applications *974*
22.6.3	Cost, Cleanliness, Scale-up, and Safety Considerations *976*
22.7	Solvent-free Microwave Extraction: Specific Effects and Proposed Mechanisms *976*
22.7.1	Effect of Operating Conditions *976*
22.7.2	Effect of the Nature of the Matrix *977*
22.7.3	Effect of Temperature *979*
22.7.4	Effect of Solubility *980*
22.7.5	Effect of Molecular Polarity *981*
22.7.6	Proposed Mechanisms *982*
22.8	Conclusions *983*
	References *983*

Index *986*

Preface

The domestic microwave oven was a serendipitous invention. Percy Spencer was working for Raytheon, a company very involved with radar during World War II, when he noticed the heat generated by a radar antenna. In 1947 an appliance called a Radarange appeared on the market for food processing. The first kitchen microwave oven was introduced by Tappan in 1955. Sales of inexpensive domestic ovens now total many billions of dollars (euros) annually.

Numerous other uses of microwaves have been discovered in the recent years – essentially drying of different types of material (paper, rubber, tobacco, leather...), treatment of elastomers and vulcanization, extraction, polymerization, and many applications in the food-processing industry.

The field of microwave-assisted organic chemistry is therefore quite young. The first pioneering publication was in 1981, the not very well known patent of Bhargava Naresh, from BASF Canada Inc., dealing with the use of microwave energy for production of plasticizer esters. Next, two famous papers from the groups of R. Gedye and R.J. Giguere appeared in 1986. These authors described several reactions completed within a few minutes when conducted in sealed vessels (glass or Teflon) in domestic microwave ovens. Although the feasibility of the procedure was thus apparent, a few explosions caused by the rapid development of high pressure in closed systems were also reported. To prevent such drawbacks, safer techniques were developed – reactions in open beakers or flasks, or solvent-free reactions, as developed since 1987 in France – in Orsay (G. Bram and A. Loupy), Caen (D. Villemin), and Rennes (J. Hamelin and F. Texier-Boullet). Combination of solvent-free procedures with microwave irradiation is an interesting and well-accepted approach within the concepts of green chemistry. This combination takes advantage both of the absence of solvent and of microwave technology under economical, efficient, and safe conditions with minimization of waste and pollution.

Approximately 2500 publications on microwave activation in organic synthesis have appeared in the literature. The spectacular growth of this technique is undoubtedly connected with the development of new, adapted, reactors enabling accurate control and reproducibility of the microwave-assisted procedures but also to the increasing involvement of industrial and pharmaceutical laboratories. This interest has been also been manifested by several reviews and books in the field including, very recently, the Wiley–VCH book by Oliver Kappe and Alexander Stadler

Microwaves in Organic Synthesis, Second edition. Edited by A. Loupy
Copyright © 2006 WILEY-VCH Verlag GmbH & Co. KGaA, Weinheim
ISBN: 3-527-31452-0

on Methods and Principles in Medicinal Chemistry. Many international congresses, all over the world, have been devoted to microwaves.

The objective of this book is to focus on the different fields of application of this technology in several aspects of organic synthesis. This second edition is a revised and enlarged version of the first. Eight new fields of application have been included. The chapters, which complement each other, are written by the most eminent scientists, all well-recognized in their fields.

After essential revision, and description of wave–material interactions, microwave technology, and equipment (Chapter 1), the development and design of reactors including scaling-up (Chapter 2) and the concepts of microwave-assisted organic chemistry in pressurized reactors (Chapter 3) are described. Special emphasis on the possible intervention of specific (not purely thermal) microwave effects is discussed in Chapter 4 and this is followed by up-to-date reviews of their effects on selectivity (Chapter 5). The next topics to be treated are coupling of microwave activation with phase-transfer catalysis (Chapter 6), ionic liquids (Chapter 7), mineral solid supports under "dry media" conditions (Chapter 8) and, more specifically, on graphite (Chapter 9). Applications in which microwave-assisted techniques have afforded spectacular results and applications are discussed extensively Chapters 10, 11, and 12 for heterocyclic chemistry, cycloadditions, and carbohydrate chemistry, respectively. Finally, the techniques have led to fruitful advances in microwave catalysis (Chapter 13) and have been applied to polymer chemistry (Chapter 14) and to organometallic chemistry using transition metal complexes (Chapter 15). Very promising techniques are now under intense development as a result of application of microwave irradiation to combinatorial chemistry (Chapter 16), multi-component reactions (Chapter 17), radiochemistry (Chapter 18), and photochemistry (Chapter 19). Other promising fields are solid-phase peptide synthesis (Chapter 20), fullerene and carbon nanotube chemistry (Chapter 21), and extraction of essential oils (Chapter 22).

I wish to thank sincerely all my colleagues and, nevertheless, (essentially), friends involved in the realization of this book. I hope to express to them, all eminent specialists, my friendly and scientific gratitude for agreeing to devote their competence and time to submitting and reviewing papers to ensure the success of this book.

Gif sur Yvette, February 22nd, 2006 *André Loupy*

About European Cooperation in COST Chemistry Programs

Microwave irradiation is a means of rapidly introducing energy into a chemical system in a manner different from the traditional methods of thermal heating. Other techniques can also be used to introduce energy into a system and, in this way, by accelerating reactions and also causing other more specific individual effects, both microwaves and ultrasonic irradiation provide means to achieve "Chemistry in High-Energy Microenvironments". Comparison of microwaves with ultrasound is the subject of COST Chemistry Action D32, which seeks to explore further uses of these methods individually, and also in combination. COST is the acronym for, in French "Co-opération Européenne dans le domaine de la Recherche Scientifique et Technique" or, in English, "European Co-operation in the Field of Scientific and Technical Research". COST started in 1992 and earlier initiatives D6 "Chemistry under Extreme and Non-Classical Conditions", and D10 "Innovative Transformations in Chemistry" involved pioneering studies of aspects of ultrasound and, more recently, microwaves in chemistry. Now D32, which started in 2004, has, as one of its objectives to take these techniques further.

Cost Actions are interdisciplinary and the objectives encourage the collaboration of chemists, biochemists and material scientists, and enable the transfer of experience between fundamentally orientated and applied research in these areas. COST Actions are split into working groups which have a cohesive theme within the overall objectives of the action. In D32 there are four microwave-based working groups involving collaboration between scientists with different expertise in modern technology:

- "Diversity oriented synthesis under (highly efficient) microwave conditions". The objectives of this working group are:
  - to apply solvent-less reaction systems and alternative reagents (those which are optimum for microwave activation) to already known reactions;
  - to perform organic reactions in aqueous media, and in this way substitute VOCs by water as the reaction media;
  - to evaluate multicomponent reactions under microwave and/or microreactor conditions to minimize waste and optimize atom-efficiency in comparison with conventional stepwise reactions under traditional heating conditions; and
  - to establish, in a combinatorial approach, novel atom economic multi-

component reactions assisted by microwave and/or microreactor technology to efficiently access pharmaceutically relevant N-heterocyclic compounds.

In such ways, microwaves can play an important role in the development of green chemistry. The use of microwaves for "green" reactions for industrial purposes can be accomplished by minimization of waste, replacement of hazardous and polluting materials, recycling of materials, and energy saving.

- "Microwave and High-intensity Ultrasound in the Synthesis of Fine Chemicals" The concern of this group is the application of high-intensity ultrasound (HIU) and microwave (MW), in combination, to the synthesis of fine chemicals, because these forms of energy can induce reactions that would be otherwise very laborious and bring about peculiar chemoselectivity. The synergic effect of these techniques can be exploited in such diverse fields as organic synthesis and sample preparation for analytical procedures. This results in challenging practicalities and technical hurdles, however, and the potential of this promising combination has yet to be investigated systematically. One objective of COST D32 is to bring experts in microwave-promoted reactions and in sonochemistry together for a joint effort. This is offering new opportunities of growth for both techniques, opening up new applications. Another important objective is to have, in each working group, as many scientists active in both fundamental and industrial research as possible, with collaboration that will lead to stimulating exchanges and innovative projects. Many important reactions shall be studied in unusual reaction media (water, ionic liquids, supercritical fluids) and under unusual conditions (sonication under pressure and HIU/MW combined).

- Finally, a significant issue that COST D32 intends to address is the challenge of scaling up results obtained in the laboratory. This is the theme of the third working group "Development and design of reactors for microwave-assisted chemistry in the laboratory and on the pilot scale". The aim of this working group is interdisciplinary comparative study of different fields of chemical technology and reactor design, to demonstrate the real advantages of microwave technology as a method for activation of chemical reactions and processes with a high radiation flow density (or high power density). The different concepts of technical microwave systems (monomode and/or multimode), different methods of temperature measurement (IR, fiber optics, metal sensors), and of energy entry (pulsed or unpulsed) will be compared and the advantages of the different options determined for some typical reactions. Safety issues (dielectric properties, temperature and power measurement, materials) will be addressed as also will risk management for the new technology. Scale-up for technical usefulness will be tested for different reactors and different types of reaction (heterogeneous catalysis, two liquid phases) and combination of microwaves with other forms of energy (UV–visible, ultrasound, plasma) will be examined, together with transfer from batch to continuous process, which is important for the economic use of microwave-assisted reactions. The energy efficiency of microwave-assisted reactions and processes will be compared with that of conventional reactions and processes.

- "Ultrasonic and Microwave-assisted Synthesis of Nanometric Particles". The objectives of this fourth working group are to use the techniques singly or in tandem to produce technologically-useful ceramics, metals, alloys, and other nanoparticulate systems for the wide range of applications for which these materials are being studied.

In summary, one set of objectives of COST D32 is to establish a firm EU base in microwave chemistry and to exploit the new opportunities provided by microwave techniques, singly or in appropriate combination, for the widest range of applications in modern chemistry.

Coventry (UK), February 22nd, 2006   *David Walton*

List of Authors

Alessia Amore
Vrije Universiteit
Department of Chemistry
Synthetic Bio-organic Chemistry
De Boelelaan 1083
1081 HV Amsterdam
The Netherlands

George Bashiardes
Université de Poitiers
40 avenue du Recteur Pineau
86022 Poitiers
France

Jean Pierre Bazureau
Université de Rennes 1
Sciences Chimiques de Rennes
UMR CNRS 6226
Bât. 10A
Campus de Beaulieu
Avenue du Général Leclerc
CS 74205
35042 Rennes cedex 1
France

Thierry Besson
Université de Rouen
UFR Médicine-Pharmacie
UMR CNRS 6014
22, bld Gambetta
76183 Rouen cedex 1
France

Tijmen de Boer
Vrije Universiteit
Department of Chemistry
Synthetic Bio-organic Chemistry
De Boelelaan 1083
1081 HV Amsterdam
The Netherlands

Dariusz Bogdal
Politechnika Krakowska
Department of Chemistry
ul. Warszawska 24
31-155 Krakow
Poland

Werner Bonrath
DSM Nutritional Products
R&D Chemical Process Technology
PO Box 3255
4002 Basel
Switzerland

Khalid Bougrin
Université Mohammed V
Faculté des Sciences
Laboratoire de Chimie des Plantes et de
Synthèse Organique et Bio-organique
Avenue Ibn Batouta
BP 1014
Rabat
Morocco

Farid Chemat
Université d'Avignon
Sécurité et Qualité
des Produits d'Origine
Végétale
33, rue Louis Pasteur
84029 Avignon cedex 1
France

Ugo Chiacchio
Università di Catania
Dipartimento di Scienze Chimiche
viale A. Doria 6
95125 Catania
Italy

List of Authors

Vladimír Církva
Academy of Sciences of the Czech Republic
Institute of Chemical Process Fundamentals
Rozvojova 135
165 02 Prague
Czech Republic

Jonathan M. Collins
CEM Corporation
PO Box 200
3100 Smith Farm Road
Matthews
NC 28106-0200
USA

Michael J. Collins
CEM Corporation
PO Box 200
3100 Smith Farm Road
Matthews
NC 28106-0200
USA

Antonino Corsaro
Università di Catania
Dipartimento di Scienze Chimiche
viale A. Doria 6
95125 Catania
Italy

Pilar de la Cruz
Universidad de Castilla-La Mancha
Departamento de Química Orgánica
Facultad de Ciencias del Medio Ambiente
45071 Toledo
Spain

Angel Díaz-Ortiz
Universidad de Castilla-La Mancha
Departamento de Química Orgánica
Facultad de Química
13071 Ciudad Real
Spain

Jacques Dubac
2 Chemin du Taur
31320 Pechbusque
France

Milan Hájek
Academy of Sciences of the Czech Republic
Institute of Chemical Process Fundamentals
Rozvojová 135
165 02 Prague 6
Suchdol
Czech Republic

Jack Hamelin
Université de Rennes 1
Sciences Chimiques de Rennes
UMR CNRS 6226
Bât. 10A
Campus de Beaulieu
Avenue du Général Leclerc
CS 74205
35042 Rennes cedex 1
France

Antonio de la Hoz
Universidad de Castilla-La Mancha
Departamento de Química Orgánica
13071 Ciudad Real
Spain

John R. Jones
University of Surrey
School of Biomedical and Molecular Sciences
Department of Chemistry
Guildford
Surrey GU2 7XH
UK

Yuhong Ju
US Environmental Protection Agency
National Risk Management Research Laboratory
Clean Processes Branch
26 West Martin Luther King Drive
MS 443
Cincinnati, OH 45268
USA

C. Oliver Kappe
Karl-Franzens-University
Institute of Chemistry
Heinrichstrasse 28
8010 Graz
Austria

Petr Klán
Masaryk University
Faculty of Science
Department of Organic Chemistry
Kotlářská 2
611 37 Brno
Czech Republic

Fernando Langa
Universidad de Castilla-La Mancha
Departamento de Química Orgánica
Facultad de Ciencias del Medio Ambiente
45071 Toledo
Spain

Mats Larhed
Uppsala University
Department of Medicinal Chemistry
Organic Pharmaceutical Chemistry
BMC
Box 574
75123 Uppsala
Sweden

Nicholas E. Leadbeater
University of Connecticut
Department of Chemistry
55 North Eagleville Road
Storrs, CT 06269-3060
USA

André Loupy
29 Allée de la Gambeauderie
91190 Gif sur Yvette
France

Shui-Yu Lu
National Institutes of Health
National Institute of Mental Health
Molecular Imaging Branch
Building 10. Room B3 C346
10 Center Drive, MSC 1003
Bethesda, MD 20892-1003
USA

Marie-Elisabeth Lucchesi
Université de la Réunion
Naturelles et de Science des Aliments
Laboratoire de Chimie des Substances
BP 7151
15 rue René Cassin
97715 Saint Denis Messageries cedex 9
France

Katarzyna Matras
Politechnika Krakowska
Department of Chemistry
ul. Warszawska 24
31-155 Krakow
Poland

Florence Mongin
Université de Rennes 1
Institut de Chimie, Synthèse and
Electrosynthèse Organiques 3
UMR 6510
Bât. 10A
Campus de Beaulieu
Avenue du Général Leclerc
CS 74205
35042 Rennes cedex 1
France

Andrés Moreno
Universidad de Castilla-La Mancha
Facultad de Química
Departamento de Química Orgánica
14071 Ciudad Real
Spain

Thach Le Ngoc
National University of Hochiminh City
University of Natural Sciences
Department of Organic Chemistry
227 Nguyen Van Cu Str., Dist. 5
Hochiminh City
Vietnam

Peter Nilsson
Biolipox
Medicinal Chemistry
Berzelius väg 3, plan 5
17165 Solna
Sweden

Kristofer Olofsson
Biolipox
Medicinal Chemistry
Berzelius väg 3, plan 5
17165 Solna
Sweden

Bernd Ondruschka
Friedrich Schiller University of Jena
Department of Technical Chemistry and
Environmental Chemistry
Lessingstr. 12
07743 Jena
Germany

Romano V.A. Orru
Vrije Universiteit
Department of Chemistry
Synthetic Bio-organic Chemistry
De Boelelaan 1083
1081 HV Amsterdam
The Netherlands

Laurence Perreux
Université Paris-Sud
ICMMO
Laboratoire des Réactions Sélectives sur
Supports
bât 410
91405 Orsay cedex
France

Alain Petit
Université Paris-Sud
ICMMO
Laboratoire des Réactions Sélectives sur Supports
bâtiment 410,
91405 Orsay cedex
France

Venerando Pistarà
Università di Catania
Dipartimento di Scienze Chimiche
viale A. Doria 6
95125 Catania
Italy

Brett A. Roberts
Monash University
ARC Special Research Centre for Green Chemistry
Clayton Campus
Victoria 3800
Australia

Giovanni Romeo
Università di Messina
Dipartimento Farmaco-Chimico
viale SS. Annunziata 6
98168 Messina
Italy

Mohamed Soufiaoui
Université Mohammed V
Faculté des Sciences
Laboratoire de Chimie des Plantes et de Synthèse Organique et Bio-organique
Avenue Ibn Batouta
BP 1014
Rabat
Morocco

Alexander Stadler
Anton Paar GmbH
Anton-Paar Strasse 20
8054 Graz
Austria

Christopher R. Strauss
CSIRO Molecular and Health Technologies
Private Bag 10
Clayton South
Victoria 3169
Australia

Didier Stuerga
University of Burgundy
GERM/LRRS and Naxagoras Technologie
BP 47870
21078 Dijon cedex
France

Françoise Texier-Boullet
Université de Rennes 1
Institut de Chimie, Synthèse and Electrosynthèse Organiques 3
UMR 6510
Bât. 10A
Campus de Beaulieu
Avenue du Général Leclerc
CS 74205
35042 Rennes cedex 1
France

Valérie Thiéry
Université de La Rochelle
UFR Sciences Fondamentales et Sciences pour l'Ingénieur, LBCB, FRE CNRS 2766
Avenue Marillac, Bâtiment Marie Curie
17042 La Rochelle cedex 1
France

Hanna M. Torenius
King's College London
Strand
London WC2R 2LS
UK

Rajender S. Varma
US Environmental Protection Agency
National Risk Management Research Laboratory
Clean Processes Branch
26 West Martin Luther King Drive
MS 443
Cincinnati, OH 45268
USA

# 1
Microwave–Material Interactions and Dielectric Properties, Key Ingredients for Mastery of Chemical Microwave Processes

Didier Stuerga

1.1
Fundamentals of Microwave–Matter Interactions

The main focus of the revised edition of this first chapter is essentially the same as the original – to explain in a chemically intelligible fashion the physical origin of microwave–matter interactions and, especially, the theory of dielectric relaxation of polar molecules. This revised version contains approximately 60% new material to scan a large range of reaction media able to be heated by microwave heating. The accounts presented are intended to be illustrative rather than exhaustive. They are planned to serve as introductions to the different aspects of interest in comprehensive microwave heating. In this sense the treatment is selective and to some extent arbitrary. Hence the bibliography contains historical papers and valuable reviews to which the reader anxious to pursue particular aspects should certainly turn.

It is the author's conviction, confirmed over many years of teaching experience, that it is much safer – at least for those who rate not trained physicists – to deal intelligently with an oversimplified model than to use sophisticated methods which require experience before becoming productive. Because of comments about the first edition, however, the author has included more technical material to enable better understanding of the concepts and ideas. These paragraphs could be omitted, depending on the level of comprehension of the reader. They are preceded by two different symbols – ✖ for TOOLS and ● for CONCEPTS.

After consideration of the history and position of microwaves in the electromagnetic spectrum, notions of polarization and dielectric loss will be examined. The orienting effects of electric fields and the physical origin of dielectric loss will be analyzed, as also will transfers between rotational states and vibrational states within condensed phases.

Dielectric relaxation and dielectric losses of pure liquids, ionic solutions, solids, polymers and colloids will be discussed. Effect of electrolytes, relaxation of defects within crystals lattices, adsorbed phases, interfacial relaxation, space charge polarization, and the Maxwell–Wagner effect will be analyzed. Next, a brief overview of thermal conversion properties, thermodynamic aspects, and athermal effects will be given.

1.1.1
Introduction

According to the famous chemistry dictionary of N. Macquer edited in 1775, "All the chemistry operations could be reduced to decomposition and combination; hence, the fire appears as an universal agent in chemistry as in nature" [1]. Heating remains the primary means of stimulating chemical reactions which proceed slowly under ambient conditions, but several other stimulating techniques – photochemical, ultrasonic, high pressure and plasma – can be used. In this book, we describe results obtained by use of microwave heating. Microwave heating or dielectric heating is an alternative to conventional conductive heating. This heating technique uses the ability of some materials (liquids and solids) to transform electromagnetic energy into heat. This "in situ" mode of energy conversion is very attractive for chemistry applications and material processing.

If the effect of the temperature on reaction rates is well known, and very easy to express, the problem is very different for effects of electromagnetic waves. What can be expected from the orienting action of electromagnetic fields at molecular levels? Are electromagnetic fields able to enhance or to modify collisions between reagents? All these questions are raised by the use of microwave energy in chemistry.

1.1.1.1 History

How it all began There is some controversy about the origins of the microwave power cavity called the magnetron – the high-power generator of microwave power. The British were particularly forward-looking in deploying radar for air defense with a system called Chain Home which began operation in 1937. Originally operating at 22 MHz, frequencies increased to 55 MHz. The superiority of still higher frequencies for radar was appreciated theoretically but a lack of suitable detectors and of high-power sources stymied the development of microwaves. Magnetrons provide staggering amounts of output power (e.g. 100 kW on a pulse basis) for radar transmitters. The earliest description of the magnetron, a diode with a cylindrical anode, was published by A.W. Hull in 1921 [2, 3]. It was developed practically by Randall and Booth at the University of Birmingham in England ca 1940 [4]. On 21 February 1940, they verified their first microwave transmissions: 500 W at 3 GHz. A prototype was brought to the United States in September of that year to define an agreement whereby United States industrial capability would undertake the development of microwave radar. In November 1940 the Radiation Laboratory was established at the Massachusetts Institute of Technology to exploit microwave radar. More than 40 types of tube would be produced, particularly in the S-band (i.e. 300 MHz). The growth of microwave radar is linked with Raytheon Company and P.L. Spencer who found the key to mass production. Microwave techniques were developed during and just before World War II when most effort was concentrated on the design and manufacture of microwave navigation and communications equipment for military use. Originally, microwaves played a leading role during the World War II, especially during the Battle of Britain when English planes could fight one against three thanks to radar. It hardly seems surprising that with all this magnetron manufacturing expertise that microwave cooking would be invented at Raytheon and that the first microwave oven would be built there.

Since these beginnings the heating capability of microwave power has been recognized by scientists and engineers but radar development had top priority. A new step began with the publication of microwave heating patents by Raytheon on 9 October 1945. Others patents followed as problems were encountered and solutions found. Probably the first announcement of a microwave oven was a magazine article describing a newly developed Radarange for airline use [5, 6]. This device, it was claimed, could bake biscuits in 29 s, cook hamburgers in 35 s, and grill frankfurters in 10 s. This name Radarange almost became generic name for microwave ovens. An early prototype picture is shown in the book of Decareau and Peterson [7]. This first commercial microwave oven was developed by P.L. Spencer from Raytheon in 1952 [8]. The legend says that P.L. Spencer, who studied high-power microwave sources for radar applications, noticed the melting of a chocolate bar put in his pocket. Another story says that P.L. Spencer had some popping corn in his pocket that began to pop as he was standing alongside a live microwave source [7].

These first oven prototypes were placed in laboratories and kitchens throughout the United States to develop microwave cooking technology. The transition from crude aircraft heater to domestic oven took almost eight years. The turning point in the story of the microwave oven was 1965, which saw the beginning of a flurry of manufacturing activity and the issue of hundreds of patents on different aspects of oven design, processes, packaging, food products, appliance and techniques. The widespread domestic use of microwave ovens occurred during the 1970s and 1980s as a result of the generation of the mass market and also of Japanese technology transfer and global marketing.

From cooking to microwave processing The first studies of the effects of microwave heating were carried out at the Massachusetts Institute of Technology's Department of Food Technology on blanching of vegetables, coffee roasting, effect of cooking, with the hope of vitamin retention [9]. Microwave and conventional freeze-drying of foods were compared by Jackson et al. [10]. The Food Research Laboratory of Raytheon have made extensive studies that led to the first microwave freeze-drying pilot plant unit [11–16].

Microwave processing began on a commercial scale in the early 1960s when Cryodry Corporation of San Ramon, California, introduced the first conveyor system for sale. The first market was the potato chip finish drying process with several systems operating in United States and Europe [17, 18]. These systems operated at 915 MHz. Several 5 to 10 kW pilot plant conveyor systems were sold during this time to food manufacturers by Raytheon and Litton Industries Atherton Division. These systems all operated at 2450 MHz. One poultry-processing system [19] had a total output of 130 kW, split between two conveyor units. This system combined microwave power and saturated steam to precook poultry parts for the institutional and restaurant food service market. This system also operated at 2450 MHz.

Food applications included microwave tempering of frozen foods, pasta drying, precooking of bacon, poultry processing, meat pie cooking, frankfurter manufacturing, drying egg yolk paste, baking, sterilization, potato processing, cocoa bean roasting, and vacuum drying [7, 20]. Curiously, microwave heating of industrial applications was initiated by the domestic oven.

Pre-historical foundations Many histories of electromagnetic waves and especially microwaves begin with publication of the *Treatise on Electricity and Magnetism* by James Clerk Maxwell in 1873. These equations were initially expressed by J.C. Maxwell in terms of quaternions. Heaviside and Gibbs would later reject quaternions in favor of a classical vector formulation to frame Maxwell's equations in the well known form. Students and users of microwave heating, perhaps benumbed by divergence, gradient, and curl, often fail to appreciate just how revolutionary this insight was. The existence of electromagnetic waves that travel at the speed of light were predicted by arbitrarily adding an extra term (the displacement current) to the equations that described all previously known electromagnetic behavior. According to Lee [21], and contrary to the standard story presented in many textbooks, Maxwell did not introduce the displacement current to resolve any outstanding conundrums but he was apparently inspired more by an aesthetic sense that nature simply should provide for the existence of electromagnetic waves. Maxwell's work was magical and arguably ranks as the most important intellectual achievement of the 19th century. According to the Nobel physicist R. Feynman, future historians would still marvel at this work, long after another event of that time – the American Civil War – has faded into merely parochial significance [21].

Maxwell died in 1879 (48 years old), and H. Von Helmholtz sponsored a prize for the first experimental evidence of Maxwell's forecasting. H. Hertz verified that Maxwell's forecasting was correct in 1888 at the Technische Hochschule in Karlsruhe. According to Lee [21] another contestant in the race, was O. Lodge, a professor at University College in Liverpool, who published his own experimental evidence one month before H. Hertz. Hertz is the German word for heart and human heart beats approximately once per second, it is perhaps all for the best that Lodge didn't win the race and that lodgian waves with frequencies measured in giga-lodges never appeared.

How was it possible to produce and detect electromagnetic waves in the 1880's? The first experiment of H. Hertz produced microwaves (frequency close to GHz). His basic transmitter-receiver is shown in Fig. 1.1. The generator is a Ruhmkorff coil or a transformer able to produce very high potential (1). This device is very similar to starter of car. The high voltage in the secondary causes spark discharge within straight wire connections to produce the desired resonant frequency (2). The detector is a ring antenna with a spark gap (3). Detection is based on induction

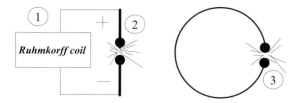

Fig. 1.1. Spark transmitter and receiver of Hertz's original experiment.

of sufficient voltage in the ring antenna to produce a visible spark. Hertz demonstrated the essential physics of wave phenomena, for example polarization and reflection. He died of blood poisoning from an infected tooth in 1894 at the age of 36. The commercial applications of wireless would be developed by G. Marconi. Many details about the whole history of microwave technology can be found elsewhere [21].

1.1.1.2 The Electromagnetic Spectrum

In the electromagnetic spectrum, microwaves occur in a transitional region between infrared and radiofrequency radiation, as shown in by Fig. 1.2. The wavelengths are between 1 cm and 1 m and frequencies between 300 GHz and 300 MHz.

The term "microwave" denotes the techniques and concepts used and a range of frequencies. Microwaves may be transmitted through hollow metallic tubes and

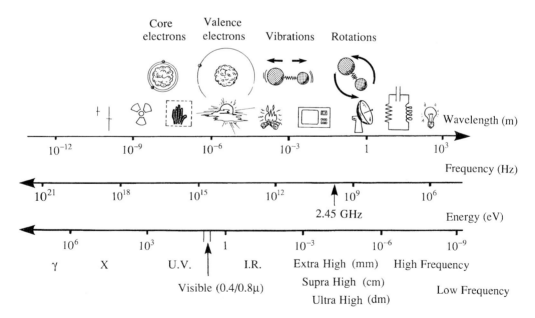

Fig. 1.2. The electromagnetic spectrum.

may be focused into beams by the use of high-gain antennas. Microwaves also change direction when traveling from one dielectric material into another, similar to the way light rays are bent (refracted) when they passed from air into water. Microwaves travel in the same manner as light waves; they are reflected by metallic objects, absorbed by some dielectric materials and transmitted without significant absorption through other dielectric materials. Water, carbon, and foods with a high water content are good microwave absorbers whereas ceramics and most thermoplastic materials absorb microwaves only slightly.

The fundamental connection between energy E, frequency, ν, wavelength, λ, and angular frequency, ω, is given by Eq. (1):

$$E = \hbar\omega = h\nu = \frac{hc}{\lambda} \qquad (1)$$

To avoid interference with telecommunications and cellular 'phone frequencies, heating applications must use ISM bands (Industrial Scientific and Medical frequencies) which are 27.12 and 915 MHz and 2.45 GHz (i.e. wavelengths of 11.05 m and 37.24 and 12.24 cm, respectively). Domestic ovens and laboratory systems usually work at 2.45 GHz. At frequencies below 100 MHz, when conventional open wire circuits are used, the technique will be referred to as radiofrequency heating. The object to be heated is placed between the two electrodes of a capacitor. At frequencies above 500 MHz, however, wired circuits cannot be used and the power is transferred to the applicator containing the material to be processed. Hence, the microwave applicator is a metallic box in which the object to be heated is placed. These operating conditions will be referred as microwave-heating processes. In the microwave band the wavelength is of the order of the size of production and transmission elements. The elements cannot, therefore, be regarded as points in comparison with the wavelength, in contrast with circuit theory. In the same way, it is impossible to consider them far larger than the wavelength, as in geometrical optics. Hence, because of the position of microwaves in the electromagnetic spectrum, both quantum mechanics (corpuscular aspects) and the Maxwell equations (wavelike aspects) will be used. Detailed analysis of these phenomena is beyond the scope of this work.

1.1.1.3 What About Chemistry? Energetic Comments

It is well known that γ or X photons have energies suitable for excitation of inner or core electrons. We can use ultraviolet and visible radiation to initiate chemical reactions (photochemistry, valence electron). Infrared radiation excites bond vibrations only, whereas microwaves excite molecular rotation.

Energy associated with chemical bonds and Brownian motion are compared in Table 1.1. The microwave photon corresponding to the frequency used in microwave heating systems, for example domestic and industrial ovens, has energy close to 0.00001 eV (2.45 GHz, 12.22 cm). According to these values, the microwave photon is not sufficiently energetic to break hydrogen bonds; it is also much smaller than that of Brownian motion and obviously cannot induce chemical reactions. If

Tab. 1.1. Brownian motion and bond energies.

	Brownian motion	Hydrogen bonds	Covalent bonds	Ionic bonds
Energy (eV)	~0.017 (200 K)	~0.04 to 0.44	~4.51 (C–H); ~3.82 (C–C)	~7.6
Energy (kJ mol^{-1})	1.64	~3.8 to 42	~435 (C–H); ~368 (C–C)	~730

no bond breaking can occur by direct absorption of electromagnetic energy, what, then, can be expected from the orienting effects of electromagnetic fields at molecular levels? Are electromagnetic fields able to enhance or to modify collisions between reagents? Do reactions proceed with the same reaction rate with and without electromagnetic irradiation at the same bulk temperature? In the following discussion the orienting effects of the electric field, the physical origin of the dielectric loss, transfers between rotational and vibrational states in condensed phases, and thermodynamic effects of electric fields on chemical equilibrium will be analyzed.

TOOLS More about energy partition of molecular systems Rotational motion of molecular systems are much slower than the vibrational motion of the relatively heavy nuclei forming chemical bonds, and even slower than the electronic motion around nuclei. These vastly differing time scales of the different types of motion lead to a natural partitioning of the discrete energy spectrum of matter into progressively smaller subsets associated with electronic, vibrational, and rotational degrees of freedom.

The Born–Oppenheimer approximation is based on this assumption and enables reduction of the mathematically intractable spectral eigenvalue problem to a set of separable spectral problems for each type of motion. According to this approximation, energy levels associated with each type of motion are proportional to the ratio of electronic mass (m_e) to the nuclear mass (M_N). This ratio, ξ, quite smaller than unity, is given by Eq. (2):

$$\zeta \propto \left(\frac{m_e}{M_N}\right)^{1/4} \tag{2}$$

The electronic energy (ΔE_{elec}) is of the order of ξ, the vibrational energy (ΔE_{Vib}) of the nuclei is of the order of ξ^2, and the rotational energy (ΔE_{Rot}) of the molecule is of the order of ξ^4. In quantum mechanics, states are described by wave functions or Hamiltonian operators, whose discrete eigenvalues define the set of energy levels and whose corresponding eigenfunctions are the basis states. Hence, the total quantum wave function Ψ for a molecule can be written in separable form as described by Eq. (3):

$$\Psi = \Psi_{Elec}(r, R_0)\Psi_{Vib}(R)\Psi_{Rot}(\varphi_i) \tag{3}$$

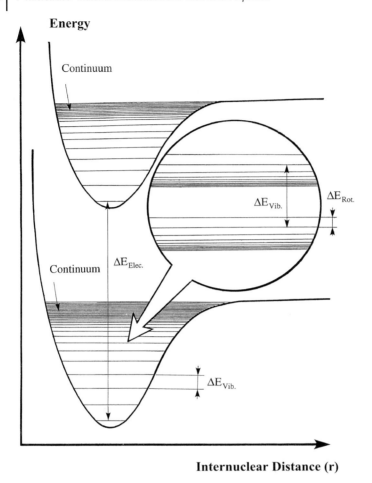

Fig. 1.3. The energy spectrum of matter.

where r is the electron coordinate, R the displacement of the nucleus from its equilibrium position, R_0, and φ_i is the Euler angle determining the orientation of the molecule in space. Figure 1.3 shows the energy spectrum of matter as it is probed on a progressively finer energy scale clearly revealing the different partition states. The fundamental and the first excited states are shown.

A resonance of a system can be produced by excitation that oscillates at a frequency close to the natural frequency of the system, unlike a relaxation, which is the restoring action of a diffusive force of thermodynamic origin. Direct resonance or a one-photon process can occur within isolated intervals of the electromagnetic spectrum from ultraviolet to visible frequencies close to 10^{15} Hz (electronic oscillator), in infrared with frequencies close to 10^{13} Hz (vibrational modes), and in the far infrared and microwave range with frequencies close to 10^{11} Hz (rotational modes).

1.1.2
The Complex Dielectric Permittivity

Insulating materials can be heated by applying high-frequency electromagnetic energy. The physical origin of this heating conversion lies with the ability of the electric field to induce polarization of charges within the heated product. This polarization cannot follow the extremely rapid reversals of the electric field and induce heating of the irradiated media.

The interaction between electromagnetic waves and matter is quantified by the two complex physical quantities – the dielectric permittivity, $\tilde{\varepsilon}$, and the magnetic susceptibility, $\tilde{\mu}$. The electric components of electromagnetic waves can induce currents of free charges (electric conduction that can be of electronic or ionic origin). It can, however, also induce local reorganization of linked charges (dipolar moments) while the magnetic component can induce structure in magnetic moments. The local reorganization of linked and free charges is the physical origin of polarization phenomena. The storage of electromagnetic energy within the irradiated medium and the thermal conversion in relation to the frequency of the electromagnetic stimulation appear as the two main points of polarization phenomena induced by the interaction between electromagnetic waves and dielectric media. These two main points of wave–matter interactions are expressed by the complex formulation of the dielectric permittivity as described by Eq. (4):

$$\tilde{\varepsilon} = \varepsilon' - j\varepsilon'' = \varepsilon_0 \varepsilon'_r - j\varepsilon_0 \varepsilon''_r \tag{4}$$

where ε_0 is the dielectric permittivity of a vacuum, ε' and ε'' are the real and imaginary parts of the complex dielectric permittivity and ε'_r and ε''_r are the real and imaginary parts of the relative complex dielectric permittivity. The storage of electromagnetic energy is expressed by the real part whereas the thermal conversion is proportional to the imaginary part.

✵ TOOLS More about polar molecules A polar molecule has a permanent electric dipole moment. The total amounts of positive and negative charges on the molecule are equal, so the molecule is electrically neutral. Distributions of the two kinds of charge are different, however, so that the positive and negative charges are centered at points separated by a distance of molecular dimensions forming an electric dipole. A dipole made up of charges $+q$ and $-q$, separated by a distance d, of magnitude qd. The dipole moment usually represented by the symbol μ is approximately 10^{-18} Coulomb (the electronic charge is of the order of 10^{-10} SI units whereas d will be of order of molecular dimensions – 10^{-10} m). The unit 10^{-18} Cb m is called the Debye (abbreviation D).

The magnitude of the dipole moment depends on the size and symmetry of the molecule. Molecules with a center of symmetry, for example methane, carbon tetrachloride, and benzene are apolar (zero dipole moment) whereas molecules with no center of symmetry are polar. Table 1.2 gives relative static dielectric permittivity

Tab. 1.2. Relative static dielectric constant, refractive index (measured at the frequency of sodium D lines) and dipole moment for a few molecules.

Molecules	ε_{Sr}	n_D^2	μ
Apolar			
n-Hexane, C_6H_{14}	1.89	1.89	–
Carbon tetrachloride, CCl_4	2.23	2.13	–
Benzene, C_6H_6	2.28	2.25	–
Polar			
Methanol, CH_3OH	33.64	1.76	1.68
Ethanol, CH_3CH_2OH	25.07	1.85	1.70
Acetone, CH_3COCH_3	21.20	1.84	2.95
Chlorobenzene, C_6H_5Cl	5.64	2.32	1.69
Water, H_2O	80.37	1.78	1.94

or dielectric constants (very low frequency or frequency close to zero), refractive indices, and dipole moments for few simple polar and apolar molecules.

From Maxwell's theory of electromagnetic waves it follows that the relative permittivity of a material is equal to the square of its refractive index measured at the same frequency. Refractive index given by Table 1.2 is measured at the frequency of the D line of sodium. Thus it gives the proportion of (electronic) polarizability still effective at very high frequencies (optical frequencies) compared with polarizability at very low frequencies given by the dielectric constant. It can be seen from Table 1.2 that the dielectric constant is equal to the square of the refractive index for apolar molecules whereas for polar molecules the difference is mainly because of the permanent dipole. In the following discussion the Clausius–Mossoti equation will be used to define supplementary terms justifying the difference between the dielectric constant and the square of the refractive index (Eq. (29); The Debye model).

The temperature dependence of the dielectric constant of polar molecules also differs from that of nonpolar molecules. Change of temperature has a small effect only for nonpolar molecules (change of density). For polar molecules, the orientation polarization falls off rapidly with increasing temperature, because thermal motion reduces the alignment of the permanent dipoles by the electric field. In the following discussion we will see that it is possible to have increasing values of dielectric permittivity with increasing temperature.

As discussed above, a molecule with a zero total charge may still have a dipole moment because molecules without a center of symmetry are polar. Similarly, a molecule may have a distribution of charge which can be regarded as two equal and opposite dipoles centered at different places. Such a distribution will have zero total charge and zero total moment but will have a quadrupole moment. Carbon disulfide, which is linear, has a quadrupole moment. Two equal and opposite quadrupole moments centered at difference places form an octupole moment. The potential arising as a result of the total charge falls off as $1/r$, that because of the dipole moment as $1/r^2$, and that because of the quadrupole moment as $1/r^3$. At large distances from the origin, the higher moments have negligible effects. The intermolecular distances in liquids and solids are not large compared with molecular dimensions, however, so quite strong interactions may arise because of higher moments.

TOOLS More about dielectrics and insulators An insulator is a material through which no steady conduction current can flow when it is submitted to an electric field. Consequently, an insulator can accumulate electric charge, hence electrostatic energy. The word dielectric, especially if it is used as adjective, covers a wide range of materials including electrolytes and even metals.

If we consider a capacitor comprising two planes parallel with S and d, the surface area of the plates and the distance between the plates, respectively (S is large compared with d so that edge effects are negligible), the vacuum capacitance is given by Eq. (5):

$$C = \varepsilon_0 \frac{S}{d} \tag{5}$$

If an alternating voltage $V = V_0 \exp(j\omega t)$ is applied to this capacitor, a charge $Q = CV$ appears on the electrode, in phase with the applied voltage. The current in the external circuit is the time-derivative of the charge Q is given by Eq. (6):

$$I = \dot{Q} = j\omega CV \tag{6}$$

This current is 90° out of phase with the applied voltage. It is a nondissipative displacement or induction current. If the volume between the electrodes is filled with a nonpolar, perfectly insulating material, the capacitor has a new capacitance, and the ratio between the vacuum and filled capacitance is the relative permittivity of the material used. The new current is larger than that above but it is still out of phase with the current. Now, if the material is either slightly conducting or polar, or both, the capacitor is not longer perfect and the current is not exactly out of phase with the voltage. Hence, there is a component of conduction in phase with the applied voltage. The origin of this current is the motion of charges. The current is composed of a displacement current and a conduction current. The loss angle is given by Eq. (7):

$$\tan\delta = \frac{\text{Dissipative term}}{\text{Capacitive term}} \tag{7}$$

The current is composed of two quantities, real and imaginary, so the dielectric permittivity will also have a complex form which will depend on the types of interaction between the electromagnetic field and matter. This discussion refers to isotropic dielectrics. Many products falls into this class but the situation is different for crystalline solids, for which the permittivity becomes a tensor quantity (values different according to crystallographic axis).

● **CONCEPTS The dielectric properties are group properties** The physical origin of polarization phenomena is the local reorganization of linked and free charges. The interaction between a dipole and an electric or magnetic field is clearly interpreted by quantum theory. For electric fields the coupling is weaker and there is such demultiplication of quantum levels that they are very close to each other. The Langevin and Boltzman theories must be used because interaction energy is continuous. Because of the weak coupling between dipole and electric field there are no quantified orientations and study of the interaction between a dipole in an electric field gives more information about the surroundings of the dipole than about the dipole itself. Dipoles are, moreover, associated with chemical bonds, and any motion of the dipole induces a correlative motion of molecular bonds whereas motion of a magnetic moment is totally independent of any molecular motion. In contrast with magnetic properties, dielectric properties are group properties and cannot be modeled by an interaction between a single dipole and electric field. A group of dipoles interacting among themselves could be considered.

The origin of confusion between the behavior of a single species and a collection, or the difference between dilute and condensed phases, is the most important problem and the source of illusions within microwave athermal effects.

1.1.2.1 Effect of the Real Part: Polarization and Storage of Electromagnetic Energy

The physical origin of polarization Polarization phenomena are expressed by the quantity $\vec{P}$ which gives contribution of matter compared with that of a vacuum. The electric field and the polarization are linked through Maxwell's equations. The constitutive equation for vacuum is Eq. (8):

$$\vec{D} = \varepsilon_0 \vec{E} \tag{8}$$

where $\vec{D}$ is the electric displacement, $\vec{E}$ the electric field. According to Eq. (8), dielectric permittivity is the ratio of the electric displacement to the electric field. For a dielectric medium characterized by $\tilde{\varepsilon}$, the constitutive equation is Eq. (9):

$$\vec{D} = \tilde{\varepsilon}\vec{E} = \varepsilon_0 \vec{E} + \vec{P} \tag{9}$$

In the global formulation of Eq. (9), we can express the term corresponding to vacuum and given by Eq. (8). Then the second and complementary term defines the contribution of matter to polarization processes or polarization $\vec{P}$. For any material, the higher the dielectric permittivity, the greater are Brownian ion processes. The polarization process described by $\vec{P}$ has its physical origin in the response of dipoles and charges to the applied field. Depending on the frequency, electromagnetic fields induce oscillation of one or more types of charge association. In any material there are a variety of types of charge association:

- inner or core electrons tightly bound to the nuclei,
- valence electrons,
- free or conduction electrons,
- bound ions in crystals,
- free ions as in electrolytes and non stoichiometric ionic crystals (for example, ionic dipoles such as OH^- have both ionic and dipolar characteristics),
- finally the multipole (mainly the quadrupole or an antiparallel association of two dipoles).

✵ TOOLS More about photon–matter interaction   Depending on the frequency, the electromagnetic field can induce oscillation of one or more types of charges association. For each configuration with its own critical frequency above which interaction with the field becomes vanishingly small, the lower the frequency the more configurations are excited. Electrons of the inner atomic shells have a critical frequency approximately that of X-rays. Consequently an electromagnetic field of wavelength more than 10^{-10} m cannot excite any vibrations, but rather induces ionization of these atoms. There is no polarizing effect on the material which has, for this frequency, the same dielectric permittivity as in vacuum. For ultraviolet radiation the energy of photons is sufficient to induce transitions of valence electrons. In the optical range an electromagnetic field can induce distortions of inner and valence electronic shells. Polarization processes, called electronic polarizability, result from a dipolar moment induced by distortion of electron shells. Electromagnetic fields in the infrared range induce atomic vibrations in molecules and crystals, and polarization processes result from the dipolar moment induced by distortion of the positions of nuclei. These polarization processes are called atomic polarization. In all these processes the charges affected by the field can be regarded as being attracted towards their central position by forces proportional to their displacement by linear elastic forces. This mechanical approach of electronic resonance is only an approximation, because electrons cannot be properly treated by classical mechanics. Quantitative treatment of these processes requires the formalism of quantum mechanics. The two types of polarization process described above can be connected together in distortion polarization.

The characteristic material response times for molecular reorientation are 10^{-12} s. Electromagnetic fields in the microwave band thus lead to rotation of polar molecules or charge redistribution; the corresponding polarization processes are called orientation polarization.

Orienting effect of a static electric field   The general problem of the orienting effect of a static electric field (orientation of polar molecules) was first considered by Debye [22, 23], Frölich [24], and more recently by Böttcher [25, 26].

A collection of molecular dipoles in thermal equilibrium is considered. It is assumed that all the molecules are identical and they can take on any orientation.

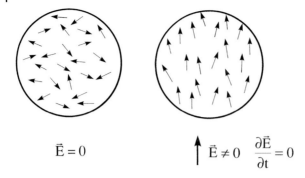

Fig. 1.4. Distribution of dipoles subjected to the effect of a static electric field.

Because of thermal energy each molecule undergoes successive collisions with the surrounding molecules. In the absence of an applied electric field, the collisions tend to maintain a perfectly isotropic statistical orientation of the molecules. This means that for each dipole pointing in one direction there is, statistically, a corresponding dipole pointing in the opposite direction as shown in Fig. 1.4.

In the presence of an applied electric field $\vec{E}$, the dipolar moment $\vec{\mu}$ of the molecule undergoes a torque $\vec{\Gamma}$. This torque tends to orientate the dipolar moment $\vec{\mu}$ parallel to the electric field. The corresponding potential energy (for a permanent or induced dipole) is minimum when the angle, θ, between the dipole and the electric field is zero. Consequently, the dipolar moment takes the same direction as the electric field. It is the same phenomena as the orientation of the compass needle in the earth's magnetic field. For molecular dipoles, however, the thermal energy counteracts this tendency and the system finally reaches a new statistical equilibrium which is depicted schematically in Fig. 1.4. In this configuration more dipoles are pointing along the field than before and the medium becomes slightly anisotropic.

The suitability of the medium to be frozen by the electric field is given by Langevin's function resulting from statistical theories which quantify competition between the orienting effect of electric field, and disorienting effects resulting from thermal agitation. The ratio of effective to maximum polarization versus the ratio of the potential interaction energy to the thermal agitation is shown in Fig. 1.5.

We can see that the Langevin function increases from 0 to 1 on increasing the strength of the electric field and/or reducing the temperature. The molecules tend to align with the field direction. For high values of the field, the orientation action dominates over the disorienting action of temperature, so that all the dipoles tend to become parallel to the applied field. Complete alignment corresponds to saturation of the induced polarization. Saturation effects only become detectable in fields of the order of 10^7 V m^{-1}. Because intermolecular distances are small in liquids and solids, however, the local field acting on molecule because of its neighbors may be very large, especially in strongly polar liquids (i.e. electric field close to

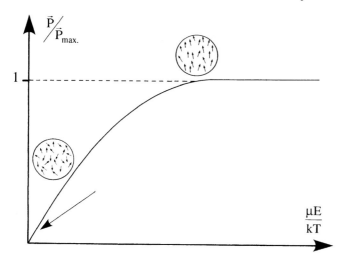

Fig. 1.5. Langevin's Function.

10^6 V m^{-1} at a distance of 5×10^{-10} m from a dipole of 1 Debye). The consequence of this should, however, be reduced, even for strongly polar liquids, because the intermolecular distance is of the same order as the molecular dimensions and it is not justified to describe the molecule as a point dipole. The situation is totally different for solids and, especially, solid surfaces for which the magnitude of the static electric field is close to saturation value. These strong static electric fields occur independently of all external electric excitation and are the physical origin of adsorption phenomena. In these circumstances adsorption can lead to freezing of molecular motion and can also induce polarization because of distortion of electronic shells. Apolar molecule can obtain polar character because of adsorption. Free molecules without dielectric losses at the operating frequency (2.45 GHz) have the capacity to be heated by microwave irradiation after adsorption on solids such as clays or alumina catalysts.

In many practical situations field strengths are well below their saturation values. The arrow on Fig. 1.5 corresponds to the usual conditions of microwave heating (no adsorption phenomena, temperature close to room temperature, 25 °C, and electric field strength close to 10^5 V m^{-1}). According to these results, the electric field strength commonly used in microwave heating is not sufficient to induce consequent freezing of media.

Calculation of the dielectric permittivity of an isotropic polar material involves the problem of the permanent dipole contribution to polarizability and the problem of calculation of the local field acting at the molecular level in terms of the macroscopic field applied. Debye's model for static permittivity considers the local field equal to the external field. This assumption is valid only for gases at low density or dilute solutions of polar molecules in nonpolar solvents. Several workers have proposed theories containing assumptions about relationships between local and external electric fields. Detailed analysis of these phenomena is beyond the scope of this section. More information can be found elsewhere [27].

1.1.2.2 Effect of Imaginary Part: Dielectric Losses

Physical origin of dielectric loss The foregoing conclusions correspond to a static description or cases for which the polarization can perfectly follow the oscillation of the electric field. Indeed, the electric field orientation depends on time with a frequency equal to 2.45 GHz (the electric field vector switches its orientation approximately every 10^{-12} s). The torque exercised by the electric field induces rotation of polar molecules, but they cannot always orient at this rate. The motion of the particles will not be sufficiently rapid to build up a time-dependent polarization $\vec{P}(t)$ that is in equilibrium with the electric field at any moment. This delay between electromagnetic stimulation and molecular response is the physical origin of the dielectric loss.

The polarization given by Eq. (9) becomes a complex quantity with the real part in phase with the excitation whereas the imaginary part has a phase lag with the excitation. This last part is the origin of the thermal conversion of electromagnetic energy within the irradiated dielectric.

◆ CONCEPTS More about delay and phase lag Matter does not respond instantaneously to stimulation induced by electromagnetic waves. In an isotropic medium this delay can be expressed by the specific formulation of polarization given by Eq. (10):

$$\vec{P} = \varepsilon_0 \int_{-\infty}^{+\infty} \chi(t-\tau) \vec{E}(\tau) \, d\tau \tag{10}$$

Because of the causality principle, Eq. (11), in which χ is the electric susceptibility, t the time, and τ the delay must be verified:

$$\chi(t-\tau) = 0, \quad t-\tau < 0 \tag{11}$$

The electric susceptibility can comprise any combination of dipolar, ionic, or electronic polarization processes. This formulation leads to relationships between the real and imaginary parts of the complex electric susceptibility, known as the Kramers–Kronig relationships [28–31] which are very similar to the frequency relations between resistance and reactance in circuit theory [30].

Consequently, after this short section on electric susceptibility, we shall always use classical elementary models which yield good results, as can be expected from correspondence principle.

Macroscopic theory of dielectric loss The main interest in dielectric theories is the frequency region at which the dispersion and absorption processes occur (the dipolar polarization can no longer change fast enough to reach equilibrium with the polarization field). When a steady electric field is applied to a dielectric, the distortion polarization (electronic and vibrational modes) will be established very quickly, essentially instantaneously compared with characteristic time of electric field. The remaining dipolar part or orientation polarization takes time to reach equilibrium state. Relaxation processes are probably the most important of the interactions between electric fields and matter. Debye [22, 23] has extended the Langevin theory of dipole orientation in a constant field to the case of a varying field. It shows that the Boltzmann factor of the Langevin theory becomes a time-dependent weighting factor. A macroscopic description, more usable, can use an exponential law with a macroscopic relaxation time, τ, or the delay in the response of the medium to the electric stimulation given by Eq. (12):

$$\vec{P}_{\text{Orientation}} = (\vec{P}_{\text{Total}} - \vec{P}_{\text{Distorsion}})\left(1 - \exp\left(-\frac{t}{\tau}\right)\right) \qquad (12)$$

Similarly, when the electric stimulation is removed, the distortion polarization falls immediately to zero whereas the distortion polarization falls exponentially. If the electric stimulation oscillates with time (angular frequency, ω; electric field strength, $\vec{E}_0$) as described by Eq. (13):

$$\vec{E} = \vec{E}_0 \exp(jwt) \qquad (13)$$

The static permittivity, ε_S, (frequency close to zero) and very high-frequency permittivity ε_∞ could be defined in term of total polarization and distortion polarization as described by Eqs (14) and (15):

$$\vec{P}_{\text{Total}} = \vec{P}_{\text{Distorsion}} + \vec{P}_{\text{Orientation}} = (\varepsilon_S - \varepsilon_0)\vec{E} \qquad (14)$$

$$\vec{P}_{\text{Distorsion}} = (\varepsilon_\infty - \varepsilon_0)\vec{E} \qquad (15)$$

According to the exponential law defined for orientation polarization (Eq. 9), the following differential equation could be defined:

$$\frac{d\vec{P}_{\text{Orientation}}}{dt} = \frac{(\vec{P}_{\text{Total}} - \vec{P}_{\text{Orientation}})}{\tau} = \frac{(\varepsilon_S - \varepsilon_\infty)\vec{E}_0 \exp(j\omega t) - \vec{P}_{\text{Orientation}}}{\tau} \qquad (16)$$

The ratio of orientation polarization to electric field becomes a complex quantity. This means that the dipolar part of the polarization is out of phase with the field, as depicted by Eq. (17):

$$\vec{P}_{\text{Total}} = \vec{P}_{\text{Orientation}} + \vec{P}_{\text{Distorsion}}$$

$$= (\varepsilon_\infty - \varepsilon_0)\vec{E}_0 \exp(j\omega t) + \frac{(\varepsilon_S - \varepsilon_0)\vec{E}_0 \exp(j\omega t)}{1 + j\omega t} \qquad (17)$$

Debye's model The Debye model could be built with these assumptions, and polarization and permittivity become complex as described by Eq. (18) where n is the refractive index and τ the relaxation time:

$$\tilde{\varepsilon} = \varepsilon' - j\varepsilon'' = n^2 + \frac{\varepsilon_s - n^2}{1 + j\omega\tau} \tag{18}$$

All polar substances have a characteristic time τ called the relaxation time (the characteristic time of reorientation of the dipolar moments in the electric field direction). The refractive index corresponding to optical frequencies or very high frequencies is given by Eq. (19):

$$\varepsilon_\infty = n^2 \tag{19}$$

whereas ε_S is the static permittivity or permittivity for static fields.

The real and imaginary parts of the dielectric permittivity of Debye's model are given by Eqs. (20) and (21):

$$\varepsilon' = n^2 + \frac{\varepsilon_s - n^2}{1 + \omega^2\tau^2} \tag{20}$$

$$\varepsilon'' = \frac{\varepsilon_s - n^2}{1 + \omega^2\tau^2} \omega\tau \tag{21}$$

Changes of ε' and ε'' with frequency are shown in Fig. 1.6. The frequency is displayed on a logarithmic scale. The dielectric dispersion covers a wide range of frequencies. The dielectric loss reaches its maximum given by Eq. (22):

$$\varepsilon''_{max} = \frac{\varepsilon_s - n^2}{2} \tag{22}$$

at a frequency given by Eq. (23):

$$\omega_{max} = \frac{1}{\tau} \tag{23}$$

This macroscopic theory justifies the complex nature of the dielectric permittivity for media with dielectric loss. The real part of the dielectric permittivity expresses the orienting effect of electric field, with the component of polarization which follows the electric field, whereas the other component of the polarization undergoes chaotic motion leading to thermal dissipation of the electromagnetic energy. This description is well adapted to gases (low density of particles). For a liquid, however, we must take into account the effect of collisions with the surroundings, and the equilibrium distribution function is no longer applicable.

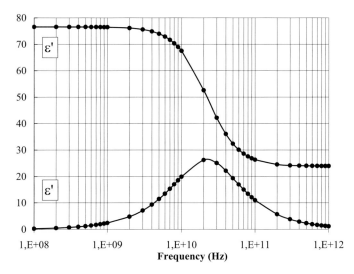

Fig. 1.6. Dependence of the complex dielectric permittivity on frequency (ε' and ε'' are, respectively, the real and imaginary parts of the dielectric loss at 25 °C; dielectric properties are: $\varepsilon_{Sr} = 78.2$, $\varepsilon_{\infty} = 5.5$, and $\tau = 6.8 \times 10^{-12}$ s).

☛ **CONCEPTS** **More about the effect of collisions on distribution functions: microscopic theory of dielectric loss** The Debye theory can define a distribution function which obeys a rotational diffusion equation. Debye [22, 23] has based his theory of dispersion on Einstein's theory of Brownian motion. He supposed that rotation of a molecule because of an applied field is constantly interrupted by collisions with neighbors, and the effect of these collisions can be described by a resistive couple proportional to the angular velocity of the molecule. This description is well adapted to liquids, but not to gases.

Molecular orientations can be specified by ϕ and θ. The fraction of molecules whose dipole moments lie in an element of solid angle $d\Omega$ is $f(\phi, \theta)\,d\theta$. The number of representative points which pass across unit length of θ in unit time is described by Eq. (24):

$$\frac{\partial q}{\partial \theta} = -K\frac{\partial f(\theta)}{\partial \theta} + f(\langle \dot{\theta} \rangle) \qquad (24)$$

where the first term describes a diffusive process with a specific constant K and the second term describes the effect of the electric field which sets the molecules in rotation with an average terminal angular velocity depending of the orientating couple and on resistive constant or damping constant of inner friction given by Eq. (25):

$$\zeta(\langle \dot{\theta} \rangle) = \frac{-\partial p\vec{E}\cos\theta}{\partial \theta} \qquad (25)$$

At equilibrium, molecular energies will be distributed according to Boltzman's law and, finally, the general formulation which defines the factor f is given by Eq. (26):

$$\frac{\partial f}{\partial \theta} = \frac{1}{\zeta \sin \theta} \frac{\partial}{\partial \theta} \left(kT \sin \theta \frac{\partial f}{\partial \theta} + fpE \sin^2 \theta \right) \tag{26}$$

The distribution function of the f factor is given by Eq. (27):

$$f = 1 + \frac{pE \cos \theta}{kT(1 + j\omega\tau)} \tag{27}$$

So the average moment in the direction of the field is given by Eq. (28) which can define the microscopic relaxation time that depends on the resistive force experienced by the individual molecules (for more details, see MacConnell [32]).

$$\langle p \cos \theta \rangle = \frac{\int_0^\pi \cos \theta \left(1 + \frac{pE \cos \theta}{kT(1 + j\omega\tau)} \right) 2\pi \sin \theta \, d\theta}{\int_0^\pi \left(1 + \frac{pE \cos \theta}{kT(1 + j\omega\tau)} \right) 2\pi \sin \theta \, d\theta} \tag{28}$$

The general equation for the complex dielectric permittivity is, then, given by Eq. (29):

$$\frac{\tilde{\varepsilon}_r - 1}{\tilde{\varepsilon}_r + 2} = \frac{\rho N}{3\varepsilon_0 M} \left(\alpha + \frac{\mu^2}{3kT(1 + j\omega\tau)} \right) \tag{29}$$

where N is the Avogadro number, M is the molar mass, ρ the specific mass, and α the atomic polarizability. The relaxation time τ is a microscopic relaxation time that depends on the average resistive force experienced by the individual molecules. In the limit of low frequency the Debye expression is obtained for the static permittivity whereas in the high frequency limit the permittivity will fall to a value which may be written as the square of the optical index (see Table 1.2, Section 1.1.2). The first term of the left side corresponds to distortion polarization whereas the other term corresponds to orientation polarization. For apolar molecules, we obtain the famous Clausius–Mosotti–Lorentz equation.

Relaxation times Debye [22, 23] suggested that a spherical or nearly spherical molecule could be treated as a sphere (radius r) rotating in a continuous viscous medium of bulk viscosity η. The relaxation time is given by Eq. (30):

$$\tau = \frac{8\pi\eta r^3}{2kT} \tag{30}$$

The relaxation time evaluated from experimental measurements is the effective time constant for the process observed in the medium studied, even for solutions.

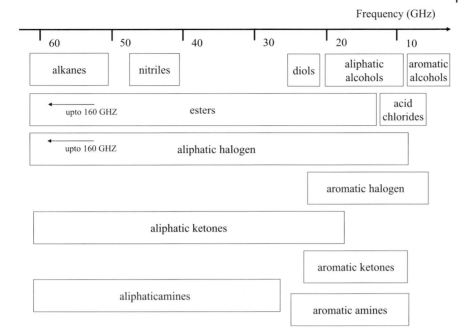

Fig. 1.7. Relaxation time range for classical organic functions.

Because of the incidence of the internal field factor, this is not the value of the molecular dipole relaxation. Depending upon the internal field assumption a variety of relationships between theoretical and effective relaxation times have been defined. Relaxation times for dipole orientation at room temperature are between 10^{-10} s for small dipoles diluted in a solvent of low viscosity and more than 10^{-4} s for large dipoles in a viscous medium such as polymers (polyethylene) or dipole relaxations in crystals (the relaxation associated with pairs of lattice vacancies).

The relaxation times of ordinary organic molecules are close to a few picoseconds. Figure 1.7 gives relaxation frequency range for classical organic functions: alkanes [33, 34], alcohols [35–39], alcohol ether [40], acid chlorides [41, 42], esters [43, 44], aliphatic [45–54] and aromatic halogens [55, 56], aliphatic [57, 58], aromatic ketones [59], nitriles [60], and aliphatic [61, 62] and aromatic amines [63].

Thus, for a frequency of 2.45 GHz these molecules can follow electric field oscillations, unlike substances which are strongly associated, for example water and alcohols, and therefore have dielectric loss at 2.45 GHz. Consequently the solvents which have dielectric loss are water, MeOH, EtOH, DMF, DMSO, and CH_2Cl_2. Dielectric loss is negligible for nonpolar solvents such as C_6H_6, CCl_4, and ethers, although addition of small amounts of alcohols can strongly increase the dielectric loss and microwave coupling of these solvents.

Effect of temperature Relaxation data for pure water play an important role in discussion of the dielectric behavior of aqueous solutions. Another practical interest is

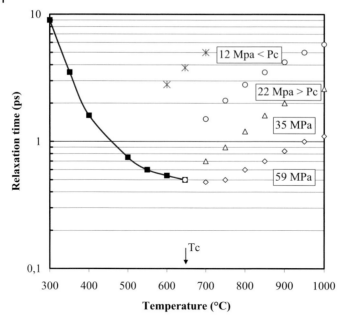

Fig. 1.8. Effect of temperature and pressure on the dielectric relaxation time of pure water. The arrow indicates the critical temperature.

the demand for dielectric reference materials suitable for calibrating and checking the performance of equipment for measurement of dipolar liquids. The design of microwave moisture measurement systems for food and other materials is another interest. The thermal and pressure dependency of the relaxation time of pure water is shown by Fig. 1.8 [64].

The critical temperature and pressure are $T_c = 647$ K and $P_c = 22.1$ MPa. In the liquid state (continuous curve) the relaxation time decreases rapidly with increasing temperature irrespective of pressure. In the gaseous state, the relaxation time is strongly pressure-dependent (positive temperature-dependence at constant pressure). The relaxation time jumps to a larger value at the boiling temperature when the pressure is lower than the critical pressure. More generally, the most relevant property at lower temperatures or higher densities is the temperature whereas density is more important at higher temperatures or lower densities. The relaxation time increases with decreasing density. The microwave field can hardly change the thermal motion of water molecules as far as they are rotating. This situation can be easily understood by considering our common experience that rapidly rotating top remains standing against gravity. The orientation of the dipolar moment may be changed when the molecule lose angular momentum, owing to collisions with other molecules. In gaseous phase, the dielectric relaxation is governed by binary collisions. Recently molecular dynamics simulations have been performed to study the dielectric properties of supercritical water. These results show that

Debye's model assumptions are not valid for supercritical water, because the dilute limit is doubted. Microscopically there are many degrees of freedom and all these motions are not totally decoupled from the others, because the eigenstate of motions are not well known the structurally disordered matter. Dielectric measurements can only probe slow dynamics which can be described by stochastic processes and classical Debye's model could be rationalized [65, 66].

The temperature and pressure dependence for methanol is quite similar to that for water. Although alcohol molecules also form hydrogen bonds, the maximum coordination is two, unlike water. The methanol molecule can form both chain and ring structures, in the same way as liquid selenium [67]. Breaking of the hydrogen-bond network occurs because of librational motion for water and stretching for methanol and most other alcohols.

At high densities, for example supercritical conditions, experimental relaxation time deviates strongly from Debye's values, owing to hydrogen bonding. In the gaseous phase free molecules are responsible for the classical dielectric relaxation and molecules incorporated within the hydrogen-bond network should be added; the general relaxation function for supercritical fluids with hydrogen bonds is given by Eq. (31):

$$\tilde{\varepsilon} = \varepsilon_\infty + (\varepsilon_S - \varepsilon_\infty)\left(\frac{1-\alpha}{1+j\omega\tau_{free}} + \frac{\alpha}{1+j\omega\tau_{bound}}\right) \tag{31}$$

where α is the fraction of bound molecules and τ_{free} and τ_{bound} and the relaxation times for free and bound molecules, respectively. The average relaxation time is given by Eq. (32):

$$\tau = (1-\alpha)\tau_{free} + \alpha\tau_{bound} \approx \tau_{free} + \alpha\tau_{bound} \quad \text{if } \alpha \ll 1 \tag{32}$$

τ_{free} can be assumed to be the binary collision time given by Eq. (33):

$$\tau_{free} = \frac{1}{4n\pi r_{eff}^2}\sqrt{\frac{m\pi}{k_B T}} \tag{33}$$

where m is the mass of the molecule, n the number density, and r_{eff} is the effective radius hard sphere diameter of molecule, equal to intramolecular distance from Raman scattering data. In the gaseous state α becomes small and eventually vanishes in the dilute limit whereas at low temperatures in the liquid state it is replaced by an enhancement factor, because of the highly correlated nature of molecular motion. Hydrogen-bonding enthalpy (vibration or stretching energy) can be obtained from Eq. (34):

$$\tau_{bound} = \tau \exp\left(\frac{\Delta H}{k_B T}\right) \tag{34}$$

For water, a liquid characterized by a discrete relaxation process in the temperature

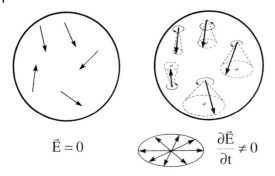

Fig. 1.9. Precession motion of the dipole of a distribution of molecules undergoing irradiation by a time dependent electric field.

range of interest, this activation enthalpy is 4.9 kcal mol^{-1}, or the hydrogen-bond energy of water.

Dynamic consequences of dielectric losses It is clear that for a substance with dielectric loss, for example water and the alcohols, the molecules do not perfectly follow the oscillations of the electric field. For media without dielectric loss, and for the same reasons as under static conditions, the strength of the electric field cannot induce rotation for all polar molecules, but statistically only for a small part (less than 1%). This means that all the molecules oscillate round an average direction (precession motion), as shown in Fig. 1.9.

The principal axis of the cone represents the component of the dipole under the influence of thermal agitation. The component of the dipole in the cone is because of the field that oscillates in its polarization plane. In this way, the dipole follows a conical orbit if Brownian movement is prevented. In reality the cone changes its direction continuously because of Brownian movement faster than the oscillation of the electric field which leads to chaotic motion. Hence the structuring effect of electric field is always negligible because of the value of the electric field strength, even more so for lossy media.

In condensed phases it is well known there are energy transfers between rotational and vibrational states. Indeed, molecular rotation does not actually occur in liquids – rotational states turn into vibrational states because of an increase in collisions. For liquids, the collision rate is close to 10^{30} collisions per second. Microwave spectroscopic studies of molecular rotation only use dilute gases to obtain pure rotational states with a sufficient lifetime. Broadening of rotational transitions induced by molecular collisions occurs because pressures are close to a few tenths of a Bar as described for Fig. 1.10.

In conclusion, for condensed phases molecular rotations have quite a short lifetime because of collisions. The oscillations eventually induced by the electric field are dissipated in liquid state leading to vibration. At densities of the collisions corresponding to those in liquids, the frequency of the collisions become comparable

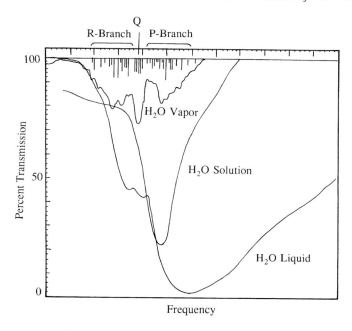

Fig. 1.10. Absorption spectrum for water (gaseous, solution, and liquid). Above the vapor band is the Mecke rotational analysis [68, 69].

with the frequency of a single rotation and, because the probability of a change in rotational state on collision is high, the time a molecule exists in a given state is small. From these remarks it is obvious that the electric field cannot induce organization in condensed phases such as the liquid state.

1.1.2.3 Thermal Dependency of Dielectric Permittivity

In contrast with Eq. (18), Eq. (29) gives the frequency behavior in relation to the microscopic characteristics of the studied medium (polarizability, dipolar moment, temperature, frequency of the field, etc). Then, for a given change of the relaxation time with temperature, we can obtain the change with frequency and temperature of the dielectric properties – the real and imaginary parts of the dielectric permittivity. In fact, for a given molecular system, it is better to use a formula containing $\tau_{\text{Inter}}(T)$, a part which depends on the temperature, and a part totally independent of the temperature, τ_{Steric}, as in Eq. (35):

$$\tau(T) = \tau_{\text{Steric}} + \tau_{\text{Inter}}(T) \tag{35}$$

Depending on the frequency of the field and the relaxation time band in relation to the temperature considered, one can observe three general changes of dielectric properties with temperature. Figure 1.11 gives the three-dimensional curves depicting the dependence dielectric properties on frequency and temperature [70].

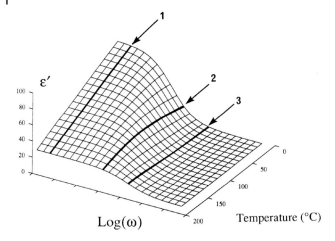

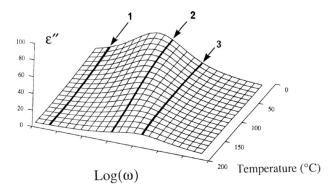

Fig. 1.11. Change of the complex dielectric permittivity with frequency and temperature (ε' is the real part and ε'' is the imaginary part of the dielectric loss) [70].

Depending on the values of the working frequency and the relaxation frequency, three general cases can be observed:

- case 1, where the real and imaginary parts of the dielectric permittivity decrease with temperature (working frequency lower than relaxation frequency);
- case 2, where the real and imaginary parts of the dielectric permittivity increase with temperature (working frequency higher than relaxation frequency); and
- case 3, where the real and/or imaginary part of the dielectric have a maximum (working frequency very close to relaxation frequency).

The two solvents most commonly used in microwave heating are ethanol and water. Values for water are given by Kaatze [71, 72] and values for ethanol by Chahine et al. [73]. Water fits most closely with case 1 because both values decrease with temperature. In contrast, for ethanol the real part increases and the dielectric loss reaches a maximum at 45 °C (case 2). For ethanol, in fact, the working frequency is higher than the relaxation frequency at room temperature. Ethanol has a single relaxation frequency close to 1 GHz at 25 °C and, furthermore, its relaxation frequency rises fairly rapidly with temperature (3 GHz at 65 °C). For water the working frequency is smaller than the relaxation frequency at all temperature (17 GHz at 20 °C and 53 GHz at 80 °C).

The pioneering work of Von Hippel [74] and his coworkers, who obtained dielectric data for organic and inorganic materials, still remains a solid basis. Study of dielectric permittivity as a function of temperature is less well developed, however, particularly for solids.

1.1.2.4 Conduction Losses

For highly conductive liquids and solids the loss term not only results from a single relaxation term as given by Eq. (21) but also from a term containing ionic conductivity, σ, as shown by Eq. (36):

$$\varepsilon'' = \frac{\varepsilon_s - n^2}{1 + \omega^2 \tau^2} \omega \tau + \frac{\sigma}{\omega} \tag{36}$$

A conducting material can be regarded as a nonconducting dielectric with resistance in parallel. The alternative graphical representation as a plot of the logarithm of dielectric losses against the logarithm of the frequency enables the a.c. conductivity associated with the relaxation dipoles to be distinguished easily from the d.c. conductivity arising from the free charges. From Eq. (36) two different ranges could be defined as shown in Fig. 1.12:

$$\text{for } \omega\tau \ll 1 \quad \varepsilon'' = \frac{\sigma}{\omega} \quad \text{range } I \tag{37}$$

$$\text{for } \omega\tau \gg 1 \quad \varepsilon'' = \frac{\varepsilon_S - \varepsilon_\infty}{\omega t} \quad \text{range } II \tag{38}$$

The second term of Eq. (36) is usually small compared with the first (typical values: $\sigma = 10^{-8}$ S, $\tau = 10^{-10}$ s, $\sigma\tau = 10^{-18}$ Ss). This is quite small compared with first term, which is of the order of 10^{-11}; it can therefore be neglected.

The hydroxide ion is a typical example of an ionic species with both ionic and dipolar characteristics. For solutions containing large amount of ionic salts this conductive loss effect can become larger than the dipolar relaxation.

For solids, conduction losses are usually very slight at room temperature but can strongly change with temperature. A typical example is alumina with very slight

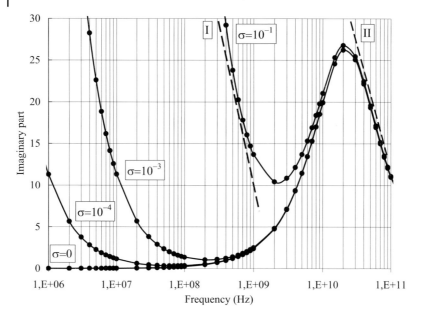

Fig. 1.12. Dependence on conductivity of change of dielectric losses.

dielectric losses at room temperature (close to 10^{-3}) and can reach fusion temperatures in several minutes in a microwave cavity [75]. This effect is a consequence of a strong increase of conduction losses associated with thermal activation of the electrons which pass from the oxygen 2p valence band to the 3s3p conduction band. In solids, moreover, conduction losses are usually enhanced by material defects which sharply reduce the energy gap between the valence and conduction bands. Because conduction losses are high for carbon black powder, the material can be used as lossy impurities or additives to induce losses within solids for which dielectric losses are too small. This trick has long been used by people using microwaves for heating applications and explains problems encountered with chocolate in microwave cookers. Chocolate contains lipid polymers with strong microwave losses. On microwave heating, chocolate degraded very quickly leading to carbon black. This increases local microwave heating. As a result microwave heating of chocolate can quickly induce a strong burning taste.

Although conductivity is usually a thermally activated process, as given by Eq. (39):

$$\sigma(T) = \sigma_0 \exp\left(-\frac{U}{k(T - T_0)}\right) \tag{39}$$

where U is the activation energy and σ_0 the conductivity at T_0, Joule heating within the sample cannot be removed quickly enough by conduction and/or convection, so the temperature of the sample increases to the fusion temperature. The temperature dependence of solid static permittivity can lead to a large increase of dielectric permittivity just below the melting point for most of crystalline or amorphous (glass) materials.

1.1.2.5 Magnetic Losses

Chemical reagents are primarily dielectric liquids or solids. Magnetic losses are, however, observed for microwave-irradiated metal oxides, for example ferrites. As for dielectrics, a complex magnetic permeability is defined as given by Eq. (40):

$$\tilde{\mu} = \mu' - j\mu'' \tag{40}$$

The real part is the magnetic permeability whereas the imaginary part is the magnetic loss. These losses are quite different from hysteresis or eddy current losses because they are induced by domain walls and electron spin resonance. These materials should be placed at the position of magnetic field maxima for optimum absorption of microwave energy. For transition metal oxides, for example those of iron, nickel, and cobalt, magnetic losses are high. These powders can thus be used as lossy impurities or additives to induce losses within solids for which dielectric losses are too small.

1.1.3
Dielectric Properties and Molecular Behavior

1.1.3.1 Dielectric Properties Within a Complex Plane

The Argand diagram Another graphical representation of substantial interest is obtained by plotting the imaginary part against the real part – the Argand diagram.

The function can be obtained by elimination of ω between Eqs (20) and (21). For simple dipole relaxation, a circle is obtained:

$$\left(\varepsilon' - \frac{\varepsilon_S - \varepsilon_\infty}{2}\right)^2 + (\varepsilon'')^2 = \left(\frac{\varepsilon_S - \varepsilon_\infty}{2}\right)^2 \tag{41}$$

The dielectric permittivity is represented by the semicircle of radius:

$$r = \frac{\varepsilon_S - \varepsilon_\infty}{2} \tag{42}$$

centered at:

$$\varepsilon' = \frac{\varepsilon_S + \varepsilon_\infty}{2} \tag{43}$$

The top of this semicircle corresponds to $\tau\omega = 1$.

30 | *1 Microwave–Material Interactions and Dielectric Properties*

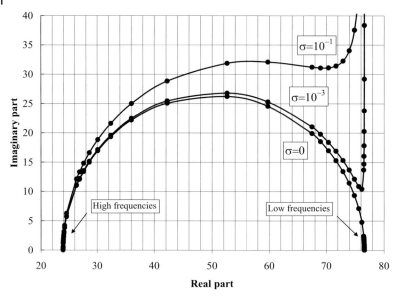

Fig. 1.13. Argand's diagram for different values of conductivity (Sm^{-1}).

This plot of experimental values is a convenient graphical test of the applicability of Debye's model. The effect of the last term on the shape of the diagram can be seen in Fig. 1.13. The larger the conductivity, the further the actual diagram departs from Debye's semicircle.

The Cole–Cole model Argand's diagram for many polar molecules in the liquid phase is actually a semicircle as predicted by Debye's model. Typical examples are pure alcohols or symmetrical molecules such as chlorobenzene in a nonpolar solvent (alkane). Many plots deviate from Argand's plot. This deviation is usually explained by assuming there is not just one relaxation time but a continuous distribution. Long molecules for which the permanent dipole moment is not aligned with the long molecular axis and polymers have broader dispersion curves and lower maximum loss than would be expected from Debye relationships. If the molecule is aligned with the field, only the longitudinal component of the dipole moment is active during the relaxation process. The molecule tends to rotate about a short molecular axis with a long relaxation time because of inertial and viscous forces. In contrast, if the molecule is perpendicular to the field the transverse component of the dipole is active and the molecule relaxes by rotating rather quickly about its long axis, because inertial and viscous forces are smaller in this configuration. If the molecules are randomly oriented relative to the field, the corresponding relaxation time is distributed between these two extreme cases. If $f(\tau)$ is the distribution function of the relaxation time between τ and $d\tau$, the corresponding Eq. (44) for dielectric loss is:

$$\varepsilon'' = \varepsilon_\infty + (\varepsilon_S - \varepsilon_\infty) \int_0^\infty \frac{f(\tau)\,d\tau}{1 + j\omega\tau} \tag{44}$$

Because this leads to circular arc centered below the axis, Cole and Cole have proposed a modified form of Debye's formula with a term h characterizing the flattening of the diagram (Eq. 45) ($h = 0$ corresponds to the classical Debye model):

$$\varepsilon'' = \varepsilon_\infty + \frac{\varepsilon_S - \varepsilon_\infty}{1 + (j\omega\tau)^{1-h}} \quad \text{with } 0 \leq h \leq 1 \tag{45}$$

The value of h found experimentally has a tendency to increase with increasing number of internal degrees of freedom in the molecules and with decreasing temperature [76]. The high relaxation is associated with group rotation and molecular tumbling. Normalized skewed arc plots give evidence of asymmetric distribution of relaxation time. The value of h increases with decreasing chain length, i.e. the distribution of relaxation time tends toward symmetric distribution with decreasing chain length. Skewed arc behavior in liquids has been reported by many workers and has been explained in terms of cooperative phenomenon and multiple relaxation processes. The molecule becomes less rigid with increasing chain length and can relax in more than one way. The different groups may rotate, as also may the whole molecule. The relaxation time for the former process is smaller than that for the latter. The intramolecular process has similar effects to the intermolecular cooperative phenomenon observed in pure polar liquids.

The Cole–Davidson model These kinds of diagram are also symmetrical or non-symmetrical and may be fairly described by an analytical relationship proposed by Davidson and Cole [77] (Eq. 46):

$$\varepsilon'' = \varepsilon_\infty + \frac{\varepsilon_S - \varepsilon_\infty}{(1 + j\omega\tau)^\alpha} \quad \text{with } 0 \leq \alpha \leq 1 \tag{46}$$

When α is close to unity this again reduces to Debye's model and for α smaller than unity an asymmetric diagram is obtained. The Cole–Cole diagram arise from symmetrical distribution of relaxation times whereas the Cole–Davidson diagram is obtained from a series of relaxation mechanisms of decreasing importance extending to the high-frequency side of the main dispersion.

Glarum's generalization Glarum [78] has suggested a mechanism which leads to a dispersion curve barely distinguishable from the empirical skewed-arc of Davidson and Cole. Glarum suggests that dipole relaxation occurs by two coexistent mechanisms. Because of lattice defects in the liquid or solid, a dipole can adapt its orientation almost instantaneously to an electric field. The presence of a hole might drastically reduce the resistance to rotation. At the same time, dipole relaxation can occur without the help of defects. Glarum believes these two processes are independent and the general correlation function is the product of the correlation function of the two mechanisms, assuming that the motion of defects is governed by a diffusion equation. If the relaxation because of arrival of a defect is rare, classical Debye relaxation is obtained ($\alpha = 1$). If defect diffusion is the dominant process, a circular arc is obtained with $\alpha = 0$. If the two processes coexist, a Cole–Davidson term $\alpha = 0.5$ is obtained. Glarum's theories have been extended by Anderson and Ulman [79]. They assumed that the orientation process is function of an environmental property called the free volume. These theories raise the possibility of deducing the rate of fluctuation of environmental conditions from dielectric measurements.

Molecules with two or more polar groups Molecules comprising of skeleton with two polar groups give two adsorptions overlapping significantly on the frequency scale. If we use subscripts 1 and 2 for the lower and higher-frequency relaxation, six quantities must be evaluated. If the processes are quite independent these can be unequivocally established from the experimental data. Coexistence of two classical Debye relaxations is described by Eqs (47) and (48):

$$\varepsilon' = \varepsilon_\infty + \frac{\varepsilon_S - \varepsilon_{\infty 1}}{1 + \omega^2 \tau_1^2} + \frac{\varepsilon_S - \varepsilon_{\infty 2}}{1 + \omega^2 \tau_2^2} \tag{47}$$

$$\varepsilon'' = \frac{\varepsilon_S - \varepsilon_{\infty 1}}{1 + \omega^2 \tau_1^2} \omega \tau_1 + \frac{\varepsilon_S - \varepsilon_{\infty 2}}{1 + \omega^2 \tau_2^2} \omega \tau_2 \tag{48}$$

Within the complex plane, two circles are obtained. The overlapping of these two circles depends on the vicinity of the relaxation time or relaxation frequency of the two polar groups. This assumption could be applied to more than two polar groups. Are there two isolated Debye's relaxations or a distribution of relaxation times for a single relaxation process? If the latter, it is better to use the Cole and Cole or Davidson and Cole models. Results from permittivity measurements are often displayed in this type of diagram. The disadvantage of these methods is that the frequency is not explicitly shown.

1.1.3.2 Dielectric Properties of Condensed Phases

In this section dipole moment values and complex dielectric permittivity are surveyed more particularly in terms of their frequency dependency for a variety of liquid and solid state systems. The varieties of dielectric phenomena encountered are described briefly. They are selected to illustrate relationships between dielectric data and the structure and behavior of molecular units.

In contrast with condensed phases, intermolecular interactions in gases are negligibly small. The dipole moment found in the gas phase at low pressure is usually accepted as the correct value for a particular isolated molecule. The molecular dipole moment calculated for pure liquid using Debye's model gives values which are usually very different from those obtained from gas measurements. Intermolecular interactions in liquids produce deviations from Debye's assumptions.

Short-range interactions lead to a strong correlation between the individual molecules and enhance the polarization. Hydrogen bonding aligns the molecules either in chain-like structures (e.g. water or alcohols) or in antiparallel arrangements (e.g. carboxylic acids). The atomic polarization increases and in the second example the orientation polarization decreases as a result of mutual cancellation of the individual molecular dipoles, leading to liquid permittivity smaller than values calculated for nonassociated liquids. If intermolecular interactions in gases are negligibly small, the Debye model enables adequate representation of the relationship between polarization and molecular dipole moment. The dipole found in the gas phase is usually accepted as the correct value [80]. The dipole moment of chlorobenzene is 1.75 D measured in the gas phase whereas it is 1.58 D in benzene, 1.68 D in dioxane and 1.51 D in carbon disulfide. In dilute solution the solution's molar polarization could be expressed as the weighted sum of the molar polarization of the individual components. Significant solute–solute effects are still present even at high dilutions.

Pure liquids – water and alcohols Water and peroxides (HO–OH) represent a limiting state of such interactions. In the liquids state, water molecules associate by hydrogen-bond formation. Despite its apparently complex molecular structure, because of its strong association, water closely followed simple Debye relaxation (at 25 °C: $\varepsilon_{Sr} = 78.2$, $\varepsilon_{\infty r} = 5.5$, $\tau = 6.8 \times 10^{-12}$ s, and $h = 0.02$, the Cole–Cole term). In such systems the molecules have usually been rigid dipoles without interaction with neighbors. The situation significantly changes when hydroxyl groups are considered. Such molecules have a dipolar group with appreciable mobility. For alcohols or phenols the hydroxyl group can rotate about the axis of the oxygen–carbon bond and can relax intramolecularly. In the liquid phase, however, hydroxyl groups of different molecules can interact forming hydrogen bonds which link molecules. The alcohols 1- and 2-propanol give almost ideal semicircles in Argand's plot. Skewed-arc is found for halogenated alkane derivatives. The skewed-arc pattern is characteristic for higher alcohols. For solutions of electrolytes in methanol and ethanol the decrease of permittivity is even more marked than that observed for water. Calculations suggest that the ionic field is probably not effective beyond the first solvation layer. Maybe for the alcohols ion solvation and its local geometrical requirements lead to proportionally greater disruption of the hydrogen-bond chains than in aqueous solutions.

Effects of electrolytes Relaxation processes are represented by several types of Argand diagram, for example the Debye diagram (I), the Cole–Cole diagram (II), the Davidson–Cole diagram (III) or a diagram with a few separated Debye regions. Original studies of aqueous salt solutions by Hasteed afforded diagrams I and II only [81, 82]. Since this pioneering work, experimental evidence suggests the occurrence of four types of diagram for ionic solutions. Unfortunately, experimental data are meager and dielectric measurements for these kinds of solution are quite limited.

Recent work has led to precise measurement of complex dielectric permittivity over a wide frequency range which covers the decimeter, centimeter, and millimeter spectral regions (7–120 GHz [83–91]). Knowledge of these data is even more crucial when moving to study of concentrated solutions and solutions at elevated temperature. In these, ion–water, water–water, and ion–ion interactions are more diverse and the corresponding relaxation processes become more complicated. The contribution to ionic losses of electric conductivity also increases, specially in the centimeter wavelength range. Increasing the temperature and concentration of the electrolytes induce an increase in relaxation frequency or a decrease in relaxation time, adding noticeable error to the determination. Experimental evidence from measurements in a wide frequency range makes it possible to describe the whole dispersion region. At high frequencies the contribution from electrical conductivity is smaller.

Figure 1.14 shows the change with frequency of the real and imaginary parts of dielectric permittivity for aqueous solutions of NaCl of high concentration. The d.c. conductivity has been avoided by the authors, who made measurements at 1 kHz using conventional methods. Addition of electrolytes usually increases conductivity, as shown in Fig. 1.15 (arrows indicate values of dielectric losses at 2.45 GHz). No resonance or relaxation processes other than the Debye rotational diffusion of water molecule occur in the high-frequency part of the millimeter wavelength range. This gives evidence that ionic losses are described by Eq. (36). The structural model of an electrolyte solution that reflects different water molecule dynamics at

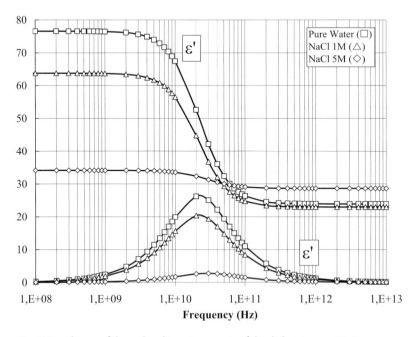

Fig. 1.14. Change of the real and imaginary parts of the dielectric permittivity.

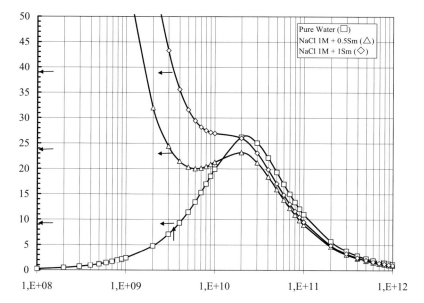

Fig. 1.15. Effect of ionic conductivity on dielectric losses.

different concentrations can be represented by the following description. In the dilute regions, water molecules can be divided into those whose state is modified by the presence of ions (hydration shell of cations and anions) and those that retain all the properties of pure water (bulk water). In highly concentrated solutions, in which bulk water is absent, the model implies the occurrence of two states of water molecules – water molecules bound by one ion, a fragment of the hydration shell, and molecules shared by cations to form ion–water clusters [92–102]. These different structural subsystems should be found within the dispersion region. LiCl and $MgCl_2$ solutions lead to two simple Debye dispersions (free and hydration-shell molecules).

First, depending on concentration ions may associate producing ion-pairs or similar solute species with an appreciable dipole moment. Such species will make their own contribution to dielectric relaxation processes. Because of their strong localized electric field, ions affect the solvent's molecular interactions. Addition of sodium chloride to water leads to a strong decrease of the real part of the dielectric permittivity equivalent to a temperature increase (e.g. 0.5 M sodium chloride at 0 °C has the same value as pure water at 30 °C). The value of relaxation time is shifted in the same sense but to a far smaller extent. The sodium chloride ions markedly change the geometric pattern of molecular interactions. According to X-ray and neutron diffraction studies, the electrostatic field neighborhood of a sodium or chloride ion is such that the interaction energy with the water molecule dipole greatly exceeds that of typical hydrogen bond between the solvent species. This means that an appreciable number of water molecules will be frozen around each ion – a change in molecular pattern from the liquid. The freezing of these water molecules forming a hydrating sheath around the ions means their dipoles are not free to re-orientate in the applied electric field. This causes the permittivity decrease. This effect has been quantitatively defined by Hasted et al. [81, 82] – for 3 M sodium chloride solution the value is one-half that for the pure solvent (e.g. nine water molecules immobilized per pair of sodium chloride ions). Neutral solute molecules such as ethyl alcohol noticeably increase the dielectric relaxation time of water [103–105]. This is frequently expressed by saying that the water tends to freeze to an ice-like configuration in the immediate neighborhood of the solute molecule. Many physicochemical data (entropies, partial vapor pressures, viscosities) agree in this respect. The situation is not simple, however, because ionized salts, for example alkyl ammonium lead to breaking down of the structure but nevertheless cause an increase in relaxation time. It is possibly contrary effects are occurring: these salts are known to form crystalline hydrate structures of the clathrate type.

Foodstuffs contain much water. Many people believe the water content is responsible for the microwave heating of food. According to Fig. 1.15, dielectric relaxation of water and corresponding dielectric losses are quite negligible for ionic solutions. Conduction losses are preponderant. Ionic species such salts (sodium chloride) induce dielectric losses in soup and microwave heating results from ionic conduction.

Intermolecular interactions and complexes   Dielectric measurements on interacting solutes in inert solvents provide information about molecule complex formation. Some such dipoles induced by intermolecular interactions and molecular complexes in benzene solution are listed in Table 1.3. The dipole moment of the complex is a function of the relative strengths of the acid and base and the intramolecular equilibrium is described by Eq. (49):

$$X\text{-}H \ldots Y \Leftrightarrow X^- \ldots H\text{-}Y^+ \tag{49}$$

Stronger acid–base complexes with proton transfer induce formation of ion-pair systems leading to high dipole moments. The OH–N interactions of pyridine in

Tab. 1.3. Dipole moments of molecular complexes in benzene solution (from Ref. [27]).

Components		μ(X–H)	μ(Y)	μ(H–Y)
CH_3COOH	C_5H_5N	1.75	2.22	2.93
$ClCH_2COOH$	C_5H_5N	2.31	2.22	4.67
$Cl_2CHCOOH$	C_5H_5N		2.22	5.24
Cl_3CCOOH	C_5H_5N		2.22	7.78
CH_3COOH	$(C_2H_5)_3N$	1.75	0.66	3.96
$s\text{-}C_6H_2(NO_3)_3OH$	$(n\text{-}Bu)_3N$	1.75	0.78	11.4

acetic and chloroacetic acids lead to a polarity increase. Thus at least 70% ionic character is expected in the trichloroacetic acid–pyridine complex. The very high dipole moment of the picric acid–base complex is indicative the predominantly ionic form.

Solutions of two nonpolar compounds often have polar properties. For iodine solutions in benzene polarization is greater than expected for nonpolar compounds. Interactions usually involve partial electron transfer from one component to another. One component has a positive charge (donor) and the other has negative charge (acceptor). In the benzene–iodine complex partial electron transfer can be envisaged between electrons of benzene π orbitals and the lowest unfilled orbital of iodine. The extent of electron transfer determines the dipole moment observed. Such systems, called charge-transfer complexes, lead to changes of other physical properties, for example magnetic properties.

The relaxation times of trihalogenated esters in solution in benzene, dioxane, and carbon tetrachloride reveal another typical type of anomalous behavior which can be explained by solute–solvent interactions. It is well known that the ester molecule is a resonance hybrid in which the carbonyl group assumes positive character. This character is further accentuated by the three electron attracting halogen group. This positive carbonyl carbon interacts strongly with the oxygen lone-pair electrons in dioxane and the π electrons of the benzene ring forming complexes with these solvents and resulting in large value of relaxation time [54].

Intermolecular interactions and hydrogen bonding Hydrogen bonding is a form of molecular orientation involving an A–H group and an electron-donor component Y. A is usually oxygen sometimes nitrogen, and less frequently carbon. Hydrogen-bond formation alters the electron distribution within the molecules, changes the polarization, and could induce a dipole moment. The polarization of N-substituted amides in solution increases with increasing concentration. Association into polar chains occurs with a dipole moment greater than that of the monomer unit. Intermolecular hydrogen bonding occurs if the NH group is at a position trans to the CO group. The two group dipole moments reinforce each other, producing an enhanced dipole. Solute–solvent interactions caused by hydrogen bonding can also increase the dipole moment. The increase is much larger than expected from the inductive effect alone – HBr has dipole moment equal of 1.08 in benzene solution compared with 2.85 in dioxane solution.

Macromolecules and polymers Because of their partial use in the electrical industry and because of the partial relevance of dielectric studies to questions of molecular mobility and relaxation time, extensive studies of polymer behavior have been performed by dielectric methods. Only the salient features can be outlined here.

Nonpolar polymers such as polyethylene, polytetrafluoroethylene, and polystyrene are especially significant because of their low loss values over the widest frequency range. For polar polymers such as poly(vinyl chloride)s, poly(vinyl acetate)s and polyacrylates dispersion is observed at lower frequencies than for the monomers, as expected. Two or more dispersion regions are commonly observed. They are referred to as α, β, and γ bands beginning with the lowest-frequency dispersion. The α dispersion is broader than a Debye process. Dielectric losses have a lower maximum and persist over a wide range of frequency. Fusoos and Kirwood [106] have successfully described this behavior. It is widely believed that α dispersion caused by Brownian motion of the polymer chain whereas the β dispersion is because of oscillatory motion or intramolecular rotation of side groups.

Small molecules adopt reasonably well defined geometric configurations and calculation of dipole moment is possible. For polymers mobile configurations with rotations about single bonds in the chain skeleton and may occur for many side groups. The measured dipole moment is the statistical average of the mobile configurations and is proportional to the square root of the number of polar groups present in the polymer. The dipole moment, μ, of a polymer is usually expressed as given by Eq. (50):

$$\langle \mu^2 \rangle = n g \mu_0^2 \qquad (50)$$

where n is the number of polar groups in the chain, μ_0 the group dipole moment of the polar unit, and g a factor depending on nature and degree of flexibility of the chain. Theoretical calculations of g have been extensively studied by Birshtein and Ptitsyn [107] for polymers with a polar group in rigid side chains and Marchal and Benoit [108] for polymers with polar groups within the chain backbone. In both polymers g is a function of the energy barrier restricting rotation within the chain backbone.

The very large dipole moment of polymers results in strong intermolecular forces in solution. Atactic and isotatic polymers have different dipole moments. The dipole moment of the atactic poly(vinyl isobutyl ether) is 10% lower than that of the isotactic form, showing that the isotactic polymer adopts a more ordered structure with group dipoles tending to align parallel to each other.

Rigid polymers have dipole moments which are proportional to the degree of polymerization. The alpha-helical form of polypeptides (e.g. gamma-benzyl glutamate) leads to a very high dipole moment, because the group dipole moments are aligned parallel. In the alpha-helix the carbonyl and amino groups are nearly parallel to the axis of the helix, which is stabilized by hydrogen bonding between these two groups. The calculated moment for this arrangement is 3.6 D for the peptide unit (2.3 D and 1.3 D, respectively, for carbonyl and amino groups) in agreement with the experimental value.

Solids and dipole relaxation of defects in crystals lattices Molecules which become locked in a solid or rigid lattice cannot contribute to orientational polarization. For polar liquids such as water, an abrupt fall in dielectric permittivity and dielectric loss occur on freezing. Ice is quite transparent at 2.45 GHz. At 273 °K, although the permittivity is very similar (water, 87.9; ice, 91.5) the relaxation times differ by a factor of 10^6 (water, 18.7×10^{-12} s; ice, 18.7×10^{-6} s). Molecular behavior in ordinary ice and a feature which may be relevant to a wide variety of solids has been further illuminated by the systematic study of the dielectric properties of the numerous phases (ice I to VIII) formed under increasing pressure. Davidson was the first to publish an exhaustive study [109]. A molecule may have equilibrium positions in a solid which correspond to potential energy minima separated by potential barrier, because of interactions with neighbors. Such molecules change their orientation either by means of small elastic displacements or by acquiring sufficient energy to jump the potential barrier. When an electric field is applied to a crystalline dipolar solid, polarization could occur by three mechanisms. The first is the distortion polarization, the second is the elastic displacement of dipoles from their equilibrium positions, and the third is a change in the relative orientations of the dipoles.

Typical lattice defects include cation vacancies; substitutional or interstitial ions are other types of more complicated structural defects. A cation vacancy behaves like a negative charge. If the temperature is high, ions are sufficiently mobile that an anion could be expelled from the lattice by the Coulomb potential of the cation vacancy. Cation and anion vacancies could form a dipole oriented along one of the six crystallographic axes. This vacancy coupling is then able to induce a crystalline dipole. Similar dipoles can also appear when an ion is substituted for the host ion. A divalent atom such as calcium substituted for a monovalent cation in alkali halides releases two electrons and becomes a doubly charged ion. The new atom of calcium has excess positive charge which can couple with a negative defect, for example an alkali metal vacancy or an interstitial halide, to create a dipole.

In the following discussion an LiF lattice with a lattice parameter containing N substitutional Mg ions per unit volume is considered. According to Langevin's model and Maxwell–Boltzmann statistics polarization arising from the dipole defects is given by Eq. (51):

$$P_{dd} = \frac{Na^2e^2}{6kT(1+j\omega\tau)} E \qquad (51)$$

An Li ion has been replaced by an Mg ion. This ion with its positive charge forms a dipole with a negative lithium vacancy sitting on one of the twelve nearest-neighbor sites normally filled with Li ions. In the absence of an electric field, these twelve positions are strictly equivalent and the lithium vacancy hops between them giving a zero average dipole moment for the defect. In the presence of an electric field the twelve sites are no longer equivalent. The twelve sites usually split into three categories in relation to the interaction energy with electric field. Hence, the value of the dipole moment is given by Eq. (52):

$$\mu = \frac{ae}{\sqrt{2}} \qquad (52)$$

Finally, the complex permittivity of the substituted lattice takes the familiar form given by Eq. (53):

$$\varepsilon'' = \varepsilon_\infty + \frac{N\mu^2}{3kT(1+j\omega\tau)} \qquad (53)$$

Equation (53) describes Debye relaxation. Magnesium and calcium-doped lithium fluorides have a characteristic Debye relaxation diagram from which the dopant concentration and the relaxation time can be deduced. Many others crystals containing mobile lattice defects have similar Debye's relaxation processes. Major understanding of the structure of color centers results from dielectric relaxation spectra. Nuclear magnetic resonance, optical and Raman spectroscopy can be used efficiently in conjunction with dielectric spectroscopy.

Solids and adsorbed phases Solid surfaces almost invariably have absorbed molecules derived from the gas or liquid medium to which the surface is exposed. The amount of such absorbed material depends upon the chemical nature of the solid surface. The absorbed layer can greatly affect the properties of the solid surface. The absorbed molecules may reorient by libratory oscillation between defined orientational positions (see *Orienting effect of a static electric field* in Section 1.1.2.1). Such a restricted process could account for the reduced effective permittivity. The extensive use of silicates (clays and soils) as catalysts in chemistry adds interest to their study.

♦ CONCEPTS More about relaxation process within solids Typical loss peaks are broader and asymmetric in solids, and frequency is often too low compared with Debye peaks. A model using hypotheses based on nearest-neighbor interactions predicts a loss peak with broader width, asymmetric shape, and lower frequency [27]. This behavior is well suited to polymeric, glassy materials and ferroelectrics. Low temperature loss peaks typically observed for polymers need many-body interactions to be obtained. Although current understanding of these processes is not yet sufficient to enable quantitative forecasting the dielectric properties of solids may offer insight into the mechanisms of many-body interactions.

Interfacial relaxation and the Maxwell–Wagner effect Dielectric absorption quantifies energy dissipation and for most systems lose energy from processes other than dielectric relaxation. These processes usually are related to the d.c. conductivity of the medium. The corresponding loss factor for higher conductivity persists at high frequencies and even in the microwave heating range. The incidence of d.c. conductivity appears for fused ionic salts as a distorting feature. In addition to d.c. conductance loss, energy dissipation can occur by scattering of the radiation at interfacial boundaries in inhomogeneous materials. In the visible region, distribution of small particles of a second material (e.g. air bubbles) in an otherwise transparent medium can render it opaque. The same feature arises in dielectric media when particles are dispersed within a matrix. The general aspect of this absorption is referred as the Maxwell–Wagner process.

If the dielectric material is not homogeneous but could be regarded as an association of several phases with different dielectric characteristic, new relaxation processes could be observed. This relaxation processes called Maxwell–Wagner processes occur within heterogeneous dielectric materials. An arrangement comprising a perfect dielectric without loss (organic solvent) and a lossy dielectric (aqueous solution) will behave exactly like a polar dielectric with a relaxation time which becomes greater as the conductivity becomes smaller. Such an arrangement, leading to a macroscopic interface between two immiscible solvents, could be extended to dispersion of slightly conducting spherical particles (radius a, permittivity and conductivity) in a non conducting medium (dielectric permittivity equal to ε_M). To calculate the effective permittivity of the dispersion, consider a sphere of radius R containing n uniformly distributed particles. These particles coalesce to form a concentric sphere of radius $\sqrt[3]{na^3}$. The effective dielectric permittivity of this heterogeneous medium is given by Eq. (54):

$$\tilde{\varepsilon} = \varepsilon_\infty + \frac{\varepsilon_S - \varepsilon_\infty}{1 + j\omega\tau} \tag{54}$$

where the static dielectric permittivity ε_S is given by Eq. (55):

$$\varepsilon_S = \varepsilon_M \left(1 + \frac{3na^3}{R^3}\right) \tag{55}$$

The system relaxes with a relaxation time given by Eq. (56):

$$\tau = \frac{\varepsilon_P + 2\varepsilon_M}{\sigma} \tag{56}$$

These interfacial polarization effects can explain strong enhancement of microwave heating rate observed by some workers with phase-transfer processes which associate organic solvents with high lossy aqueous solutions or with a dispersion of solids within liquids which are essentially nonpolar. The Maxwell–Wagner loss will not occur in homogeneous liquid systems. For supercooled liquids in particular, for example glasses or amorphous solids, the three loss processes d.c. conductivity, Maxwell–Wagner effect, and dipolar absorption occur simultaneously. Consequently, even if dielectric loss for these media is low at room temperature, slight heating will lead to a increase of dielectric loss resulting in fusion. For example, an empty drinking glass placed within a domestic microwave oven can melt easily, depending on oven power. Obviously, if a wave-guide or cavity is used, fusion can result for an empty test tube, for a test tube partially filled with lossy products, for or a test tube filled with products without dielectric loss. For products with slight dielectric loss it is better to avoid glass and use a test tube made from quartz or silica without dielectric loss at 2.45 GHz.

Colloids Colloidal solutions are the most difficult systems to measure and analyze dielectrically. If the solute particles have dipole moments, the solution should show anomalous dispersion and loss at low frequencies if orientation of the dipole involves orientation of the whole particle. If there are substantial differences among the dielectric constants or conductivity of the particles and the matrix, interfacial polarization can cause dielectric loss and frequency belongs to the low frequency range. In aqueous colloidal solutions the presence of electrolytes even in small amounts in the water should commonly cause sufficient conductivity difference to result in interfacial polarization. Another effect is that of the electric double layer at the particle surface, even in the absence of the electric field. The effect of these factors must be taken into account. Errera [110] reported the first results of large apparent permittivity for vanadium pentoxide, and Schwan [111] was the first to provide experimental evidence of the large increase of static permittivity for a suspension of polystyrene spheres in water. Values of 10 000 for diameters close to a micron have been reached for static permittivity. The dynamics of charge distribution within these systems lead to interesting dielectric properties with occur over a timescale determined by different characteristics of the system. The properties of colloids have been studied at radiofrequency but very little attention has been devoted to their properties at microwave frequencies. This is probably because the charge dynamics for particles several micrometers in diameter are too slow to result in significant losses at microwave frequencies.

Charged colloids typically consist of charged particles suspended in an electrolyte. Surface charges attract counterions leading to a double layer charge. When a microwave field is applied to a charged particle, the tangential component of the electric field around the particle surface causes azimuthal transport of the double layer ions across the particle. This results in asymmetric charge distribution within the double layer around the particle. This charge redistribution induces a change in the dipole strength, leading to a resultant electric field around the particle which opposes the applied field [112–114]. O'Brien [114] showed that flow of counterions result in a high frequency, low-amplitude relaxation with a relaxation time τ_1 given by Eq. (57):

$$\tau_1 \approx \frac{1}{\kappa^2 D} \tag{57}$$

where κ is the reciprocal of the double layer thickness and D the ion diffusivity [112]. After this charge redistribution of the double layer, a slower relaxation process occurs within the electrolyte. The low-frequency relaxation time τ_2 is given by Eq. (58):

$$\tau_2 = \frac{R^2}{D} \tag{58}$$

where R is the radius of the particle. The size dependence of this dissipative process is of particular interest for microwave heating. Hussain et al. [115] have recently shown that polystyrene particles with surface charge resulting from sulfate groups suspended in pure distilled water (diameter between 20 and 200 nm) had relaxation beyond 10 GHz associated with the dipolar relaxation of water. The dielectric permittivity and dielectric loss induced by colloids over this region are slightly lower than those produced by water. Dielectric losses are in agreement with additional dc. static conductivity of the colloid given by Eq. (36). More recently, Bonincontro and Cametti [116] and Buchner et al. [117] have described how dielectric measurements can be used to obtain information about electric polarization mechanisms occurring on different time scales for ionic and nonionic micellar solutions, liposomes, and suspensions of biological cells. Jönsson et al. have [118] discussed the effect of interparticle dipolar interaction on magnetic relaxation for magnetic nanoparticles. It seems that superparamagnetic behavior could be replaced by spin-glass-like dynamics for those systems.

Space charge polarization and relaxation Accumulation of charges in the vicinity of electrodes unable to discharge the ions arriving on them can induce relaxing space-charge. The ion behavior balances the effect of the field to accumulate charge on interfaces whereas thermal diffusion tends to avoid them (i.e. interfaces between two liquids with dielectric and conductivity differences). The layer can contain charge density and is therefore equivalent to a large dipole. Dipole reversal is a sluggish relaxation process. Unlike the classical Debye's model, it is impossible to separate dielectric permittivity into real and imaginary parts. Argand's diagram can be calculated and are found to be exactly semicircular and slightly flattened.

1.2
Key Ingredients for Mastery of Chemical Microwave Processes

After a brief overview of the thermal conversion data (thermal dependence of dielectric properties, electric field focusing effects, hydrodynamics) thermodynamic aspects and athermal effects of electric fields will be examined. Next, the effects of thermal path and hot spots induced by microwave heating will be analyzed in term of kinetic effects.

1.2.1
Systemic Approach

The tasks of chemical engineers are the design and operation of chemical reactors for converting specific feed material or reactants into marketable products. They must have knowledge of the rates of chemical reactions involved, the nature of the physical processes interacting with the chemical reactions, and conditions which affect the process. The rates of the physical processes (mass and heat transfer) involved in commonly used chemical reactors can often be estimated adequately from the properties of the reactants, the flow characteristics, and the configuration of the reaction vessel. Chemical process rate data for most industrially important reactions cannot, however, be estimated reliably from theory and must be determined experimentally.

In contrast, design of chemical microwave processes, especially scale-up of operating conditions, involves knowledge and control of several nonlinear feedback

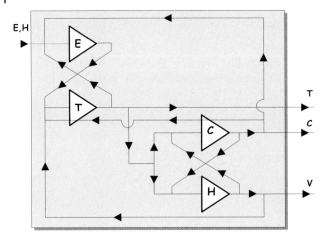

Fig. 1.16. Schematic diagram of the chemical microwave process [119].

loops. Figure 1.16 depicts chemical microwave processes in accordance with concepts of systemic theory, or the theory of systems, in cybernetic modeling. This theory studies systems in relation to feedback and coupling loops.

The microwave process includes the microwave oven, the reactors and the reactants. The inputs of the system are microwave energy (E, H) and reactants whereas the outputs are heat (T), products (C), and convection motions of fluid solutions (V). A processor is associated with each process. They are represented schematically by triangles in Fig. 1.16. In systemic theory, the meaning of a processor is a function linking input and output. These processors can be expressed by partial differential equations. The two first processors to be considered are electromagnetic (E) and the thermal (T) processors, leading to heating. The heating processor can supply the chemical (C) and hydrodynamic (H) processors. The chemistry can modify hydrodynamic conditions (viscosity, superficial tension, etc) whereas convective motion affects yield of reactions. Convective motion and chemical reactions can also modify thermal and dielectric properties. Hence, four coupling loops can be defined. Most of these coupling loops have highly nonlinear character. Since 1990, the author has studied all these couplings and separately now the general problem of a chemical microwave process can be solved with optimized devices and plants.

The importance of these phenomena cannot be overemphasized, because they have significant effects on the yields and the quality of the products. The characteristics and limitations of these systems which affect the reactor performance should be well understood to ensure successful design and operation of the plant. Consequently, a microwave plant could be optimized for several given compositions (i.e. dielectric properties and thermal dependency). An optimized microwave plant cannot be a versatile device in terms of dielectric properties or, in other words, chemical composition.

1.2.2
The Thermal Dependence of Dielectric Loss

According to Poynting's formulation, the time averaged dissipated power density $P_{Diss}(r)$ at any position r within lossy liquids or solids is given by Eq. (59):

$$P_{Diss}(r) = \frac{\omega}{2}\left(\varepsilon_0\varepsilon'' \vec{E}(r)^2 + \mu_0\mu'' \vec{H}(r)^2 + \frac{\sigma}{\omega}\vec{E}(r)^2\right) \qquad (59)$$

where ω is the angular frequency, ε_0 the dielectric permittivity of a vacuum, $\varepsilon''(r)$ the dielectric loss, $\vec{E}(r)$ the electric field amplitude, μ_0 the magnetic permeability of a vacuum, μ'' the magnetic losses, $\vec{H}(r)$ the magnetic field amplitude, and σ the ionic conductivity.

According to dielectric and/or magnetic losses and electric and magnetic field strength, the dissipation of electromagnetic energy leads to heating of the irradiated medium. Hence estimation of dissipated power density within the heated object depends directly on the electric and magnetic fields, and dielectric and magnetic loss distributions. Maxwell's equations can be used to describe the electromagnetic fields in the lossy medium and an energy balance can be solved to provide the temperature profiles within the heated reactor.

The specificity of microwave heating results from the thermal dependence of dielectric properties. The complex dielectric permittivity is very dependent on temperature and the dynamic behavior of microwave heating is governed by this thermal change. The electric field amplitude depends on the real and imaginary parts of the dielectric permittivity, which themselves depend on temperature as described by Eq. (60):

$$P_{Diss}(r, T) = \frac{\omega\varepsilon_0\varepsilon_r''(T)}{2}\vec{E}(r, \varepsilon_r'''(T), \varepsilon_r''(T))^2 \qquad (60)$$

Figure 1.17 shows the thermal feedback induced by the thermal dependence of dielectric properties. The microwave applied energy results in dissipated microwave energy depending on dielectric properties. The heating rate depends on thermal properties (thermal diffusivity, specific heat). The thermal dependence of the thermal properties is, however, very slight compared with the thermal dependence of the dielectric properties. In contrast with conventional heating techniques, the heated medium acts as an energy converter. Consequently, thermal change of dielectric properties causes changes in the dissipated energy during heating. Depending on the nature on the thermal changes, this may result in thermal runaway or, occasionally, reduced material heating. Thermal runaway is a catastrophic phenomenon in which a slight change of microwave power causes the temperature to increase rapidly. The electric field depends on spatial location in relation to the wavelength within the heated material and thus inhomogeneous heating results in the deleterious effect of inhomogeneous material properties (e.g. densification

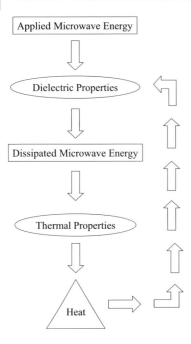

Fig. 1.17. Schematic diagram of dielectric thermal feedback [119].

and sintering in ceramics, or polymer curing and reticulation). A precise knowledge of the dielectric properties is therefore essential for any study of microwave heating or design of a microwave applicator. The thermal behavior of the heated material is, however, usually dependent not only on the dependence of dielectric losses but on the strength of the electric field applied. Both of these factors have to be known to optimize the operating conditions and microwave thermal processes. Consequently, basic understanding of microwave heating processes remains somewhat empirical and speculative, because of its highly nonlinear character.

1.2.3
The Electric Field Effects

The electric field is a crucial condition in microwave heating and the design of microwave ovens. If electric field distributions within empty microwave ovens are well known, the problem is totally different if loaded microwave ovens are considered. Perturbation theory can be used if the sample is very small. In fact, the magnitude of the perturbation is proportional to reactor-to-applicator volume ratio. The perturbation could be negligible if this ratio is close to 10^{-4} and most laboratory and industrial devices have higher ratios [120].

1.2.3.1 Penetration and Skin Depths

The wave equation of electromagnetic fields in the z direction is given by Maxwell's equations. The electric field takes the form given by Eq. (61):

$$E = E_0 \exp(j\omega t - \gamma z) \tag{61}$$

where γ is the propagation constant. Dielectric media without loss lead to a pure imaginary propagation constant given by Eq. (62):

$$\gamma = j\omega(\varepsilon_0 \varepsilon_r \mu_0 \mu_r)^{1/2} \tag{62}$$

The problem is totally different for dielectric loss. The wave is attenuated as it traverses the medium and, therefore, the power dissipated is reduced to an even larger extent. Consequently, the propagation constant becomes complex as described by Eq. (63):

$$\gamma = \alpha + j\beta \tag{63}$$

The real and imaginary parts of the complex propagation constant are called attenuation factor (Np m^{-1}) and the phase factor (Rad m^{-1}), respectively. They are given by Eqs (64) and (65):

$$\alpha = \sqrt{\frac{\omega^2 \varepsilon_0 \mu_0 \varepsilon_r'}{2}} \sqrt{\sqrt{1 + \left(\frac{\varepsilon_r''}{\varepsilon_r'}\right)^2} - 1} \tag{64}$$

$$\beta = \sqrt{\frac{\omega^2 \varepsilon_0 \mu_0 \varepsilon_r'}{2}} \sqrt{\sqrt{1 + \left(\frac{\varepsilon_r''}{\varepsilon_r'}\right)^2} + 1} \tag{65}$$

The simplified version of the attenuation factor α for highly lossy media, for example metallic or ionic conductors, is given by Eq. (66):

$$\alpha \approx \sqrt{\frac{\omega^2 \varepsilon_0 \mu_0 \varepsilon_r''}{2}} \tag{66}$$

Hence, the attenuation factor α could be defined by a distance δ which is the skin depth as described by Eq. (67):

$$\delta = \frac{1}{\alpha} \tag{67}$$

The attenuation factor leads to attenuation of electric field and power during propagation along the z coordinate as described by Eqs. (68) and (69):

$$E = E_0 \exp(-\alpha z) \exp j(wt - \beta z) \tag{68}$$

$$P = P_0 \exp(-2\alpha z) \tag{69}$$

The penetration depth P_d is defined as the distance from the surface of a lossy dielectric material at which the incident power drops to 37% (1/e). Skin depth is, in fact, equal to twice the penetration depth.

The field or power penetration increases with decreasing frequency. Penetration depths at frequencies below 100 MHz are close to meters. Penetration depths within pure water at 2.45 GHz are close to centimeters (7 mm at 3 °C and 2 cm at 40 °C). Consequently, the penetration depth within a microwave reactor filled with lossy media is obviously smaller. Microwave heating could results in unacceptable hot spots and vigorous stirring is often necessary.

Field penetration within metals is substantially smaller. Skin depth within the metallic enclosure of a microwave oven is close to several microns, depending on the nature of the metal. Although losses are larger for stainless steel than for most metals it is widely used as the internal wall of industrial microwave ovens because of its surface hardness and resistance to corrosion. Stainless steel should not be used in high-level devices, however. Skin depth within metals explains why sparks observe during microwave heating of plates with plating. The thickness of the plating is close to the skin depth and microwave heating induces plating vaporization leading to a plasma with colored sparks. This is a test enabling verification of whether the plating is made with aluminum, copper, gold, or silver!

1.2.3.2 Dimensional Resonances

According to the conclusions of the previous paragraph, microwave heating of high lossy media could be difficult in terms of thermal uniformity. Obviously, strong stirring could settle this problem. The aspect of energy profiles within lossy media is more complicated. Resonant devices and, especially, dimensional resonance according to the author's terminology can induce electric field-focusing effects. The Fig. 1.18 shows the dependence of temperature on electric field radial distribution within a water pipe of diameter 4 cm. Different radial distributions within air and water are observed and the focusing effect is observed at 10 °C, at which temperature dielectric losses are greater.

Electromagnetic field distributions within spherical and cylindrical bodies irradiated by a plane wave have been examined by several workers in relation to safety problems of radar and for medical purposes [122–126]. For spherical shapes it was found that the heating potential was higher than the value calculated by use of the average cross section model by a factor approximately ten under resonance conditions. Many common food items for example potatoes and tomatoes have diameters and dielectric properties that fall within the range for maximum core heating. Strong focusing effects are exhibited by geometrical shape of mushroom hat. Ohlsson and Risman [127] have shown that core heating releases steam which induces an energy impulse of sufficient power to cause mechanical stress within the heated material. This deleterious phenomenon is optimum for an egg of diameter close to 40 mm. This fact has been used as the well known effect that microwaves cook from inside. Accordingly, spheres and cylinders behave as dielectric resonators with strong focusing effects of electromagnetic energy. These effects are very dependent on temperature in relation to the balance between electric field distribu-

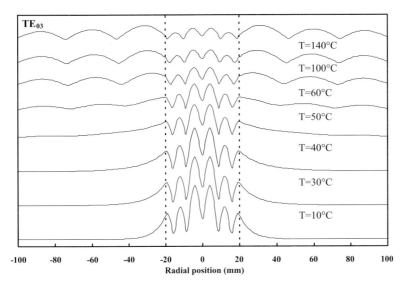

Fig. 1.18. Dependence of temperature on the radial electric field profile within a water pipe [121].

tions and dielectric properties. The same reagents placed within flasks of different shape can lead to different heating rate for the same microwave operating conditions.

In classical descriptions, thermal runaway is attributed to a strong increase of dielectric losses because of heating. So, the energy provided by microwave irradiation increases with temperature. The authors have shown it is possible to achieve thermal runaway with dielectric losses decreasing with temperature as a result of dimensional resonance or focusing effects of an electromagnetic field within the dielectric sample [128, 129].

1.2.4
Hydrodynamic Aspects

In many instances of heat transfer involving liquids, convection is an important factor. In most circumstances of heat-transfer within a conventional reactor, heat is being transferred from one fluid through a solid wall to another fluid and natural convection occurs. Density differences provide the body force required to move the fluid. As a result of thermal gradients induced by heating, vaporization and boiling can occur. Fluid motion results from different hydrodynamic instability. The author has exhaustively studied the microwave hydrodynamic behavior of water and ethanol which are two classical solvents in chemistry [130–133]. A brief overview of hydrodynamic instability (convective patterns for Rayleigh–Benard, Marangoni, and Hickman instabilities) and the design of experimental devices for heating of very polar liquids, for example water, can be found elsewhere [130], as can experimental results relating to the hydrodynamic behavior of water and

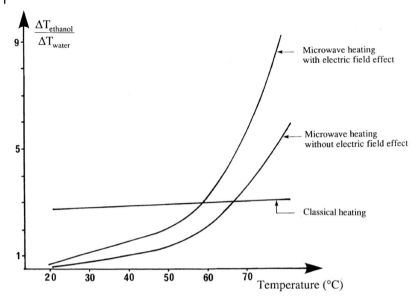

Fig. 1.19. Comparison of the effect of temperature on the heating expected with conventional and microwave heating (with and without the electric field effect) for water and ethanol [133].

ethanol under microwave heating at reduced pressure [131]. Coupling of hydrodynamic instability with linear stability analysis of experimental results is discussed in Ref. [132]. Microwave heating seems to be a tool which enables control of the spatial thermal profile in relation to the intrinsic properties of the liquid. Finally, modeling of microwave heating is described in Refs. [120] and [133].

The heating rates expected for water and ethanol as a result of conventional and microwave heating are compared in Fig. 1.19. For identical energy density and conventional heating, the ratio of the induced heating rates in water and ethanol does not change during heating. If no significant changes are observed for the temperature range, a significant difference appears when the dielectric loss effect is taken into account. In a third step, with the electric field correction, the difference is significantly amplified. Hence, we observe for temperatures below 50 °C that microwave heating is preponderant in water and for temperatures above 50 °C microwave heating is preponderant in ethanol. Thus this figure obviously proves the selectivity of classical heating and microwave heating in relation to the thermal dependence of the real and imaginary parts of the dielectric permittivity. For water or other liquids with dielectric losses decreasing with temperature, thermal conversion of electromagnetic energy is preponderant in cold areas. Thus, temperature fluctuations are eliminated by microwave heating. The intrinsic ability of microwave heating is to eliminate the thermal gradient between bulk and surface for an evaporating column. For ethanol or other liquids with dielectric losses increasing with temperature, in contrast with water, and for the same reasons, the thermal conversion of the electromagnetic energy is preponderant in hot areas. Microwave heating amplifies temperature fluctuations. Thus, we can understand why interfacial instabilities are preponderant in microwave evaporation of ethanol whereas bulk instabilities are preponderant in microwave evaporation of water.

The geometrical shape of the temperature profiles can be controlled in relation to the thermal dependence of the dielectric properties, and moreover, the absence of warming walls, as in classical heating, gives new dimensions to microwave heating. These two characteristics can make it possible to obtain specific and promising hydrodynamic effects as a result of thermal treatment of liquids.

1.2.5
Thermodynamic and Other Effects of Electric Fields

The thermodynamic effects of electric fields exist and are well known. The application of an electric field to a solution can induce an effect on chemical equilibrium. For example, the equilibrium of Eq. (18) where C has a large dipolar moment while B has a small dipole is shifted toward C under the action of an electric field:

$$A + B \rightleftharpoons C \tag{70}$$

Typical examples are the conversion of the neutral form of an amino acid into its zwitterionic form, the helix–coil transitions in polypeptides and polynucleotides, or other conformational changes in biopolymers. Reactions of higher molecularity where reactants and products have different dipole moments are subject to the same effect (association of carboxylic acids to form hydrogen-bonded dimers). Equilibrium involving ions are often more sensitive to the application of an electric field; the field induces a shift toward producing more ions. This is known as the dissociation field effect (DFE) or the "second Wien's effect" [134].

In principle the effect of an electric field on chemical equilibria can be described by the thermodynamic relationship described by Eq. (71):

$$\left(\frac{\partial \ln K_{\text{Eq.}}}{\partial |\vec{E}|}\right)_{P,T} = \frac{\Delta \mu}{RT} \tag{71}$$

where K_{eq} is the equilibrium constant, $|\vec{E}|$ the field strength (V m^{-1}), and $\Delta \mu$ the molar change of the macroscopic electric moment or the molar polarization for nonionic systems. For ionic equilibria it must be pointed out that one can never reach a true thermodynamic equilibrium because of the field-induced flow of the ions. DFE theory has been developed by Onsager [135]. The most notable result is that $\Delta \mu$ is proportional to the field strength E. Hence, according to Eq. (72):

$$\int_0^E d \ln K_{\text{Eq.}} = \frac{1}{RT} \int_0^E \Delta \mu \, dE \propto E^2 \tag{72}$$

which results from integration of Eq. (71), the change in equilibrium constant is proportional to the square of the electric field strength and the effect on the equilibrium constant is noticeable only at high field strengths. In practice electric field strengths up to 10^7 V m^{-1} are required to produce a measurable effect on normal chemical reactions. For water at 25 °C, K_{eq} changes by approximately 14% if a field of 100 kV m^{-1} is applied. Smaller fields are required to achieve a comparable shift in less polar solvents. Nonionic equilibria can also be perturbed by the DFE if they are coupled to a rapid ionic equilibrium. A possible example is depicted by Eq. (73):

$$A^- + H^+ \rightleftharpoons AH \rightleftharpoons BH \tag{73}$$

in which the slow equilibrium is coupled to an acid–base equilibrium. This is the same principle as coupling a temperature-independent equilibrium to a strongly dependent one. Such a scheme has been studied for the helix–coil transition of the poly-alpha,L-glutamic acid by Yasunaga et al. [136], in which dissociation of protons from the side chains increases the electric charge of the polypeptide which in turn induces a transition from the helix to the coil form, for the dissociation of acetic acids by Eigen and DeMayer [137], and for dissociation of water by Eigen and Demayer [138].

Hence, if thermodynamic effects of an electric effect exist, the electric field strengths necessary are too high compared with the ordinary operating conditions of microwave heating.

1.2.6
The Athermal and Specific Effects of Electric Fields

A chemical reaction is characterized by a change of free energy between the reagents and the products. According to thermodynamics, the reaction is feasible only if the change of free energy is negative. The more negative the change of free energy, the more feasible the reaction. This change of the free energy for the reaction is the balance between broken and created chemical bonds. This thermodynamic condition is not sufficient to achieve chemical reaction in a short time (or with a significant reaction rate), however. Kinetic conditions must also be satisfied to achieve the reaction. The free energy of activation depends on the enthalpy of activation which expresses the height of the energy barrier to surmount. This energy condition is only a necessary condition and is not sufficient to ensure transformation of the reagents. The relative orientation of the molecules which react is crucial, and this condition is expressed by the entropy of activation. This entropic term expresses the need for a geometrical approach to ensure effective collisions between reagents.

Thus the essential question raised by the assumption of "athermal" or "specific" effects of microwaves is the change of these characteristic terms (free energy of reaction and of activation) for the reaction studied [139]. Hence, in relation to the previous conclusions, five criteria or arguments (in a mathematical sense) relating to the existence of microwave athermal effects have been formulated by the author

[140]. More details can be found in this comprehensive paper which analyses and quantifies the likelihood of nonthermal effects of microwaves. This paper provides some guidelines for clear definition what should characterize nonthermal effects.

Hence, according to these five criteria there is no doubt an electric field cannot have any molecular effect for solutions. First, the orienting effect of the electric field is small compared with thermal agitation, because of the weakness of the electric field amplitude. Even if the electric field amplitude were sufficient, the presence of dielectric loss results in a delay of dipole moment oscillations in comparison with electric field oscillations. Heating the medium expresses the stochastic character of molecular motion induced by dissipation of the electromagnetic wave. The third limitation is the annihilation of molecular rotation in condensed phases, for example the liquid state. According to our demonstration, under usual operating conditions it will be proved that the frequently propounded idea that microwaves rotates dipolar groups is, mildly speaking, misleading.

If molecular effects of the electric field are irrelevant in microwave heating of solutions, this assumption could be envisaged for operating conditions very far from current conditions. On one hand, it will be necessary to use a stronger electric field amplitude, or to reduce the temperature according to the Langevin function. This last solution is obviously antinomic with conventional chemical kinetics, and the first solution is, currently, technologically impossible. On the other hand, it will be necessary to avoid reaction media with dielectric loss. Molecular effects of the microwave electric field could be observed paradoxically for a medium which does not heat under microwave irradiation.

In conclusion, the interaction between a dipole and an electric field is clearly interpreted by quantum theory. The coupling is weaker than with magnetic fields, and when a dipole population is subjected to an electric field there is a such demultiplication of quantum levels that they are very close to each other. The interaction energy is continuous, and we have to use Boltzmann or Langevin theories. Because of the weak coupling between dipole and electric field, and the lack of quantified orientations, the study of electric dipole behavior gives less information about the dipole itself than about its surroundings. Indeed electric dipoles are associated with molecular bonds (the electric dipole moment results from the distribution of positive and negative charges on the molecule studied; if they are centered at different points the molecule has a permanent dipole and the molecule is polar). Any motion of electric dipoles, induced for example by interaction with an electric field, leads to correlative motion of molecular bonds, whereas motions of the magnetic moment are totally independent of any molecular motion. Consequently, studies of dielectric properties must be studies of "group properties". Those properties cannot be modeled by a single dipole; a group of dipoles interacting among themselves would be a key aspect of these models. The origin of the confusion between the behavior of a single dipole and a collection of dipoles (in other words differences between dilute and condensed phases) is the most important problem, and the source of illusions for people claiming microwave effects resulting from the orientating effect of the electric field.

In conclusion, is it necessary to obtain a microwave athermal effect to justify microwave chemistry? Obviously no, it is not necessary to present microwaves effects in a scientific disguise. There are many examples where microwave heating gives particular time–temperature histories and gradients which cannot be achieved by other means especially with solid materials. Hence, rather than claiming nonthermal effects it is better to claim a means or a tool to induce a specific thermal history.

1.2.7
The Thermal Path Effect: Anisothermal Conditions

According to the author, before claiming microwave heating effects in preference to collisional or mechanistic terms, it is necessary to estimate the effect of strong heating rates induced by microwave heating. The energy density used in a domestic oven is sufficient to raise temperature from ambient to 200 °C in less than 1 min, and so cause the total reaction time to be reduced by a factor close to 10^3.

This natural tendency for thermal racing is accentuated by these implicated energy densities which are usually different from those used with a water or oil bath, particularly for sealed vessels and autoclaves as used by organic chemists. The temperature and, especially, the thermal path $T(t)$ seems to be a crucial variable. The authors have shown theoretically and experimentally that strong heating rates (up to $5° s^{-1}$) can induce selectivity or inversion between two competitive reactions [141].

The first case, depicted by Fig. 1.20, is called "induced selectivity". Numerical results correspond to a values of kinetic terms which give an optimum selectivity effect; exact values could be found elsewhere [141]. Under classical heating conditions, or a very slow heating rate, a mixture of the two products P_1 and P_2 is obtained. The ratio of the concentrations of P_1 and P_2 could be controlled by adjusting the rate of heating. Hence, under microwave heating conditions, pure P_1 or P_1

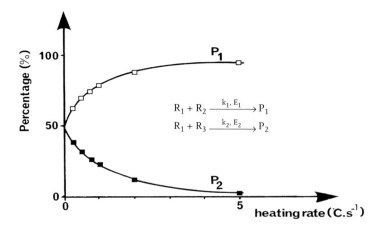

Fig. 1.20. Dependence of the amount (%) of P_1 and P_2 on heating rate. From Ref. [141].

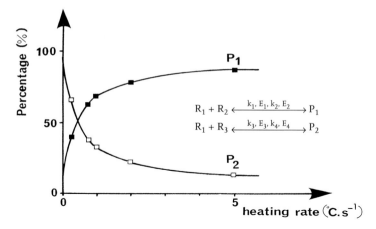

Fig. 1.21. Dependence of the amount (%) of P_1 and P_2 on heating rate from [141].

slightly contaminated by P_2 can be obtained. Chemical operating conditions (reactant concentrations) which are not selective conditions, because of side reaction leading to P_2, can become very selective when the rate of heating is very fast.

The second case, called "Isomeric inversion", is illustrated by Fig. 1.21. Under classical conditions P_1 alone is obtained whereas rapid heating or microwave heating leads to P_2. This situation is more interesting than the first because it forecasts reactivity changes induced by heating rate despite the same chemical operating conditions (reagent concentrations).

The author has illustrated these selectivity effects experimentally with a very classical reaction of much industrial interest – sulfonation of naphthalene. 2-Naphthalene sulfonic acid is a raw material used in the manufacture of pharmaceuticals, dyestuffs, and polymers. Sulfonation of naphthalene results in a mixture of 1- and 2-naphthalene sulfonic, di, tri, and tetrasulfonic acids and sulfones, depending on operating conditions. The reaction is first order with respect to naphthalene in concentrated aqueous sulfuric acid and the ratio of 1- to 2-sulfonic acid decreases slightly with increasing sulfuric acid concentration and temperature – the isomer ratio is 6 for 75% sulfuric acid and 4 for 95% sulfuric acid at 25 °C. The detailed operating conditions (concentrations, microwave applicator, heating rates, and analysis) can be found elsewhere [141]. Figure 1.22 shows the dependence of the percentage 1- and 2-sulfonic acids after reaction on the microwave power used. The isomer ratio can be controlled independently of the chemical operating conditions. This change of selectivity results from heating rate induced. At high temperature (close to 130 °C) 2-sulfonic acid is the major product rather than 1-sulfonic acid. The higher the heating rate, the more the ratio of 1- to 2-sulfonic acid changes.

To observe such kinetic effects of microwave heating, however, it is necessary to have reactions with reaction times close to heating time. Under most operating conditions reactions times are close to several tenths of minutes and anisothermal

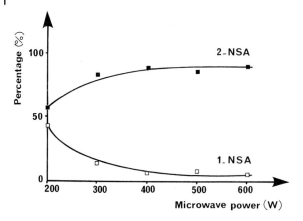

Fig. 1.22. Dependence of the percentage of P_1 and P_2 on heating rate from [141].

effects can be neglected. The use of sealed vessels and autoclaves should enable synthetic chemists to take advantage of the strong heating rate induced by microwave heating (heating times close to reaction times). Thus microwave heating seems to be an easy way of imposing very rapid heating rates for kinetic control of reactions.

1.2.8
Hot Spots and Heterogeneous Kinetics

Effects of temperature upon reaction rate and the heterogeneous character of microwave heating are well known. The wavelength of microwaves used is 12.2 cm (2.45 GHz). Within a dielectric medium, the wavelength is, to a first approximation, equal to the wavelength in air divided by the square root of real part of relative dielectric permittivity. Therefore, for very polar solvents, for example alcoholic or aqueous solutions, several local electric field maxima could be obtained within the heated sample [120, 142]. Baghurst and Mingos [143] have shown that boiling temperature of solvents under the action of microwave heating could be higher than under classical heating conditions. This temperature is referred to the nucleation-limited boiling temperature (NLBT). For most organic solvents the NLBT is within the range of a few degrees to thirty degrees above the conventional boiling point. Vigorous stirring or addition of nucleating materials (e.g. antibumping granules) can easily reduce this localized superheating. For solid reagents or materials, microwave heating can rapidly lead to fusion, depending on the thermal dependence of dielectric properties and, especially, on increase of conductivity with temperature (Section 1.1.2.4 and Ref. [75]). Consequently, most local thermal fluctuations can be amplified and temperatures close to 1000 °C can easily be reached in a few seconds. The association of organic reagents with inorganic solids as alumina, silica, and clays strongly enhances the capacity to absorb microwaves. Loupy et al. [144]

have shown that association of potassium acetate with alumina or silica leads to a reactive medium which strongly absorbs microwaves – the temperature is close to 600 °C after heating for 1 min. The author has made real-time infrared videos of powders (alumina, oxides, zeolites, water gels, etc.) under the action of microwave heating. Evidence of strong thermal gradients has been obtained (for ferrites, $50° \text{ mm}^{-1}$). The author has shown that these hot spots or areas could induce localized reaction rate enhancement [145]. These results have shown that a very small density of superheating areas is sufficient to induce a consequent rate enhancement (2% of hot spots are sufficient to increase yield by 60%), even if their effects on averaged temperatures are not detectable.

In conclusion, microwave heating offers the possibility of realizing high power densities because of core thermal conversion of electromagnetic energy. There is, therefore, increasing industrial application of microwave energy for heating, drying, curing, and sintering of materials. Microwave applicators (traveling wave, multimode, and single-mode cavities) are designed by a trial and error procedure, however. Industrial in-line processing calls for the design of specific applicators which enable high-power densities and hence rapid rates of heating to be achieved. The development of electromagnetic models for such applicators would improve our understanding of microwave processing and heating and enable the rapid design of optimized microwave devices.

References

1 P. Macquer, Dictionnaire de Chymie contenant la théorie et la pratique de cette science, Lacombe, Paris, **1766**, II, 503.
2 A.W. Hull, Phys. Rev. **1921**, 18, 31.
3 The Magnetron, AIEE J, **1921**, 40, 715–718.
4 U.S. patent 2,123,728. This patent is based on an earlier German application, filed in 1935 and described that year in Hollmann's book, Physik und Technik der Ultrakurzen Wellen, Erster band (Physics and Technology of Ultrashort Waves). This classic reference had much more influence on wartime technological developments in the UK, and the UDS than in Germany.
5 Anon. Electronics **1946**, 19, 11, 178.
6 Anon. Elec. Eng. **1946**, 65, 12, 591.
7 R.V. Decareau and R.A Peterson, Microwave processing and engineering, VCH Weinheim, **1986**.
8 P.L. Spencer, US patent 1950, 2,495,429; 1952 No. 2,593,067; 1952, US Patent No. 2,605,383.
9 B.E. Proctor and S.A. Goldblith, Electromagnetic radiation fundamentals and their applications in food technology, Adv. Food Res. **1951**, 3, 120–196.
10 S.M. Jackson, S.L. Rickter and C.O. Chischester, Food Technol. **1957**, 11, 9, 468–470.
11 D.A. Copson and R.V. Decareau, Food Res. **1957**, 22, 402–403.
12 D.A. Copson, Food Technol. **1954**, 8, 9, 397–399.
13 D.A. Copson, IRE Trans. PGME-4 **1956**, 27–35.
14 D.A. Copson, Food Technol. **1958**, 12, 6, 270–272.
15 D.A. Copson, Microwave heating. 2nd Edn., Westport, Connecticut, AVI Publishing Co., USA, **1975**.
16 D.A. Copson and E.Z. Krajewski, **1958**, US Patent no. 3,020,645.

17 S.P. Lipoma and H.E. Watkins, **1968**, US Patent no. 3,365,301.
18 J.P. O'Meara, J. Microwave Power, **1973**, 8, 167–172.
19 D.P. Smith, Microwave Energy Appl. Newsl. **1969**, 2, 3, 9–10.
20 R.V. Decareau, Microwaves in the food processing industry, New York, Academic press, USA, **1985**.
21 T.H. Lee, Planar Microwave Engineering, Cambridge University Press, Cambridge, UK, **2004**.
22 P. Debye, The collected papers of Peter J. W. Debye, Ed. Interscience Publishers, New York, USA, **1913**.
23 P. Debye, Polar molecules, Ed. Dover, New York, USA, **1947**.
24 H. Frölich, Theory of dielectric, dielectric constant and dielectric loss, Ed. Clarendon Press, Oxford, UK, **1958**.
25 C.J. Böttcher, Theory of electric polarization. vol 1: Dielectric in static fields, Ed. Elsevier, New York, USA, **1973**.
26 C.J. Böttcher, Theory of electric polarization. vol 2: Dielectric in time-dependent fields, Ed. Elsevier, Amsterdam, Holland, **1978**.
27 N.E. Hill, W.E. Vaughan, A.H. Price and M. Davies, Dielectric properties and molecular behavior, Van Nostrand Reinhold Company, London, UK, **1969**.
28 L.D. Landau and E.M. Lifshitz, Electrodynamics of continuous media (translated by J.B. Sykes and J.S. Bell), Pergamon, Elmsford, New York, USA, **1960**.
29 J.D. Jackson, Classical electrodynamics, 2nd Edn., Wiley, New York, USA, **1975**.
30 S. Ramo, J.R. Whinnery, and T. Van Duzer, Fields and Waves in communication, 2nd End. Wiley, New York, USA, **1984**.
31 R. Coelho, Physics of dielectrics for the engineer, Fundamental studies in engineering 1, Elsevier Scientific Publishing Company, The Netherlands, **1978**.
32 J. MacConnell, Rotational Brownian motion and dielectric theory, Ed. Academic press, London, England, **1980**.
33 O. Göttmann and U. Stumper, Chem. Phys. Lett., **1973**, 22, 2 387–389.
34 G.W.F. Pardoe, Trans. Faraday Soc., **1970**, 66, 2699–2709.
35 R.S. Sharma and G.Q. Sofi, Acta Licencia Indica, **1976**, 2, 4, 364–368.
36 J. Crossley, Adv. Mol. Relax. Interact. Process., **1979**, 14, 115–120.
37 L. Koszoris and G. Masszi, Acta Biochim. et Biophys. Acad. Sci. Hung., **1982**, 17 (3/4), 237–249.
38 F.F. Hanna and A.M. Bishai, Z. Phys. Chemie, Leipzig **1976**, 257, 6, 1241–1248.
39 F.F. Hanna and A.M. Bishai, Z. Phys. Chemie, Leipzig **1977**, 258, 4, 609–614.
40 H.D. Purohit and R.J. Sengwa, Bull. Chem. Soc., Japan, **1991**, 64, 6, 2030–2031.
41 H.D. Purohit and H.S. Sharma, Indian J. Pure Appl. Phys., **1973**, 11, 664–665.
42 H.D. Purohit and H.S. Sharma, Bull. Chem. Soc., Japan, **1977**, 50, 10, 2606–2608.
43 Y. Koga, H. Takahashi and K. Higasi, Bull. Chem. Soc., Japan, **1974**, 47, 1, 84–87.
44 G.P. Srivastava, P.C. Mathur and M. Krishna, Can. J. Phys., **1972**, 50, 1449–1452.
45 A.A. Anthony and C.P. Smyth, J. Am. Chem. Soc., **1964**, 86, 152–158.
46 S. Chandra and J. Prakash, Journal of Chemical Physics, **1971**, 54, 12, 5366–5371.
47 J. Crossley and C.P. Smyth, J. Am. Chem. Soc., **1969**, 91, 10, 2482–2487.
48 S.K. Srivastava and S.L. Srivastava, Indian J. Phys., **1977**, 51B, 26–32.
49 S.K. Roy, K. Sengupta and S.B. Roy, Indian J. Phys., **1977**, 51B, 42–49.
50 H. Purohit, S. Kumar and A.D. Vyas, Adv. Mol. Relax. Interact. Process., **1980**, 16, 166–173.
51 H. Purohit, H.S. Sharma and A.D. Vyas, Bull. Chem. Soc., Japan, **1975**, 48, 10, 2785–2788.
54 H.D. Purohit, H.S. Sharma and A.D. Vyas, Can. J. Phys., **1977**, 55, 1902–1905.
55 M.L. Arrawatia, M.L. Sisodia, P.C.

Gupta and S.C. Kabra, J. Chem. Soc., Faraday Trans. 2, **1991**, 77, 169–180.

56 Gerhard Klages, Z. Naturforsch, **1988**, 43a, 1–13.

57 J. Crossley, J. Chem. Phys., **1972**, 56, 6, 2549–2552.

58 M.P. Madan, Can. J. Phys., **1975**, 53, 23–27.

59 D.B. Farmer, P.F. Mountain and S. Walker, J. Phys. Chem., **1973**, 77, 5, 714–717.

60 P. Firman, A. Marchetti, M. Xu, E.M. Eyring and S. Petrucci, J. Phys. Chem., **1991**, 95, 7055–7061.

61 F.J. Arcega Solsona and J.M. Fornies-Marquina, J. Phys. D. Appl. Phys., **1982**, 15, 1783–1793.

62 K. Sengupta, S.K. Roy and S.B. Roy, Indian J. Phys., **1977**, vol. 51B, 34–41.

63 J.K. Vij, I. Krishan and K.K. Srivastana, Bull. Chem. Soc., Japan, **1973**, 46, 17–20.

64 M. Yao and Y. Hiejima, J. Mol. Liq., **2002**, 96–97, 207–220.

65 M.S. Skaf and D. Laria, J. Chem. Phys., **2000**, 113, 3499–3502.

66 C.N. Yang and H.J. Kim, J. Chem. Phys., **2000**, 113, 6025–6032.

67 A.H. Narten and A. Habenshuss, J. Chem. Phys., **1984**, 80, 3387–3393.

68 E.L. Kinsey and J.W. Ellis, Phys. Rev., **1937**, 51, 1074–1078.

69 W. West, J. Chem. Phys., **1939**, 7, 795–801.

70 D. Stuerga and P. Gaillard, J. Microwave Power Elec. Energy, **1996**, 31, 2, 101–113.

71 U. Kaatze and V. Uhlendorf, Z. Phys. Chem. Neue Folge **1981**, 126, 151–165.

72 U. Kaatze, J. Chem. Eng. Data **1989**, 34, 371–374.

73 R. Chahine, T.K. Bose, C. Akyel and R.G. Bosisio, J. Microwave Power Elec. Energy, **1984**, 19, 2, 127–134.

74 A.R. Von Hippel, Dielectric materials and applications, MIT press, USA, **1954**.

75 A.J. Berteaud and J.C. Badot, J. Microwave Power Elec. Energy, **1976**, 11, 315–318.

76 K.S. Cole and R.H. Cole, J. Chem. Phys., **1949**, 9, 341–345.

77 D.W. Davidson and R.H. Cole, J. Chem. Phys., **1951**, 18, 1417–1422.

78 S.H. Glarum, J. Chem. Phys., **1960**, 33, 1371–1375.

79 J.E. Anderson and R. Ullman, J. Chem. Phys., **1967**, 47, 2178–2184.

80 A.L. McClellan, Tables of experimental dipole moments, W.H. Freeman and Company, San Francisco; USA, **1967**.

81 J.B. Hasted, D.M. Ritson and C.H. Collie, J. Chem. Phys., **1948**, 16, 1, 1–11.

82 J.B. Hasted, D.M. Ritson and C.H. Collie, J. Chem. Phys., **1948**, 16, 1, 12–21.

83 A.K. Lyashchenko and A.Y. Zasetsky, J. Mol. Liq., **1998**, 77, 61–75.

84 A.A. Asheko and K.E. Nemchenko, J. Mol. Liq., **2003**, 105, 2–3, 295–298.

85 A.S. Lileev, Z.A. Filimonova and A.K. Lyashchenko, J. Mol. Liq., **2003**, 103–104, 299–308.

86 J. Barthel, J. Mol. Liq., **1995**, 65/66, 177–185.

87 M. Nakahara and C. Wakai, J. Mol. Liq., **1995**, 65/66, 149–155.

88 D.G. Frood and T.J. Gallagher, J. Mol. Liq., **1996**, 69, 183–200.

89 J. Barthel, R. Buchner, P.N. Eberspächer, M. Münsterer, J. Stauber, and B. Wurn, J. Mol. Liq., **1998**, 78, 83–109.

90 T. Tassaing, Y. Danten and M. Besnard, J. Mol. Liq., **2002**, 101/1–3, 149–158.

91 J.K. Vij, D.R.J. Simpson, and O.E. Panarina, J. Mol. Liq., **2004**, 112, 125–135.

92 G. Sutmann and R. Vallauri, J. Mol. Liq. **2002**, 98–99, 213–224.

93 A.A. Chialvo, P.T. Cummings, and J.M. Simonson, J. Mol. Liq., **2003**, 103–104, 235–248.

94 A. Geiger, M. Kleene, D. Paschek and A. Rehtanz, J. Mol. Liq., **2003**, 106/2–3, 131–146.

95 G.G. Malenkov, D.L. Tytik, and E.A. Zheligovskaya, J. Mol. Liq., **2003**, 106/2–3, 179–198.

96 A.K. Lyashchenko and V.S. Dunyashev, J. Mol. Liq., **2003**, 106/2–3, 199–213.

97 V.I. Lobyshev, A.B. Solovey, and N.A. Bulienkov, J. Mol. Liq., **2003**, 106/2–3, 277–297.
98 A. Botti, F. Bruni, S. Imberti, M.A. Ricci, and A.K. Soper, J. Mol. Liq., **2005**, 117, 77–79.
99 A. Botti, F. Bruni, S. Imberti, M.A. Ricci, and A.K. Soper, J. Mol. Liq., **2005**, 117, 81–84.
100 R.L. Smith, S.B. Lee, H. Homori and K. Arai, Fluid Phase Equilib., **1998**, 144, 315–322.
101 L. de Maeyer and G. Kessling, J. Mol. Liq., **1995**, 67, 193–210.
102 K. Ando and J.T. Hynes, J. Mol. Liq., **1995**, 64, 25–37.
103 E. Guardia, J. Marti, J.A. Padro, L. Saiz, and A.V. Komolkin, J. Mol. Liq., **2002**, 96–97, 3–17.
104 A.A. Atams, N.A. Atamas, and L.A. Bulavin, J. Mol. Liq., **2005**, 120, 15–17.
105 T. Sato and R. Buchner, J. Mol. Liq., **2005**, 117, 23–31.
106 R.M. Fuoss and J.G. Kirwood, J. Am. Chem. Soc., **1941**, 63, 385.
107 T.M. Birshtein and O.B. Ptitsyn, Zh. Fiz. Khim., **1954**, 24, 1998, 217–222.
108 J. Marchal and H. Benoit, J. Chem. Phys., **1955**, 52, 818–820.
109 D.W. Davidson, Molecular relaxation processes, Chemical Society Publ., London, UK 1966 no. 20–33.
110 J. Errera, Colloid Science Vol 2, Chemical Catalog Co., New York, USA, **1926**.
111 H.P. Schwan, Adv. Biol. Med. Phys., **1957**, 5, 147–149.
112 J. Lyklema and H.P. Van Leeuwen, Adv. Colloid Interface Sci., **1999**, 83, 33–46.
113 J.E. Hinch, J. Chem. Soc. Faraday Trans. II, **1984**, 80, 535–543.
114 R.W. O'Brien, J. Colloid Interface Sci., **1986**, 113, 81–88.
115 S. Hussain, I.J. Youngs, and I.J. Ford, J. of Physics D: Applied Physics, **2004**, 37, 318–325.
116 A. Bonincontro and C. Cametti, Colloids Surf. A: Physicochem. Eng. Aspects, **2004**, 246, 115–120.
117 R. Buchner, C. Baar, P. Fernandez, S. Schrödle, and W. Kunz, J. Mol. Liq., **2005**, 118, 179–187.
118 P.E. Jönsson, J.L. Garcia-Palacios, M.F. Hansen, and P. Nordblad, J. Mol. Liq., **2004**, 114, 131–135.
119 D. Stuerga, The future of thermal microwaves: Challenges of thermal uniformity, dimensional resonances, chemical and hydrodynamic selectivities HDR dissertation (in French), University of Burgundy, July 13, **1994**.
120 D. Stuerga and M. Lallemant, J. Microwave Power Electromag. Energy, **1993**, 28, 2, 73–83.
121 C. Lohr, P. Pribetich and D. Stuerga, Microwave Optical Technol. Lett., **2004**, 42, 1, 46–50.
122 D.E. Livesay and K. Chen, IEEE Trans. MTT, **1974**, 22, 1273–1274.
123 A. Taflove and M.E. Brodwin, IEEE Trans. MTT, **1975**, 23, 888–891.
124 H.N. Kritikos and H.P. Schwan, IEEE Trans. BE, **1975**, 22, 457–459.
125 M.J. Haymann, O.P. Gandhi, J.A. d'Andrea and I. Chattergee, IEEE Trans. MTT, **1979**, 27, 809–811.
126 R.J. Spiegel, IEEE Trans. MTT, **1984**, 32, 730–732.
127 T. Ohlsson and P.O. Risman, J. of Microwave Power, **1978**, 13–4, 303–305.
128 D. Stuerga, I. Zahreddine, and M. Lallemant, C.R. Acad. Sci. Paris, **1992**, 315, II, 1319–1324.
129 D. Stuerga, I. Zahreddine, C. More and M. Lallemant, C.R. Acad. Sci. Paris, **1993**, 316, II, 901–906.
130 D. Stuerga and M. Lallemant, J. Microwave Power Electromag. Energy, **1993**, 28, 4, 206–218.
131 D. Stuerga and M. Lallemant, J. Microwave Power Electromag. Energy, **1993**, 28, 4, 219–233.
132 D. Stuerga, A. Steinchen-Sanfeld, and M. Lallemant, J. Microwave Power Electromag. Energy, **1994**, 29, 1, 3–19.
133 D. Stuerga and M. Lallemant, J. Microwave Power Electromag. Energy, **1994**, 29, 1, 20–30.
134 M. Wien, Phys. Z,. **1931**, 32, 545.
135 L. Onsager, J. Chem. Phys., **1934**, 2, 599–615.
136 T. Yasunaga, T. Sano, K. Takahashi, H. Takenaka, and S. Ito, Chem. Lett., **1973**, 405–408.

137 M. Eigen and L. Demayer, in Techniques of Organic Chemistry Vol. VIII Eds. S.L. Friess, E.S. Lewis and A. Weissberger, Wiley Interscience, New York, USA, **1963**.
138 M. Eigen and L. Demayer, Z. Electrochem., **1955**, 59, 10, 986–993.
139 L. Perreux and A. Loupy, Tetrahedron, **2001**, 57, 9199–9223.
140 D. Stuerga and P. Gaillard, J. Microwave Power Electromag. Energy, **1996**, 31, 2, 87–113. It is better to read the guest editorial by P.O. Risman of this issue of the journal.
141 D. Stuerga, K. Gonon, and M. Lallemant, Tetrahedron, **1993**, 49, 28, 6229–6234.
142 C. Lohr, P. Pribetich and D. Stuerga, Microwave Optical Technol. Lett., **2004**, 42, 5, 365–369.
143 D.R. Baghurst and M.D. Mingos, J. Chem. Soc. Chem. Commun., **1992**, 674–677.
144 G. Bram, A. Loupy, M. Majdoub, E. Gutierrez, and E. Ruiz-Hitzky, Tetrahedron, **1990**, 46, 15, 5167–5176.
145 D. Stuerga, P. Gaillard and M. Lallemant, Tetrahedron, **1996**, 52, 15, 5505–5510.

2
Development and Design of Laboratory and Pilot Scale Reactors for Microwave-assisted Chemistry

Bernd Ondruschka, Werner Bonrath, and Didier Stuerga

2.1
Introduction

The chemical industry is concerned with production of chemicals with very simple to very complex structures [1]. When dealing with relatively simple structures, there does not usually seem to be any need for deeper understanding of chemistry than that to which an engineer is normally exposed. Most reaction engineering textbooks are designed with this basic assumption [2]. Catalysis, which is invariably an integral part of the reaction engineer's basic knowledge, has been connected to the production of large-volume chemicals which are often relatively simple in structure. During the last years, new catalytic processes have been established in the fine chemical and pharmaceutical industries, which deal with more complex structures [3]. Increasing attempts by chemists to extend the use of catalysis to the production of medium and small-volume chemicals has resulted in a change in perspective requiring closer liaison and better mutual understanding between chemists and engineers. Another change is the increasing role of process intensification. This is nowhere more evident than in the production of organic chemicals. Process intensification consist of the development of novel apparatus and techniques which, compared with those commonly used today, are expected to bring dramatic improvements in manufacturing and processing, substantially reducing equipment-size/production capacity ratio, energy consumption and waste production, and ultimately resulting in cheaper, sustainable techniques [4]. A common approach is reaction-rate enhancement by extending known or emerging laboratory techniques to industrial production. This technology can be chemistry and engineering-intensive, or both. Nowadays, attractive examples are the use of high frequencies (microwave-assisted chemistry [5, 6]), ultrasound (sonochemistry) [7], photons (photochemistry) [8], enzymes (biotechnology) [9], immiscible phases (phase-transfer catalysis) [10], microparticles (microphase engineering) [11], membranes (membrane reactor engineering) [12], combinations of reactions with different separation technology (multifunctional or combinatorial reactor engineering or reactive separations) and mixing [13]. Their use in the production of medium and small-volume chemicals such as pesticides, drugs, pharmaceuticals, perfumery chemicals, and other consumer products is being increasingly explored both by academia and industry [14]. Some of this technology has developed little beyond the laboratory stage, although it has been a part of the synthetic organic methods for several years. This overview explores microwave technology for the reaction rate and/or yield and/or selectivity enhancement. The application of microwave technology demonstrates the generalization of known reaction engineering principles for homogeneous and heterogeneous reaction systems.

2.2
Basic Concepts for Reactions and Reactors in Organic Synthesis

Two essential questions about chemical syntheses are important to chemists and engineers engaged in research, reaction design, or operation involving chemical reactions [2]:

- What are the reaction equilibrium conditions?
- How rapidly is it possible to attain a desirable approach to equilibrium conditions?

Proper answers are rather complex, because different properties and conditions of a chemical system affect both equilibrium and reaction rate. Although the questions are related, no unified quantitative treatment yet exists, and to a large extent they are handled separately by the sciences of thermodynamics and reaction kinetics. Fortunately, with the help of thermodynamic and kinetics, the questions can be answered for many reactions with the aid of data and generalizations obtained by thermal, spectroscopic, and chromatographic measurements, and/or experimental computer chemistry, and the estimation methods of Benson [15].

Chemical reaction rate may be particularly affected by factors such as flow conditions, phase boundaries, and the presence of foreign substances; well-known factors such as temperature, pressure, and relative amounts of the reactants also strongly affect the equilibrium. The ultimate objective of the engineer working in this field is to design processes and equipments for conducting reactions on a larger scale or to modify, as needed, existing equipment or designs. Rate data may be obtained with either batch or continuous equipment. With the former, the reactants are charged in bulk to a stirred vessel and observations are made during the course of the reaction. With the latter, reactants are charged continuously at measured rates through a tube reactor. The tubular-flow reactor is either the differential type or integral. The former is so short only a small, though necessarily measurable, amount of conversion occurs. This affords direct evaluation of the instantaneous reaction rate. In the latter comparatively large conversion may occur. Both types have their utility, also for microwave-assisted reactions.

Reactions can be classified as reversible, irreversible, parallel, and consecutive. With regard to operating conditions they are isothermal at constant volume, isothermal at constant pressure, nonisothermal, adiabatic, and polytropic. Reactions are also classified according to the phases involved:

1. homogeneous (gaseous, liquid, or solid) phase and
2. heterogeneous (controlled by diffusive mass transfer or by chemical resistance).

It is, furthermore, often very important to distinguish between:

3. noncatalyzed, and
4. catalyzed reactions.

Ideal and real equipment types are just as much a basis for a further differentiation:

5. stirred tank or tank battery,
6. single or multiple-tubular reactor, and
7. reactor filled with solid particles, inert or catalytic (fixed/moving/fluidized bed).

There are these well-known types of reaction:

8. batch,
9. semi-batch, and
10. continuous.

Clearly, these classifications are also useful for reactions under the action of microwave irradiation [6]. From the engineering perspective, the principal distinctions are drawn between homogeneous and heterogeneous liquid and/or gas-phase reactions, solvent-free reaction conditions, and between batch, semi-batch and continuous reactions. The highest influence has the necessary equipment and operating conditions on the one side and the design, synthetic, and analytical methods on the other side.

2.3
Methods for Enhancing the Rates of Organic Reactions

According to Scheme 2.1 the principles of green chemistry can be condensed to the word "productively" [16]. Two main goals are to eliminate or minimize the use of volatile organic solvents in modern syntheses and to reduce energy inputs. Development of new synthetic solvent-free methods with microwave assistance is an important topic of research with growing popularity, because solvent-free reactions reduce solvents usage, simplify synthesis and separation procedures, prevent waste, and avoid the hazards and toxicity associated with the use of solvents [135].

Green chemistry must be combined with more environmentally friendly technology if step-change improvements are necessary in chemical manufacturing processes. Synthetic chemists have traditionally not been adventurous in their choice of reactor type: the familiar round-bottomed flask with (mechanical, magnetic) stirrer remains the automatic choice for most, even when the chemistry they plan

P	Prevent waste.
R	Renewable materials.
O	Omit derivatization steps.
D	Degradable chemical products.
U	Use safe synthetic methods.
C	Catalytic reagents.
T	Temperature, pressure ambient.
I	In-process monitoring.
V	Very few auxiliary substances.
E	Factor, maximize feed in product.
L	Low toxicity of chemical products.
Y	Yes, it is safe!

Scheme 2.1. Condensed principles of green chemistry.

to use is innovative, e.g. the use of a nonvolatile solvent (ionic liquids) or a heterogeneous catalyst as an alternative to a soluble reagent. An increasing number of research articles describing green chemical (engineering) reactions are based on alternative reactors, however.

Energy questions have often been (somewhat) neglected in the calculations of resource utilization for a chemical process. Batch processes based on large reaction vessels can run for many hours or even days to maximize yield and often suffer from poor mixing and heat-transfer characteristics. As the cost of energy increases and greater efforts are made to control emissions associated with generating energy, the use of energy will become one of the most important parts green chemistry calculations. Future trends will be not only better-designed reactors such as those described below but alternative energy sources. Table 2.1 compares general and specific methods for the stimulation of reactions, and system requirements.

Microwave-assisted chemistry is based on the use of intensive directed irradiation and its purposeful application can have several advantages:

Table 2.1. Methods for chemical activation.

General energy methods	System requires
Piezo chemistry	High pressure
Thermo chemistry	Heating
Sono chemistry	Sound source and liquid
Specific energy methods	**System requires**
Electro chemistry	Conducting media
Photo chemistry	Chromophore(s)
Microwave chemistry	**Polar media or species/ions**

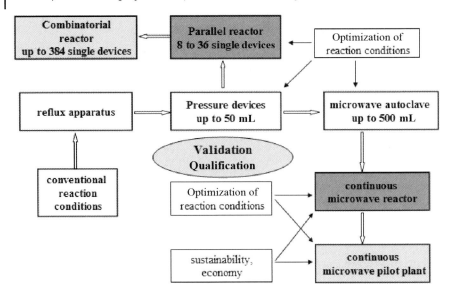

Fig. 2.1. Development concepts for microwave-assisted reactions and processes.

- short heating times,
- wide usable ranges of temperature and pressure,
- accumulation of energy added to the system,
- inversion of heating flux in the reactor,
- high(er) energy efficiency,
- sophisticated measurement and safety technology, and
- modular systems which enable changing from mg to kg scale.

Toward development of microwave-initiated reactions and processes the flow-chart in Fig. 2.1 was designed in Jena and its versatility successfully tested, Fig. 2.2 [17–20].

2.4
Microwave-assisted Organic Syntheses (MAOS)

During the last two decades there has been interest in the potential of microwave energy as a means of rate enhancement in synthesis. Since pioneering reports in *Tetrahedron Letters* – both in 1986 (Gedye et al. and Giguere et al., but see also the "pre" pioneering report of N. Bhargava, Can. Pat. 353929, 1980) – a plethora of reactions have been conducted using microwave irradiation. Many reviews, overviews, and some books or monographs have been published [5, 6, 21–29].

Nowadays it is accepted that under the action of microwave irradiation chemical reactions are affected by:

2.4 Microwave-assisted Organic Syntheses (MAOS)

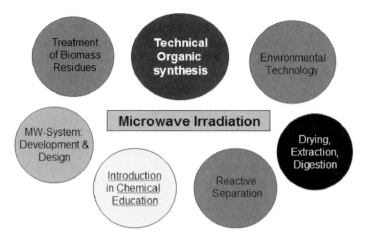

Fig. 2.2. Microwave-assisted investigations at the Institute of Technical Chemistry and Environmental Chemistry, Friedrich Schiller University of Jena.

- overheating,
- hot-spot formation,
- polarization,
- dielectric properties,
- solvent sensitivity to microwaves,
- spin alignment, and (partly)
- nuclear spin rotation.

This framework represents the major effects of microwaves, which are not always equally important.

The microwave region of the electromagnetic spectrum corresponds to wavelengths from 1 cm to 1 m. The most used wavelength is 12.24 cm (2.45 GHz). The heating effect produced by microwaves is especially affected by the dielectric constant, ε'_r, of the medium, which is a measure of the ability of the molecules to be polarized by an external electrical field. Another important factor is the dielectric loss, ε''_r, which is a measure of the efficiency with which electromagnetic energy can be converted into heat. The dielectric loss for a molecule goes through a maximum as the dielectric constant is reduced. The ratio of dielectric loss to dielectric constant is defined as the dielectric loss tangent, $\tan \delta$, which is an important number for characterizing microwave heating.

2.4.1
Microwave Ovens and Reactors – Background

This part is written as a practical tool for people interested in laboratory or industrial equipment. Exhaustive coverage of microwave oven design would require more space than is available in this chapter. Excellent coverage of applicator theory [30–32] and the basis of electromagnetic waveguides and cavities [33] can be found elsewhere.

2.4.1.1 Applicators, Waveguides, and Cavities

The microwave applicator is the component of a processing system in which energy is applied to the product to be heated. The equipment designer's objective is to ensure that this is accomplished in the most efficient manner possible. The packaging of the reagents (powder, liquids, pellets) coupled with dielectric characteristics enable the designer to literally mold a process in terms of applying the required quantity of energy where it is needed. A wide variety of applicators are available or have been patented to cover almost any conceivable applications of microwave power [30, 31]. In chemistry applications the goal is to achieve the desirable reaction or products.

High-power microwaves are generated by vacuum tubes. The magnetron and klystron are the most commonly used tubes for generation of continuous waves of power suitable for microwave processing. Power is normally launched from the microwave tube into a transmission line or waveguide, where it travels to a load or termination such as an antenna or a microwave heating applicator.

Lumped circuits with capacitors and inductors used at lower radio frequencies are not usable at microwave frequencies. Open transmission lines are not used at microwave frequencies because the radiation would be excessive. This has led to the use of waveguides as transmission media and to the use of resonant cavities as applicators. Waveguides are metallic tubes of circular or rectangular cross-section. Resonant cavities are metallic boxes (parallelepipedic or circular). Voltage and currents are not the fundamental concerns. In fact, power is considered to travel in the transmission line by means of electromagnetic waves which consist of alternating electric and magnetic fields. When energy is launched into waveguide, many modes may be excited. For a waveguide, an infinite number of modes or wave configurations can exist. Only those modes above cutoff will attenuate in a short distance. Cutoff conditions are defined by the dimensions of the waveguide cross-section. Discontinuities within the waveguide may excite higher-order modes, resulting in higher energy storage at the discontinuity. In the same way as for waveguides, an infinite number of modes or wave configurations can exist for microwave cavities. According to the geometry and dimensions of the boxes, single-mode or multimode cavities can be obtained. Domestic ovens belong to this last category. Empty domestic ovens have approximately two-hundred modes [34].

When microwaves travel along a waveguide terminated by the microwave heating application (for example a resonant cavity loaded by the object to be heated) a reflected wave travels back toward the source. The wave traveling towards the termination is called the incident wave and the wave traveling back to the magnetron is the reflected wave. The goal of microwave oven design is to ensure that all the incident power is absorbed by the load. In other words, the resonant frequency of the loaded oven (and not the empty oven) should be close to frequency of the magnetron (i.e. 2.45 GHz). If too much energy is reflected back to the source, the magnetron may be damaged. This is why it is not advisable to run empty domestic ovens. Most commercial ovens are protected by an automatic thermal cut-off in the event of poor matching between magnetron and oven.

2.4.1.2 Single-mode or Multi-mode?

The electric field pattern produced by the standing waves within the cavity may be extremely complex. Some areas may receive large amounts of energy whereas others may receive little energy [35–37]. To minimize this, domestic ovens use mode stirrers or fan-shaped paddles and turntables. These devices are designed for typical domestic loads. Consequently, the use of domestic ovens for laboratory purposes with small loads and poorly lossy media can lead to bad operating conditions, specially difficult to reproduce. The use of single-mode or "quasi"-single-mode cavities enables definition of precise positions within the cavities where the electric field strength is maximum. The electric field strength is, moreover, much higher than in multi-mode devices. The effective cavity power is three orders of magnitude higher.

Use of single-mode cavities is more complex, however, because insertion of the lossy sample changes the frequency resonance of the device. A multi-mode cavity is more versatile than a single-mode device, because of the number of modes. A movable plunger can enable a tuning of the cavity to ensure good matching with the source. Because the dielectric properties of materials are very dependent on temperature, real-time tuning is necessary during the heating process. This is usually achieved by use of a computer. Use of a circulator which directs the reflected energy into a dummy water load is needed to avoid damage of the magnetron in the event of poor matching. Another reason for irreproducible operating conditions in the domestic oven is the variable power. The variable power of domestic ovens is the consequence of periodic switching of the magnetron power. Large switching periods are undesirable in chemistry because of the cooling period between switching steps.

In conclusion, use of multi-mode systems for laboratory purposes requiring reproducible operating conditions implies the same geometry, volume and position for the samples to be heated and switching cycles smaller than the characteristic time of the chemical kinetics of the reaction studied. The use of single-mode cavities seems to be of particular importance for enabling efficient application of microwave energy. One mistake is to believe that single-mode cavities imply small operating volume close to several cubic centimeters. An efficient temperature-control system enabling feedback control during heating would be an asset in the event of thermal runaway and/or exothermic reactions. Thermal runaway is a consequence of the thermal dependence of dielectric losses. Above a threshold value of temperature, the rate of heating becomes very high.

2.4.1.3 Limits of Domestic Ovens

Most microwave-promoted organic synthesis has been performed in multi-mode domestic ovens. In these ovens, despite the power level which commonly fluctuates as a result of the on–off cycles and heterogeneous energy, there are other problems, relating to safety. Heating organic solvents in open vessels can lead to violent explosion induced by electric arcs inside the cavity or sparking as a result of switching of the magnetron. Conventional chemical reflux can be used if the water condenser is outside the microwave cavity. To do this it is necessary to connect the reaction vessel to the condenser through a port that ensures microwave leakage is within safe limits. Mingos et al. [38] described this kind of domestic oven modification for atmospheric pressure operation.

Another efficient way of using microwaves safely in organic synthesis is the use of solvent-free procedures (Ref. [81] and Chapter 8 in this book).

High pressure operating conditions induced by microwave heating are very attractive for chemists. Microwave irradiation leads to core heating, so microwave heating of autoclaves has advantages compared with conventional heating modes. The closed vessel must be transparent to electromagnetic waves and must sustain the pressures induced by vaporization of solvents. The vessel material should have very low dielectric losses. This material must also be chemically inert and able to accommodate the rate of pressure increase induced by microwave heating. This depends on microwave power level and also on the media heated. The Paar instrument company [39] has designed this kind of vessel. They are made from Teflon and polyetherimide and they can support pressures up to 80 atm and temperatures up to 300 °C (Fig. 2.3) [40, 41]. The reagent volume is approximately 20 cm^3. Obviously, for safe use of this system, a pressure-release system enables avoidance of violent explosion of the vessels. For safety and reproducible operating conditions, however, we recommend use of systems with real time monitoring of pressure.

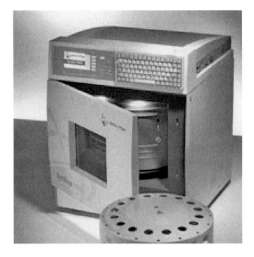

Fig. 2.3. Microwave setup "Synthos 3000" (Paar).

2.4.1.4 Temperature Measurement Limits

Another problem with domestic ovens is that of temperature measurement. Classical temperature sensors should be avoided because of the power levels – temperature measurements would be distorted by strong electric currents induced inside metal wires used to connect the temperature sensor. The technological solution is optical fiber thermometers [42–45], although measurements are limited to below 250 °C. For higher values, surface temperature can be estimated by use of an infrared camera or a pyrometer [46, 47]. Because of the volumic nature of microwave heating, however, surface temperatures are often less than core temperatures.

2.4.1.5 Design Principles of Microwave Applicators

The classical industrial design of microwave applicators and, specifically, the choice of geometrical shape are based on a simple similarity principle between wave propagation and spatial distribution within the empty and the loaded microwave applicators. The dielectric load of the reactor defined by chemical vessels and the reactants to be transformed. This theoretical approach is approximate. The spatial distribution of electromagnetic fields within applicators strongly depends on geometrical shape and the dielectric properties of the load [48–52]. Hence, this design method will be valid only if the dielectric perturbation induced by the reactor is negligible. In fact, the magnitude of the perturbation is proportional to reactor-to-applicator volume ratio. The perturbation would be negligible if this ratio were close to 10^{-4} but for most laboratory and industrial devices the ratio is higher [53].

It is, therefore, more efficient but also more complicated to be guided by a geometrical matching principle. According to this principle the microwave applicator designer wishes to ensure a good match between electric field spatial distribution and the geometrical shape of the chemical vessel used. This geometrical matching principle is easier to apply for mono-mode applicators because of knowledge of wave-propagation directions and spatial distribution. The limit of this design method is that it requires a knowledge of both empty applicator modes and loaded applicator modes. A general schematic diagram of this design principle is given in Fig. 2.4.

In the following text the different laboratory, experimental and industrial devices will be described according to the geometric shape of microwave applicators and reactors. Two fundamental transverse cross-sections of microwave applicators will be distinguished – rectangular and circular – whereas reactor geometrical shape can be cylindrical or egg-shaped. The geometrical shapes of microwave applicators and chemical reactors and the physical nature of the reactant phases (solid, liquid, or gas) are the most important aspects of the following description of laboratory, experimental, and industrial microwave reactors.

First, laboratory and experimental reactors will be described. The vessel containing the reactants or their supports are made of convenient dielectric materials (cylindrical or egg-shaped reactor). Original microwave reactors will be described. The first is a metallic cylindrical reactor which is also the microwave applicator. It enables high pressures to be achieved. The other microwave reactor is egg-shaped, leading to high focusing of the microwave power.

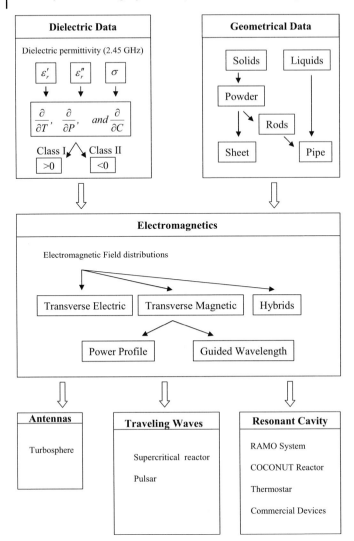

Fig. 2.4. Design principles of microwave applicators.

Second, industrial microwave reactors will be described. Most industrial applicators are made of rectangular wave guides [54]. Reactants are contained within a pipe or placed on a simple conveyer belt.

2.4.2
Scale-up of Microwave Cavities

As often within reactor design, reaction field homogeneity within the microwave applicator is an essential criterion (cf. the phenomenon of field distributions under sonochemical or photochemical conditions). The irradiation field obtained in domestic microwave reactors is clearly not homogeneous [136, 137]. For this reason the most important problems in scale-up from microwave cavities are:

- energy efficiency,
- uniformity of heating in the irradiated zone, and
- safe monitoring and control of the reaction.

2.4.3
Efficiency of Energy and Power

The power transferred to a reaction mixture by microwaves is given by:

$$P_d = (\lambda + 2\pi v \varepsilon'_r)|E|^2$$

where P_d is the power transferred in an unit volume of material with conductivity λ and dielectric constant ε'_r. E is the local electric field intensity of the microwaves of frequency v. Typical applications for which microwaves have been used so far are dielectric materials for which conductive heat transfer is negligible. In reactions involving catalysts, however, conductive heat transfer may be significant [55, 56]. In general, the power transferred in a microwave oven is used to heat many materials besides providing the heat of reaction; the total power supplied, P_t, can be divided into two components:

$$P_t = P_d + \Sigma(\Delta T/\Delta t)c_p\rho$$

The summation in the second term indicates that many materials are heated by microwaves. The main focus of microwave engineers is to minimize the second term. This can be done, for example, by appropriate choice of insulation and other materials. Materials with lower heat capacity and density have lower absorption of microwave energy [57].

2.4.4
Field Homogeneity and Penetration Depth

The spatial nonuniformities are a major concern in the scale-up of microwave reactions. Traditionally, turntables and (mono- or multi-)mode mixers in domestic microwave ovens have been used to reduce spatial nonuniformity [58]. There are two types of nonuniformity in the microwave irradiation process. The first is known as the "standing wave effect", which is the repeating field intensity variation within the microwave applicator. This intensity typically follows a half sine-wave pattern. Because of this "standing wave effect", the field intensity can change from the maximum to zero in a distance of one-quarter of the operating wavelength (12.24 cm). The second type of inhomogeneity is related to the penetration depth of microwaves. An approximate relationship for the penetration depth D_p for small dielectric heat loss is:

$$D_p \approx \lambda_{mw}(\varepsilon'_r/\varepsilon''_r)^{1/2}$$

where λ_{mw} is the wavelength of the microwave irradiation. From this expression one can expect that the depth to which microwaves can penetrate depends largely on the material properties and the wavelength used. Hence a reaction conducted in a (for example, Erlenmeyer) flask in, for example, a modified domestic or commercial microwave oven may not proceed to any appreciable extent when performed in a larger microwave reactor. The classical problem of reaction scale-up has yet to be seriously addressed for microwave reactions. The problem of the spatial nonuniformity of microwave irradiation can be partly overcome by using mechanical (or magnetic) stirrers to mix its contents. Another and/or additional way of overcoming this limitation is to use microwaves of different wavelengths (cf. pioneering work by Linn. [59] or by using radio-wave systems). The preferred method of overcoming this problem is use of continuous reaction mode combined with suitable reactor diameter.

2.4.5
Continuous Tube Reactors

The objective of this section is to give a comprehensive and critical overview on the concept of continuous-flow processes including microwave-assistance. The first continuous microwave system was described by Strauss et al. [24]. Only a small volume of the reaction solution was exposed to microwave irradiation and the power needed for heating was low. The design overcomes the penetration depth problem. Liquids flowing through a small tube are likely to be heated uniformly in the radical direction while passing through the microwave field. The main factor in the design of such reactors is the length of the reaction section, which depends on the necessary average hydrodynamic residence time for attractive yields and selectivity. A variety of continuous flow systems have been proposed and tested [60–74].

2.4.6
MAOS – An Interdisciplinary Field

It is well known that a wide variety of organic reactions are accelerated substantially by microwave irradiation in sealed tubes. These rate enhancements can be attributed to superheating of the solvent, because of the increased pressure generated when the reactions are performed in the a.m. manner. Furthermore several reports have described increased reaction rates for reactions conducted under the action of microwave irradiation at atmospheric pressure, suggesting specific or nonthermal activation by microwaves. Some of these re-studied reactions occur at similar rates under the action of microwave and conventional heating under identical conditions, e.g. temperature.

There is, nevertheless, experimental evidence that some reactions occur faster than conventionally heated reactions at the same temperature. Although this suggests that nonthermal effects are involved, these rate enhancements could also be attributable to localized superheating or hot spots. In fact, most reports of substantial rate enhancements under the action of microwaves seem to be in heterogeneous reactions, for example under dry media conditions or when solid catalysts are used [75, 76]. In these circumstances localized heating of a microwave-absorbing solid support or catalyst could lead to an increased reaction rate [136, 137].

For many reactions with efficient mechanical stirring and strictly similar conditions, important specific not purely thermal microwave effects were observed when compared with conventional heating (Ref. [138] and Chapter 4 in this book). They can be attributed to polarity increases during the progress of a reaction depending strongly on medium (avoiding polar solvents) and mechanisms (transition states more polar than ground states).

The question of reproducibility and scale-up will always imply the question about reaction conditions. In addition, the reaction medium (phase) plays a much more important role for this kind of power input compared with classical reactions. Besides the molecular mass, reaction mixture polarity is essential for absorption of microwave power. Because dielectric constants are known for a few compounds only and, moreover, at near room temperature, more problems are predictable and require close contact with neighboring disciplines, for example with electrical engineering. The primary literature reflects the incomplete nature of results from microwave-assisted reactions and processes, as it does for conventional syntheses. The dependence of reaction engineering on technical considerations is, however, greater for microwave-assisted reactions, so improved description of reaction conditions is crucial.

Here we would like to remark that correct and/or complete description of the setup used, the method of temperature measurement, the reaction size, and the energy input are especially important in a microwave procedure [20, 76].

2.5
Commercial Microwave Reactors – Market Overview

In the 1990s, with use of microwaves in synthetic chemistry, the market for suitable equipment grew. At the beginning of the 21st century it is expected that the relevant market for equipment for chemical synthesis will overtake the market for chemical analysis. The three leading players (CEM Corporation, Biotage, and Milestone (MLS)) manufacture equipment which is suitable for laboratory use. The target customers are academic working groups and laboratories in the chemical, pharmaceutical and biochemical industries. The intention of all producers and users is to extend the dimensions of product capacity.

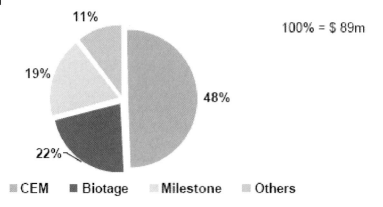

Fig. 2.5. Microwave market in 2003.

The market for installations used in microwave chemistry can be estimated to about 90 million dollars (2003), cf. Fig. 2.4 and Ref. [28].

By contacting the homepages of these enterprises [41, 58, 77, 78], one can follow recent developments and the availability of microwave equipment. A recent state-of-the-art-overview of this subject has been written by Kappe and Stadler [6b].

On September 22, 2005, the CEM Corporation gave notice of the formation of a new division within the company that will focus on developing systems for scaling up reactions performed using microwaves. Investment of at least $10 million over the next five years will be devoted to developing systems for synthesis of multikilogram to manufacturing-scale quantities of materials. "The formation of this new division will enable CEM to offer a complete microwave synthesis solution from discovery phase to full production scale and will have a major impact on companies specializing in pharmaceutical sciences, biosciences, specialty chemicals and polymer synthesis" (M.J. Collins [77] and Chapter 20 in this book).

The relatively low cost of domestic microwave ovens was, and remains, an attractive option for performing reactions on a laboratory scale. A common modification is addition of a port to enable insertion of reaction tubes or (open or closed) flasks into the cavity. Some published papers describe microwave ovens for safer operation [6, 18].

Reactions performed under atmospheric conditions at the boiling point of the reaction mixture are the mainstay of traditional synthetic methods. Traditional reflux is easily adapted to a wide range of reaction volumes, glass components are easy to configure, and reaction homogeneity is achieved by constant of the boiling solution or by stirring.

Reflux conditions under the action of microwaves gives chemists the opportunity to study the unique effects of microwaves on reaction rates and mechanisms. Microwave enhancement of reflux reactions depends entirely on relative absorption of microwaves and differential and localized heating of the different components of the reaction mixture.

2.5.1
Prolabo's Products

This first laboratory equipment is no longer commercially available since the closure of the French Prolabo company. The products are now supported by CEM [77]. Prolabo's systems are described here, however, because many laboratories use these devices [79]. The Synthewave™ 402 and 1000 are systems devoted to laboratory synthesis [80, 81, 127]. They have a closed rectangular waveguide section as the cavity. The magnitude of microwave power available is 300 W. They can be used with cylindrical tubes of several different diameters. A condenser can be connected to the tube. The originality of this arrangement is to enable measurement of temperature by use of an infrared pyrometer. Synthewave™ systems enable measurements at the bottom of the heated tube and are very suitable for laboratory experiments. Users should be aware of the problem of temperature measurement. It is possible to calibrate the IR emissivity factor for each medium to be tested. This setup could be combined with a computer for regulation of the temperature of the reagents. The Synthewave™ 402 and 1000 are illustrated in Fig. 2.6.

The Soxwave™ 100 is a variant which has been designed for extractions, for example Soxhlet, which are laborious. The extraction tube is capped with a cooling column. Optional temperature control during extraction was available. Maxidigest™ (one microwave unit, 15 to 250 W) and Microdigest™ (3 to 6 independent digestions at one time with integrated magnetic stirrers) are other variants for digestion.

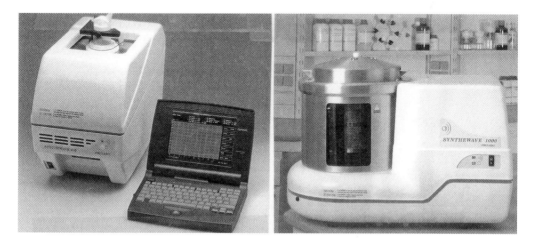

Fig. 2.6. General view of the Synthewave™ 402 (left) and the Synthewave™ 1000 (right).

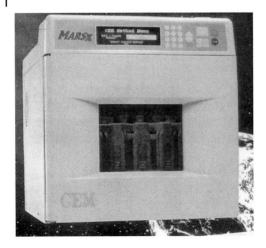

Fig. 2.7. General view of the MARS™.

2.5.2
CEM's products [82–85]

The MARS™ (Fig. 2.7) has a multimode cavity very similar to that of a domestic oven but with safety precautions (15-mL vessels up to 0.5-L round-bottomed flasks, magnetic stirring, temperature control). The microwave power available is 300 W. The optical temperature sensor is immersed in the reaction vessel for rapid response up to 250 °C. Ceiling mounting is available to enable connection with a conventional reflux system located outside the cavity or to facilitate addition of reactants. These ports are provided with a ground choke to prevent microwave leakage. It is also possible to use a turntable for small vessels with volumes from 0.1 mL to 15 mL (120 positions for 15 mL vessels). Pressure vessels are available (33 bar monitored, 20 bar controlled).

The Discover™ system is a new device designed for synthesis. CEM claims a single-mode applicator and focused microwaves. According to technical data the applicator comprises two concentric cavities with an aperture ensuring coupling (Fig. 2.8).

Discover™ enables microwave syntheses to be performed both in pressurized reaction vials (with volumes up to 80 mL) and at atmospheric pressure in standard glassware (with volumes up to 125 mL) with reflux condensers, addition funnels, or inert atmosphere work. It has a patented circular waveguide capable of self-tuning. This applicator features multiple points for the microwave energy to enter the cavity, compensating for variations in the coupling characteristics and physical size of the sample and the geometrical placement of the sample in the cavity. It ensures that reactions, irrespective of volume, always receive the optimum amount of energy. One can work at temperatures up to 300 °C with standard IR thermometry or optional fiber-optic temperature measurement.

2.5 Commercial Microwave Reactors – Market Overview

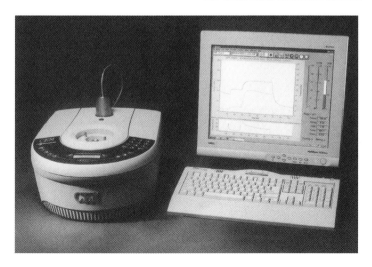

Fig. 2.8. General view of Discover™.

For pressure feedback control, the patented IntelliVent Pressure System is designed to vent automatically should the pressure from the reaction rise above the normal operating limit; the IntelliVent sensor system will maintain a completely contained system while it safely releases the excess pressure. In-situ variable speed stirring and cooling are available.

Finally, integrated software enables editing of reaction conditions in real-time and simplifies data management. Discover™ can accommodate modules for automation, HTS (Fig. 2.9), low-temperature reactions (Fig. 2.10), scale up (Fig. 2.11),

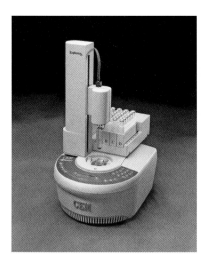

Fig. 2.9. The Explorer PLS™ system.

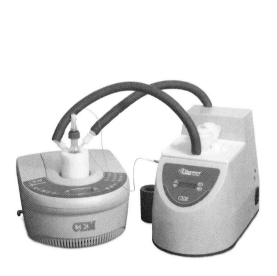
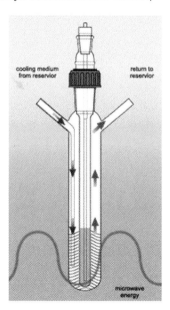

Fig. 2.10. The Coolmate system.

peptide synthesis (Chapter 20 in this book), and *in situ* spectroscopic analysis (Fig. 2.12).

2.5.3
Milestone's Products

Milestone (MLS) has developed new setups under multi-mode conditions for both scale-down and scale-up [86, 87]. All apparatus developments are based on the

Fig. 2.11. The Voyager™ system.

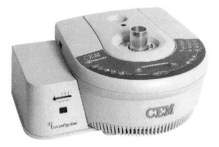

Fig. 2.12. The InvestigatorT System.

Ethos philosophy [58]. The newest device is the "multiSYNTH" batch system, Fig. 2.13, which is a "hybrid" microwave platform for synthesis under GMP (good management practice) conditions. The unique feature of this installation is its capability to utilize both a mono or multi-modal microwave input in one and the same cavity. Further characteristics are pulsed and unpulsed power supply (800 W), stirring by oscillating/rotating the sample(s) and the vessel(s), automatic temperature (IR, fiber optic) and pressure control, rapid cooling of vessels, suitability for a single test vessel or a 12-position carousel, high flexibility of sample size (1.5 to 10 mL, 0.25 to 5 mL, single reactor up to 1 000 mL (1 bar) and pressure vessels up to 70 mL (20 bar). Automatic, real-time monitoring and feedback control of many data for permanent process control is another advantage of the multiSYNTH system.

Other microwave batch apparatus, for example "rotaPREP" or "microCLAVE"

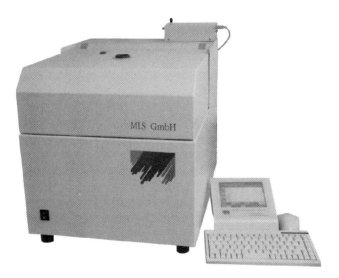

Fig. 2.13. The Ethos multiSYNTH.

Fig. 2.14. The Ethos rotaPREP.

("μCLAVE") (Figs. 2.14 and 2.15), has been successfully used in a variety of research projects. Additional background information is given elsewhere [17–20].

Another attractive microwave device for numerous batch applications is the Milestone/MLS component "ultraCLAVE", Fig. 2.16.

Fig. 2.15. The Ethos microCLAVE.

Fig. 2.16. The ultraCLAVE.

This Milestone-patented ultraCLAVE combines microwave irradiation with high-pressure vessel technology, enabling researchers to conduct batch reactions at pressures and temperatures up to 200 bar and 280 °C. The ultraCLAVE was designed specifically for semi-automated batch processing of samples, ideal for high-throughput or large-volume applications. Loading and unloading of sample containers is the only manual operation required. All other functions (raising/lowering the vessel cover, inert gas pressurization and venting, etc.) is performed automatically under computer control.

The 3.5-L reaction chamber of the ultraCLAVE enables processing of several samples (rotors with as many as 50 positions are available) or a single large reaction mixture. Reactions can be scaled from microliters to 3 L under identical conditions using the same reaction system. Continuous, unpulsed delivery of microwave power enables the most precise possible control over reaction conditions at millisecond frequencies.

The ultraCLAVE is designed for safe processing of hazardous samples, for example waste, toxic chemicals, drugs, explosive, and pyrophoric materials. Semi-automated operation and complete computer control enable the nitrogen pressure compensation within the ultraCLAVE.

The Ethos MR™ contains a multimode cavity very similar to that of a domestic oven but with safety precautions. It can use standard glass (420 mL up to 2.5 bar) or polymer reactors (375 mL up to 200 °C and 30 bar) with magnetic stirring. The microwave power available is 1 kW. The optical temperature sensor is immersed in the reaction vessel for rapid response up to 250 °C. An infrared sensor is also available. A ceiling mounting is available to enable connection with a conventional reflux system located outside the cavity or for addition of reactants. The Ethos CFR™ is illustrated in Fig. 2.17. It is a continuous-flow variant of the Ethos MR™.

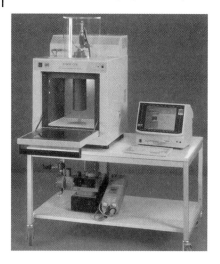

Fig. 2.17. Ethos CFR™.

2.5.4
Biotage's Products

The product portfolio of Biotage (formerly "Personal Chemistry") consists of mono-mode cavity microwave synthesizers. Emrys™, the logo for a series of microwave synthesizers, is Biotage's flagship product [5, 6, 78]. The Emrys™ Creator is a microwave system for rapid testing of synthesis research proposals; the Emrys™ Optimizer is an automated microwave synthesizer for (medicinal) working groups; and the Emrys™ Advancer is a microwave synthesizer for preparing compounds on a multigram scale safely and rapidly (minutes), Fig. 2.18 (cf. Ref. [5]).

The former SmithSynthesizer™ and SmithCreator™ are systems devoted to laboratory synthesis. They contain a closed rectangular waveguide section as the cavity. They can use specific cylindrical tubes. Pressure and temperature sensors enable real-time monitoring and control of operating conditions. This system was a good solution for laboratory experiments. The SmithSynthesizer™ is shown in Fig. 2.19.

Chemspeed (August, Switzerland) has recently developed an innovative microwave synthesizer, using a state-of-the-art Biotage microwave synthesizer, Fig. 2.20. The powerful hardware and software is organized in the pathway reaction preparation – capping and crimpling – transfer to microwave synthesizer – microwave irradiation – addition of reagents for further steps (from case to case recovery to the MW reactor) – work-up (SPE, filtration, online analysis ...). The advantage is the robotic solid/liquid and MW gravimetric solid handling (in other words, accelerator technology).

Fig. 2.18. Biotage's scale-up synthesizer "Adventure".

2.6
Selected Equipment and Applications

Based on the practical experience of the working group in Jena in recent years, this chapter is mainly devoted to a selection of our own results. Figure 2.21 illustrates step-by-step scale-up for synthesis of the innovative compound family "ionic liquids" [88, 89].

Fig. 2.19. The SmithSynthesizer™.

Fig. 2.20. The Microwave Synthesizer "SWAVE".

Today's organic chemists are discovering the advantages of controlled microwaves for heating synthetic reactions. There is, as a result, a growing demand for the ability to scale up synthetic reactions using microwaves from gram to kilogram scale. To address these needs some producers have developed a new flow-through system. This continuous-flow microwave reactor is the link between laboratory-scale reactions and industrial-scale synthetic production. Milestone/MLS and the working group in Jena have together developed a microwave-based miniplant setup named "Ethos Pilot", Figs. 2.22 and 2.23. The system "Pilot 4000" is characterized by [90, 91]:

- on-line coupling with a rapid HPLC system (Bischoff, Leonberg),
- temperatures up to 240°,
- pressures up to 60 bar,
- reactor length (for example 70 cm),
- feed flow ca. 25 L h^{-1}, by use of different pumping systems, and
- 4000 W microwave energy.

The Pilot miniplant system consists of a microwave labstation fitted with a vertical reactor. Reagents are pumped from the bottom of the reactor and the reaction products flow from the top (scale-up from grams to kilograms) into a water-cooled heat exchanger. In-line sensors enable continuous monitoring of reaction temperature. Homogeneity of temperature along the entire length of the reactor is ensured by use of a unique rotating microwave diffuser. The microwave system is controlled by an external touch screen control terminal. Intuitive software enables the

2.6 Selected Equipment and Applications | 87

Laboratory microwave equipment

0.15 mol | **2 mol** | **>3 mol h⁻¹**

Picture 1: Microwave parallel reactor | **Picture 2: Microwave autoclave** | **Picture 3: Continuous microwave system**

6-fold (10-fold) rotor system HPR 1000/6 (10), Fa. MLS GmbH Leutkirch/Allgäu, Germany
→ 6 (10) identical Teflon pressure vessels
→ reaction volume up to 60 mL per vessel
→ fiber-optic temperature control
→ magnetic agitation

"μClave", Fa. MLS GmbH Leutkirch/Allgäu, Germany
→ autoclave reactor
→ reaction volume up to 500 mL
→ up to 260°C and 60 bar
→ temperature and pressure sensor
→ mechanical or magnetic agitation

"contFlow", Fa. MLS GmbH Leutkirch/Allgäu, Germany
→ flow reactor
→ variable reactors
→ up to 10 h⁻¹
→ up to 180°C and 60 bar
→ temperature sensor and pressure sensor

Fig. 2.21. Synthesis of ionic liquids under the action of microwave irradiation.

Fig. 2.22. The Miniplant "Pilot 4000".

user to control everything from process conditions (temperature, pressure, etc.) to hardware operating conditions (pump speed, back pressure valve, and safety operating limits). The user simply defines the desired temperature–time profile, and the Pilot will automatically carry out the programmed cycle, monitoring the pro-

Fig. 2.23. The Ethos Pilot 4000.

cess conditions to ensure precise, repeatable adherence to the given method. A high-performance polymer shield and a computer-controlled pressure-control valve help to ensure safe conditions in the flow-through vessel at all times. Practical experience and conclusions are given elsewhere [91]. The Pilot system was successfully tested in research projects sponsored by industry or academia. Esterification and transesterification reactions (biodiesel production), Fischer glycosidations and sugar derivatization, hydrolyses of starch and proteins, and Maillard reactions were investigated in the fullest detail [92, 93].

2.6.1
Heterogeneous Catalysis

This section briefly reviews microwave-assisted heterogeneous gas-phase catalysis [94]. This special means of nonclassical energy input by use of microwave radiation is still a fringe area of catalysis research and alternative reaction engineering in chemistry and chemical engineering. Microwave-assisted heterogeneous gas-phase catalysis is, however, expected to gain significant popularity in academia and industry in the near future.

Since 1995 the number of research groups and publications (approx. 20) on this subject has significantly increased. Although metal catalysts were often investigated originally, the focus today is almost entirely on perovskite-type mixed oxides, because of to their catalytic and microwave-absorbing properties. With few exceptions, so-called monomode microwave applicators operating at 2.45 GHz were used to activate the catalysts. In those applicators a standing wave (mode) is generated. Multimode microwave applicators have so far been used in catalyst testing to a very limited extent only.

Oxidative coupling of methane with the formation of new C–C bonds is by far the most investigated reaction in recent years. Almost the same interest is devoted to the partial oxidation of short-chain hydrocarbons to give syngas. Catalysis of automobile exhaust fumes is expected to be a prosperous research and development field in the future. High heating rates and quick power control using microwave technology enables to rapid adaption to changing pollutant concentrations.

The work of Perry [121] on the exothermic CO oxidation on Pt/Al_2O_3 and Pd/Al_2O_3 catalysts proved there are only minimal measurable differences between catalysis under the action of microwaves and under classical conditions. Because of imprecise temperature measurement, these differences diminish even further if experimental conditions are improved. For exothermic reactions, formation of hot spots can therefore be excluded for metal-containing supported catalysts in continuous microwave fields at low frequencies [139]. Mingos et al. [140] found hot spots of diameter 0.09–1 mm on a 20% MoS_2/γ–Al_2O_3 catalyst used for endothermic decomposition of H_2S. Resulting differences between catalyst bed temperatures was approx. 100–200 K.

The appropriate catalyst to choose for heterogeneous (or immobilized homogeneous) gas-phase catalysis under the action of microwaves unclear. The choice is difficult because catalysts must have appropriate catalytic properties and suitable

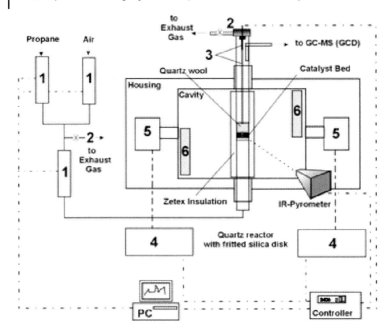

Fig. 2.24. Schematic diagram of the microwave-assisted gas-phase (oxidation) apparatus.

microwave properties (dielectric/magnetic properties, conductivity). One possibility for creation of such catalysts is to add microwave-active components (e.g. graphite, cf. Chapter 9 in this book) to a thermally-optimized catalyst and test it in screening experiments. The advantages of microwave technology can be better exploited if the catalytically-active component and the microwave active component are deposited on a microwave-transparent support. The goal is the formation of hot spots inside or on the surface of the catalyst only and, consequently, reduction of the temperatures of the catalyst bed and the gas phase, which inhibits undesired secondary reactions.

An apparatus suitable for gas phase catalysis has been developed and successfully tested in Jena (Figs. 2.24 and 2.25). A Panasonic NE-1846-type microwave oven was modified by Fricke and Mallah Microwave Technology (Hannover) [95, 96].

2.6.2
Hyphenated Techniques in Combination with Microwaves

2.6.2.1 Microwave Oven Cascade

Bodgal et al. developed and tested a continuous reactor (consisting of six domestic microwave ovens) with a rotating quartz tube (Fig. 2.26). This installation was designed for study of microwave irradiation for chemical recycling of some polymeric materials, especially polyurethane [97].

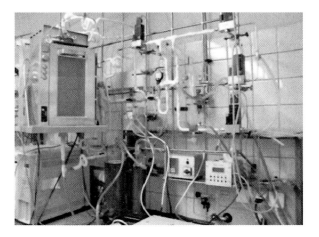

Fig. 2.25. Microwave setup for gas-phase reactions.

2.6.2.2 Photoconversions by Use of Microwave–UV

In recent years, some working groups have demonstrated experimentally that stimulation of selected reactions by microwave assistance in combination with photochemistry is (partly) advantageous. Special investigations with combined microwave–photochemistry experiments in Brno (Klan), Prague (Hajek), and Jena (Ondruschka) should be mentioned here. One setup of the working group in Jena is shown in Fig. 2.27, cf. Refs [98–101].

The detailed background of this impressive application of microwaves is given elsewhere [102].

Conscious choice of the lamp filling (element combination, pressure), envelope

Figure 2.26. Continuous microwave reactor with rotating quartz tube (Politechnika Krakowska).

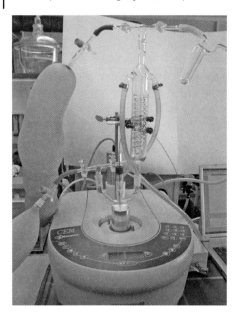

Figure 2.27. Microwave–UV studies based on the Discover™ system.

material, and optimum temperature can lead, under microwave conditions, to a dramatic effect of the spin dynamics of radicals formed. This can be regarded as an archetype of a nonthermal microwave effect. Activities are planned within the European COST D 32 project. Also worthy of mention in this connection is that the (constructive and destructive) photoconversions observed clearly demonstrated the power of this alternative form of energy. Electrodeless lamp devices nevertheless certainly require rapid and innovative further development.

2.6.2.3 Microwaves–Ultrasound

Patent and other literature shows that some working groups concentrate their interest on interactions between these two alternative forms of energy. Examples include the investigations of Chemat et al. [103], Song et al. [104, 105], and Cravotto et al. [106, 107]. Figures 2.28 and 2.29 are from the working group in Torino (Turin). Results obtained from use of high-intensity ultrasound and microwaves, alone or combined, to promote Pd/C-catalyzed aryl–aryl coupling give the hope that further studies will lead to better yields and selectivity in these synthetic challenges.

2.6.3
Combination of Microwave Irradiation with Pressure Setup

Milestone/MLS has developed special arrangements for microwave irradiation (2.45 GHz) which can work under high pressure and, if necessary, also under flow conditions. One example is a research project on remediation of organic-

Figure 2.28. Rearrangement of microwave oven with ultrasound device.

polluted waste water, Fig. 2.30 [108]. The apparatus used, the Ethos Pilot CFR HP, consists of a high-pressure (up to 600 bar and 400 °C) ceramic tube reactor.

Another important application is microwave-assisted extraction of natural products in combination with subcritical and supercritical-fluid (carbon dioxide) extraction. This complex installation (Fig. 2.31) was designed, built, and tested at the

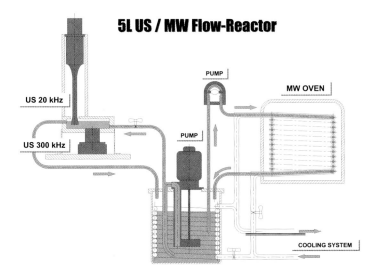

Figure 2.29. Schematic diagram of the combination of different alternative energies.

Figure 2.30. Combination of microwave irradiation and a high-pressure device.

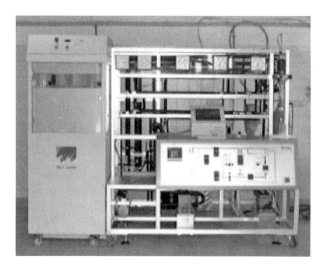

Figure 2.31. Modified ultraCLAVE II (Milestone/MLS) in combination with an SCF apparatus (XO1-500 AF, SITEC, Zurich).

Institute of Nonclassical Chemistry of the University of Leipzig [109–111]. Similar extractions [112] and apparatus [113] have been reported by Paré et al.

2.6.3.1 The RAMO System

Some years ago D. Stuerga designed a microwave reactor, called the RAMO (reacteur autoclave microonde), which is not a commercial device. The microwave applicator and the reactor are original. The resonant frequency of the cavity can be controlled by varying the position of a plunger. The effective cavity power can be increased by three orders of magnitude. The autoclave is made of polymeric materials, which are microwave transparent, chemically inert, and sufficiently strong to accommodate the pressures induced. The reactants are placed in a Teflon flask inserted within a polyetherimide flask. A fiber-optic thermometry system, a pressure transducer, and a manometer enable simultaneous measurement of temperature and pressure within the reactor. The system is controlled by pressure. The reactor is shown in Fig. 2.32.

The microwave power can be adjusted to enable constant pressure within the vessel. A incorporated pressure-release valve enables use of this experimental device routinely and safely. An inert gas, for example argon, can be introduced into the reactor to avoid sparking risk with flammable solvents. This experimental device can raise the temperature from ambient to 200 °C in less than 20 s (pressure approx. 1.2 MPa and heating rate approx. $7° \ s^{-1}$). The RAMO system has been designed for nanoparticle growth and characterization [114–117]. The RAMO system is a batch system which could be easily converted to a continuous process on an industrial scale (several hundred kilograms per second).

Figure 2.32. General view of the RAMO system.

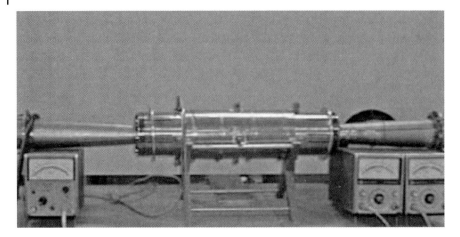

Figure 2.33. Supercritical microwave reactor applicator.
(The two cones are cylindrical waveguides for the microwave power supply.)

2.6.3.2 Supercritical Microwave Reactor

Several years ago Delmotte et al. have designed a microwave reactor for high pressure chemistry (Fig. 2.33) [118].

The microwave applicator and reactor are identical to accommodate the mechanical constraints induced by high pressure within liquids. This is the main interest of this device. The metallic cylindrical pipe is simultaneously a waveguide and the reactor. The cylindrical reactor–applicator has a steel wall approximately 30 mm thick. This thickness enables use of internal pressures above 30 MPa, which are above the critical point of water. The internal diameter of the reactor is 50 mm and its length is 500 mm. The system is powered simultaneously by two 6-kW generators placed at the ends of the reactor. This simultaneous supply is necessary to overcome the penetration depth for water, defined as the distance from the surface of the material at which the power drops to $1/e$ of its value at the surface. The penetration depth of microwaves is 15 mm for water at 20 °C. Electromagnetic energy transfer is ensured by use of matched alumina windows. The mode of propagation within the reactor is theoretically the TE_{11} mode. The system is of interest for performing very specific chemical reactions, for example oxidation in aqueous media under critical conditions.

2.6.3.3 Coconut Reactor

Stuerga and Pribetich have designed an egg-shaped microwave reactor. Its name has been chosen because of its appearance (a black egg which reveals, after opening, a white core). The coconut reactor is illustrated in Fig. 2.34.

The origin of this project is the classical observation of egg explosion during microwave cooking. Spheroidal objects act as dielectric lenses focusing electromagnetic energy. The difference between a sphere and an egg is the amount of focus-

Figure 2.34. General view of the coconut reactor.

ing. The authors do not wish to focus all the microwave power within a small area. The most important point is to control the shape and volume of the focusing spot. This focusing spot is called a caustic curve in geometrical optic. The focusing effect obtained enables heating rates five times higher than those obtained with the RAMO system (close to $35°\ s^{-1}$).

2.6.4
Synthesis of Laurydone®

The target of this ADEME (agency for environment and energy control)-supported project was the development of a new microwave-assisted pilot method for synthesis of laurydone® (the lauric ester of the pyroglutamic acid) [119], a product of particular interest in the production of lipsticks. This synthesis is usually conducted in toluene in the presence of *p*-toluenesulfonic acid as catalyst. The relatively long processing time, particularly removal of toluene from the reaction mixture, has been overcome. The objectives achieved by use of the microwave-assisted reaction were reaction without toluene and catalyst and significant shortening of the reaction time. The enterprises Sairem, Bioeurope, and de Dietrich together developed an attractive industrial process after obtaining successful laboratory results. Because the product was highly microwave-absorbing and flammable, direct treatment in a tank was not considered. An external continuous reactor (recirculation loop, Fig. 2.35) was tested and later preferred. The power used is 6 kW (2.45 GHz). Temperature control (150 °C) is based on a sensor, with a regulator to adjust the microwave power. The energy balance shows that energy consumption is as low as 1 kW kg^{-1} ester.

This example shows microwave-assisted reaction control is suitable for production of special compounds.

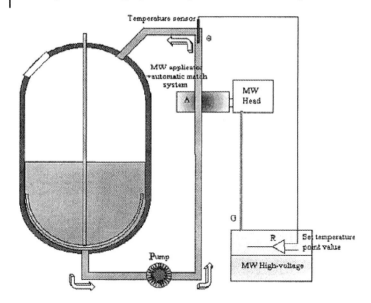

Figure 2.35. Installation for microwave-initiated synthesis of laurydone®.

2.6.5
Industrial Equipment: Batch or Continuous Flow?

As already mentioned, industrial equipment is classically divided into batch reactors and continuous flow systems. Typical batch reactors for industrial heterogeneous or homogeneous reactions accommodate, approximately, several cubic meters of liquid reagents. The design of microwave applicators capable of heating these volumes meets scientific and technological limits of microwave penetration within high lossy media. Because the penetration depth of microwaves is less than decimeters for classical solvents such as water, alcohols, etc., microwave treatment of a cubic meter of solvent is usually technologically impossible. Heating several cubic meters of liquid reagents inside batch reactors would, moreover, require microwave power close to one megawatt. This microwave power is out of range of the classical devices of microwave technology, typically approximately one-hundred kilowatts at 915 MHz or 2.45 GHz frequencies.

We therefore believe that industrial scale technological management of microwave-assisted chemical reactions is not compatible with batch reactors coupled with multimode applicators. Typical processes with systematic reduction of the dielectric losses of reactant under consideration, for example filtration and drying of mineral or pharmaceutical powders are compatible with multimode applicators.

2.6.5.1 Turbosphere
As far as we are aware, the only industrial batch microwave device is the microwave variant of the Turbosphere (*"all in one solution"* mixer/granulator/dryer

Figure 2.36. General view of the microwave variant of the Turbosphere.

designed by Moritz), shown in Fig. 2.36. This industrial equipment is sold by P. Guérin, a company well known in the pharmaceutical industry [120].

In contrast, several industrial continuous-flow systems are commercially available. These systems enable spatial positioning of a small quantity of matter (from grams to one kilogram) in front of a microwave source of several kilowatts power (typically close to 2 or 6 kW). According to the flow and heating rates expected, several modular power units should be associated and microwave power magnitude could be typically approximately 10 kW. The industrial continuous flows systems described below have been designed for the food industry and for drying, and not specially for the chemical industry. These devices could be used for chemical processes, however.

2.6.5.2 Pulsar System

This first industrial device was designed by MES [122] for drying. It could be used for solid-state reactions with powder reactants. The reactor cannot, therefore, be a classical chemical vessel or a classical chemical reactor with stirrer and other associated technical devices but a container able to enclose a reactant powder layer. The geometric shape of the microwave applicator is parallelepipedical box and the reactants are supported by a dielectric conveyor belt with edges, as shown by Fig. 2.37.

Each applicator is connected to one or two microwave generators on both sides of the equipment (courtesy of M. Delmotte and MES company). The conveyor belt passes through several large open parallelepiped applicators. The geometric structure of the waveguides is derived from a standard waveguide adjusted to the TE_{01} mode at a frequency of 2450 MHz [122]. After interface matching of the layer of

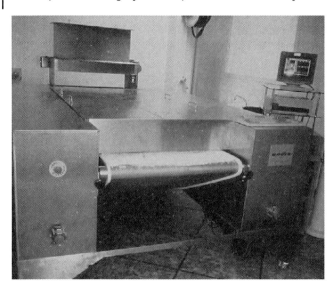

Figure 2.37. General view of the Pulsar system.

material, each open parallelepiped applicator transfers the electromagnetic power supplied by the generators to the reactive bed. The power is dissipated as heat and the temperature of reactant gradually increases.

This equipment could be used for chemical reactions based on strong solid–gas interaction with gas adsorbed on powder, for example limited air oxidation, or with gas release (water, ammonia), for example esterification. The oversized applicator structure enables design of a dielectric pipe to manage such matter transfer. This equipment can be also used for many reactions on solid supports. A typical unit is powered by microwave generators units of 2 or 6 kW for a total microwave power close to 20 or 60 kW.

2.6.5.3 The Thermostar System

This industrial equipment has been designed by MES [122–124]. The Thermostar system comprises a cylindrical vessel associated with a parallelepipedic applicator. Circular pipes are very classical geometrical shapes for industrial reactors. The Thermostar device contains parallelepipedic microwave applicators crossing a dielectric pipe. Two variants of this device have been designed, for reactants in the liquid or solid states.

The first variant is designed for liquid phases. Figure 2.38 shows the microwave applicator; the vertical or horizontal pipes are arranged according to nature of the product to be treated.

Six successive units are visible and are set to achieve heating of the reactants in the glass pipe. Each applicator is a parallelepiped metallic box ($200 \times 200 \times 300$ mm^3). The reactor is a dielectric pipe with diameter between 70 and 100 mm.

Figure 2.38. General view of the Thermostar system.

This pipe passes through the box following the electric field direction within the excitation waveguide. This waveguide can be seen in Fig. 2.38. This equipment is specified for heating liquid phases. The liquid moves from the bottom to top, with the higher temperature in the upper part of the pipe. The reactants can only be liquid phases or gas and liquid phases with a gas release system in the bottom part of the pipe (for example a bubble-blowing system).

The second variant is designed for solid state reactants with the exclusion of liquid or gas. This powder variant of Thermostar is illustrated in Fig. 2.38. The microwave applicator is the same as for the device for heating liquids but reactant transport is ensured by a metal screw set within the dielectric pipe. This specific traveling metallic screw crosses all the microwave applicators. The collinearity of this metal screw with the electric field is ensured because the major electric field direction is parallel to the major direction and perpendicular to the local curving of the screw. A typical industrial unit for solid or liquid reactants is powered with microwave generator units of 2 or 6 kW for total microwave power close to 20 or 60 kW.

2.7
Qualification and Validation of Reactors and Results

Qualification of technical devices and validation of reactions and processes are fundamental aspects of the introduction of new techniques, technology, and methods in the laboratory and in industry. In the pharmaceutical industry, product development and production is unimaginable without qualification and validation. The definition and reproducibility of system and reaction conditions are pivotal elements in addition to the ability to keep exact records of reactions and processes. Unfortunately, quality management is often neglected in basic chemistry research and process development in academia. Often it is also completely omitted from chemistry education. Consideration of both a.m. criteria is, in our opinion, the best guarantor of emotion-free discussions on the existence of microwave effects (cf. Refs 76 and 138 and Chapter 4 of this book).

The recent literature on microwave-assisted chemistry has reported a multitude of different effects in chemical reactions and processes and attributed them to microwave radiation. Some of these published results cannot be reproduced, however, because the household microwave ovens employed often have serious technical shortcomings. Published experimental procedures are often insufficient and do not enable reproduction of the results obtained. Important factors required for qualification and validation, for example exact records, reproducibility, and transparency of reactions/processes, are commonly not reported, which poses a serious drawback in the industrial development of microwave-assisted reactions and processes for synthesis of fine chemicals, intermediates, and pharmaceuticals. Technical microwave devices for synthetic chemistry have been on the market for a while (cf. a.m. explanations) and should enable comparative investigations to be conducted under set conditions. These investigations would enable better assessment of the observed effects. It is, furthermore, possible to obtain a better insight into the often discussed (nonthermal) microwave effects from these experiments (Ref. [138] and Chapter 4 of this book). Technical microwave systems are an important first step toward the use of microwave energy for technical synthesis. The actual scale-up of chemical reactions in the microwave is, however, still to be undertaken. Comparisons between microwave systems with different technical specifications should provide a measure for qualification of the systems employed, which in turn is important for validation of reactions and processes performed in such commercial systems.

2.8
Conclusion and Future

The permanent development of chemical methods available at the beginning of the 21st century creates the impression that molecules can be prepared irrespective of their structural complexity [125–128]. It seems to be only a question of sufficient manpower and budget to achieve this synthetic aim in a reasonable time. This impression, however, denies the fact that many synthetic routes and chemical methods are far from being highly efficient, particularly when multistep sequences are envisaged. Despite tremendous efforts in automation and in the development of new chemical methods in the past decade there is still an overall deficiency in new chemical technology [129]. Flow microwave-assisted processes can be regarded as a significant breakthrough towards more efficient syntheses, including multistep, parallel, and/or combinatorial sequences [130–132].

Indeed, flow processes are already combined with functionalized solid phases, with ionic liquids, supercritical fluids such as carbon dioxide, and microwave irradiation. Nevertheless, numerous examples of above mentioned processes with standard and more and less innovative laboratory equipment take advantage of this new approach. Whatever chemists and chemical engineers require – synthesis of few milligrams of compounds in drug chemistry, synthesis of building blocks on a multigram scale for parallel synthesis, or even the kilogram production of fine chemicals – flow processes are a helpful tool and a crucial link between differently scaled reactions.

The potential of microwave methods will be very dependent on the development of instrumentation, on more efficient control of process conditions, and on reduction of the cost of equipment. Flow microwave equipment seems to be optimum devices for the design of classical and automated systems. Undoubtedly, with the advent of new technology and materials, novel engineering solutions for implementation of microwave heating will also arise (new ways to supply the radiation, new design of vessels and their materials, new flow systems, automated injection units and types of coupling with instrumental (spectroscopic and/or chromatographic real-time) determination) and new ways of controlling the reaction conditions (temperature, pressure, safety) and "good management practice" feedback will appear.

Furthermore, consolidation of the theoretical basis of microwave pretreatment is a topical goal [133]. New studies on this subject should probably include in-depth investigation of the interaction of microwaves with substances of different chemical nature (solvents, analytical, structural components of microwave systems), research into the effect of physical properties of microwaves (frequency and intensity) on the pathway of physicochemical processes in solution and in heterogeneous systems (cf. Ref. [134]).

References

1 L. K. DORAISWAMY, *Organic Synthesis Engineering*, Oxford University Press, Inc., New York **2001**.
2 O. LEVENSPIEL, *Chemical Reaction Engineering*, 3rd edn, J. WILEY and SONS, HOBOKEN (NJ) **1999**.
3 W. BONRATH, T. NETSCHER, *Appl. Catal. A: General* **2005**, *280*, 55–73.
4 A. STANKIEWICZ, *Chem. Eng. Sci.* **2001**, *56*, 359–364.
5 J. P. TIERNEY, P. LIDSTRÖM, *Microwave Assisted Organic Synthesis*, Blackwell Publ. Ltd., Oxford (UK) **2005**.
6 (a) C. O. KAPPE, A. STADLER, *Microwaves in Organic and Medicinal Chemistry*, Wiley–VCH, Weinheim **2005**; (b) Chapter 3, p. 29–55.
7 W. BONRATH, *Ultrasonics Sonochem.* **2005**, *12*, 103–106.
8 (a) D. WÖHRLE, M. W. TAUSCH, W.-D. STOHRER, *Photochemie: Konzepte, Methoden, Experimente*, Wiley–VCH, Weinheim **1998**; (b) J. MATTAY, A. GRIESBECK (eds.), *Photochemical Key Steps in Organic Synthesis*, VCH, Weinheim **1994**.
9 See for example K. NAKAMURA, R. YAMANAKA, T. MATSUDA, T. HARADA,

Tetrahedron Asymmetry **2003**, *14*, 2659–2681.
10 (a) C. M. STARKS, M. HALPER, *Phase-transfer Catalysis: Fundamentals, Applications, and Industrial Applications*, Springer, Berlin 1994; (b) A. G. VOLKOV, *Interfacial Catalysis*, CRC Press, Boca Raton **2002**.
11 cf. [1], p. 744, 758; (b) B. ROZENBERG, *Heterophase Networks Polymers: Synthesis, Characterization, and Properties*, CRC Press, Boca Raton **2003**.
12 J. G. SANCHEZ MARCANO, T. T. TSOTSIS, *Catalytic membranes & Catalytic Membrane Reactors*, Wiley–VCH, Weinheim **2002**.
13 C. A. M. AFONSO, J. G. CRESPO, *Green Separation Processes – Fundamentals and Applications*, Wiley–VCH **2005**.
14 J. CLARK, D. MACQUARRIE (eds.), *Handbook of Green Chemistry & Technology*, Blackwell Sci. Ltd., London **2002**.
15 S. W. BENSON, *Thermochemical Kinetics*, Wiley, New York **1976**.
16 S. L. Y. TANG, R. L. SMITH, M. POLIAKOFF, *Green Chem.* **2005**, *7*, 761–762.
17 M. NÜCHTER, B. ONDRUSCHKA, U. MÜLLER, T. DITTMAR, in: *Innovative Energieträger in der Verfahrenstechnik (Dortmund 2000)*, Shaker, Aachen **2000**, 85–101.
18 M. NÜCHTER, B. ONDRUSCHKA, A. TIED, W. LAUTENSCHLÄGER, in: *Innovative Energieträger in der Verfahrenstechnik II (Dortmund 2001)*, Shaker, Aachen **2001**, 85–97.
19 M. NÜCHTER, U. MÜLLER, B. ONDRUSCHKA, A. TIED, W. LAUTENSCHLÄGER, *Chem. Eng. Tech.* **2003**, *26*, 1207–1216.
20 M. NÜCHTER, B. ONDRUSCHKA, W. BONRATH, A. GUM, *Green Chem.* **2004**, *6*, 128–141.
21 A. LOUPY (ed.), *Microwaves in Organic Synthesis*, Wiley–VCH, Weinheim **2002**.
22 B. L. HAYES, *Microwave Synthesis: Chemistry at the Speed of Light*, CEM Publ. Matthews (NC) **2002**.
23 C. R. STRAUSS, *Aust. J. Chem.* **1999**, *52*, 83–96.
24 B. A. ROBERTS, C. R. STRAUSS, *Acc. Chem. Res.* **2005**, *38*, 653–661.
25 B. L. HAYES, *Aldrichim. Acta* **2004**, *37*, 66–77.
26 C. O. KAPPE, *Angew. Chem., Int. Ed.* **2004**, *43*, 6250–6284.
27 A. DE LA HOZ, A. DIAZ-ORTIZ, A. MORENO, *Chem. Soc. Rev.* **2005**, *34*, 164–178.
28 www.evaluserve.com (Developments in Microwave Chemistry, 2005).
29 www.uni-graz.at/~kappeco
30 R. V. DECAREAU, R. A. PETERSON, *Microwave processing and engineering*, VCH Weinheim **1986**.
31 R. V. DECAREAU, *Microwaves in the food processing industry*, Academic Press, New York **1985**.
32 A. C. METAXAS, R. J. MEREDITH, *Industrial microwave heating*, Peregrinus Ltd, London **1983**.
33 C. A. BALANIS, *Advanced engineering electromagnetics*, Wiley & sons, New York **1989**.
34 J. E. GERLING, *J. Microwave Power Electromag. Ener.* **1987**, *22*, 199–207.
35 M. WATANABE, M. SUZUKI, S. OHKAWA, *J. Microwave Power* **1978**, *13*, 173–181.
36 S. WASHISU, I. FUKAI, *J. Microwave Power* **1980**, *15*, 59–61.
37 S. Z. CHU and H. K. CHEN, *J. Microwave Power Electromag. Ener.* **1988**, *23*, 139–143.
38 D. MICHAEL, P. MINGOS, D. R. BAGHURST, *Chem. Soc. Rev.* **1991**, *20*, 1–47.
39 *Paar instrument company*, Moline, Illinois 61265-9984 (USA), www.paarinst.com
40 US Pat. 4882128 (**1989**).
41 (a) www.anton-paar.com; (b) J. M. KREMSER, C. O. KAPPE, *Eur. J. Org. Chem.* **2005**, 3672–3679.
42 *Luxtron corporation*, Santa Clara (CA), www.luxtron.com
43 *Fiso technologies Inc.*, Sainte-Foy (Quebec), www.fiso.com
44 *Ircon, Inc Corporate*, Illinois (USA), www.ircon.com
45 *Land Infrared*, Newtown (PA), www.landinst.com
46 *Mikron Instruments Company*, Hancock (Michigan), www.mikroninst.com

47 *Raytek corporation*, Santa Cruz (CA), www.raytek.com
48 A. CALMELS, D. STUERGA, P. LEPAGE, P. PRIBETICH, *Microwave Opt. Techn. Lett.* **1999**, *21*, 477–482.
49 D. STUERGA, A. CALMELS, P. PRIBETICH, *Microwave Opt. Techn. Lett.* **2001**, *30*, 193–195.
50 P. O. RISMAN, T. OHLSSON, *J. Microwave Power Electromag. Ener.* **1991**, *22*, 193–198.
51 C. LOHR, P. PRIBETICH, D. STUERGA, *Microwave Opt. Techn. Lett.* **2004**, *42*, 46–50.
52 C. LOHR, P. PRIBETICH, D. STUERGA, *Microwave Opt. Techn. Lett.* **2004**, *42*, 365–369.
53 D. STUERGA and M. LALLEMANT, *J. Microwave Power Electromag. Ener.* **1993**, *28(2)*, 73–83.
54 J. M. OSEPCHUK, *IEEE Trans. MTT* **2002**, *50*, 975–985.
55 H. WILL, P. SCHOLZ, B. ONDRUSCHKA, *Chem. Eng. Tech.* **2004**, *27*, 1–10.
56 H. WILL, *Mikrowellenassisterte heterogene Gasphasenkatalyse: Konzeption, Aufbau und Austestung eines multimoden Mikrowellenreaktor neuen Typs, PhD thesis*, Jena **2002**.
57 H. WILL, P. SCHOLZ, B. ONDRUSCHKA, W. BURCKHARDT, *Chem. Eng. Tech.* **2003**, *26*, 1146–1149.
58 www.milestonesrl.com and www.mls-mikrowellen.de
59 www.linn.de
60 N. S. WILSON, C. R. SARKO, G. P. ROTH, *Org. Proc. Res. Dev.* **2004**, *8*, 535–538.
61 M. C. BAGLEY, R. L. JENKINS, M. C. LUBINU, C. MASON, R. WOOD, *J. Org. Chem.* **2005**, *70*, 7003–7006.
62 G. JAS, A. KIRSCHNING, *Chem. Eur. J.* **2003**, *9*, 5708–5723.
63 B. M. KHADILKAR, V. R. MADYAR, *Org. Proc. Res. Dev.* **2001**, *5*, 452–455.
64 S. SAABY, K. R. KNUDSEN, M. LADLOW, S. V. LEY, *Chem Commun.* **2005**, 2909–2911.
65 P. WATTS, S. J. HASWELL, *Chem. Soc. Rev.* **2005**, *34*, 235–246.
66 E. COMER. M. G. ORGAN, *J. Am. Chem. Soc.* **2005**, *127*, 8160–8167.
67 PING HE, S. J. HASWELL, P. D. I. FLETCHER, *Appl. Catal. A: General* **2004**, *274*, 111–114.
68 PING HE, S. J. HASWELL, P. D. I. FLETCHER, *Sens. Actuators B* **2005**, *105*, 516–520.
69 PING HE, S. J. HASWELL, P. D. I. FLETCHER, *Lab on a chip* **2004**, *4*, 38–41.
70 E. ESVELD, F. CHEMAT, J. VAN HAVEREN, *Chem. Eng. Tech.* **2000**, *23*, 279–283.
71 E. ESVELD, F. CHEMAT, J. VAN HAVEREN, *Chem. Eng. Tech.* **2000**, *23*, 429–435.
72 S. SAABY, I. R. BAXENDALE, S. V. LEY, *Org. Biomol. Chem.* **2005**, *3*, 3365–3368
73 U. R. PILLAI, E. SHALE-DEMSSIE, R. S. VARMA, *Green Chem.* **2004**, *6*, 295–298.
74 R. J. J. JACHUCK, D. K. SELVARAJ, R. S. VARMA, *Green Chem.* **2006**, *8*, 29–33.
75 H. WILL, P. SCHOLZ, B. ONDRUSCHKA, *Top. Catal.* **2004**, *29*, 175–162.
76 M. NÜCHTER, B. ONDRUSCHKA, D. WEISS, R. BECKERT, W. BONRATH, A. GUM, *Chem. Eng. Tech.* **2005**, *28*, 871–881.
77 (a) www.cem.com; (b) J. D. FERGUSON, *Mol. Diversity*, **2003**, *7*, 281–286.
78 (a) www.biotage.com; (b) J.-S. SCHANCHE, *Mol. Diversity*, **2003**, *7*, 293–300.
79 A. LOUPY, P. PIGEON, M. RAMDANI, *Tetrahedron* **1996**, *52*, 6705–6721.
80 C. LIMOUSIN, J. CLÉOPHAX, A. PETIT, A. LOUPY, G. LUKACS, *J. Carbohydr. Chem.* **1997**, *16*, 327–342.
81 A. LOUPY, A. PETIT, J. HAMELIN, F. TEXIER-BOULLET, P. JACQUAULT, D. MATHÉ, *Synthesis* **1998**, 1213–1234.
82 *CEM Corporation*, Matthews (NC), www.cemsynthesis.com
83 EU Pat. 0455513 (**1991**).
84 EU Pat. 0604970 (**1994**).
85 US Pat. 5796080 (**1998**).
86 L. FAVRETTO, *Mol. Diversity* **2003**, *7*, 287–290.
87 M. NÜCHTER. B. ONDRUSCHKA, *Mol. Diversity*, **2003**. *7*, 253–264.
88 A. STARK. R. BIERBAUM, B. ONDRUSCHKA, *unpublished results*, Jena **2003/04**.
89 M. NÜCHTER, B. ONDRUSCHKA, J. HOFFMANN, A. TIED, *Studies in Surface*

Science and Catalysis (S.-E. PARK, J.-S. CHANG, K.-W. LEE, eds.), Elsevier, Amsterdam **2004**, *153*, 131–136.
90 R. BIERBAUM, M. NÜCHTER, B. ONDRUSCHKA, *Chem. Eng. Tech.* **2005**, *28*, 427–431.
91 R. BIERBAUM, Th. DIMMIG, B. ONDRUSCHKA, *Prerequisites for the Application of Microwave-assisted Chemical Reaction Engineering*, Proceed. Ampere 2005, Modena, **2005**; cf.: R. BIERBAUM, *Reaktionstechnische Rahmenbedingungen für den Einsatz einer mikrowellenassistierten Reaktionsführung*, PhD thesis, Bergakademie Freiberg **2005**.
92 M. NÜCHTER, U. NÜCHTER, B. ONDRUSCHKA, G. KERNS, K. FISCHER, in: *Mikrowelleneinsatz in den Materialwissenschaften, der chemischen Verfahrenstechnik und in der Festkörperchemie* (Hrsg. M. Willert-Porada), (Dortmund 1998), Shaker, Aachen **1998**, 219–230.
93 www.ituc.uni-jena.de
94 M. HAJEK, *Microwave Catalysis in Organic Chemistry*, in: [10], chapter 10, 345–378 and Chapter 13 in this book.
95 H. WILL, P. SCHOLZ, B. ONDRUSCHKA, *Chem. Eng. Tech.* **2004**, *27*, 113–122.
96 S. FÄLSCH, *Beitrag zur mikrowellenassistierten heterogenen Gasphasenkatalyse*, PhD thesis, Jena **2005**.
97 D. BOGDAL (Krakow), e-mail: pcbogdal@cyf-kr.edu.pl
98 M. NÜCHTER, B. ONDRUSCHKA, A. JUNGNICKEL, U. MÜLLER, *J. Phys. Org. Chem.* **2000**, *13*, 579–586.
99 M. NÜCHTER, B. ONDRUSCHKA, U. SUNDERMEIER, *Kombination von Ultraviolett- und Mikrowellenstrahlung zum Abbau von Xenobiotika im Problemwässern*, Proceed. 4. GVC-Kongress, Bremen **1999**, 1357–1371.
100 U. SUNDERMEIER, Diploma thesis, Jena **1999**.
101 A. JUNGNICKEL, Diploma thesis, Jena **1999**.
102 P. KLAN, M. HAJEK, *Microwave Photochemistry*, in: [10], chapter 14, 463–486 and Chapter 19 in this book.
103 A. LAGHA, S. CHEMAT, P. V. BARTELS, F. CHEMAT, *Analusis* **1999**, *37*, 452–447.
104 Y. PENG, R. DOU, G. SONG, J. JIANG, *Synlett* **2005**, 2245–2247.
105 Y. PENG, G. SONG, *Green Chem.* **2003**, *5*, 704–706.
106 G. CRAVOTTO, M. BEGGIATO, A. PENONI, G. PALMISANO, S. TOLLARI, J.-M. LÉVÊQUE, W. BONRATH, *Tetrahedron Lett.* **2005**, *46*, 2267–2271.
107 B. TOUKONIITTY, J.-P. MIKKOLA, D. YU. MURZIN, T. SALMI, *Appl. Catal. A: General* **2005**, *279*, 1–22.
108 Fraunhofer Institute of Chemical Technology (Pfinztal), Synthesis Report Contract No. BRPR-CT 98-0639 (Supercritical Fluids to extract and/or degrade organic waste materials), cf. www.ict.fraunhofer.de and Ref. [58].
109 www.uni-leipzig.de/~lnc
110 H. SONNENSCHEIN, J. GERMANUS, P. HARTING, in: *Innovative Energieträger in der Verfahrenstechnik* (Dortmund 2001), Shaker, Aachen **2001**, 39–47.
111 H. SONNENSCHEIN, J. GERMANUS, P. HARTING, *Chem. Ing. Tech.* **2002**, *74*, 270–274.
112 J. R. J. PARÉ, J. M. R. BÉLANGER, *TrAC, Trends in Analytical Chemistry* **1994**, *13*, 176–184.
113 J. R. J. PARÉ, J. M. R. BÉLANGER, *Microwave-assisted Process (MAP): Principles and Applications, Techniques and Instrumentation in Analytical Chemistry* **1997**, *18*, 395–420.
114 K. BELLON, P. RIGNEAU, D. STUERGA, *Euro. Phys. J. Appl. Phys.* **1999**, *AP7*, 41–43.
115 K. BELLON, D. CHAUMONT, D. STUERGA, *J. Mat. Res.* **2001**, *16*, 2619–2622.
116 E. MICHEL, D. STUERGA, D. CHAUMONT, *J. Mat. Sci. Lett.* **2001**, *20*, 1593–1595.
117 T. CAILLOT, D. AYMES, D. STUERGA, N. VIART, G. POURROY, *J. Mat. Sci.*, **2002**, *37*, 5153–5158.
118 French Pat. 99 03749 (**1999**).
119 (a) J. F. ROCHAS, J. P. BERNARD, L. LASSALLE, *Synthesis of Laurydone Under Microwave Field*, Preprint of SAIREM (Neyron, France, **2005**); (b) French Pat. 2833260 (**2003**). *"Procedure for preparation of an ester by microwave heating of a heterogeneous mixture of acid and alcohol"*
120 Pierre Guérin SA, Mauzé sur le

Mignon (France), www.pierreguerin.com
121 W. L. Perry, D.W. Cooke, J. D. Katz, *J. Catal.* **1999**, *171*, 431–438.
122 *MES SA*, Villejuif (France), www.m-e-s.net
123 French Pat. 82 04398 (**1982**).
124 French Pat. 86 01104 (**1986**).
125 B. M. Khadilkar, G. L. Rebeiro, *Org. Proc. Res. Dev.* **2002**, *6*, 826–828.
126 K. T. J. Loones, B. U. W. Maes, G. Rombouts, S. Hostyn, G. Diels, *Tetrahedron* **2005**, *61*, 10338–10348.
127 J. Cléophax, M. Liagre, A. Loupy, A. Petit, *Org. Proc. Res. Dev.* **2000**, *4*, 498–504.
128 N. G. Anderson, *Org. Proc. Res. Dev.* **2001**, *5*, 613–621.
129 W. Bonrath, *Ultrasonics Sonochem.* **2004**, *11*, 1–4.
130 A. Stadler, B. H. Yousefi, D. Dallinger, P. Walla, E. van den Eycken, N. Kaval, C. O. Kappe, *Org. Proc. Res. Dev.* **2003**, *7*, 707–716.
131 H. S. Ku, E. Siores, A. Taube, J. A. R. Ball, *Comput. Ind. Eng.* **2002**, *12*, 281–290.
132 I. V. Kubrakova, *Russ. Chem. Rev.* **2002**, *71*, 283–294.
133 D. Bathen, *Dimensionen von Mikrowellen- und Ultraschall-Anlagen im technischen Massstab – eine Einführung*, in: *Innivative Energieträger in der Verfahrenstechnik* (Dortmund 2001), Shaker, Aachen **2001**, 9–14.
134 M. Karches, H. Takashima, Y. Kanno, *Ind. Eng. Chem. Res.* **2004**, *43*, 8200–8206.
135 T. Tanaka, *Solvent-free Organic Synthesis*, Wiley–VCH, Weinheim, **2003**.
136 D. Stuerga, P. Gaillard, M. Lallement, *Tetrahedron* **1996**, *52*, 5505–5510.
137 (a) D. Bogdal, M. Lukasiewicz, J. Pielichowski, A. Miciak, S. Bednarz, *Tetrahedron* **2003**, *59*, 9136–9139. (b) M. Lukasiewicz, D. Bogdal, J. Pielichowski, *Adv. Synth. Catal.* **2003**, *345*, 1269–1272.
138 (a) L. Perreux, A. Loupy, *Tetrahedron* **2001**, *57*, 9199–9223. (b) A. Loupy, F. Maurel, A. Sabatié-Gogova, *Tetrahedron* **2004**, *60*, 1683–1691.
139 J. R. Thomas, Jr., *Catal. Lett.* **1997**, *49*, 137–141.
140 X. Zhang, D. O. Hayard, D. M. P. Mingos, *Chem. Commun.* **1999**, 975–976.

3
Roles of Pressurized Microwave Reactors in the Development of Microwave-assisted Organic Chemistry

Thach Le Ngoc, Brett A. Roberts, and Christopher R. Strauss

3.1
Introduction

It is generally accepted that the field of microwave-assisted organic synthesis began with the 1986 disclosures of the groups of Gedye [1] and Giguere [2]. Those researchers reported the use of domestic microwave ovens and simple sealed microwave-transparent vessels for relatively routine organic transformations. Although their reaction times were impressively short, vessel deformation and explosions created risks and safety hazards [1–3]. Both groups recognized that the development of high pressure was the underlying cause of the problems they had encountered. Individually, they described different procedures for trying to manage the technique more safely, but not for controlling it. Although the workers did not say so themselves, others interpreted their reports to suggest that heating of solvents in domestic microwave ovens was inherently dangerous and that organic solvents were incompatible with microwave energy. Although the latter interpretation was subsequently shown to be overly simplistic, early investigators recognized that eliminating, or at least minimizing, the risks would be necessary before tangible practical benefits of microwave chemistry could be realized.

For more than ten years, the most common and popular strategy was to employ domestic microwave ovens under solvent-free conditions [4–6]. The use of organic reactants adsorbed on to a solid support such as silica gel or alumina, typically containing inorganic reagents such as CsF, CrO_3 or MnO_2, reduced the possibility of uncontrolled exothermal processes [4–6]. "Dry media" reactions, as they came to be known, could also be performed in open vessels at atmospheric pressure. The pioneers of that approach introduced new techniques and ways of thinking about organic synthesis. They also produced many results several of these, although beyond the scope of this chapter, have been discussed by leading exponents in other chapters of this book.

Significantly, advocates of solvent-free approaches promoted as major advantages that unmodified domestic microwave ovens could be used and that sample mixing and temperature measurement were unnecessary during reactions. Our group, however, began with two main alternative suppositions:

Microwaves in Organic Synthesis, Second edition. Edited by A. Loupy
Copyright © 2006 WILEY-VCH Verlag GmbH & Co. KGaA, Weinheim
ISBN: 3-527-31452-0

1. that temperature measurement would be essential to enable other organic chemists to reproduce reaction conditions; and
2. because solvents had fundamental, long-established roles in synthesis, chemists would prefer to perform microwave-assisted organic chemistry in the presence of organic solvents if it were possible and advantageous to do so.

We also considered that if reactions could be performed routinely with lower-boiling point solvents under moderate pressure, at temperatures of approximately 200 °C, such processes could, perhaps, proceed more rapidly and conveniently than by conventional means.

3.2
Toward Dedicated Microwave Reactors

Investigation of these propositions required dedicated microwave reactors. It was essential that such systems operated reliably, safely, and routinely with volatile compounds, including organic solvents at high temperatures and at moderate pressures. To attract the attention and maintain the interest of chemists, microwave technology would need to have advantages that were unavailable with the conventional chemical process equipment of the day.

Since the mid-1970s, microwave heating with pressure vessels had been used in analytical laboratories to speed the rate of digestion and dissolution of solid samples, for example ores, hair, and foodstuffs. Digestion is a degradative process, typically involving treatment of a small sample with excess of a strong oxidizing acid such as nitric, perchloric, or sulfuric. The objective of digestion is to obtain a clear solution usually for quantitative analysis of mineral components, without charring and physical loss of material. Although it is essential that digestion is standardized for high precision, conditions such as temperature, reaction time, sample stirring, and cooling rate are not essential to the outcome and so are not actively monitored.

Conversely, synthesis is a constructive process. Reactants that are not particularly stable to strong acids and bases may be required and in much larger amounts than those used for digestion. They may be expensive and highly reactive, even on the laboratory-scale. The ability to define and control conditions, including temperature, time, sample stirring, addition or withdrawal of materials, and post-reaction cooling, is almost always vital for satisfactory outcomes and for reproducibility.

Existing techniques and equipment for microwave digestion were deemed unsatisfactory for synthetic applications. Fabrication of appropriate reactors for synthesis with solvents involved overcoming several technical problems that had not been recognized previously. Not surprisingly, proof of principle took longer to achieve than for solvent-free microwave processes. With safety as the highest priority, means of measuring and controlling temperature (difficult and/or costly to implement in a microwave environment in the late 1980s and early 1990s) were necessary. Precise control of pressure and microwave power input was also required. Technology for rapid cooling of reactions were needed, as were effective, automatic fail-safe emergency procedures in the event of unexpected difficulties. Additional major issues included scale, vessel design and construction, including materials, means for stirring reaction mixtures, computer control, uniform heating, and sample withdrawal. To emphasize the breadth of these technical challenges, it seems that emergency shut-down procedures, stirring, withdrawal of samples during reactions, and rapid cooling, either concurrently or post-reaction, had not been described (or perhaps even contemplated) before our reactors were developed.

3.2.1
The Continuous Microwave Reactor (CMR)

The continuous microwave reactor (CMR), built in 1988, but not reported in peer-reviewed open literature until 1994, was the first microwave system designed for chemical processing with organic solvents [7]. Commercial systems, developed subsequently under license, typically comprised a microwave cavity fitted with a vessel of microwave-transparent, inert material (Fig. 3.1). Plumbing in the microwave zone was connected to a metering pump and pressure gauge at the inlet end and to a heat exchanger and a pressure-regulating valve at the outlet. The heat exchanger enabled rapid cooling of the effluent, under pressure, immediately after it left the irradiation zone. Temperature was monitored immediately before and after cooling. Variables such as internal volume of the plumbing within the microwave zone, flow rate and control of the applied microwave power enabled flexible operation. To withstand corrosive acids and bases within the plumbing, contact was avoided between metal surfaces and reaction mixtures. Feed-back microprocessor control enabled setting of pump rates and temperatures for heating and cooling

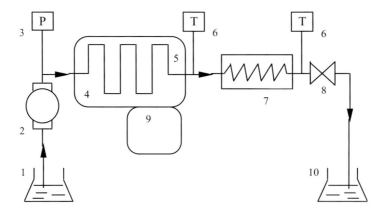

Fig. 3.1. Schematic diagram of the CMR: 1, reactants for processing; 2, metering pump; 3, pressure transducer; 4, microwave cavity; 5, reaction coil; 6, temperature sensor; 7, heat exchanger; 8, pressure regulator; 9, microprocessor controller; 10, product vessel.

Fig. 3.2. Early example (ca 1997) of a CMR manufactured by Milestone mls.

of reactions. Fail-safe measures were instituted to shut down the system if the temperature exceeded the maximum allowable by 10 °C, or in the event of blockage, leakage, or rupture.

Commercial units (Fig. 3.2) had volumes up to 120 mL within the microwave zone, 80 mL within the cooling zone, and could be operated at flow rates up to 100 mL min^{-1}. Under such arrangements, residence times in the microwave zone were typically 2–10 min.

3.2.2
Microwave Batch Reactors

Complementary laboratory-scale microwave batch reactors (MBRs) were built for synthesis and for kinetics studies [8]. The embodiment shown in Fig. 3.3 had a capacity of up to 200 mL. Reactions could be heated to 260 °C and although the system could accommodate internal pressures up to 10 MPa (100 atmospheres) it was found pressures exceeding 5 MPa were rarely required. The main features included: rapid heating capability (1.2 kW microwave output); infinitely variable control of microwave power; measurement of absorbed and reflected microwave energy; a load-matching device to maximize heating efficiency; direct measurement of the reaction temperature and pressure; a stirrer for mixing and to ensure uniform temperature within the sample; valving and plumbing to facilitate sample introduction and withdrawal during the heating period; chemically inert wettable surfaces and fittings; rapid cooling post-reaction, and a facility for conducting reactions under an atmosphere of inert gas. Microwave power input was computer-controlled. Heating could be performed at high or low rates as required and designated temperatures could be maintained for hours, if desired.

Fig. 3.3. Dr Ulf Kreher at work with the MBR.

3.3
Applications of the New Reactors

3.3.1
With Low-boiling Solvents

Immediately after their development, both microwave reactor systems were applied to common synthetic processes, for example rearrangement, addition, substitution, condensation, and elimination. Examples included esterification, amidation, transesterification, acetalization, hydrolysis of esters and amides, isomerization, decarboxylation, oxidation, etherification, and formation of aminoreductones [7, 8]. Many organic solvents were employed without mishap, including acetone, chloroform, methanol, ethyl acetate, ethanol, butanone, acetonitrile, isopropanol, toluene, tetrahydrofuran, pyridine, acetic acid, chlorobenzene, propionic acid, dimethylformamide, and dimethyl sulfoxide. Excellent yields were often obtained with short reaction times and at temperatures more than 150 °C higher than the boiling point of the solvent under atmospheric pressure. Volatile reactants or gases such as ammonia, dimethylamine, formaldehyde, and carbon dioxide were processed without difficulty. When gases were formed during reactions, over-pressure situations rarely occurred. With the MBR, heating could be maintained within a narrow temperature range for hours if necessary.

Typically, reaction times were up to three orders of magnitude shorter than for literature preparations performed conventionally at lower temperatures under atmospheric pressure. Thus, procedures requiring hours or days at reflux or ambient temperature and pressure often were complete within minutes and with high selectivity on use of microwave heating. Other technologies or methods compatible with these systems, included solvent-free conditions with "neat" starting material, tandem or cascade, catalyzed or uncatalyzed reactions, use of aqueous media at high temperature, and non-extractive techniques for product isolation.

The use of low-boiling solvents for moderately high temperature reactions is particularly strongly appealing when such solvents are acceptable environmentally. Ethanol, for example, is industrially produced by fermentation. It is readily recyclable and is both a renewable and biodegradable resource [9]. In many societies it is used in cooking and consumed in beverages. Although ethanol had been widely used as an organic solvent at low to reflux temperatures, reports of its applications in synthesis at higher temperature were limited. The supercritical temperature is 243 °C, so in working with ethanol an upper temperature limit of 230 °C was employed to avoid problems and risks associated with excessive pressure, diffusion of supercritical fluids through polymeric materials, and poor microwave absorption.

We have previously reported the intramolecular aldol reaction of hexane-2,5-dione to afford 3-methylcyclopent-2-en-1-one in highly dilute aqueous base [10]. We now reveal that the reaction also can be conducted with excellent conversion in ethanol containing 0.01–0.07% by weight NaOH. Under the latter conditions, when the dione was treated at 200 °C for 15 min, with 0.02% w/w ethanolic NaOH, all the starting material was consumed and the yield was greater than 90% (Scheme 3.1). This result is indicative of the opportunities still available for structured research into the use of low-boiling solvents for synthesis at high temperatures.

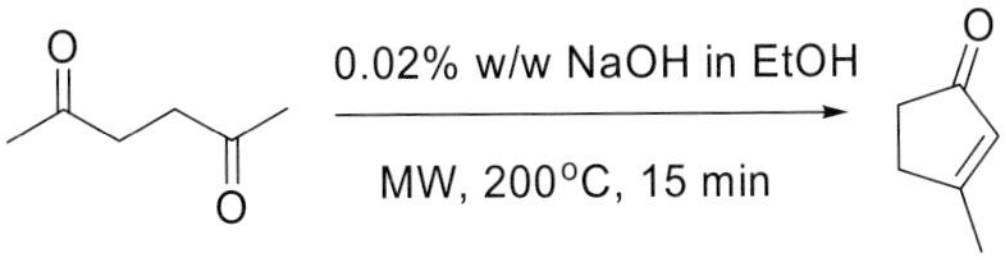

Scheme 3.1. Intramolecular aldol reaction of hexane-2,5-dione to afford 3-methylcyclopent-2-en-1-one in ethanol containing highly dilute base.

3.3.2
Kinetics Measurements

Development of the field of microwave-assisted organic chemistry, be it with "dry media" or with solvents in open or sealed vessels, was accompanied by claims of reactions proceeding faster in a microwave environment than under conventional conditions at the same temperature. In such reports, which were common in the 1990s, yields were usually quoted after a given time under apparently comparable conditions. Reaction kinetics, which are essential to meaningful discussions about reaction rates were not determined. A major reason for that appeared to be the lack of appropriate equipment for conducting kinetics studies. For the first time, MBRs, through their capacity for temperature measurement and for stirring of samples, enabled complete histories of thermally homogeneous reactions to be obtained. Accordingly, the reactors were used for comparative kinetics measurements for microwave-heated and for conventionally heated reactions, which were performed in oil baths. Examples we examined all involved thermally homogeneous reaction mixtures and the rates of microwave and conventionally heated reactions were found to be the same within experimental error. This applied also to reactions for which other workers had previously reported the occurrence of non-thermal microwave effects [11]. Our work was not the last word in this subject, however, and discussion continues about the effects of microwave energy on reaction rates.

3.3.3
Core Enabling Technology

The CMR and MBR were enabling technology that could stand alone or be integrated with other technologies and methods [12]. In our laboratories they became core technologies toward the development of new tools for preparative organic chemistry. The objective was to be able to select appropriate combinations to solve specific problems in environmentally acceptable ways. The strategy was employed to establish several improved or new environmentally benign chemical processes. Now, by out-scaling with multi-vessel carousels, others have broadened the scope to include reactions conducted in parallel, an approach that has proven effective in teaching laboratories and for scaling-up syntheses [13, 14].

3.4
Commercial Release of MBRs and CMRs

By mid-1995, only 220 papers had been published on microwave-assisted organic chemistry [15]. Until then the vast majority of the papers had covered "dry media" reactions and for approximately the next five years this trend continued. By the turn of the 21st Century, however, the applicability of dedicated microwave reactors and associated chemistry had stimulated other researchers and encouraged commercial manufacturers to construct systems for laboratory and pilot-scale studies [13, 14].

There followed a phase of rapid development coinciding with the commercial release of batch and continuous microwave reactors employing the essential design criteria and operational principles underpinning the MBR and CMR. Now, microwave reactions may be performed in commercially available systems, with or without magnetic stirring, in vessels of glass, quartz, ceramic, or polymeric materials. Reactions may be cooled externally or more directly by contact with a cold-finger. Microwave units may have the capability for "simultaneous heating and cooling" as promoted by one manufacturer or for concurrent heating and cooling as we originally termed it [15]. Measurement of temperature may be conducted remotely by means of an infrared pyrometer or directly by use of an optical fiber, gas expansion thermometer, or grounded thermocouple. Notwithstanding relatively minor variations in methodology and terminology, manufacture of microwave reactors designed for synthesis with organic solvents under pressure is performed in several

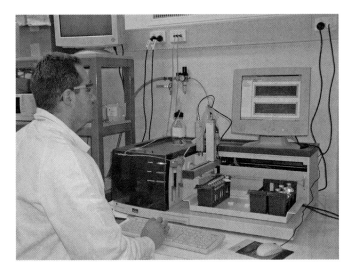

Fig. 3.4. Dr Brett Roberts with the Biotage reactor.

countries in Europe and Asia and in the United States. Distribution is global. Reactors include robotically operated units with capacities low enough to accommodate post-reaction cooling by blowing of air across the outside surface of the vessel (Fig. 3.4).

Batch reactors can process mixtures ranging in scale from multigram or milligram to one or two kilogram and the microwave energy may be applied through monomodal (for low-scale reactions) or multi-modal cavities [13, 14]. Although substantial emphasis is sometimes placed upon it, the modality of microwave applicators usually does not greatly impinge upon the outcome of reactions conducted above the multi-milligram scale in appropriately designed systems. One reason for this is that dielectric properties of materials change with temperature and chemical composition. Consequently, as reactions progress, regular and frequent tuning of either multi-modal or monomodal cavities will be necessary to ensure maximum absorption of the input energy, a procedure that is usually automated in computer-managed units.

Researchers in synthetic chemistry, particularly those engaged in bioactive compound discovery, tend to work with reaction volumes in the range 0.1–10 mL. Commercial CMR may be used for process optimization, small scale manufacture for pilot-scale studies, or to produce high-value and low-volume chemical compounds. Typical throughputs are up to 100 mL min^{-1}.

Temperature and pressure limits vary according to manufacturers' specifications and usually depend on the physical and chemical properties of materials employed in the construction of reaction vessels. In batch systems the recommended upper temperature is usually between 200 °C (vessels fabricated from polymeric materials such as PTFE, for example) and 300 °C (quartz vessels). Most commercial systems operate at pressures up to 2 MPa, but two manufacturers have recently developed equipment capable of working in the 8 MPa to 20 MPa range. Another manufacturer has estimated that 500 000 reactions are conducted annually with its systems.

From the foregoing discussion, the technology for microwave-assisted organic synthesis with organic solvents (under pressure if necessary) is now employed extensively and routinely in chemical discovery, organic synthesis, and medicinal chemistry [16, 17]. Microwave reactors have found *niche* applications in the production of intermediates, flavors and fragrances, specialty chemicals, and pilot-scale manufacture. Units capable of parallel or sequential operation, some of which may be operated robotically and remotely, are employed around the clock in industrial chemical discovery.

So, not only these microwave systems can generate reaction times that are orders of magnitude shorter than those obtained by traditional methods, but through automation they can considerably extend the working day for synthetic chemists on the basis of time alone.

3.5
Advantages of Pressurized Microwave Reactors

Significant advantages of microwave reactors designed for operating with low-boiling solvents at elevated temperature and pressure are summarized below:

1. Rapid response compared with conventional heating.
2. Input microwave energy is adsorbed directly by the sample, not by the vessel.
3. When power is turned on or off, energy input starts or ceases immediately; an important safety consideration.
4. The power can be controlled to match that required.
5. Reactor systems can be automated and operated remotely.
6. Thermally unstable products can be cooled rapidly, after the heating step.
7. Stirring of the sample limits thermal gradients: the temperature of the material on the inner walls of the reaction vessel is not significantly different from that in the body of the mixture, hence unwanted pyrolysis can be minimized or avoided.
8. Heating rates can be varied as desired and designated temperatures can be maintained for lengthy periods.
9. Reactions known to require high temperatures and higher boiling solvents can be performed under pressure at these temperatures, but in lower-boiling solvents, facilitating work-up.
10. The available temperature range for many solvents is increased dramatically, in many cases by more than 100 °C.
11. Low boiling reactants can be heated to high temperatures and rapidly cooled under pressure. Losses of volatiles are minimal.
12. Reactions can be sampled for analysis while material is being processed. With

the CMR, reaction mixtures can be subjected to multiple passes if required, or the conditions can be altered during a run.
13. Reaction conditions can be measured and reproduced.
14. Reactions performed on a laboratory scale are amenable to scale up, because the conditions are defined.
15. Moderate to high temperature reactions can be conducted in vessels fabricated from inert materials such as PFA Teflon, PTFE, or quartz. This is beneficial where strongly acidic or basic reactants or products are incompatible with metals or borosilicate glass.
16. In multi-phase systems, selective heating is possible.

3.6 Applications

3.6.1 General Synthesis

Microwave systems have promoted thermal preparation of heat-labile compounds, reactions that require high temperature, and they have aided optimization of established reactions known to require elevated temperature. Improved conditions obtained in comparison with literature methods have typically involved a combination of increased convenience, time savings, greater yields, greater selectivity, the need for less catalyst, or use of a more environmentally benign solvent or reaction medium [7, 8, 15].

The use of pressurized microwave systems to conduct Heck, Suzuki, and Stille reactions in solution was first reported by Hallberg and Larhed in 1996 [18]. This topic has been the subject of active investigation ever since [13, 16, 17, 19, 20]. It seems that most metal-catalyzed reactions can be performed within minutes by microwave "flash heating" in pressurized systems, sometimes with high regio and enantioselectivity [19]. Hallberg and Larhed, in collaboration with Curran and co-workers also introduced fluorous tin reagents for Stille reactions under microwave conditions [21]. They not only improved the synthetic procedure, but integrated it with isolation and purification of the products. Reactions that usually required about 24 h at 80 °C under reflux conditions, were completed within minutes under the action of microwave heating at higher temperature Problems associated with the separation of residual tin salts (which are toxic and normally difficult to remove) were precluded. Some other recent examples of microwave-assisted organic chemistry, summarized below, reveal the convenience, capability, and versatility of microwave reactors.

In a continuation of their extensive work on microwave-assisted Heck coupling [18, 19], the Larhed group performed regioselective, chelation-controlled double β-arylations of terminal alkenes, employing vinyl ethers as chelating alkenes and aryl bromides as coupling partners, with Herrmann's palladacycle as the palladium source (Scheme 3.2) [22].

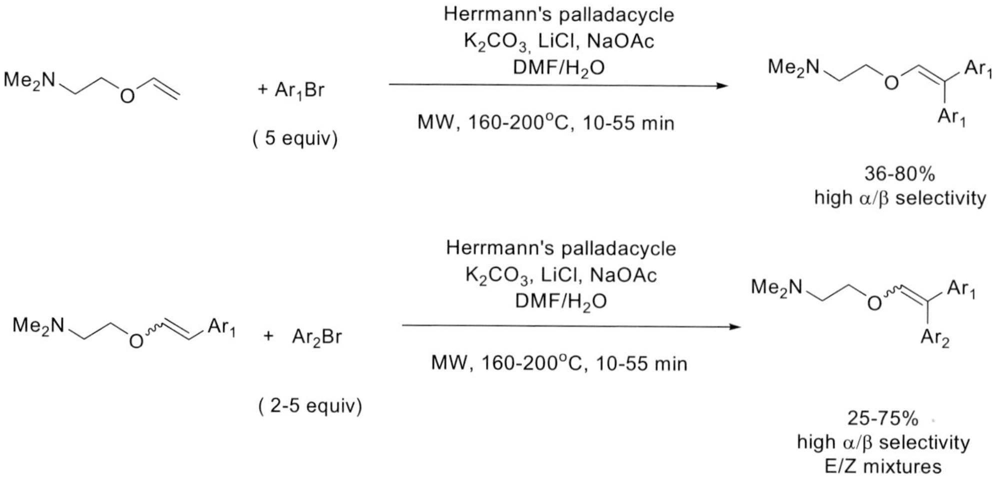

Ar_1 = phenyl
Ar_2 = phenyl, 2-naphthyl

Herrmann's palladacycle = [*trans*-di(μ-acetato)bis[*o*-(di-*o*-tolyl-phosphino)benzyl]dipalladium (II)]

Scheme 3.2. Microwave-assisted double-Heck arylations of vinyl ethers.

In work by Romera et al. [23], the selective monoalkylation of functionalized anilines with alkyl halides and tosylates was achieved in good yields (Scheme 3.3). Only minor quantities of the over-alkylated products were formed.

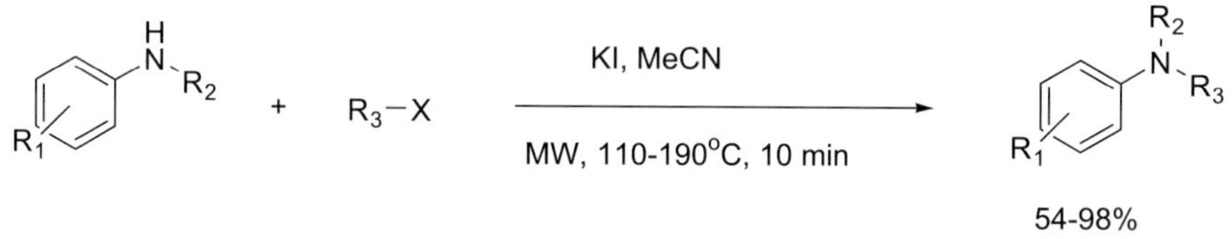

Scheme 3.3. *N*-alkylation of anilines.

Steven Ley and his group at Cambridge performed a microwave-assisted Claisen rearrangement of an allylaryl ether, to give the corresponding 2-allylphenolic deriv-

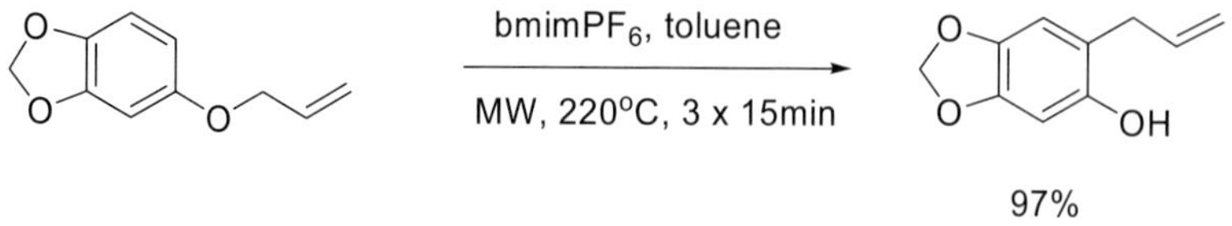

bminPF_6 = 1-butyl-3-methylimidazolium hexafluorophosphate

Scheme 3.4. Claisen rearrangement of an allylaryl ether.

ative in 97% yield by serially heating the starting material to 220 °C [24]. With toluene employed as solvent, its microwave susceptibility was increased by doping with a small quantity of an ionic liquid (Scheme 3.4).

A similar approach employing an ionic liquid as a susceptor was adopted by Leadbeater and Torenius for base-catalyzed Michael addition of imidazole to methyl acrylate [25]. At 200 °C after 5 min and with triethylamine (TEA) in equimolar quantities, a 75% yield was obtained (Scheme 3.5).

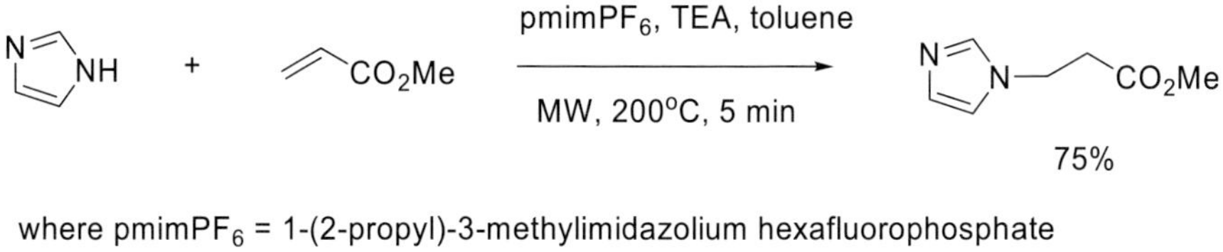

where pmimPF$_6$ = 1-(2-propyl)-3-methylimidazolium hexafluorophosphate

Scheme 3.5. Microwave-assisted Michael addition.

Tye and Whittaker pursued Ugi-type three-component condensations to prepare a range of pyrrole derivatives [26]. Levulinic acid was reacted with two isonitriles and four amines to afford eight different examples. Under conventional conditions, the reactions required 48 h to return moderate yields, but microwave heating in methanol at 100 °C afforded improved yields in only 30 min (i.e. in two orders of magnitude less time; Scheme 3.6).

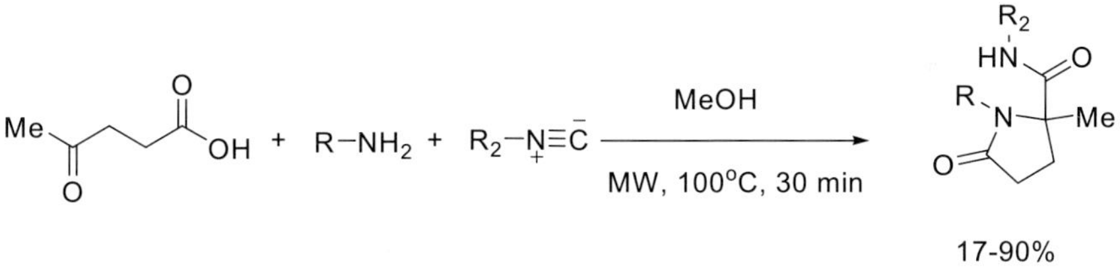

Scheme 3.6. Ugi-type three-component condensation.

Reaction mixtures involving one of six different derivatives of 2-aminophenol and eight separate derivatives of benzoyl chloride were heated in dioxane or xylene at temperatures between 210 °C and 250 °C to give a library of 48 benzoxazoles in good to excellent yields (Scheme 3.7) [27]. This simple method precluded requirements for Lewis acid or base catalysts.

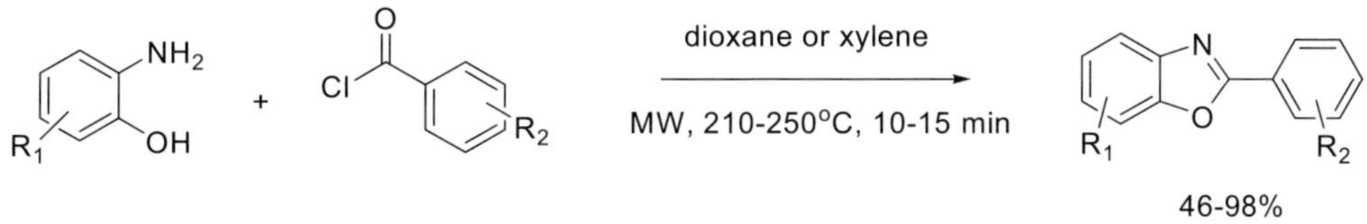

Scheme 3.7. Microwave-assisted benzoxazole synthesis.

Kappe and Stadler in extensive studies of the Biginelli reaction, used microwave conditions to develop members of a library of compounds within 20 min [28]. Their process was superior to that involving conventional reflux conditions with regard to both yield and time (Scheme 3.8).

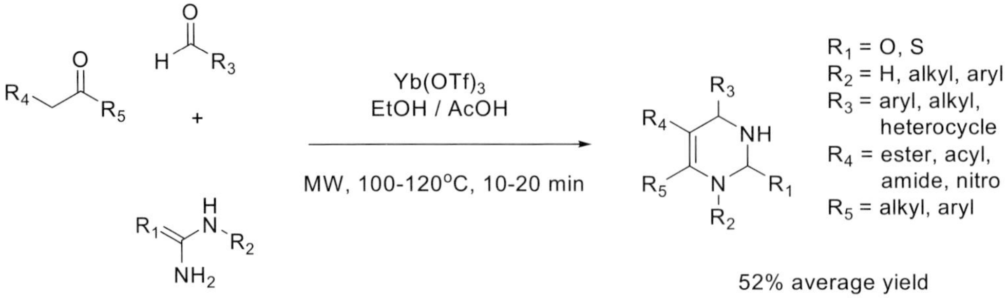

Scheme 3.8. Microwave-assisted Biginelli reaction.

Lehmann and co-workers performed Mannich reactions with paraformaldehyde, secondary amines as their hydrochloride salts, and substituted acetophenones as the starting constituents [29]. With dioxane as solvent they performed reactions for 10 min at 180 °C to give the desired β-amino ketones in moderate to good yields (Scheme 3.9). The syntheses were scaled-up twentyfold. A monomodal reactor was used for small-scale preparations and a multimodal system for scale-up.

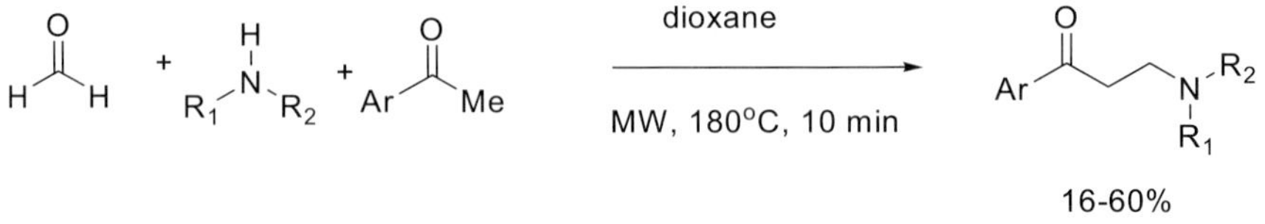

Scheme 3.9. Microwave-assisted Mannich reaction.

Wu et al. developed a rapid, efficient synthesis of 4-chloro-5-bromopyraxolopyrimidine by microwave-assisted electrophilic aromatic substitution [30]. With N-bromosuccinimide in acetonitrile for 10 min at 100 °C the starting chloro derivative was converted to the desired product in excellent yield and regioselectivity (Scheme 3.10).

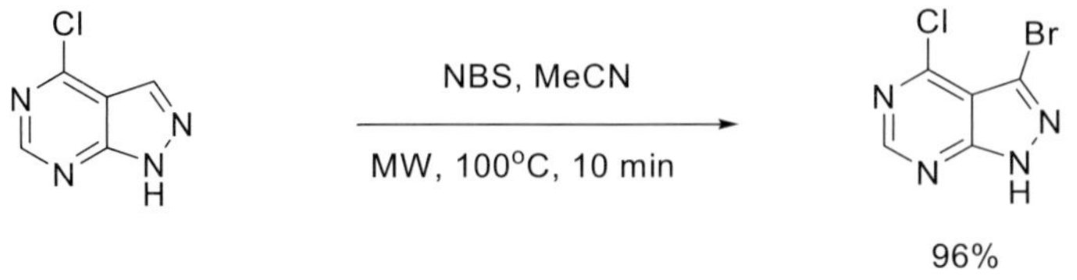

Scheme 3.10. Microwave-assisted electrophilic aromatic substitution.

3.6.2
Kinetic Products

In complex reactions, kinetic products may be formed first but may not necessarily be the most stable. Depending upon the particular pathway, the thermodynamically most stable products may or may not arise by further reaction of kinetic intermediates. Because of their potentially transient nature, some kinetic products may prove difficult to prepare and can therefore present interesting and significant synthetic challenges.

Because MBR and CMR enable rapid heating and cooling of reactions, in principle they could be usefully applied to the preparation of kinetic products. A successful example with important industrial applicability is the acid-catalyzed hydrolysis of cellulose to afford practical conversions to glucose and oligosaccharides [12, 15]. Also, treatment of (S)-(+)-carvone in water for 10 min at temperatures between 180 and 250 °C, afforded 8-hydroxy-p-6-menthen-2-one as an intermediate on the pathway to carvacrol [31]. Addition of water to the 8,9 olefinic bond of carvone was regioselective and proceeded at a lower temperature than did isoaromatization (Scheme 3.11). The kinetic product was readily isolated by differential extraction from the recyclable starting ketone, affording a preparative route despite the low conversion.

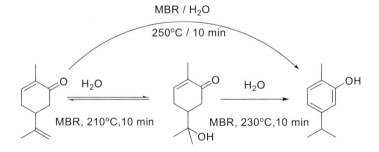

Scheme 3.11. Kinetic and thermodynamic products from carvone in water.

3.6.3
Selective Synthesis of O-Glycofuranosides and Pyranosides from Unprotected Sugars

The Fischer–Helferich synthesis of alkyl-O-glycosides involves acid catalyzed condensation of a sugar with an alcohol [32]. More than forty years ago, Bishop and co-workers showed that for a range of sugars, α and β O-glycofuranosides were formed kinetically and α and β O-glycopyranosides were thermodynamic products [33–35]. These findings were of interest to us owing to the high significance now attached to carbohydrates in bio-organic chemistry. Because equilibria affecting ring size and anomeric ratios are difficult to control, preparation of individual isomers often involves exhaustive chromatography with or without protection and subsequent deprotection of the hydroxyl groups of the sugar and activation of the glycosyl donor.

For methyl glycosides of D-allose, Evans and Angyal found that addition of $CaCl_2$ and $SrCl_2$ affected the isomer ratio, enabling the furanosides to accumulate [36]. They required an order of magnitude more acid, however, to compensate for a significantly slower reaction rate. In an extension of that approach, Lubineau and Fischer obtained a mixture of methyl α and β O-glucofuranosides in the ratio 32:68 and in 75% yield by leaving glucose and MeOH to stand in the presence of $FeCl_3$ at room temperature for 7 days [37]. The analogous reaction with galactose required heating to 60 °C for 24 h followed by heating under reflux for 4 h. Ferrieres et al. [38] adapted Lubineau's method to include solvents such as THF and 1,4-dioxane. In some instances reaction times were reduced from days to several hours by sonication.

Nuchter et al. [39] recently studied the direct glycosylation of glucose, mannose, and galactose by microwave heating with alcohols in the presence catalytic amounts of HCl. Furanoside derivatives were unstable under these conditions and with glucose, the workers obtained the methyl O-α-glucopyranoside exclusively and quantitatively after heating at 140 °C for 20 min. Independently, in unpublished work, we also applied the microwave systems to the glycosidation of glucose, galactose, and mannose with low-molecular-weight alcohols including methanol, ethanol, and isopropanol, but with the objective of obtaining the kinetic products that had eluded Nuchter and co-workers. For reactions of glucose with methanol reported herein, acidic mixtures of the sugar and the alcohol were heated in the MBR at temperatures between 70 °C and 150 °C, for 1–15 min, then cooled rapidly and neutralized with $NaHCO_3$. At 80 °C after 2 min (with Amberlyst 15 as catalyst), the methyl β and α-O-glucofuranosides were the only glycosidic products, formed in 19 and 14% yield, respectively. After 15 min at 70 °C, these glucosides comprised 34 and 20%, respectively, of the product mixture, but smaller amounts of the methyl β- and α-O-glucopyranosides (7 and 5% respectively) were also present, with unreacted glucose. At 150 °C after 2 min the thermodynamically most stable products, methyl β- and α-O-glucopyranosides were obtained in 39 and 61% yield, respectively. These results indicated that the rate of conversion of glucose to the methyl O-glucofuranosides was comparable with that for the equilibration of the glucofuranosides to the pyranosides. They also confirmed that although vigorous conditions can readily afford pyranosides (as reported by Nuchter et al. [39]), the furanosides can be produced selectively and directly.

Finally, by employing the CMR, with sulfuric acid as catalyst (2.5% w/v relative to MeOH), at 80 °C and with a residence time of 6 min in the microwave zone, the methyl β and α-O-glucofuranosides comprised nearly 80% of the product mixture (52 and 27% respectively). These conditions were suitable for large-scale preparation of the furanosides. When the residence time was extended to 12 min, the selectivity was reversed, with the methyl β and α-O-glucopyranosides being obtained in 23 and 54% yield, respectively (Scheme 3.12).

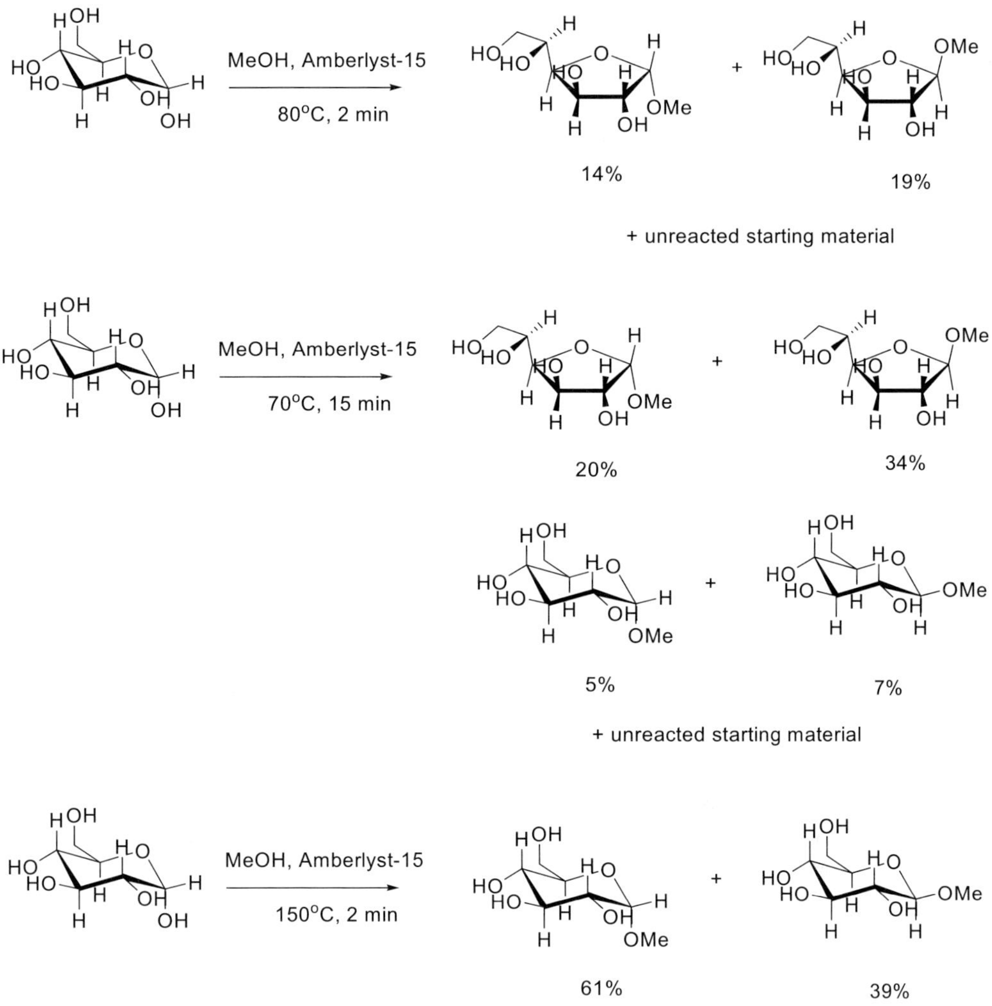

Scheme 3.12. Microwave-assisted glycosylation.

3.6.4
Synthesis in High-temperature Water

In studies, with microwave reactors, with the objective of discovering environmentally benign media for organic synthesis, high-temperature water was found to be useful within the range 100 to 260 °C [31]. Until that work, exploration of this temperature range had usually been neglected for synthesis. The rationale for employing water under those conditions now is widely understood [15]. The ionic product increases one-thousandfold between 25 °C and 240 °C, making water a stronger acid and base at high temperature. In addition, as shown in Fig. 3.5, the dielectric

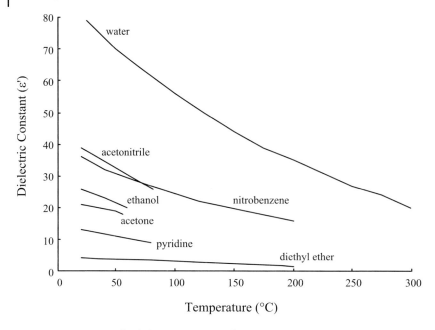

Fig. 3.5. Decreases in the dielectric constants of water and organic solvents with increasing temperature (reproduced from Ref. [15] with permission).

constant decreases, indicating that the polarity is reduced as the temperature is increased. Owing to these apparently contradictory properties, the roles of water can vary among those of solvent, reactant, catalyst, and medium, or a combination of all four, in organic reactions.

When addition of acid or base was found to be necessary, typically, less agent was needed for high-temperature processes than for those at and below 100 °C, and reactions often proceeded more selectively [10, 31]. Often the requirement was orders of magnitude lower. Neutralization of such reaction mixtures afforded small amounts of salt, thereby avoiding production of excessive waste [12]. Apart from polluting soil and ground water, salts can lower the pH of atmospheric moisture and contribute to acid dew or acid rain. Also, hydrophobic resins or adsorbents were employed for concentration and isolation of reaction products from water. Advantages accruing from avoidance of extraction with organic solvents include ease of use, potential for high throughput, and low waste generation owing to ready recyclability of the resin and the solvent used for desorption [10].

A diverse range of reactions were studied and high selectivity was obtained with seemingly minor variations in the conditions [31]. Scale-up was also demonstrated, including to continuous operation with the CMR [15]. This indicated the potential of aqueous high-temperature media for future development of clean processes. Obvious advantages of water include low cost, negligible toxicity and safe handling and disposal. Some examples of organic reactions in water from our laboratories are summarized below.

3.6.5
Biomimetic Reactions

Naturally occurring monoterpene alcohols were heated in water without prior derivatization with typical biological water-solubilizing groups, for example phosphates or glycosidic units. Biomimetic reactions that normally would be acid-catalyzed, proceeded on the underivatized compounds in the absence of added acidulant. Cooling of the mixtures rendered the products insoluble and readily isolable, and the aqueous phase did not require neutralization before work-up.

Geraniol, nerol, and linalool are almost insoluble in water at ambient temperature. Although acid-labile, they do not readily react in water at moderate temperature and neutral pH. In unacidified water at 220 °C in the MBR, they reacted within minutes. Geraniol rearranged to α-terpineol (18%) and linalool (16%), predominantly. Smaller amounts of the monoterpene hydrocarbons were also obtained, including myrcene, α-terpinene (10%), limonene (11%), γ-terpinene, the ocimenes, α-terpinolene, and alloocimenes (Scheme 3.13) [31].

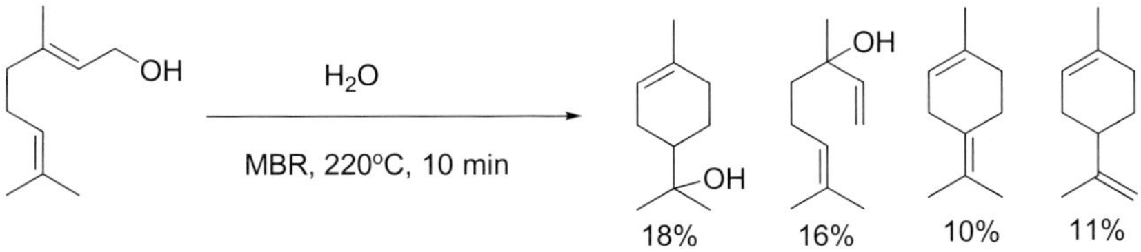

Scheme 3.13. Reaction of geraniol in water at 220 °C in the MBR.

Nerol and linalool underwent considerably more elimination than did geraniol, to give the same hydrocarbons. The major products and their relative proportions were consistent with those for carbocationic rearrangement of derivatives of linalool, nerol, and geraniol under acidic conditions.

α-Ionones and β-ionones can be converted to ionene, catalyzed by HI with small amounts of phosphorus. A cleaner cyclization occurred on heating β-ionone in water at 250 °C in the MBR. In the work-up the usual exhaustive washing procedures were unnecessary. Similarly, as mentioned above, carvacrol was prepared almost quantitatively, by isoaromatization of carvone in water at 250 °C for 10 min. A conventional, literature method under acidic conditions took longer and proceeded with lower conversion. The above examples show that elevated temperatures under neutral pH conditions can have advantages over acidic (or basic) reagents at lower temperatures.

3.6.6
Indoles

In the first example of water as the reaction medium for Fischer indole synthesis, 2,3-dimethylindole was obtained in 67% yield from phenylhydrazine and butan-2-one, at 220 °C for 30 min (Scheme 3.14). Neither a preformed hydrazone nor addition of acid was required [31].

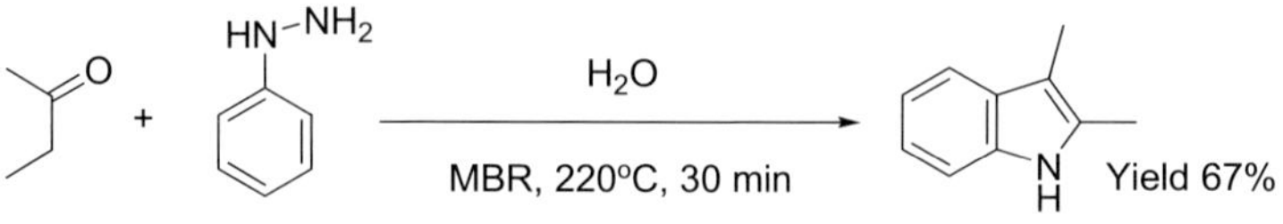

Scheme 3.14. Fischer indole synthesis in water.

Interestingly, the Fischer indole synthesis does not readily proceed from acetaldehyde to afford indole. Usually, indole-2-carboxylic acid is prepared from phenylhydrazine with pyruvic acid or with a pyruvate ester, followed by hydrolysis. Traditional methods for decarboxylation of indole-2-carboxylic acid to form indole are not environmentally benign. They include pyrolysis or heating with copper-bronze powder, copper(I) chloride, "copper" chromite, "copper" acetate, or copper(II) oxide in, for example, heat-transfer oils, glycerol, quinoline, or 2-benzylpyridine. Decomposition of the product during lengthy thermolysis or purification affects the yields.

In water at 255 °C, decarboxylation of indole-2-carboxylic acid was quantitative within 20 min in the MBR (Scheme 3.15) [31].

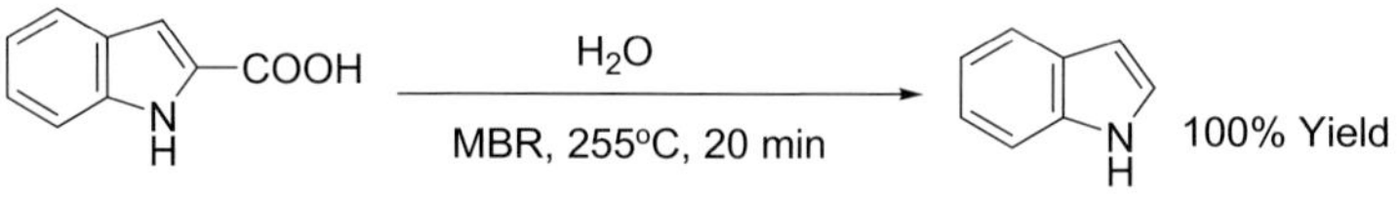

Scheme 3.15. Decarboxylation in water.

A semi-systematic study of the hydrolysis of ethyl indole-2-carboxylate in aqueous media at high temperature indicated that decarboxylation of indole-2-carboxylic acid proceeded by an arenium ion mechanism and was inhibited by base. Because base facilitated hydrolysis of the ester, it was possible to obtain either the acid or indole from the ester merely by manipulating the number of equivalents of base present.

Recently, through the agency of a microwave reactor housing thick quartz vessels that could withstand pressures of 8 MPa and heat water to 300 °C, Kremsner and Kappe have conducted more interesting research into several aspects of the topics described above [40].

3.7
Effect of the Properties of Microwave Heating on the Scale-up of Methods in Pressurized Reactors

After approximately two decades of microwave-assisted organic synthesis, the field can no longer be regarded as in its infancy. Hence, at first glance it may seem surprising that despite various and significant advantages, few laboratory-scale microwave procedures have been further developed into larger-scale industrial processes. Considerations of scale-up, however, require an appreciation of crucial factors including mechanisms of heat transfer by microwaves, effects and potential effects of microwaves on materials, and cost of plant. This understanding may well be required on a case by case basis and cannot be superficial.

First, to avoid potential interference to communications and radar bands, the entire microwave spectrum is not readily available for chemistry. By international convention the frequencies 915 ± 25 MHz, 2450 ± 13 MHz, 5800 ± 75 MHz, and 22125 ± 125 MHz have been assigned for industrial and scientific microwave heating and drying applications. For synthetic chemistry, equipment operating at 2450 MHz, corresponding to a wavelength of 12.2 cm, is used almost exclusively.

Also, the depth of penetration of electromagnetic radiation increases with increasing wavelength (or decreasing frequency) and can vary with temperature. With pure water the depth of penetration at 2450 MHz, increases from 6 mm at 3 °C to approximately 40 mm at 60 °C. Thus the capacity for absorption of microwave energy by water varies inversely with temperature. This trend is observed for a range of organic solvents also. Variations in susceptibility to microwaves can be attenuated (in either direction, however) if, as reactions proceed, the dielectric properties of the product mixtures differ substantially from those of the starting materials [15].

Briefly, the primary mechanism of microwave heating is dielectric loss. The dielectric loss factor (loss factor; ε'') and the dielectric constant (ε') of a material determine the efficiency of heat transfer to the sample. Their quotient ($\varepsilon''/\varepsilon'$) is the dissipation factor (tan δ). High values indicate ready susceptibility to microwave energy. Materials dissipate microwave energy mainly by dipole rotation and ionic conduction. Dipole rotation affects the alignment of molecules with permanent or induced dipoles, with the electric field component of the radiation.

At 2450 MHz, the field oscillates 4.9×10^9 times per second and sympathetic agitation of the molecules generates heat. The quantity of heat produced by dipole rotation is dictated by the dielectric relaxation time of the sample, which in turn, is affected by temperature and viscosity. Ionic conduction on the other hand, occurs by migration of dissolved ions with the oscillating electric field. Heat is generated by frictional losses that depend on the size, charge and conductivity of the ions and on their interactions with the solvent.

If mixing is adequate, temperature gradients within the sample will be minimized by microwave bulk heating and the extent of conduction and convection will be low. Thus differences between the means of energy transfer in conventional and microwave heating have significant implications for the performance of thermochemical reactions. They are reflected in the methods, monitoring, types of equipment, and vessels employed. Also, as mentioned above, the samples themselves have a substantial effect. Dimensions, volume, shape, composition, and physical and chemical properties are important considerations, as are the reactions in question, the media used, and other safety issues. Obviously, these factors become increasingly important with scale [14, 15].

Although high rates of microwave heating are usually regarded as desirable, difficulties can arise if the dielectric loss of a material increases with temperature. The material then will absorb microwave energy with increasing efficiency as temperatures rise, and thermal runaway could result. This characteristic of microwave heating can occasionally be beneficial, particularly for selective heating of catalysts. Usually, however, microwave-driven thermal runaways should be avoided by careful monitoring and control of temperature and power input. In contrast, as already mentioned, the dielectric constants of solvents usually decrease with increasing temperature. Their efficiency at absorbing microwave energy will diminish with increasing temperature, which can lead to poor matching of the input and absorbed microwave energy. This phenomenon becomes pronounced as liquids approach the supercritical fluid state. Solvents and reaction temperatures should be chosen with these considerations in mind, particularly as excess input microwave energy can lead to the build up of electric charge within the sample, followed by discharge by arcing.

Composition can affect heating rates, particularly when ionic conduction becomes possible as a result of addition or formation of salts. For compounds of low molecular weight, the dielectric loss contributed by dipole rotation decreases with increasing temperature, but that resulting from ionic conduction increases. When an ionic sample is microwave-irradiated, heat is initially produced predominantly as a result of dielectric loss by dipole rotation, and the contribution from ionic conduction becomes more significant with temperature rise.

As a mixture gains complexity, owing to the conversion of starting materials during a reaction, an increasing tendency for microwave absorption would be expected, unless polar components (e.g. ethanol and formic acid) were undergoing condensation to form a considerably less polar product (ethyl formate). In that circumstance the reaction temperature may not be sustainable without increasing the power, and arcing could occur within the sample unless the microwave load is matched.

From the foregoing discussion, the propensity of a sample to undergo microwave heating is related to its dielectric and physical properties. Compounds with high dielectric constants (e.g. ethanol and dimethylformamide) tend to absorb microwave irradiation readily whereas less polar substances (for example aromatic and aliphatic hydrocarbons) or compounds with no net dipole moment (e.g. carbon dioxide, dioxane, and tetrachloromethane) and highly ordered crystalline materials, are poorly absorbing.

As mentioned above, monomodal cavities have been used for microwave chemistry on the sub-gram scale. Occasionally special benefits have been claimed for so-called "focused" microwaves. From the foregoing discussion, however, the dielectric properties of a sample can alter substantially with temperature and/or with changing chemical composition. Hence, irrespective of whether multi-modal and unimodal cavities are employed, frequent tuning may be necessary if heating efficiency is to be retained. When tuning is performed frequently, transfer of microwave conditions between monomodal and multi-modal cavities is usually facile. With the MBR (which had a tunable multi-modal cavity), Cablewski et al. performed five reactions that had been conducted earlier on the gram scale or below with "focused" microwaves [41]. These were scaled-up between 40 and 60 fold and reaction conditions and yields were comparable with those on the smaller scale. The reactions studied were diverse. They comprised Suzuki coupling, a Hantzsch reaction, benzoylation of diacetone glucose catalyzed by polymer-bound dimethylaminopyridine, synthesis of hydantoin from phenylisothiocyanate, and the Pd-catalysed asymmetric alkylation of 4-methoxyphenol with cyclohex-2-enyl carbonate in the presence of Trost's catalyst [41].

Results of Kappe et al. for scale-up of the hydrolysis of benzamide, from 50 mg to 5 g were consistent with these findings (Scheme 3.16) [42].

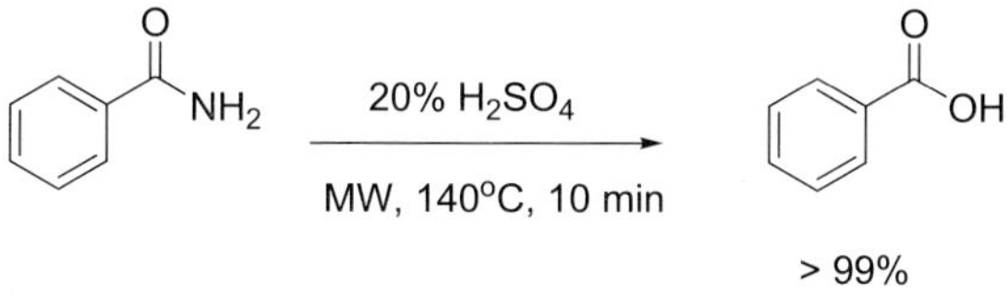

Scheme 3.16. Hydrolysis of benzamide.

Other examples of scale-up involved a triphenylphosphine-free one-pot Wittig olefination, a one-step three-component synthesis of imidazo annulated pyridine and a metal-catalyzed Suzuki coupling. Kappe and co-workers also recently transferred conditions for reactions originally performed on a small scale with a monomodal system, to scale-up by parallel synthesis in a multimodal batch reactor [13]. Typically, the scale-up was 100-fold, from 1 mmol; examples included Biginelli condensations, Heck and Negishi couplings, and Diels–Alder cycloadditions with gaseous reactants.

3.8
Software Technology for Translation of Reaction Conditions

Traditional methods for organic synthesis mostly involve glass reaction vessels operated at atmospheric pressure or below. Routine use of such equipment has meant that for the past one and a half centuries, modest temperatures and lengthy times have become the normal reaction conditions for established processes. Statistics obtained by literature searching for reaction conditions support that contention [43]. Work with the microwave reactors, however, has demonstrated clearly that if reactions can be performed at significantly higher temperatures than normal, considerable savings in time and energy can be realized. These and other advances in hardware for synthesis are in the process of revolutionizing approaches to organic chemistry, if they have not done so already.

Yet despite having undergone rapid growth, particularly since 2000, microwave technology is not yet employed routinely in all synthetic laboratories. To those familiar with the microwave technique, this may appear somewhat surprising in view of significant potential advantages with regard to yields and time. The cost of commercial dedicated microwave equipment, although a substantive issue for some, does not seem to be the only mitigating factor. It seemed to us that a major impediment to uptake was associated with a lack of experience regarding where to start. Seemingly, many synthetic chemists had perceived that considerable empirical work could be required before they could adapt established conditions into appropriate alternatives employing higher temperatures.

If that interpretation were correct, to utilize the microwave technology conveniently, those accustomed to traditional methods would require answers to questions such as "In what time could I expect to perform, with comparable results, the same reaction at 180 °C instead of in 7 h at 78 °C?" or "If I wanted to complete the reaction in twice the yield and in 8 min instead of in 10 h at 100 °C, what temperature would I need?" In any event, it was clear that chemists need to be able to transpose established conditions from ambient to ca 120 °C, predictably into alternatives employing higher temperatures, without first conducting extensive empirical investigations. In a significant departure from the use of the Arrhenius equation, along with reaction temperature and time, the new conditions also needed to take yields or conversions into account. The last aspect, to the best of our knowledge, had not previously been rigorously considered in that context.

We have now developed software-based technology for estimating optimum reaction conditions [43, 44]. The program employs a strategy that enables the chemist to focus on the desired conditions rather than using more painstaking mapping of the contours on the reaction surface. For a wide range of reactions and conditions examined we have found the number of required confirmatory experiments to be, typically, only one or two, irrespective of the process under consideration. A first estimate produces a set of conditions that, when tested experimentally (the *first iteration*), usually affords a yield or conversion close to that desired. Experimental results from the first iteration may be used to obtain a second estimate that could also be tested by experiment (the *second iteration*). Although rarely required, further iterations can be performed in a similar manner.

To elaborate, by starting with data from only a single experiment (*reference data*) for the reaction in question (e.g. 43% yield in 3 h at 78 °C), the developed software could estimate the time expected to give any specified yield, between 5 and 95%, for that reaction at the same temperature. Alternatively, different conditions of time (between 0.1 and 10^5 min) and temperature (between 0 and 270 °C) that would be expected to return any nominated yield or conversion between 5 and 95% could be calculated. Starting with the aforementioned set of reference data for the subject reaction (43% yield in 3 h at 78 °C), it could be estimated, for example, that the reaction could proceed to 79% yield in 34 min at 150 °C or in 87% yield after 5 min at 204 °C. For both estimates, all of the variables of time, temperature, and yield or conversion differ from those in the reference data set. Such estimates are not produced at random, but are based on data for the conditions requested and sought by the investigator.

For any specific reaction, a calculation requires one set of experimentally determined reference data with regard to temperature, time, yield or conversion, along with the desired reaction temperature or time and yield or conversion: a total of three experimentally established and two requested items of data [43]. The software can then estimate the one remaining unknown, from time, temperature, yield, or conversion.

3.9 Conclusion

Microwave-assisted organic chemistry has developed into a significant field and continues to experience exponential growth and to command its own dedicated, international conferences. The number of papers in peer-reviewed journals began with two in 1986, grew to 220 by mid-1995, was well over 1000 by 2002 and probably has trebled since then. Commercial microwave reactors based upon the principles of the CMR and MBR have found applications in the research and manufacture of high-value low-volume chemicals. They satisfy criteria concerning measurement and control of reaction conditions, failsafe procedures, advantages unavailable with conventional equipment, and capabilities for scaling-up and scaling-down chemical processes.

Pressurized microwave reactors, through a capability of accommodating organic solvents safely, have significantly extended the useful operating temperature range for low-boiling organic solvents. They have enabled new chemical processes that require moderate temperatures.

Economic and safety considerations encourage minimum stockpiling of chemicals and avoidance of transport of hazardous substances. These increasing demands present opportunities for microwave chemistry in the development of environmentally benign methods for preparation of intermediates, specialty, chemicals and pharmaceutical products. Within the next few years individual chemical reactors may be required to perform diverse tasks and to be easily relocated as community pressure mounts to restrict movement of chemicals by road, rail, air, and sea. Future requirements could include quick start up and shut down, capability to produce high yields with minimum development, low holding capacity, almost complete elimination of waste, with the potential for accidents during transportation and storage being avoided by just-in-time, point-of-use production.

The MBR and CMR are portable, multi-purpose and self-contained, features that could become increasingly important in this regard. Their rapid throughput capabilities and the materials of construction enable easy cleaning for re-use and promote short turnaround times. Safety advantages include control and method of energy input, low volumes undergoing reaction at one time and opportunities for remote, programmable operation.

Acknowledgments

CSIRO is thanked for hosting Professor Thach Le Ngoc on sabbatical leave for one year and for granting a Post-doctoral Fellowship to Dr Brett Roberts. Kodak (Australasia), Milestone and Personal Chemistry (now Biotage) are acknowledged for commercial collaboration. The authors are indebted to the following co-workers who were engaged on aspects of the work and whose contributions are also acknowledged through co-authorship of cited references: A. F. Faux, T. Cablewski, R. W. Trainor, K. D. Raner, J. S. Thorn, L. Bagnell, U. Kreher, L. Mokbel, F. Vyskoc, and M. Ballard.

References

1 GEDYE, R., SMITH, F., WESTAWAY, K., ALI, H., BALDISERA, L., LABERGE L., ROUSELL, J., *Tetrahedron Lett.*, **1986**, *27*, 279.
2 GIGUERE, R. J., BRAY, T. L., DUNCAN, S. M., MAJETICH, G., *Tetrahedron Lett.*, **1986**, *27*, 4945.
3 GEDYE, R. N.; SMITH, F. E.; WESTAWAY, K. C., *Can. J. Chem.*, **1988**, *66*, 17.
4 LOUPY, A., PETIT, A., HAMELIN, J., TEXIER-BOULLET, F., JACQUAULT, P., MATHÉ, D., *Synthesis*, **1998**, 1213.
5 LANGA, F., DE LA CRUZ, P., DE LA HOZ, A., DIAZ-ORTIZ, A., DIEZ-BARRA, E., *Contemp. Org. Synthesis*, **1997**, *4*, 373.
6 VARMA, R. S., *Green Chem.*, **1999**, *1*, 43.
7 CABLEWSKI, T., FAUX, A. F., STRAUSS, C. R., *J. Org. Chem.*, **1994**, *59*, 3408.
8 RANER, K. D., STRAUSS, C. R., TRAINOR, R. W., THORN, J. S., *J. Org. Chem.*, **1995**, *60*, 2456.
9 DANNER, H., BRAUN, R., *Chem. Soc. Rev.*, **1999**, *28*, 395.
10 BAGNELL, L., BLIESE, M., CABLEWSKI, T., STRAUSS, C. R., TSANAKTSIDIS, J., *Aust. J. Chem.*, **1997**, *50*, 921.
11 RANER, K. D., STRAUSS, C. R., VYSKOC, F., MOKBEL, L., *J. Org. Chem.*, **1993**, *58*, 950.
12 STRAUSS, C. R., *Aust. J. Chem.*, **1999**, *52*, 83.
13 KAPPE, C. O., STADLER, A., *Microwaves in Organic and Medicinal Chemistry*, **2005**, Wiley-VCH, Weinheim.
14 ROBERTS, B. A., STRAUSS, C. R., in *Microwave Assisted Organic Synthesis*, TIERNEY, J., LIDSTRÖM, P., eds., **2005**, Blackwell, Oxford, Ch. 9, p. 237.
15 STRAUSS, C. R., TRAINOR, R. W., *Aust. J. Chem.*, **1995**, *48*, 1665.
16 LIDSTRÖM, P., TIERNEY, J., WATHEY, B., WESTMAN, J., *Tetrahedron*, **2001**, *57*, 9225.
17 KAPPE, C. O., *Angew. Chem. Int. Ed.*, **2004**, *43*, 6250.
18 LARHED, M., HALLBERG, A., *J. Org. Chem.*, **1996**, *61*, 9582.
19 LARHED, M., MOBERG, C., HALLBERG, A., *Acct. Chem. Res.*, **2002**, *35*, 717.
20 ERSMARK, K., LARHED, M., WANNBERG, J., *Curr. Opinion Drug Discovery & Dev.*, **2004**, *7*, 417.
21 OLOFSSON, K., KIM, S-Y., LARHED, M., CURRAN, D. P., HALLBERG, A., *J. Org. Chem.*, **1999**, *64*, 4539.
22 SVENNEBRING, A., NILSSON, P., LARHED, M., *J. Org. Chem.*, **2004**, *69*, 3345.
23 ROMERA, J. L., CID, J. M., TRABANCO, A. A., *Tetrahedron Lett.*, **2004**, *45*, 8797.
24 BAXENDALE, I. R., LEE, A.-I., LEY, S. V., *J. Chem. Soc., Perkin Trans.*, **2002**, 1850.

25 Leadbeater, N. E., Torenius, H. M., *J. Org. Chem.*, **2002**, *67*, 3145.
26 Tye, H., Whittaker, M., *Org. Biomol. Chem.*, **2004**, *2*, 813.
27 Pottorf, R. S., Chadha, N. K., Katkevics, M., Ozola, V., Suna, E., Ghane, H., Regber, T., Player, M. R., *Tetrahedron Lett.*, **2003**, *44*, 175.
28 Kappe, C. O., Stadler, A., *J. Comb. Chem.*, **2001**, *3*, 624.
29 Lehmann, F., Pilotti, Å., Luthman, K., *Mol. Diversity*, **2003**, *7*, 145.
30 Wu, T. Y. H., Schultz, P. G., Ding, S., *Org. Lett.*, **2003**, *5*, 3587.
31 An, J., Bagnell, L., Cablewski, T., Strauss, C. R., Trainor, R. W., *J. Org. Chem.*, **1997**, *62*, 2505.
32 Schmidt, R. R., *Angew. Chem. Int. Ed.*, **1986**, *25*, 212.
33 Bishop, C. T., Cooper, F. P., *Can. J. Chem.*, **1962**, *40*, 224.
34 Bishop, C. T., Cooper, F. P., *Can. J. Chem.*, **1963**, *41*, 2743.
35 Smirnyagin, V., Bishop, C. T., *Can. J. Chem.*, **1968**, *46*, 3085.
36 Evans, M. E., Angyal, S., *J. Carbohydrate Res.*, **1972**, *25*, 43.
37 Lubineau, A., Fischer, J.-C., *Synth. Commun.*, **1991**, *21*, 815.
38 Ferrieres, V., Bertho, J.-N., Plusquellec, D., *Tetrahedron Lett.*, **1995**, *36*, 2749.
39 Nuchter, M., Ondruschka, B., Lautenschlager, W., *Synth. Commun.*, **2001**, *31*, 1277.
40 Kremsner, J. M., Kappe, C. O., *Eur. J. Org. Chem.*, **2005**, *17*, 3672.
41 Cablewski, T., Hellman, B., Pilotti, P., Thorn, J., Strauss, C. R., *Personal communication*. See www.biotage.com for more information.
42 Horeis, G., Pichler, S., Stadler, A., Gössler, W., Kappe, C. O., *Fifth International Electronic Conference on Synthetic Organic Chemistry (ECSOC-5)* E0000.
43 Roberts, B. A., Strauss, C. R., *Acct. Chem Res.*, **2005**, *38*, 653.
44 Roberts, B. A., Strauss, C. R., *Molecules*, **2004**, *6*, 459.

4
Nonthermal Effects of Microwaves in Organic Synthesis

Laurence Perreux and André Loupy

4.1
Introduction

Microwave (MW) activation has become widely accepted and very popular unconventional technology in organic chemistry as an alternative to, and often improvement on, conventional heating. This statement is clearly evident from the annual number of publications on MW-assisted organic chemistry as growing rapidly with more than 2500 publications up to the end of 2005 since the pioneering work of Gedye [1] and Giguere [2] in 1986 (Fig. 4.1).

The main results are compiled in exhaustive reviews [3–7] and books [8–11]. Most of these publications describe important acceleration for a wide range of organic reactions, especially when conducted under solvent-free conditions. The combination of solvent-free reaction conditions and microwave irradiation leads to large reductions in reaction times, enhancements in conversions, and, sometimes [4, 12], in selectivity with several advantages of the eco-friendly approach, termed "Green Chemistry".

A substantial number of these reports are however based on inaccurate or unfounded comparisons with classical conditions which do not enable unequivocal conclusions to be made about microwave effects. Very often, for this reason, some apparent contradictions and controversies have appeared in the literature [13–17]. To try to rationalize all these results, it is necessary to propose a plausible interpretation of the effects based on accurate and reliable data resulting from strict comparisons between reactions conducted with microwave irradiation or conventional heating and, otherwise, similar conditions (reaction medium, temperature, time, pressure) [4, 18–20]. A monomode microwave reactor which gives wave focusing should preferably be used (for reliable homogeneity in the electric field) and accurate control of the temperature (using an optical fiber or by infrared detection) throughout the reaction [4, 21]. This gives the possibility of operating with rather similar profiles of temperature increases in both kinds of activation. Based on such strict comparisons, it then becomes possible to make an educated judgment about the suitability, or otherwise, of using microwave irradiation according to reaction type and experimental conditions.

Microwaves in Organic Synthesis, Second edition. Edited by A. Loupy
Copyright © 2006 WILEY-VCH Verlag GmbH & Co. KGaA, Weinheim
ISBN: 3-527-31452-0

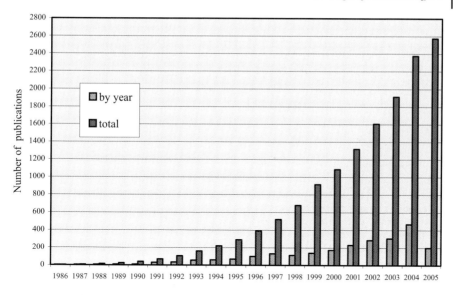

Fig. 4.1. Number of publications dealing with microwave irradiation in organic synthesis for the period 1986–2005.

4.2
Origin of Microwave Effects

Microwave (MW) irradiation consists in electromagnetic waves in the range 0.3 to 300 GHz (i.e. wavelengths from 1 cm to 1 m). All standard equipment (domestic ovens or more specifically dedicated reactors for chemical synthesis, cf. Chapters 1 and 2) operate at a frequency of $v = 2.45$ GHz (corresponding to $\lambda = 12.2$ cm) to avoid interferences with radio and radar frequencies. When applying Planck's law ($E = hc/\lambda$), the energy involved in material–MW interactions is much to low (approx. 1 J mol^{-1}) to induce chemical activation. It is even much lower than the energy of Brownian motion. Thus, absorption of MW by materials is mainly connected with electromagnetic interactions which are specific to polar molecules (dipoles) and increasingly important when the polarity of materials is increased.

MW-enhanced chemistry is based on the efficiency of interactions of molecules with waves by "microwave dielectric heating" effects. This phenomenon depends on the ability of materials to absorb MW radiation and convert it into heat. The electric component of the electromagnetic field has been shown to be the most important [22–24]. It results in two main mechanisms – dipolar polarization and ionic conduction. Irradiation of polar molecules at MW frequencies results in orientation of the dipoles or ions in the applied electric field (Scheme 4.1) [25].

One can, consequently, thus distinguish two main effects:

- *electrostatic polar effects* which lead to dipole–dipole-type interactions between the dipolar molecules and the charges of the electric field (Scheme 4.1, b). These re-

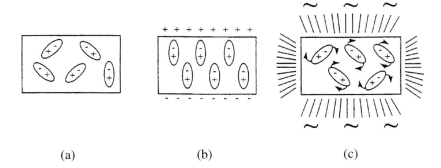

Scheme 4.1. Effects of the surrounding electric field on the mutual orientation of dipoles (a) without any constraint, (b) submitted to a continuous electric field, and (c) submitted to an alternating electric field of high frequency.

sult in energy stabilization which is electrostatic in nature. This phenomenon could be the origin of *specific nonthermal effects* and is expected to be rather similar to dipole–dipole interactions induced by a polar solvent [26] (ex: DMSO).

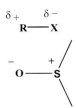

- *thermal effects* (dielectric heating) result from the tendency of the dipoles to follow the inversion of the alternating electric field and induce energy dissipation in the form of heat through molecular friction and dielectric loss (Scheme 4.1, c). This energy dissipation in the core of materials enables much more regular temperature distribution compared with conventional heating. Classical thermal phenomena (conduction, convection, radiation, etc.) only play a secondary role in the a posteriori equilibration of temperature.

In this range of frequency, the *charge space polarization* [22, 23, 27] can also intervene and can be of prime importance with semiconductors, because it concerns materials which contain free conduction electrons. This phenomenon is essential in the heating of more or less magnetic solid particles, for example a variety of mineral oxides or metallic species.

The acceleration of reactions by microwave exposure results from material–wave interactions leading to thermal effects (which may be easily estimated by temperature measurement) and specific (not purely thermal) effects. Clearly, a combination of these two contributions can be responsible for the observed effect, discussed in

Tab. 4.1. Boiling points (°C) of some polar solvents under MW irradiation in the absence or presence of a nucleation regulator.

Solvent	Boiling point	Microwave exposure		
		Multimode [11, 12]	Monomode (100 W)	Monomode + boiling chips
Water	100	105	100	100
1-Heptanol	176	208	180	173
Ethyl acetate	77	102	92	77
Chloroform	61	89	85	62
Cyclohexanone	155	186	168	155

terms of thermal and nonthermal specific effects [5, 28], which are clearly shown to be dependent on medium and reaction mechanisms (see below).

For liquid products (example solvents), only polar molecules selectively absorb microwaves – nonpolar molecules are inert to microwave dielectric loss. In this context of efficient microwave absorption, it has also been shown that higher boiling points could be observed when solvents are submitted to microwave irradiation conditions compared with conventional heating. This effect, called the "superheating effect" [29, 30] has been attributed to retardation of nucleation in microwave heating (Table 4.1).

It is clearly connected to the effect of stirring and the presence of a nucleation regulator [31]. It is also related to the microwave power. This effect has been shown to be eliminated when the experiments are performed with well-stirred mixtures [32–35] using low microwave power. It could, however, be connected with the absence of stirring, i.e. in closed vessels inside a domestic microwave oven.

4.3
Specific Nonthermal Microwave Effects

The effect of microwave irradiation in chemical reactions can be attributed to a combination of the thermal and nonthermal effects, i.e. overheating, hot spots, and selective heating, and nonthermal effects of the highly polarizing field, in addition to effects on the mobility and diffusion that may increase the probabilities of effective collisions. These effects can be rationalized considering the Arrhenius law [36, 37] and result from modification of each of the terms of the equation $k = A \exp(-\Delta G^{\ddagger}/RT)$

1. *Increase in the pre-exponential factor, A*, which is representative of the probability of molecular impacts. The collision efficiency can be effectively influenced by

mutual orientation of polar molecules involved in the reaction. Because this factor depends on the vibration frequency of the atoms at the reaction interface, it could be postulated that the microwave field might affect this one. Binner et al. [38] explained the increased reaction rates observed during the microwave synthesis of titanium carbide in this way:

$$TiO_2 + 3C \longrightarrow TiC + 2CO\uparrow$$

Calculations have shown that the faster diffusion rates might be explained by an increase in the factor A with no change in activation energy. Miklavc [39], by analyzing the rotational dependence of the reaction $O + HCl \rightarrow OH + Cl$, concluded that marked acceleration may occur as a result of the effects of rotational excitation on collision geometry.

2. *Intervention of localized high temperatures:*
   - either at a microscopic level [20, 40] as postulated in sonochemistry to justify the sonochemical effect [36]. There is an unavoidable lack of experimental evidence as we can, necessarily, only have access to macroscopic temperature. It was suggested that, in some examples, MW activation could originate from hot spots generated by dielectric relaxation on a molecular scale;
   - or at a macroscopic level, because it is essential when strong MW-absorbing solid catalysts, for example graphite [41] or Magtrieve (an oxidant) [42], are involved, where the (high) temperatures are not homogeneously distributed. Some controversy also exists about the effects of MW irradiation in heterogeneous catalysis, however, because modification of the catalyst's electronic properties, and even structure, has also been proposed [43].
3. *Decrease in the activation energy* $\Delta G^\ddagger$ is certainly the main effect. When considering the contribution of enthalpy and entropy to the value of $\Delta G^\ddagger$ [$\Delta G^\ddagger = \Delta H^\ddagger - T\Delta S^\ddagger$], it may be predicted that the magnitude of the $-T\Delta S^\ddagger$ term would decrease in a microwave-induced reaction, this being more organized when compared with classical heating as a consequence of dipolar polarization.

Lewis et al. [44] presented experimental evidence for such an assumption after measurement of the dependence of rate constants on temperature for the unimolecular imidization of a polyamic acid (Eq. (1), Fig. 4.2, Table 4.2).

$$\text{(structure with } O_2S\text{-substituted aryl amide and carboxylic acid)} \xrightarrow{NMP} \text{(imide product)} \qquad (1)$$

where NMP = N-methylpyrrolidinone.

The apparent activation energy is substantially reduced. The same evidence, i.e. a decrease in $\Delta G^\ddagger$, was also observed for the decomposition of sodium hydrogen

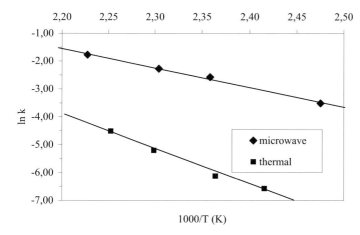

Fig. 4.2. First-order kinetic plots for microwave (MW) and thermal (Δ) activation of the imidization reaction.

carbonate in aqueous solution [45]:

$$2NaHCO_3 \rightarrow Na_2CO_3 + H_2O + CO_2$$

To justify this reduction in the energy of activation one can assume greater stabilization by the MW of the transition state (TS) of the reaction when compared with its ground state (GS) (Fig. 4.3).

This is presumably true when the polarity increases during the course of the reaction, i.e. if the TS is more polar than the GS. It will be true for a reaction in which the dipole moment is enhanced from GS to TS [5].

The effects of the electric field, E, have been determined, by *ab initio* calculations, by evaluating the saving in the standard enthalpy of formation (ΔH_f) of different ion-pairs M^+X^- on application of electric fields (regarded as dipolar, uniform, and continuous) of different types [46]. Calculations were based on the effect of the electric field E (with optimum effects when considering E to be collinear with the dipole moment) according to ion-pair dissociation by considering the stabilization (decreases in ΔH_f) as a function of the interatomic distance, d, between Li^+ and F^- and the nature of ions at their equilibrium distance for several

Tab. 4.2. Results from the Arrhenius plots in Fig. 4.2.

Activation mode	$\Delta G^{\ddagger}$ (kJ mol^{-1})	log A
MW	57 ± 5	13 ± 1
Δ	105 ± 14	24 ± 4

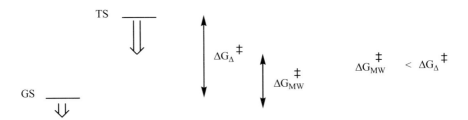

Fig. 4.3. Relative stabilization of TS and GS and the effect on $\Delta G^{\ddagger}$.

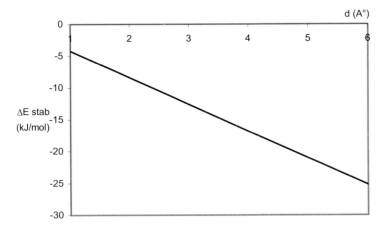

Fig. 4.4. Dependence of the energy of stabilization by E (norm = 5×10^8 V m^{-1}) on the interionic distance d (Li···F).

Tab. 4.3. Effect of the electric field E (norm = 5×10^8 V m^{-1}) on the equilibrium distance of ion-pairs.

Salt	$d_{equil.}$ (Å)	$\Delta E_{stab.}$ (kJ mol^{-1})[a]	μ (Debye)
LiCl	2.07	−7.93	7.6
NaCl	2.40	−9.88	9.5
KCl	2.80	−11.99	11.5

[a] Decrease in the enthalpy of formation

M$^+$ (Li, Na, K) and X$^-$ (F, Br, Cl). The main results are illustrated in Fig. 4.4 and Table 4.3.

It is obvious from this theoretical approach that the energetic stabilization of ion-pairs induced by interaction with the electric field E becomes increasingly important as the size of the ions and their dipole moments (μ) increase. The more polar ion-pairs are more stabilized by E, clearly increasing from tight to loose ion-pairs, i.e. with their dissociation and polarity.

4.4
Effects of the Medium

Microwave effects can be treated according to the reaction medium. Solvent effects are of particular importance [47, 48].

4.4.1
Polar Solvents

If polar solvents are used, either protic (e.g. alcohols) or aprotic (e.g. DMF, CH$_3$CN, DMSO, etc.), the main interaction might occur between MW and polar molecules of the solvent. Energy transfer is from the solvent molecules (present in large excess) to the reaction mixtures and the reactants, and it would be expected that any specific effects of MW on the reactants would be masked by solvent absorption of the field. The reaction rates should therefore be nearly the same as those observed under conventional heating (Δ).

This is essentially true, as evidenced by the rates of esterification in alcoholic media of propan-1-ol with ethanoic acid [49] (Fig. 4.5) or of propan-2-ol with mesi-

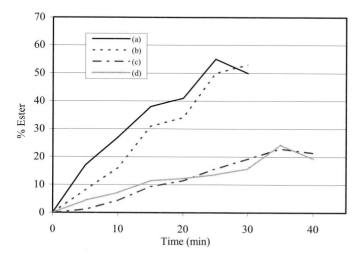

Fig. 4.5. Comparative conversions under the action of MW and Δ during esterification of propanol with ethanoic acid. (a) H$_2$SO$_4$ catalyst, Δ; (b) H$_2$SO$_4$ catalyst, MW; (c) silica catalyst, Δ; (d) silica catalyst, MW.

toic acid [50]. The absence of a specific MW effect became apparent from several experiments carefully conducted in alcohols or in DMF with similar conditions under MW or classical heating [15].

More recently [51], the MW-mediated Biginelli dihydropyrimidine synthesis (Eq. 2) was reinvestigated using a purpose-built commercial microwave reactor with on-line temperature, pressure, and microwave power control. When transformations were performed with MW heating at atmospheric pressure in ethanol solution there was no increase in either rate or yield when the temperature was identical with that used for conventional heating. The only significant rate and yield enhancements were obtained when the reaction was performed under solvent-free conditions in an open system.

$$ArCHO + \underset{H_2N}{\overset{H_2N}{>}}=O + \underset{Me}{\overset{EtO_2C}{>}}=O \xrightarrow{EtOH} \underset{Me}{\overset{EtO_2C}{\underset{N}{\bigwedge}}}\overset{Ar}{\underset{H}{\bigvee}}\overset{H}{\underset{O}{}} \quad (2)$$

A rapid and efficient procedure for flash heating by microwave irradiation has been described for attachment of aromatic and aliphatic carboxylic acids to chloromethylated polystyrene resins via their cesium salts (Eq. 3) [35]:

$$\text{(P)-O-C}_6\text{H}_4\text{-CH}_2\text{Cl} \xrightarrow[\text{NMP, MW}]{RCO_2H, CsCO_3} \text{(P)-O-C}_6\text{H}_4\text{-CH}_2\text{-O-C(O)-R} \quad (3)$$

Significant rate accelerations and higher loadings are observed when the microwave-assisted procedure was compared with the conventional thermal procedure. Reactions times were reduced from 12 to 48 h with conventional heating at 80 °C to 5–15 min with MW flash heating in NMP at temperatures up to 200 °C. Finally, careful kinetic comparison studies have shown that the observed rate enhancements can be attributed to the rapid direct heating of the solvent (NMP) rather than to a specific nonthermal microwave effect [35].

The synthesis of β-lactams from diazoketones and imines can be realized not only by using photochemical reaction conditions but also under the action of microwave irradiation. When the reaction was performed in o-dichlorobenzene at 180 °C, however, the rates of the thermal and microwave-assisted formation of β-lactams were shown to be identical within the limits of experimental error (80–85% conversion after 5 min) [52].

As described above, however, some rather small differences could appear, taking into account the superheating effect of the solvent under the action of microwaves in the absence of any stirring. This probably occurs in the isomerization of safrole and eugenol in ethanol under reflux (MW 1 h compared with Δ 5 h to obtain equivalent yields) [53].

4.4 Effects of the Medium

It was suggested superheating of the solvent was responsible for the observed rate enhancement under the action of microwave irradiation of the synthesis of 3,5-disubstituted 4-amino-1,2,4-triazoles in 1,2-ethyleneglycol as (polar) solvent (Eq. 4) [54]:

$$Ar-C\equiv N \xrightarrow[\substack{H_2N\text{-}NH_2,\ 2HCl \\ H_2N\text{-}NH_2,\ H_2O}]{HOCH_2CH_2OH} \begin{array}{c} Ar-\underset{\underset{H}{N-N}}{\overset{N-N}{\diagdown\diagup}}-Ar \\ \downarrow \\ Ar-\underset{N-N}{\overset{NH_2}{\diagdown\diagup}}-Ar \end{array} \tag{4}$$

The yields obtained by use of MW and Δ were clearly different for very short reaction times and became similar after 15 min at 130 °C (Fig. 4.6).

4.4.2
Nonpolar Solvents

More striking is the use of nonpolar solvents (e.g. xylene, toluene, carbon tetrachloride, hydrocarbons), because these are transparent to, and only weakly absorb microwaves. They can, therefore, be used to enable specific absorption by the reactants. If these reactants are polar, energy transfer from the reactants to the solvent occurs and the results may be different under the action of MW and Δ. This effect seems to be clearly dependent on the reaction and was therefore the subject of controversy. In xylene under reflux, for example, no MW-specific effect was observed

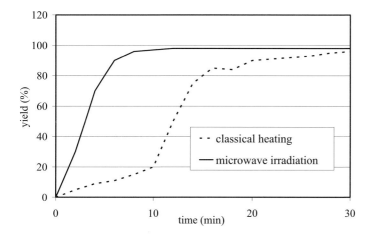

Fig. 4.6. Dependence of on time for 3,5-diphenyl-4-amino-triazole synthesis at 130 °C.

for the Diels–Alder [13] reaction depicted by Eq. (5), whereas important specific effects were described for aryldiazepinone synthesis [55, 56] (Eq. 6):

$$k_{MW} \equiv k_\Delta \quad (5)$$

10 min 140°C MW = 80-95% Δ= 10-30% (6)

These examples will be discussed and explained later during discussion of the dependence of MW effects on reaction mechanism.

The effect of solvent clearly seems to be very important with regard to the possibility of MW-specific effects. These decrease when the polarity of the solvent is increased. This effect was shown in at least two studies by Berlan et al. [33] and, later, by Bogdal [57]. In the first study, the acceleration under the action of MW was much more apparent in xylene ($\mu = 1$ Debye) than in the more polar dibutyl ether ($\mu = 4$ Debye) for the Diels–Alder reaction of 2,3-dimethylbutadiene with methylvinylketone (Fig. 4.7).

In the second investigation [57], involving synthesis of a coumarin by Knoevenagel condensation, supported by rate constant measurements and activation energy calculations, it was found that the effect of MW was more important when the reaction was conducted in xylene whereas it was noticeably reduced in ethanol (Eq. (7), Table 4.4).

Tab. 4.4. Rate constants ($\times 10^3$) for the reaction depicted in Eq. (7).

T (°C)	Xylene k_1 (mol L s^{-1})		Ethanol k_1 (mol L s^{-1})	
	MW	Δ	MW	Δ
60	5.7	2.2	6.9	4.9
80	12.2	3.7	12.9	8.6

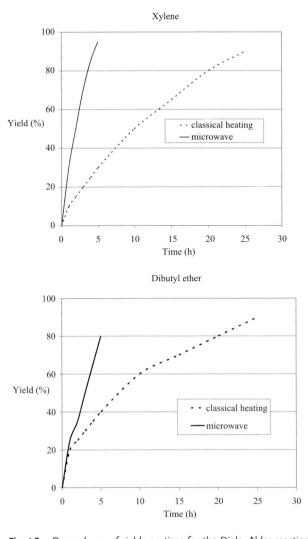

Fig. 4.7. Dependence of yields on time for the Diels–Alder reaction at 95 °C.

4.4.3
Solvent-free Reactions

Microwave effects are mostly likely to be observed in solvent-free reactions [4]. In addition to the preparative interest of these methods in terms of use, separation, and economical, safe, and clean procedures (green chemistry), absorption of microwave radiation should now be limited to the reactive species only. The possible specific effects will therefore be optimum, because they are not moderated or impeded by solvents. They can be accomplished by to following three methods [4, 58]:

1. reactions between the neat reagents, in quasi-equivalent amounts, requiring at least one liquid phase in heterogeneous media and leading to interfacial reactions [58–61];
2. solid–liquid phase-transfer catalysis (PTC) conditions, in anionic reactions using the liquid electrophile as both reactant and organic phase and a catalytic amount of tetraalkylammonium salts as the transfer agent [62]; and
3. reactions using impregnated reagents on solid mineral supports (aluminas, silicas, clays) in dry media [4, 12, 63].

Some other specific effects with different origins could, however, arise from the supports. The mineral supports are usually poor heat conductors, i.e. significant temperature gradients can develop inside the vessels under the action of conventional heating. In contrast, they behave as efficient absorbers of microwave energy with consequently good homogeneity in temperature (Scheme 4.2).

Finally, because the interpretation proposed for specific MW nonthermal effects is rather similar to that for solvent effects (Hughes-Ingold theory), one can expect that aprotic polar solvents can be removed and replaced by MW activation requiring operation with a nonpolar solvent or, much better, under solvent-free conditions. In this situation we add also advantages of green chemistry conditions. In

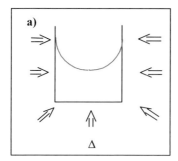 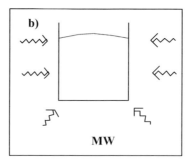

Scheme 4.2. Temperature gradients in materials submitted either to traditional heating (a) or to microwave irradiation (b).

Tab. 4.5. Results from the phenacylation of 1,2,4-triazole.

Conditions	Mode	N_1	N_4	$N_{1,4}$
DMF	Δ or MW	95	5	0
o-xylene	Δ	32	28	40
	MW	100	0	0
No solvent	Δ	33	29	38
	MW	100	0	0

important work this analogy has been verified in an example involving regioselectivity [106]. The phenacylation of 1,2,4-triazole (Eq. (8), Table 4.5) was shown to be solvent-dependent under the action of conventional heating. When using MW irradiation in either xylene or solvent-free conditions, the same selectivity is obtained as is observed classically in DMF. A very important specific MW effect is apparent from the regioselectivity which is similar to that when polar solvents are used. This finding is good support of our hypothesis and tentative of rationalization of polar MW effects.

(8)

4.5
Effects Depending on Reaction Mechanisms

The effects of MW result from material–wave interactions and, because of the dipolar polarization phenomenon, the greater the polarity of a molecule (for example the solvent) the more pronounced the MW effect when temperature is increased

[64]. In terms of reactivity and kinetics, a specific effect has therefore been considered; this is related to the reaction mechanism and, in particular, how the polarity of the system is altered during the progress of the reaction. These assumptions are evidently connected with the Hughes–Ingold model [65], universally adopted to explain solvent effects [26] and, especially, the intervention of aprotic dipolar solvents.

Specific microwave effects can be expected for polar mechanisms, when the polarity is increased during the reaction from the ground state to the transition state (as more or less implied by Abramovich in the conclusion of his review in 1991 [66]). The outcome essentially depends on the medium and the reaction mechanism. If stabilization of the transition state (TS) is more effective than stabilization of the ground state (GS), the result is enhancement of reactivity by reduction of the activation energy (Fig. 4.3), because of dipole–dipole type electrostatic interactions of polar molecules with the electric field.

4.5.1
Isopolar Transition-state Reactions

Isopolar activated complexes differ very little or not at all in charge separation or charge distribution from the corresponding initial reactants. These complexes are implied in pericyclic reactions such as Diels–Alder cycloadditions and the Cope rearrangement.

Ground and transition states have a priori identical polarity, because no charges are developed during the reaction path. Following this rule, specific microwave effects would not be expected in these reactions (Fig. 4.8), as has been verified when the reactions were performed in a nonpolar solvent [13, 14]. Solvent effects in these reactions are also small, or negligible, for the same reasons [67].

Such a conclusion is, nevertheless, connected with the synchronous character of the mechanism. If a stepwise process is involved (non-simultaneous formation of the two new bonds), as is the situation when dienes and (or) dienophiles are unsymmetrical or in hetero Diels–Alder reactions, a specific microwave effect could intervene as charges are developed in the transition state. This could certainly be true of several cycloadditions [68, 69] and, particularly, for 1,3-dipolar cycloaddi-

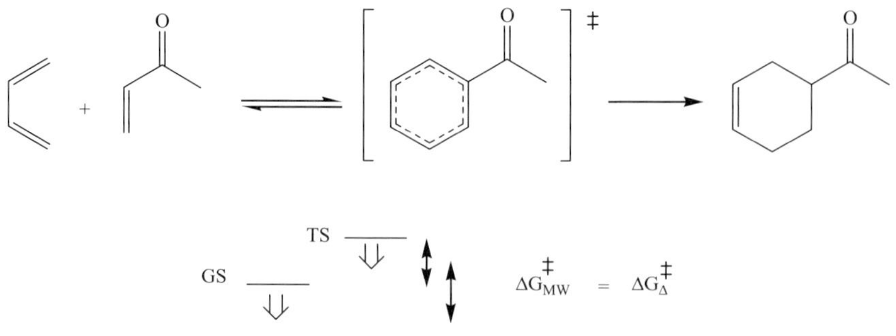

$$\Delta G^\ddagger_{MW} = \Delta G^\ddagger_{\Delta}$$

Fig. 4.8. Similar stabilization of isopolar TS and GS (concerted synchronous mechanism).

4.5 Effects Depending on Reaction Mechanisms

tions [70]. Such an assumption was verified experimentally and justified by considering theoretical calculations predicting an asynchronous mechanism in the cycloaddition of N-methylazomethine ylide to C_{70} fullerene [48] (Eq. 9).

$$C_{70} \xrightarrow[\text{MW or } \Delta]{CH_3NHCH_2CO_2H/HCHO} C_{70}\text{-fused pyrrolidine-N-CH}_3 \quad (9)$$

During the course of a study of the [3+2]cycloaddition of azidomethyldiethylphosphonate to acetylenes and enamines leading to alkyltriazoles under solvent-free conditions, we observed that specific effects can be involved, depending on the nature of the substituents on the dipolarophiles [71] (Eq. 10, Table 4.6). They were explained by considering the nonsynchronous character of the mechanism.

$$(EtO)_2P(O)CH_2N_3 + R_1\text{-}{\equiv}\text{-}R_2 \longrightarrow \text{triazole product} \quad (10)$$

The synthesis of biologically significant fluorinated heterocyclic compounds was accomplished by 1,3-dipolar cycloaddition of nitrones to fluorinated dipolarophiles

Tab. 4.6. Thermal or microwave activation for the cycloaddition depicted in Eq. (10).

R_1	R_2	Activation	Conditions		Yield (%)[a]
			t (min)	T (°C)	
CH_3	$P(O)(OEt)_2$	Δ	20	90	5
		MW	20	90	78
H	CO_2Et	Δ	5	100	70
		MW	5	100	92
C_6H_5	CO_2Et	Δ	10	160	>98
		MW	10	160	>98
H	C_6H_5	Δ	30	120	40
		MW	30	120	>98
H	CH_2OH	Δ	30	100	40
		MW	30	100	>98

[a] The ratio of the two isomers formed remained identical under both conditions of activation

Tab. 4.7. Thermal or microwave activation for the cycloaddition depicted in Eq. (11).

Activation	Conditions			Yield (%)
	Solvent	Time	Temp. (°C)	
Δ	toluene	24 h	110	65
MW	none	3 min	119	98
Δ	none	3 min	119	64
Δ	none	30 min	119	98

[72]. This reaction was noticeably improved under solvent-free conditions and using microwave irradiation (Eq. (11), Table 4.7).

$$\underset{Ph}{\overset{H}{\rightthreetimes}}\!\!=\!\!\underset{O^-}{\overset{Me}{N^+}} + \underset{F_3C}{\overset{HO}{\rightthreetimes}}\!\!=\!\!\underset{H}{\overset{CO_2Et}{\rightthreetimes}} \longrightarrow \underset{Me}{\underset{N}{\overset{F_3C}{\underset{HO}{\bigvee}}\!\!\overset{CO_2Et}{\underset{O}{\bigvee}}\!\!Ph}} \quad (11)$$

It is apparent there is a definite advantage to operating under solvent-free conditions. The specific microwave effect here is of rather low magnitude but evident, because after 3 min the yield increases from 64 to 98%. Prolongation of the reaction time with classical heating led to an equivalent result. The microwave effect is rather limited here, because of a near-synchronous mechanism.

Loupy et al. [73] described the reaction of 1-ethoxycarbonylcyclohexadiene (**I**) and 3-ethoxycarbonyl-α-pyrone (**II**) under solvent-free conditions – an irreversible Diels–Alder cycloadditions with an acetylenic compound. Because a specific microwave effect was apparent for compound **II** only (Fig. 4.9), it was concluded that higher yields are related to variation of the dipole moment from the ground state GS to the transition state TS. These conclusions are supported by results from *ab initio* calculations (Table 4.8).

From these calculations, it is obvious to conclude that the first reaction is synchronous with similar bond lengths formed whereas the second one is asynchronous with quite different bond lengths formed. Furthermore, the dipole moments remained the same for GS and TS in the first reaction whereas they are noticeably increased in the TS for the second. All these conclusions strongly support the evidence and interpretation of important specific not purely thermal MW effects when asynchronous mechanisms are involved.

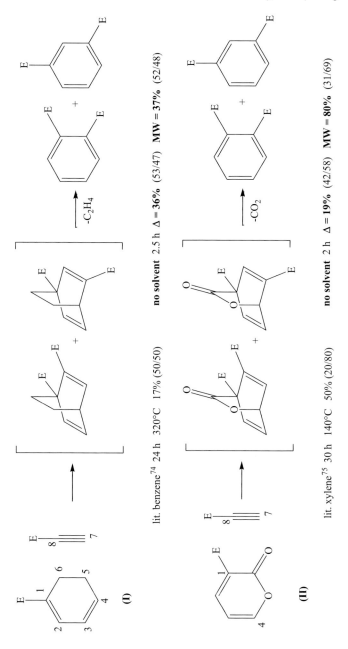

Fig. 4.9. Irreversible Diels–Alder cycloadditions with ethyl propiolate (E = CO$_2$Et).

Tab. 4.8. Lengths of the bonds formed and dipole moments of the reagents (GS = ground state and TS = transition state) performed at the HF/6.31G(d) level.

Reaction	Dipole moments (Debye)		Bond lengths in TS (Å)	
	GS	TS	1–8	4–7
I	2.2 and 2.4	1.9	2.16	2.16
II	2.2 and 2.3	4.8	2.38	2.03

MW effects can therefore act as a useful tool enabling appreciation of the asynchronicity of TS. This is especially true of 1,3-dipolar cycloadditions (cf. Chapter 11) [76]. Because these reactions are well-known to proceed via asynchronous mechanisms, MW-specific effects are expected, and were often apparent when MW and Δ were seriously checked under similar sets of conditions [77, 78]. A specific case is illustrated here (Fig. 4.10) [79].

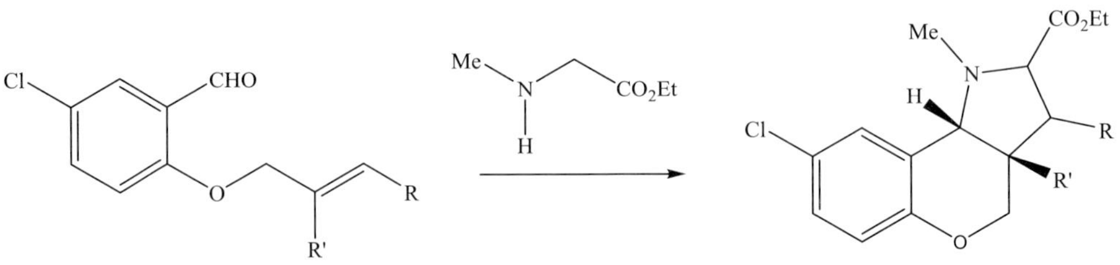

	Toluene reflux Dean-Stark		Solvent-free	
			Δ	MW
R' = H R = H	24 h	88%	1.5 h 92%	15 min 93%
R' = Me R = H	24 h	72%	9 h 72%	30 min 98%

Fig. 4.10. Synthesis of pyrrolidines by condensation of the O-allylic compounds with ethyl sarcosinate.

4.5.2
Bimolecular Reactions Between Neutral Reactants Leading to Charged Products

Typical reactions are amine and phosphine alkylations or additions to a carbonyl group (Scheme 4.3). In these examples, because of the development of dipoles in

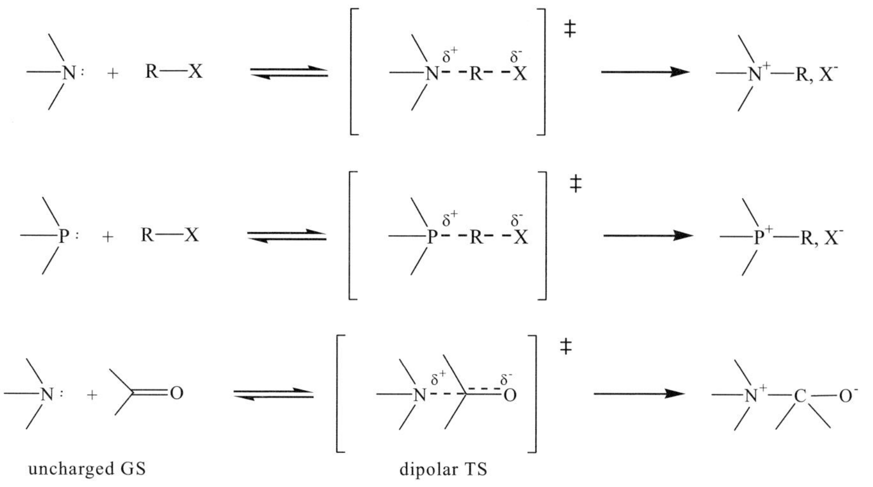

Scheme 4.3

uncharged GS → dipolar TS

R = alkyl group   X = halide

Scheme 4.3

the TS, we are concerned with a polarity increase during the course of the reaction starting from the GS toward the TS. Favorable microwave effects are consequently expected.

The magnitude of these effects might be related to the nature of substituents α to N or P and to the structure of the leaving group, as exemplified by several observations that will be described and discussed below.

4.5.3
Anionic Bimolecular Reactions Involving Neutral Electrophiles

These reactions comprise nucleophilic S_N2 substitutions, β-eliminations, and nucleophilic additions to carbonyl compounds or activated double bonds, etc. They involve the reactivity of anionic species Nu^- associated with counter-ions M^+ to form ion-pairs with several possible structures [80] (Scheme 4.4).

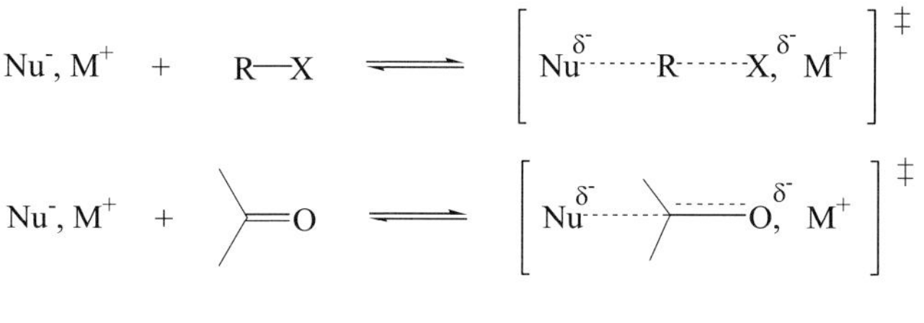

Scheme 4.4

The transition states loose ion-pairs in so far as they involve charge-delocalized anions, thereby enhancing polarity compared with the ground states (in which the ion-pairs are tighter), because of an increase in anionic dissociation as the more bulky product anion is formed. As a consequence, *specific microwave effects, directly connected to polarity enhancement, should depend on the structure of reactive ion-pairs in the GS*:

- If tight ion-pairs (between two hard ions) are involved in the reaction, the microwave accelerating effect then becomes more important, because of enhancement of ionic dissociation during the course of the reaction as tight-ion-pairs (GS) are transformed into more polar loose ion-pairs (TS).
- If, on the other hand, loose ion-pairs (between soft ions) are involved, microwave acceleration is limited, because ionic interactions are only slightly modified from GS to TS.

This duality in the behavior of some S_N2 reactions can be foreseen and observed (*vide infra*) by comparing reactions involving hard or soft nucleophilic anionic reagents according to the cation and the leaving group.

4.5.4
Unimolecular Reactions

Entropic contributions to the acceleration of first-order reactions by microwaves should be negligible ($\Delta S^\ddagger = 0$). When ionization (S_N1 or E_1) or intramolecular addition (cyclization) processes are involved a microwave effect could be viewed as resulting from a polarity increase from the GS to the TS, because of the development of dipolar intermediates (Scheme 4.5).

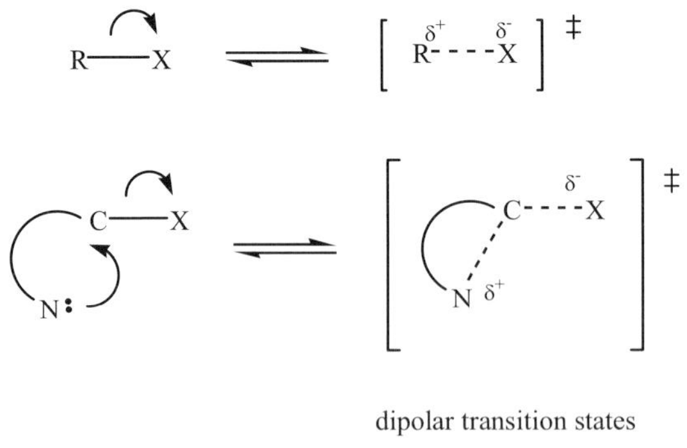

dipolar transition states

R = alkyl group   X = halide

Scheme 4.5

4.6
Effects Depending on the Position of the Transition State Along the Reaction Coordinates

The position of the transition state along the reaction coordinates in relation to the well-known Hammond postulate [81] will be considered as it is of important influence to interpret any medium and structural effects.

If the activation energy $\Delta G^{\ddagger}$ of a reaction is only small, the TS looks like the GS (it is depicted as a "reactant-like transition state"). Consequently, the polarity is only slightly modified between the GS and TS during the course of the reaction and only weak specific microwave effects can be foreseen under these conditions (exothermic reactions).

By way of contrast, a more difficult reaction implies a higher activation energy. The TS therefore occurs later along the reaction path (endothermic reactions) and, consequently, the effect of polarity effects might be significantly larger. *It might be assumed that a microwave effect should be more pronounced when the TS occurs later along the reaction coordinates* (depicted more as a "product-like transition state") and is, therefore, more prone to develop increased polarity (Scheme 4.6).

This conclusion is in agreement with a remark of Lewis who stated that "slower reacting systems tend to show a greater effect under microwave radiation than faster reacting ones" [82]. In this way, during solvent-free Wittig olefination with phosphoranes, it was shown that the benefit of MW irradiation increases with less reactive systems. The best stabilized phosphoranes do not react at all in the solid state with aldehydes or ketones under conventional heating but necessitate MW irradiation [83].

Consequently, a microwave effect can be important when steric effects are involved in a reaction, as exemplified by the increased magnitude of the effect for saponifications of hindered mesitoic esters relative to benzoic esters [84] (*vide supra*).

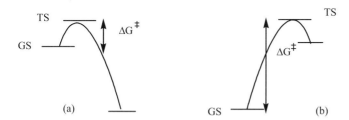

Scheme 4.6. (a) small $\Delta G^{\ddagger}$ → early TS → little change in polarity TS/GS → weak microwave effect; (b) large $\Delta G^{\ddagger}$ → late TS → important change in polarity TS/GS → large microwave effect.

4.7
Effects on Selectivity

A few examples of increased selectivity can be found in the literature [85–90] where the steric course, and the chemo- or regioselectivity of reactions can be altered under the action of microwave irradiation compared with conventional heating. They have been listed and discussed in a recent review by De la Hoz et al. [91] and are discussed in Chapter 5 of this book. The main problem when trying to attribute accurately any MW effect lies in the fact that the *kinetic control* of reactions is not ensured. Of course, the only serious conclusions need to eliminate the thermodynamic control of reactions which is, unfortunately, highly probable when high temperatures are concerned.

Related to this issue, which could be extremely interesting, is that simultaneous external cooling of reaction mixture while heating with MW leads to enhancement of the overall process. This is now possible using a Coolmate system from CEM. It enables greater MW power to be applied directly to the reaction mixture and prevents overheating. Some results appeared [92, 93] but are limited to savings of reaction time whereas promising results could be expected for selectivity, particularly in asymmetric induction which usually needs low temperatures.

As another consequence of the assumptions above, it might be foreseen that specific nonthermal MW effects could be important in determining the selectivity of some reactions. When competitive reactions are involved, the GS is common for both processes. The mechanism occurring via the more polar TS could, therefore, be favored by use of microwave radiation (Scheme 4.7).

Langa et al. [48, 94, 95], while performing cycloaddition of N-methylazomethine ylide with C_{70} fullerene, proposed a rather similar approach. Theoretical calculations predict an asynchronous mechanism, suggesting that this phenomenon can be explained by considering that, under kinetic control, "microwave irradiation will favor the more polar path corresponding to the hardest transition state".

Two representative examples from our laboratory will be described below. The first concerns the aromatic nucleophilic substitution (S_NAr) of potassium methoxide with either activated (*p*-nitrophenyl) or nonactivated (α-naphthyl) aromatic

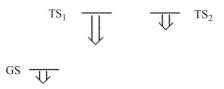

Scheme 4.7. The more polar TS1 is more stabilized by dipole–dipole interactions with the electric field and, therefore, is more prone to microwave effects.

chlorides under solid–liquid solvent-free phase-transfer catalysis [96]. Under similar sets of conditions and identical profiles of increasing temperature, yields were largely improved under the action of MW irradiation compared with conventional heating (Δ), because of the disappearance of many by-products resulting from radical reactions (Fig. 4.11).

This observation is a striking example of the effect of MW on selectivity, because the less polar mechanism involving radical reactions is now disappearing in comparison with the more polar S_NAr mechanism.

A second example is the competition between Diels–Alder cycloaddition and Michael addition during the reaction of 2-methoxythiophene **III** with DMAD under solvent-free conditions [73] (Fig. 4.12).

Microwaves were shown to affect both reactivity and selectivity. The effect on yield is rather limited in the Diels–Alder reaction and found to be higher in the Michael addition. This process was even more favored by use of acetic acid as the solvent. These results can be explained by the natural assumption that the transition state leading to Michael addition (**M**) much more polar than that leading to Diels–Alder cycloaddition (**DA**).

These assumptions were confirmed by calculations performed on the common GS and each TS, and consideration of the dipole moments. When one considers the relative dipole moments of the TS, the polarity is higher for that leading to Michael addition. Consequently, under MW activation, one can expect enhancement in **M** compared with **DA**. The most important factor affecting the selectivity is the solvent; a protic solvent stabilizes the more polar TS, thus strengthening the tendency revealed in its absence.

4.8
Some Illustrative Examples

To illustrate these trends, we now present some typical illustrative examples. These were selected because strict comparisons of microwave and classical heating activation were made under similar conditions (time, temperature, pressure, etc . . .) for the same reaction medium and using, preferably, a single mode system equipped with stirring. They mostly involve reactions performed under solvent-free conditions or, occasionally, in a nonpolar solvent, because these conditions also favor observation of microwave effects.

4.8.1
Bimolecular Reactions Between Neutral Reactants

These reactions are among the most propitious for revealing specific microwave effects because the polarity is evidently increased during the course of the reaction from a neutral ground state to a dipolar transition state.

158 | 4 Nonthermal Effects of Microwaves in Organic Synthesis

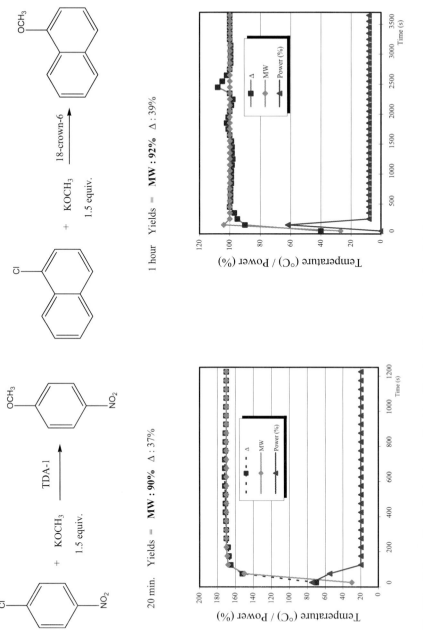

Fig. 4.11. S_NAr reactions of potassium methoxide with solvent-free phase-transfer catalysis.

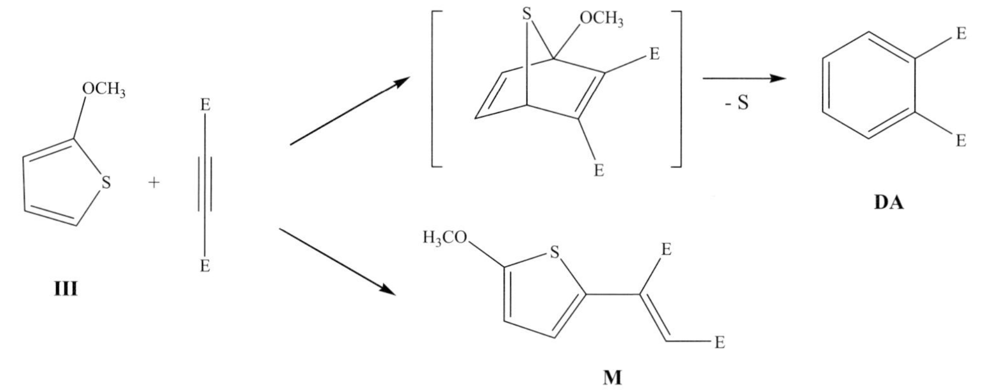

4.8 Some Illustrative Examples

2 h 140°C	no solvent	**M/Δ** =	MW : 65/6	Δ : 40/7
2 h 100°C	acetic acid	**M/Δ** =	MW : 10/49	Δ : 6/4

Dipole moments (Debye) **III** = 1.8 , DMAD = 2.8, Transition State for **DA** = 5.4 and for **M** = 6.1

Fig. 4.12. Reaction of thiophene **III** with dimethylacetylenedicarboxylate (DMAD) E = CO_2Me.

4.8.1.1 Nucleophilic Additions to Carbonyl Compounds

The most typical situation is addition of an amine to a carbonyl group (Eq. 12):

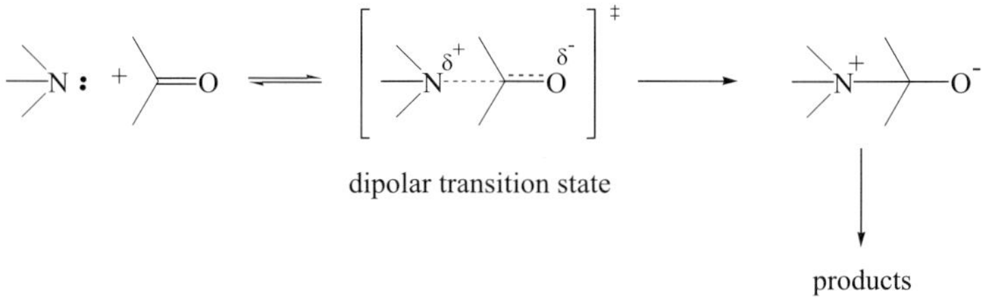

(12)

This example covers very classical processes such as the syntheses of a wide variety of compounds including imines, enamines, amides, oxazolines, hydrazones etc ...

Amines

Imine or enamine synthesis   It has been shown by Varma et al. [97] that reaction of primary and secondary amines with aldehydes and ketones is substantially accelerated by microwaves under solvent-free conditions in the presence of montmorillonite K10 clay, affording high yields of imines and enamines (Eq. 13):

$$\text{(13)}$$

A more elaborate example is the Niementowski reaction to give access to quinazolinones and quinolines [98]. The determining step is the reaction of anthranilic acid with some amides or ketones (Eq. 14):

$$\text{(14)}$$

It has been shown that a mixture of indoloquinazoline and anthranilic acid, when adsorbed on graphite, leads to cyclization in good yields after 30 min at 140 °C in noticeably less time than for the purely thermal procedure under similar conditions, in which a very poor yield was obtained even after 24 h [99].

Comparative results have been obtained in the synthesis of 3*H*-quinazolin-4-one [100]. The fusion of anthranilic acid with formamide led to the formation of an *o*-amidine intermediate and usually proceeded by intramolecular cyclization (Eq. 15):

$$\text{(15)}$$

20 min   MW = 92%   6 h   Δ = 59%

A large specific MW effect was observed in the solvent-free synthesis of *N*-sulfonylimines, a similar type of reaction [101] (Eq. 16):

$$\text{(16)}$$

Ar = C_6H_5 Ar' = pCH$_3$-C_6H_4 6 min   190 °C   MW = 91%   Δ = 40%

Intermolecular hydroacylation of 1-alkenes with aldehydes catalyzed by Wilkinson complex [Rh(PPh$_3$)$_3$Cl] was performed classically in toluene at 150 °C for 24 h to give access to a variety of ketones in good yields (70–90%). It was then shown by Jun, Loupy et al. [102] that this reaction (under the action of homogeneous catalysis) can be achieved better under solvent-free conditions with microwave activation and taking advantage of large specific MW effects. The authors attributed these effects to formation of an aldimine by the *in situ* previous condensation of aldehyde and aminopicoline giving an intermediate aldimine (Eq. 17):

$$R_2 = nC_8H_{17}$$

	R_1 = CF$_3$	10 min	160°C	MW = 91%	Δ = 52%
	R_1 = OMe	10 min	100°C	MW = 90%	Δ = 48%

With the same type of molecule, nonthermal specific microwave effects were apparent in the Wilkinson complex-catalyzed orthoalkylation of ketimines [103], which occurred via initial transimination (Eq. 18):

30 min   200°C   MW = 96%   Δ = 67%

To explain this result, it may be assumed that the transimination by reaction of the starting imine, **1**, with 2-amino-3-picoline, **2**, might be the rate-determining step. In such a situation, the transition state is expected to be more polar and, therefore, sensitive to MW dipole–dipole stabilization. The qualitative polarities between ground and transition states have been determined by PM3 computations, and clearly showed the enhancement of the polarity during the process (Eq. 19):

162 | 4 Nonthermal Effects of Microwaves in Organic Synthesis

$$\text{(19)}$$

	1	**2**	**3**
	[GS]		[Dipolar TS]
μ (Debye)	1.23	1.96	6.36

Hydrazone synthesis In a typical example, a mixture of benzophenone and hydrazine hydrate in toluene resulted in 95% yield of the hydrazone within 20 min [104] (Eq. 20):

$$\text{(20)}$$

The hydrazone was subsequently treated with KOH under the action of MW irradiation to undergo Wolff–Kishner reduction (leading to $PhCH_2Ph$) within 25–30 min in excellent yield (95%). As an extension, the reaction of neat 5- or 8-oxobenzopyran-2(1H)-ones with a variety of aromatic and heteroaromatic hydrazines is remarkably accelerated by irradiation in the absence of any catalyst, solid support, or solvent [105] (Eq. 21). Kinetic considerations for the reaction between two solids below their melting points have been explained by the formation of a eutectic melt during the reaction:

$$\text{(21)}$$

Ex: R = 4-nitrophenyl   MW 9 min 130 °C 98%
                          Δ   1 h   130 °C 80%

Amidation of carboxylic acids Uncatalyzed amidations of acids have been realized under solvent-free conditions and a very important microwave effect was observed [106, 107]. The best results were obtained by use of a slight excess of either amine or acid (1.5 equiv.). The reaction involves thermolysis of the previously formed ammonium salt (acid–base equilibrium), and is promoted by nucleophilic attack of the amine on the carbonyl moiety of the acid and removal of water at high temperature. The large difference between yields (MW $\gg \Delta$) might be a consequence of interaction of the polar TS with the electric field (Eq. (22), Table 4.9):

$$PhCH_2NH_3^+, RCO_2^- \rightleftharpoons \left[\text{tetrahedral intermediate} \right]^{\ddagger} \xrightarrow{-H_2O} PhCH_2NHC(=O)R \tag{22}$$

Considering that water can be removed at 150 °C equally for both types of activation, the noticeable difference in yields is clearly indicative of improved nucleophilic addition of the amine to the carbonyl group when performed under the action of microwave irradiation.

During an extended study of the synthesis and polymerization of chiral methacrylamide, Iannelli and co-workers [108] conducted careful comparisons when per-

Tab. 4.9. Reaction of benzylamine with carboxylic acids at 150 °C for 30 min.

R	Amine/acid	Yield (%)	
		MW	Δ
Ph	1:1	10	10
	1.5:1	75	17
	1:1.5	80	8
$PhCH_2$	1:1	80	63
	1.5:1	93	40
	1:1.5	92	72
$CH_3(CH_2)_8$	1:1	85	49

forming the reaction of methacrylic acid and (R)-1-phenylethylamine. Kinetic comparisons of reactions performed under the action of MW or classical heating revealed the greater selectivity of the MW-accelerated reaction. The use of differential scanning calorimetry (DSC) enabled reproduction of almost the same heating profile as observed under the action of microwaves (Eq. 23). Yields obtained under the action of MW were by far better than those from simple thermal activation (Fig. 4.13).

$$\text{methacrylic acid} + H_2N\text{-CH(CH}_3\text{)-Ph} \xrightarrow{-H_2O} \text{amide} \qquad (23)$$

Under the MW conditions applied, important accelerations were observed together with better selectivity as the desired amidation was clearly preferred to the Michael side-reactions observed under the action of conventional heating. Although the debate about the existence of a so-called "specific microwave effect" remains open, it is the authors' opinion that their results cannot be attributed solely to the exceptionally strong heating effects of MW. They therefore strongly understand (as we do) that highly polar intermediates (zwitterions and salts) interact directly on a molecular level with the electromagnetic field associated with the microwave.

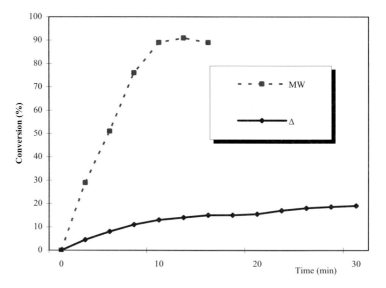

Fig. 4.13. Comparative kinetic plots under the action of MW irradiation and in oil bath.

The preparation of aliphatic, aromatic, or functionalized tartramides directly from tartaric acid and amines under solvent-free conditions with MW irradiation was also described (Eq. 24) [109]. Under identical profiles of temperature increase (Fig. 4.14), the yields under the action of MW irradiation were far greater than with conventional heating.

$$\underset{\substack{HO \\ HO}}{\overset{CO_2H}{\underset{CO_2H}{\diagdown}}} \xrightarrow[\text{MW 12 min.}]{\text{RNH}_2 \text{ (2.8 equiv.)}} \underset{\substack{HO \\ HO}}{\overset{CONHR}{\underset{CONHR}{\diagdown}}} \quad (24)$$

R = PhCH$_2$, n-C$_6$H$_{13}$, n-C$_4$H$_9$, Ph   71-83%

R = PhCH$_2$ 180°C   12 min   MW = 80%   16 h   Δ = 68%

Synthesis of 2-oxazolines Oxazolines can be readily synthesized by means of a noncatalyzed solvent-free procedure by two successive nucleophilic additions on a carbonyl group with the formation of an amide as an intermediate [110, 111] (Eq. 25):

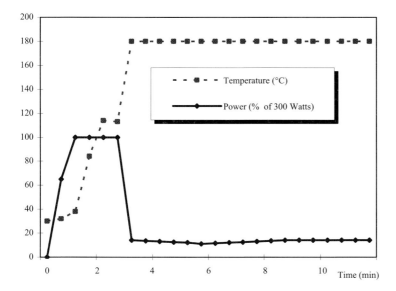

Fig. 4.14. Power and temperature evolutions during MW irradiation in the amidation of tartaric acid.

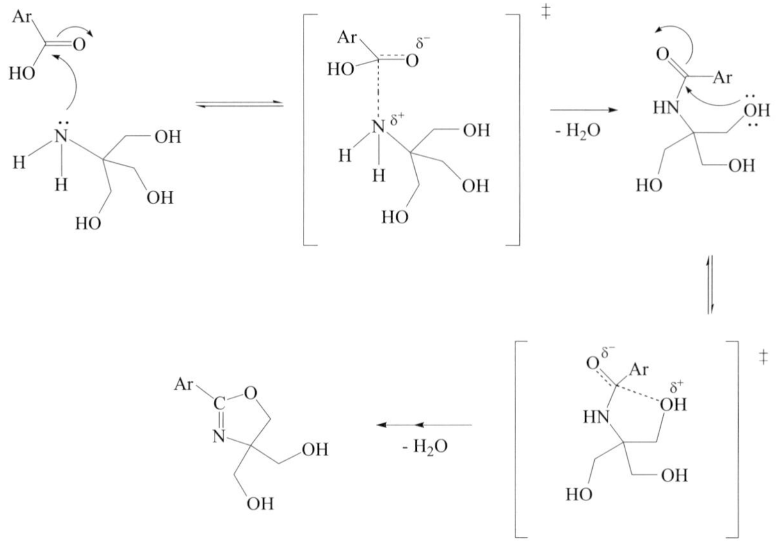

10 min 200 °C MW 80-95% Δ < 5%

(25)

When considering the overall mechanism, two consecutive steps leading to the development of more polar TS than the GS can both be accelerated by specific MW effects. The first is nucleophilic attack of the amine on the carbonyl moiety (leading to an intermediate amide); the second is nucleophilic attack of the hydroxyl function on the carbonyl to achieve cyclization.

Facile syntheses of a variety of 2-substituted 2-oxazolines were obtained with N-acylbenzotriazoles under mild conditions and short reaction times under MW irradiation (Eq. 26) [112]:

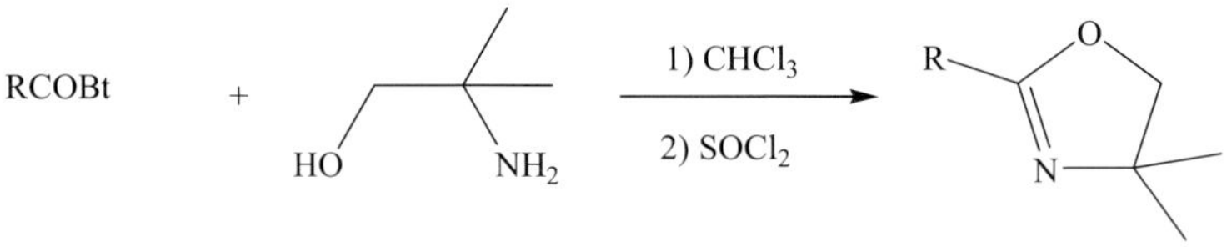

(26)

Bt = benzotriazolyl

MW	10 min	80°C	Yield = 98%
Δ	30 min	80°C	Conversion = 60–70% + by-products

With the same kind of molecule and the same method of synthesis, 4-arylidene-2-phenyl-5(4H)-oxazolones were synthesized by the Erlenmeyer reaction from aromatic aldehydes and hippuric acid using calcium acetate under solvent-free conditions (Eq. 27) [113]:

$$\text{ArCHO} + \underset{\text{Ph}}{\underset{|}{\text{C}}}(=O)-\underset{H}{N}-CH_2-COOH \xrightarrow{Ac_2O/Ca(OAc)_2}$$

(oxazolone product, Ar-CH= with N=C-Ph) 70-99% (27)

Ex: Ar = Ph   48-50°C   5 min.   Yields : MW = 97%   Δ = 50%

Finally, in a similar procedure, synthesis of long chain 2-imidazolines was realized by condensation of aminoethylethanolamine with several fatty acids under solvent-free MW conditions using CaO as support (Eq. 28) [114]:

$$H_2N-CH_2CH_2-NH-CH_2CH_2-OH + R-COOH \longrightarrow \text{(2-imidazoline)-CH}_2CH_2OH$$ (28)

R = n-$C_{17}H_{35}$ 150°C   7.5 min   Yields :   MW = 92%   Δ = traces.

Synthesis of aminotoluenesulfonamides These compounds were prepared by reaction of aromatic aldehydes with sulfonamides under the action of microwaves in the presence of a few drops of DMF to enable better energy transfer [115] (Eq. (29), Table 4.10):

Tab. 4.10. Reaction of 5-amino-2-toluenesulfonamide with aromatic aldehydes.

Ar	Time (min)	Yield (%)		
		MW (no DMF)	MW (ε DMF)	Δ (ε DMF)
p-$NO_2C_6H_4$	1	40	98	5
o-ClC_6H_4	2	22	90	6
5-NO_2 2-furyl	2	20	96	5

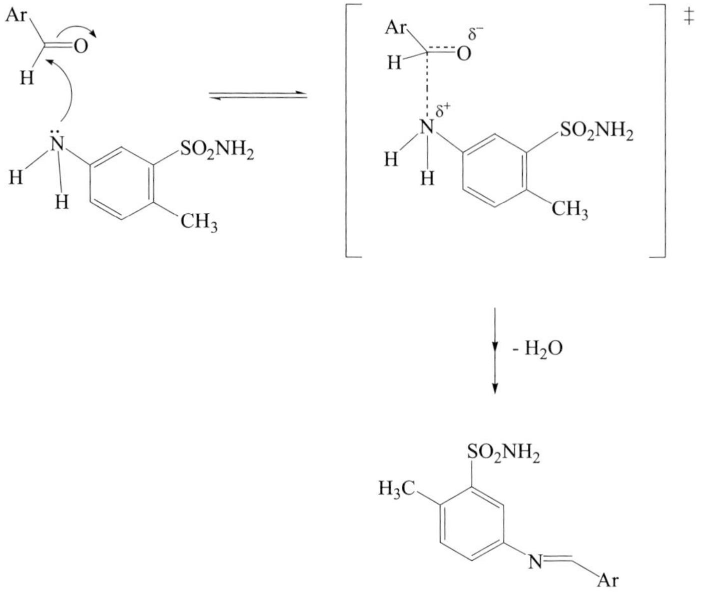

[Scheme for Eq. (29): ArCHO + aniline derivative (with SO$_2$NH$_2$ and CH$_3$ substituents) → dipolar transition state → imine product with loss of H$_2$O]

(29)

Thionation of carbonyl compounds A series of 1,4-dithiocarbonyl piperazines has been synthesized from aldehydes, piperazine and elemental sulfur under the action of microwave irradiation and solvent-free conditions. An important non-thermal effect of the radiation was revealed (Eq. 30) [116]:

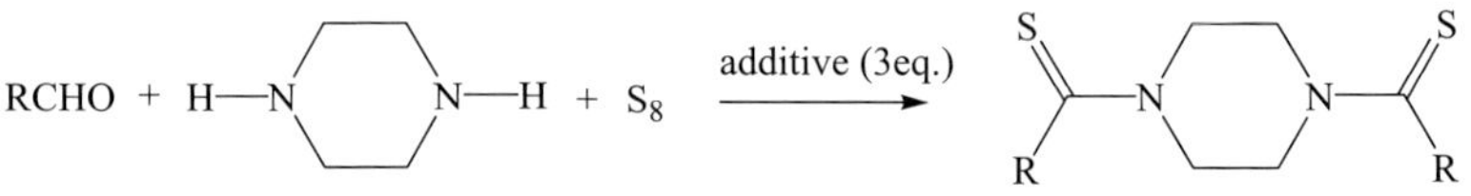

$$\text{RCHO} + \text{H-N}\underset{}{\overset{}{\bigcirc}}\text{N-H} + S_8 \xrightarrow{\text{additive (3eq.)}} \text{R-C(=S)-N}\underset{}{\overset{}{\bigcirc}}\text{N-C(=S)-R}$$

(30)

	time	temp	MW	Δ
DMF	5min	100–102°C	85%	48%
EG	5min	105–107°C	80%	40%

This is consistent with nucleophilic attack of a neutral molecule (amine or sulfur) on the carbonyl compound, leading to a dipolar TS.

A series of γ-thionolactones has been synthesized, with good yields, using a new combination of Lawesson's reagent (LR) and hexamethyldisiloxane (HMDO) in solvent-free conditions under the action of MW irradiation (Eq. 31). An important MW-specific effect was observed when the reaction yield was carefully compared with that from traditional heating under similar conditions [189]:

4.8 Some Illustrative Examples

[Scheme showing conversion of γ-butyrolactone derivative with C=O to C=S using LR/HMDO, solvent-free] (31)

$$LR = H_3CO-\text{C}_6H_4-P(=S)(S)_2P(=S)-C_6H_4-OCH_3$$
(Lawesson's reagent, dimeric structure with P-S-P-S four-membered ring)

Ex : R = H R' = n-C$_8$H$_{17}$ 5 min. 120°C   Yields MW : 85% Δ : 27%

Similarly, from the same laboratory, a new and practical procedure for synthesis of isothiocyanates was described. They were readily obtained from the corresponding isocyanates using Lawesson's reagent under the action of MW irradiation and solvent-free conditions [190] (Eq. 32). Important specific not purely thermal MW effects were measured accurately here. Because thionation has been described to occur via a polar transition state reaction [191], it is expected to be accelerated under the action of MW irradiation; this could explain the significant rate enhancement.

$$R-N=C=O \xrightarrow[\text{solvent-free}]{\text{LR}} R-N=C=S$$

(32)

R = phenyl 4 min 140°C Yields MW : 73% Δ : 23%

  cyclohexyl ibid   MW : 94% Δ : 42%

  n-hexyl   ibid   MW : 97% Δ : 35% [80% in 30 min].

Condensation of urea with carbonyl compounds A rapid and efficient MW-assisted synthesis of hydantoins and thiohydantoins was described by Muccioli et al. [117]. The most straightforward conditions for synthesis of phenytoin are the base-catalyzed condensation using benzil and urea, known as the Biltz synthesis (Eq. 33). MW activation of the Biltz synthesis of phenytoin improved both yield and reaction time. The first step consists in MW activation of the reaction of benzil with (thio)urea; the second includes the conversion of the resulting 2-(thio)hydantoin to hydantoin using hydrogen peroxide. When reactions were performed at the same temperature under both reaction conditions, yields were by far better under the action of MW and emphasized the evident specific MW effects. These are perfectly expected when one considers the polar TS involved in the first step (nucleophilic addition of neutral NH$_2$ group on carbonyl moiety).

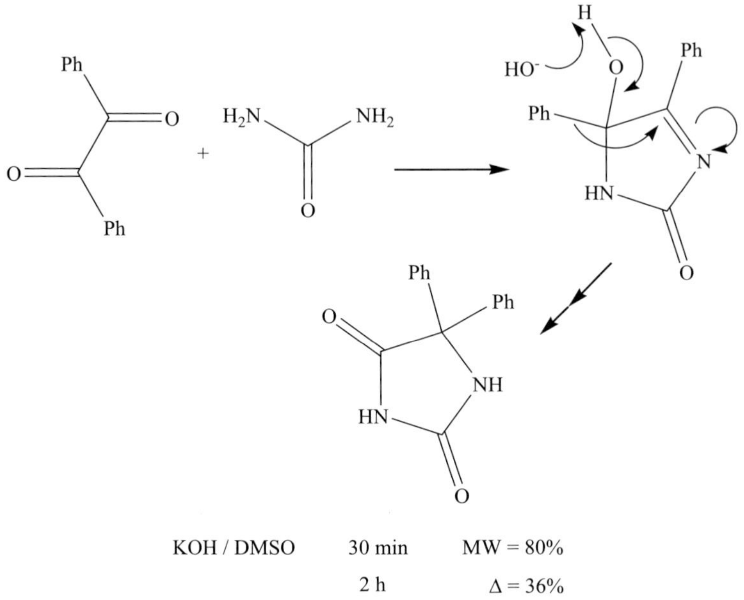

[Scheme for Eq. (33): reaction of diphenyl diketone (PhCOCOPh) with urea (H$_2$N-CO-NH$_2$) via hydroxide-mediated mechanism to give the hydantoin product]

KOH / DMSO   30 min   MW = 80%
              2 h   Δ = 36%

(33)

The synthesis of metallophthalocyanines [118] from phthalic anhydride and urea is another interesting example of a mechanism with charge development in the transition state (Eq. 34) and in which a large specific microwave effect was apparent.

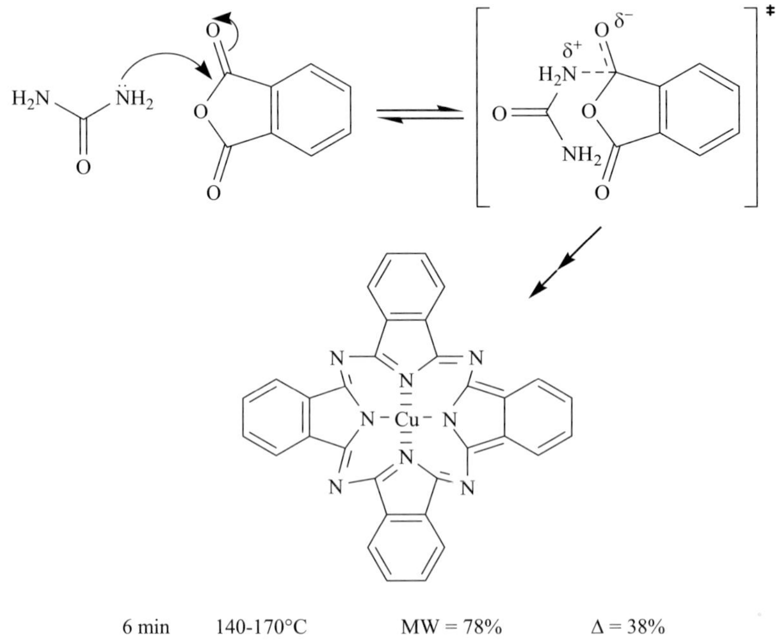

[Scheme for Eq. (34): urea + phthalic anhydride → charged transition state → Cu-phthalocyanine]

6 min   140–170 °C   MW = 78%   Δ = 38%

(34)

Base-free ester aminolysis An efficient method of amide synthesis by solvent-free ester aminolysis under the action of microwave activation has been described by Toma et al. [119] (Eq. 35):

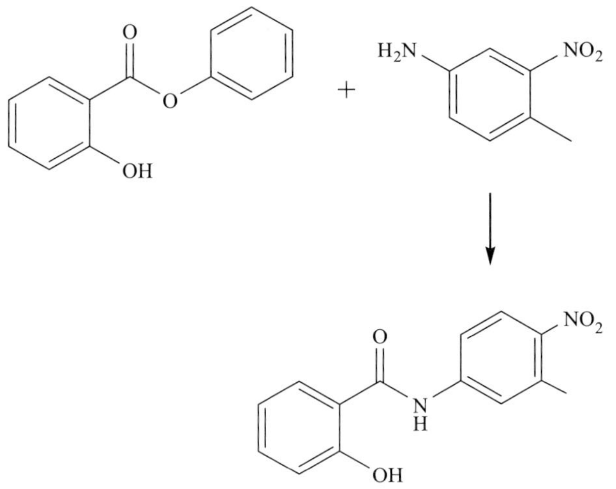

(35)

5 min   162°C   MW = 98%   Δ = 0%

The extraordinary specific microwave effects can be explained by important enhancement of the polarity during the process, because of the involvement of a dipolar transition state (Scheme 4.8).

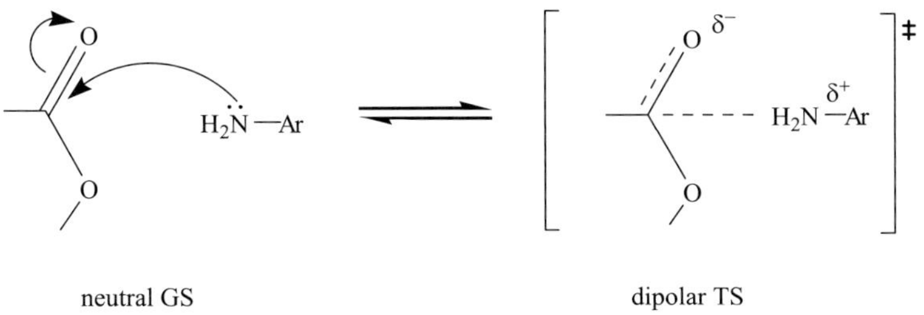

neutral GS   dipolar TS

Scheme 4.8

Alcohols

Solvent-free esterification of fusel oil Fusel oil basically comprises a mixture of alcohols such as isopentanol and isobutanol. Synthesis of isopentyl stearate has been performed using both MW irradiation and conventional heating under solvent-free conditions (Eq. 36) [120]:

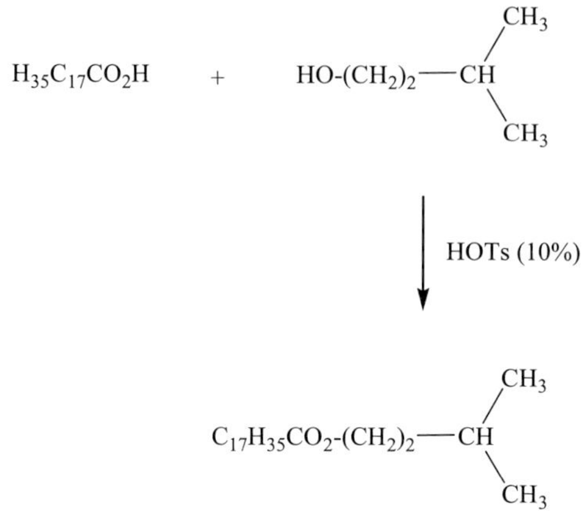

$$H_{35}C_{17}CO_2H \;+\; HO\text{-}(CH_2)_2\text{---}CH(CH_3)_2$$

$$\downarrow \text{HOTs (10\%)}$$

$$C_{17}H_{35}CO_2\text{-}(CH_2)_2\text{---}CH(CH_3)_2 \tag{36}$$

80 s	90 °C	MW	98%
80 s	120 °C	Δ	13%
10 min	120 °C	Δ	57%

The very important specific microwave effect is consistent with a mechanism which involves the formation of a dipolar TS from neutral molecules (Scheme 4.9).

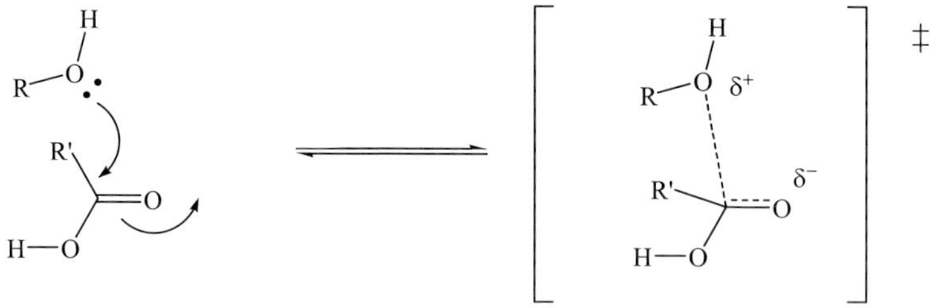

Scheme 4.9

Synthesis of alkyl p-toluenesulfinates The reactions of aliphatic alcohols with *p*-toluenesulfinic acid are accelerated by microwave irradiation under solvent-free conditions in the presence of silica gel, affording a high-yielding synthesis of *p*-toluenesulfinate esters [121] (Eq. 37):

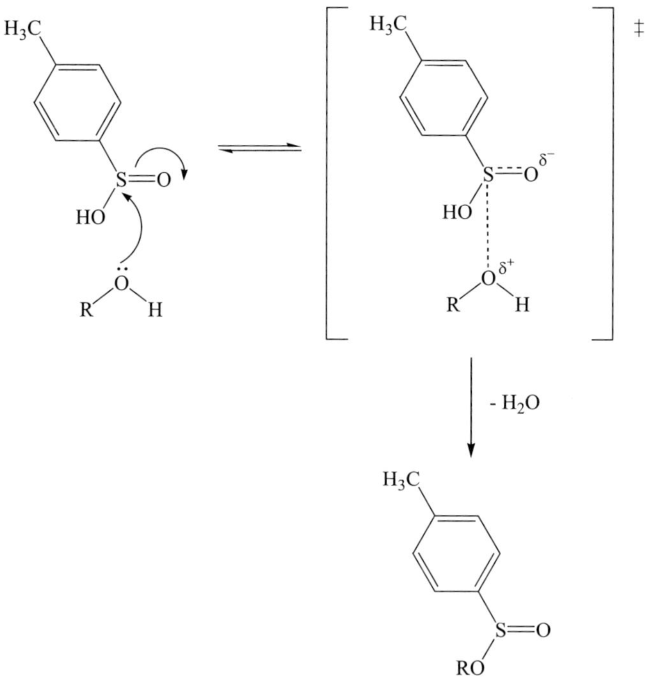

(37)

R = (CH$_3$)$_2$CHCH$_2$ 70 °C   MW   1.5 min   95%   Δ   30 min   10%

Synthesis of aminocoumarins by the Pechmann reaction Efficient synthesis of 7-aminocoumarins has been performed by the Pechmann reaction between *m*-aminophenols and β-ketonic esters. A comparative study of this procedure showed that use of microwave irradiation reduced the reaction time from several hours, if conventional heating was used, to a few minutes only (Eq. 38) [122]:

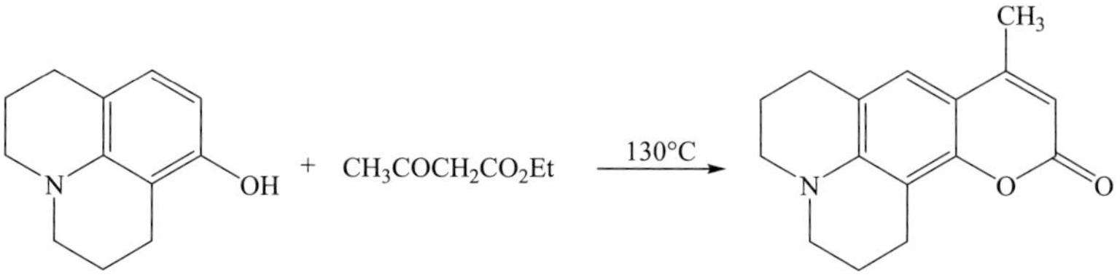

(38)

MW   12 min   62%
Δ   390 min   62%

Synthesis of cyclic acetals Cyclic ketals (potential cosmetics ingredients) have been obtained in excellent yields from a cineole ketone under the action of microwave in solvent-free conditions or in toluene. The results reported compared very favorably with those obtained by use of conventional heating (Eq. (39), Table 4.11) [123].

Tab. 4.11. Synthesis of ketal from cineole ketone and propylene glycol ($R_1 = CH_3$, $R_2 = H$).

Method	Activation	Time (min)	Yield (%)
Alumina	MW	30	78
	Δ	300	27
Toluene	MW	15	90
	Δ	360	30

$$(39)$$

Similarly, some cyclic ketals derived from 2-adamantanone were obtained in excellent yields thanks to microwave activation under solvent-free conditions [124] (Eq. 40). The important specific MW effect observed is consistent with mechanism considerations; the dipolar transition state is more polar, and therefore more stabilized by the field than the ground state.

$$(40)$$

140°C 15 min MW = 94%
       6 h   Δ = 60%

4.8.1.2 Michael Additions

Imidazole has been condensed via a 1,4 Michael addition with ethyl acrylate by use of basic clays (Li^+ and Cs^+ montmorillonites) under solvent-free conditions with microwave irradiation [125] (Eq. 41):

$$(41)$$

montmorillonite Li^+ 1 min 40 °C MW 40% Δ 0%
                        5 min 75 °C MW 72% Δ 27%

It was shown that MW irradiation accelerated the 1,4 Michael addition of primary and cyclic secondary amines to acrylic esters leading to several β-amino acid derivatives in good yields within short reaction times [126] (Eq. 42):

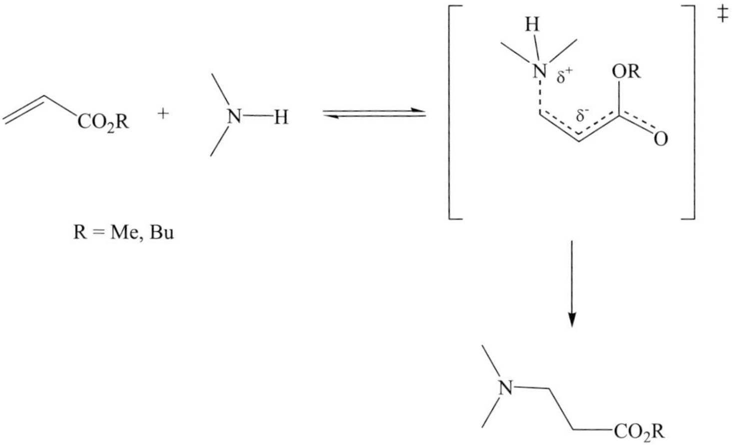

$$\text{CH}_2=\text{CH–CO}_2\text{R} + \text{R}_2\text{N–H} \rightleftharpoons \left[\begin{array}{c} \text{H} \\ \text{N}^{\delta+} \cdots \text{OR} \\ \cdots \cdots \delta^- \cdots \text{O} \end{array} \right]^{\ddagger} \longrightarrow \text{R}_2\text{N–CH}_2\text{CH}_2\text{–CO}_2\text{R} \quad (42)$$

R = Me, Bu

1,2 Asymmetric induction with up to 76% diastereoisomeric excess was observed in reactions of several amines with β-substituted acrylic acid esters from D-(+)-mannitol, in the absence of solvent, after exposure to microwaves for 12 min [127].

Microwave activation coupled with dry media technique as a green chemistry procedure has been applied to the synthesis of a series of 1,5-benzothiazepines [128–130]. These compounds are well known cardiovascular drugs acting as calcium-channel blockers. The reaction using montmorillonite KSF as an inorganic solid was noticeably accelerated under the action of MW activation when results were compared with those obtained by classical heating under similar conditions (Eq. 43):

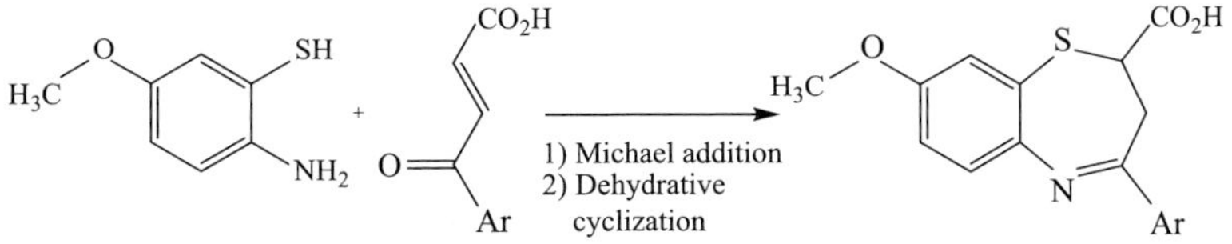

(43)

Ex: Ar = 2-methyl,4-fluorophenyl

13 min	140 °C	Montmorillonite KSF	MW = 71%	Δ = 22%
		Neat + DMF (2-3 drops)	MW = 66%	Δ = traces

The rate-determining step is assumed to be Michael addition of the benzenethiol moiety to the carbon–carbon double bond of α,β-unsaturated carbonyl compound. Specific MW effects can be thus expected when considering evolution of the polarity during the reaction progress. The transition state is more polar than the ground state and consequently more stabilized by the electromagnetic field (Scheme 4.10) resulting in a decrease in the activation energy.

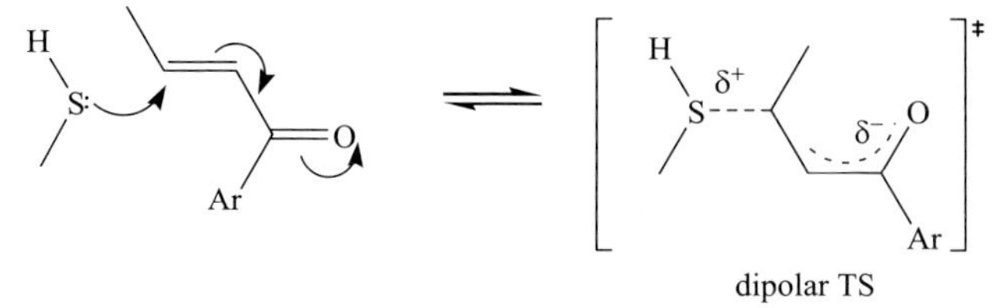

Scheme 4.10

The same explanation of the microwave effect can be advanced for the solvent-free synthesis of 4-aryl substituted 5-alkoxycarbonyl-6-methyl-3,4-dihydropyridones [131] by condensation of Meldrum's acid, methyl acetoacetate, and benzaldehyde derivatives (Eq. 44):

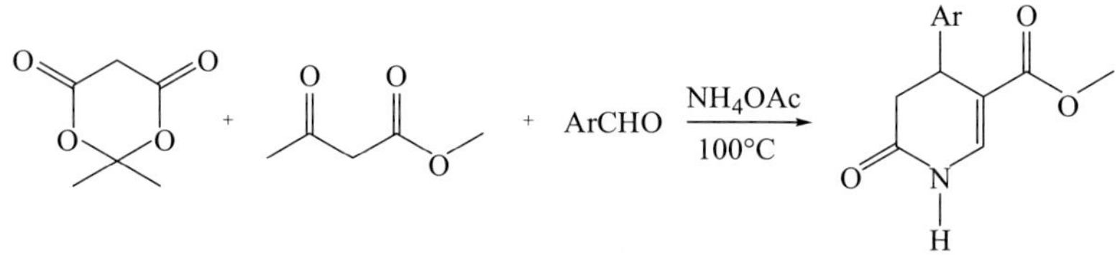

(44)

| Ar = C_6H_5 | 15 min | MW = 86% | 120 min | Δ = 20% |
| Ar = 3-NO_2-C_6H_4 | 10 min | MW = 82% | 120 min | Δ = 17% |

Nitrocyclohexanols were synthesized by a double and diastereoselective Michael addition followed by ring closure (Eq. 45). When performed in the presence of KF-alumina under solvent-free conditions, an important microwave effect was observed:

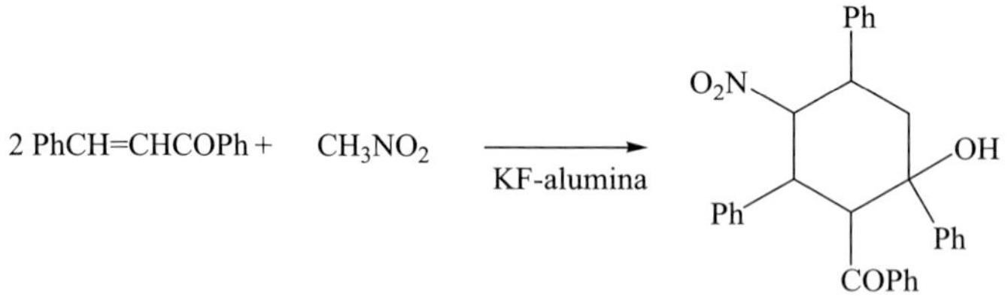

$$2\ PhCH=CHCOPh + CH_3NO_2 \xrightarrow{KF\text{-alumina}}$$

(45)

| Monomode reactor | 90°C | 15 min | MW = 60% |
| | | | Δ = traces |

4.8.1.3 S_N2 Reactions

Reaction of pyrazole with phenethyl bromide In the absence of base, the phenethylation of pyrazole under solvent-free conditions is more rapid by far under the action of microwaves (8 min at 145 °C) compared with Δ, which requires 48 h

[132] (Eq. 46):

$$R = C_6H_5CH_2CH_2 \quad (46)$$

Ring opening of epoxides by amines Microwave-assisted ring opening of (R)-styrene oxide by pyrazole and imidazole leads to corresponding (R)-1-phenyl-2-azolylethanols. With pyrazole, use of microwave irradiation increases both chemo- and regioselectivity compared with conventional heating [133] (Eq. 47):

$$(47)$$

via dipolar TS

The synthesis of anti amino alcohols by aminolysis of vinyl epoxides is greatly improved by MW activation (using a focused MW reactor at 20–30 W) because of simplified handling, short reaction times, and high yielding reactions also with sterically hindered substrates (Eq. 48) [134]:

$$(48)$$

R_1 = H, Bn or BnO; R_2 = H or Me; R_3 = H or CH_2OBn ; R_4 = H or CH_2OBn

8-15 min   MW = 84-100%

1-4.5 h   125-170°C   Δ = 79-83%

N-alkylation of 2-halopyridines A microwave-assisted procedure (focused waves) for N-alkylation of 2-halopyridines has been described; the noticeable microwave effect was indicative of a polar TS [135] (Eq. 49):

$$ \text{(49)} $$

R = CO$_2$Et	165 - 170 °C	MW	40 min	80%
		Δ	23 h	46%
R = CN	165 - 170 °C	MW	40 min	56%
		Δ	50 h	10%

Synthesis of phosphonium salts Using a domestic oven it was shown that the reaction of triphenylphosphine and an organic halide is very rapid under the action of microwave irradiation. The reaction times were reduced to only few minutes in contrast with conventional heating, which requires from 30 min to 14 days [136]. Temperatures were not measured, however, so conclusions cannot be certain.

Nucleophilic substitutions of benzyl halides as electrophiles with Ph$_3$P or Bu$_3$P as nucleophiles have been conducted under solvent-free conditions with accurate control of the power and temperature using a monomode reactor (Synthewave S402). The results were carefully compared under similar conditions with either MW or Δ activation [137] (Eq. (50), Table 4.12):

Tab. 4.12. Solvent-free benzylation of triphenyl(butyl)phosphine.

R	X	Temp. (°C)	Time (min)	Yield (%)	
				MW	Δ
Ph	Br	100	2	99	98
	Cl	100	10	78	24
	Cl	150	10	94	91
n-Bu	Br	100	0.75	94	92
	Cl	100	0.75	91	64
	Cl	100	10	92	87

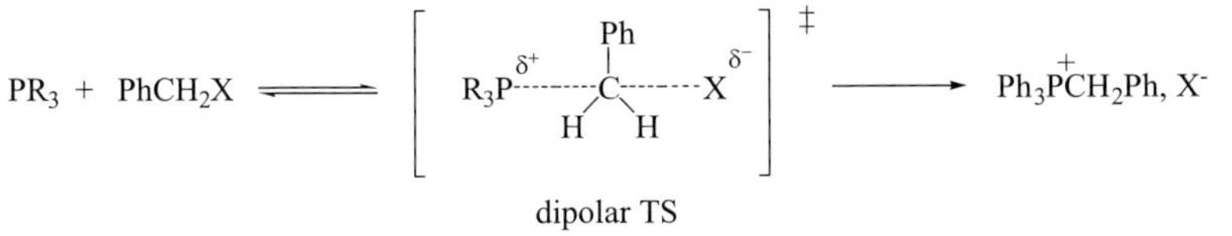

$$PR_3 + PhCH_2X \rightleftharpoons \left[R_3\overset{\delta^+}{P}\cdots\underset{\underset{H}{|}}{\overset{\overset{Ph}{|}}{C}}\cdots\overset{\delta^-}{X} \right]^{\ddagger} \longrightarrow Ph_3\overset{+}{P}CH_2Ph, X^- \quad (50)$$

dipolar TS

R = Ph, n-Bu   X = Br, Cl

Whereas no effect was observed with benzyl bromide, important specific not purely thermal effects were observed when benzyl chloride was used, evidence, therefore, that MW effects are very dependent on the leaving group. This result can be justified when one considers that the TS involving C–Cl bond-breaking will occur later along the reaction coordinates than that involving C–Br, because of the poorer electrophilicity of the first bond (for instance, the bond energies are 66 and 79 kcal mol^{-1} respectively for C–Br and C–Cl). With a similar interpretation, the increased MW effect in the reaction of benzyl chloride with PPh$_3$ compared with PBu$_3$ can be justified because the first phosphine is less reactive, i.e. with a TS later along the reaction coordinates.

Although the MW effect is not appreciable at 150 °C, in the reaction of PPh$_3$ with PhCH$_2$Cl, it becomes clearly apparent when the temperature is reduced to 100 °C. When delineating MW effects, careful attention must be paid to the temperature. If it is too high, the MW effect will be masked and the temperature must be minimized to start from a low yield under Δ and therefore make it possible to observe MW enhancements.

This conclusion is in agreement with kinetic results obtained by Radoiu et al. [138] for the transformation of 2- and 4-^{t}butyl phenols in the liquid phase in the presence of montmorillonite KSF as catalyst under the action of either MW or Δ (Eq. (51), Table 4.13):

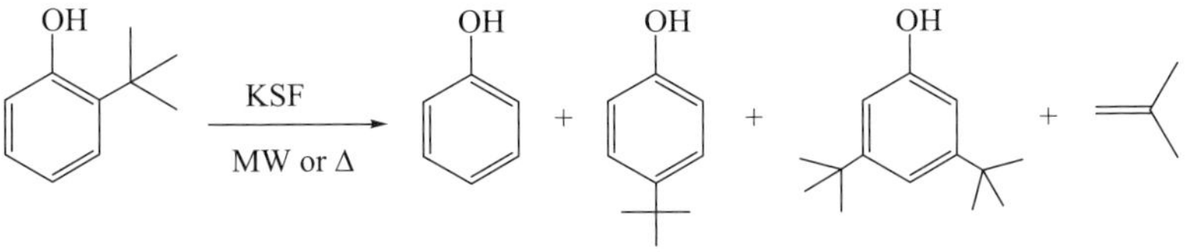

[Scheme 51: 2-tert-butylphenol → phenol + 4-tert-butylphenol + 3,5-di-tert-butylphenol + isobutylene, with KSF, MW or Δ] (51)

Tab. 4.13. Dependence on temperature of rate constants (r°) for transformation of 2-^{t}butyl phenols under the action of MW or Δ.

Temp (°C)	r°$_{MW}$ (10^3 s^{-1})	r°$_\Delta$ (10^3 s^{-1})	r°$_{MW/\Delta}$
22	1.5	0.07	21.6
75	3.2	1.2	2.7
105	10.3	7.1	1.5
198	21.0	20.0	1.1

Alkylation of amines The N-alkylation is the first step for the synthesis of ionic liquids (cf. Chapter 7). The mechanism is especially suitable for observation of specific not purely thermal effects occurring as a result of a dipolar TS. In our laboratory, we have checked the dependence of MW effects on the leaving group (Cl or Br) and on the nature of the amine (Eq. 52) [139]. The main results are given in Table 4.14.

$$H_3C-N\underset{(I)}{\overbrace{}}N + R-X \longrightarrow H_3C-N\underset{}{\overbrace{}}\overset{+}{N}-R \quad X^-$$

$$\underset{(II)}{\text{pyridine}} + R-X \longrightarrow \underset{R}{\overset{+}{N}}\text{-pyridinium} \quad X^- \tag{52}$$

From this table, it is clearly apparent that specific not purely thermal MW effects occur for all compounds and are more pronounced with chlorides than bromides and with pyridine **II** than with N-methylimidazole **I**. These observations are once more explained by considering that MW effects increase as the TS lies later along the reaction coordinates (more difficult reactions). These are well supported by theoretical approaches giving access to the dipole moments of both starting materials and transition states. It was shown that the experimental magnitude

Tab. 4.14. Solvent-free N-alkylation of amines **I** and **II** by n-alkyl halides.

Amine	Alkyl halide	Conditions		Yield (%)	
		Temp. (°C)	Time (min)	MW	Δ
I	n-Butyl Br	90	8	89	73
	n-Butyl Cl	150	15	95	38
	n-Octyl Br	120	8	91	67
	n-Octyl Cl	160	15	98	41
	n-Hexadecyl Br	140	15	98	51
	n-Hexadecyl Cl	160	30	98	29
II	n-Butyl Br	130	10	94	54
	n-Butyl Cl	180	40	62	27
	n-Octyl Br	130	10	92	40
	n-Octyl Cl	180	60	82	56

Tab. 4.15. Semi-empirical calculations for amine alkylations: enhancements of dipole moments from GS to TS, energy of activation evaluated in the gas phase.

Reaction	Yields MW/Δ	Δμ (Debye) TS/GS	ΔG‡ (kcal mol^{-1})
I + n-BuBr	89/73	8.77	28.8
I + n-BuCl	95/38	9.76	43.5
II + n-BuBr	94/54	9.15	32.5
II + n-BuCl	62/27	10.01	47.7

of MW effect is strictly related to enhancement in dipole moments from the GS to the TS (Table 4.15) and to the increase of the energy of activation.

Similar results and conclusions were drawn during the solvent-free preparation of chiral ionic liquids from (−)-N-ephedrine (Eq. 53) [140]:

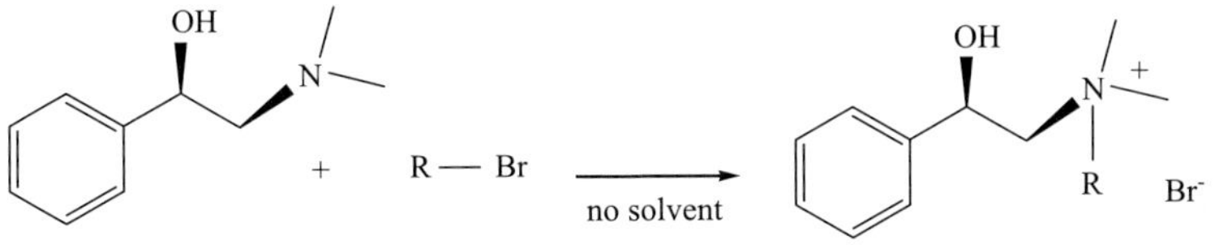

Ex: R = n-C$_8$H$_{17}$ 30 min 95°C   Yields : MW = 85%   Δ = 6%

(53)

An efficient and clean synthesis of N-aryl azacycloalkanes from dihalides and aniline derivatives has recently been achieved using MW irradiation in an aqueous potassium carbonate medium (added to trap the HX formed) [192]. When compared with conventional heating under similar conditions, large specific MW effects were apparent. The observations are evidently consistent with mechanistic considerations as the polar transition state is favored by MW irradiation (Eq. 54):

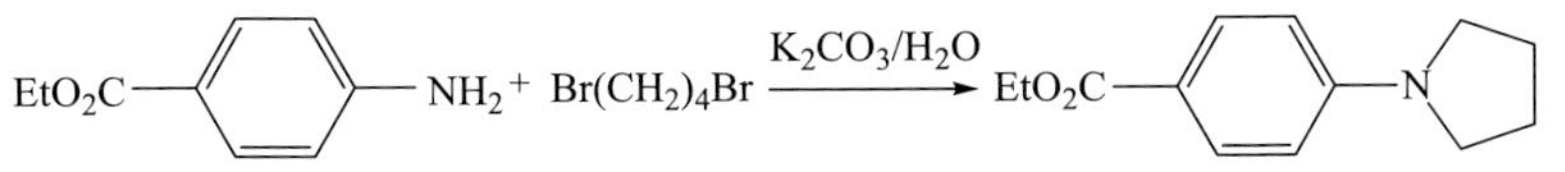

EtO$_2$C—⟨aryl⟩—NH$_2$ + Br(CH$_2$)$_4$Br $\xrightarrow{\text{K}_2\text{CO}_3/\text{H}_2\text{O}}$ EtO$_2$C—⟨aryl⟩—N⟨pyrrolidine⟩

120°C   20 min.   Yields MW : 91%
                        Δ   : traces
         8 h   Δ   : 58%

(54)

4.8.1.4 Vinylic and Aromatic Nucleophilic Substitutions

Aromatic nucleophilic substitution   An expeditious microwave-assisted S$_N$Ar reaction with cyclic amines has been reported for activated aromatic substrates [141] (Eq. 55):

Tab. 4.16. S$_N$Ar reaction between *p*-chlorotoluene and piperidine (X = CH$_2$, R = CH$_3$).

	Activation mode	Conditions		Yield (%)
K$_2$CO$_3$/EtOH	Δ	16 h	Reflux	60
	MW	6 min	Reflux	70
Basic alumina	MW	75 s	—[a]	92

[a] Not determined, but certainly very high, because the vessel was placed inside an alumina bath (prone to microwave absorption)

$$R-\underset{}{\bigcirc}-Cl + \underset{H}{\overset{X}{\bigcirc}}_{N} \rightleftharpoons \left[\underset{R}{\overset{X}{\bigcirc}}_{\overset{NH}{\underset{\delta^-}{\bigcirc}}\overset{\delta^+}{Cl}} \right]^{\ddagger} \longrightarrow \underset{R}{\overset{X}{\bigcirc}}_{N}-\underset{}{\bigcirc} + HCl \quad (55)$$

R = CHO, NO$_2$ X = CH$_2$, O

The reactions were performed in a heterogeneous medium using K$_2$CO$_3$ in ethanol (MW or Δ) or basic alumina in dry media to trap the hydrochloric acid formed. Because of the formation of a dipole in the TS, the microwave effect depicted in Table 4.16 was observed.

Synthesis of aromatic ethers has been performed under solvent-free phase-transfer catalysis conditions by reaction of several aryl halides with potassium methoxide or phenoxide in the presence of a catalytic amount of 18-crown-6. The specific MW effects were shown to be very dependent on the nucleophile and on the structure of the aromatic compound (activated or nonactivated, chloride or fluoride) (Eq. (56), Table 4.17) [96, 142].

$$ArX + RO^-,K^+ \xrightarrow[\text{no solvent}]{\text{18-crown-6}} ArOR + K^+,X^- \quad (56)$$

Ar = O$_2$N—⟨⟩—, naphthyl, pyridyl   X = F, Cl, Br

It is clearly apparent that in the first series the specific MW effect is clearly increasing as the conditions become more harsh, i.e. with later TS along the reaction coordinates. The same conclusions can also be drawn from comparative experiments with 3-halopyridines and the sequence of the MW effect was shown to be related to the energy of activation as evaluated by *ab initio* calculations (Table 4.17).

Tab. 4.17. Etherification of aromatic halides under solvent-free PTC conditions.

R	Ar	X	Time (min)	Temp. (°C)	%MW	%Δ	$\Delta G^\ddagger$ (kcal mol^{-1})
CH$_3$O [96]	p-NO$_2$C$_6$H$_4$	F	3.5	80	100	81	
	β-Naphthyl	F	60	80	94	52	
		Cl	60	100	71	27	
CH$_3$O [142]	3-Pyridyl	F	20	130	98	95	69.3
		F	20	180	57	15	150.7
		Cl	45	200	55	1	176.7

The coupling of microwave activation and the basic system potassium fluoride/potassium carbonate was shown to be an efficient method for preparing 2-nitrophenylamines [142] by nucleophilic aromatic substitution. Within few minutes, excellent results were obtained compared with classical heating (Eq. 57):

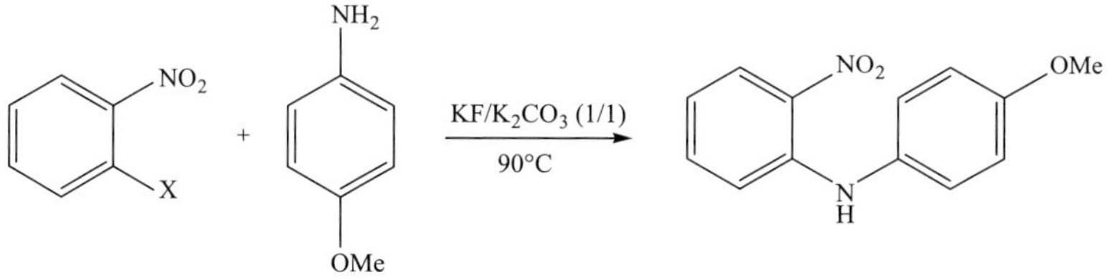

(57)

| X = F | 8 min | MW = 93% | Δ = 23% |
| X = Cl | 10 min | MW = 90% | Δ = 0% |

The nucleophilic substitution reaction installing a benzylic amine on monochlorotriazine derivatives [143] was performed under the action of MW irradiation, enabling aminotriazine compounds to be obtained very quickly with good yields in very short reaction times compared with conventional heating (15 min instead of 2 days) (Eq. 58):

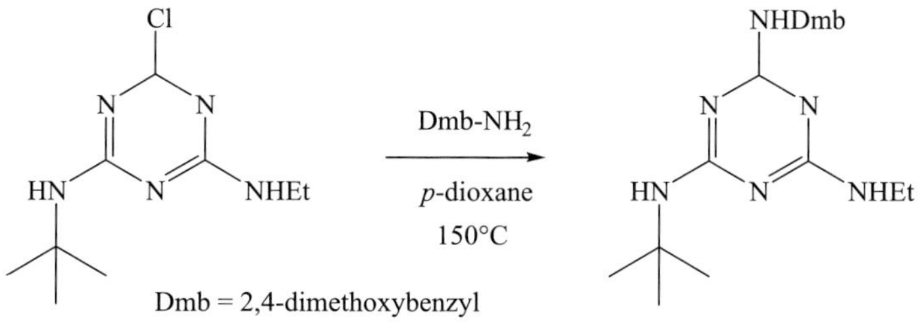

(58)

Dmb = 2,4-dimethoxybenzyl

| 15 min | MW = 97% |
| 2 days | Δ = 86% |

Vinylic nucleophilic substitutions Cherng and co-workers [145] have shown that microwave irradiation can greatly facilitate the synthesis of a variety of substituted uracils by vinylic substitution (Eq. 59):

$$\text{Uracil} + NH_2CH_2Ph \longrightarrow \text{6-benzylamino uracil derivative} \quad (59)$$

20 min   150 °C   MW = 88%   Δ = 34%

A similar result was obtained in the synthesis of 6-benzylaminouracil (15 min, 130 °C, MW = 94%/Δ = 41%).

Microwave-assisted solvent-free reactions have been used by Jenekhe [146] to synthesize quinoline derivatives. An important specific nonthermal microwave effect has been observed compared with conventional heating (Eq. 60). This MW effect is consistent with mechanistic considerations, because the rate-determining step is the internal cyclization depicted in Eq. (60) resulting from nucleophilic attack of the enamine on the carbonyl moiety occurring via a dipolar transition state.

DPP = diphenylphosphonate

$$\text{PhCOCH}_3 + \text{2-aminobenzaldehyde} \xrightarrow[4 \text{ min}/108°C]{\text{DPP}} \text{intermediate} \xrightarrow{-H_2O} \text{2-phenylquinoline} \quad (60)$$

1 equiv. DPP	MW = 78%	Δ = 24%
0.5 equiv. DPP	MW = 78%	Δ = 15%

4.8.2
Bimolecular Reactions with One Charged Reactant

The TS for anionic S_N2 reactions involves loose ion-pairs as in a charge-delocalized (soft) anion. On the another hand, the GS could involve a neutral electrophile and

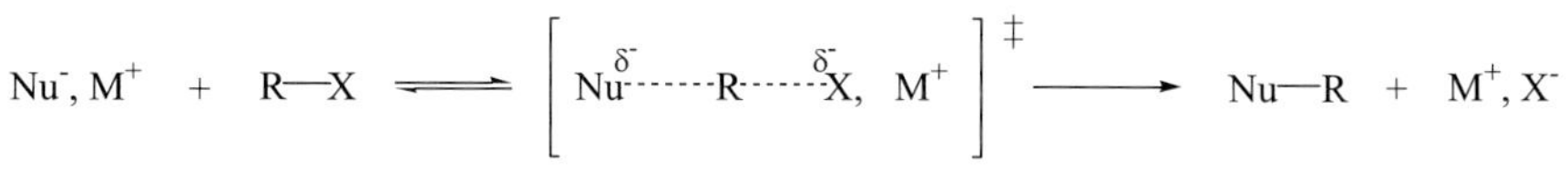

$$\text{Nu}^-, \text{M}^+ + \text{R}{-}\text{X} \rightleftharpoons \left[\text{Nu}^{\delta^-}{\text{-----}}\text{R}{\text{-----}}\text{X}^{\delta^-}, \text{M}^+ \right]^{\ddagger} \longrightarrow \text{Nu}{-}\text{R} + \text{M}^+, \text{X}^-$$

Scheme 4.11

either tight or loose ion-pairs, depending on the anion structure (hard or soft) (Scheme 4.11).

4.8.2.1 Anionic S_N2 Reactions Involving Charge-localized Anions

In this case, the anion being hard and with a high charge density, the reactions are concerned with tight-ion-pairs. During the course of the reaction, ionic dissociation is increased and, hence, polarity is enhanced from the GS towards the TS. Specific microwave effects should be expected.

Selective dealkylation of aromatic alkoxylated compounds Selective de-ethylation of 2-ethoxyanisole is achieved by use of KO^{t}Bu as the reagent in the presence of 18-crown-6 as the phase-transfer agent (PTA). With addition of ethylene glycol (E.G.), the selectivity is reversed and demethylation occurs (Eq. (61), Table 4.18). Although involvement of microwaves is favorable in both examples, the second reaction was shown to be more strongly accelerated than the first [147].

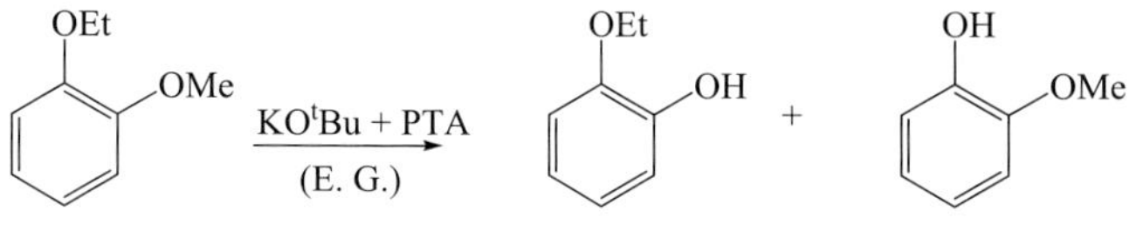

(61)

Demethylation results from the S_N2 reaction whereas de-ethylation occurred via the E2 mechanism (Scheme 4.12).

The microwave-specific effect is more apparent in the demethylation (S_N2). Microwave acceleration clearly is more pronounced with the difficulty of the reaction,

Tab. 4.18. Reaction of KO^{t}Bu with 2-ethoxyanisole in the presence of 18-crown-6 and, optionally, ethylene glycol.

Additive	Time	Activation	Temp (°C)	1 (%)	2 (%)	3 (%)
–	20 min	MW	120	7	0	90
–	20 min	Δ	120	48	0	50
E.G.	1 h	MW	180	0	72	23
E.G.	1 h	Δ	180	98	0	0

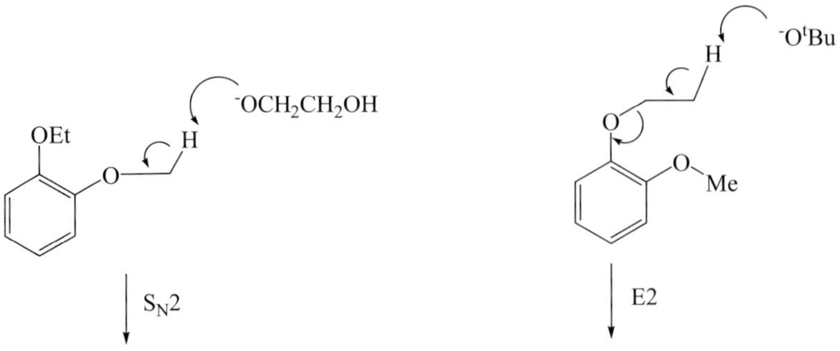

Scheme 4.12. Mechanisms for demethylation and de-ethylation of ethoxyanisole.

thus constituting a clear example of an increased microwave effect with a more difficult reaction, indicative of a later TS position along the reaction coordinates. The microwave effect may also be connected to the more localized charge in the S_N2 transition state (three centers) when compared with that of β-E_2 (charge developed over five centers).

Alkylation of dianhydrohexitols under the action of Phase-transfer Catalysis (PTC) conditions Dianhydrohexitols, important by-products of biomass (Scheme 4.13) derived from corn starch, were dialkylated under PTC conditions in the presence of a small amount of xylene.

Attempts were made to use dialkylations as model reactions before subsequent polymerizations and revealed very important specific microwave effects [148] (Eq. (62), Table 4.19):

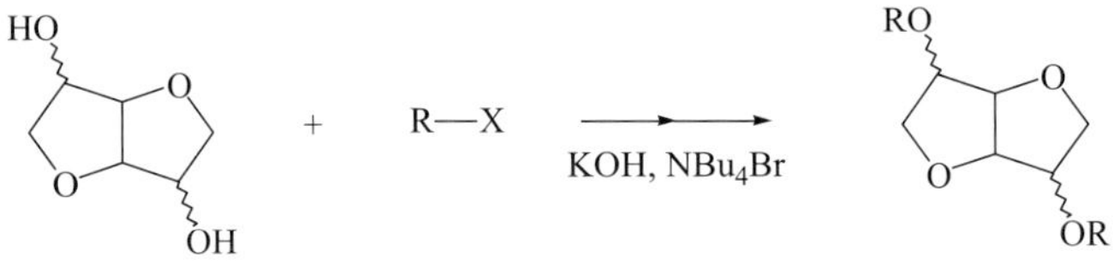

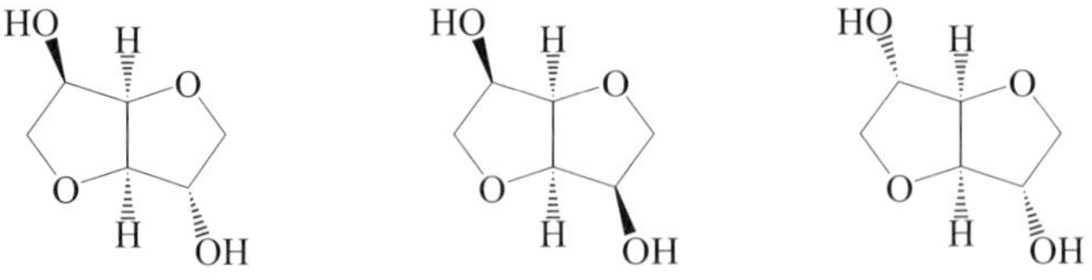

Scheme 4.13. Structure of 1,4:3,6-dianhydrohexitols (A = isosorbide, B = isomannide, C = isoidide).

Tab. 4.19. Dialkylation of dianhydrohexitols under PTC conditions.

RX	t (min)	T °C	Yield (%)					
			A		B		C	
			MW	Δ	MW	Δ	MW	Δ
PhCH$_2$Cl	5	125	98	13	98	15	97	20
nC$_8$H$_{17}$Br	5	140	96	10	74	10	95	10

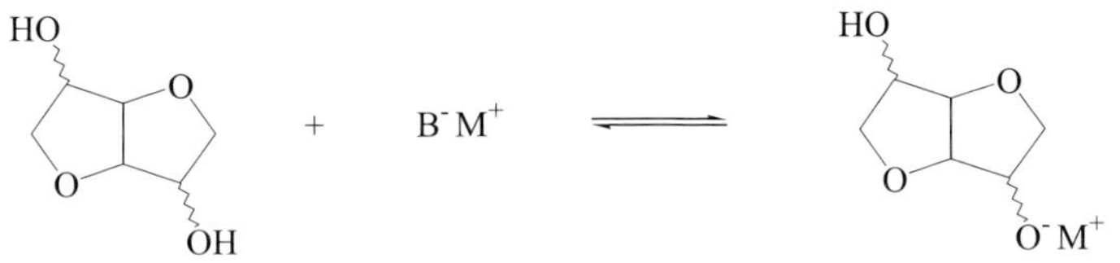

Scheme 4.14

These observations are consistent with the reactive species being formed from tight ion-pairs between cations and the alkoxide anions resulting from abstraction of hydrogen atoms in A, B, and C (Scheme 4.14).

The reaction of monobenzylated isosorbide D with ditosylates (Eq. 63) is more subtle; the microwave-specific effect (Table 4.20) appeared when the temperature was reduced to 80 °C (modulated by the presence of cyclohexane) whereas it was

Tab. 4.20. Yields from reaction of monobenzylated isosorbide D with ditosylates for 15 min.

R	Yield (% E)			
	T 110 °C (xylene)		T 80 °C (cyclohexane)	
	MW	Δ	MW	Δ
(CH$_2$)$_8$	95	91	96	39
(CH$_2$)$_6$	91	90	96	45
CH$_2$CH$_2$OCH$_2$CH$_2$	92	92	91	36

masked at the higher temperature of 110 °C (maintained by use of toluene) [149].

$$\underset{D}{\text{HO-furofuran-OCH}_2\text{Ph}} \xrightarrow[\text{TsO-R-OTs}]{\text{KOH, NBu}_4\text{Br}} \underset{E}{\text{PhH}_2\text{CO-furofuran-O-R-O-furofuran-OCH}_2\text{Ph}} \tag{63}$$

The Krapcho reaction Dealkoxycarbonylation of activated esters occurs classically under drastic thermal conditions [150]. It constitutes a typical example of a very slow-reacting system (with a late TS along the reaction coordinates) and is, therefore, prone to a microwave effect. The rate-determining step involves a nucleophilic attack by a halide anion and requires anionic activation, which can be provided by solvent-free PTC conditions under the action of microwave irradiation [151, 155]. These results illustrate the difficult example of cyclic β-ketoesters with a quaternary carbon atom in the α position relative to each carbonyl group (Eq. 64), which classically gave only 20% yield using $CaCl_2$ in DMSO under reflux for 3 h. Some typical results are summarized in Table 4.21.

$$\text{cyclohexanone-C(R)(COOEt)} \xrightarrow[\text{H}_2\text{O (2 eq)}]{\text{LiBr (2 eq), NBu}_4\text{Br (0.1 eq)}} \text{cyclohexanone-CHR} \tag{64}$$

$$\text{cyclohexanone-C(R)-C(=O)-O-Et} \quad Br^-, M^+ \quad (S_N2)$$

Tab. 4.21. Krapcho reaction under solvent-free PTC conditions.

R	Reaction conditions		Yield (%)	
	t (min)	Temp. (°C)	MW	Δ
H	8	138	96	<2
C_2H_5	15	160	94	<2
$n\text{-}C_4H_9$	20	167	89	<2
$n\text{-}C_6H_{13}$	20	186	87	<2
	60	186		22
	180	186		60

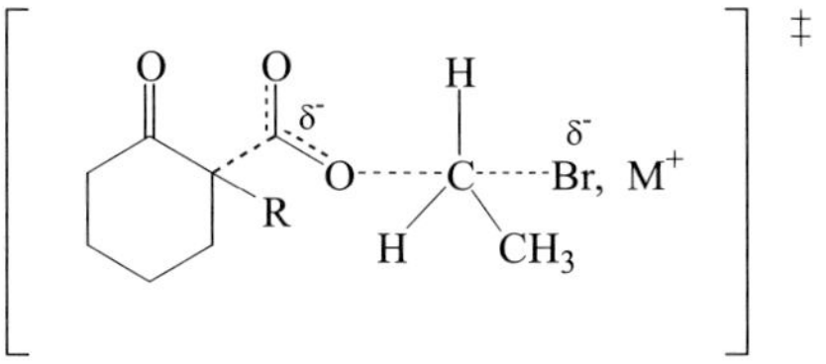

Scheme 4.15. Transition state for the Krapcho reaction.

A definite microwave effect is involved when strict comparisons of MW and Δ activations are considered and is compatible with the mechanistic assumption that a very polar TS is developed (Scheme 4.15).

Anionic β-elimination Ketene acetal synthesis by β-elimination of haloacids from halogenated acetals under solvent-free PTC conditions under well controlled conditions using thermal activation (Δ), ultrasound (US), or microwave irradiation [152] (MW) has been described. Mechanistically, as the TS is more charge delocalized than the GS and the polarity is enhanced during the course of the reaction, a favorable microwave effect is expected, and is actually observed (Eqs. (65) and (66), Scheme 4.16).

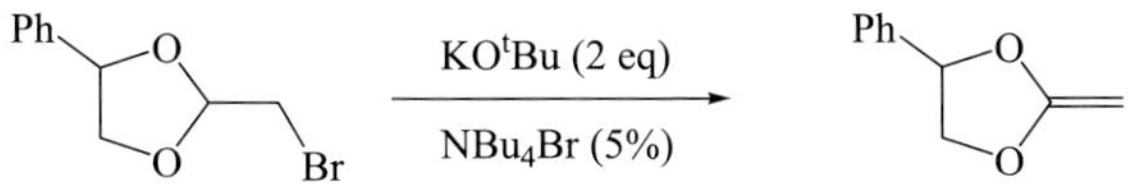

5 min   75 °C   Δ: 36%   US: 55%   MW: 87%   (65)

25 min   90 °C   Δ: 15%   US: 45%   MW: 79%   (66)

tight ion pairs (GS)   loose ion pairs (TS)

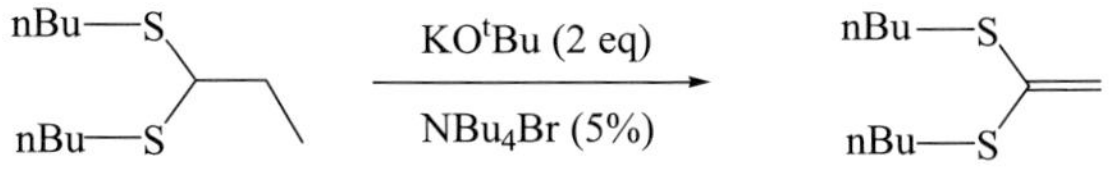

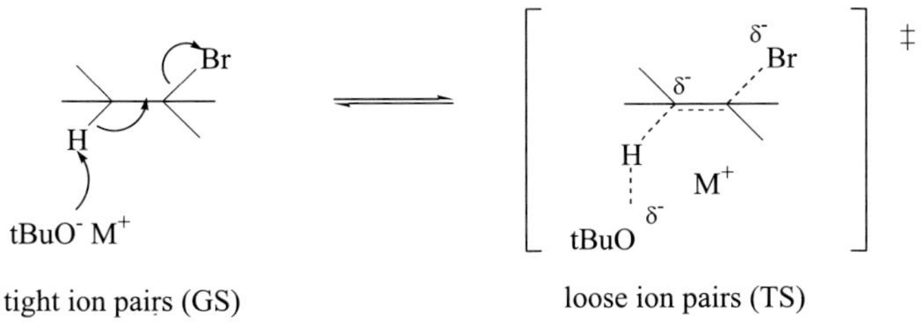

Scheme 4.16. Evolution of the polarity in β-elimination.

4.8.2.2 Anionic S_N2 Reactions Involving Charge-delocalized Anions

Weak or nonexistent microwave effects are expected for these reactions, because the GS and TS have rather similar polarities and both involve loose ion-pairs.

Alkylation of potassium benzoate Alkylation of several substituted benzoic acid salts with *n*-octyl bromide was performed under solvent-free PTC conditions with excellent yields ($\geq 95\%$) with a very short reaction time (2–7 min) [153]. Oil bath heating (Δ) led to yields equivalent to those produced under microwave irradiation, which thus revealed only thermal effects were important in the temperature range used, 145–202 °C (Eq. 67).

$$Z\text{-C}_6H_4\text{-CO}_2H + nC_8H_{17}Br \xrightarrow[\text{NBu}_4\text{Br}]{\text{K}_2\text{CO}_3} Z\text{-C}_6H_4\text{-CO}_2nC_8H_{17}$$

Z = H	2.5 min	145 °C	Δ = MW	99 %	(67)
Z = OCH$_3$	2 min	145 °C		98 %	
Z = CN	2 min	202 °C		95 %	

In contrast, when *n*-octyl bromide was used with the less reactive terephthalate species, which constitutes a "slow-reacting system", the yield was improved from 20 to 84% when microwaves were used instead of Δ; this can be attributed to a later TS along the reaction coordinates (Eq. 68):

$$HO_2C\text{-C}_6H_4\text{-CO}_2H + RBr \xrightarrow[\text{NBu}_4\text{Br}]{\text{K}_2\text{CO}_3} RO_2C\text{-C}_6H_4\text{-CO}_2R \quad (68)$$

R = *n*-C$_8$H$_{17}$ 6 min   175 °C   MW 84% Δ 20%

Pyrazole alkylation in basic media [122] A very important microwave-specific effect was apparent in the absence of a base in the reaction of pyrazole with phenethyl bromide (reaction times: MW = 8 min, Δ = 48 h, Eq. 69). When the same reaction was performed in the presence of KOH, the microwave effect disappeared (*vide supra* Eq. 46).

$$\text{pyrazole-H} + \text{KOH} \underset{-H_2O}{\rightleftharpoons} \text{pyrazole}^- K^+ \text{ (loose ion pair)} \xrightarrow{\text{PhCH}_2\text{CH}_2\text{Br}} \text{N-CH}_2\text{CH}_2\text{Ph pyrazole} \quad (69)$$

8 min   145 °C   MW 64 % Δ 61 %

This effect could be predicted when considering the weak evolution of polarity between the GS and TS, because the reactive species consist of loose ion-pairs (involving a soft anion).

Selective alkylation of β-naphthol in basic media Alkylations in dry media of the ambident 2-naphthoxide anion were performed under focused microwave activation. Whereas the yields were identical to those obtained with classical heating for benzylation, they were significantly improved under microwave irradiation conditions for the more difficult *n*-octylation (less reactive electrophilic reagent). No change in selectivity was observed, however, indicating the lack of effect of ionic polarization [154]. The absence or weakness of the microwave effect was assumed to be related to loose ion-pairs involving the soft naphthoxide anion in the GS and a small change in polarity in an early TS. When the TS occurred later along the reaction coordinates (e.g. for *n*-octylation requiring a higher temperature), more polarity is developed and, consequently, the microwave effect could appear (Eq. (70), Table 4.22; limited here to the lithiated base).

$$\text{β-naphthol-OH} + \text{Li}^+\text{B}^- + \text{R Br} \longrightarrow \underset{\text{Mono alkylation}}{\text{R-substituted naphthol-OH}} + \underset{\text{Di alkylation}}{\text{R,R-disubstituted naphthalenone}} \quad (70)$$

Alkylation of malonate anion Hernandez et al. [155, 156] have developed a rapid and easy procedure for the synthesis of ^{14}C-labeled esters using a combination of solvent-free phase-transfer catalysis conditions and microwave technology. A detailed study of the first step, which consists in malonic alkylation, did not reveal any advantages of microwave activation compared with conventional heating (Eq. 71):

Tab. 4.22. C-alkylation of β-naphthol in the presence of lithiated base under solvent-free conditions.

RX	Li$^+$ B$^-$	t (min)	Temp (°C)	Yield (%)			
				MW Mono C	MW Di C	Δ Mono C	Δ Di C
PhCH$_2$Br	LiOH	4	190	98	2	97	2
	LiO^{t}Bu	4	137	9	91	8	88
nC$_8$H$_{17}$Br	LiOH	9	240	92	1	62	–
	LiO^{t}Bu	10	200	27	56	7	20

$$\text{CH}_2(\text{CO}_2\text{Et})_2 + \text{R-Br} \xrightarrow[\text{2 min / 130°C}]{\text{KOtBu / Aliquat 336}} \text{R-CH}(\text{CO}_2\text{Et})_2$$

R = $n\text{C}_{10}\text{H}_{21}$ MW = 69%   Δ = 60%

R = $n\text{C}_{18}\text{H}_{37}$ MW = 65%   Δ = 62%

$$R = n\text{C}_{10}\text{H}_{21} \quad \text{MW} = 69\% \quad \Delta = 60\% \tag{71}$$

$$R = n\text{C}_{18}\text{H}_{37} \quad \text{MW} = 65\% \quad \Delta = 62\%$$

The specific microwave effect was absent because the difference between the polarity of the GS and TS is very small. The reactive species in GS comprise very loose ion-pairs, because of delocalization of the negative charge in the malonate ion (Eq. 72):

$$\underset{[\text{GS}]}{\overset{\text{EtO}_2\text{C}}{\underset{\text{EtO}_2\text{C}}{>}}\!\!\!\!\!\!-\,\,\,^+\text{NR}_4} + \text{R-Br} \rightleftharpoons \left[\underset{[\text{TS}]}{\overset{\text{EtO}_2\text{C}}{\underset{\text{EtO}_2\text{C}}{>}}\cdots\text{R}\cdots\text{Br}^{\delta-}\,\,^+\text{NR}_4}\right]^{\ddagger} \tag{72}$$

4.8.2.3 Nucleophilic Additions to Carbonyl Compounds

Saponification of hindered aromatic esters   This is a typical example of an enhanced microwave-specific effect related to the difficulty of the reaction, which presumably proceeds via later and later TS. Whereas essentially thermal effects are observed (approx. 200 °C) with methyl and octyl benzoate, a microwave-specific effect is increasingly apparent with hindered esters and becomes optimum with mesitoyl octanoate (Eq. (73), Table 4.23) [157].

$$\text{RCO}_2\text{R}' \xrightarrow[\substack{\text{1) NaOH (2 eq), Aliquat 336 (10\%)}\\ \text{2) HCl (2N)}}]{200\,°\text{C, 5 min}} \text{RCO}_2\text{H} \tag{73}$$

Tab. 4.23.   Solvent-free PTC saponifications of aromatic esters.

R	R′	Yield (%)	
		MW	Δ
Ph	Me	92	73
Ph	nOct	98	86
mesityl	Me	90	48
mesityl	nOct	97	39

4.8 Some Illustrative Examples

PTC transesterification in basic medium The microwave-assisted PTC transesterification of several carbohydrates in basic medium with methyl benzoate or laurate has been studied [158]. Small amounts of DMF were necessary to provide good yields within 15 min at 160 °C. Rate enhancements were compared with conventional heating (Δ) under the same conditions and specific microwave activation was mostly seen when the less reactive fatty compounds were involved (Eq. (74), Table 4.24).

$$\text{Carbohydrate-OH} \xrightarrow[\varepsilon \text{ DMF, RCO}_2\text{CH}_3]{\text{K}_2\text{CO}_3,\ \text{NBu}_4\text{Br (5\%)}} \text{Carbohydrate-OCOR} + \text{CH}_3\text{OH} \quad (74)$$

15 min, 160 °C

The reactive species under these conditions consist of tight-ion-pairs involving the alkoxide anion from the carbohydrate (charge localized anion). The less reactive long-chain methyl laurate leads to a later TS along the reaction coordinates and the magnitude of the microwave effect is therefore increased.

Tab. 4.24. Transesterification in 15 min at 160 °C in basic medium with methyl benzoate and laurate (monomode reactor, relative amounts 1:2:2).

R	Yield (%)	
	MW	Δ
Ph	96	21
$CH_3(CH_2)_{10}$	88	0

Ester aminolysis in basic medium Ester aminolysis usually occurs under harsh conditions – high temperatures and extended reaction periods or use of strong alkali metal catalysts. An efficient solid-state synthesis of amides from nonenolizable esters and amines using KO^{t}Bu under microwave irradiation [159] has been described. The reaction of esters with octylamine was extensively studied to identify possible microwave effects [160] (Eq. (75), Table 4.25):

$$RCO_2Et + CH_3(CH_2)_7NH_2 \xrightarrow[150\ °C]{10\ \text{min}} RCONH(CH_2)_7CH_3 + EtOH \quad (75)$$

Tab. 4.25. Ester aminolysis with *n*-octylamine at 150 °C for 10 min.

R	Base	Yield (%)	
		MW	Δ
Ph	–	0	0
	KO^{t}Bu	80	22
	KO^{t}Bu + Aliquat 336	87	70
PhCH$_2$	–	63	6
	KO^{t}Bu + Aliquat 336	63	36

The microwave-specific effect is increased when the reaction is performed in the absence of a phase-transfer catalyst, showing that the nature of the reactive species is of great importance in ionic dissociation (Eq. 76):

$$CH_3(CH_2)_7NH_2 + KO^tBu \rightleftharpoons \underset{\text{tighter ion pair}}{CH_3(CH_2)_7NH^-, K^+} + {}^tBuOH \uparrow$$

$$CH_3(CH_2)_7NH^-, K^+ + R_4N^+, Cl^- \rightleftharpoons \underset{\text{looser ion pair}}{CH_3(CH_2)_7NH^-, R_4N^+} + K^+, Cl^- \tag{76}$$

As expected, larger microwave effect is observed for the tighter ion-pair (RNH$^-$, K$^+$). As an extension of this work, and in order to gain further insight, the effect of amine substituents was studied [161] in the reaction with ethyl benzoate (Eq. (77), Table 4.26):

$$PhCO_2Et + RNH_2 \xrightarrow[10\ \text{min}/150\ °C]{{}^tBuOK} PhCONHR + EtOH\uparrow \tag{77}$$

The microwave effects are clearly substituent-dependent and, as in the former example, disappear on adding a phase-transfer agent. When the substituent R can delocalize the negative charge on the amide anion (R = Ph), the ion-pairs RNH$^-$, M$^+$ exist in a looser association. Consequently, a decrease in microwave effect is expected as the evolution from the GS towards the TS occurs with only a slight modification of polarity in the ion-pairs. Conversely, the microwave effect is optimum with the tighter ion-pair (nC$_8$H$_{17}$N$^-$, K$^+$).

Tab. 4.26. Ethyl benzoate aminolysis with different amines at 150 °C for 10 min (relative amounts PhCO$_2$Et:RNH$_2$: KO^{t}Bu = 1.5:1:2).

R	Yield (%)		Yield (%) with NR$_4$Cl	
	MW	Δ	MW	Δ
Ph	88	73	90	83
PhCH$_2$	84	42	98	85
nC$_8$H$_{17}$	80	22	87	70

Tighter ion-pair effects were also claimed for the same reaction by other authors [185]. Unfortunately, unadapted comparisons were conducted, for example solvent-free MW reactions and conventional heating in DMSO as a solvent, making accurate conclusions difficult to obtain.

In a parallel study [162] it was shown that formamide and primary and secondary amines react with esters in the presence of potassium *tert*-butoxide under microwave irradiation. Substituted amides are formed in yields (generally more than 70%) much higher than under the action of conventional heating (Eq. (78), Table 4.27):

$$RCO_2R' + HCONH_2 \xrightarrow[MW]{^tBuOK} RCONH_2 + R'CO_2H$$

$$RCO_2R' + R_1R_2NH_2 \xrightarrow[MW]{^tBuOK} RCONR_1R_2 + R'OH \qquad (78)$$

Tab. 4.27. Aminolysis of esters with different amines under the action of microwave irradiation for 3 min and with conventional heating.

Ester	Amine	T (°C)	Yield (%)	
			MW	Δ
CH$_3$CO$_2$Et	BuNH$_2$	95	70	25 (2d)
	Et$_2$NH	129	74	30 (1d)
PhCH$_2$CO$_2$Me	BuNH$_2$	105	97	36 (1d)
	Et$_2$NH	204	70	25 (6 h)

d = days; h = hours

4.8.2.4 Reactions Involving Positively Charged Reactants

Relatively few comparative MW/Δ studies are currently available on this topic.

Friedel-Crafts acylation of aromatic ethers Solvent-free benzoylation of aromatic ethers has been performed under the action of microwave irradiation in the presence of a metallic catalyst, $FeCl_3$ being one of the most efficient [163]. With careful control of the temperature and other conditions, nonthermal microwave effects have not been observed either in terms of yield or isomeric ratios of the obtained products (Eq. 79):

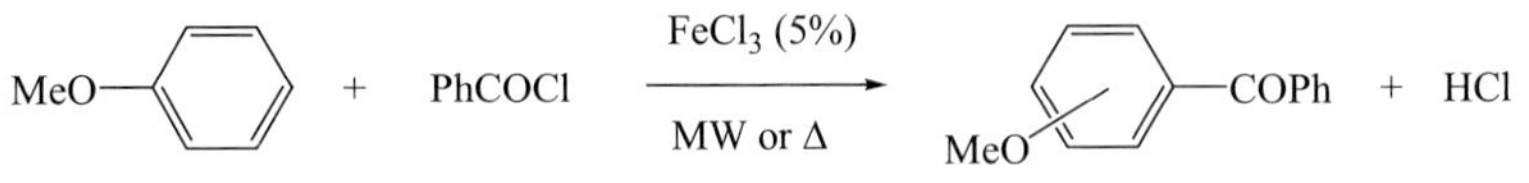

$$MeO\text{-}C_6H_5 + PhCOCl \xrightarrow[\text{MW or }\Delta]{FeCl_3\ (5\%)} MeO\text{-}C_6H_4\text{-}COPh + HCl \quad (79)$$

1 min   165 °C   yield = 95%   para/ortho = 94/6

The reactive species is the acylium ion resulting from abstraction of a chloride anion from benzoyl chloride (Scheme 4.17). This reagent comprises an ion-pair formed between two large (soft) ions which are therefore associated as loose ion-pairs. According to these assumptions, the absence of a microwave effect should be expected as the polarity evolution is very weak between the GS and TS (two loose ion-pairs of similar polarities).

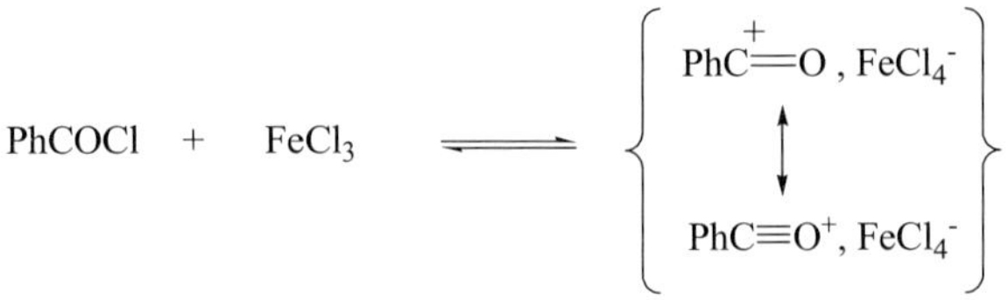

$$PhCOCl + FeCl_3 \rightleftharpoons \left\{ \begin{array}{c} Ph\overset{+}{C}=O\ ,\ FeCl_4^- \\ \updownarrow \\ PhC\equiv O^+,\ FeCl_4^- \end{array} \right\}$$

Scheme 4.17

A much more environmentally sound procedure was advocated by Paul et al. [164], who used Zn powder as a catalyst for Friedel–Crafts acylation of aromatic compounds. Zinc is a nontoxic, safe, and inexpensive metal which can be used in solvent-free conditions. It was shown to have a remarkably high activity in the acylation of a series of aromatic compounds with acetyl and benzoyl chlorides when performed under microwave irradiation. Yields were much more better than when using conventional thermal heating (Eq. (80), Table 4.28). The Zn powder could also be re-used up to six times after simple washing with diethyl ether and dilute HCl.

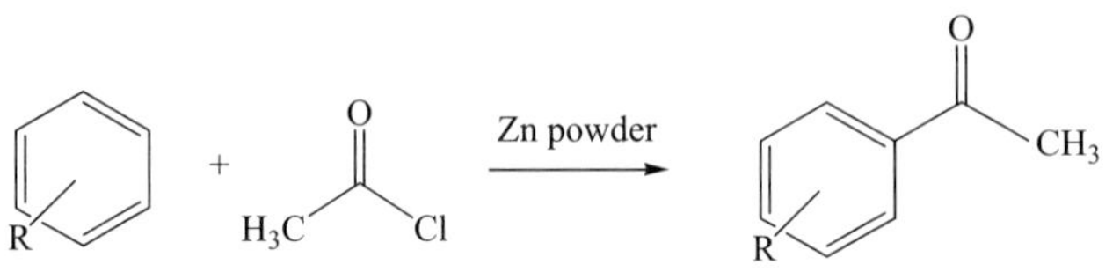

$$R\text{-}C_6H_5 + H_3C\text{-}COCl \xrightarrow{Zn\ powder} R\text{-}C_6H_4\text{-}CO\text{-}CH_3 \quad (80)$$

Tab. 4.28. Friedel–Crafts acylation of aromatic compounds by the use of Zn powder.

R	Conditions		Yields (%)	
H	62 °C	30 s	MW = 95%	Δ = 20%
CH_3	53 °C	40 s	MW = 70%	Δ = 8%
NO_2	84 °C	3 min	MW = 79%	Δ = 10%

In contrast, MW effects were absent in the sulfonylation of mesitylene in the presence of metallic catalysts such as $FeCl_3$ (5%) (Eq. 81) [165]. A plausible explanation could be that the reactive species $[PhSO_2]^+$, $[FeCl_4]^-$ are a very loose ion-pair in the GS and therefore susceptible to a significant MW effect.

$$PhSO_2Cl + FeCl_3 \rightleftharpoons [PhSO_2]^+ [FeCl_4]^- \longrightarrow \text{mesityl-}SO_2Ph + HCl \quad (81)$$

30 min   110 °C   MW = 80%   Δ = 78%

Formylation using the Vilsmeier reagent [166] Substituted acetophenones were irradiated in a domestic microwave oven with $POCl_3$–DMF/SiO_2 to give β-chlorovinylaldehydes in 2 min with yields of 75–88%. Under the same conditions, the use of conventional heating led to only 30–40% (Eq. 82):

$$Cl\text{-}C_6H_4\text{-}COCH_3 \xrightarrow[SiO_2]{POCl_3\text{-}DMF} Cl\text{-}C_6H_4\text{-}C(CHO)=CH\text{Et} \quad (82)$$

2 min   75–78 °C   MW = 79%   Δ = 35%

The specific microwave effect can be attributed to two different facts:

- improvement in the formation of chloroiminium species (so-called Vilsmeier reagent)

- enhancement of subsequent electrophilic aromatic substitution

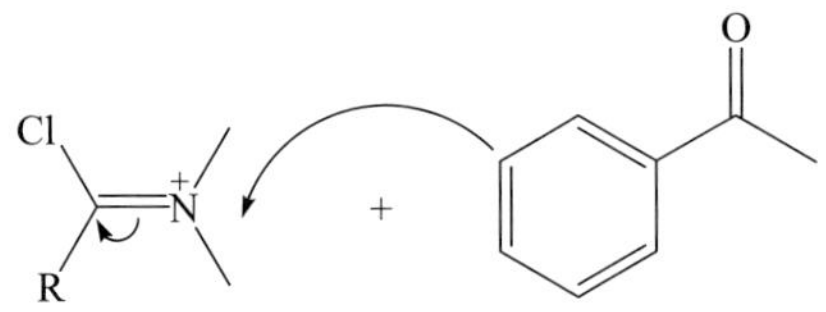

S$_N$2 reactions with tetralkylammonium salts The S$_N$2 reaction of triphenylphosphine with triethylbenzylammonium chloride (the tertiary amine is thus the leaving group) has been studied under solvent-free microwave conditions. The reaction occurred only under MW [137] irradiation (Eq. 83):

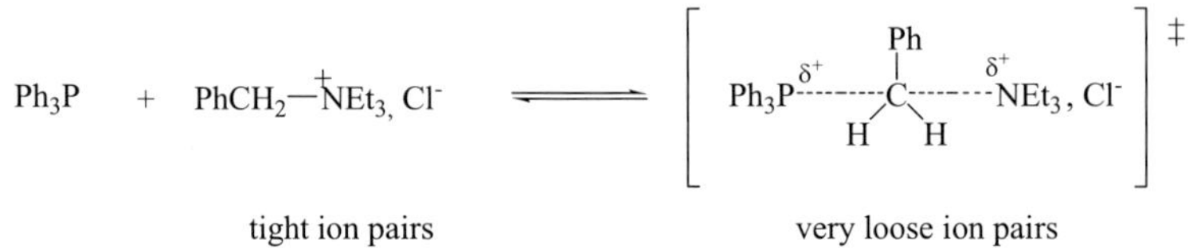

$$Ph_3P + PhCH_2-\overset{+}{N}Et_3, Cl^- \rightleftharpoons \left[Ph_3P^{\delta+}\cdots\underset{H\ H}{\overset{Ph}{C}}\cdots NEt_3, Cl^- \right]^{\ddagger}$$

tight ion pairs   very loose ion pairs

150 °C   10 min   MW = 70%   Δ = 0%

(83)

This effect is readily attributable to the very loose structure of the ion-pairs in the TS (also involving delocalization in the phenyl groups of the phosphine) and which are therefore far more polar than the initial ion-pairs in the GS. Furthermore, because tetraalkylammonium is a poor leaving group reacting under rather harsh conditions (high energy of activation), we are certainly concerned with a late TS very prone to MW effects.

4.8.3
Unimolecular Reactions

4.8.3.1 Imidization Reaction of a Polyamic Acid [44]

The polyamic acid (EQ) precursor (Fig. 4.15) was prepared by adding stoichiometric amounts of 3,3,4,4-benzophenonetetracarboxylic acid dianhydride (BTDA) and diaminodiphenylsulfone (DDS). The solution highly concentrated in NMP was then submitted to either thermal or MW activation with accurate monitoring of the temperature (Scheme 4.18).

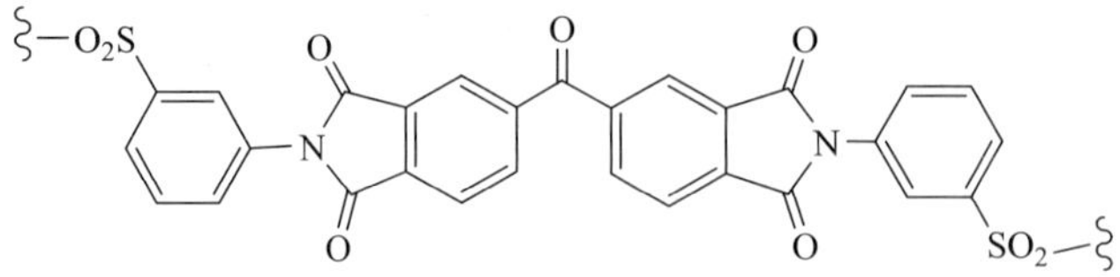

Fig. 4.15. Polyamic acid structure.

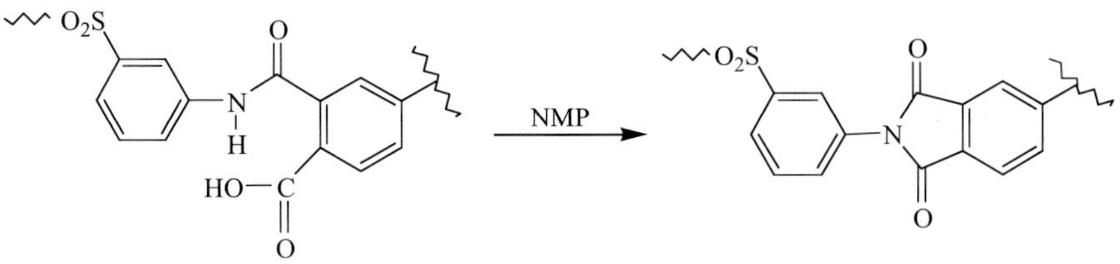

Scheme 4.18

Analysis of kinetic data showed that the apparent activation energy for the reaction was reduced from 105 to 57 kJ mol^{-1}. This observation was consistent with the polar mechanism of this reaction, implying the development of a dipole in the transition state (Scheme 4.19).

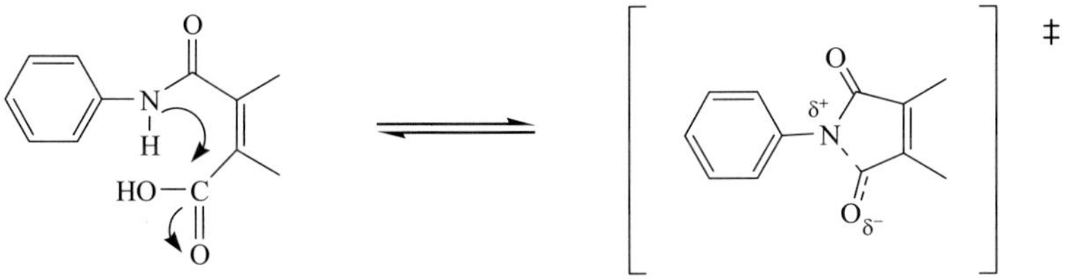

Scheme 4.19. Mechanism for the imidization reaction.

4.8.3.2 Cyclization Reactions

Cyclocondensation of *N*-(trifluoroacetamido)-*o*-arylenediamines led to a series of 2-trifluoromethylarylimidazoles in good yield on montmorillonite K10 in dry media under the action of microwave irradiation within 2 min. Using conventional heating under the same conditions, no reaction was observed [167] (Eq. (84), Table 4.29).

Tab. 4.29. Cyclization of monotrifluoroacetylated *o*-arylenediamines.

R_1	R_2	Temp (°C)	Yield (%)	
			MW (2 min)	Δ (20 h)
H	H	125	87	23
H	CH$_3$	127	84	19
NO$_2$	H	134	95	28

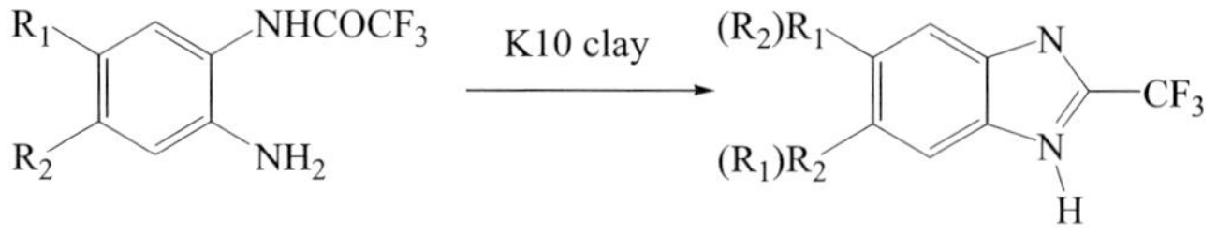

4 Nonthermal Effects of Microwaves in Organic Synthesis

[Reaction scheme 84: substituted aniline with NHCOCF$_3$ and NH$_2$ groups, with K10 clay, gives benzimidazole with CF$_3$ substituent]

(84)

This observation is consistent with the assumptions of the authors, who predicted microwave effects when the polarity is enhanced in a dipolar TS. The kinetic rate-determining step consists in intramolecular attack of the nitrogen lone pair on the carbon atom of the carbonyl moiety (Scheme 4.20).

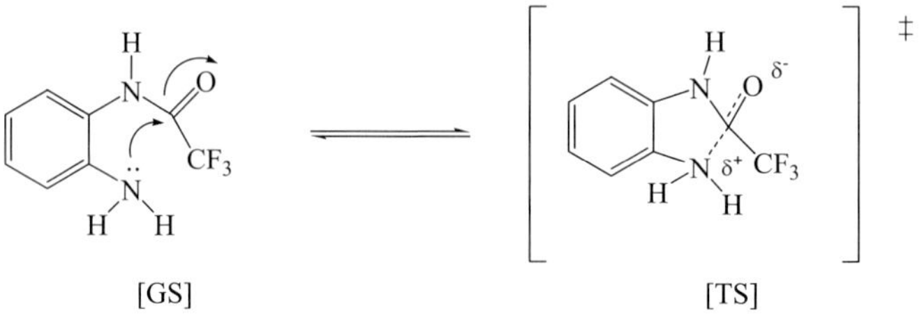

[Scheme showing ground state [GS] and transition state [TS] with δ+ and δ− charges]

Scheme 4.20. Mechanism of cyclocondensation.

A similar conclusion can be drawn in the Paal–Knorr cyclization [168] between an amido-1,4-diketone and a large variety of primary amines, performed under microwave activation (Eq. 85), the rate-determining step being the nucleophilic attack of the amine on the carbonyl moiety via a dipolar TS.

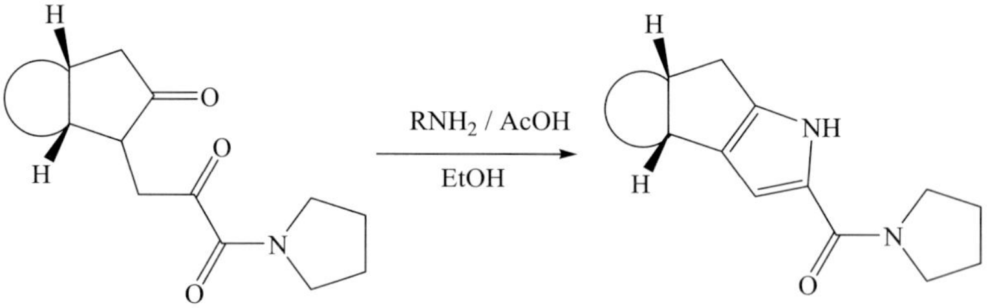

[Reaction scheme 85: amido-1,4-diketone + RNH$_2$ / AcOH in EtOH gives pyrrole product with pyrrolidine amide]

(85)

R = CH$_2$Ph	5 min	MW = 75%	20 min	Δ = 55%
R = CH$_2$CH$_2$OMe	5 min	MW = 77%	15 min	Δ = 58%

The synthesis of phthalocyanines [118] under the action of microwave irradiation can be performed easily in less time by using phthalonitrile as the substrate. Solvent-free conditions led to by far the best results and an important specific microwave effect was apparent (Eq. 86):

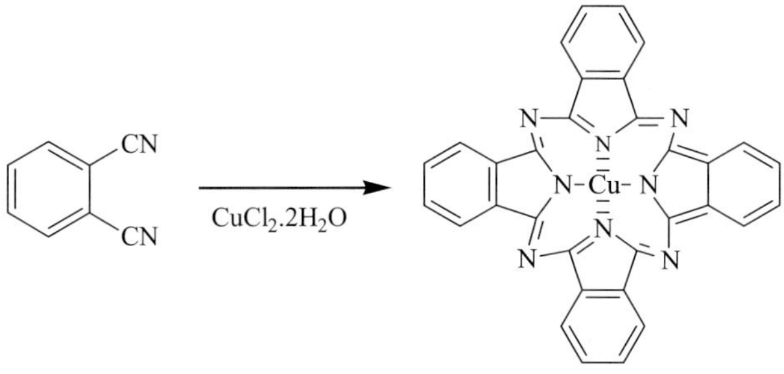

(86)

15 min   200 °C   MW = 88%   Δ = 10%

This effect can be explained when one considers the enhancement in the polarity of the system when the reaction is in progress, because of a well-fitting mechanism.

4.8.3.3 Intramolecular Nucleophilic Aromatic Substitution

Pyrazolo[3,4b]quinolines and pyrazolo[3,4b]pyrazoles have been synthesized by reacting β-chlorovinylaldehydes with hydrazine or phenylhydrazine using a catalytic amount of *p*-toluenesulfonic acid (PTSA) under the action of microwave irradiation. The yields are much better than under the action of Δ using the same conditions [169] (Eq. 87):

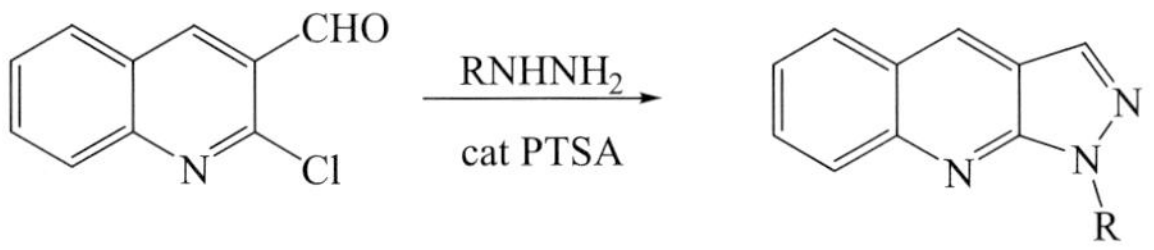

(87)

R = H   1.5 min   127–130 °C   MW = 97%   Δ = 30%

The noticeable enhancement of reaction rate because of an MW-specific effect is consistent with a reaction mechanism in which the kinetic rate-determining step is nucleophilic attack of an amino group on the chloroquinoline ring (Scheme 4.21).

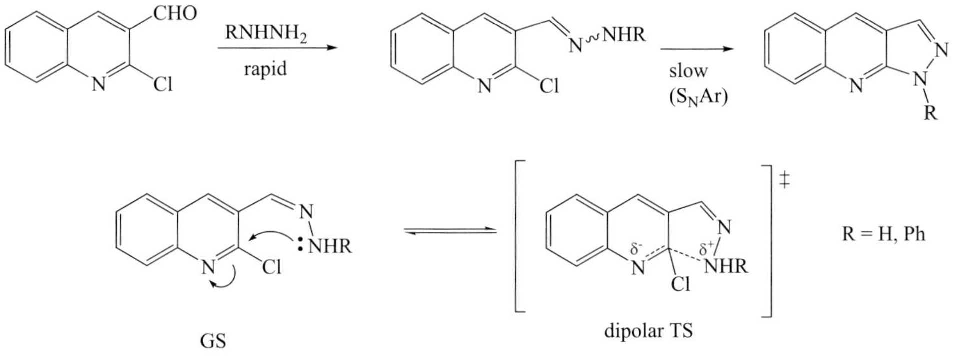

R = H, Ph

GS   dipolar TS

Scheme 4.21. Mechanism of internal S_NAr.

4.8.3.4 Intramolecular Michael Additions

ortho-Aminochalcones have been cyclized to tetrahydroquinolones in dry media using K10 clay as the support under the action of MW irradiation. The role of microwaves in accelerating the process was apparent because a relatively extended reaction time was required under the action of conventional heating to obtain similar yields [170] (Eq. (88), Scheme 4.22).

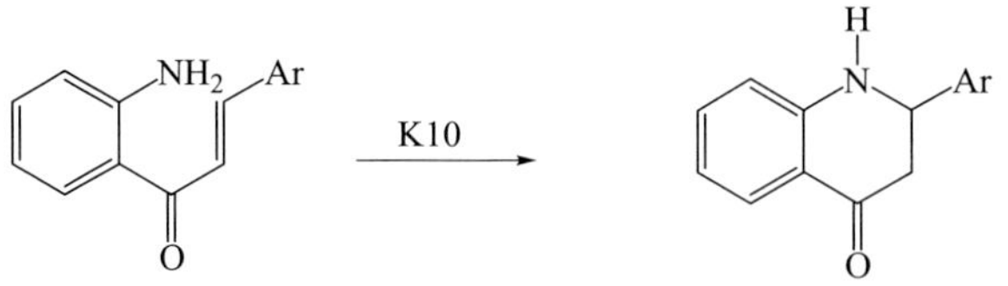

$$\text{(88)}$$

MW   1.5-2 min   110 °C   70-80%

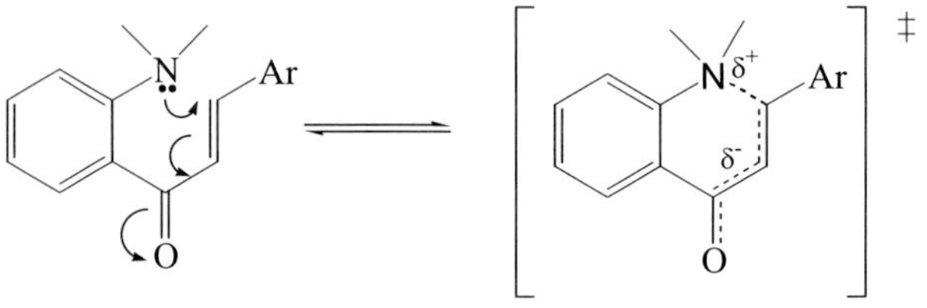

Scheme 4.22. Mechanism of intramolecular Michael addition of an amino group.

A similar study of *ortho*-hydroxychalcones in dry media using silica gel has been reported [171]. Conventional thermal cyclization, under the same conditions as for microwave irradiation, required a much longer reaction time (Eq. (89), Table 4.30, Scheme 4.23).

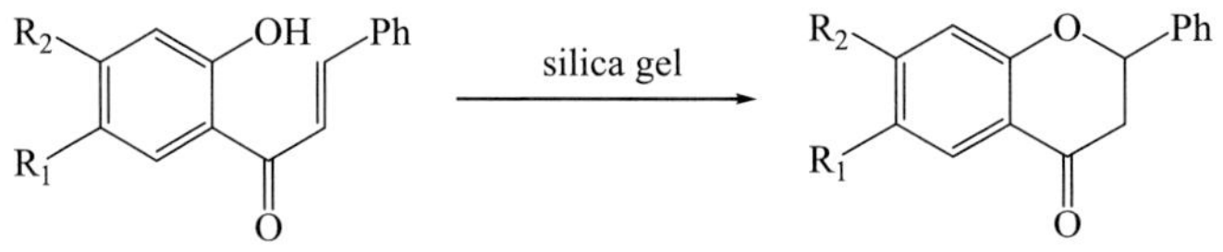

$$\text{(89)}$$

Tab. 4.30. Intramolecular Michael addition of o-hydroxychalcones for 20 min at 140 °C.

R_1	R_2	Mode	Yield (%)
H	H	MW	82
		Δ	44
CH_3O	H	MW	61
		Δ	22

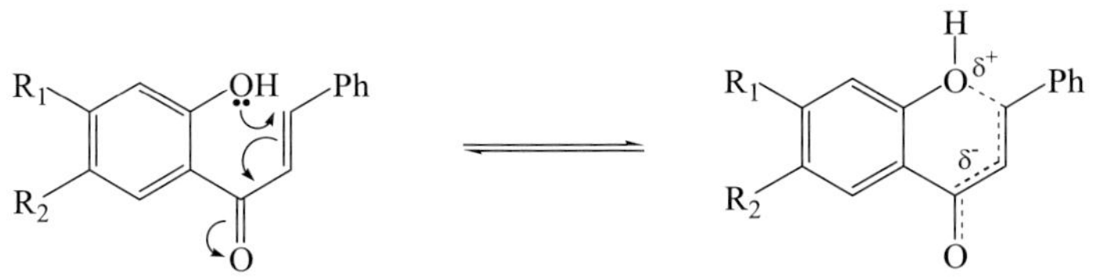

Scheme 4.23. Mechanism of intramolecular Michael addition of an hydroxyl group.

4.8.3.5 Deprotection of Allyl Esters

Carboxylic acids are regenerated from their corresponding substituted allyl esters on montmorillonite K10 by use of microwave irradiation under solvent-free conditions to afford enhanced yields and reduced reaction times compared with thermal conditions [172] (Eq. 90):

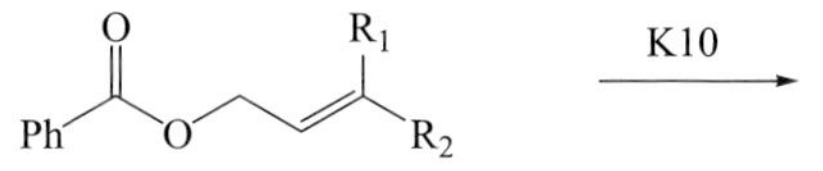

$$\text{(90)}$$

R_1 = H	R_2 = Ph	110 °C	MW : 20 min 96%	Δ : 6 h	96%
R_1 = R_2 = H		110 °C	MW : 20 min 98%	Δ : 5 h	94%

This effect can also get again be rationalized by a mechanism with intervention of a polar TS (Scheme 4.24).

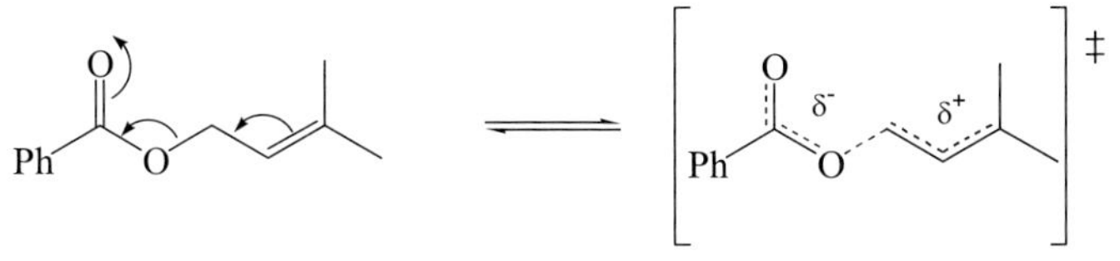

Scheme 4.24. Mechanism of deprotection of allyl esters.

4.8.3.6 Synthesis of Pyrido-fused Ring Systems

Significant improvements in the synthesis of pyrido-fused heterocycles were achieved when the reaction was performed under solvent-free conditions and applying the *tert*-amino effect as the key ring-closure method [183]. It was shown, as depicted in Eq. (91), that MW can improve the reaction yield under similar sets of conditions compared with Δ heating. When extending the reaction time, as often quoted, nearly quantitative yields were obtained, irrespective of the mode of activation. The specific MW effect observed here is consistent with the proposed polar mechanism for the cyclization [184].

[Scheme for Eq. (91): solvent-free, 180°C]

5 min   MW = 58%   Δ = traces

22 min   MW = 94%   Δ = 94%

4.8.3.7 Ring-closing Olefin Metathesis

An efficient method for ring-closing metathesis [186] was established between solvent-free conditions under the action of MW in a monomode reactor (Eq. 92):

[Scheme for Eq. (92): Grubbs catalyst, no solvent]

Grubbs catalyst = $\text{Cl}_2(\text{PCy}_3)_2\text{Ru}=\text{CHPh}$

X = Cl   5 min. 50°C   MW : 67%   Δ : 48%

X = CF$_3$ 3 min. 50°C   MW : 90%   Δ : 74%

Nonthermal MW-specific effects were revealed and shown to be more important with the less reactive systems. They can be related to the polarity increases as the reaction progresses involving polar metallacyclobutanes with polar carbon–metal bonds.

4.8.4
Some Illustrative Examples of Effects on Selectivity

Very few results are available on selectivity effects, and in which kinetic details of reactions have been described, because of a lack of strict comparisons between MW and Δ activation. Reports of the effect of MW on selectivity have been reviewed by De la Hoz et al. [89, 91] and a discussed in Chapter 5 of this book.

4.8.4.1 Benzylation of 2-Pyridone

Regiospecific *N*- or *C*-benzylations of 2-pyridone have been conducted under solvent-free conditions in the absence of base. The regioselectivity was controlled by the activation method (MW or Δ) or, when using microwaves, by the emitted power level or the leaving group of the benzyl halides [90] (Eq. 93):

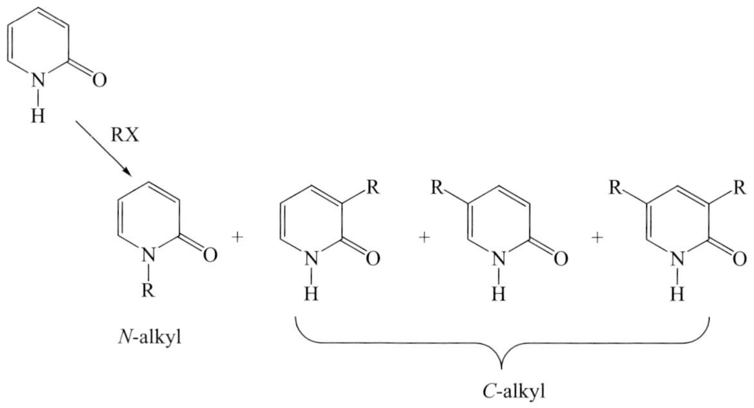

N-alkyl   *C*-alkyl   (93)

PhCH$_2$Br   5 min   196 °C   MW : > 98% *C*-alkyl   Δ : > 98% *N*-alkyl

PhCH$_2$I   5 min   180 °C   MW : > 98% *C*-alkyl   Δ : traces

To explain these results it can be assumed that the TS leading to *C*-alkylation is more polar than that responsible for *N*-alkylation. This assumption, however, assumes the occurrence of kinetic control (which is not known for certain in this example).

4.8.4.2 Addition of Vinylpyrazoles to Imine Systems

On microwave irradiation, vinylpyrazoles react with *N*-trichloroethylidene carbamate, undergoing addition to the imine system through the conjugated vinyl group [173] (Eq. 94):

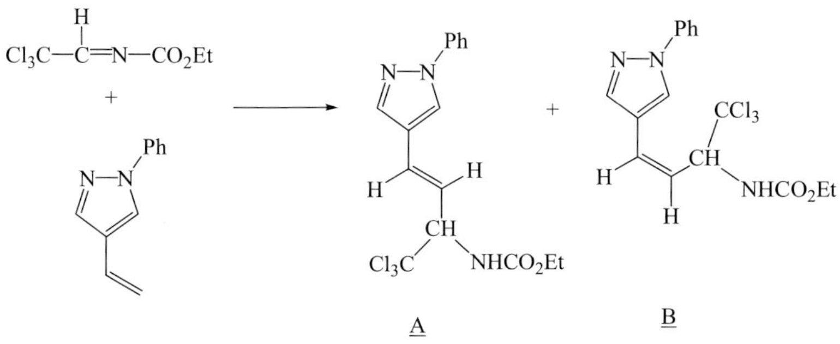

A   B   (94)

MW   15 min   125 °C   A : 70%   B : 15%

Use of microwave irradiation as an energy source is crucial to performing the reaction and avoiding decomposition or dimerization of the starting pyrazole, which are observed in the absence of MW irradiation.

A similar conclusion was drawn during an examination of the Diels–Alder reaction of 6-demethoxy-β-dihydrothebaine with methyl vinyl ketone under the action of microwave irradiation [174]. When performed under conventional heating conditions, extensive polymerization of the dienophile was observed whereas reaction is much cleaner under MW activation (Eq. 95):

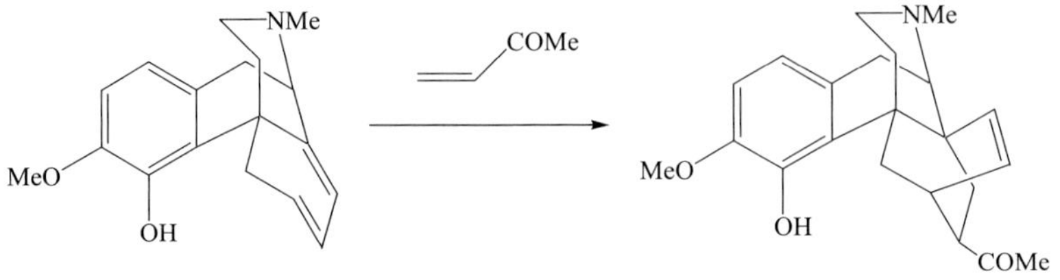

(95)

4.8.4.3 Stereo Control of β-Lactam Formation [175, 176]

Formation of β-lactams by reaction of an acid chloride, a Schiff base, and a tertiary amine (Eq. 96) seems to involve many pathways, some of which are very fast at higher temperatures. When conducted in open vessels in unmodified microwave ovens, high-level irradiation led to preferential formation of the *trans*-β-lactams (55%) whereas at low power, the *cis* isomer was obtained as the only product (84%). The failure of the *cis* isomers to isomerize to the *trans* compounds is an example of induced selectivity. It can therefore be stated that the transition state leading to the *trans* isomer seems more polar than that giving the cis compound.

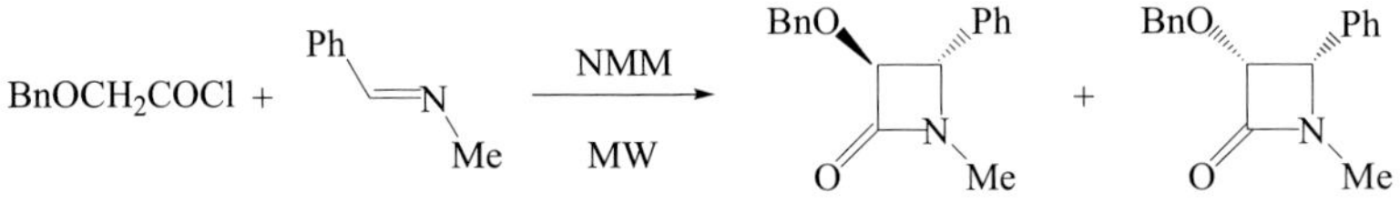

(96)

NMM= *N*-methylmorpholine

4.8.4.4 Cycloaddition to C_{70} Fullerene

It has been claimed that the regioselectivity of cycloaddition of *N*-methylazomethine ylide to C_{70} is slightly affected by using microwave irradiation as the source. By choosing an appropriate solvent (*o*-dichlorobenzene, ODCB) and emitted microwave power, the ratio 1a:1b is modified from 50:50 to 45:55 [48, 94] (Eq. (97), Table 4.31).

Tab. 4.31. Yields of monoadducts and isomer distribution for cycloaddition (Eq. 97) in ODCB at 180 °C.

Mode		Time (min)	Yield (%)	% 1a	% 1b	% 1c
MW	120 W	30	39	50	50	–
	180 W	30	37	45	55	–
	300 W	15	37	47	53	–
Δ		120	32	46	46	8

$$C_{70} \xrightarrow[\text{MW or }\Delta]{CH_3NHCH_2CO_2H/HCHO} \text{products} \quad (97)$$

1a (1-2 isomer)   **1b** (5-6)   **1c** (7-21)

Theoretical calculations predict an asynchronous mechanism and suggest that modification of the regiochemical outcome is related to the energies and hardness of the TS involved. Perhaps, in other words, the more polar TS are favored under the action of MW.

4.8.4.5 Selective Alkylation of 1,2,4-Triazole

Bu using MW irradiation under solvent-free conditions it was possible to achieve regiospecific benzylation in position 1 of 1,2,4-triazole whereas only the 1,4-dialkylated product was obtained in poor yield under the action of conventional heating [177] (Eq. 98):

$$\text{triazole-H} + PhCH_2Br \longrightarrow N_1 + N_{1,4} \quad (98)$$

molar ratio 1:1

5 min   160°C   MW = 70%   100% N_1
              Δ = 14%   100% $N_{1,4}$

Tab. 4.32. Phenacylations of 1,2,4-triazole.

Ar	X	Time (min)	Temp. (°C)	Mode	Global yield (%)	N_1	N_4	$N_{1,4}$
phenyl	Cl	20	140	MW	90	100	–	–
				Δ	98	33	29	38
3,5-dichlorophenyl	Cl	25	140	MW	95	100	–	–
				Δ	98	38	27	35
4-bromophenyl	Br	25	170	MW	90	100	–	–
				Δ	98	38	28	34

This observation may be explained by the increased efficiency of the first benzylation ($S_N 2$ reaction between two neutral reagents proceeding via a dipolar TS) under microwave conditions.

For extension to the preparation of azolic fungicides, phenacylation was next examined. Under the action of microwave irradiation, exclusive reaction in position 1 (or equivalent 2) occurred whereas mixtures of N_1, N_4, and $N_{1,4}$ products were obtained by use of Δ under the same conditions [106, 178] (Eq. (99), Table 4.32).

$$\text{triazole-H} + \text{ArCOCH}_2\text{X} \xrightarrow{\text{no solvent}} N_1 + N_4 + N_{1,4} \quad (99)$$

Because kinetic control was certain, this clear microwave effect could possibly be because of a difference in polarity of the transition state, with a more polar TS apparently being formed on π attack by the nitrogen atom in position 2. Theoretical calculations are currently in progress to try to confirm this assumption.

This example is decisive proof of the analogy between polar solvent and solvent-free MW-assisted procedures. This behavior undeniably confirms the effect of polar dipole–dipole type interactions with the electric field (Section 4.4).

4.8.4.6 Rearrangement of Ammonium Ylides

Ammonium ylides can isomerize to (1,2) rearrangement products (Stevens rearrangement) or to (2,3) shift products (Sommelet–Hauser sigmatropic rearrangement) when allyl or benzyl are located on the nitrogen atom. A strong microwave effect has been noticed (Eq. 100) [179]:

$$\text{Ph-N}^+\text{-CN} \xrightarrow{2 \text{ min}} \text{A (1,2)} + \text{B (2,3)} \quad (100)$$

| MW | yield 69% | A : B = 70 : 30 | Δ | yield 56% | A : B = 10 : 90 |

Under similar profiles of temperature increase it has been shown that selectivity favoring (1,2) Stevens rearrangement occurs under the action of microwaves. A tentative explanation can be to consider that, under the action of MW, the more polar mechanism (1,2 ionic shift) is favored compared with the less polar mechanism (2,3 radical shift). This result may be indicative of competition between ionic and radical pathways.

4.9
Concerning the Absence of Microwave Effects

The absence of microwave effects can have at least three different origins:

1. *A transition state of polarity similar to that of the ground state.* This is the situation for synchronous mechanisms in some pericyclic reactions when performed in nonpolar solvents [13, 14] or in neat liquids (Eqs 101 and 102):

$$\text{anthracene} + \text{CH}_2=\text{C(CO}_2\text{Et)}_2 \xrightarrow[\text{MW or } \Delta]{\text{toluene}} \text{adduct} \quad (101)$$

$$\text{citronellal} \xrightarrow{\text{neat liquid}} \text{isopulegol} \quad (102)$$

2. *A very early transition state along the reaction coordinates* (cf Hammond postulate) which cannot enable development of polarity between the GS and TS (reactant-

like). This will occur when the reactions are rather easy and do not require classically harsh conditions (exothermic reactions). This origin has been shown for phthalimide syntheses by reaction of phthalic anhydride with amino compounds [180] and for chalcone syntheses by reaction of aromatic aldehydes and acetophenones [181], etc ... Slight differences can appear when performing the reaction in the presence of a solvent, because of a superheating effect if no stirring is used.

3. *Temperature which is too high*, which may produce good yields in short reaction times under the action of conventional heating. To find evidence of microwave effects, it is necessary to reduce the temperature under conventional conditions to start from a rather poor yield (<30–40%) to discover possible microwave activation. Example of this have been revealed in studies in which a microwave effect occurred at relatively low temperatures but was masked at higher temperatures and in which elevated yields were obtained from conventionally heated reactions [137, 138, 149, 182].

4.10
Conclusions: Suitable Conditions for Observation of Specific MW Effects

We have proposed in this review a rationalization of microwave effects in organic synthesis based on the effect of the medium and on mechanistic considerations. The most suitable conditions in which there is any chance of checking specific not purely thermal effects are:

1. *Polar mechanisms* via polar transition states. If the polarity of a system is enhanced from the ground state to the transition state, acceleration can result from an increase in material–wave interactions during the course of the reaction. The most frequently encountered examples are unimolecular or bimolecular reactions between neutral molecules (because dipoles are developed in the TS) and anionic reactions of tight ion-pairs, i.e. involving charge-localized anions (leading to ionic dissociation in the TS).
2. *Slower reacting systems*, necessitating a high energy of activation and consequently with late product-like transition states along the reaction coordinates, in agreement with the Hammond postulate.
3. *Nonpolar solvents or, even better, solvent-free conditions* (green chemistry) to avoid that microwave effects will be masked or limited by solvent effects.
4. *Appropriate temperatures*. Δ and MW yields (generally more than 90%) can be almost the same and any MW effects masked if temperatures which are too high (or reaction times which are too long) are used. Use of less harsh conditions can enable the discovery of specific MW effects, which can lead to smaller amounts of by-products, because product decomposition occurs at higher temperatures. A characteristic example, from the many available, is given in Eq. (103) [149], in which MW specific effects were apparent at 80 °C but were masked at 110 °C.

4.10 Conclusions: Suitable Conditions for Observation of Specific MW Effects

[Scheme (103): dimerization via Br-(CH$_2$)$_8$-Br, KOH, NBu$_4$Br]

Solvent	Time	Temp	MW	Δ
Toluene	10 min.	110 °C	95%	91%
Cyclohexane	10 min.	80 °C	96%	39%

A striking example in which all these conditions are fulfilled may be the Leuckart reductive amination of carbonyl compounds. This reaction is well known but, unfortunately, using classical procedures, is only possible under very harsh conditions (240 °C, sealed containers, and extended reaction times) and gives modest yields (≤ 30%) [187]. These difficulties are a good challenge to check the effectiveness of microwave irradiation, because the mechanism develops a dipolar transition state (Eq. (104), Table 4.33) [188], and this should also favor involvement of a microwave effect because the TS certainly occurs very late along the reaction coordinates when one considers the harsh usual thermal conditions:

$$R_1R_2C{=}O + HCONH_2 + HCO_2H \longrightarrow R_1R_2C{=}NHCHO$$

[Mechanism showing carbonyl attack by formamide nitrogen through a dipolar transition state with δ^- on O and δ^+ on N, leading to products] (104)

Tab. 4.33. Microwave-mediated Leuckart reductive amination of carbonyl compounds within 30 min.

R$_1$	R$_2$	T (°C)	Activation	Yield (%)
Ph	Ph	202	MW	>98
		202	Δ	2
p-CH$_3$OC$_6$H$_4$	p-CH$_3$OC$_6$H$_4$	193	MW	95
		193	Δ	3
Ph	CH$_2$Ph	210	MW	95
		210	Δ	12

This is a distinctive example of a pronounced microwave effect for a reaction occurring with a very late dipolar transition state. It could happen as a miraculous MW effect which can be, finally, easily justified.

Many types of carefully controlled experiment must be performed, however, to evaluate the reality and limitations of these approaches, if valid comparisons are to be made.

References

1 R. N. Gedye, F. Smith, K. Westaway, H. Ali, L. Baldisera, L. Laberge and J. Rousell, *Tetrahedron Lett.*, **1986**, *27*, 279–282.
2 R. J. Giguere, T. L. Bray, S. M. Duncan and G. Majetich, *Tetrahedron Lett.*, **1986**, *27*, 4945–4958.
3 S. Caddick, *Tetrahedron*, **1995**, *51*, 10403–10432.
4 A. Loupy, A. Petit, J. Hamelin, F. Texier-Boullet, P. Jacquault and D. Mathé, *Synthesis*, **1998**, 1213–1234.
5 L. Perreux and A. Loupy, *Tetrahedron*, **2001**, *57*, 9199–9223.
6 P. Lidström, J. Tierney, B. Wathey and J. Westman, *Tetrahedron*, **2001**, *57*, 9225–9283.
7 C. O. Kappe, *Angew. Chem. Int. Ed.*, **2004**, *43*, 6250–6284.
8 *Microwaves in Organic Synthesis* (Ed: A. Loupy), Wiley–VCH, Weinheim, **2002**.
9 B. L. Hayes, *Microwave Synthesis: Chemistry at the speed of Light*, CEM Publishing, Matthews NC, **2002**.
10 C. O. Kappe and A. Stadler, *Microwaves in Organic and Medicinal Chemistry*, Vol. 25 Methods and Principles in Medicinal Chemistry, Wiley–VCH, Weinheim, **2005**.
11 *Microwave-Assisted Organic Synthesis* (Eds.: P. Lidström, J. P. Tierney), Blackwell, Oxford, **2005**.
12 R. S. Varma, *Green Chem.*, **1999**, *1*, 43–55.
13 K. D. Raner, C. R. Strauss, F. Vyskoc and J. Mokbel, *J. Org. Chem.*, **1993**, *58*, 950–953.
14 R. Laurent, A. Laporterie, J. Dubac, S. Lefeuvre and M. Audhuy, *J. Org. Chem.*, **1992**, *57*, 7099–7102.
15 K. C. Westaway and R. N. Gedye, *J. Microwave Power Electromagnetic Energy*, **1995**, *30*, 219–229.

16 N. Kuhnert, *Angew. Chem.* **2002**, *114*, 1943–1946.
17 C. R. Strauss, *Angew. Chem. Int. Ed.* **2002**, *41*, 3589–3590.
18 G. Bram, A. Loupy, M. Majdoub, E. Guttierez and E. Ruiz-Hitzky, *Tetrahedron*, **1990**, *46*, 5167–5176.
19 (a) A. Petit, A. Loupy, P. Maillard and M. Momenteau, *Synthetic Commun.*, **1992**, *22*, 1137–1142; (b) A. Loupy and Le Ngoc Thach, *Synth. Commun.*, **1993**, *23*, 2571–2577.
20 K. G. Kabza, B. R. Chapados, J. E. Gestwicki and J. L. McGrath, *J. Org. Chem.*, **2000**, *65*, 1210–1214.
21 P. Goncalo, C. Roussel, J. M. Melot and J. Vebrel, *J. Chem. Soc. Perkin Trans. 2*, **1993**, 2111–2115.
22 D. M. P. Mingos and D. R. Baghurst, *Microwave-Enhanced Chemistry* (Eds.: H. M. Kingston, S. J. Haswell), Am. Chem. Soc., Washington, **1997**, 4–7.
23 C. Gabriel, S. Gabriel, E. H. Grant, B. S. Halstead and D. M. P. Mingos, *Chem. Soc. Rev.*, **1998**, *27*, 213–223.
24 S. Kalhori, B. Minaev, S. Stone-Elander and N. Elander, *J. Phys. Chem. A*, **2002**, *106*, 8516–8524.
25 P. Zenatti, M. Forgeat, C. Marchand and P. Rabette, *Technologie et Stratégie, Bulletin de l'OTS*, **1992**, *55*, 4–8.
26 C. Reichardt, *Solvents and solvent effects in Organic Chemistry*, 3rd Edition, Wiley–VCH, Weinheim, **2003**, 5–56.
27 J. Thuéry, *Microwaves: industrial, scientific and medicinal applications*, Artech House, **1992**.
28 A. de la Hoz, A. Diaz-Ortiz and A. Moreno, *Chem. Soc. Rev.*, **2005**, *34*, 164–178.
29 G. Bond, R. B. Moyes, J. D. Pollington and D. A. Whan, *Chem. and Ind.*, **1991**, 686–687.
30 D. R. Baghurst and D. M. P. Mingos, *J. Chem. Soc., Chem. Commun.*, **1992**, 674–677.
31 R. Saillard, M. Poux, J. Berlan and M. Audhuy-Peaudecerf, *Tetrahedron*, **1995**, *51*, 4033–4042.
32 S. Rault, A. G. Gillard, M. P. Foloppe and M. Robba, *Tetrahedron Lett.*, **1995**, *36*, 6673–6674.
33 F. Chemat, E. Esweld, *Chem. Eng. Technol.*, **2001**, *7*, 735–744.
34 A. Stadler and C. O. Kappe, *Tetrahedron*, **2001**, *57*, 3915–3920.
35 A. Stadler and C. O. Kappe, *Eur. J. Org. Chem.*, **2001**, 919–925.
36 J. Berlan, P. Giboreau, S. Lefeuvre and C. Marchand, *Tetrahedron Lett.*, **1991**, *32*, 2363–2366.
37 S. C. Jullien, M. Delmotte, A. Loupy and H. Jullien, *Symposium Microwave and High Frequency*, Nice (France), October 1991, vol. II, 397–400.
38 J. G. P. Binner, N. Hassine and T. E. Cross, *J. Mat. Sci.*, **1995**, *30*, 5389–5393.
39 A. Miklavc, *Chem. Phys. Chem.*, **2001**, *2*, 552–555.
40 D. Stuerga and P. Gaillard, *J. Microwave Power Electromagnetic Energy*, **1996**, *31*, 87–100.
41 A. Laporterie, J. Marquié and J. Dubac, *Microwaves in Organic Synthesis* (Ed.: A. Loupy), Wiley–VCH, Weinheim, **2002**, chapter 7, 219–252.
42 (a) D. Bogdal, M. Lukasiewicz, J. Pielichowski, A. Miciak and S. Bednarz, *Tetrahedron*, **2003**, *59*, 649–653; (b) M. Lukasiewicz, D. Bogdal and J. Pielichowski, *Adv. Synth. Catal.*, **2003**, *345*, 1–4.
43 H. Will, P. Scholz and B. Ondruschka, *Chem. Ing. Tech.*, **2002**, *74*, 1057–1067.
44 D. A. Lewis, J. D. Summers, T. C. Ward and J. E. McGrath, *J. Polym. Sci.*, **1992**, *30A*, 1647–1653.
45 C. Shibata, T. Hashima and K. Ohuchi, *Jpn J. Appl. Phys.*, **1995**, *35*, 316–319.
46 S. Marque, F. Maurel and A. Loupy, *International Symposium on Microwave Science*, Takamatsu (Japan), July 27th **2004**.
47 A. Loupy, *International Conference of Microwave Chemistry*, Prague (Czech Republic), 6–11 Sept, **1998**, plenary lecture PL2.
48 F. Langa, P. de la Cruz, A. de la Hoz, E. Espildora, F. P. Cossio and B. Lecea, *J. Org. Chem.*, **2000**, *65*, 2499–2507.

49 S. D. POLLINGTON, G. BOND, R. B. MOYES, D. A. WHAN, J. P. CANDLIN and J. R. JENNINGS, *J. Org. Chem.*, **1991**, *56*, 1313–1314.
50 K. D. RANER and C. R. STRAUSS, *J. Org. Chem.*, **1992**, *57*, 6231–6234.
51 A. STADLER and C. O. KAPPE, *J. Chem. Soc., Perkin Trans. 2*, **2000**, 1363–1368.
52 M. R. LINDER and J. PODLECH, *Org. Lett.*, **2001**, *3*, 1849–1851.
53 G. V. SALMORIA, E. L. DALL'OGLIO and C. ZUCCO, *Synthetic Commun.*, **1997**, *27*, 4335–4340.
54 F. BENTISS, M. LAGRENÉE and D. BARBRY, *Tetrahedron Lett.*, **2000**, *41*, 1539–1542.
55 K. BOUGRIN, A. K. BENNANI, S. FKIH-TETOUANI and M. SOUFIAOUI, *Tetrahedron Lett.*, **1994**, *35*, 8373–8376.
56 A. CHANDRA SHEKER REDDY, P. SHANTAN RAO and R. V. VENKATARATNAM, *Tetrahedron Lett.*, **1996**, *37*, 2845–2848.
57 D. BOGDAL, *Monogr. Politech. Krakow*, **1999**, *248*, 1–134; *Chem. Abstr.*, **2000**, *133*, 58369.
58 A. LOUPY, *Modern Solvents in Organic Synthesis, Topics Curr. Chem.*, **1999**, *206*, 155–207.
59 K. TANAKA and F. TODA, *Chem. Rev.*, **2000**, *100*, 1025–1074.
60 G. W. V. CAVE, C. L. RASTON and J. L. SCOTT, *J. Chem. Soc., Chem. Commun.*, **2001**, 2159–2169.
61 K. TANAKA, *Solvent-Free Organic Synthesis*, Wiley–VCH, Weinheim, **2003**.
62 S. DESHAYES, M. LIAGRE, A. LOUPY, J. L. LUCHE and A. PETIT, *Tetrahedron*, **1999**, *55*, 10851–10870.
63 G. BRAM, A. LOUPY and D. VILLEMIN, *Solid Supports and Catalysts in Organic Synthesis* (Ed.: K. SMITH) Ellis Horwood, PTR Prentice Hall, *Organic Chemistry Series*, Chichester (U.K.), **1992**, chapter 12, 302–326.
64 R. N. GEDYE, F. E. SMITH and K. C. WESTAWAY, *Can. J. Chem.*, **1998**, *66*, 17–26.
65 E. D. HUGHES and C. K. INGOLD, *J. Chem. Soc.*, **1935**, *23*, 244.
66 R. ABRAMOVICH, *Org. Prep. Proc. Int.*, **1991**, *23*, 683–711.
67 G. MAJETICH and K. WHELESS, *Microwave Enhanced Chemistry* (Ed.: H. M. KINGSTON and S. J. HASWELL), Amer. Chem. Soc., Washington, **1997**, 455–505.
68 J. R. CARRILLO, P. DE LA CRUZ, A. DIAZ-ORTIZ, M. J. GOMEZ-ESCALONILLA, A. DE LA HOZ, F. LANGA, A. MORENO and P. PRIETO, *Recent Res. Dev. Org. and Bioorg. Chem*, **1997**, *1*, 68–84 and references cited therein.
69 J. R. CARRILLO, A. DIAZ-ORTIZ, F. P. COSSIO, M. J. GOMEZ-ESCALONILLA, A. DE LA HOZ, A. MORENO and P. PRIETO, *Tetrahedron*, **2000**, *56*, 1569–1577.
70 S. RIGOLET, P. GONCALO, J. M. MELOT and J. VEBREL, *J. Chem. Res. (S)*, **1998**, 686–687 and *(M)*, **1998**, 2813–2833.
71 F. LOUERAT, K. BOUGRIN, A. LOUPY, A. M. OCHOA DE RETANA, J. PAGALDAY and F. PALACIOS, *Heterocycles*, **1998**, *48*, 161–169.
72 A. LOUPY, A. PETIT and D. BONNET-DELPON, *J. Fluorine Chem.*, **1995**, *75*, 215–217.
73 A. LOUPY, F. MAUREL and A. SABATIÉ-GOGOVA, *Tetrahedron*, **2004**, *60*, 1683–1691.
74 C. M. WYNN and P. S. KLEIN, *J. Org. Chem.*, **1966**, *31*, 4251–4252.
75 J. A. REED, C. L. SCHILLING JR., R. F. TARWIN, T. A. RETTIG and J. K. STILLE, *J. Org. Chem.*, **1969**, *34*, 2188–2191.
76 G. BASHIARDES, C. CANO and B. MAUZÉ, *Synlett*, **2005**, 1425–1428.
77 J. AZIZIAN, A. R. KARIMI, A. A. MOHAMMADI and M. R. MOHAMMADDIZADEH, *Synthesis*, **2004**, 2263–2265.
78 M. A. CHIACCHIO, L. BORRELLO, G. DI PASQUALE, A. POLLICINO, F. A. BOTTINO and A. RESCIFINA, *Tetrahedron*, **2005**, *61*, 7986–7993.
79 G. BASHIARDES, I. SAFIR, A. S. MOHAMED, F. BARBOT and J. LADURANTY, *Org. Lett.*, **2003**, 4915–4918.
80 M. SZWARC, *Ions Pairs in Organic Reactions*, Wiley–Interscience, New York, **1972/1974**, vol. 1 and 2.
81 G. S. HAMMOND, *J. Amer. Chem. Soc.*, **1935**, *77*, 334–338.
82 D. A. LEWIS, *Mat. Res. Soc. Symp. Proc.*, **1992**, *269*, 21–31.
83 T. THIEMANN, M. WATANABE, Y.

Tanaka and S. Matatka, *New J. Chem.*, **2004**, *28*, 578–584.
84. A. Loupy, P. Pigeon, M. Ramdani and P. Jacquault, *Synthetic Commun.*, **1994**, *24*, 159–165.
85. I. Forfar, P. Cabildo, R. M. Claramunt and J. Elguero, *Chem. Lett.*, **1994**, 2079–2080.
86. B. Herradon, A. Morcuende and S. Valverde, *Synlett*, **1995**, 455–458.
87. J. R. Carrillo-Munoz, D. Bouvet, E. Guibé-Jampel, A. Loupy and A. Petit, *J. Org. Chem.*, **1996**, *61*, 7746–7749.
88. J. A. Vega, S. Cueto, A. Ramos, J. J. Vaquero, J. L. Garcia-Navio, J. Alvarez-Builla and J. Ezquerra, *Tetrahedron Lett.*, **1996**, *37*, 6413–6416.
89. F. Langa, P. de la Cruz, A. de la Hoz, A. Diaz-Ortiz and E. Diez-Barra, *Contemporary Org. Synth.*, **1997**, 373–386.
90. I. Almena, A. Diaz-Ortiz, E. Diez-Barra, A. de la Hoz and A. Loupy, *Chem Lett.*, **1996**, 333–334.
91. A. De la Hoz, A. Diaz-Ortiz and A. Moreno, *Curr. Org. Chem.*, **2004**, *8*, 903–918.
92. J. J. Chen and S. V. Deshpande, *Tetrahedron Lett.*, **2003**, *44*, 8873–8876.
93. N. E. Leadbeater, S. J. Pillsbury and V. A. Williams, *Tetrahedron*, **2005**, *61*, 3565–3585.
94. F. Langa, P. de la Cruz, A. De la Hoz, E. Espildora, F. P. Cossio and B. Lecea, *International Conference on Microwave Chemistry*, Antibes (France), 4–7 Sept **2000**, 33–36.
95. A. De la Hoz, A. Diaz-Ortiz, A. Moreno and F. Langa, *Eur. J. Org. Chem.*, **2000**, 3659–3673.
96. M. Chaouchi, A. Loupy, S. Marque and A. Petit, *Eur. J. Org. Chem.*, **2002**, 1278–1283.
97. R. S. Varma, R. Dahiya and S. Kumar, *Tetrahedron Lett.*, **1997**, *38*, 2039–2042.
98. M. S. Khajavi, P. Asfani and R. M. Kourosh, *Iran J. Chem., Chem. Eng.*, **1998**, *17*, 29–32.
99. L. Domon, C. Le Coeur, A. Grelard and T. Besson, *Tetrahedron Lett.*, **2001**, *38*, 2039–2042.
100. F. R. Alexandre, A. Berecibar and T. Besson, *Tetrahedron Lett.*, **2002**, *43*, 3911–3913.
101. A. Vass, J. Dudas and R. S. Varma, *Tetrahedron Lett.*, **1999**, *34*, 4951–4954.
102. A. Loupy, S. Chatti, S. Delamare, D. Y. Lee, J. H. Chung and C. H. Jun, *J. Chem. Perkin Trans. 1*, **2002**, 1280–1285.
103. G. Vo-Thanh, H. Lahrache, A. Loupy, I. J. Kim, D. H. Chang and C. H. Jun, *Tetrahedron*, **2004**, *60*, 5539–5543.
104. S. Gadhwal, M. Baruah and J. S. Sandhu, *Synlett*, **1999**, 1573–1574.
105. (a) M. Jeselnik, R. S. Varma, S. Polanc and M. Kocevar, *Chem. Commun.*, **2001**, 1716–1717; (b) M. Jeselnik, R. S. Varma, S. Polanc and M. Kocevar, *Green Chem.*, **2002**, *4*, 35–38.
106. A. Loupy, L. Perreux, M. Liagre, K. Burle and M. Moneuse, *Pure Appl. Chem.*, **2001**, *73*, 161–166.
107. L. Perreux, A. Loupy and F. Volatron, *Tetrahedron*, **2002**, *58*, 2155–2162.
108. M. Iannelli, V. Alupei and H. Ritter, *Tetrahedron*, **2005**, *61*, 1509–1515.
109. F. Massicot, R. Plantier-Royon, C. Portella, D. Saleur, A. V. L. Sudha, *Synthesis*, **2001**, 2441–2444.
110. A. L. Marrero-Terrero and A. Loupy, *Synlett*, **1996**, 245–246.
111. F. Garcia-Tellado, A. Loupy, A. Petit and A. L. Marrero-Terrero, *Eur. J. Org. Chem.*, **2003**, 4387–4391.
112. A. R. Katritzky, C. Cai, K. Suzuki, S. K. Singh, *J. Org. Chem.*, **2004**, *69*, 811–814.
113. S. Paul, P. Nanda, R. Gupta and A. Loupy, *Tetrahedron Lett.*, **2004**, *45*, 425–427.
114. R. Martinez-Palou, G. de Paz, J. Marin-Cruz and G. Zepeda, *Synlett*, **2003**, 1847–1849.
115. R. Perez, E. R. Perez, M. Suarez, L. Gonzalez, A. Loupy, M. L. Jimeno and C. Ochoa, *Org. Prep. Proc. Int.*, **1997**, *29*, 671–677.
116. M. Gupta, S. Paul and R. Gupta, *Synthetic Commun.*, **2001**, *31*, 53–59.
117. G. C. Muccioli, J. H. Poupaert, J. Wouters, B. Norberg, W. Pooppitz,

G. K. E. Scriba and D. M. Lambert, Tetrahedron, **2003**, *59*, 1301–1307.

118 A. Burczyk, A. Loupy, D. Bogdal and A. Petit, Tetrahedron, **2005**, *61*, 179–188.

119 E. Veverkova, M. Meciarova, S. Toma and J. Balko, Monatsh Chem., **2003**, *134*, 1215–1219.

120 E. R. Perez, N. C. Carnevalli, P. J. Cordeiro, U. P. Rodrigues-Filho and D. W. Franco, Org. Prep. Proc. Int., **2001**, *33*, 395–400.

121 A. R. Hajipour, J. E. Mallakpour and A. Afrousheh, Tetrahedron, **1999**, *55*, 2311–2316.

122 S. Frère, V. Thiéry and T. Besson, Tetrahedron Lett., **2001**, *42*, 2791–2794.

123 M. T. Genta, C. Villa, E. Mariani, A. Loupy, A. Petit, R. Rizzetto, A. Mascarotti, F. Morini and M. Ferro, Int. J. Pharmaceutics, **2002**, *231*, 11–20.

124 M. T. Genta, C. Villa, E. Mariani, M. Longobardi and A. Loupy, Int. J. Cosmetic. Sc., **2002**, *24*, 257–262.

125 R. M. Martin-Aranda, M. A. Vicente-Rodriguez, J. M. Lopez-Pestana, A. J. Lopez-Peinado, A. Jerez, J. D. Lopez-Gonzalez and M. A. Banares-Munoz, J. Mol. Cat., **1997**, *124 A*, 115–121.

126 N. N. Romanova, A. G. Gravis, G. M. Shaidullina, I. F. Leshcheva and Y. G. Bundel, Mendeleev Commun., **1997**, 235–236.

127 N. N. Romanova, A. G. Gravis, I. F. Leshcheva and Y. G. Bundel, Mendeleev Commun., **1998**, 147–148.

128 A. Dandia, M. Sati and A. Loupy, Green Chem., **2002**, *4*, 599–602.

129 A. Dandia, M. Sati, K. Arya and A. Loupy, Heterocycles, **2003**, *60*, 563–569.

130 A. Dandia, M. Sati, K. Arya, R. Sharma and A. Loupy, Chem. Pharm. Bull., **2003**, *51*, 1137–1141.

131 H. Rodriguez, M. Suarez, R. Perez, A. Petit and A. Loupy, Tetrahedron Lett., **2003**, *44*, 3709–3712.

132 O. Correc, K. Guillou, J. Hamelin, L. Paquin, F. Texier-Boullet and L. Toupet, Tetrahedron Lett., **2004**, *45*, 391–395.

132 I. Almena, E. Diez-Barra, A. de la Hoz, J. Ruiz and A. Sanchez-Migallon, J. Heterocycl. Chem., **1998**, *35*, 1263–1268.

133 H. Glas and W. R. Thiel, Tetrahedron Lett., **1998**, *39*, 5509–5510.

134 B. Olofsson and P. Somfai, J. Org. Chem., **2002**, *37*, 8574–8583.

135 J. A. Vega, J. J. Vaquero, J. Alvarez-Builla, J. Ezquerra and C. Hamdouchi, Tetrahedron, **1999**, *55*, 2317–2326.

136 J. J. Kiddle, Tetrahedron Lett., **2000**, *41*, 1339–1341.

137 J. Cvengros, S. Toma, S. Marque and A. Loupy, Can. J. Chem., **2004**, *82*, 1365–1371.

138 M. T. Radoiu, J. Kurfurstova and M. Hajek, J. Mol. Cat., **2000**, *160*, 383–392.

139 B. Pégot, G. Vo-Thanh, F. Maurel and A. Loupy, unpublished results.

140 G. Vo-Thanh, B. Pégot and A. Loupy, Eur. J. Org. Chem., **2004**, 1112–1116.

141 M. Kidwai, P. Sapra and B. Dave, Synthetic Commun., **2000**, *34*, 4479–4488.

142 M. Lloung, A. Loupy, S. Marque and A. Petit, Heterocycles, **2004**, *63*, 297–308.

143 Z. B. Xu, Y. Lu and Z. R. Guo, Synlett, **2003**, *4*, 564–566.

144 E. Hollink, E. E. Simanek and D. E. Bergbreiter, Tetrahedron Lett., **2005**, *46*, 2005–2008.

145 W. P. Fang, Y. T. Cheng, Y. R. Cheng and Y. J. Cherng, Tetrahedron, **2005**, *61*, 3107–3113.

146 S. J. Song, S. J. Cho, D. K. Park, T. W. Kwon and S. A. Jenekhe, Tetrahedron Lett., **2003**, *44*, 255–257.

147 A. Oussaid, Le Ngoc Thach and A. Loupy, Tetrahedron Lett., **1997**, *38*, 2451–2454.

148 S. Chatti, M. Bortolussi and A. Loupy, Tetrahedron Lett., **2000**, *41*, 3367–3370.

149 S. Chatti, M. Bortolussi and A. Loupy, Tetrahedron, **2000**, *56*, 5877–5883.

150 A. P. Krapcho, Synthesis, **1982**, 805–822 and 893–914.

151 J. P. Barnier, A. Loupy, P. Pigeon, M. Ramdani and P. Jacquault,

J. Chem. Soc. Perkin Trans 1, **1993**, 397–398.

152 A. Diaz-Ortiz, P. Prieto, A. Loupy and D. Abenhaim, Tetrahedron Lett., **1996**, 37, 1695–1698.

153 A. Loupy, P. Pigeon and M. Ramdani, Tetrahedron, **1996**, 52, 6705–6712.

154 A. Oussaid, E. Pentek and A. Loupy, New J. Chem., **1997**, 21, 1339–1345.

155 L. Hernandez, E. Casanova, A. Loupy and A. Petit, Czech. J. Phys., **2003**, 53, 751–754.

156 L. Hernandez, E. Casanova, A. Loupy and A. Petit, J. Label. Compd. Radiopharm., **2003**, 46, 151–156.

157 L. Perreux and A. Loupy, Org. Prep. Proc. Int., **2003**, 35, 361–368.

158 C. Limousin, J. Cléophax, A. Loupy and A. Petit, Tetrahedron, **1998**, 54, 13567–13578.

159 R. S. Varma and K. P. Naicker, Tetrahedron Lett., **1999**, 40, 6177–6180.

160 A. Loupy, L. Perreux and A. Petit, Ceramic Trans., **2001**, 111, 163–172.

161 L. Perreux, A. Loupy and M. Delmotte, Tetrahedron, **2003**, 59, 2185–2189.

162 F. Z. Zradni, J. Hamelin and A. Derdour, Synthetic Commun., **2002**, 32, 3525–3531.

163 C. Laporte, J. Marquié, A. Laporterie, J. R. Desmurs and J. Dubac, C. R. Acad. Sci. Paris, t.2, **1999**, serie IIc, 455–465.

164 S. Paul, P. Nanda, R. Gupta and A. Loupy, Synthesis, **2003**, 18, 2877–2881.

165 J. Marquié, A. Laporterie and J. Dubac, J. Org. Chem., **2001**, 66, 421–425.

166 S. Paul, M. Gupta and R. Gupta, Synlett, **2000**, 1115–1118.

167 K. Bougrin, A. Loupy, A. Petit, B. Daou and M. Soufiaoui, Tetrahedron, **2000**, 57, 163–168.

168 S. Werner and P. S. Iyer, Synlett, **2005**, 9, 1405–1408.

169 S. Paul, M. Gupta, R. Gupta and A. Loupy, Tetrahedron Lett., **2001**, 42, 3827–3829.

170 R. S. Varma and R. K. Saini, Synlett, **1997**, 857–858.

171 T. Patonay, R. S. Varma, A. Vass, A. Levai and J. Dudas, Tetrahedron Lett., **2001**, 42, 1403–1406.

172 S. Gajare, N. S. Shaid, B. K. Bonde and V. H. Deshpande, J. Chem. Soc. Perkin Trans 1, **2000**, 639–640.

173 J. R. Carrillo, A. Diaz-Ortiz, A. de la Hoz, M. J. Gomez-Escalonilla, A. Moreno and P. Prieto, Tetrahedron, **1999**, 55, 9623–9630.

174 J. T. M. Linders, J. P. Kokje, M. Overhand, T. S. Lie and L. Maat, Rec. Trav. Chim. Pays Bas, **1988**, 107, 449–454.

175 A. K. Bose, B. K. Banik and M. S. Manhas, Tetrahedron Lett., **1995**, 36, 213–216.

176 M. S. Manhas, B. K. Banik, A. Mathur, J. E. Vincent and A. K. Bose, Tetrahedron, **2000**, 56, 5587–5601.

177 D. Abenhaim, E. Diez-Barra, A. de la Hoz, A. Loupy and A. Sanchez-Migallon, Heterocycles, **1994**, 38, 793–802.

178 E. Perez, A. Loupy, M. Liagre, A. M. de Guzzi Plepis and P. J. Cordeiro, Tetrahedron, **2003**, 59, 865–870.

179 S. Torchy, G. Cordonnier and D. Barbry, International Electronic Conferences on Synthetic Organic Chemistry, 5th, 6th **2001** and **2002**, 7th, 8th, **2003** and 2004, 1839–1843.

180 T. Vidal, A. Petit, A. Loupy and R. N. Gedye, Tetrahedron, **2000**, 56, 5473–5478.

181 E. Le Gall, F. Texier-Boullet and J. Hamelin, Synthetic Commun., **1999**, 29, 3651–3657.

182 C. H. Jun, J. H. Chung, D. Y. Lee, A. Loupy and S. Chatti, Tetrahedron Lett., **2001**, 42, 4803–4805.

183 N. Kawal, B. Halasz-Dajka, G. Vo Thanh, W. Dehaen, J. Van der Eycken, P. Màtyus, A. Loupy and E. Van der Eycken, Tetrahedron, **2005**, 61, 9052–9057.

184 W. Verboom, M. R. J. Hamzink, D. N. Reinhoudt and R. Visser, Tetrahedron Lett., **1984**, 25, 4309–4312.

185 V. Polshettiwar and M. P. Kaushik, Indian J. Chem., **2005**, 44B, 773–777.

186 G. Vo-Thanh and A. Loupy, Tetrahedron Lett., **2003**, 44, 9091–9094.

187 A. W. Ingersoll, J. H. Brown, C. K. Kim, W. D. Beauchamp and G. Jennings, *J. Amer. Chem. Soc.*, **1936**, *58*, 1808–1811.
188 A. Loupy, D. Monteux, A. Petit, J. M. Aizpurua, E. Dominguez and C. Palomo, *Tetrahedron Lett.*, **1996**, *37*, 8177–8180.
189 J. J. Filippi, X. Fernandez, L. Lizzani-Cuvelier and A. M. Loiseau, *Tetrahedron Lett.*, **2003**, *44*, 6647–6650.
190 L. Valette, S. Poulain, X. Fernandez and L. Lizzani-Cuvelier, *J. Sulfur Chem.*, **2005**, *26*, 155–161.
191 M. Jesberger, T. P. Davis and L. Barner, *Synthesis*, **2003**, 1929–1958.
192 Y. Ju and R. S. Varma, *Org. Lett.*, **2005**, *7*, 2409–2411.

5
Selectivity Under the Action of Microwave Irradiation

Antonio de la Hoz, Angel Díaz-Ortiz, and Andrés Moreno

5.1
Introduction

Microwave radiation is an alternative to conventional heating as a method of introducing energy into reactions. Microwave heating exploits the ability of some compounds (liquids or solids) to transform electromagnetic energy into heat. The use of microwaves as a mode of heating *in situ* has many attractions in chemistry because, in contrast to conventional heating, its magnitude depends on the dielectric properties of the molecules. As a guide, liquid compounds with high dielectric constants tend to absorb microwave radiation whereas less polar substances and highly ordered crystalline materials are poor absorbers. In this way absorption of the radiation and heating may be performed selectively. The use of microwave irradiation has led to the introduction of new concepts in chemistry because the absorption and transmission of the energy is completely different from that in the conventional mode of heating. The shape and size of the sample can also have an effect and these factors can affect the scale-up of some reactions.

Transfer of energy with microwaves is not by conduction or convection but by dielectric loss. The propensity of a sample to undergo microwave heating depends on its dielectric properties, the dielectric loss factor (ε'') and the dielectric constant (ε'). The dielectric constant represents the ability of a substance to absorb microwaves while the dielectric loss factor represents the ability of a substance to transform this energy into heat. A high dissipation factor ($\tan \delta = \varepsilon''/\varepsilon'$) results in a high susceptibility to microwave energy. Dielectric factors relevant to microwave heating have been reviewed [1].

The wavelength of radiation in the microwave region enables entire bulk quantities of a given material to be heated simultaneously without any major temperature gradient and, moreover, heating depends on the dielectric properties of the material and the specific heat capacity, the emissivity, the geometry, the volume (or mass), and the strength of the applied field. This type of heating mechanism can only be achieved by using electromagnetic waves in the microwave and RF region; the penetration depth of other forms of electromagnetic radiation (e.g. infrared) is too small and thermal conductivity is the limiting factor.

Microwaves in Organic Synthesis, Second edition. Edited by A. Loupy
Copyright © 2006 WILEY-VCH Verlag GmbH & Co. KGaA, Weinheim
ISBN: 3-527-31452-0

Instrumentation can also have a significant effect on heating pattern and power densities and, consequently, on the absorption of the energy. A comprehensive survey of microwave instrumentation can be found elsewhere [1, 2; see also Chapters 1 and 2].

Microwave irradiation has been successfully applied in chemistry since 1975 and many examples in organic synthesis have been described [3, 4]. Several reviews have been published on the application of this technique to solvent-free reactions [5], cycloaddition reactions [6], synthesis of radioisotopes [7], fullerene chemistry [8] and advanced materials [9], polymers [10], heterocyclic chemistry [11], carbohydrates [12], homogeneous [13] and heterogeneous catalysis [14], medicinal and combinatorial chemistry [15], and green chemistry [16]. All these applications are described elsewhere in this book.

Control of selectivity (chemo, regio, stereo, and enantioselectivity) is among the most important objective in organic synthesis. The efficient use of reaction conditions (temperature, time, solvent, etc.), kinetic or thermodynamic control, protecting or activating groups (for example chiral auxiliaries), and catalysts (including chiral catalysts) have all been used to obtain the desired isomer.

The objective of this review is not to compile examples of the synthetic applications of microwave irradiation in which only better yields and reduced reaction times have been achieved, but to summarize examples in which microwave radiation has resulted in chemo, regio, or stereoselectivity that differ from those obtained by use of conventional heating. Possible explanations for this behavior will be given throughout the text. Some examples will be discussed separately in Chapter 4 (Section 4.8.4) as an extension of previous original considerations in terms of specific electrostatic effects and of the relative stabilization of the more polar transition states [17a].

5.2
Selective Heating

5.2.1
Solvents

The most notable characteristic of microwave-assisted reactions is the spectacular acceleration often obtained. This effect is particularly important in the synthesis of short-lived radioisotopes [7], combinatorial chemistry [15], and catalysis [13, 14]. The effect has been used to avoid the decomposition of products and reagents and, in this way, to dramatically improve yields and to perform reactions that do not occur when conventional heating is used [18].

Several authors have postulated the existence of a so-called "microwave effect" to explain results that cannot be explained solely by the effect of rapid heating. Hence, rate acceleration or changes in reactivity and selectivity could be explained in terms of a specific radiation effect and not merely by a thermal effect [17]. The existence of such a "microwave effect" is still a controversial issue [17, 19] and is

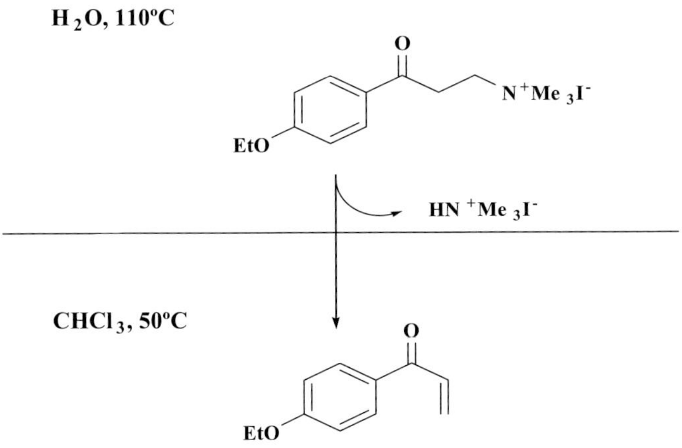

Fig. 5.1. Selective heating of water–chloroform mixtures.

beyond the scope of this chapter (see Chapter 4). It is clear, however, that microwave irradiation is a selective mode of heating. Characteristically, microwaves generate rapid and intense heating of polar substances whereas apolar substances do not absorb the radiation and are not heated. This selective effect was elegantly exploited by Strauss [20] in a Hofmann elimination reaction using a two-phase water–chloroform system (Fig. 5.1). The temperatures of the aqueous and organic phases were 110 and 50 °C, respectively, because of differences between the dielectric properties of the solvents. As a result of this difference decomposition of the final product is avoided. Comparable conditions would be difficult to obtain using traditional heating methods.

A similar effect was observed by Hallberg in the preparation of β,β-diarylated aldehydes by hydrolysis of enol ethers in a two-phase (toluene–$HCl_{(aq)}$) system [21]. Overheating of polar liquids is another effect that can be exploited to advantage. Mingos [22] detected this effect on applying microwaves to polar liquids and found that overheating to between 13–26 °C above the normal boiling point may occur (Fig. 5.2). This effect can be explained by the "inverted heat transfer" effect (from the irradiated medium toward the exterior), because boiling nuclei are formed at the surface of the liquid.

This effect could account for the enhancement in reaction rates observed in organic and organometallic chemistry, and modification of the selectivity. It is, nevertheless, limited to reactions conducted in closed vessels without any stirring or nucleation regulator, and disappears in the presence of boiling chips [23]. Chemat [23] studied the origin, effect, and application of superheating of organic liquids. The factors considered included the effect on the super-heating of adding boiling nuclei, and of microwave power, volume, pressure, and chemical composition. It was shown how the induced superheating phenomenon could be exploited to accelerate homogeneous reactions, because under microwave conditions as chemical processes occur at temperatures tens of degrees higher than when classical heating

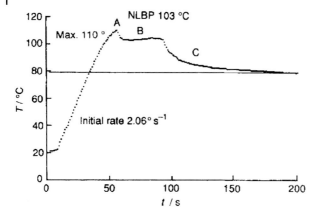

Fig. 5.2. Heating profile of ethanol under microwave irradiation.

is used (Fig. 5.3). This approach also enables reactions to be performed at atmospheric pressure with the same yields and reaction times as when conducted in closed reactors with classical heating.

Marken et al. [24] showed that the effect of 2.45 GHz microwave radiation on electroorganic processes in microwave-absorbing (organic) media can be dramatic but was predominantly thermal in nature. The process studied was the oxidation of 2 mM ferrocene in acetonitrile (0.1 M NBu$_4$PF$_6$) with a Pt electrode. Sigmoidal steady-state responses were detected and, as expected, an increase in microwave power led to an increase in the limiting current. This effect was qualitatively attributed to the formation of a "hot spot" in close proximity to the electrode surface. Focusing of microwaves at the end of the metal electrode is responsible for this highly localized thermal effect. When the microwave power was switched off, the voltammetric characteristics observed at room temperature returned immediately.

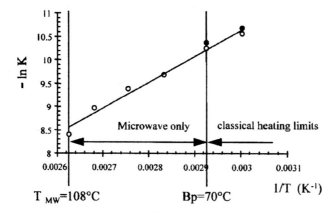

Fig. 5.3. Kinetics of esterification reaction: (○) microwave heating; (●) classical heating.

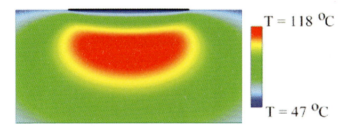

Fig. 5.4. Thermography of an electroorganic process in acetonitrile under microwave irradiation.

The temperature was seen to increase away from the electrode surface with a "hot spot" region at a distance of approximately 40 µm. The "hot spot" temperature (Fig. 5.4) was 118 °C and thus substantially higher than the boiling point of acetonitrile (81.6 °C) and also much higher than the temperature of the electrode (47 °C). Under these conditions the velocity of acetonitrile convection through the "hot spot" region is 0.1 cm s^{-1} and, therefore, the solvent typically passes through the high-temperature region in less than 100 ms.

5.2.2
Catalysts

The characteristics of microwave heating have been exploited efficiently in catalysis. In homogeneous catalysis [13] several advantages have been observed:

- the combination of metal catalysts under air and with water as the solvent;
- the use of milder and less toxic reagents at high temperatures; and
- the possibility of integrating efficient synthesis with non-chromatographic purifications.

In heterogeneous catalysis [14b] benefits arise from the possibility that heating can be concentrated on the catalyst's mass while the surroundings remain relatively cool, a situation that can enhance selectivity. Bogdal [25], for example, described the oxidation of alcohols using Magtrieve (DuPont's trademark for the oxidant based on tetravalent chromium dioxide CrO_2) (Scheme 5.1). Irradiation of Magtrieve led to rapid heating of the material up to 360 °C within 2 min. When toluene was introduced into the reaction vessel, the temperature of Magtrieve reached ca. 140 °C within 2 min and was more uniformly distributed (Fig. 5.5). This experi-

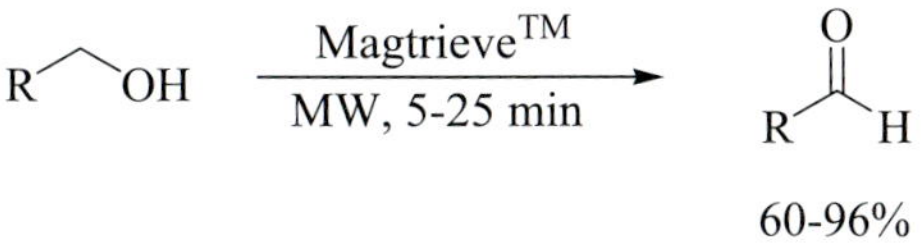

$$R\text{—OH} \xrightarrow[\text{MW, 5-25 min}]{\text{Magtrieve}^{\text{TM}}} R\text{—CHO}$$

60-96%

Scheme 5.1

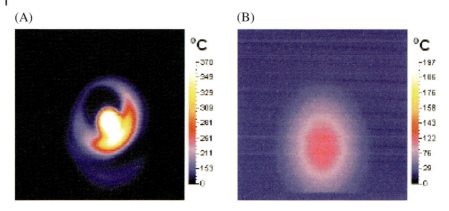

Fig. 5.5. Temperature profile for Magtrieve (2 min): (A) without solvent; (B) in toluene solution.

ment shows that the temperature of the catalyst can be higher than the temperature of the bulk solvent, which implies that such a process might be more energy-efficient than conventional processes.

The same effect in Zeolite–guest systems was demonstrated by Auerbach [26] by equilibrium molecular dynamics and non-equilibrium molecular dynamics after experimental work by Conner [27]. The energy distributions obtained in Zeolite and Zeolite-Na are shown in Fig. 5.6A. At equilibrium, all the atoms in the system are at the same temperature. When Na-Y Zeolite is exposed to microwave energy, however, the effective steady-state temperature of Na atoms is substantially higher than that of the rest of the framework; this is indicative of athermal energy distribution. The steady-state temperatures for binary methanol/benzene mixtures in

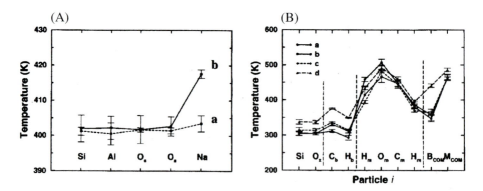

Fig. 5.6. (A) Energy distributions in NaY at (a) thermal equilibrium and (b) nonequilibrium, with the external field. (B) Steady-state energy distributions for binary mixtures in siliceous-Y (a) 1:1, (b) 2:2, (c) 4:4, and (d) 8:8 methanol–benzene per unit.

$$H_2S(g) \xrightarrow{\gamma\text{-}Al_2O_3 \text{ or } MoS_2/\gamma\text{-}Al_2O_3} H_2(g) + 1/2\, S_2(g)$$

Scheme 5.2

both siliceous Zeolites are shown in Fig. 5.6B. Statistically different temperatures for each component were found and $T_{methanol} \gg T_{benzene} > T_{zeolite}$. These results suggest that methanol dissipates energy to benzene, although this process is much too slow to lead to thermal equilibrium under steady-state conditions.

Several authors have detected or postulated the presence of "hot spots" in samples irradiated with microwaves. This effect arises as a consequence of the inhomogeneity of the applied field, meaning that the temperature in certain zones within the sample is much greater than the macroscopic temperature and is, therefore, not representative of the reaction conditions as a whole. This overheating effect has been demonstrated by Mingos in the decomposition of H_2S over γ-Al_2O_3 and MoS_2–γ-Al_2O_3 (Scheme 5.2) [28]. The conversion efficiency under microwave and conventional thermal conditions is compared in Fig. 5.7. The higher conversion under microwave irradiation was attributed to the presence of hot spots. The temperature in the hot spots was estimated to be approximately 100–200 °C above the bulk temperature on the evidence of calculations and several transformations that occurred. These transformations included the transition from γ- to α-alumina and the melting of MoS_2, which occurs at temperatures much higher than the measured bulk temperature. The size of the hot spots was estimated to be as large as 100 μm.

Hot spots may be created by the difference between the dielectric properties of

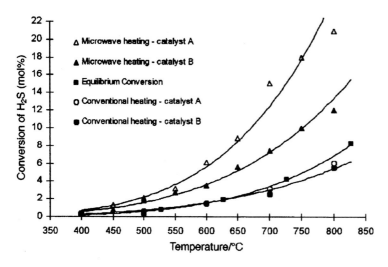

Fig. 5.7. Dependence of H_2S conversion on temperature with mechanically mixed catalyst A and impregnated catalyst B.

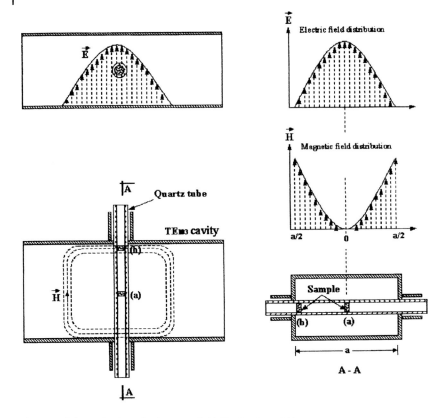

Fig. 5.8. Schematic view of microwave field distribution within the TE$_{103}$ single-mode microwave cavity.

materials, by the uneven distribution of electromagnetic field strength, or by volumetric dielectric heating under microwave conditions [29]. These effects have also been explained in terms of temperature gradients within a solid, meaning that the hot spots could not be directly measured [25].

Cheng [30] emphasized the importance of the magnetic field component in microwave heating. In his experiments a single-mode cavity with a cross-section of 86 mm by 43 mm was used; this works in TE$_{103}$ single mode (Fig. 5.8). The maximum electric field (E) is in the center of the cross section, where the magnetic field (H) is at its minimum. The maximum magnetic field is near the wall, where the electric field is at its minimum. In this way a sample can be placed within the cavity with a maximum electric or magnetic field as desired; as an example, the behavior of powder-compact samples of FeO, Fe$_2$O$_3$, and Fe$_3$O$_4$ is shown in Fig. 5.9.

Conductive samples, for example metal powders and carbide samples, can be heated much more efficiently in the magnetic field. In contrast, for pure ceramic samples, which are insulators with little conductivity, much higher heating rates were obtained in the pure electric field. It was concluded that it is not possible to

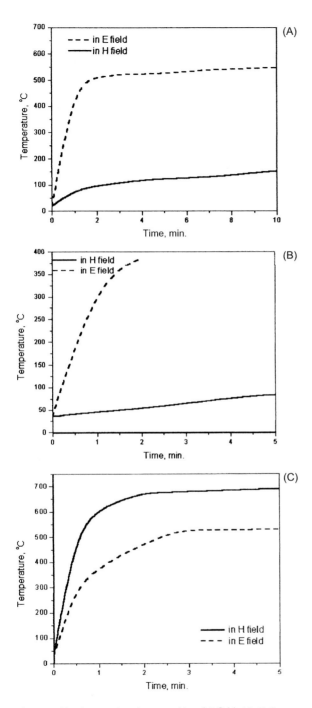

Fig. 5.9. Heating rate in microwave H and E field. (A) FeO powder-compact sample, (B) Fe_2O_3 powder-compact sample, (C) Fe_3O_4 powder-compact sample.

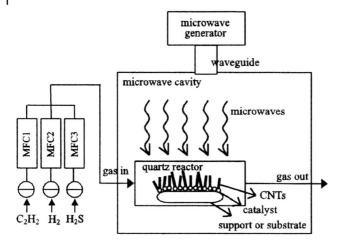

Fig. 5.10. Schematic view of the experiments.

ignore the effect of the magnetic component, especially for conductor and semiconductor materials. The contribution of the magnetic loss mechanisms could be hysteresis, eddy currents, magnetic resonance, and domain wall oscillations.

Selective heating has been used in the synthesis and purification of carbon nanotubes (CNT). For example, Lee [31a] described the preparation of carbon nanotubes in a microwave oven (Fig. 5.10). The support material (carbon black, a microwave absorber, or SiO_2, a microwave insulator) was placed in a quartz reactor with a flow of acetylene (C_2H_2) gas without a catalyst and was irradiated by microwaves. Identical experiments were repeated with carbon black loaded with Co, Ni, and Fe catalysts.

The catalyst particles only are heated to the temperature necessary for CNT synthesis, without raising the temperature of the substrate on which the catalyst lies. Co is believed to be the most effective catalyst, although Ni and Fe are almost as good. The decomposition of acetylene is initiated on the heated catalyst. When the reaction occurs, formation of carbon species continues, because of the continuous supply of energy from the exothermic reaction and from the synthesized carbon forms, which can absorb additional microwave energy. The nature of the support also has a strong effect on the reaction; SiO_2, a microwave insulator, does not absorb microwave energy and dissipates the energy absorbed by the catalyst particles, meaning that CNT cannot be synthesized with Co–SiO_2 (Fig. 5.11a). In contrast, carbon is a microwave absorber and can be heated up to 1000 °C. The catalyst particles are, therefore, easily heated with carbon supports (Fig. 5.11b).

The same strategy has been used to prepare flexible carbon nanotubes on a polymer support in a monomode reactor within 2 s [31b]. This technique enables the preparation to be performed without preheating of the catalyst film, under atmospheric operating conditions, with rapid synthesis and the ability to synthesize CNT on polymer substrates.

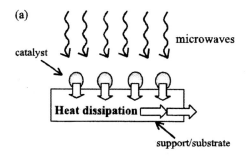
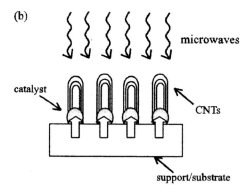

Fig. 5.11. Mechanism of CNT synthesis using microwave heating. (a) microwave-insulating substrate, (b) microwave-absorbing substrate.

Purification of carbon nanotubes has been performed in multimode [31] and monomode [32] systems. Prato purified HIPCO carbon nanotubes in a multimode oven. Common impurities are amorphous carbon and iron particles. The raw material was soaked in diethyl ether to obtain a more compact material. After evaporation of the solvent the flask was subjected to microwave heating (80 W) and an immediate weight loss occurred (5 s). This process was then repeated. Results from iron analysis were 16% (w/w) for the first run and 9% (w/w) after the second heating cycle. It is clear in this instance that the microwave heating is selectively directed to the iron particles. It can be seen from Fig. 5.12 that, although the quality of the tubes remained similar to the original material, most of the iron spots had disappeared.

Similar selective heating of the catalyst particles was observed by Harutyunyan [33] in the first step of a general purification procedure for single-wall CNT produced by the arc-discharge technique with an Ni–Y catalyst added to the electrodes. Microwave heating as an initial step for metal removal was compared with "selective oxidation", which refers to selective oxidation of undesirable minority phases (Figs. 5.13 and 5.14).

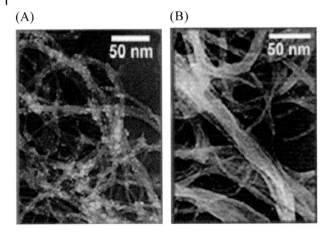

Fig. 5.12. TEM image of (A) raw HIPCO tubes, (B) after MW and acid washing treatment.

5.2.3
Reagents; Molecular Radiators

Larhed [34] described the molybdenum-catalyzed allylic alkylation of (E)-3-phenyl-2-propenyl acetate. The reaction occurs with good reproducibility, complete conversion, high yields, and excellent enantiomer excess (ee) in only a few minutes (Scheme 5.3). In the standard solvent (THF), and with an irradiation power of 250 W, a yield of 87% was obtained and high regioselectivity and ee (98%) were achieved. Regioselectivity was somewhat lower (17–19:1) than in the previously re-

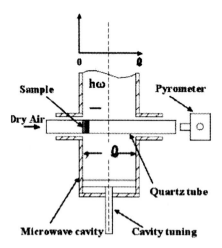

Fig. 5.13. A schematic diagram of the microwave system.

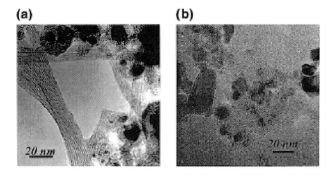

Fig. 5.14. TEM and HRTEM micrographs of (A) as-prepared SWNT material; (B) SWNT material after microwave heating at 500 °C.

ported two-step method (32–49:1). Alkylation also worked on polymer-supported reagents and, consequently, can be applied in combinatorial chemistry.

The high temperatures obtained (220 °C) are not only because of increased boiling points at elevated pressure but also because of a significant contribution from sustained overheating. The yields from the oil-bath experiments were lower than those for the corresponding microwave-heated reactions. In pure, microwave-transparent (non polar) solvents, the added substances, whether ionic or non-ionic, must therefore contribute to the overall temperature profile when the reaction is performed. It seems reasonable that when the substrates act as

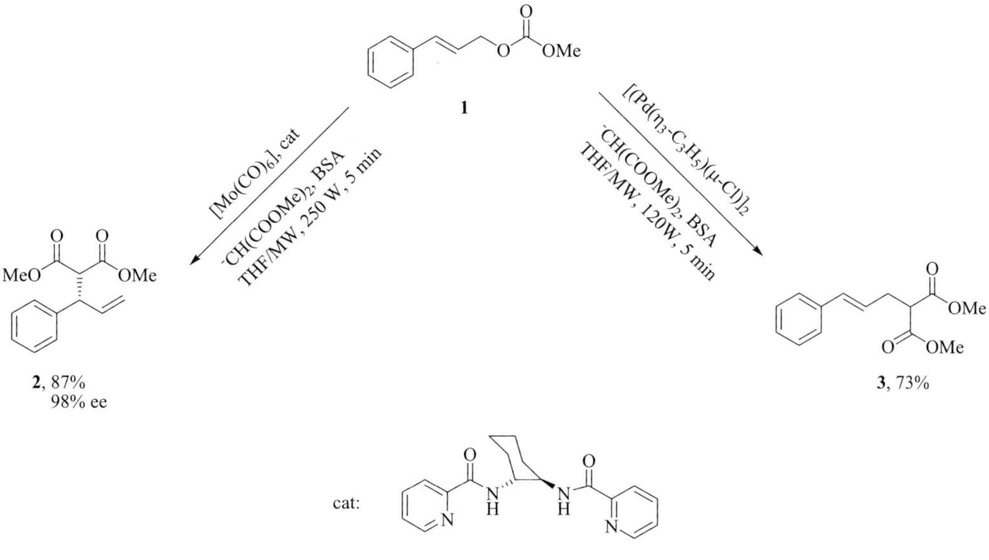

Scheme 5.3

"molecular radiators" in channeling energy from microwave radiation to bulk heat, their reactivity might be enhanced. The concept and advantages of "molecular radiators" have also been described by other authors [35].

5.2.4
Susceptors

A susceptor can be used when the reagents and solvents do not absorb microwave radiation. Susceptors are inert compounds that efficiently absorb microwave radiation and transfer the thermal energy to another compound that is a poor radiation absorber. This method has an interesting advantage – if the susceptor is a catalyst the energy can be focused on the surface of the susceptor where the reaction occurs. In this way, thermal decomposition of sensitive compounds can be avoided. In contrast, transmission of the energy occurs through conventional mechanisms.

In solvent-free or heterogeneous conditions graphite has been used as a susceptor (Chapter 9 in this book). For example, Garrigues [36] described the cyclization of (+)-citronellal to (−)-isopulegol and (+)-neoisopulegol on graphite. The stereoselectivity of the cyclization can be altered under the action of microwave irradiation.

Ionic liquids have been used both in solution and under homogeneous conditions. For example, Ley [37] described the preparation of thioamides from amides. Although the reaction under classical conditions occurs in excellent yield, the reaction time can be shortened by using MW irradiation (Scheme 5.4). The reaction was performed in toluene and, because this is not an optimum solvent for absorption and dissipation of MW energy, a small amount of polar solvent was added to the reaction mixture to ensure efficient heat distribution.

In this regard, Leadbeater (Ref. [38] and Chapter 7 in this book) studied the use of ionic liquids as aids for microwave heating of a nonpolar solvent (Table 5.1). It was shown that apolar solvents can, in a very short time, be heated to temperatures way above their boiling points in sealed vessels using a small quantity of an ionic

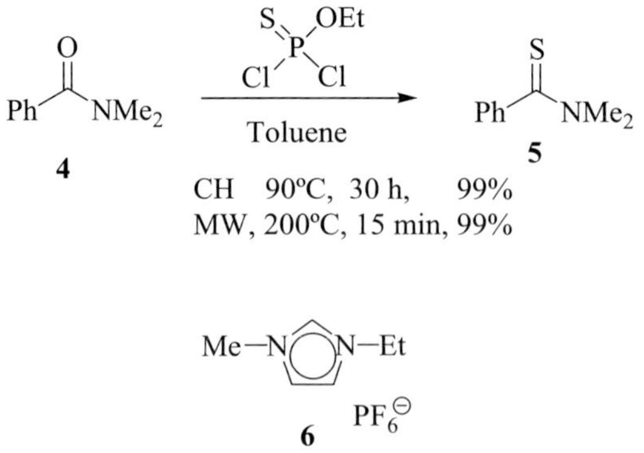

Scheme 5.4

Tab. 5.1. The microwave heating effects of adding a small quantity of ionic liquids, **8** and **9**, to hexane, toluene, THF, and dioxane.

[Structures of ionic liquids **8** (imidazolium with I⁻) and **9** (imidazolium with Br⁻)]

Solvent	IL[a]	T IL (°C)	t (s)	T (°C)[b]	b.p. (°C)
Hexane	8	217	10	46	69
	9	228	15		
Toluene	8	195	150	109	111
	9	130	150		
THF	8	268	70	112	66
	9	242	60		
Dioxane	8	264	90	76	101
	9	248	90		

[a] Ionic liquid 1 mmol mL^{-1} solvent
[b] Temperature reached without ionic liquid

liquid. It was found that 0.2 mmol ionic liquid was the optimum amount to heat 2 mL solvent. These solvent mixtures were tested with model reactions, for example as Diels–Alder cycloadditions, Michael additions, and alkylation reactions.

5.3
Modification of Chemoselectivity and Regioselectivity

5.3.1
Protection and Deprotection of Alcohols

Protection and deprotection of alcohols are important steps in organic synthesis and can be used to obtain selectivity – particularly in the chemistry of carbohydrates (for extended considerations, see Chapter 12), where the presence of several hydroxyl groups makes it difficult to obtain the desired selectivity. Microwave irradiation has been used for selective protection and deprotection of alcohols in several systems.

Herradón [39] conducted a study on the selective benzoylation of polyols by microwave irradiation and excellent results were obtained. In the example shown, the reaction conducted under the action of radiation and in the presence of dibutyltin oxide led exclusively to product **10**, which is benzoylated in the 2-position. This product is formed via a dibutyltin acetal, which catalyzes and controls the direction

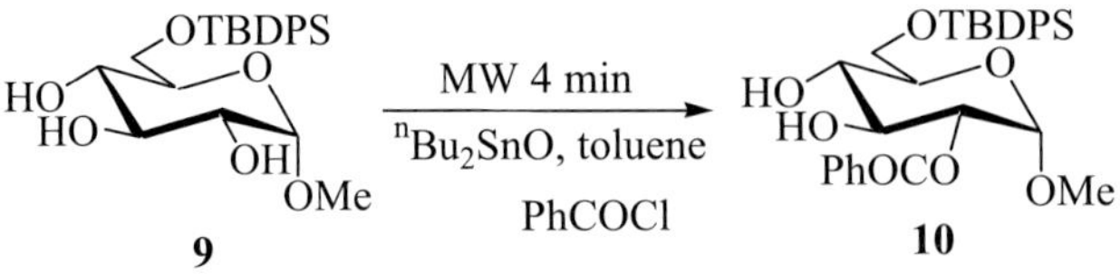

Scheme 5.5

of the reaction (Scheme 5.5). Under classical conditions, the non-catalyzed acylation is as fast as the tin-catalyzed reaction, so there is no advantage in using the tin species.

A similar strategy was used by Ballel [40] in the tin-mediated regioselective 3-O-alkylation of lactose and galabiose derivatives. The corresponding reactions under conventional heating did not yield any product.

Herradón also described the acylation of polyols [41] and amino alcohols (Scheme 5.6) [42] catalyzed by dibutyltin oxide. The chemoselectivity of the reaction depends on the power applied during irradiation.

More recently, Caddick et al. [43] studied the selective benzoylation of primary hydroxyl groups using dibutyltin oxide as a catalyst and triethylamine as the base. These experiments again show that the stoichiometry and the mode of heating have a significant effect on the selectivity of the reaction.

Ley [44] performed deprotection of pivalic esters on alumina under the action of microwave activation in the absence of solvent. It was found that selective deprotection of the 6-position could be achieved without migration of groups or isomerization of the anomeric center (Scheme 5.7). Yields of the order of 90% were obtained; such levels cannot be achieved by classical heating.

Lardy [45] studied the selective acetylation of sterols in the semi-solid state. It was found that the reaction is both chemoselective and regioselective under the action of microwave irradiation. For example, thermal heating of progesterone (16) led to enolization at the 3 and 20-carbonyls to give a 2:1 mixture of mono- and dienolacetates. Under the action of microwave irradiation, the 3-enol acetate was the major product (95% conversion) (Scheme 5.8) and the 3,5-diene-3-acetate (17) was also isolated with excellent selectivity.

Varma [46] studied the selective deprotection of diacetate 18 in the absence

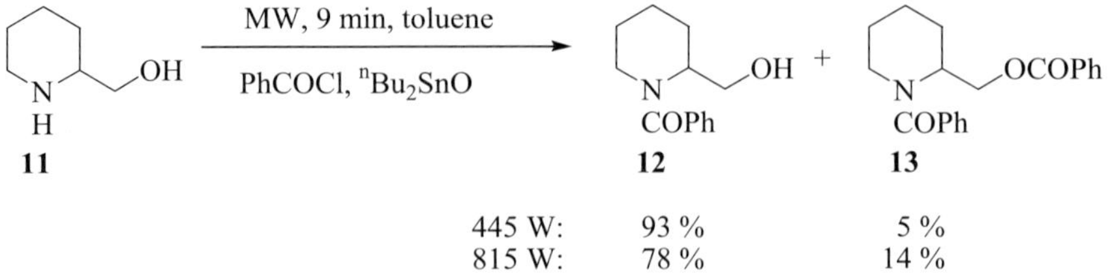

Scheme 5.6

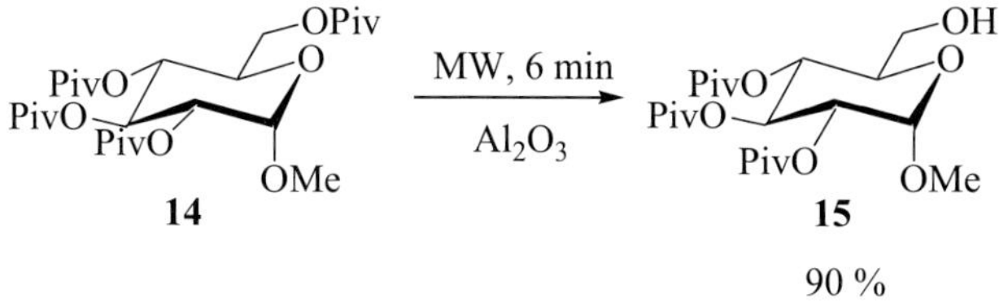

5.3 Modification of Chemoselectivity and Regioselectivity | 235

Scheme 5.7

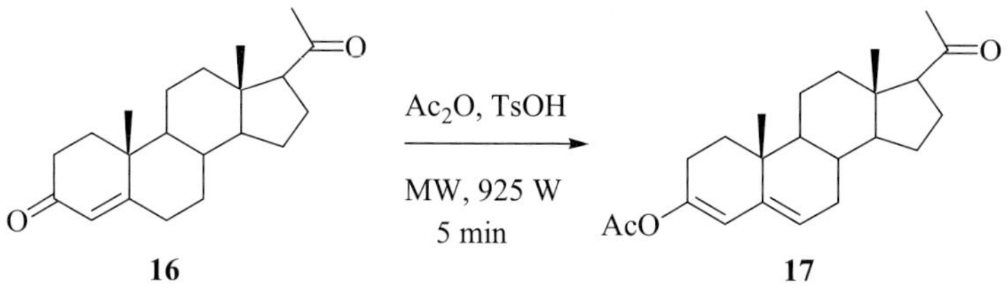

Scheme 5.8

of solvent. Mono or di-deprotection could be controlled by adjusting the reaction time, as outlined in Scheme 5.9. Mono-deprotection of this system is not possible by use of conventional heating.

Similarly, Das [47] used ammonium formate on silica gel to achieve selective and eco-friendly deprotection of aryl acetates under microwave irradiation.

Dealkylation of 2-ethoxyanisole has been described by Loupy [48]. Deethylation was the dominant reaction with potassium *tert*-butoxide without any change in selectivity under the action of microwave irradiation. Demethylation became the dominant reaction when ethylene glycol was added, however; under these conditions no reaction was observed when conventional heating was used.

Sarma [49] studied the monotetrahydropyranylation of symmetrical diols catalyzed by iodine. Monoprotected diols were obtained in 75% yield within 3 min on irradiation in a microwave oven. When the same reaction was performed under reflux, however, the conversion was quantitative within 30 min now the selectivity was very poor (mono:diether yield 43%:51%) (Fig. 5.15). Even at room temperature, the reaction proceeded slowly with no selectivity (mono:diether ratio 1:1) [49].

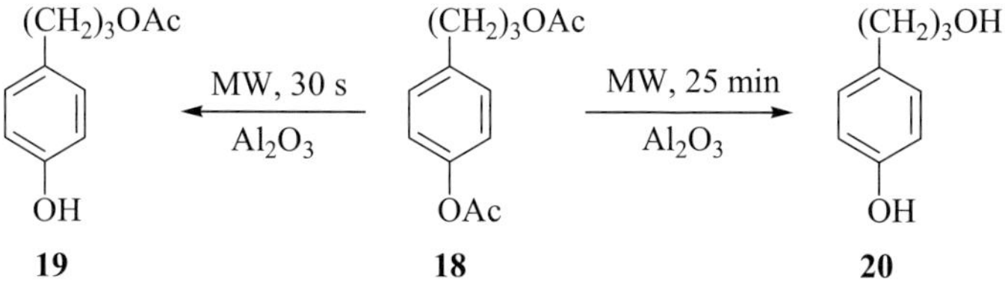

Scheme 5.9

HO−CH$_2$CH$_2$-OH $\longrightarrow$ HO−CH$_2$CH$_2$-OTHP + THPO−CH$_2$CH$_2$-OTHP
       21   **22**   **23**

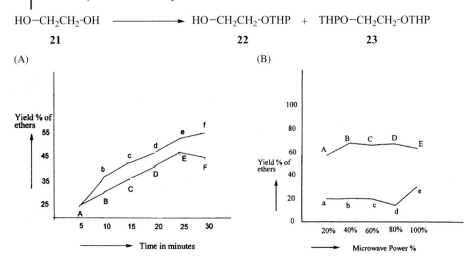

Fig. 5.15. Reaction of ethane-1,2-diols with DHP. Curve Adcdef is for diether **23** and ABCDEF is for monoether **22**.
(A) Dependence of yield on reaction time under reflux in THF.
(B) Dependence of yield on microwave power for 120 s.

5.3.2
Electrophilic Aromatic Substitution

Electrophilic aromatic substitution is the most traditional method for introducing functional groups into an aromatic ring. Lack of complete regioselectivity produces mixtures of compounds, however. Microwave irradiation has been used to modify and invert the selectivity of these reactions. In the sulfonation of naphthalene (**24**) under microwave irradiation, Stuerga [50] showed that the ratio of 1- and 2-naphthalenesulfonic acids (1- and 2-NSA, **25** and **26**) obtained is a function of the applied power (Fig. 5.16).

It is believed that when competing reactions occur, one reaction can be favored over another by controlling the rate of heating. If we consider a kinetic model for two competitive reactions based on Arrhenius' Law, reduction of the reaction time enables two special situations to be envisaged:

- The first situation is so-called *induced selectivity*. This is represented in Fig. 5.17(A), which shows the concentrations of product 1 (P1) and product 2 (P2) as a function of heating rate. Under the action of classical heating (slow heating) a mixture of P1 and P2 is obtained. By modifying the heating rate it is possible to obtain P1 as the principal product.
- The second situation is represented in Fig. 5.17(B) and is described as *inversion*. Under classical conditions P2 is the major product, whereas the use of microwaves or very rapid heating gives P1 as essentially the only product. This situa-

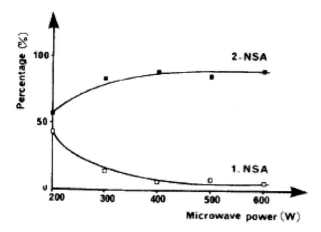

Fig. 5.16. Dependence of the amount (%) of 1- and 2-NSA **25** and **26** on microwave power.

tion is particularly interesting, because changes in the reactivity induced by heating rate can be envisaged.

This possibility led to new ways of accessing kinetically controlled products and demonstrates that microwave heating can have very important applications in the control of chemo, regio, and stereoselectivity.

In the same context, Claramunt [51] reported complete inversion of selectivity in the reaction between 1-bromoadamantane (**28**) and pyrazole (**27**). When the reaction is performed in an autoclave at 230 °C the exclusive product is 4-(1-adamantyl)pyrazole (**29**), but when the reaction is conducted in a microwave reactor the corresponding 3-substituted isomer **30** (44%) is obtained in high purity (Scheme 5.10).

Similarly, in the reaction of 2-pyridone (**31**) with benzyl halides **32** in the absence of solvent [52], the classical route gives N-alkylation whereas microwave irradiation

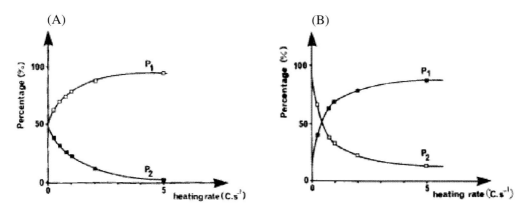

Fig. 5.17. Dependence of the amount (%) of P1 and P2 on heating rate.

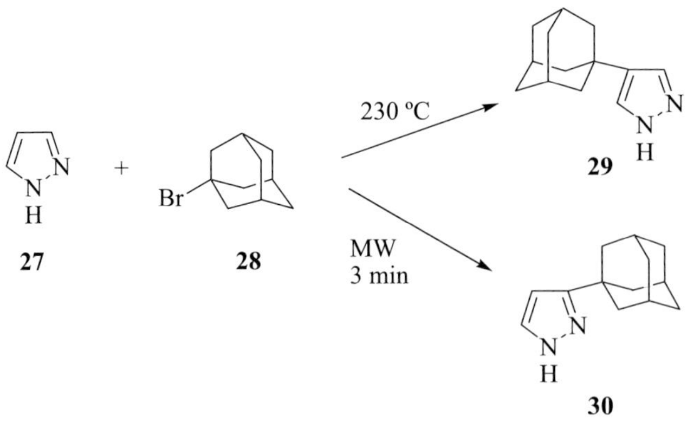

Scheme 5.10

leads to *C*-alkylation. Use of microwaves enables the selectivity to be controlled by altering the irradiation power and the nature of the leaving group, meaning that the aforementioned phenomenon of induced selectivity can be observed. At low power, alkylation occurs mainly at C-5 whereas higher irradiation power leads to a slight excess of the C-3 product. *C*-alkylation is, moreover, observed when benzyl bromide is used whereas use of benzyl chloride results in *N*-alkylation (Scheme 5.11, Table 5.2).

The Fries rearrangement has been studied by several groups, who have described the reaction on solid supports and under the action of microwave irradiation. Paul and Gupta [53] indicated that in the reaction catalyzed by zinc powder, the *ortho/para* selectivity can occasionally be modified by using microwave heating. For instance, *o*-hydroxyacetophenone (**38**) was obtained exclusively from phenylacetate (**37**) under the action of MW irradiation whereas *p*-hydroxyacetophenone (**39**) was obtained exclusively by conventional heating (Scheme 5.12). A similar result was obtained by Khadilkar [54] in the reaction catalyzed by aluminum trichloride, although in this reaction the change was less dramatic. Kad [55] reported that mod-

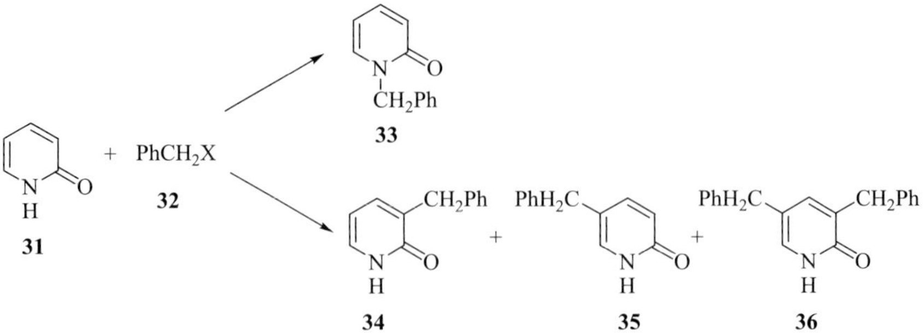

Scheme 5.11

Tab. 5.2. Benzylation of 2-pyridone (31). Product distribution.

X	Reaction conditions	T (min)	T (°C)	N/C (33/34 + 35 + 36)
Cl (32a)	MW, 780 W	5	198	100/0
	CH	5	176	100/0
Br (32b)	MW, 150 W	5	81	100/0
	MW, 450 W	2,5	180	0/100
	CH	5	196	100/0
I (32c)	MW, 150 W	5	146	0/100
	CH	5	180	traces/0

ification of the selectivity was not achieved in the same rearrangement catalyzed by Montmorillonite K-10.

Klán [56] described the photo-Fries rearrangement of phenyl acetate (37) under the action of microwaves and when irradiated with an electrodeless discharge lamp (EDL) (Chapter 19 of this book). The reaction provides two main products, 2- and 4-hydroxyacetophenone (Scheme 5.13; 38 and 39, respectively). Product distributions are given in Table 5.3.

ortho/para selectivity in microwave experiments was slightly different from that obtained by use of conventional heating (CH). These differences can be attributed to superheating effects in the microwave field for all solvents and were measured directly with a fiber-optic thermometer or estimated on the assumption that the temperature-dependence of the product ratio is linear.

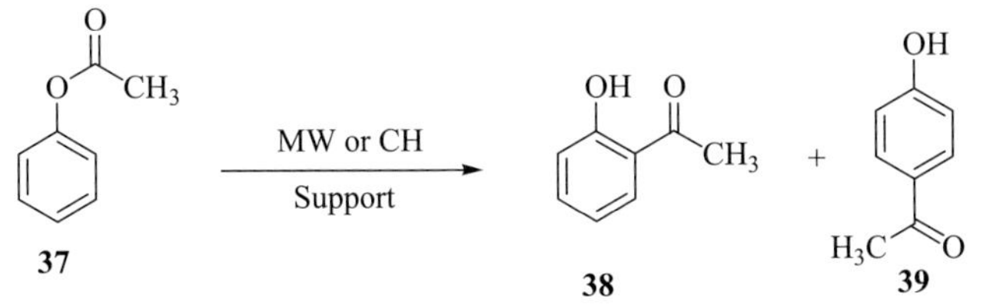

Support	ortho/para 38/39	Yield
Zn powder (MW, 3h)	100/0	75%
Zn powder (CH, 6h)	0/100	69%
$AlCl_3$ (MW, 3 min)	73/23	70%
$AlCl_3$ (CH, 5 min)	58/42	43%
K-10 Clay (MW, 4 min)	90/10	90%
K-10 Clay (CH, 3h)	90/10	80%

Scheme 5.12

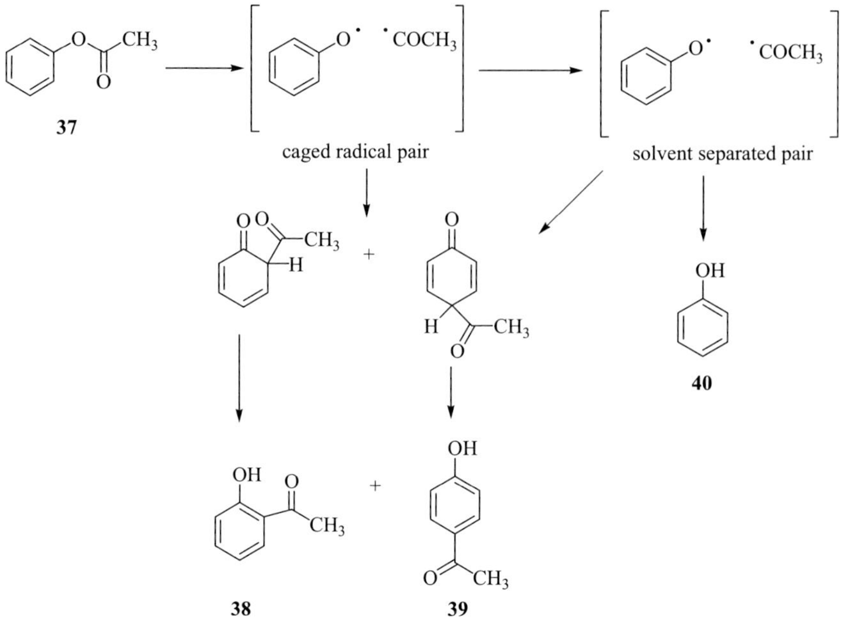

5 Selectivity Under the Action of Microwave Irradiation

Scheme 5.13

Hájek [57] studied the transformation of *tert*-butylphenols catalyzed by KSF and observed differences in reaction rate and product distributions when comparing microwave irradiation and conventional heating (Fig. 5.18). Several conditions were studied in the isomerization and transalkylation, including temperature and solvent effects, and it was concluded that different reaction rates and selectivity are a consequence of "microwave-induced polarization". In this way, the absorbed 2-*tert*-butylphenol molecules are affected to a greater extent by microwave irradiation

Tab. 5.3. Photo-Fries reaction under the action of MW and when irradiated at >254 nm.

Solvent	Reaction conditions	40/38 + 39	ortho/para	T (°C)	Superheating effect (°C)
Methanol	CH	0.21	1.18	20	–
	CH	0.32	0.95	65	–
	MW	0.35	0.98	71	12
Acetonitrile	CH	0.25	1.65	20	–
	CH	0.38	1.08	81	–
	MW	0.41	0.96	90	14

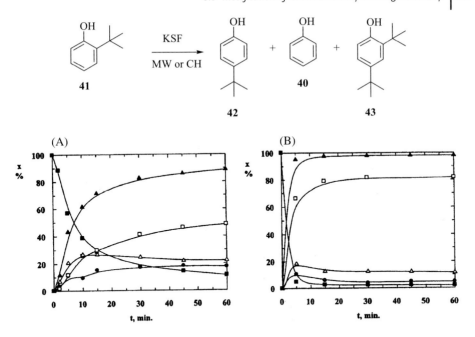

Fig. 5.18. Transformation of 2-*tert*-butylphenol (**41**) (2-TBP, ■) catalyzed by Montmorillonite KSF catalyst under the action of conventional heating (CH) (A) and microwave (MW) conditions (B); ▲, total conversion (**42** + **43** + **40**); ● selectivity for phenol (**40**); □ selectivity for 4-TBP (**42**); △ selectivity for 2,4-DTBP (**43**) at 75 °C.

than by conventional heating alone. This interaction can result in marked changes in both reaction rates and selectivity.

5.3.3
Synthesis and Reactivity of Heterocyclic Compounds

Heterocyclic compounds have a wide range of applications and are also extensively distributed in nature. These compounds are also important intermediates in organic synthesis. Several examples involving modification of selectivity in the preparation and reactivity of heterocyclic compounds have been reported. The degradation of ethyl indole-2-carboxylate (**44**) with 0.2 M NaOH has been reported by Strauss [20]. This reaction leads to the formation of indole (**46**) if the power input enables a temperature of 255 °C to be achieved or to indol-2-carboxylic acid (**45**) if the temperature is limited to 200 °C (Scheme 5.14).

In the alkylation of 1,2,4-triazole (**47**) [58] with benzyl chloride in the absence of base and solvent, the use of microwaves enables the pure N-1 alkylated product **48** to be obtained whereas the classical route leads exclusively to either quaternization or decomposition (Scheme 5.15).

Similarly, Loupy [59] undertook the preparation of 2,4-dichlorophenacylazoles by

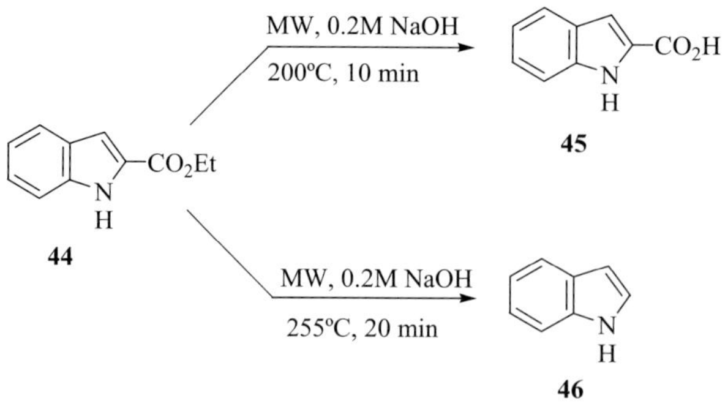

Scheme 5.14

reaction of the appropriate azole with 2,2′,4′-trichloroacetophenone in the absence of base and solvent. It was found that significantly increased regioselectivity was obtained on use of microwave irradiation. The not purely thermal specific microwave effects were shown to be very important in this reaction when solvent-free conditions or a non-polar solvent is used whereas they are masked when the reactions were performed in a polar solvent (DMF) [59b]. This point is extensively discussed in Chapter 4.

Cardillo [60] described the regioselective rearrangement of aziridine **51** in toluene, a good solvent to observe the effect of microwave irradiation (Scheme 5.16); under these conditions temperatures were 54–56 °C. Several Lewis acids were tested as catalysts and the best results were obtained with $BF_3.Et_2O$. Under the action of microwave irradiation, the yield and regioselectivity were greater than 99%, whereas with conventional heating only 65% yield was obtained and a poorer selectivity (85:15) was observed together with the presence of different by-products.

Echevarría [61] conducted reactions between 5-amino-1,3-dimethylpyrazole (**55**) and benzaldehydes **56** (Scheme 5.17). The reaction with benzaldehyde (**56a**) in the absence of solvent at room temperature produced the desired imine **57a** in low yield (10%). Reaction in the solid state with silica gel as catalyst at room tem-

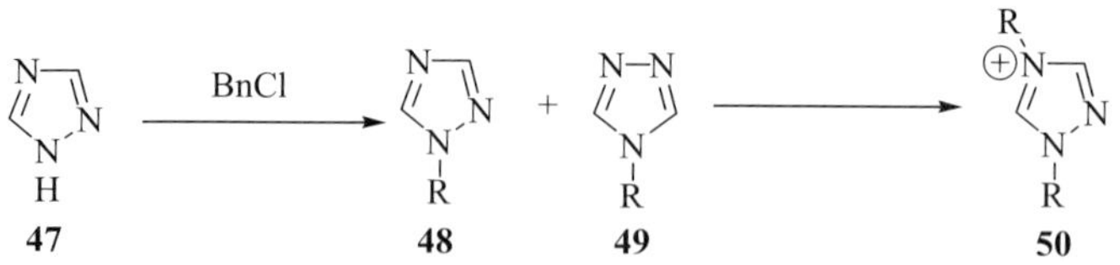

		48/49/50	Yield (%)
CH	5 min, 165°C	0/0/100	13
CH	1h, 120°C	decomp.	--
MW	450W, 5 min, 165°C	100/0/0	70

Scheme 5.15

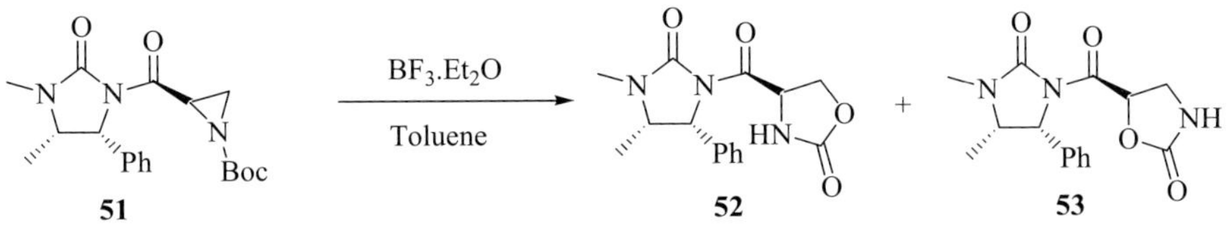

Scheme 5.16

MW, 240W, 54–56 °C, 99%, >99:1
CH 50 °C, 65%, 85:15

perature for 120 min furnished compound **60a** in 12% yield. When the reaction was performed in a microwave oven yields were greatly improved and selectivity depended on the nature of the substituent on the benzaldehyde. Reaction with benzaldehyde (**56a**) produced **60a** in 85% yield, but with tolualdehyde (**56b**) the tricyclic compound **61b** was obtained in 62% yield and, finally, with *p*-nitrobenzaldehyde (**56c**) compound **59c** was obtained in 99% yield.

The temperature obtained when using microwaves depends on the dielectric constant of the reagents and, therefore, the relative permittivity of the three benzaldehydes should be different. Microwave irradiation not only affords better yields and cleaner reactions than conventional heating, but even leads to different compounds – evidence of a change not only of reactivity but also of selectivity.

The desulfonylation of *N*-sulfonyl tetrahydroisoquinolines **63** with potassium fluoride on alumina under the action of microwave irradiation enables the selective synthesis of 3,4-dihydroisoquinolines **64** and isoquinolines **65** (Scheme 5.18) [62].

Microwave irradiation (490 W, microwave oven) of the N-sulfonyl heterocycles resulted in good yields of the corresponding 3,4-dihydroisoquinolines **64** in 10–20 s. Interestingly, increasing the irradiation time completely transformed the starting materials into the corresponding isoquinolines **65**, providing a highly selective and convenient strategy enabling access both classes of compound. Conventional heating, on the other hand, led to complete consumption of the starting material only after 48 h under reflux in toluene. Under these conditions, the 3,4-dihydroisoquinoline **63** was the only reaction product.

Hamelin [63] reported that pyridazinone **68** and heterobicyclic **69** compounds were formed in a ratio of 7:93 in the reaction of glyoxal monophenylhydrazones **66** with *β*-ketoesters **67** in the absence of solvent under the action of classical heating. Use of microwaves in the same reaction led to inversion of the reactivity and resulted in a product ratio of 85:15 (Scheme 5.19).

In the same way, reaction of glyoxal mono-1,1-dimethylhydrazone (**70**) with *β*-ketoesters **67** in the absence of solvent [64] gave 1-aminopyrroles **73** and other nitrogen heterocycles **71** and **72**. The selectivity was found to depend on a variety of conditions, for example temperature, time, and mode of heating (Scheme 5.20).

Uchida et al. [65] reported the preparation of *cis* and *trans*-2,4,5-triarylimidazolines from aromatic aldehydes. Microwave irradiation of a mixture of benzaldehyde (**56a**) and hexamethyldisilazane on silica gel in the absence of sol-

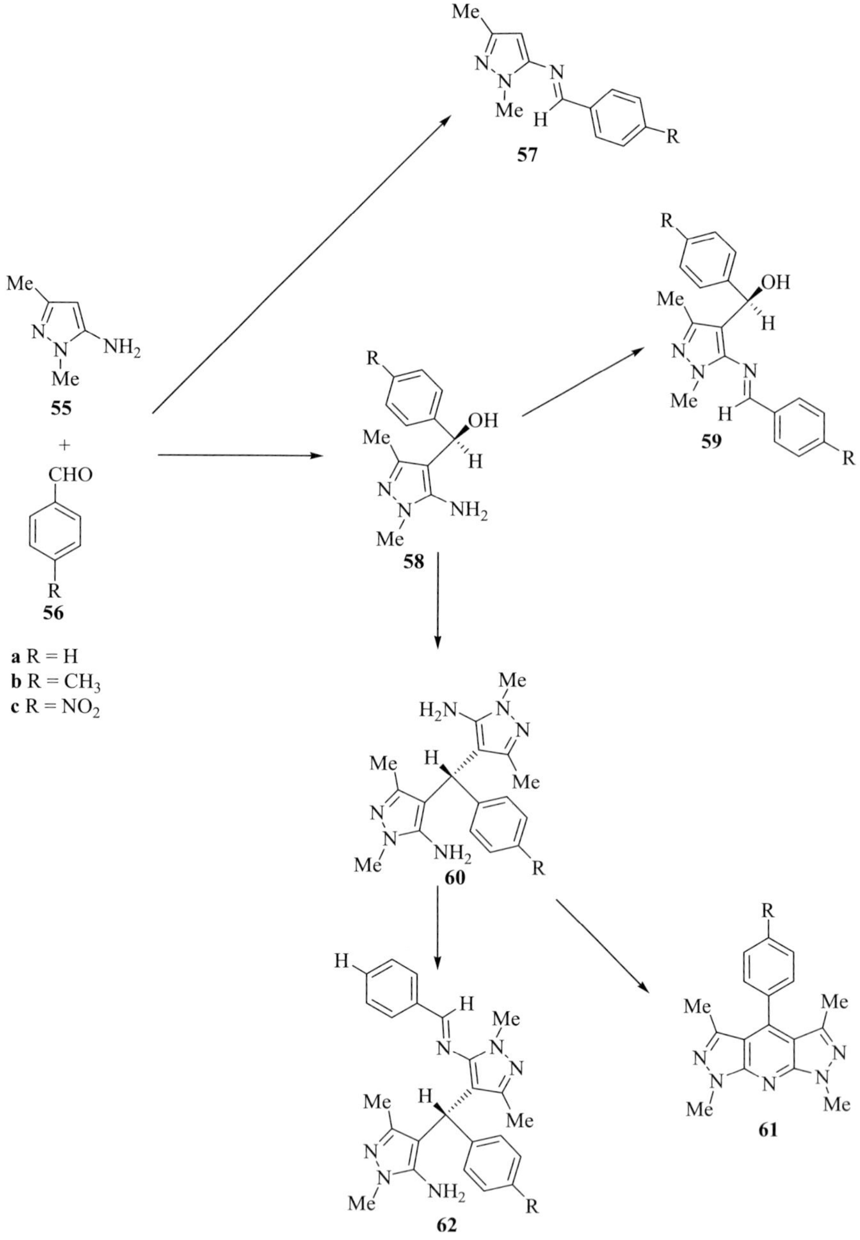

a R = H
b R = CH₃
c R = NO₂

Scheme 5.17

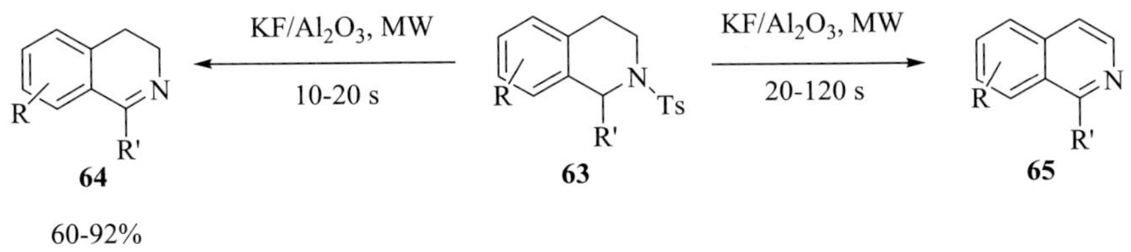

5.3 Modification of Chemoselectivity and Regioselectivity | 245

Scheme 5.18

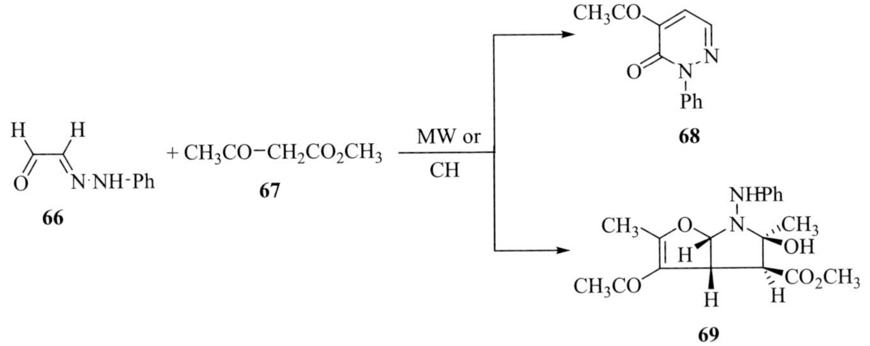

Scheme 5.19

vent for 5 min gave the bis imine **74** in 79% yield. In contrast, the cis imidazoline **75** was obtained directly by conventional heating. The methano diimines were cyclized to a mixture of cis and trans imidazolines by irradiation with one equivalent of a base for 5 min [65].

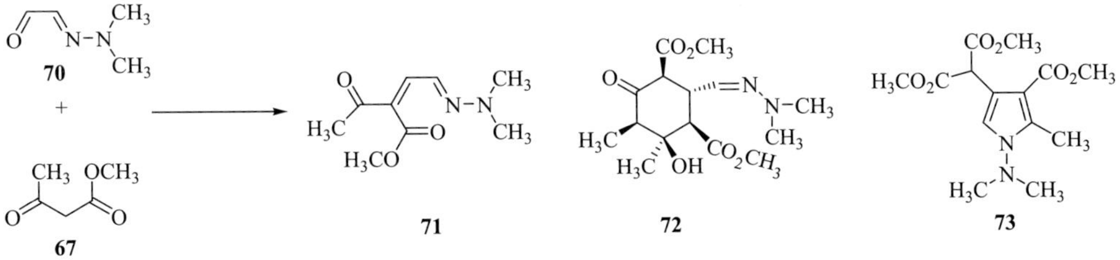

T(°C)	t	P(W)	71	72	73
20	48 h[a]	---	0	50	50
20	7 d[b]	---	9	0	91
100	4 min[a]	30	0	67	33
78	4 min[b]	30	50	0	50

[a] piperidine 0.15 mL
[b] piperidine 0.01 mL

Scheme 5.20

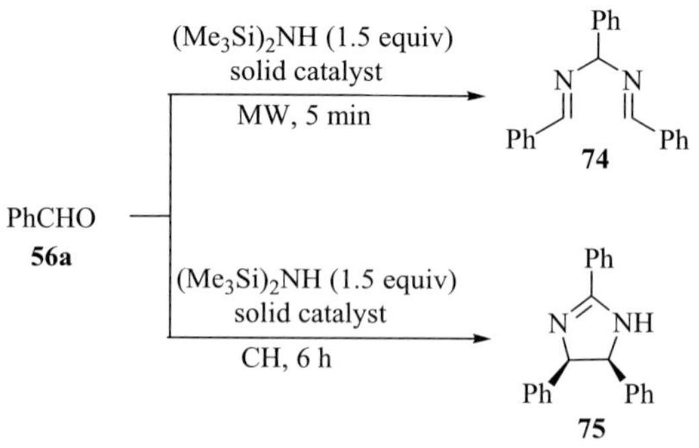

Scheme 5.21

The condensation of 2,5,6-triaminopyrimidin-4-one (**76**) with unsymmetrical α,β-dicarbonyl compounds **77** led to the substituted pterins, with preferential formation of the unwanted 7-isomer rather than the 6-isomer. A one-pot synthesis of 6-methylpterin **78** involved condensation with methylglyoxal (**77**) at a controlled temperature (0–5 °C). Sodium bisulfate was used to mask the more reactive aldehyde function. Under the action of microwave irradiation, however, the 6-isomer can be obtained with total regioselectivity without addition of sodium bisulfate or hydrazine hydrate (Scheme 5.22) [66].

The intramolecular cyclization of δ-iminoacetylenes to pyrazino[1,2-a]indoles described by Abbiati [67] is a new example of modification of selectivity. When 1-propargyl indoles **79** were treated in a sealed tube at 100 °C with 2 M ammonia in methanol, the corresponding pyrazino indoles **80** and **81** were obtained in good yields. Differences between the relative ratio of **80** and **81** were related to:

1. the relative stability of pyrazinoindoles and dihydropyrazinoindoles; and
2. the different reaction times required by the different substituted substrates.

Formation of dihydropyrazinoindoles is a kinetically controlled process whereas pyrazinoindoles are the thermodynamically controlled products – it is well known

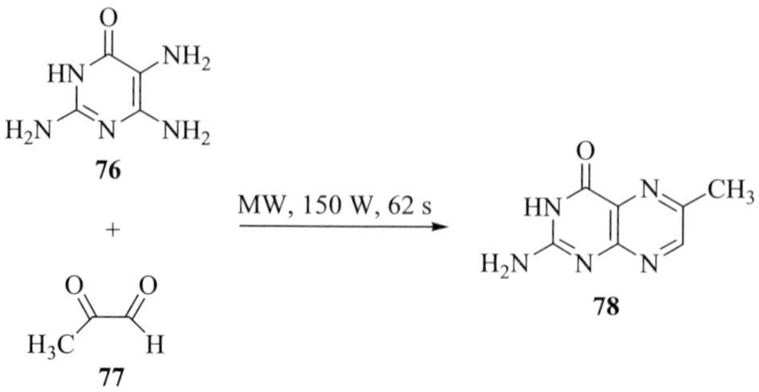

Scheme 5.22

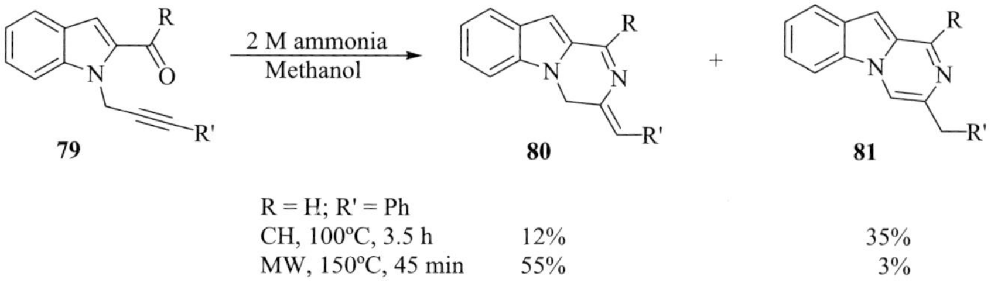

5.3 Modification of Chemoselectivity and Regioselectivity

Scheme 5.23

R = H; R' = Ph
CH, 100°C, 3.5 h — 12% / 35%
MW, 150°C, 45 min — 55% / 3%

that prolonged reaction times promote the formation of the latter. The microwave-assisted reaction was performed in a multimode oven at 150 °C in 2 M ammonia in methanol. The reaction was quicker than that with conventional heating and overall yields were increased by 11–36%. The ratio of dihydropyrazinoindoles to pyrazinoindoles was shifted toward the fully conjugated system.

5.3.4
Cycloaddition Reactions

Cycloaddition reactions have been performed with great success with the aid of microwave irradiation (Chapter 11). All the problems associated with these reactions have been conveniently solved by the rapid heating achieved with microwave irradiation, a situation not accessible by classical methods [4a, 6]. In some examples the selectivity of the reaction has also been modified. Langa described the cycloaddition of N-methylazomethine ylides to C_{70} to give three regioisomers (**83a–c**) by attack at the 1–2, 5–6, and 7–21 bonds (Scheme 5.24) [68]. Under the action of conventional heating the 7–21 **83c** isomer was formed in only a low proportion

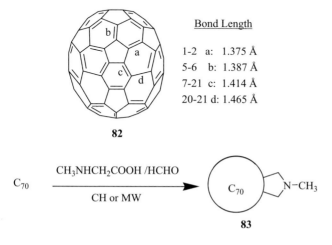

Bond Length
1-2 a: 1.375 Å
5-6 b: 1.387 Å
7-21 c: 1.414 Å
20-21 d: 1.465 Å

82

83

Scheme 5.24

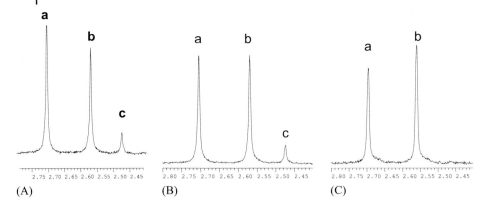

Fig. 5.19. ^{1}H NMR spectra (methyl groups) of adducts **83a**, **83b**, and **83c** (Scheme 5.24). From left to right: (A) classical heating in toluene; (B) classical heating in o-DCB; (C) microwave irradiation at 180 W in o-DCB.

and the 1–2 **83a** isomer was found to predominate. Use of microwave irradiation in conjunction with o-DCB (o-dichlorobenzene), which absorbs microwaves efficiently, gave rise to significant changes in reactivity. In contrast with classical conditions, the isomer **83c** was not formed under the action of microwave irradiation, irrespective of the irradiation power, and isomer **83b** was predominant at higher power (Scheme 5.24 and Fig. 5.19).

A computational study on the mode of cycloaddition showed that reaction is stepwise, with the first step consisting of a nucleophilic attack on the azomethine ylide. The most negative charge on the fullerene moiety in transition states **83a** and **83b** is located on the carbon adjacent to the carbon–carbon bond being formed. In transition state **c**, however, the negative charge is delocalized throughout the C_{70} subunit. The relative ratio of isomers **83a–c** is related to the degree of hardness, and the formation of **83b** should be favored under the action of microwave irradiation. It is noticeable that purely thermal arguments predict the predominance of **83a** under the action of microwave irradiation, which is in marked contrast with the result found experimentally. This system can be used as a predictive model in competitive reactions with a non-concerted mechanism in which at least one polar transition state is involved and in which the transition with the harder transition state will be favored by use of microwave irradiation.

This model was used by Díaz-Ortiz [69] in the preparation of nitroproline esters **86** by 1,3-dipolar cycloaddition of imines **84**, derived from α-aminoesters, with β-nitrostyrenes **85**, in the absence of solvent (Scheme 5.25). Conventional heating produced the expected isomers **86a** and **86b** by the *endo* and *exo* approaches. Under the action of microwave irradiation, however, a new compound, isomer **86c**, was obtained. The authors showed that this isomer arises by thermal isomerization of the imine by rotation in the carboxyl part of the ylide. Isomer **86c** is then produced by an *endo* approach. The exclusive formation of the second dipole

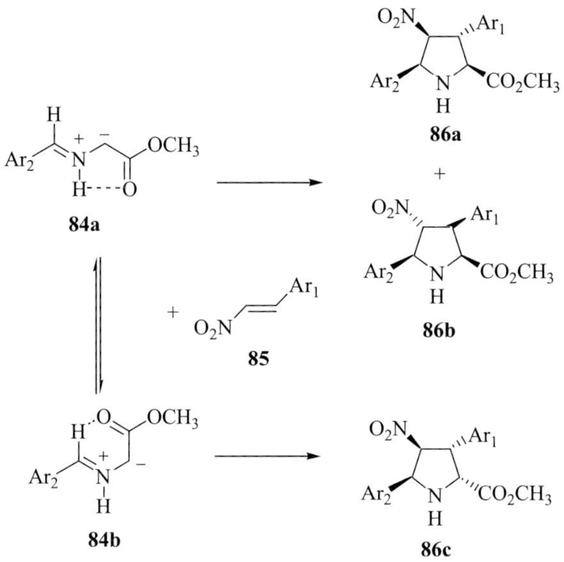

5.3 Modification of Chemoselectivity and Regioselectivity | 249

Scheme 5.25

under the action of microwave irradiation is most probably related to the higher polarity, hardness, and lower polarizability in comparison with the first dipole.

Hong [70] described the cycloaddition reactions of fulvenes **87** with quinones **88** and several activated alkenes and alkynes in an attempt to provide new examples of the microwave effect. Two examples warrant particular attention. Reaction of 6,6-dimethylfulvene (**87**) with *p*-benzoquinone (**88**) produced the [4+2] cycloaddition when the reaction was performed with conventional heating. Use of microwave heating, however, gave rise to the [6+4] cycloaddition product (Scheme 5.26).

Unfortunately, no control experiments in DMSO with conventional heating and in C_6H_6 with microwave activation were performed to enable unambiguous conclusions.

In contrast, reaction with dimethyl maleate (**91**) gave only the [4+2] product both by conventional heating and with microwave irradiation. Under the action of

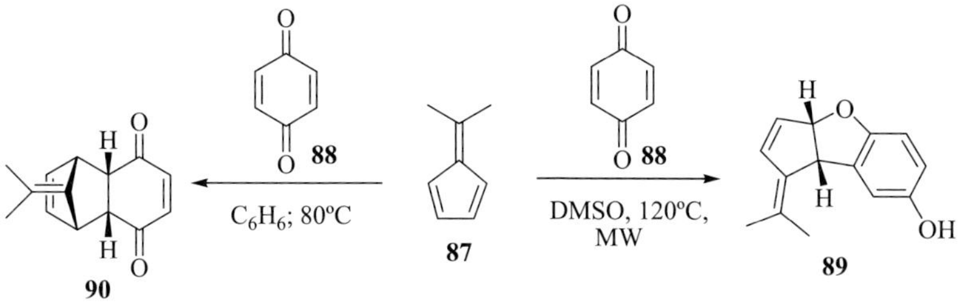

Scheme 5.26

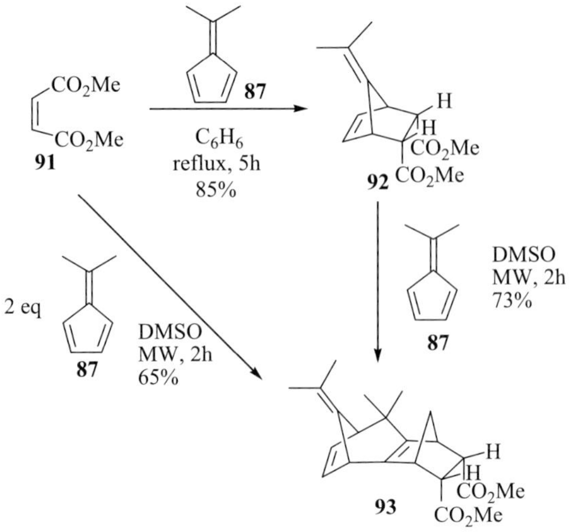

Scheme 5.27

microwaves, however, addition of a second equivalent of 6,6-dimethylfulvene (**87**) was observed. Formation of this new product can be explained by a [4+2] cycloaddition with a first equivalent of 6,6-dimethylfulvene (**87**) then protonation, a 1,2-alkyl shift, and [4+3] cycloaddition with a second equivalent of 6,6-dimethylfulvene (**87**). Alternatively, this product can be obtained by dimerization of 6,6-dimethylfulvene (**87**) via [4+2] or a [6+4] cycloaddition then [4+2] cycloaddition with dimethyl maleate (**91**) (Scheme 5.27) [70].

Wagner [71] performed 1,3-dipolar cycloaddition of nitrones **95** with *trans*-[PtCl$_2$(PhCN)$_2$] (**94**) and showed that microwave irradiation enhances the rate of the cycloaddition substantially and also favors selectivity towards the monocycloadduct **96**, compared with thermal conditions (Scheme 5.28). Both cycloadditions can be performed at room temperature overnight, indicating that the difference between reactivity is not dramatic, although it is sufficient to achieve high selectivity. The progress of the reaction with time is shown in Fig. 5.20. The nitrone

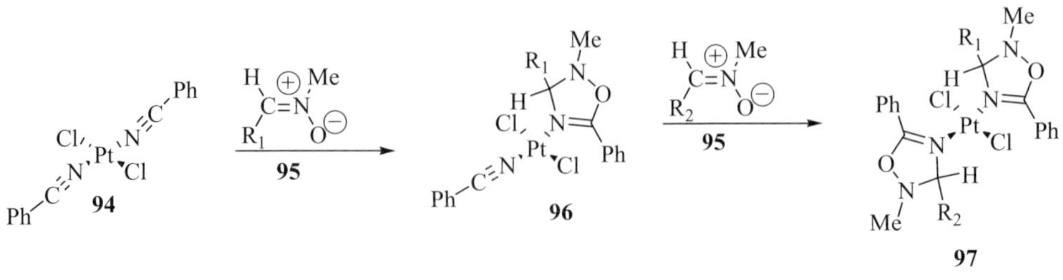

Scheme 5.28

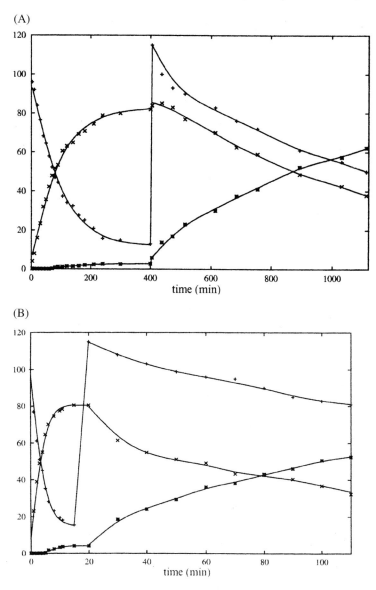

Fig. 5.20. Reaction of trans-[PtCl$_2$(PCN)$_2$] **94** and nitrone **95** under (A) thermal conditions (B) the action of microwave irradiation (+**95**; ×**96**; ∗**97**).

is consumed at the same rate the mono-cycloaddition product is formed. The sample can be used to study the rate of the second cycloaddition – when the nitrone was almost consumed another equivalent was added and the reaction was again followed by ^{1}H NMR spectroscopy. The thermal reaction is highly selective, the second reaction is slower than the first cycloaddition and the half-lives differ by a factor of approximately 6. Under the action of microwave irradiation, both cycloaddition steps were significantly accelerated compared with under thermal conditions. Whereas the first cycloaddition is accelerated by a factor of 25, however, the second is accelerated by a factor of 7 only.

5.3.5
Polymerization

Microwave irradiation has been successfully applied in polymer chemistry (Ref. [10] and Chapter 14 of this book) – for the synthesis and processing of polymers, e.g. for modification of the surface and cross-linking, and also in the degradation of polymers. Microwave plasmas also have been used in the polymerization and surface modification of materials. The enhanced reaction rates have been attributed to thermal effects – although for some reactions it seems the advantages arise from the selective excitation of one of the educts involved. Shifts in selectivity have also been observed.

Loupy [72] described the synthesis of polyethers derived from isoidide and isosorbide by phase-transfer catalysis under the action of conventional heating and microwave irradiation (Scheme 5.29). Although similar temperature profiles were observed in these reactions (Fig. 5.21), noticeably different results were obtained under the action of microwave irradiation:

1. reaction times were markedly reduced;
2. higher molecular weight polymers were obtained with better homogeneity; and
3. the structures of polymers were quite different, with significant differences in chain termination.

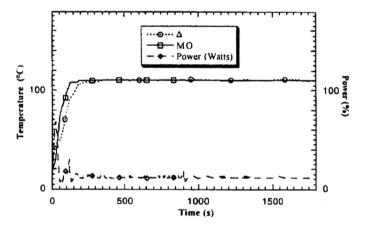

Fig. 5.21. Profile of the increase in temperature under the action of microwave irradiation and conventional heating for the reaction of **98** with **99**.

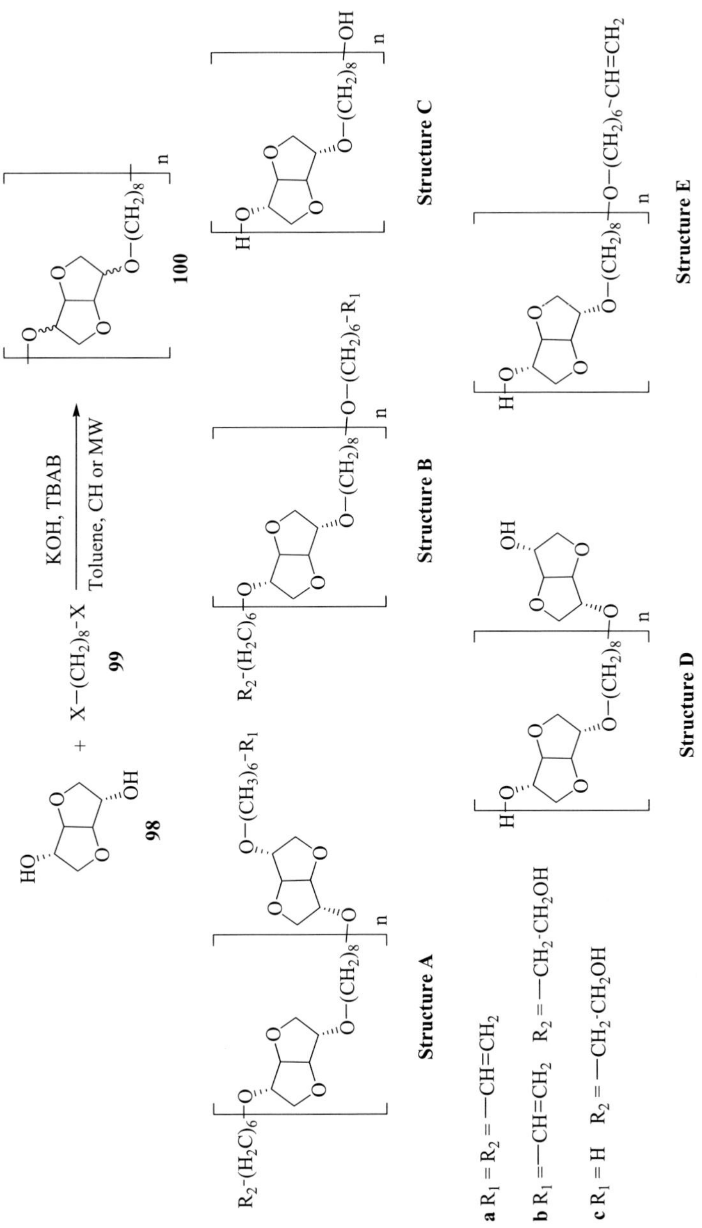

5.3 Modification of Chemoselectivity and Regioselectivity

Scheme 5.29

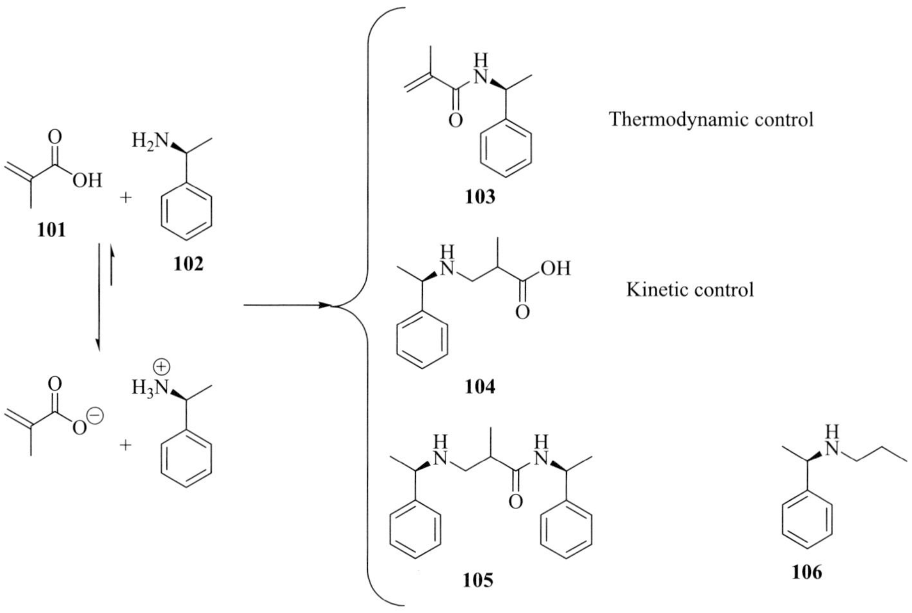

Scheme 5.30

Under the action of microwave irradiation, terminal ethylenic groups were formed rather rapidly (structures A_1, A_2, A_3, B_1 and B_2). With conventional heating, however, the terminal units essentially consisted of hydroxyl functions (Structures A_1, C, D, and E). The authors believe the formation of ethylenic units can be foreseen as an example of a specific microwave effect on selectivity on the basis of an enhanced stabilization of the more polar transition state (E2 rather than S_N2) under the action of microwave irradiation [17a].

Ritter [73] described the synthesis and polymerization of methacrylic acid (**101**) [73a] and acrylic acid [73b] with (R)-1-phenylethylamine (**102**) (Scheme 5.30). Reactions under the action of conventional heating were performed in a differential scanning calorimeter (DSC); this enabled the heating profile observed under microwave conditions to be reproduced. After 15 min of microwave irradiation, conversion to the amide was 93% whereas with thermal heating it was only 12% after the same period of time. A detailed study of both reactions led to the identification of two side-products, 2-methyl-3-(1-phenylethylamino)propionic acid (**104**) and 2-methyl-N-(1-phenylethyl)-3-(1-phenylethylamino)propionamide (**105**), as a result of a Michael addition reaction between the starting acid and the amide product. In the reaction performed by thermal heating, it was possible to identify the presence of (1-phenylethyl)propylamine (**106**), which was not detected after the reaction under the action of microwave irradiation. From kinetic data (Fig. 5.22) obtained by GC–MS it was possible to observe that both sets of reaction conditions led to **104** as the main Michael addition compound formed during the first 5 min. After this

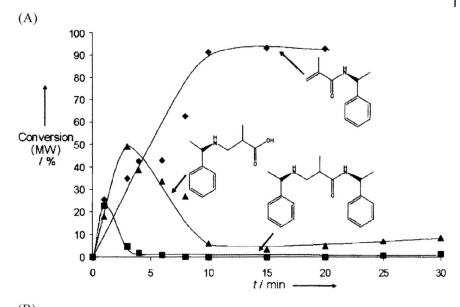

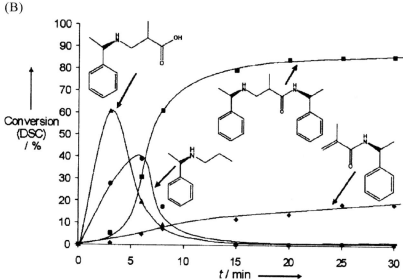

Fig. 5.22. Kinetic plots of the reactions shown in Scheme 5.30: (A) under the action of microwave irradiation at 180 °C (IR pyrometer); (B) conventional thermal heating at 200 °C (DSC).

time, the reversible nature of the reaction led to a decrease in the amount of this compound and, when microwave irradiation was used, the desired amide **103** became the main product. In contrast, classical thermal conditions give rise to competition between the appearance of compounds **103** and **105**. The authors believe that the high stability of amide **103** is the force driving rapid decomposition of product **104** by a retro-Michael reaction and this can be related to the use of MW irradiation, because when competitive reactions are involved, the mechanism occurring via the hardest, more polar transition state should be favored [18].

Zhang [74] performed cationic chain polymerization of **107** to produce a bisaliphatic epoxy resin and discovered some modification in the selectivity under the action of microwave irradiation. The extent of polymerization was followed by DSC (Fig. 5.23). Two exothermic peaks (peaks 1 and 2) were related to the exothermic polymerization of the epoxide **107** (Fig. 5.23A). Under thermal conditions the polymerization at peak 1 occurred first as the temperature gradually increased. As the temperature continued to increase, polymerization at peak 2 occurred. In the microwave field (Fig. 5.23B), however, the order was reversed and it was peak 2 that disappeared first and peak 1 second. The authors believe microwave energy will affect the internal energy of a chemical with a permanent dipole moment. This effect is too small to be significant, however. Microwave energy will also substantially affect the Gibbs free energy of a chemical with a permanent dipole moment. As a consequence, if two reactions are possible for the same system, one reaction could occur to a greater extent than the other in microwave fields.

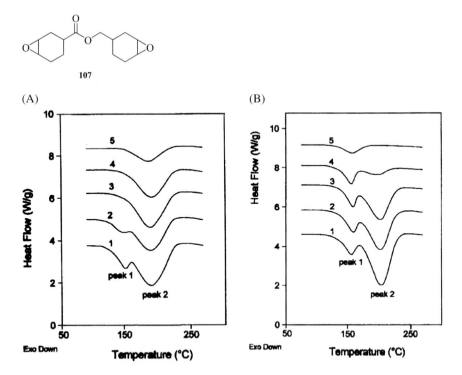

Fig. 5.23. DSC curves for the epoxide **107**: (A) under the action of conventional heating at 150 °C (1), 160 °C (2), 164 °C (3), 179 °C (4) and 199 °C (5); (B) under the action of microwave irradiation for 24 s (1), 28 s (2), 32 s (3), 38 s (4),

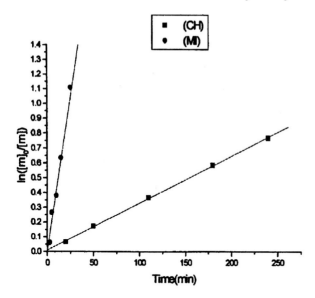

Fig. 5.24. Kinetics of atom-transfer radical polymerization of methyl methacrylate (MMA) under the action of microwave irradiation (MI) and conventional heating (CH). Conditions: $T = 72\ °C$, $[MMA]_0 = 9.46$ M; $[MMA]_0/[DClX]_0/[CuCl]_0/[DETA]_0 = 400:1:1:3$.

Zhu [75] described the atom-transfer radical polymerization of methyl methacrylate using α,α'-dichloroxylene/CuCl/N,N,N',N'',N''-pentamethyldiethylenetriamine as initiation system. The apparent k_p under the action of microwave irradiation ($7.6 \times 10^{-4}\ s^{-1}$) is much higher than that with conventional heating ($5.3 \times 10^{-5}\ s^{-1}$), which indicates that application of microwave irradiation enhances the rate of polymerization (Fig. 5.24).

The authors determined the concentration of Cu in solution for this heterogeneous system and showed that under the action of microwave irradiation it is higher than that with conventional heating within the same period of time (Fig. 5.25). As a consequence, microwave irradiation increases the dissolution of CuCl in the system and increases the efficiency of the initiator. Molecular weight distributions in the microwave process range from 1.2 to 1.5, whereas for conventional heating they are in the range 1.2 to 1.6. This is an example of selective heating that enhances both the rate of polymerization and initiator efficiency.

5.3.6
Miscellaneous

Microwave irradiation has been used to modify selectivity in rearrangement and coupling reactions. Indeed, the photo-Fries rearrangement described above could also be classified in this section [56]. Sudrik described the Wolff rearrangement of α-diazoketones under conventional heating and microwave irradiation [76]. The

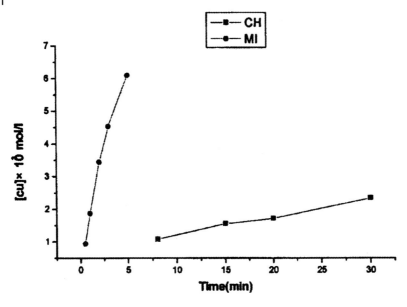

Fig. 5.25. Dependence of [Cu] on time in solution. Conditions: $[MMA]_0 = 0.047$ M; $[MMA]_0/[CuCl]_0/[PMDETA]_0 = 1200:1:1:3$.

results obtained show the superiority of the microwave-promoted rearrangement over the conventional heating method. Conformationally restricted substrates such as 3-diazocamphor (**108**) were used to prove the existence of a non-thermal microwave effect. Compound **108**, on thermolysis or transition metal catalysis, is known to undergo intramolecular C–H insertion to give the tricyclic ketone **110** (Scheme 5.31). Photolysis in methanol yields the methyl ester of the Wolff rearrangement product. Microwave irradiation in benzylamine (**109**) gave exclusively a diastereomeric mixture of the ring-contracted bicyclic amide **111** in 73% yield. Microwave irradiation of 3-diazocamphor (**108**) in the presence of water also produced the tricyclic ketone **110** as the principal product along with several side-products. The author attributes this microwave-specific behavior to the effective dielectric coupling of the 3-diazocamphor with microwaves.

Klán et al. described the Norrish type-II reaction of valerophenones in microwave photochemistry (Scheme 5.32) [56]. Equimolecular mixtures of both ketones were irradiated at ≥ 280 nm in a variety of solvents; such an experimental arrangement guaranteed identical photochemical conditions for the two compounds. The fragmentation to cyclization ratio varied from 5 to 8 and was characteristic for given reaction conditions (Table 5.4). The photochemical efficiency R ($R = [\mathbf{112a}] + [\mathbf{114a}]/[\mathbf{113b}] + [\mathbf{114b}]$) was found to be temperature-dependent and the magnitude is most probably related to the solvent basicity. The authors consider that superheating by microwave irradiation is probably responsible for the observed changes in selectivity. A linear dependence of R with temperature was observed by considering the estimated overheating (Table 5.4 and Fig. 5.26).

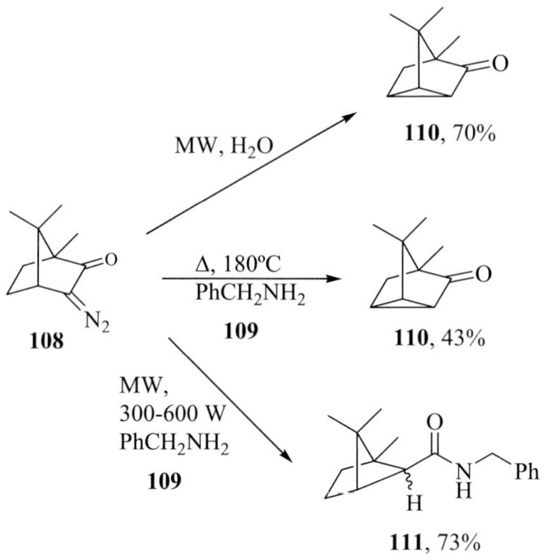

5.3 Modification of Chemoselectivity and Regioselectivity | 259

Scheme 5.31

The authors consider these systems to be a photochemical thermometer for estimating superheating effects in microwave-assisted reactions.

Larhed reported a MW-assisted enantioselective Heck reaction between cyclopentene **116** and phenyltriflate **115** using Pd_2dba_3 as a catalyst in conjunction with

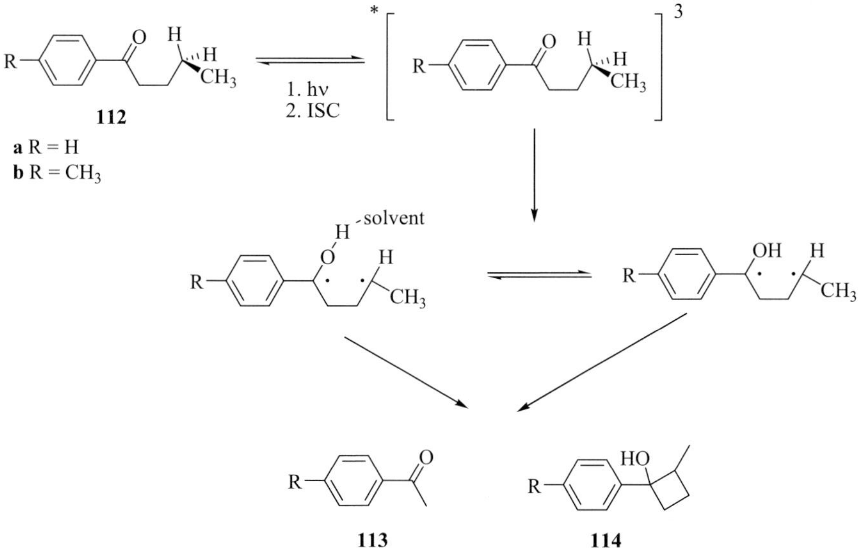

a R = H
b R = CH_3

Scheme 5.32

Tab. 5.4. Norrish type-II reaction in the MW field irradiated at >280 nm.

Solvent	Reaction conditions	R	T (°C)	Superheating effect (°C)
Methanol	CH	2.25	20	–
	CH	1.52	65	–
	MW	1.34	75	11
Acetonitrile	CH	2.12	20	–
	CH	1.12	81	–
	MW	0.98	90	9

R = [112a] + [114a]/[113b] + [114b]

proton sponge (Scheme 5.33) [77]. Under the action of classical heating (70 °C, 5 days), this reaction in THF furnished compound **117** in 80% yield and with 86% ee. A mixture of compounds **118** and **119** was also obtained (20%). Microwave arylation for only 4 h at 140 °C resulted exclusively in the formation of **117** in 78% yield and only traces (1–2%) of isomers **118** and **119**. Only moderate stereoselectivity was achieved in this reaction, however.

NiCl$_2$-catalyzed hydrophosphinylation of triple bonds in acetonitrile gave excellent yields and an impressive reduction in the reaction time from 1–38 h to 1–10 min [78]. Stereoselective cis addition to the triple bond was always observed but on using compounds with an asymmetrically substituted triple bond, for example phenylacetylene (**120**), two regioisomers were obtained **122** and **123**. In this reaction the selectivity was enhanced by use of microwave irradiation (Scheme 5.34).

It is remarkable that under the action of microwave irradiation the regioselectiv-

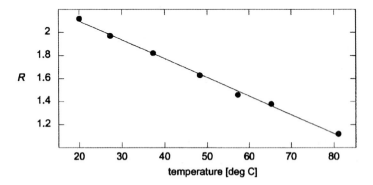

Fig. 5.26. Linear temperature dependence of Norrish type-II photochemistry system in acetonitrile (R = [112a] + [114a]/ [113b] + [114b]).

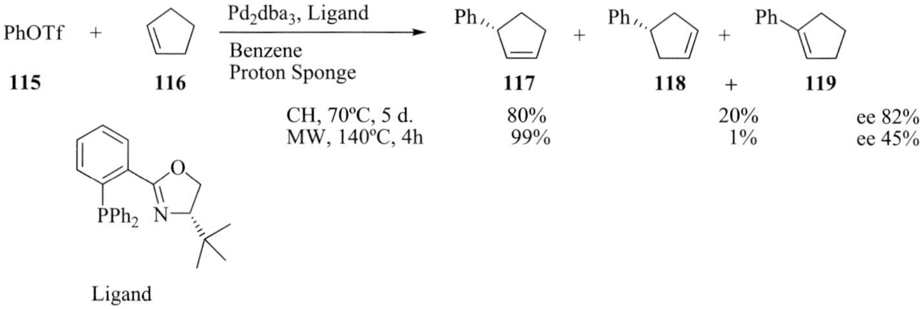

Scheme 5.33

PhOTf	+	(cyclopentene)	→ Pd₂dba₃, Ligand / Benzene / Proton Sponge	Ph⋯(cyclopentene)	+ Ph⋯(cyclopentene)	+ Ph(cyclopentene)
115		**116**	CH, 70°C, 5 d.	**117**	**118**	**119**
			MW, 140°C, 4h	80%	20%	ee 82%
				99%	1%	ee 45%

Ligand: 2-(oxazolinyl)phenyl-PPh₂ with tert-butyl substituent

ity increased at lower temperature but decreased when conventional heating was used, when the reaction was performed in acetonitrile under reflux.

Hegedus [79] described the complexation of 1,8-pyrazine-capped 5,12-dioxocyclams with the aim of obtaining coordination oligomers containing multiple communicating metal centers with potentially useful electronic, optical, or catalytic properties. Treatment of **124** with copper(II) tetrafluoroborate in methanol under reflux gave the desired pyrazine-capped dioxocyclam copper(II) **125** complex in good yield as a green crystalline solid (Scheme 5.35). In an attempt to reduce the reaction time required for complexation the ligand was irradiated for 2 min in a domestic microwave oven in the presence of a fivefold excess of copper(II) tetrafluoroborate and sodium carbonate in methanol in a pressure tube. The resulting solution was filtered through Celite and left to stand for 2 days. Dark blue crystals of **126** were deposited (Scheme 5.35). This complex was only formed under the action of microwave irradiation. Prolonged heating of the starting materials only produced the complex **125**.

Romanova [80] studied the effect of microwave irradiation on the direction and stereochemistry of the Rodionov reaction. Microwave irradiation of a mixture of equimolecular amounts of benzylammonium acetate **128**, monoethyl malonate **130**, and 2-phenylpropanal **127** resulted in the formation of β-amino ester **132** in 38% yield and ethyl-4-phenyl-2-pentenoate **133** in 60% yield, exclusively as the trans isomer (Scheme 5.36). By contrast, thermal activation does not lead to forma-

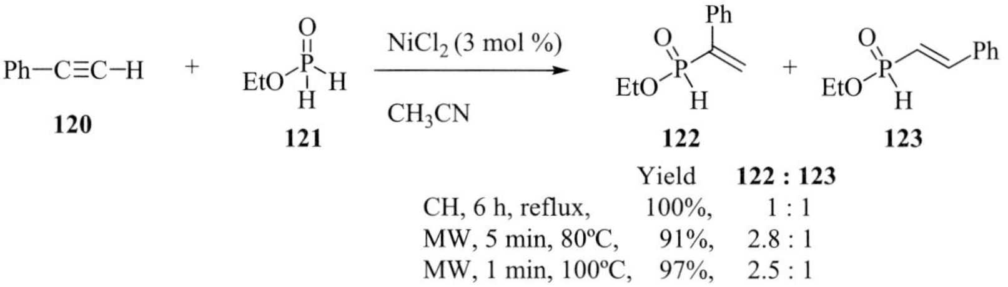

Scheme 5.34

Ph—C≡C—H + EtO-P(=O)(H)H → NiCl₂ (3 mol %), CH₃CN → EtO-P(=O)(H)-C(=CH₂)Ph + EtO-P(=O)(H)-CH=CH-Ph

120 + **121** → **122** + **123**

	Yield	122 : 123
CH, 6 h, reflux,	100%,	1 : 1
MW, 5 min, 80°C,	91%,	2.8 : 1
MW, 1 min, 100°C,	97%,	2.5 : 1

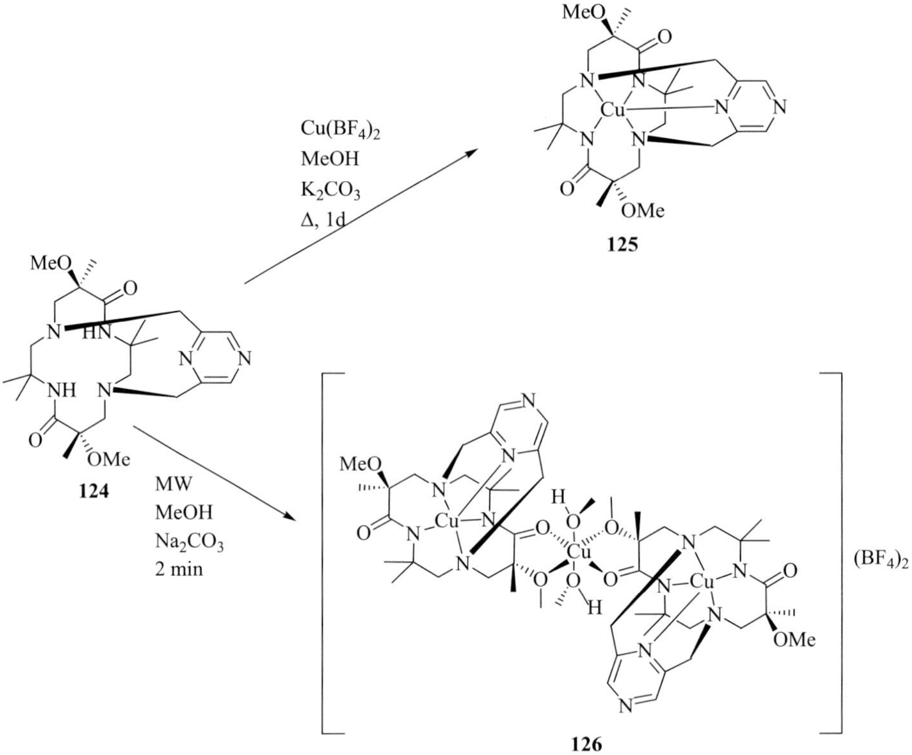

Scheme 5.35

tion of β-amino ester **132**. In this reaction, the product is **133**, which was obtained in 20% yield and with a *trans/cis* ratio of 11:1. In addition, partial transformation of benzylammonium acetate into N-benzylacetamide **134** (22%) also occurred (Scheme 5.36). The author concluded that the main factors determining the course of the Rodionov reaction are the mode of activation and the acidity of the medium.

Kad [81] described the selective oxidation of methylallyl groups with SeO_2 over silica gel under solvent-free conditions. Selective oxidation to the aldehyde (**138**) was easily achieved under the action of microwaves whereas a mixture of the alcohol (**137**) and aldehyde (**138**) was observed on conventional heating (Scheme 5.37).

The direct transformation of amines to ketones, catalyzed by Pd/C in water under the action of microwave irradiation, was performed by Miyazawa [82]. The reaction strongly depended on the structure of the diol, particularly the number of hydrogen atoms on the carbon adjacent to the nitrogen. *sec*-Butylamine (**139**) was completely converted to 2-butanone (**140**) within 1 h on irradiation at 50 W in a glass ampoule (Scheme 5.38). The reaction using a preheated oil bath at 170 °C for 1 h gave the desired 2-butanone in 29% yield without by-products but a longer reaction time increased the amount of di-*sec*-butylamine (**142**), presumably formed

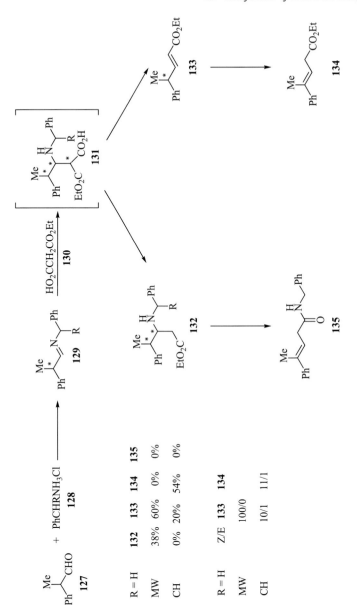

Scheme 5.36

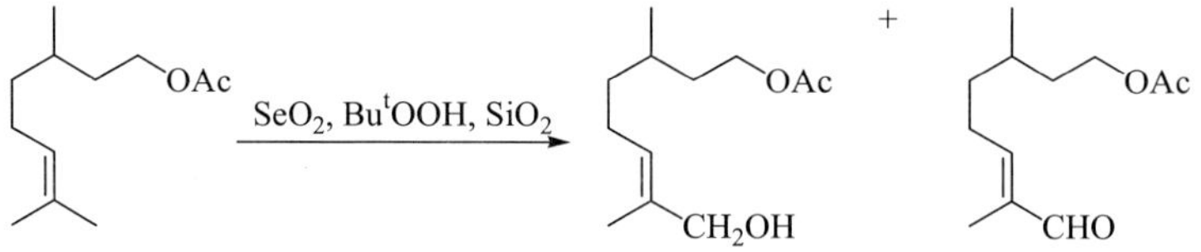

Scheme 5.37

	136	137	138
MW, 640 W, 10 min, 85%		0	100
CH, 24 h		75%	80	20

(Note: MW gives 137:138 = 0:100; CH gives 75% yield with 137:138 = 80:20)

by the reductive amination of 2-butanone with *sec*-butylamine. Thus, microwave irradiation in a sealed vessel effectively enhances selectivity and accelerates the reaction.

5.4
Modification of Stereo and Enantioselectivity

The most important aspect of the synthesis of organic molecules that contain one or more stereogenic elements is usually that of stereochemical control. Indeed, such control is an essential aspect of the design of a good synthesis and, in addition, affects selection of the most appropriate method – including the choice of a particular route [83]. Several reports have described how microwave irradiation can be used to modify stereoselectivity in some reactions. The possibility of stereochemical modification by simply changing the mode of activation is a very attractive prospect. Slight modification of the stereoselectivity in the Rodionov reaction has been reported by Romanova [80].

Kuang [84] described the stereoselective synthesis of (*Z*) and (*E*)-1-bromoalkenes by elimination from 1,2-dibromoalkanes **143** under the action of microwave irradiation (Scheme 5.39). (*Z*)-1-Bromoalkenes **144a** were obtained in 0.2–1 min, in the presence of triethylamine, in excellent yields (∼90%) and with high *Z/E* stereoselectivity (>99:1) [84a]. The corresponding (*E*)-1-bromoalkenes **144b** were obtained in 0.5–3 min by using silver acetate in acetic acid, again in excellent yields (∼90%)

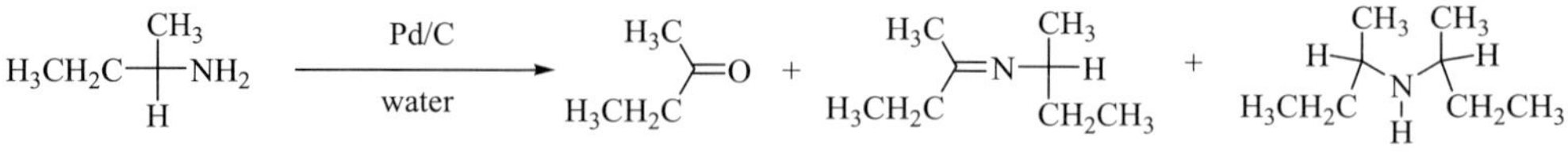

	139	**140**	**141**	**142**
MW, 50 W, 1 h, Conversion 100%		100	0	0
CH, 170°C, 1 h, Conversion 29%		100	0	0
CH, 170°C, 9 h, Conversion 76%		59	Trace	41

Scheme 5.38

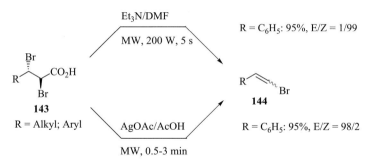

Scheme 5.39

and with high Z/E stereoselectivity ($>3:97$) [84b]. The authors indicated that yields and stereoselectivity obtained with these methods are substantially higher than those from previous procedures.

(Z)-1-Bromoalkenes **144a** can be prepared *in situ* as intermediates for the synthesis of terminal alkynes **145** by elimination in the presence of DBU or for the preparation of (Z)-enines **146** by palladium-catalyzed coupling with alkynes (Scheme 5.40) [84c]. Both reactions were performed with microwave irradiation and gave good to excellent yields.

A variety of substrates were examined by Chanda [85] for aziridination with chloramine-T **150** and Bromamine-T **148** in the presence of a copper catalyst. It is interesting to note that aziridination did not occur with the less reactive cinnamates **147** in the presence of $CuCl_2$ and $CuBr_2$ (Scheme 5.41). The reaction did, however, proceed under the action of microwave irradiation with $CuBr_2$ to yield the aziridine **149**, albeit in poor yield.

Aziridination under the action of ultrasound irradiation with bromamine-T **148** resulted in the selective formation of the trans aziridine **149b** (Scheme 5.41). This is in contrast with results from the same reaction under microwave irradiation – a

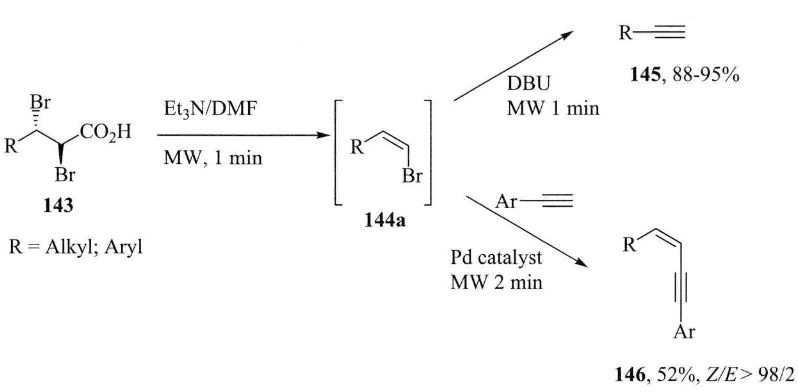

Scheme 5.40

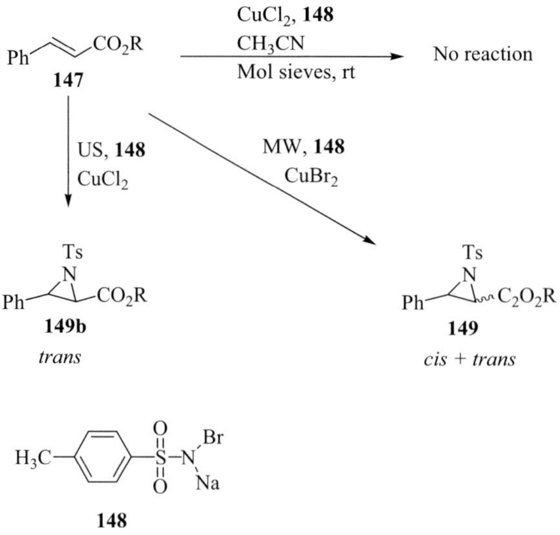

Scheme 5.41

[Scheme 5.41 shows alkene 147 (Ph-CH=CH-CO₂R) reacting under three conditions: with CuCl₂/148 in CH₃CN with molecular sieves at rt giving no reaction; with US, 148, CuCl₂ giving trans-aziridine 149b (Ts-N, Ph, CO₂R); with MW, 148, CuBr₂ giving cis + trans aziridine 149. Compound 148 is shown as H₃C-C₆H₄-S(O)₂-N(Br)(Na).]

mixture of *cis* and *trans* isomers of the aziridine was obtained. Such results were attributed by the authors to the difference between the activation mechanisms of the two processes.

Bose has described reactions between acid chlorides **151** and Schiff bases **152** in which the stereoselectivity depends on the order of addition of the reagents (Scheme 5.42) [86]. When the condensation was conducted by a "normal addition" sequence (i.e. acid chloride last), only the *cis* β-lactam **153a** was formed. If, however, the "inverse addition" technique (triethylamine last) was used, 30% *cis* **153a** and 70% *trans* **153b** β-lactams were obtained under the same conditions. When the reaction was conducted in a microwave oven with chlorobenzene as the solvent, the ratio of *trans* **153b** to *cis* **153a** β-lactams was 90:10, irrespective of the order of addition, and isomerization to the thermodynamically more stable *trans* β-lactam **153b** did not occur.

This effect was explained by Cossío, who postulated that under the action of microwave irradiation the route involving direct reaction between the acyl chloride

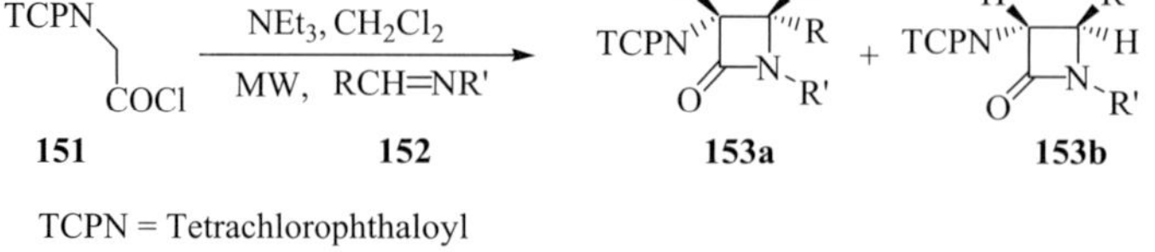

Scheme 5.42

[Scheme 5.42: TCPN-CH(COCl) 151 reacts with RCH=NR' 152 in the presence of NEt₃, CH₂Cl₂, MW to give cis β-lactam 153a and trans β-lactam 153b. TCPN = Tetrachlorophthaloyl.]

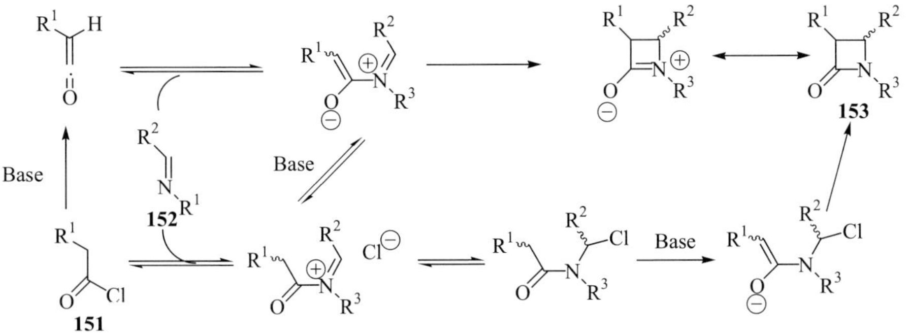

Scheme 5.43

and the imine competes efficiently with the ketene-imine reaction pathway (Scheme 5.43) [87].

The Diels–Alder reaction between cycloalkanones (**155**) and cyclic dienes (**154**) in toluene and catalyzed by AlCl$_3$ under the action of microwave irradiation was reported by Reddy [88]. Reactions were performed in a domestic microwave oven and gave adducts in good yields within 2 min (Scheme 5.44). Interestingly, it was also observed that microwave irradiation increased selectivity for the *endo* product. This result is in contrast with that reported by Gedye [89] in the cycloaddition of cyclopentadiene with methyl acrylate in methanol. In this reaction, significant modification of *endo/exo* selectivity, in comparison with previously reported results, was not observed.

A study of the mutarotation of α-D-glucose **157a** to β-D-glucose **157b** (Fig. 5.27) has been described by Pagnota [90]. It was found that in EtOH–H$_2$O, 1:1, apart from more rapid equilibration with microwaves in comparison with conventional heating, microwaves led to a modification of the equilibrium position such that a greater amount of the α-D-glucose (**157a**) was obtained than was obtained by classical heating. This extraordinary effect cannot be explained by a classical heating effect and is the clearest example of a possible specific reaction induced by a microwave radiation field.

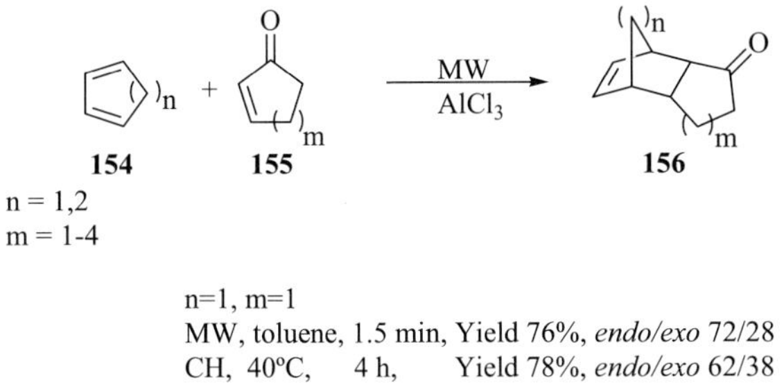

n = 1,2
m = 1-4

n=1, m=1
MW, toluene, 1.5 min, Yield 76%, *endo/exo* 72/28
CH, 40°C, 4 h, Yield 78%, *endo/exo* 62/38

Scheme 5.44

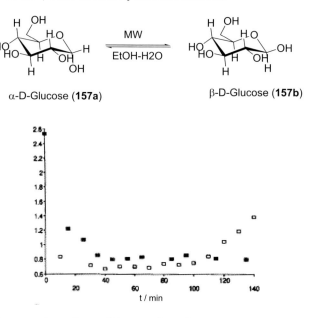

Fig. 5.27. Dependence of the ratio of α-D-Glucose **157a** to β-D-Glucose **157b** on time for (■) the conventionally heated and (□) the microwave-irradiated reaction.

The stereoselectivity in the cyclization of citronellal **158** to isopulegol **159** and neoiso-pulegol **160** on graphite can be altered by using microwave irradiation (Scheme 5.45) [36]. Isopulegol **159** is always the main diastereoisomer, irrespective of the method of heating, but use of microwaves increases the amount of neoiso-pulegol to 30%.

Jacob [91] described the same reaction using silica-supported $ZnCl_2$ (SiO_2/$ZnCl_2$, 10%). This catalyst promotes the selective cyclization of citronellal within 1.5 min under MW irradiation and gives quantitative yield (100%) with a good ratio of (+)-neoisopulegol (**160**) to isopulegol (**159**) (76:24). In contrast, the cyclization products were obtained in only 72% yield, together with by-products and loss of selectivity, when the reaction was heated at 58 °C using an oil bath until all of the citronellal (**158**) was consumed (1 h).

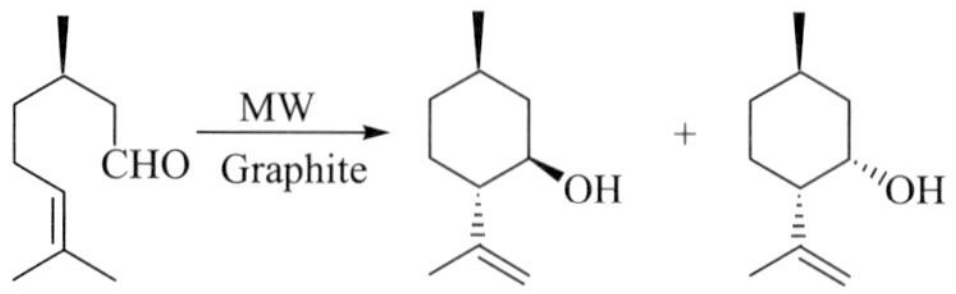

Scheme 5.45

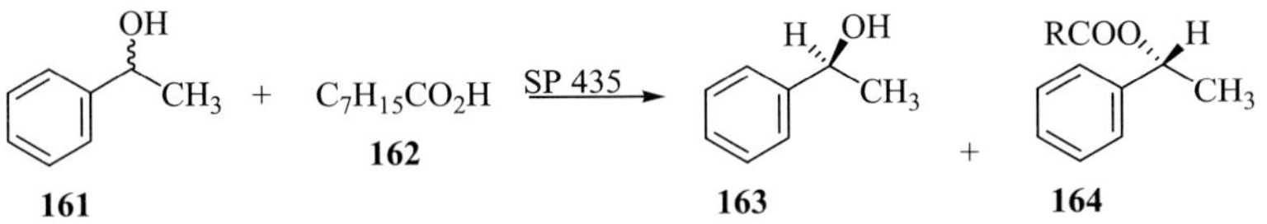

Scheme 5.46

		%	ee **163** (S)
Conventional Heating	10 min 78 °C	48 %	62
Microwaves (300 + 80 W)	1 + min 95 °C	47 %	86
Microwaves (60 + 20 W)	5 + 5 min 78 °C	52 %	93

Guibé-Jampel and Loupy [92] showed that yield and stereoselectivity in the acylation of racemic 1-phenylethanol (**161**) catalyzed by supported enzymes can be enhanced by use of microwave irradiation (Scheme 5.46). The authors suggest that the specificity of the reaction can be attributed to an improvement in the reversibility of the reaction, because of better elimination of water and/or reduction in $\Delta H^{\ddagger}$ and $\Delta S^{\ddagger}$.

Synergism between microwave irradiation and enzyme catalysis has been ascribed to reduction of the Michaelis constants by use of microwaves, without any associated change in the form of the rate equation [93].

The preparation of ruthenium bis(diimine)sulfoxide complexes by reaction of cis-[Ru(bipyridine)$_2$(Cl)$_2$] (**165**) with enantiomerically pure chiral sulfoxides **166** was described by Aït-Haddou [94] as a new concept in the preparation of optically active octahedral ruthenium complexes (Scheme 5.47). The reaction produces two diastereomeric complexes **167** and **168** and the microwave-irradiated reactions resulted in excellent yields and high reaction rates with a notable increase in the observed diastereomeric excess.

Moberg [95] described the Mo(0)-catalyzed allylic alkylation of dimethyl malonate with 3-arylprop-2-enyl carbonates **169** using the enantiomerically pure chiral diamine **172** as ligand (Scheme 5.48). The reaction produced two regioisomers

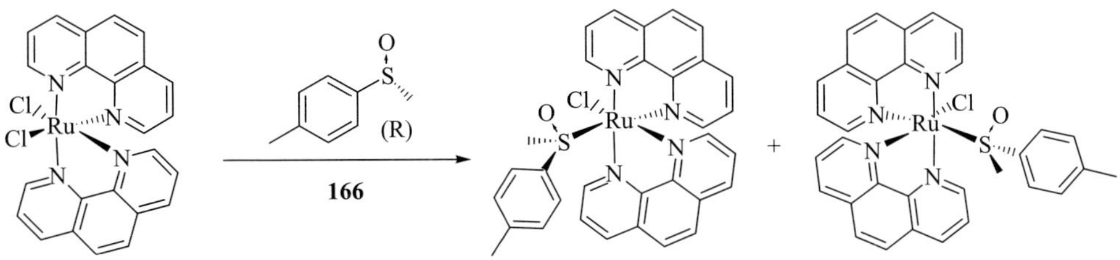

375 W, 2 min. Yield 97%, de 73.7 %

Scheme 5.47

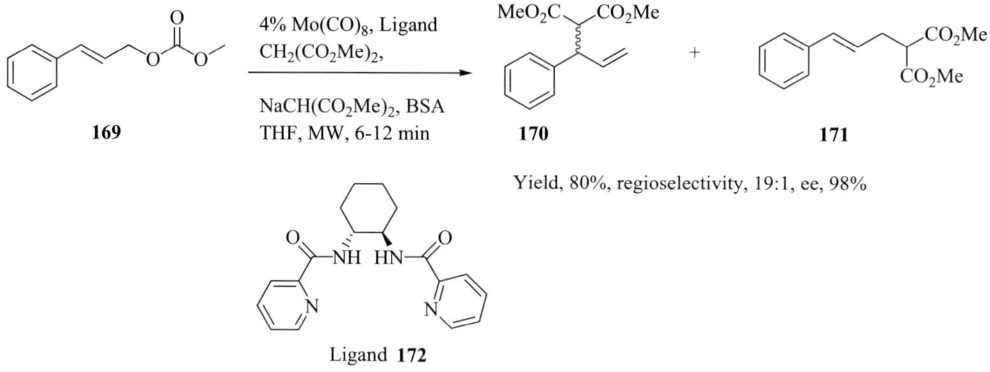

270 | 5 Selectivity Under the Action of Microwave Irradiation

Scheme 5.48

Yield, 80%, regioselectivity, 19:1, ee, 98%

(branched **170** and linear **171**) and in the first compound a new chiral center was created. Microwave-activated reactions occurred in 6 min with high regioselectivity (10:1 to 69:1); the branched isomer **170** was the predominant product. A high enantiomer excess was also observed in the branched isomer **170** (74–98 ee). The use of a solid-supported ligand led to a branched-to-linear ratio of 35:1 and an enantiomer excess of 97% [95b]. The resin could be recovered and used at least seven times without a significant change in the outcome of the reaction.

Tanaka [96] described the preparation of helical aromatic compounds by electrophilic aromatic substitution of *p*-phenylenediamines with carboxylic acids, catalyzed by $ZnCl_2$ (Scheme 5.49). Microwave irradiation led to a reduction in the reaction time from 9 h to 5 min, and a small increase in yield (10%) – except when low-boiling carboxylic acids such as acetic and propionic acid were used. On starting with optically pure (S)-**174**, the conventional heating reaction gave a racemic

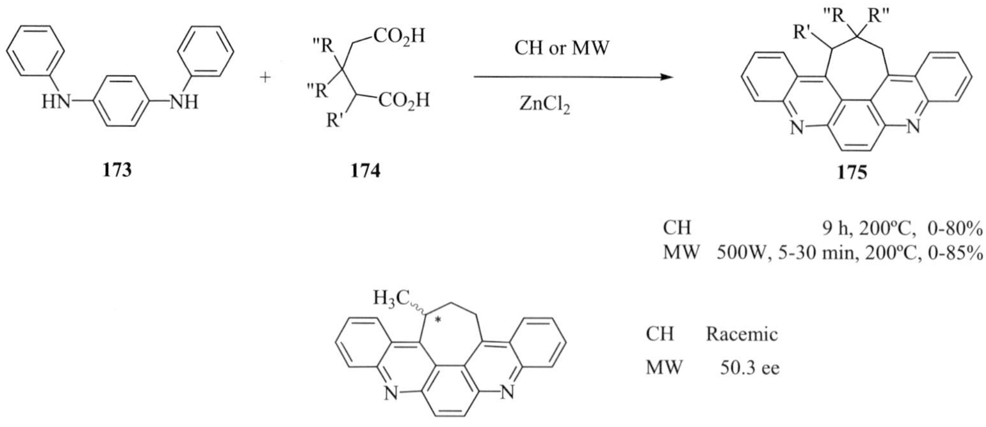

CH	9 h, 200°C, 0-80%
MW	500W, 5-30 min, 200°C, 0-85%

CH	Racemic
MW	50.3 ee

Scheme 5.49

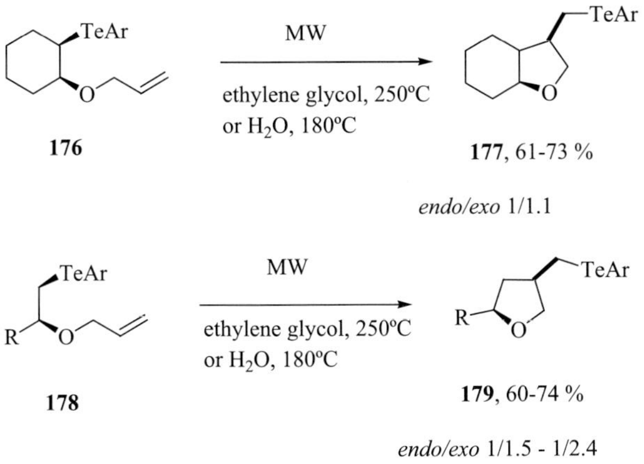

Scheme 5.50

176 (cyclohexane with TeAr and O-allyl) → MW, ethylene glycol, 250°C or H$_2$O, 180°C → **177**, 61–73 %

endo/exo 1/1.1

178 (R-CH(TeAr)-O-allyl) → MW, ethylene glycol, 250°C or H$_2$O, 180°C → **179**, 60–74 %

endo/exo 1/1.5 - 1/2.4

mixture of **175** as did use of racemic **174**. Racemization of the sp^3 chiral carbon adjacent to the carbonyl group or in the reaction intermediates seemed to occur through an enol form at elevated temperature. In contrast, 50.3% ee was achieved on using microwave irradiation. The rapid increase in temperature led the reactive starting materials and intermediates to give the final products without racemization of the sp^3 carbon, thus resulting in higher stereoselectivity.

The microwave-assisted group-transfer cyclization of organotellurium compounds was reported by Engman [97a]. Microwave irradiation led to reduction of the reaction time from a few hours to minutes (Scheme 5.50). Group-transfer cyclization of primary and secondary alkyl aryltellurides could be induced to occur without additives in an environmentally benign solvent such as water. Group-transfer cyclization products were always formed as mixtures of *exo/endo* or *cis/trans* isomers. In comparison with tin-promoted, light-induced group transfer, the selectivity was significantly reduced [97b]. The predominant formation of exo and trans isomers is consistent with the Beckwith–Houk model for ring-closure of 5-hexenyl radicals assuming a chair-like transition state.

Ruthenium-catalyzed asymmetric reductions of aromatic ketones **180** can be performed under microwave irradiation. Moberg [98] described this reaction using a monomode microwave reactor and ruthenium complexes **182** with enantiomerically pure chiral diamines **181** (Scheme 5.51). The reaction is very fast and efficient; even sterically hindered *tert*-butylphenylketone, which is normally quite unreactive, was reduced in almost quantitative yield in 3 minutes. The enantioselectivity was, however, lower than that obtained under standard conditions similarly to that described by Larhed [77] in the enantioselective Heck reaction between cyclopentene **115** and phenyl triflate **116** (Scheme 5.33).

A similar reduction in stereoselectivity in ethyl alcohol was observed by Toukoniitty [99] in the hydrogenation of ethyl pyruvate to ethyl lactate, while similar stereoselectivity was observed in toluene. The authors consider that this result is a con-

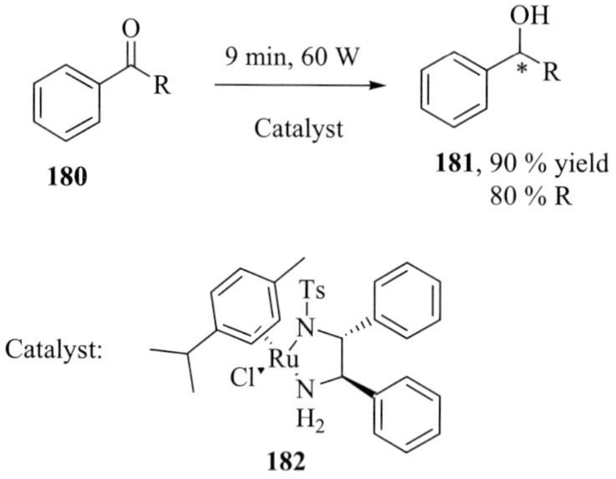

5 Selectivity Under the Action of Microwave Irradiation

Scheme 5.51

sequence of local superheating or polar ethyl alcohol in the cavity, which is not possible in non-polar toluene.

Chen [100] reported the selective oxidation of glycosyl sulfides **183** to sulfoxides **184** by use of magnesium monoperoxyphthalate under the action of microwave irradiation (Scheme 5.52). Sulfoxides **184** were obtained as a mixture of R and S isomers and microwave irradiation reduced the reaction time from 10 to 0.7 h. Interestingly, the R and S isomer ratios are not similar to those reported previously by the same authors [101] and they concluded that these modifications of the selectivity may have been associated with a "microwave effect" or the oxidant used.

5.5
Conclusions

In conclusion, it has been extensively shown that microwave irradiation is a selective mode of activation. Radiation is selectively absorbed by polar molecules in the presence of apolar molecules, a property that leads to selective thermal gradients.

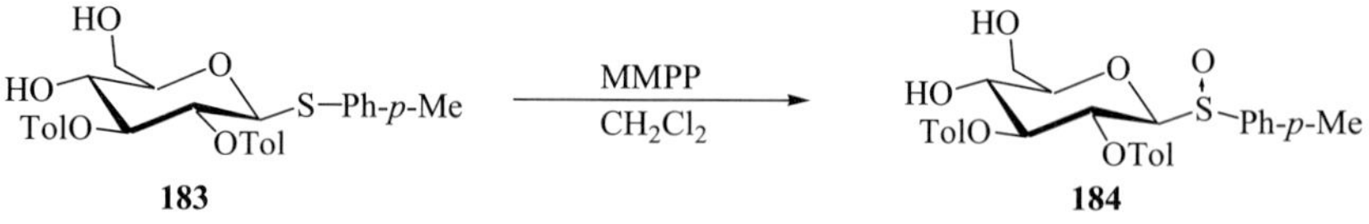

	Oxidant	Yield	R:S ratio
MW, 20 min	MMPP,	87%	5:1
CH, 33 h	Oxone/SiO$_2$	82%	10:1
CH, 42 h	t-BuOOH/SiO$_2$	85%	9:1

Scheme 5.52

Overheating of polar substances and the presence of hot spots has also been demonstrated; the latter is particularly important in heterogeneous systems. All of these effects can be used to significantly improve reactions and even to perform reactions that do not occur under the action of conventional heating.

More importantly, these effects can be used to modify the chemo, regio, and stereoselectivity of a given reaction, sometimes leading to complete inversion of the selectivity simply by changing the mode of heating between conventional heating (conductive heating) and microwave heating (dielectric heating).

Several authors also postulate the occurrence of a non-thermal effect that originates from the polarizing electromagnetic field. In this way the most polar transition state, i.e. the harder transition state, will be favored under the action of microwave irradiation (Chapter 4). These results have been supported by computational calculations. The experimental and theoretical results described here now need to be supported with further examples. If these results are confirmed, however, such systems could be used as a predictive tool to show which reactions can be improved or have modified selectivity under the action of microwave irradiation and even to predict the result of the reaction.

Some other characteristic examples are described in Chapter 4, essentially in connection with medium and mechanistic effects and understood in terms of specific effects derived from enhancement of polarity during the course of the reactions (i.e. between the ground and the transition state) [102].

Acknowledgments

Financial support from the DGICYT of Spain through project BQU2001-1095 and from the Consejería de Ciencia y Tecnología JCCM through project PAI-05-019 is gratefully acknowledged.

References

1 (a) D. M. P. MINGOS, A. G. WHITTAKER, *Microwave Dielectric Heating Effects in Chemical Synthesis in Chemistry under Extreme or non Classical Conditions*, R. VAN ELDIK, C. D. HUBBARD, Eds., John Wiley and Sons, **1997**, 479–545; (b) C. GABRIEL, S. GABRIEL, E. H. GRANT, B. S. J. HALSTEAD, D. M. P. MINGOS, *Chem. Soc. Rev.* **1998**, *27*, 213–223; (c) D. STUERGA, M. DELMOTTE, *Wave–Material Interactions, Microwave Technology and Equipment.* In *Microwaves in Organic Synthesis*; ed. A. LOUPY; Wiley–VCH: Weinheim, **2002**.

2 (a) J. D. FERGUSON, *Molecular Diversity* **2003**, *7*, 281–286 (http://www.cem.com); (b) L. FAVRETTO, *Molecular Diversity* **2003**, *7*, 287–291 (http://www.milestonesci.com); (c) J.-S. SCHANCHE, *Molecular Diversity* **2003**, *7*, 293–300 (http://www.biotagedcg.com).

3 (a) A. K. BOSE, M. J. MANHAS, B. K. BANIK, E. W. ROBB, *Res. Chem. Intermed.* **1994**, *20*, 1–11; (b) S. CADDICK, *Tetrahedron* **1995**, *52*, 10403–10432; (c) S. A. GALEMA, *Chem. Soc. Rev.* **1997**, *25*, 233–238; (d) P. LIDSTRÖM, J. TIERNEY, B. WATHEY, J.

WESTMAN, *Tetrahedron* **2001**, *57*, 9225–9283; (e) C. O. KAPPE, *Angew. Chem. Int. Ed.* **2004**, *43*, 6250–6284; (f) B. L. HAYES, *Aldrichimica Acta* **2004**, *37*, 66–77.

4 (a) *Microwaves in Organic Synthesis*; ed. A. LOUPY: Wiley–VCH: Weinheim, **2002**; (b) B. L. HAYES, *Microwave Synthesis: Chemistry at the Speed of Light*, CEM Publishing, Matthews, NC, **2002**; (c) *Microwave-Assisted Organic Synthesis*, Eds. P. LIDSTÖM, J. P. TIERNEY, Blackwell Scientific, **2005**; (d) C. O. KAPPE, A. STADLER, *Microwaves in Organic and Medicinal Chemistry*, in *Methods and Principles in Medicinal Chemistry (Volume 25)*, Eds. R. MANNHOLD, H. KUBINYI, G. FOLKERS: Wiley, **2005**.

5 (a) A. LOUPY, G. BRAM, J. SANSOULET, *New J. Chem.* **1992**, *16*, 233–242; (b) R. S. VARMA, *Tetrahedron* **2002**, *58*, 1235–1255.

6 A. DÍAZ-ORTIZ, F. LANGA, A. DE LA HOZ, A. MORENO, *Eur. J. Org. Chem.* **2000**, *4*, 3659–3673.

7 (a) N. ELANDER, J. R. JONES, S. Y. LU, S. STONE-ELANDER, *Chem. Soc. Rev.* **2000**, *29*, 239–249; (b) S. STONE-ELANDER, N. ELANDER, *J. Lab. Comp. Radiopharm.* **2002**, *45*, 715–746.

8 (a) F. LANGA, P. DE LA CRUZ, E. ESPÍLDORA, J. J. GARCÍA, J. J. GARCÍA, M. C. PÉREZ, A. DE LA HOZ, *Carbon* **2000**, *38*, 1641–1645; (b) F. LANGA, P. DE LA CRUZ, E. ESPÍLDORA, A. DE LA HOZ, *Applications of Microwave Irradiation to Fullerene Chemistry*, in *Fullerenes*, Vol. 9, The Electrochemical Society, New York, **2000**, pp. 168–178.

9 S. BARLOW, S. R. MARDER, *Adv. Funct. Mater.* **2003**, *13*, 517–518.

10 (a) L. ZONG, S. ZHOU, N. SGRICCIA, M. C. HAWLEY, L. C. KEMPEL, *J. Microwave Power Electromagn. Energy* **2003**, *38*, 49–74; (b) F. WIESBROCK, R. HOOGENBOOM, U.S. Schubert, *Macromol. Rapid. Commun.* **2004**, *25*, 1739–764.

11 (a) Y. XU, Q.-X. GUO, *Heterocycles* **2004**, *63*, 903–974; (b) N. N. ROMANOVA, P. V. KUDAN, A. G. GRAVIS, Y. G. BUNDEL, *Chem. Heterocycl. Comp.* **2000**, *36*, 1130–1140;

(c) A. R. KATRITZKY, S. K. SINGH, *Arkivoc* **2003**, xiii, 68–86.

12 (a) S. K. DAS, *Synlett* **2004**, 915–932; (b) A. CORSARO, U. CHIACCHIO, V. PISTARA, G. ROMEO, *Current Org. Chem.* **2004**, *8*, 511–538.

13 M. LARHED, C. MOBERG, A. HALLBERG, *Acc. Chem. Res.* **2002**, *35*, 717–727.

14 (a) H. WILL, P. SCHOLZ, B. ONDRUSCHKA, *Chem. Ing. Tech.* **2002**, *74*, 1057–1067; (b) R. B. MOYES, G. BOND, *Microwave Heating in Catalysis*, in *Handbook of Heterocyclic Catalysis* 8.5. Wiley, **1997**.

15 (a) C. O. KAPPE, *Curr. Opin. Chem. Biol.* **2002**, *6*, 314–320; (b) M. LARHED, A. HALLBERG, *Drug Discovery Today* **2001**, *6*, 406–416; (c) A. LEW, P. O. KRUTZIK, M. E. HART, A. R. CHAMBERLIN, *J. Comb. Chem.* **2002**, *4*, 95–105; (d) H. E. BLACKWELL, *Org. Biomol. Chem.* **2003**, *1*, 1251–1255; (e) V. SANTAGADA, E. PERISSUTTI, G. CALIENDO, *Current. Med. Chem.* **2002**, *9*, 1251–1283; (f) V. SANTAGADA, F. FRENCENTESE, E. PERISSUTTI, L. FAVRETTO, G. CALIENDO, *QSAR Comb. Sci.* **2004**, *239*, 919–946.

16 (a) R. S. VARMA, *Clean Products and Processes* **1999**, *1*, 132; (b) R. S. VARMA, in *Advances in Green Chemistry: Chemical Syntheses Using Microwave Irradiation*, AstraZeneca Research Foundation, Kavitha Printers, Bangalore, India, **2002**; (c) A. K. BOSE, M. S. MANHAS, S. N. GANGULY, A. H. SHARMA, B. K. BANIK, *Synthesis* **2002**, 1578–1591; (d) M. NÜCHTER, B. ONDRUSCHKA, W. BONRATH, A. GUM, *Green Chem.* **2004**, *6*, 128–141; (e) B. A. ROBERTS, C. R. STRAUSS, *Acc. Chem. Res.* **2005**, 651–653; (f) N. E. LEADBEATER, *Chem. Commun.* **2005**, 2881–2902.

17 (a) L. PERREUX, A. LOUPY, *Tetrahedron* **2001**, *57*, 9199–9223; (b) A. DÍAZ-ORTIZ, A. DE LA HOZ, A. MORENO, *Chem. Soc. Rev.* **2005**, 157–168.

18 F. LANGA, P. DE LA CRUZ, A. DE LA HOZ, A. DÍAZ-ORTIZ, E. DÍEZ-BARRA, *Contemp. Org. Synth.* **1997**, *4*, 373–386.

19 N. KUHNERT, *Angew. Chem. Int. Ed.* **2002**, *41*, 1863–1866.

20 C. R. Strauss, R. W. Trainor, *Aust. J. Chem.* **1995**, *48*, 1665–1692.
21 P. Nilsson, M. Larhed, A. Hallberg, *J. Am. Chem. Soc.* **2001**, *123*, 8217–8225.
22 D. R. Baghurst, D. M. P. Mingos, *J. Chem. Soc., Chem. Commun.* **1992**, 674–675.
23 F. Chemat, E. Esveld, *Chem. Eng. Technol.* **2001**, *24*, 735–744.
24 Y. C. Tsai, B. A. Coles, R. G. Compton, F. Marken, *J. Am. Chem. Soc.* **2002**, *124*, 9784–9788.
25 (a) D. Bogdal, M. Lukasiewicz, J. Pielichowski, A. Miciak, S. Bednarz, *Tetrahedron* **2003**, *59*, 649–653; (b) M. Lukasiewicz, D. Bogdal, J. Pielichowski, *Adv. Synth. Catal.* **2003**, *345*, 1269–1272.
26 C. Blanco, S. M. Auerbach, *J. Am. Chem. Soc.* **2002**, *124*, 6250–6251.
27 M. D. Turner, R. L. Laurence, W. C. Conner, K. S. Yngvesson, *AIChE J.* **2000**, *46*, 758–768.
28 X. Zhang, D. O. Hayward, D. M. P. Mingos, *Chem. Commun.* **1999**, 975–976.
29 X. Zhang, D. O. Hayward, D. M. P. Mingos, *Catal. Lett.* **2003**, *88*, 33–38.
30 J. Cheng, R. Roy, D. Agrawal, *Mat. Res. Innovat.* **2002**, *5*, 170–177.
31 (a) E. H. Hong, K.-H. Lee, S. H. Oh, C.-G. Park, *Adv. Funct. Mater.* **2003**, *13*, 961–966; (b) B.-J. Yoon, E. H. Hong, S. E. Jee, D.-M. Yoon, D.-S. Shim, G.-Y. Son, Y. J. Lee, K.-H. Lee, H. S. Kim, C. G. Park, *J. Am. Chem. Soc.* **2005**, *127*, 8234–8235.
32 E. Vázquez, V. Georgakilas, M. Prato, *Chem. Commun.* **2002**, 2308–309.
33 A. R. Harutyunyan, B. K. Pradhan, J. Chang, G. Chen, P. C. Eklund, *J. Phys. Chem. B* **2002**, *106*, 8671–8675.
34 N. F. K. Kaiser, U. Bremberg, M. Larhed, C. Moberg and A. Hallberg, *Angew. Chem. Int. Ed. Engl.* **2000**, *39*, 3595–3598.
35 A. Steinber, A. Stadlet, S. F. Mayer, K. Faber and C. O. Kappe, *Tetrahedron Lett.* **2001**, *42*, 6283–6286.
36 B. Garrigues, R. Laurent, C. Laporte, A. Laporterie, J. Dubac, *Liebigs Ann.* **1996**, 743–744.
37 S. V. Ley, A. G. Leach, R. I. Storer, *J. Chem. Soc., Perkin 1* **2001**, 358–361.
38 N. E. Leadbeater, H. M. Torrenius, *J. Org. Chem.* **2002**, *67*, 3145–148.
39 A. Morcuende, S. Valverde, B. Herradón, *Synlett* **1994**, 89–91.
40 L. Ballel, J. A. F. Joosten, F. Ait el Maate, R. M. J. Liskamp, R. J. Pieters, *Tetrahedron Lett.* **2004**, *45*, 6685–6687.
41 B. Herradón, A. Morcuende, S. Valverde, *Synlett* **1995**, 455–458.
42 A. Morcuende, M. Ors, S. Valverde, B. Herradón, *J. Org. Chem.* **1996**, *61*, 5264–5270.
43 S. Caddick, A. J. McCarroll, D. A. Sandham, *Tetrahedron* **2001**, *57*, 6305–6310.
44 S. V. Ley, D. M. Mynett, *Synlett* **1993**, 793–794.
45 P. Marwah, A. Marwah, H. A. Lardy, *Tetrahedron* **2003**, *59*, 2273–2287.
46 R. S. Varma, M. Varma, A. K. Chatterjee, *J. Chem. Soc., Perkin Trans. 1* **1993**, 999–1000.
47 C. Ramesh, G. Mahender, N. Ravindranath, B. Das, *Green Chem.* **2003**, *5*, 68–70.
48 A. Oussaid, Le Ngoc Thach, A. Loupy, *Tetrahedron Lett.* **1997**, *38*, 2451–2454.
49 N. Deka, J. C. Sarma, *J. Org. Chem.* **2001**, *66*, 1947–1948.
50 D. Stuerga, K. Gonon, M. Lallemant, *Tetrahedron* **1993**, *49*, 6229–6234.
51 (a) I. Forfar, P. Cabildo, R. M. Claramunt, J. Elguero, *Chem. Lett.* **1994**, 2079–2080; (b) P. Cabildo, R. M. Claramunt, I. Forfar, C. Foces-Foces, A. L. Llamas-Saiz, *Heterocycles* **1994**, *37*, 1623–1636.
52 I. Almena, A. Díaz-Ortiz, E. Díez Barra, A. de la Hoz, A. Loupy, *Chem. Lett.* **1996**, 333–334.
53 S. Paul, M. Gupta, *Synthesis* **2004**, 1789–1792.
54 B. M. Khadilkar, V. R. Madyar, *Synth. Commun.* **1999**, *29*, 1195–1200.
55 G. L. Kad, I. R. Trehan, J. Naur, S. Nayyar, A. Arora, J. S. Bar, *Indian J. Chem* **1996**, *35B*, 734–36.
56 P. Klán, J. Literák, S. Relich,

J. Photochem. Photobiol. A 2001, 143, 49–57.
57 M. Hájek, M. T. Radioiu, J. Mol. Catal. A 2000, 160, 383–92.
58 D. Abenhaïm, E. Díez-Barra, A. de la Hoz, A. Loupy, A. Sánchez-Migallón, Heterocycles 1994, 38, 793–802.
59 (a) E. Pérez, A. Loupy, M. Liagre, A. M. Guzzi Plepis, P. J. Cordeiro, Tetrahedron 2003, 59, 865–870; (b) A. Loupy, L. Perreux, M. Liagre, K. Burle and M. Moneuse, Pure Appl. Chem. 2001, 73, 161–166.
60 G. Cardillo, L. Gentilucci, M. Gianotti, A. Tolomelli, Synlett 2000, 1309–1311.
61 A. Esteves-Souza, A. Echevarría, I. Vencato, M. L. Jimeno, J. Elguero, Tetrahedron 2001, 57, 6147–6153.
62 C. C. Silveira, C. R. Bernardi, A. L. Braga, T. S. Kaufman, Synlett 2002, 907–910.
63 S. Jolivet-Fouchet, L. Toupet, F. Texier-Boullet, J. Hamelin, Tetrahedron 1996, 52, 5819–5832.
64 S. Jolivet-Fouchet, J. Hamelin, F. Texier-Boullet, L. Toupet, P. Jacquault, Tetrahedron 1998, 54, 4561–4578.
65 H. Uchida, H. Tanikoshi, S. Nakamura, P. T. Reddy, T. Toru, Synlett 2003, 1117–1120.
66 S. Goswami, A. K. Adak, Tetrahedron Lett. 2002, 43, 8371–8373.
67 G. Abbiati, A. Arcadi, A. Bellinazi, E. Beccalli, E. Rossi, S. Zanzola, J. Org. Chem. 2005, 70, 4088–4095.
68 F. Langa, P. de la Cruz, A. de la Hoz, E. Espíldora, F. P. Cossío, B. Lecea, J. Org. Chem. 2000, 65, 2499–2507.
69 A. Díaz-Ortiz, A. de la Hoz, M. A. Herrero, P. Prieto, A. Sánchez-Migallón, F. P. Cossío, A. Arrieta, S. Vivanco, C. Foces-Foces, Molecular Diversity 2004, 7, 175–180.
70 B. C. Hong, Y. J. Shr, J. H. Liao, Org. Lett. 2002, 4, 663–666.
71 B. Desai, T. N. Danks, G. Wagner, J. Chem. Soc., Dalton Trans. 2004, 166–171.
72 (a) S. Chatti, M. Bortolussi, A. Loupy, J. C. Blais, D. Bogdal, M. Majdoub, Eur. Polym. J. 2002, 38, 1851–1861; (b) S. Chatti, M. Bortolussi, A. Loupy, J. C. Blais, D. Bogdal, P. Roger, J. Appl. Polymer Sci. 2003, 90, 1255–1266; (c) S. Chatti, M. Bortolussi, A. Loupy, J. C. Blais, Eur. Polym. J. 2004, 40, 561–577.
73 (a) M. Iannelli, V. Alupei, H. Ritter, Tetrahedron 2005, 61, 1509–1515; (b) M. Iannelli, H. Ritter, Macromol. Chem. Phys. 2005, 206, 349–353.
74 (a) J. V. Crivello, D. Zhang, J. O. Stoffer, Polym. Mat. Sci. Eng. 1999, 81, 118–119; (b) D. Zhang, J. V. Crivello, J. O. Stoffer, J. Polym. Sci. B 2004, 42, 4230–4246.
75 X. Li, X. Zhu, Z. Cheng, W. Xu, G. Chen, J. Appl. Polymer Sci. 2004, 92, 2189–2195.
76 S. G. Sudrik, S. P. Chavan, K. R. S. Chandrakumar, S. Pal, S. K. Date, S. P. Chavan, H. R. Sonawane, J. Org. Chem. 2002, 67, 1574–1579.
77 P. Nilsson, H. Gold, M. Larhed, A. Hallberg, Synthesis 2002, 1611–1614.
78 P. Ribière, K. Bravo-Altamirano, M. I. Sntczak, J. D. Hawkins, J.-L. Mantchamp, J. Org. Chem. 2005, 70, 4064–4072.
79 L. S. Hegedus, M. J. Sundermann, P. K. Dorhout, Inorg. Chem. 2003, 42, 4346–4354.
80 N. N. Romanova, A. G. Gravis, P. V. Kudan, I. F. Lescheva, N. V. Zyk, Russian J. Org. Chem. 2003, 39, 692–697.
81 J. Singh, M. Sharma, G. L. Kad, B. R. Chhabra, J. Chem. Res. (S) 1997, 264–265.
82 A. Miyazawa, K. Tanaka, T. Sakakura, M. Tashiro, H. Tashiro, G. K. S. Prakash, G. A. Olah, Chem. Commun. 2005, 2104–2106.
83 E. L. Eliel, S. H. Wilen, Stereochemistry of Organic Compounds, John Wiley and Sons, New York 1994.
84 (a) C. Kuang, H. Senboku, M. Tokuda, Tetrahedron Lett. 2001, 42, 3893–3896; (b) C. Kuang, H. Senboku, M. Tokuda, Tetrahedron 2005, 61, 637–642; (c) C. Kuang, Q. Yang, H. Senboku, M. Tokuda, Tetrahedron 2005, 61, 4043–4052.

85 B. M. Chanda, R. Vyas, A. V. Bedekar, *J. Org. Chem.* **2001**, *66*, 30–34.
86 (a) A. K. Bose, B. K. Banik, M. S. Manhas, *Tetrahedron Lett.* **1995**, *36*, 213–216; (b) A. K. Bose, M. Jayaraman, A. Okawa, S. S. Bari, E. W. Robb, M. S. Manhas, *Tetrahedron Lett.* **1996**, *37*, 6989–6992.
87 A. Arrieta, B. Lecea, F. P. Cossío, *J. Org. Chem.* **1998**, *63*, 5869–5876.
88 M. Karthikeyan, R. Kamakshi, V. Sridar, B. S. R. Reddy, *Synth. Commun.* **2003**, *33*, 4199–4204.
89 R. N. Gedye, W. Rank, K. C. Westaway, *Can. J. Chem.* **1991**, *69*, 706–711.
90 M. Pagnota, C. L. F. Pooley, B. Gurland, M. Choi, *J. Phys. Org. Chem.* **1993**, *6*, 407–411.
91 R. G. Jacob, G. Perin, L. N. Loi, C. S. Pinno, E. J. Lenardao, *Tetrahedron Lett.* **2003**, *44*, 3605–3608.
92 J. R. Carrillo-Muñoz, D. Bouvet, E. Guibé-Jampel, A. Loupy, A. Petit, *J. Org. Chem.* **1996**, *61*, 7746–7749.
93 G. D. Yadav, P. S. Lathi, *J. Mol. Catal. A* **2004**, *223*, 51–56.
94 F. Pezet, J.-C. Daran, I. Sasaki, H. Aït-Haddou, G. G. A. Balavoine, *Organometallics* **2000**, *19*, 4008–4015.
95 (a) O. Belda, C. Moberg, *Synthesis* **2002**, 1601–1606; (b) O. Belda, S. Lundgren, C. Moberg, *Org. Lett* **2003**, *5*, 2275–2278; (c) O. Belda, C. Moberg, *Acc. Chem. Res.* **2004**, *37*, 159–167.
96 M. Watanabe, H. Suzuki, Y. Tanaka, T. Ishida, T. Oshikawa, A. Tori, *J. Org. Chem.* **2004**, *69*, 7794–7801.
97 (a) C. Ericsson, L. Engman, *J. Org. Chem.* **2004**, *69*, 5143–5146; (b) L. Engman, V. Gupta, *J. Org. Chem.* **1997**, *62*, 157–173.
98 S. Lutsenko, C. Moberg, *Tetrahedron: Asymmetry* **2001**, *12*, 2529–2532.
99 B. Toukoniitty, J.-P. Mikkola, T. Salmi, D. Yu Murzin, *Proceedings of the 10th International Conference on Microwave and High Frequency Heating*, Modena (Italy), **2005**.
100 M. Y. Chen, L. N. Patkar, C. C. Lin, *J. Org. Chem.* **2004**, *69*, 2884–2887.
101 M. Y. Chen, L. N. Patkar, H. T. Chen, C. C. Lin, *Carbohydr. Res.* **2003**, *338*, 1327–1332.
102 A. Loupy, F. Maurel and A. Sabatié-Gogova, *Tetrahedron* **2004**, *60*, 1683–1691.

6
Microwaves and Phase-transfer Catalysis

André Loupy, Alain Petit, and Dariusz Bogdal

6.1
Phase-transfer Catalysis

In the last three decades there have been many reports of organic reactions performed under phase-transfer catalytic (PTC) conditions [1–4]. This technique has found numerous applications in essentially all fields of organic synthesis, industrial chemistry, biotechnology, and material science. It is encountered in the manufacture of advanced pharmaceuticals, fragrances, crop-protection chemicals, highly advanced engineering plastics, materials for semiconductors, and electro-optical and data storage devices [5]. It is worth remarking that in the 1990s sales of products made by use of phase-transfer catalysis exceeded 10 USD billion year^{-1}.

The concept of PTC was developed mainly because of the work of independent research groups led by Makosza [1] and, later, Starks [2], Brandström [3], and Dehmlow [4]. They introduced techniques in which the reactants were situated in two separate phases, i.e. liquid–liquid or solid–liquid. Because the phases were non-miscible, ionic reagents (i.e. salts, bases, or acids) were dissolved in the aqueous phase while the substrate remained in the organic phase (liquid–liquid PTC). In solid–liquid PTC, on the other hand, ionic reagents can be used in their solid state as a suspension in the organic medium. Transport of the anions from the aqueous or solid phase to the organic phase, where the reaction occurred, was ensured by use of catalytic amounts of lipophilic agents, usually quaternary onium salts or cation-complexing agents (e.g., crown ethers or cryptates). Because the reactions proceeded very slowly, or not at all, in the absence of a catalyst, phase-transfer catalysts were found to be of utmost importance in the extraction of reaction species between phases so that the reaction could proceed, thus increasing yields and rates of reactions substantially.

In general, under PTC conditions, three types of catalytic procedure can be considered (Scheme 6.1):

1. Liquid–liquid PTC in which the inorganic anions or anionic species generated from relatively strong organic acids are located in the aqueous phase and react

Microwaves in Organic Synthesis, Second edition. Edited by A. Loupy
Copyright © 2006 WILEY-VCH Verlag GmbH & Co. KGaA, Weinheim
ISBN: 3-527-31452-0

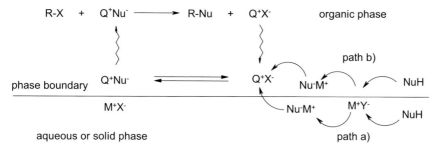

Scheme 6.1

with organic compounds in a liquid–liquid two-phase system. They are transferred the organic phase as loose lipophilic ion pairs, because of the intervention of the phase-transfer catalyst which continuously transfers the anionic species from the aqueous phase to the organic phase, in which the reaction occurs (Scheme 6.1, path a).

2. Liquid–liquid PTC conditions in which weak organic acids react in the presence of concentrated aqueous sodium or potassium hydroxide, which is in contact with the organic phase containing an anion precursor and organic reactants. The anions are created on the phase boundary and continuously introduced, associated with the cations of the catalyst, into the organic phase, in which further reactions occur (Scheme 6.1, path b).

3. Solid–liquid PTC conditions in which the nucleophilic salts (organic or mineral) are transferred from the solid state (as they are insoluble) to the organic phase by means of a phase-transfer agent. Most often, the organic nucleophilic species can be formed by reaction of their conjugated acids with solid bases (sodium or potassium hydroxides, or potassium carbonate) (Scheme 6.1, path b). Another proposed mechanism suggests that interfacial reactions occur as a result of absorption of the liquid phase on the surface of the solid.

The organic phase can be a non-polar organic solvent (benzene, toluene, hexane, dichloromethane, chloroform, etc.) or a neat liquid substrate, usually the electrophilic reagent, which acts both as a reactive substrate and the liquid phase.

In chemical syntheses under the action of microwave (MW) irradiation, the most successful applications are necessarily found to be use of solvent-free systems [6]. In these systems, microwaves interact directly with the reagents and can, therefore, drive chemical reactions more efficiently. The possible acceleration of such reactions might be optimum, because they are not moderated or impeded by solvents. Reactions on solid mineral supports and, in turn, the interaction of microwaves with the reagents on the solid-phase boundary, which can substantially increase the rate of the reactions, are of particular interest [7]. PTC reactions are perfectly tailored for microwave activation, and the combination of solid–liquid PTC and microwave irradiation gives the best results [8]:

1. after ion-pair exchange with the catalyst, the nucleophilic ion pair [Q$^+$Nu$^-$] is a highly polar species especially prone to interaction with microwaves;
2. the extraction processes, during the reaction, in which a PTC mechanism is mainly involved, can be accelerated at the phase boundary, and an increase of reaction rate might be expected; and
3. reactions under solid–liquid PTC-like reactions on solid mineral supports under "dry media", solvent-free conditions often suffer from difficulties with heat transfer through the reaction medium and homogenous heating under conventional conditions. Because microwave irradiation is a means of volumetric heating of materials, temperature is more uniform under microwave conditions. Improvement of temperature homogeneity and heating rates implies faster reactions and less degradation of the final products.

Similarly to classical PTC reaction conditions, under solid–liquid PTC conditions with use of microwaves, the role of catalyst is very important. It has usually been found that in the absence of a catalyst the reactions proceed very slowly or not at all. The need to use a phase-transfer catalyst implies also the application of at least one liquid component (i.e. the electrophilic reagent or the solvent). It has been shown [9] that ion-pair exchange between the catalyst and nucleophilic anions proceeds efficiently only in the presence of a liquid phase.

During investigation of the formation of tetrabutylammonium benzoate from potassium benzoate and tetrabutylammonium bromide, and thermal effects related to this under the action of MW irradiation, it was shown that potassium benzoate did not absorb MW significantly (Fig. 6.1, curves a and b). Even in the pres-

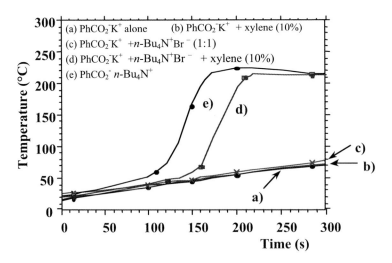

Fig. 6.1. Thermal behavior induced by microwave irradiation of PhCO$_2$K under different conditions (monomode reactor, 180 W).

ence of tetrabutylammonium bromide (TBAB), the temperature increase for solid potassium benzoate was very modest (Fig. 6.1, curve c). In the presence of small amounts of xylene (Fig. 6.1, curve d), a non-polar solvent (i.e. inert to MW irradiation), a large temperature increase provides evidence of the formation of tetrabutylammonium benzoate in the liquid phase, from which the positive thermal effect results (Fig. 6.1, curve d).

It must be stressed that a liquid component can be substituted with an efficient absorber of MW irradiation together with a low-melting component. The use of most typical PTC solvents (non-polar aromatic or aliphatic hydrocarbons, or highly chlorinated hydrocarbons) is most interesting for microwave activation, because such solvents are transparent or absorb microwaves only weakly. They can, therefore, enable specific absorption of MW irradiation by the reagents, and the results or product distributions might be different under microwave and conventional conditions [7]. *Hence, by coupling microwave technology and solid–liquid phase-transfer conditions*, we create *a clean, selective and efficient method* for performing organic reactions, with substantial improvements in terms of conditions and simplicity of operating procedures. This is essentially useful for poorly reactive systems involving, for instance, hindered electrophiles or long-chain halides.

Numerous reactions in organic synthesis can be achieved under solid–liquid PTC and with microwave irradiation in the absence of solvent, usually under normal pressure in open vessels. Increased amounts of reactants can be used to ensure better compatibility between the in-depth penetrability of materials and the radiation wavelength.

Because microwave activation is still a new technique, the number of examples of its combination with PTC might seem to be limited; it has, however, increased substantially from the first edition of the book in 2002. As a close recent development (largely described in Chapter 7 of this book), the use of ionic liquids can often be regarded as equivalent to use of PTC. They consist of salts comprising loose ion pairs between large cations (most often imidazolium or pyridinium) and large charge-delocalized anions. They behave as a very polar solvent or as species able to absorb strongly MW irradiation and therefore able to enhance reaction temperature.

6.2
Synthetic Applications of Phase-transfer Processes

6.2.1
O-Alkylations

In conventional methods, PTC has provided interesting procedures for O-alkylation, and coupling PTC conditions with MW activation has proved quite fruitful for such reactions.

Tab. 6.1. Alkylation of $CH_3COO^-K^+$ under MW + PTC conditions (domestic oven, 600 W).

RX	t (min)	Final temperature (°C)	Yield (%)
$n\text{-}C_8H_{17}Br$	1	187	98
$n\text{-}C_8H_{17}Cl$	1	162	98
$n\text{-}C_8H_{17}I$	2	165	92
$n\text{-}C_{16}H_{33}Br$	1	169	98

6.2.1.1 Synthesis of Esters

Alkyl acetates Potassium acetate can be readily alkylated in a domestic microwave oven by use of equivalent amounts of salt and alkylating agent in the presence of Aliquat 336 (10 mol%). Some important results, exemplified in Eq. (1), are given in Table 6.1 [10, 11].

$$CH_3COO^-K^+ + R-X \longrightarrow CH_3COOR + K^+X^- \quad (1)$$

Yields are always almost quantitative within 1–2 min, irrespective of chain length and the nature of the halide leaving groups. This procedure was scaled up from 50 mmol to the 2 mol scale (i.e. from 15.6 to 622.4 g total starting materials) in a larger batch reactor (Synthewave 1000 from Prolabo) [12]. Yields were equivalent to those obtained under similar conditions (5 min, 160 °C) in laboratory-scale experiment (Synthewave 402) with nearly equivalent set of conditions of reaction time, temperature, and emitted electric power (Table 6.2).

Xu et al. have obtained similar results with n-butyl bromide using TBAB (10 mol%) and alumina (4:1 w/w) as the catalyst [13]. Benzyl acetate was also conveniently prepared from sodium acetate and benzyl halide by use of MW irradiation and PTC in synergy [14].

Tab. 6.2. Synthesis of n-octyl acetate under MW + PTC conditions (Synthewave, 5 min, 160 °C).

Reactor	Amounts of material [g (mol)]			Total amount (g)	Yield (%)
	CH_3COOK	n-OctBr	Aliquat		
MW S402	4.9 (0.05)	9.7 (0.05)	1 (0.0025)	15.6	98
MW S1000	196 (2)	386 (2)	40.4 (0.1)	622.4	98

Tab. 6.3. Reaction of benzyl halides with hexanoic acid under MW + PTC conditions (10 min, 560 W, 10% PhCH$_2$N$^+$Me$_3$Cl$^-$).

ArCH$_2$X	Yield (%)
Benzyl bromide	72
Benzyl iodide	90
α-Bromo-*p*-xylene	76
α-Iodo-*p*-xylene	92
α-Bromomesitylene	81

Long-chain esters   As a generalization of the above method, stearyl stearate was synthesized within 1 min in quantitative yield (Eq. 2) [11].

$$C_{17}H_{35}COO^-K^+ \; + \; n\text{-}C_{18}H_{37}Br \; \xrightarrow[\text{MW (600W) 1 min}]{\text{Aliquat 10\%}} \; C_{17}H_{35}COO \; n\text{-}C_{18}H_{37} \quad\quad (2)$$
$$99\%$$

It has been shown that reaction of carboxylic acids with benzyl halides, which does not occur when heated conventionally, could be performed efficiently under the action of MW irradiation in the presence of a quaternary ammonium salt as a catalyst (Eq. 3) [15]. Typical results are given in Table 6.3.

$$\text{Ar}-\text{CH}_2\text{X} \; + \; n\text{-}C_5H_{11}\text{COOH} \; \xrightarrow{Q^+X^-} \; \text{Ar}-\text{CH}_2-\text{O}-\text{CO}-\text{C}_5\text{H}_{11} \; + \; \text{HX} \quad\quad (3)$$

Aromatic esters   It is possible to alkylate benzoic acids directly, without the need to prepare reactive potassium salts in a separate step, because they can be generated *in situ* by reacting the acid with a base (potassium carbonate or hydroxide) in the presence of a phase-transfer catalyst. As an illustration of this principle, a volatile polar molecule is a by-product, eliminated as a result of exposure to MW (Eq. 4), and the equilibrium is shifted to completion. The second effect of irradiation is activation of the alkylation step itself (Eq. 5). All the reagents can be used in the theoretical stoichiometry. Some indicative results are given in Table 6.4 [9].

$$\text{ArCOOH} \; + \; \text{KOH} \; \xrightleftharpoons{\text{Aliquat 10\%}} \; \text{ArCOO}^-K^+ \; + \; H_2O\nearrow \quad\quad (4)$$

$$\text{ArCOO}^-K^+ \; + \; R-X \; \xrightarrow{\text{Aliquat 10\%}} \; \text{ArCOOR} \; + \; K^+X^- \quad\quad (5)$$

A striking example in this series is the alkylation of terephthalic acid (Eq. 6). The specific effect of microwaves is clearly apparent in this example, because, other fac-

Tab. 6.4. Alkylation of potassium 4-Z-benzoate under MW + PTC conditions (domestic oven, 600 W, 10% Aliquat).

Z	t (min)	Yield (%)	
		Preformed salt	Salt *in situ*
H	2.5	99	99
NMe$_2$	3	97	100
OMe	2	82	98
CN	3	80	95
NO$_2$	2	81	95

tors being equal, the yields are unambiguously much higher. It is noteworthy that the specific not purely thermal MW effects are important in the more difficult reaction and this point is discussed in Chapter 4 of this book, on effects depending on the position of the transition state along the reaction coordinates.

$$\text{terephthalic acid} + 2\ n\text{-OctBr} \xrightarrow[\text{Aliquat 10\%}]{2\ K_2CO_3} \text{di-}n\text{-octyl terephthalate} \quad (6)$$

MW (600W)	7 min	227 °C	87%
Classical heating	7 min	227 °C	20%

Solid–liquid solvent-free PTC was applied, with noticeable improvement and simplification compared with classical procedures, in a green chemistry context, for synthesis of some aromatic esters useful as cosmetic ingredients – 3-methylbutyl 4-methoxycinnamate, 2-ethylhexyl 4-methoxycinnamate, 2-ethylhexyl 4-(dimethylamino)benzoate, and 2-ethylhexyl salicylate, all well-known ultraviolet B sunscreen filters, 4-isopropylbenzyl salicylate, a UV absorber and cutaneous antilipoperoxidant, the parabens propyl 4-hydroxybenzoate and butyl 4-hydroxybenzoate (Eq. 7), and antimicrobial agents [16].

$$\text{ex}:\ HO\text{-}C_6H_4\text{-}COOH + nC_4H_9Br \xrightarrow[\text{Aliquat 336}]{K_2CO_3}_{\text{10 min 160°C}} HO\text{-}C_6H_4\text{-}COOnC_4H_9 \quad 95\% \quad (7)$$

O-Alkylation reactions of carboxylic acids such as aryloxyacetic acids (unsubstituted furoic acids and benzofuroic acids) with (un)substituted ω-haloacetophenones in dry media under the action of MW with PTC have been described [17].

Tab. 6.5. Synthesis of ethers under MW + PTC conditions (domestic oven, 560 W).

R	R'X	t (min)	Yield (%)
Et	$PhCH_2Cl$	5	85
n-Bu	$PhCH_2Cl$	10	78
n-Oct	$PhCH_2Cl$	10	88
n-Oct	n-BuBr	10	78
$PhCH_2$	n-BuBr	10	92

6.2.1.2 Ether Synthesis

Aliphatic ethers Yuan et al. studied two types of conditions for this reaction, using either the alcohols or the corresponding halides as starting materials [18, 19]. In the presence of quaternary ammonium salts, the reactions shown in (Eq. 8) were complete within a few minutes. Typical results are given in Table 6.5.

$$ROH + R'X \xrightarrow{PhCH_2N^+Me_3Cl^-} R-O-R' + HX \qquad (8)$$

More recently, this method has been extensively applied to a wide range of Williamson syntheses in dry media with K_2CO_3 and KOH as bases, TBAB as phase-transfer agent, and a variety of aliphatic alcohols (e.g. n-octanol and n-decanol, yields 75–92%) [20].

The direct O-alkylation of 2-bromo-3-pyridinol, via the standard Williamson reaction (Eq. 9) leads to competitive reactions owing to the drastic experimental conditions required (basic medium, high temperatures, long reaction times). Very short reactions times of 45 to 60 s proved to be sufficient to achieve O-alkylation almost quantitatively (Table 6.6) under MW + PTC conditions. Attempts to perform the reaction in the absence of catalyst (TBAB) failed [21].

Tab. 6.6. Synthesis of 3-alkoxy-2-bromopyridine under MW + PTC conditions (domestic oven, 300 W).

R	t (s)	Yield (%)
$CH_3(CH_2)_5$	45	86
$CH_3(CH_2)_7$	50	84
$CH_3(CH_2)_9$	60	85
$CH_3(CH_2)_{11}$	60	82

$$\text{pyridine-OH derivative} + \text{R-Br} \xrightarrow[\text{TBAB}]{\text{KOH}} \text{pyridine-OR derivative} \quad (9)$$

Furanic diethers [22] A new family of furanic diethers has been obtained by alkylation of 2,5-furandimethanol (Eq. 10) under the action of microwaves under PTC solvent-free conditions.

$$\text{HOH}_2\text{C-furan-CH}_2\text{OH} + 2\ \text{R-X} \xrightarrow[\text{Aliquat}]{\text{KOH}} \text{ROH}_2\text{C-furan-CH}_2\text{OR} \quad (10)$$

Microwaves Monomode Reactor (30W)	5 min	180°C	93%
Conventional heating (oil bath)	5 min	180°C	41%

The diethers were synthesized in high yields with short reaction times. When compared with classical heating, under otherwise comparable conditions, reaction times were substantially improved by use of microwave activation.

Diethers from dianhydrohexitols A series of new ethers has been obtained by alkylation of dianhydrohexitols (isosorbide, isoidide, isomannide) under the action of MW irradiation and PTC conditions. Yields exceeded 90%, a dramatic improvement compared with those from conventional heating with similar temperature profiles. The best yields, for example from isosorbide, were obtained in the presence of a small amount of xylene and TBAB as catalyst at 140 °C (Eq. (11), Table 6.7) [23].

Tab. 6.7. Reaction of isosorbide with several alkylating agents in the presence of KOH and TBAB (relative amounts 1:3:3:0.1) under the action of MW irradiation in xylene solution.

R-X	t (min)	T (°C)	Yield (%)	
			MW	Δ
n-$C_8H_{17}Br$	5	140	96	10
$C_6H_5CH_2Cl$	5	125	98	13
3-Cl-$C_6H_4CH_2Cl$	5	125	95	15
4-Cl-$C_6H_4CH_2Cl$	5	125	96	14
3-F-$C_6H_4CH_2Cl$	5	125	95	15
$CH_3CH_2OCH_2CH_2Br$	30	100	78	45

6.2 Synthetic Applications of Phase-transfer Processes

$$\text{HO-[isosorbide]-OH} + 2\ R\text{-}X \xrightarrow[\text{Xylene 5 min}]{\text{KOH / TBAB}} \text{RO-[isosorbide]-OR} \quad (11)$$

R = n-Oct, ArCH$_2$, CH$_3$CH$_2$OCH$_2$CH$_2$
X = Cl, Br

Later, several polyethers were synthesized from isosorbide and isoidide by means of a MW-assisted PTC method (Eq. 12). In addition to increasing the rate of the reaction, the MW affected the structure of the polymers (determined by MALDI–TOF mass spectrometry) (Table 6.8).

$$\text{HO-[isoidide]-OH} + X\text{-}(CH_2)_8\text{-}X \xrightarrow[\substack{\text{Toluene}\\ \text{30 min 110°C}}]{\text{KOH / TBAB}} \left[\text{O-[isoidide]-O-}(CH_2)_8\right]_n \quad (12)$$

MW-assisted synthesis proceeded more rapidly than with conventional heating, and reaction time was reduced to 30 min with the yield of approximately 67–71% of the high-molecular-weight fraction (FP MeOH), which is more important here. Under conventional conditions this fraction was afforded in ca 10–12% yield within 30 min. Under conventional conditions similar yields of the polyethers

Tab. 6.8. Polyethers from isoidide and 1,8-dimesyloctane. Effect of reaction time on the yields of high-molecular-weight fraction (FP MeOH), low molecular weight fraction (FP Hex), and molecular weight distribution. Mn and Mw are, respectively, the number average and weight average molecular weights of the FP MeOH fraction and the ratio Mw/Mn is the polydispersity index (monomode microwave reactor, 300 W).

t	Activation mode[a]	FP MeOH (%)	FP hex (%)	Total yield (%)	Mn (g mol^{-1})	Mw (g mol^{-1})	Mw/Mn
30 min	MW	67	18	85	3500	4700	1.34
60 min	MW	71	19	90	5100	6900	1.35
30 min	Δ	12	81	93	2600	3400	1.31
1 day	Δ	64	25	89	3200	4100	1.28
1 week	Δ	83	5	88	3900	5100	1.31
1 month	Δ	91	0	91	4100	5000	1.22

[a] MW, microwave irradiation; Δ, conventional thermal heating

were obtained when the reaction time was extended to 24 h. These yields remained practically unchanged even though the synthesis was performed for another 7 days (Table 6.8). It was also found that the mechanism of chain termination is different under microwave and conventional conditions. The polyethers prepared with conventional heating have shorter chains with terminal hydroxyl groups whereas under MW irradiation the polymer chains were longer with terminal ethylenic ends. In fact, under the action of microwaves, terminal ethylenic ends were formed rapidly and set up a hindrance to further polymer growth. In contrast, under conventional conditions, terminations essentially consisted of hydroxyl functions [24].

Phenolic ethers Gram quantities of aryl-2-(N,N-diethylamino)ethyl ethers, compounds of biological interest, have been prepared in a household microwave oven, with potassium hydroxide, and glyme as the transfer agent, according to Eq. (13)] [25]:

$$\text{ArOH} \xrightarrow[\text{ClCH}_2\text{CH}_2\text{NEt}_2]{\text{KOH, glyme}} \text{Ar-O-CH}_2\text{CH}_2\text{NEt}_2 \quad 92\text{-}99\% \tag{13}$$
MW (600W) 2 min

Under the action of MW irradiation several phenols react remarkably fast in dry media with primary alkyl halides to give aromatic ethers (Eqs. 14 and 15) [26, 27].

$$\text{ArOH} + \text{R-X} \xrightarrow[\text{MW (300W) 25-65 sec.}]{\text{K}_2\text{CO}_3 / \text{KOH, TBAB 10\%}} \text{Ar-OR} \tag{14}$$

ex : Y = H   R-X = PhCH$_2$Cl   50 sec.   78%
               Me$_2$SO$_4$ 25 sec.   87%
Y = p NH$_2$ PhCH$_2$Cl   40 sec.   89%

$$\text{Cl-C}_6\text{H}_4\text{-OH} + \text{ClCH}_2\text{COOH} \xrightarrow[\text{H}_2\text{O}]{\text{NaOH, Et}_3\text{N}^+\text{C}_{12}\text{H}_{25}\text{ Br}^-} \text{Cl-C}_6\text{H}_4\text{-OCH}_2\text{COOH}$$

MW   0.5 h   92%
Δ   5 h   96%

$$\tag{15}$$

2′-Benzyloxyacetophenone is an important intermediate in the manufacture of drugs used as diuretics, antihypertensives, platelet anti-aggregants, lipoxygenase analgesics, and prostaglandins, and for treatment of metabolic disorders. It was elegantly formed by synergistic combination of solid–liquid PTC and MW irradiation (Eq. 16) [28]. The rates of reaction are increased by orders of magnitude and the reaction is selective at 80 °C, in comparison with liquid–liquid PTC which is slow and produces by-products. The same system has been used with a series of phenolic compounds (Eq. 17) [29, 30].

6.2 Synthetic Applications of Phase-transfer Processes | 289

$$\text{2-hydroxyacetophenone} + \text{benzyl chloride} \xrightarrow[\text{Toluene}]{\text{NaOH, TBAB}} \text{O-benzylated product} \quad (16)$$

$$\text{o-cresol} + \text{benzyl chloride} \xrightarrow[\text{Toluene, 90°C}]{\text{NaOH, TBAB}} \text{O-benzylated product} \quad (17)$$

		Δ	MW
10 min	conversion	4%	24%
3 h	conversion	84%	85%
	selectivity (O-alkylated product)	89.6%	98.8%

Catechol has been reacted with β-methallyl chloride under the action of MW and PTC conditions. Yields of 2-methallyloxyphenols varied from 59 to 68% under liquid–liquid conditions (Table 6.9), whereas no reaction was observed in a solid–liquid PTC procedure (Eq. 18) [31].

$$\text{catechol} + H_2C=C(CH_3)Cl \xrightarrow[\text{H}_2\text{O, MW, 1-2 min}]{\text{base, PTC}} \text{mono} + \text{di} \quad (18)$$

A similar reaction is the methylenation of 3,4-dihydroxybenzaldehyde in the presence of a phase-transfer catalyst on a benign calcium carbonate surface [32]. Presumably, bonding of the vicinal hydroxyl groups is low, thereby enhancing the reaction with the alkylating agent under the action of solvent-free MW irradiation (Eq. 19).

Tab. 6.9. Reaction of catechol with β-methallyl chloride (8 mmol) in the presence of a base (2 mmol) and phase-transfer catalyst (0.2 mmol) under the action of microwave irradiation.

Base	PTC	Yield (%)	Mono/di ratio
NaOH	PEG-400	64	6.1
KOH	TBAB	59	5.1
K_2CO_3	TBAB	68	6.4

Tab. 6.10. Alkylation of 3,4-dihydroxybenzaldehyde with long-chain alkyl bromide under the action of MW + PTC conditions.

R	$C_{12}H_{25}$	$C_{14}H_{29}$	$C_{16}H_{33}$	$C_{18}H_{37}$
Yield (%)	86 (Δ18)	80	81	80

$$\text{HO-C}_6\text{H}_3(\text{OH})\text{-CHO} \xrightarrow[\text{Support / MW}]{\text{ICH}_2\text{I / TBAB / CaCO}_3} \text{methylenedioxybenzaldehyde} \quad 3 \text{ min } 157°\text{C } 95\% \tag{19}$$

With the aim of improvement and simplification of the synthesis of potential cosmetic compounds, alkylations of mono- and di-hydroxybenzaldehydes with long-chain halides were efficiently realized under solvent-free MW + PTC conditions (Eqs. (20) and (21), Table 6.10) [33, 34].

$$\text{OHC-C}_6\text{H}_4\text{-OH} + n\text{-C}_{16}\text{H}_{33}\text{Br} \xrightarrow[\text{15 min } 140°\text{C}]{\text{KOH / Aliquat}} \text{OHC-C}_6\text{H}_4\text{-O-}n\text{C}_{16}\text{H}_{33} \quad \text{MW } 96\% \quad \Delta \; 22\% \tag{20}$$

$$\text{OHC-C}_6\text{H}_3(\text{OH})\text{-OH} + 2\,\text{RBr} \xrightarrow[\text{TBAB} \atop \text{MW, 10 min } 130°\text{C}]{\text{KOH + K}_2\text{CO}_3} \text{OHC-C}_6\text{H}_3(\text{OR})\text{-OR} \tag{21}$$

The synthesis of 8-quinolinyl ethers from 8-hydroxyquinolines and organic halides under PTC conditions and microwave irradiation has also recently been reported (Eq. 22) [35].

$$\text{Y-quinoline-OH} + \text{R-X} \xrightarrow[\text{DMF} \atop \text{MW (750W) 3-15 min}]{\text{NaOH, PEG 400}} \text{Y-quinoline-OR} \quad 76\text{-}91\% \tag{22}$$

Ex : Y = H   R-X = PhCH$_2$Cl   7 min   87%
               PhCOCH$_2$Br   13 min   91%

Desyl ethers are key intermediates in the synthesis of biologically active furanopyrones. MW-assisted synthesis of these compounds has been performed in an open vessel under PTC solvent-free conditions (Eq. 23) by alkylation of 7-hydroxy-4-methyl coumarin with desyl chloride [36].

$$\text{H}_3\text{CO} \underset{\text{OH}}{\overset{\text{O} \quad \text{O}}{\bigotimes}} + \underset{\text{Cl}-\text{CH}-\text{COPh}}{\overset{\text{Ph}}{|}} \xrightarrow[\text{MW 2 min}]{\text{K}_2\text{CO}_3 / \text{NBu}_4\text{HSO}_4} \text{H}_3\text{CO} \underset{98\%}{\overset{\text{O} \quad \text{O}}{\bigotimes}} \underset{\text{O}-\text{CH}-\text{COPh}}{\overset{\text{Ph}}{|}} \quad (23)$$

Alkylations of phenols with epichlorhydrin under PTC conditions and MW irradiation were described twice in 1998. Subsequently, ring-opening reactions of the epoxide group were also performed using microwaves (Eqs. 24 and 25) [37, 38]. In the first publication, catalytic synthesis of chiral glycerol sulfide ethers was described [37]. The second reported the preparation of biologically active amino ethers [38].

$$\text{ArOH} + \text{Cl}\diagdown\!\!\!\triangle\text{O} \xrightarrow[\text{MW}]{\text{NaOH, TBAB}} \text{ArO}\diagdown\!\!\!\triangle\text{O} \xrightarrow[\substack{\text{NaOH, PEG 400}\\ \text{DMF}\\ \text{MW (750W) 8 min}}]{\text{Ar'SH}} \underset{67\text{-}92\%}{\text{ArO}\diagdown\!\!\underset{\text{OH}}{|}\!\!\diagup\text{SAr'}} \quad (24)$$

$$\text{ArOH} + \text{Cl}\diagdown\!\!\!\triangle\text{O} \xrightarrow[\text{MW}]{\text{K}_2\text{CO}_3/\text{KOH, TBAB}} \underset{67\text{-}96\%}{\text{ArO}\diagdown\!\!\!\triangle\text{O}}$$

$$\text{ArO}\diagdown\!\!\!\triangle\text{O} \xrightarrow[\substack{\text{K}_2\text{CO}_3/\text{NaOH, TBAB}\\ \text{Silica gel}\\ \text{MW 10-40 min}}]{\text{RNH}_2} \underset{67\text{-}89\%}{\text{ArO}\diagdown\!\!\underset{\text{OH}}{|}\!\!\diagup\text{NHR}} \quad (25)$$

Condensation of salicylaldehyde and its derivatives with a variety of esters of chloroacetic acids in the presence of TBAB led to the synthesis of benzo[b]furans by means of a solid–liquid PTC reaction under the action of microwave irradiation [39]. This was a modification of one of the most popular routes to substituted benzo[b]furans, i.e. O-alkylation of O-hydroxylated aromatic carbonyl compounds with α-halogenated carbonyl compounds then intramolecular condensation. The mixture of aldehydes and chloroacetic acid esters were absorbed on potassium carbonate then irradiated in an open vessel in a domestic MW oven for 8–10 min (Eq. 26).

$$\underset{\underset{Y}{X}}{\overset{\text{CHO}}{\bigodot\!\!\text{OH}}} + \text{Cl}\diagdown\!\!\diagup\text{CO}_2\text{R} \xrightarrow[\text{8 - 10 min}]{\text{K}_2\text{CO}_3, \text{TBAB}} \underset{\underset{Y}{X}\ (1)}{\overset{\text{CHO}}{\bigodot\!\!\text{O}\diagdown\!\!\diagup\text{CO}_2\text{R}}} \longrightarrow \underset{\underset{Y}{X}\ (2)}{\bigodot\!\!\bigodot\!\!\text{CO}_2\text{R}} \quad (26)$$

It was found that the first step was rate-determining. When, moreover, the reaction was performed with the same reaction temperature profiles under both conventional heating (oil bath) and MW (monomode cavity) conditions, different dis-

Tab. 6.11. Distribution of the intermediate (**1**) and final product (**2**) in the synthesis of benzo[b]furans under both conventional and MW conditions.

T (°C)	t (min)	Solvent/amount (mL)	Yield (%) 1	Yield (%) 2
110	10/MW	–	12	66
110	20/MW	–	–	100
110	20/Δ	–	4	96
85	20/MW	–	15	85
85	20/Δ	–	90	10
85	30/MW	Cyclohexane/1.5	62	38
85	30/MW	EtOH(100%)/1.5	0	100
85	30/MW	EtOH(96%)/1.5	0	100
85	30/Δ	EtOH(100%)/1.5	47	53

tributions of the intermediate (**1**) and final (**2**) products were obtained (Table 6.11). Indeed, the product distribution was strongly affected by microwaves when the reaction was run at 85 °C rather than 110 °C, and addition of a small amount of a polar or a non-polar solvent also affected the distribution of the products. In this work two solvents capable of extensive coupling (i.e. ethanol) and not coupling (i.e. cyclohexane) with microwaves were used. Addition of ethanol strongly shifted the product distribution towards the final product (**2**), whereas addition of cyclohexane resulted in much lower yield of **2** [40].

Phenolic polyethers Polycondensations of 3,3-bis(chloromethyl)oxetane and a variety of bisphenols have been studied using the MW–PTC technique (Eq. 27) [41]. The results obtained showed the advantages of microwaves in terms of the molecular weights of the crystalline polymer, which were reflected in higher values of the transition temperatures (T_g) and melting points (T_m); reaction times were also reduced for all types of structure.

	Microwave	Conventional thermal
Yield	72%	50%
T_g	117°C	103°C
T_m	256°C	232°C

(27)

Tab. 6.12. N-Alkylation of saccharin with bromides under MW + PTC conditions (750 W).

R	Catalyst	t (min)	Yield (%)
PhCH$_2$	None	4	8
	SiO$_2$	4	75
	SiO$_2$ + TEBA	4	92
n-C$_{16}$H$_{33}$	SiO$_2$	6	82
	SiO$_2$ + TEBA	6	97

6.2.2
N-Alkylations

6.2.2.1 Saccharin [42]

Rapid N-alkylation of saccharin sodium salt (Eq. 28) by a series of halides has been performed on silica gel in a domestic MW oven. TEBA was shown to provide a useful co-catalytic effect (Table 6.12).

$$\text{Saccharin-N}^{\ominus}\text{Na}^{\oplus} + \text{R-Br} \xrightarrow[\text{TEBA}]{\text{Silica gel}} \text{Saccharin-N-R} \qquad (28)$$

6.2.2.2 Benzoxazinones and Benzothiazinones

Rapid N-alkylation of these compounds was performed under the action of PTC with MW irradiation (Eq. 29) [43].

$$\text{Y = O, S} + \text{R-X} \xrightarrow[\substack{\text{Silica gel} \\ \text{MW 8–10 min}}]{\text{NaOH, TBAB}} \text{product, 72–90\%} \qquad (29)$$

Y = O	RX = MeI	8 min	90%
	PhCH$_2$Br	10 min	84%
Y = S	RX = MeI	8 min	86%
	PhCH$_2$Br	10 min	80%

Michael reaction of the same compounds with acrylonitrile and methyl acrylate was promoted under the same conditions within 9–10 min [44].

6.2.2.3 Barbitone [45]

Under solid–liquid PTC conditions, 5,5-diethylbarbituric acid was N,N-dialkylated in a good yield in the presence of a lipophilic ammonium salt and potassium carbonate when reaction mixtures were irradiated in a household MW oven (Eq. 30).

$$\text{Barbituric acid} + 2\ R-X \xrightarrow[\text{MW (400W) 6-18 min}]{K_2CO_3,\ Me_3N^+C_{16}H_{33},\ Br^-} \text{N,N-dialkyl product} \qquad (30)$$

Ex : PhCH$_2$Br, 10 min, 99%   86-99%

6.2.2.4 Amides and Lactams

The N-alkylation of acetanilide under the action of MW irradiation in the presence of a phase-transfer catalyst has been reported (Eq. 31) [46].

$$H_3C-C(O)-NHPh + R-X \xrightarrow[\text{MW}]{\text{NaOH, TBAB}} H_3C-C(O)-N(R)-Ph \qquad (31)$$

Ex : R-X = nBuI   80 sec   88%

N-Substituted amides and lactams can be rapidly N-alkylated under solid–liquid PTC conditions with MW irradiation. The reactions were performed simply by mixing an amide with a 50% excess of an alkyl halide and a catalytic amount of TBAB. These mixtures were absorbed on a mixture of potassium carbonate and potassium hydroxide [47] and then irradiated in an open vessel in a domestic MW oven for 55–150 s (Eq. 32).

$$R-X + R^1-NH-C(O)-R^2 \xrightarrow[\text{MW, 55-120 sec}]{K_2CO_3,\ KOH,\ TBAB} R^1-N(R)-C(O)-R^2$$

$$R-X + (H_2C)_n\text{-lactam(NH)} \xrightarrow[\text{MW, 80-150 sec}]{K_2CO_3,\ KOH,\ TBAB} (H_2C)_n\text{-lactam(N-R)} \qquad (32)$$

Microwaves   55-150 sec   45-92%

The starting reagents in the Gabriel amine synthesis, N-alkylphthalimides, have been obtained under the action of MW irradiation in a solid–liquid PTC system. The reactions were conducted with high yields (50–90%) simply by mixing phthalimide with 50% excess alkyl halide and catalytic TBAB, which were later absorbed on potassium carbonate. Irradiation of the reaction mixtures in a domestic oven led to the desired phthalimide derivatives (Eq. 33) within short reaction times (4–10 min) [48].

$$\text{Phthalimide-NH} + \text{R-X} \xrightarrow[\text{MW (450W) 4-10 min}]{\text{K}_2\text{CO}_3,\ \text{TBAB}} \text{Phthalimide-NR} \quad 49\text{-}95\% \tag{33}$$

Ex : R-X = PhCH$_2$Cl   4 min   93%
n-BuBr   4 min   73%
n-C$_{12}$H$_{25}$I   10 min   95%

The same reaction (R–X = n-OctBr) was studied using TBAB and several basic supports. Na$_2$SO$_4$ and CaCO$_3$ (yields 93 and 95%, respectively, within 3 min of MW irradiation) were shown to be more efficient than K$_2$CO$_3$ (71%) [49].

Syntheses of N-alkyl phthalimides and N-alkyl succinimides by alkylation of potassium salts have been performed on silica gel in dry media under MW + PTC conditions with fairly high yields [50].

Melatonin has been prepared from phthalimide by successive N and C-alkylation, cyclization, hydrolysis, decarboxylation, and acetylation. The four one-pot reactions were performed under MW irradiation in good yields with short reaction times (Eq. 34) [51].

$$\text{Phthalimide-NH} \xrightarrow[\substack{2\text{-}\ \text{CH}_3\text{COCH}_2\text{COOC}_2\text{H}_5 \\ \text{K}_2\text{CO}_3\ /\ \text{MWI}}]{\substack{1\text{-}\ \text{Br(CH}_2)_3\text{Br}\ /\ \text{K}_2\text{CO}_3\ /\ \text{TEBA} \\ \text{CH}_3\text{CN}\ /\ \text{MWI}}} \text{Phthalimide-N-(CH}_2)_3\text{-CH(COOC}_2\text{H}_5)\text{-COCH}_3 \tag{34}$$

6.2.2.5 Aromatic Amines

N-Ethylaniline has been alkylated by reaction with benzyl chloride under liquid–liquid PTC conditions in the presence of 30% sodium hydroxide solution and CTAB as catalyst (Eq. 35). Microwave irradiation (25 min) of the reaction mixture in a sealed vessel afforded N-benzyl-N-ethylaniline in 90% yield, compared with 16 h of conventional heating (oil bath) [27].

$$\text{PhNHEt} + \text{PhCH}_2\text{Cl} \xrightarrow[\text{H}_2\text{O}]{\text{NaOH / CTAB}} \text{Ph-N(Et)-CH}_2\text{Ph} \tag{35}$$

MW   25 min   90%
Δ   16 h   90%

6.2.2.6 N-Phenylpyrrolidino[60]fullerene [52]

A facile synthesis of a series of N-alkylpyrrolidino[60]fullerenes by solvent-free PTC under the action of microwaves has been described (Eq. 36).

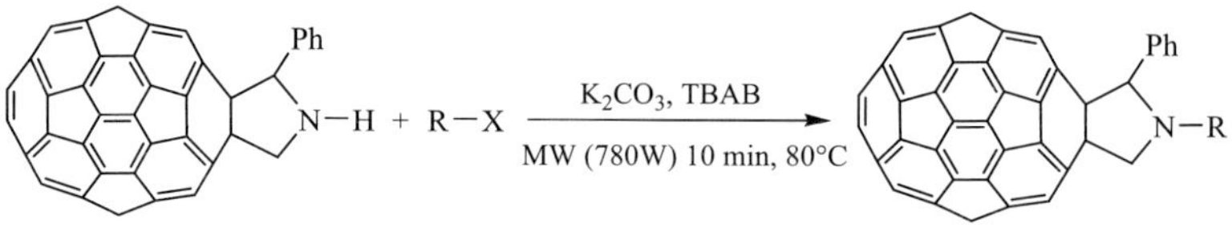

$$\text{[C}_{60}\text{-pyrrolidine]N-H} + \text{R-X} \xrightarrow[\text{MW (780W) 10 min, 80°C}]{\text{K}_2\text{CO}_3\text{, TBAB}} \text{[C}_{60}\text{-pyrrolidine]N-R} \tag{36}$$

Ex : R-X = PhCH$_2$Br   70%
CH$_2$=CH-CH$_2$Br   73%
n-C$_8$H$_{17}$Br   31%

The synergy between the dry media and the MW irradiation was convincingly demonstrated in this work. For example, with the allylic compound the yield is only 16% after 24 h in toluene under reflux whereas no reaction occurs after 10 min at 100 °C by classical heating, thus revealing an important specific microwave effect.

6.2.2.7 Five-membered Nitrogen Heterocycles

Under the action of MW irradiation, several azaheterocycles (pyrrole, imidazole, indole, and carbazole) can react remarkably quickly with alkyl halides to give N-alkyl derivatives (Eqs. 37 and 38) [53, 54]. Such reactions have been performed simply by mixing the azaheterocyclic compound with 50% excess alkyl halide and a catalytic amount of TBAB. The reactants were absorbed either on a mixture of potassium carbonate and potassium hydroxide or on potassium carbonate alone and then irradiated in a domestic MW oven for 30 s–10 min.

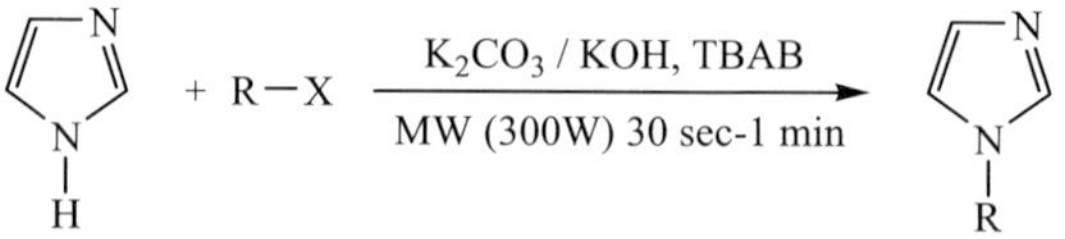

$$\text{imidazole-NH} + \text{R-X} \xrightarrow[\text{MW (300W) 30 sec-1 min}]{\text{K}_2\text{CO}_3 \text{ / KOH, TBAB}} \text{imidazole-NR} \quad 73\text{-}89\% \tag{37}$$

Ex : PhCH$_2$Cl, 40 sec, 89%

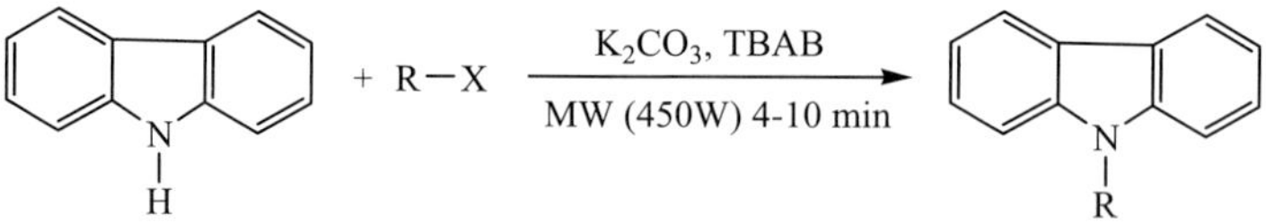

$$\text{carbazole-NH} + \text{R-X} \xrightarrow[\text{MW (450W) 4-10 min}]{\text{K}_2\text{CO}_3\text{, TBAB}} \text{carbazole-NR} \quad 79\text{-}95\% \tag{38}$$

Ex : R-X = PhCH$_2$Cl   4 min   70%
n-BuBr   4 min   73%
n-C$_{10}$H$_{21}$Cl   10 min   95%

6.2.2.8 Pyrimidine and Purine Derivatives [55]

Selective N-alkylation of 6-amino-2-thiouracil with different halides has been performed efficiently by use of MW-assisted methods in the presence of small

Tab. 6.13. Alkylation of 6-amino-2-thiouracil under PTC + MW conditions.

RX	t (min)	T (°C)	Yield (%)
$p\text{-}NO_2C_6H_4CH_2Cl$	6	70–75	91
$p\text{-}BrC_6H_4COCH_2Br$	6	70–75	95
$Br\text{-}(CH_2)_5\text{-}Br$	10	80–85	87
$n\text{-}C_{16}H_{33}Br$	9	105–110	91

amounts of DMF to improve energy transfer (Eq. (39a), Table 6.13) [55a]. No reaction was observed under the same conditions in a thermoregulated oil bath.

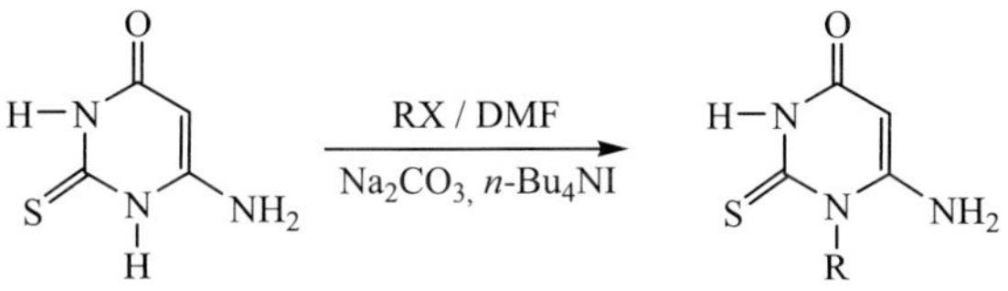

(39a)

Microwave-assisted ring opening of epoxides with pyrimidine nucleobases provided rapid entry into C-nucleoside synthesis. The use of microwaves led to different selectivity from that resulting from conventional heating (Eq. 39b) [55b].

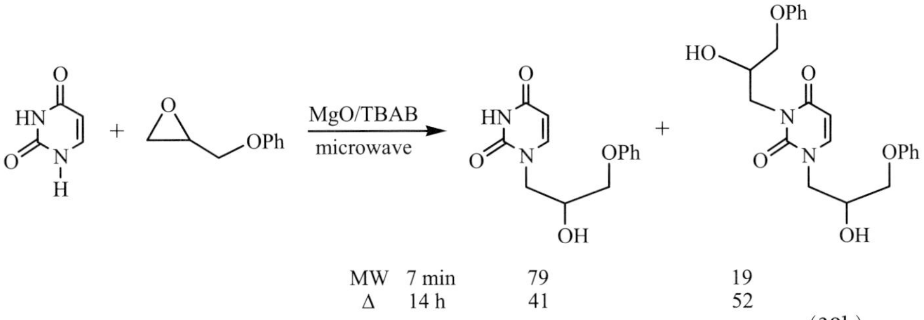

MW	7 min	79	19
Δ	14 h	41	52

(39b)

6.2.3
C-Alkylations of Active Methylene Groups [56–65]

Several monoalkylations of functionalized acetates (Eq. 40) have been described in a series of papers. The reactions were performed on potassium carbonate, either pure or mixed with potassium hydroxide. Some significant results are given in Table 6.14.

Tab. 6.14. Monoalkylation of functionalized acetates in a microwave oven (650 W).

R	R'X	t (min)	Yield (%)	Refs.
$PhSO_2$	$PhCH_2Cl$	3	76	57
	n-BuBr	3	83	
	n-OctBr	3	79	
PhCH=N	$PhCH_2Cl$	1	63	58
	n-BuBr	2	55	
PhS	$PhCH_2Cl$	4.5	83	59, 60
	n-BuBr	4.5	59	
CH_3CO	$PhCH_2Cl$	3	81	61
	n-BuBr	4.5	61	
COOEt	$PhCH_2Cl$	2	72	62
	n-BuBr	2	86	
	AllylBr	2	75	
p-NO_2-C_6H_4	Ph-$(CH_2)_3$-I	3.5	55	63
	Ph-$(CH_2)_8$-I	7	79	
p-NO_2-C_6H_4	Ph-S-$(CH_2)_6$-Br	7	50	63
	Ph-S-$(CH_2)_8$-Br	7	59	

$$R\diagdown COOEt + R'-X + Base \xrightarrow{n\text{-}Bu_4N^+X^-} R\underset{R'}{\diagdown COOEt} \qquad (40)$$

Rapid monoalkylations are achieved in good yields compared with classical methods. Of particular interest is the synthesis of α-amino acids by alkylation of aldimines under the action of MW activation. Subsequent acidic hydrolysis of the alkylated imine provided leucine, serine, or phenylalanine in preparatively useful yields within 1–5 min [58].

A rapid and novel one-pot radiochemical synthesis of ^{14}C-labeled esters has been developed by use a combination of solvent-free conditions and MW irradiation (Eq. 41) [64].

$$^{14}CH_2(COOEt)_2 \xrightarrow[R-X \quad 130°C]{KOtBu, Aliquat 336} R^{14}CH(COOEt)_2$$

R—X			
	$nC_{10}H_{21}Br$	2 min	74% *
	$nC_{10}H_{21}Cl$	10 min	20%
	$nC_{18}H_{37}Br$	2 min	65%

* classical synthesis NaH, DMF 12 h 100°C 69%

(41)

Cyclopropane derivatives can be obtained by reacting active methylene compounds and 1,2-dibromoethane using a solvent-free PTC + MW technique (Eq. 42) [65].

$$\underset{CO_2Et}{\overset{CN}{>}} \xrightarrow[\text{MW, 10 min}]{BrCH_2CH_2Br, K_2CO_3, \text{Aliquat}} \underset{\underset{86\%}{CO_2Et}}{\overset{CN}{\triangleright\!\!\!<}} \quad (42)$$

Alkylation of phenylacetonitrile has been performed by solid–liquid PTC in 1–3 min under the action of MW irradiation (Eq. (43), Table 6.15). The nitriles so obtained can subsequently be quickly hydrolyzed in a MW oven to yield the corresponding amides or acids [66].

$$PhCH_2CN + R-X \xrightarrow[H_2O]{NaOH, TEBACl} Ph-\underset{R}{\overset{H}{\underset{|}{C}}}-CN + Ph-\underset{R}{\overset{R}{\underset{|}{C}}}-CN \quad (43)$$

The condensation of phenylacetonitrile with benzaldehydes has been performed using K_2CO_3 in the presence of TBAB for 3 min with MW activation. By extending the reaction time to 10%, four different products were obtained from phenyl or nitrile group migration (Eq. 44) [67a]. These results have been extended to a series of arylacetonitriles with several aliphatic and aromatic aldehydes [67b].

	1	2	3	4
K_2CO_3 Aliquat 3 min	90	4	0	0
10 min	4	22	16	21

(44)

Functional groups were selectively introduced at the C-2 position of isophorone via epoxide ring-opening with several nucleophiles obtained from active methylene groups. Different behavior was observed depending on the reaction conditions and the nature of nucleophilic agents [68]. The best experimental systems involved

Tab. 6.15. Alkylation of phenylacetonitrile under MW + PTC conditions (NaOH 6 equiv., TEBA 15%).

R	PhCH$_2$CN/RX	t (min)	Power (W)	Yield (%)
PhCH$_2$Br	1.5	1	350	62/32
	1.5	3	160	58/35
n-HexBr	1.5	1	350	70/0
AllylBr	1.5	3	160	48/10
	0.5	2	160	96/2

PTC or KF-alumina under solvent-free conditions and MW irradiation (Eq. (45), Table 6.16).

$$\text{epoxyisophorone} + H_2C\overset{X}{\underset{Y}{\diagdown}} \xrightarrow{K_2CO_3 / \text{Aliquat}} \text{intermediate-OH} \xrightarrow{-H_2O} \text{product} \quad (45)$$

6.2.4
Alkylations with Dihalogenoalkanes

6.2.4.1 O-Alkylations

Ethers from alkylation of furfuryl alcohol with dihalides have been obtained under solvent PTC + MW conditions with quasi-quantitative yields (Eq. 46) [22].

$$\text{furfuryl-CH}_2\text{OH} + X-R-X \xrightarrow[\text{Aliquat}]{\text{KOH}} \text{furfuryl-CH}_2-O-R-O-CH}_2-\text{furfuryl} \quad (46)$$

Tab. 6.16. Epoxyisophorone-ring opening by reaction with several active methylene compounds under MW + PTC conditions.

X	Y	T (°C)	t	Yield (%)	
				MW	Δ
COOEt	COOEt	95	2 min	83	40
CN	Ph	70	40 s	91	51
CN	CN	58	20 s	94	64
COMe	COOEt	85	2 min	78	40
CN	CONH$_2$	90	1 min	91	54

Tab. 6.17. Reaction of monobenzyloxy isosorbide with X-R-X under the action of MW irradiation.

X	R	Solvent	T (°C)	Yield (%)	
				MW	Δ
Br	$(CH_2)_8$	Toluene	110	63	
	$(CH_2)_6$	Toluene	110	68	
	$(CH_2)_4$	Toluene	110	60	
OTs	$(CH_2)_8$	Cyclohexane	80	96	39
	$(CH_2)_6$	Cyclohexane	80	96	40
	$(CH_2)_4$	Cyclohexane	80	96	45
	$CH_2CH_2OCH_2CH_2$	Cyclohexane	80	91	36

In an analogous manner, the selectively obtained exo mono-benzylated isosorbide [69] was alkylated by use of alkyl dibromides to afford new ethers from dianhydrohexitols moieties separated by ether functions (Eq. (47), Table 6.17) [70]. Tosylates seemed to be a better leaving group for minimizing competitive β-elimination [71].

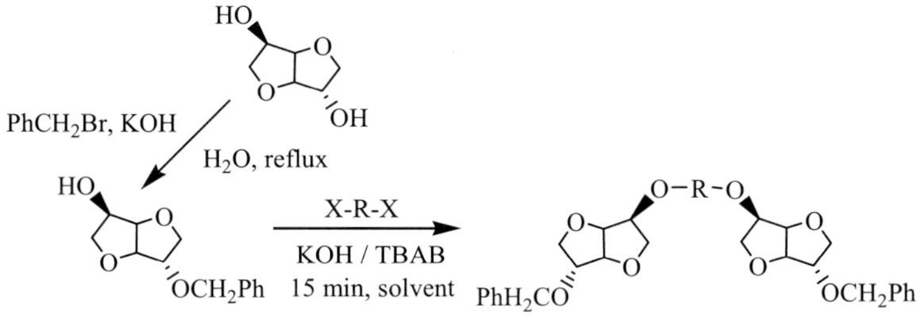

(47)

6.2.4.2 S-Alkylations

6-Mercaptobenzimidazo[1,2-c]quinazoline reacts with dibromo derivatives under PTC conditions (K_2CO_3/TBAB). Microwave irradiation enabled a striking reduction in reaction times (12 h compared with 15 min) with rather similar yields (86 and 81% respectively) when compared with conventional heating (Eq. 48) [72].

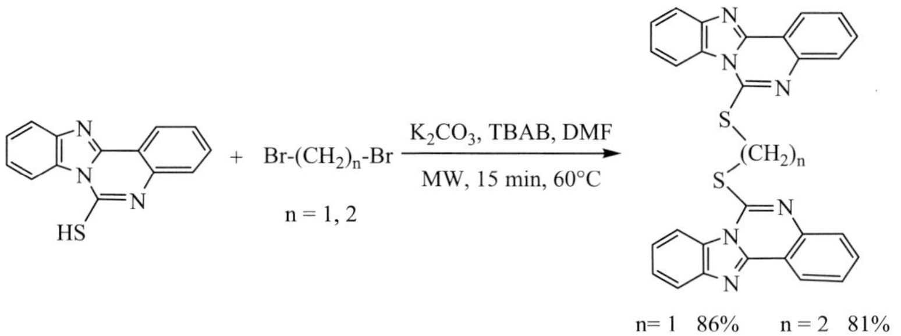

(48)

n = 1 86%   n = 2 81%

6.2.5
Nucleophilic Addition to Carbonyl Compounds

6.2.5.1 Aldol Condensation

Jasminaldehyde can be obtained classically from heptanal and benzaldehyde in 70% yield within 3 days at room temperature (Eq. 49). By use of a 600-W domestic MW oven, however, an enhanced yield of 82% was achieved in only 1 min. The amount of side-products (self condensation of n-heptanal) decreased from 30 to 18% when this technique was used [73]. α-Hexylcinnamic aldehyde and chalcone have been synthesized, by the same method, from octanal and acetophenone, respectively [74].

$$\text{PhCHO} + \text{CH}_3(\text{CH}_2)_5\text{CHO} \xrightarrow[\text{MW, 1 min, 82\%}]{\text{KOH, Aliquat}} \underset{\text{H}}{\overset{\text{Ph}}{>}}=\underset{\text{C}_5\text{H}_{11}}{\overset{\text{CHO}}{<}} \quad (49)$$

A second example of aldolization (Eq. 50) is the "dry" reaction of ferrocene carbaldehyde with carbonyl compounds in the presence of potassium hydroxide, and Aliquat as a catalyst [75]. Reactions which are too slow at room temperature are efficiently accelerated by use of microwaves, giving very good yields within a few minutes.

$$\text{Fc-CHO} + \text{R}_1-\text{H}_2\text{C}-\text{CO-R}_2 \xrightarrow[-\text{H}_2\text{O}]{\text{KOH, Aliquat}} \text{Fc-CR}_1=\text{C(R}_2\text{)=O} \quad (50)$$

Microwave activation and solvent-free PTC has been shown to be of prime efficiency for the synthesis of new benzylidene cineole derivatives (UV sunscreens) by the Knoevenagel reaction. When performed classically by use of KOH in ethanol at room temperature for 12 h (Eqs. 51 and 52) the yield was 30%. This was improved to 90–94% within 2–6 min under PTC + MW conditions (Tables 6.18 and 6.19) [33, 34].

$$\text{cineole-ketone} + \text{H-CO-C}_6\text{H}_4\text{-R} \xrightarrow[\text{no solvent}]{\text{KOH (1.5 equiv.)}\atop\text{Aliquat (5\%)}} \text{cineole=CH-C}_6\text{H}_4\text{-R} \quad (51)$$

$$\text{cineole-ketone} + \text{H-CO-C}_6\text{H}_3(\text{OR})_2 \xrightarrow[\text{TBAB}]{\text{K}_2\text{CO}_3 + \text{KOH}} \text{cineole=CH-C}_6\text{H}_3(\text{OR})_2 \quad (52)$$

Tab. 6.18. Benzylidene derivatives of 2-cineolylol from p-substituted benzaldehydes [33].

R	t (min)	T (°C)	Yield (%)	
			MW	Δ
CH_3	2	180	90	20
$N(CH_3)_2$	2	180	94	40
$OC_{16}H_{33}$	6	200	94	12

Tab. 6.19. Benzylidene derivatives of 2-cineolylol from 3,4-dialkoxybenzaldehydes [34].

R	t (min)	T (°C)	Yield (%)	
			MW	Δ
$C_{12}H_{25}$	60	140	80	13
$C_{14}H_{29}$	60	140	78	
$C_{16}H_{33}$	60	140	81	
$C_{18}H_{37}$	60	140	87	

6.2.5.2 Ester Saponification

Esters are easily saponified in a few minutes by use of powdered potassium hydroxide (2 mol equiv.) and Aliquat (10 mol%) in the absence of solvent (Eq. (53), Table 6.20) [76].

$$R-COOR' \xrightarrow[\text{2. HCl}]{\text{1. KOH, Aliquat}} R-COOH \qquad (53)$$

From the few examples summarized in Table 6.20, three important conclusions can be drawn:

- rapid and easy reactions occur, even with the most hindered mesitoic esters, which are otherwise practically nonsaponifiable under classical conditions [77];
- the advantage of using a monomode MW reactor rather than a domestic oven is evident; and

Tab. 6.20. Ester saponifications under MW + PTC conditions.

R	R'	Multimode oven (250 W)		Monomode reactor (90 W)			Classical heating		
		t (min)	Yield (%)	t (min)	T (°C)	Yield (%)	t (min)	T (°C)	Yield (%)
Ph	Me	0.5	87	1	205	96	1	205	90
	n-Oct	1	83	2	210	94	2	210	72
2,4,6-Me$_3$C$_6$H$_2$	Me	2	75	2	140	84	2	140	38
	n-Oct	2	57	4	223	82	4	223	0

- more interesting fundamentally is the very strong specific non-thermal MW effect, as is apparent from comparison with classical heating. This effect grows as ester reactivity falls.

In a more detailed study, it was shown that MW effects are very dependent on the temperature and the nature of the cation associated with hydroxyl anion [78] (for example, Eq. (54) and Table 6.21).

$$\text{Ph-CO}_2\text{CH}_3 + M^+OH^- \xrightarrow[\text{then HCl}]{\text{Aliquat}} \text{Ph-CO}_2H + CH_3OH \quad (54)$$

With NaOH, the MW effect is more important than when the reaction is performed with KOH. This is consistent with tighter ion pairs in the previous case

Tab. 6.21. Saponification of methyl benzoate under MW + PTC conditions.

M$^+$	Base (equiv.)	T (°C)	t (min)	Yield (%)			
				Without aliquat		With aliquat	
				MW	Δ	MW	Δ
K$^+$	2	200	5	94	90	98	98
	1	70	60	77	42	67	65
Na$^+$	2	200	5	92	77	90	73

Tab. 6.22. Ester aminolysis under MW + PTC conditions (KO*t*Bu, 10 min, 150 °C).

R	R'	Aliquat	Yield (%) MW	Δ
C_6H_5	nC_8H_{17}	−	80	22
		+	87	70
	$C_6H_5CH_2$	−	84	42
		+	98	85
$C_6H_5CH_2$	C_6H_5	+	98	33

(cation hardness: $Na^+ > K^+$), thus leading to less polar species. The importance of these effects is very dependent upon the reactivity of the systems and becomes more critical when the temperature is reduced (see extended discussion in Chapter 4 of this book).

6.2.5.3 Ester Aminolysis

Solvent-free ester aminolysis has been studied under the action of MW irradiation in the presence of KOtBu with or without a phase-transfer agent (Eq. 55) [79].

$$RCO_2Et + R'NH_2 \xrightarrow[\text{MW or }\Delta \quad 150°C]{\text{KO}t\text{Bu (Aliquat 336)}} RCONHR' \qquad (55)$$

$R = C_6H_5, C_6H_5CH_2, nC_5H_{11}$
$R' = C_6H_5, C_6H_5CH_2, nC_8H_{17}$

Within 10 min at 150 °C, excellent yields were obtained under the action of MW (80–98%) whereas they were limited when conventional heating was used under similar conditions (Table 6.22).

6.2.5.4 Base-catalyzed Transesterifications

Transesterifications of methyl esters with high boiling alcohols, as shown in (Eq. 56), occurred readily under MW irradiation because of the displacement by evaporation of the polar volatile methanol (Table 6.23) [11].

$$ArCOOMe + n\text{-}C_8H_{17}OH \underset{\text{Aliquat}}{\overset{\text{Base}}{\rightleftharpoons}} ArCOOn\text{-}C_8H_{17} + MeOH \qquad (56)$$

This study was next extended to the synthesis of benzoyl and dodecanoyl derivatives from protected carbohydrates [80]. MW-assisted PTC transesterifications with

Tab. 6.23. Base-catalyzed transesterifications under MW + PTC conditions (600 W).

Ar	Base	t (min)	Yield (%)
C_6H_5	KOH	3	40
	KOt-Bu	3	42
	K_2CO_3	2.5	90
2,4,6-$Me_3C_6H_2$	KOH	10	64
	KOt-Bu	10	68
	K_2CO_3	10	89

methyl benzoate or dodecanoate were studied for several carbohydrates. Small amounts of dimethylformamide (DMF) were shown to be necessary to provide good yields (76–96%) within 15 min. Rate enhancements compared to conventional heating (Δ) and specific MW activation were especially noticeable when less reactive fatty compounds were involved (Eq. 57).

$$\text{carbohydrate-OH} + RCO_2CH_3 \xrightarrow[15 \text{ min, } 160°C]{K_2CO_3, \text{ TBAB}} \text{carbohydrate-OR} \quad (57)$$

R = Ph		MW	44%		
	DMF	MW	96%	Δ	21%
R = $CH_3(CH_2)_{10}$	DMF	MW	88%	Δ	0% (id. 12h)

6.2.5.5 Synthesis of Acyl Isothiocyanates

The reaction of acyl chloride with potassium thiocyanate under PTC conditions (with PEG 400 as catalyst) is a good method for preparation of acyl isothiocyanate (Eq. 58). It has been conducted efficiently under solvent-free conditions and MW activation [81, 82].

$$PhCOCl + KSCN \xrightarrow[\text{MW solvent-free}]{\text{PEG 400}} Ph-\overset{O}{\underset{\parallel}{C}}-N=C=S \quad (58)$$

Subsequently, by reaction with amine (Eq. 59), N-aryl and N'-aroyl thioureas were prepared in excellent yields (86–98%) using this two-step procedure.

$$Ph-\overset{O}{\underset{\parallel}{C}}-N=C=S + RNH_2 \xrightarrow[\text{MW solvent-free}]{} Ph-\overset{O}{\underset{\parallel}{C}}-NH-\overset{S}{\underset{\parallel}{C}}-NHR \quad (59)$$

In other work [83, 84], thiosemicarbazides were obtained in a two-step procedure, the second being the reaction of acyl isothiocyanates with aryloxyacetic acid hydrazide (Eq. 60).

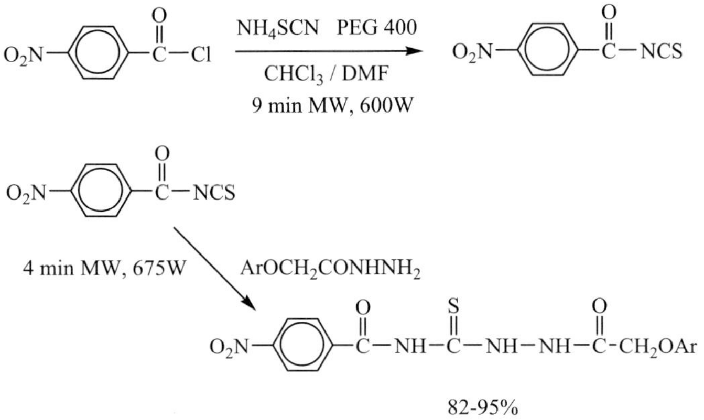

$$O_2N-C_6H_4-C(O)-Cl \xrightarrow[\text{CHCl}_3 / \text{DMF} \\ \text{9 min MW, 600W}]{\text{NH}_4\text{SCN PEG 400}} O_2N-C_6H_4-C(O)-NCS$$

$$O_2N-C_6H_4-C(O)-NCS \xrightarrow[\text{4 min MW, 675W}]{\text{ArOCH}_2\text{CONHNH}_2} O_2N-C_6H_4-C(O)-NH-C(S)-NH-NH-C(O)-CH_2OAr \quad (60)$$

82-95%

Phenylacetyl (and benzoyl) phenyl thioureas were prepared by similar procedures [85, 86] by reacting a series of acyl chlorides with ammonium thiocyanate in the first step (Eq. 61).

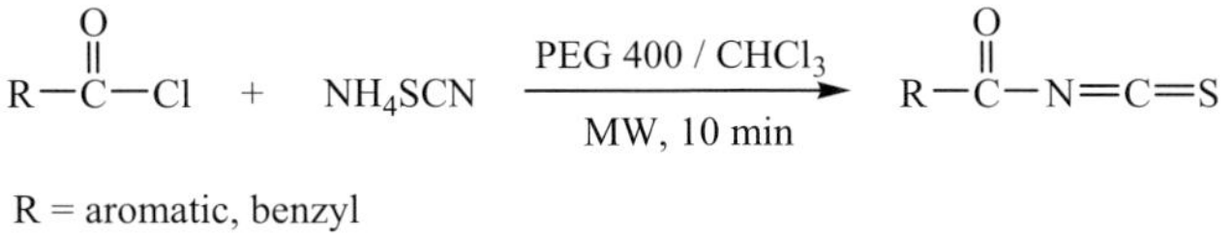

$$R-C(O)-Cl + NH_4SCN \xrightarrow[\text{MW, 10 min}]{\text{PEG 400 / CHCl}_3} R-C(O)-N=C=S \quad (61)$$

R = aromatic, benzyl

6.2.6
Deprotonations

6.2.6.1 Base-catalyzed Isomerization of Allylic Aromatic Compounds [87]

Eugenol is a natural product available from a variety of essential oils (cinnamon-tree or pimentos leaves). Its isomerization (Eq. 62) into isoeugenol, the starting material for synthetic vanillin, is rather difficult and proceeds in modest yields under relatively harsh conditions. It can, however, be very efficiently prepared by use of 2.2 molar equivalents of base and catalytic (5%) amounts of Aliquat in the absence of solvent.

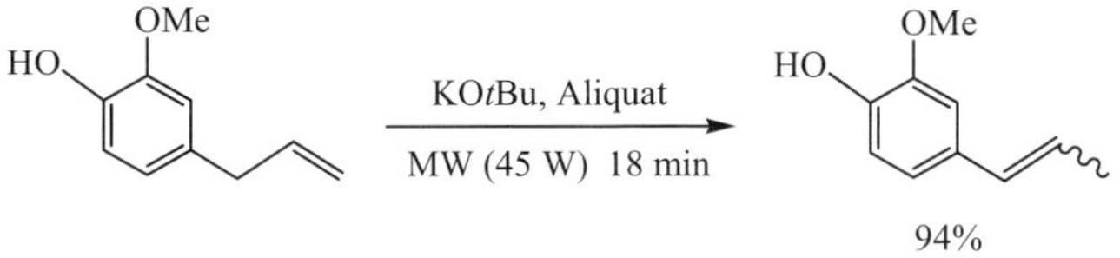

$$\text{eugenol} \xrightarrow[\text{MW (45 W) 18 min}]{\text{KO}t\text{Bu, Aliquat}} \text{isoeugenol} \quad (62)$$

94%

Isomerization of safrole to isosafrole was realized in the same way by use of KO*t*Bu/Aliquat under solvent-free conditions and exposure to MW (Eq. 63) [88].

$$\text{safrole} \xrightarrow[\text{150W 8 min 117°C}]{\substack{\text{KO}t\text{Bu - Aliquat} \\ \text{monomode MW}}} \text{isosafrole} \quad 99\% \tag{63}$$

6.2.6.2 Carbene Generation (α-Elimination) [89]

Dichlorocarbene has been generated under solid–liquid conditions by use of microwaves. Heating a mixture of $CHCl_3$, powdered NaOH, and trace amounts of CTAB under reflux in cyclohexene under the action of MW irradiation afforded 90% dichloronorcarane (Eq. 64) within 20 min, compared with 81% in 60 min without MW or only 12% in 90 min without the catalyst [89a].

$$CHCl_3 \xrightarrow[\text{CTAB}]{\text{NaOH}} [\ddot{C}Cl_2] \xrightarrow{\text{cyclohexene}} \text{dichloronorcarane} \tag{64}$$

Generation of dichlorocarbene was also reported when ammonium salts were supported on a non-cross-linked polystyrene matrix [89b]. Then the reaction involved mixing of chloroform, polymer-supported PTC catalyst, and styrene in the presence of a sodium hydroxide solution. The product, 1,1-dichloro-2-phenylcyclopropane was afforded in a very good yield (94%) after 10 min of MW irradiation at 65 W.

6.2.6.3 β-Elimination

Bromoacetals in basic media can be converted to cyclic ketene acetals (Eq. 65). These β-eliminations, previously performed under solid–liquid PTC without solvent and with sonication [90], were further improved by MW irradiation (Table 6.24) [91].

$$\text{Br-acetal} \xrightarrow[\text{TBAB}]{\text{Base}} \text{ketene acetal} \tag{65}$$

With potassium *t*-butoxide and TBAB, higher yields were obtained more rapidly than under sonication conditions or with conventional heating.

Tab. 6.24. β-Elimination from bromoacetals in a monomode reactor (75 W). Comparison with sonochemical conditions and classical heating.

Compound	Base	Ultrasound conditions			Microwave conditions			Classical heating		
		t (h)	T (°C)	Yield (%)	t (min)	T (°C)	Yield (%)	t (min)	T (°C)	Yield (%)
Ph-dioxolane-Br	KOH, TBAB	1	75	81	10	130	81			
	KOt-Bu, TBAB	1	35	70	5	75	87	5	75	36
dioxane-Br	KOH, TBAB	1	75	81	10		28			
	KOt-Bu, TBAB	1	45	70	5	60	95	5	64	41

6.2.7
Miscellaneous Reactions

6.2.7.1 Aromatic Nucleophilic Substitution (S_NAr)

Irradiation of a mixture of *ortho* or *para*-nitrochlorobenzene and ethanol in the presence of sodium hydroxide and a phase-transfer agent yields the corresponding ethoxy aromatic compounds within a few minutes (Eq. 66) [92]. The same procedure was subsequently applied to 2-chlorophenol [93]. In both reactions, PEG 400 was shown to be the most efficient catalyst (Table 6.25).

$$Z\text{-}C_6H_4\text{-}Cl + EtOH + NaOH \xrightarrow[-NaCl, -H_2O]{\text{Catalyst 10\%}} Z\text{-}C_6H_4\text{-}OEt \quad (66)$$

Tab. 6.25. Aromatic nucleophilic substitution under MW + PTC conditions (420–700 W).

Z	Catalyst	t (min)	Yield (%)	Ref.
4-NO$_2$	None	2	14	92
	PhCH$_2$N$^+$Me$_3$Cl$^-$	2	51	
	PEG 400	2	99	
2-NO$_2$	PEG 400	2	99	92
2-OH	None	2	63	93
	PhCH$_2$N$^+$Me$_3$Cl$^-$	2	70	
	PEG 400	2	82	

Solvent-free S_NAr reactions under solid–liquid PTC conditions were realized by use of methoxide or phenoxide as nucleophiles. The main results and comparison with those from classical heating are indicated in Table 6.26 for activated (e.g. 4-nitrohalobenzenes) or nonactivated (e.g. β-naphthyl halides) substrates [94].

Diaryl ethers were easily obtained from reactions of phenols with nitroarylfluorides under solvent-free + MW conditions by use of a KF/alumina + Aliquat system (Eq. 67) [95].

$$R\text{-}C_6H_4\text{-}OH + O_2N\text{-}C_6H_4\text{-}F \xrightarrow[\text{no solvent } 10 \text{ min MW}]{\text{KF-alumina / Aliquat 336}} R\text{-}C_6H_4\text{-}O\text{-}C_6H_4\text{-}NO_2 \qquad (67)$$

R = (o+p) H, CH$_3$, OCH$_3$, tBu, NO$_2$ 72–92%

Etherifications of heterocyclic compounds have been performed with good efficiency by solid–liquid PTC coupled with MW irradiation (Eq. (68), Table 6.27) [96]. Yields and conditions involved here are a noticeable improvement over classical methods from the standpoint of green chemistry.

$$\text{2-X-pyridine} + RO^- K^+ \xrightarrow[\text{MW}]{\text{18-crown-6}} \text{2-OR-pyridine} \qquad (68)$$

Tab. 6.26. Solvent-free PTC S_NAr reactions under MW + PTC conditions (potassium salts).

	Catalyst	t (min)	T (°C)	Yield (%)	
				MW	Δ
Cl–C$_6$H$_4$–NO$_2$ + PhO$^-$	18-Crown-6	5	150	93	86
F–C$_6$H$_4$–NO$_2$ + PhO$^-$	–	30	150	98	73
Cl–C$_6$H$_4$–NO$_2$ + CH$_3$O$^-$	18-Crown-6	20	170	90	37
F–C$_6$H$_4$–NO$_2$ + CH$_3$O$^-$	18-Crown-6	3.5	80	100	81
Cl-naphthyl + CH$_3$O$^-$	18-Crown-6	60	100	71	27
F-naphthyl + CH$_3$O$^-$	18-Crown-6	60	80	94	57

Tab. 6.27. Solvent-free PTC S_NAr of pyridine halides.

X	RO⁻	Conditions	Yield (%)
3-F	PhO⁻	30 min 180 °C	85
3-Cl	PhO⁻	45 min 180 °C	55
2-Cl	PhO⁻	20 min 150 °C	95
3-F	MeO⁻	20 min 100 °C	100
3-Cl	MeO⁻	20 min 130 °C	92

Selective and efficient fluorinations of chlorodiazines were obtained under solvent-free MW-assisted PTC conditions (Eq. 69) [97].

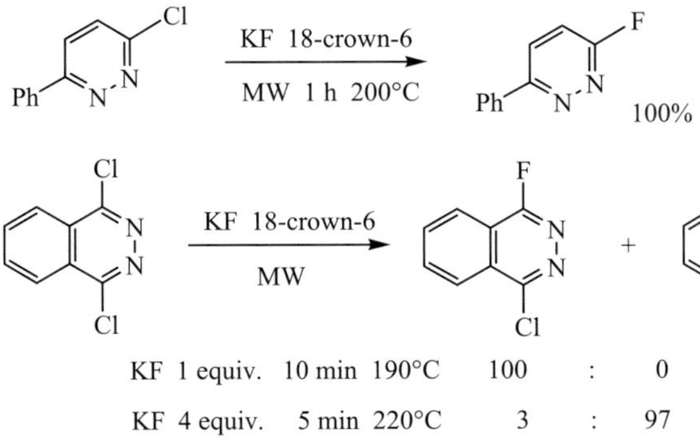

(69)

6.2.7.2 Dealkoxycarbonylations of Activated Esters (Krapcho Reaction)

This type of reaction is usually best performed in DMSO solution under reflux. A simpler procedure has been proposed which uses anionic activation and MW irradiation, with a metallic salt as the reagent and a PTC in the absence of solvent [98]. This procedure was applied to the striking example of cyclic β-ketoesters with substantial improvements (Eq. (70) and Table 6.28) which are readily apparent when the maximum yields obtained under classical Krapcho conditions (<20% when R = H) [99].

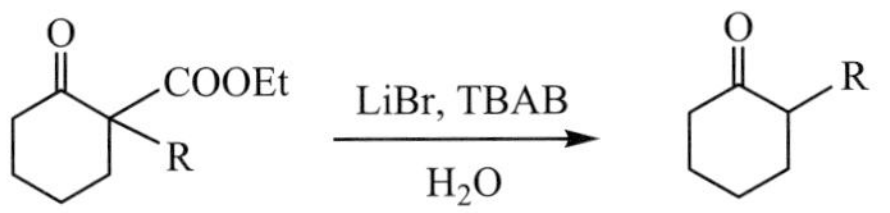

(70)

To assess the specific not purely thermal MW effects, the second experiment (R = Et) was performed using conventional oil bath heating under the same set of conditions of time and temperature (15 min, 160 °C). No reaction was observed. Further heating for 3 h led to total conversion but the yield (60%) was limited by

Tab. 6.28. Dealkoxycarbonylation of cyclic β-ketoesters in a monomode reactor.

R	Microwave Conditions		T (°C)	Yield (%)
	t (min)	Power (W)		
H	8	30	138	96
Et	15	30	160	94
n-Bu	20	45	167	89
n-Hex	20	90	186	87

product degradation. Clearly, compared with conventional heating, MW activation results in a large reduction in time, simplified experimental conditions, and prevention of product degradation at high temperature.

6.2.7.3 [1,3]-Dipolar Cycloaddition of Diphenylnitrilimine [100]

Diphenylnitrilimine (DNPI) can be subjected to 1,3-dipolar cycloaddition with activated double bonds as dipolarophiles (Eq. 71). It can be generated *in situ* by reaction of hydrazonoyl chloride with a base.

$$\text{Ph-C} \equiv \overset{\oplus}{\text{N}} - \overset{\ominus}{\text{NPh}}$$

(71)

The cycloaddition has been performed almost quantitatively within 6 min under the action of microwaves, using KF as base and Aliquat as the phase-transfer agent. For illustration, results obtained for substituted chalcones are listed in Table 6.29. When the same reactions are performed all other factors being equal (time, temperature), no reaction occurs under classical thermal conditions. This behavior once again confirms a specific radiation effect.

6.2.7.4 Synthesis of β-Lactams [101]

With KF in the presence of a phase-transfer agent (18-crown-6), silyl ketene acetals react with aldimines to give β-lactams within a few minutes under the action of microwave irradiation in closed Teflon vessels (Eq. 72).

Tab. 6.29. 1,3-Dipolar cycloaddition of DNPI to chalcones in a monomode reactor (30 W).

Z	t (min)	T (°C)	Yield (%)
H	6	170	90
Br	5	170	93
Cl	6	174	95
Me	6	168	87
OMe	6	175	89

$$\text{Ph-CH=N-Ph} + \text{Me(OSiMe}_3\text{)C=CH(OMe)} \xrightarrow[\text{MW (300 W) 7 min}]{\text{KF, 18-crown-6}} \text{β-lactam} \quad (72)$$

6.2.7.5 Selective Dealkylations of Aromatic Ethers

Ethylisoeugenol and ethoxyanisole have been selectively demethylated or deethylated using potassium t-butoxide in the presence of 18-crown-6 (Eq. 73) [102].

$$\text{ArOEt(OMe)} \xrightarrow[\text{(ethylene glycol)}]{\text{KOtBu + 18-crown-6}} \text{ArOEt(OH)} + \text{ArOH(OMe)} \quad (73)$$

R = CH=CH-CH$_3$, H

Under solvent-free conditions, only deethylation is observed whereas in the presence of ethylene glycol (EG) the selectivity is totally reversed and demethylation becomes the major process. In both, considerable increases in reaction rate were observed under the action of MW irradiation when compared with classical heating (Δ) (Table 6.30).

Tab. 6.30. Selective dealkylations of 2-ethoxyanisole under MW + PTC conditions (R = H).

	Conditions	t (min)	T (°C)	Starting material (%)	1 (%)	2 (%)
–	MW	20	120	7	–	90
–	Δ	20	120	48	–	50
EG	MW	75	180	–	72	23
EG	Δ	75	180	98	–	–

Tab. 6.31. Deethylation of 2-ethoxyanisole within 20 min under MW + PTC conditions (KO*t*Bu 2 equiv.; TDA-1 10%).

Reactor	T (°C)	Amounts of materials [g (mmol)]			Total amount (g)	Yield (%)
		Ethoxyanisole	KO*t*Bu	TDA-1		
MW S402	120	0.76 (5)	1.12 (10)	0.324 (1)	2.206	90
MW S1000	140	37.24 (245)	54.98 (490)	15.9 (49)	108.12	82

This procedure was extended to the hundred-gram scale with success because, under rather similar conditions, yields remained excellent (82%) requiring only a slight modification in temperature (140 instead of 120 °C) (Table 6.31) [12].

6.2.7.6 Synthesis of Dibenzyl Diselenides

A simple, rapid and efficient method has been reported for synthesis of dibenzyl diselenides under the action of MW irradiation. Benzyl halides were reacted with selenium powder in the presence of a base and a phase-transfer agent (Eq. (74), Table 6.32) [103]. The reactions were performed either in THF or in $C_6H_6-H_2O$.

$$R\text{-}C_6H_4\text{-}CH_2X + 2\,Se \xrightarrow[C_6H_6 - H_2O,\ MW]{NaOH,\ PEG\ 400} R\text{-}C_6H_4\text{-}CH_2\text{-}Se\text{-}CH_2\text{-}C_6H_4\text{-}R \quad (74)$$

The same authors have more recently described the synthesis of dibenzoyl diselenides by reaction of selenium with sodium hydroxide under the action of PTC and MW irradiation conditions to afford sodium diselenides which reacted further with benzoyl chloride at 0 °C [104].

Tab. 6.32. Synthesis of dibenzyl diselenides under MW + PTC conditions (15 min, 750 W; NaOH 15 equiv.; PEG-400 5%).

R	X	Yield (%)
H	Cl	75
H	Br	95
4-Br	Br	85
4-CH_3	Br	91
4-NO_2	Br	72

Tab. 6.33. MW-promoted hydrolysis of nitriles by use of the NaOH–PEG system.

R	t (s)	Yield (%)
Ph_2CH	90	94
Ph	45	83
$PhCH_2$	90	71
$C_{17}H_{35}$	60	78
2-Pyridyl	40	52

6.2.7.7 Selective Hydrolysis of Nitriles to Amides [105]

As already mentioned (Section 6.3), nitriles can be hydrolyzed to amides or acids [67]. Nitriles have been efficiently converted into the corresponding amides in the presence of PEG-400 and aqueous NaOH system under the action of MW irradiation (Eq. (75), Table 6.33).

$$R-C \equiv N \xrightarrow[\text{MW}]{\text{aq NaOH - PEG 400}} R-\overset{O}{\underset{\|}{C}}-NH_2 \quad (75)$$

6.2.7.8 Synthesis of Diaryl-α-tetralones [106]

One-pot synthesis of diaryl-α-tetralones via Michael condensation and subsequent Robinson annulation reactions of isophorone with chalcones was performed efficiently in a solvent-free PTC system under the action of MW irradiation. Compared with conventional heating, substantial rate enhancements were observed, within very short reaction times, by use of microwaves (Eq. (76), Table 6.34). They were far better than those achieved by the classical method (NaOEt in EtOH under reflux for 24 h; 40–56%).

Tab. 6.34. Synthesis of diaryl-α-tetralones from isophorone and chalcone under MW + PTC conditions.

Ar	t (min)	T (°C)	Yield (%)	
			MW	Δ
C_6H_5	5	110	86	51
p-CH_3O-C_6H_4	5	108	86	51
p-CH_3-C_6H_4	6	120	89	52
o-Cl-C_6H_4	5	110	88	54
p-Cl-C_6H_4	6	110	86	56

$$\text{(scheme)} \quad \xrightarrow[\text{Aliquat 2\%}]{\text{NaOEt (2 equiv)}} \quad [\text{enolate } M^+] \quad \xrightarrow{\text{Ar-CH=CH-CO-Ph}} \quad \text{product} \qquad (76)$$

6.2.7.9 Intramolecular Cyclization

Malonic acid allylic esters undergo intramolecular cyclization under solid–liquid PTC conditions in the presence of Aliquat 336, potassium carbonate, and iodine (Eq. 77) [107]. Application of MW irradiation to this procedure enabled 2–3 fold reduction in reaction time compared with conventional conditions. It was found that use of MW affected the exo/endo diastereoisomer ratio (a linear correlation between MW power and exo isomer concentration was observed [108]).

$$\text{allylic malonate} \xrightarrow[\text{Aliquat 336, toluene}]{K_2CO_3,\ I_2} \text{bicyclic product} \qquad (77)$$

Microwave (boiling toluene)	10–16 min	65–67%
Conventional heating (boiling toluene)	25–30 min	65–94%
Room temperature (toluene)	6–13 h	65–94%

6.2.7.10 Heck, Suzuki and Sonogashira Cross-coupling Reactions

Reaction of organic halides with alkenes catalyzed by palladium compounds (Heck type reaction) is known to be a useful method for carbon–carbon bond formation at unsubstituted vinylic positions. The first reports of the application of MW methodology to this type of reaction were published by Hallberg et al. in 1996 [109] and by Diaz-Ortiz et al. in 1997 [110]; both used in triethylamine solutions. Later, Villemin et al. studied the possibility of Heck coupling of iodoarenes with methyl acrylate in aqueous solution under pressurized conditions [111]. The reactions were conducted in a Teflon autoclave under the action of MW irradiation in the presence of palladium acetate, different phosphine ligands, and tetrabutylammonium hydrogen sulfate (TBAHS) as PTC catalyst, to afford the desired coupling products in 40 to 90% yield.

The palladium catalyzed Heck coupling reaction induced by MW irradiation under solvent-free liquid–liquid PTC conditions, in the presence of potassium carbonate and a small amount of $[Pd(PPh_3)_2Cl_2]$-TBAB as catalyst, has recently been reported [112]. The arylation of alkenes with aryl iodides proceeded smoothly to afford exclusively trans products in high yields (86–93%) (Eq. 78).

$$R\text{-}C_6H_4\text{-}I + CH_2=CHY \xrightarrow[\text{TBAB, H}_2\text{O}]{[Pd(PPh_3)_2Cl_2],\ K_2CO_3} R\text{-}C_6H_4\text{-}CH=CH\text{-}Y \qquad (78)$$

Microwaves	10 min	86–93%
Conventional heating	10 min	5–15%
	3–7 h	54–90%

More recently, Najera et al. performed a number of Heck reactions of deactivated aryl halides and styrenes under phosphane-free conditions using oxime-derived palladacycles or palladium acetate as catalyst [113]. Coupling can be performed either with dicyclohexylmethylamine as base and TBAB as PTC catalyst or in neat water with triethylamine in N,N-dimethylacetamide (DMA) solutions under the action of MW irradiation.

A series of Pd-catalyzed intramolecular Heck reactions of N-(2-halophenyl)-substituted enaminones was performed by Pombo-Villar et al. [114]. The reactions were performed in the presence of palladium acetate with a phosphonium ligand as catalyst, sodium acetate, and tetrabutylammonium chloride (TBACl) as the PTC catalyst in DMF solution to afford desired coupling product in 25 to 99% yields (Eq. 79).

$$\text{Ar-X with enaminone} \xrightarrow[\text{DMF}]{\text{Pd(OAc)}_2 \text{ P-ligand} \atop \text{Bu}_4\text{NCl, AcONa}} \text{cyclized product} \quad (79)$$

X = Br, I

Reaction of aryl halides with boronic acids catalyzed by palladium compounds (Suzuki reaction) is one of the most versatile reactions for selective formation of carbon–carbon bonds. One of the first reports of the application of microwaves to this type of reaction was published by Larhed et al. in 1996 [115]. Subsequently, Varma et al. described the Suzuki-type coupling of boronic acids and aryl halides (Eq. 80) in the presence of palladium chloride and poly(ethylene glycol) (PEG-400) under the action of MW irradiation [116]. The reactions were performed at 100 °C to give the desired coupling products in 50 to 90% yield within 50 s. The coupling reaction can be also conducted under conventional conditions (oil bath, 100 °C), but to achieve similar yields a longer reaction time was needed (15 min). It was found that addition of KF affords better yields.

$$\text{R-C}_6\text{H}_4\text{-X} + (\text{HO})_2\text{B-C}_6\text{H}_4\text{-X} \xrightarrow[\text{MW, 50 sec}]{\text{PdCl}_2, \text{KF, PEG 400}} \text{R-biaryl} \quad (80)$$

MW (240 W)	50 sec	50-90%
Classical heating	15 min	50-90%

Leadbeater et al. recently showed that the PTC procedure with MW irradiation can be used to prepare of biaryls using water, palladium acetate, and TBAB as solvent, catalyst, and phase-transfer agent, respectively [117]. The desired coupling products were obtained in good yield (60 to 90%). The reaction can, however, be performed equally well using MW and conventional heating methods. Although it was later reported that Suzuki-type coupling of boronic acids and aryl halides was possible without the need for a transition-metal catalyst, further reassessment of the reaction showed that ultralow palladium contaminants (ca. 50 ppb) either in the commercially available bases used [118] or in reaction vessels [119] promoted the reaction.

Application of immobilized palladium catalyst under the action of MW, with phase-transfer agents, has also been reported to promote the Suzuki reactions of a range of aryl halides and triflates in solvents such as water and ethanol under MW conditions [120, 121]. Similar results were obtained when aryl halides were attached to a PEG matrix as para-substituted benzoates [122]. It was found that conventional thermal conditions induced up to 45% cleavage of the benzoates whereas this side reaction was suppressed when MW conditions were employed.

Related to Suzuki coupling reaction involving organoboron reagents is the so-called Hiyama coupling of organosilanes with halides and triflates to form unsymmetrical biaryl compounds. DeShong et al. described rapid, MW-promoted Hiyama coupling of bis(catechol)silicates and aryl bromides (Eq. 81) [123]. Suitable reaction conditions were use of 5% palladium catalyst, a phosphonium ligand, and tetrabutylammonium fluoride (TBAF) as PTC catalyst in THF solution. Exposure of the reaction mixture to MW irradiation at 110 °C for 10–15 min provided the desired coupling products in very good yields of 80 to 100% (Eq. 81).

$$\text{bis(catechol)silicate} \cdot \text{HNEt}_3 + \text{ArX}(R_1) \xrightarrow[\text{MW (50W), 10 min, 120°C}]{\text{Pd(dba)}_2, \text{TBAF, THF}} \text{biaryl} \quad (81)$$

MW (50 W)	10 min	80–90%
Classical heating	10 min	< 40%

Reaction of terminal acetylenes with aryl and vinyl halides catalyzed by palladium/copper compounds (the Sonogashira reaction) is known to be a useful method for the preparation of unsymmetrical alkynes. The first procedure for application of microwave methodology in the Sonogashira reaction was reported by Erdelyi et al. [124], and reports of Sonagashira coupling induced by MW irradiation under PTC conditions were recently published by Leadbeater et al. [125] and Van der Eycken [126]. In the same way as for Suzuki reaction [118], both reports claimed it was possible to perform transition-metal free Sonogashira coupling in the presence of TBAB and NaOH [125] or PEG and Na_2CO_3 [126] as PTC catalyst and base, respectively. The typical reaction conditions involved irradiation of the reaction mixture in aqueous solution to ca 170–175 °C for 5 to 25 min in a sealed vessel (Eq. 82).

$$\text{Naphthyl-Br} + \text{HC≡C-Ph} \xrightarrow[\text{MW, 15 min, 175°C}]{\text{TBAB, Na}_2\text{CO}_3, \text{H}_2\text{O}} \text{Naphthyl-C≡C-Ph} \quad (82)$$

MW (150 W)	15 min	40–80%
Classical heating	720 min	0%

More recently, Pombo-Villar et al. demonstrated the palladium-catalyzed synthesis of diaryl acetylenes by a one-pot direct coupling of 1-aryl-2-trimethylsilylacetylenes and aryl halides in the presence of palladium acetate, sodium acetate, and TBACl as PTC catalyst [127]. This method was used to couple a range of aryl halides and arylsilylacetylenes and does not involve use of a copper catalyst. It was possible to introduce an aldehyde moiety and an amino group to the phenylacetylene substrate and obtain the desired coupling products in 15 min with 60 and 81% yield, respectively. Thus, neither of these electron-withdrawing and electron-donating functional groups seems to significantly affect the reactivity of the substrate (Eq. 83) [127]. The reaction yields for experiments run under the same set of conditions were similar.

$$\text{Ph-X} + \text{Me}_3\text{Si-C≡C-Ph} \xrightarrow[\text{MW (450W), 15 min, 100°C}]{\text{Pd(OAc)}_2, (o\text{-Tol})_3\text{P, TBACl}} \text{Ph-C≡C-Ph} \quad (83)$$

X = Br, I

MW (450 W)	15 min	74%
Classical heating	15 min	62%

The Sonogashira coupling reaction of aryl halides were attached to a PEG matrix as para-substituted benzoates was also reported [128]. It was shown that MW irradiation can reduce the reaction time from 3 h under conventional conditions to 1.5 min, and that yields of the products were approximately the same. Similarly, the cleavage proceeded faster under MW conditions (1 min compared with 5 h at 50 °C under conventional conditions).

6.2.7.11 Oxidation Reactions

By a modification of Noyori's procedure of alcohol oxidation by hydrogen peroxide [129], primary and secondary alcohols have been oxidized to the equivalent carboxylic acids and ketones within 20–30 min under the action of MW irradiation [130]. The reactions were performed under liquid–liquid PTC conditions using 30% aqueous H_2O_2 in the presence of sodium tungstate and tetrabutylammonium hydrogen sulfate (TBAHS) as catalyst. The experimental procedure involves simple mixing of an alcohol, $Na_2WO_4 \cdot H_2O$ and TBAHS then addition of 30% aqueous H_2O_2 in 25:1:1:125 molar ratios for primary alcohols and 25:1:1:40 molar ratios for secondary alcohols, in an open vessel. The best results were obtained when the temperatures of reaction mixtures were 90 °C and 100 °C for primary and secondary alcohols, respectively (Eq. 84).

$$\underset{R}{\overset{}{\frown}}OH + H_2O_2 \xrightarrow[MW, 20\ min]{Na_2WO_4,\ TBAHS} \underset{R}{\overset{O}{\frown}}OH$$

$$\underset{R_1\ \ R_2}{\overset{OH}{\frown}} + H_2O_2 \xrightarrow[MW, 10\ min]{Na_2WO_4,\ TBAHS} \underset{R_1\ \ R_2}{\overset{O}{\frown}} \quad (84)$$

$$\underset{OH}{\overset{OH}{\frown}} + H_2O_2 \xrightarrow[MW, 10\ min]{Na_2WO_4,\ TBAHS} \underset{OH}{\overset{O}{\frown}}$$

Microwaves	10–20 min	60–97%
Conventional heating at 90°C	4 h	83–96%

The oxidation of secondary alcohols under the action of MW irradiation can be performed when hydrogen peroxide is substituted by hydrogen peroxide urea adduct (UHP) [131]. The microwave-assisted reaction results in appropriate carbonyl compounds, higher yield, and shortening of the reaction time compared with conventional conditions. Typical results are given in Table 6.35 [131].

Hydrogen peroxide has also been used under the action of MW irradiation for epoxidation of simple or cyclic alkenes. The reactions were accomplished under liquid–liquid PTC conditions in ethylene chloride solution in the presence of Na_2WO_4 and Aliquat 336 as catalysts. The best results were obtained at 70 °C

Tab. 6.35. Oxidation of secondary alcohols by means of hydrogen peroxide urea adduct (UHP) under MW + PTC conditions (multimode reactor, 600 W).

Alcohol	Reaction time (min)	Yield (%)		Product
		Microwave	Conventional	
Octan-2-ol	120	88	66	Octan-2-one
Cyclohexanol	60	98	62	Cyclohexanone
1-Phenyl-ethanol	60	98	56	1-Phenylethanone
2-Ethylhexane-1,3-diol	120	77	65	3-Hydroxymethylheptan-4-one
1-Octen-3-ol	120	85	66	1-Octen-3-one

when the concentration of hydrogen peroxide was set to 8% and the pH of aqueous phase was kept below 2 (Eq. 85) [132].

$$\text{CH}_2=\text{CH-(CH}_2)_n\text{-CH=CH}_2 + \text{H}_2\text{O}_2 \xrightarrow[\text{ethylene chloride}]{\text{Na}_2\text{WO}_4,\ \text{Aliquat 336}} \text{epoxide product} \qquad (85)$$

Microwaves	100 min	91-98%
Conventional heating	100 min	54-65%

6.2.7.12 S-Alkylation of n-Octyl bromide [49]

The synthesis of octylthiocyanate, by reaction of n-octyl bromide with KSCN, and its subsequent isomerization to isothiocyanate, has been realized by use of TBAB under the action of MW irradiation. The effect of inorganic solid supports (SiO_2, K10, graphite, NaCl) has been studied (Eq. 86).

$$\text{R-Br} \xrightarrow[\text{4 min - MW}]{\text{KSCN / TBAB}} \text{R-SCN} \longrightarrow \text{R-NCS} \qquad (86)$$

6.2.7.13 Reductive Decyanation of Alkyldiphenylmethanes [133]

The reaction was performed in aqueous NaOH with PEG-400 as phase-transfer agent (Eq. 87).

$$\text{Ph}_2\text{C(CN)R} \xrightarrow[\text{2 min - MW}]{\text{NaOH / PEG 400}} \text{Ph}_2\text{CHR} \qquad (87)$$

R = H, $(CH_2)_n CH_3$, $CH_2CH_2CH_2NR_2$ (R = H 94%)

6.2.7.14 Synthesis of Symmetrical Disulfides

A rapid and general method for the synthesis of symmetrical disulfides involves reaction of sulfur with NaOH under PTC + MW irradiation conditions to give sodium disulfide, which reacts with alkyl halides to afford the disulfides in good to excellent yields (Eq. 88) [134].

$$2\ \text{RX} + 2\ \text{S} \xrightarrow[\text{DMF, MW, 6-7 min}]{\text{NaOH   PEG 400}} \text{RSSR} \qquad (88)$$

yields: tBuCl 90% $nC_5H_{11}Cl$ 88% $PhCH_2Cl$ 91%

6.2.7.15 The Hantzsch Reaction [135]

The synthesis of a variety of substituted 1,4-dihydropyridines has been achieved by reaction of aldehydes, ethyl/methyl acetoacetates, and ammonium acetate in water using a phase-transfer catalyst under the action of MW irradiation. Compared with classical Hantzsch's reaction conditions, this new method consistently has the advantage of good yields and short reaction times. Bifunctional compounds containing two units have been synthesized in good yields using dialdehyde as precursor (Eq. 89).

$$R_1CHO + Me\overset{O}{\underset{}{\diagdown}}\overset{O}{\underset{}{\diagup}}R_2 + NH_4OAc \xrightarrow[MW\ 5-10\ min]{TBAB\ /\ H_2O} \underset{Me\ \ N\ \ Me}{\underset{H}{R_2\diagdown\diagup R_1 \diagdown\diagup R_2}} \qquad (89)$$

72 - 92%

6.3
Conclusion

The use of MW irradiation to provide the energy for the activation of chemical species certainly leads to faster and cleaner reactions compared with conventional heating. The coupling of microwave technology with solvent-free solid–liquid PTC conditions is one of a group of particularly efficient, powerful, and attractive methods.

Significant improvements of yields and/or reaction conditions can be achieved, with substantial simplification in operating procedures. The powerful synergistic combination of PTC and MW techniques has certainly enabled an ever increasing number of reactions to be conducted under cleaner and milder conditions. The inherent simplicity of the method can, furthermore, be allied with all the advantages of green chemistry procedures in terms of reactivity, selectivity, atom economy, safety, and ease of manipulation.

References

1 (a) MAKOSZA, M.; FEDORYNSKI, M. in *"Handbook of Phase-Transfer Catalysis"*, Blackie Academic & Professional, London, New York, **1997**, chapter 4, 135–17; (b) MAKOSZA, M.; FEDORYNSKI, M. *Polish J. Chem.* **1996**, *70*, 1093; (c) MAKOSZA, M. *Survey Prog. Chem.* **1980**, *9*, 1.

2 STARKS, C.M.; LIOTTA, C.L.; HALPERN, M. in *"Phase-transfer catalysis"*, Chapman & Hall, New York, London, **1994**.

3 BRANDSTRÖM, A. in *"Preparative Ion Pair Extraction, an Introduction to Theory and Practice"*, Apotekarsocieteten-Hassle Lakemedel, Stockholm, **1974**.

4 DEHMLOW, E.V.; DEHMLOW, S.S. in *"Phase-transfer catalysis"*, 3rd edn, Verlag Chemie, Weinheim, **1993**.

5 SASSON, Y.; NEUMANN, R. in *"Handbook of Phase-transfer catalysis"*, Chapman and Hall, London, **1997**.

6 LOUPY, A.; PETIT, A., HAMELIN, J.; TEXIER-BOULLET, F.; JACQUAULT, P., MATHÉ, D. *Synthesis* **1998**, 1213–1234.

7 PERREUX, L.; LOUPY, A. *Tetrahedron* **2001**, *57*, 9199–9223.

8 DESHAYES, S; LIAGRE, M.; LOUPY, A.; LUCHE, J.L., PETIT, A. *Tetrahedron* **1999**, *55*, 10851–10870.

9 LOUPY, A.; PIGEON, P.; RAMDANI, M. *Tetrahedron* **1996**, *52*, 6705–6712.

10 Bram, G.; Loupy, A.; Majdoub, M. *Synth. Commun.* **1990**, *20*, 125–129.

11 Loupy, A.; Petit, A.; Ramdani, M.; Yvanaef, C., Majdoub, M.; Labiad, B., Villemin, D. *Can. J. Chem.* **1993**, *71*, 90–95.

12 Cléophax, J.; Liagre, M.; Loupy, A.; Petit, A. *Org. Proc. Res. Develop.* **2000**, *4*, 498–504.

13 Xu, W.G.; Liu, F.A.; Yu, Y.X.; Jin, S.X.; Liu, J.; Jin, Q.H. *Chem. Res. Chin. Univ.* **1992**, *8*, 324–326. *Chem. Abstr.* **1993**, *118*, 233448x.

14 Jiang, Y.L.; Yuan, Y.C.; Sun, Y.H. *Chem. Res. Chin. Univ.* **1994**, *10*, 159–162. *Chem. Abstr.* **1994**, *121*, 300548g.

15 Yuncheng, Y.; Julin, Y.; Dabin, G. *Synth. Commun.* **1992**, *22*, 3109–3114.

16 Villa, C.; Baldarassi, S.; Gambaro, R.; Mariani, E.; Loupy, A. *Int. J. Cosmetic Sci.* **2005**, *27*, 11–16.

17 Li, Z.; Quan, Z.J.; Wang, X.C. *Chemical Papers* **2004**, *58*, 256–259.

18 Yuan, Y.C.; Jiang, Y.L.; Gao, D.B. *Chinese Chem. Lett.* **1992**, *3*, 613–614. *Chem. Abstr.* **1993**, *118*, 38524s.

19 Yuan, Y.C.; Jiang, Y.L.; Pang, J.; Zhang, X.H.; Yang, C.G. *Gazz. Chim. Ital.* **1993**, *123*, 519–520.

20 Bogdal, D.; Pielichowski, J.; Jaskot, K. *Org. Prep. Proc. Int.* **1998**, *30*, 427–432.

21 Matondo, H.; Baboulène, M.; Rico-Lattes, I. *Appl. Organometal. Chem.* **2003**, *17*, 239–243.

22 Majdoub, M.; Loupy, A.; Petit, A.; Roudesli, S. *Tetrahedron* **1996**, *52*, 617–628.

23 (a) Chatti, S.; Bortolussi, M.; Loupy, A. *Tetrahedron Lett.* **2000**, *41*, 3367–3370; (b) Chatti, S.; Bortolussi, M.; Loupy, A. *Tetrahedron* **2001**, *57*, 4365–4370.

24 (a) Chatti, S.; Bortolussi, M.; Loupy, A.; Blais, J.C.; Bogdal, D.; Majdoub, M. *Eur. Polym. J.* **2002**, *38*, 1851–1861; (b) Chatti, S.; Bortolussi, M.; Loupy, A.; Blais, J.C.; Bogdal, D.; Roger, P. *J. Appl. Polym. Sci.* **2003**, *90*, 1255–1266; (c) Chatti, S.; Bortolussi, M.; Bogdal, D.; Blais, J.C.; Loupy, A. *Eur. Polym. J.* **2004**, *38*, 561–577.

25 Campbell, L.J.; Borges, L.F.; Heldrich, F.J. *Biomed. Chem. Lett.* **1994**, *4*, 2627–2630.

26 (a) Bogdal, D.; Pielichowski, J.; Boron, A. *Synth. Commun.* **1998**, *28*, 3029–3039; (b) Bratulescu, G.; Le Bigot, Y., Delmas, M.; Pogany, I. *Rev. Roum. Chim.* **1998**, *43*, 321–326.

27 Jiang, Y.L.; Hu, Y.Q.; Pang, J.; Yuan, Y.C. *J. Amer. Oil Chem. Soc.* **1996**, *73*, 847–850.

28 Yadav, G.D.; Bisht, P.M. *J. Mol. Catal.* **2004**, *221*, 59–69.

29 Yadav, G.D.; Bisht, P.M. *Synth. Commun.* **2004**, *34*, 2885–2892.

30 Yadav, G.D.; Bisht, P.M. *Catalysis Commun.* **2004**, *5*, 259–263.

31 Li, J.; Pang, J.; Cao, G.; Xi, Z. *Synth. Commun.* **2000**, *30*, 1337–1342.

32 Varma, R.S. *Green Chemistry* **1999**, *1*, 43–55.

33 Mariani, E.; Genta, M.T.; Bargagna, A.; Neuhoff, C.; Loupy, A.; Petit, A. in "Application of the Microwave Technology to Synthesis and Material Processing", Mucchi Ed. **2000**, 157–167.

34 Villa, C.; Genta, M.T.; Bargagna, A.; Mariani, E.; Loupy, A. *Green Chemistry* **2001**, *3*, 196–200.

35 Wang, J.X; Zhang, M.; Hu, Y. *Synth. Commun.* **1998**, *28*, 2407–2413.

36 Reddy, Y.T., Rao, M.K.; Rajitha, B. *Indian J. Heterocyclic Chem.* **2000**, *10*, 73–74.

37 Wang, J.X; Zhang, M.; Huang, D., Hu, Y. *J. Chem. Res. (S)* **1998**, 216–217.

38 Pchelka, B., Plenkiewicz, J. *Org. Prep. Proced. Int.* **1998**, *30*, 87–90.

39 Bogdal, D.; Warzala, M. *Tetrahedron* **2000**, *56*, 8769–8773.

40 Bogdal, D.; Bednarz, S. to be published.

41 Hurduc, N.; Abdelylah, D.; Buisine, J.M.; Decock, P.; Surpateanu, G. *Eur. Polym. J.* **1997**, *33*, 187–190.

42 Ding, J.; Gu, H.; Wen, J.; Lin, C. *Synth. Commun.* **1994**, *24*, 301–303.

43 Huang, Z.Z.; Zu, L.S. *Org. Prep. Proced. Int.* **1996**, *28*, 121–123.

44 Wu, L.L.; Huang, X. *Hecheng Huaxue* **1997**, *5*, 179–181. *Chem. Abstr.* **1998**, *128*, 204852a.

45 Ding, J.; Yang, J.; Fu, M. *Hecheng Huaxue* **1997**, *5*, 309–310. *Chem. Abstr.* **1998**, *128*, 230343e.

46 Zhang, X.H.; You, Y.E.; Guo, M. *Hecheng Huaxue* **1998**, *2*, 220–222. *Chem. Abstr.* **1998**, *129*, 230514g.

47 Bogdal, D. *Molecules* **1999**, *3*, 333.

48 Bogdal, D.; Pielichowski, J.; Boron, A. *Synlett* **1996**, 873–874.

49 Vass, A.; Toth, J.; Pallai-Varsanyi, E. in *"Effect of inorganic solid support for microwave assisted organic reactions"* OR 19, presented at the International Conference on Microwave Chemistry, Prague, Czech Republic, Sept. 6–11, **1998**.

50 Hekmatshoar, R.; Heravi, M.M.; Baghernejad, B.; Asadolah, K. *Phosphorus, Sulfur and Silicon and the Related Elements* **2004**, *179*, 1611–1614.

51 He, L.; Li, J.L.; Zhang, J.J.; Su, P.; Zheng, S.L. *Synth. Commun.* **2003**, *5*, 741–747.

52 De la Cruz, P.; De la Hoz, A.; Font, L.M.; Langa, F.; Pérez-Rodriguez, M.C. *Tetrahedron Lett.* **1998**, *39*, 6053–6056.

53 Bogdal, D.; Pielichowski, J.; Jaskot, K. *Heterocycles* **1997**, *45*, 715–722.

54 Bogdal, D.; Pielichowski, J.; Jaskot, K. *Synth. Commun.* **1997**, *27*, 1553–1560.

55 Rodriguez, H.; Pérez, R.; Suarez, M.; Lam, A.; Cabrales, N.; Loupy, A. *Heterocycles* **2001**, *55*, 291–301.

56 Jiang, Y.; Wang, Y.; Deng, R.; Mi, A. *A.C.S. Symp. Ser.* **1997**, *659*, 203–213.

57 Wang, Y.L.; Jiang, Y.Z. *Synth. Commun.* **1992**, *22*, 2287–2291.

58 Deng, R.H.; Mi, A.Q.; Jiang, Y.Z. *Chinese Chem. Lett.* **1993**, *4*, 381–384. *Chem. Abstr.* **1993**, *119*, 271670s.

59 Deng, R.H.; Jiang, Y.Z. *Hecheng Huaxue* **1994**, *2*, 83–85. *Chem. Abstr.* **1994**, *121*, 108127c.

60 Deng, R.H.; Wang, Y.L.; Jiang, Y.Z. *Synth. Commun.* **1994**, *24*, 1917–1921.

61 Deng, R.H.; Wang, Y.L.; Jiang, Y.Z. *Synth. Commun.* **1994**, *24*, 111–115.

62 Wang, Y.; Deng, R.; Mi, A.; Jiang, Y. *Synth. Commun.* **1995**, *25*, 1761–1764.

63 Abramovitch, R.A.; Shi, Q.; Bogdal, D. *Synth. Commun.* **1995**, *25*, 1–7.

64 Hernandez, L.; Casanova, E.; Loupy, A.; Petit, A. *J. Label. Comp. & Radiopharm.* **2003**, *46*, 151–158.

65 Gumaste, V.K.; Khan, A.J.; Bhawal, B.M.; Deshmukh, A.R.A.S. *Indian J. Chem., Section B* **2004**, *43*, 420–422.

66 Barbry, D.; Pasquier, C.; Faven, C. *Synth. Commun.* **1995**, *25*, 3007–3013.

67 (a) Loupy, A.; Pellet, M.; Petit, A.; Vo-Thanh, G. *Org. Biomol. Chem.* **2005**, *3*, 1534–1540; (b) Guillot, R.; Loupy, A.; Meddour, A.; Pellet, M.; Petit, A. *Tetrahedron* **2005**, *61*, 10129–10137.

68 Rissafi, B.; Rachiqi, N.; El Louzi, A.; Loupy, A.; Petit, A.; Fkih-Tetouani, S. *Tetrahedron* **2001**, *57*, 2761–2768.

69 Abenhaim, D.; Loupy, A.; Munnier, L.; Tamion, R.; Marsais, F.; Quéguiner, G. *Carbohydr. Res.* **1994**, *261*, 255–266.

70 Chatti, S.; Bortolussi, M.; Loupy, A. *Tetrahedron* **2000**, *56*, 5877–5883.

71 Chatti, S.; Bortolussi, M.; Loupy, A. *Tetrahedron* **2001**, *57*, 4365–4370.

72 Soukri, M.; Guillaumet, G.; Besson, T.; Aziane, D.; Aadil, M.; El Essassi, M.; Aksira, M. *Tetrahedron Lett.* **2000**, *41*, 5857–5860.

73 Abenhaim, D.; Chu Pham Ngoc Son; Loupy, A.; Nguyen Ba Hiep *Synth. Commun.* **1994**, *24*, 1199–1205.

74 Su, G.F.; Wan, H.J.; Mou, H.T.; Huang, S.W. *Jingxi Huahong* **2002**, *19*, 15–17. *Chem. Abstr.* **2002**, *137*, 95477.

75 Villemin, D.; Martin, B.; Puciova, M.; Toma, S. *J. Organomet. Chem.* **1994**, *484*, 27–31.

76 Loupy, A.; Pigeon, P.; Ramdani, M.; Jacquault, P. *Synth. Commun.* **1994**, *24*, 159–165.

77 Dietrich, B.; Lehn, J.M. *Tetrahedron Lett.* **1973**, *14*, 1225–1228.

78 Perreux, L.; Loupy, A. *Org. Prep. Proced. Int.* **2003**, *35*, 361–368.

79 Perreux, L.; Loupy, A.; Delmotte, M. *Tetrahedron* **2003**, *59*, 2185–2189.

80 Limousin, C.; Cléophax, J.; Loupy, A.; Petit, A. *Tetrahedron* **1998**, *54*, 13567–13578.

81 Wei, T.B.; Lin, Q.; Zhang, Y.M.; Wei, W. *Synth. Commun.* **2004**, *34*, 181–186.

82 LIN, Q.; ZHANG, Y.M.; WEI, T.B.; WANG, H. *J. Chem. Res. (S)* **2004**, 298–299.
83 LI, Z.; WANG, X.C. *Synth. Commun.* **2002**, *32*, 3087–3092.
84 LI, Z.; WANG, X.C. *Chinese Chem. Lett.* **2002**, *13*, 717–720.
85 BAI, L.; LI, K.; LI, S.; WANG, J.X. *Synth. Commun.* **2002**, *32*, 1001–1007.
86 BAI, L.; LI, S.; WANG, J.X.; CHEN, M. *Synth. Commun.* **2002**, *32*, 127–132.
87 LOUPY, A.; LE NGOC, T. *Synth. Commun.* **1993**, *23*, 2571–2577.
88 LE NGOC, T.; TRAN, H.A.; NGUYEN, A.K.; TRAN, K.P. *Tap chi Hoa hoc* **1999**, *37*, 92–94.
89 (a) CHEN, X.; HONG, P.J.; DAI, S. *Huaxue Tongbao* **1993**, *11*, 29–30. *Chem. Abstr.* **1994**, *121*, 56753g; (b) CHEN, Z.-X.; XU, G.-Y.; YANG, G.-C.; WANG, W. *React. Funct. Polym.* **2004**, *61*, 139–146.
90 DIAZ-ORTIZ, A.; DIEZ-BARRA, E.; DE LA HOZ, A.; PRIETO, P. *Synth. Commun.* **1993**, *23*, 1935–1942.
91 DIAZ-ORTIZ, A.; PRIETO, P.; ABENHAIM, A.; LOUPY, A. *Tetrahedron Lett.* **1996**, *37*, 1695–1698.
92 YUAN, Y.C.; GAO, D.B.; JIANG, Y.L. *Synth. Commun.* **1992**, *22*, 2117–2119.
93 PANG, J.; XI, Z.; CAO, G. *Synth. Commun.* **1996**, *26*, 3425–3429.
94 CHAOUCHI, M.; LOUPY, A.; MARQUE, S.; PETIT, A. *Eur. J. Org. Chem.* **2002**, *7*, 1278–1283.
95 LEE, J.C.; CHOI, J.H.; LEE, J.S. *Bull. Korean Chem. Soc.* **2004**, *25*, 1117–1118.
96 LLOUNG, M.; LOUPY, A.; MARQUE, S.; PETIT, A. *Heterocycles* **2004**, *63*, 297–308.
97 MARQUE, S.; SNOUSSI, H.; LOUPY, A.; PLÉ, N.; TURCK, A. *J. Fluorine Chem.* **2004**, *125*, 1847–1851.
98 LOUPY, A.; PIGEON, P.; RAMDANI, M.; JACQUAULT, P. *J. Chem. Res. (S)* **1993**, 36–37.
99 BARNIER, J.P.; LOUPY, A.; PIGEON, P.; RAMDANI, M.; JACQUAULT, P. *J. Chem. Soc. Perkin Trans I* **1993**, 397–398.
100 BOUGRIN, K.; SOUFIAOUI, M.; LOUPY, A.; JACQUAULT, P. *New J. Chem.* **1995**, *19*, 213–219.
101 TEXIER-BOULLET, F.; LATOUCHE, R.; HAMELIN, J. *Tetrahedron Lett.* **1993**, *34*, 2123–2126.
102 OUSSAID, A.; LE NGOC, T.; LOUPY, A. *Tetrahedron Lett.* **1996**, *52*, 2451–2454.
103 WANG, J.X.; BAI, L.; LI, W.; HU, Y. *Synth. Commun.* **2000**, *30*, 325–332.
104 WANG, J.X.; BAI, L.; LIU, Z. *Synth. Commun.* **2000**, *30*, 971–977.
105 BENDALE, P.M.; KHADILKAR, B.M. *Synth. Commun.* **2000**, *30*, 1713–1718.
106 RISSAFI, B.; EL LOUZI, A.; LOUPY, A.; PETIT, A.; SOUFIAOUI, M.; FKIH-TÉTOUANI, S. *Eur. J. Org. Chem.* **2002**, *15*, 2518–2523.
107 TOKE, L.; HELL, G.T.; SZABO, G.; TOTH, G.; BIHARI, M.; ROCKENBAUER, A. *Tetrahedron* **1993**, *49*, 5133–5146.
108 FINTA, Z.; HELL, L.; TOKE, L. *J. Chem. Res. (S)* **2000**, 242–244.
109 LARHED, M.; LINDENBERG, G.; HALLBERG, A. *Tetrahedron Lett.* **1996**, *37*, 8219–8222.
110 DIAZ-ORTIZ, A.; PRIETO, E. *Synlett* **1997**, 269–270.
111 VILLEMIN, D., NECHAB, B. 2nd International Electronic Conference on Synthetic Organic Chemistry (ECSOC-2), http://www.mdpi.org/ecsoc/, September 1–30, 1998.
112 WANG, J.X., LIU, Z.; HU, Y.; WIE, B.; BAI, L. *J. Chem. Res. (S)* **2000**, 484–485.
113 BOTELLA, L.; NAJERA, C. *Tetrahedron* **2004**, *60*, 5563–5570.
114 SORENSEN, U.; POMBO-VILLAR, E. *Helv. Chim. Acta* **2004**, *87*, 82–89.
115 LARHED, M.; HALLBERG, A. *J. Org. Chem.* **1996**, *61*, 9582–9584.
116 NAMBOODIRI, V.V.; VARMA, R.S. *Green Chem.* **2001**, *3*, 146–148.
117 LEADBEATER, N.E.; MARCO, M. *J. Org. Chem.* **2003**, *68*, 888–892.
118 ARVELA, R.K.; LEADBEATER, N.E.; SANGI, M.S.; WILLIAMS, V.A.; GRANADOS, P.; SINGER, R.D. *J. Org. Chem.* **2004**, *70*, 161–168.
119 ONDRUSCHKA, B; NUCHTER, M. to be published.
120 WANG, Y.; SAUER, D.R. *Org. Lett.* **2004**, *6*, 2793–2796.
121 SOLODENKO, W.; SCHON, U.; MESSINGER, J.; GLINSCHERT, A.; KIRSCHNING, A. *Synlett* **2004**, 1699–1702.

122 Blettner, C.G.; Konig, W.A.; Stenzel, W.; Schotten, T. *J. Org. Chem.* **1999**, *64*, 3885–3890.
123 Seganish, W.M.; DeShong, P. *Org. Lett.* **2004**, *6*, 4379–4381.
124 Erdelyi, M.; Gogoll, A. *J. Org. Chem.* **2001**, *66*, 4165–4169.
125 Leadbeater, N.E.; Marco, M.; Tominack, B. *J. Org. Lett.* **2003**, *5*, 3919–3922.
126 Appukkuttan, P.; Dehaen, W.; Van der Eycken, E. *Eur. J. Org. Chem.* **2003**, 4713–4716.
127 Sorensen, U.; Pombo-Villar, E. *Tetrahedron* **2005**, *61*, 2697–2703.
128 Xia, M.; Wang, Y.G. *Chinese Chem. Lett.* **2002**, *13*, 1–2.
129 Sato, K.; Aoki, M.; Takagi, R.; Noyori, R. *J. Amer. Chem. Soc.* **1997**, *119*, 12386–12387.
130 Bogdal, D.; Lukasiewicz, M. *Synlett* **2000**, 143–145.
131 Lukasiewicz, M.; Bogdal, D.; Pielichowski, J. 8th International Electronic Conference on Synthetic Organic Chemistry (ECSOC-2), http://www.mdpi.org/ecsoc/, November 1–30, **2004**.
132 Bogdal, D.; Lukasiewicz, M.; Pielichowski, J.; Bednarz, S. *Synth. Commun.* **2005**, in preparation.
133 Khadilkar, B.M.; Bendale, P.M. in "*Microwave assisted reductive decyanation of alkyldiphenylmethanes*" OR 24, presented at the International Conference on Microwave Chemistry, Prague, Czech Republic, Sept. 6–11, **1998**.
134 Wang, J.X.; Gao, L.; Huang, D. *Synth. Commun.* **2002**, *32*, 963–969.
135 Salehi, H.; Guo, Q.X. *Synth. Commun.* **2004**, *34*, 4349–4357.

7
Microwaves and Ionic Liquids

Nicholas E. Leadbeater and Hanna M. Torenius

7.1
Introduction

There is a growing demand for faster, cleaner and more efficient methods for organic synthesis because of increasing environmental awareness, and health and safety issues. In this regard, microwave activation has became an increasingly popular tool as evidenced by the number of recent reviews and books on the topic [1, 2]. The disadvantage of traditional heat sources, for example oil baths and electrical heaters, is that they are rather slow, and long reflux at high temperatures may lead to byproduct formation and product decomposition. Microwave irradiation is a rapid and direct heating technique which often dramatically reduces reaction times from hours to minutes, increases product yields, and enhances product purity [3]. In drug discovery, in which small quantities of many compounds are needed quickly, use of microwaves has proved to be a valuable tool for rapid synthesis of large libraries of compounds [4–6].

There are two mechanisms by which microwaves interact with reaction mixtures [7]. Polarization of dielectric material arises when the distribution of an electron cloud is distorted or physical rotation of molecular dipoles occurs. For generation of heat on irradiation with microwaves, at least one component of a reaction mixture must have a dipole moment. Compounds with high dipole moments also have large dielectric constants, ε. The selectivity of microwave irradiation is clear when comparing the heating of water and hexane. Water, a polar solvent, has a high dielectric constant and therefore heats rapidly on microwave irradiation whereas hexane, a nonpolar solvent, heats very slowly.

Microwave activation also occurs via a conduction mechanism. In a solution that contains ionic material, the ions start to move under the influence of the electric field of the microwave irradiation. This results in expenditure of energy because of an increased collision rate, thus converting kinetic energy into heat. This effect is particularly important at higher temperatures.

Issues arise when chemists want to use nonpolar reagents in conjunction with less polar solvents such as hexane ($\varepsilon = 1.88$), toluene ($\varepsilon = 2.38$), tetrahydrofuran ($\varepsilon = 7.58$), and dioxane ($\varepsilon = 2.21$). Because of their lower dielectric constants, these

Microwaves in Organic Synthesis, Second edition. Edited by A. Loupy
Copyright © 2006 WILEY-VCH Verlag GmbH & Co. KGaA, Weinheim
ISBN: 3-527-31452-0

solvents are not usually suitable for microwave chemistry except when highly polar reagents are used. In other cases, a highly polar reaction medium is desirable. Use of ionic liquids in conjunction with microwave heating can help address both these situations.

Ionic liquids are interesting materials which have found new functions in chemistry [8]. Their unique properties and ionic structure has made them useful in synthesis and catalysis [9]. Because of their low toxicity, low vapor pressure and recyclability these compounds have a "green" reputation.

The study of ionic liquids started with the attempt to develop more efficient electrolytes for batteries. High melting-point molten salts were already widely used, but the high melting points of these substances caused problems inside the batteries. To find better electrolytes, first inorganic $NaCl-AlCl_3$ molten salt, melting point 107 °C, was studied. This was followed by study of mixtures of 1-ethylpyridinium halides and $AlCl_3$. The alkylpyridinium cations were relatively easy to reduce, so more resistant ionic liquids were synthesized. Alkylimidazolium salts afforded the most promising ionic liquids [10].

Ionic liquids differ from classical molten salts by being liquids at room temperature, or their melting points are below 100 °C. They have an ionic structure and usually consist of an organic cation and an inorganic or organic anion. The most common cations found in ionic liquids are the tetraalkylammonium, tetraalkylphosphonium, N-alkylpyridinium, 1,3-dialkylimidazolium, and trialkylsulfonium moieties [11]. Some common ionic liquid cations and anions are presented in Fig. 7.1.

The nature of both the cation and the anion determine the melting point of the ionic liquid. For example, if 1,3-dialkylimidazolium-based ionic liquids are to be liquids at room temperature the cation should be unsymmetrical [11]. Of all the ionic liquids, 1,3-dialkylimidazolium-based ionic liquids are most widely used in synthesis and catalysis.

Halide-based ionic liquids are simple to prepare. Conventionally, they are synthesized by reacting heterocyclic starting materials, for example N-methylimidazole, pyridine, or thiazole, with an appropriate alkyl halide in a nonpolar solvent, for example toluene (Scheme 7.1). The reactions can be also performed using a large excess of the alkyl halide as reagent and solvent. The reactions usually take between 3 and 72 h to reach completion. Symmetrical N,N'-disubstituted imidazolium halide

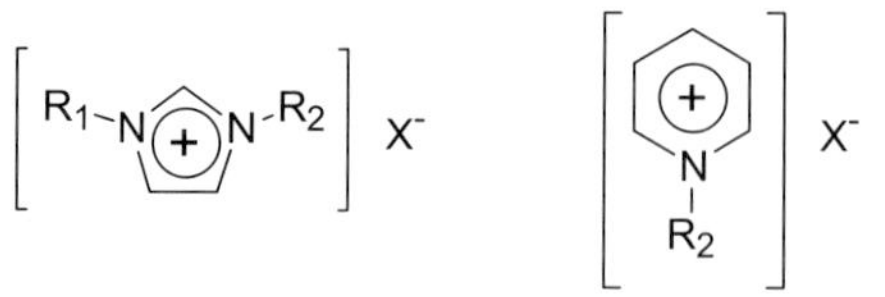

$R_1 = C_nH_{2n+1}$, mostly n = 1-2; $R_2 = C_nH_{2n+1}$, n = 1-8

$X = I^-, Br^-, Cl^-, PF_6^-, BF_4^-, AlCl_4^-, SbF_6^-$

Fig. 7.1. Structure of ionic liquids.

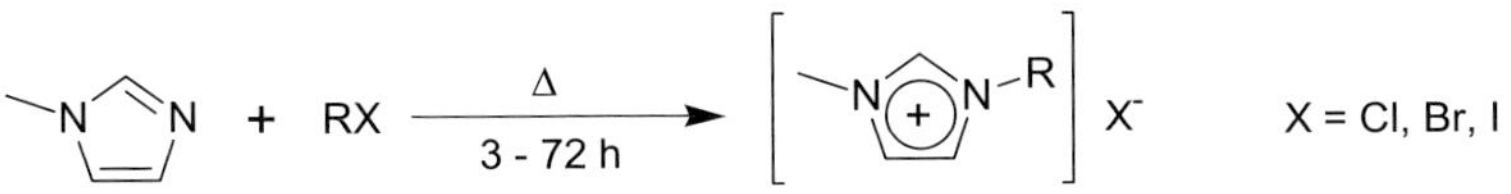

Scheme 7.1. Synthes is of 1,3-dialkylimidazolium halide ionic liquids.

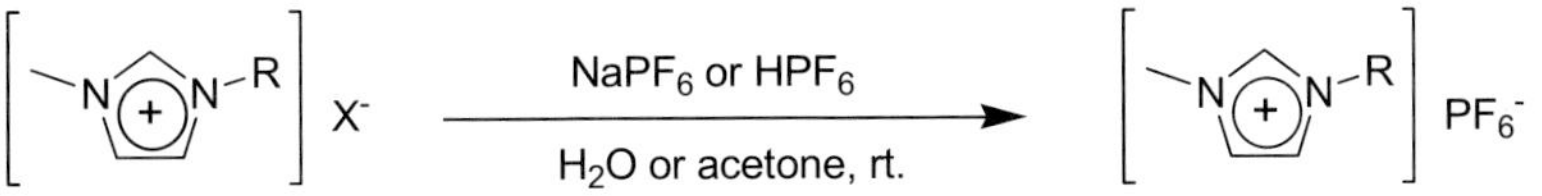

Scheme 7.2. Preparation of fluorophosphate-based ionic liquids.

ionic liquids can be synthesized from N-trimethylsilylimidazole and 2 or 3 equiv. alkyl and benzyl halides [12]. This reaction is also typically performed in toluene, under reflux, for up to 24 h.

Halide-based ionic liquids can function as starting materials for the preparation of other ionic liquids. For example, the common hexafluorophosphate PF_6 and tetrafluoroborate BF_4 ionic liquids can be prepared by simple anion exchange (Scheme 7.2).

Ionic liquids have several interesting and beneficial properties, because of their ionic structure and low melting points. They have negligible vapor pressure so do not evaporate. They are stable at high temperatures, are extremely good at dissolving diverse organic compounds and catalysts, but in many cases are relatively inert toward reactions. They are poorly miscible in common nonpolar solvents, so can be used for extraction. Their properties can also be tuned. Their solubility in water depends on the anion, the temperature, and the substituents. Although imidazolium-based halide ionic liquids are usually water-soluble, their PF_6 derivatives are insoluble in water. The presence of impurities can alter the physical and chemical properties of ionic liquids. They are usually hydroscopic. Water is difficult to remove, because of intermolecular hydrogen bonding. To a general approximation, they are more polar than acetonitrile, but less polar than methanol. Changing the counter anion does not change the polarity of the medium significantly and neither does the presence of a small amount of water.

Ionic liquids usually have good thermal stability, although decomposition can occur at high temperatures. Vacuum pyrolysis studies of 1,3-dialkylimidazolium halide-based ionic liquids showed that the primary decomposition pathway was dealkylation (Scheme 7.3) [13]. This can be prevented by using non-nucleophilic

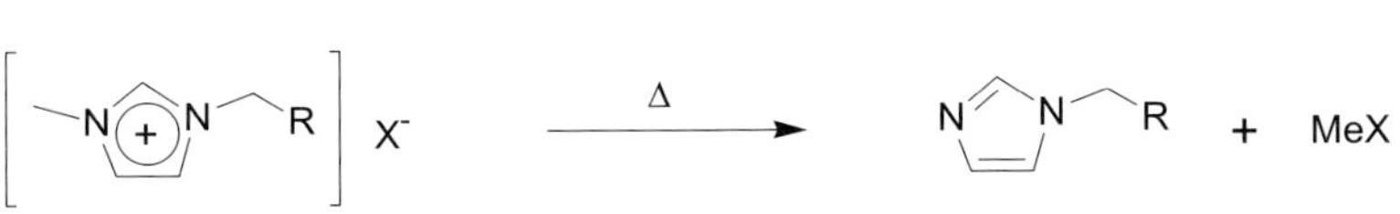

Scheme 7.3. Decomposition of 1,3-dialkylimidazolium halide-based ionic liquids.

anions. For BF_4 and PF_6-based ionic liquids the primary decomposition pathway is attack by traces of HF generated by reaction of the anion with any residual water present in the ionic liquid.

The thermal properties and decomposition of imidazolium, tetraethylammonium, and tetrabutylammonium-based ionic liquids have been studied in detail by use of calorimetry [14]. The freezing points of samples were much lower than their melting points. The greatest supercooling was observed for ionic liquids bearing the 1-propyl-3-methylimidazolium cation. Thermal decomposition was dependent on the structure of the liquid. For imidazole-based ionic liquids, increasing the substitution of the ring increased the thermal stability, because of removal of the ring hydrogen atoms. Removal of the hydrogen in the 2 position had the largest effect. Attachment of an isopropyl group to the nitrogen reduced the thermal stability. Imidazolium cations were usually more stable than the tetraalkylammonium cations. The order of anion stability was:

$$PF_6 > \text{bisperfluoro(ethylsulfonyl)imide} > \text{bis(trifluoromethylsulfonyl)imide}$$
$$\approx BF_4 > \text{tris(trifluoromethylsulfonyl)methide} \approx AsF_6 \gg I, Br, Cl.$$

7.2
Ionic Liquids in Conjunction with Microwave Activation

Use of ionic liquids in conjunction with microwave irradiation in synthetic chemistry can be classified into three topics:

- use of microwave heating in the preparation of ionic liquids;
- use of ionic liquids as solvents and reactants for microwave-promoted synthesis, and
- use of ionic liquids as aids for heating nonpolar reaction mixtures using microwave irradiation.

7.2.1
Synthesis of Ionic Liquids Using Microwave Heating

As mentioned in the introduction, ionic liquids are conventionally prepared by heating heterocyclic starting materials, for example N-methylimidazole, pyridine, or thiazole with an appropriate alkyl halide under reflux in a nonpolar solvent, for example toluene or in a large excess of the alkyl halide. Conventional synthesis of ionic liquids has several disadvantages. First, long reflux times are often needed to obtain reasonable yields. Second, purification of the ionic liquid is often problematic and large quantities of organic solvents are required to extract impurities. This is especially true with halide-based ionic liquids – excess halide is difficult to remove completely. This has prompted the use of microwave-mediated approaches for the preparation ionic liquids. The first report was by Varma and coworkers who, using a domestic microwave oven, prepared a range of ionic liquids from

1-methylimidazole and n-butyl halides [15]. After irradiating N-methylimidazole and n-butyl bromide for 2 min in the domestic oven, 1-butyl-3-methylimidazolium bromide [BMIM]Br was obtained in 78% yield. When the ionic liquid starts to form, the polarity of the reaction medium increases and, therefore, microwave absorption increases. The observed reactivity order was I > Br > Cl. The yields of ionic liquid improved when sequential irradiation was used. For example, irradiation for 30 s then 3 × 15 s irradiation intervals is better than one period of 75 s.

This method is not only limited to halide-based ionic liquids. Synthesis of ionic liquids containing a tetrafluoroborate anion and imidazolium cation has been performed successfully [16] by use of a modified domestic microwave oven with pulsed irradiation (5 × 30 s). Approximately 90% yields of the desired ionic liquids were usually obtained compared with 36% after the same reaction time using conventional heating. Microwave irradiation has been used in the synthesis of 1-ethyl-3-methylimidazolium benzoate and dialkyl imidazolium tetrachloroaluminate ionic liquids [17, 18]. These reactions were again performed in a domestic microwave oven using pulsed irradiation.

Although reports of ionic liquid synthesis in domestic microwave ovens are promising, synthesis in these systems has several disadvantages. For chloroaluminates there were problems with the reproducibility of the synthesis. In domestic microwave ovens, control of the reaction conditions is difficult and overheating of the reaction mixtures can easily occur. A modified procedure was published, again using a domestic oven, but placing the reaction vessel in a beaker of water [19]. This water moderation reduced the amount of irradiation reaching the sample and thus reduced the effects of overheating. It also functioned as a heat sink. Other problems with domestic ovens in the synthesis of ionic liquids are that volatile alkyl halides are released inside microwave cavity, which can be hazardous, and large-scale synthesis is difficult because of poor control, possible contamination, and product decomposition.

Safe and clean large-scale synthesis of ionic liquids can be achieved by using modern accurate microwave equipment. Khaldilkar and Rebeiro studied the synthesis of imidazolium, pyridine, and 2,6-lutidine-based ionic liquids using alkyl halides as reagents [20]. Excellent yields were reported after irradiation for 20–55 min, the length of the alkyl chain having an effect on the reaction time required. Different reaction temperatures were required, depending on the substrates used. At 100 °C no product was ever observed. At 150 °C the imidazolium-based ionic liquids could be prepared. For ionic liquids bearing pyridinium cations a reaction temperature of 200 °C was required. The reactions could be scaled up to 50 g and no large excess of alkyl halide was needed. Deetlefs and Seddon studied the synthesis of pyrazolium, thiazolium, imidazolium, and pyridine-based ionic liquids [21]. Selected results are summarized in Table 7.1. The authors reported that the reactions are up to 72 times faster than when using conventional heating. They also found that if the microwave irradiation is prolonged, decomposition of the ionic liquid occurs. Imidazolium halide based ionic liquids could be prepared in 150, 200, 250, and 300 mmol quantities using the same reaction conditions. Synthesis of pyrazolium and thiazolium chloride ionic liquids was not successful.

Tab. 7.1. Preparation of ionic liquids using microwave irradiation.

Ionic liquid	Reaction temp. (°C)	Irradiation time (min)	Yield (%)
[methylimidazolium butyl] I⁻	165	4	92
[methylimidazolium butyl] Br⁻	80	6	93
[methylimidazolium butyl] Cl⁻	150	20	96
[pyridinium butyl] I⁻	165	10	95
[pyridinium butyl] Br⁻	130	30	95
[pyridinium butyl] Cl⁻	120	40	90
[methylthiazolium butyl] I⁻	135	35	96
[methylthiazolium butyl] Br⁻	130	65	94

Chiral ionic liquids have been prepared by Loupy and coworkers using controlled microwave irradiation (Scheme 7.4) [22]. They developed a solvent-free direct alkylation technique using a range of alkyl bromides and (1R,2S)-N-methylephedrine as reagents, the latter being obtained by reductive amination of (1R,2S)-ephedrine. They prepared a series of halide-based ionic liquids bearing alkyl chains of different lengths. Depending on the boiling point of the alkyl halide the reactions were either performed in 10 mL sealed vessels or in open 100 mL flasks. The temperature was accurately measured and monitored during the course of microwave irradiation. Product yields were carefully compared with those obtained using an oil

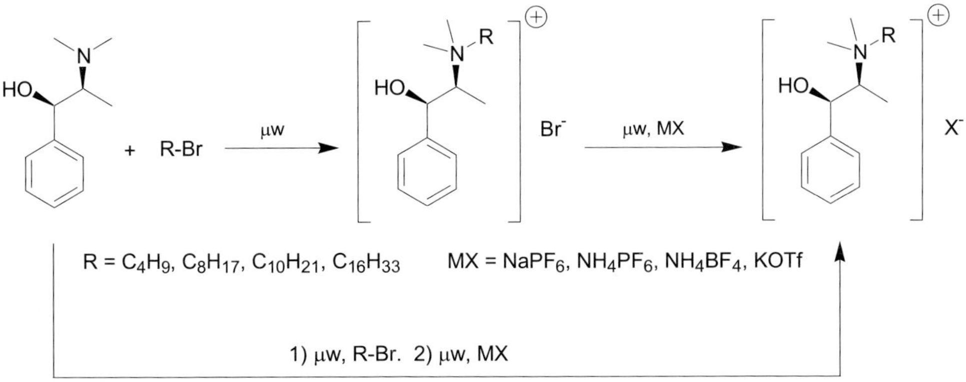

Scheme 7.4. Preparation of chiral ionic liquids using microwave activation.

bath, running the reactions for the same time at the same temperature with similar profiles of temperature increase. Using conventional heating, yields were noticeably lower than when using microwave assistance and the authors suggest the existence of a nonthermal microwave specific effect. The halide-based ionic liquids were then converted into $^-PF_6$, $^-BF_4$, and ^-OTf analogs, again using microwave irradiation. The quaternization and anion metathesis steps could be performed in a two-step, one-pot method, all crude products from reaction of alkyl bromides and (1R,2S)-N-methylephedrine being directly submitted to the anion-exchange step. Ionic liquids prepared this way were then used successfully in asymmetric Baylis–Hillman reactions [23].

7.2.2
Reactions Using Microwave Irradiation and Ionic Liquids as Solvents and Reagents

The widest application of ionic liquids in microwave chemistry has been as polar solvents. They heat very rapidly under the action of microwave irradiation and have excellent solvating properties. Many chemical reactions benefit from this combination of rapid heating and an ionic reaction medium.

7.2.2.1 Hydrogenation
Catalytic hydrogenation reactions are very important in organic synthesis [24]. Two-phase hydrogenation reactions using ionic liquids and rhodium catalysts have been reported [25, 26]. The most common ionic liquids used in hydrogenations are [BMIM]PF_6, [BMIM]SbF_6, and [BMIM]BF_4. Although methods using hydrogen gas in conjunction with microwave heating have not been reported, hydrogenation reactions are possible under microwave conditions by use of catalytic transfer hydrogenation [27]. Transfer hydrogenation of alkenes, alkynes, and nitro compounds have been reported using 10% Pd–C as a catalyst, formic acid salts as a source of hydrogen and [BMIM]PF_6 as the solvent (Scheme 7.5) [28]. When

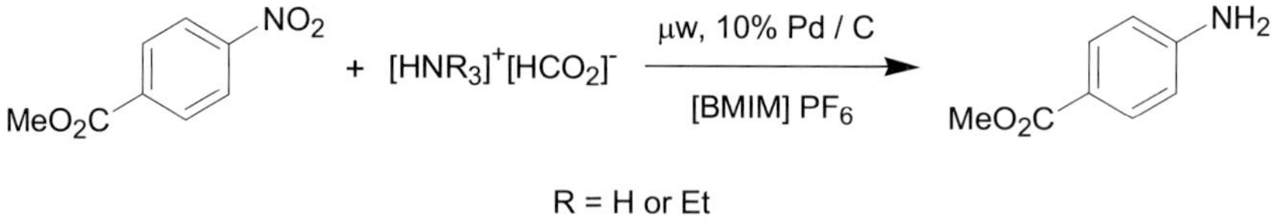

![Scheme 7.5 reaction]

Scheme 7.5. Ionic liquids and microwave heating in transfer hydrogenation reactions.

4-nitrobenzoic acid methyl ester was used as starting material, 4-aminobenzoic acid methyl ester was formed in 70% yield after 7 min irradiation. After optimization, heating for 70 min at 150 °C was found to lead to complete conversion. The authors reported that monitoring the pressure of the reaction gave a clear indication of its progress. When the reaction reached constant pressure the hydrogenation was complete. The catalyst could be recycled but, after five cycles, 40% loss of activity was observed. This could be restored by adding a corresponding amount of fresh catalyst.

7.2.2.2 Ruthenium-catalyzed Olefin Metathesis

Ring-closing metathesis (RCM) is one of the most important methods of synthesis of versatile heterocyclic compounds important in both medicinal chemistry and natural product synthesis and was the topic of the 2005 Nobel Prize in Chemistry [29]. Ruthenium-based "Grubbs" catalysts are commonly used in these reactions, because of their stability and high reactivity (Fig. 7.2) [30]. Removal of the catalysts from the product is not simple, however, and recycling of the catalysts is not possible using conventional methods. Ionic liquids such as [BMIM]PF_6 have been used as solvents in an attempt to avoid some of these issues [31]. Using 5 mol% of the Grubbs type 1 catalyst, a range of RCM reactions could be performed by heating at 50 °C for 20 h. Product isolation was simple but the recyclability of the catalyst was poor, because of decomposition in the ionic liquid. N-heterocyclic carbene-based ruthenium complexes have also been used for RCM reactions in conjunction with ionic liquids as solvents [32, 33]. When RCM reactions are performed using microwave irradiation and ionic liquids as solvents, excellent yields and very fast reactions have been reported [34]. When [BMIM]PF_6 was used as solvent and Grubbs type II as catalyst, for many substrates conversion to product was complete after

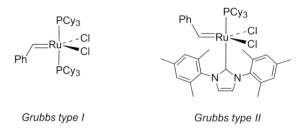

Grubbs type I   *Grubbs type II*

Fig. 7.2. Grubbs catalysts types I and II.

Tab. 7.2. Microwave ring-closing metathesis using ionic liquids as solvents.

$$\text{CH}_2=\text{CH-CH}_2-X-\text{CH}_2-\text{CH}=\text{CH}_2 \xrightarrow[\substack{2\text{-}3 \text{ mol \% catalyst} \\ [\text{BMIM}]\text{PF}_6}]{\mu w, 15 \text{ s}} \text{cyclopentene with X}$$

Substrate	Product	[BMIM]PF$_6$		CH$_2$Cl$_2$	
		μw yield (%)	Δ yield (%)	μw yield (%)	Δ yield (%)
EtO$_2$C, CO$_2$Et (diallyl)	EtO$_2$C, CO$_2$Et cyclopentene	100	3	100	21
EtO$_2$C, CO$_2$Et (allyl/methallyl)	EtO$_2$C, CO$_2$Et methylcyclopentene	100	0	79	9
Ts-N(allyl)$_2$	Ts-N pyrroline	100	0	91	45
diallyl ether	dihydrofuran	0	2	85	4

microwave irradiation for only 15 s. Representative results of the study are collected in Table 7.2. Conventional heating using the same temperature and time did not lead to significant product formation. For allyl ether, no product was formed when using ionic liquids as solvents. The authors believe this is because of the relatively low boiling point of the ether. Using microwave irradiation and running the reactions for 120 s, similar yields could be obtained using dichloromethane as a solvent. The authors suggest a potential non-thermal microwave effect to explain the enhanced reaction rates they observe both in ionic liquids and in dichloromethane.

The reactions have also been performed under solvent-free conditions with considerable success and almost quantitative yields of the products [35a]. For example, the last entry in Table 7.2 was improved to 100% when irradiation was performed for 3 min. at 50 °C. Under solvent-free conditions, the existence of nonthermal specific microwave effects has been proposed on the basis of comparison with control reactions performed using conventional heating.

Kappe and coworkers studied the same RCM reactions, and dichloromethane was used as a solvent to probe more thoroughly the existence, or not, of a specific nonthermal microwave effect [35b]. Quartz reaction vessels were used in the study to minimize microwave absorption and hence indirect heating from normal borosilicate glass. It was found that when the power was 100 W the dienes used in the reactions absorbed significantly whereas absorption by the Grubbs catalyst was negligible. Comparisons of microwave irradiation and conventional heating were performed using a preheated oil bath. A solution of the diene was equilibrated before the catalyst was added. Under these conditions complete conversion was observed in the same time as when using microwave irradiation and thus the idea of a nonthermal microwave effect as the cause of the rapid reactions was considered improbable when a polar solvent is used.

7.2.2.3 Synthesis of Nitriles from Aryl Halides

Nitriles are versatile and important components of a range of dyes, natural products, and pharmaceuticals. Aryl nitriles can be synthesized from aryl halides by direct reaction between aryl halides and copper cyanide, known as the Rosemund von Braun reaction [36]. These cyanation reactions can have several disadvantages, in particular the long reaction times required. Ren and coworkers showed that 1,3-dialkylimidazolium halide-based ionic liquids can be used as solvents in the Rosemund von Braun reaction [37]. Complete conversion, based on GC–MS analysis, was achieved after 24 h at 90 °C. When using microwave irradiation and ionic liquid as a solvent, Leadbeater and coworkers showed the reaction times could be reduced to between 3 and 10 min [38]. Under the optimized reaction conditions, 2 equiv. CuCN and 1 equiv. aryl halide were rapidly heated to 200 °C in [i-PrMIM]Br as solvent. Representative results are collected in Table 7.3. The microwave method works as well as the conventional method for a range of aryl iodide and aryl bro-

Tab. 7.3. Rosemund von Braun reaction using ionic liquids and microwave heating.

Entry	Aryl halide	Time (min)	Isolated yield (%)
1	4-Iodotoluene	3	75
2	4-Bromotoluene	10	63
3	4-Chlorotoluene	10	0
4	4-Iodoanisole	3	65
5	4-Bromoanisole	10	40
6	4-Bromoacetophenone	10	64
7	4-Iodonitrobenzene	3	55
8	4-bromonitrobenzene	10	16
9	4,4′-Dibromobiphenyl	10	19

mide starting materials. Whereas 3 min irradiation was sufficient for aryl iodide substrates, bromides required longer (10 min) and aryl chlorides failed to react. Extraction of product from the ionic liquid proved problematic after both conventional and microwave heating, and although GC–MS results showed product formation to be quantitative, actual isolated yields were substantially lower. An alternative microwave method is to use $Ni(CN)_2$ as a cyanating agent and NMP as a solvent; higher product recoveries are possible with this method [39]. This highlights a limitation of ionic liquids as solvents.

7.2.2.4 Palladium-catalyzed Heck Reactions

The Heck reaction is an important tool in organic synthesis and drug discovery [40]. Typically, Heck reactions are performed in polar solvents, for example DMF or NMP, and a variety of palladium catalysts are used. Addition of an ionic salt to the reaction mixture can stabilize the palladium catalyst and Heck reactions have been performed, for example, using molten tetrabutylammonium bromide as solvent and a palladacycle as catalyst [41, 42]. Using these conditions, aryl chlorides could be coupled with styrene using 0.5 mol% of the palladacycle. Microwave-assisted Heck reactions of chloroarene substrates have been reported with tetrabutylammonium bromide as nonaqueous ionic liquid and an Mg–Al layered double hydroxide-supported palladium catalyst (Fig. 7.3) [43]. The yields of the Heck reaction using microwave irradiation were similar to those obtained using conventional methods, but the reaction time could be reduced from 15–40 h to 30–60 min (Scheme 7.6). The double hydroxide-supported catalyst was a better choice than homogenous $PdCl_2$, especially for electron-neutral and electron-rich substrates.

Heck couplings of aryl halides with alkenes in room temperature ionic liquids have been reported [44]. Dialkylimidazolium and n-hexylpyridinium hexafluorophosphate and tetrafluoroborate ionic liquids were used as solvents, $Pd(OAc)_2$ and $PdCl_2$ as catalysts, and an amine as base. Addition of ligands such as triphenylphosphine, tri-o-tolylphosphine, and triphenylarsine usually had a negative affect on product yield. Addition of cosolvents such as DMF was not necessary. Palladium-catalyzed Heck reactions involving aryl bromides have been performed

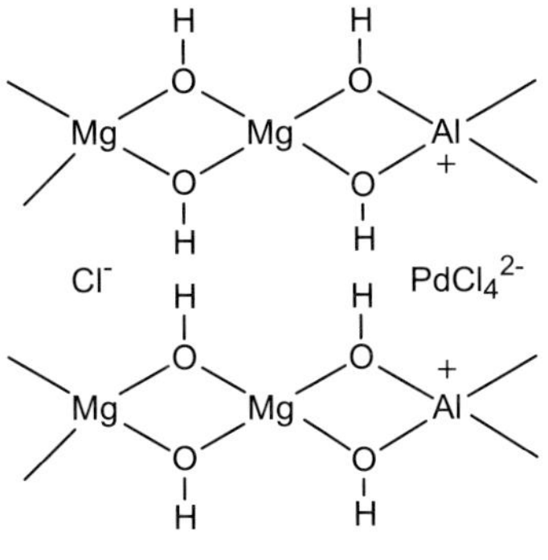

Fig. 7.3. Mg–Al layered double hydroxide-supported palladium catalyst.

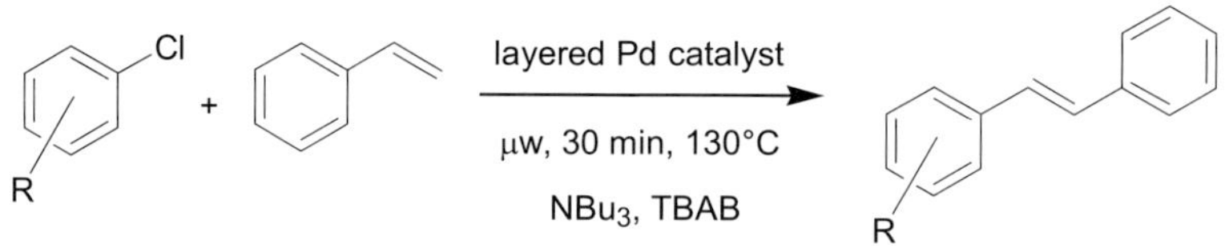

Scheme 7.6. Layered palladium-catalyzed microwave and TBAB-assisted Heck reaction.

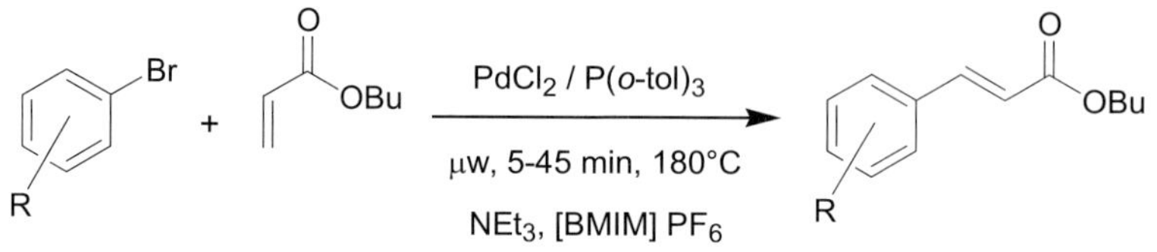

Scheme 7.7. Heck reactions using ionic liquids and microwave heating.

in [BMIM]PF$_6$ as solvent using microwave heating (Scheme 7.7) [45]. Different catalysts and ligands were screened using bromoanisole and butyl acrylate as substrates. In [BMIM]PF$_6$ as solvent PdCl$_2$ gave better results than Pd(OAc)$_2$ and, in this study, addition of P(o-tol)$_3$ as ligand proved beneficial for the reaction. The catalyst remained in the ionic liquid at the end of the reaction and could be recycled up to five times. Reaction of butyl vinyl ether with 1-bromonaphthalene in [BMIM]PF$_6$ was also studied and different conditions were required. It was necessary to use Pd(OAc)$_2$ as the catalyst and DPPP (1,3-bis(diphenylphosphino)-propane) as ligand. The optimum yield was obtained after heating at 130 °C for 120 min. If the reaction temperature was increased, PdCl$_2$ was used, or no DPPP was added, the regioselectivity of the reaction suffered.

Heck reactions catalyzed by palladium on carbon have been performed using microwave heating in open vessels with 1-octanyl-3-methylimidazolium tetrafluoroborate [OMIM]BF$_4$ as the solvent (Scheme 7.8) [46]. Short reaction times were needed and for aryl iodides the yields varied from 35% for o-iodobenzoic acid methyl ester to 86% for iodobenzene. With aryl bromides yields varied from 89% for p-bromonitrobenzene to 33% for p-bromotoluene. When aryl chlorides

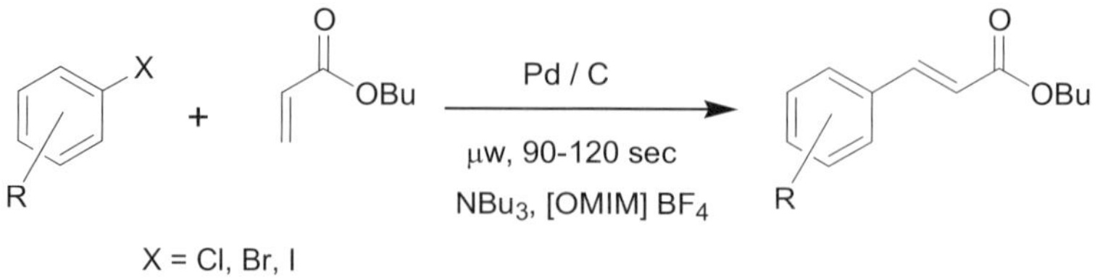

Scheme 7.8. Heck reactions in an open vessel using ionic liquids and microwave heating.

were used, poor yields were obtained. The catalyst was reported to remain in the ionic liquid at the end of the reaction and could therefore be reused.

7.2.2.5 Benzylation in Ionic Liquids Using Thermal and Microwave Conditions

The DABCO (1,4-diazabicyclo[2.2.2]octane)-catalyzed N-benzylation of heterocycles using dibenzyl carbonate as an alkylating agent benefits from the use of ionic liquids either as additives or as solvents [47]. They have effects on both the yield and the rate of the reaction. Several imidazolium and ammonium-based ionic liquids have been screened. Best results were achieved when tetraoctylammonium chloride and tetrabutylammonium chloride (TBAC) were used in stoichiometric quantities. Using 10–30 mol% DABCO, 1 equiv. TBAC as additive, and dimethylacetamide (DMA) or acetonitrile as solvent, a range of substrates were screened using both conventional and microwave heating. Representative examples are shown in Table 7.4 with results from conventional and microwave activation. Conventional heating reactions were performed at 135 °C for between 30 min and 2 h. For the microwave reactions a continuous-flow reactor was employed and the reactions were run at 160 °C with residence times of between 6 and 18 min. Yields were slightly lower than with conventional heating. Of the imidazolium-based ionic liquids screened, [HMIM]Cl gave similar yields to TBAC but only when used as solvent rather than stoichiometric additive.

7.2.2.6 Phosgenation

Phosgenation is used widely in industry for preparation of carbonates and isocyanates [48, 49]. Phosgene is highly toxic making it difficult to handle. Ionic liquids have been used in alcohol phosgenation using triphosgene as a phosgene equivalent. The reaction has been performed using microwave irradiation in a specially designed apparatus [50]. A glass reactor consisting of two chambers (a developing chamber for decomposition of triphosgene and a reaction chamber) was placed in a microwave oven. Benzyltriethylammonium chloride, phenanthroline monohydrate, and 1-ethyl-2,3-dimethylimidazolium-bis-trifluoromethanesulfonoimide were employed in the decomposition of the triphosgene. When an ionic liquid was used as a solvent a 94% yield of dibutyl carbonate was obtained from n-butanol after a reaction time of 30 min. The catalyst and ionic liquid could be recycled.

7.2.2.7 Microwave and Ionic Liquid-assisted Beckmann Rearrangement

The Beckmann rearrangement is an efficient way of obtaining substituted amides from oximes using phosphorus reagents such as PCl_5 and P_2O_5–methanesulfonic acid, or Brønsted acids such as sulfuric acid, phosphoric acid, formic acid, or silica gel. Under solvent-free conditions, adsorption of reagents on acidic supports, for example $HCOOH–SiO_2$ or montmorillonite clays, microwave and conventional heating methods have been compared (Scheme 7.9) [51, 52]. Good yields of amides were be obtained after 7–10 min microwave irradiation at between 112–152 °C in open vessels. In a preheated sand bath using the same reaction times yields were sometimes substantially lower. Classical heating led to large differences between reaction yields whereas microwave heating was more consistent. The

Tab. 7.4. DABCO catalyzed *N*-benzylation using thermal and microwave irradiation.

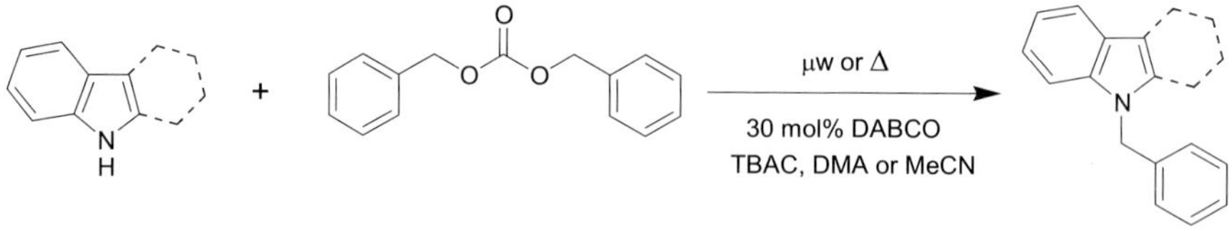

Starting material	Product	Conventional heating		Microwave heating	
		Time (min)	Yield (%)	Time (min)	Yield (%)
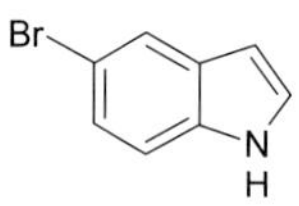 5-Br-indole	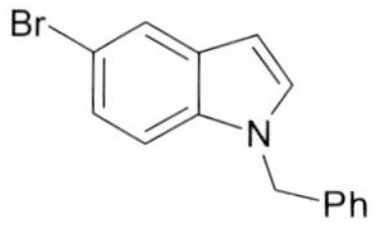 5-Br-N-benzylindole	30	83	6	76
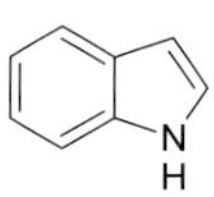 indole	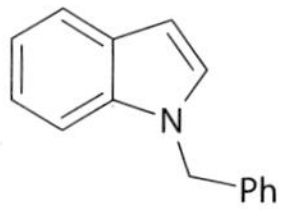 N-benzylindole	120	80	12	70
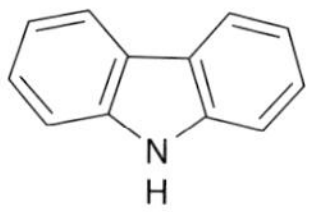 carbazole	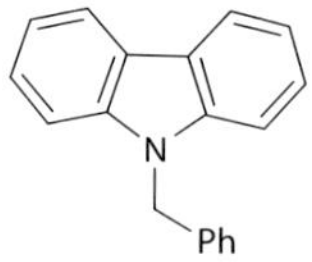 N-benzylcarbazole	180	89	18	82
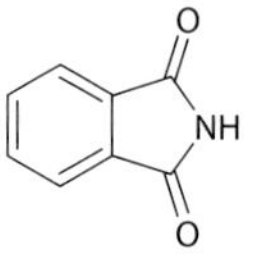 phthalimide	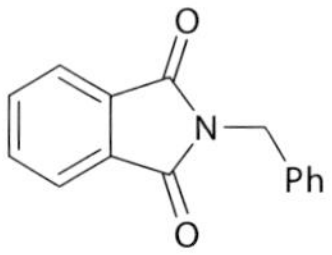 N-benzylphthalimide	60	87	6	84
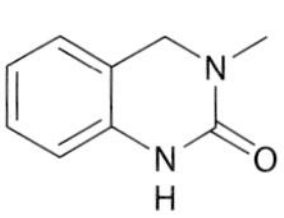 3-methyl-3,4-dihydroquinazolin-2(1H)-one	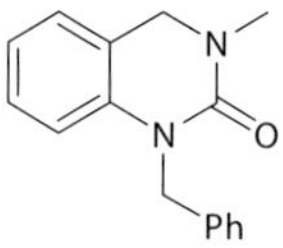 N-benzyl derivative	120	71	12	41

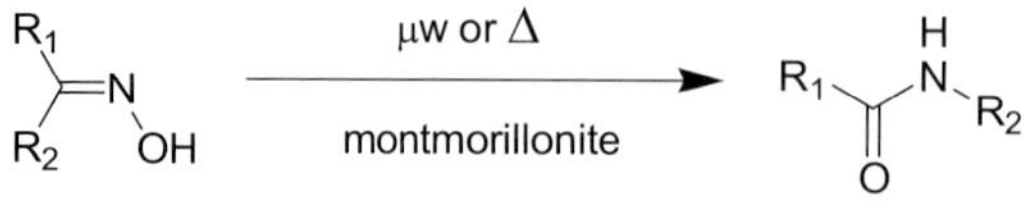

Scheme 7.9. Solvent-free Beckmann rearrangements.

Tab. 7.5. Microwave-assisted Beckmann rearrangement using ionic liquids.

$$\underset{R_2}{\overset{R_1}{>}}=N_{OH} \xrightarrow[H_2SO_4, \text{ ionic liquid}]{\mu w} R_1 \underset{O}{\overset{H}{\underset{\|}{\overset{|}{N}}}} R_2$$

R_1	R_2	Ionic liquid	Time (s)	H_2SO_4 (mol%)	Yield (%)
C_6H_5	C_2H_5	[BMIM]PF_6	40	25	99
C_6H_5	C_2H_5	[BMIM]PF_6	60	5	99
C_6H_5	C_2H_5	[BMIM]BF_4	60	25	99
C_6H_5	C_2H_5	[BMIM]SbF_6	60	25	99
4-Cl-C_6H_4	C_2H_5	[BMIM]PF_6	60	5	81
4-Me-C_6H_4	C_2H_5	[BMIM]PF_6	40	5	91
C_6H_5	C_6H_5	[BMIM]PF_6	60	5	99
4-MeO-C_6H_4	C_6H_5	[BMIM]PF_6	60	5	99

authors discussed the possibility of nonthermal microwave effects to explain the different yields. $ZnCl_2$ was found to be the best acidic support for microwave-promoted Beckmann rearrangement of aldoximes [53] whereas for oximes, $BiCl_3$ was best [54].

The Beckmann rearrangement can be performed under mild conditions using ionic liquids as solvents and PCl_5, P_2O_5, and $POCl_3$ as catalysts [55]. The best ionic liquid was found to be [N-BuPy]BF_4, which led to high product yields and selectivity. The Beckmann rearrangement of cyclohexanone oxime in [BMIM]PF_6 and [BMIM]BF_4 has also been reported [56]. In these reactions P_2O_5 and P_2O_5–methanesulfonic acid were used as catalysts. Reaction times varied from 16–24 h and excellent yields were usually obtained when [BMIM]PF_6 was used. No product was formed if [BMIM]BF_4 was used; this was attributed to the higher water content. Lee and coworkers have studied the ionic liquid-mediated Beckmann rearrangement using microwave heating [57]. They screened several ionic liquids and the amount of sulfuric acid needed as catalyst. Representative results are collected in Table 7.5. The microwave method was found to be superior to analogous conventional heating. All the ionic liquids tested worked well under the action of microwaves and only 5 mol% sulfuric acid was needed. The stability of the ionic liquids used was assayed by use of thermogravimetric analysis. It was found that all the ionic liquids tested ([BMIM]PF_6, BF_4, SbF_6, and OTf) were thermally stable and decomposition only occurred between 380–400 °C. In the Beckmann rearrangement reactions sequential irradiation cycles of 20 s were used. The total reaction time varied between 40 s and 120 s.

7.2.2.8 Conversion of Alkyl Alcohols to Halides

Ren and Wu showed that alcohols can be converted to alkyl halides by using 1,3-dialkylimidazolium halide-based ionic liquids as both solvent and nucleophile in

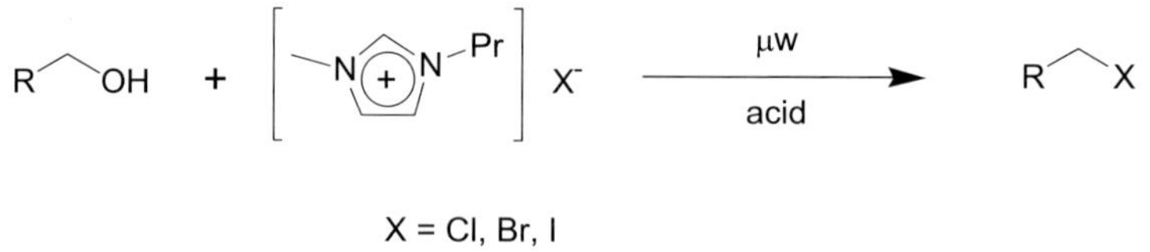

Scheme 7.10. Conversion of alcohols into halides by use of microwaves with ionic liquids as reagents and solvents.

the presence of either an organic or inorganic acid [58]. The reactions generally take between 24–48 h to reach completion at room temperature. The reaction can be greatly accelerated by use of microwave irradiation, the mixture being held at 200 °C for between 30 s and 10 min (Scheme 7.10) [58]. The effect of toluene as cosolvent was also studied and found to be beneficial when using allylic or benzylic alcohols as substrates. By use of [PMIM]I, primary alcohols were successfully converted to the corresponding iodides, very rapidly and with good yields. In contrast with the conventional method, use of an organic acid instead of H_2SO_4 was beneficial in the microwave method. The reaction time was found to be very important. When using [PMIM]I, if the reaction mixture is heated for longer than 30 s, significant decomposition is observed. Attempts to use secondary and tertiary alcohols were not successful. Long-alkyl chain ionic liquids, for example [OMIM]Br or [OMIM]I, can be used in the conversion of fatty alcohols to the corresponding alkyl halides by use of the same method [59].

7.2.2.9 Synthesis of Triazines

The combination of [BMIM]PF_6 as solvent and microwave irradiation can be used for synthesis of 6-aryl-2,4-diaminotriazines [60]. The KOH-catalyzed condensation between benzonitrile and dicyandiamide in [BMIM]PF_6 was studied to optimize the reaction conditions. The conditions were then applied to a range of substrates. Reaction times of 10–15 min were required, compared with 15–24 h when conventional heating was used (Scheme 7.11). Product yields were high.

7.2.2.10 1,3-Dipolar Cycloaddition Reactions

Several ionic liquids have been screened for use in 1,3-dipolar cycloaddition reactions [61]. Reaction times and product yields can be enhanced using [EMIM]BF_4,

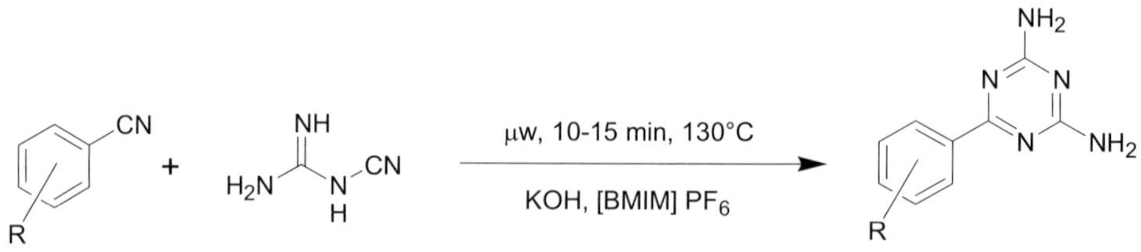

Scheme 7.11. Synthesis of 6-aryl-2,4-diaminotriazines using ionic liquids and microwave heating.

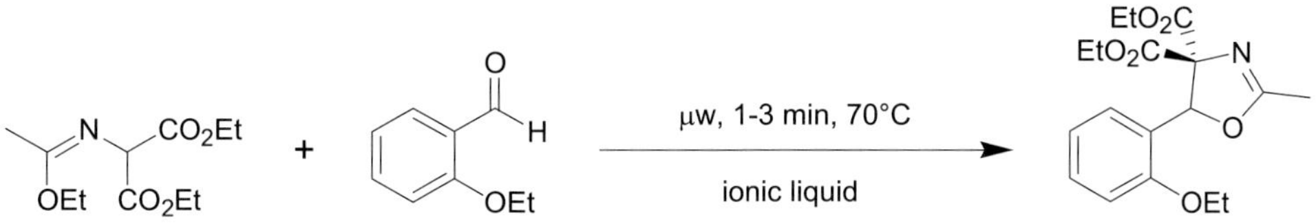

Scheme 7.12 1,3-dipolar cycloaddition reactions using ionic liquids and microwave activation.

PF_6, or ONf as solvents. When reaction of 2-methoxybenzaldehyde with an imidate derived from diethylaminomalonate was used as a test, reaction times and product yields could be enhanced by use of [EMIM]BF_4, PF_6, or ONf as solvents rather than conventional organic media. When reactions in [EMIM]BF_4 and [EMIM]PF_6 were performed with microwave irradiation excellent conversion to product was achieved after 1–3 min. Addition of 5% glacial acetic acid further increases the reaction rate (Scheme 7.12).

The same authors have also reported 1,3-dipolar cycloadditions using 2-hydroxy and 3-hydroxybenzaldehydes grafted on a soluble ionic liquid support [62]. New benzaldehyde-supported ionic liquids were prepared via two different routes. In the first approach the synthesis started from an N-alkylimidazole and 2-chloroethanol, thermolysis of which, followed by anion exchange to form the BF_4 or PF_6 ionic liquid, gave the desired supports. After esterification with an acid-functionalized 2-hydroxybenzaldehyde, excellent yields of the benzaldehyde-supported ionic liquids were obtained. The synthetic approach is shown in Scheme 7.13.

The second synthetic approach involved the reaction of 1-imidazole, chloroethanol, and sodium ethoxide to form (2-hydroxyethyl)imidazole. Treatment of this with dicyclohexylcarbodiimide (DCC) to form the N,N'-dicyclohexyl isourea followed by reaction with hydroxybenzaldehyde and quaternization with an alkyl iodide gave the iodide-based supported benzaldehyde. Anion exchange then was performed to generate the desired supported ionic liquids (Scheme 7.14).

For application in 1,3-dipolar cycloaddition reactions, diversity could be introduced by first reacting the benzaldehyde-grafted ionic liquid with different alkyl amines using microwave heating. This was followed by conventional heating with an imidate to give the desired supported product. Treatment of the bound cycloadduct with NaOMe in methanol resulted in cleavage from the ionic liquid phase (Scheme 7.15).

7.2.2.11 Knoevenagel Reactions

The ionic liquid grafted benzaldehydes prepared for the 1,3-dipolar cycloaddition reactions have also been used successfully as substrates for Knoevenagel reactions using microwave irradiation (Scheme 7.16). One equivalent of different malonate derivatives with a variety of electron-withdrawing groups was added to the ionic liquid phase and piperidine was used as catalyst. Reaction times varied from 15–60 min. The product was cleaved from the ionic liquid phase by using the NaOMe

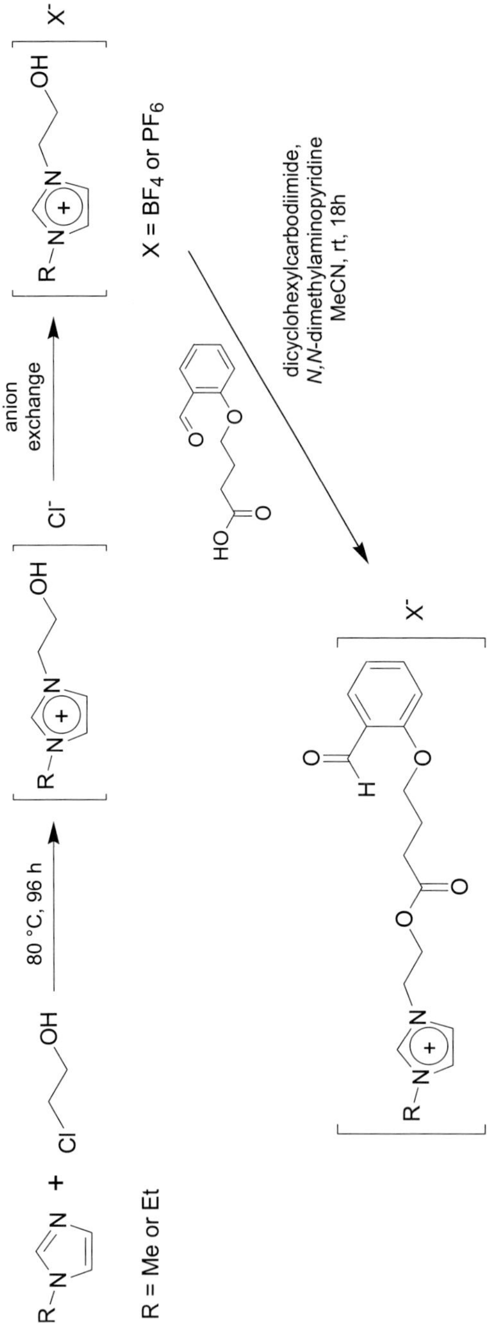

Scheme 7.13. Synthesis of 2-hydroxybenzaldehyde-grafted ionic liquids.

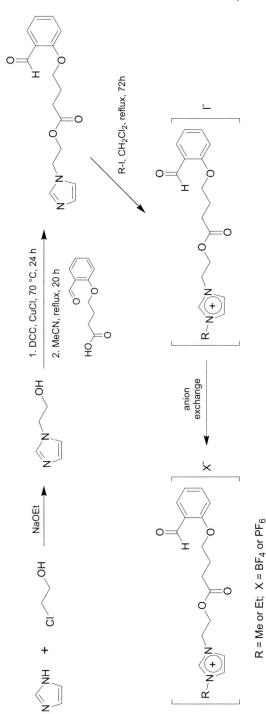

Scheme 7.14. Alternative synthesis of 2-hydroxybenzaldehyde-grafted ionic liquids.

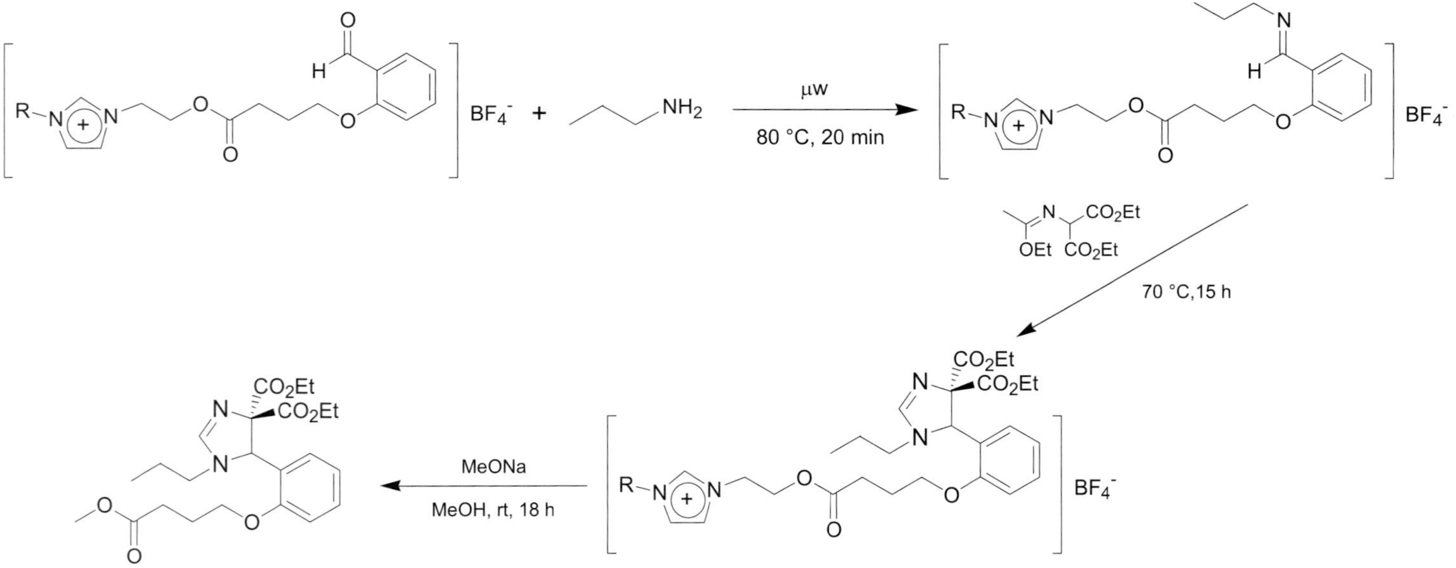

Scheme 7.15. 1,3-Dipolar cycloaddition reactions using ionic liquid-supported benzaldehydes.

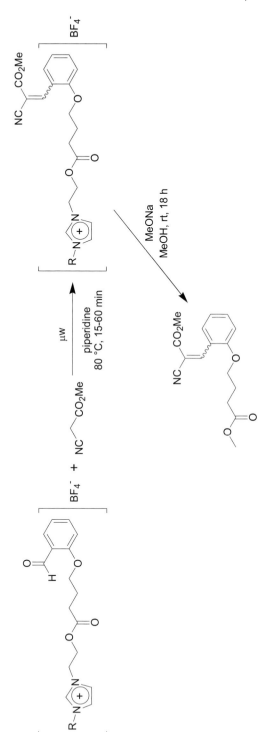

Scheme 7.16. Solvent-free Knoevenagel reactions using ionic liquid-supported benzaldehydes.

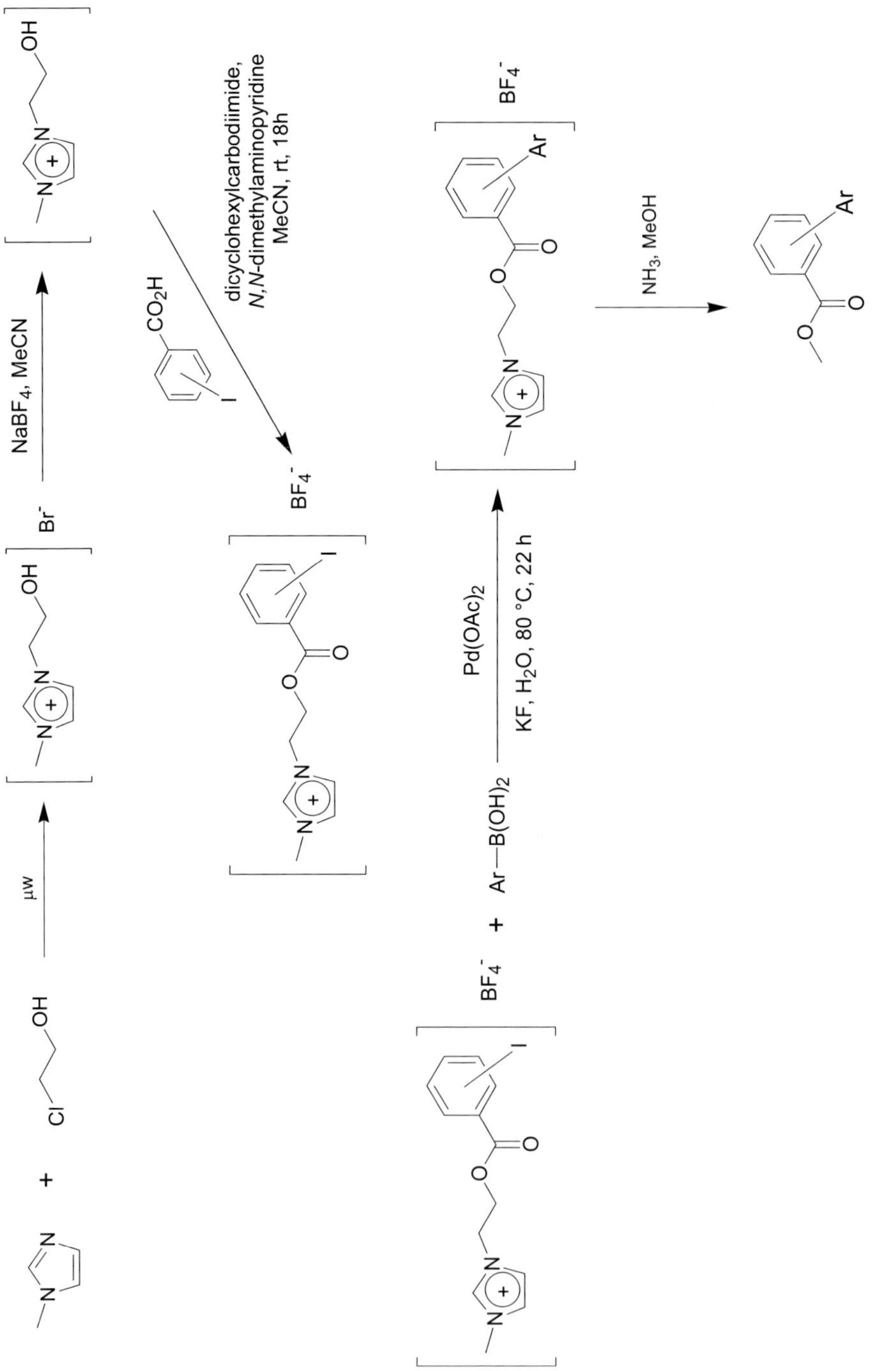

Scheme 7.17. Ionic liquid-supported Suzuki reactions.

in methanol procedure. The final products were obtained without the need for further purification.

7.2.2.12 Ionic Liquids as Soluble Supports for the Suzuki Coupling Reaction

Soluble ionic liquid supports have been developed for use in Suzuki coupling reactions (Scheme 7.17) [63]. The basic support was prepared by reaction of N-methylimidazole and 2-bromoethanol using microwave irradiation followed by anion exchange. Suitably functionalized aryl iodides were then attached to the support using standard coupling procedures. The resulting soluble ionic liquid-supported iodide was coupled to a range of boronic acids using palladium acetate as a catalyst, cesium fluoride as base, and water as solvent. The biaryl product was then cleaved from the support by use of ammonia in methanol. Yields varying from 55 to 83% were reported.

7.2.3
Use of Ionic Liquids and Microwaves in Multicomponent Reactions

7.2.3.1 The Kabachnik–Fields Reaction in Ionic Liquids Using Microwave Heating

The Kabachnik–Fields reaction is an effective means of preparing biologically active α-amino phosphonates [64]. It involves the three component reaction of an aromatic aldehyde, an aniline, and diethylphosphite. The reaction has recently been performed using microwave irradiation with [BMIM]PF$_6$, [BMIM]SbF$_6$, [BMIM]BF$_4$, and DMF as solvents and lanthanide triflates as catalysts (Scheme 7.18) [65]. The reactions were performed using a domestic microwave oven and pulsed irradiation. Catalyst activity in the ionic liquids was found to be higher than or comparable with that in DMF. It was also found that catalyst activity varied depending on the ionic liquid used. For example, Yb(OTf)$_3$ was very active in [BMIM]BF$_4$ but Sc(OTf)$_3$ was more active in [BMIM]PF$_6$. Excellent product yields were obtained.

7.2.3.2 Multicomponent Synthesis of Propargylamines

A variety of propargylamines can be prepared by three-component Mannich condensation of acetylenes, aldehydes, and secondary amines (Scheme 7.19) [66, 67]. These reactions were performed using CuCl to activate the acetylene component. Reaction times varied from 3–36 h. Silver iodide has recently been used to catalyze the reaction [68].

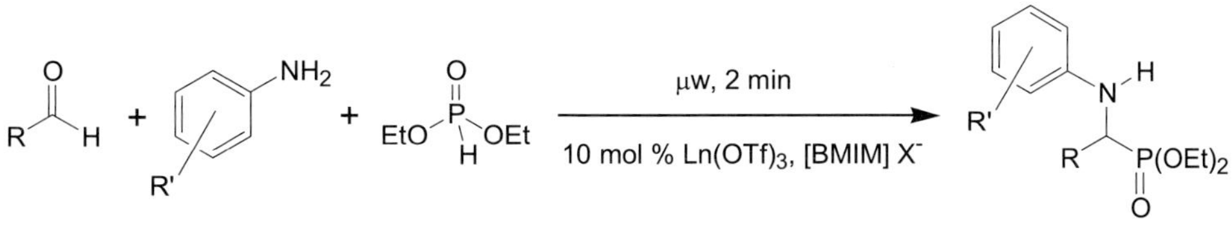

Scheme 7.18. The Kabachnik–Fields reaction using ionic liquids and microwave heating.

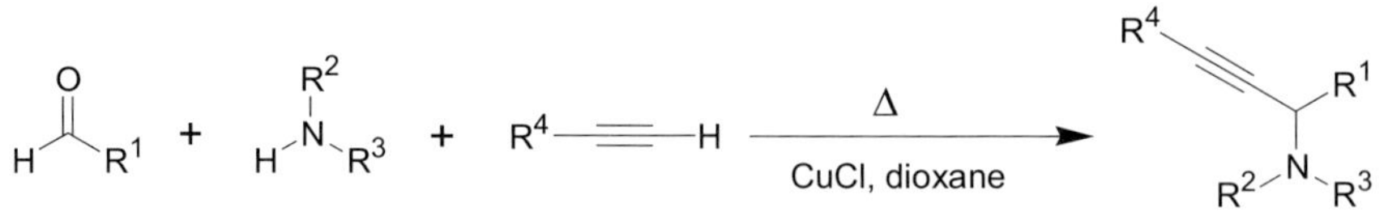

Scheme 7.19. Mannich-type multicomponent synthesis of propargylamines.

Mannich-type multicomponent reactions have been performed using ionic liquids with microwave irradiation [69]. Optimum reaction conditions were screened using piperidine as the amine, phenylacetylene as the alkyne, and benzaldehyde or 4-chlorobenzaldehyde as the aldehyde component. The effects of solvent and irradiation time were studied. Reactions were best run in dioxane with addition of a small quantity of ionic liquid. If only ionic liquid was used significant decomposition was observed. When the optimum conditions had been found, several substrates were screened. Selected examples are shown in Table 7.6. For aromatic or aliphatic aldehydes with cyclic amines and aromatic alkynes, optimum conditions were 6 min microwave irradiation at 150 °C. For noncyclic amines slightly longer irradiation times were required and for aliphatic alkynes 10 min heating at 150 °C gave the best yields of product. Conventional heating methods gave lower yields of the desired products.

7.2.3.3 One-pot Pictet–Spengler Reactions

The Pictet–Spengler reaction is an efficient way of forming tetrahydro-β-carboline rings, functionality that is abundant in nature [70]. These reactions have been performed solvent free using microwave irradiation with the reagents adsorbed on a silica support [71]. Parallel synthesis of 1,2,3,4-tetrahydro-β-carbolines has also been reported [72]. Enhanced product yields compared with those obtained using conventional heating were observed in both instances. When microwave irradiation was used, Yb(OTf)$_3$ was shown to be an effective catalyst for the reaction [73]. Occasionally, however, for example reaction of tryptamine, addition of 50 mol% of a chloroaluminate salt, for example [BMIM]Cl-AlCl$_3$, was necessary. Excellent yields (85–96%) were obtained after microwave heating at 100–120 °C for 30–60 min using dichloromethane as a solvent (Scheme 7.20).

7.2.3.4 Multicomponent Reactions Using Soluble Ionic Liquid Supports and Microwave Heating

Microwave irradiation has been used in the synthesis of a range of hydrophilic poly(ethylene glycol)-functionalized ionic liquids based on 1,3-disubstituted imidazolium cations and fluorinated anions (Scheme 7.21) and these have been used as supports in multicomponent reactions. Starting from either N-methylimidazole or N-butylimidazole, ethylene glycol functionality was introduced by reaction with chloroalcohols. Ionic liquids bearing one, two, or three ethylene glycol repeating units were prepared this way. Microwave irradiation was used to facilitate the reaction and, to avoid decomposition because of overheating as the imidazolium salts

7.2 Ionic Liquids in Conjunction with Microwave Activation | 351

Tab. 7.6. Multicomponent synthesis of propargylamines using an ionic liquid and microwave heating.

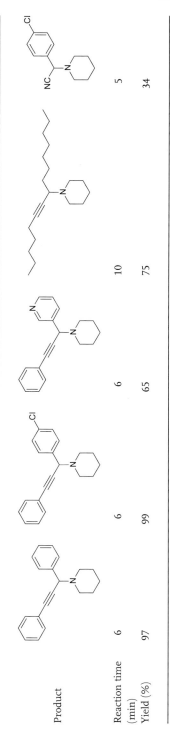

Product				
Reaction time (min)	6	6	6	10	5
Yield (%)	97	99	65	75	34

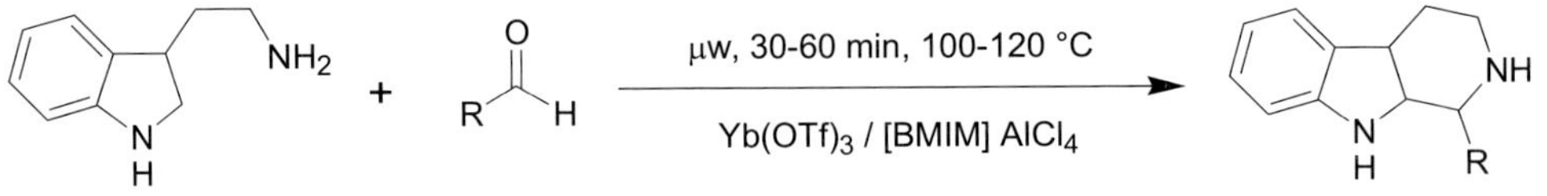

Scheme 7.20. The microwave-promoted Pictet–Spengler reaction using Yb(OTf)$_3$ and [BMIM]Cl-AlCl$_3$.

are formed, pulses of low-power irradiation were used. The temperature reached and the time of the pulses varied from reaction to reaction. Having made the halide-based materials, anion exchange was then used to give the desired final supports, termed [PEG$_n$-RMIM]X ionic liquids ($n = 1, 2$ or 3; R = Me or Bu; X = BF$_4^-$, PF$_6^-$, or NTf$_2^-$).

Synthesis of thiazolidinones [PEG$_n$-RMIM]X ionic liquids have been used for rapid synthesis of a small library of amido 4-thiazolidinones from amine, aldehyde, and mercaptoacid components (Scheme 7.22) [74]. In an initial feasibility study, acid-functionalized benzaldehydes were first coupled to the [PEG$_n$-RMIM]X ionic liquids. Imines were formed by reaction of the supported aldehydes with primary amines. The reactions were run in open vessels. Optimum results were obtained by irradiating the reaction mixture with low power at 100 °C for 20 min. The imines were then condensed with mercaptoacids to give the desired thiazolidinones which were then cleaved from the ionic liquid support by amide formation. Microwave irradiation was again used in this cleavage step. The procedure entailed addition of a small amount of solid potassium *tert*-butoxide to a premixed mixture of the amine and supported thiazolidinone and microwave exposure for 10–20 min at 100 or 150 °C depending on the amine used. In another study, a series of one-

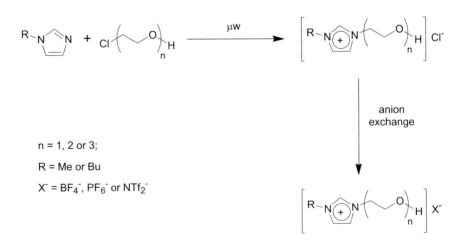

Scheme 7.21. Synthesis of [PEG$_n$-RMIM]X ionic liquids.

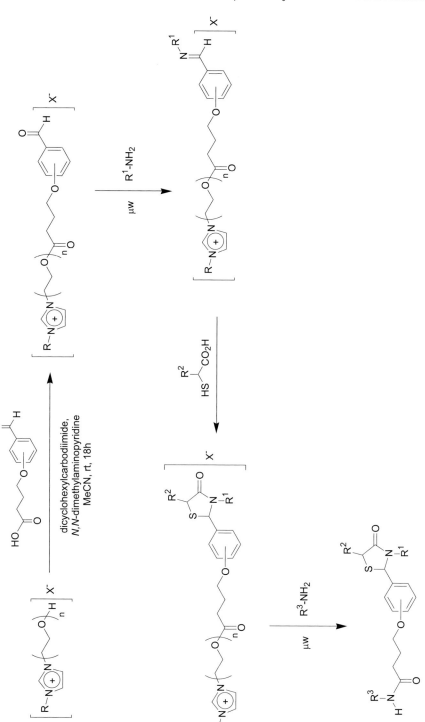

Scheme 7.22. Synthesis of thiazolidinones using [PEG$_n$-RMIM]X ionic liquids and microwave heating.

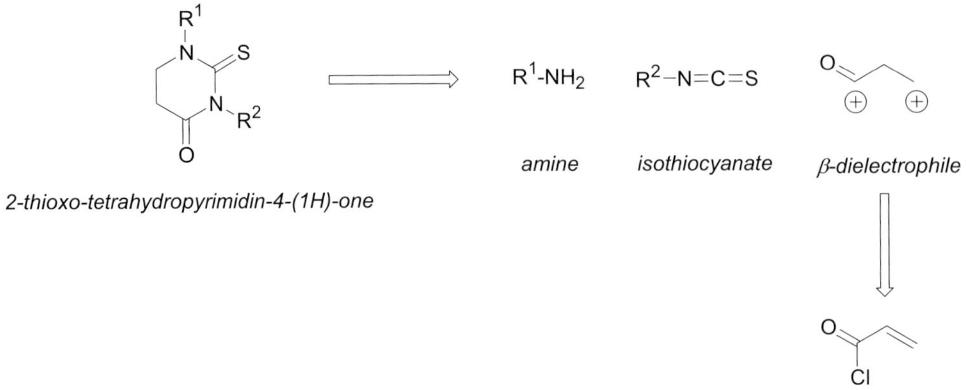

Scheme 7.23. Construction of 2-thioxotetrahydropyrimidin-4-(1H)-ones.

pot three-component condensations were undertaken using a stoichiometry ratio of [PEG$_n$-RMIM]X bound aldehyde:amine:mercaptoacid of 1:1:1. Here, microwave irradiation with low power for 1–2 h was required. Again, cleavage was by formation of amides. Using this strategy, a small library of 4-thiazolidinones could be prepared.

Synthesis of 2-thioxotetrahydropyrimidin-4-(1H)-ones   Libraries of 2-thioxotetrahydropyrimidin-4-(1H)-ones can be prepared by multicomponent reaction of amine, isothiocyanate, and β-dielectrophile components (Scheme 7.23). [PEG$_n$-RMIM]X ionic liquids have been used in this reaction with microwave irradiation (Scheme 7.24) [75]. In a step-by-step approach, treatment of [PEG$_n$-RMIM]X with acryloyl chloride provided a supported acryloyl ionic liquid phase which was reacted with an amine and then isothiocyanate to give a thiourea. Cyclization–cleavage from the ionic liquid support to give the desired 2-thioxotetrahydropyrimidin-4-(1H)-one was achieved by using low-power microwaves at 120 °C for 15–45 min in the presence of diethylamine.

7.2.4
Use of Ionic Liquids as Heating Aids

In their microwave-promoted synthesis of thiocarbonyls using a polymer-supported thionating reagent, Ley and coworkers found that addition of a small quantity of an ionic liquid to a toluene solution can greatly increase the rate and yields of the reaction [76]. It has subsequently been shown that, by use of microwave irradiation, nonpolar solvents such as hexane, toluene, THF, and dioxane can be heated using microwave irradiation to high temperatures in sealed vessels using a small quantity of an ionic liquid additive [77]. This enables their use as solvents in synthesis with nonpolar reactants. In the microwave cavity, with efficient stirring, the ionic liquid forms small droplets that heat the nonpolar medium.

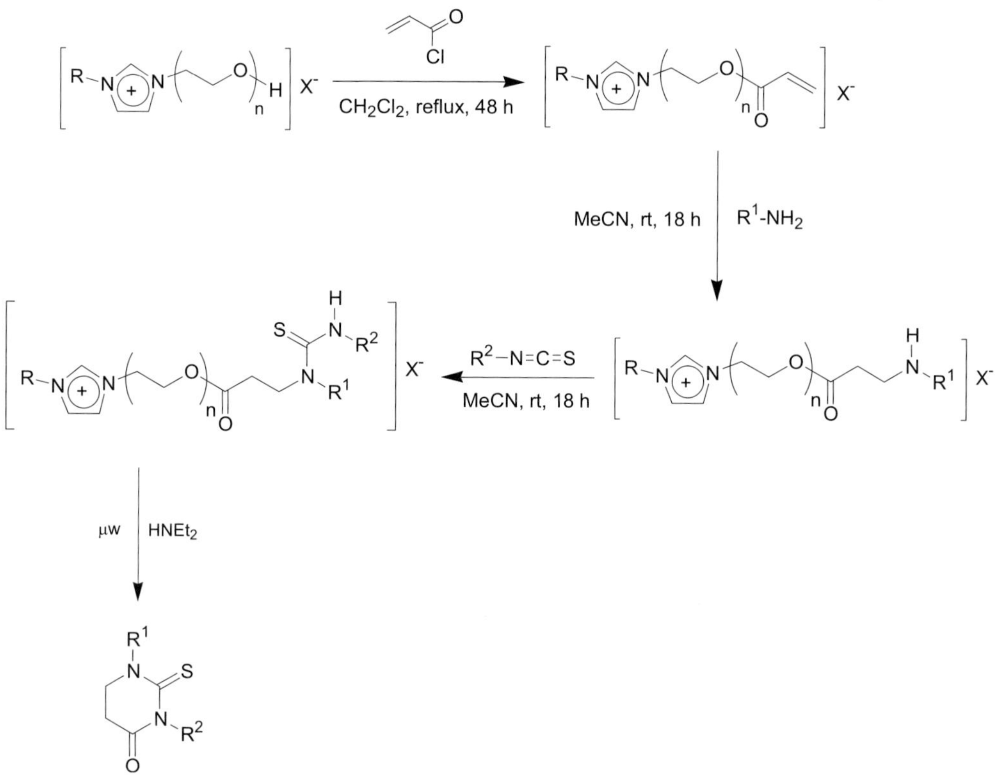

Scheme 7.24. Synthesis of 2-thioxotetrahydropyrimidin-4-(1H)-ones using [PEG$_n$-RMIM]X ionic liquids and microwave activation.

A range of 1,3-dialkylimidazolium halides have been screened for their ability to heat nonpolar solvents on microwave irradiation. Although high temperatures were reached very rapidly, the halide-based ionic liquids decomposed and contaminated the solvent. This problem was overcome by using PF$_6$-based ionic liquids as the heating aids. To show the benefits of the ionic liquids as heating aids, three reactions were screened (Scheme 7.25). The first was the Diels–Alder reaction between 2,3-dimethylbutadiene and methyl acrylate. This is traditionally performed in toluene or xylene and takes between 18 and 24 h to reach completion, giving yields of product varying from 9–90% depending on the solvent used and the temperature at which the reaction is conducted [78–80]. Even when using catalysts such as AlCl$_3$, the reaction still takes 5 h to reach 73% yield [81]. Using a mixture of toluene and ionic liquid, the authors were able to prepare the Diels–Alder adduct in 80% yield in 5 min. During the reaction, a microwave power of 100 W was used and the temperature of the mixture was held at 200 °C. In a control experiment, in which the reaction was repeated but in the absence of the ionic liquid, no product

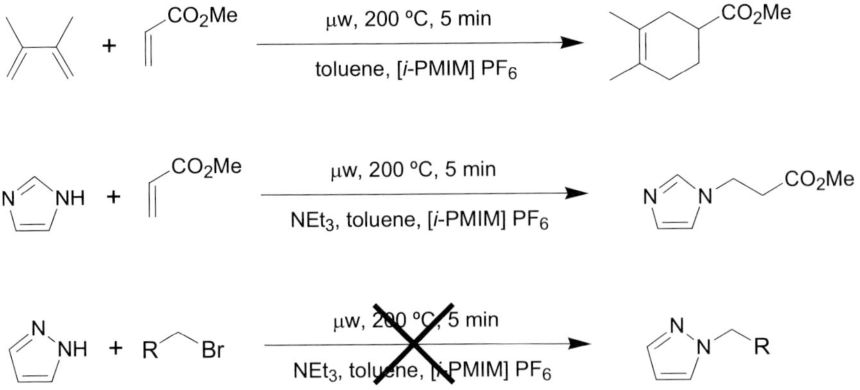

Scheme 7.25. Reactions in which ionic liquids have been shown to function as heating aids.

was formed after an identical time. Interestingly, this same reaction has been reported previously using microwave irradiation with xylene as solvent but the time taken to reach the same level of product yield was well in excess of 3 h, presumably because the reaction mixture was heated to 95 °C only [82]. The authors do not say whether this is the maximum temperature obtainable because of the use of a non-polar solvent or if this was the temperature chosen for the study.

The second reaction chosen for study was the base-catalyzed Michael addition of imidazole to methyl acrylate. This was chosen as a test for whether the ionic liquid would act solely as a heating agent in the reaction rather than becoming chemically involved, because there would be scope for reaction between the ionic liquid and the base or acrylate during the course of the reaction. This reaction had already been performed using microwave irradiation, with basic clays (Li$^+$ and Cs$^+$ montmorillonites) as catalysts [83]. Using the ionic liquid heating technique, the authors replaced the clays with triethylamine and found that use of toluene as solvent resulted in a good yield of the desired product, again after microwave heating for just 5 min at 200 °C.

The third reaction studied was the alkylation of pyrazoles with alkyl halides. This reaction was not possible using the ionic liquid heating technique, because the aryl halide reacted with the ionic liquid at the elevated temperatures used in the reaction. This highlights one of the limitations of this method.

The heating behavior of ionic liquids combined with organic solvents under microwave irradiation has also been the subject of a recent paper by Ondruschka and coworkers [84]. Kappe et al. have used microwaves for inter and intramolecular hetero Diels–Alder reactions of 2(1H)-pyrazinones using dichloroethane as a solvent with small amounts of ionic liquid as heating aids (Scheme 7.26) [85]. The conventional reaction can take up to 2 days to reach completion. With the microwave heating the reaction time could be reduced to 50 min in the absence of the ionic liquid and only 18 min in its presence. Product yields were reported to be similar to those obtained by use of conventional heating. Interestingly, when the

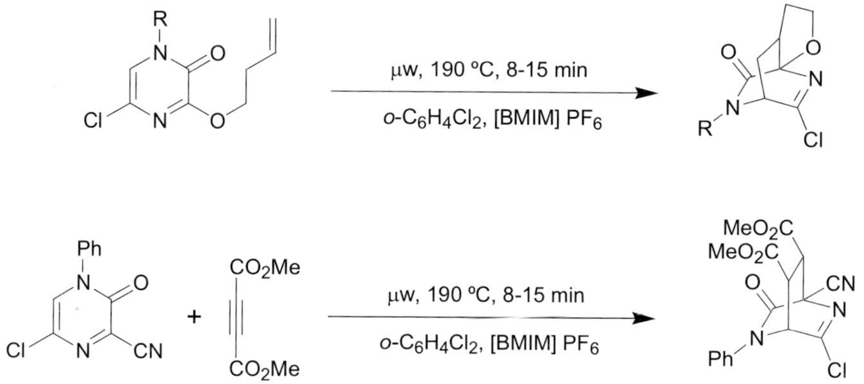

Scheme 7.26. Hetero Diels–Alder reactions using an ionic liquid as a heating aid in conjunction with microwave heating.

reaction was performed with ethene as the alkene coupling partner and an ionic liquid was used as heating aid the reaction temperature rose too rapidly and significant decomposition was observed.

7.3 Conclusions

Microwave irradiation as a tool for synthetic organic chemistry has found increasing use since the first publications in 1986. The development of the microwave apparatus has enabled scientists to perform reliable, safe, and reproducible reactions. The main advantage of microwave irradiation is rapid reaction compared with conventional heating. Debate on the possible existence of nonthermal microwave effects is still ongoing (see Chapter 4 of this book for an extended discussion). Because of their ionic structure, low vapor pressure, good thermal stability, and excellent solvation properties, ionic liquids are very versatile when used in conjunction with microwave chemistry. They can be used as solvents in a variety of reactions, and as heating aids. Microwave methods can also be used to advantage in the preparation of ionic liquids, essentially under solvent-free conditions.

Abbreviations

[BMIM]Cl-AlCl$_3$	1-Butyl-3-methylimidazolium tetrachloroaluminate
[BMIM]Cl	1-Butyl-3-methylimidazolium chloride
[BMIM]Br	1-Butyl-3-methylimidazolium bromide
[BMIM]BF$_4$	1-Butyl-3-methylimidazolium tetrafluoroborate
[BMIM]NO$_3$	1-Butyl-3-methylimidazolium nitrate

[BMIM]NTf$_3$	1-Butyl-3-methylimidazolium bis(trifluoromethane)sulfonimide
[BMIM]PF$_6$	1-Butyl-3-methylimidazolium hexafluorophosphate
[BMIM]OTf	1-Butyl-3-methylimidazolium trifluoromethanesulfonate
[BMIM]SbF$_6$	1-Butyl-3-methylimidazolium hexafluoroantimonate
[EMIM]BF$_4$	1-Ethyl-3-methylimidazolium tetrafluoroborate
[EMIM]NTf$_3$	1-Ethyl-3-methylimidazolium bis(trifluoromethane)sulfonimide
[EMIM]ONf$_6$	1-Ethyl-3-methylimidazolium nonafluorobutansulfonate
[EMIM]PF$_6$	1-Ethyl-3-methylimidazolium hexafluorophosphate
[HMIM]Cl	1-Hexyl-3-methylimidazolium chloride
[PMIM]Br	1-Propyl-3-methylimidazolium bromide
[PMIM]I	1-Propyl-3-methylimidazolium iodide
[i-PMIM]Br	1-Isopropyl-3-methylimidazolium bromide
[N-BuPy]BF$_4$	N-Butylpyridinium tetrafluoroborate
[OMIM]Br	1-Octyl-3-methylimidazolium bromide
[OMIM]I	1-Octyl-3-methylimidazolium iodide
[OMIM]BF$_4$	1-Octyl-3-methylimidazolium tetrafluoroborate
[OMIM]PF$_6$	1-Octyl-3-methylimidazolium hexafluorophosphate
[PMIM]I	1-Propyl-3-methylimidazolium iodide
[PMIM]Br	1-Propyl-3-methylimidazolium bromide
[PMIM]BF$_4$	1-Propyl-3-methylimidazolium tetrafluoroborate
[PMIM]PF$_6$	1-Propyl-3-methylimidazolium hexafluorophosphate
TBAB	Tetrabutylammonium bromide
TBAC	Tetrabutylammonium chloride

References

1 For books on microwave activation in synthesis: (a) Kappe, O.; Stadler, A. *Microwaves in Organic and Medicinal Chemistry*, 2005, Wiley–VCH, Weinheim; (b) Tierney, J. P.; Lidström, P. Eds. *Microwave Assisted Organic Synthesis*, 2005, Blackwell, Oxford; (c) Loupy, A. Ed. *Microwaves in Organic Synthesis*, 2002, Wiley–VCH, Weinheim; (d) Hayes, B. L., *Microwave synthesis: Chemistry at the Speed of Light*, 2002, CEM Publishing, Matthews.

2 For reviews on microwave activation in synthesis: (a) Kappe, C. O. *Angew. Chem. Int. Ed.* 2004, 43, 6250–6284; (b) Loupy, A. *C. R. Chimie* 2004, 7, 103–112; (c) Lidström, P.; Tierney, J.; Wathey, B.; Westman, J. *Tetrahedron*, 2001, 57, 9225–9283; (d) Loupy, A.; Petit, A.; Hamelin, J.; Texier-Boullet, F.; Jacquault, P.; Mathé, D. *Synthesis* 1998, 1213–1235; (e) Caddick, S., *Tetrahedron*, 1995, 38, 10403–10432; (f) Strauss, C. R., *Aust. J. Chem.*, 1999, 52, 83–96.

3 For examples see: (a) Westman, J. *Org. Lett.*, 2001, 3, 3745–3747; (b) Kuhnert, N.; Danks, T. N. *Green Chem.*, 2001, 3, 68–70; (c) Perreux, L.; Loupy, A. *Tetrahedron.*, 2001, 57, 9199–9123.

4 Kappe, C. O. *Current Opinion in Chemical Biology*, 2002, 6, 314–320.

5 Lew, A.; Krutzik, P. O.; Hart, M. E., Chamberlin, A. R. *J. Comb. Chem.*, 2002, 4, 95–105.

6 (a) Larhed, M.; Hallberg, A. *Drug Discovery Today*, 2001, 6, 406–416; (b) Wathey, B.; Tierney, J.; Lindström, P.; Westman, J. *Drug Discovery Today*, 2002, 7, 373–380.

7 For a review see: GABRIEL, C.; GABRIEL, S.; GRANT, E. H.; HALSTEAD, B. S. J.; MINGOS, D. M. P. Chem. Soc. Rev., **1998**, *27*, 213–223.
8 For an introduction to ionic liquids see: WASSERSCHEID, P.; WELTON, T. *Ionic liquids in Synthesis*, **2003**, Wiley VCH, Weinheim.
9 For the general information and use of ionic liquids see for example: (a) WELTON, T. Chem. Rev., **1999**, *99*, 2071–2083; (b) WASSERSCHEID, P.; KEIM. Angew. Chem. Int. Ed., **2000**, *39*, 3772–3789; (c) OLIVIER-BOURBIGOU, H.; MAGNA, L. J. Mol. Catal. A: Chem, **2002**, *348*, 1–19.
10 WILKES, J. S.; LEVISKY, J. A.; WILSON, R.; HUSSEY, C. L. Inorg. Chem., **1982**, *21*, 1263–1264.
11 SHELDON, R. Chem. Commun., **2001**, 2399–2407.
12 HARLOW, K.; HILL, A. F.; WELTON, T. Synthesis, **1996**, 697–698.
13 (a) BEGG, C. G.; GRIMMETT, M. R.; WETHEY, P. D. Aust. J. Chem., **1973**, *26*, 2435–2438; (b) CHAN, B. K. M., CHANG, N.-H.; GRIMMETT, M. R. Aust. J. Chem., **1977**, *30*, 2005–2013.
14 NGO, H.; LE COMPTE, K.; HARGENS, L.; MCEWAN, A. B. Thermochimica Acta, **2000**, *357–358*, 97–102.
15 VARMA, R. S.; NAMBOODIRI, V. V. Chem. Commun., **2001**, 643–644.
16 NAMBOODIRI, V. V.; VARMA, R. S. Tetrahedron Lett., **2002**, *43*, 5381–5383.
17 MURUGESAN, S.; KARST, N.; ISLAM, T.; WIENCEK, J. M.; LINHARDT, R. J. Synlett, **2003**, 1283–1286.
18 NAMBOODIRI, V. V.; VARMA, R. S. Chem. Commun, **2002**, 342–343.
19 LAW, M. S.; WONG, K. Y.; CHAN, T. H. Green Chem., **2002**, *4*, 328–330.
20 KHADILKAR, B. M.; REBEIRO, G. L. Org. Proc. Res. Dev., **2002**, *6*, 826–828.
21 DEETLEFS, M.; SEDDON, K. R. Green Chemistry, **2003**, *5*, 181–186.
22 VO THANH, G.; PEGOT, B.; LOUPY, A. Eur. J. Org. Chem., **2004**, 1112–1116.
23 PEGOT, B.; VO THANH, G.; GORI, D.; LOUPY, A. Tetrahedron Lett., **2004**, *45*, 6425–6428.
24 TANG, W.; ZHANG, X. Chem. Rev., **2003**, *103*, 3029–3069.
25 CHAUVIN, Y.; MUSSMANN, L.; OLIVIER, H. Angew. Chem. Int. Ed., **1995**, *34*, 2698–2700.
26 DYSON, P. J.; ELLIS, D. J.; PARKER, D. G.; WELTON, T. Chem. Commun., **1999**, 25–26.
27 BANIK, B. K.; BARAKAT, K. J.; WAGLE, D. R.; MANHAS, M. S.; BOSE, A. K. J. Org. Chem., **1999**, *64*, 5764–5753.
28 BERTHOLD, H.; SCHOTTEN, T.; HÖNIG, H. Synthesis, **2002**, 1607–1610.
29 GRUBBS, R. H. Ed. *Handbook of Metathesis*, **2003**, Wiley–VCH, Weinheim.
30 See for example: (a) NICOLAOU, K. C.; BULGER, P. G.; SARLAH, D. Angew. Chem. Int. Ed., **2005**, *44*, 4490–4527; (b) ASTRUC, D. New. J. Chem., **2005**, *29*, 42–56; (c) SCHROCK, R. R. J. Mol. Catal. A: Chem., **2004**, *213*, 21–30; (d) HOVEYDA, A. H.; GILLINGHAM, D. G.; VAN VELDHUIZEN, J. J.; KATAOKA, O.; GARBER, S. B.; KINGSBURY, J. S.; HARRITY, J. P. A. Org. Biol. Chem., **2004**, *2*, 8–24; (e) CONNON, S. J.; BLECHERT, S., Angew. Chem. Int. Ed., **2003**, *42*, 1900–1923; (f) FURSTNER, A., Angew. Chem. Int. Ed., **2000**, *39*, 3012–3043; (g) BUCHMEISER, M. R. Chem. Rev. **2000**, *100*, 1565–1604.
31 BUIJSMAN, R. C.; VAN VUUREN, E.; STERRENBURG, J. G. Org. Lett., **2001**, *3*, 3785–3787.
32 AUDIC, N.; CLAVIER, H.; MAUDUIT, M.; GUILLEMIN, J.-C. J. Am. Chem. Soc., **2003**, *125*, 9248–9249.
33 YAO, Q.; ZHANG, Y. Angew. Chem. Int. Ed., **2003**, *42*, 3395–3398.
34 MAYO, K. G.; NEARHOOF, E. H.; KIDDLE, J. J. Org. Lett., **2002**, *4*, 1567–1570.
35 (a) VO THANH, G.; LOUPY, A. Tetrahedron Lett. **2003**, *44*, 9091–9094; (b) GARBACIA, S.; DESAI, B.; LAVASTRE, O.; KAPPE, C. O. J. Org. Chem., **2003**, *68*, 9136–9139.
36 For review on the cyanation see: (a) ELLIS, G. A.; ROMNEY-ALEXANDER, T. M. Chem. Rev., **1987**, *87*, 779–794; (b) GRUSHIN, V. V.; ALPER, H. Chem. Rev., **1994**, *94*, 1047–1062.
37 WU, J. X.; BECK, B.; REN, R. X. Tetrahedron Lett., **2002**, *43*, 387–389.
38 LEADBEATER, N. E.; TORENIUS, H. M.; TYE, H. Tetrahedron, **2003**, *59*, 2253–2258.

39 Arvela, R. K.; Leadbeater, N. E. *J. Org. Chem.*, **2003**, *68*, 9122–9125.
40 See for example: Beletskaya, I.; Cheprakov, A. V. *Chem. Rev.*, **2000**, *100*, 3009–3066.
41 Jeffery, T., *Tetrahedron Lett.*, **1985**, *26*, 2667–2670.
42 Herrmann, W. A.; Böhm, V. P. W. *J. Organomet. Chem.*, **1999**, *572*, 141–145.
43 Choudary, B. M.; Madhi, S.; Chowdari, N. S.; Kantam, M. L.; Sreedhar, B. *J. Am. Chem. Soc.*, **2002**, *124*, 14127–14136.
44 Carmichael, A. J.; Earle, M. J.; Holbrey, J. D.; McCormac, P. B.; Seddon, K. R. *Org. Lett.*, **1999**, *1*, 997–1000.
45 Vallin, K. S. A.; Emilsson, P.; Larhed, M.; Hallberg, A. *J. Org. Chem.*, **2002**, *67*, 6243–6246.
46 Xie, X.; Lu, J.; Chen, B.; Han, J.; She, X.; Pan, X. *Tetrahedron Lett.*, **2004**, *45*, 809–811.
47 Shieh, W.-C.; Lozanov, M.; Repič, O. *Tetrahedron Lett.*, **2003**, *44*, 6943–6945.
48 Delledonne, D.; Rivetti, F.; Romano, U. *Appl. Catal.*, **2001**, *221*, 241–251.
49 Wegener, G.; Brandt, M.; Duda, L.; Hofmann, J.; Klesczewski, B.; Koch, D.; Kumpf, R.-J.; Orzesek, H.; Pirkl, H.-G.; Six, C.; Steinlein, C.; Weisbeck, M. *Appl. Catal.*, **2001**, *221*, 303–335.
50 Trotzki, R.; Nüchter, M.; Ondruschka, B. *Green Chem.*, **2003**, *5*, 285–290.
51 Zhou, J.-F.; Tu, S.-J.; Feng, J.-C. *Synth. Commun.*, **2002**, *32*, 959–962.
52 Bosch, A. I. de la Cruz, P.; Diez-Parra, E.; Loupy, A.; Langa, F. *Synlett*, **1995**, 1259–1260.
53 Loupy, A.; Régnier, S., *Tetrahedron Lett.*, **1999**, *40*, 6221–6224.
54 Thakur, A. J.; Boruah, A.; Prajapati, D.; Sandhu, J. S. *Synth. Commun.*, **2000**, *30*, 2105–2111.
55 Peng, J.; Deng, Y., *Tetrahedron Lett.*, **2001**, *42*, 403–405.
56 Ren, R. X.; Zueva, L. D.; Ou, W. *Tetrahedron Lett.*, **2001**, *42*, 8441–8443.
57 Lee, J. K.; Kim, D.-C.; Song, C. E.; Lee, S.-G. *Synth. Commun.*, **2003**, *33*, 2301–2307.
58 Ren, R. X.; Wu, J. X. *Org. Lett.*, **2001**, *3*, 3727–3728.
59 Nguyen, H.-P.; Matondo, H.; Baboulene, M. *Green Chem.* **2003**, *5*, 303–305.
60 Peng, Y.; Song, G., *Tetrahedron Lett.* **2004**, *45*, 5313–5316.
61 Fraga-Dubreuil, J.; Bazureau, J. P. *Tetrahedron Lett.*, **2000**, *41*, 7351–7355.
62 Fraga-Dubreuil, J.; Bazureau, J. P. *Tetrahedron Lett.*, **2001**, *42*, 6097–6100.
63 Miao, W.; Chan, T. H. *Org. Lett.*, **2003**, *5*, 5003–5005.
64 Qian, C.; Huang, T. *J. Org. Chem.*, **1998**, *63*, 4125–4128.
65 Lee, S.; Lee, J. K.; Song, C. E.; Kim, D. *Bull. Kor. Chem. Soc.*, **2002**, *23*, 667–668.
66 Dyatkin A. B.; Rivero, R. A. *Tetrahedron Lett.*, **1998**, *39*, 3647–3650.
67 McNally, J. J.; Youngman, M. A.; Dax, S. L. *Tetrahedron Lett.*, **1998**, *39*, 967–970.
68 Wei, C.; Li, Z.; Li, C.-J. *Org. Lett.*, **2003**, *5*, 4473–4475.
69 Leadbeater, N. E.; Torenius, H. M.; Tye, H. *Mol. Div.*, **2003**, *7*, 135–144.
70 Cox, E. D.; Cook, J. M. *Chem. Rev.*, **1995**, *95*, 1797–1842.
71 Pal, B.; Jaisankar, P.; Giri, V. S. *Synth. Commun.*, **2003**, *33*, 2339–2348.
72 Wu, C.-Y.; Sun, C-M. *Synlett*, **2002**, 1709–1711.
73 Srinivasan, N.; Ganesan, A. *Chem. Commun.*, **2003**, 916–917.
74 Fraga-Dubreuil, J.; Bazureau, J. P. *Tetrahedron*, **2003**, *59*, 6121–6130.
75 Hakkou, H.; Van den Eynde, J. J.; Hamelin, J.; Bazureau, J. P. *Tetrahedron*, **2004**, *60*, 3745–3753.
76 Ley, S. V.; Leach, A. G.; Storer, R. I. *J. Chem. Soc., Perkin Trans. 1*, **2001**, 358–361.
77 Leadbeater, N. E.; Torenius, H. M. *J. Org. Chem.*, **2002**, *67*, 3145–3148.
78 Monin, J. *Helv. Chim. Acta.*, **1958**, *41*, 2112–2120.
79 Vedejis, E.; Gadwood, R. C. *J. Org. Chem.*, **1978**, *43*, 376–378.
80 Fang, X.; Warner, B. P.; Watkin, J. G. *Synth. Commun.*, **2000**, *30*, 2669–2676.

81 INUKAI, T.; KASAI, M. *J. Org. Chem.*, **1965**, *30*, 3567–3569.
82 BERLAN, J.; GIBOREAU, P.; LEFEUVRE, S. MARCHAND, C. *Tetrahedron Lett.*, **1991**, *32*, 2363–2366.
83 MARTÍN-ARANDA, R. M.; VINCENTE-RODRÍGUEZ, M. A.; LÓPEZ-PESTAÑA, J. M.; LÓPEZ-PEINADO, A. J.; JEREZ, A.; LÓPEZ-GONZÁLEZ, J. DE D.; BAÑARES-MUÑOZ, M. A. *J. Mol. Catal. A.*, **1997**, *124*, 115–121.
84 HOFFMANN, J.; NÜCHTER, M.; ONDRUSCHKA, B.; WASSERSCHEID, P. *Green Chem.*, **2003**, *5*, 296–299.
85 VAN DER EYCKEN, E.; APPUKKUTTAN, P.; DE BORGGRAEVE, W.; DEHAEN, W.; DALLINGER, D.; KAPPE, C. O. *J. Org. Chem.*, **2002**, *67*, 7904–7907.

8
Organic Synthesis Using Microwaves and Supported Reagents

Rajender S. Varma and Yuhong Ju

8.1
Introduction

In the electromagnetic radiation region, microwaves (0.3–300 GHz) lie between radiowave (Rf) and infrared (IR) frequencies with relatively large wave lengths (1 mm–1 m). Microwaves, nonionizing radiation incapable of any molecular activation or breaking of bonds, are a form of energy that manifest as heat by interaction with the medium or materials wherein they can be reflected (metals), transmitted (good insulators that will not heat), or absorbed (reducing the available microwave energy and rapidly heating the sample). This unconventional microwave (MW) energy source has been used for heating food materials for almost 50 years [1] and is now being used for a variety of chemical applications including organic synthesis [2–11] wherein chemical reactions are accelerated by selective absorption of MW radiation by polar molecules, nonpolar molecules being inert to MW dielectric loss [12]. Initial experiments with microwave heating exploited the use of high dielectric solvents, for example dimethyl sulfoxide (DMSO) and dimethylformamide (DMF) in a household kitchen MW oven. The rate enhancements in such reactions are believed to be because of rapid superheating of the polar solvents and pressure effects [12]. In these solution-phase reactions, however, development of high pressures and the use of specialized sealed or Teflon vessels are limitations, although they have been circumvented by introduction of commercial MW instruments with appropriate temperature and pressure controls.

Heterogeneous reactions facilitated by supported reagents on inorganic oxide surfaces have received attention in recent years, both in the laboratory and in industry. Although the first description of the surface-mediated chemical transformation dates back to 1924 [13], it was not until almost half a century later that this technique received extensive attention with the appearance of several reviews, books, and account articles [14–22].

A related development that has had profound effect on heterogeneous reactions is the use of microwave (MW) irradiation techniques for the acceleration of organic reactions. Since the appearance of initial reports on the application of microwaves for chemical synthesis in polar solvents [11], the approach has blossomed into a

Microwaves in Organic Synthesis, Second edition. Edited by A. Loupy
Copyright © 2006 WILEY-VCH Verlag GmbH & Co. KGaA, Weinheim
ISBN: 3-527-31452-0

useful technique for a variety of applications in organic synthesis and functional group transformations, as is testified by a large number of publications and review articles on this theme [2–10, 22–25]. Although reactions in conventional organic solvents [23], ionic liquids [26–28] and aqueous media [29] have grown as a result of the availability of new, commercial, MW systems, the focus has shifted to less cumbersome solvent-free methods wherein the neat reactants, often in the presence of mineral oxides or supported catalysts [2–6, 8, 10, 22, 25b], undergo facile reactions to provide high yields of pure products, thus eliminating or minimizing the use of organic solvents. The application of microwave irradiation with the use of catalysts or mineral-supported reagents, under solvent-free conditions, enables organic reactions to occur expeditiously at ambient pressure [2–6, 8, 10, 18, 22, 24], thus providing unique chemical processes with special attributes such as enhanced reaction rates, higher yields, improved purity of final products, and associated ease of manipulation. These reactions are performed with the reagents immobilized on the porous solid supports and have advantages over conventional solution-phase reactions, because of the good dispersion of active reagent sites, associated selectivity, and easier work-up.

The ready availability of inexpensive household MW ovens that can be safely used for solvent-free reactions and an opportunity to work with open vessels, thus avoiding the risk of high-pressure development, are the main reasons for the worldwide popularity of this approach. The bulk temperature attained in these solvent-free reactions is relatively low although higher localized temperatures may be reached during microwave irradiation. Unfortunately, in many of the reactions reported, accurate temperature measurement has not been attempted. Although there is relatively poor understanding of the reasons for the dramatic rate acceleration, and some researchers are skeptical about reproducibility, this MW strategy has been the most widely practiced approach in laboratories around the globe. The recyclability of some of these solid supports makes these processes truly eco-friendly green procedures. Previous cartography of the oven using a cobalt chloride aqueous solution to determine the best location for placement of vessels (high electric field density) enables accurate and reproducible MW experiments [30].

8.2
Microwave-accelerated Solvent-free Organic Reactions

The initial laboratory-scale feasibility of microwave-promoted solvent-free procedures [24] has been illustrated for a wide variety of useful chemical transformations such as protection/deprotection (cleavage), condensation, rearrangement, oxidation, and reduction, and in the synthesis of several heterocyclic compounds on mineral supports [4, 8, 22]. A range of industrially significant chemical entities and precursors, for example imines, enamines, enones, nitroalkenes, sulfur compounds, and heterocycles have been synthesized in a relatively environmentally benign manner by using MW [2–6, 8, 10, 22, 25]. A vast majority of these solvent-free reactions have been performed in open glass containers such as test tubes, beakers, and round-bottomed flasks, etc., using neat reactants in an unmodified household MW oven or commercial MW equipment, usually operating at 2450 MHz. The general procedure involves simple mixing of neat reactants with the catalyst, their adsorption on mineral or "doped" supports, and exposure of the reaction mixture to irradiation in a microwave oven.

A reaction accelerated by microwave irradiation has often been compared with the same reaction in an oil bath at the same bulk temperature. Unfortunately, there have been quite a few reports in the chemical literature that have not been conducted with such proper control of conditions and, consequently, fair comparison is not often possible. Nevertheless, using this MW approach, the problems associated with waste disposal of solvents that are used several times in chemical reactions, and use of an excess of reagents are avoided or minimized. Discussion of the preparation of supported reagents or catalysts has not been included in this chapter because numerous review articles are available on this theme [14–22].

8.2.1
Protection–Deprotection Reactions

Protection–deprotection reaction sequences are an integral part of organic syntheses such as the preparation of monomers, fine chemicals, and reaction intermediates or precursors for pharmaceuticals. These reactions often involve use of acidic, basic, or hazardous reagents and toxic metal salts [31]. The solvent-free MW-accelerated protection/deprotection of functional groups, developed during the last decade, is an attractive alternative to conventional cleavage reactions.

8.2.1.1 Formation of Acetals and Dioxolanes

Within the framework of nonalimentary preparation of products from biomass, Loupy et al. prepared acetals of L-galactono-1,4-lactone (an important byproduct from the sugar beet industry) in excellent yields [32] by adsorbing the lactone and a long-chain aldehyde on montmorillonite K10 or KSF clay then exposing the reaction mixture to MW irradiation (Scheme 8.1). Improvements over the conventional method are substantial (DMF, H_2SO_4, 24 h at 40 °C, yields less than 20–25%).

Hamelin and coworkers protected aldehydes and ketones as acetals and dioxolanes by using orthoformates, 1,2-ethanedithiol, or 2,2-dimethyl-1,3-dioxolane. This acid-catalyzed reaction proceeds in the presence of p-toluenesulfonic acid (PTSA)

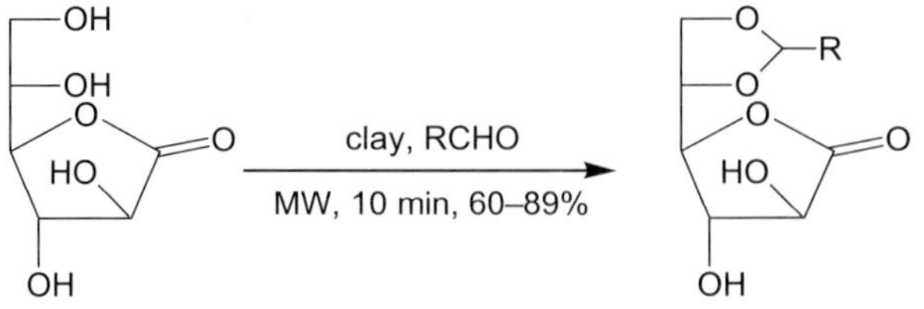

Scheme 8.1. Formation of acetal derivatives of L-galactono-1,4-lactone.

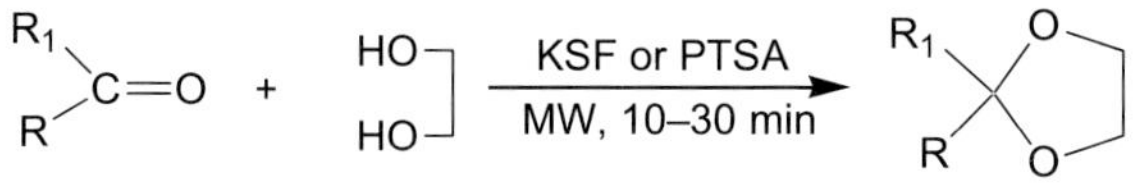

Scheme 8.2. Formation of dioxolanes.

or KSF clay under solvent-free conditions (Scheme 8.2). The yields obtained with the microwave method are better than those obtained using conventional heating in an oil bath [33].

Thioacetals have been prepared using an essentially similar technique [34]. The active methylene compounds are adsorbed on KF–alumina, mixed with methanesulfonothioate, and irradiated in a microwave oven to produce thioacetals in good yields (Scheme 8.3).

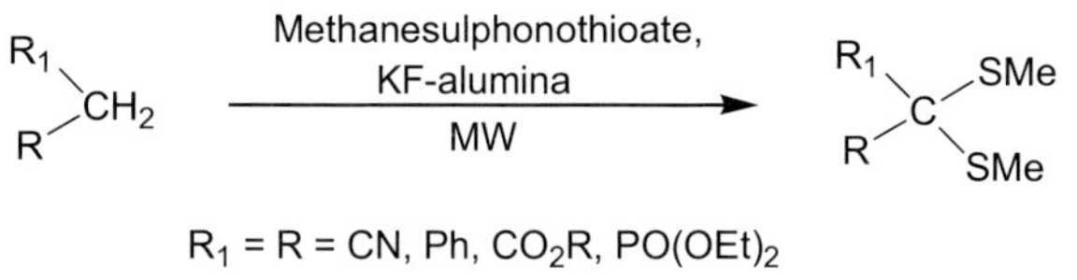

$R_1 = R = CN, Ph, CO_2R, PO(OEt)_2$

Scheme 8.3. Formation of thioacetals.

Acetic anhydride–pyridine on basic alumina has been used to conduct acetylations of hydroxy, thiol and amino groups under microwave irradiation conditions [35]. This rapid, safe, and eco-friendly technique can be applied for a broad variety of N, O, and S acetylations.

8.2.1.2 N-Alkylation Reactions

Several solvent-free N-alkylation reactions have been reported which involve use of tetrabutylammonium bromide (TBAB), as a phase-transfer agent, under microwave irradiation conditions, an approach that is developed in Chapter 6 [36]. An experimentally simple microwave-assisted solvent-free N-arylation of primary amines with sodium tetraphenylborate or arylboronic acids, promoted by inexpensive cupric acetate on the surface of KF–alumina, has been reported. The reaction is selective for mono-N-arylation and a variety of functional groups are tolerated in the process (Scheme 8.4) [37].

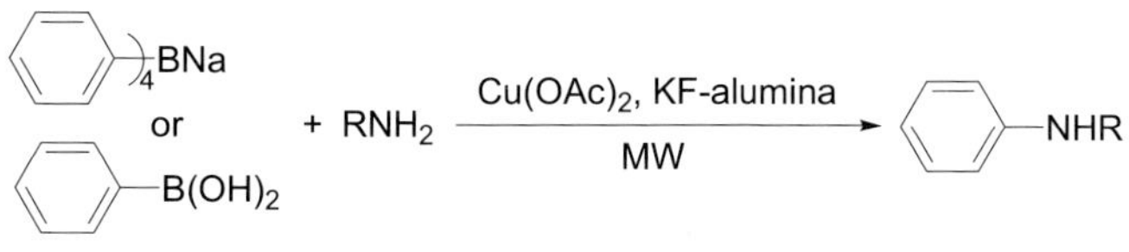

Scheme 8.4. MW-assisted solvent-free N-arylation.

8.2.1.3 Deacylation Reactions

The first report on the utility of recyclable alumina as a viable support surface for deacylation reaction is credited to Varma and his colleagues [38]. This high-school science project demonstrated that the orthogonal deprotection of alcohols is possible under solvent-free conditions on a neutral alumina surface using MW irradiation (Scheme 8.5). Interestingly, chemoselectivity between alcoholic and phenolic groups in the same molecule has been achieved simply by varying the reaction time – the phenolic acetates are deacetylated faster than alcoholic analogs [38].

In an extremely simple approach, an unmodified household microwave oven has been used in this study with excellent results. The generation of higher temperatures is avoided simply by intermittent irradiation [38].

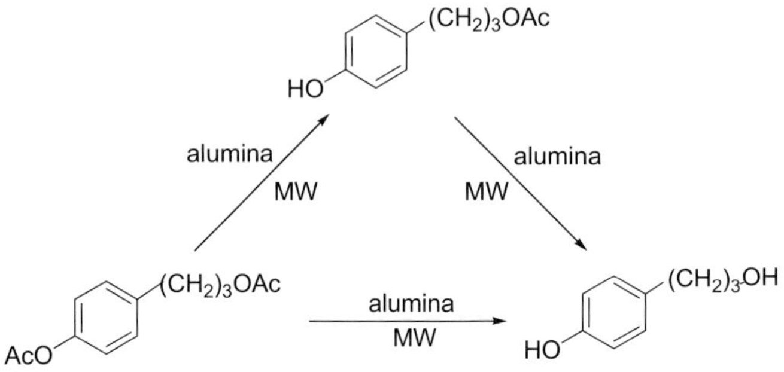

Scheme 8.5. Deacylation of alcohols and phenols on alumina.

8.2.1.4 Cleavage of Aldehyde Diacetates

Brief exposure of the diacetate derivatives of aromatic aldehydes to MW irradiation on a neutral alumina surface enables rapid regeneration of aldehydes (Scheme 8.6) [39]. Selectivity in these deprotection reactions is achievable merely by adjusting the time of irradiation. As an example, for molecules bearing acetoxy functionality (R = OAc), the aldehyde diacetate is selectively removed in 30 s whereas an extended period of 2 min is required to cleave both the diacetate and ester groups. The yields obtained are better than those possible by conventional heating methods and the procedure is applicable to compounds, for example cinnamaldehyde diacetate, bearing olefinic moieties [39].

Interestingly, acylal formation has been accomplished with acetic anhydride [40] on K10 clay (75–98%) as well as deacylation [41].

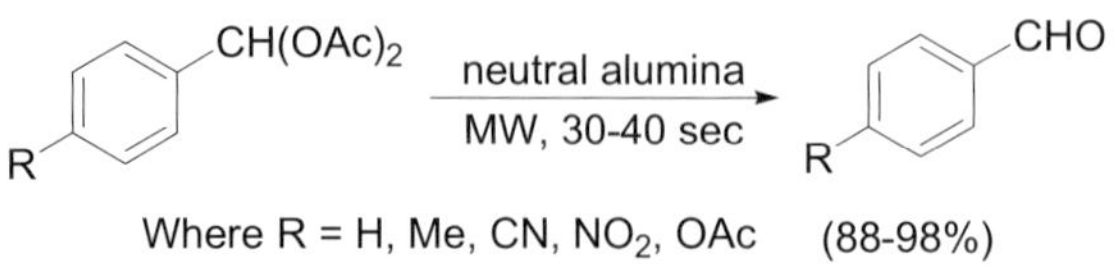

Where R = H, Me, CN, NO_2, OAc (88-98%)

Scheme 8.6. Cleavage of aldehyde diacetates on alumina.

8.2.1.5 Cleavage of Carboxylic Esters on a Solid Support

An efficient procedure for debenzylation of esters under solvent-free conditions has been reported by Varma et al. (Scheme 8.7) [42]. By altering the surface characteristics of the solid support cleavage of the 9-fluorenylmethoxycarbonyl (Fmoc) group and related protected amines can be achieved in a similar fashion. The optimum conditions for cleavage of N-protected moieties are use of basic alumina and an irradiation time of 12–13 min at ∼130–140 °C.

Hydrolytic deprotection of carboxylic acids from their correspondent allyl esters, under "dry conditions" on montmorillonite K10 clay, under the action of MW irradiation, has also been reported (Scheme 8.8) [43].

Microwave-enhanced hydrolysis of esters utilizing potassium fluoride-doped alumina in the absence of solvents has been developed by Kabalka and coworkers [44]. Carboxylic acids are produced in excellent yields with the corresponding alcohols, thus providing a reliable, rapid and practical procedure for the deprotection of esters (Scheme 8.9).

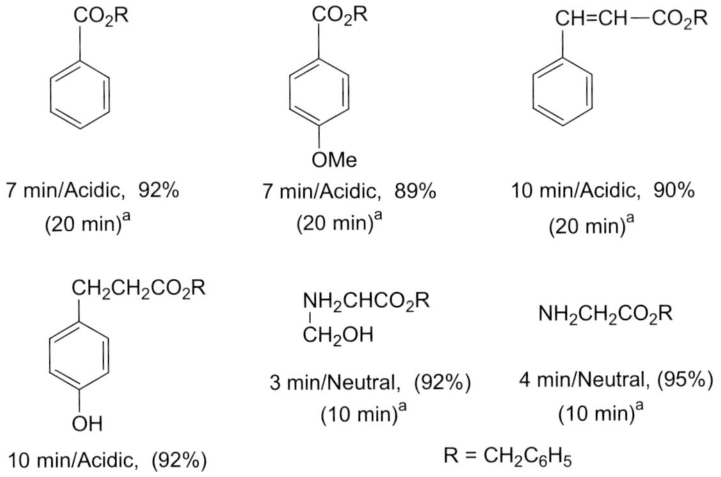

[a] Time in parentheses refer to deprotection in oil bath at the same temperature

Scheme 8.7. Debenzylation of carboxylic esters on alumina.

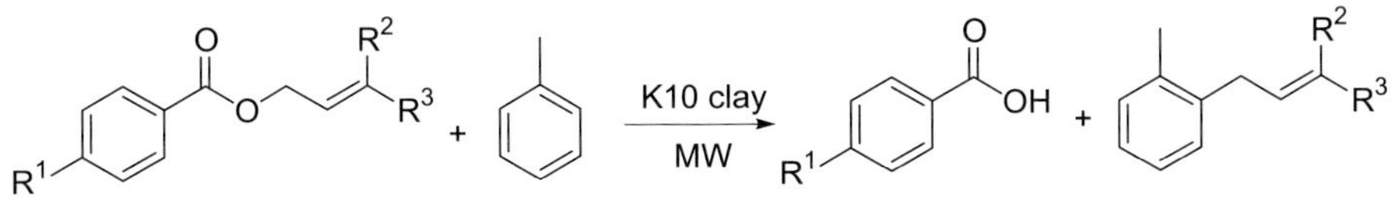

R^1 = H, CH_3, OCH_3, NO_2, Cl; R^2, R^3 = H, CH_3, Ph.

Scheme 8.8. MW deprotection of allyl esters on clay.

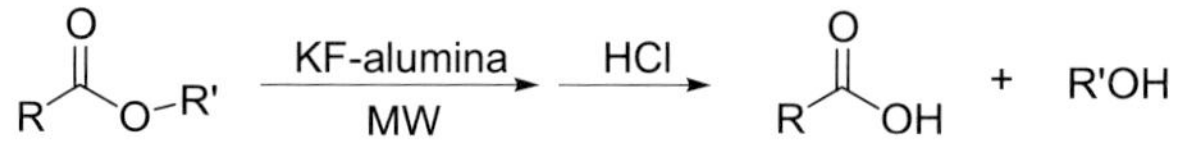

Scheme 8.9. MW deprotection (hydrolysis) of esters on alumina.

8.2.1.6 Selective Cleavage of N-tert-Butoxycarbonyl Groups

This approach may find application in peptide bond formation that would eliminate the use of irritating and corrosive chemicals such as trifluoroacetic acid and piperidine, as has been demonstrated for the deprotection of N-boc groups (Scheme 8.10). Solvent-free deprotection of N-tert-butoxycarbonyl groups occurs on exposure to microwave irradiation in the presence of neutral alumina "doped" with aluminum chloride (Scheme 8.10) [45].

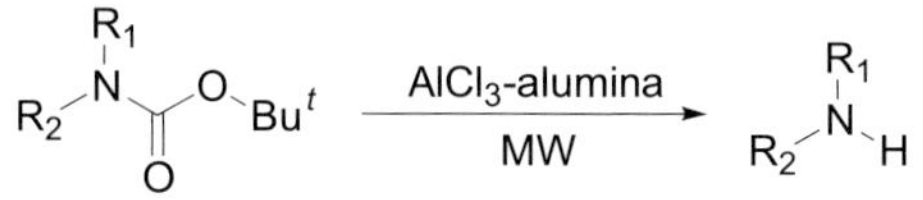

Scheme 8.10. Deprotection of N-tert-butoxycarbonyl group on alumina.

8.2.1.7 Desilylation Reactions

A variety of alcohols protected as the t-butyldimethylsilyl (TBDMS) ether derivatives can be rapidly regenerated to the corresponding hydroxy compounds on an alumina surface using MW irradiation (Scheme 8.11) [46]. This approach avoids use of the corrosive fluoride ions normally used for cleavage of the silyl protecting groups.

Deprotection of trimethylsilyl ether has also been accomplished (88–100%) on K10 clay [47] or oxidative cleavage (70–95%) in the presence of clay and iron(III) nitrate [48]. Another oxidative deprotection of trimethylsilyl ethers using supported potassium ferrate, K_2FeO_4 and MW has been reported [49].

8.2.1.8 Dethioacetalization Reaction

Thioacetals and ketals are important protecting groups used in organic manipulations. The regeneration of carbonyl compounds by cleavage of acid and base-stable thioacetals and thioketals is a challenging task. Cleavage of thioacetals normally requires use of toxic heavy metals such as Ti^{4+}, Cd^{2+}, Hg^{2+}, Ag^{2+}, Tl^{3+}, or uncommon reagents such as benzeneseleninic anhydride [50]. A high-yielding solid-state dethioacetalization reaction has been reported by Varma et al. using Clayfen (clay supported Fe(III) nitrate) (Scheme 8.12) [50]. The reaction is quite general and is devoid of byproducts formation except for substrates bearing free phenolic groups, where ring nitration may compete with dethioacetalization.

A report on the deprotection of thioacetals with Clayan (clay supported ammonium nitrate) (80–89%) soon followed [51].

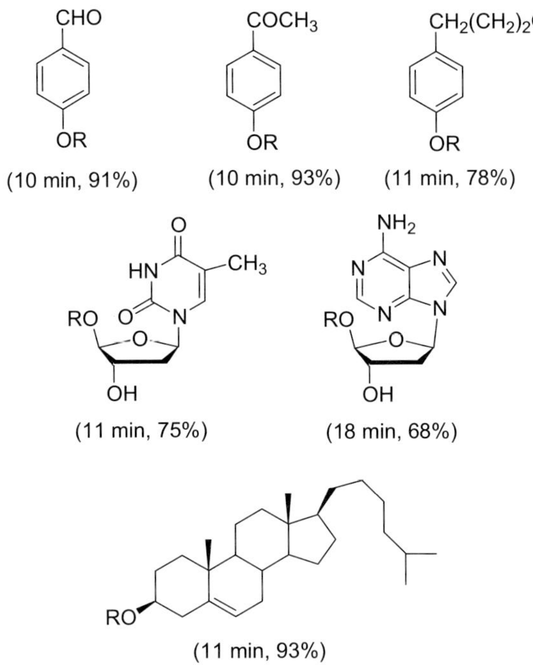

8.2 Microwave-accelerated Solvent-free Organic Reactions | 369

(10 min, 91%)   (10 min, 93%)   (11 min, 78%)

(11 min, 75%)   (18 min, 68%)

(11 min, 93%)

Where R = *tert*-butyldimethylsilyl (TBDMS)

Scheme 8.11. Desilylation reactions on alumina.

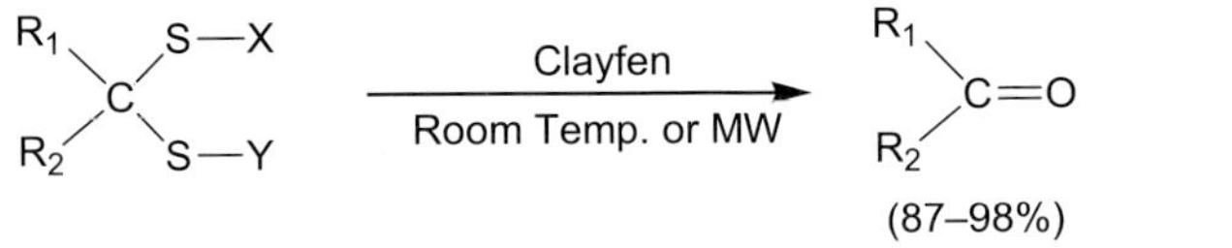

(87–98%)

Where R_1 = C_6H_5, *p*-CH_3 C_6H_4, *p*-$NO_2C_6H_4$; R_2 = H ; X–Y = –(CH$_2$)$_2$–

R_1 = R_2 = C_2H_5 ; X–Y = –(CH$_2$)$_2$– ; R_1 = R_2 = C_6H_5 ; X = Y = C_2H_5

R_1 = C_6H_5 ; R_2 = CH_3 ; X–Y = –(CH$_2$)$_2$–

R_1–R_2 = isoflavanolyl , 2-methylcyclohexyl; X–Y = –(CH$_2$)$_2$–

Scheme 8.12. Dethioacetalization reactions using Clayfen.

8.2.1.9 Deoximation Reactions

Oximes have been used as protecting groups for carbonyl compounds, owing to their hydrolytic stability. Consequently, the development of newer deoximation reagents has continued with the availability of a wide range of such agents, namely Raney nickel, pyridinium chlorochromate, pyridinium chlorochromate–H_2O_2, triethylammonium chlorochromate, dinitrogen tetroxide, trimethylsilyl chlorochromate, Dowex-50, dimethyl dioxirane, H_2O_2 over titanium silicalite-1, zirconium sulfophenyl phosphonate, *N*-haloamides, and bismuth chloride [52, 53].

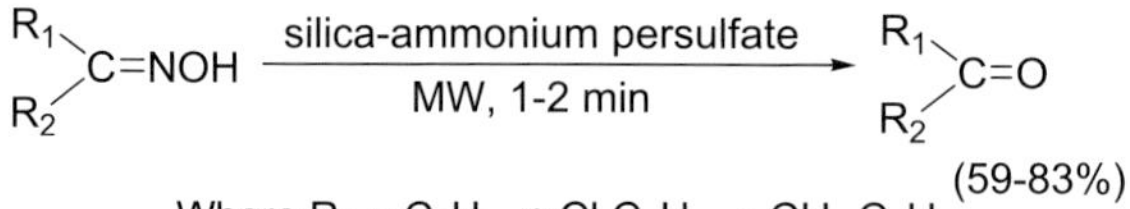

$R_1R_2C=NOH$ →[silica-ammonium persulfate / MW, 1-2 min] $R_1R_2C=O$

(59-83%)

Where R_1 = C_6H_5, p-Cl C_6H_4, p-CH_3 C_6H_4, p-CH_3O C_6H_4; R_2 = CH_3
R_1 = 2-thienyl, 1-naphthyl, C_6H_5, p-$NO_2C_6H_4$, m, p-$(CH_3O)_2C_6H_3$; R_2 = H
and R_1 = R_2 = cyclohexyl

Scheme 8.13. Deoximation of carbonyl compounds by silica-supported ammonium persulfate.

The quest for a solvent-free deprotection procedure has led to the use of relatively benign reagent, ammonium persulfate on silica, for regeneration of carbonyl compounds (Scheme 8.13) [53]. Neat oximes are simply mixed with solid supported reagent and the contents are irradiated in a MW oven to regenerate free aldehydes or ketones in a process that is applicable to both aldoximes and ketoximes. The critical role of the surface must be emphasized, because the same reagent supported on clay surface delivers predominantly the Beckmann rearrangement products – the amides [54].

A facile deoximation procedure with sodium periodate impregnated on moist silica (Scheme 8.14) has also been introduced that is applicable exclusively to ketoximes [55]. Aldehydes have been regenerated from the corresponding bisulfites (85–98%) on KSF clay surface [56].

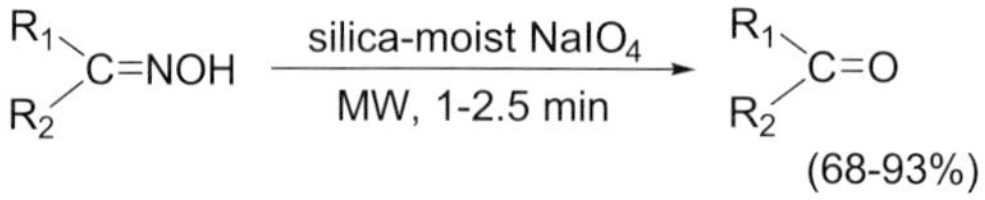

$R_1R_2C=NOH$ →[silica-moist $NaIO_4$ / MW, 1-2.5 min] $R_1R_2C=O$

(68-93%)

Where R_1 = CH_3 ; R_2 = C_6H_5, p-Cl C_6H_4, p-Br C_6H_4,
p-$CH_3C_6H_4$, p-$CH_3OC_6H_4$, p-NH_2 C_6H_4.
R_1 = R_2 = C_6H_5; R_1 = R_2 = cyclohexyl, tetrahydronaphthyl
and R_1 = C_2H_5, R_2 = n-C_4H_9

Scheme 8.14. Deoximation of ketoximes with silica-supported periodate.

Heravi et al. have used zeolite-supported thallium(III) nitrate to convert oximes into the parent carbonyl compounds in high yields (Scheme 8.15) [57]. Silica-supported ceric ammonium nitrate has also been used under the action of MW

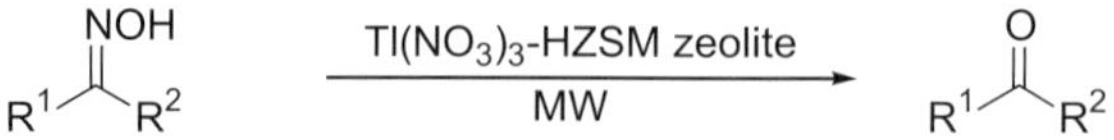

$R^1-C(=NOH)-R^2$ →[$Tl(NO_3)_3$-HZSM zeolite / MW] $R^1-C(=O)-R^2$

R^1 = aryl alkyl; R^2 = H, aryl, alkyl

Scheme 8.15. MW-assisted deoximation reactions.

irradiation to regenerate carbonyl compounds from oximes, semicarbazones, and phenylhydrazones efficiently [58].

8.2.1.10 Cleavage of Semicarbazones and Phenylhydrazones

Carbonyl compounds are also rapidly regenerated from the corresponding semicarbazone and phenylhydrazone derivatives by use of montmorillonite K10 clay impregnated with ammonium persulfate (Scheme 8.16) [59]. Interestingly, the microwave or ultrasound irradiation techniques can be used in these solvent-free procedures. Microwave exposure achieves deprotection in minutes whereas ultrasound-promoted reactions require 1–3 h for regeneration of carbonyl compounds [59].

Regeneration of carbonyl compounds from hydrazones (75–98%) [60] and semicarbazones (55–90%) [61] has also been achieved with bismuth trichloride.

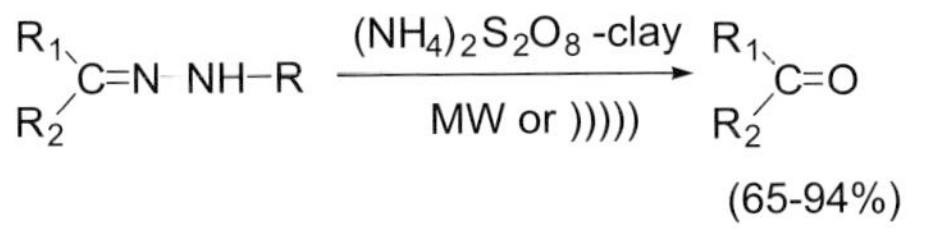

$$\underset{R_2}{\overset{R_1}{>}}C=N-NH-R \xrightarrow[\text{MW or)))))}]{(NH_4)_2S_2O_8\text{-clay}} \underset{R_2}{\overset{R_1}{>}}C=O$$

(65–94%)

$R_1 = n\text{-}C_4H_9, C_6H_5, p\text{-Cl } C_6H_4, p\text{-}CH_3 C_6H_4, p\text{-OH } C_6H_4$
$R_2 = CH_3, C_2H_5$ and $R = CNNH_2, C_6H_5$

Scheme 8.16. Regeneration of carbonyls from semicarbazone and phenylhydrazone derivatives.

8.2.1.11 Dethiocarbonylation

Dethiocarbonylation, transformation of thiocarbonyls to carbonyls, has been accomplished with several reagents, for example trifluoroacetic anhydride, CuCl/MeOH/NaOH, tetrabutylammonium hydrogen sulfate/NaOH, clay/ferric nitrate, NOBF$_4$, bromate and iodide solutions, alkaline hydrogen peroxide, sodium peroxide, thiophosgene, trimethyloxonium fluoroborate, tellurium-based oxidants, dimethyl selenoxide, benzeneseleninic anhydride, benzoyl peroxide, and halogen-catalyzed alkoxides under phase-transfer conditions [62]. These methods have limitations, however, for example the use of the stoichiometric amounts of the oxidants that are often inherently toxic or require longer reaction time or involve tedious procedures. In a process accelerated by MW irradiation, Varma et al. demonstrated efficient dethiocarbonylation process under solvent-free conditions using Clayfen or Clayan (Schemes 8.17 and 8.18) [62].

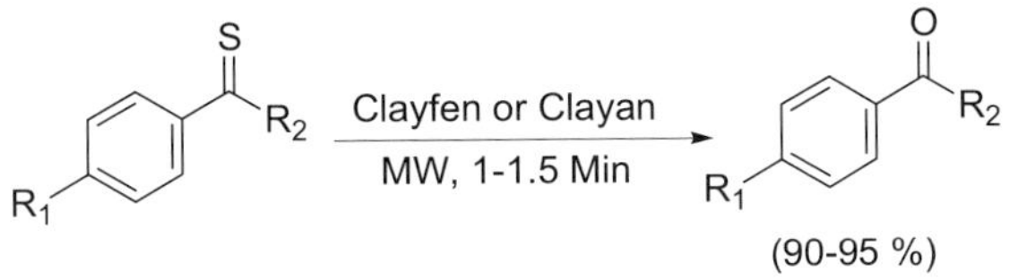

(90–95 %)

Where $R_1 = H, R_2 = CH_3$; $R_1 = H, Br, CH_3, R_2 = C_6H_5$

Scheme 8.17. Solvent-free dethiocarbonylation using Clayfen or Clayan.

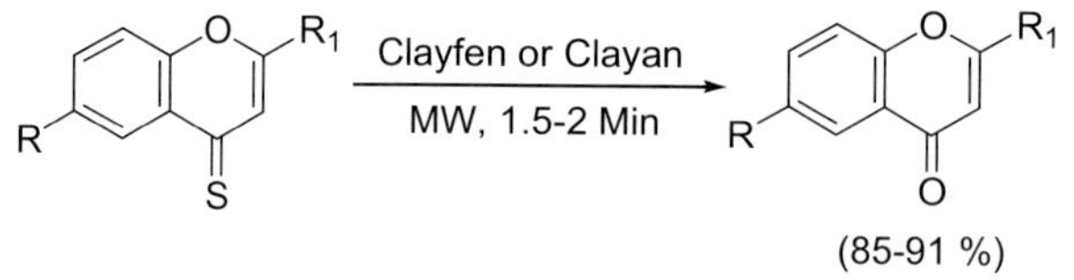

Where R = H; $R_1 = C_6H_5$, p-MeC_6H_4, p-$MeOC_6H_4$
R =OCH_3; R_2 = p-MeC_6H_4

Scheme 8.18. Transformation of thiocarbonyl derivatives of flavonoids with Clayfen and Clayan.

8.2.1.12 Cleavage of Methoxyphenyl Methyl (MPM) and Tetrahydropyranyl (THP) Ethers

Using clay-supported ammonium nitrate (Clayan), selective deprotection of MPM ether has been achieved using microwave irradiation under solvent-free conditions (Scheme 8.19) [63]. The same reagent has been used for cleavage of THP ethers. A similar selective preparation and cleavage of THP ethers has been achieved under the action of microwave irradiation catalyzed by iodine [64] or neat reaction in an ionic liquid [28].

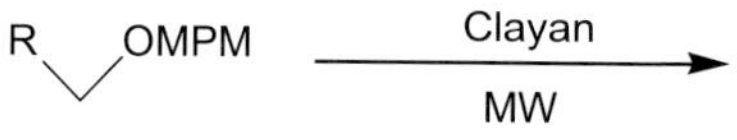

R = alkyl, alkenyl, alkynyl, aryl, acetate, ester, benzyl or silyl ether groups

Scheme 8.19. Cleavage of methoxyphenyl methyl (MPM) ethers using Clayan.

Alcohols and amines have been regenerated by MW-promoted cleavage of sulfonates (83–90%) and sulfonamides (76–85%), respectively, on a basic KF–alumina surface (Scheme 8.20) [65].

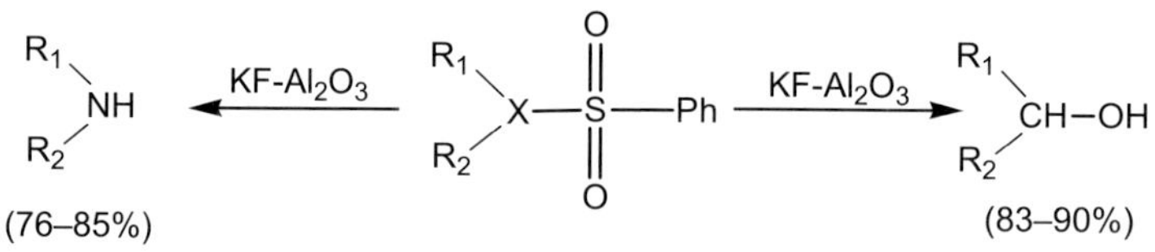

Where X = N, CH-O-

Scheme 8.20. Cleavage of sulfonates and sulfonamides on basic KF–alumina surface.

8.2.2 Condensation Reactions

A wide variety of MW-assisted aldol [66, 67] and Knoevenagel condensation reactions have been accomplished using relatively benign reagents such as ammonium acetate [68]; these include the Gabriel synthesis of phthalides with potassium acetate [69].

8.2.2.1 Wittig Olefination Reactions

Some difficult Wittig reactions of stable phosphorus ylides with ketones have been accelerated by use of MW irradiation (Scheme 8.21) [70]. Compared with the conventional method an improved yield was achieved in a shorter time by use of MW irradiation in the absence of solvent.

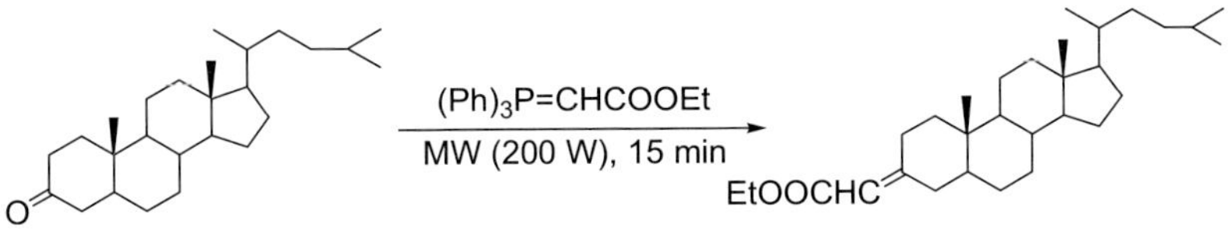

Scheme 8.21. MW-assisted Wittig olefination reaction.

Additional reports on olefination reactions have appeared [71] including the preparation of several phosphonium salts using a domestic MW oven wherein the rate of the reaction of neat triphenylphosphine and organic halide was remarkably enhanced in a pressure tube with a threaded Teflon cap [72].

8.2.2.2 Knoevenagel Condensation Reactions – Synthesis of Coumarins

Knoevenagel condensation reaction of creatinine with aldehydes occurs rapidly under solvent-free reaction conditions at 160–170 °C under the action of focused microwave irradiation (Scheme 8.22) [73].

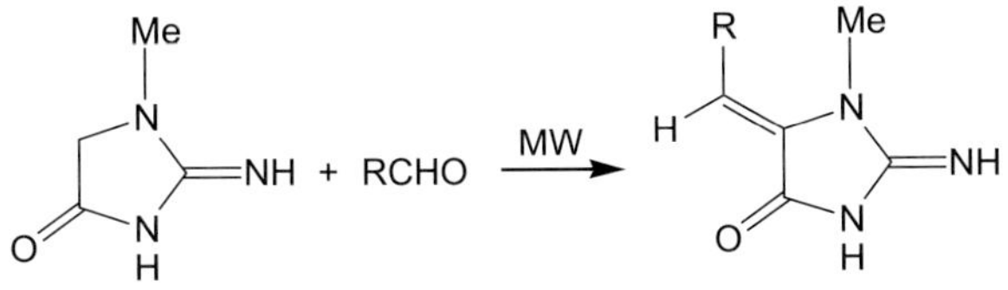

Scheme 8.22. Knoevenagel condensation reaction of creatinine with aldehydes.

5-Nitrofurfurylidines have been prepared by condensation of 5-nitrofurfuraldehyde with active methylene compounds, under the action of microwave, irradiation using $ZnCl_2$ and K10 as catalysts [74].

The classical Pechmann approach for synthesis of coumarins via the microwave-promoted reaction [75] has been extended to a solvent-free system in which salicylaldehydes undergo Knoevenagel condensation with a variety of ethyl acetate derivatives in the presence of piperidine to afford coumarins (Scheme 8.23) [76].

8.2.2.3 Synthesis of Imines, Enamines, Nitroalkenes, and N-Sulfonylimines

The preparation of imines, enamines, nitroalkenes and N-sulfonylimines proceeds via azeotropic removal of water from the intermediate in reactions that are nor-

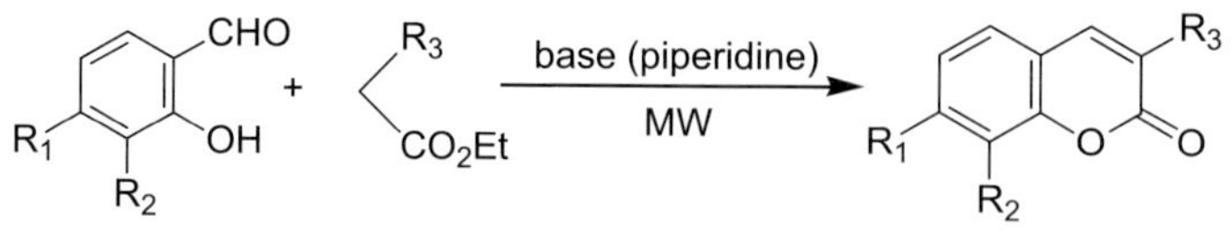

Scheme 8.23. MW-assisted synthesis of coumarins.

mally catalyzed by *p*-toluenesulfonic acid, titanium(IV) chloride, or montmorillonite K10 clay. A Dean–Stark apparatus is traditionally used, and requires a large excess of aromatic hydrocarbons such as benzene or toluene for azeotropic water elimination.

MW-expedited dehydration reactions using montmorillonite K10 clay [77] or Envirocat reagent [74], EPZG® (Schemes 8.24 and 8.25) have been demonstrated in a facile preparation of imines and enamines by reaction of primary and secondary amines with aldehydes and ketones, respectively. The generation of polar transition state intermediates in such reactions, and their enhanced coupling with microwaves, is possibly responsible for these rapid imine or enamine-forming reactions. To prevent the loss of low boiling reactants, use of microwave oven at lower power levels or intermittent heating has been used [77, 78].

Formation of benzil diimines by MW-assisted solvent-free reaction of benzil with aromatic amines on an alumina surface for 4 min has recently been reported to afford 1,2,3,4-tetraaryl-1,4-diaza-1,3-butadienes in good yields [79].

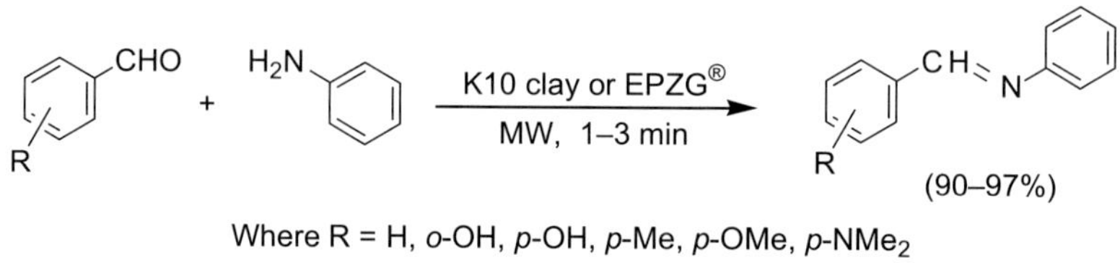

Where R = H, *o*-OH, *p*-OH, *p*-Me, *p*-OMe, *p*-NMe$_2$

Scheme 8.24. Clay-catalyzed formation of imines under solvent-free conditions.

$n_1 = 1$; $n_2 = 2$; Y = CH$_2$; $n_1 = 1$; $n_2 = 2$; Y = O
$n_1 = 2$; $n_2 = 1$; Y = CH$_2$; $n_1 = 2$; $n_2 = 2$; Y = CH$_2$
$n_1 = 2$; $n_2 = 2$; Y = O

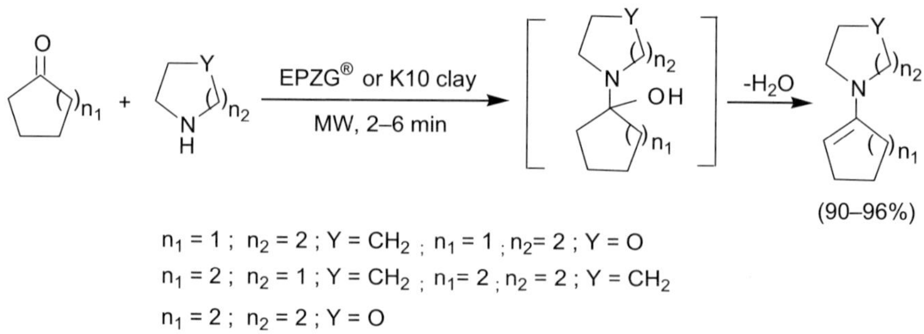

Scheme 8.25. Clay-catalyzed formation of enamines under solvent-free conditions.

8.2 Microwave-accelerated Solvent-free Organic Reactions | 375

The preparation of β-enamino ketone and esters on solid montmorillonite K10 clay coupled with MW irradiation applies to cyclic, acyclic and α-chloro-substituted β-dicarbonyl compounds with amines or their corresponding ammonium acetates (Scheme 8.26) [80].

The condensation of neat carbonyl compounds with nitroalkanes to afford nitroalkenes (Henry reaction) also proceeds rapidly via this MW approach in the presence of only catalytic amounts of ammonium acetate, thus avoiding use of a large excess of polluting nitrohydrocarbons normally employed (Scheme 8.27) [81].

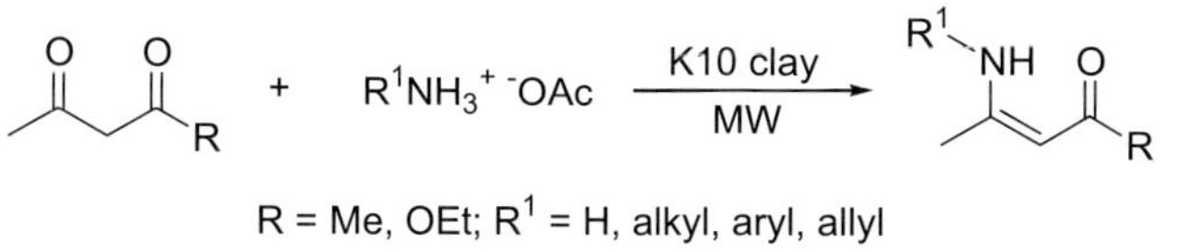

R = Me, OEt; R^1 = H, alkyl, aryl, allyl

Scheme 8.26. MW synthesis of β-enamino carbonyl compounds.

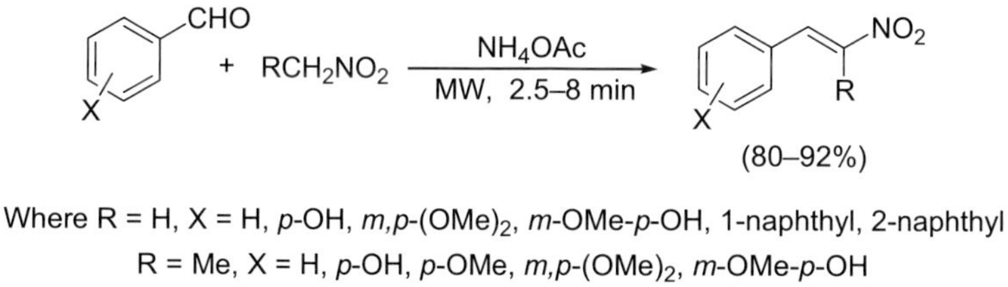

(80–92%)

Where R = H, X = H, p-OH, m,p-(OMe)$_2$, m-OMe-p-OH, 1-naphthyl, 2-naphthyl
R = Me, X = H, p-OH, p-OMe, m,p-(OMe)$_2$, m-OMe-p-OH

Scheme 8.27. MW-assisted preparation of α,β-unsaturated nitroalkenes.

The cycloaddition, reduction, and oxidation reactions of α,β-unsaturated nitroalkenes provide easy access to a vast array of functionality that includes nitroalkanes, N-substituted hydroxylamines, amines, ketones, oximes, and α-substituted oximes and ketones [82–84]. Consequently, there are numerous possibilities of using these *in situ* generated nitroalkenes for preparation of valuable building blocks and synthetic precursors.

Expeditious preparation of N-sulfonylimines has been optimized for one-pot solvent-free operation that involves microwave thermolysis of aldehydes and sulfonamides in presence of relatively benign reagents, calcium carbonate and montmorillonite K10 clay (Scheme 8.28) [85].

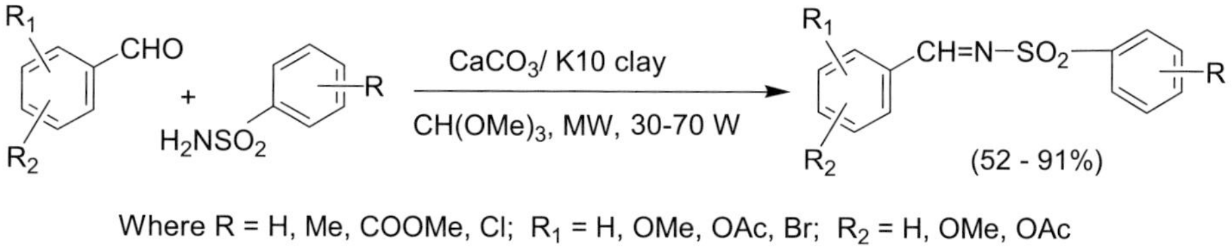

(52 - 91%)

Where R = H, Me, COOMe, Cl; R$_1$ = H, OMe, OAc, Br; R$_2$ = H, OMe, OAc

Scheme 8.28. One-pot solvent-free preparation of N-sulfonylimines.

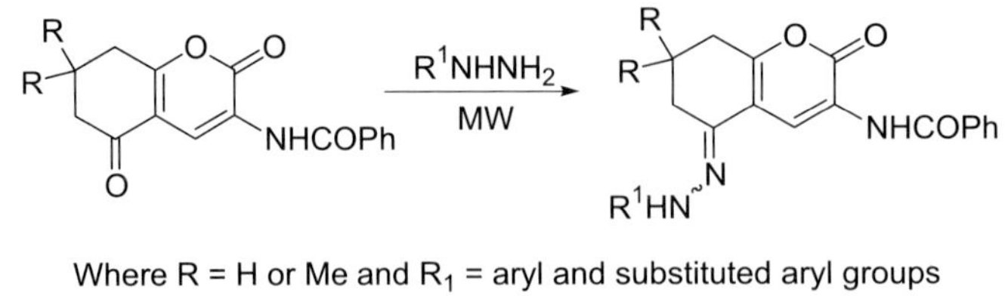

Scheme 8.29. Formation of hydrazones under solvent-free and catalyst-free conditions.

Bis(indolyl)nitroethanes have been obtained readily in 7–10 min in high yields (70–86%) on fine TLC-grade silica gel (5–40 μm) via Michael reaction of 3-(2′-nitrovinyl) indole with indoles. The same reaction is reported to require 8–14 h for completion at room temperature [86]. Several functionalized resins have been prepared from Merrifield resins via a MW-assisted procedure that used mixed solvent system to facilitate swelling of the resins and coupling with microwaves [87]. These resins can function as solid supports or polymeric scavengers in solid-phase synthesis.

The formation of hydrazones from the corresponding carbonyl compounds has been accomplished initially in toluene [88]. Treatment of hydrazone with alkali (KOH) accomplishes Wolff–Kichner reduction that proceeds in good yield under MW irradiation conditions [89]. Varma and Kocevar's group have shown that solvent-free and catalyst-free reaction of hydrazines with carbonyl compounds is possible on MW irradiation (Scheme 8.29) [90]. Interestingly, the general reaction proceeds smoothly even for solid reactants and is completed below the melting points of the two reactants, possibly via the formation of a eutectic. The reactions have been conducted in a household MW oven and the control experiments were conducted concurrently in separate open beakers; the reactions can be essentially followed by visual observation when a melt is obtained [91].

Pyrazolo[3,4-b]quinolines and pyrazolo[3,4-c]pyrazoles have been synthesized by microwave irradiation of β-chlorovinylaldehydes and hydrazines in the presence of p-toluenesulfonic acid (PTSA) under solvent-free conditions (Scheme 8.30) [92] via hydrazone formation and sequential cyclization.

An interesting solid-state synthesis of amides, using potassium *tert*-butoxide and accessible reagents, non-enolizable esters and amines, under the action of MW irradiation, has also been reported [93].

The kinetics of the acid-catalyzed esterification reaction of 2,4,6-trimethylbenzoic acid in i-PrOH under the action of microwave irradiation have been investigated

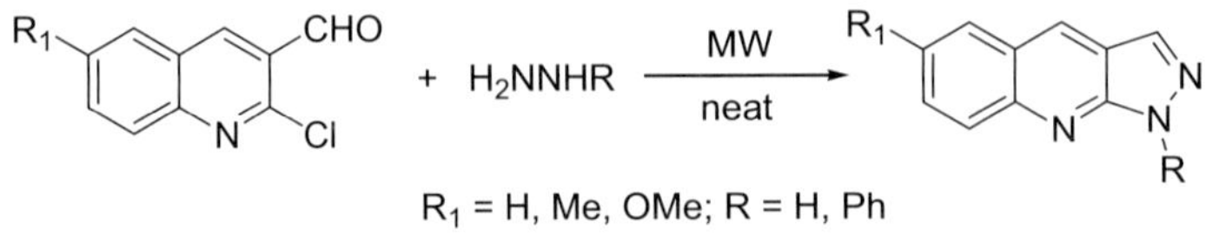

Scheme 8.30. MW-assisted solvent-free synthesis of pyrazolo[3,4-b]quinolines.

[94]. A simple and practical technique for MW-assisted synthesis of esters has been reported wherein the reactions are conducted either on solid mineral supports or by using a phase-transfer catalyst (PTC) in the absence of organic solvents [95]. The esterification of enols with acetic anhydride and iodine has also been recorded [96].

A detailed account of condensation reactions used in heterocyclic chemistry can be found in Section 8.2.6, in Chapter 10 [97], and, for cycloaddition reactions, in Chapter 11 [98]. A previously unknown class of compounds, spiro[3H-indole-3,2'-[4H] pyrido[3,2-e]-1,3-thiazine]-2,4'(1H) diones, can be synthesized by reaction of in situ-generated 3-indolylimine with 2-mercaptonicotinic acid under the action of MW in the absence of solvent. Both neat reactions and reactions on solid supports such as silica gel, alumina etc., effectively promote the reaction whereas reactions under thermal heating conditions failed to proceed (Scheme 8.31) [99].

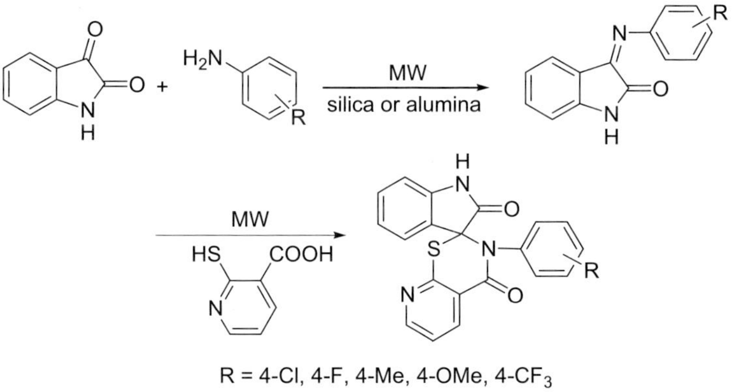

Scheme 8.31. MW-promoted synthesis of spiro[indole-pyrido] thiazines.

8.2.2.4 MW-assisted Michael-addition Reactions

Solvent-free Michael addition between diethyl ethoxymethylenemalonate (EMME) and a variety of O, S, N nucleophiles either neat or on an alumina support, under the action of MW irradiation, has been explored [100] and found to be a useful procedure for nucleophilic addition to EMME (Scheme 8.32).

Another solvent-free MW-accelerated conjugate addition of aldehydes to α,β-unsaturated ketones has been demonstrated by Yadav and coworkers [101]; in the

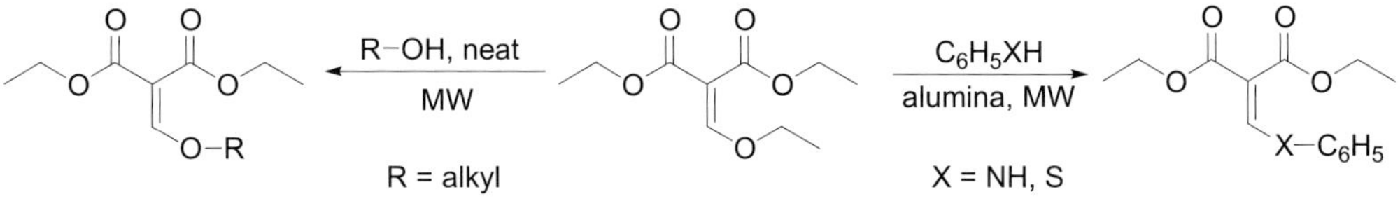

Scheme 8.32. Solvent-free nucleophilic Michael addition to EMME.

presence of 5-(2-hydroxyethyl)-1,3-thiazolium halides and DBU adsorbed on the basic alumina surface, 1,4-diketones are obtained in enhanced yields and reduced reaction times compared with conventional methods (Scheme 8.33).

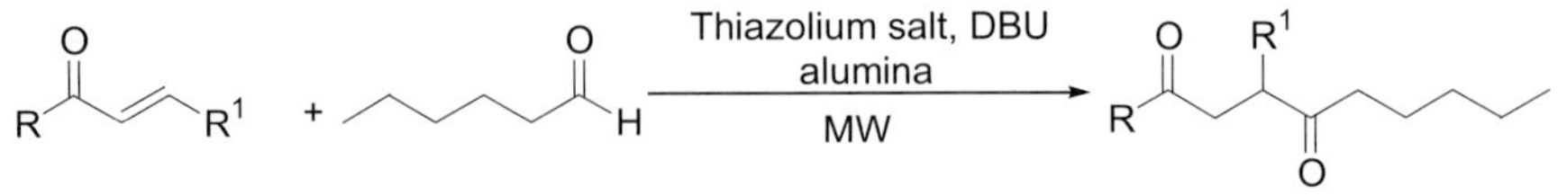

Scheme 8.33. MW synthesis of 1,4-diketones.

A series of nitrocyclohexanol derivatives has been synthesized by using MW irradiation starting from nitromethane and unsaturated ketones in the presence of alumina-supported potassium fluoride under solvent-free conditions (Scheme 8.34). The reaction involves a double and diastereoselective Michael addition followed by ring closure [102].

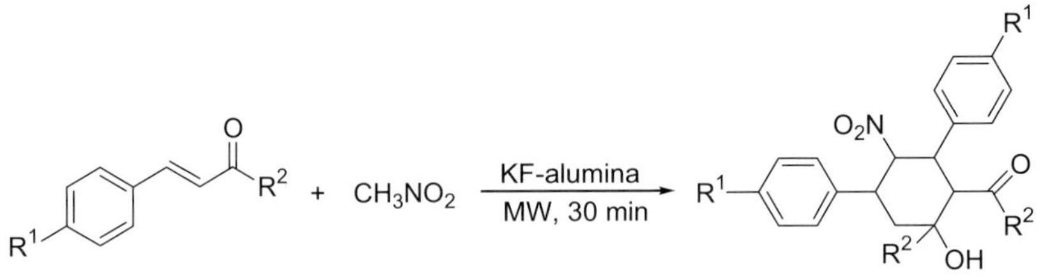

Scheme 8.34. MW synthesis of nitrocyclohexanol.

8.2.2.5 MW-assisted Solid Mineral-promoted Miscellaneous Condensation Reactions

Microwave-assisted solvent-free synthesis of a quinoline-3,4-dicarboximide library on inorganic solid supports has recently been reported [103]. Wet clay K10 was shown to be the best medium for the condensation reaction between 2-methylquinoline -3,4-dicarboxylic anhydride and several primary amines. Microwave irradiation is essential for rapid and complete formation of imides (Scheme 8.35).

Facile synthesis of cyclic ethers from dihalo compounds on alumina under solvent-free conditions has been accomplished in good yields by Mihara et al. (Scheme 8.36) [104].

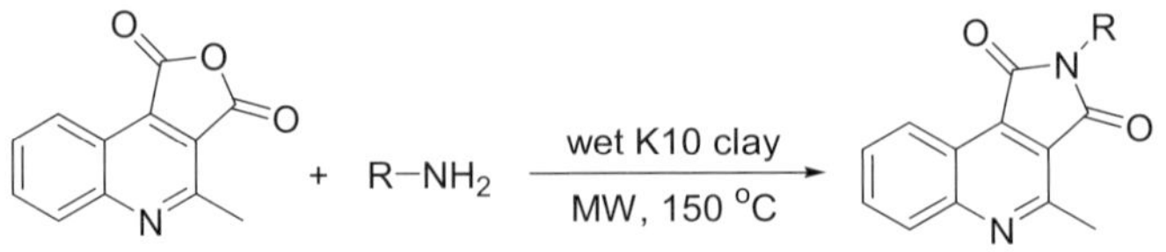

Scheme 8.35. MW synthesis of quinoline-3,4-dicarboximides.

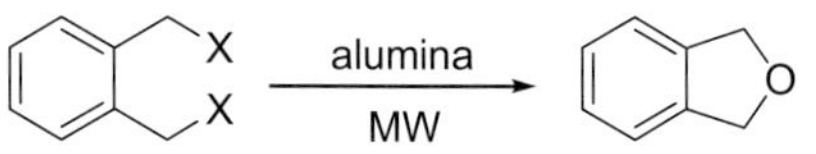

Scheme 8.36. Solid-supported synthesis of cyclic ethers.

An expedient montmorillonite K10 clay-catalyzed cycloisomerization of salicylaldehyde 4-(β-D-ribo- or β-D-2′-deoxyribofuranosyl)semicarbazones yields benzoxazinone nucleosides, 4-hydrazino-3,4-dihydro-3-(β-D-ribo- or β-D-2′-deoxyribofuranosyl)-2H-benz[e]-1,3-oxazin-2-ones, which readily undergo reductive dehydrazination on alumina-supported copper(II) sulfate to furnish 3,4-dihydro-3-(β-D-ribo- or β-D-2′-deoxyribofuranosyl)-2H-benz[e]-1,3-oxazin-2-ones under the action of microwave irradiation (Scheme 8.37) [105]. The reaction is solvent-free.

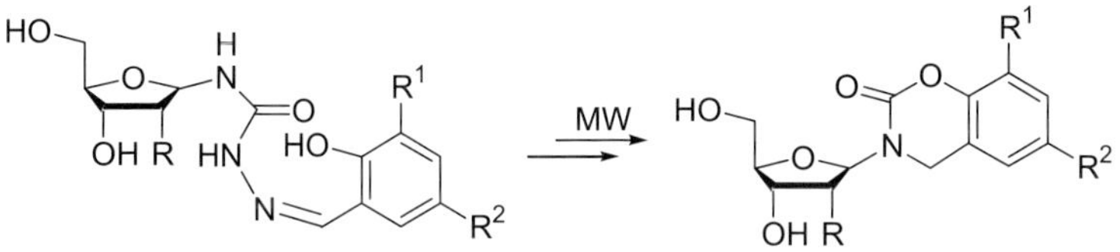

Scheme 8.37. Synthesis of benzoxazinone nucleosides.

Practical access to 1,6-anhydro-β-D-hexopyranoses by a solid-supported solvent-free procedure has been demonstrated by Bailliez et al. [106]. Microwave irradiation of 6-O-tosyl or 2,6-di-O-tosyl peracetylated hexopyranoses absorbed on basic alumina furnished the corresponding 1,6-anhydro-β-D-hexopyranoses. Direct access to 1,6:3,4-dianhydro-β-D-altro-pyranose from D-glucose is also described (Scheme 8.38).

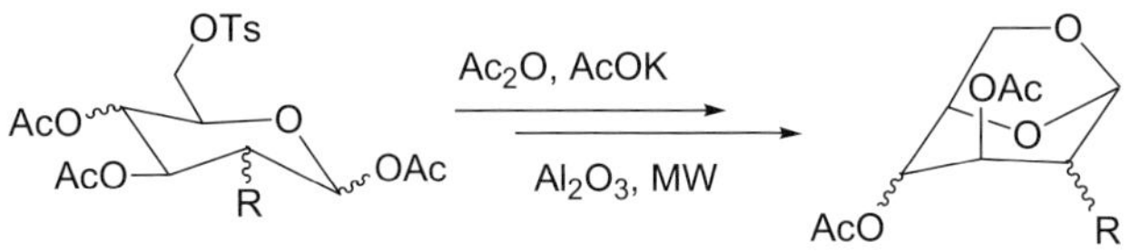

Scheme 8.38. Solid-supported solvent-free approach for preparation of 1,6-anhydro-β-D-hexopyranoses.

8.2.3
Isomerization and Rearrangement Reactions

Numerous rearrangement and isomerization reactions have been reported using MW irradiation. Some reactions are performed in the solution phase whereas others proceed on a graphite or mineral support surface often "doped" with Lewis acids. Sometimes the reactions proceed on heating the neat reactants. Notable examples are the benzil–benzilic acid rearrangement [107], solvent-free Beckmann rearrangement on K10 clay [54], Fries rearrangement on K10 clay that affords mixture of ortho and para products [108], and thia-Fries rearrangement of arylsulfonates using aluminum and zinc chloride on silica gel [109].

8.2.3.1 Eugenol–Isoeugenol Isomerization
Eugenol undergoes MW-assisted isomerization to isoeugenol under solvent-free conditions in the presence of potassium *tert*-butoxide and a catalytic amount of phase-transfer reagent [36].

8.2.3.2 Pinacol–Pinacolone Rearrangement
An example of solvent-free pinacol–pinacolone rearrangement using MW irradiation has been reported. The process involves irradiation of the gem diols with Al^{3+}-montmorillonite K10 clay for 15 min to afford the rearrangement product in excellent yields (Scheme 8.39) [24a]. Comparative studies performed by conventional heating in an oil bath showed that reaction times are too long (15 h). When using KSF clay, Villemin observed the similar rearrangement and the Meyer–Schuster acidic rearrangement (Scheme 8.40) [24b].

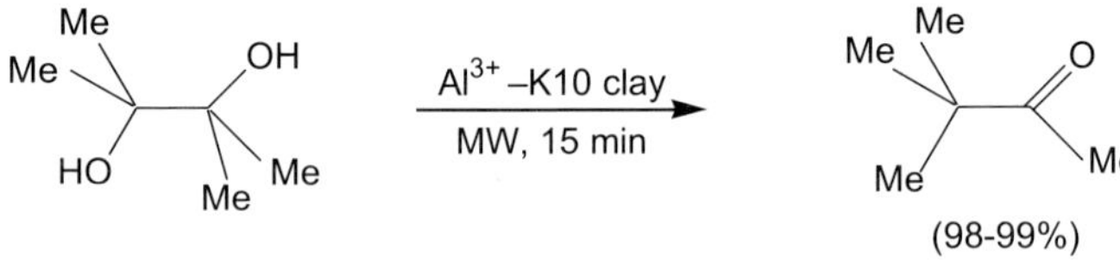

Scheme 8.39. Pinacol–pinacolone rearrangement on Al^{3+}–montmorillonite K10 clay.

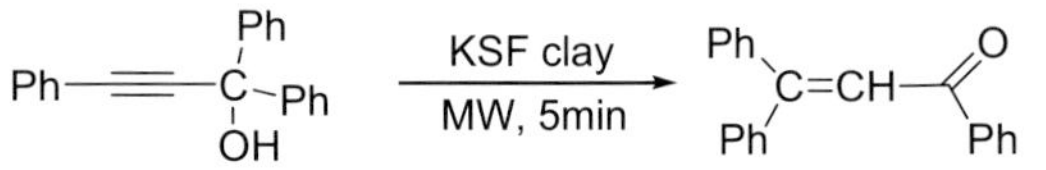

Scheme 8.40. Meyer–Schuster rearrangement on KSF clay.

An efficient ring-expansion transformation has also been described under solvent-free conditions (Scheme 8.41) [110]. This microwave procedure is superior to the same reactions conducted in traditional methanolic solution.

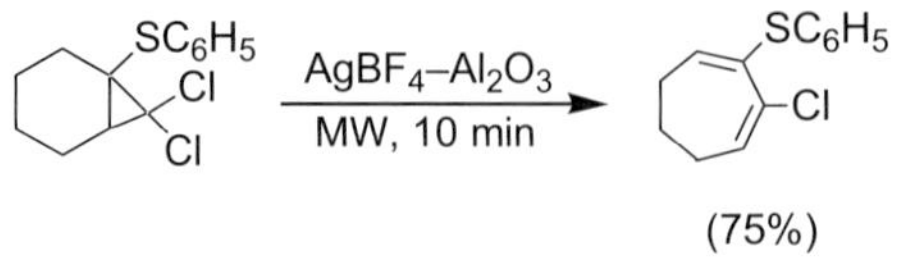

Scheme 8.41. MW-assisted ring expansion reaction on alumina.

8.2.3.3 Beckmann Rearrangement

A simple montmorillonite K10 clay surface is one among numerous acidic supports that have been explored for Beckmann rearrangement of oximes (Scheme 8.42) [54]. The conditions are not, however, adaptable for aldoximes that are readily dehydrated to the corresponding nitriles under solvent-free conditions. Zinc chloride has been used in this rearrangement for benzaldehyde and 2-hydroxyacetophenone, the latter being adapted for the synthesis of benzoxazoles.

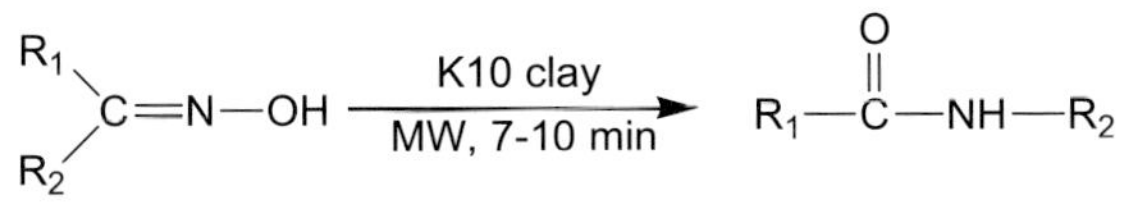

Where R_1 = CH_3 or C_6H_5 and R_2 = C_6H_5 or substituted phenyl

Scheme 8.42. Beckmann rearrangement of oximes on clay.

8.2.3.4 Claisen Rearrangement

A few solvent-free examples of Claisen rearrangement reactions under the action of microwave irradiation have been described. One involves the double Claisen rearrangement of bis(4-allyloxyphenyl)sulfone into bis(3-allyl-4-hydroxyphenyl)sulfone (Scheme 8.43) [111]. Similarly, 3'-allyl-2'-hydroxyacetophenone has been obtained in quantitative yield from 2'-allyloxyacetophenone by Bennett et al. [112].

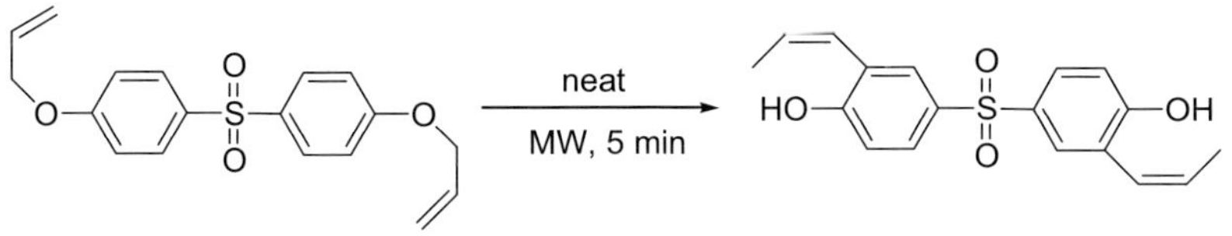

Scheme 8.43. Solvent-free Claisen rearrangement.

Among numerous other studies, Ferrier rearrangement is noticeably improved because it proceeds well (72–83%) on irradiation of the neat reactants (Scheme 8.44) [113].

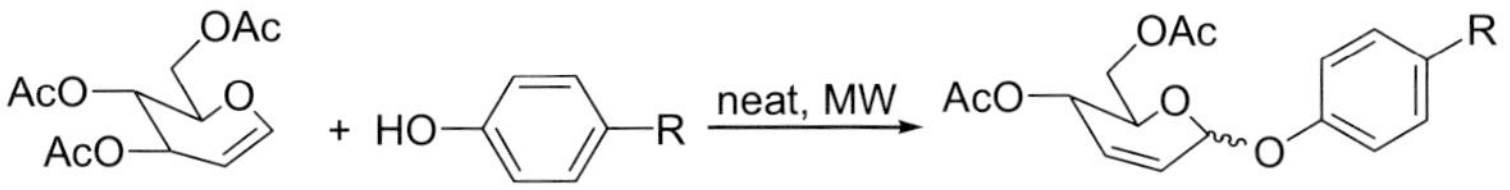

Scheme 8.44. Solvent-free Ferrier rearrangement.

8.2.4
Oxidation Reactions – Oxidation of Alcohols and Sulfides

Metal-based reagents have been extensively used in organic synthesis. Peracids, peroxides, potassium permanganate ($KMnO_4$), manganese dioxide (MnO_2), chromium trioxide (CrO_3), potassium dichromate ($K_2Cr_2O_7$) and potassium chromate (K_2CrO_4) are some of the common oxidizing reagents used for organic functional groups [114, 115].

The utility of such reagents in the oxidation processes is compromised by their inherent toxicity, cumbersome preparation, potential danger in handling of metal complexes, difficulties encountered in product isolation, and waste disposal problems. Immobilization of metallic reagents on solid supports has circumvented some of these drawbacks and provided an attractive alternative in organic synthesis, because of the selectivity and associated ease of manipulation. Localization of metals on the mineral oxide surfaces also reduces the possibility of their leaching into the environment.

8.2.4.1 Activated Manganese Dioxide–Silica

Manganese dioxide (MnO_2) supported on silica gel provides an expeditious and high-yield route to carbonyl compounds. Benzyl alcohols are thus selectively oxidized to carbonyl compounds using 35% MnO_2 "doped" silica under MW irradiation conditions (Scheme 8.45) [116].

Manganese dioxide on bentonite clay has also been used for oxidation of phenols to quinones (30–100%) [117] and MnO_2 on silica effects the dehydrogenation of pyrrolodines (58–96%) [118].

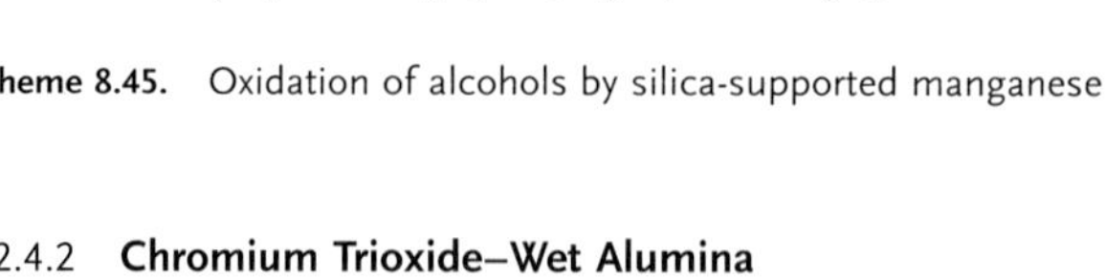

$$\begin{array}{c} R_1 \\ \diagdown \\ CH-OH \\ \diagup \\ R_2 \end{array} \xrightarrow[\text{MW, 20-60 sec}]{MnO_2\text{-silica}} \begin{array}{c} R_1 \\ \diagdown \\ C=O \\ \diagup \\ R_2 \end{array}$$

(67-96%)

Where R_1 = H; R_2 = C_6H_5, p-MeC_6H_4, p-$MeOC_6H_4$, $C_6H_5CH=CH$
R_1 = Et, C_6H_5, C_6H_5CO; R_2 = C_6H_5; R_1 = R_2 = hydroquinone
R_1 = p-$MeOC_6H_4CO$; R_2 = p-$MeOC_6H_4$

Scheme 8.45. Oxidation of alcohols by silica-supported manganese dioxide.

8.2.4.2 Chromium Trioxide–Wet Alumina

Use of chromium(VI) reagents for oxidative transformation is compromised by their inherent toxicity, involved preparation of the various complex forms (with pyridine or acetic acid), and cumbersome work-up procedures. Chromium trioxide (CrO_3) immobilized on premoistened alumina enables efficient oxidation of benzyl alcohols to carbonyl compounds by simple mixing with different substrates

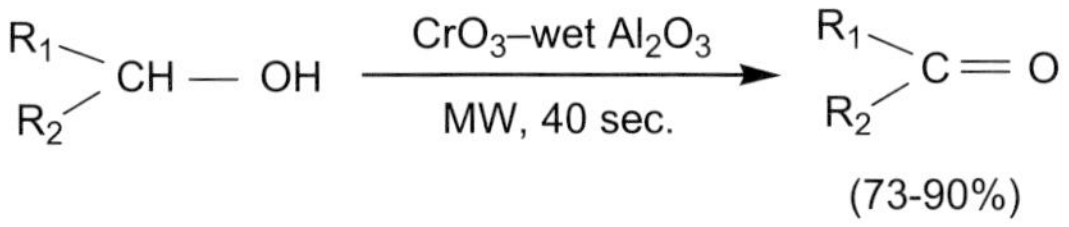

$$\underset{R_2}{\overset{R_1}{>}}CH-OH \xrightarrow[\text{MW, 40 sec.}]{\text{CrO}_3\text{–wet Al}_2\text{O}_3} \underset{R_2}{\overset{R_1}{>}}C=O$$

(73-90%)

R_1 = C_6H_5, p-MeC$_6$H$_4$, p-MeOC$_6$H$_4$, p-NO$_2$C$_6$H$_4$; R_2 = H
R_1 = C_6H_5 ; R_2 = Me, C_6H_5, C_6H_5CO; R_1 = R_2 = ⌬, ⌬⌬

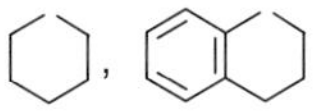

Scheme 8.46. Oxidation of alcohols by chromium trioxide supported on premoistened alumina.

(Scheme 8.46). Interestingly, no overoxidation to carboxylic acids is observed and products are devoid of tar contaminants, a typical occurrence in many CrO$_3$ oxidations [119].

8.2.4.3 Selective Solvent-free Oxidation with Clayfen

A facile method for the oxidation of alcohols to carbonyl compounds has been reported by Varma et al. using montmorillonite K10 clay-supported iron(III) nitrate (Clayfen) under solvent-free conditions [120]. This MW-expedited reaction presumably proceeds via intermediate nitrosonium ions. Interestingly, no carboxylic acids are formed in the oxidation of primary alcohols. The simple solvent-free experimental procedure involves mixing of neat substrates with Clayfen and brief exposure of the reaction mixture to irradiation in a MW oven for 15–60 s. This rapid, manipulatively simple, inexpensive, and selective procedure avoids use of excess solvents and toxic oxidants (Scheme 8.47) [120]. Solid-state use of Clayfen has afforded higher yields and the amounts used are half that used by Laszlo et al. [17, 19].

A ground mixture of iron(III) nitrate and HZSM-5 zeolite, termed "zeofen", has also been used, both in dichloromethane solution and in the solid state under MW irradiation conditions [121]. It has been suggested that the zeolite aids the reproducibility of the reaction but any other aluminosilicate support would probably be equally effective. Recent studies point out attractive alternatives that do not employ any solid supports in such oxidations with nitrate salts [122].

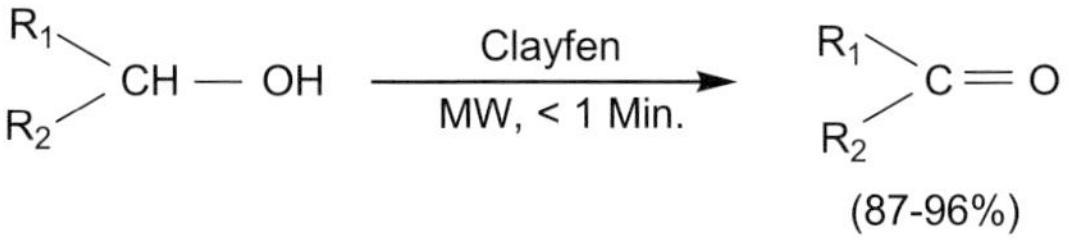

$$\underset{R_2}{\overset{R_1}{>}}CH-OH \xrightarrow[\text{MW, < 1 Min.}]{\text{Clayfen}} \underset{R_2}{\overset{R_1}{>}}C=O$$

(87-96%)

Where R_1 = H; R_2 = C_6H_5, p-MeC$_6$H$_4$, p-MeOC$_6$H$_4$, 2-tetrahydrofuranyl
R_1 = Et, C_6H_5CO; R_2 = C_6H_5; R_1, R_2 = cyclohexyl
R_1 = p-MeOC$_6$H$_4$CO; R_2 = p-MeOC$_6$H$_4$

Scheme 8.47. Solvent-free selective oxidation of alcohols with Clayfen.

8.2.4.4 Oxidations with Claycop–Hydrogen Peroxide

Metal ions play an important role in several of these oxidative reactions and in biological dioxygen metabolism. As an example, copper(II) acetate and hydrogen peroxide have been used to produce a stable oxidizing agent, hydroperoxy copper(II) compound. The same oxidation system is also obtained from copper(II) nitrate and hydrogen peroxide (Eq. 1) [123] but requires neutralization of the ensuing nitric acid by potassium bicarbonate to maintain a pH ~5.

$$2\,Cu\,(NO_3)_2 + H_2O_2 + 2\,H_2O \rightarrow 2\,CuO_2H + 4\,HNO_3 \qquad (1)$$

The copper(II) nitrate immobilized on K10 clay (Claycop)–hydrogen peroxide system has been shown to be an effective oxidant for a variety of substrates and provides excellent yields (Scheme 8.48) [124]; maintenance of the pH of the reaction mixture is not required.

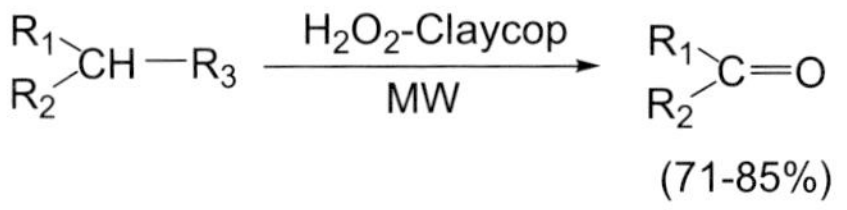

$$\underset{R_2}{\overset{R_1}{>}}CH-R_3 \quad \xrightarrow[MW]{H_2O_2\text{-Claycop}} \quad \underset{R_2}{\overset{R_1}{>}}C=O$$

(71-85%)

$R_1 = C_6H_5, p\text{-}NO_2C_6H_4$; $R_2 = H, C_6H_5$; $R_3 = H, Br, CN, NH_2, COOH$

Scheme 8.48. Oxidation reactions with Claycop and hydrogen peroxide.

8.2.4.5 Other Metallic Oxidants – Copper Sulfate or Oxone®–Alumina

Symmetrical and unsymmetrical benzoins have been rapidly oxidized to benzils in high yields using solid reagent systems, copper(II) sulfate–alumina [125] or Oxone®–wet alumina [125, 126] under the influence of microwaves (Scheme 8.49). Conventionally, the oxidative transformation of α-hydroxyketones to 1,2-diketones is accomplished by reagents such as nitric acid, Fehling's solution, thallium(III) nitrate (TTN), ytterbium(III) nitrate, ammonium chlorochromate–alumina and Clayfen. In addition to the extended reaction time, most of these processes suffer from drawbacks such as the use of corrosive acids and toxic metals that generate undesirable waste products.

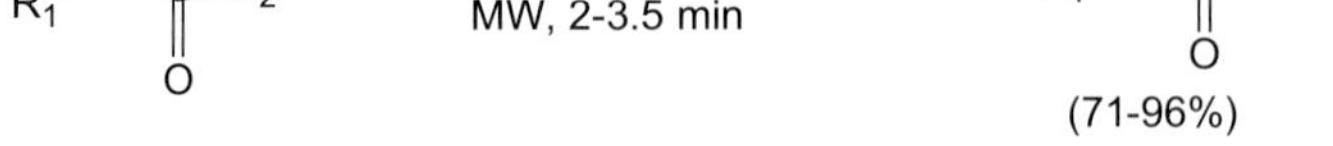

$$R_1\underset{O}{\overset{OH}{\underset{|}{\text{C}}}}R_2 \quad \xrightarrow[MW,\,2\text{-}3.5\,min]{Oxone^®\text{-}Al_2O_3 \text{ or } CuSO_4\text{-}Al_2O_3} \quad R_1\underset{O}{\overset{O}{\text{C}}}\underset{O}{\overset{}{\text{C}}}R_2$$

(71-96%)

Where $R_1 = R_2 = C_6H_5, p\text{-}MeC_6H_4, p\text{-}MeOC_6H_4, p\text{-}ClC_6H_4$, (furyl)

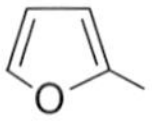

$R_1 = C_6H_5$; $R_2 = p\text{-}MeC_6H_4, p\text{-}MeOC_6H_4$

and $R_1 = Me$; $R_2 = C_6H_5$

Scheme 8.49. Oxidation of α-hydroxyketones with alumina-supported copper sulfate or Oxone®.

Under these solvent-free conditions, oxidation of primary alcohols (e.g. benzyl alcohol) and secondary alcohols (e.g. 1-phenyl-1-propanol) is rather sluggish and poor, and is of little practical utility. Consequently, the process is applicable only to α-hydroxyketones as exemplified by a variety of examples including a mixed benzylic/aliphatic α-hydroxyketone, 2-hydroxypropiophenone, that furnishes the corresponding vicinal diketone [126, 127].

8.2.4.6 Nonmetallic Oxidants – Alumina Impregnated with Iodobenzene Diacetate (IBD)

Several organohypervalent iodine reagents have been used for oxidation of alcohols and phenols such as iodoxybenzene, o-iodoxybenzoic acid (IBX), bis(trifluoroacetoxy)iodobenzene (BTI), and Dess–Martin periodinane, etc. Use of inexpensive iodobenzene diacetate (IBD) as an oxidant, however, has not been fully exploited. Most of these reactions are conducted in high-boiling DMSO or toxic acetonitrile, media that result in increased burden on the environment.

Varma and coworkers explored the use of hypervalent iodine compounds on solid supports for the first time and developed a facile oxidative procedure that rapidly converts alcohols into the corresponding carbonyl compounds, in almost quantitative yields, using alumina-supported IBD under solvent-free conditions and MW irradiation [128]. Use of alumina as a support improved yields markedly compared with neat IBD (Scheme 8.50). 1,2-Benzenedimethanol, under these conditions, undergoes cyclization to afford 1(3H)-isobenzofuranone.

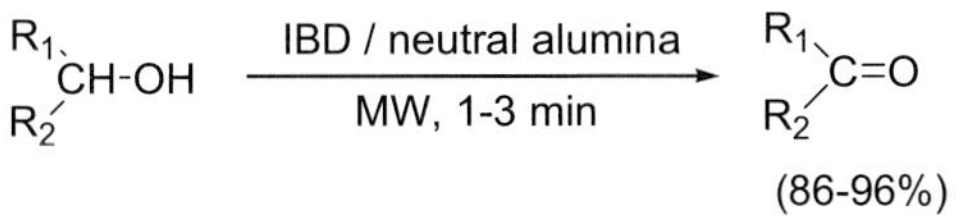

$$\begin{array}{c} R_1 \\ R_2 \end{array}\!\!\!\text{CH-OH} \quad \xrightarrow[\text{MW, 1-3 min}]{\text{IBD / neutral alumina}} \quad \begin{array}{c} R_1 \\ R_2 \end{array}\!\!\!\text{C=O}$$

(86-96%)

Where R_1 = C_6H_5, substituted phenyl and R_2 = H, C_2H_5, COC_6H_5

Scheme 8.50. Oxidation of alcohols using alumina-supported iodobenzene diacetate.

8.2.4.7 Oxidation of Sulfides to Sulfoxides and Sulfones – Sodium Periodate–Silica

Oxidation of sulfides to the corresponding sulfoxides and sulfones proceeds under rather strenuous conditions requiring strong oxidants such as nitric acid, hydrogen peroxide, chromic acid, peracids, and periodate. By use of MW irradiation this oxidation can be achieved under solvent-free conditions and with the desired selectivity to either sulfoxides or sulfones using 10% sodium periodate on silica (Scheme 8.51) [129]. A smaller amount of the active oxidizing agent is used, which is safer and easier to handle.

Several refractory thiophenes, that are often not reductively removable by conventional refining processes, can be oxidized under these conditions, e.g. benzothiophenes are oxidized to the corresponding sulfoxides and sulfones using ultrasonic and microwave irradiation, respectively, in the presence of $NaIO_4$–silica [129]. A noteworthy feature of the procedure is its applicability to long-chain fatty

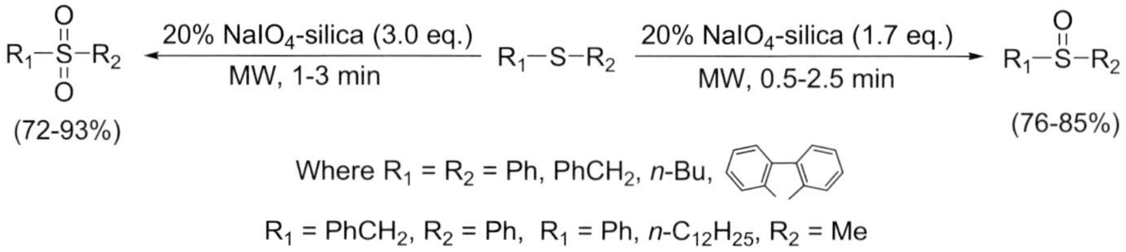

Scheme 8.51. Oxidation of sulfides to sulfoxides and sulfones by silica-supported NaIO$_4$.

$$R_1-S(=O)_2-R_2 \xleftarrow[\text{MW, 1-3 min}]{\text{20\% NaIO}_4\text{-silica (3.0 eq.)}} R_1-S-R_2 \xrightarrow[\text{MW, 0.5-2.5 min}]{\text{20\% NaIO}_4\text{-silica (1.7 eq.)}} R_1-S(=O)-R_2$$

(72-93%)   (76-85%)

Where $R_1 = R_2 =$ Ph, PhCH$_2$, n-Bu, (biphenyl)

$R_1 =$ PhCH$_2$, $R_2 =$ Ph; $R_1 =$ Ph, n-C$_{12}$H$_{25}$, $R_2 =$ Me

sulfides that are insoluble in most solvents and are consequently difficult to oxidize.

8.2.4.8 Oxidation of Sulfides to Sulfoxides – Iodobenzene Diacetate–Alumina

As described above (Section 8.2.4.6), the solid reagent system, IBD–alumina is a useful oxidizing agent and its use has been extended to rapid, high-yield, and selective oxidation of alkyl, aryl and cyclic sulfides to the corresponding sulfoxides on microwave activation (Scheme 8.52) [130].

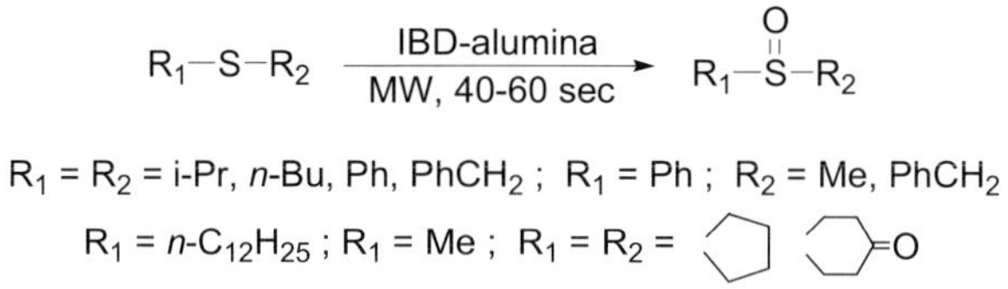

$$R_1-S-R_2 \xrightarrow[\text{MW, 40-60 sec}]{\text{IBD-alumina}} R_1-S(=O)-R_2$$

$R_1 = R_2 =$ i-Pr, n-Bu, Ph, PhCH$_2$; $R_1 =$ Ph ; $R_2 =$ Me, PhCH$_2$
$R_1 =$ n-C$_{12}$H$_{25}$; $R_1 =$ Me ; $R_1 = R_2 =$ (cyclopentyl) (cyclohexanone)

Scheme 8.52. Oxidation of sulfides to sulfoxides by alumina-supported IBD.

8.2.4.9 Oxidation of Arenes and Enamines – Potassium Permanganate–Alumina

KMnO$_4$-impregnated alumina oxidizes arenes to ketones within 10–30 min under solvent-free conditions under the action of focused microwaves [131]. β,β-Disubstituted enamines have been successfully oxidized to carbonyl compounds with KMnO$_4$–Al$_2$O$_3$ in domestic (255 W, 82 °C) and focused (330 W, 140 °C) microwave ovens, under solvent-free conditions, by Hamelin et al. [132]. The yields are better if focused ovens are used. When the same reactions are conducted in an oil bath at 140 °C, no carbonyl compound formation is observed (Scheme 8.53).

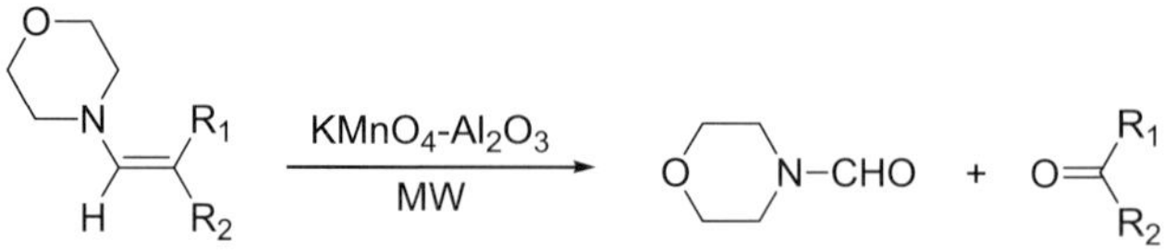

$$\text{(morpholino-enamine)} \xrightarrow[\text{MW}]{\text{KMnO}_4\text{-Al}_2\text{O}_3} \text{O=N-CHO (morpholine-formamide)} + O=C(R_1)(R_2)$$

Scheme 8.53. Oxidation of enamines with alumina-supported potassium permanganate.

Interestingly, 2,5-disubstituted 1,3,4-oxadiazoles have been synthesized by oxidation of 1-aroyl-2-arylidene hydrazines with potassium permanganate on silica, alumina, or montmorillonite K10 clay surfaces under MW irradiation conditions [133]. K10 clay seemed to be the most appropriate support.

8.2.4.10 Oxidation Using [Hydroxyl(tosyloxy)iodo]benzene (HTIB)

The oxidation of benzylic alcohols to carbonyl compounds is a fundamental transformation in organic synthesis. An efficient method for oxidation of benzylic alcohols with HTIB, Koser's reagent, under solvent-free MW irradiation conditions has been described by Lee et al. (Scheme 8.54) [134]. The salient environmentally friendlier features of the solvent-free reaction are rapid reaction kinetics, experimental simplicity, and higher product yields, although the purification process still requires traditional flash column chromatography. Several biologically useful α-ketoesters have been synthesized in high yields from the corresponding α-hydroxyl esters.

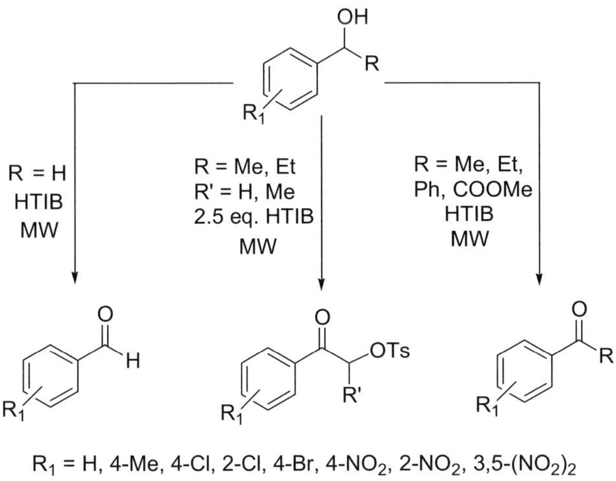

R_1 = H, 4-Me, 4-Cl, 2-Cl, 4-Br, 4-NO_2, 2-NO_2, 3,5-$(NO_2)_2$

Scheme 8.54. Efficient oxidations of benzylic alcohol with HTIB.

8.2.4.11 Other Oxidation Reactions

A rapid self-coupling reaction of β-naphthols occurs in presence of iron(III) chloride, $FeCl_3 \cdot 6H_2O$, using a focused MW oven under solvent-free conditions, and is far superior to classical heating mode [135].

Further examples of oxidative procedures with microwave irradiation include the oxidation of benzylic bromides to the corresponding aldehydes with pyridine N-oxides (15–95%) [136]. The catalyst system, V_2O_5/TiO_2 and forms of this catalyst modified by addition of an effective MW coupling dielectric, for example MoO_3, WO_3, Nb_2O_5, or Ta_2O_5, have been explored for selective oxidation of toluene to benzoic acid under the action of MW irradiation. In the conventional heating process the additives have no positive effect on catalyst performance [137].

8.2.5
Reduction Reactions

In the domain of MW-assisted chemistry, reduction reactions were the last to appear on the scene. Use of ammonium formate and catalytic transfer hydrogenation were the initial examples [23b].

8.2.5.1 Reduction of Carbonyl Compounds with Aluminum Alkoxides

The pioneering work of Posner on the reduction of carbonyl compounds with isopropyl alcohol and alumina [138] has now been adapted to give an expeditious solvent-free reduction procedure that utilizes aluminum alkoxides under MW irradiation conditions (Scheme 8.55) [139].

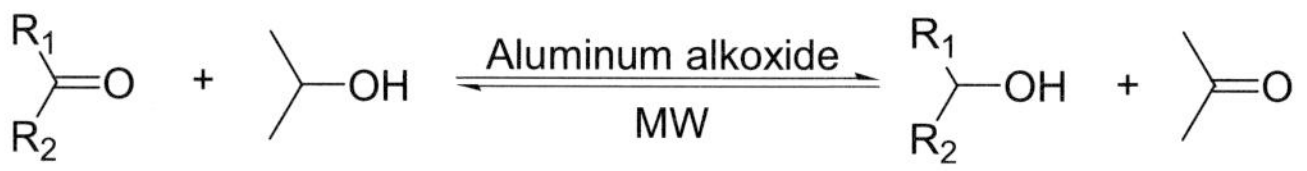

$$\underset{R_2}{\overset{R_1}{>}}\!\!=\!\!O \;+\; \rangle\!\!-\!OH \;\underset{MW}{\overset{\text{Aluminum alkoxide}}{\rightleftarrows}}\; \underset{R_2}{\overset{R_1}{>}}\!\!-\!OH \;+\; \rangle\!\!=\!\!O$$

Scheme 8.55. Solvent-free reduction of carbonyl compounds using aluminum alkoxides.

8.2.5.2 Reduction of Carbonyl Compounds to Alcohols – Sodium Borohydride–Alumina

Inexpensive and safe sodium borohydride ($NaBH_4$) has been extensively used as a reducing agent because of its compatibility with protic solvents. Solid-state reduction of carbonyl compounds has been achieved by mixing substrates with $NaBH_4$ and storing the reaction mixture in a dry box for five days. The disadvantage of heterogeneous reduction with $NaBH_4$ is that the solvent slows the reaction rate whereas in the solid-state reactions the required time (5 days) is too long for it to be of any practical utility [140].

Bram and Loupy developed efficient and selective solid-supported reducing agents, alkaline borohydride on "Fontainebleau sand" (a pure nonhydrated silica gel) or alumina [141] for regioselective reduction of carbonyl compounds including α,β-ethylenic ketones under the action of conventional heating. Varma and coworkers reported, for the first time, however, a simple method for expeditious reduction of aldehydes and ketones that uses alumina-supported $NaBH_4$ and proceeds in the solid state using microwaves [142]. The process, in its entirety, involves the mixing of carbonyl compounds with (10%) $NaBH_4$–alumina and irradiating the reaction mixture in a household MW oven for 0.5–2 min (Scheme 8.56).

The useful chemoselective feature of the reaction is apparent from the reduction of *trans*-cinnamaldehyde (cinnamaldehyde/$NaBH_4$–alumina, 1:1 mol equiv.); the olefinic moiety remains intact and only the aldehyde functionality is reduced, in a facile reaction.

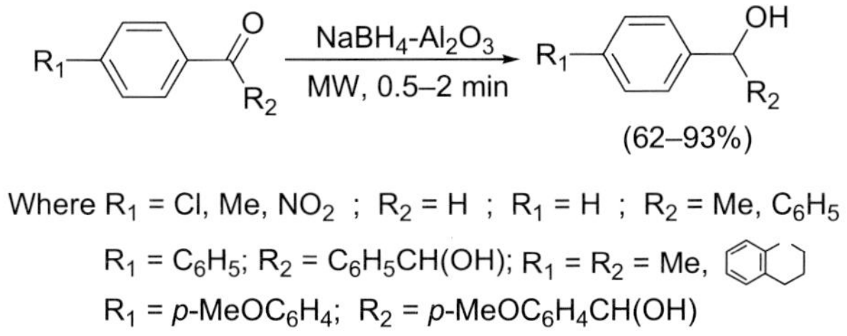

8.2 Microwave-accelerated Solvent-free Organic Reactions

$$R_1\text{-}C_6H_4\text{-}C(=O)R_2 \xrightarrow[\text{MW, 0.5–2 min}]{\text{NaBH}_4\text{-Al}_2O_3} R_1\text{-}C_6H_4\text{-}CH(OH)R_2$$

(62–93%)

Where R_1 = Cl, Me, NO_2 ; R_2 = H ; R_1 = H ; R_2 = Me, C_6H_5
R_1 = C_6H_5; R_2 = $C_6H_5CH(OH)$; R_1 = R_2 = Me, (naphthyl)
R_1 = p-MeOC$_6$H$_4$; R_2 = p-MeOC$_6$H$_4$CH(OH)

Scheme 8.56. Reduction of carbonyl compounds using alumina-supported NaBH$_4$.

No side-products are formed and the reaction does not proceed in the absence of alumina. Further, the reaction rate improves in the presence of moisture. The moisture is absorbed by alumina during recovery of the product. The alumina support can be recycled and reused for subsequent reduction, repeatedly, by mixing with fresh borohydride without loss of activity. In terms of safety, the air used for cooling the magnetron ventilates the microwave cavity, thus preventing any ensuing hydrogen from reaching explosive concentrations. The technique has been elegantly used for MW-enhanced solid-state deuterations using sodium borodeuteride-impregnated alumina [143]. Further extension of this work to the specific labeling of molecules has been explored [144] and is discussed elsewhere in this book [145].

8.2.5.3 Reductive Amination of Carbonyl Compounds

Sodium cyanoborohydride [146], sodium triacetoxyborohydride [147], or NaBH$_4$ coupled with sulfuric acid [148] are common agents used for reductive amination of carbonyl compounds. These reagents either generate waste or involve the use of corrosive acids. The environmentally friendlier procedures developed by Varma and coworkers have been extended to a solvent-free reductive amination procedure for carbonyl compounds, using moist montmorillonite K10 clay-supported sodium borohydride, that is facilitated by microwave irradiation (Scheme 8.57) [149].

Some practical applications of NaBH$_4$ reductions on mineral surfaces of *in situ*-generated Schiff's bases have been successfully demonstrated. The solid-state re-

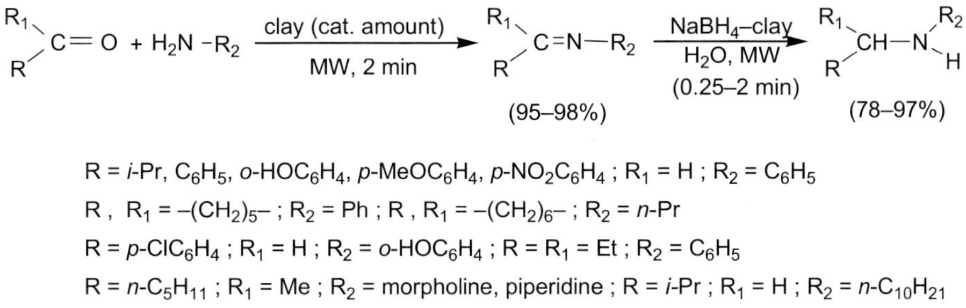

$$R\text{-}C(R_1)\text{=}O + H_2N\text{-}R_2 \xrightarrow[\text{MW, 2 min}]{\text{clay (cat. amount)}} R\text{-}C(R_1)\text{=}N\text{-}R_2 \xrightarrow[\text{H}_2\text{O, MW (0.25–2 min)}]{\text{NaBH}_4\text{-clay}} R\text{-}CH(R_1)\text{-}N(R_2)H$$

(95–98%) (78–97%)

R = i-Pr, C$_6$H$_5$, o-HOC$_6$H$_4$, p-MeOC$_6$H$_4$, p-NO$_2$C$_6$H$_4$; R_1 = H ; R_2 = C$_6$H$_5$
R , R_1 = –(CH$_2$)$_5$– ; R_2 = Ph ; R , R_1 = –(CH$_2$)$_6$– ; R_2 = n-Pr
R = p-ClC$_6$H$_4$; R_1 = H ; R_2 = o-HOC$_6$H$_4$; R = R_1 = Et ; R_2 = C$_6$H$_5$
R = n-C$_5$H$_{11}$; R_1 = Me ; R_2 = morpholine, piperidine ; R = i-Pr ; R_1 = H ; R_2 = n-C$_{10}$H$_{21}$

Scheme 8.57. Reductive amination of carbonyl compounds using clay-supported NaBH$_4$.

ductive amination of carbonyl compounds on inorganic solid supports such as alumina, clay, and silica, etc., and especially on the surface of K10 clay, rapidly afford secondary and tertiary amines [149]. Clay behaves as a Brönsted acid and also provides water from its interlayers thus enhancing the reducing properties of $NaBH_4$ [22].

$NaBH_4$ supported on alumina under solvent-free conditions has been used for hydrochalcogenation of methyl propiolate derivatives with phenylchalcogenolate anion generated *in situ* from the respective diphenyl dichalcogenide (S, Se, Te) [150]. This improved method is general and furnishes the (Z)-β-phenylchalcogeno-α,β-unsaturated esters in good yields and with selectivity comparable with procedures using organic solvents under an inert atmosphere (Scheme 8.58).

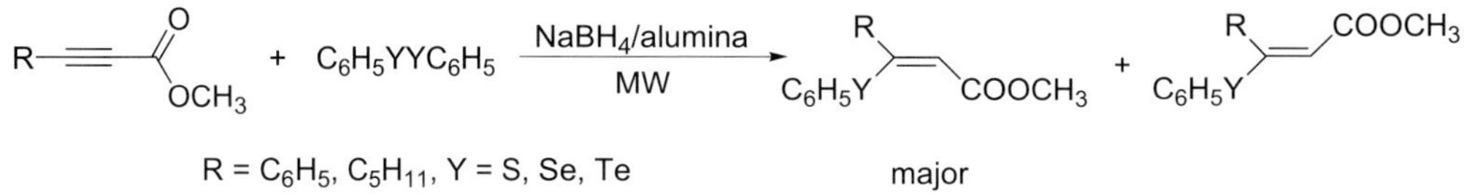

$$R-\!\!\equiv\!\!-COOCH_3 \;+\; C_6H_5YYC_6H_5 \;\xrightarrow[MW]{NaBH_4/alumina}\; \underset{C_6H_5Y}{\overset{R}{\diagdown}}\!\!=\!\!\underset{COOCH_3}{\overset{}{\diagup}} \;+\; \underset{C_6H_5Y}{\overset{R}{\diagdown}}\!\!=\!\!\underset{}{\overset{COOCH_3}{\diagup}}$$

R = C_6H_5, C_5H_{11}, Y = S, Se, Te   major

Scheme 8.58. MW-assisted hydrochalcogenation of methyl propiolate derivatives.

8.2.5.4 Solid-state Cannizzaro Reaction

The Cannizzaro reaction, the disproportionation of an aldehyde to an equimolar mixture of primary alcohol and carboxylic salt [151, 152], is restricted to aldehydes without an α hydrogen atom and which cannot therefore undergo aldol condensation. This oxidation–reduction reaction is usually conducted under strongly basic conditions and suffers from the disadvantages of low yields of the desired products and extended reaction times [153, 154]. Use of the crossed-Cannizzaro reaction [153], using a scavenger and inexpensive paraformaldehyde, has, however, afforded improved yields of alcohols before the introduction of hydride reducing agents. The popularity of the Cannizzaro reaction in synthetic organic chemistry dropped substantially after the discovery of lithium aluminum hydride, $LiAlH_4$, in 1946.

In numerous reactions involving arylaldehydes on a variety of mineral oxide surfaces, formation of Cannizzaro-derived alcoholic contaminants has been consistently observed [155]. Interestingly, no useful product is formed under solvent-free MW irradiation conditions on an alumina surface with calcium hydroxide or in the presence of a strong base such as sodium hydroxide. The reaction remains incomplete with concomitant formation of several unidentified products, reminiscent of earlier described observations on basic alumina surface [155]. Finally, Varma et al. discovered that the reaction proceeds rapidly on a barium hydroxide, $Ba(OH)_2 \cdot 8H_2O$, surface, which constitutes the first application of this reagent in a solvent-free crossed-Cannizzaro reaction [156]. In a typical experimental procedure, a mixture of benzaldehyde (1 mmol) and paraformaldehyde (2 mmol) is mixed with barium hydroxide octahydrate (2 mmol) and irradiated in an MW oven (100–110 °C) or heated in an oil bath (100–110 °C) (Scheme 8.59). Arylaldehydes bearing an electron-withdrawing substituent undergo reaction at a much

$$RCHO + (CH_2O)_n \xrightarrow[\text{MW or oil bath}]{\text{Ba(OH)}_2 \cdot 8H_2O} RCH_2OH + RCOOH$$
$$\phantom{RCHO + (CH_2O)_n \xrightarrow[\text{MW or oil bath}]{\text{Ba(OH)}_2 \cdot 8H_2O}} \text{(80-99\%)} \quad \text{(1-20\%)}$$

Scheme 8.59. Solvent-free crossed-Cannizzaro reaction using paraformaldehyde.

faster rate than aldehydes with electron-releasing groups [156]. Soon thereafter, additional research groups started exploring the general utility of this reaction using microwaves [157, 158].

8.2.5.5 Reduction of Aromatic Nitro Compounds to Amines with Alumina-supported Hydrazine

Varma and his coworkers have described a simple and efficient procedure wherein aromatic nitro compounds are readily reduced to the corresponding amino compounds in good yields with hydrazine hydrate supported on alumina in the presence of iron(III) chloride ($FeCl_3 \cdot 6H_2O$), Fe(III) hydroxide, or Fe(III) oxides (Scheme 8.60) [159].

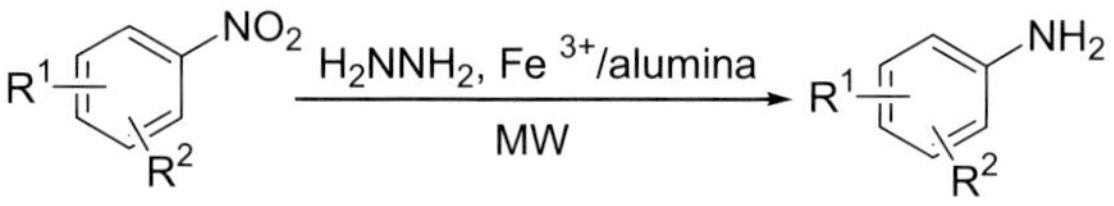

$$R^1\text{-Ar}(R^2)\text{-NO}_2 \xrightarrow[\text{MW}]{H_2NNH_2,\ Fe^{3+}/\text{alumina}} R^1\text{-Ar}(R^2)\text{-NH}_2$$

Scheme 8.60. Reduction of nitro compounds to amines with alumina-supported hydrazine.

8.2.6
Synthesis of Heterocyclic Compounds

Heterocyclic chemistry has benefited substantially from MW-expedited processes developed over the last decade. An exhaustive overview is provided elsewhere in this book [97, 98]; limited solvent-free chemistry utilizing mineral-supported reagents is covered in this section.

8.2.6.1 Flavones

Naturally occurring flavonoids are oxygen heterocyclic compounds widely distributed in the plant kingdom; the most abundant are the flavones. Members of this class have a wide variety of biological activity and have been useful in the treatment of a variety of diseases [160, 161]. Flavones have been prepared by a variety of methods, for example Allan–Robinson synthesis and synthesis from chalcones via an intramolecular Wittig strategy [162]. The commonly followed approach, however, involves the Baker–Venkataraman rearrangement, wherein *o*-hydroxyacetophenone is benzoylated to form the benzoyl ester which is then treated with base (KOH/pyridine) to effect acyl migration, forming a 1,3-diketone

[163, 164]. This diketone is then cyclized under strongly acidic conditions with sulfuric acid and acetic acid to afford a flavone.

A solvent-free synthesis of flavones has been achieved that simply involves the MW irradiation of *o*-hydroxydibenzoylmethanes adsorbed on montmorillonite K10 clay for 1–1.5 min. Rapid and exclusive formation of flavones occurred in good yields (Scheme 8.61) [165]. The intramolecular Michael addition of *o*-hydroxychalcones on a silica gel surface has also been reported [166].

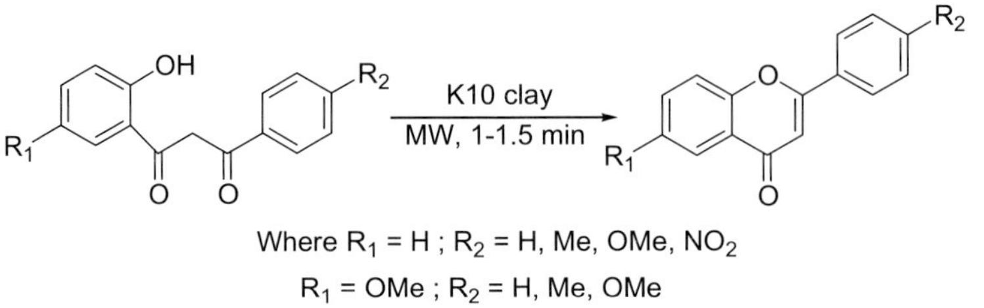
OH
R2
K10 clay
MW, 1-1.5 min
R1
O
O
O
R2
R1
O
Where R1 = H ; R2 = H, Me, OMe, NO2
R1 = OMe ; R2 = H, Me, OMe

Scheme 8.61. Formation of flavones by cyclization of *o*-hydroxydibenzoylmethanes on K10 clay.

8.2.6.2 **2-Amino-substituted Isoflav-3-enes**

The estrogenic properties of isoflav-3-enes are well known and consequently, several derivatives of these chromene heterocycles have been the target of medicinal chemists. Varma and coworkers uncovered a useful enamine-mediated pathway to this class of compound [167–169]. The group also discovered a facile and general method for MW-expedited synthesis of isoflav-3-enes substituted with basic moieties at the 2-position (Scheme 8.62) [170]. These promising results are especially appealing in view of the convergent one-pot approach to 2-substituted isoflav-3-enes wherein *in situ*-generated enamine derivatives have been subsequently reacted with *o*-hydroxyaldehydes in the same pot (Scheme 8.62) [170].

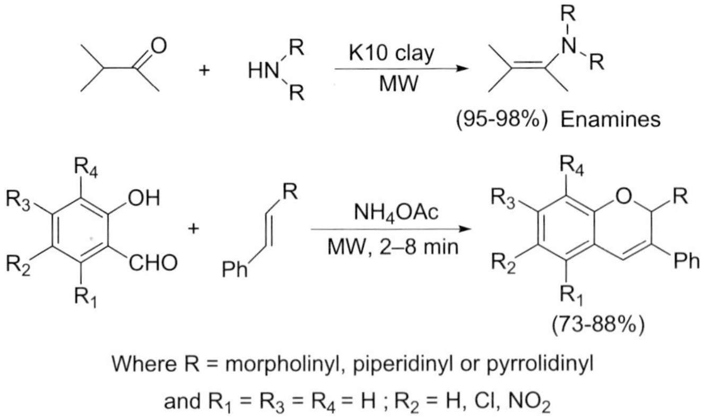
O
+
HN
R
R
K10 clay
MW
R
N
R
(95-98%) Enamines
R4
R3
OH
R2
CHO
R1
+
R
Ph
NH4OAc
MW, 2–8 min
R4
R3
O
R
R2
Ph
R1
(73-88%)
Where R = morpholinyl, piperidinyl or pyrrolidinyl
and R1 = R3 = R4 = H ; R2 = H, Cl, NO2

Scheme 8.62. One-pot synthesis of 2-substituted isoflav-3-enes from *in situ*-generated enamines.

8.2.6.3 Substituted Thiazoles, Benzothiazepines, and Thiiranes

Thiazole and its derivatives are conventionally prepared from lachrymatory α-haloketones and thioureas (or thioamides) by the Hantzsch procedure [171]. In a marked improvement, Varma et al. synthesized the title compounds by simple reaction of α-tosyl oxyketones, generated *in situ* from arylmethyl ketones and [hydroxy(tosyloxy)iodo]benzene (HTIB), with thioamides in the presence of K10 clay using microwave irradiation (Scheme 8.63). The process is solvent-free in both the steps [172].

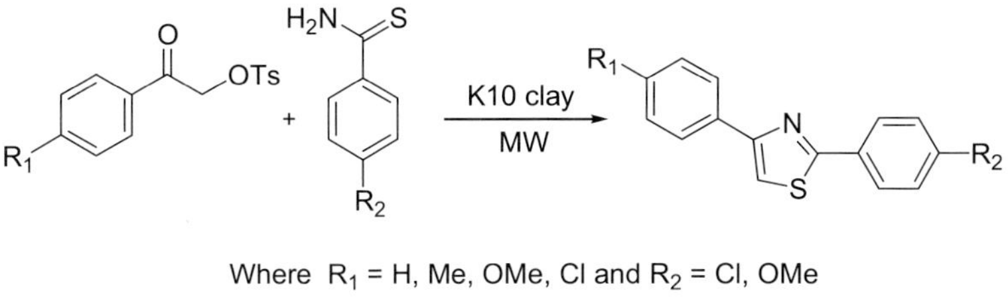

Where R_1 = H, Me, OMe, Cl and R_2 = Cl, OMe

Scheme 8.63. Synthesis of substituted thiazoles from *in situ*-generated α-tosyl oxyketones.

The corresponding bridgehead heterocycles are difficult to obtain by conventional heating methods, because of reaction of α-tosyl oxyketones with ethylenethioureas. The MW-assisted process, on the other hand, is complete within a short time (Scheme 8.64) [172, 173].

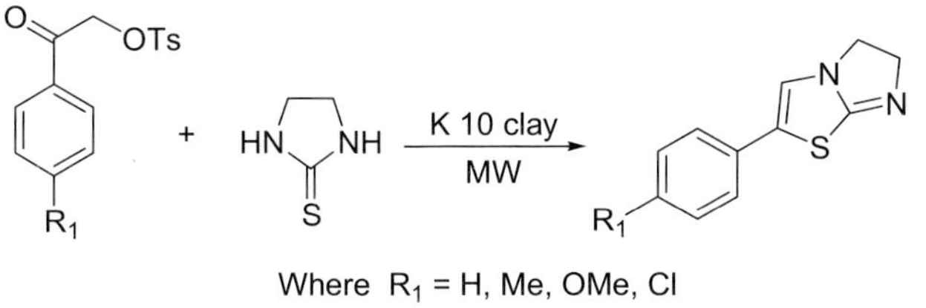

Where R_1 = H, Me, OMe, Cl

Scheme 8.64. Synthesis of bridgehead thiazoles from α-tosyl oxyketones.

Solvent-free cyclization of converting *N*-aryl iminodithiazoles into 2-cyano benzothiazoles under the action of MW has been investigated to develop environmentally friendly procedures for synthesis of heterocyclic molecules for which traditional methods failed or are less attractive [174].

Several 1,4-dicarbonyl compounds have been successfully subjected to MW-mediated ring closure with Lawesson's reagent to afford useful *S*-heterocycle-containing liquid crystalline targets. The new method of ring-closure has proven useful in improving the yield of an earlier synthesized liquid crystal entity and several newer reactions have been scaled up to several grams without compromising the yield (Scheme 8.65) [175].

A solvent-free MW-promoted one-pot synthesis of fluorinated 2,3-dihydro-1,5-benzothiazepines by Michael addition followed by dehydrative cyclization has

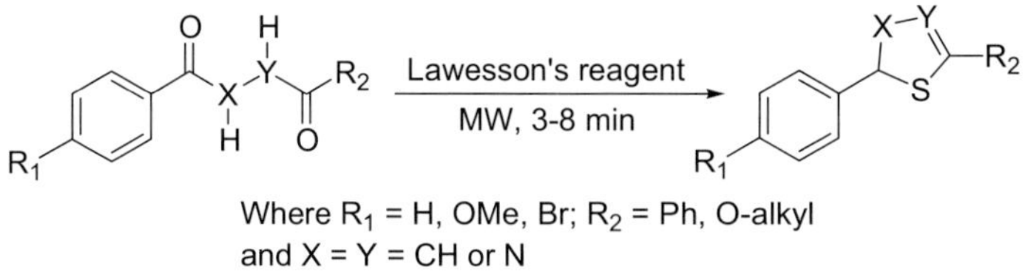

Scheme 8.65. MW-assisted synthesis of S-heterocycle-containing liquid-crystalline compounds.

been described by Dandia et al. [176]. This efficient procedure demonstrates that MW activation combined with recoverable catalysts is superior to conventional conditions in improving reaction yields and simplifying the work-up procedure (Scheme 8.66). The synthesis of more complicated tetracyclic 1,5-benzothiazepines soon followed (Scheme 8.67) [177].

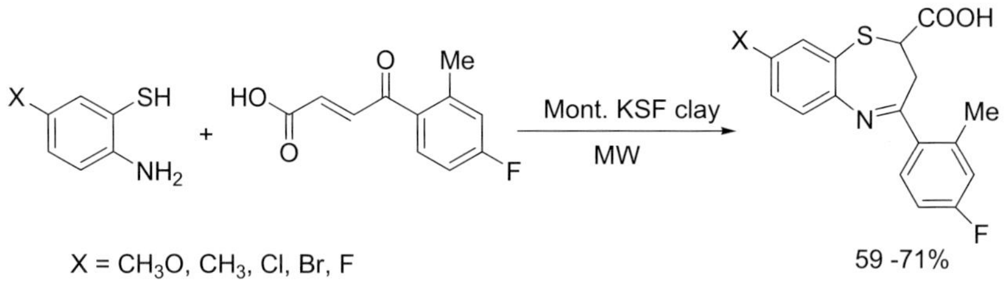

Scheme 8.66. MW-assisted synthesis of dihydrobenzothiazepines.

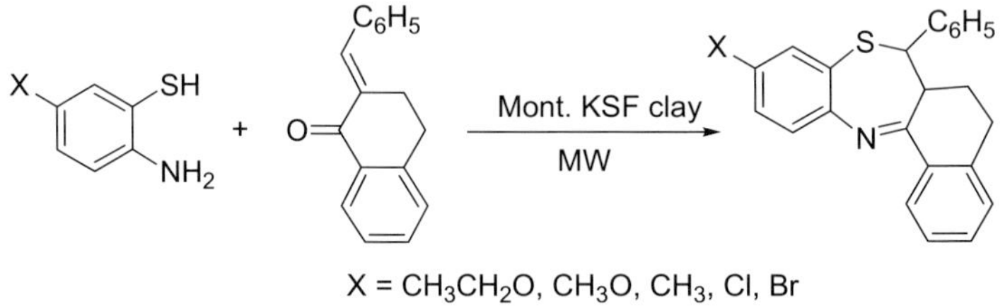

Scheme 8.67. MW-assisted tetracyclic benzothiazepines synthesis.

A simple method has been developed for synthesis of thiiranes from epoxides via a one-pot reaction of epoxides with diethylphosphite in the presence of ammonium acetate or ammonium hydrogen carbonate, sulfur, and acidic alumina under solvent-free conditions using microwave irradiation (Scheme 8.68) [178].

8.2.6.4 Synthesis of 2-Aroylbenzofurans

Pharmacologically important natural 2-aroylbenzofurans are easily obtainable under basic solvent-free conditions from α-tosyl oxyketones and salicylaldehydes

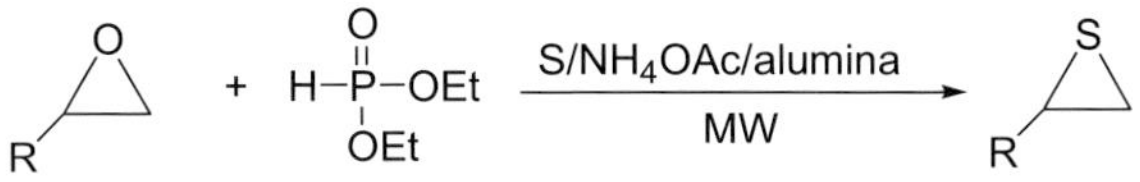

Scheme 8.68. Synthesis of thiiranes from epoxides.

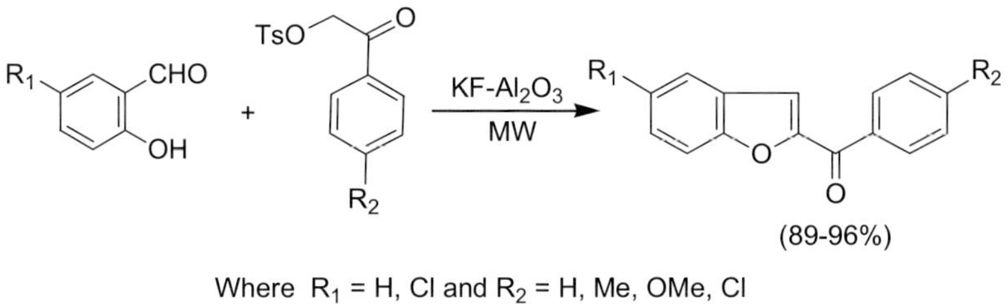

Scheme 8.69. Synthesis of 2-aroylbenzofurans on potassium fluoride-doped alumina.

in the presence of potassium fluoride-doped alumina using microwave irradiation (Scheme 8.69) [172, 173].

8.2.6.5 Synthesis of Quinolones and other Nitrogen Heterocycles

2'-Aminochalcones, which are readily available, provide easy access to 2-aryl-1,2,3,4-tetrahydro-4-quinolones in yet another solvent-free cyclization reaction (internal Michael addition) using K10 clay under microwave irradiation conditions [179]; the products are valuable precursors for medicinally important quinolones (Scheme 8.70).

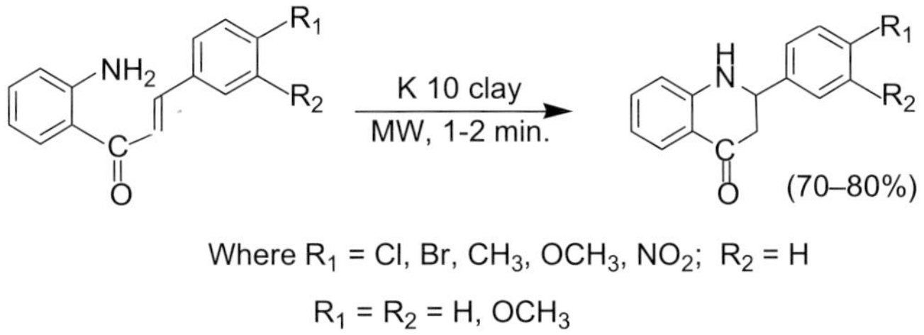

Scheme 8.70. Synthesis of 2-aryl-1,2,3,4-tetrahydro-4-quinolones by cyclization on clay.

A one-pot MW-accelerated synthesis of selective glycine receptor antagonists 3-aryl-4-hydroxyquinolin-2(1H)-ones has been developed via amidation of malonic ester derivatives with anilines and subsequent cyclization of the intermediate, malondianilides under solvent-free conditions (Scheme 8.71) [180].

Reaction of anthranilic acid derivatives with formamide on silica gel, acidic alumina, or montmorillonite K10 under the action of MW irradiation generates quinazolines in good yields [181]. Another one-step MW method for preparation of sub-

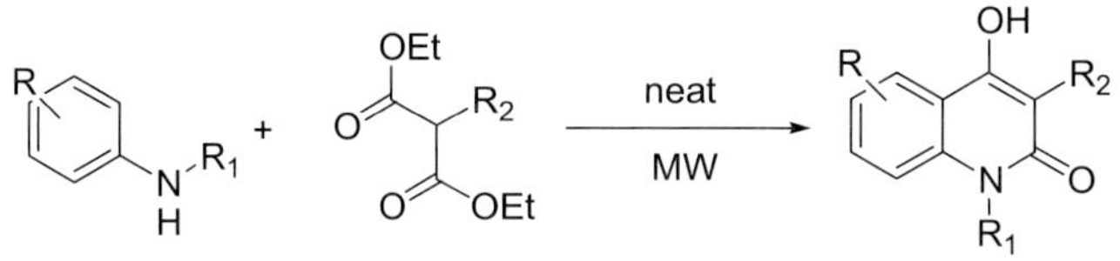

Scheme 8.71. MW-assisted synthesis of 3-aryl-4-hydroxyquinolin-2(1H)-ones.

stituted 7,7′-bis-indolizines, based on 1,3-dipolar cycloaddition of 4,4′-bipyridinium ylides, generated *in situ* with activated alkynes, uses KF/alumina under solvent-free conditions [182].

Fischer indolization, a general method for synthesis of 2-heteroaryl-5-methoxyindoles, has been reported to proceed under controlled microwave conditions on a solid support (Scheme 8.72). This route uses 2-acetylpyridine as a model and 3-acetyl derivatives of cycloalkeno[c]fused pyridines as the synthetic building blocks to assemble 5-methoxy-2-(2-pyridyl)indoles, a key step in the total synthesis of new 9-methoxyindolo[2,3-a]quinolizine alkaloids [183].

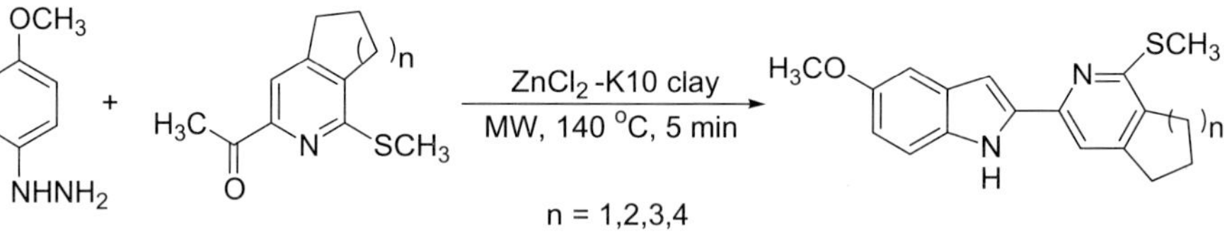

Scheme 8.72. MW-assisted Fisher indole synthesis on clay.

The synthesis of 2,4,5-triarylimidazoline derivatives from aldehydes and alumina-supported ammonium acetate under solvent-free conditions using MW irradiation has been explored by Kaboudin and Saadati [184]. A similar solid-supported route from benzoin, aldehydes, and ammonium acetate, possibly involving air as an oxidant, had been reported earlier (Scheme 8.73) [185].

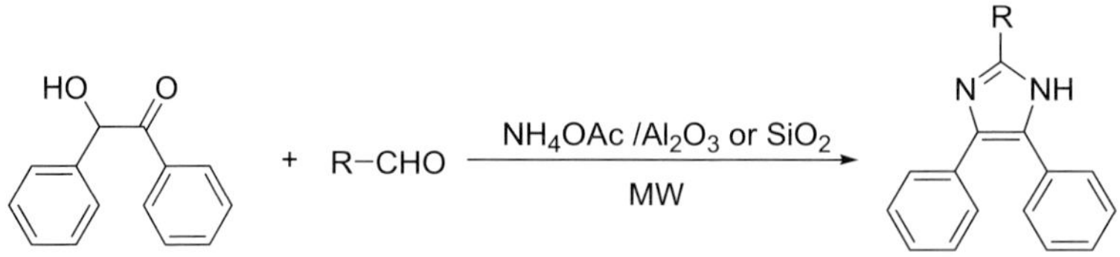

Scheme 8.73. MW synthesis of 2,4,5-substituted imidazolines.

Collman and Decréau have developed a modified solvent-free approach for preparation of new free base tris-aryl and tris-pyrimidyl corroles using MW irradiation [186]. Compared with conventional heating, the MW technique afforded higher yields and led to noticeably cleaner reaction products (Scheme 8.74).

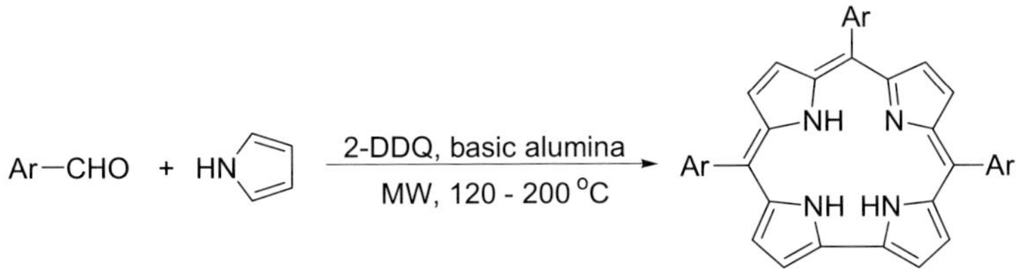

Scheme 8.74. Microwave-assisted synthesis of corroles on basic alumina.

8.2.6.6 Solvent-free Assembly of Pyrido Fused-ring Systems

Microwave-assisted solvent-free synthesis of pharmacologically important pyrido fused-ring systems has recently been accomplished and is an improved method for assembly of a variety of pyridopyridazine and quinoline derivatives. Benz-1,3-oxazine formation has also been investigated in dry media using Al_2O_3–KF as a solid base (Scheme 8.75) [187].

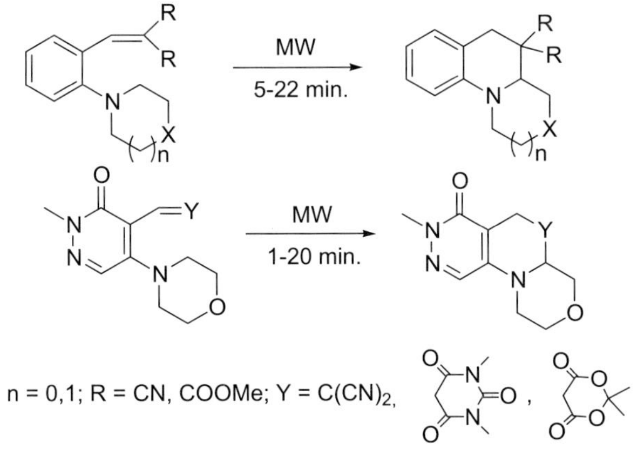

n = 0,1; R = CN, COOMe; Y = C(CN)$_2$,

Scheme 8.75. MW-assisted synthesis of pyrido fused-ring systems.

8.2.6.7 Synthesis of Uracils

Solvent-free condensation of malonic acid or cyanoacetic acid and ureas in the presence of acetic anhydride under MW irradiation conditions affords functionalized uracil derivatives such as 6-hydroxy or 6-amino uracils in high to excellent yields (Scheme 8.76) [188].

8.2.6.8 MW Synthesis of Benzoxazinones

A concise preparation of pharmaceutically useful benzoxazin-2-ones in a few steps from salicylaldehydes and 2-hydroxyacetophenone, on montmorillonite K10 clay and alumina-supported copper sulfate, under solvent-free MW irradiation conditions, has been reported [189]. This high-yielding, expeditious, and eco-friendly

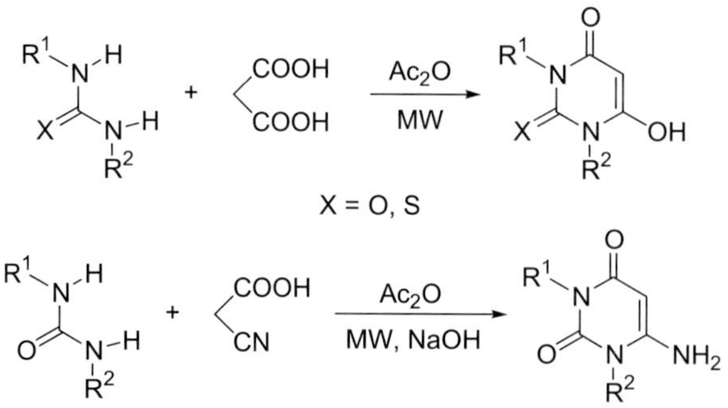

Scheme 8.76. MW-assisted synthesis of uracils.

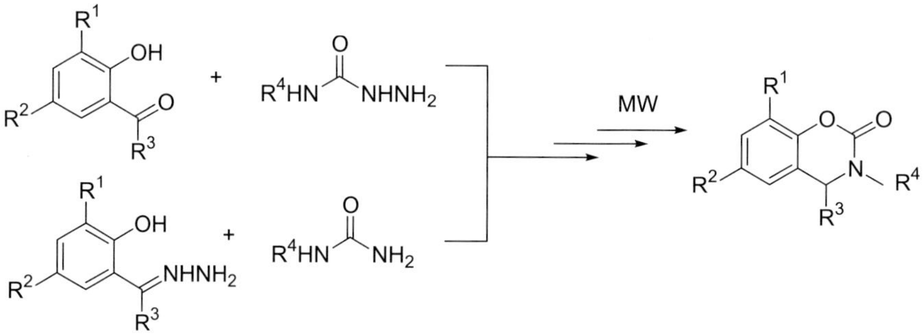

Scheme 8.77. MW-assisted syntheses of benzoxazin-2-ones.

process that leads to synthetically manipulable products may find application in the generation of libraries of this class of compound (Scheme 8.77).

8.2.6.9 Multi-component Reactions

The multiple-component condensation (MCC) approach is attracting attention because diversity can be conveniently achieved in a single step simply by varying the reacting components. The generation of small-molecule libraries requires the development of efficient procedures with special emphasis on the ease of reaction manipulation. This approach is reviewed in detail in Chapter 17. Varma and Kumar have developed such a facile procedure which is amenable to the generation of libraries of imidazo[1,2-a]pyridines, imidazo[1,2-a]pyrazines and imidazo[1,2-a]pyrimidines under solvent-free conditions using MW irradiation (Scheme 8.78) [190]. This is a marked improvement on the conventional two-component synthesis that requires lachrymatory α-haloketones and restricts the generation of a diverse library of these molecules.

Aldehydes and the corresponding 2-aminopyridine, pyrazine, or pyrimidine are mixed in the presence of a catalytic amount of clay (50 mg) to generate iminium

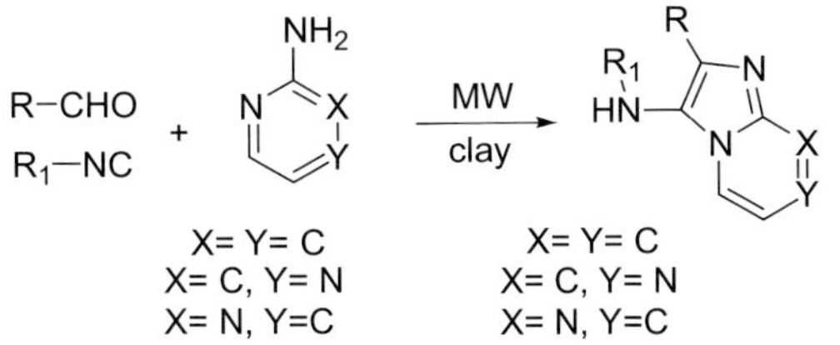

Scheme 8.78. Synthesis of imidazo[1,2-a]annulated N-heterocycles by the Ugi reaction.

intermediate. Isocyanides are subsequently added to the same container and the reactants are further exposed to MW to afford the corresponding imidazo[1,2-a]pyridines, imidazo[1,2-a]pyrazines, and imidazo[1,2-a]pyrimidines (Scheme 8.78). The process is general and convenient for all the three reaction components, e.g. aldehydes (aliphatic, aromatic, and vinylic), isocyanides (aliphatic, aromatic, and cyclic), and amines (2-aminopyridine, 2-aminopyrazine, and 2-aminopyrimidine). A library of imidazo[1,2-a]pyridines, imidazo[1,2-a]pyrazines, and imidazo[1,2-a]pyrimidines can be readily obtained by varying the three components [190].

Other workers have described convenient syntheses of highly substituted pyrroles (60–72%) on silica gel using readily available α,β-unsaturated carbonyl compounds, amines, and nitroalkanes under the action of MW irradiation [191]. The neat reactants have been used under solvent-free conditions to generate Biginelli and Hantzsch reaction products with enhanced yields and shortened reaction times (Scheme 8.79) [192].

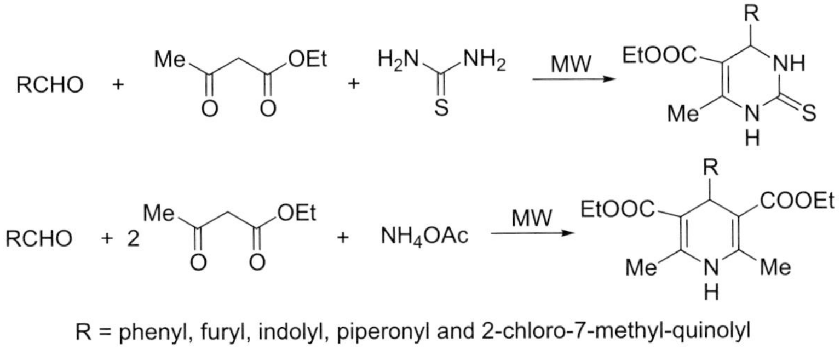

R = phenyl, furyl, indolyl, piperonyl and 2-chloro-7-methyl-quinolyl

Scheme 8.79. MW-assisted one-pot synthesis of N-heterocycles.

Several three-component condensations, catalyzed by iodine–alumina, for the synthesis of 3,4-dihydropyrimidin-2(1H)-ones under solvent-free conditions have been reported. The reaction proceeds via condensation of an aldehyde, ethyl acetoacetate, and a urea or thiourea under MW irradiation conditions in the presence of 10% iodine adsorbed on neutral alumina to afford substituted 3,4-dihydropyrimidin-2(1H)-ones in excellent yields (Scheme 8.80) [193].

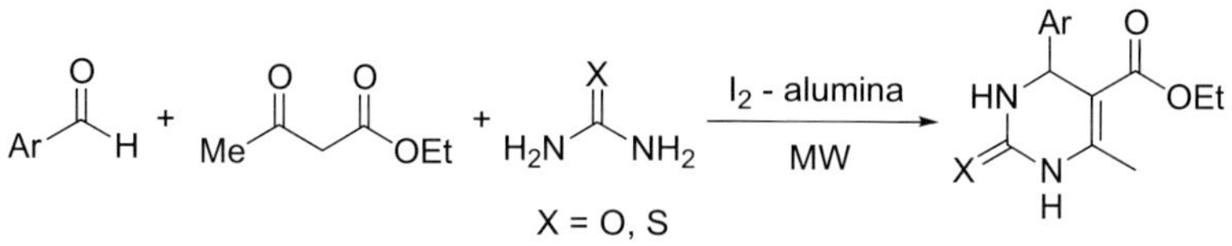

Scheme 8.80. MW synthesis of 3,4-dihydropyrimidin-2(1H)-ones.

A three-component expeditious synthesis of 3,6-diaryl-5-mercaptoperhydro-2-thioxo-1,3-thiazin-5-ones from 2-methyl-2-phenyl-1,3-oxathiolan-5-one, an aromatic aldehyde, and an N-aryldithiocarbamic acid has recently appeared [194]. The synthesis is diastereoselective and involves tandem Knoevenagel, Michael, and ring transformation reactions occurring under solvent-free MW irradiation conditions in a single pot (Scheme 8.81).

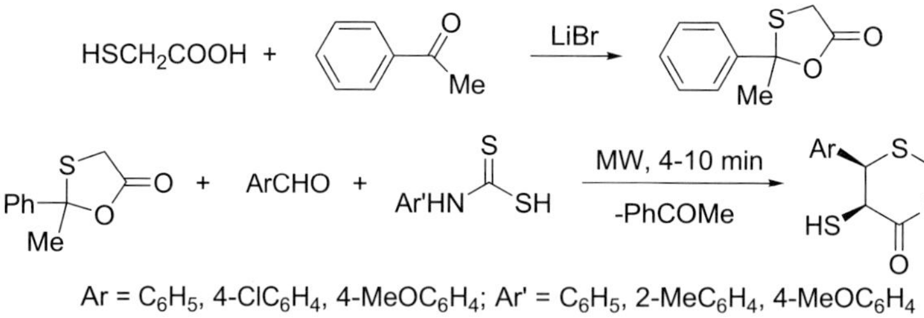

Ar = C_6H_5, 4-ClC_6H_4, 4-$MeOC_6H_4$; Ar' = C_6H_5, 2-MeC_6H_4, 4-$MeOC_6H_4$

Scheme 8.81. MW-assisted solvent-free synthesis of 1,3-thiazines.

A multicomponent reaction has been described for assembly of quinoxaline derivatives using montmorillonite K10 clay as catalyst under MW irradiation conditions (Scheme 8.82) [195].

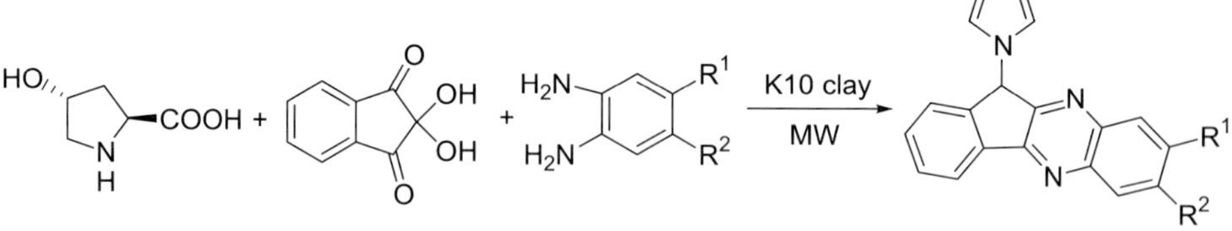

Scheme 8.82. Three-component coupling approach to quinoxaline derivatives on clay.

A similar strategy was originally used for the Biginelli condensation reaction to synthesize a set of pyrimidinones (65–95%) in a household MW oven [196], an approach that has been successfully applied to combinatorial synthesis [197]. Yet another example is the convenient synthesis of pyrroles (60–72%) on silica gel using readily available enones, amines, and nitro compounds [198]. Three-component condensation of an aldehyde, urea or thiourea, and a dicarbonyl com-

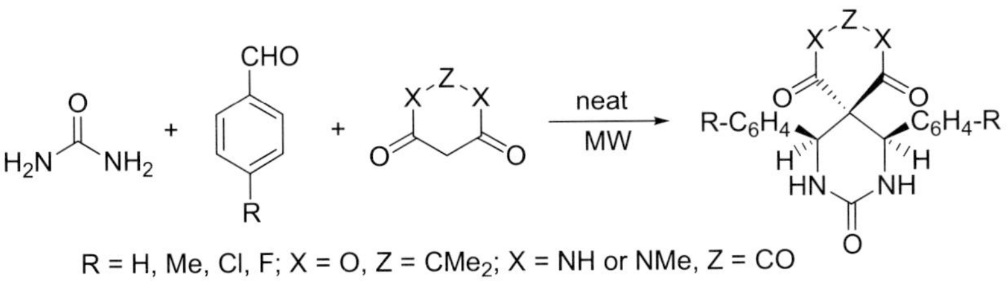

Scheme 8.83. Solvent-free synthesis of spiro-fused heterocyles.

pound generating 3,4-dihydropyrimidin-2(1H)-one under solvent-free conditions, expeditiously generating spiro-fused heterocycles in higher yields and with a simplified purification step, has recently been demonstrated (Scheme 8.83) [199].

8.2.7
Miscellaneous Reactions

8.2.7.1 Transformation of Arylaldehydes to Nitriles

Arylaldehyde to nitrile conversion is an important chemical transformation [200]. The reaction usually proceeds via an aldoxime intermediate that is subsequently dehydrated using a wide variety of reagents, for example chloramine/base [201], O,N-bis-(trifluoroacetyl) hydroxylamine or trifluoroacetohydroximic acid [202], p-chlorophenyl chlorothionoformate/pyridine [203], triethylamine/dialkyl hydrogen phosphinates [204], $TiCl_4$/pyridine [205], triethylamine/phosphonitrilic chloride [206], and 1,1′-dicarbonylbiimidazole [207]. The dehydration of aldoxime is a time-consuming process even for one-pot reactions [208]. The application of hydroxylamine hydrochloride-impregnated clay developed by Varma et al. reduces the entire operation to a one-pot synthesis using microwaves wherein arylaldehydes are rapidly converted into nitriles in good yields (89–95%) in the absence of solvent [127, 209]. In this general reaction, a variety of aldehydes undergo this facile conversion to the corresponding nitriles in a short time (1–1.5 min) on MW irradiation (Scheme 8.84). With aliphatic aldehydes, however, only poor yields of nitriles (10–15%) are obtained with concomitant formation of undesirable products.

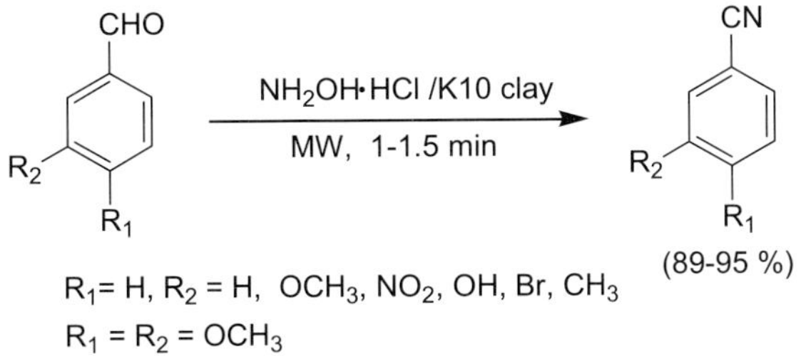

Scheme 8.84. Transformation of arylaldehydes to nitriles by use of hydroxylamine hydrochloride–clay.

Several variations of the aforementioned transformation, on silica surfaces, soon followed [210–212]. The nitriles have also been obtained from carboxylic acids (20–93%) on alumina support [213] and from the corresponding amides (80–95%) in toluene [214].

8.2.7.2 Nitration of Styrenes – Preparation of β-Nitrostyrenes

Solid-state synthesis of β-nitrostyrenes has been reported by Varma et al. in a process that uses readily available styrene and its substituted derivatives and, Clayfen and Clayan, inexpensive clay-supported nitrate salts (Scheme 8.85) [215]. In a simple experiment a mixture of styrene with Clayfen or Clayan is irradiated in a MW oven (~100–110 °C, 3 min) or heated in an oil bath (~100–110 °C, 15 min). With Clayan, intermittent heating is recommended with 30-s intervals to maintain the temperature below 60–70 °C. Remarkably, the reaction proceeds only in solid state; it results in the formation of polymeric products in organic solvents.

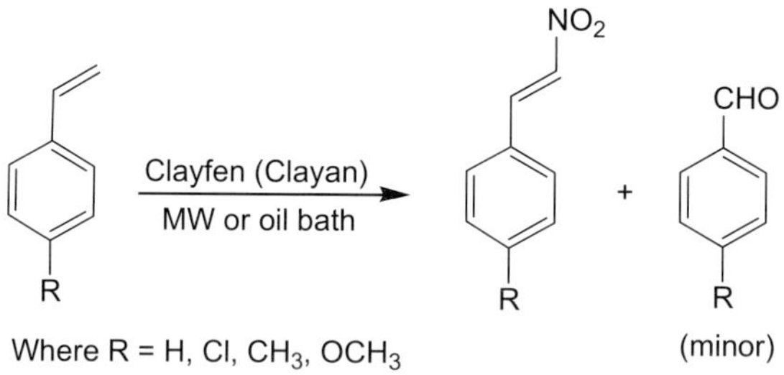

Where R = H, Cl, CH_3, OCH_3

Scheme 8.85. Preparation of β-nitrostyrenes using clay-supported nitrate salts.

8.2.7.3 Bromination of Alkanones Using Microwaves

The synthesis of α-bromo and α,α-dibromoalkanones using dioxane–dibromide and silica gel under solvent-free MW irradiation conditions has been developed (Scheme 8.86) [216].

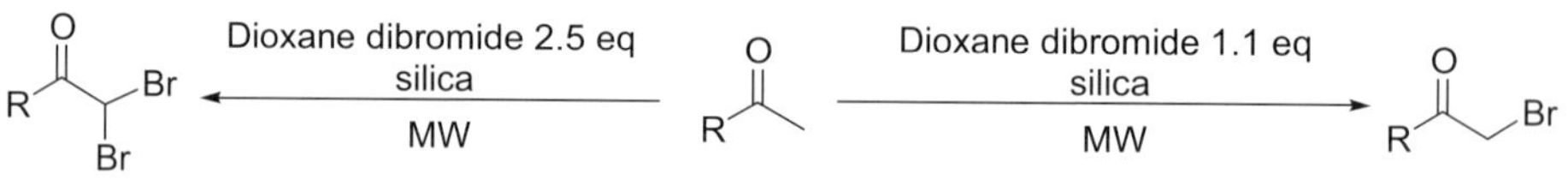

Scheme 8.86. Solvent-free synthesis of bromoalkanones.

8.2.7.4 MW-assisted Elimination Reactions

Solvent-free dehydration of hydroxypyrrolidines to pyrrolines under MW conditions has been reported by Collina et al. (Scheme 8.87) [217]. This high-yielding dehydration procedure retains the configuration of the adjacent carbon without racemization.

Microwave-assisted elimination of trans-4-(4-fluorophenyl)-3-chloromethyl-1-

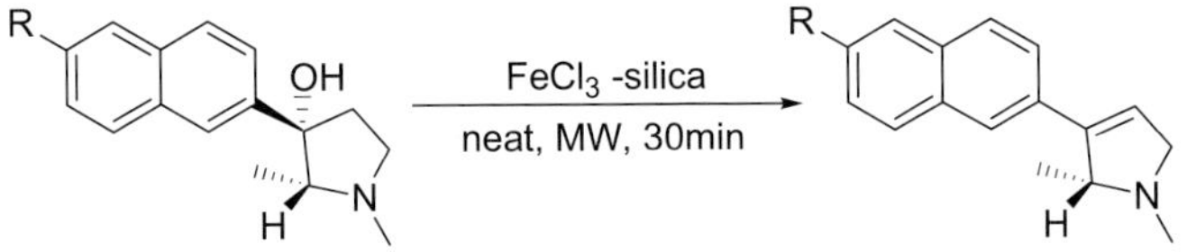

Scheme 8.87. Solvent-free dehydration of hydroxypyrrolidines to pyrrolines.

methylpiperidine on KF–alumina under solvent-free conditions has been reported to afford higher yields of 4-(4-fluorophenyl)-3-methylene-1-methylpiperidine (65.5–71%) in considerably shorter reaction times (20–40 min) than conventional heating [218].

8.2.7.5 Organometallic Reactions (Carbon–Carbon Bond-forming Reactions)

Palladium catalyzed reaction of aryl halides with olefins is a useful synthetic method for C–C bond formation [219, 220]. The most commonly used catalyst is palladium acetate, although other palladium complexes have also been used. Solvent-free Heck reactions with excellent yields have been performed in a household MW oven with palladium acetate as catalyst and triethylamine as base (Scheme 8.88) [221]. A comparison study revealed that the longer reaction times and deployment of high pressures, typical of the classical heating method, are avoided using this MW procedure.

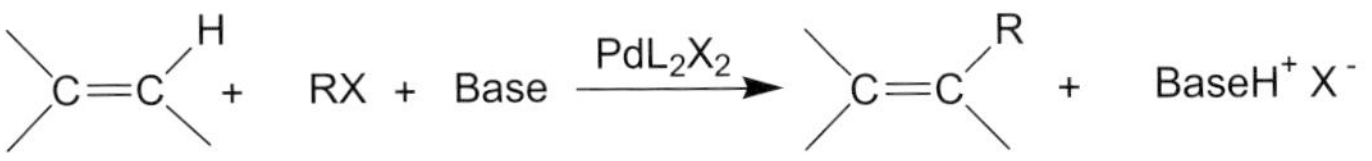

Scheme 8.88. Palladium-catalyzed carbon–carbon bond-forming reactions using microwaves.

A rapid MW-assisted palladium-catalyzed coupling of heteroaryl and aryl boronic acids with iodo and bromo-substituted benzoic acids, anchored on TentaGel has been achieved [222]. An environmentally friendly Suzuki cross-coupling reaction has been developed that uses polyethylene glycol (PEG) as the reaction medium and palladium chloride as a catalyst [223]. A solvent-free Suzuki coupling has also been reported on palladium-doped alumina in the presence of potassium fluoride as base [224]. This approach has been extended to the Sonogashira coupling reaction wherein terminal alkynes couple readily with aryl or alkenyl iodides on palladium-doped alumina in the presence of triphenylphosphine and cuprous iodide (Scheme 8.89) [225].

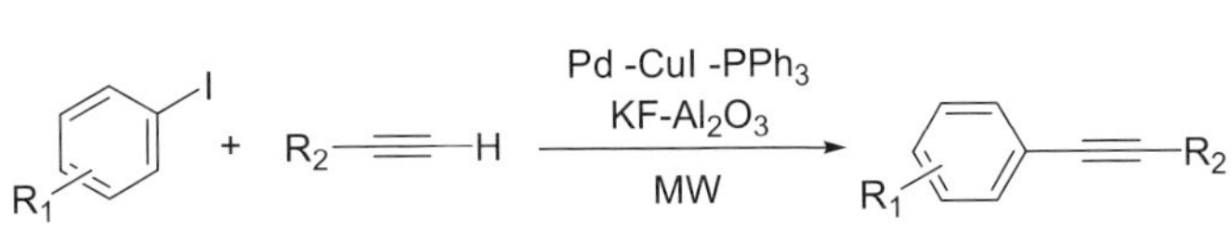

Scheme 8.89. MW-expedited Sonogashira coupling on palladium-doped alumina.

A rapid and efficient molybdenum-catalyzed, MW-accelerated asymmetric allylic alkylation under noninert conditions has been reported [226]. Intermolecular hydroacylation of 1-alkenes with aldehydes has been presented as a greener alternative to the classical approach using a homogeneous catalyst in toluene.

8.2.7.6 Synthesis of Radiolabeled Compounds – Exchange Reactions

The already described, MW-expedited borohydride reduction [142] has been adapted for efficient deuteration and tritiation procedures using MW irradiation and solid hydrogen/deuterium/tritium donors with minimal radioactive waste generation, in contrast with classical tritiation efforts [227]. Jones and coworkers have circumvented the traditional disadvantages of tritium-labeling techniques, as demonstrated in deuterated and tritiated borohydride reductions [143, 144] based on similar MW-expedited reduction executed on an alumina surface [142]. Hydrogen-exchange reactions that require prolonged reaction time (24 h) [228, 229] and elevated temperatures are the primary beneficiaries of this microwave approach (Scheme 8.90) [143, 145]. The high purity of labeled materials, efficient insertion and excellent regioselectivity are some of the advantages of this emerging technology.

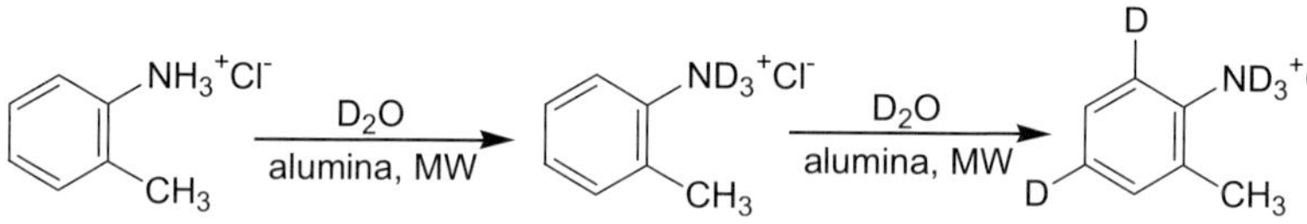

Scheme 8.90. Deuteration reactions using MW irradiation.

In faster, selective, and cleaner applications of the microwave-accelerated reactions, Stone-Elander et al. synthesized a variety of radiolabeled (^{3}H, ^{11}C, and ^{19}F) organic compounds by nucleophilic aromatic and aliphatic substitution reactions, esterifications, condensations, hydrolysis, and complexation reactions using monomodal MW cavities on a microscale [144]. Substantially less radioactive waste is generated in these procedures, which are discussed, at length, in Chapter 18 [145].

Hydrogenation reactions in which $H_2/D_2/T_2$ gases are replaced by different formates proceed very rapidly under MW irradiation conditions (Scheme 8.91) [230]. The pattern of labeling can be easily modified and the advantages are especially noteworthy for tritium, because high specific activity tritiated water is hazardous to use.

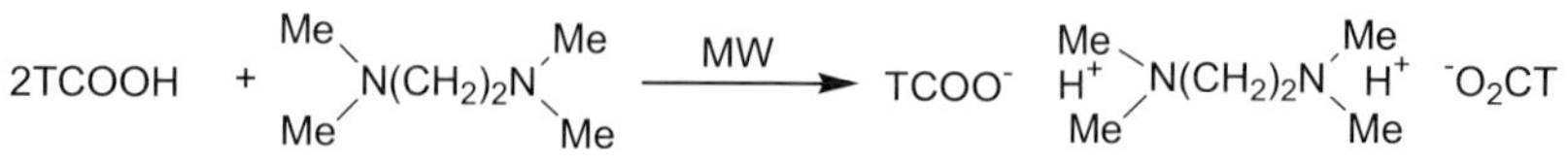

Scheme 8.91. Hydrogenation reactions using formates.

8.2.7.7 Enzyme-catalyzed Reactions

In conventional synthetic transformations, enzymes are normally used in aqueous or organic solvent at moderate temperatures to preserve the activity of enzymes. Consequently, some of these reactions require longer reaction times. In newer developments enzymes are immobilized on solid supports [231] where they are amenable to relatively higher temperature reaction with adequate pH control. The application of MW irradiation has been explored with two enzyme systems – *Pseudomonas* lipase dispersed in Hyflo Super Cell and commercially available SP 435 Novozym (*Candida antarctica* lipase grafted on an acrylic resin).

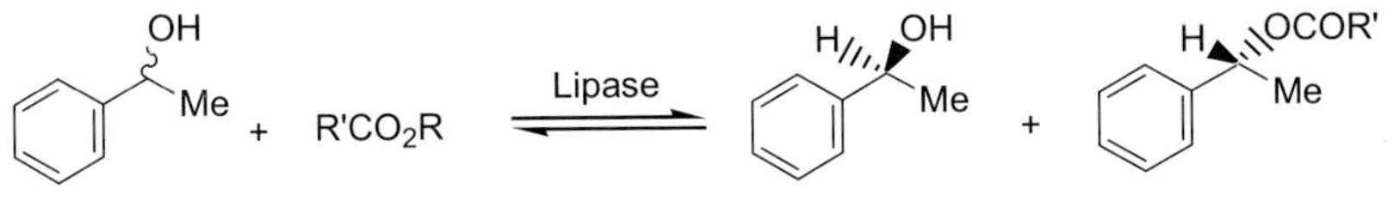

Scheme 8.92. MW-assisted resolution of racemic 1-phenylethanol via transesterification.

Resolution of racemic 1-phenylethanol has been achieved under solvent-free microwave irradiation conditions by transesterification using the enzymes mentioned above (Scheme 8.92) [232]. Comparison of the MW-assisted reaction with conventional heating revealed the former to be more enantioselective, presumably because of efficient removal of low molecular weight alcohols or water on exposure to microwaves or, alternatively, an entropic effect because of dipolar polarization that induces prior organization of the system. Thermostable enzymes, for example a crude homogenate of *Sulfolobus solfataricus* and recombinant β-glucosidase from *Pyrococcus furiosus* have been successfully applied to transglycosylation reactions in which recycling of the biocatalyst is feasible.

8.2.7.8 Solvent-free Synthesis of Ionic Liquids

Room-temperature ionic liquids (RTIL), most often consisting of N-alkylimidazolium cations and different large anions [233], have received wide attention because of their potential in a variety of commercial applications, especially as substitutes for traditional volatile organic solvents [234, 235]. They are polar in nature but consist of poorly coordinating ions and are a polar alternative for biphasic systems. Other important attributes of these ionic liquids include negligible vapor pressure, potential for recycling, compatibility with a variety of organic compounds and organometallic catalysts, and ease of separation of products from reactions [236]. Unfortunately, most conventional preparative procedures for synthesis of ionic liquids involve several hours of heating in solvents under reflux and use of a large excess of alkyl halides/organic solvents that diminish their true potential as "greener" solvents. Their use and coupling with MW irradiation is reviewed in Chapter 7.

Ionic liquids, being polar and ionic in character, couple with MW irradiation very efficiently and are, therefore, ideal microwave-absorbing candidates for expediting chemical reactions. The first efficient preparation of 1,3-dialkylimidazolium halides via microwave irradiation has been described by Varma et al. The reaction time was

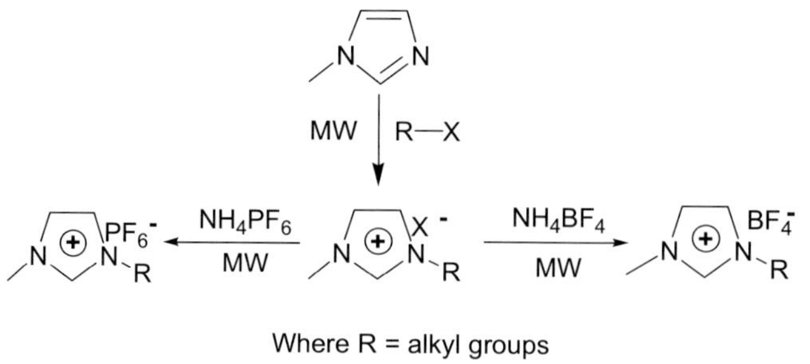

Scheme 8.93. MW-assisted preparation of ionic liquids.

reduced from several hours to minutes and the reaction avoids the use of a large excess of alkyl halides/organic solvents as the reaction medium (Scheme 8.93) [26–28].

This approach avoids the use of volatile organic solvents, and is much faster, efficient, and eco-friendly. Significant rate enhancements are reported in the 1,3-dipolar cycloaddition reactions, including use of covalently grafted dipolarophiles on the ionic liquids [237].

1-Alkyl-3-methylimidazolium tetrachloroindate(III), [Rmim][InCl$_4$] and 1-butyl-3-methylimidazolium tetrachlorogallate, [bmim][GaCl$_4$] have recently been prepared by Varma and Kim using a simple solvent-free MW procedure. These ionic liquids have been successfully used as recyclable catalysts for efficient and eco-friendly protection of alcohols to form tetrahydropyranyl (THP) ethers and for efficient acetalization of aldehydes under mild conditions, respectively (Scheme 8.94) [238, 239].

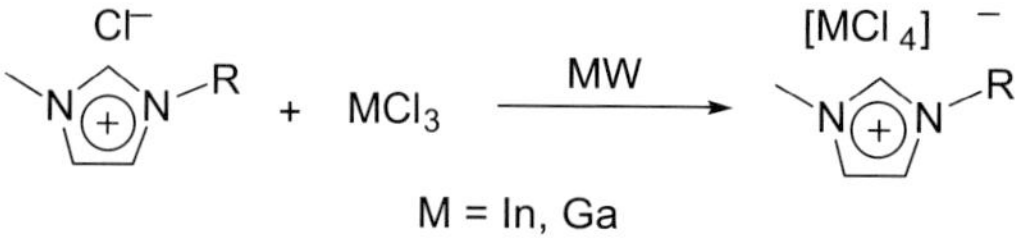

Scheme 8.94. MW-assisted solvent-free synthesis of indium and gallium-based ionic liquids.

A tetrahalideindate(III)-based ionic liquid has recently been used as a recyclable catalyst in the coupling of carbon dioxide and epoxides to provide ready access to cyclic carbonates (Scheme 8.95) [240]. Mechanistic details have been described.

8.3
Conclusions

Microwave heating, being specific and instantaneous, is unique and has found a place in chemical synthesis. Specifically, the solvent-free reactions are convenient to perform and have clear advantages over conventional heating procedures, as

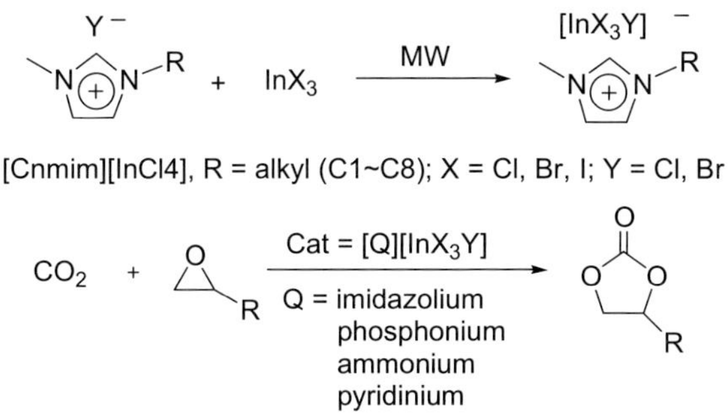

Scheme 8.95. Ionic liquid-catalyzed coupling of epoxides with carbon dioxide.

has been summarized in this chapter. Numerous selective organic functional group transformations have been accomplished more efficiently and expeditiously using a variety of reagents supported on mineral oxides as catalysts. Although much work has been performed around the world using unmodified household microwave ovens (multimode applicators), more recent work does emphasize the advantages of using commercial systems which not only enable improved temperature and power control but make it possible to conduct relatively large-scale reactions [241, 242], with the additional option of continuous operation. Engineering and scale-up aspects of chemical process development have also been discussed [243].

There are distinct advantages of these solvent-free procedures in instances in which catalytic amounts of reagents or supported agents are used, because they enable reduction or elimination of solvents, thus preventing pollution "at source". Although not delineated completely, reaction rate enhancements achieved by use of these methods may be ascribed to nonthermal effects. Rationalization of microwave effects and mechanistic considerations are discussed in detail elsewhere in this book [25, 244]. There has been an increase in the number of publications [23c, 244, 245] and patents [246–256], and increasing interest in the pharmaceutical industry [257–259], with special emphasis on combinatorial chemistry and even polymerization reactions [260–263], and environmental chemistry [264]. The development of newer microwave systems for solid-state reaction [265], and introduction of the concepts of process intensification [266], may help realization of the full potential of microwave-enhanced chemical syntheses under solvent-free conditions.

References

1 C. R. BUFFLER, *Microwave Cooking and Processing*, Van Nostrand Reinhold, New York, **1993**, pp 1–68.
2 R. S. VARMA in *ACS Symposium Series No. 767/Green Chemical Syntheses and Processes* (P. T. ANASTAS, L. HEINE, T. WILLIAMSON, Eds.), Chapter 23, pp 292–313, American Chemical Society, Washington DC, **2000**.
3 R. S. VARMA in *Green Chemistry:*

Challenging Perspectives (P. TUNDO, P. T. ANASTAS, Eds.), pp 221–244, Oxford University Press, Oxford, **2000**.
4 R. S. VARMA, *Green Chem.*, **1999**, *1*, 43.
5 R. S. VARMA, *Clean Products and Processes*, **1999**, *1*, 132.
6 R. S. VARMA in *Microwaves: Theory and Application in Material Processing IV*, (D. E. CLARK, W. H. SUTTON, D. A. LEWIS, eds.), pp 357–365, American Ceramic Society, Westerville, Ohio, **1997**.
7 (a) R. A. ABRAMOVICH, *Org. Prep. Proc. Int.*, **1991**, *23*, 683; (b) S. CADDICK, *Tetrahedron*, **1995**, *51*, 10403.
8 (a) R. S. VARMA, *Pure Appl. Chem.*, **2001**, *73*, 193; (b) A. LOUPY, L. PERREUX, M. LIAGRE, K. BURLE, M. MONEUSE, *Pure Appl. Chem.*, **2001**, *73*, 161.
9 (a) F. LANGA, P. DE LA CRUZ, A. DE LA HOZ, A. DÍAZ-ORTIZ, E. DÍEZ-BARRA, *Contemp. Org. Chem.*, **1997**, *4*, 373; (b) A. DE LA HOZ, A. DÍAZ-ORTIZ, A. MORENO, *Chem. Soc. Rev.*, **2005**, *34*, 164.
10 A. LOUPY, A. PETIT, J. HAMELIN, F. TEXIER-BOULLET, P. JACQUAULT, D. MATHÉ, *Synthesis*, **1998**, 1213.
11 (a) R. GEDYE, F. SMITH, K. WESTAWAY, H. ALI, L. BALDISERA, L. LABERGE, J. ROUSELL, *Tetrahedron Lett.*, **1986**, *27*, 279; (b) R. J. GIGUERE, T. L. BRAY, S. M. DUNCAN, G. MAJETICH, *Tetrahedron Lett.*, **1986**, *27*, 4945.
12 (a) C. GABRIEL, S. GABRIEL, E. H. GRANT, B. S. J. HALSTEAD, D. M. P. MINGOS, *Chem. Soc. Rev.*, **1998**, *27*, 213; (b) D. M. P. MINGOS in *Microwave-assisted Organic Synthesis* (Eds.: P. LIDSTRÖM, J. P. TIERNEY), Blackwell, Oxford, **2005**, Chapter 1.
13 Using chemical reagents on porous carriers, *Akt.-Ges. Fur Chemiewerte. Brit. Pat.*, **1924**, *231*, 901. (*Chem. Abstr.* **1925**, *19*, 3571).
14 A. MCKILLOP, K. W. YOUNG, *Synthesis*, **1979**, 401 and 481.
15 G. H. POSNER, *Angew Chem. Int. Ed. Engl.*, **1978**, *17*, 487.
16 A. CORNELIS, P. LASZLO, *Synthesis*, **1985**, 909.
17 (a) P. LASZLO, *Preparative Chemistry Using Supported Reagents*, Academic Press, Inc. San Diego, **1987**; (b) LASZLO, P. *Science*, **1987**, *235*, 1473.
18 G. BRAM, A. LOUPY, D. VILLEMIN, in *Solid Supports and Catalyst in Organic Synthesis*, K. SMITH, Ed. Ellis Horwood, Chichester, **1992**, Chapter 12.
19 M. BALOGH, P. LASZLO, *Organic Chemistry Using Clays*, Springer, Berlin, **1993**.
20 (a) J. H. CLARK, *Catalysis of Organic Reactions by Supported Inorganic Reagents*, VCH publisher, Inc., New York, 1994; (b) J. H. CLARK, D. J. MACQUARRIE, *Chem. Commun.*, **1998**, 853.
21 G. W. KABALKA, R. M. PAGNI, *Tetrahedron*, **1997**, *53*, 7999.
22 R. S. VARMA, *Tetrahedron*, **2002**, *58*, 1235.
23 (a) C. R. STRAUSS, R. W. TRAINOR, *Aust. J. Chem.*, **1995**, *48*, 1665; (b) A. K. BOSE, B. K. BANIK, N. LAVLINSKAIA, M. JAYARAMAN, M. S. MANHAS, *Chemtech*, **1997**, *27*, 18; (c) P. LINDSTRÖM, J. TIERNEY, B. WATHEY, J. WESTMAN, *Tetrahedron*, **2001**, *57*, 9225.
24 (a) E. GUTIÉRREZ, A. LOUPY, G. BRAM, E. RUIZ-HITZKY, *Tetrahedron Lett.*, **1989**, *30*, 945; (b) A. BEN ALLOUM, B. LABIAD, D. VILLEMIN, *Chem. Commun.*, **1989**, 386; (c) G. BRAM, A. LOUPY, M. MAJDOUB, *Synth. Commun.*, **1990**, *20*, 125; (d) G. BRAM, A. LOUPY, M. MAJDOUB, E. GUTIÉRREZ, E. RUIZ-HITZKY, *Tetrahedron*, **1990**, *46*, 5167.
25 (a) C. O. KAPPE, *Angew. Chem. Int. Ed.*, **2004**, *43*, 6250; (b) L. PERREUX, A. LOUPY, *Tetrahedron*, **2001**, *57*, 9199.
26 R. S. VARMA, V. V. NAMBOODIRI, *Chem. Commun.*, **2001**, 643.
27 R. S. VARMA, V. V. NAMBOODIRI, *Pure Appl. Chem.*, **2001**, *73*, 1307.
28 V. V. NAMBOODIRI, R. S. VARMA, *Chem. Commun.*, **2002**, 342.
29 (a) THACH LE NGOC, B. A. ROBERTS, C. R. STRAUSS, this book, Chapter 3; (b) C. J. LI, T. H. CHAN, *Organic Reactions in Aqueous Media*. John Wiley and Sons: New York, 1997; (c) C. J. LI, *Chem. Rev.*, **2005**, *105*, 3095; (d) S. KOBAYASHI, K. MANABE, *Acc. Chem. Res.*, **2002**, *35*, 209; (e) W. WEI, C. C. K. KEH, C. J. LI, R. S. VARMA, *Clean Tech. Environ. Policy* **2005**, *7*, 62;

(f) S. Narayan, J. Muldoon, M. G. Finn, V. V. Fokin, H. C. Kolb, K. B. Sharpless, *Angew. Chem., Int. Ed.*, **2005**, *44*, 3275; (g) J. E. Klijn, J. B. F. N. Engberts, *Nature*, **2005**, *435*, 746; (h) Y. Ju, R. S. Varma, *Org. Lett.*, **2005**, *7*, 2409; (i) Y. Ju, R. S. Varma, *J. Org. Chem.*, **2006**, *71*, (in press).

30 D. Villemin, F. Thibault-Starzyk, *J. Chem. Educ.*, **1991**, *68*, 346.

31 T. W Greene, P. G. M. Wuts, *Protective Groups in Organic Synthesis*, 2nd Edition, John Wiley and Sons, New York, N.Y. **1991**.

32 M. Csiba, J. Cléophax, A. Loupy, J. Malthête, S. D. Gero, *Tetrahedron Lett.*, **1993**, *34*, 1787.

33 B. Perio, M. J. Dozias, P. Jacquault, J. Hamelin, *Tetrahedron Lett.*, **1997**, *38*, 7867.

34 D. Villemin, A. Ben Alloum, F. Thibault-Starzyk, *Synth. Commun.*, **1992**, *22*, 1359.

35 S. Paul, P. Nanda, R. Gupta, A. Loupy, *Tetrahedron Lett.*, **2002**, *43*, 4261.

36 A. Loupy, A Petit, D. Bogdal, this book, Chapter 6.

37 P. Das, B. Basu, *Synth. Commun.*, **2004**, *34*, 2177.

38 R. S. Varma, M. Varma, A. K. Chatterjee, *J. Chem. Soc, Perkin Trans 1*, **1993**, 999.

39 R. S. Varma, A. K. Chatterjee, M. Varma, *Tetrahedron Lett.*, **1993**, *34*, 3207.

40 D. Kamarkar, D. Prajapati, J. S. Sandhu, *J. Chem. Res. (S)*, **1998**, 382.

41 E. R. Pérez, A. L. Marrero, R. Pérez, M. A. Autie, *Tetrahedron Lett.*, **1995**, *36*, 1779.

42 R. S. Varma, A. K. Chatterjee, M. Varma, *Tetrahedron Lett.*, **1993**, *34*, 4603.

43 A. S. Gajare, N. S. Shaikh, B. K. Bonde, V. H. Deshpande, *J. Chem. Soc., Perkin Trans 1*, **2000**, 639.

44 G. W. Kabalka, L. Wang, R. M. Pagni, *Green Chem.*, **2001**, *3*, 261.

45 D. S. Bose, V. Lakshminarayana, *Tetrahedron Lett.*, **1998**, *39*, 5631.

46 R. S. Varma, J. B. Lamture, M. Varma, *Tetrahedron Lett.*, **1993**, *34*, 3029.

47 M. M. Mojtahedi, M. R. Saidi, M. M. Heravi, M. Bolourtchian, *Monatsh. Chem.*, **1999**, *130*, 1175.

48 M. M. Mojtahedi, M. R. Saidi, M. Bolourtchian, M. M. Heravi, *Synth. Commun.*, **1999**, *29*, 3283.

49 M. Tajbakhsh, M. Heravi, S. Habibzadeh, *Phosphorus, Sulfur and Silicon*, **2003**, *178*, 361.

50 R. S. Varma, R. K. Saini, *Tetrahedron Lett.*, **1997**, *38*, 2623 and references cited therein.

51 H. M. Meshram, G. S. Reddy, G. Sumitra, J. S. Yadav, *Synth. Commun.*, **1999**, *29*, 1113.

52 A. Corsaro, U. Chiacchio, V. Pistaria, *Synthesis*, **2001**, 1903.

53 R. S. Varma, H. M. Meshram, *Tetrahedron Lett.*, **1997**, *38*, 5427 and references cited therein.

54 A. I. Bosch, P. de la Cruz, E. Díez-Barra, A. Loupy, F. Langa, *Synlett*, **1995**, 1259.

55 R. S. Varma, R. Dahiya, R. K. Saini, *Tetrahedron Lett.*, **1997**, *38*, 8819.

56 A. K. Mitra, A. De, N. Karchaudhuri, *J. Chem. Res. (S)*, **1999**, 560.

57 M. D. Heravi, M. Ghassemzadeh, *Phosphorus, Sulfur and Silicon*, **2003**, *178*, 119.

58 K. V. N. S. Srinivas, B. Das, *J. Chem. Res. (S)*, **2002**, 556.

59 R. S. Varma, H. M. Meshram, *Tetrahedron Lett.*, **1997**, *38*, 7973.

60 A. Boruah, B. Boruah, D. Prajapati, J. S. Sandhu, *Synlett*, **1997**, 1251.

61 M. Baruah, D. Prajapati, J. S. Sandhu, *Synth. Commun.*, **1998**, *28*, 4157.

62 R. S. Varma, D. Kumar, *Synth. Commun.*, **1999**, *29*, 1333 and references cited therein.

63 J. S. Yadav, H. M. Meshram, G. Sudershan, G. Sumitra, *Tetrahedron Lett.*, **1998**, *39*, 3043.

64 N. Deka, J. C. Sarma, *J. Org. Chem.*, **2001**, *66*, 1947.

65 G. Sabitha, S. Abraham, B. V. S. Reddy, J. S. Yadav, *Synlett*, **1999**, 1745.

66 J. W. Elder, *J. Chem. Ed.*, **1994**, *71*, A142.

67 D. Abenhaïm, C. P. N. Son, A.

Loupy, N. B. Hiep, *Synth. Commun.*, **1994**, *24*, 1199.
68. A. K. Mitra, A. De, N. Karchaudhuri, *Synth. Commun.*, **1999**, *29*, 2731.
69. M. Lacova, J. Chovancova, E. Ververkova, S. Toma, *Tetrahedron*, **1996**, *52*, 14995.
70. A. Spinella, T. Fortunati, A. Soriente, *Synlett*, **1997**, 93.
71. Y. Lakhrissi, C. Taillefumier, M. Lakhrissi, Y. Chapleur, *Tetrahedron Asymmetry*, **2000**, *11*, 417.
72. J. J. Kiddle, *Tetrahedron Lett.*, **2000**, *41*, 1339.
73. D. Villemin, B. Martin, *Synth. Commun.*, **1995**, *25*, 3135.
74. D. Villemin, B. Martin, *J. Chem. Res. (S)*, **1994**, 146.
75. V. Singh, J. Singh, P. Kaur, G. L. Kad, *J. Chem. Res. (S)*, **1997**, 58.
76. D. Bogdal, *J. Chem. Res. (S)*, **1998**, 468.
77. R. S. Varma, R. Dahiya, S. Kumar, *Tetrahedron Lett.*, **1997**, *38*, 2039.
78. R. S. Varma, R. Dahiya, *Synlett*, **1997**, 1245.
79. G. S. Singh, D. S. Mahajan, *J. Chem. Res. (S)*, **2004**, 410.
80. H. T. S. Braibante, M. E. F. Braibante, G. B. Rosso, D. A. Oriques, *J. Braz. Chem. Soc.*, **2003**, *14*, 994.
81. R. S. Varma, R. Dahiya, S. Kumar, *Tetrahedron Lett.*, **1997**, *38*, 5131.
82. R. S. Varma, G. W. Kabalka, *Heterocycles*, **1986**, *24*, 2645.
83. G. W. Kabalka, R. S. Varma, *Org. Prep. Proc. Int.*, **1987**, *19*, 283.
84. G. W. Kabalka, L. H. M. Guindi, R. S. Varma, *Tetrahedron*, **1990**, *46*, 7443.
85. A. Vass, J. Dudas, R. S. Varma, *Tetrahedron Lett.*, **1999**, *40*, 4951.
86. M. Chakrabarty, R. Basak, N. Ghosh, *Tetrahedron Lett.*, **2001**, *42*, 3913.
87. H. Yang, Y. Peng, G. Song, X. Qian, *Tetrahedron Lett.*, **2001**, *42*, 9043.
88. S. Gadhwal, M. Boruah, J. S. Sandhu, *Synlett*, **1999**, 1573.
89. E. Parquet, Q. Lin, *J. Chem. Educ.*, **1997**, *74*, 1225.
90. M. Ješelnik, R. S. Varma, S. Polanc, M. Kočevar, *Chem. Commun.*, **2001**, 1716.
91. M. Ješelnik, R. S. Varma, S. Polanc, M. Kočevar, *Green Chem.*, **2002**, *4*, 35.
92. S. Paul, M. Gupta, R. Gupta, A. Loupy, *Tetrahedron Lett.*, **2001**, *42*, 3827.
93. R. S. Varma, K. P. Naicker, *Tetrahedron Lett.*, **1999**, *40*, 6177.
94. K. D. Raner, C. R. Strauss, *J. Org. Chem.*, **1992**, *57*, 6231.
95. A. Loupy, A. Petit, M. Ramdani, C. Yvanaeff, B. Labiad, D. Villemin, *Can. J. Chem.*, **1993**, *71*, 90.
96. D. J. Kalita, R. Borah, J. C. Sarma, *J. Chem. Res. (S)*, **1999**, 404.
97. J. P. Bazureau, J. Hamelin, F. Mongin, F. Texier-Boullet, this book, Chapter 10.
98. K. Bougrin, M. Soufiaoui, G. Bashiardes, this book, Chapter 11.
99. A. Dandia, K. Arya, M. Sati, S. Gautam, *Tetrahedron*, **2004**, *60*, 5253.
100. A. Loupy, S. J. Song, S. J. Cho, D. K. Park, T. W. Kwon, *Synth. Commun.*, **2005**, *35*, 79.
101. J. S. Yadav, K. Anuradha, B. V. Subba Reddy, B. Eeshwaraiah, *Tetrahedron Lett.*, **2003**, *44*, 8959.
102. O. Correc, K. Guillou, J. Hamelin, L. Paquin, F. Texier-Boullet, L. Toupet, *Tetrahedron Lett.*, **2004**, *45*, 391.
103. A. Mortoni, M. Martinelli, U. Piarulli, N. Regalia, S. Gagliardi, *Tetrahedron Lett.* **2004**, *45*, 6623.
104. M. Mihara, Y. Ishino, S. Minakata, M. Komatsu, *Synlett*, **2002**, 1526.
105. L. D. S. Yadav, B. S. Yadav, V. K. Rai, *Tetrahedron Lett.*, **2004**, *45*, 5351.
106. V. Bailliez, R. M. de Figueiredo, A. Olesker, J. Cléophax, *Synthesis*, **2003**, 1025.
107. H.-M. Yu, S.-T. Chen, M.-J. Tseng, K.-T. Wang, *J. Chem. Res. (S)*, **1999**, 62.
108. G. L. Kad, I. R. Trehan, J. Kaur, S. Nayyar, A. Arora, J. S. Brar, *Ind. J. Chem.*, **1999**, *35B*, 734.
109. F. M. Moghaddam, M. G. Dakamin, *Tetrahedron Lett.*, **2000**, *41*, 3479.
110. D. Villemin, B. Labiad, *Synth. Commun.*, **1992**, *22*, 2043.
111. T. Yamamoto, Y. Wada, H. Enokida, M. Fujimoto, K. Nakamura, S. Yanagida, *Green Chem.*, **2003**, *5*, 690.

112 C. J. BENNETT, S. T. CALDWELL, D. B. MCPHAIL, P. C. MORRICE, G. G. DUTHIE, R. C. HARTLEY, *Bioorg. Med. Chem.*, **2004**, *12*, 2079.

113 S. SOWMYA, K. K. BALASUBRAMANIAN, *Synth. Commun.*, **1994**, *24*, 2097.

114 A. J. FATIADI in *Organic Synthesis by Oxidation with Metal Compounds*, W. J. MIJS, C. R. H. I. DEJONGE (Eds.) Plenum Press, New York, **1986**, pp 119–260.

115 B. M. TROST, (Ed.), *Comprehensive Organic Synthesis (Oxidation)*, Pergamon, New York, **1991**, Vol 7.

116 R. S. VARMA, R. K. SAINI, R. DAHIYA, *Tetrahedron Lett.*, **1997**, *38*, 7823.

117 J. GÓMEZ-LARA, R. GUTIÉRREZ-PÉREZ, G. PENIERES-CARILLO, J. G. LÓPEZ-CORTÉS, A. ESCUDERO-SALAS, C. ALVAREZ-TOLEDANO, *Synth. Commun.*, **2000**, *30*, 2713.

118 B. OUSSAID, B. GARRIGUES, M. SOUFIAOUI, *Can. J. Chem.*, **1994**, *72*, 2483.

119 R. S. VARMA, R. K. SAINI, *Tetrahedron Lett.*, **1998**, *39*, 1481.

120 R. S. VARMA, R. DAHIYA, *Tetrahedron Lett.*, **1997**, *38*, 2043.

121 M. M. HERAVI, D. AJAMI, K. AGHAPOOR, M. GHASSEMZADEH, *Chem. Commun.*, **1999**, 833.

122 R. S. VARMA, V. V. NAMBOODIRI, *Solvent-free Oxidation of Alcohols Using Iron(III) nitrate Nonahydrate*, IUPAC CHEMRAWN XIV World Conference on "Green Chemistry: Toward Environmentally Benign Processes and Products," University of Colorado, Boulder, June 9–13, **2001**.

123 P. CAPDEVIELLE, M. MAUMY, *Tetrahedron Lett.*, **1990**, *31*, 3891.

124 R. S. VARMA, R. DAHIYA, *Tetrahedron Lett.*, **1998**, *39*, 1307.

125 R. S. VARMA, D. KUMAR and R. DAHIYA, *J. Chem. Res. (S)*, **1998**, 324.

126 R. S. VARMA, R. DAHIYA and D. KUMAR, *Molecules Online*, **1998**, *2*, 82.

127 R. S. VARMA, K. P. NAICKER, D. KUMAR, R. DAHIYA, P. J. LIESEN, *J. Microwave Power Electromag. Energy*, **1999**, *34*, 113.

128 R. S. VARMA, R. DAHIYA, R. K. SAINI, *Tetrahedron Lett.*, **1997**, *38*, 7029.

129 R. S. VARMA, R. K. SAINI, H. M. MESHRAM, *Tetrahedron Lett.*, **1997**, *38*, 6525.

130 R. S. VARMA, R. K. SAINI, R. DAHIYA, *J. Chem. Res. (S)*, **1998**, 120.

131 A. OUSSAID, A. LOUPY, *J. Chem. Res. (S)*, **1997**, 342.

132 H. BENHALILIBA, A. DERDOUR, J.-P. BAZUREAU, F. TEXIER-BOULLET, J. HAMELIN, *Tetrahedron Lett.*, **1998**, *39*, 541.

133 S. ROSTAMIZADEH, S. A. G. HOUSAINI, *Tetrahedron Lett.* **2004**, *45*, 8753.

134 J. C. LEE, J. Y. LEE, S. J. LEE, *Tetrahedron Lett.*, **2004**, *45*, 4939.

135 D. VILLEMIN, F. SAUVAGET, *Synlett*, **1994**, 435.

136 D. BARBRY, P. CHAMPAGNE, *Tetrahedron Lett.*, **1996**, *37*, 7725.

137 Y. LIU, Y. LU, P. LIU, R. X. GAO, Y. Q. YIN, *Appl. Catal. A General*, **1998**, *170*, 207.

138 G. H. POSNER, A. W. RUNQUIST, M. J. CHAPDELAINE, *J. Org. Chem.*, **1977**, *42*, 1202 and references cited therein.

139 D. BARBRY, S. TORCHY, *Tetrahedron Lett.*, **1997**, *38*, 2959.

140 F. TODA, K. KIYOSHIGE, M. YOGI, *Angew. Chem. Int. Ed. Engl.*, **1989**, *28*, 320.

141 G. BRAM. E. D'LNCAN, A. LOUPY. D., *J. Chem. Soc., Chem. Commun.*, **1981**, 1066.

142 R. S. VARMA, R. K. SAINI, *Tetrahedron Lett.*, **1997**, *38*, 4337.

143 W. T. ERB, J. R. JONES, S. Y. LU, *J. Chem. Res (S)*, **1999**, 728.

144 N. ELANDER, J. R. JONES, S. Y. LU, S. STONE-ELANDER, *Chem. Soc. Rev.*, **2000**, *29*, 239.

145 J. R. JONES, S. Y. LU, this book, Chapter 18.

146 R. F. BORCH, M. D. BERSTEIN, H. D. DURST, *J. Am. Chem. Soc.*, **1971**, *93*, 289.

147 A. F. ABDEL-MAGID, K. G. CARSON, B. D. HARRIS, C. A. MARYANOFF, R. D. SHAH, *J. Org. Chem.*, **1996**, *61*, 3849 and references cited therein.

148 V. GIANCARLO, A. G. GIUMANINI, P. STRAZZOLINI, M. POIANA, *Synthesis*, **1993**, 121.

149 R. S. VARMA, R. DAHIYA, *Tetrahedron*, **1998**, *54*, 6293.

150 G. PERIN, R. G. JACOB, F. DE

Azambuja, G. V. Botteselle, G. M. Siqueira, R. A. Freitag, E. J. Lenardão, *Tetrahedron Lett.*, **2005**, *46*, 1679.

151 S. Cannizzaro, *Ann.*, **1853**, *88*, 129.
152 T. A. Geissman, *Organic Reactions*, **1944**, *II*, 94.
153 C. G. Swain, A. L. Powell, W. A. Sheppard, C. R. Morgan, *J. Am. Chem. Soc.*, **1979**, *101*, 3576.
154 E. C. Ashby, D. T. Coleman III, M. P. Gamasa, *Tetrahedron Lett.*, **1983**, *24*, 851.
155 R. S. Varma, G. W. Kabalka, L. T. Evans, R. M. Pagni, *Synth. Commun.*, **1985**, *15*, 279.
156 R. S. Varma, K. P. Naicker, P. J. Liesen, *Tetrahedron Lett.*, **1998**, *39*, 8437.
157 J. A. Thakuria, M. Baruah, J. S. Sandhu, *Chem. Lett.*, **1999**, *9*, 995.
158 A. Sharifi, M. M. Mojtahedi, M. R. Saidi, *Tetrahedron Lett.*, **1999**, *40*, 1179.
159 A. Vass, J. Dudás, J. Tóth, R. S. Varma, *Tetrahedron Lett.*, **2001**, *42*, 5347.
160 R. S. Varma, *Nutrition*, **1996**, *12*, 643.
161 J. F. M. Post, R. S. Varma, *Cancer Lett.*, **1992**, *67*, 207.
162 Y. LeFloc'h, M. LeFeuvre, *Tetrahedron Lett.*, **1986**, *27*, 2751.
163 W. Baker, *J. Chem. Soc.*, **1933**, 1381.
164 H. S. Mahal, K. Venkataraman, *J. Chem. Soc.*, **1934**, 1767.
165 R. S. Varma, R. K. Saini, D. Kumar, *J. Chem. Res (S)*, **1998**, 348.
166 T. Patonay, R. S. Varma, A. Vass, A. Levai, J. Dudas, *Tetrahedron Lett.*, **2001**, *42*, 1403.
167 F. M. Dean, R. S. Varma, *J. Chem. Soc. Perkin Trans 1*, **1982**, 1193.
168 F. M. Dean, M. Varma, R. S. Varma, *J. Chem. Soc. Perkin Trans 1*, **1982**, 2771.
169 F. M. Dean, R. S. Varma, *Tetrahedron Lett.*, **1981**, *22*, 2113.
170 R. S. Varma, R. Dahiya, *J. Org. Chem.*, **1998**, *63*, 8038.
171 A. Hantzsch, *Liebigs Ann.*, **1885**, *250*, 262.
172 R. S. Varma, D. Kumar, P. J. Liesen, *J. Chem. Soc., Perkin Trans 1*, **1999**, 4093.
173 R. S. Varma, *J. Heterocyclic Chem.*, **1999**, *35*, 1565.
174 S. Frère, V, Thiéry, T. Besson, *Synth. Commun.*, **2003**, *33*, 3795.
175 A. A. Kiryanov, P. Sampson, A. J. Seed, *J. Org. Chem.*, **2001**, *66*, 7925.
176 A. Dandia, M. Sati, A. Loupy, *Green Chem.*, **2002**, *4*, 599.
177 (a) A. Dandia, M. Sati, K, Arya, A. Loupy, *Heterocycles*, **2003**, *60*, 563; (b) A. Dandia, M. Sati, K, Arya, R. Sharma, A. Loupy, *Chem. Pharm. Bull.*, **2003**, *51*, 1137.
178 B. Kaboudin, H. Norouzi, *Tetrahedron Lett.*, **2004**, *45*, 1283.
179 R. S. Varma, R. K. Saini, *Synlett*, **1997**, 857.
180 J. H. M. Lange, P. C. Verveer, S. J. M. Osnabrug, G. M. Visser, *Tetrahedron Lett.*, **2001**, *42*, 1367.
181 S. Balalaie, A. Sharifi, B. Ahangarian, E. Kowsari, *Heterocycl. Commun.*, **2001**, *7*, 337.
182 R. M. Dinica, C. Pettinari, *Heterocycl. Commun.*, **2001**, *7*, 381.
183 (a) T. Lipinska, E. Guibé-Jampel, A. Petit, A. Loupy, *Synth. Commun.*, **1999**, *29*, 1349; (b) T. Lipinska, *Tetrahedron Lett.*, **2004**, *45*, 8831.
184 B. Kaboudin, F. Saadati, *Heterocycles*, **2005**, *65*, 353.
185 Y. Xu, L. F. Wang, H. Salehi, W. Deng, Q. X. Guo, *Heterocycles*, **2004**, *63*, 1613.
186 J. P. Collman, R. A. Decréau, *Tetrahedron Lett.*, **2003**, *44*, 1207.
187 N. Kaval, B. Halasz-Dajka, G. Vo-Thanh, W. Dehaen, J. Van der Eycken, P. Màtyus, A. Loupy, E. Van Der Eycken, *Tetrahedron*, **2005**, *61*, 9052.
188 I. Devi, P. J. Bhuyan, *Tetrahedron Lett.*, **2005**, *46*, 5727.
189 L. D. S. Yadav, R. Kapoor, *J. Org. Chem.*, **2004**, *69*, 8118.
190 R. S. Varma, D. Kumar, *Tetrahedron Lett.*, **1999**, *40*, 7665.
191 B. C. Ranu, A. Hajra, U. Jana, *Synlett*, **2000**, 75.
192 M. Kidwai, S. Saxena, R. Mohan, R. Venkataraman, *J. Chem. Soc., Perkin Trans 1*, **2002**, 1845.
193 I. Saxena, D. C. Borah, J. C. Sarma, *Tetrahedron Lett.*, **2005**, *46*, 1159.

194 L. D. S. Yadav, S. Yadav, V. K. Rai, *Tetrahedron*, **2005**, *61*, 10013.
195 J. Azizian, A. R. Karimi, Z. Kazemizadeh, A. A. Mohammadi, M. R. Mohammadizadeh, *Tetrahedron Lett.*, **2005**, *46*, 6155.
196 C. O. Kappe, D. Kumar, R. S. Varma, *Synthesis*, **1999**, 1799.
197 C. O. Kappe, A. Stadler, this book, Chapter 16.
198 B. C. Ranu, A. Hazra, U. Jana, *Synlett*, **2000**, 75.
199 A. Shaabani and A. Bazgir, *Tetrahedron Lett.*, **2004**, *45*, 2575.
200 M. Miller, G. Loudon, *J. Org. Chem.*, **1975**, *40*, 126.
201 D. T. Mowry, *Chem. Rev.*, **1948**, *42*, 250.
202 J. H. Pomeroy, C. A. Craig, *J. Am. Chem. Soc.*, **1959**, *81*, 6340.
203 D. L. Clive, *J. Chem. Soc., Chem. Commun.*, **1970**, 1014.
204 P. J. Foley, *J. Org. Chem.*, **1969**, *34*, 2805.
205 W. Lehnert, *Tetrahedron Lett.*, **1971**, *6*, 559.
206 G. Rosini, G. Baccolini, S. Cacchi, *J. Org. Chem.*, **1973**, *38*, 1060.
207 H. G. Foley, D. R. Dalton, *J. Chem. Soc., Chem. Commun.*, **1973**, 628.
208 D. Villemin, M. Lalaoui, A. Ben Alloum, *Chem. Ind. (London)*, **1991**, 176.
209 R. S. Varma, K. P. Naicker, *Molecules Online*, **1998**, *2*, 94.
210 J.-C. Feng, B. Liu, N.-S. Bian, *Synth. Commun.*, **1998**, *28*, 3765.
211 B. Das, P. Madhusudhan, B. Venkataiah, *Synlett*, **1999**, 1569.
212 A. K. Chakraborti, G. Kaur, *Tetrahedron*, **1999**, *55*, 13265.
213 J.-C. Feng, B. Liu, Y. Liu, *Synth. Commun.*, **1996**, *26*, 4545.
214 D. S. Bose, B. Jayalakshmi, *J. Org. Chem.*, **1999**, *64*, 1713.
215 R. S. Varma, K. P. Naicker, P. J. Liesen, *Tetrahedron Lett.*, **1998**, *39*, 3977.
216 S. Paul, V. Gupta, R. Gupta, A. Loupy, *Tetrahedron Lett.*, **2003**, *44*, 439.
217 S. Collina, G. Loddo, A. Barbieri, L. Linati, S. Alcaro, P. Chimenti, O. Azzolina, *Tetrahedron Asymmetry*, **2004**, *15*, 3601.
218 H. Navratilova, Z. Krit, M. Potacek, *Synth. Commun.*, **2004**, *34*, 2101.
219 R. F. Heck, *Org. React.*, **1982**, *27*, 345.
220 R. S. Varma, K. P. Naicker, P. J. Liesen, *Tetrahedron Lett.*, **1999**, *40*, 2075.
221 A. Díaz-Ortiz, P. Prieto, E. Vázquez, *Synlett*, **1997**, 269.
222 M. Larhed, G. Lindeberg, A. Hallberg, *Tetrahedron Lett.*, **1996**, *37*, 8219.
223 V. V. Namboodiri, R. S. Varma, *Green Chem.*, **2001**, *3*, 146 and references cited therein.
224 G. W. Kabalka, R. M. Pagni, V. V. Namboodiri, C. M. Hair, *Green Chem.*, **2000**, *2*, 120.
225 G. W. Kabalka, L. Wang, V. V. Namboodiri, R. M. Pagni, *Tetrahedron Lett.*, **2000**, *41*, 5151.
226 N.-F. K. Kaiser, U. Bremberg, M. Larhed, C. Moberg, A. Hallberg, *Angew Chem. Int. Ed. Engl.*, **2000**, *39*, 3596.
227 K. E. Wilzbach, *J. Am. Chem. Soc.*, **1957**, *79*, 1013.
228 H. Andres, H. Morimoto, P. G. Williams, E. M. Zippi, *Synthesis and Applications of Isotopically Labelled Compounds*, J. Allen, R. Voges (Eds.), Wiley, Chichester, 1995, p 83.
229 N. H. Werstiuk, in *Isotopes in the Physical and Biological Sciences*, E. Buncel, J. R. Jones, (Eds.), Vol. 1, Labelled Compounds (Part A) Elsevier, Amsterdam, **1987**, pp 124–155.
230 M. H. Al-Qahtani, N. Cleator, T. N. Danks, R. N. Garman, J. R. Jones, S. Stefaniak, A. D. Morgan, A. J. Simmonds, *J. Chem. Res. (S)*, **1998**, 400.
231 E. Guibé-Jampel, G. Rousseau, *Tetrahedron Lett.*, **1987**, *28*, 3563.
232 J. R. Carrillo-Muñoz, D. Bouvet, E. Guibé-Jampel, A. Loupy, A. Petit, *J. Org. Chem.*, **1996**, *61*, 7746.
233 T. Welton, *Chem. Rev.*, **1999**, *99*, 2701.
234 J. S. Wilkes, J. A. Levinsky, R. A. Wilson, C. L. Hussey, *Inorg. Chem.*, **1982**, *21*, 1263.
235 J. D. Holbrey, K. R. Seddon, *Clean Products and Processes*, **1999**, *1*, 223.

236 J. G. Huddleston, A. E. Visser, W. M. Reichert, H. D. Willauer, G. A. Brocker, R. D. Rogers, *Green Chem.*, **2001**, *3*, 156.
237 J. Fraga-Dubreuil, J. P. Bazureau, *Tetrahedron Lett.*, **2000**, *41*, 7351.
238 Y. J. Kim, R. S. Varma, *Tetrahedron Lett.*, **2005**, *46*, 1467.
239 Y. J. Kim, R. S. Varma, *Tetrahedron Lett.*, **2005**, *46*, 7447.
240 Y. J. Kim, R. S. Varma, *J. Org. Chem.*, **2005**, *70*, 7882.
241 B. Pério, M. J. Dozias, J. Hamelin, *Org. Proc. Res. Dev.*, **1998**, *2*, 428.
242 (a) J. Cléophax, M. Liagre, A. Loupy, A. Petit, *Org. Proc. Res Dev.*, **2000**, *4*, 498; (b) A. Stadler, B. H. Yousefi, D. Dallinger, P. Walla, E. Van der Eycken, N. Kaval, C. O. Kappe, *Org. Proc. Res Dev.*, **2003**, *7*, 701.
243 M. Mehdizadeh, *Res. Chem. Intermed.*, **1994**, *20*, 79.
244 L. Perreux, A. Loupy, this book, Chapter 4.
245 M. Nüchter, B. Ondruschka, W. Bonrath, A. Gum, *Green Chem.*, **2004**, 6, 128.
246 B. Herzog. (Celanese GmbH). Catalyst Based on Pd. Au and Alkali Metals for the Preparation of Vinyl Acetate. *Eur. Patent Application* EP 922,491, **1999** (DE Application 19,754,992, 11 Dec 1997); *Chem. Abstr.*, **1999**, *131*, 45225s.
247 J.-R. Desmurs, J. Dubac, A. Laporterie, C. Laporte, J. Marquié (Rhodia Chimie). Method for Acylation or Sulfonylation of an Aromatic Compound. *PCT International Application* WO 40,339, **1998** (FR Application 97/2,917, 12 Mar 1997); *Chem. Abstr.*, **1998**, *129*, 244928g.
248 A Pöppl, S. Witt, B. Zimmermann (Henkel KgaA). Solid Perfumed Deodorant Composition Containing Alunite. *PCT International Application* WO 23, 197 (DE Application 19,548,067, 26 Jun 1997); *Chem. Abstr.*, **1997**, *127*, 70625x.
249 O. Rhode, S. Witt, I. Hardacker (Henkel K.A.). Method for Preparing of Alkyl-Glycosides Using Microwave Irradiation. *PCT International Application* WO 3,869, **1999** (DE Application 19,730,836, 21 Jan 1999); *Chem. Abstr.*, **1999**, *130*, 110555v.
250 I. T. Badejo (Bayer Corporation). Microwave Syntheses of Quinacridones, 6,13-Dihydroquinacridones and 6,13-Quinacridonequinones at Moderate Temperatures. *Eur. Patent Application* EP 905,199, **1999** (U.S. Patent Application 63,128, 20 Apr 1998); *Chem. Abstr.*, **1999**, *130*, 253670q.
251 P. Coe, T. Waring, C. Mercier (Rhone–Poulenc Chemicals). Preparation of Fluoro Compounds by Treatment of the Corresponding Amines with Hydrogen Fluoride and a Nitrosating Agent Under Ultrasound or Microwave Irradiation. *PCT International Application* WO 41,083, **1997** (GB Application 96/9, 154, 1 May 1996); *Chem. Abstr.*, **1998**, *128*, 13127h.
252 G. Forat, J.-M. Mas, L. Saint-Jalmes (Rhone–Poulenc Chimie). Method for Grafting a Substituted Difluoromethyl Group to a Compound Containing an Electrophilic Group with Microwave Irradiation. *PCT International Application* WO 5, 609, **1998** (FR Application 96/9,754, 1 Aug 1996); *Chem. Abstr.*, **1998**, *128*, 166999u.
253 D. Séméria, M. Philippe (L'Oréal). High-Yield Preparation of Ceramides by Conducting the Aminoalcohol Amidation in the Presence of Microwave Irradiation. *Eur. Patent Application* EP 884,305, **1998** (FR Application 97/7,240, 11 Jun 1997); *Chem. Abstr.*, **1999**, *130*, 52677y.
254 G. Lindeberg, M. Larhed, A. Hallberg (Labwell AB). Method for Organic Reactions – Transition Metal Catalyzed Organic Reactions. *PCT International Application* WO 43,230, **1997** (SE Application 96/3,913, 25 Oct 1996); *Chem. Abstr.*, **1998**, *128*, 34382c.
255 J. Burkhardt, P. Schubert. Process for the preparation of an organozinc derivatives and its use as a nutritional supplement. *Eur. Pat. Appl.* EP 1529774, **2005**.
256 R. Xiong, S. D. Pastor, P. Bujard. Process preparation of metal oxide coated organic material by microwave

deposition. *PCT International Application* WO 2004111298, **2004**.

257 F.-R. ALEXANDRE, L. DOMON, S. FRÈRE, A. TESTARD, V. THIÉRY, T. BESSON, *Molecular Diversity*, **2003**, *7*, 273.

258 (a) G. A. STROHMEIER, C. O. KAPPE, *J. Comb. Chem.* **2002**, *4*, 154; (b) C. O. KAPPE, A. STADLER, *Microwaves in Organic and Medicinal Chemistry*, Vol. 25. Methods and Principles in Medicinal Chemistry (Eds, R. MANNHOLD, H. KUBINYI, G. FOLKERS). Wiley–VCH, Weinheim, **2005**.

259 G. CALIENDO, E. PERISSUTTI, V. SANTAGADA, F. FIORINO, B. SEVERINO, D. CIRILLO, R. D'EMMANUELE DI VILLA BIANCE, L. LIPPOLIS, A. PINTO, R. SORRENTINO, *Eur. J. Med. Chem.*, **2004**, *39*, 825.

260 X. FANG, R. HUTCHENON, D. A. SCOLA, *J. Polym. Sci.: Part A: Polym. Chem.*, **2000**, *38*, 1379.

261 S. VELMATHI, R. NAGAHATA, J. SUGIYAMA, K. TAKEUCHI, in *Microwave 2004, Proceedings of International Symposium on Microwave Science and Its Application to Related Science*, Takamatsu, Japan, **2004**, pp. 91–93.

262 W. ZHANG, J. GAO, C. WU, *Macromolecules*, **1997**, *30*, 6388.

263 K. FAGHIHI, K. ZAMANI, A. MIRSAMIE, S. MALLAKPOUR, *J. Appl. Poly. Sc.*, **2004**, *91*, 516.

264 W. VWTER, D. JANUSSEN, *Environ. Sci. Technol.*, **2005**, *39*, 3889.

265 *Solid Phase Microwave Reactor (SPMR)*, www.milestone.srl.com.

266 R. J. J. JACHUCK, D. K. SELVARAJ, R. S. VARMA, *Green Chem.*, **2006**, *8*, 29.

9
Microwave-assisted Reactions on Graphite

Thierry Besson, Valérie Thiéry, and Jacques Dubac

9.1
Introduction

During the last 15 years numerous papers dealing with the use of microwave (MW) irradiation, rather than conventional heating, in organic and inorganic chemistry have reported dramatic reductions of reaction time and significant enhancement of yields and purity of the products. Despite the possibility of operating with pressurized reactors [1], however, MW irradiation of chemicals reactions involving low boiling reagents and/or products can involve serious safety problems. Consequently, MW-assisted solvent-free reactions ("dry media") have been widely investigated in organic synthesis [2]. Among the materials most often used as supports are alumina, silica, clays, and zeolites, which are sometimes also used as catalysts. When properly dried, however, these materials are good-to-moderate MW absorbers and poor thermal conductors. For reactions which require high temperatures, the idea of using a reaction support which takes advantage both of efficient MW coupling and strong adsorption of organic molecules has stimulated great interest. Because most organic compounds do not interact appreciably with MW radiation, such a support could be an ideal "sensitizer", able to absorb, convert, and transfer energy provided by a MW source to the chemical reagents.

Most forms of carbon, except diamond, which are renowned as supports for precious metal catalysts in some applications [3], interact strongly with MW [4]. Amorphous carbon and graphite, in their powdered form, irradiated at 2.45 GHz, rapidly (within 1 min) reach very high temperatures (>1300 K). This property has been used to explain MW-assisted syntheses of inorganic solids [5]. In these syntheses, carbon is either a "secondary susceptor" which assists the initial heating but does not react with other reactants, or is one of the reactants, e.g. in the synthesis of metal carbides. MW–carbon coupling has also been widely developed:

- by Wan and coworkers for gas-phase reactions; for example, in the synthesis of hydrogen cyanide from ammonia and carbon or methane [6], in the MW-induced catalytic reaction of water and carbon [7], and in the removal and/or destruction of acid gaseous pollutants such as SO_2 and NO_x [7, 8]; and

Microwaves in Organic Synthesis, Second edition. Edited by A. Loupy
Copyright © 2006 WILEY-VCH Verlag GmbH & Co. KGaA, Weinheim
ISBN: 3-527-31452-0

- for processing of polymers and composites in which carbon black or graphite particles or fibers are induced in the material [9].

The MW-promoted cracking of organic compounds in the presence of silica-supported graphite [10a] or activated charcoal [10b] has also been reported.

Graphite, the most stable of the three allotropic forms of carbon, has two structures, α (hexagonal form), and β (rhombohedral form), which interconvert easily [11]. In a graphite layer each carbon atom is strongly bonded to three other carbon atoms in a planar configuration (sp^2 hybridization); the remaining p electrons (one per carbon) are delocalized. The resulting carbon–carbon bonds are very strong (477 kJ mol^{-1}). The interlayer bonds, in contrast, are weak (17 kJ mol^{-1}), giving rise to the mechanical (lubricant) and chemical (intercalation) properties of graphite. Electronically, graphite is a semimetal of high electrical and thermal conductivity [11]. As for the other semimetallic materials, the electronic current (σ) is the main factor in graphite–MW interaction [5]. The rate of heating of a MW-irradiated material has been estimated to be $\Delta T/t = \sigma \cdot |E|^2/\rho c$, where E is the electric field, ρ is the density, and C is the specific heat capacity of material [5]. Compared with other dielectric solids, graphite has an unusually high thermal conductivity (a weak C, 0.63 kJ kg^{-1} K^{-1} at room temperature) [11]. This thermal conductivity, which decreases exponentially with increasing temperature, is a determining factor in the high rate of heating of graphite on MW irradiation, although other types of MW interaction, e.g. the excitation of weak interlayer bonds and, especially in graphite powder, eddy currents or localized plasma effects, can also lead to very rapid dissipation of energy in graphite [5].

Because of its strong coupling with MW, its good adsorbent properties towards organic molecules [12], and its layer structure which enables it to form intercalated compounds [13], graphite has great potential in MW-assisted synthetic applications in organic chemistry, despite its weak fractal dimension ($D \approx 2$) [14].

Papers on the use of graphite in organic synthesis are quite recent. Studies in this field have increased since the work of Laurent (Sections 9.2.1, 9.2.2, 9.2.4, 9.2.7, 9.3.2, and 9.3.3) [15], Bond (Section 9.2.3) [16], and Villemin (Section 9.2.4) [17] and their coworkers.

These applications are presented here in two parts. In the first part of this chapter, graphite behaves as an energy converter (or "sensitizer") capable of conveying the energy carried by MW radiation to the chemical reagents. The objective of the authors was to review recent developments in microwave-assisted synthesis of heteroaromatic compounds under conditions that include the use of graphite and microwave irradiation in the quinazoline and thiazole ring-forming step (Sections 9.2.5.1–9.2.5.3). As for the preceding review [18], acylation of aromatic compounds (Section 9.3.2), acylative cleavage of ethers (Section 9.3.3), and ketocarboxylation of carboxylic diacids (Section 9.3.4) will also be included. The second part of this chapter reveals the surprising catalytic activity of some metal inclusions of graphite. Results are abundantly described and discussed in this review. Description of microwave-assisted chemistry is followed by general notes (Section 9.4) on graphite properties, MW apparatus, typical procedures, and safety measures.

9.2
Graphite as a Sensitizer

Owing to particularly strong interaction with MW radiation and high thermal conductivity, graphite is an efficient sensitizer. It is capable of converting radiation energy to thermal, which is then transmitted instantaneously to the same reactions as performed by classical heating. Two reaction types take advantage of this graphite–MW coupling:

1. reactions which require a high temperature, and
2. reactions involving chemical compounds, for example organic compounds, which have low dielectric loss and are not heated sufficiently under the action of MW irradiation.

Its inert behavior toward numerous chemical compounds and its adsorbent properties (responsible for the retention of volatile or sublimable organic compounds) make graphite the support of choice for thermal reactions. Among its impurities magnetite has been revealed to be an active catalyst and some reactions can be performed without any added catalyst (see Section 7.3 in the preceding review [18]). Two processes are then possible, the graphite-supported reaction ("dry" process) and reaction in the presence of a small amount of graphite (solid–liquid medium).

This section covers reactions in which graphite is a sensitizer, without participation of its metal inclusions as possible catalysts, although a catalyst can be added to the graphite. The amount of graphite can be varied. It is usually at least equal to and most often greater than that of the reagents, resulting in a graphite-supported–microwave (GS–MW) process. Occasionally, optimization of processes have shown that a "catalytic amount" (10% or less than 10% by weight) of graphite is sufficient to induce rapid and strong heating of the mixture. Some novel examples are described.

9.2.1
Diels–Alder Reactions

Many Diels–Alder (DA) cycloadditions have been studied under the action of MW irradiation [19]. The use of a "dry process", as in GS–MW coupling, is of great interest for difficult reactions which need high temperatures, particularly those involving poor MW-absorbing reagents. Some reactions which normally require use of an autoclave can, moreover, occur in an open reactor, owing to retention of a possibly volatile reagent by the graphite.

Among the dienes known to be weakly reactive are anthracene (Scheme 9.1, **1**), metacrolein dimethylhydrazone (**2**), and 3,6-diphenyl-1,2,4,5-tetrazine (**3**). DA cycloadditions with these dienes require long reaction times under classical heating conditions (Table 9.1).

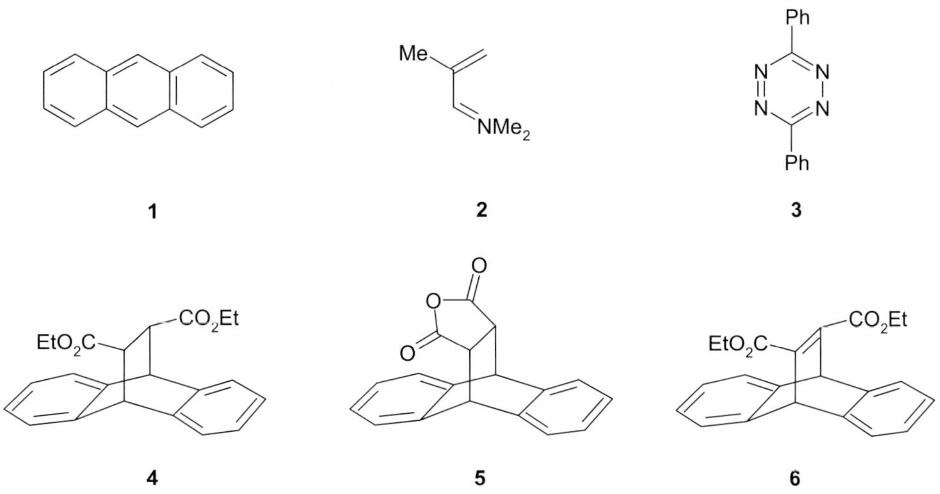

Scheme 9.1

Tab. 9.1. Diels–Alder reactions of dienes **1–3** using the GS–MW process [15].

Entry	Adduct	SG–MW irradiation conditions[a] (T_{max})	Yield (%)[b]	Conventional heating conditions	Yield (%)
1	4	120 W; 1 min[c] (370 °C)	92	Dioxane, reflux, 60 h [20]	90
2	4	30 W; 1 min × 3[d] (147 °C)	92	–	–
3	5	30 W; 1 min × 3[d] (155 °C)	75	p-Xylene, reflux, 10 min [21]	90
4	6	30 W; 1 min × 3[d] (130 °C)	97	–[e] [20]	Quant.
5	7	30 W; 1 min × 10[d] (171 °C)	50	No reaction [21, 22]	–
6	8	30 W; 1 min × 10[d] (157 °C)	62	$CHCl_3$, reflux, 120 h[f]	70
7	9	30 W; 1 min × 5[d] (168 °C)	72	CCl_4, 60 °C, 3 h [23]	–[e]
8	10b	30 W; 1 min × 5[d] (154 °C)	93	75 °C, 30 min [24]	–[e]
9	11	30 W; 1 min × 20[d] (160 °C)	60	Toluene, reflux, 50 h [25]	94

[a] See typical procedure; reagents used in equimolar amounts (entries 1–5) or excess dienophile, 5:1 (entries 6, 7, 9) or 2:1 (entry 8)
[b] Yield of isolated product relative to the minor reagent
[c] Continuous MW irradiation (CMWI); applied incident power; irradiation time; maximum temperature indicated by IR pyrometer
[d] Sequential MW irradiation (SMWI); applied incident power; time and number of irradiations; interval between two irradiations: 2 min (entries 2–5), 1 min (entries 6–9)
[e] Not given
[f] This work

Three reactions of **1**, with diethyl fumarate, maleic anhydride, and dimethyl acetylenedicarboxylate (DMAD), are highly representative of the variety of experimental conditions used in the GS–MW process [26, 27]. Continuous MW irradiation (CMWI) with an incident power of 120 W for 1 min led to a large increase in temperature ($T_{max} > 300\ °C$). Adduct **4** was obtained almost quantitatively (Table 9.1, entry 1) whereas only traces of adducts **5** and **6** were detected. When the incident power was reduced (30 W) and sequential MW irradiation (SMWI) was used, adducts **5** and **6** were obtained in good yield (Table 9.1, entries 3 and 4). This controlled irradiation enabled the temperature to be limited ($T_{max} < 200\ °C$) and avoided the retro-DA reaction. In the reaction between **1** and diethyl fumarate similar SMWI conditions also gave the adduct **4** in high yield (Table 9.1, entry 2).

Other DA reactions of **1** (and some of its derivatives) in SMWI processes have been reported [28]. Under powerful irradiation ($T_{max} > 300\ °C$), all products decomposed by the retro-DA reaction.

The hetero-DA reaction with azadienes, a well known synthetic method for obtaining nitrogen heterocycles, suffers from difficulties, because of the low reactivity of the diene. For example, azadiene **2** did not react with DMAD under the action of conventional heating [22]. Sequential exposure to MW irradiation (30 W) for 10 min on a graphite support ($T_{max} = 171\ °C$) led to the adduct **7** (Scheme 9.2) with 60% conversion (50% in isolated product) [26, 27].

An equivalent yield was obtained by ultrasonic irradiation of the neat reaction mixture at 50 °C for 50 h [29].

The DA reaction of tetrazines such as **3** was also studied by use of the GS–MW process [26, 27]. The expected adduct, however, decomposed, with elimination of nitrogen followed by dehydrogenation, giving a pyridazine or a dihydropyridazine [23–25]. With 2,3-dimethylbutadiene and cyclopentadiene as dienophiles, SMWI

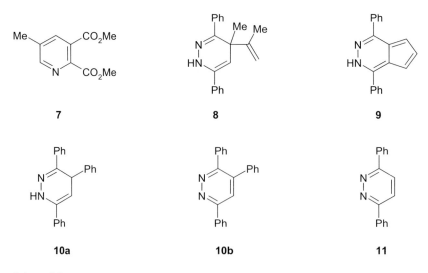

Scheme 9.2

gave dihydropyridazines **8** and **9**, as under classical heating [23] (Table 9.1, entries 6 and 7).

Under classical conditions, the reaction between **3** and styrene required 50 h of heating at 110 °C, and gave the dihydropyridazine adduct **10a** [24]. After SMWI with 30 W incident power for 5 min (T_{max} = 154 °C), the adduct **10a** was not detected whereas the totally dehydrogenated product, pyridazine **10b**, was isolated in almost quantitative yield (Table 9.1, entry 8). Ethyl vinyl ether and **3** gave the same product, pyridazine **11**, under both classical heating [25] and MW irradiation conditions (Table 9.1, entry 9). In this instance the DA adduct lost nitrogen and ethanol.

Synthesis of these adducts was realized in very short times compared with the same reactions under the action of classical heating. The efficiency of the MW process is all the more noteworthy because the three dienophiles (dimethylbutadiene, cyclopentadiene, and ethyl vinyl ether) are volatile. Although an excess of these reagents relative to **3** was used, the adsorption power of graphite was responsible of their retention, because the temperature of the reaction mixture exceeded their boiling points of approximately 120–130 °C.

This retention of the reagents by the graphite support is known from a series of experiments involving volatile dienes such as 2,3-dimethylbutadienes (**12**) and isoprene (**13**) (Table 9.2) [30, 31]. Reaction of **12** with diethyl mesoxalate gave **14** (Scheme 9.3) in 75% yield after SMW irradiation at low power (30 W) for 20 × 1 min only (Table 9.2, entry 1).

When conventional heating was used, a satisfactory yield was obtained after 4 h at 135 °C in a sealed tube [32]. Ethyl glyoxylate is a weaker carbonyl dienophile than diethyl mesoxalate, and a catalyst ($ZnCl_2$) was added to afford the expected DA adducts with dienes **12** and **13** in goods yields (Table 9.2, entries 2 and 7). For reaction with **13**, the catalyst $SnCl_4$ [34] was substituted by $ZnCl_2$ to prevent formation of the ene reaction product (Table 9.2, entry and 7). Although adduct **16** was previously prepared from **15** [32], its direct DA synthesis from **12** and glyoxylic acid could be performed under the action of MW and without a catalyst (Table 9.2, entry 3).

Reactions performed with methyl vinyl ketone and metacrolein as ethylenic dienophiles also revealed the clear advantage of SMWI conditions over conventional heating (Table 9.2, entries 4–6) [31]. In the reaction of isoprene with methyl vinyl ketone (Table 9.2, entry 6), selectivity for the para adduct (54%) was much better than when conventional heating was used (26%), probably owing to the reduction of the reaction time.

Another example of the retention of volatile DA reagents is that of cyclopentadiene in a tandem retro-DA–DE "prime" reaction [15, 38]. This reaction type is the thermal decomposition of a DA adduct (A) and the generation of a diene (generally the initial diene) which is trapped *in situ* by a dienophile leading to a new adduct (B) [39]. Cyclopentadiene (**22**) (b.p. 42 °C) is generated by thermolysis of its dimer at approximately 160 °C [40]. An equimolar mixture of commercial crude dicyclopentadiene (**21**) and dimethyl maleate was irradiated in accordance with the GS–MW process, in an open reactor, under 60 W incident power, for 4 min

Tab. 9.2. Diels–Alder reactions with 2,3-dimethylbutadiene (**12**) and isoprene (**13**) by use of the GS–MW process [27, 30, 31].

Entry	Adduct	MW irradiation reaction time (min)[a]	Yield (%)[b]	Conventional heating conditions	Yield (%)
1	14	30 W; 1 min × 20[c] (165 °C)	75	MeCN, 135 °C[d], 4 h [32]	86
2	15	30 W; 1 min × 10[c,e] (165 °C)	87	140 °C, 6 h [33]	65
3	16	30 W; 1 min × 10[c] (249 °C)	54	–	–
4	17	30 W; 1 min × 10[c] (126 °C)	89	Xylene, 95 °C, 80 h [35]	–[f]
5	18	30 W; 1 min × 10[c] (133 °C)	77	Benzene, 150 °C[d] [36]	65
6	19	30 W; 1 min × 2[c,e] (89 °C)	54	CH_2Cl_2 ($AlCl_3$), 20 °C, 24 h [37]	26
7	20	30 W; 1 min × 10[c,e] (146 °C)	73	CH_2Cl_2 ($SnCl_4$), 20 °C, 18 h [34]	10

[a] Reagents used in equimolar amounts (entries 4–6) or in excess diene, 5:1 (entries 2, 3), 3:1 (entry 7), 2:1 (entry 1)
[b] Yield of isolated product relative to the minor reagent.
[c] Sequential MW irradiation (SMWI); applied incident power; time and number of irradiations; interval between two irradiations: 1 min (entries 1–3), 3 min (entries 4–7)
[d] Reaction in sealed tube
[e] Reactions performed in the presence of a catalyst: $ZnCl_2$ (entries 2, 7), $AlCl_3$ (entry 6)
[f] Not given

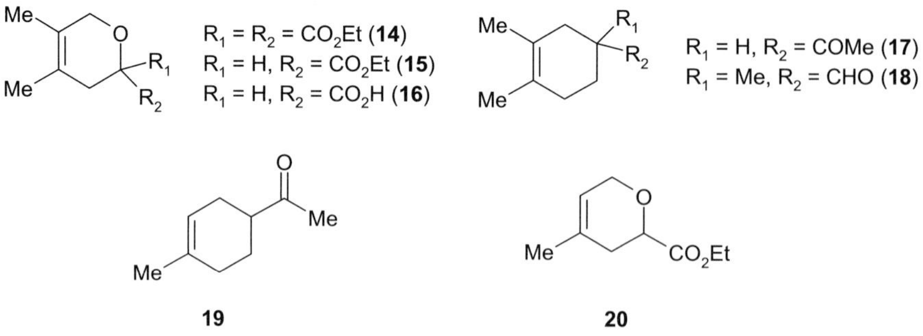

$R_1 = R_2 = CO_2Et$ (**14**)
$R_1 = H, R_2 = CO_2Et$ (**15**)
$R_1 = H, R_2 = CO_2H$ (**16**)

$R_1 = H, R_2 = COMe$ (**17**)
$R_1 = Me, R_2 = CHO$ (**18**)

19

20

Scheme 9.3

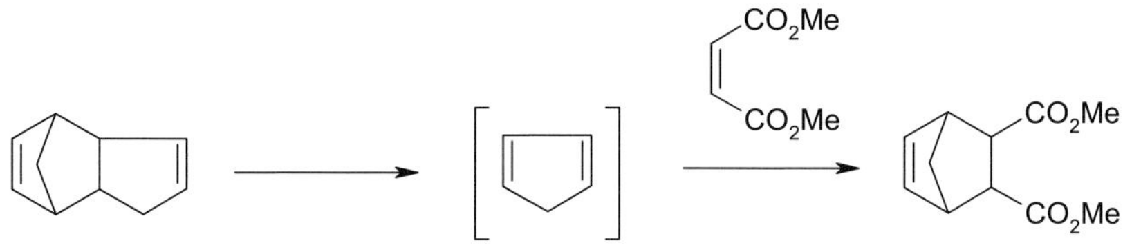

Scheme 9.4

(8×30 s). The expected adduct **23** was isolated in 40% yield (Scheme 9.4). The isomeric composition of **23** (endo–endo:exo–exo = 65:35) was identical with that obtained under classical conditions from **22** and methyl maleate [41]. The overall yield of this tandem reaction can be increased from pure dimer **21** (61%) and the same tandem reaction has also been reported using ethyl maleate as dienophile [31].

The main advantages of the GS–MW process are the rapid increase of temperature, the retention of organic molecules, and the possibility of performing some reactions under one-pot conditions.

9.2.2
Ene Reactions

The ene reaction (or Alder reaction) is a cycloaddition which requires an activation energy higher than that of the Diels–Alder reaction [41]. Without catalyst it usually occurs under pressure and/or at high temperature. The reaction with an alkene (ene) is much easier if the latter is more substituted (high HOMO) and the enophile is more electron-poor (low LUMO).

The alkene 1-decene (**24**) was poorly reactive in the carbonyl–ene reaction with ethyl mesoxalate and required a temperature up to 170 °C for a very long time (5 h) [42]. When performed in a homogeneous liquid medium at the same temperature, but under the action of MW irradiation the reaction gave a similar result. Reaction time was appreciably reduced by the use of GS–MW coupling [30]. Thus, irradiation at 60 W for only 10 min led to the ene adduct **25** in 50% yield (Scheme 9.5). Under these conditions a maximum temperature of 230 °C was measured. To obtain the same yield with conventional heating at 170° C, reaction for 1 h is required.

The stereoselectivity of the reaction was not related to the mode of heating. Better conversion of **24** was obtained by increasing the irradiation (incident power >60 W), but the occurrence of side products made isolation of **25** more difficult.

(−)-β-Pinene (**26**), a more reactive ene than **24**, reacted with ethyl mesoxalate under CCl_4 reflux in 90% yield after 5 h conventional or MW heating [42]. The reaction supported on graphite occurred in only 2×1 min of MW irradiation with an incident power of 60 W. The adduct **27** was obtained in 67% isolated yield (Scheme 9.6) [30]. This yield was obtained after 2 h reaction under CCl_4 reflux.

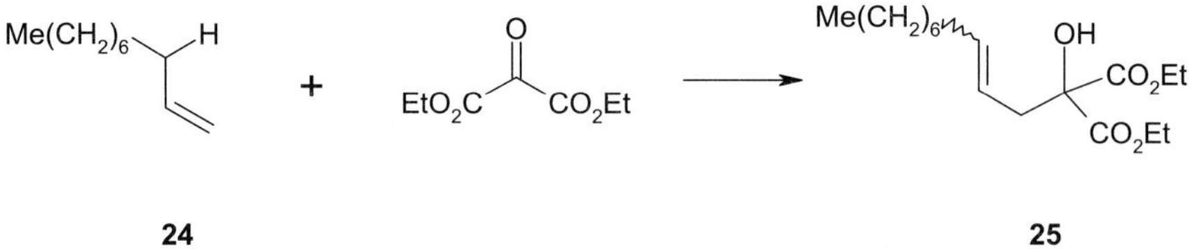

Me(CH₂)₆—CH=CH₂ + EtO₂C–C(=O)–CO₂Et → Me(CH₂)₆–CH=CH–CH₂–C(OH)(CO₂Et)(CO₂Et)

24 → **25**

Thermal heating: 170°C, 1 h   50%
                    5 h   quant.(c/t = 26/74) [42]

GS/MW: 60W, 1min x 10   50% (c/t = 30/70) [30]

Scheme 9.5

The intramolecular cyclization of (+)-citronellal (**28**) leads to a mixture of isomeric pulegols (**29**) and, particularly, to (−)-isopulegol (**29a**) which is an intermediate in the synthesis of (−)-menthol [43]. The reaction can be performed by heating in the absence [44] or presence [43, 45] of a catalyst, including zeolite [46] and montmorillonite [42] under the action of MW. Cyclization of **28**, when performed by the GS–MW process −4 min (8 × 30 s) irradiation at 30 W incident power – resulted in 88% conversion (80% isolated yield) to the pulegols (**29**) (Scheme 9.7).

The same reaction under conventional heating or MW heating at a constant temperature of 180 °C yielded only 50% conversion after 4 h [42]. Considering that the maximum temperature measured during GS–MW irradiation was only 210 °C, and that the reaction was not catalyzed, the acceleration was clearly very strong. It has been observed that, in contrast with clays such as montmorillonite, graphite in catalytic amounts does not accelerate the cyclization of **28** from room temperature to 180 °C [42]. The authors believe that under the heterogeneous conditions of the GS–MW process the temperature shown by IR pyrometer is too low and gives no indication of probable "hot spots" produced at the surface of graphite grains (Section 9.4.2). The stereoselectivity was, moreover, somewhat different from that obtained by classical heating – although the amount of (+)-neoisopulegol (**29b**) was increased (−)-isopulegol (**29a**) remained the main diastereoisomer (68%).

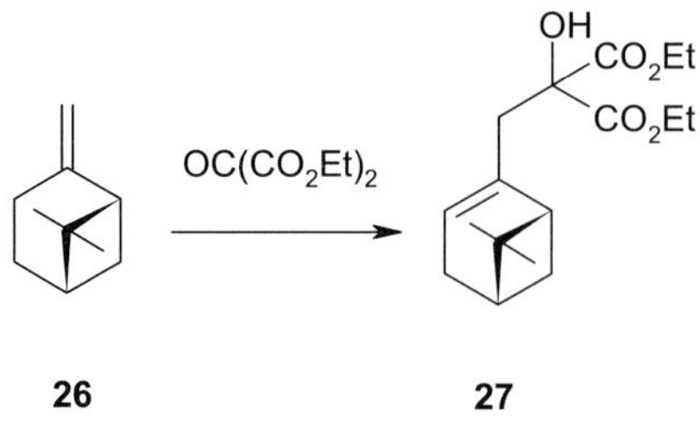

26 —OC(CO₂Et)₂→ **27**

Thermal heating: CCl₄ reflux, 2 h   65%
                       5 h   92% [42]

GS/MW: 60W, 1min x 2   67% [30]

Scheme 9.6

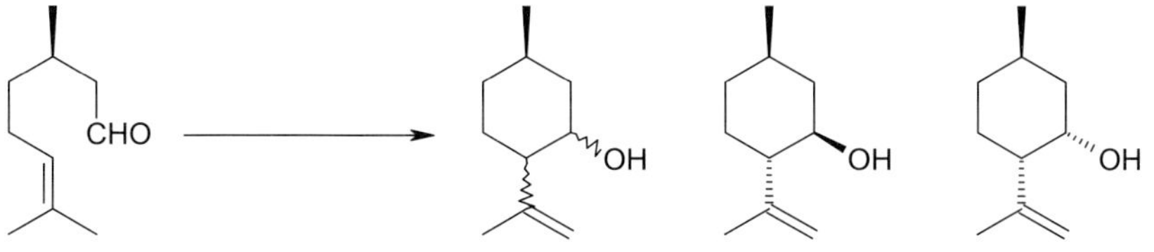

Scheme 9.7

Thermal heating: 180°C, 30 h   90% (**29a/29b**/other = 71/16/13) [44]

GS/MW: 30W, 30 s x 8   80% (**29a/29b**/other = 68/30/2) [30]

Scheme 9.7

9.2.3
Oxidation of Propan-2-ol

Reaction of an alcohol over basic catalysts favors dehydrogenation to give the corresponding carbonylated derivative. This reaction has been studied for propan-2-ol over a series of alkaline carbon catalysts, under the action of conventional heating and MW irradiation (Scheme 9.8) [16].

The main effect of MW irradiation on the graphite and charcoal-supported catalysts is to reduce the average temperature required for reaction to occur. The authors believe this is the result of "hot spots" formed within the catalyst bed (Section 9.4.2). Graphite-supported catalysts seem, moreover, to be more selective than the equivalent charcoal-supported catalysts, especially under the action of MW irradiation – 3.6–97.7% compared with 68.4–86.3%. This might be because of the hydrophobic nature of the graphite which directs the reaction away from the production of water by dehydration of the alcohol.

MeCHOHMe → (1% M / C, N_2) → MeCOMe

M: Li, Na, K
C: graphite or charcoal
T: 222-345°C (MW)
     260-530°C (Δ)

Scheme 9.8

9.2.4
Thermolysis of Esters

The thermolysis of esters is a much used reaction in organic [47] and organometallic [48] syntheses, generally for the creation of a carbon–carbon double bond. The

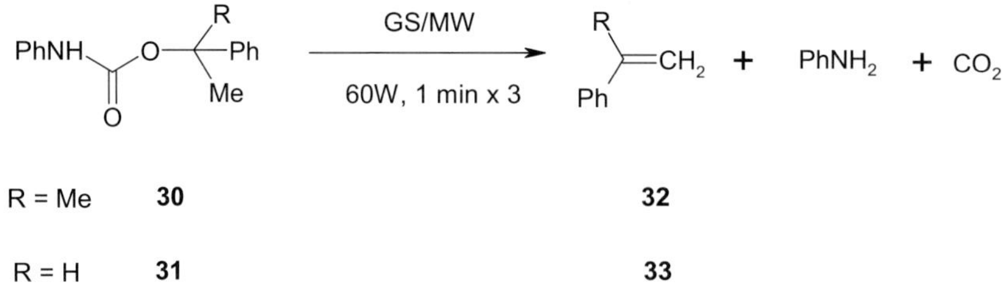

PhNH–C(=O)–O–C(R)(Ph)(Me) →[GS/MW, 60W, 1 min × 3] Ph–C(R)=CH$_2$ + PhNH$_2$ + CO$_2$

R = Me   **30**   **32**

R = H   **31**   **33**

Scheme 9.9

mechanism is often like that of the retro-ene reaction, requiring high temperatures. Among the esters, those of carbamic acid are more easily decomposed than those of carboxylic acids.

The high temperatures reached in the GS–MW process have been used to achieve thermal decomposition of O-alkylcarbamates (Scheme 9.9) [15]. The latter are prepared from the corresponding alcohols and phenyl isocyanate in the presence of stannous octanoate [49].

With 1,1-dimethylbenzyl phenylcarbamate (**30**), a tertiary carbamate:

- Heating in *p*-xylene under reflux for 10 h led to no decomposition;
- MW irradiation (150 W) of pure crystalline **30** led to 45% 2-methylstyrene (**32**) after 60 min (6 × 10 min); and
- MW irradiation of graphite powder impregnated with **30** gave 90% **32** after 3 min (3 × 1 min) under 60 W incident power ($T_{max} = 320\ °C$).

This example shows that the GS–MW process can be used to accomplish thermal decomposition which cannot be performed efficiently by use of MW irradiation alone, because of weak MW absorption by starting compound (**30**) of probable low dielectric loss.

The decomposition of a secondary carbamate, 1-methylbenzyl phenylcarbamate (**31**), was more difficult, and only 60% of styrene (**33**) was obtained under the same conditions ($T_{max} = 340\ °C$). Attempts to decompose a primary carbamate, 1-octyl phenylcarbamate, failed because its sublimation preceded its decomposition [15]. All these reactions have been performed in an open reactor by the above procedure (Section 9.2.1).

Some esters not having an aliphatic hydrocarbon chain are liable to thermal rearrangement. This is observed for O-arylthiocarbamates, for which rearrangement into S-arylthiocarbamates has been studied by Villemin et al. on different supports and under the action of MW irradiation (Scheme 9.10) [17].

No rearrangement was observed for the pure compound **34a**, adsorbed or not on KF–Al$_2$O$_3$, probably owing to its low dielectric loss. By using supports known to convert MW energy into thermal energy, the authors achieved a conversion rate of

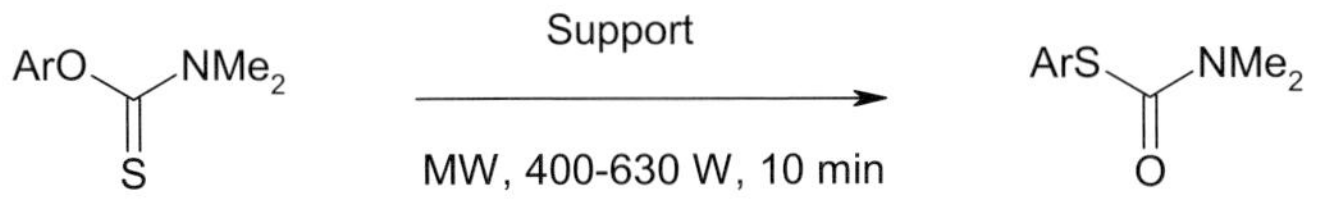

ArO_C(S)_NMe₂ → ArS_C(O)_NMe₂

34 → **35**

Conditions: Support, MW, 400–630 W, 10 min

Ar	Support	Conversion (%)
Ph (a)	alumina[a]	15
Ph (a)	silica[a]	20
Ph (a)	vermiculite[a]	30
Ph (a)	silicon carbide[a]	40
Ph (a)	graphite[b]	70
4-methylphenyl (b)	graphite[b]	45
4-ntrophenyl (c)	graphite[b]	90
4-carboxymethylphenyl (d)	graphite[b]	70
2-naphthyl (e)	graphite[b]	30

[a] MW irradiation: 630 W, 10 min; [b] MW irradiation: 400 W, 10 min; T_{end} = 240–270 °C

Scheme 9.10

15 to 90% for **34**. The best yields, 30 (**35e**) to 90% (**35c**), were obtained on graphite powder.

9.2.5
Thermal Reactions in Heterocyclic Syntheses

For more than a century, heterocycles have constituted one of the largest area of research in organic chemistry. The presence of heterocycles in all kinds of organic compound of interest in biology, pharmacology, optics, electronics, material sciences, and so on is very well known. Sometimes, the preparation of these heterocyclic systems by conventional routes is hard work that implies many synthetic steps and extensive starting material [50a]. The recent availability of commercial microwave systems specific for synthesis offers improved opportunities for reproducibility, efficient synthesis, rapid reaction optimization, and the potential discovery of new chemistries. Microwaves also have an advantage where processes involve sensitive reagents or when products may decompose under prolonged reaction conditions.

9.2.5.1 Synthesis of Quinazolines and Derivatives

Many reactions in heterocyclic multistep syntheses involve thermal condensations. Among these, the Niementowski reaction is the most common method for synthesis of the 3*H*-quinazolin-4-one ring. It involves the fusion of anthranilic acid (or a derivative, e.g. 2-aminobenzonitrile) with formamides or thioamides (or their *S*-methyl derivatives) and usually needs high temperatures and requires lengthy and tedious conditions. Recently, Besson and coworkers studied the possibilities offered by this reaction and explored the preparation of novel bioactive heterocycles (e.g. **38**, **39**, and **40** in Scheme 9.11) in which the quinazoline skeleton is fused with thiazole, indole or benzimidazole rings [50a–c].

They first investigated thermal heating of benzimidazo or indolo[1,2-*c*]quinazolines (**36** and **37**) and anthranilic acids [50b, c] (Scheme 9.11). As this

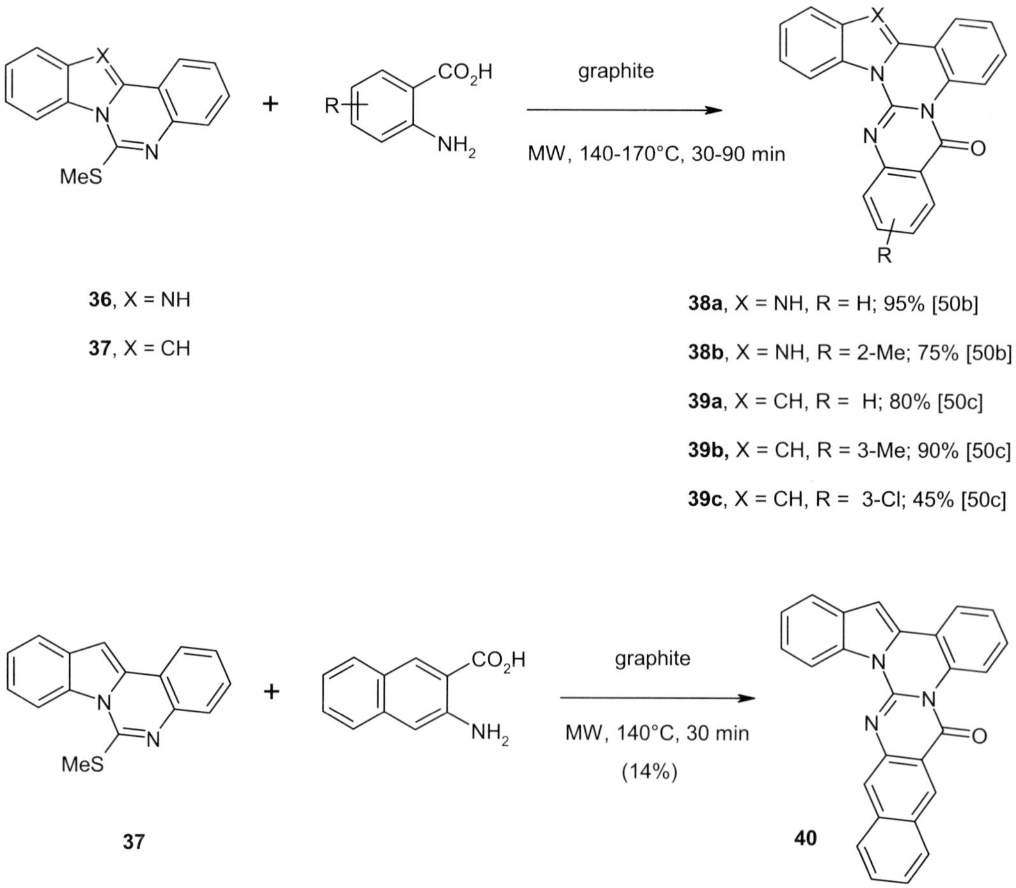
X
N
N
MeS
+
R
CO_2H
NH_2
graphite
MW, 140-170°C, 30-90 min
X
N
N
N
O
R
36, X = NH
37, X = CH
38a, X = NH, R = H; 95% [50b]
38b, X = NH, R = 2-Me; 75% [50b]
39a, X = CH, R = H; 80% [50c]
39b, X = CH, R = 3-Me; 90% [50c]
39c, X = CH, R = 3-Cl; 45% [50c]
N
N
MeS
+
CO_2H
NH_2
graphite
MW, 140°C, 30 min
(14%)
N
N
N
O
37
40

Scheme 9.11. Synthesis of triaza or tetraazabenzo[*a*]-benzimidazo (**38**) or indeno[1,2-*c*]quinazolines (**39**).

group has shown, the yield of this reaction depends on the mode of activation. Thus, thermal heating of the two starting material compounds, neat at 120 °C or 140 °C or in butanol under reflux for 48 h, did not give more than 50% of final product [50b]. In contrast, MW irradiation enabled a striking reduction of reaction time accompanied by a real improvement in the yield. Then, benzimidazole (**38**) and indole (**39, 40**) polycyclic derivatives (Scheme 9.11) are obtained in interesting yields (45–95%). Longer heating periods resulted in degradation products [50c].

By following the same strategy, this research group also achieved the efficient microwave-assisted synthesis of original 8*H*-quinazolino[4,3-*b*]quinazolin-8-ones [50d]. Homogeneous and heterogeneous conditions were studied and it was observed that microwave irradiation (150 °C, 60 W) of a mixture 4-(thiomethyl)quinazoline (**41**) and excess anthranilic acid, absorbed on graphite, led to the cyclized products (**42a–g**) (Scheme 9.12). The yields are good (21–79%) and reaction time is shorter than for purely thermal procedures (oil bath or metal bath). During this study Besson and coworkers have also shown that although this graphite-supported method can be regarded as an interesting solvent-free procedure, a more productive (with better yields: 41–85%) microwave-assisted homogeneous method (using acetic acid as solvent) was achieved.

The three studies described above confirmed that the ratio of the quantity of ma-

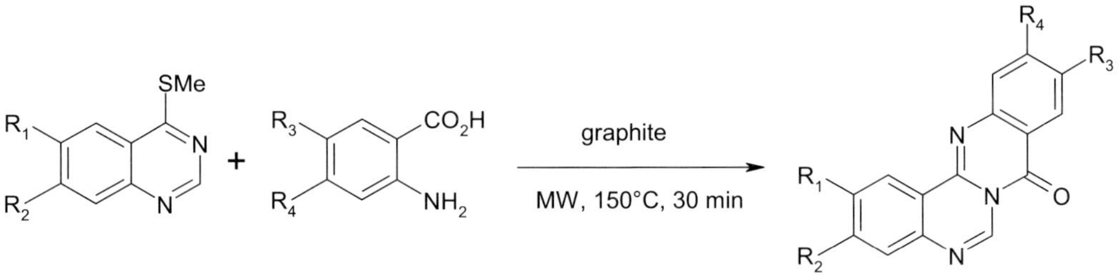

(a) $R_1 = R_2 = R_3 = R_4 = H$; 79%

(b) $R_1 = Me, R_2 = R_3 = R_4 = H$; 58%

(c) $R_1 = Br, R_2 = R_3 = R_4 = H$; 21%

(d) $R_1 = OMe, R_2 = OMe, R_3 = R_4 = H$; 34%

(e) $R_1 = R_2 = R_4 = H, R_3 = Me$; 53%

(f) $R_1 = R_2 = R_4 = H, R_3 = Br$; 50%

(g) $R_1 = R_2 = H, R_3 = R_4 = OMe$; 29%

Scheme 9.12

terial to the quantity of support may be very important. The adsorption of organic molecules by graphite was very useful in procedures in which anthranilic derivatives can sublime during the reaction, changing the ratio of the reactants and leading to byproducts, necessitating complex work-up. Use of excess graphite in these experiments prevented substantial sublimation of the anthranilic acids and led to the good yields observed. Reactions performed with smaller quantities of graphite gave worse results (mainly low yields). It is also important to notice that all these microwave-assisted reactions were realized at atmospheric pressure in monomode microwave reactors.

Pursuing their efforts on the Niementowski reaction and its possibilities, Besson and coworkers have recently extended the family of fused quinazolinones which can be obtained via the microwave-assisted Niementowski reaction from the starting amidines (**43**). The authors described rapid and convenient access to pentacyclic 6,7-dihydro-5a,7a,13,14-tetraazapentaphene-5,8-diones (**44**) structurally related to well studied terrestrial alkaloids (e.g. rutaecarpine and luotonine A) [50e]. The strong thermal effect, because of graphite–microwaves interaction, was particularly efficient in these reactions, in which the quinazolinone and the piperazine rings are fused. Only 10% by weight of graphite was used, in pressurized monomode reactors (Scheme 9.13).

During investigation of the biological potential of sulfur-containing heterocycles, the strategy applied above was extended to the preparation of novel 2,3-condensed thieno[2,3-*d*]pyrimidinones [50f] from **43** or **45**, themselves obtained by condensation of ethylenediamine on imino-1,2,3-dithiazoles. Formation of the 5,6-dihydro-3-thia-4a,6a,12,13-tetraazaindeno[5,6-*a*]anthracene-4,7-dione (**46**) was realized in the presence of a "catalytic amount" (approx. 10% by weight) of graphite. Here again the synergic effect of pressure and microwaves afforded very rapid heating with lack of sublimation of anthranilic acid and its derivatives into the vials (Scheme 9.14).

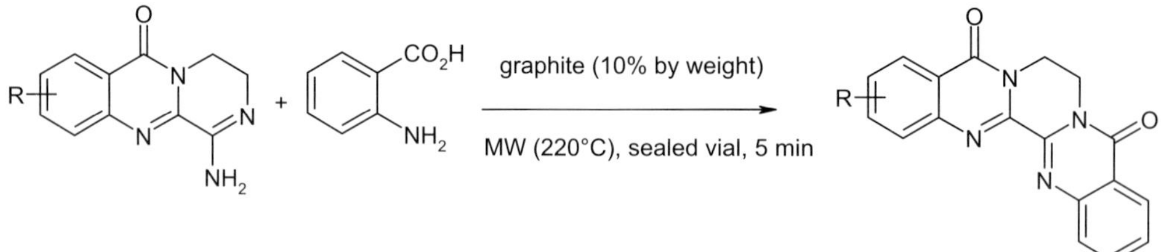

(a) R = H; 52%

(b) R = 4-Cl, 68%

(c) R = 5-Br; 65%

(d) 4,5-diOMe; 36%

Scheme 9.13

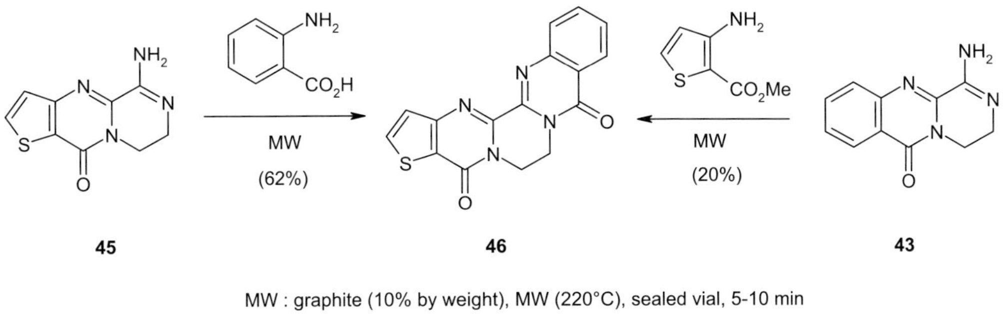

NH2
CO2H
MW
(62%)
NH2
CO2Me
MW
(20%)
45
46
43
MW : graphite (10% by weight), MW (220°C), sealed vial, 5-10 min

Scheme 9.14

The most important drawback of these procedures is the quantity of product available at the end of the reaction but the high reproducibility of the process is important and enables, in adapted reactors, interesting extension of the scale (multigram).

9.2.5.2 **Benzothiazoles and Derivatives**

Benzothiazoles are important heterocycles and continue to be interesting synthetic targets because several classes of annulated thiazoles and thiazolyl (hetero)arenes have a diverse array of biological activity.

One of the most studied microwave-assisted formations of the benzothiazole ring is initiated by an addition–elimination reaction. Treatment of aniline derivatives with 4,5-dichloro-1,2,3-dithiazolium chloride (commonly called Appel's salt) gave 4-chloro-5*H*-1,2,3-iminodithiazoles (**47**) which are very versatile intermediates in the synthesis of a variety of heterocycles (Scheme 9.15). Besson and coworkers

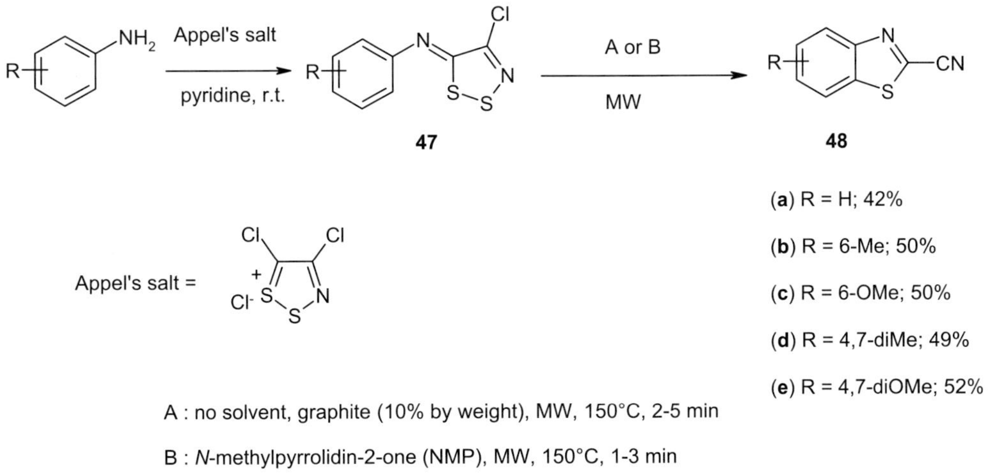

Appel's salt
pyridine, r.t.
A or B
MW
47
48
(a) R = H; 42%
(b) R = 6-Me; 50%
(c) R = 6-OMe; 50%
(d) R = 4,7-diMe; 49%
(e) R = 4,7-diOMe; 52%
Appel's salt =
A : no solvent, graphite (10% by weight), MW, 150°C, 2-5 min
B : N-methylpyrrolidin-2-one (NMP), MW, 150°C, 1-3 min

Scheme 9.15

demonstrated that heating of these compounds at elevated temperature gave benzothiazoles (**48**) in good yields and in very short times [50g]. Depending on the nature of the substituents present on the aromatic ring, this procedure provided a number of benzothiazoles (**48**) in 30–70% yields.

By varying the ratio of the quantities of reactant and support (graphite) the authors discovered that short (2–5 min) microwave irradiation (150 W) of the starting imino-1,2,3-dithiazoles (**47**) at 150 °C in the presence of a small amount of graphite (10% by weight) afforded the required 2-cyanobenzothiazoles (Scheme 9.15). Under similar experimental conditions (with same quantity of starting material, graphite, and the same reaction time), conventional heating afforded the products after a very long time.

It may be noticed that the first approach described in this work consisted in adsorbing the starting material on excess graphite (3 equiv. by weight) and exposure of the powder obtained to microwaves for different times and at different power. Under these conditions, 1 g **47** may be heated (150 °C) in one step by use of this process. Although this may avoid the presence of carbonaceous compounds, scale up of such a procedure was rapidly limited. These results may be linked with the difficulty of obtaining good temperature control at the surface of the solid phase. In some of the experiments described in this work [50g], it was observed that large quantity of graphite may lead to technical problems (hazardous electric arcs and important local elevation of temperature) with regard to controllability and reproducibility.

The last example of a reaction performed between the conditions described above is the synthesis of novel benzothiazolyl indolo (**49**) or benzimidazo[1,2-c]quinazolines (**50**), which were obtained by condensation of the 2-cyanobenzothiazoles (**48**) 2-(2-aminophenyl)indole or benzimidazole [50h]. Microwave irradiation (220 °C) of the two reactants at atmospheric pressure in the presence of graphite as sensitizer (10% by weight) afforded the expected products in good yields and in short times (Scheme 9.16).

Here again, use of a small amount of graphite was sufficient to induce strong heating and enable easy work-up.

9.2.5.3 Synthesis of 2H-Benzopyrans (Coumarins)

Among polyheterocyclic systems, coumarins are synthesized by many routes, including the Pechmann reaction [51], which involves condensation of phenols with β-ketoesters. This reaction, which has been the most widely applied method, has recently been studied under the action of MW irradiation by several authors [52], including the GS–MW process for 4-substituted 7-aminocoumarins [53]. Synthesis of methyl 7-aminocoumarin-4-carboxylates (**51** and **52**) by the Pechmann reaction involves heating a mixture of *m*-aminophenol and dimethyl oxalate at 130 °C (Scheme 9.17). Under such conditions, however, the yield of the reaction is variable, usually low (36%). Use of graphite as a support led to the expected lactone in slightly better yield (44%).

Addition of a solid acid catalyst such as Montmorillonite K10 increased the yield significantly under the action of either thermal heating or MW irradiation (e.g.

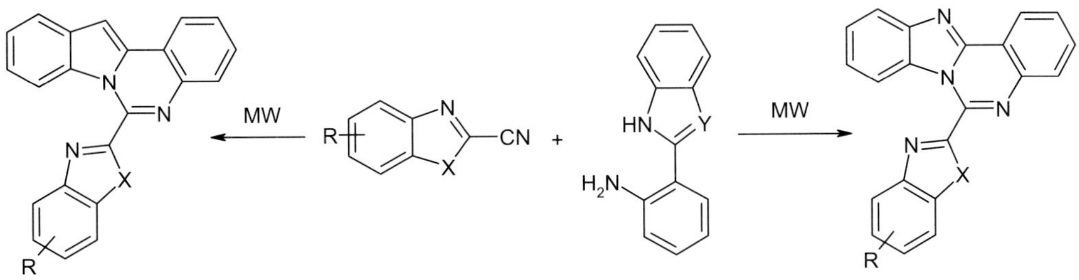

49

MW : graphite (10 % by weight), μw (150°C, 150 W), 30-240 min.

50

X = S or O
Y = CH, N

(a) X = S, R = H; 56%

(b) X = S, R = 6-F; 53%

(c) X = S, R = 6-Me; 60%

(d) X = S, R = 6-OMe; 68%

(e) X = S, 4,7-diMe; 61%

(f) X = S, R = 4,7-diOMe; 54%

(g) X = S, R = 6-NO$_2$; 12%

(a) X = S, R = H; 70%

(b) X = S, R = 6-F; 56%

(c) X = S, R = 6-Me; 71%

(d) X = S, R = 6-OMe; 73%

(e) X = S, 4,7-diMe; 65%

(f) X = S, R = 4,7-diOMe; 58%

(g) X = S, R = 6-NO$_2$; 18%

Scheme 9.16

51a: 64% (Δ) and 66% (MW)). Under the latter conditions the reaction time was noticeably reduced. Comparable results were obtained in the synthesis of N-substituted aminocoumarins (**52**) [53]. In this part of the work it was observed that association of solid and/or liquid reactants on graphite as support may involve uncontrolled reactions and are generally worse than comparable thermal reactions. In these circumstances simple fusion of the reactants may lead to more convenient procedures.

9.2.6
Decomplexation of Metal Complexes

Decomplexation of some metal complexes calls for drastic conditions. This is true for (η-arene)(η-cyclopentadienyl)iron(II) hexafluorophosphates, [FeAr(Cp)][PF$_6$] [54, 55]. Although their chemical decomplexation is known [55a], the most widely used method is pyrolytic sublimation at high temperatures (>200 °C) [55b]. To evaluate MW irradiation as the method of decomplexation of such iron complexes, Roberts and coworkers performed the reaction in the presence of graphite [54]. They discovered that the yield of the free ligand from the [Fe(η-N-phenylcarbazole)(η-Cp) [PF$_6$] complex (**53**) depended on the kind (flakes or pow-

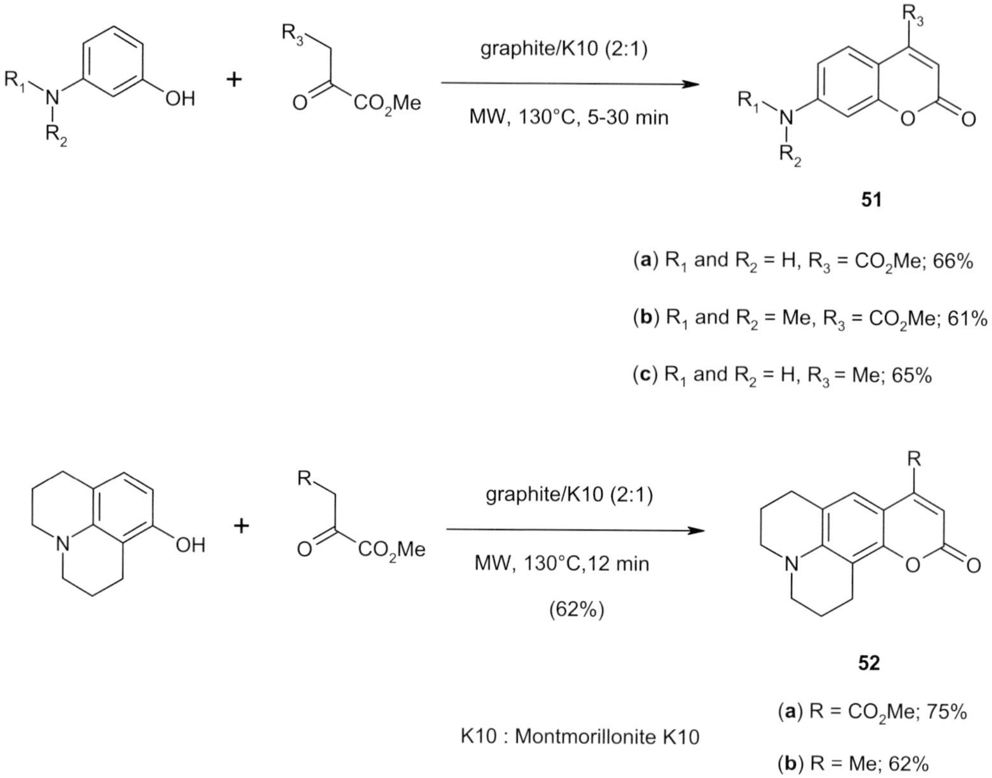

Scheme 9.17

Reaction 1: graphite/K10 (2:1), MW, 130°C, 5-30 min → **51**
- (a) R₁ and R₂ = H, R₃ = CO₂Me; 66%
- (b) R₁ and R₂ = Me, R₃ = CO₂Me; 61%
- (c) R₁ and R₂ = H, R₃ = Me; 65%

Reaction 2: graphite/K10 (2:1), MW, 130°C, 12 min (62%) → **52**
- (a) R = CO₂Me; 75%
- (b) R = Me; 62%

K10 : Montmorillonite K10

der) and amount of graphite, and on the irradiation time. In an optimized experiment, 96% N-phenylcarbazole was obtained from **53** (1 g) by use of 1 g graphite flakes and 2 min MW irradiation. Experiments were performed using a conventional domestic MW oven (the temperature of the reaction mixture at the end of the irradiation was not given). Pentamethylbenzene, N-phenylpyrrole, and 1,2-diphenylindole have been decomplexed from the corresponding iron(II) complexes by use of the same GS–MW process; yields were 94%, 62%, and 71% respectively. The results showed this to be a rapid and efficient method of decomposition.

9.2.7
Redistribution Reactions between Tetraalkyl or Tetraarylgermanes, and Germanium Tetrahalides

Synthesis of alkyl (or aryl) halogermanes (**56**) from a germanium tetrahalide (**54**) occurs in two steps (Scheme 9.18) [56]. The most difficult to realize, and the least selective, is the second, i.e. the redistribution between alkyl (or aryl) and halide substituents of R₄Ge and GeX₄ compounds. Depending on the ratio of these two

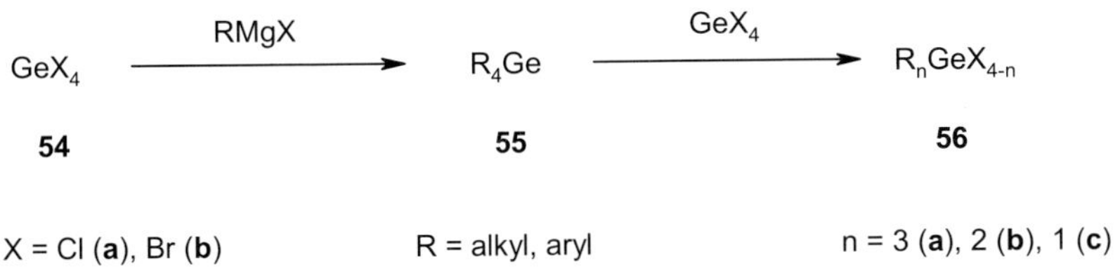

$$GeX_4 \xrightarrow{RMgX} R_4Ge \xrightarrow{GeX_4} R_nGeX_{4-n}$$

    54   55   56

X = Cl (**a**), Br (**b**)   R = alkyl, aryl   n = 3 (**a**), 2 (**b**), 1 (**c**)

Scheme 9.18

compounds, the reaction gives alkyl (or aryl) halogermanes, **56a**, **56b**, or **56c**. This requires a catalyst, the most frequently used being the corresponding aluminum halide, and its amount must be relatively high (approx. 20 mol%) [56]. The experimental conditions are, moreover, rather drastic – heating in a sealed tube between 120 °C (arylated series) and 200 °C (alkylated series) for several hours.

These redistribution reactions are possible at atmospheric pressure under the action of MW. Irradiation is performed for a few minutes in the presence of the same catalyst [57]. These reactions with the less volatile germanium tetrabromide (**54b**) (b.p. 184 °C) have also been performed by use of the GS–MW process, without added catalyst (Table 9.3, entries 1 and 3) [15]. In this instance, despite the use of weaker incident power, the temperature reached 420 °C, very much higher than that obtained under the action of MW irradiation of a reaction mixture containing $AlBr_3$ (200 to 250 °C) (Table 9.3, entries 2 and 4); The presence of this catalyst substantially favors redistribution towards the dibrominated products (**56b**) (84% for R = ^{n}Bu, 85% for R = Ph) relative to the monobrominated compounds (**56a**), which are the major products of the GS–MW process (78% and 43% respectively).

Tab. 9.3. Redistribution reactions between germanium tetrabromide (**54b**) and tetrabutyl or tetraphenylgermanes (**55**) under the action of MW irradiation [15].

Entry	R	$GeBr_4$[e]	R_4Ge[e]	$AlBr_3$[e]	Exp. conditions	R_4Ge[c]	56 (a:b:c)[c]	Yield[d]
1[a]	^{n}Bu	5.20	5.75		90 W, 3 min[e]	17	83 (78:3:19)	80
2[b]	^{n}Bu	5.20	6.10	1.0	300 W, 8 min[f]		100 (8:84:8)	87
3[a]	Ph	5.75	5.20		90 W, 3 min[e]	9	91 (43:32:25)	85
4[b]	Ph	5.70	5.20	1.0	210 W, 8 min[f]	3	97 (9:85:6)	96

[a] Reaction performed by the GS–MW process with 6 g graphite as support
[b] Ref. [57]
[c] Products (%). Products were analyzed by GC and ^{1}H NMR after alkylation [57]
[d] Recovered Ge products
[e] Continuous irradiation
[f] Sequential irradiation, 2 min × 4

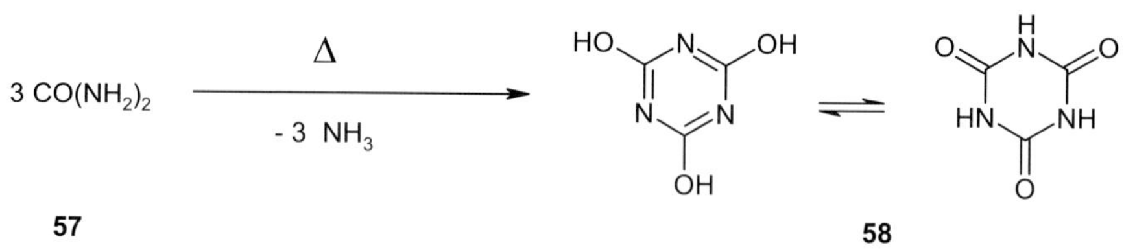

$$3\ CO(NH_2)_2 \xrightarrow[-3\ NH_3]{\Delta} \mathbf{57} \longrightarrow \mathbf{58}$$

57 **58**

Scheme 9.19

The tribrominated product (**56c**), the most difficult to prepare, have been obtained with a rather high selectivity (73 to 80%) by use of the catalytic process under the action of MW [57]. In this reaction, therefore, the GS–MW process seems less effective than the MW-assisted and AlX_3-catalyzed process.

9.2.8
Pyrolysis of Urea

The reaction usually used to produce cyanuric acid (**58**) is the thermolysis of urea (**57**) between 180 °C and 300 °C (Scheme 9.19) [58]. The reaction occurs with formation of ammonia, which itself can react with **58** to give secondary products. It is, therefore, necessary to eliminate NH_3 and to operate with an open reactor.

This reaction has been studied with classical and MW heating under homogeneous and heterogeneous conditions [59]. Table 9.4 summarizes the results.

Tab. 9.4. MW-assisted thermolysis of urea (**57**) under solvent-free conditions (Scheme 9.19) [59].

Entry	Conditions[a,b]	T_{max} (°C)	Time (min)	Yield (%)	Selectivity (%)	Reaction rate ($10^3\ s^{-1}$)
1	MW[c]	200	1	5.2	20.2	8.8
2	Δ[c]	200	1	4.5	30.3	8.7
3	MW[c]	200	30	68.4	73.6	8.8
4	Δ[c]	200	30	37.9	72.2	8.7
5	MW[d]	200	1	9.9	56.3	21.9
6	Δ[d]	200	1	4.6	33.6	8.7
7	MW[d]	300	3	4.6	93.5	12028
8	Δ[d]	300	3	61.2	45.6	7156

[a] Microwave (MW) or conventional (Δ) heating
[b] The incident MW power was adjusted to furnish the maximum temperature (T_{max})
[c] Homogeneous phase; from 20 g urea (mp 133–135 °C)
[d] Heterogeneous phase; from 20 g urea +5 g graphite

When the reaction was conducted in the homogeneous phase at 200 °C (Table 9.4, entries 1–4), identical reaction rates and similar yields and selectivity were obtained for both heating modes. Kinetic data for the first-order equation were similar: Ea (MW) = 159 ± 3 kJ mol^{-1}, Ea (Δ) = 160 ± 3 kJ mol^{-1}; ln A (MW) 35 ± 1, ln A (Δ) 34 ± 1. In contrast, in the presence of graphite (**57**:graphite = 4:1, w/w), improved yield and selectivity were obtained under the action of MW irradiation compared with conventional heating (Table 9.4, entries 5–8 at the same bulk temperature). Chemat and Poux ascribed this phenomenon to localized superheating ("hot spots") on the graphite surface (Section 9.4.2).

9.2.9
Esterification of Stearic Acid with *n*-Butanol

Esterification of stearic acid (**59**) with butanols has been studied under homogeneous and heterogeneous conditions [60]. The yield of butyl stearates (Me(CH$_2$)$_{14}$CO$_2$R (**60**), R = ^{n}Bu (**a**), ^{s}Bu (**b**), ^{t}Bu (**c**)) depended on the catalysts, on the isomeric form of the butanol, and on the mode of heating (conventional heating and MW irradiation). The esterification yield was similar under homogeneous conditions, irrespective of the mode of activation. In contrast, under heterogeneous conditions (e.g. with iron(III) sulfate or potassium fluoride as catalyst), after 2 h at the same bulk temperature (140 °C), the MW-irradiated reaction resulted in a higher (1.2 to 1.4 fold) yield of **60a** than did conventionally heating. This difference was more evident when graphite was added to the reaction mixture – similar yields (75–95% **60a**) were obtained after only 5 min MW irradiation. Despite the small amount of graphite added (approx. 1:10, w/w, relative to the reagents), the reaction mixture rapidly reached a much higher temperature (250 to 300 °C) than in the absence of graphite.

9.3
Graphite as Sensitizer and Catalyst

As a support for chemical reactions, graphite has often been "modified" by addition of a variety of substances which can be intercalated between the carbon layers (GIC–graphite intercalation compounds) or dispersed on the graphite surface, depending on the preparation conditions [13]. The resulting compounds, especially the GIC, have been used as reagents and as catalysts in numerous reactions, particularly in organic transformations [13, 61]. Depending on the intercalating guest, the carbon lattice behaves as an electron acceptor (e.g. with metals–C$_8$K) or as an electron donor (e.g. with halogens–C$_8$Br). The electron power does not, however, seem to give Lewis-type catalytic activity to the graphite itself. As long ago as 1994 [15] Dubac, Laurent and coworkers reported that reactions known to require a Lewis acid catalyst can be conducted in the presence of unmodified graphite. The authors showed that the catalytic ability of graphite is a result of metal impurities, not the carbon lattice [66]. Most reported graphite-catalyzed reactions have been performed under the action of conventional heating, in the presence of a solvent and a small amount of graphite [62–64].

The use of graphite-supported methodology has been reported for three types of reaction – Friedel–Crafts acylation [15, 27, 66], the acylative cleavage of ethers [15], and the ketodecarboxylation of carboxylic diacids [67, 68], either with conventional heating (GS/d) or MW irradiation (GS–MW coupling).

First, however, the authors describe the analysis of two commercial graphites of different purity which are used for these experiments.

9.3.1
Analysis of Two Synthetic Commercial Graphites

The two synthetic, unmodified graphites used in the experiments were:

- graphite **A**, Aldrich 28,286-3; purity 99.1%; particle size 1–2 μm; and
- graphite **B**, Fluka 50870; purity 99.9%; similar granulometry.

Elemental analysis of graphites **A** and **B** by X-ray fluorescence and optical emission (ICPMS) spectrometry showed the presence of approximately a dozen elements. Graphite **A** contained more iron and aluminum than graphite **B** (**A**, Fe 0.41%, Al 0.02%; **B**, Fe 0.007%, Al 0.002%, by weight). Other elements (Ca, K, Si, Ti, V, Mn) were also present at lower concentrations. A careful study by transmission electron microscopy (TEM) of graphite **A** revealed small crystallites; the EDX spectrum of these showed that iron and oxygen were the main elements (Fig. 7.1 in the preceding review [18]). From these crystallites it was possible to obtain X-ray diffraction patterns in which, in addition to ring-shaped spots from graphite, heavy spots corresponding to magnetite (Fe_3O_4) were observed [66]. In a sample of graphite **B**, these particles were not noticeable owing to their small density.

The authors were therefore using two graphites of sharply different purity. The high iron content (0.41%) of graphite **A** leads one to expect catalytic action, either directly by the oxide (Fe_3O_4) or after a possible transformation on contact with MW radiation or a chemical reagent.

9.3.2
Acylation of Aromatic Compounds

Acylation of aromatic compounds (Friedel–Crafts, FC, acylation), of great industrial interest, suffers from an important catalysis problem [69]. Most of the Lewis acids used as catalysts (traditionally metal chlorides such as $AlCl_3$) complex preferentially with the ketone produced instead of with the acylating agent [70] (Scheme 9.20). Except for bismuth (III) chloride with acid chlorides, rarely does a metal chloride complex preferentially with the acylating agent [71, 72].

A stoichiometric amount of promoter, at least, is required for the reaction to proceed, leading to an environmentally hostile process with gaseous effluents and mineral wastes. With some metal salts, however, an increase in reaction tempera-

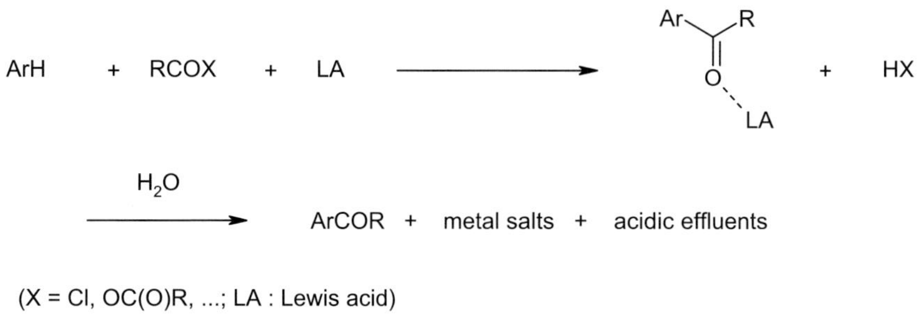

9.3 Graphite as Sensitizer and Catalyst | 439

ArH + RCOX + LA ⟶ Ar–C(=O)–R · · · LA + HX

⟶ (H₂O) ArCOR + metal salts + acidic effluents

(X = Cl, OC(O)R, …; LA : Lewis acid)

Scheme 9.20

ture sets them free from their complex with the ketone, and a true catalytic reaction becomes possible [73]; this is observed for iron(III) chloride [74] and some metal triflates [72, 75], including their use under the action of MW irradiation [76].

In 1994 preliminary results revealed, surprisingly, that FC acylation could be achieved in the presence of graphite **A**, under solvent-free conditions, under the

Tab. 9.5. Graphite-supported acylation of anisole (**61**) by use of a variety of acylating reagents (RCOX) under the action of MW irradiation[a] [27, 66].

Entry	RCOX[b]	Conditions[c] (T_{max})	Product[d]	Conversion and yield (%)[e]
1	MeCOCl	45 W; 1 min × 8, (230 °C)	**68**[g]	62 (54)
2	ⁱPrCOCl	60 W; 1 min × 8, (260 °C)	**69**[g]	80 (74)
3	UndCOCl[f]	90 W; 2 min × 4, (410 °C)	**70**[g]	82 (74)
4	PhCOCl	150 W; 20 s × 15, (330 °C)	**71**[g]	100 (91)
5	FuCOCl[f]	90 W; 2 min × 5, (430 °C)	**72**[h]	60 (50)
6	(MeCO)₂O	+FeCl₃[i], 60 W; 2 min × 6, (310 °C)	**68**[g]	65 (56)
7	(ⁱPrCO)₂O	+FeCl₃[i], 90–30 W; 10 min[j], (260 °C)	**69**[g]	69 (63)

[a] Graphite **A**, 5 g; **61**, 10 to 20 mmol
[b] **51**:RCOX (mol) = 2:1 (entries 1, 2), 4:1 (entries 3–7)
[c] Applied incident power; sequential irradiation time; period between two irradiations: 2 min, except entry 4, 1 min 40 s
[d] Products: Me(C₆H₄)COR, R = Me (**68**), ⁱPr (**69**), Und (**70**), Ph (**71**), Fu (**72**)
[e] Conversion determined by GC; yield of isolated product (in brackets) relative to the minor reagent
[f] Und = undecyl; Fu = 2-furyl
[g] p:o ≥ 95:5
[h] p:o = 82:18
[i] FeCl₃:(RCO)₂O = 1:10 (mol)
[j] Continuous MW irradiation with degressive power

action of either classical heating or MW irradiation [15]. More recently the same reaction has been reported in the presence of a small amount of the same graphite and using a solvent [62]. We have explained this "catalytic effect" of graphite [66].

The procedure entailed MW irradiation, at atmospheric pressure, of graphite powder A impregnated with reagents. The first experiments were performed with an activated aromatic, anisole (**61**), using a variety of acylating reagents (Table 9.5). With volatile acid chlorides such as acetyl or isobutyryl, the reaction occurred in convenient yields (entries 1 and 2), and boiling was delayed as a result of graphite adsorption. With nonvolatile benzoyl chloride conversion was quantitative (entry 4). Good yields were also obtained for a long-chain derivative, known to resinify in the presence of a Lewis acid (entries 3 and 5). The method was also applied to the benzoylation of other aromatic compounds (Table 9.6). The benzoylation of benzene itself, volatile and less reactive, seemed more difficult to perform (Table 9.6, entry 4). Silyl-substituted aromatics reacted by ipso Si substitution [77], and were less volatile. With trimethylsilylbenzene, benzoylation occurred with an overall

Tab. 9.6. Graphite-supported benzoylation of aromatic compounds under the action of MW irradiation[a] [27, 66].

Entry	ArH[b]	Conditions[c] (T_{max})	Product[d]	Conversion and yield (%)[e]
1	62	90 W; 2 min × 5, (390 °C)	73	73 (64)
2	63	60 W; 2 min × 7, (350 °C)	74[f]	54 (45)
3	64	90 W; 2 min × 6, (450 °C)	75[g]	85 (76)
4	65	60 W; 2 min × 7, (370 °C)	76	23 (15)
5	66	150 W; 20 s × 15, (328 °C)	76 77	42 (25) (8)
6	67	90 W; 2 min × 4, (380 °C)	71[h]	60 (50)

[a] Graphite **A**, 5 g; **61**, 10 to 20 mmol
[b] ArH: veratrole (**62**), toluene (**63**), naphthalene (**64**), benzene (**65**), Me_3SiPh (**66**), p-$Me_3Si(C_6H_4)OMe$ (**67**); ArH:PhCOCl (mol) = 4:1 (entries 1–5), 1:1.2 (entry 6)
[c] Applied incident power; sequential irradiation time; interval between two irradiations: 2 min, except entry 5, 1 min 40 s
[d] Products: 2,4-$MeO(C_6H_3)COPh$ (**73**), $Me(C_6H_4)COPh$ (**74**), $C_{10}H_8COPh$ (**75**), PhCOPh (**76**), p-$Me_3Si(C_6H_4)COPh$ (**77**)
[e] Conversion determined by GC; yield of isolated product (in brackets) relative to the minor reagent
[f] p:o = 85:15
[g] α:β = 55:45
[h] p:o = 95:5

Tab. 9.7. Acylation of aromatic compounds in the presence of a small amount of graphite and under the action of MW irradiation [27, 66].

Entry	ArH[a]	RCOX[b]	Graphite[c]	Conditions[d] (T_{max})	Product[a]	Conversion and yield (%)[e]
1	61	MeCOCl	A	45 W; 1 min × 8 (120 °C)	68[f]	27 (20)
2	61	iPrCOCl	A	60 W; 1 min × 8 (120 °C)	69[f]	90 (80)
3	61	PhCOCl	A	300 W; 1 min (165 °C)	71[f]	85 (76)
4	61	PhCOCl	C	300 W; 1 min (159 °C)	71[f]	95 (85)
5	61	PhCOCl	A	60 W; 2 min × 7 (178 °C)	71[f]	100 (98)
6	63	PhCOCl	A	150 W; 20 s × 60 (120 °C)	74[g]	36 (29)
7	64	PhCOCl	A	300 W; 1 min (160 °C)	75[h]	40 (35)

[a] ArH: anisole (**61**), toluene (**63**), naphthalene (**64**)
[b] ArH:RCOX (mol) = 4:1 (entries 3–6), 2:1 (entries 1, 2, 7)
[c] Graphite C: iron–Graphimet (Alfa 89 654)
[d] Continuous (entries 3, 4, 7) or sequential (entries 1, 2, 5, 6) MW irradiation; interval between two irradiations: 1 min (entry 5), 2 min (entries 1, 2), 40 s (entry 6)
[e] Conversion determined by GC; yield of isolated product (in brackets) relative to the minor reagent
[f] $p{:}o \geq 93{:}7$
[g] $p{:}m{:}o = 82{:}2{:}16$
[h] $\alpha{:}\beta = 70{:}30$

yield higher than for benzene, but competitive hydrogen substitution was also observed (entry 5).

If graphite **A** has catalytic activity in these reactions, the amount of graphite could be reduced, and 0.5 g (instead of 5 g) of graphite **A** was, indeed, sufficient to promote these reactions (Table 9.7) [66]. In the process in which a small amount of graphite was used:

- the temperature gradient was lower than for the GS–MW process; and
- vaporization of the reactants was not delayed.

Consequently:

- sequential MW irradiation was preferable to continuous (compare Table 9.7, entries 3 and 5); and
- the MW power must be reduced and the reaction time increased (compare Table 9.5, entry 4, and Table 9.7, entry 5).

Comparative attempts at graphite-supported acylation of anisole, toluene, and naphthalene using classical heating afforded interesting results [66]. With nonvolatile reactants the yields were almost identical with those obtained under the action of MW. In contrast, if at least one reactant was volatile (MeCOCl, ^{i}PrCOCl) or sublimable (naphthalene), the yield obtained under the action of MW was higher. SMWI enables control and limitation of these phenomena.

With regard to a mechanistic hypothesis, the catalytic effect of graphite itself behaving as a Lewis acid was excluded – the use of graphite **B** instead of graphite **A** resulted in no reaction, or sometimes only a trace of the acylation product [27, 66]. The presence of relatively large amounts of Fe_3O_4 in graphite **A** (Section 9.3.1) led us to believe this impurity was responsible for the catalytic activity observed. Several further experiments supported this.

Graphites combined with a variety of metals ("Graphimets") are known for their catalytic properties [13]. The iron–graphite compound in which the presence of Fe_3O_4 crystallites has been shown [78] proved very efficient for the acylation of anisole (Table 9.7, entry 4). Because the iron content (5%) was much higher than that of graphite **A**, the Graphimet could be a convenient material for GS–MW experiments, but its cost, especially relative to that graphite **A**, limits its use.

For the benzoylation of anisole (Table 9.5, entry 4), graphite **A** was replaced by graphite **B** doped with Fe_3O_4 (28 mg for 5 g graphite, the same iron content as in graphite **A**). Although unachievable with graphite **B** alone, benzoylation of anisole was now possible, but with lower yield than for use of graphite **A** (50% instead of 100%). This showed the activity of Fe_3O_4 became much stronger when subjected to the high graphitization temperature.

The catalytic effect of graphite **A** thus depends on iron impurities, e.g. Fe_3O_4, and probably also on iron sulfides or sulfates, because sulfur is also present in this graphite, and all these iron compounds are known catalysts of FC acylation [69, 73, 74]. In this respect it seems that $FeCl_3$ could be the true catalyst generated *in situ* by the reaction of the different iron compounds with acid chloride and hydrogen chloride. In the absence of a chlorinating agent, for example using an acid anhydride as the reagent and an iron oxide (Fe_2O_3 or Fe_3O_4) as the catalyst, acylation does not occur. The authors have effectively shown that the GS–MW process using acid anhydrides as reagents is efficient only after addition of a catalytic amount of $FeCl_3$ (Table 9.5, entries 6 and 7).

Finally, a sample of graphite **A** was analyzed after acylation using an acid chloride as reagent. No Fe_3O_4 crystallites were observed and an EDX spectrum revealed small deposits containing iron, chlorine, and oxygen. Thus, formation of $FeCl_3$ from Fe_3O_4 crystallites is highly probable. Loupy and coworkers have shown that α-Fe_2O_3 can be generated from Fe_3O_4 under the action of MW at high temperature [79]; the formation of $FeCl_3$ would be a result of chlorination of Fe_3O_4 and/or Fe_2O_3. Because Fe_3O_4 interacts strongly with MW [4], the presence of hot spots in the region of Fe_3O_4 crystallites could also lead to increased catalytic activity.

Compared with previous FC acylations, these processes are clean, without aqueous workup, and therefore without effluents ("green chemistry"). The graphite is, moreover, inexpensive and can be safely stored or discarded. Its activity is, however, limited to activated aromatic compounds.

The process which seems to have the most possibilities for a scale-up development is that using a amount of graphite for which the desorption treatment can be totally suppressed in a continuous flow system. We recently proposed use of such a process to perform FC acylations under the action of MW with FeCl₃ as catalyst [76d]. Replacement of FeCl₃ by a graphite bed is quite conceivable in the same continuous flow apparatus.

9.3.3
Acylative Cleavage of Ethers

Preparation of esters using an acid chloride as reagent is usually performed from alcohols and rarely from ethers [80]. Protection of hydroxyl groups as ether derivatives is, however, a widely used method in organic synthesis [81]. The acylative cleavage of ethers is one possible deprotection process. It occurs with an acid halide or anhydride in the presence of a catalyst, usually a Lewis acid. In 1998, Kodomari and coworkers reported the cleavage of some ethers (benzylic, allylic, *tert*-butylic, and cyclic) with acid halides in the presence of graphite **A** [63]. The reaction was performed under reflux in dichloromethane. Under these conditions, however, cleavage of primary or secondary alkyl ethers did not occur.

By use of the solvent-free GS–MW process, which enables reaction at high temperatures, cleavage of these alkyl ethers became possible (Scheme 9.21) [15]. An equimolar (10 mmol) mixture of benzoyl chloride and *n*-butyl oxide adsorbed on 5 g graphite **A** was sequentially irradiated with 90 W incident power. The conversion reached 80% (yield of isolated *n*-butyl benzoate (**79**): 62%). With ethyl oxide, the yield of ethyl benzoate (**78**) was lower, but noteworthy considering the volatility of this oxide and the significant retentive power of graphite towards organic compounds. These preliminary results have not yet been expanded, but it is certain that more reactive ethers, for example those substituted with *sec* or *tert*-alkyl, benzylic or allylic groups, will be cleavable in the same way.

With regard to the mechanism of this reaction and the nature of the catalyst, the authors do not believe the graphite itself is the catalyst. In fact, diethyl and di-*n*-butyl ethers are inert toward benzoyl chloride in the presence of graphite **B**. It is also known that metal chlorides, especially FeCl₃ [82], are catalysts for this reac-

PhCOCl + R₂O →(Graphite **A**) RCl + PhCO₂R

R = Et   60 W; 2 min x 7   **78** (50%)*
          (T_{max} = 350°C)

R = ⁿBu   90 W; 2 min x 4   **79** (62%)*
          (T_{max} = 450°C)

* Not maximized yields

Scheme 9.21

tion. After careful mechanistic study of FC acylations (Sections 9.3.1 and 9.3.2), the authors proposed that the catalyst of the graphite-assisted acylative cleavage of ethers is $FeCl_3$ generated *in situ* from Fe_3O_4 (and/or Fe_2O_3) and the acid chloride. Then, C–O bond cleavage would involve the O-acylation of the ether ($[R_2O-COR']^+$) followed by the nucleophilic displacement (S_N1 or S_N2) of one of the two hydrocarbon groups (R) of ether by the anionic part of the reagent (Cl^-), as with the $FeCl_3$–Ac_2O system [82]. It is interesting to note here that phenolic ethers, for example anisole or veratrole, preferentially afford the acylation of the aromatic nuclei rather than the cleavage of the ether group (Section 9.3.2).

9.3.4
Ketodecarboxylation of Carboxylic Diacids

Thermal decomposition of carboxylates gave ketones by a decarboxylation mechanism [83]. The same ketones were obtained directly from the corresponding carboxylic acids by a decarboxylation–dehydration process in the presence of several catalysts, for example thorium [84] or manganous [85] oxides. This catalytic route enables reduction of the reaction temperature, which still remains high (250–350 °C). Although the method is of little use for synthesis of aliphatic ketones, it is an important route for preparing cyclic ketones from carboxylic diacids [83, 86]. Among these ketones, cyclopentanone is an important industrial compound [87]. For its synthesis from adipic acid, typical procedures using a variety of catalysts (barium hydroxide [88a], metal oxides [83], carbonates [88b], potassium fluoride [88c]) have been described. The search for an efficient and eco-friendly industrial process is still the subject of current interest, however [89, 90]. In this respect, with a view to reducing energy and raw material consumption, a new approach has been undertaken using graphite-supported chemistry [67, 68].

Catalytic cyclization of a diacid requires two contradictory thermal conditions – a temperature high enough to have a convenient reaction rate but low enough to avoid vaporization of the diacid. For adipic acid, the limiting temperature is approximately 290–295 °C [83, 86, 88]. Because of its properties of rapid conversion of MW energy and retention of organic molecules, graphite could enable a high reaction temperature to be reached rapidly, although it delays the vaporization of the diacid. Because, moreover, magnetite (Fe_3O_4) is a catalyst for the decarboxylation of acids [91], no added catalyst would be necessary.

To determine approximately the optimum temperature of graphite-supported cyclization of adipic acid, a series of experiments was performed with classical heating. Using the two graphites **A** and **B** (Section 9.3.1), no significant vaporization of adipic acid (**80**) was observed up to 450 °C at atmospheric pressure. Graphite **A** proved to be the more efficient, giving 90% yield of cyclopentanone (**84**) after 30 min irradiation (Table 9.8, entry 1). Graphite **B** gave a lower yield under the same conditions (entry 2).

The retentive power of graphite towards adipic acid and the catalytic effect of the magnetite, especially present in **A**, are obvious. TEM examination of a graphite **A** before and after reaction showed that crystallites of Fe_3O_4 seemed to be smaller

Tab. 9.8. Graphite-supported thermal ketodecarboxylation of diacids **80–83**[a,b] [67, 68].

Entry	Diacid	Graphite	Ketone yield (%)[c]
1	80	A	74, 90 (85)
2	80	B	74, 22
3	81	A	75, 60
4	81	A	75, 80 (74)
5	82	A	76, 80 (72)
6	83	A	77, 17

[a] Temperature of the electrical oven: 450 °C; reaction time: 30 min (entries 1–4), 60 min (entries 5, 6); pressure: atm. p. (entries 1–3), 300 mm Hg (entries 4–6)
[b] The optimized diacid:graphite ratio was 15 mmol:5 g
[c] Conversion determined by GC and isolated yield (in brackets) from four experiments

after reaction. The same graphite sample was reused for three successive reactions without significant loss in yield, however. When applied to the synthesis of other cyclic ketones (Scheme 9.22), less volatile than **84**, it was observed that pressure had an effect on recovery of product (Table 9.8, entries 3 and 4). A slightly reduced pressure (300 mm Hg) was necessary to obtain 3-methylcyclopentanone (**85**) or cyclohexanone (**86**) in convenient yield (Table 9.8, entries 4 and 5). For cyclization of suberic acid (**83**), a less favorable structure, the yield in cycloheptanone (**87**) remained low (Table 9.8, entry 6).

Some MW-promoted decarboxylations have been reported in the literature [92], even the decarboxylation of magnesium, calcium and barium salts of alkanoic

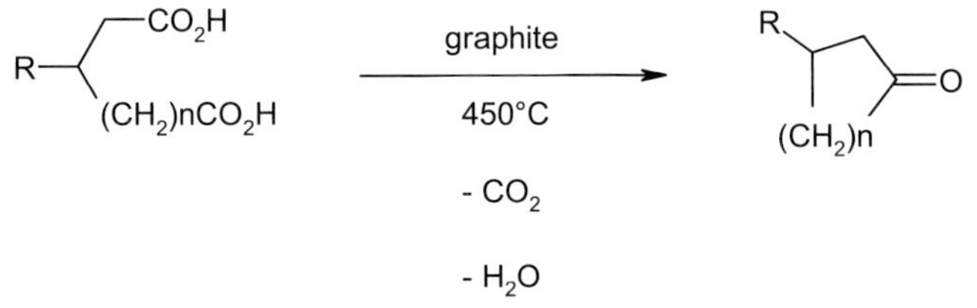

$$R-\overset{-CO_2H}{\underset{(CH_2)nCO_2H}{\bigg\langle}} \xrightarrow[\substack{450°C \\ -CO_2 \\ -H_2O}]{graphite} \underset{(CH_2)n}{R}\!\!\!\!\!\bigg\rangle\!\!=\!\!O$$

n = 2; R = H	80	84
n = 2; R = Me	81	85
n = 3; R = H	82	86
n = 4; R = H	83	87

Scheme 9.22

Tab. 9.9. Graphite-supported ketodecarboxylation of adipic acid (**80**) under the action of MW irradiation[a] [67, 68].

Entry	Graphite and added catalyst	MW conditions[b,c]	Yield (%)[d]
1	**A**; none	90 W; 2 min × 2	60
2	**A**; none	90 W; 2 min × 2 + 75 W; 2 min × 4	90
3	**B**; none	90 W; 2 min × 2	19
4	**B**; none	90 W; 2 min × 2 + 75 W; 2 min × 4	33
5	**B**; Fe_3O_4 (28 mg)	90 W; 2 min × 2	51
6	**B**; Fe_2O_3 (29 mg)	90 W; 2 min × 2	51
7	**B**; FeO (26 mg)	90 W; 2 min × 2	35
8	**B**; Al_2O_3 (28 mg)	90 W; 2 min × 2	15
9	**B**; Bi_2O_3 (28 mg)	90 W; 2 min × 2	16
10	**B**; KF (21 mg)	90 W; 2 min × 2	14
11	**B**; Na_2CO_3 (80 mg)	90 W; 2 min × 2	29
12	**B**; Cs_2CO_3 (244 mg)	90 W; 2 min × 2	26
13	Fe_3O_4 (3.47 g) without graphite	90 W; 2 min × 2	10

[a] Mass of **80**: 2.19 g (15 mmol); mass of graphite: 5 g
[b] Sequential MW irradiation controlled to a maximum temperature of 450 °C
[c] Applied incident power and irradiation time; interval between two irradiations: 2 min
[d] Yield of cyclopentanone (**84**) from GC analysis

acids [92a]. The authors have shown for FC acylations and acylative cleavage of ethers (Sections 9.3.2 and 9.3.3) that the graphite-supported process takes advantage of MW, because graphite and magnetite are among the solids with the most efficient MW absorbing power [4], providing an elevated temperature in the core. Consequently, cyclization of **80** was achieved under GS–MW conditions. A SMWI mode was optimized and coupled with a limitation of the reaction temperature to 450 °C, using the two graphites **A** and **B** (Table 9.9, entries 1–4).

Graphite **A** was again superior, giving, under these optimized conditions (entry 2), a 90% yield in cyclopentanone (**84**) after only 6 × 2 min irradiation. Under the same conditions, graphite **B** gave only a 33% yield (entry 4).

To compare their activities further, a variety of catalysts were added to graphite **B**, and the results were analyzed by comparison with the reference experiment (Table 9.9, entry 3) for which the yield was low (19%). When doped with Fe_3O_4, graphite **B** gave a 51% conversion of **80** (entry 5), almost as much as graphite **A** alone (entry 1). The two other iron oxides, Fe_2O_3 and FeO, seemed to be active but the other catalysts, Al_2O_3, Bi_2O_3, and KF, were inactive.

Because Fe_3O_4 itself strongly absorbs MW [4] and is good catalyst for decarboxylations [91], is the graphite necessary? When Fe_3O_4 was used in the absence of graphite, the yield of ketone **84** decreased dramatically (10%) (Table 9.9, entry 13).

Adipic acid was recovered almost completely on the walls of the reactor and on the cold finger. Consequently, the presence of graphite as support, able to adsorb and retain the adipic acid, and then enable cyclization before vaporization, is essential.

Comparison of reaction times revealed a shortening under the action of MW irradiation (Table 9.9, entry 2 – overall reaction time 22 min) compared with conventional heating (Table 9.8, entry 1, 30 min), for the same maximum temperature. This can be explained by a higher temperature gradient and the presence of "hot spots" on the graphite surface under MW conditions.

A reaction mechanism with Fe_3O_4 as catalyst has been proposed [68], in agreement with previous work on the decarboxylation of acids in the presence of a metal oxide [83]. After the transient formation of iron (II) and iron (III) carboxylates from the diacid and Fe_3O_4 (with elimination of water), the thermal decarboxylation of these salts should give the cyclic ketone and regeneration of the catalyst.

This "dry", solvent-free GS–MW process, which rapidly induces high temperatures, is very useful for decarboxylation of diacids. This has several advantages for example:

- medium-grade commercial graphite, of low cost, can be used;
- the diacid is confined to the graphite which prevents its vaporization; and
- the volatile ketone produced is recovered by distillation as the only organic compound.

A large-scale process could be devised by using a continuous supply of diacid (as a solid or in the molten state) on a graphite bed.

9.4
Notes

9.4.1
MW Apparatus, Typical Procedures, and Safety Measures

Many interesting reports of studies employing domestic microwave instruments have appeared in the literature. There is, however, some debate over the safety and reproducibility of nondedicated microwave instruments in the laboratory environment. Custom-designed microwave instruments for laboratory use are now commercially available and enable superior control of the reaction conditions and enhanced reproducibility whilst also having a key safety advantage [93]. Wherever possible, this review will focus on chemistry performed using the latter type of instrumentation. All the apparatus used in the work described in this chapter has an IR pyrometer enabling continuous recording of reaction mixture temperature, a MW power modulator, open vessels enabling reactions at normal or reduced pressure, and a rotating reactor. Some reactor equipment also includes a dry ice condenser, with easy rotation of the whole arrangement.

Graphite reflects MW like a metal and its heating is very dependent on particle size. With large particles (flakes), electric discharges were observed under MW irradiation conditions whereas with fine particles (powder) a rapid increase of temperature was achieved. Although quartz reactors are highly preferable for reaction with graphite, Pyrex glasses are most commonly used and the temperature of the reaction mixture must be controlled to avoid melting of the reactor.

Typical procedures for GS–MW processes have been described in the literature cited above. Deposition of nonvolatile reactants on graphite has been performed by using a volatile and inert solvent (usually diethyl ether or dichloromethane), easily removed under reduced pressure before MW irradiation. When one of reactants was volatile, it was added neat to graphite powder previously prepared with the nonvolatile reactant as above. The solvent used for impregnation of graphite must be totally evaporated before MW irradiation; if not it can occasionally react with one of the other reagents, owing to high temperatures reached. For example, formation of the ester RCO_2Et by acylative cleavage of Et_2O by RCOCl (Section 9.3.3) has been observed in FC acylations [15, 27]. After irradiation, desorption of the reaction product(s) has been achieved by washing the graphite with a proper solvent. Use of an ultrasonic bath can be advantageous.

9.4.2
Temperature Measurement

Knowledge of the temperature of the reaction media is fundamental. Under MW irradiation conditions this measurement poses some problems [94]. In stirred homogeneous media measurement with an IR pyrometer is satisfactory, although an optical-fiber thermometer is preferable, but limited to high temperatures. Consequently, for measurement of the temperature of graphite-supported reactions under MW irradiation an IR pyrometer was used. This enables measurement of the temperature of the lower reactor walls, and not inside the reaction mixture. Stuerga and Gaillard showed that the temperature of a microwave-irradiated heterogeneous (especially solid) medium was not uniform, but there were "hot spots" [95a]. Similar conclusions were reported by Bogdal and coworkers who studied a rapid and efficient method for selective oxidation of aliphatic and benzylic alcohols to their corresponding carbonyl compounds [95b]. This group used the solid and magnetically retrievable oxidant Magtrieve, which is based on tetravalent chromium dioxide (CrO_2). Microwave irradiation of Magtrieve at continuous power in an open vessel led to rapid heating of the material to 360 °C within 2 min. Temperature heterogeneity was recorded with a thermo-vision camera during the irradiation; the highest value was observed in the center of the reaction vessel. Working in a weak microwave absorber (e.g. toluene as solvent) enabled use of Magtrieve at a more uniform temperature (140 °C).

Dubac and coworkers have observed, by using a thermocouple at the end of a reaction, that the temperature inside the graphite powder was twenty to fifty degrees higher than that indicated by an IR pyrometer. During MW irradiation of C-supported catalysts in the absence of nitrogen carrier, Bond and coworkers [16] observed very small bright spots within the catalyst bed. They proposed that these bright emissions, from continually fluctuating positions, could be because of the generation of plasmas. MW-absorbing impurities, for example Fe_3O_4 crystallites, on the graphite surface could also induce local superheating.

9.4.3
Mechanism of Retention of Reactants on Graphite

The phenomenon of adsorption of organic molecules by graphite is well known, molecular arrangements on surfaces have been recorded and thermodynamic data have been determined [12]. The specific surface area of graphite **A** is 13 m^2 g^{-1} [15]. Adsorption of 2 mL of an organic compound by 5 g of graphite would therefore give a layer 300 Å thick. Van der Walls interactions, responsible for the adsorption, are not effective for such large thicknesses. Other dynamic phenomena must be considered, for example intercalation and diffusion between the carbon layers. Molecules of appreciable size, for example aromatic hydrocarbons and C_{60} fullerene, have been intercalated [96]. In view of the high temperatures in the GS–MW process, operation of these different phenomena (adsorption, diffusion, and intercalation) seems likely. The interlayer spaces could behave as micro reactors (or micro "pressure cookers") in which reactants, or at least a part of them, should be confined. After a reaction, however, we have not observed (by XRD) any change in the interlayer distance of graphite. Consequently, at the end of the reaction, and at room temperature, the interlayer distance should revert to its normal value (3.354 Å) [11], after ejecting the reaction product to the surface. It must be noted here that the intercalation of organic compounds (pyridine derivatives) into a layered mineral oxide (α-VO(PO$_4$).2H$_2$O) occurs more rapidly under the action of MW than with conventional thermal methods [97]. Because the graphite used is of low grain size (1–2 µm), intergrain confinement because of capillary action is also possible.

9.4.4
Graphite or Amorphous Carbon for C/MW Coupling

Because amorphous carbon, as graphite, heats rapidly under the action of MW irradiation [4], its use as a sensitizer has been widely reported [5–10]. MW-assisted esterification of carboxylic acids with alcohols was performed on activated carbon in good yields (71–96%) [98]. When Dubac and coworkers used charcoal powder as a support they had difficulty desorbing the reaction products [15]. Even with a continuous extractor, desorption was never quantitative. Desorption of reaction products from graphite powder is much easier than from amorphous carbon. Charcoal and different carbon blacks have highly variable structures and properties, which depend on the carbonaceous starting material and the preparation conditions [3, 11]. The graphitization of carbon, which is required to achieve high corrosion resistance, leads to materials of more homogeneous structure and properties, enabling highly reproducible reactions. Graphite is, therefore, preferred in supported reactions, although use of intercalated reagents remains possible.

9.5
Conclusion

For the last four years, the main published work on this topic has been thermal reactions in heterocyclic chemistry and, in particular, synthesis of fused quinazolines and thiazoles (Sections 9.3.1 and 9.3.2) in which graphite behaves as an energy converter (or "sensitizer") capable of conveying the energy carried by the MW radiation to the chemical reagents. Compared with the first review [18], this chapter is mainly an update on recent developments in the microwave-assisted synthesis of heteroaromatic compounds under conditions that use both graphite and microwave irradiation in the ring-forming step. While preparing this review we observed that use of graphite in thermal reactions is not highly developed in organic chemistry. The examples described in this chapter demonstrate that under the action of a microwave field graphite behaves efficiently as an energy converter (or "sensitizer") capable of conveying the energy carried by the MW radiation to chemical reagents. Thus, graphite-assisted microwave synthesis enables easy and rapid access to a variety of aromatic heterocyclic compounds which may have interesting pharmaceutical potential. These aromatic heterocyclic compounds have been synthesized mainly in the single-mode focused microwave synthesizers now commercially available. The procedures described are reproducible and can be developed in a combinatorial chemistry strategy.

The work described in this review is also shows that performing microwave-assisted reactions in the presence of graphite should be considered with special attention. A few of these considerations can be applied generally in conducting microwave heated reactions and include the following:

1. The ratio of the quantities of material and support (e.g. graphite) is very important. Use of excess graphite may avoid excessive sublimation of the reactants or the products, and the reactions can be performed at atmospheric pressure, whereas use of a "catalytic amount" of graphite (less than 10% by weight) enables working in sealed vials and use of the synergic effect of pressure with rapid microwave heating of graphite.
2. For solid starting materials, the use of solid supports can offer operational, economical, and environmental advantages over conventional methods. The main aspect of the experiments described in this review is the use of very small quantities of graphite. In some examples the authors have re-used recycled graphite four times without changing the yields of the reactions. Association of liquid/solid reactants on a solid support like graphite may, however, lead to uncontrolled reactions, which may lead to worse results than the conventional thermal reactions. In these circumstances, simple fusion of the products or addition of an appropriate solvent may lead to more convenient mixtures or solutions for microwave-assisted reactions [99].

The strategies explored and defined in the various examples presented here open the door to wider application of microwave chemistry in industry. The most important problem for chemists today (drug-discovery chemists in particular) is to scale-up microwave chemistry reactions for a large variety of synthetic reactions with minimal optimization of the procedures for scale-up. There is currently a growing demand for industry to scale-up microwave-assisted chemical reactions, which is pushing the major suppliers of microwave reactors to develop new systems [100]. In the next few years, these new systems will evolve to enable reproducible and routine kilogram-scale microwave-assisted synthesis.

References

1 C. R. STRAUSS, *Microwave-assisted Organic Chemistry in Pressurized Reactors*, in: Microwaves in Organic Synthesis, A. LOUPY, (ed.), first edition, **2002**, Chapter 2, p 35.
2 (a) C. O. KAPPE, *Angew. Chem. Int. Ed.* **2004**, *43*, 6250; (b) L. PERREUX, A. LOUPY, M. DELMOTTE, *Tetrahedron* **2003**, *59*, 2185; (c) L. PERREUX, A. LOUPY, F. VOLATRON, *Tetrahedron* **2002**, *58*, 2155; (d) L. PERREUX, A. LOUPY, *Tetrahedron* **2001**, *57*, 9199; (e) P. LIDSTRÖM, J. TIERNEY, B. WATHEY, J. WESTMAN, *Tetrahedron* **2001**, *57*, 9225; (f) A. LOUPY, A. PETIT, J. HAMELIN, F. TEXIER-BOULET, P. JACQUAULT, D. MATHÉ, *Synthesis* **1998**, 1213.
3 D. S. CAMERON, S. J. COOPER, I. L. DODGSON, B. HARRISON, J. W. JENKINS, *Catal. Today* **1990**, *7*, 113.
4 (a) J. W. WALKIEWICZ, G. KAZONICH, S. L. MCGILL, *Miner. Metal. Proc.* **1988**, *5*, 39; (b) D. M. P. MINGOS, D. R. BAGHURST, *Chem. Soc. Rev.* **1991**, *20*, 1; (c) D. M. P. MINGOS, *Chem. Ind.* **1994**, 596.
5 K. J. RAO, B. VAIDHYANATHAN, M. GANGULLI, P. A. RAMAKRISHAN, *Chem. Mater.* **1999**, *11*, 882, and references cited therein.
6 J. K. S. WAN, T. A. KOCH, *Res. Chem. Intermed.* **1994**, *20*, 29.
7 J. K. S. WAN, *Res. Chem. Intermed.* **1993**, *19*, 147.
8 M. Y. TSE, M. C. DEPEW, J. K. S. WAN, *Res. Chem. Intermed.* **1990**, *13*, 221.
9 (a) D. A. LEWIS, *Mater. Res. Soc. Symp. Proc.* **1992**, 269 (Microwave Processing of Materials III), 21; (b) M. C. HAWLEY, J. WEI, V. ADEGBITE, *Mater. Res. Soc. Symp. Proc.* **1994**, 347 (Microwave Processing of Materials IV), 669, and references cited therein.
10 M. S. IOFFE, E. A. GRIGORYAN, *Neftekhimiya* **1993**, *33*, 557; *Chem. Abstr.* **1994**, *120*, 109907y; (b) D. D. TANNER, Q. DING, P. KANDANARACH-CHI, J. A. FRANZ, *Prep. Symp.-Am. Chem. Soc., Div. Fuel Chem.* **1999**, *44*, 133; *Chem. Abstr.* **1999**, *130*, 324766p.
11 KIRK–OTHMER, *Encyclopedia of Chemical Technology*, 3rd edn., Vol. 4. Wiley–Interscience, New York, **1978**, p. 556, and references cited therein.
12 (a) D. P. E. SMITH, J. K. H. HÖRBER, G. BINNING, H. NEJOH, *Nature* **1990**, *344*, 641; (b) J. P. RABE, S. BUCHHOLZ, *Sciences* **1991**, *253*, 424; (c) J. P. RABE, *Ultramicroscopy* **1992**, 42, and references cited therein. For analysis of volatile organic compounds by trapping in a carbonaceous adsorbent and by thermal desorption using MW, see (d) M. ALMARCHA, J. ROVIRA, *Tec. Lab.* **1991**, *13*, 322. For adsorption of two organic solvents, methyl isobutyl ketone and methyl isobutyl carbinol, on to graphite as a function of evaporation temperature, see (e) D. S. MARTIN, P. WEIGHTMAN, *Surf. Sci.* **1999**, *441*, 549.
13 (a) M. A. M. BOERSMA, *Catal. Rev. Sci. Eng.* **1974**, *10*, 243; (b) H. B. KAGAN, *Chemtech* **1976**, 510; (c) H. B. KAGAN, *Pure Appl. Chem.* **1976**, *46*, 177; (d) H. SELIG, L. B. EBERT, *Adv. Inorg. Chem. Radiochem.* **1980**, *23*, 281; (e) R. SETTON, in: Preparative Chemistry Using Support Reagents. P. LASZLO,

(ed.), Academic Press, London, **1987**, Chapter 15, p. 255; (f) R. Czuk, B. I. Glänzer, A. Fürstner, *Adv. Organomet. Chem.* **1988**, *28*, 85.

14 D. Avnir, D. Farin, P. Pfeiffer, *J. Chem. Phys.* **1983**, *79*, 3566.

15 (a) R. Laurent, Thèse, Université Paul Sabatier, Toulouse, **1994**; (b) M. Audhuy-Peaudecerf, J. Berlan, J. Dubac, A. Laporterie, R. Laurent, S. Lefeuvre, French Patent, 1994, no. 94.09073 (FR. Appl. 20 July **1994**).

16 G. Bond, R. B. Moyes, I. Theaker, D. A. Whan, *Top. Catal.* **1994**, *1*, 177.

17 (a) D. Villemin, M. Hachemi, M. Lalaoui, *Synth. Commun.* **1996**, *26*, 2461; (b) A. Ben Alloum, Thèse, Université de Caen, **1991**.

18 A. Laporterie, J. Marquié, J. Dubac, in: *Microwaves in organic synthesis*, A. Loupy (ed.), Wiley–VCH, Weinheim, **2002**, Chapter 7, p. 219.

19 (a) A. de la Hoz, A. Diaz-Ortiz, F. Langa, in: *Microwaves in organic synthesis*, A. Loupy, ed., Wiley–VCH, Weinheim, **2002**, Chapter 9, p 295; (b) K. Bougrin, M. Soufiaoui, G. Bashiardes, this book, Chapter 11.

20 O. Diels, K. Alder, *Liebigs Ann.* **1931**, *486*, 191.

21 R. J. Giguere, T. L. Bray, S. Duncan, G. Majetich, *Tetrahedron Lett.* **1986**, *27*, 4945.

22 L. Ghosez, B. Serckx-Poncin, M. Rivera, P. Bayard, F. Sainte, A. Demoulin, A. M. Frisque-Hesbain, A. Mockel, L. Munoz, C. Bernad-Henriet, *Lect. Heterocyclic Chem.* **1985**, *8*, 69.

23 J. C. Martin, D. R. Bloch, *J. Am. Chem. Soc.* **1971**, *93*, 451.

24 R. A. Carboni, R. V. Linisey Jr, *J. Am. Chem. Soc.* **1959**, *81*, 4342.

25 J. Sauer, A. Mielert, D. Lang, D. Peter, *Chem. Ber.* **1965**, *98*, 1435.

26 B. Garrigues, C. Laporte, R. Laurent, A. Laporterie, J. Dubac, *Liebigs Ann.* **1996**, 739.

27 C. Laporte, Thèse, Université Paul-Sabatier, Toulouse, **1997**.

28 P. Garrigues, B. Garrigues, *C. R. Acad. Sci. Paris*, t. 1, Sér. IIc **1998**, 545.

29 M. Villacampa, J. M. Pérez, C. Avendano, J. C. Menédez, *Tetrahedron* **1994**, *50*, 10047.

30 B. Garrigues, R. Laurent, C. Laporte, A. Laporterie, J. Dubac, *Liebigs Ann.* **1996**, 743.

31 C. Laporte, A. Oussaid, B. Garrigues, *C. R. Acad. Sci. Paris*, Sér. IIc **2000**, 321.

32 R. Bonjouklian, R. A. Ruden, *J. Org. Chem.* **1977**, *42*, 4095.

33 A. Stambouli, M. Chastrette, M. Soufiaoui, *Tetrahedron Lett.* **1991**, *32*, 1723.

34 E. I. Klimova, E. G. Treshchova, Y. A. Arbuzoz, *Zh. Org. Khim.* **1970**, *6*, 413.

35 J. Berlan, P. Giboreau, S. Lefeuvre, C. Marchand, *Tetrahedron Lett.* **1991**, *32*, 2363.

36 J. E. Baldwin, M. J. Lusch, *J. Org. Chem.* **1979**, *44*, 1923.

37 L. Eklund, A. K. Axelsson, A. Nordahl, R. Carlson, *Acta Chim. Scand.* **1993**, *47*, 581.

38 M. Audhuy-Peaudecerf, J. Marquié, C. Laporte, A. Laporterie, *Int. Conf. Microwave Chemistry*, Sept. 6–11, **1998**, Prague, Czech Republic.

39 T. L. Ho, *Tandem Organic Reactions*. Wiley, New York, **1992**, p. 144.

40 R. B. Moffet, *Org. Synth.* Coll. Vol. IV, **1963**, 238.

41 (a) H. M. R. Hoffmann, *Angew. Chem. Int. Ed. Engl.* **1969**, *8*, 556; (b) W. Oppolzer, V. Snieckus, *Angew. Chem. Int. Ed. Engl.* **1978**, *17*, 476; (c) J. Dubac, A. Laporterie, *Chem. Rev.* **1987**, *87*, 319; (d) W. Oppolzer, *Angew. Chem. Int. Ed. Engl.* **1989**, *101*, 38; (e) K. Mikami, M. Shimizu, *Chem. Rev.* **1992**, *92*, 1021.

42 R. Laurent, A. Laporterie, J. Dubac, J. Berlan, S. Lefeuvre, M. Audhuy, *J. Org. Chem.* **1992**, *57*, 7099.

43 S. Akutagawain, in: *Chirality in Industry*. A. N. Collins, G. N. Sheldrake, J. Crosby, (eds.), Wiley, New York, **1992**, Chapter 16, p. 313, and references cited therein.

44 K. H. Schulte-Elte, G. Ohloff, *Helv. Chim. Acta* **1967**, *50*, 153.

45 (a) Y. Nakatami, K. Kawashina, *Synthesis* **1978**, 827; (b) S. Sakane, K.

Maruoka, H. Yamamoto, *Tetrahedron Lett.* **1986**, *42*, 2203.
46 J. Ipaktschi, M. Brück, *Chem. Ber.* **1990**, *87*, 1591.
47 (a) M. B. Smith, J. March, *Advanced Organic Chemistry; Reactions, Mechanisms, and Structure*, 5th edn. Wiley–Interscience, New York, **2001**, p. 1329; (b) C. H. De Puy, R. W. King, *Chem. Rev.* **1960**, *60*, 431.
48 J. Dubac, A. Laporterie, G. Manuel, *Chem. Rev.* **1990**, *90*, 215.
49 F. Hostettler, E. F. Cox, *Ind. Eng. Chem.* **1960**, *52*, 609.
50 (a) F. R. Alexandre, L. Domon, S. Frère, A. Testard, V. Thiéry, T. Besson, *Molecular Diversity* **2003**, *7*, 273; (b) M. Soukri, G. Guillaumet, T. Besson, D. Aziane, M. Aadil, El-M. Essassi, M. Akssira, *Tetrahedron Lett.* **2000**, *41*, 5847; (c) L. Domon, C. Le Coeur, A. Grelard, V. Thiéry, T. Besson, *Tetrahedron Lett.* **2001**, *42*, 6671; (d) M. de F. Pereira, L. Picot, J. Guillon, J. M. Léger, C. Jarry, V. Thiéry, T. Besson, *Tetrahedron Lett.* **2005**, *46*, 3445; (e) F. R. Alexandre, A. Berecibar, R. Wrigglesworth, T. Besson, *Tetrahedron* **2003**, *59*, 1413; (f) M. de F. Pereira, L. Picot, J. Guillon, J. M. Léger, C. Jarry, V. Thiéry, T. Besson, *Tetrahedron Lett.* **2005**, *46*, 3445; (g) M. de F. Pereira, V. Thiéry, T. Besson, *J. Sulfur Chem.* **2006**, *27*, 49; (h) S. Frère, V. Thiéry, T. Besson, *Synth. Commun* **2003**, *33*, 3789; (i) S. Frère, V. Thiéry, C. Bailly, T. Besson, *Tetrahedron* **2003**, *59*, 773.
51 (a) A. Russel, J. R. Frye, *Org. Synth.* **1941**, *21*, 22; (b) Sethna, R. Phadke, *Org. React. (N. Y.)* **1953**, *7*, 1.
52 (a) V. Singh, J. Singh, K. P. Kaur, G. L. Kad, *J. Chem. Res. (S)* **1997**, 58; (b) J. Singh, J. Kaur, S. Nayyar, G. L. Kad, *J. Chem. Res. (S)* **1998**, 280; (c) A. de La Hoz, A. Moreno, E. Vasquez, *Synlett* **1999**, 608.
53 S. Frère, V. Thiéry, T. Besson, *Tetrahedron Lett.* **2001**, *42*, 2791.
54 Q. Dabirmanesh, S. I. S. Fernando, R. M. G. Roberts, *J. Chem. Soc., Perkin Trans 1* **1995**, 743.
55 (a) R. A. Brown, S. I. S. Fernando, R. M. G. Roberts, *J. Chem. Soc., Perkin Trans 1* **1994**, 197; (b) R. G. Sutherland, A. S. Abd-El-Aziz, A. Piorko, C. C. Lee, *Synth. Commun.* **1988**, *18*, 291.
56 M. Lesbre, P. Mazerolles, J. Satgé, *Organic Compounds of Germanium.* Wiley, London, **1991**.
57 R. Laurent, A. Laporterie, J. Dubac, J. Berlan, *Organometallics* **1994**, *13*, 2493.
58 (a) V. V. Dragalov, S. V. Karachinsky, O. Y. Peshkova, V. P. Kirpichev, *J. Anal. Appl. Pyrolysis* **1993**, *25*, 311.
59 F. Chemat, M. Poux, *Tetrahedron Lett.* **2001**, *42*, 3693.
60 F. Chemat, M. Poux, S. A. Galema, *J. Chem. Soc., Perkin Trans. 1* **1997**, 2371.
61 J. M. Lalancette, M. J. Fournier-Breault, R. Thiffault, *Can. J. Chem.* **1974**, *52*, 589; (b) G. A. Olah, J. Kaspi, *J. Org. Chem.* **1977**, *42*, 3046; (c) G. A. Olah, J. Kaspi, J. Bukala, *J. Org. Chem.* **1977**, *42*, 4187; (d) K. Laali, J. Sommer, *Nouv. J. Chim.* **1981**, *5*, 469; (e) G. Gondos, I. Kapocsi, *J. Phys. Chem. Solids* **1996**, *57*, 855; (f) A. A. Slinkin, Yu. N. Novikov, N. A. Ptibytkova, L. I. Leznover, A. M. Rubinshtein, M. E. Vol'pin, *Kinet. Katal.* **1973**, *14*, 633; (g) H. B. Kagan, T. Yamagashi, J. C. Motte, R. Setton, *Isr. J. Chem.* **1978**, *17*, 274; (h) Y. N. Novikov, M. E. Vol'pin, *Physica Ser. B, C* (Amsterdam) **1981**, *105*, 471; (i) R. Setton, F. Beguin, S. Piroelle, *Synth. Met.* **1982**, *4*, 299.
62 M. Kodomari, Y. Suzuki, Y. Kouji, *J. Chem. Soc., Chem. Commun.* **1997**, 1567.
63 Y. Suzuki, M. Matsushima, M. Kodomari, *Chem. Lett.* **1998**, 319.
64 Yu. A. Lapin, I. H. Sanchez (Great Lakes Chemical) US Patent 6,147,226; *Chem. Abstr.* **2000**, *133*, 350130h.
65 M. Nagai, *Shokubai* **1998**, *40*, 631; *Chem. Abstr.* **1999**, *130*, 81220z.
66 C. Laporte, P. Baulès, A. Laporterie, J.-R. Desmurs, J. Dubac, *C. R. Acad. Sci. Paris, t. 1, Sér. IIc*, **1998**, 141.

67 J. Marquié, Thèse, Université Paul Sabatier, Toulouse, **2000**.

68 J. Marquié, A. Laporterie, J. Dubac, N. Roques, *Synlett* **2001**, 493.

69 (a) G. A. Olah, *Friedel–Crafts and Related Reaction*, Vols. I–IV. Wiley–Interscience, New York, **1963–1965**; (b) G. A. Olah, *Friedel–Crafts Chemistry*, Wiley–Interscience, New York, **1973**; (c) H. Heaney, in: *Comprehensive Organic Synthesis*, Vol. 2. B. M. Trost, (ed.), Pergamon Press, Oxford, **1991**, Chapter 3.2, p. 733; (d) G. A. Olah, V. P. Reddy, G. K. S. Prakash, in: *Encyclopedia of Chemical Technology*, 4th edn., Vol. 11. Wiley, New York, **1994**, p. 1042; (e) S. B. Mahato, *J. Indian Chem. Soc.* **2000**, *77*, 175.

70 R. Ashforth, J.-R. Desmurs, *Ind. Chem. Libr.*, 8 (The roots of development). J.-R. Desmurs, S. Ratton, (eds.). Elsevier, Amsterdam, **1996**, p. 3.

71 J.-R. Desmurs, M. Labrouillère, J. Dubac, A. Laporterie, H. Gaspard, F. Metz, *Ind. Chem. Libr.*, 8 (The roots of development). J.-R. Desmurs, S. Ratton, (eds.). Elsevier, Amsterdam, **1996**, p. 15.

72 C. Le Roux, J. Dubac, *Synlett* **2002**, 181.

73 D. E. Pearson, C. A. Buehler, *Synthesis* **1972**, 533.

74 J. J. Scheele, *Electrophilic Aromatic Acylation*. Tech. Hogesch, Delft, The Netherlands, **1991**; *Chem. Abstr.* **1992**, *117*, 130844y.

75 (a) A. Kawada, S. Mitamura, S. Kobayashi, *J. Chem. Soc., Chem. Commun.* **1993**, 1157; (b) I. Hachiya, M. Moriwaki, S. Kobayashi, *Tetrahedron Lett.* **1995**, *36*, 409; (c) A. Kawada, S. Mitamura, S. Kobayashi, *J. Chem. Soc., Chem. Commun.* **1996**, 183; (d) J. Izumi, T. Mukaiyama, *Chem. Lett.* **1996**, 739; (e) J.-R. Desmurs, M. Labrouillère, C. Le Roux, H. Gaspard, A. Laporterie, J. Dubac, *Tetrahedron Lett.* **1997**, *38*, 8871; (f) S. Répichet, C. Le Roux, J. Dubac, J.-R. Desmurs, *Eur. J. Org. Chem.* **1998**, 2743; (g) S. Kobayashi, S. Iwamoto, *Tetrahedron Lett.* **1998**, *39*, 4697; (h) J. Matsuo, K. Odashima, S. Kobayashi, *Synlett* **2000**, 403; (i) S. Kobayashi, I. Komoto, *Tetrahedron* **2000**, *56*, 6463; (j) C. J. Chapman, C. G. Frost, J. P. Hartley, A. J. Whittle, *Tetrahedron Lett.* **2001**, *42*, 773; (k) S. Kobayashi, I. Komoto, J. Matsuo, *Adv. Synth. Catal.* **2001**, *343*, 71; (l) R. P. Singh, R. M. Kamble, K. L. Chandra, P. Saravanan, V. K. Singh, *Tetrahedron* **2001**, *57*, 241.

76 (a) J.-R. Desmurs, J. Dubac, A. Laporterie, C. Laporte, J. Marquié, (Rhodia Chimie, Fr.) PCT Int. Appl. WO 9840,339 (FR Appl. 97/2.917. 12 Mar 1997); *Chem. Abstr.* **1998**, *129*, 244928g; (b) C. Laporte, J. Marquié, A. Laporterie, J.-R. Desmurs, J. Dubac, *C. R. Acad. Sci. Paris*, t. 2, Sér. IIc, **1999**, 455; (c) J. Marquié, C. Laporte, A. Laporterie, J. Dubac, J.-R. Desmurs, N. Roques, *Ind. Eng. Chem. Res.* **2000**, *39*, 1124; (d) J. Marquié, G. Salmoria, M. Poux, A. Laporterie, J. Dubac, N. Roques, *Ind. Eng. Chem. Res.* **2001**, *40*, 4485.

77 B. Benneteau, J. Dunoguès, *Synlett* **1993**, 171.

78 D. J. Smith, R. M. Fisher, L. A. Freeman, *J. Catal.* **1981**, *72*, 51.

79 (a) M. Gasnier, L. Albert, J. Derouet, L. Beaury, A. Loupy, A. Petit, P. Jacquault, *J. Alloys Compd.* **1993**, *198*, 173; (b) M. Gasnier, A. Loupy, A. Petit, H. Jullien, *J. Alloys Compd.* **1994**, *204*, 165.

80 M. B. Smith, J. March, *Advanced Organic Chemistry; Reactions, Mechanisms, and Structure*, 5th edn. Wiley–Interscience, New York, **2001**, p. 482.

81 T. W. Greene, P. G. M. Wuts, *Protective Groups in Organic Synthesis*, 2nd edn. J. Wiley, New York, **1991**, Chapter 2, p. 10.

82 B. Ganem, V. R. Small, Jr., *J. Org. Chem.* **1974**, *39*, 3728.

83 H. Kwart, K. King, in: *The Chemistry of Carboxylic Acids and Esters*. S. Patai, (ed.), Wiley–Interscience, London, **1969**, Chapter 8, p. 341.

84 J.-B. Senderens, *Compt. Rend.* **1909**, t. *148*, 297.

85 P. Sabatier, A. Mailhe, *Compt. Rend.* **1914**, t. *158*, 985.

86 (a) L. Ruzicka, W. Brugger, M. Pfeiffer, H. Schinz, M. Stoll, *Helv. Chim. Acta* **1926**, *9*, 499; (b) L. Ruzicka, W. Brugger, C. F. Seidel, H. Schinz, *Helv. Chim. Acta* **1928**, *11*, 496; (c) L. Ruzicka, M. Stoll, H. Schinz, *Helv. Chim. Acta* **1928**, *11*, 670; (d) L. Ruzicka, H. Schinz, W. Brugger, *Helv. Chim. Acta* **1928**, *11*, 686.

87 H. Siegel, E. Eggersdorfer, in: *Ullman's Encyclopedia of Industrial Chemistry*, Vol. A15. W. Gerhartz, (ed.), VCH, Weinheim, **1990**, p. 77.

88 (a) J. F. Thorpe, G. A. R. Kon, *Org. Synth.* Vol. I, **1941**, 192; (b) A. L. Liberman, T. V. Vasina, *Izv. Akad. Nauk. SSSR, Ser. Khim.* **1968**, *3*, 632; (c) L. Rand, W. Wagner, P. O. Wagner, L. R. Kovac, *J. Org. Chem.* **1962**, *27*, 1034.

89 (a) M. Alas, M. Crochemore (Rhône–Poulenc Chimie), *Eur. Patent Appl.* EP 626,364 (FR Appl. 93/6,477, 28 May 1993); *Chem. Abstr.* **1995**, *122*, 105296s; (b) M. Alas, M. Crochemore (Rhône–Poulenc Chimie), *Eur. Patent Appl.* EP 626,363 (FR Appl. 93/6,476, 28 May 1993); *Chem. Abstr.* **1995**, *122*, 105297t.

90 (a) S. Liang, R. Fisher, F. Stein, J. Wulff-Doring (BASF Aktiengesllschaft) *PCT Int. Appl.* WO 9961,402 (DE Appl. 19,823,835, 28 May 1998); *Chem. Abstr.* **1999**, *131*, 352841y; (b) R. Fisher, S. Liang, R. Pinkos, F. Stein (BASF AG) *Ger. Offen.* DE 19,739,441; *Chem. Abstr.* **1999**, *130*, 224607u.

91 A. Mailhe, *Compt. Rend.* **1913**, *157*, 219.

92 (a) V. Gareev, V. V. Zorin, S. I. Maslennlkov, D. L. Rakhmankulov, *Bashk. Khim. Zh.* **1998**, *5*, 33; *Chem. Abstr.* **1999**, *130*, 311519k; (b) H. M. Sampath Kumar, B. V. Subbaredy, S. Anjaneyulu, J. S. Yadav, *Synth. Commun.* **1998**, *28*, 3811; (c) C. Afloroaei, M. Vlassa, A. Becze, P. Brouant, J. Barbe, *Heterocycl. Commun.* **1999**, *5*, 249.

93 (a) J. Cléophax, M. Liagre, A. Loupy, A. Petit, *Org. Process Res. Dev.* **2000**, *4*, 498, and references cited therein; (b) Detailed descriptions of microwave reactors with integrated robotics were recently published: (a) J. D. Ferguson, *Mol. Diversity* **2003**, *7*, 281; (b) L. Favreto, *Mol. Diversity* **2003**, *7*, 287; (c) J.-S. Schanche, *Mol. Diversity* **2003**, *7*, 293.

94 (a) M. Delmotte, D. Stuerga, *Microwave Technology, Wave-Materials Interactions and Equipment*, in: Microwaves in organic synthesis, A. Loupy, (ed.), Wiley–VCH, Weinheim, **2002**, Chapter 1, 1–33; (b) D. Stuerga, this book, Chapter 1.

95 (a) D. Stuerga, P. Gaillard, *Tetrahedron* **1996**, *52*, 5505; (b) D. Bogdal, M. Lukasiewicz, J. Pielichowski, A. Miciak, Sz. Bednarz, *Tetrahedron* **2003**, *59*, 649.

96 B. A. Averill, T. E. Sutto, J. M. Fabre, *Mol. Cryst. Liq. Cryst.* **1994**, *244*, 77.

97 K. Chatakondu, M. L. Green, D. M. P. Mingos, S. M. Reynolds, *J. Chem. Soc., Chem. Commun.* **1989**, 1515.

98 Fan, K. Yuan, C. Hao, N. Li, G. Tan, X. Yu, *Org. Prep. Proc. Int.* **2000**, *32*, 287; *Chem. Abstr.* **2000**, *133*, 176808y.

99 T. Vidal, A. Petit, A. Loupy, R. Gedye, *Tetrahedron* **2000**, *56*, 5473.

100 (a) T. Cablewski, A. F. Faux, C. R. Strauss, *J. Org. Chem.* **1994**, *59*, 3408; (b) F. Chemat, M. Poux, J. L. Di Martino, J. Berlan, *Chem. Eng. Technol.* **1996**, *19*, 420; (c) F. Chemat, E. Esveld, M. Poux, J. L. Di Martino, *J. Microwave Power Electromagn. Energy* **1998**, *33*, 88; (d) E. Esveld, F. Chemat, J. Van Haveren, *J. Chem. Eng. Technol.* **2000**, *23*, 279 and 429.

10
Microwaves in Heterocyclic Chemistry

Jean Pierre Bazureau, Jack Hamelin, Florence Mongin, and Françoise Texier-Boullet

10.1
Introduction

This chapter will deal with applications of microwave irradiation in the synthesis of heterocycles by a variety of means. It is not intended to be exhaustive, but rather emphasizes significant examples of the use of microwave heating.

We have chosen to report reactions in solvents first, then reactions without solvent. Each section is grouped in accordance with the main types heterocycle, in order of increasing complexity, commencing with three, four, and five-membered ring systems containing one heteroatom, and their benzo analogues, followed by five-membered systems with more than one heteroatom, and then the analogous higher-membered ring systems. Reactions involving more complex and polyheterocyclic systems are presented at the end of each section.

10.2
Microwave-assisted Reactions in Solvents

In this section we will examine examples of organic transformations performed by using microwave energy to activate organic mixtures. In all the examples included in this section the chemical reagents are dissolved in a polar solvent that couples effectively with microwaves and generates the heat energy required to promote the transformations.

10.2.1
Three, Four, and Five-membered Systems with One Heteroatom

10.2.1.1 Synthesis of Aziridines

Aziridines, the smallest heterocycles, are an important class of compound in organic chemistry. Interesting access to aziridines by using Bromamine-T as a source of nitrogen in the Cu(II)-catalyzed aziridination of olefins in MeCN was recently

Microwaves in Organic Synthesis, Second edition. Edited by A. Loupy
Copyright © 2006 WILEY-VCH Verlag GmbH & Co. KGaA, Weinheim
ISBN: 3-527-31452-0

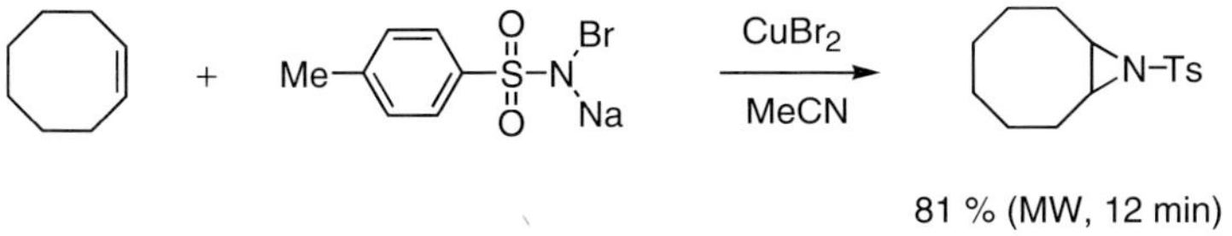

Scheme 10.1

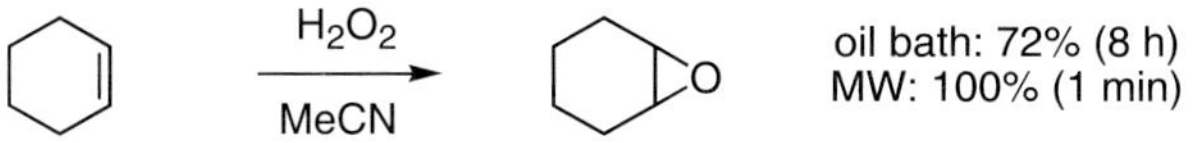

cyclooctene + Me-C6H4-S(O)2-N(Br)(Na) → [CuBr2, MeCN] → bicyclic N-Ts aziridine
81 % (MW, 12 min)

Scheme 10.2

cyclohexene → [H2O2, MeCN] → cyclohexene oxide
oil bath: 72% (8 h)
MW: 100% (1 min)

reported (Scheme 10.1) [1]. Application of microwaves has resulted in enhanced yields of aziridines in short reaction times [2].

10.2.1.2 Synthesis of Oxiranes

Varma and coworkers described an efficient microwave-assisted epoxidation of olefins (Scheme 10.2). This approach significantly minimizes the longer reaction times required in conventional heating of olefins [3].

10.2.1.3 β-Lactam Chemistry

Bose and coworkers have described hydrogenation reactions using ammonium formate as hydrogen donor and Pd/C as catalyst for selective transformations of β-lactams **1**, as shown in Scheme 10.3 and Table 10.1 [4]. The authors report reaction times similar to those achieved with a preheated oil bath at 130 °C on a small scale; on a larger scale, however, microwave-assisted reactions seem to proceed more rapidly.

The same authors have also studied the preparation of substituted vinyl β-lactams **5**, with efficient stereocontrol, by use of limited amounts of solvent (chlorobenzene) (Scheme 10.4) [5]. Microwave oven-induced reaction enhancement (MORE) chemistry techniques have been used to reduce pollution at the source and to increase atom economy.

A comparative study of the reactivity of the oxalimide **6** in a variety of solvents (xylene, chlorobenzene, toluene) and of methylphosphinite **7** was performed with the Synthewave 402 focused microwave reactor (Merck Eurolab, div. Prolabo,

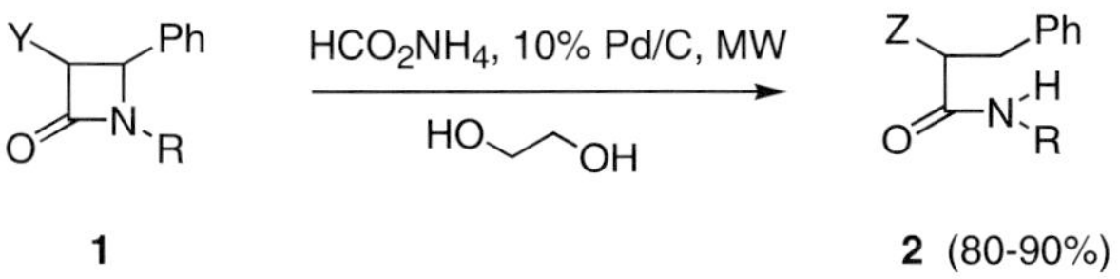

Scheme 10.3

β-lactam **1** (Y, Ph, N-R) → [HCO2NH4, 10% Pd/C, MW, HOCH2CH2OH] → **2** (Z, Ph, H, N-R) (80–90%)

Tab. 10.1. Transformations of β-lactams **1** by hydrogenation under the action of microwaves.

Product 1	Y	R	Product 2	Z	Yield of 2 (%)
1a	CH=CH$_2$	An	2a	CH$_2$CH$_3$	83
1b	MeC(=CH$_2$)	An	2b	MeCH(Me)	80
1c	CH=CH$_2$	Ph	2c	CH$_2$CH$_3$	90
1d	O-CH$_2$CH=CH$_2$	An	2d	OCH$_2$CH$_2$CH$_3$	87
1e	OBn	An	2e	OH	80
1f	OPh	An	2f	OPh	84
1g	CH=CH$_2$	Bn	2g	CH$_2$CH$_3$	85
1h	OBn	Bn	2h	OH	83

An = 4-MeOC$_6$H$_4$, Bn = CH$_2$Ph

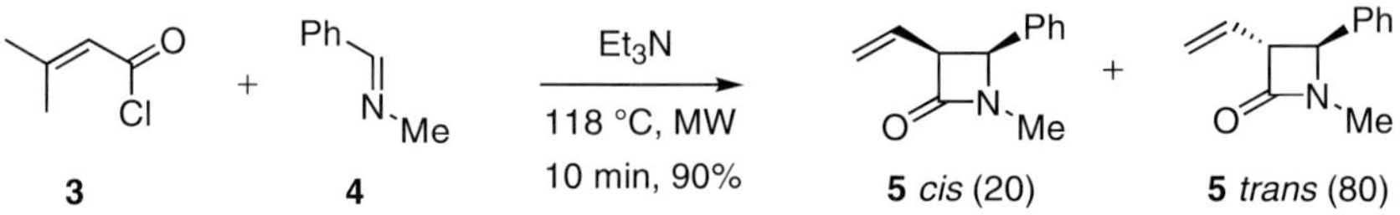

Scheme 10.4

France), using different conditions of power and exposure time (Scheme 10.5) [6]. In all experiments yields were better than those of previous procedures with classical heating (Table 10.2), and the authors wrote "it is clear that microwave technique is applicable to highly functionalized compounds containing stereogenic centers without appreciable modification of these centers".

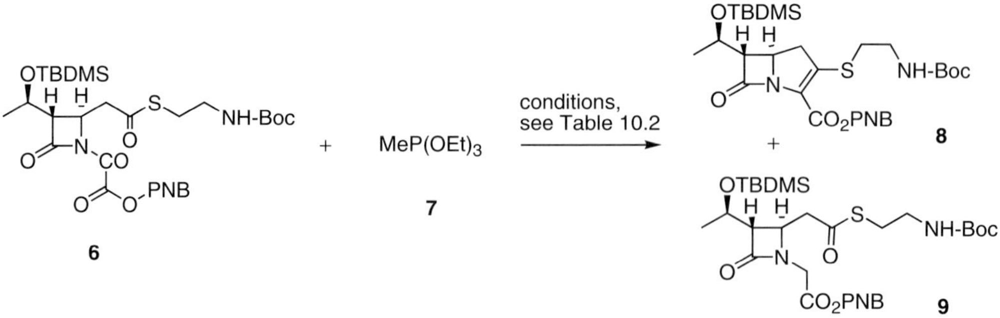

Scheme 10.5

Tab. 10.2. Cyclization of **6** under the action of microwave irradiation.

Solvent	Time (min)	Power (W)	Final temp. (°C)	Yield of 8 (%)
Xylene	10	300	140	50
Chlorobenzene	2	300	150	35
Neat	10	300	80	0

10.2.1.4 Pyrrolidine Chemistry

Esterification of carboxylic acids with dimethyl carbonate (DMC) is strongly accelerated by employing DBU (1 equiv.) as catalyst and microwaves as the energy source. Notable advantages of this strategy are:

1. use of a cheaper and "green" methylating reagent (DMC);
2. efficiency with and applicability to sterically hindered acids;
3. compatibility with commonly used amino acids and protecting groups, for example Boc and Cbz; and
4. high yields (Scheme 10.6) [7].

10.2.1.5 Synthesis of Pyrroles and Related Compounds

Kilburn's group efficiently performed the radical cyclization of alkenyl and alkynyl isocyanides with thiophenol, 2-mercaptoethanol, and ethanethiol in the presence of a radical initiator under the action of microwaves at 130 °C. Direct comparison with results obtained using conventional heating showed the advantage of using microwaves in terms of faster reactions and better yields in the preparation of functionalized pyrrolines and pyroglutamates [8]. Wilson and coworkers have used a single-mode microwave oven at 60–80 W to improve the ring-closing metathesis (RCM) of diolefin substrates in 5–10 min in the synthesis of dihydropyrroles, dihydrofurans, and cyclopentenes [9]. Electron-deficient and sterically congested double bonds have also been used in RCM reactions with microwave irradiation in the synthesis of Δ^3-pyrrolines and Δ^3-tetrahydropyridines [10]. Kappe, Lavastre and coworkers also used a similar approach with standard Grubbs type II and a cationic ruthenium allenylidene catalyst in neat and ionic liquid-doped dichloromethane under sealed-vessel conditions. The same authors, in care-

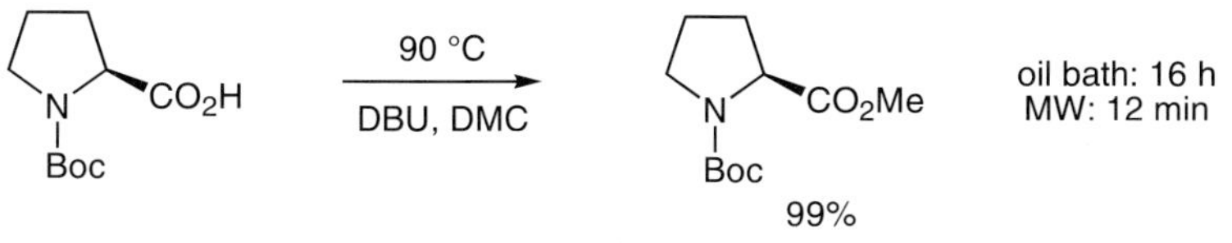

Scheme 10.6

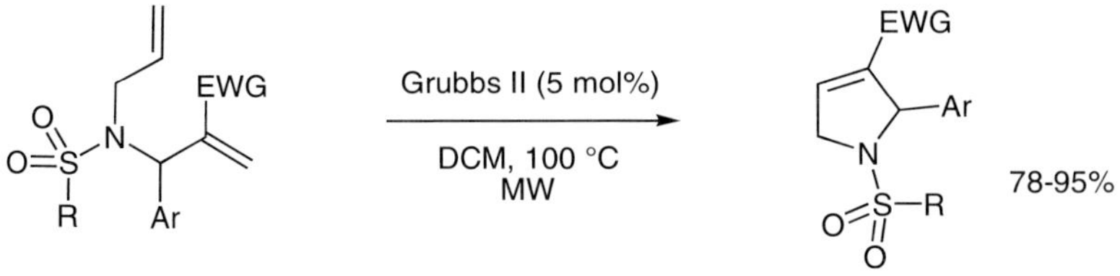

Scheme 10.7

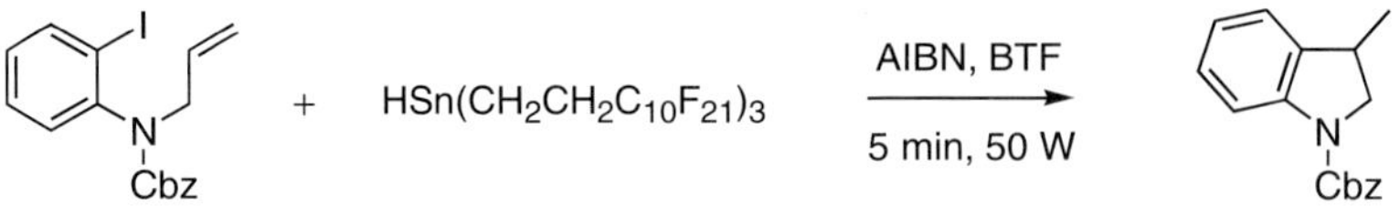

Scheme 10.8

ful comparison studies, showed that the observed rate enhancements are not the results of a nonthermal microwave effect [11]. A similar approach was used by Balan and Adolfsson to prepare 2,5-dihydropyrroles, the starting diene substrates being obtained from aza-Baylis–Hillman adducts (Scheme 10.7) [12].

Schobert et al. have demonstrated [3,3]-sigmatropic rearrangements of allyl tetronates and allyl tetramates to 3-allyltetronic and tetramic acids, respectively, within 20–60 min at 130–190 °C (300 W) [13].

A single-mode microwave oven was used at 50 W to improve radical-mediated reduction and cyclization of halide **10** with $HSn(CH_2CH_2C_{10}F_{21})_3$, **11**, in benzotrifluoride (BTF) to enable the preparation of **12** in high yield (93%) after 5 min, as illustrated in Scheme 10.8 [14].

Rapid and efficient Suzuki coupling of protoxygencol has been developed using polymer-supported palladium catalysts under microwave conditions at 110 °C within 10 min. in the synthesis of various benzolidines [15]. Radical cyclization of resin-bound N-(2-bromophenyl)acrylamides using Bu_3SnH proceeded smoothly in DMF or toluene as solvent under the action of microwave irradiation for preparation of 2-oxindoles as illustrated in Scheme 10.9. This method is superior to conventional solution synthesis [16].

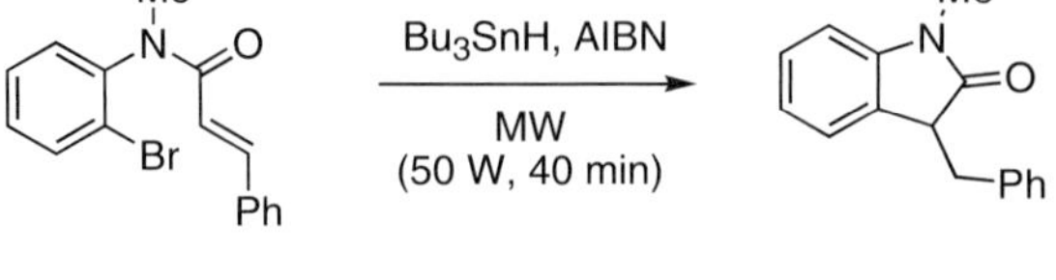

Scheme 10.9

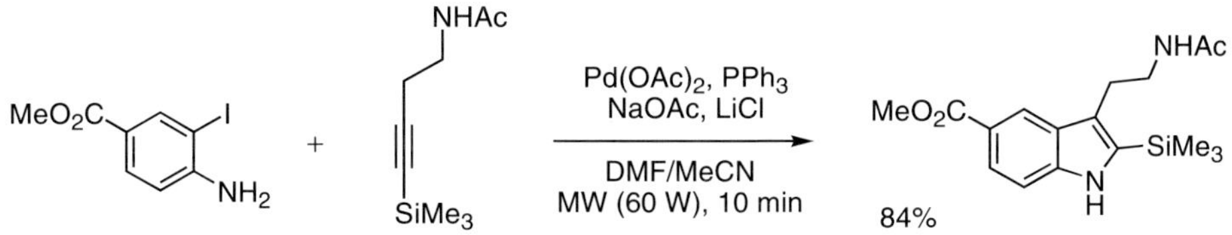

Scheme 10.10

Studer and coworkers have conducted extensive investigations leading to a small library of α,β-unsaturated oxindoles. A one-pot sequence comprising homolytic aromatic substitution followed by an ionic Horner–Wadsworth–Emmons olefination proved to be efficient, in short reaction time, under the action of microwave irradiation [17]. Pd-catalyzed reactions of aryl bromides with bis(pinacolato)diboron in DMSO, DME, or dioxane could be dramatically accelerated by using microwaves. The yields obtained are higher and superior to those described previously [18]. Elegant one-pot synthesis of polyarylpyrroles from but-2-ene and but-2-yne-1,4-diones in PEG 200 has been described by Rao's group; the procedure entails hydrogenation of the carbon–carbon double or triple bonds then amination–cyclization using ammonium and alkylammonium formates in the presence of palladium under the action of microwave irradiation (2–3 min) [19]. The synthesis of the indole core of melatonin analogs by Pd-mediated heteroannulation of a silylated alkyne with 2-iodoanilines has been achieved rapidly (11 min), in 84% yield, under the action of microwave irradiation (60 W) to speed drug-discovery processes (Scheme 10.10) [20].

Pyrrole is one of the most prominent heterocycles in several natural products and synthetic pharmaceuticals. The most common approach to pyrroles is the Paal–Knorr reaction. Taddei and coworkers have investigated a rapid and versatile synthesis of tetrasubstituted pyrroles in few highly efficient steps:

- functional homologation of a β-ketoester with an aldehyde;
- oxidation to give a series of differently substituted 1,4-dicarbonyl compounds; and
- cyclization with the Paal–Knorr procedure under the action of microwave irradiation (Scheme 10.11) [21].

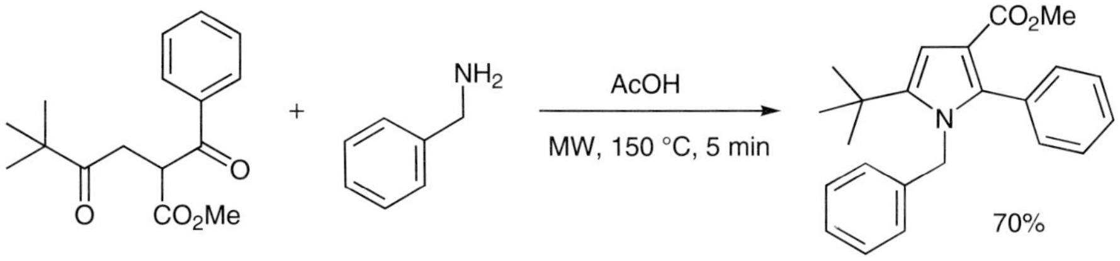

Scheme 10.11

10.2.1.6 Synthesis of Tetrahydrofurans

A series of primary and secondary alkyl aryl tellurides has been found to undergo rapid (3–10 min) group-transfer cyclization to afford tetrahydrofuran derivatives in good yields (60–74%) under microwave heating conditions at 250 °C in ethylene glycol or at 180 °C in water, the only drawback of the process being the loss in diasteroselectivity as a consequence of the higher reaction temperature [22]. Li and coworkers have developed a successful method for synthesis of vinyl cycloethers by direct addition of THF and 1,4-dioxane to alkynes; reaction occurs at 200 °C in 40–180 min under the action of microwave irradiation (300 W) [23].

10.2.1.7 Synthesis of Furans and Related Compounds

Rao et al. have shown that but-2-ene-1,4-diones, and but-2-yne-1,4-diones can be converted to furan derivatives conveniently by using HCO_2H and catalytic Pd/C in PEG 200 with microwave irradiation. The reaction takes less than 5 min, is amenable to scale-up, and works well for preparation of both 2,5-diaryl and 2,3,5-triphenylfuran derivatives [24]. Because of the usefulness of phthalides in the synthesis of biologically active substances, a rapid carbonylation–lactonization reaction utilizing in-situ-generated CO has been developed by Alterman and coworkers for preparation of phthalides, dihydroisocoumarin, dihydroisoindone, and phthalimide from aryl bromide precursors [25]. Chromium carbene-mediated Dötz benzannulation of a variety of alkynes has been shown to proceed remarkably rapidly under microwave-assisted conditions [26]. A microwave process is highly beneficial in enhancing the rates of tandem cyclization–Claisen rearrangement. The authors claimed that the method enables rapid assembly (15–30 min) of a variety of cycloheptane-containing polycyclic structures in yields ranging from good to excellent (Scheme 10.12) [27].

10.2.1.8 Synthesis of Thiophenes and Related Compounds

Leadbeater and Marco reported an elegant ligand-free Pd-catalyzed Suzuki reaction in water using microwave heating. This method is useful for coupling boronic acids and 2-bromothiophene (150 °C, 5 min, 70%) [28]. Zhang et al. have also developed a similar strategy combined with easy fluorous separation for synthesis of thiophene and hydantoïn derivatives [29]. Mutule and Suna intensively studied a

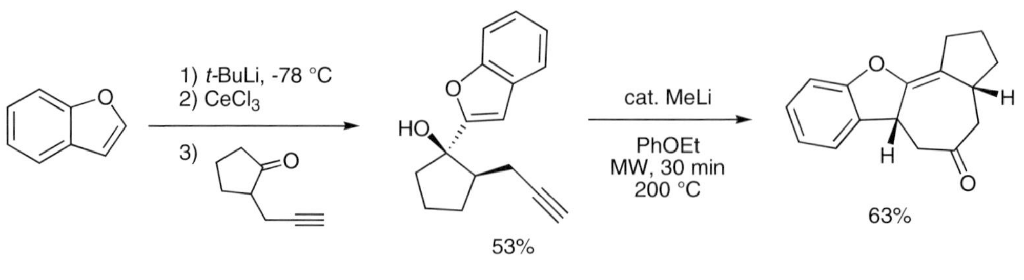

Scheme 10.12

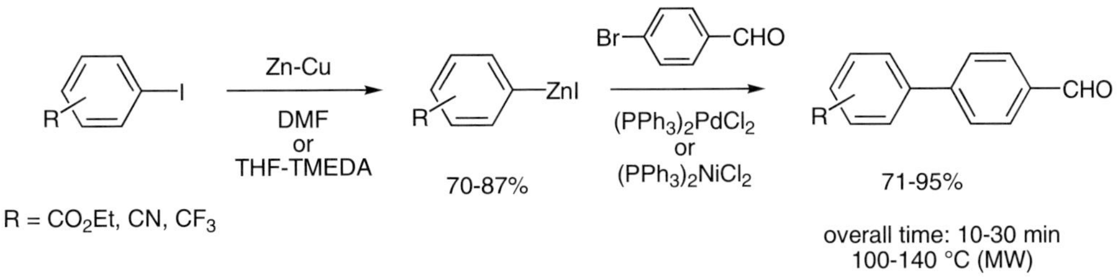

[Scheme 10.13 depicts: Ar-I (R = CO₂Et, CN, CF₃) with Zn-Cu in DMF or THF-TMEDA → Ar-ZnI (70–87%) → coupling with Br-C₆H₄-CHO using (PPh₃)₂PdCl₂ or (PPh₃)₂NiCl₂ → biaryl-CHO product (71–95%). Overall time: 10–30 min, 100–140 °C (MW).]

Scheme 10.13

sequential arylzinc formation–Negishi cross-coupling procedure in a microwave environment (Scheme 10.13) for access to thiophene derivatives [30].

The classical decarboxylation of benzo[b]thiophene-2-carboxylic acids involves a Cu-mediated reaction in quinoline and a subsequent problematic work-up. A new homogeneous method was developed and adapted under the action of microwave irradiation by using an organic base (DBU) and a high-boiling polar solvent (dimethylacetamide). All the decarboxylations (65–100%) were performed at 200 °C for 45 min in a sealed microwave vessel at 600 W (Scheme 10.14) [31].

10.2.2
Five-membered Systems with Two Heteroatoms

10.2.2.1 Synthesis of Cyclic Ureas

Cyclic ureas have many applications as intermediates in the preparation of biologically active molecules. The conventional methods involve cyclization of 1,2-diamines with phosgene or oxidative carbonylation of diamines. Varma and coworkers developed a direct synthesis of cyclic ureas from urea and diamines in the presence of ZnO using microwaves. The major advantage of the method is that the reaction is accelerated by exposure to microwave irradiation; the byproducts were, moreover, easily eliminated compared with traditional methods [32].

10.2.2.2 Hydantoin, Creatinine, and Thiohydanthoin Chemistry

A successful combination of microwave technology with PEG 6000 as liquid phase has been applied to the synthesis of 1,3-disubstituted hydantoins in four steps with overall yield ranging from 70 to 93% as described in Scheme 10.15 [33].

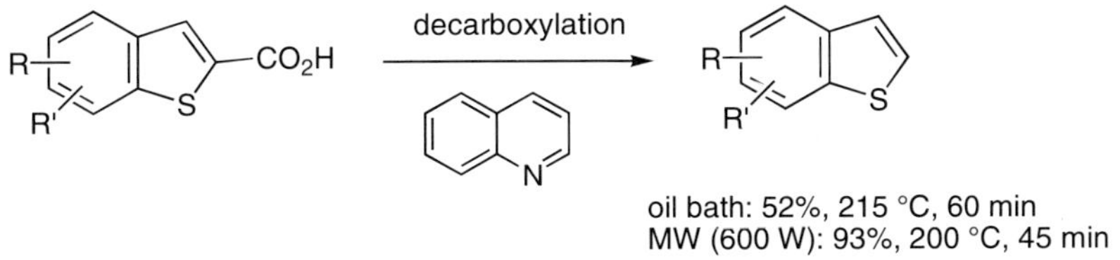

[Scheme 10.14 depicts: benzo[b]thiophene-2-carboxylic acid (R, R' substituted) undergoes decarboxylation in quinoline to give the corresponding benzo[b]thiophene. oil bath: 52%, 215 °C, 60 min; MW (600 W): 93%, 200 °C, 45 min.]

Scheme 10.14

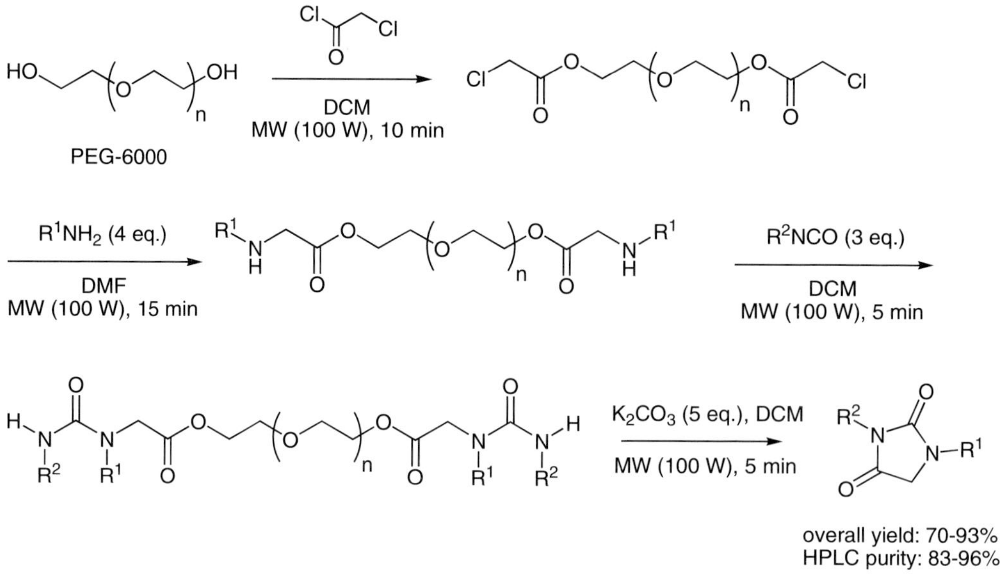

Scheme 10.15

The significance of this work is that the cyclization–cleavage step involved mild, basic conditions and microwave flash heating.

Microwave-mediated condensation can be used to prepare heterobicycles as illustrated in Scheme 10.16 [34]. The imino derivatives **15** were obtained directly from 3-formylchromone [35] **13** and creatinine [36] **14**, with DMSO as solvent and boric acid as catalyst, and with microwave irradiation and classical heating (Table 10.3).

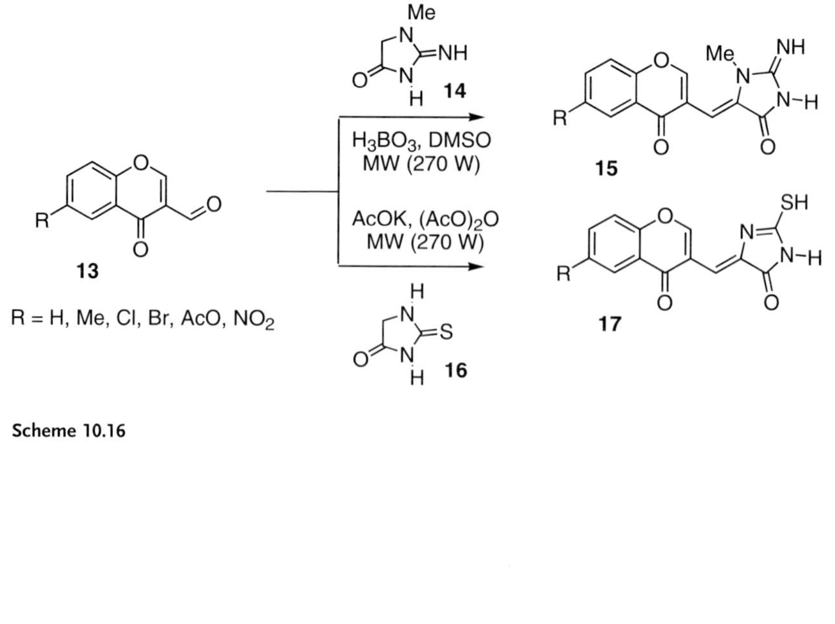

Scheme 10.16

Tab. 10.3. Condensation of 3-formylchromones **13** with creatinine **14** and thiohydantoin **16**.

Product	R	Conventional heating		Microwave irradiation	
		Reaction time (min)	Yield (%)	Reaction time (min)	Yield (%)
15a	H	60	71	3	75
15b	Cl	60	72	2	76
17a	H	60	72	8	79
17b	Br	60	88	9	96

The authors have also investigated the preparation of the 2-thioxo-5-imidazolidin-4-ones **17** from thiohydantoin **16** in the presence of potassium acetate, with acetic anhydride as solvent, using both methods. Here, the effect of microwave irradiation was reduction of reaction times and a small increase in the yields.

10.2.2.3 Synthesis of Pyrazoles and Related Compounds

Kidwai et al. have demonstrated that formic acid can be used to catalyze cyclocondensation of hydrazones to give new fungicidal pyrazoles, as shown in Scheme 10.17 [37]. The significance of their work, clearly exploiting nonthermal specific microwave effects, is noticeable when one considers that in the classical approach cyclocondensation of hydrazones requires 30–35 h with constant heating at 100–120 °C whereas the same reaction is complete in 4–7 min with improved yield when performed under the action of microwave irradiation.

An extremely simple one-pot synthesis of pyrazoles, pyrimidines, and isoxazoles has been realized by reacting enamino ketones, formed *in situ* with the appropriate bidentate nucleophile, under the action of microwaves [38]. Another approach to pyrazole from 4-alkoxy-1,1,1-trichloro-3-alken-2-ones and hydrazines, with toluene as solvent, is also possible under microwave conditions [39]. The Ullman coupling of (S)-[1-(3-bromophenyl)ethyl]ethylamine with N–H-containing heteroarenes such as pyrazole in N-methylpyrrolidone afforded the N-arylated compounds in high yields under microwave heating conditions at 198 °C [40].

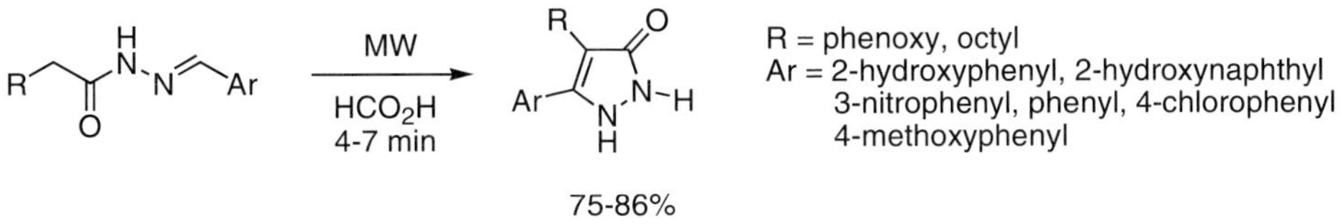

Scheme 10.17

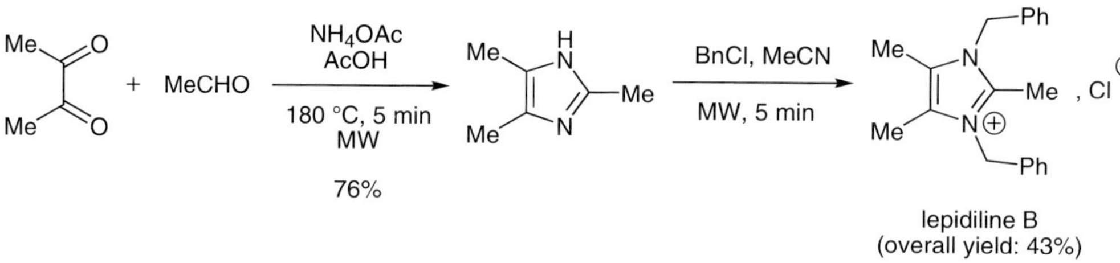

Scheme 10.18

10.2.2.4 Synthesis of Imidazoles and Related Compounds

Combs et al. have described an unprecedented *in situ* thermal reduction of the N–O bond on microwave irradiation at 200 °C for 20 min in the synthesis of 2,4,5-triarylimidazoles from the parent keto-oximes and aldehydes [41]. A similar method has been extended to a short synthesis of lepidiline B (Scheme 10.18) and trifenagrel illustrating the utility of microwave technology [42].

The preparation of important active metabolite EXP-3174 of losartan by oxidation with manganese dioxide in water has been improved under the action of microwave irradiation. Conventional heating (100 h, reflux, 9%) and microwave irradiation (50 min, 200–300 °C, 64%) have also been compared by an Italian group [43]. Bergman and coworkers have shown that microwave irradiation strongly accelerates the Rh-catalyzed intramolecular coupling of a benzimidazole C–H bond to pendant alkenes [44]. The cyclic products were formed in moderate to excellent yields with reaction times less than 20 min in 1,2-dichlorobenzene–acetone (3:1) as solvent. New benzimidazoloquinones substituted at the 2-position by a sulfonyl group have been synthesized in seven steps and 50% overall yield; the final step is microwave-assisted and uses 2-chloromethyl-1,5,6-trimethylbenzimidazole-4,7-dione as starting material. The authors claimed that microwave irradiation enables easy and rapid access to original heterocyclic quinones with potential antitumor activity [45].

10.2.2.5 Synthesis of Oxazolidines

Straightforward access to N-protected 5-oxazolidinones using amino acids, paraformaldehyde, and PTSA in a minimum amount of toluene is accelerated by exposure to microwaves (3 min). This approach is not only simple and efficient but can be conveniently scaled-up to large quantities [46].

10.2.2.6 Synthesis of Oxazoles and Related Compounds

Player and coworkers have developed a facile parallel synthesis of a 48-membered library of benzoxazoles using two efficient methods in solution. Both methods involve the use of 2-aminophenol, acid chloride, with dioxane (or xylene) as solvent in a sealed reaction vessel for a reaction time of 15 min at 210 °C (or 250 °C). All benzoxazoles were obtained in good to excellent yields (46–98%) [47]. The use of

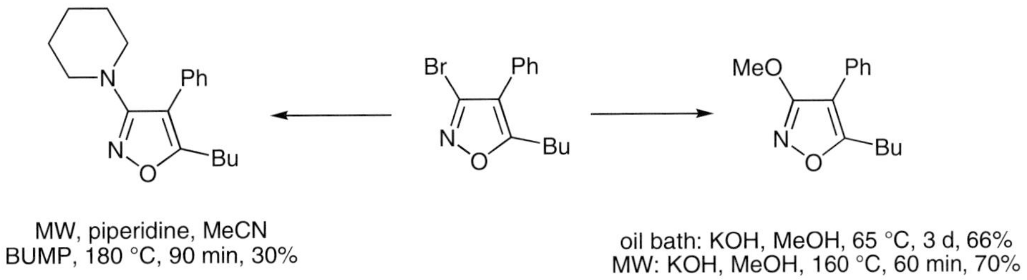

Scheme 10.19

MW, piperidine, MeCN
BUMP, 180 °C, 90 min, 30%

oil bath: KOH, MeOH, 65 °C, 3 d, 66%
MW: KOH, MeOH, 160 °C, 60 min, 70%

phosphazene bases under microwave irradiation facilitates the amination (S_NAr) process and enables the corresponding 3-aminoisoxazoles to be obtained in moderate yield (Scheme 10.19). Under thermal conditions the 3-bromoisoxazoles were found to be inert to substitution [48].

The cyclocondensation of 4-alkoxy-1,1,1-trichloro-3-alken-2-ones with hydroxylamine using toluene as solvent is more efficient under the action of microwave irradiation than using the classical method (the average time ratio for the two methods is 1:160) [49]. Katritzky et al. have demonstrated the applicability of microwaves to the synthesis of a variety of 2-oxazolines from readily available N-acylbenzotriazoles and 2-amino-2-methyl-1-propanol under mild conditions (80 °C) with short reaction times (12 min). Use of N-acyl benzotriazoles also avoids some complications in microwave reactions, for example dimerization or the exclusive formation of amides from carboxylic acids [50].

10.2.2.7 Synthesis of Thiazolidines

Under microwave conditions, reaction of benzylidene-anilines and mercaptoacetic acid in benzene gives 1,3-thiazolidin-4-ones in high yield (65–90%) whereas the same reaction performed conventionally (reflux) requires longer reaction times and gives lower yields (25–69%). This difference seems to be because of the formation of intermediates and byproducts during the conventional reaction [51]. 4-Thiazolodinones were also obtained in high yields (68–91%) and short reaction times under the action of microwave irradiation via a three-component one-pot condensation using the environmentally benign solvent ethanol in open vessels at atmospheric pressure [52].

10.2.2.8 Synthesis of Thiazoles and Related Compounds

Click chemistry has been used to prepare a series of 1,4-disubstituted-1,2,3-triazoles by a Cu(I)-catalyzed three-component reaction of alkyl halides, sodium azide, and alkynes. The method avoids isolation and handling of potentially unstable small organic azides and provides triazoles in the pure form and high yields (81–89%). Microwave irradiation dramatically reduces reaction times from hours to minutes [50].

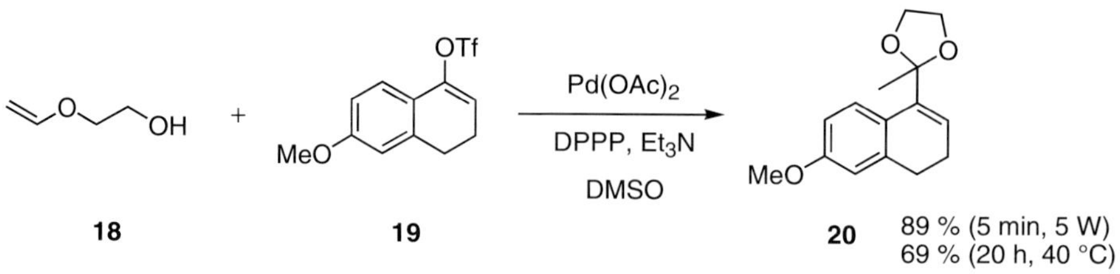

Scheme 10.20

10.2.2.9 Dioxolan Synthesis

A regiocontrolled Heck vinylation of commercially available 2-hydroxyethyl vinyl ether **18** in dry DMSO has been reported (Scheme 10.20) [53]. Flash heating by microwave irradiation (5 min, 5 W), as a complement to the standard thermal heating (20 h, 40 °C), was used to reduce reaction times drastically, with better yields of isolated product **20**.

10.2.3
Five-membered Systems with More than Two Heteroatoms

10.2.3.1 Synthesis of Triazoles and Related Compounds

Barbry and coworkers conducted extensive investigations leading to the development of a range of microwave-assisted syntheses of 3,5-disubstituted 4-amino-1,2,4-triazoles as potentially good corrosion inhibitors [54]. The group found, for example, that 4-amino-1,2,4-triazole **23** is quickly prepared (5 min) by reaction of 2-cyanopyridine **21** with hydrazine dihydrochloride in the presence of excess hydrazine hydrate in ethylene glycol under the action of microwave irradiation (95% yield). The analogous reaction performed with conventional heating proceeds in 85% yield and requires a longer reaction time (ca. 45 min) (Scheme 10.21).

10.2.3.2 Synthesis of Oxadiazoles and Related Compounds

In the preparation of novel 1,3,4-oxadiazoles **25** from 1,2-diacylhydrazines **24**, Brain et al. used a highly efficient cyclodehydration assisted by use of microwaves,

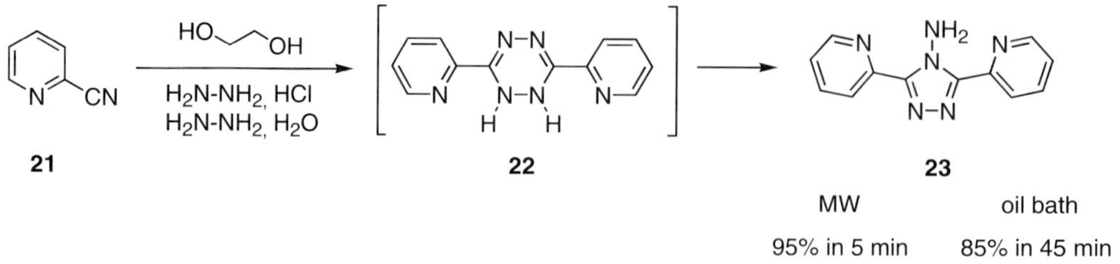

Scheme 10.21

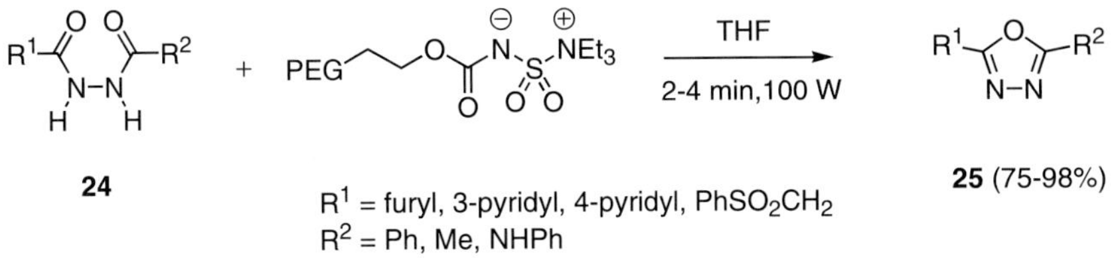

Scheme 10.22

with THF as solvent, using polymer-supported Burgess reagent (Scheme 10.22) [55].

Microwave irradiation coupled with poly(ethylene glycol)-supported Burgess reagent reduced the reaction time to 2–4 min and led to improved yields. Use of harsh reagents (e.g. $SOCl_2$, $POCl_3$, and polyphosphoric acid) for the cyclodehydration was avoided.

10.2.3.3 Synthesis of Tetrazoles and Related Compounds

The tetrazole functional group is of particular interest in medicinal chemistry, because of its potential role as a bioisostere of the carboxyl group. In this context, Schulz et al. have demonstrated the synthetic utility of the cyano group of arylnitrile boronates as a source of tetrazole derivatives under microwave conditions. The reaction is conducted with azidotrimethylsilane and dibutyltin oxide as catalyst to provide aryltetrazole boronates in yields ranging from 60 to 93% [56]. In the same manner, microwaves may assist successful conversion of sterically hindered nitriles into tetrazoles [57].

10.2.4
Six-membered Systems with One Heteroatom

10.2.4.1 Piperidine Chemistry

Piperidine, morpholine, pyrrolidine, N-alkylpiperazines, and anilines have been converted to N-aryl derivatives by use of aryl triflates under the action of microwave irradiation in NMP [58]. The amination reaction proceeded efficiently in the absence of base and catalyst, and was not affected by traces of water (Scheme 10.23).

10.2.4.2 Synthesis of Pyridines and Related Compounds

In 1998, Khmelnitsky and coworkers examined microwave-assisted parallel Hantzsch pyridine synthesis [59]. They demonstrated the benefits of microwave ir-

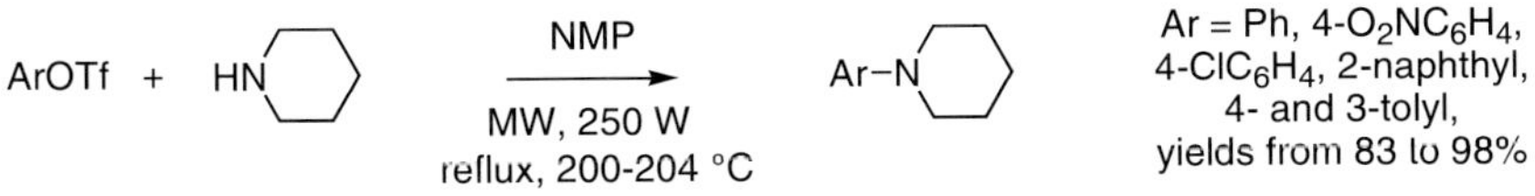

Scheme 10.23

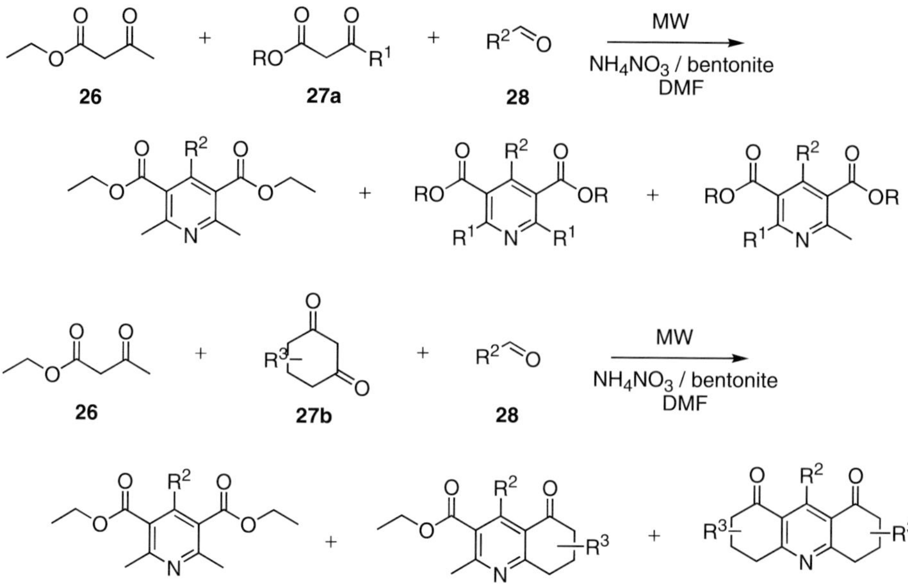

Scheme 10.24

radiation in a 96-well plate reactor for high throughput, automated production of a pyridine combinatorial library. In each reaction, ethyl acetoacetate (26) was used as one of the components of the Hantzsch synthesis, whereas the second 1,3-dicarbonyl compound 27a (or 27b) and the aldehyde 28 were used in all possible combinations (one unique combination per well) (Scheme 10.24). Each well plate containing ammonium nitrate (on bentonite) was impregnated with N,N-dimethylformamide as energy-transfer medium. The 96-well reactor placed in a microwave oven (1300 W) was irradiated for 5 min at 70% power level.

More recently, Bagley and coworkers prepared tri or tetrasubstituted pyridines by microwave irradiation of ethyl β-aminocrotonate and a variety of alkynones in a single synthetic step and with total control of regiochemistry (Scheme 10.25) [60].

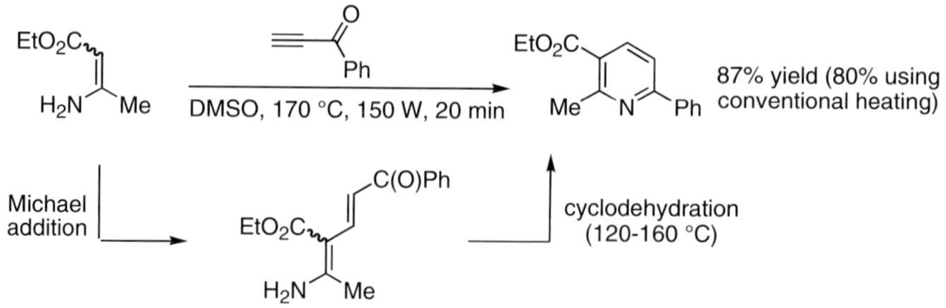

Scheme 10.25

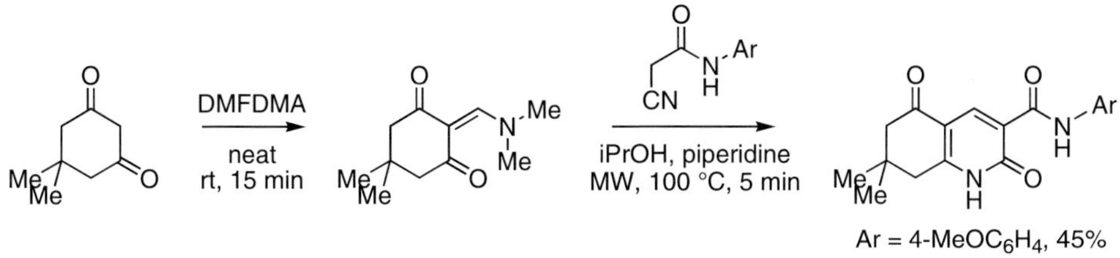

Scheme 10.26

This new one-pot Bohlmann–Rahtz procedure, conducted at 170 °C in a self-tunable microwave synthesizer, gives superior yields to similar experiments conducted using conductive-heating techniques in a sealed tube, and can be performed in the presence of a Brönsted or Lewis acid catalyst.

2-Pyridone and 2-quinolone analogues are well known biologically active heterocyclic scaffolds. In 2004 Kappe and coworkers generated libraries of 3,5,6-substituted 2-pyridone derivatives by rapid microwave-assisted solution-phase methods using a one-pot, two-step procedure [61]. This three-component condensation of CH-acidic carbonyl compounds, N,N-dimethylformamide dimethylacetal (DMFDMA), and methylene-active nitriles led to 2-pyridones and fused analogues in moderate to good yields (Scheme 10.26).

The quinoline scaffold and derivatives occur in a large number of natural products and drug-like compounds. A method for microwave-assisted synthesis of 2-aminoquinolines has been described by Wilson et al. [62]. The process involves rapid microwave irradiation of secondary amines and aldehydes to form enamines, then addition of 2-azidobenzophenones with subsequent irradiation to produce the 2-aminoquinoline derivatives (Scheme 10.27). Purification of the products was accomplished in a streamlined manner by using solid-phase extraction techniques to produce the desired compounds in high yields and purity. Direct comparison of the reaction under thermal and microwave conditions, using identical stoichiometry and sealed reaction vessels, showed the latter resulted in improved yield.

A new type of N-hydroxylacridinedione derivative was recently obtained in excel-

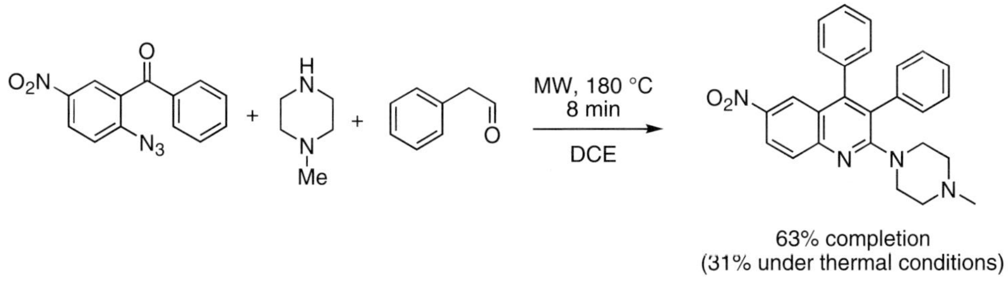

Scheme 10.27

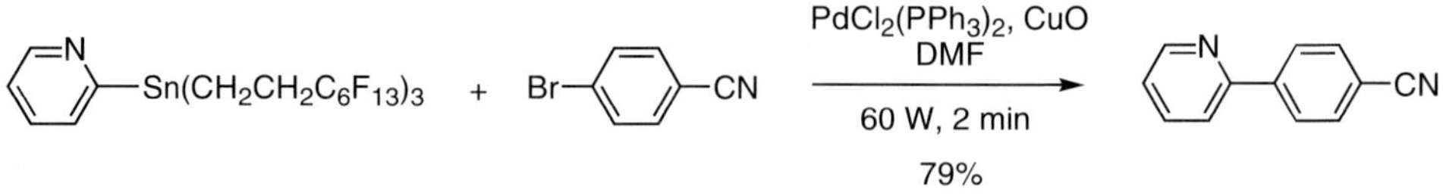

Scheme 10.28

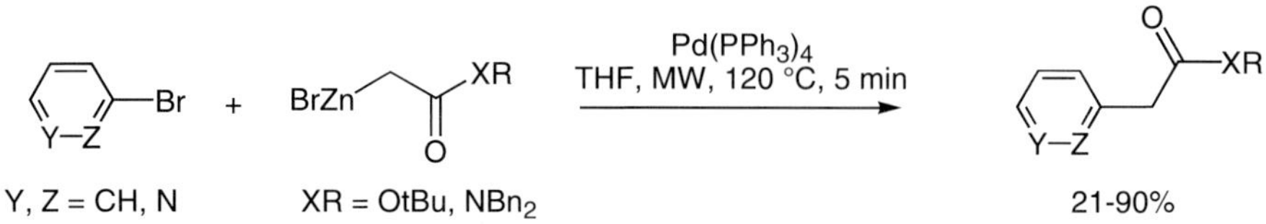

Scheme 10.29

lent yields (80–95%) within a short reaction time (4 to 8 min) by sequential addition, elimination, and cyclization reactions when a mixture of aldoxime and dimedone in glycol was subjected to microwave irradiation [63].

The efficiency of fluorous Stille coupling reactions is enhanced by use of microwave irradiation [64]. The reaction indicated in Scheme 10.28 proceeds in 79% yield after 2 min with DMF as the microwave-active solvent.

Functionalization of pyridines from the corresponding bromo substrates has been achieved by use of Reformatsky reagents and a microwave-accelerated reaction procedure [65]. This approach utilizes Pd(0)-catalyzed α-arylation of esters and amides (Scheme 10.29). The Reformatsky reagent was prepared by microwave irradiation of *tert*-butyl bromoacetate or dibenzyl bromoacetamide with Zn in THF for 5 min at 100 °C.

Negishi cross-coupling was used for synthesis of pyridylpyrimidines using classical thermal or microwave-assisted conditions [66]. Distribution and yield of desired compounds and possible by-products proved to be very dependent on the type of energy input. In contrast with thermal conditions, the microwave-assisted method enabled efficient access to dicoupled compounds (Scheme 10.30).

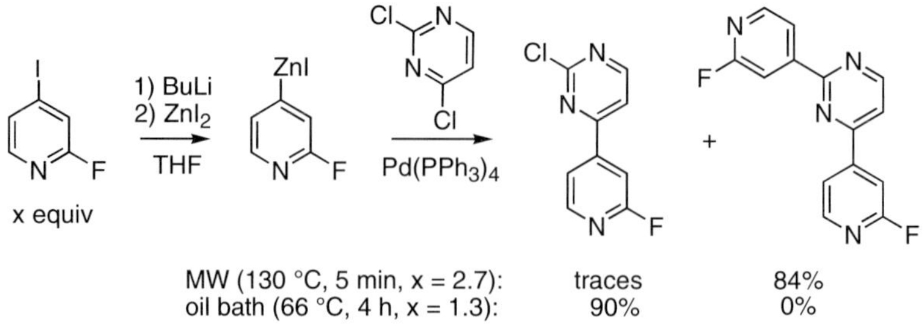

Scheme 10.30

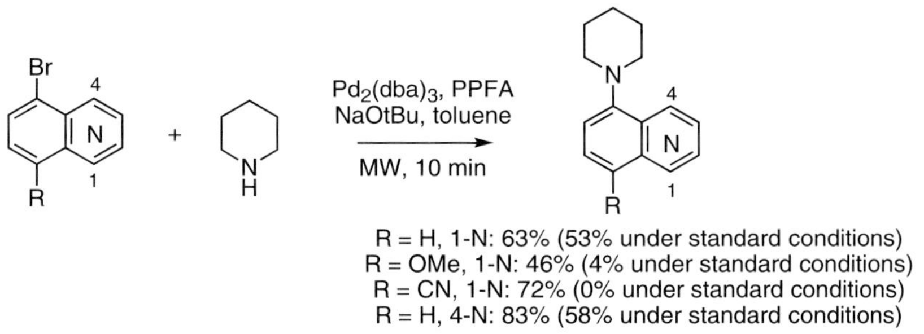

Scheme 10.31

R = H, 1-N: 63% (53% under standard conditions)
R = OMe, 1-N: 46% (4% under standard conditions)
R = CN, 1-N: 72% (0% under standard conditions)
R = H, 4-N: 83% (58% under standard conditions)

In 2004, Walla and Kappe described a more straightforward means of access to arylzinc halides – microwave-accelerated Rieke Zn insertion into aryl halides [67].

Cu-catalyzed cross-coupling of aryl iodides with 2- and 3-pyridylacetylenes under the action of microwave irradiation has been described [68]. Use of 10 mol% CuI and 2 equiv. Cs_2CO_3 results in 79% yield.

Hamann and coworkers described a rapid synthesis of 5- and 8-aminoquinolines, in good yields, from the corresponding heterocyclic bromides by Pd-catalyzed aryl amination under microwave conditions (Scheme 10.31) [69]. Yields were consistently improved compared with those obtained under standard conditions. When 5-bromo-8-cyanoquinoline was used as substrate, no desired products were obtained under standard conditions with several different primary and secondary amines whereas microwave conditions provided the desired products in good to excellent yields.

Similar reactions were reported with activated and unactivated azaheteroaryl chlorides using temperature-controlled microwave heating [70]. Good yields were obtained in short reaction times.

N-Arylsulfonamides are an important class of therapeutic agents in medicinal chemistry. Traditional preparation involves reaction of sulfonyl chlorides with amines, but the sulfonation of weakly nucleophilic amines is difficult. Pd catalysis proved to be effective for N-Arylation of sulfonamides with 4-chloroquinoline (Scheme 10.32) [71].

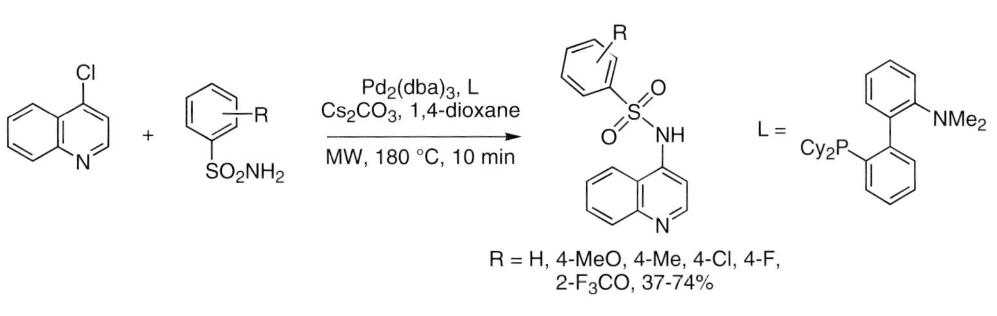

R = H, 4-MeO, 4-Me, 4-Cl, 4-F, 2-F_3CO, 37–74%

Scheme 10.32

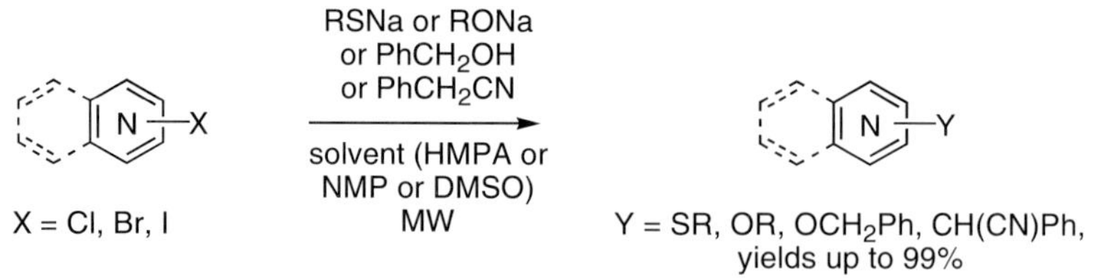

Scheme 10.33

Similarly, N-arylsulfoximines and related species have been prepared by Pd-catalyzed coupling of a sulfoximine with aryl chlorides under the influence of microwave irradiation [72]. Appropriately functionalized systems gave rise to benzothiazines directly in a one pot process.

Syntheses of substituted pyridines [73], quinolines [74], and isoquinolines [74] by S_NAr reactions of haloheterocycles with sulfur, oxygen, and carbon nucleophiles under the action of focused microwave irradiation were developed by Cherng using aprotic dipolar solvents (Scheme 10.33). The reactions were complete within several minutes with yields up to 99%. The method using microwave irradiation is superior to those conducted with conventional heating. This method is especially important for the preparation of 3-pyridyl ethers, thioethers and acetonitriles, because these compounds are not easily accessible by conventional procedures. It was next extended to the pyrimidine and pyrazine series [75].

In medicinal chemistry nitriles are very useful because they can be transformed into biologically important structures such as tetrazoles, triazoles, oxazoles, thiazoles, oxazolidinones ... A variety of aryl nitriles has been prepared in excellent yields by Pd-catalyzed coupling of aryl halides with $Zn(CN)_2$ using polymer-supported triphenylphosphine as the ligand and DMF as the solvent under microwave irradiation conditions (Scheme 10.34) [76]. Products were obtained in high yields and excellent purity without the need for purification.

Trabanco and coworkers developed an operationally simple method for the α-allylation of aryl alkanones with allyl alcohol under microwave irradiation conditions (Scheme 10.35) [77]. The corresponding α-allyl ketones were obtained in moderate to good yields with minor quantities of diallylation by-products.

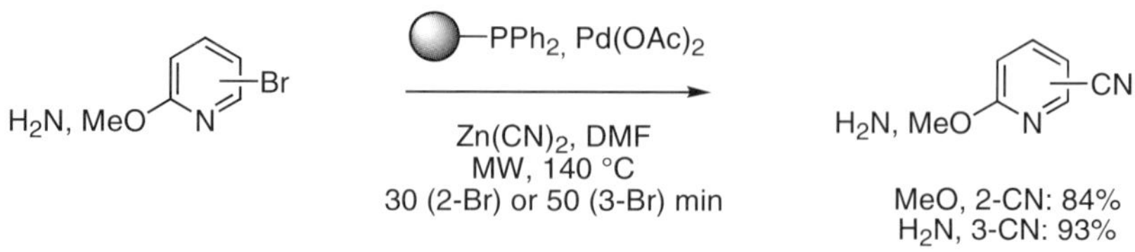

Scheme 10.34

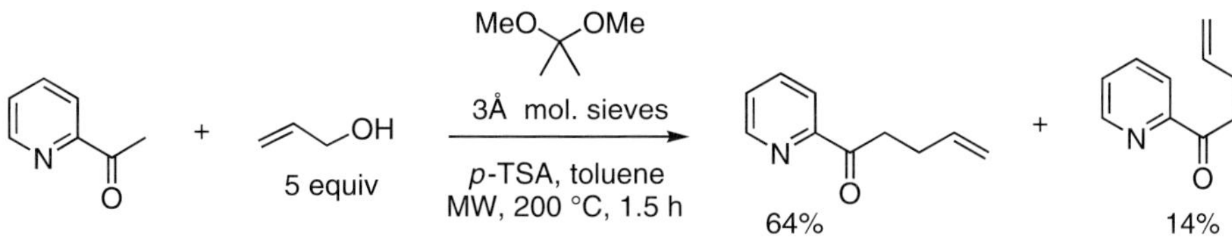

Scheme 10.35

The 2-pyridone core structure is present in a wide range of compounds with antibacterial, antifungal, and antitumor activity; members of this family also play an important role in Alzheimer's disease. By employing microwave-assisted organic synthesis, efficient conditions have been established for introduction of aminomethylene substituents in highly substituted bicyclic 2-pyridones. To incorporate tertiary aminomethylene substituents in the 2-pyridone framework, a microwave-assisted Mannich reaction using preformed imminium salts proved to be effective. Primary amino methylene substituents were introduced via cyanodehalogenation then borane dimethyl sulfide reduction of the afforded nitrile (Scheme 10.36) [78]. Microwave irradiation proved superior to traditional conditions for these transformations.

The perfluorooctylsulfonyl group was introduced by Zhang and coworkers as a traceless tag for solution-phase Pd-catalyzed deoxygenation reactions of phenols. The synthetic efficiency is improved by use of microwave irradiation for the reaction and fluorous solid phase extraction for separation [79].

Alvarez-Builla and coworkers have used microwave irradiation to improve Hantzsch 1,4-dihydropyridine synthesis [80]. Reduced reaction times and improved yields are generally associated with this procedure as exemplified in Scheme 10.37.

Khadilkar and Madyar have developed a large scale continuous synthesis of Hantzsch 1,4-dihydropyridine-3,5-dicarboxylates in aqueous hydrotope solution, using a modified domestic microwave oven [81]. The authors used novel reusable aqueous hydrotope solution as a safe alternative to inflammable organic solutions, in a microwave cavity, for synthesis of commercially important calcium blockers such as nifedipine, nitrendipine, and a variety of other 1,4-dihydropyridines (DHP) (Scheme 10.38). Nitrendipine (R = 3-NO_2, R^1 = Me) has been obtained in 94% yield (50 g) after 24 min by microwave irradiation of the reaction mixture (final temperature 86 °C) at a flow rate of 100 mL min^{-1}. The reaction mixture was circulated through the microwave cavity in four cycles of 6 min each; a 2-min gap between each cycle was imposed to avoid excessive heating.

It should be noted that the Hantzsch 1,4-DHP synthesis could be conducted with a soluble polymer. Successful derivatization of poly(styrene–co-allyl alcohol) under the action of microwave irradiation with a variety of ethyl oxopropanoates and cthyl 3-aminobut-2-enoates was reported by Vanden Eynde's group [82].

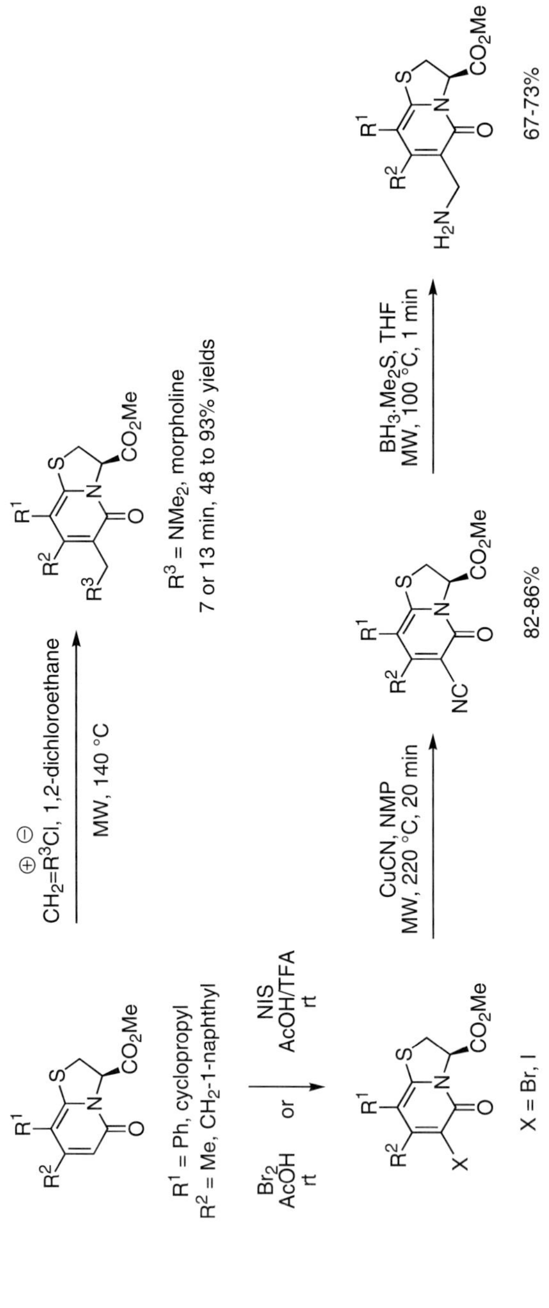

476 | *10 Microwaves in Heterocyclic Chemistry*

Scheme 10.36

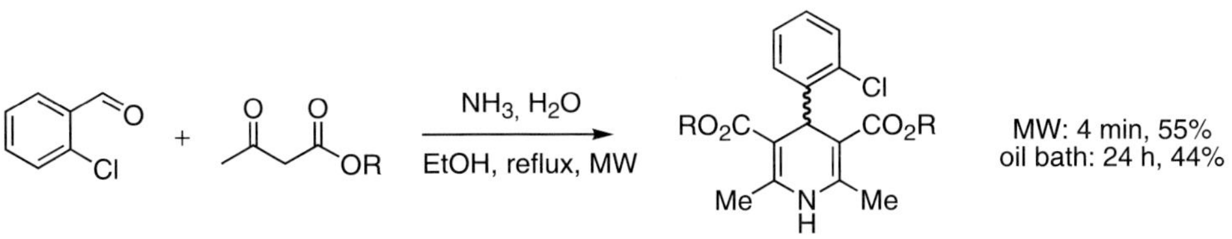

Scheme 10.37

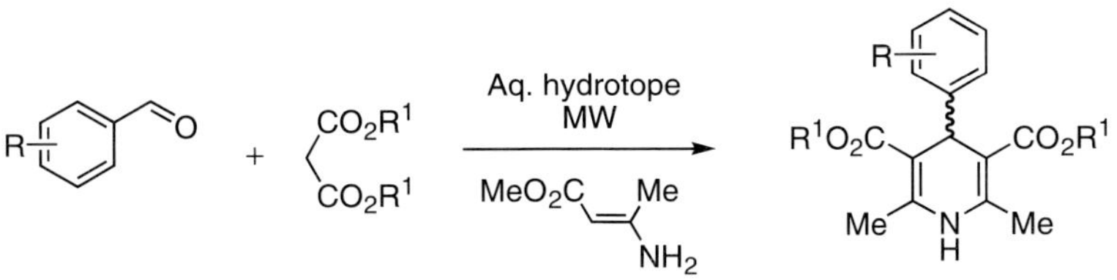

Scheme 10.38

10.2.4.3 Synthesis of Tetrahydropyrans

In 2004 Percy and coworkers observed unexpected selective formation and reactions of epoxycyclooctenones under microwave-mediated conditions [83]. Topologically mobile difluorinated cyclooctenones undergo epoxidations with methyl(trifluoromethyl)dioxirane. The epoxides resist conventional hydrolysis but react smoothly in basic media under microwave irradiation to afford unique hemiacetals and hemiaminals in good yields (Scheme 10.39).

10.2.4.4 Coumarin Chemistry

For preparation of allyl coumarins and dihydrofuranocoumarins by tandem Claisen rearrangement–cyclization, conventional procedures require vigorous reaction conditions, tedious work-up, and long reaction times and lead to low yields. The rearrangement of allyloxycoumarins **29** to dihydrofuranocoumarins **30** in good yields has been optimized in a sealed Teflon container with BF_3–etherate in N-methylformamide (NMF) [84]; the short reaction time (10 min) showed the

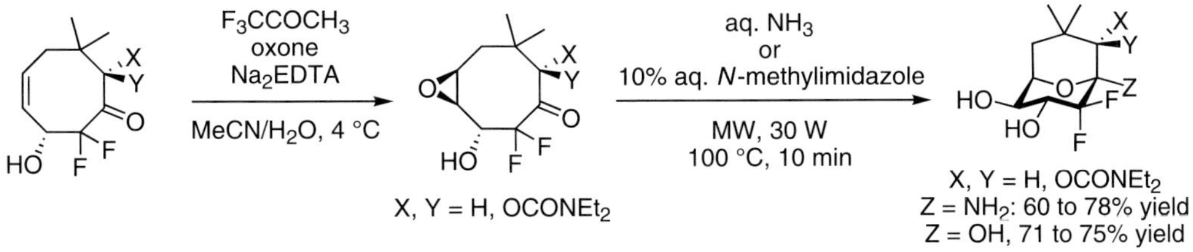

Scheme 10.39

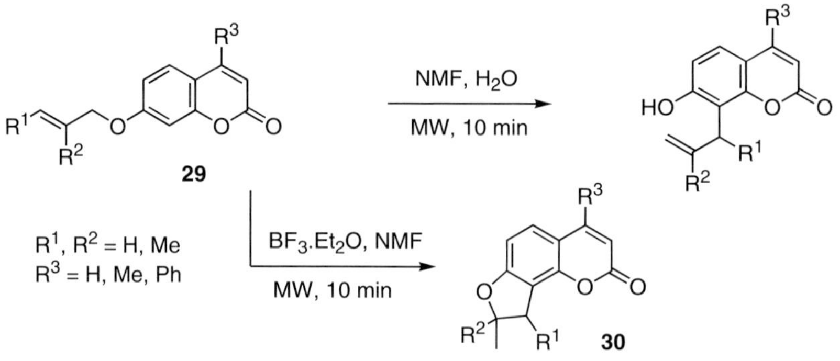

Scheme 10.40

microwave-assisted procedure is the best means of preparation of these compounds, as shown in Scheme 10.40.

In 2002 Raghunathan and coworkers described intramolecular domino Knoevenagel hetero Diels–Alder reactions between 4-hydroxycoumarin and its benzo analogues and O-prenylated aromatic aldehydes or citronellal, to afford pyrano fused polycyclic frameworks. High chemoselectivity was achieved by application of microwave irradiation (Scheme 10.41) [85]. The reaction described is promising because it enables easy access to skeletons that occur in natural products.

Methods to form fluorescein and rhodamine dyes typically feature high-temperature condensation reactions that are not readily adapted to form small libraries of derivatives. Burgess and co-workers developed borylation, Suzuki, and Sonogashira reactions involving bromo derivatives of fluorescein and rhodamine [86]. Organoboron compounds, biaryls, and alkynes were readily synthesized by

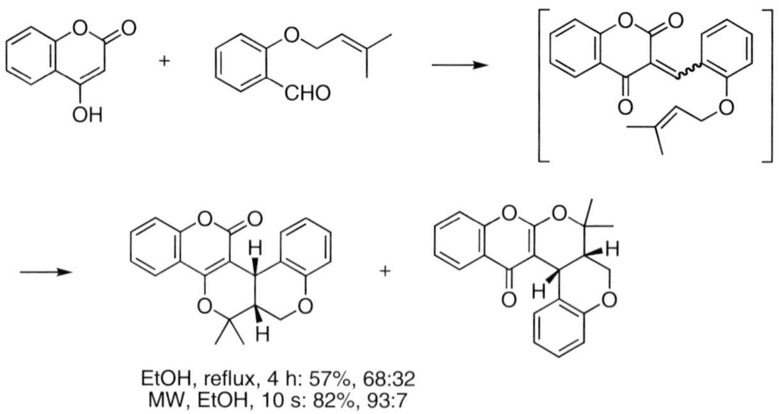

Scheme 10.41

using bursts of microwave irradiation to give high temperatures for relatively short times.

10.2.5
Six-membered Systems with More than One Heteroatom

10.2.5.1 Piperazine Chemistry
2,5-Diketopiperazines can be regarded as the smallest cyclic peptides, derived from the head-to-tail folding of a linear unprotected dipeptide. Microwave irradiation has been applied in the synthesis of 2,5-diketopiperazines using the dimerization of an active peptide compound as the key reaction (Scheme 10.42) [87]. Conventional heating and microwave irradiation were compared; synthesis using microwave irradiation gave the desired compounds in higher yields and with shorter reaction times than those obtained by conventional heating.

N-aryl-2-piperazinones are of great interest as farnesyl-transferase inhibitors, and both N-aryl-2-piperazinones and N-aryl-2,5-piperazinediones are synthetic precursors of N-arylpiperazines which are key elements in monoamine receptor-active drugs. Lange et al. enhanced the Goldberg reaction with microwaves for synthesis of N-arylpiperazinones, N-arylpiperazinediones, and N-aryl-3,4-dihydroquinolines [88]. Microwave irradiation greatly accelerates the reaction when NMP is used as solvent.

10.2.5.2 Synthesis of Tetrahydropyrimidine Derivatives
Isothioureas and cyclic guanidines are frequently used in bioactive compounds, either to modulate solubility or to pick up electrostatic interactions. Their use as central scaffolds in drug design is, however, very limited because of the lack of general applicable and mild synthetic methods or straightforward procedures. Wellner and coworkers reported the construction of cyclic isothioureas and related guanidine derivatives using solely microwave-assisted chemistry, without any need for activating agents or protecting group manipulations (Scheme 10.43) [89]. Formation of substituted guanidines from the corresponding isothiouronium salts was controlled by the nucleophilicity of the counterion and affected by the reaction temperature.

Yeh and Sun have combined the advantages of microwave technology with liquid phase combinatorial chemistry to facilitate the synthesis of thioxotetrahydropyrimidinones [90]. The reactions were significantly accelerated.

10.2.5.3 Synthesis of Pyrimidines and Related Compounds
Ethynyl ketones are valuable synthetic intermediates in the preparation of a wide range of simple nitrogen-containing heteroaromatic molecules. Bagley and coworkers described syntheses of 2,4-disubstituted and 2,4,6-trisubstituted pyrimidines in high yields using microwave irradiation of a mixture of amidine and alkynone in acetonitrile at 120 °C (Scheme 10.44) [91]. This procedure is more efficient than other reported conditions using traditional heating methods, and often does not require further purification.

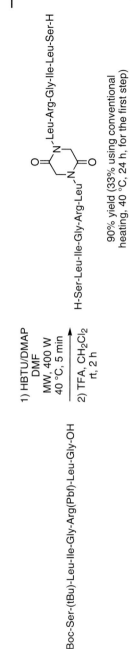

Scheme 10.42

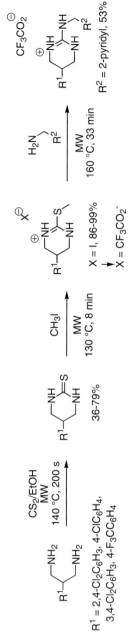

Scheme 10.43

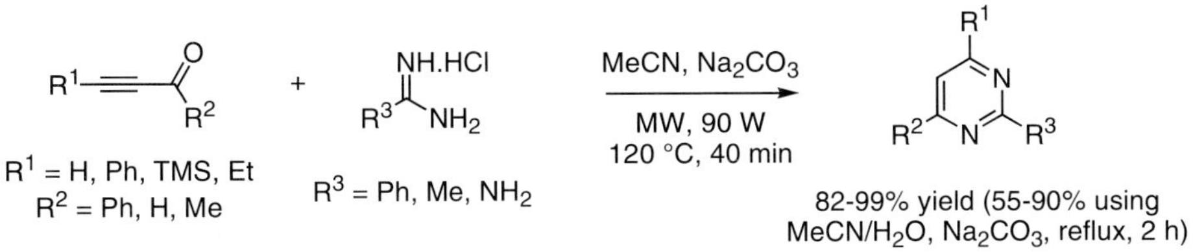

Scheme 10.44

R^1 —≡— C(O)—R^2 + R^3—C(=NH·HCl)NH$_2$ → pyrimidine

R^1 = H, Ph, TMS, Et
R^2 = Ph, H, Me
R^3 = Ph, Me, NH$_2$

Conditions: MeCN, Na$_2$CO$_3$, MW, 90 W, 120 °C, 40 min

82–99% yield (55–90% using MeCN/H$_2$O, Na$_2$CO$_3$, reflux, 2 h)

Hydroxymethylation of uracil, 5-hydroxymethyl-2′,3′-O-isopropylideneuridine, 5′-O-*tert*-butyldiphenylsilyl-2′,3′-O-isopropylideneuridine, 2′,3′-O-isopropylidenecytidine, and 2′,3′-O-isopropylidene-5′-O-tritylcytidine was efficiently achieved in very high yields by El Ashry and coworkers with paraformaldehyde in alkaline medium under the action of microwave irradiation [92].

Paolini et al. has reported regioselective microwave-assisted iodination of pyrimidinones, uracils, cytosine, and their unprotected nucleosides, using N-iodosuccinimide to give the corresponding 5-iodo derivatives [93].

The effect of monomode microwave irradiation in the Pd-catalyzed phenylation of 5-iodouracil with nontoxic sodium tetraphenylborate as phenyl reagent has been examined (Scheme 10.45) [94]. The authors showed that the use of monomethylformamide (MMF) as solvent increases the yield of the coupling (70%), because MMF has a high boiling point (180 °C) and is more polar ($\varepsilon = 182.5$) than other amides used in the microwave-activated reactions.

4-Aminoquinazolines are useful as fungicides and as anti-inflammatory, antimicrobial, and antihypertensive agents. In particular, they are potent and highly selective inhibitors of tyrosine kinase. Han and coworkers have developed an efficient method for their synthesis using microwave irradiation reactions between N-(2-cyanophenyl)-N,N-dimethylformamidine derivatives and amines (Scheme 10.46) [95]. This procedure is straightforward and high yielding, an advantage over alternate routes which often require several steps.

The Niementowski synthesis of the 3H-quinazolin-4-one core was re-investigated by Besson and coworkers using microwave irradiation [96]. It enabled significant rate enhancements and good yields compared with conventional reaction conditions.

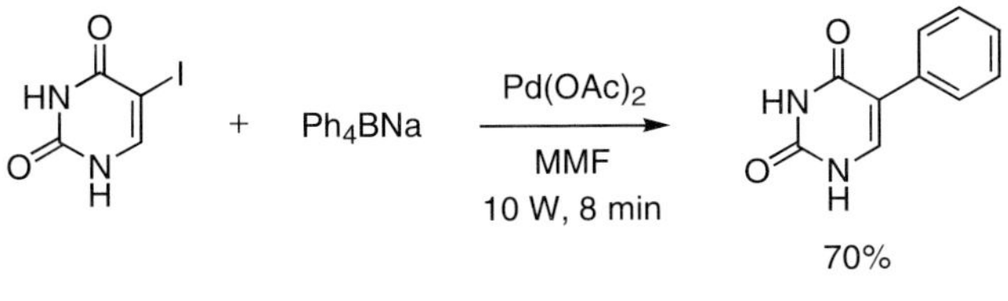

Scheme 10.45

5-iodouracil + Ph$_4$BNa → 5-phenyluracil

Conditions: Pd(OAc)$_2$, MMF, 10 W, 8 min, 70%

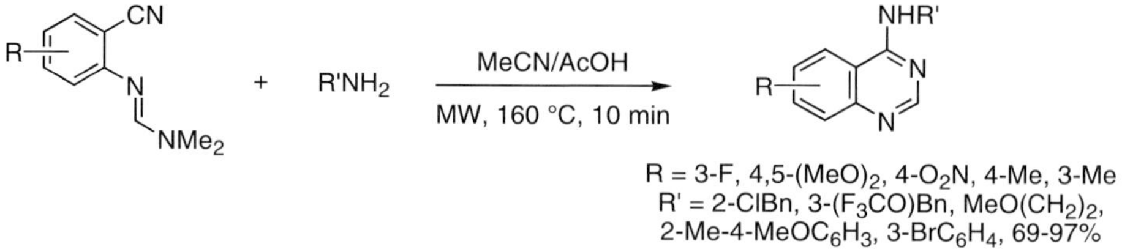

CN
R
N
NMe2
+ R'NH2
MeCN/AcOH
MW, 160 °C, 10 min
NHR'
N
N
R
R = 3-F, 4,5-(MeO)2, 4-O2N, 4-Me, 3-Me
R' = 2-ClBn, 3-(F3CO)Bn, MeO(CH2)2,
2-Me-4-MeOC6H3, 3-BrC6H4, 69-97%

Scheme 10.46

Wu and coworkers developed a microwave-assisted reaction to facilitate the construction of 4,5-disubstituted pyrazolopyrimidines. This one-pot two-step process involves sequential S_NAr displacement of the C4 chloro substituent with a variety of anilines and amines, then Suzuki coupling with different boronic acids (Scheme 10.47) [97]. Use of microwave irradiation leads to high product conversion, low side product formation, and shorter reaction times.

Dihydropyrimidinone (DHPM) derivatives have received substantial attention in recent years because of to their activity as calcium-channel blockers, antihypertensive, antibacterial, antitumor, and anti-inflammatory agents. The introduction of fluoro or trifluoromethyl groups into an organic molecule frequently provides compounds of pharmacological interest, compared with their nonfluorinated analogs. Microwaves have been found to greatly accelerate the Biginelli synthesis of fluorine-containing ethyl 4-aryl-6-methyl-1,2,3,4-tetrahydropyrimidin-2-one/thione-5-carboxylates in unsealed vessels with ethanol as energy-transfer medium and a small amount of concentrated HCl as catalyst (Scheme 10.48) [98]. Short reaction

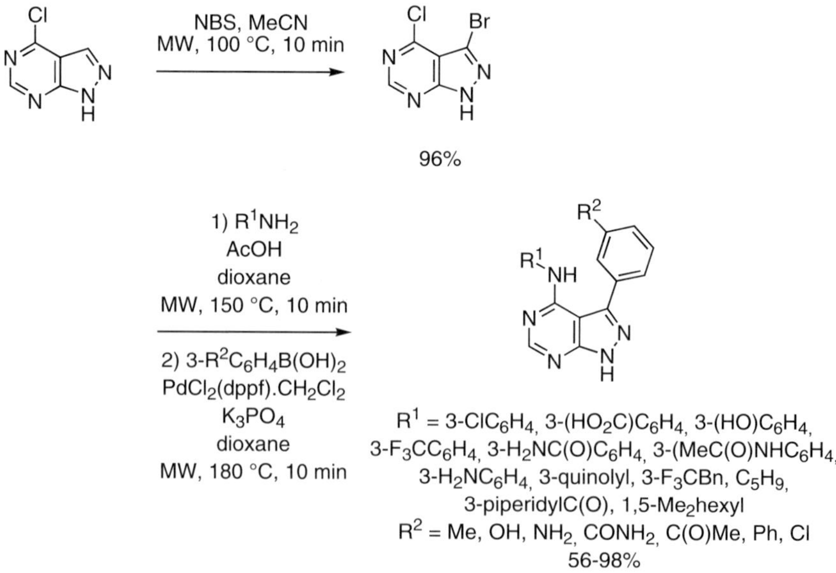

Cl
N
N
N
N
H
NBS, MeCN
MW, 100 °C, 10 min
Cl
Br
96%
1) R1NH2
AcOH
dioxane
MW, 150 °C, 10 min
2) 3-R2C6H4B(OH)2
PdCl2(dppf).CH2Cl2
K3PO4
dioxane
MW, 180 °C, 10 min
R2
R1
NH
R1 = 3-ClC6H4, 3-(HO2C)C6H4, 3-(HO)C6H4,
3-F3CC6H4, 3-H2NC(O)C6H4, 3-(MeC(O)NHC6H4,
3-H2NC6H4, 3-quinolyl, 3-F3CBn, C5H9,
3-piperidylC(O), 1,5-Me2hexyl
R2 = Me, OH, NH2, CONH2, C(O)Me, Ph, Cl
56-98%

Scheme 10.47

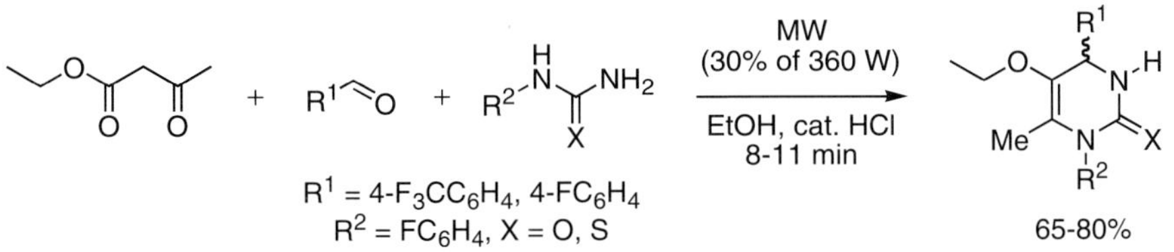

Scheme 10.48

times, enhanced yields, and easy workup are the main advantages observed by Dandia et al.

N-Bromosuccinimide has recently been used as a catalyst in the preparation of DHPM and the corresponding thio derivatives under the action of microwave irradiation [99]. Good yields, short reaction times, and mild reaction conditions make this procedure complementary to existing methods. Among these, use of polyphosphate ester as a reaction mediator has been successfully attempted for Biginelli condensation under microwave conditions [100].

The development of a support/linker system for the microwave-assisted synthesis of dihydropyrimidine test libraries, and methods for solid-phase scale-up on cellulose were recently described [101].

10.2.5.4 Synthesis of Pyrazines and Related Compounds

Zhang and Tempest have efficiently incorporated microwave and fluorous technologies in a Ugi–de-Boc cyclization strategy for construction of quinoxalinones and

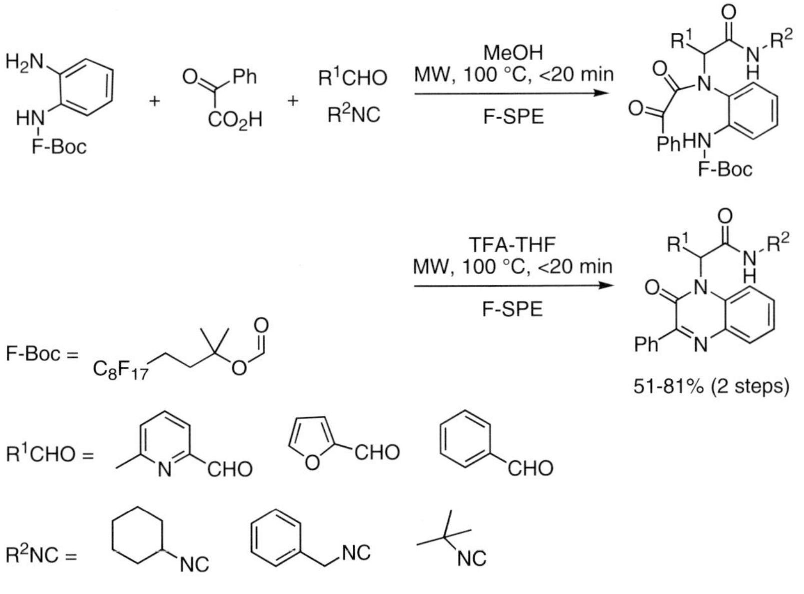

Scheme 10.49

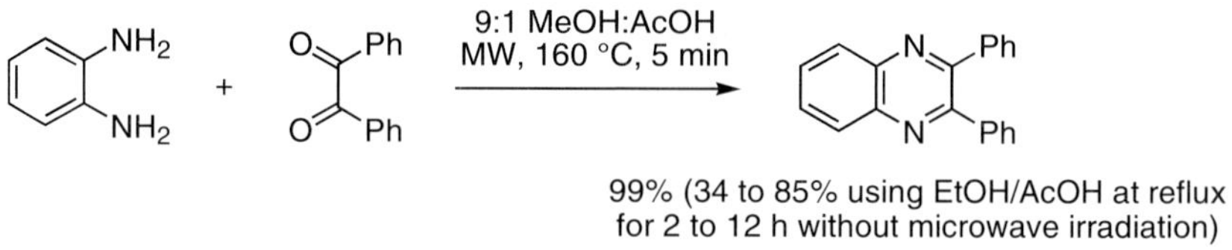

Scheme 10.50

benzimidazoles [102]. For instance, in the synthesis of substituted quinoxalinones, a fluorous-Boc-protected diamine was used for the Ugi reactions. Both the Ugi and the post-condensation reaction proceeded rapidly under the action of microwave irradiation, and the reaction mixtures were purified by solid-phase extraction on FluoroFlash cartridges (F-SPE) (Scheme 10.49).

Compared with the original Ugi–de-Boc cyclization procedures, which take 1–2 days, the fluorous/microwave approach has more favorable reaction and purification conditions – less than 20 min for each reaction and no need for the double scavenging step.

In 2002 Isay-type condensations for synthesis of 6-substituted pterins, including pterin sugar derivatives and 2-substituted quinoxalines, were developed under microwave irradiation conditions [103]. Interestingly, the isomer-free 6-substituted pterins desired were obtained in moderate to good yields whereas mixtures of both 6- and 7-isomers (major) are usually formed when conventional Isay type condensations are used.

Lindsley and coworkers prepared functionalized quinoxalines and heterocyclic pyrazines in excellent yields from 1,2-diketone intermediates under the action of microwave irradiation (Scheme 10.50) [104]. In addition to being a general method for synthesis of a variety of aryl/heteroaryl 1,2-diamines and 1,2-diketones, this procedure suppresses the formation of the polymeric species characteristic of traditional thermal conditions.

In both reactions of scheme 10.51, the thermal instability of Ru(II)-catalyst systems during prolonged heating at 88 °C in ring-closing metathesis reactions of dienyne and triyne substrates has been circumvented by use of a microwave-induced temperature increase to 160 °C when the reaction times were reduced to minutes (Scheme 10.51) [105].

Single mode microwave-assisted combinatorial synthesis of biologically interesting quinoxalinones has been described by Tung and Sun [106]. It enabled chiral libraries of quinoxalinones to be assembled by use of $S_N Ar$ reactions, reduction, and concomitant cyclization under the action of microwave irradiation.

10.2.5.5 Synthesis of Triazines

The triazine ring system is a component of commercial dyes, herbicides, insecticides, and even pharmaceutical compounds. By application of microwave technology, a general procedure has been developed for rapid synthesis of diverse 3,5,6-trisubstituted 1,2,4-triazines from a variety of 1,2-diketones and acyl hydrazides in excellent yields and purities (Scheme 10.52) [107].

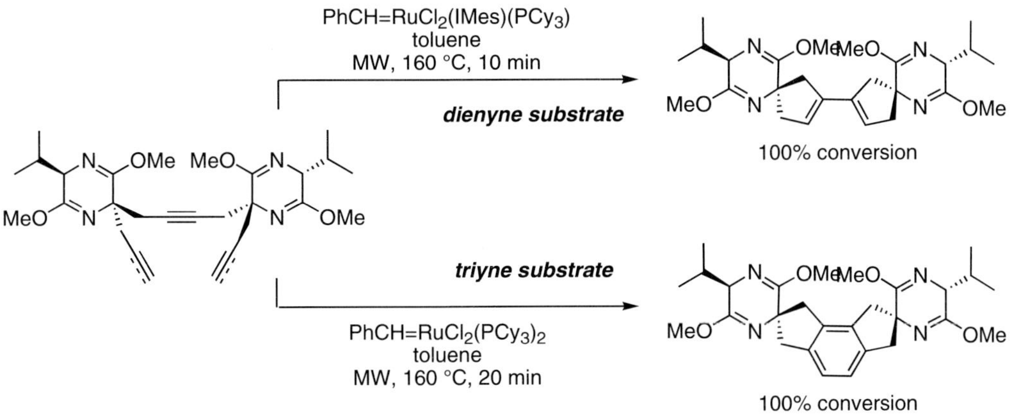

Scheme 10.51

A library of twenty phenyldihydrotriazines was prepared and compared with an identical library generated by conventional parallel synthesis. Microwave synthesis reduced reaction times from an average of 22 h to 35 min, and compounds generated using microwave irradiation were purer. Isolated yields of all the compounds were comparable for both methods [108].

10.2.6
Synthesis of More Complex or Polyheterocyclic Systems

The purpose of this section is to highlight the applications of microwave irradiation to syntheses of more complex or polyheterocyclic systems. When conventional thermal procedures (metal or oil bath) fail microwave irradiation can be used as an alternative to classical methods enabling development of easy and rapid access to new heterocycles irrespective of the conditions needed in the homogeneous phase.

Besson and coworkers reported an original approach for the synthesis of the rare thiazolo[5,4-*f*]quinazoline **31** in seven steps from commercially available 2-amino-5-nitrobenzonitrile (**32**) (Scheme 10.53) [109]. The authors studied the conversion

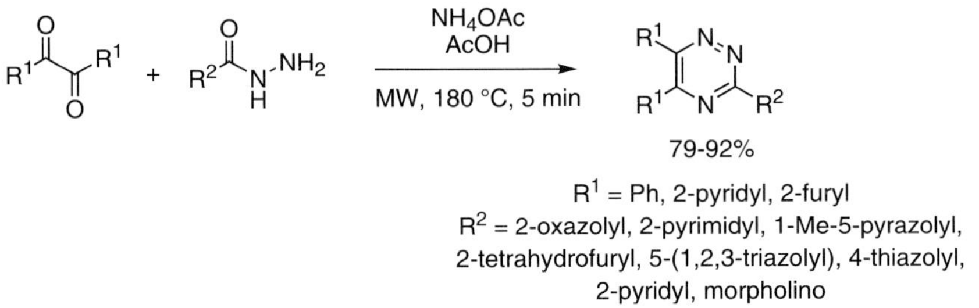

R^1 = Ph, 2-pyridyl, 2-furyl
R^2 = 2-oxazolyl, 2-pyrimidyl, 1-Me-5-pyrazolyl, 2-tetrahydrofuryl, 5-(1,2,3-triazolyl), 4-thiazolyl, 2-pyridyl, morpholino

Scheme 10.52

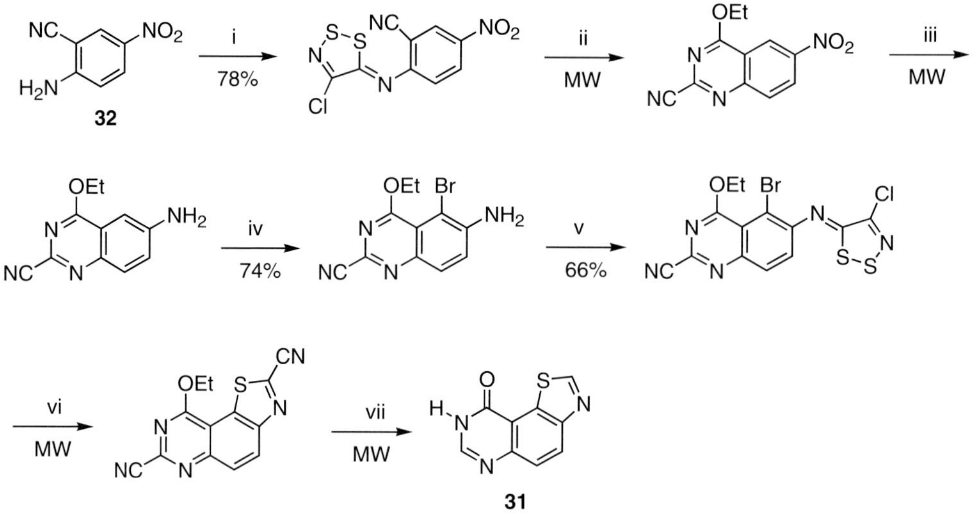

Scheme 10.53. Reagents and conditions (for reaction times and yields of steps ii, iii, vi and vii, see Table 10.4). (i) 4,5-dichloro-1,2,3-dithiazolium chloride, pyridine, rt, 10 h; (ii) NaH, EtOH, reflux; (iii) SnCl$_2 \cdot$ 2H$_2$O, EtOH, 70 °C; (iv) Br$_2$, AcOH, rt, 24 h; (v) 4,5-dichloro-1,2,3-dithiazolium chloride, rt, 4 h; (vi) CuCN, pyridine, reflux; (vii) HCl, reflux.

of all four steps in the synthesis of thiazoloquinazoline **31** to microwave conditions, with the same concentration of starting material and volume of solvent, and found that yields of the desired compounds were always better than those obtained with conventional heating (Table 10.4). The overall time for the synthesis of **31** was substantially reduced and the overall yield was enhanced.

A new means of access to 5a,10,14b,15-tetraazabenzo[a]indeno[1,2-c]anthracen-5-ones **33** and to benzimidazo[1,2-c]quinazoline **34** was reported in 2000 [110]. The preparation of 6-mercapto-benzimidazo[1,2-c]quinazoline (**34**) was easily ac-

Tab. 10.4. Comparison of conventional heating and microwave irradiation for the different steps of Scheme 10.53.

Step	Starting material	Product	Conventional heating		Microwave irradiation	
			Reaction time (min)	Yield (%)	Reaction time (min)	Yield (%)
ii	23	24	640	37	80	61
iii	24	25	60	72	10	94
vi	27	28	90	50	20	53
vii	28	29	60	49	10	50

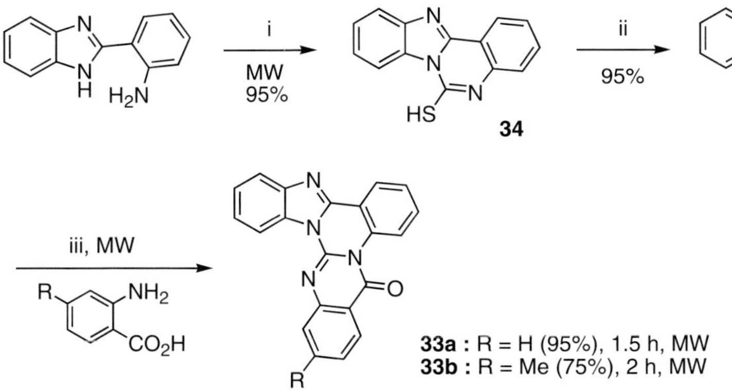

Scheme 10.54. Reagents and conditions: (i) CS_2, MeOH, reflux, MW, 55 min; (ii) MeI, NaH, DMF, 25 °C, 5 min; (iii) graphite, MW.

complished at 60 °C in 55 min, under the action of focused microwave irradiation, by reaction of starting 2-(2-aminophenyl)benzimidazole with CS_2 in the presence of KOH and a protic solvent (MeOH) (Scheme 10.54). Under similar conditions, conventional heating led to the same yield but with a significantly extended reaction time of 24 h.

The benzodiazepine nucleus is a well studied traditional pharmacophoric scaffold that has emerged as a core structural unit of a variety of sedative–hypnotic, muscle relaxant, anxiolytic, antihistaminic and anticonvulsant agents. The literature on this subject, although very copious, is mainly patents. An original investigation of the preparation of 1,5-benzodiazepin-5-ones **35** (Scheme 10.55) was described by Santagada et al. [111]. Noticeable improvements were achieved for

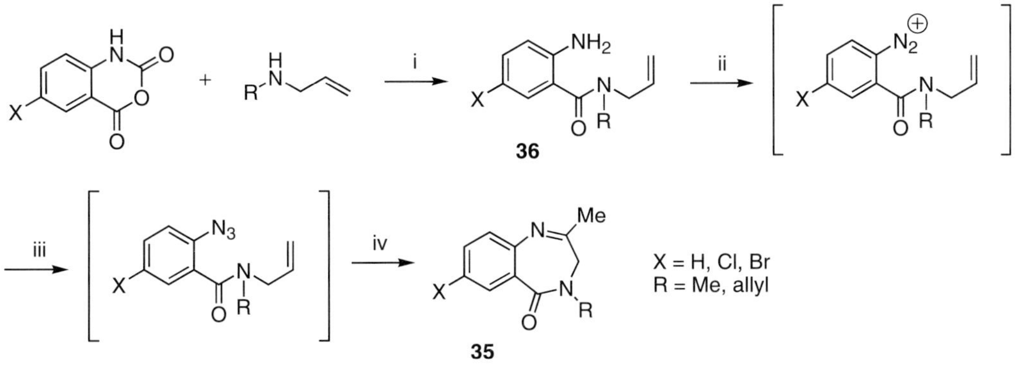

Scheme 10.55. Reagents and conditions: (i) DMF, reflux, 1 h, MW; (ii) HCl M, $NaNO_2$, AcOH, 30 min; (iii) NaN_3, Et_2O, 40 min; (iv) DMF, MW.

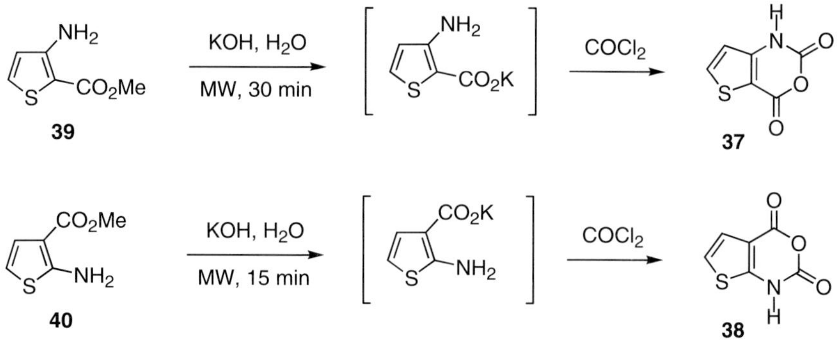

Scheme 10.56

steps (i) and (iv) when performed under the action of microwave irradiation with the same concentration of starting material and volume of DMF. For step (i), the product **36a** (R = Me, X = H) was obtained in 91% yield after 5 min under microwave conditions at 150 W, compared with only 65% using conventional heating in oil bath, under reflux in DMF for 180 min. For step (iv), the product **35a** was isolated in 60% yield after a reduced reaction time of 5 min at 80 °C under the action of microwave irradiation (150 W), compared with 40% using conventional heating conditions (60 min, reflux).

Thiaisatoic acid is a starting material of great interest in the field of heterocyclic chemistry, as a potential new pharmacological scaffold. Rault and coworkers studied the synthesis of 2-thiaisatoic anhydride **37** and 3-thiaisatoic anhydride **38** using the alkaline hydrolysis of ortho aminoesters **39** and **40** as a key step (Scheme 10.56) [112]. The authors showed it is possible to perform this reaction in a multimode microwave oven in a few minutes on a large scale in water containing a slight excess of potassium hydroxide but without cosolvent, avoiding decarboxylation reaction observed under the action of long classical heating [113]. After microwave heating, the potassium carboxylates could be treated directly by bubbling phosgene through the aqueous solution to yield the anhydrides **37** and **38** in 85 and 67% yields, respectively, with purity >90%.

Under the action of conventional heating the rearrangement of pyrrolo[2,1-c]-benzodiazepine **41**, in boiling phosphorus oxychloride, into benzo[h][1,6]-naphthyridine **42** required a long reaction time (5 h). Under microwave heating conditions, compound **42** is obtained with reduced reaction time (1.5 h) and better purity. The chlorine displacement with N-methylpiperazine was next achieved in DMF in a sealed glass vessel. After 1 h under microwave irradiation the derivative **43** was obtained in 68% yield (Scheme 10.57) [114].

5-Deaza-5,8-dihydropterins were prepared from 2,6-diaminopyrimidin-4-one, 1,3-dicarbonyl compounds, or benzoylacetonitrile, and aromatic or aliphatic aldehydes [115]. This three-component cyclocondensation proceeds under microwave-assisted conditions in good yield using Zn(II) bromide as a Lewis acid catalyst, and with total control of regiochemistry.

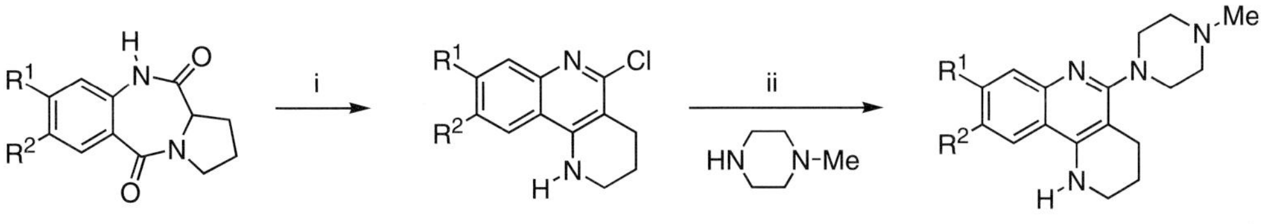

Scheme 10.57. Reagents and reaction conditions: (i) POCl$_3$, pyridine, reflux, MW (700 W), 1 h 45 min; (ii) DMF, 240 °C, 6 bar, MW, 1 h.

The importance of pyrimido[1,2-*a*]pyrimidines is well recognized by biological chemists. A one-pot access to these structures was realized by condensation of 2-aminopyrimidine, aromatic aldehydes and active methylene reagents under the action of microwave irradiation [116].

Many nucleoside analogues substituted at the 5 position of the heterocycle, especially those in the 2′-deoxyuridine series, are known to have potent biological properties, and have been investigated as antiviral and anticancer agents. Pyrimidinones, uracils, and uridine iodinated at C5 were converted to the corresponding alkynyl derivatives by direct microwave-enhanced Sonogashira coupling using PdCl$_2$(PPh$_3$)$_2$, Et$_3$N and CuI (Scheme 10.58) [117]. The use of microwave irradiation afforded the coupling products in a few minutes and in good yields.

Hopkins and Collar used microwave activation to improve the synthesis of 6-substituted 5*H*-pyrrolo[2,3-*b*]pyrazines. The reaction is a Pd-catalyzed heteroannulation process followed by deprotection (Scheme 10.59) [118].

A variety of fused 3-aminoimidazoles has been synthesized by a microwave-assisted Ugi three-component coupling reaction catalyzed by scandium triflate in methanol as solvent. The reactions of heterocyclic amidines with aldehydes and isocyanides were performed in 33 to 93% yields within only 10 min of microwave irradiation using a simple one-stage procedure [119].

In 2002 Raboisson et al. reported a method for synthesis of 4,7-dihydro-1*H*-imidazo[4,5-*e*]-1,2,4-triazepin-8-one derivatives by use of microwave methodology. The reaction involves aminoimidazole carboxhydrazide and ortho esters, and the subsequent alkylation of the 1,2,4-triazepin-8-one obtained occurs regioselectively at C1 [120].

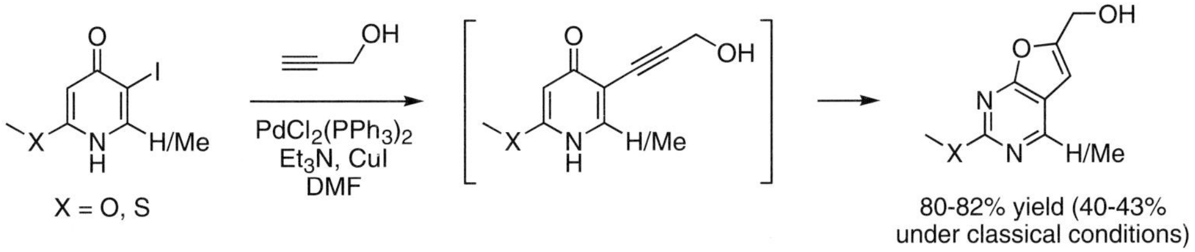

Scheme 10.58

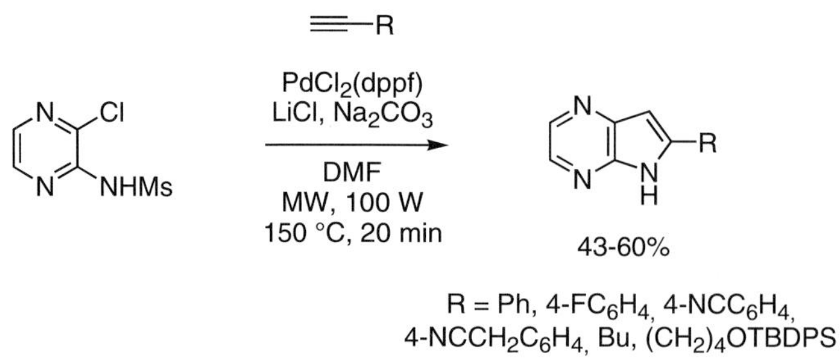

Scheme 10.59

Indolizino[1,2-*b*]quinoline is a constituent of the skeleton of camptothecin, an antitumor agent, and of mappicine, an antiviral agent. When prepared by means of a Friedländer reaction, indolizino[1,2-*b*]quinolines were obtained in moderate or low yields by conventional heating. Starting from unstable 2-aminobenzaldehydes or imine and indolizinones, and using acetic acid as solvent, use of microwave irradiation resulted in yields of 57–91%, compared with 31–48% by the conventional method (Scheme 10.60) [121]. This improvement may be because of shortened reaction times, which limit any degradation of unstable reactants. Solvents can, moreover, be heated (or superheated) to a high temperature, known as the nucleation limiting boiling point (when the reactions are performed in a closed vessel without any stirring), and such superheating may increase overall reaction rates. Finally, another additional specific (non thermal) mechanism of microwave irradiation leading to the rate and yield enhancements must evidently be taken into account (see Chapter 4 of this book for an extended discussion).

The canthines are a tetracyclic subclass of β-carboline alkaloids bearing an additional D-ring. Members of the canthine family have been shown to have pharmacological activity, including antifungal, antiviral, and antitumor properties. In 2003 Lindsley and coworkers described a microwave-mediated procedure for one-pot synthesis of the basic canthine skeleton [122]. The key step is an inverse-electron demand Diels–Alder reaction and subsequent chelotropic expulsion of N_2, a reaction that can be achieved after a three-component condensation of an acyl hydrazide-tethered indole with a 1,2-diketone and excess ammonium acetate to form a triazine (Scheme 10.61).

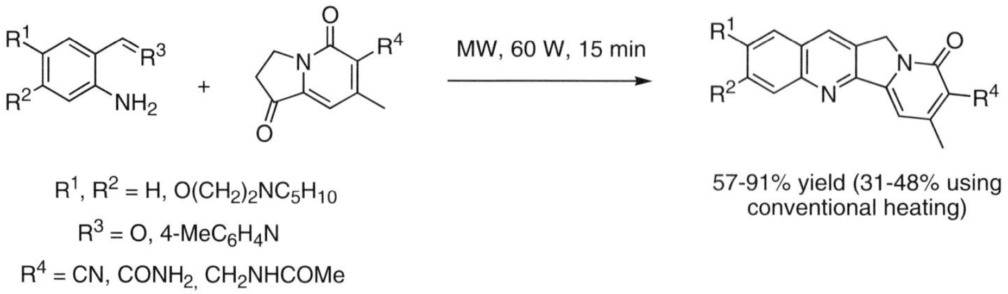

Scheme 10.60

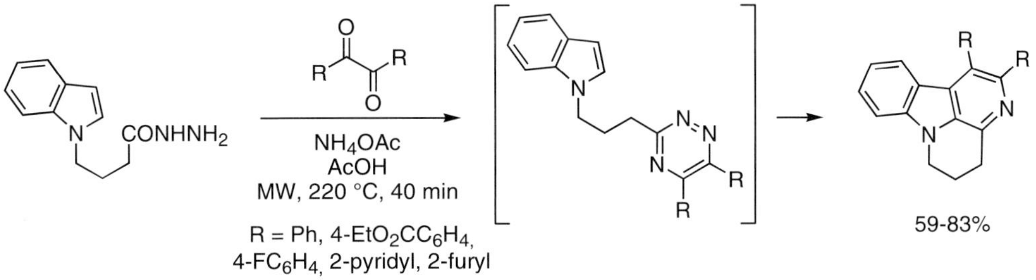

Scheme 10.61

The reaction was conducted in a single-mode microwave synthesizer, with reaction times reduced 10 to 700-fold over conventional thermal methods.

The 1,2,3,4-tetrahydro-β-carboline pharmacophore is an important structural element in several tryptophan-derived natural product alkaloids. A soluble polymer supported synthesis of 1,2,3,4-tetrahydro-β-carboline derivatives was reported by Wu and Sung. The one-pot condensation of Fmoc-protected tryptophan anchored to MeO-PEG-OH via an ester linkage has been performed with a variety of aldehydes and ketones under the action of microwave irradiation to provide immobilized 1,2,3,4-tetrahydro-β-carboline derivatives; these were liberated from the soluble matrix in good yield and high purity [123].

More recently, Chu and coworkers prepared diastereoisomers of 1,3-disubstituted 1,2,3,4-tetrahydro-β-carbolines in short times (0.5 to 4 h) with good yields (50–98%) using the Pictet–Spengler reactions of tryptophan with aldehydes under acidic conditions at ambient temperature [124]. Although reaction rates are intrinsically slow, ketone reactions can be accelerated (from days to minutes) by using microwaves with open vessels; isolated yields are high (67–99%).

Microwave irradiation has been used to accelerate the conversion of isatins to their thiosemicarbazones, which were cyclized into 1,2,4-triazino[5,6-b]indole-3-thiols. Selective alkylation of these tricyclic systems with different alkylating agents has also been achieved under microwave irradiation conditions [125].

The development of a cost-effective microwave-assisted procedure for synthesis of pyrido fused-ring systems, by applying the *tert*-amino effect, has been described by Van der Eycken and coworkers (Scheme 10.62) [126].

Reactions that required hours or even days under conventional heating conditions could be completed within 3–42 min under the action of microwave irradiation, with minimum energy demands. The isolated yields obtained with the

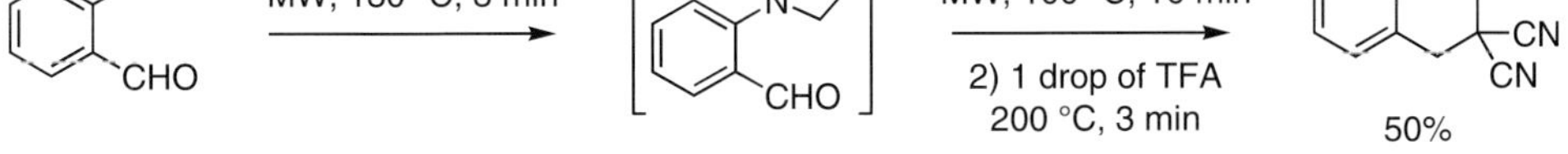

Scheme 10.62

microwave-assisted procedures have been found to be as good as those obtained under conventional heating conditions, or even superior.

10.3
Solvent-free Synthesis

Although the initial reason of the development of solvent-free conditions for microwave irradiation was safety, it soon became apparent that use of these conditions had many other benefits – simplicity, efficiency, easy work-up, very often higher yields, and enhanced reaction rates. The absence of solvent is, furthermore, time and moneysaving and often enables elimination of waste treatment. This class of synthesis is now clearly seen as a basis of "green chemistry".

Solvent-free synthesis can be realized under a variety of conditions; we will give selected results and, when available, comparisons with the same solvent-free conditions but with classical heating.

10.3.1
Three, Four and Five-membered Systems with One Heteroatom

10.3.1.1 Synthesis of Aziridines

A dry media technique using microwave irradiation for synthesis of aziridines appeared in the literature in 1996 (Scheme 10.63) [2]. The reaction was performed under the action of focused microwaves and conventional heating and it is noteworthy that elimination is more efficient than Michael addition (leading to aziridines) under the action of irradiation, the conditions being the same.

10.3.1.2 Synthesis of Thiiranes

An expeditious one-pot synthesis of thiiranes has been described by Yadav and coworkers (Scheme 10.64) [127]. The reaction proceeds with high diasteroselectiv-

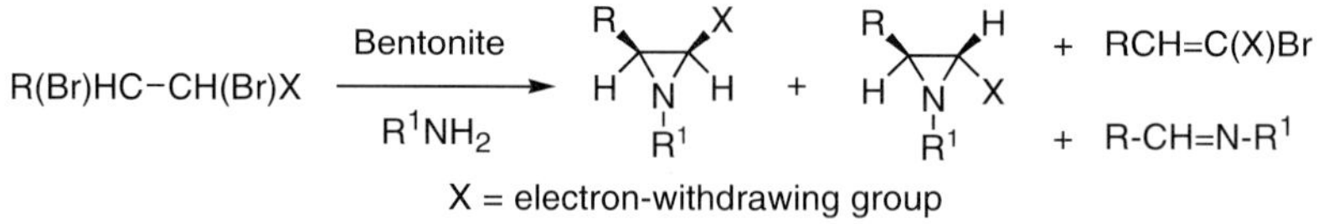

Scheme 10.63

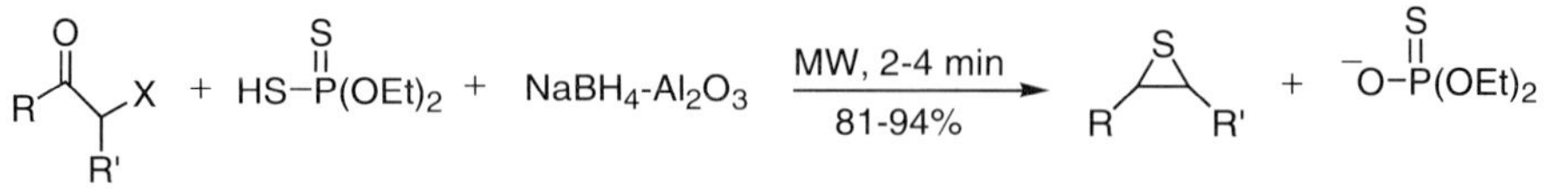

Scheme 10.64

ity. Yields under microwave irradiation conditions (from 81 to 94%) are usually twice those obtained by use of an oil bath.

10.3.1.3 Synthesis of Lactams and Imides

In 1979 Olah and coworkers reported one step conversion of alicyclic ketones into lactams by reaction of hydroxylamine O-sulfonic acid and formic acid under reflux for a few hours [128]. More recently, this reaction has been achieved under solvent-free conditions with silica gel as inorganic support and focused irradiation, as exemplified in Scheme 10.65 for caprolactam synthesis [129].

In a multimode oven, the power is too high and cannot be modulated, so only decomposition products are obtained.

The reaction of silyl ketone acetals with imines under the action of irradiation has been explored. The versatility of the microwave approach is illustrated in Scheme 10.66 [130]. When the reactions were performed by adsorption of the reagents on to K_{10} montmorillonite clay, the aminoester is formed. Conversely, when the reaction was performed by mixing the neat reagents with KF and 18-crown-6 and irradiation in a closed vessel, β-lactams were isolated in moderate to good yields.

The same group reported the formation of heterobicyclic lactams as illustrated below (Scheme 10.67) [131]. The reaction was performed by supporting the reagents on alumina and irradiating in an open vessel.

Solvent-free microwave reactions between phthalic anhydride and amino compounds were carefully re-examined by Gedye, Loupy, and coworkers, who showed

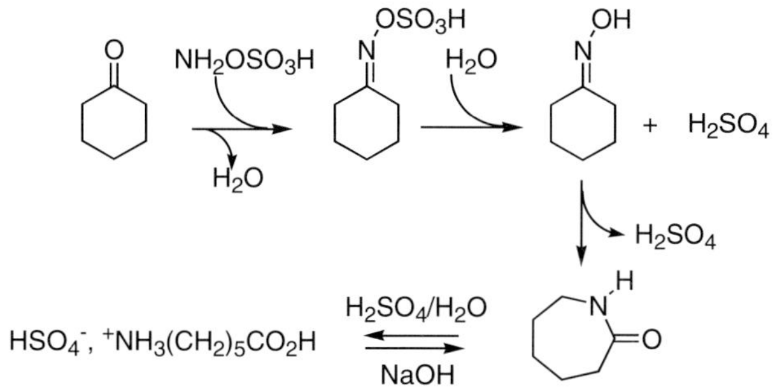

Ketone	Lactam	Time^a (min)	Yield (%)
Cyclopentanone	Valerolactam	15	60
Cyclohexanone	Caprolactam	10	86
Cycloheptanone	2-Azacyclooctanone	20	72
Cyclooctanone	2-Azacyclononanone	15	65
Cycloundecanone	2-Azacyclododecanone	15	72
Cyclododecanone	2-Azacyclotridecanone	20	82

^{a} Irradiation at 30 W, the temperatures reached by the reaction mixture are in the range 100–120 °C, depending on the nature of the ketone.

Scheme 10.65

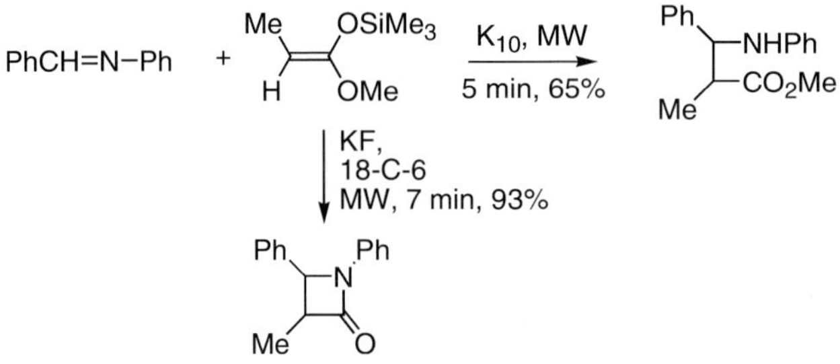

Scheme 10.66

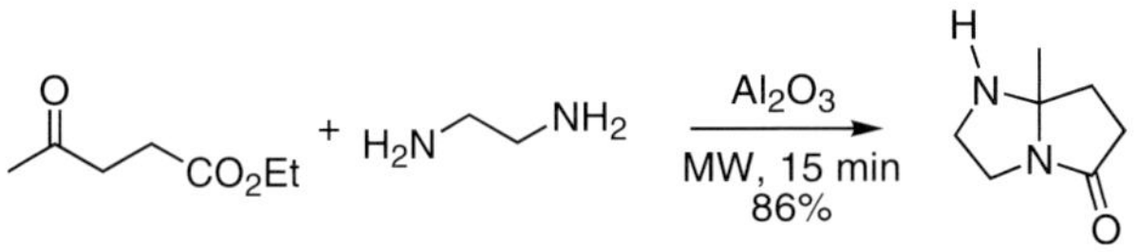

Scheme 10.67

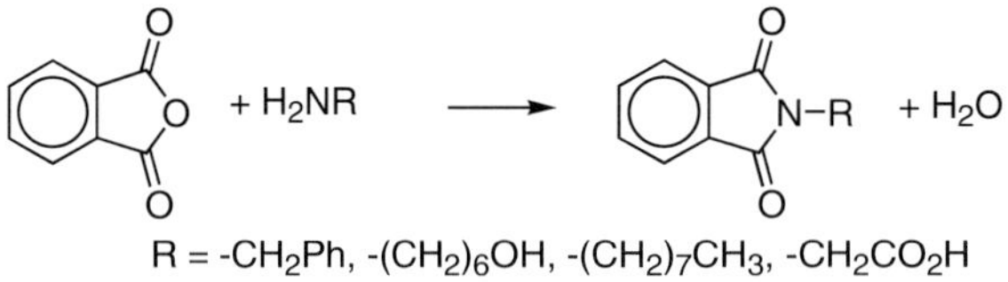

Scheme 10.68

that the reaction needs at least one liquid phase [132]. The reaction occurs after melting of phthalic anhydride and subsequent dissolution in the amine. It was concluded that reactions between two solids might not occur and that a high-boiling-point solvent might be necessary. Excellent yields (>90%) were always obtained within short reaction times (5–10 min) (Scheme 10.68).

More recently, this procedure has been used for a new synthesis of thalidomide, now a "rehabilitated drug" (Scheme 10.69) [133].

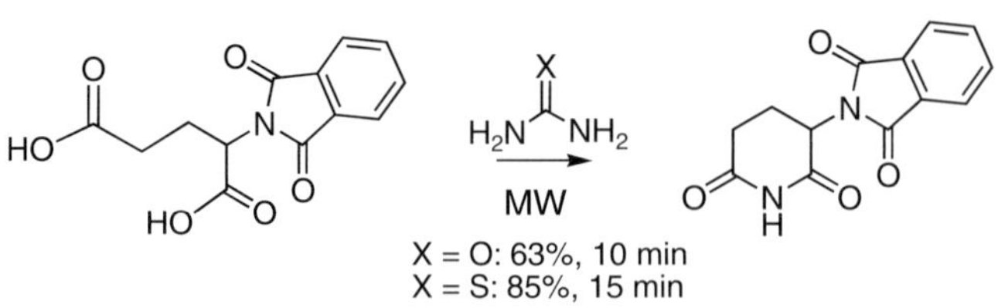

Scheme 10.69

10.3.1.4 Synthesis of Pyrroles and Related Compounds

Efficient syntheses of highly substituted alkylpyrroles and fused pyrroles have been achieved by three-component coupling of:

- an α,β-unsaturated aldehyde or ketone, an amine, and a nitroalkene on the surface of silica gel;
- an α,β-unsaturated nitroalkene, an aldehyde or ketone, and an amine on the surface of alumina without solvent under microwave irradiation conditions.

The different routes are as shown in Scheme 10.70 [134]. A variety of starting compounds was used and yields of isolated products ranged from 60 to 86% for irradiation times of 5 to 15 min. This green procedure avoiding solvents constitutes by far a better and more practical alternative to existing methods.

Different heterocycles, e.g. furan, pyrrole, N-benzylpyrrole, indole, and pyrazole react with methyl α-acetamidoacrylate to give α-amino acid precursors under irradiation with silica-supported Lewis acids as catalysts [135]. In homogeneous catalysis, long reaction times were required. The reaction of vinylpyrazoles with imines has also been realized [136].

A one-pot synthesis of indolizines via a three-component reaction has been reported by Boruah and coworkers (Scheme 10.71) [137].

A parallel synthesis of a 28-member library of phthalimides was reported in 2002 [138]. After chromatography the library members were obtained in good yields (Scheme 10.72).

In the same way, a library of quinoline-3,4-dicarboxamides was obtained under the action of microwave irradiation: several primary amines react with quinoline-3,4-dicarboxylic imides. Among a variety of supports, clay K_{10} proved to be the best and irradiation gave much higher yields [139]. Collina et al. set up an interest-

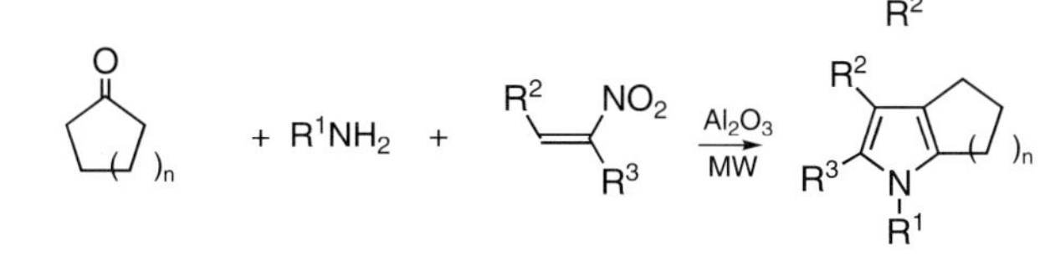

Scheme 10.70

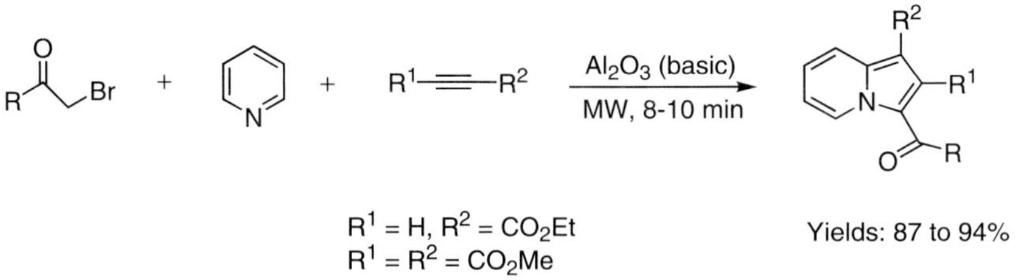

Scheme 10.71

R¹ = H, R² = CO₂Et
R¹ = R² = CO₂Me

Yields: 87 to 94%

ing high yield regioselective procedure to prepare racemic and enantiomeric 1,2-dimethyl-3-[2-(6-substituted naphthyl)]-2H,5H-pyrrolines by dehydration of the corresponding pyrrolidines under the action of microwave irradiation (over FCl_3–SiO_2) [140].

A general procedure for preparation of 2-heteroaryl-5-methoxyindoles was reported by Lipinska, who used a microwave-induced solid-supported clay $K_{10}/ZnCl_2$ Fischer indolization [141].

A variety of bis(indolyl)nitroethanes has been obtained in high yields (70–86%) by reacting 3-(2-nitrovinyl)indole with indole and 1- or 2-alkylindoles on silica gel (TLC-grade) under microwave irradiation conditions (Scheme 10.73) [142].

A simple and rapid synthesis of tetrapyrrole macrocycles has been achieved under dry media conditions with microwave activation [143]. Pyrrole and benzaldehyde adsorbed on silica gel afford tetraphenylporphyrin within 10 min (Scheme 10.74), whereas with conventional methods (e.g. acetic acid in the presence of pyridine) 24 h were necessary. This procedure has recently been applied to the synthesis of corroles [144].

In the Prolabo Synthewave 402 monomode reactor (at 135 W), a yield of 9.5% was obtained compared with 4% in a multimode domestic oven. The advantages of the method are its rapidity and more straightforward workup; yields suffer from the same limitations as under classical conditions. It was later shown that the yield increases to 31% when the irradiation was performed in the presence of toluene [145].

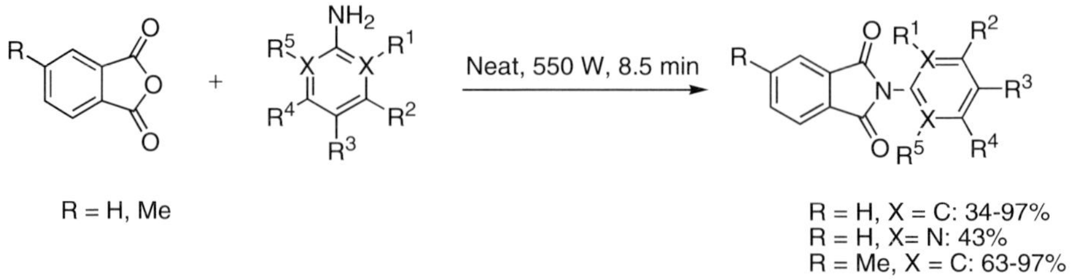

R = H, Me

Neat, 550 W, 8.5 min

R = H, X = C: 34-97%
R = H, X= N: 43%
R = Me, X = C: 63-97%

Scheme 10.72

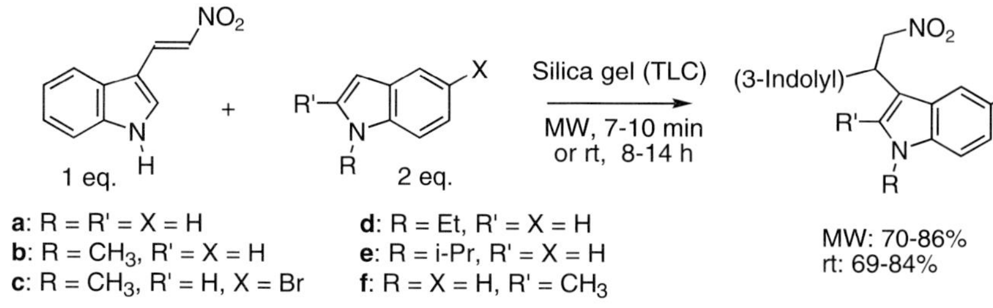

Scheme 10.73

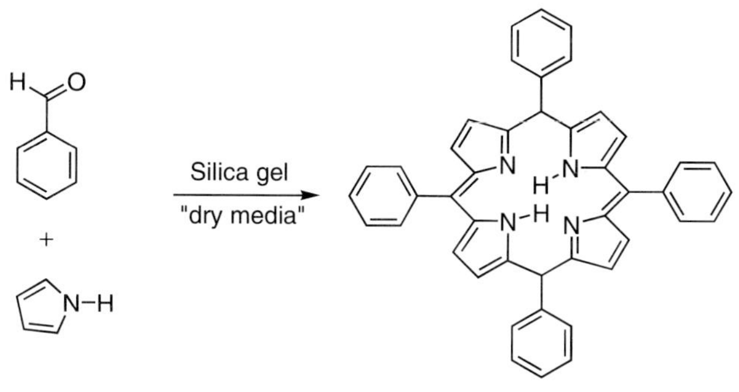

Scheme 10.74

10.3.1.5 Tetrahydrofuran Chemistry

An interesting ring opening of a fatty oxirane has been reported by Loupy and co-workers [146]. Diethyl acetamidomalonate opens the oxirane ring according to Scheme 10.75. Addition of lithium chloride is necessary for the reaction to proceed. Whereas no reaction was observed under salt-free conditions, the lactone was obtained in 90% yield within 5 min in the presence of on alumina impregnated with LiCl and KF (Al_2O_3–LiCl–KF = 4:1:1 by weight) under the action of focused microwave irradiation at 150 W. With conventional heating no reaction occurred.

Loupy and coworkers have synthesized new diethers by alkylation of isosorbide under the action of irradiation with PTC conditions [147]. The yields were very good (>90%) within a few minutes. Even when similar temperature profiles were

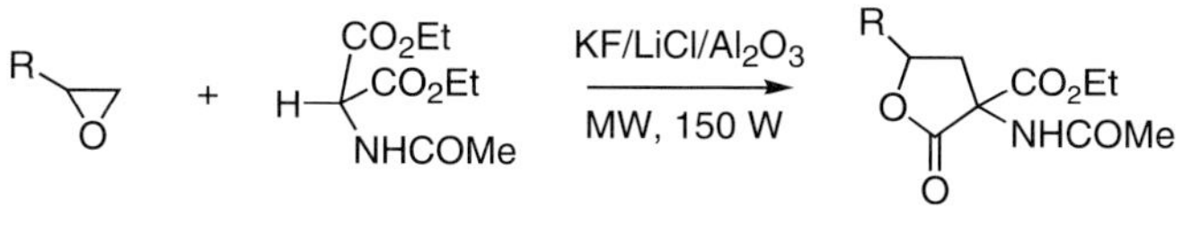

Scheme 10.75

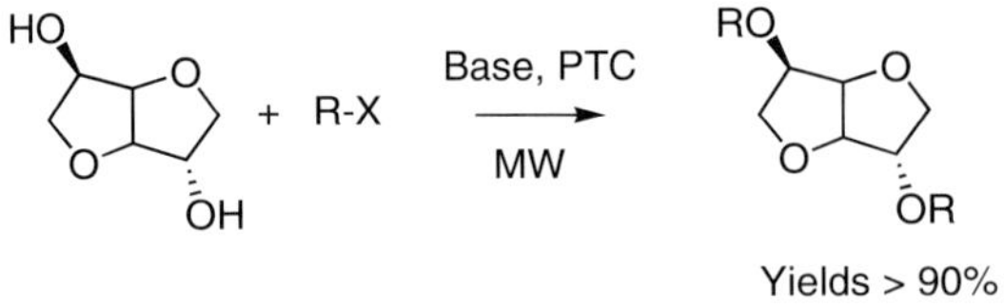

Scheme 10.76

used, the yields were much lower under the action of conventional heating (Scheme 10.76).

In the same way, new diols have been obtained from dianhydrohexitol ethers [148].

More recently, Kotha et al. realized the synthesis of a variety of bis-alkyl ketones from the corresponding alkyloxy compounds under microwave irradiation conditions on silica without solvent [149].

A series of γ-thionolactones has been prepared in good yields by a combination of Lawesson's reagent and hexamethyldisiloxane between solvent-free conditions with irradiation [150].

10.3.1.6 Synthesis of Furans and Related Compounds

Tokuda and coworkers discovered easy conditions for the preparation of furans starting from fused-ring alkylidenecyclopropanes under solvent-free conditions with irradiation (Scheme 10.77) [151].

Benzo[b]furans have been formed by condensation of salicylaldehydes with a variety of esters of chloroacetic acids, in the presence of tetrabutylammonium bromide (TBAB), under the action of microwaves without solvent (Scheme 10.78) [152]. Potassium carbonate (20 mmol), TBAB (0.5 mmol), and salicylaldehyde derivatives (5.0 mmol) were mixed, and a chloroacetic ester (10 mmol) was then added dropwise. The mixture was stirred and then irradiated for 8 to 10 min. After extraction with CH_2Cl_2 and evaporation, the products were purified by column chromatography. If an oil bath is used, full conversion of an aldehyde usually requires 3 h.

Vo-Thanh and Loupy reported a ruthenium-catalyzed ring closure metathesis of olefins under solvent-free irradiation conditions to give a variety of heterocyclic compounds [153].

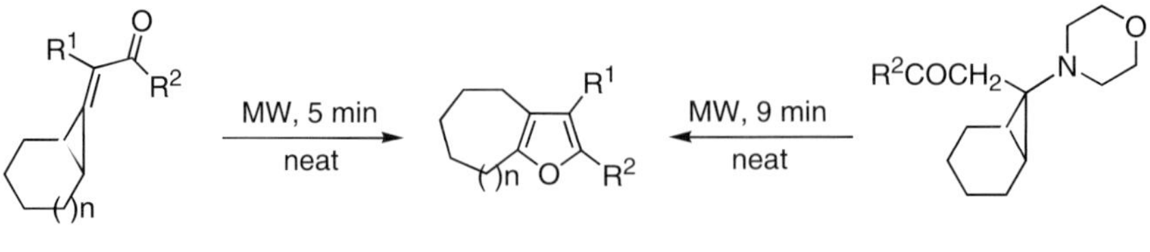

Scheme 10.77

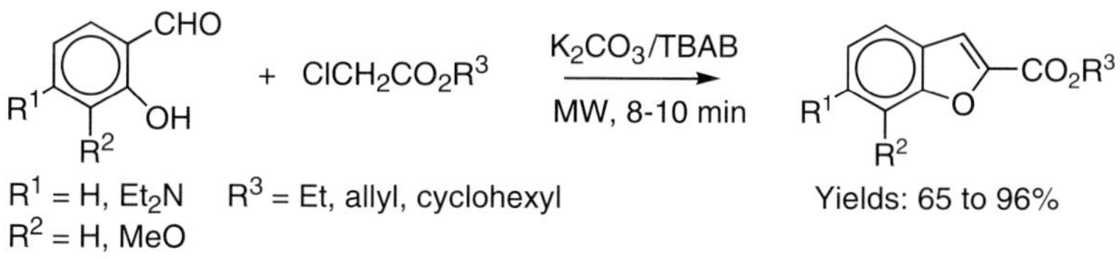

Scheme 10.78

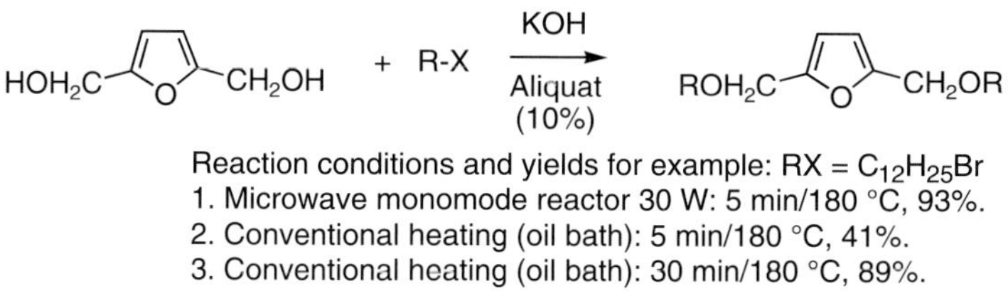

R^1 = H, Et$_2$N R^3 = Et, allyl, cyclohexyl
R^2 = H, MeO

Aromatic aldehyde with OH + ClCH$_2$CO$_2$R^3 → benzofuran-CO$_2$R^3, K$_2$CO$_3$/TBAB, MW, 8–10 min, Yields: 65 to 96%

Scheme 10.79

HOH$_2$C–(furan)–CH$_2$OH + R-X → ROH$_2$C–(furan)–CH$_2$OR, KOH, Aliquat (10%)

Reaction conditions and yields for example: RX = C$_{12}$H$_{25}$Br
1. Microwave monomode reactor 30 W: 5 min/180 °C, 93%.
2. Conventional heating (oil bath): 5 min/180 °C, 41%.
3. Conventional heating (oil bath): 30 min/180 °C, 89%.

Among the various derivatives of biomass, furan compounds obtained from furfural are important (200 000 t year^{-1}). A new family of furan diethers has been obtained by alkylation of 2,5-furandimethanol or furfuryl alcohol under the action of microwave irradiation with PTC solvent-free conditions (Scheme 10.79) [154]. Reaction times were improved by use of microwave irradiation, and the same conditions were extrapolated to the synthesis of a series of new furan diethers by alkylation of furfuryl alcohol by dihalides. 1,4-Diketones substituted by furans or thiophenes were synthesized by conjugate addition of aldehydes to α,β-unsaturated ketones by irradiation without solvent in the presence of thiazolium halides and DBU adsorbed on alumina [155].

10.3.1.7 Synthesis of Thiophenes and Related Compounds

Barbarella and coworkers achieved rapid, efficient, and clean synthesis of highly pure thiophene oligomers by solvent-free microwave-assisted coupling of thienyl boronic acids and esters with thienyl bromides using Al$_2$O$_3$ as support [156]. This procedure is a general and very rapid route to the preparation of soluble thiophene oligomers. The α-halogenation of 2-acrylthiophene using magnesium halides under solvent-free microwave irradiation was achieved by Lee et al. [157].

10.3.2
Five-membered Systems with Two Heteroatoms

10.3.2.1 Synthesis of Cyclic Ureas

Reaction of strong CH-acidic such as Meldrum's acid or barbituric acid derivatives, with aldehydes and urea was studied by Shaabani et al.; this led to an efficient solvent-free synthesis of spiro-fused heterocycles by use of microwave irradiation [158].

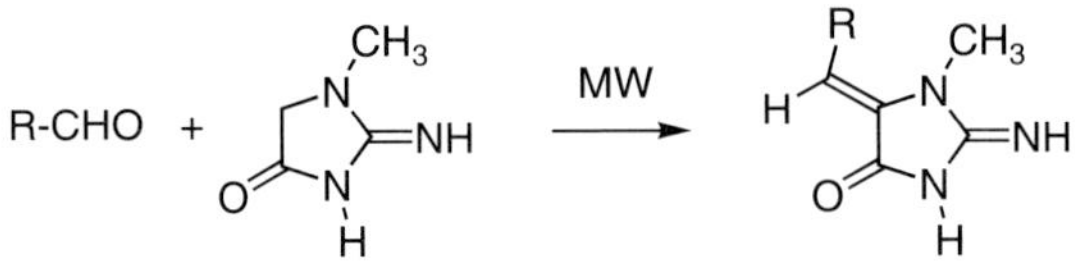

[Scheme 10.80 reaction diagram]

Scheme 10.80

10.3.2.2 Hydantoin and Creatinine Chemistry

Knoevenagel condensation under the action of microwave irradiation has been widely studied. It has also been used for the synthesis of heterocycles in accordance with Scheme 10.80 [36]. At 160–170 °C under the action of focused irradiation (40–60 W), condensation occurs within 45 s to 4 min, without solvent, without base, and without catalyst. Unfortunately, no comparisons were made with conventional heating and classical methods.

More recently, Loupy et al. set up an efficient synthesis of 1,5-disubstituted hydantoins and thiohydantoins under solvent-free conditions under the action of microwaves [159]. This method can be extended toward parallel synthesis.

10.3.2.3 Synthesis of Pyrazoles and Related Compounds

Condensation of aromatic acyl compounds with N,N-dimethylformamide dimethyl acetal in a pressure tube under the action of microwave irradiation affords easy access to 1-aryl-3-dimethylaminoprop-2-enones in almost quantitative yields after 6 min. These intermediates can be reacted with hydrazine hydrate under conventional reflux in ethanol to form the corresponding 3-substituted pyrazoles (Scheme 10.81) [160].

1,3,5-Trisubstituted pyrazolines were readily oxidized to the corresponding pyrazoles by 1,3-dibromo-5,5-dimethylhydantoin (DBH) which reduces the time and enhances the yield [161].

10.3.2.4 Synthesis of Imidazoles and Related Compounds

In 1996, the condensation of orthoesters with *ortho*-phenylenediamine leading to benzimidazoles was achieved in toluene under reflux or without solvent by use of focused microwaves (Scheme 10.82) [162].

A more striking example was reported more recently by Soufiaoui and

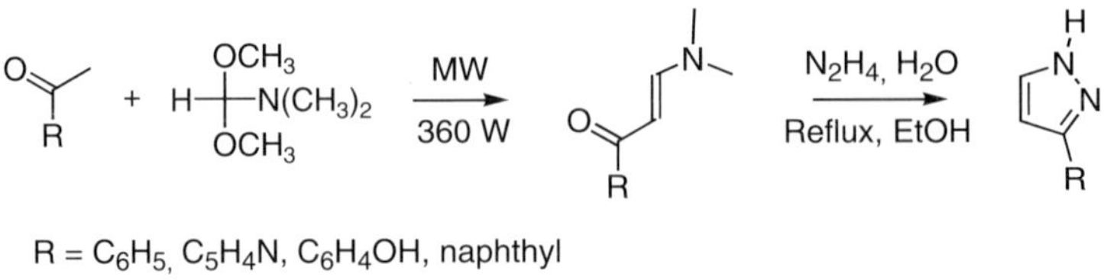

[Scheme 10.81 reaction diagram]

R = C_6H_5, C_5H_4N, C_6H_4OH, naphthyl

Scheme 10.81

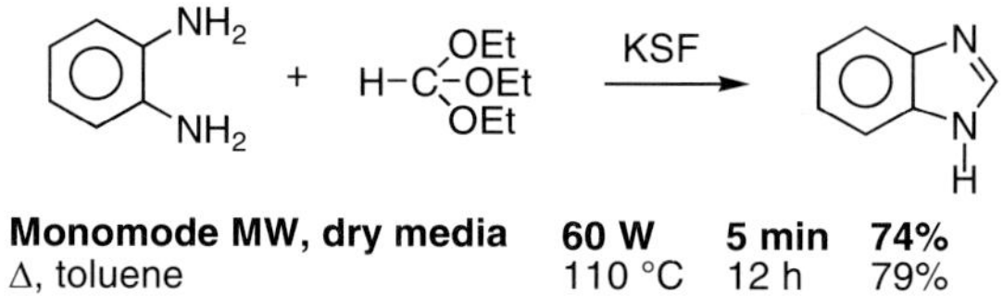

10.3 Solvent-free Synthesis | 501

Scheme 10.82

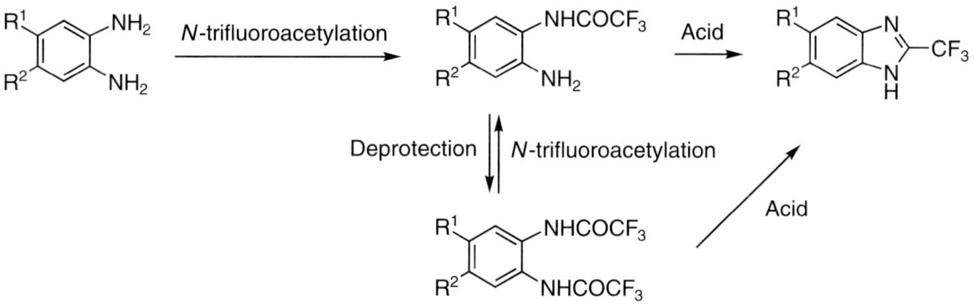

Monomode MW, dry media	60 W, 5 min, 74%
Δ, toluene	110 °C, 12 h, 79%

Scheme 10.83

coworkers who studied the cyclocondensation of N-(carbotrifluoromethyl)-ortho-arylenediamines on K_{10} clay in dry media under the action of irradiation for 2 min in a domestic oven (Scheme 10.83) [163]. It is worth noting that with classical heating under the same conditions only traces of heterocycles were observed. The difference is discussed in terms of enhancement of the polarity of the transition state compared with the initial state (see Chapter 4 of this book).

Imidazole has been condensed with ethyl acrylate by using two basic clays (Li^+ and Cs^+ montmorillonites) as catalysts in a microwave oven [164]. The role of alkali promoters (Li^+ and Cs^+) was studied and it was found that the greater the basicity and the irradiation time and power, the higher were the conversions. The yield of N-substituted imidazole is optimum for 0.1 g Cs^+ montmorillonite at 850 W after irradiation for 5 min (Scheme 10.84). The reaction proceeds in high yields and selectivity under the action of irradiation and is more efficient than with conventional heating.

An expeditious solvent-free synthesis of imidazoline derivatives, using basic or neutral alumina and microwave irradiation, has been reported [165]. The reaction

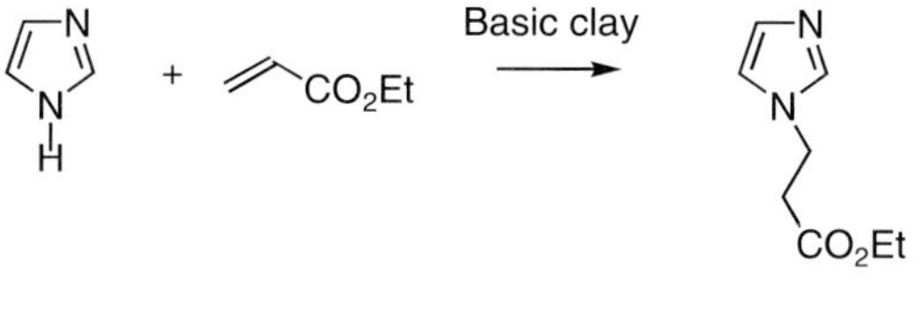

Scheme 10.84

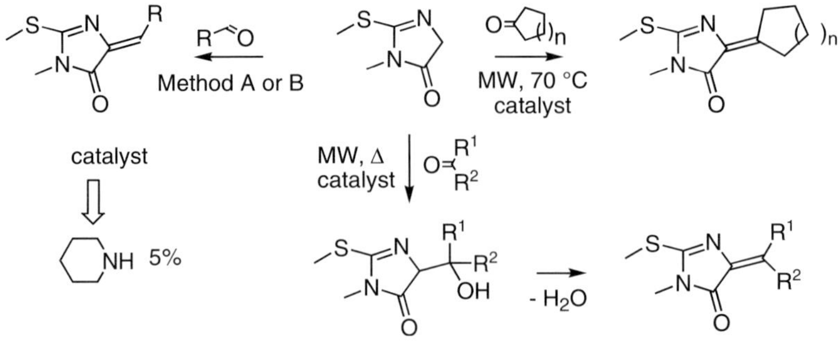

Scheme 10.85

time was reduced from hours to minutes with improved yield compared with conventional heating.

An efficient solvent-free synthesis of imidazolones via the Knoevenagel reaction using microwave irradiation has been documented (Scheme 10.85) [166]. In method A, the reaction is performed in CH_2Cl_2 under reflux. Method B is without solvent under the action of microwave irradiation. It has been shown that yields are much higher and reaction times much lower for method B. This method was applied to the synthesis of the precursor of leucettamine B (mediator of inflammation) [167].

β-Hydrazinoacrylates derived from benzimidazole, benzoxazole, and benzothiazole were readily prepared in good yields by transamination of the corresponding 3-dimethylamino acrylates with a variety of hydrazines (Scheme 10.86) [168]. These hydrazino acrylates can then be cyclized to the corresponding 1,2-dihydropyrazol-3-ones either under the action of irradiation or by heating conventionally. These last compounds have also been prepared directly from the starting acrylates.

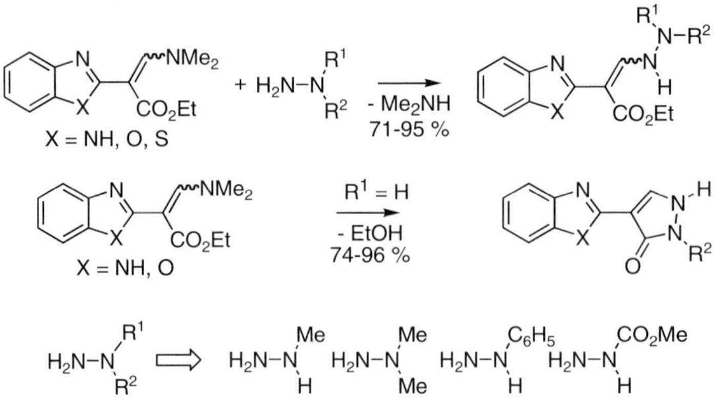

Scheme 10.86

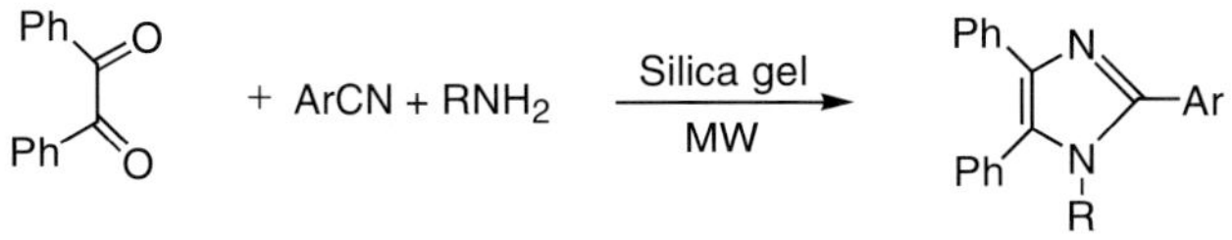

Scheme 10.87

Tetrasubstituted imidazoles have been prepared in high yield by one-pot three-component condensation of benzil, benzonitrile derivatives, and primary amines adsorbed on silica gel without solvent (Scheme 10.87) [169]. Under irradiation conditions yields range from 58 to 92% (8 min).

Long chain 2-alkyl-1-(2-hydroxyethyl)-2-imidazolines (at 150 °C) and their amide precursors at 120 °C were readily obtained by condensation of aminomethylethanolamine with fatty acids using CaO as inorganic support [170]. A nonthermal effect is implicated. A one-pot selective synthesis of cis- and trans-2,4,5-triarylimidazolines from aromatic aldehydes was achieved under irradiation conditions via methadiamines and selected bases [171].

10.3.2.5 Synthesis of Oxazoles

Rather good yields of polysubstituted oxazoles are obtained starting from a variety of carbonyl compounds using sequential treatment with HDNIB [hydroxy-(2,4-dinitrobenzenesulfonyloxy)iodo]benzene and amides under the action of irradiation according to Scheme 10.88 [172].

10.3.2.6 Synthesis of Dioxolanes

Acetals of L-galactono-1,4-lactone (an abundant by-product of the sugar beet industry) have thermotropic liquid crystalline properties. Acetalization in DMF containing H_2SO_4 in the presence of anhydrous $CuSO_4$ gave (after 12 to 24 h at 40 °C) yields of 20–35% (Scheme 10.89) [173].

The use of montmorillonites KSF or K_{10} as acidic supports and irradiation at 60 W (monomode oven, final temperature of 155 °C) led to yields threefold higher in 10 min. A mixture of an aldehyde or a ketone, ethylene glycol (EG), and p-toluenesulfonic acid (pTSA) led, after irradiation, to the corresponding dioxolane (Scheme 10.90) [174].

Unfortunately, the comparison is not reliable because the temperature was not measured in the domestic oven [175].

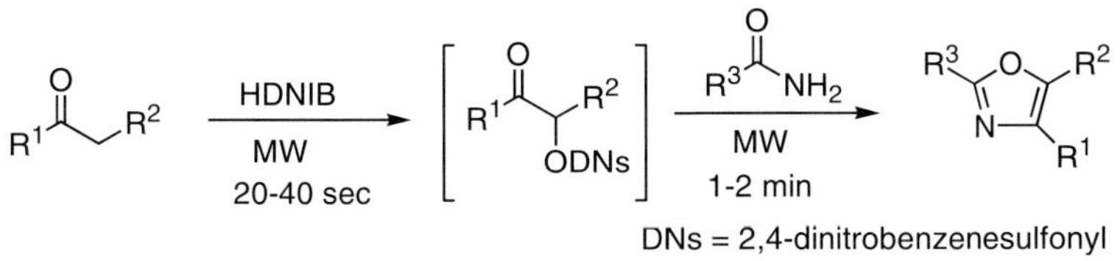

Scheme 10.88

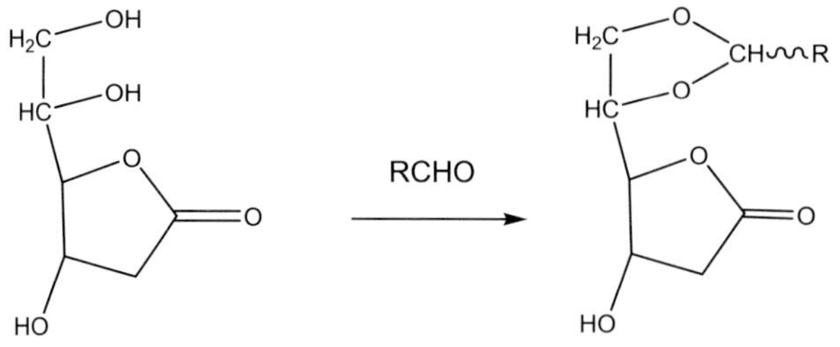

Scheme 10.89

Example : R = CH$_3$-(CH$_2$)$_{11}$

DMF, H$_2$SO$_4$, CuSO$_4$ 24 h.   40 °C   25%
KSF, "dry media",   10 min.   60 W   66%
K$_{10}$, "dry media",   10 min.   60 W   89%

Scheme 10.89

Dioxolane formation by acid-catalyzed exchange between 2,2-dimethyl-1,3-dioxolane (DMD) and a ketone in an inert (nonpolar) solvent, or simply in excess DMD, requires 4 to 7 h under classical conditions [176]. This reaction is readily achieved under microwave irradiation in high yields within 4 to 30 min (Scheme 10.91) [174]. It has been shown that yields are higher under the action of microwaves. This dioxolane exchange was subsequently scaled up to 250 g in the Synthewave 1000 with the same yields [177] and this type of carbonyl protection was extended to dithiolanes and oxathiolanes [178].

Ketene acetals and dithioacetals have been prepared in a solvent-free procedure by base-catalyzed β-elimination under PTC conditions. The yields obtained by use of irradiation are much higher than those obtained by use of ultrasound or conventional heating under the same conditions [179] (Scheme 10.92). These results are obviously indicative of a specific effect of irradiation.

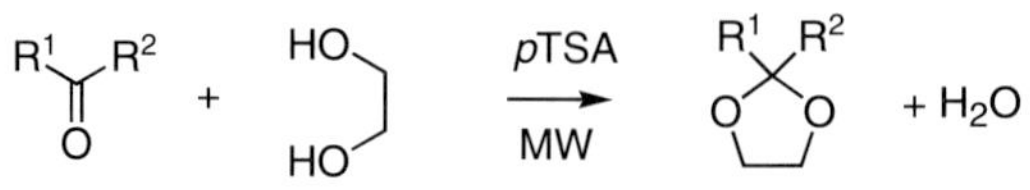

Example: R^1 = Ph; R^2 = H			
Focused microwave oven, 1.5 eq. EG	15 min	80 °C	80%
Domestic oven, 10 to 15 eq. EG	2 min	650 W	81%
Example: R^1 = Ph, R^2 = CH$_3$			
Focused microwave oven, EG 3 eq.	30 min	120 °C	90%
Domestic oven, 10 to 15 eq. EG	2 min	650 W	71%

Scheme 10.90

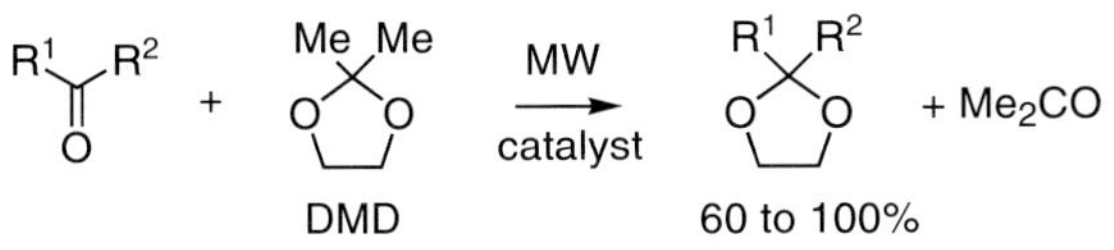

10.3 Solvent-free Synthesis

R^1	R^2	DMD	Catalyst	Power (W)	Time (min)	T max (°C)	Yield (%)
Ph	H	1.5	KSF	300	4	80	90
Ph	Me	3	pTSA	150	30	110	95
Ph	Ph	3	pTSA	150	30	120	60
-(CH$_2$)$_5$-	-(CH$_2$)$_5$-	3	pTSA	150	30	120	100
Me	CH$_2$COMe	3	pTSA	150	30	110	100
Me	CH$_2$COMe	3	pTSA	150	30	120	95

Scheme 10.91

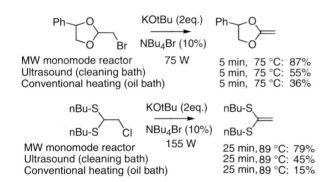

Scheme 10.92

More recently, the synthesis of 5-phenyl-1,3-dioxolan-4-ones starting from mandelic acid (racemic and optically pure) and aldehydes and ketones catalyzed by copper sulfate was achieved in good yields [180].

10.3.2.7 Synthesis of Thiazoles and Related Compounds

Benzothiazoles can be synthesized by condensation of carboxylic acids with 2-aminothiophenol (Scheme 10.93) [181].

Acetylations of thiazoles have been reported by Gupta et al. [182].

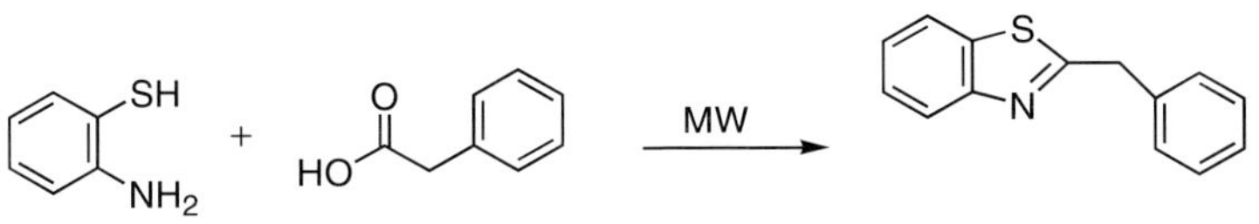

Scheme 10.93

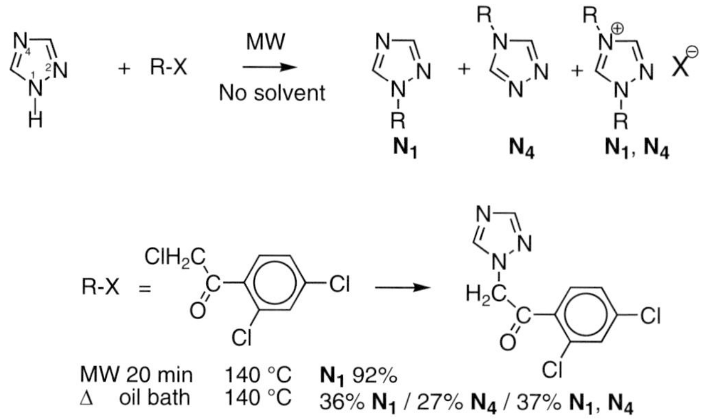

Scheme 10.94

10.3.3
Five-membered Systems with More than Two Heteroatoms – Synthesis of Triazoles and Related Compounds

Selective alkylation of 1,2,4-triazole in position 1 is of primary interest for the synthesis of biologically active molecules such as fungicides (fluconazole, flutriafole, …). Direct alkylation with classical heating gave a mixture of 1- and 4-alkylated triazoles together with quaternary salts resulting from alkylations at both the 1 and 4 positions. Interestingly, it has been shown that benzylation and phenacylation occurred selectively at position 1 without any base under the action of irradiation and under solvent-free conditions (Scheme 10.94) [183]. The reaction with (2,4-dichloro)phenacyl chloride was studied in particular depth (Chapter 4). This reaction has been scaled-up to more than 100 g by use of a Synthewave 1000 oven.

Benzylation of 1,2,4-triazole and benzotriazole has been achieved by de la Hoz, Loupy and coworkers [184].

10.3.4
Six-membered Systems with One Heteroatom

10.3.4.1 Synthesis of Pyridines and Related Compounds
The condensation of β-formylenamides with cyanomethylenes under the action of irradiation is catalyzed by basic alumina to afford fused pyridines in excellent yields and within very short times (Scheme 10.95) [185]. This procedure avoids the use of harsh conditions.

Recently, the group of Boruah achieved the synthesis of annelated pyridines from β-formyl enamides via a Henry reaction in a one-pot process [186]. Under the action of microwave irradiation, in the presence of a base, for example pyrrolidine, morpholine or piperidine, yields from 80 to 90% were obtained after 8 to 10 min (Scheme 10.96).

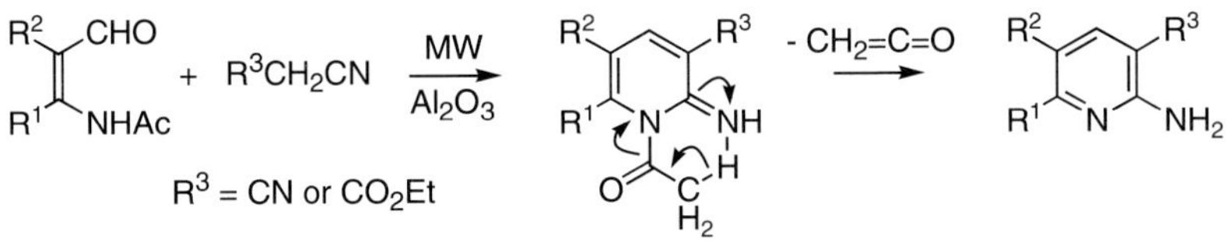

Scheme 10.95

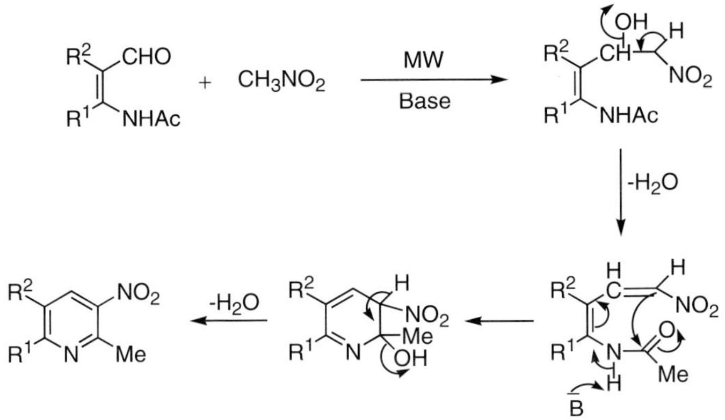

Scheme 10.96

A one-pot synthesis of N-substituted 4-aryl-1,4-dihydropyridines using a microwave-assisted procedure was described in 2001 (Scheme 10.97) [187]. Among a variety of solid supports (K_{10}, acidic alumina, zeolite HY, silica gel), silica gel was proved to be the most efficient. After irradiation for 4 min, the reactions led to yields ranging from 62 to 94%.

Loupy et al. have reported a specific non-thermal microwave effect in the synthesis of 4-aryl substituted 5-alkoxycarbonyl-6-methyl-3,4-dihydropyridones (81–91% under the action of microwaves compared with 17–28% with conventional heating) [188].

It is of primary interest to avoid corrosive mineral acids in synthetic processes. This can easily be achieved by use of acidic solid supports coupled with microwave irradiation. The technique has been applied to the preparation of quinolines (Scheme 10.98) [189]. This procedure is a safe, green alternative to the use of H_2SO_4 at more than 150 °C.

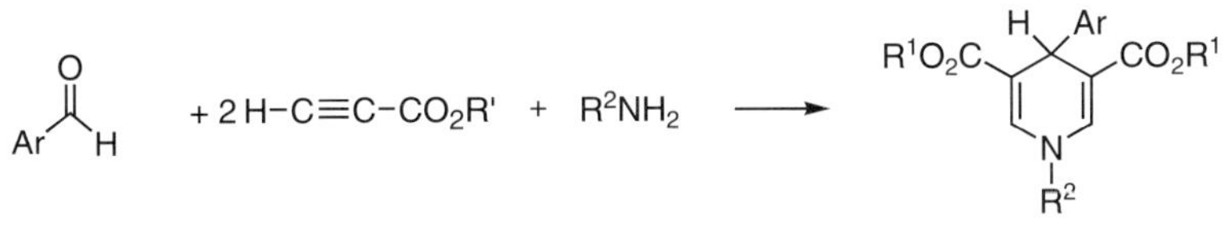

Scheme 10.97

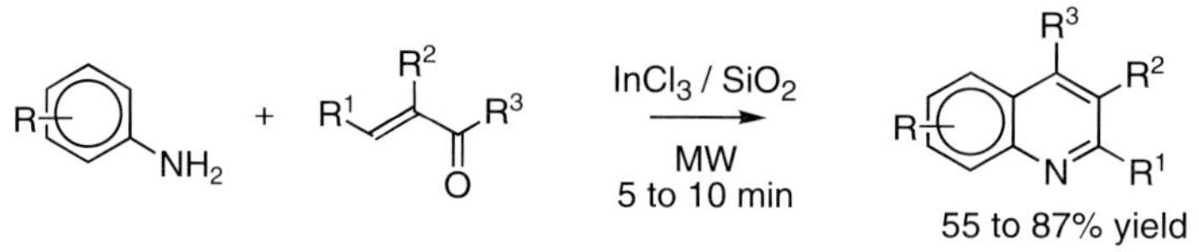

Scheme 10.98

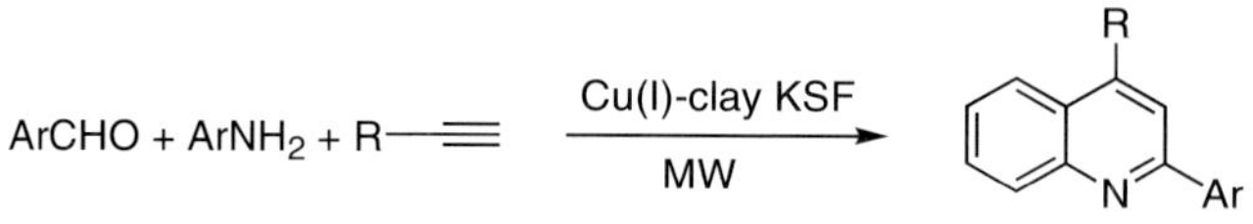

Aryl amine + α,β-unsaturated carbonyl R^1, R^2, R^3 → quinoline via InCl$_3$/SiO$_2$, MW, 5 to 10 min, 55 to 87% yield.

Scheme 10.99

ArCHO + ArNH$_2$ + R−≡ —Cu(I)-clay KSF, MW→ 2,4-disubstituted quinoline.

A minilibrary of twelve quinoline derivatives has been prepared by Friedlander coupling condensation between a variety of acetophenones and 2-aminoacetophenone or benzophenone, in the presence of diphenylphosphate as catalyst, after 4 min under irradiation conditions [190].

An expeditious preparation of 2,4-disubstituted quinoline (Scheme 10.99) has been reported by Yadav et al. [191]. Although it has been mentioned that microwaves enhance rates and the yields compared with conventional heating, the comparison is not safe, because the temperatures for the microwave experiments are not reported.

Almost the same procedure has been used for the synthesis of 2-pentafluorophenylquinoline derivatives [192].

For the Bernthsen reaction, an important route to acridines, it has recently been shown that under microwave irradiation it is possible to reduce the amount of catalyst compared with the classical route yet with noticeable yield enhancement (Scheme 10.100) [193].

Depending on irradiation time, it is possible to prepare either dihydroisoquinolines or isoquinolines in satisfactory yields by irradiation of N-sulfonyl tetrahydroisoquinoline derivatives in the presence of KF on alumina [194].

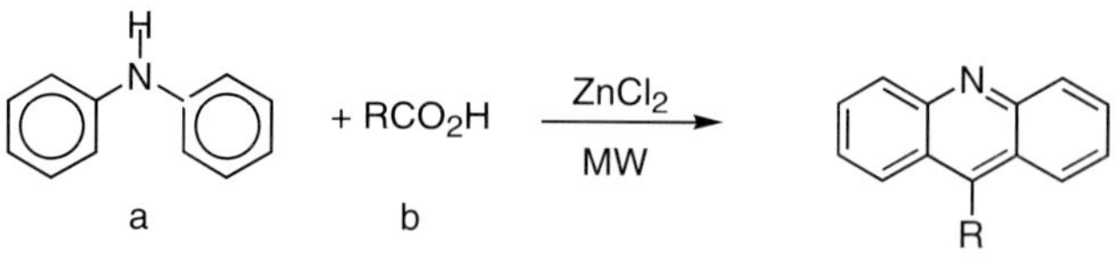

Scheme 10.100

Diphenylamine (a) + RCO$_2$H (b) —ZnCl$_2$, MW→ acridine.

MW a: ZnCl$_2$: b = 1 : 1 : 1 Yields: 57 to 98%
conventional a: ZnCl$_2$: b = 1 : 5 : 1 Yields: 18 to 53%

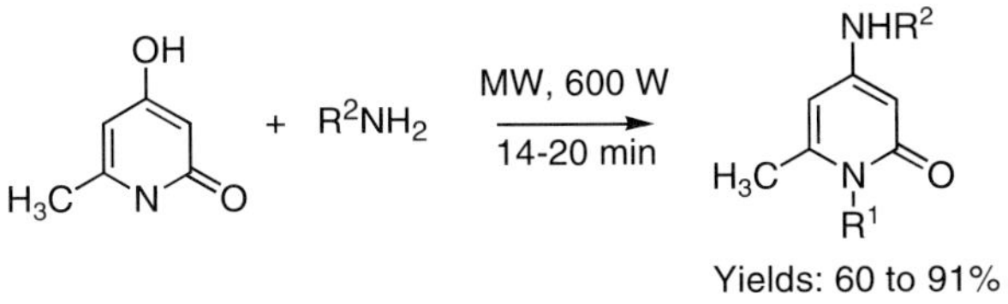

Scheme 10.101

A survey of microwave activation in the chemistry of Hantzsch 1,4-dihydropyridines (1,4-DHP) was reported in 2003 [195]. The experimental method proposed more than a century ago remains the most widely used for synthesis of these heterocycles. Since 1992 the process has been adapted to microwave irradiation under a variety of conditions to reduce the reaction time and enhance the yield. Among these experiments, Zhang reported a solvent-free process starting from 3-aminocrotonate (20 mmol), methyl acetoacetate (20 mmol), and aromatic aldehydes (20 mmol) in a domestic oven [196]. Yields from 59 to 77% were reported for 10-min reaction. A variety of conditions (solution, dry media, solvent-free) has been used for microwave-assisted synthesis of Hantzsch 1,4-DHP. Only procedures involving solvent-free conditions under the action of microwave irradiation led to the aromatized pyridine derivatives.

2-Pyridones were studied for N- and C-alkylation reactions by de la Hoz and coworkers. As already mentioned for 1,2,4-triazoles, the selectivity of the alkylation is highly dependent on the activation technique (microwave or conventional heating) [197] (see Chapter 5).

The microwave-assisted nucleophilic substitution of 4-hydroxy-6-methyl-2(1H)-pyridones has also been studied (Scheme 10.101) [198].

The Hantzsch dihydropyridine synthesis has been performed in a single-mode microwave cavity [199]. In comparison with both conventional methods and microwave-assisted reactions performed in a domestic oven, reaction times were shorter and yields were higher (Scheme 10.102). The improved yields under microwave conditions enabled the synthesis of a small library.

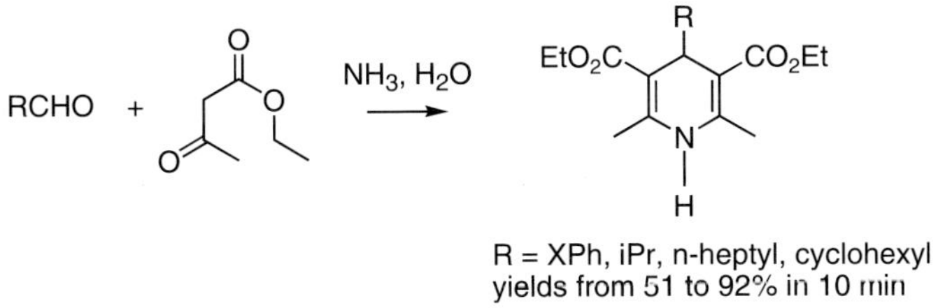

Scheme 10.102

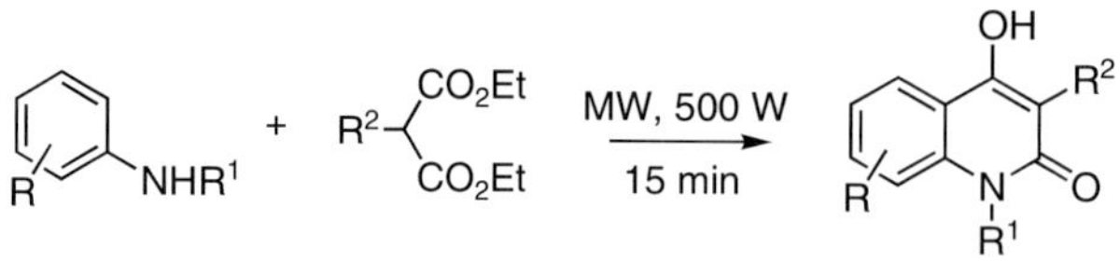

Scheme 10.103

3-Aryl-4-hydroxyquinolin-2(1H)-ones, which are of pharmaceutical interest, were readily obtained in a one-pot procedure by formal amidation of malonic ester derivatives with an aniline then cyclization of the intermediate malondianilides (Scheme 10.103) [200]. The synthesis of Merck's glycine NMDA receptor or antagonist L-701,324 is illustrative. It can be prepared in one step by use of this procedure whereas the previously reported synthetic procedure comprises several reaction steps.

Among various polyaryls, Basu et al. [201] reported "microwave-assisted Suzuki coupling" on a KF–alumina surface starting from 1,6-dibromopyridine to give 1,6-diphenylpyridine.

A transition metal-free direct amination of halo(pyridine or pyrimidine) has been developed in good yields under the action of computer-controlled microwave irradiation [202]. This solvent-free reaction is useful for coupling of halo(pyridine and pyrimidine) with pyrrolidine and piperidine derivatives by nucleophilic aromatic substitution (S_NAr).

Loupy and coworkers have shown that Zn powder is an inexpensive, green, and efficient catalyst for Friedel–Crafts acylation without solvent under the action of microwave irradiation, and that pyridine could also be acylated in these conditions [203].

A variety of carboxylic acids, including nicotinic and quinaldic acid undergo amidation with urea in the presence of imidazole under microwave irradiation conditions without solvent [204].

10.3.4.2 Synthesis of 1,8-Cineole Derivatives

1,8-Cineole derivatives, which are of interest as potential cosmetic products, have been obtained by a "green chemistry procedure", without solvent and with microwave activation according to Scheme 10.104 [205]. The yields are significantly better than for a previously reported procedure (90–98% instead of 50–70%) and the

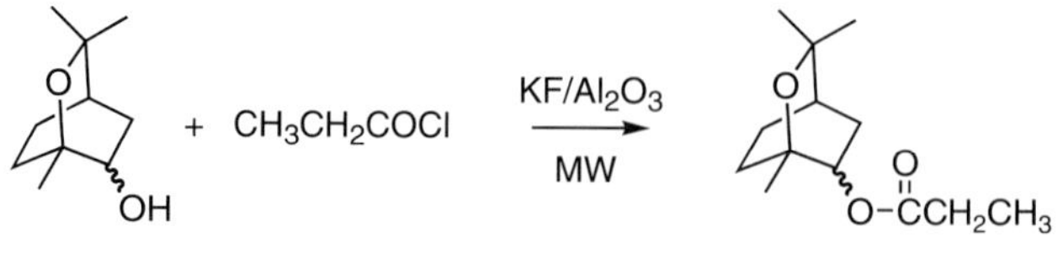

Scheme 10.104

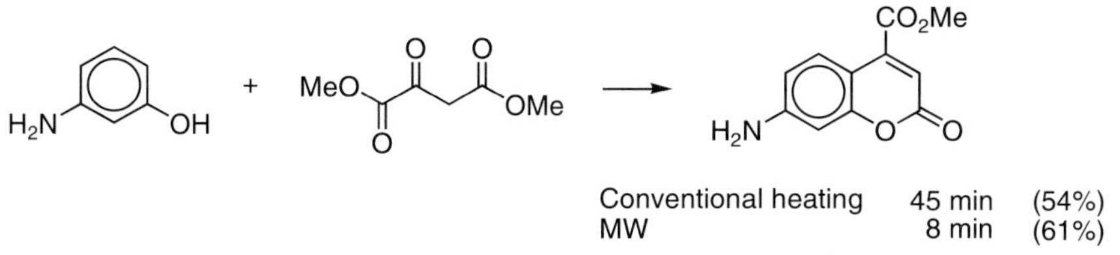

Scheme 10.105

	Conventional heating	45 min	(54%)
	MW	8 min	(61%)

reaction times are considerably reduced. These improvements are connected "to the intervention of highly polar reactants and then consequently prove to develop strong interactions with microwaves" (see Chapter 4).

10.3.4.3 Synthesis of Coumarins and Related Compounds

Coumarins are important compounds with applications as pharmaceuticals and agrochemicals. Some 7-aminocoumarin-4-carboxylates have been synthesized by use of the Pechmann reaction, with microwave irradiation of the reactants on solid supports (graphite–K_{10}) (Scheme 10.105) [206]. The synthesis of unsubstituted coumarins (C-4 position) has also been reported [207].

As mentioned by Varma et al., the reactions of 5- or 8-oxobenzopyran-2-ones with a variety of aromatic and heteroaromatic hydrazines are accelerated under the action of microwave irradiation without catalyst support or solvent [208].

Sodium bromide-catalyzed three-component cyclocondensation of aryl aldehydes, alkylnitriles, and dimedone to give excellent yields of tetrahydrobenzo[*b*]pyrans [209].

10.3.5
Six-membered Systems with More than One Heteroatom

10.3.5.1 Synthesis of Pyrimidines and Related Compounds

The Biginelli reaction was revisited by Kappe and Stadler in 2000. This synthesis was reinvestigated under the action of microwave irradiation under a variety of different conditions [210]. Under atmospheric pressure in ethanol solution, there is no difference from conventional heating. Under pressure, the yield is reduced and by-products are formed. In an open system, rate and yield enhancements are significant and this is rationalized by the rapid evaporation of the solvent which means that this is in fact a solvent-free reaction. This was confirmed by performing the reaction without solvent under the action of microwaves or with thermal heating (Scheme 10.106).

This was later extended to the synthesis of novel pyrimido[1,3-*a*]pyrimidines under solvent-free conditions. Ethyl 2-amino-4-aryl-1,4-dihydro-6-phenylpyrimidine-5-carboxylates reacted regioselectively with 3-formylchromone or diethyl (ethoxymethylene)malonate, without solvent, to afford pyrimido[1,3-*a*]pyrimidines (Scheme 10.107) [211].

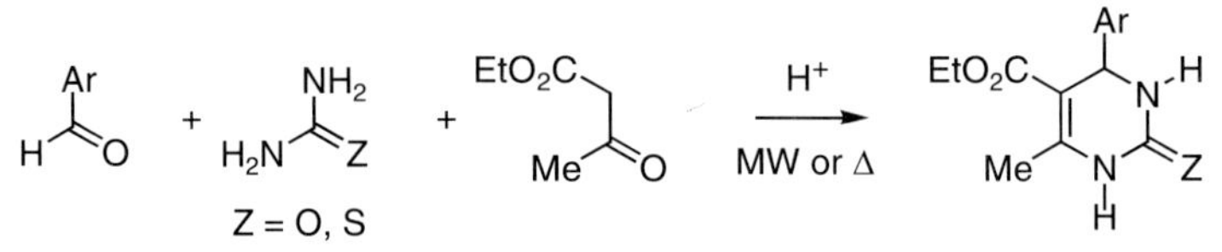

Scheme 10.106

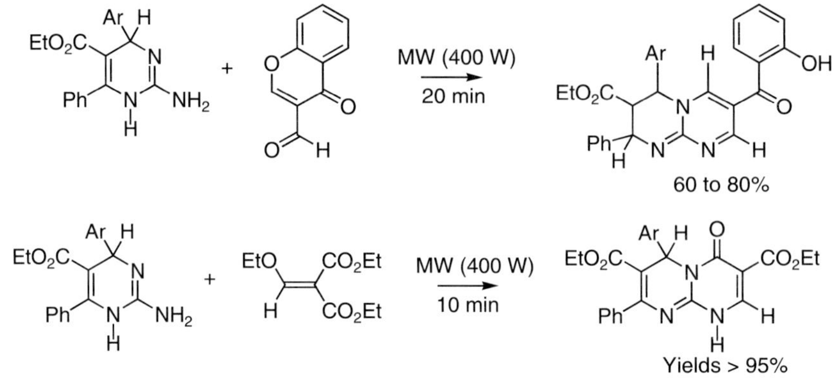

Scheme 10.107

This Biginelli reaction was further exemplified by Prajapati et al. for the synthesis of 3,4-dihydropyrimidine-2(3H)-ones [212].

Quite recently, it has been shown that reduction of nitro and azido arenes to N-arylformamides using $ZnHCO_2NH_4$ under microwave conditions proceeds well. In the absence of irradiation the reaction leads to amines. This interesting procedure was extended to the synthesis of 4-(3H)-quinozolinones and pyrrolo[2,1-c][1,4]benzodiazepines [213].

10.3.5.2 Synthesis of Pyrazines and Related Compounds

Quinoxaline-2,3-diones have been prepared by use of single-mode irradiation [214]. Previous attempts in solution led to explosions, but the authors successfully used solvent-free conditions with acidic supports or catalysts (the best being p-toluenesulfonic acid) and irradiation times of 3 min (Scheme 10.108).

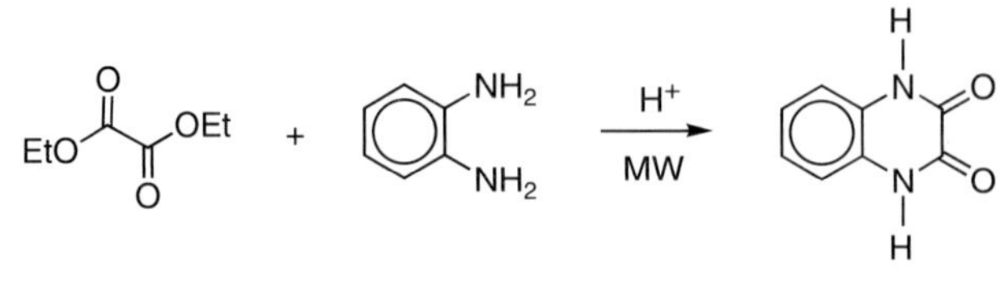

Scheme 10.108

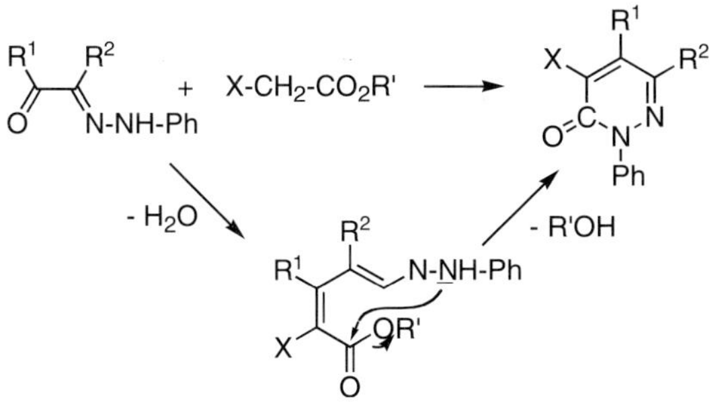

44: X = CN, R^1 = Me, R^2 = H

Scheme 10.109

p-Toluenesulfonic acid also proved to be the best catalyst for formation of pyrazolo[3,4-*b*]quinolines from 2-chloro-3-formylquinolines and hydrazine hydrate–phenyl hydrazine under microwave irradiation [215].

Several 3-(2*H*)-pyridazinones have been prepared from monophenyl hydrazones of 1,2-dicarbonyl compounds and a variety of active methylene compounds within 1–20 min without solvent under focused irradiation conditions in the presence of carefully adjusted amounts of piperidine or solid potassium *tert*-butoxide (isolated yields 50–89%), in accordance with Scheme 10.109 [216]. In the synthesis of the pyridazinone **44**, microwave irradiation has no specific effect, because the result (72%) was identical with that obtained by use of classical heating under the same conditions. With the dry media procedure it was possible to isolate the intermediate alkene, which was not obtained in the previously reported procedure. When the active methylene compound is a keto ester, the reaction follows a different pathway [216b].

10.3.5.3 *N*-Formylmorpholine Chemistry

Oxidations of *β*,*β*-disubstituted enamines have been performed over $KMnO_4$–Al_2O_3 (Scheme 10.110) and the results compared with those obtained by use of a domestic oven, a focused oven, and classical heating [217].

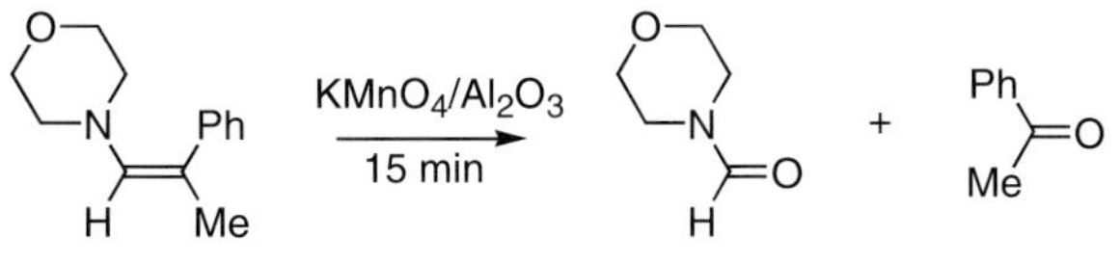

Domestic oven	255 W		73%
Monomode reactor	300 W	140 °C	83%
Classical heating		140 °C	< 2%

Scheme 10.110

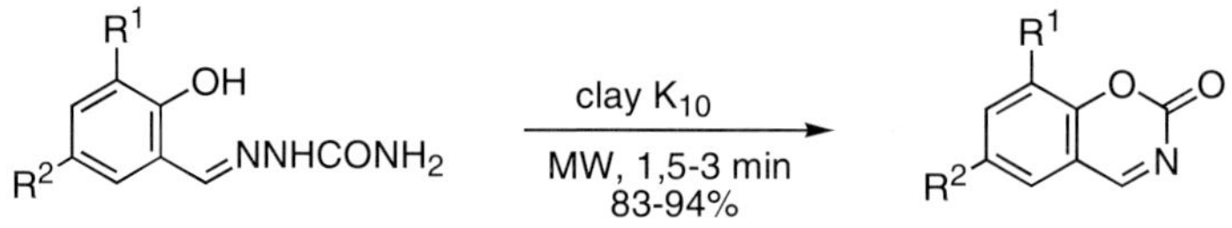

Scheme 10.111

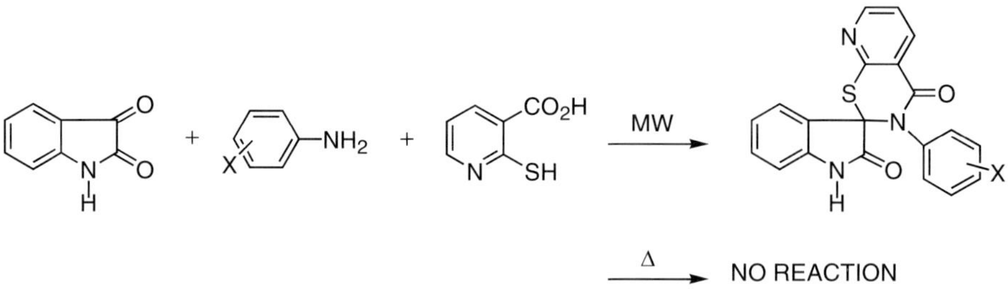

Scheme 10.112

10.3.5.4 Synthesis of Benzoxazines and Related Compounds

Good yields of 2H-benz[e]-1,3-oxazin-2-ones were obtained by cyclization of salicylaldehyde semi-carbazones catalyzed by clay K_{10} under irradiation conditions (Scheme 10.111) [218].

More recently, the same group reported an analogous cyclization for the synthesis of 2H-benz[e]-1,3-oxazine-2-thiones [219] and functionalized 1,3-thiazines [220].

Previously unknown pyridothiazines have been prepared by treatment of in situ generated 3-indolylimine derivatives with 2-mercaptonicotinic acid under irradiation conditions, according to Scheme 10.112 [221].

This is clearly indicative of the specific, not thermal, microwave effect (for a discussion, see Chapter 4 of this book).

10.3.5.5 Synthesis of Triazines

Melamines have been synthesized by an efficient procedure according to Scheme 10.113 [222]. The scope of the reaction in relation to the amine used has been studied and it has been shown that the procedure is valid when moderately bulk amines are used.

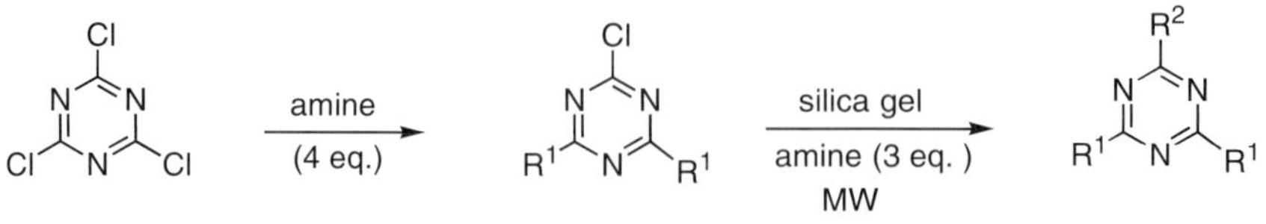

Scheme 10.113

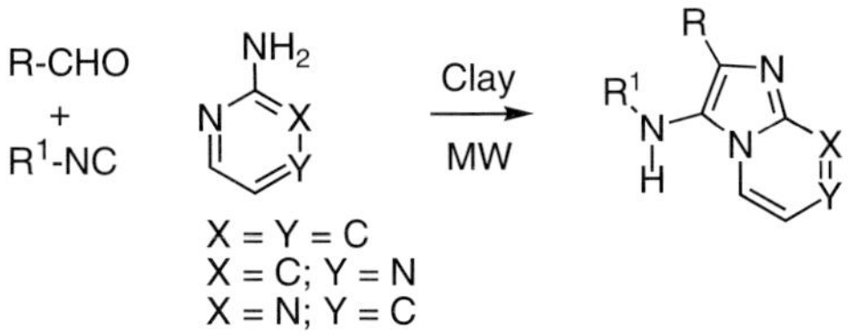

Scheme 10.114

10.3.6
Synthesis of More Complex or Polyheterocyclic Systems

A rapid one-pot synthesis of imidazo[1,2-a]pyridines, pyrazines and pyrimidines was described in 1999 by Varma and Kumar, who used recyclable montmorillonite clay K_{10} under solvent-free conditions with microwave irradiation (Scheme 10.114) [223]. The whole process, formation of the iminium ion by condensation of the aldehyde with an amine (1 min at 900 W in a domestic oven) then nucleophilic attack of the isocyanide (1 min at 450 W and 1 min cooling) took 3 min and gave the pure products in yields ranging from 56 to 88%.

Also of interest is a new easy means of preparing 3-substituted (S)-2,3,6,7,12,12a-hexahydropyrazino[1′,2′:1,6]pyrido[3,4-b]indole-1,4-diones (Scheme 10.115) [224]. This process is much more efficient than the classical procedure with TFA in CH_2Cl_2 for 2 h at 30 °C, which leads to 60% yield instead of 95% after 5 min under irradiation conditions.

Focused irradiation of N-acylimidates mixed with imidazolidine ketene aminals provided a new means of access to 2,3-dihydroimidazo[1,2-c]pyrimidines (Scheme 10.116) [225].

Dandia et al. described an efficient and clean procedure for synthesis of a series of 8-substituted 2-carboxy-2,3-dihydro-1,5-benzothiazepines in solvent-free conditions under the action of irradiation [226]. It is noteworthy that, under thermal conditions, the benzothiazepines were obtained along with other products.

It has been reported by Nemes and coworkers that α,β-unsaturated aldehydes and ketones react with 1-cyanomethylene-6,7-dimethoxy-1,2,3,4-tetrahydroisoqui-

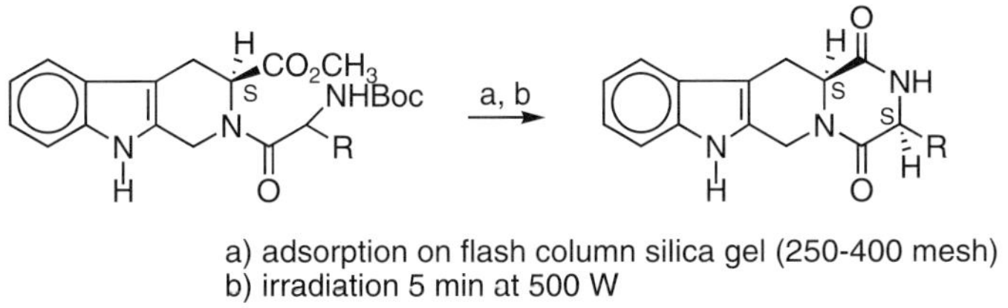

a) adsorption on flash column silica gel (250-400 mesh)
b) irradiation 5 min at 500 W

Scheme 10.115

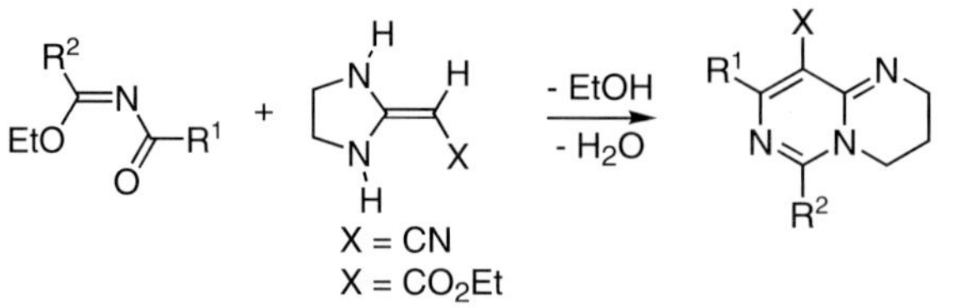

Scheme 10.116

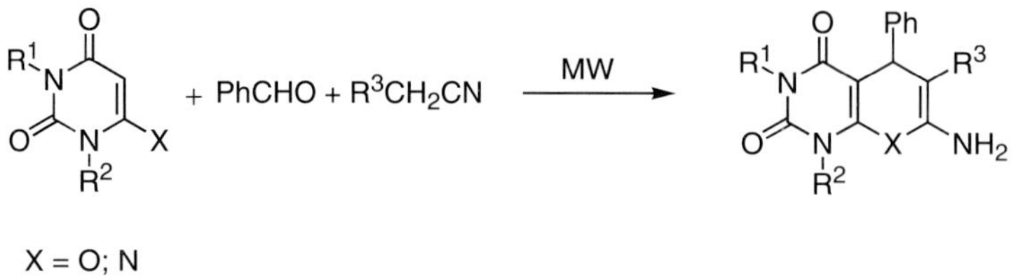

Scheme 10.117

X = O; N

noline to give, under the action of irradiation, 6,7-dihydro-4*H*-benzo[a]quinolizines and 2*H*-quinolizines respectively [227]. 6-Arylbenzimidazo[1,2-c]quinazolines were obtained by irradiation of 2-benzimidazoylbenzamides supported on SiO$_2$–MnO$_2$ (95:5) [228].

Pyrano- and pyridopyrimidines have been synthesized in a three-component, one-pot procedure according to Scheme 10.117 [229].

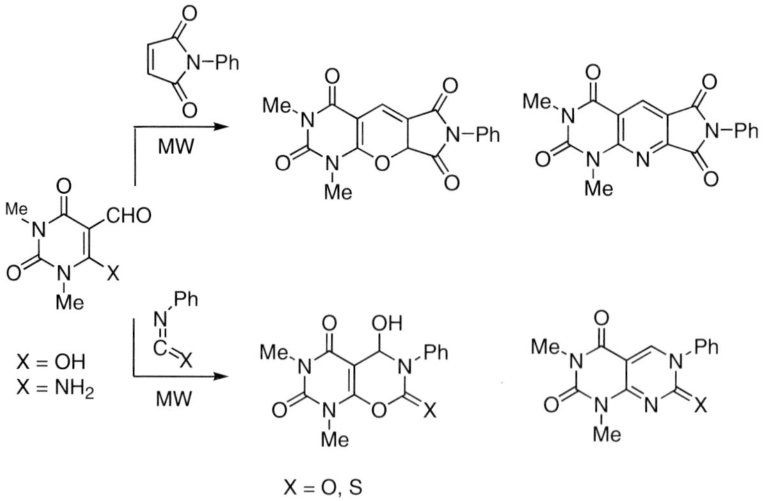

Scheme 10.118

The same type of heterocycle was prepared by Bhuyan et al. by a similar procedure using triethyl orthoformate and alkylnitriles in the presence of acetic anhydride under the action of irradiation [230]. More recently, the same group reported the preparation of uracils by microwave irradiation in the solid state, as summarized by Scheme 10.118 [231].

Polyheterocyclic systems have been efficiently synthesized by cyclocondensation between anthranilic acid and lactim ethers derived from 2,5-piperazinediones. This reaction leads to pyrazino[2,1-b]quinazoline-3,6-diones. When applied to bis(lactim) ethers the procedure again works better under the action of microwaves, with reaction times reduced from hours to minutes and enhancement of yields [232].

10.4
Conclusion

Again, in this new edition, microwave irradiation seems a very powerful tool owing to versatility which enables the use of a wide variety of experimental conditions and very often leads to enhanced yields in very short reaction times. For this chapter in the previous edition, which was published at the end of 2002, we collected 113 references. Despite the fact that some of these were withdrawn, three years later we found 232 references (and most probably missed some); that is twice as many in three years! This means the technique has become very popular and demonstrates its usefulness, as was foreseen in previous reports.

References and Notes

1 CHANDA, B.M.; VYAS, R.; BEDEKAR, A.V. *J. Org. Chem.* **2001**, *66*, 30.

2 SAOUDI, A.; HAMELIN, J.; BENHAOUA, H. *J. Chem. Res. (S)* **1996**, 492.

3 PILLAI, U.R.; SAHLE-DEMESSIE, E.; VARMA, R.S. *Tetrahedron Lett.* **2002**, *43*, 2909.

4 BOSE, A.K.; BANIK, B.K.; BARAKAT, K.J.; MANHAS, M.S. *Synlett* **1993**, 575.

5 (a) BOSE, A.K.; BARIK, B.K.; MATHUR, C.; WAGLE, D.R.; MANHAS, M. *Tetrahedron* **2000**, *56*, 5603; (b) BOSE, A.K.; BARIK, B.K.; MATHUR, C.; WAGLE, D.R.; MANHAS, M. *Tetrahedron* **2000**, *56*, 5587.

6 PANUNZIO, M.; CAMPANA, E.; VICENNATI, P.; BREVIGLIERI, G. in "Fifth International Electronic Conference on Synthetic Organic Chemistry (ECSOC-5), 1–30 September 2001", http://www.mdpi.org/ecsoc-5/e0010/e0010.htm.

7 SHIEH, W.C.; DELL, S.; REPIC, O. *Tetrahedron Lett.* **2002**, *43*, 5607.

8 LAMBERTO, M.; CORBETT, D.F.; KILBURN, J.D. *Tetrahedron Lett.* **2003**, *44*, 1347.

9 YANG, C.; MURRAY, W.V.; WILSON, L.J. *Tetrahedron Lett.* **2003**, *44*, 1783.

10 GRIGG, R.; MARTIN, W.; MORRIS, J.; SRIDHARAN, V. *Tetrahedron Lett.* **2003**, *44*, 4899.

11 GARBACIA, S.; DESAI, B.; LAVASTRE, O.; KAPPE, C.O. *J. Org. Chem.* **2003**, *68*, 9136.

12 BALAN, D.; ADOLFSSON, H. *Tetrahedron Lett.* **2004**, *45*, 3089.

13 SCHOBERT, R.; GORDON, G.J.; MULLEN, G.; STEHLE, R. *Tetrahedron Lett.* **2004**, *45*, 1121.

14 NAMBOODIRI, V.V.; VARMA, R.S. *Green Chem.* **2001**, *3*, 146, and references cited therein.

15 Wang, Y.; Sauer, D.R. *Org. Lett.* **2004**, *6*, 2793.
16 Akamatsu, H.; Fukase, K. *Synlett* **2004**, 1049.
17 Teichert, A.; Jantos, K.; Harms, K.; Studer, A. *Org. Lett.* **2004**, *6*, 3477.
18 Appukkuttan, P.; Van der Eycken, E.; Dehaen, W. *Synlett.* **2003**, 1204.
19 Rao, H.S.P.; Jothilingam, S.; Scheeren, H.W. *Tetrahedron* **2004**, *60*, 1625.
20 Fînaru, A.; Berthault, A.; Besson, T.; Guillaumet, G.; Berteina-Raboin, S. *Tetrahedron Lett.* **2002**, *43*, 787.
21 Minetto, G.; Raveglia, L.F.; Taddei, M. *Org. Lett.* **2004**, *6*, 389.
22 Ericsson, C.; Engman, L. *J. Org. Chem.* **2004**, *69*, 5143.
23 Zhang, Y.; Li, C.J. *Tetrahedron Lett.* **2004**, *45*, 7581.
24 Rao, H.S.P.; Jothilingam, S. *J. Org. Chem.* **2003**, *68*, 5392.
25 Wu, X.; Mahalingam, A.K.; Wan, Y.; Alterman, M. *Tetrahedron Lett.* **2004**, *45*, 4635.
26 Hutchinson, E.J.; Kerr, W.J.; Magennis, E.J. *Chem. Commun.* **2002**, 2262.
27 McIntosh, C.E.; Martínez, I.; Ovaska, T.V. *Synlett* **2004**, 2579.
28 Leadbeater, N.E.; Marco, M. *Org. Lett.* **2002**, *4*, 2973.
29 Zhang, W.; Chen, C.H.T.; Lu, Y.; Nagashima, T. *Org. Lett.* **2004**, *6*, 1473.
30 Mutule, I.; Suna, E. *Tetrahedron Lett.* **2004**, *45*, 3909.
31 Allen, D.; Callaghan, O.; Cordier, F.L.; Dobson, D.R.; Harris, J.R.; Hotten, T.M.; Owton, W.M.; Rathmell, R.E.; Wood, V.A. *Tetrahedron Lett.* **2004**, *45*, 9645.
32 Kim, Y.J.; Varma, R.S. *Tetrahedron Lett.* **2004**, *45*, 7205.
33 Lee, M.J.; Sun, C.M. *Tetrahedron Lett.* **2004**, *45*, 437.
34 Lacova, M.; Gasparova, R.; Loos, D.; Liptay, T.; Pronayova, N. *Molecules* **2000**, *5*, 167.
35 (a) Nohara, A.; Sugihara, H.; Ukawa, K. *Japan Kokai Tokyo Koho* **1978**, 111,070; *Chem. Abstr.* **1979**, *90*, 5482z; (b) Klutchko, S.; Kaminski, D.; Von Strandtmann, M. *U.S. Patent* **1977**, 4,008,252; *Chem. Abstr.* **1977**, *87*, 5808a.
36 Villemin, D.; Martin, B. *Synth. Commun.* **1995**, *25*, 3135.
37 Kidwai, M.; Bhushan, K.R.; Misra, P. *Indian J. Chem.* **2000**, *39B*, 458.
38 Molteni, V.; Hamilton, M.M.; Mao, L.; Crane, C.M.; Termin, A.P.; Wilson, D.M. *Synthesis* **2002**, 1669.
39 Martins, M.A.P.; Pereira, C.M.P.; Beck, P.; Machado, P.; Moura, S.; Teixeira, M.V.M.; Bonacorso, H.G.; Zanatta, N. *Tetrahedron Lett.* **2003**, *44*, 6669.
40 Wu, Y.-J.; He, H.; L'Heureux, A. *Tetrahedron Lett.* **2003**, *44*, 4217.
41 Sparks, R.B.; Combs, A.P. *Org. Lett* **2004**, *6*, 2473.
42 Wolkenberg, S.E.; Wisnoski, D.D.; Leister, W.H.; Wang, Y.; Zhao, Z.; Lindsley, C.W. *Org. Lett* **2004**, *6*, 1453.
43 Santagada, V.; Fiorino, F.; Perissutti, E.; Severino, B.; Terracciano, S.; Teixeira, C.E.; Caliendo, G. *Tetrahedron Lett.* **2003**, *44*, 1149.
44 Tan, K.L.; Vasudevan, A.; Bergman, R.G. *Org. Lett.* **2003**, *5*, 2131.
45 Boufatah, N.; Gellis, A.; Maldonado, J.; Vanelle, P. *Tetrahedron* **2004**, *60*, 9131.
46 Tantry, S.J.; Babu, K.; Babu, V.V.S. *Tetrahedron Lett.* **2002**, *43*, 9461.
47 Pottorf, R.S.; Chadha, N.K.; Katkevics, M.; Ozola, V.; Suna, E.; Ghane, H.; Regberg, T.; Player, M.R. *Tetrahedron Lett.* **2003**, *44*, 175.
48 Moore, J.E.; Spinks, D.; Harrity, J.P.A. *Tetrahedron Lett.* **2004**, *45*, 3189.
49 Martins, M.A.P.; Beck, P.; Cunido, W.; Pereira, C.M.P.; Sinhorin, A.P.; Blanco, R.F.; Peres, R.; Bonacorso, H.G.; Zanatta, N. *Tetrahedron Lett.* **2002**, *43*, 7005.
50 Katritzky, A.R.; Cai, C.; Suzuki, K.; Singh, S.K. *J. Org. Chem.* **2004**, *69*, 811.
51 Bolognese, A.; Correale, G.; Manfra, M.; Lavecchia, A.; Novellino, E.; Barone, V. *Org. Biomol. Chem.* **2004**, *2*, 2809.
52 Gududuru, V.; Nguyen, V.; Dalton, J.T.; Miller, D.D. *Synlett* **2004**, 2357.

53 Vallin, K.S.A.; Larhed, M.; Johansson, K.; Hallberg, A. *J. Org. Chem.* **2000**, *65*, 4537.
54 Bentiss, F.; Lagrenée, M.; Barbry, D. *Tetrahedron Lett.* **2000**, *41*, 1539.
55 Brain, C.T.; Paul, J.M.; Loang, Y.; Oakley, P.J. *Tetrahedron Lett.* **1999**, *40*, 3275.
56 Schulz, M.J.; Coats, S.J.; Hlasta, D.J. *Org. Lett.* **2004**, *6*, 3265.
57 Bliznets, I.V.; Vasil'ev, A.A.; Shorshnev, S.V.; Stepanov, A.E.; Lukyanov, S.M. *Tetrahedron Lett.* **2004**, *45*, 2571.
58 Xu, G.; Wang, Y.G. *Org. Lett.* **2004**, *6*, 985.
59 Cotterill, I.C.; Usyatinsky, A.Y.; Arnold, J.M.; Clark, D.S.; Dordik, J.S.; Michels, P.C.; Khmelnitsky, Y.L. *Tetrahedron Lett.* **1998**, *39*, 1117.
60 Bagley, M.C.; Lunn, R.; Xiong, X. *Tetrahedron Lett.* **2002**, *43*, 8331.
61 Gorobets, N.Y.; Yousefi, B.H.; Belaj, F.; Kappe, C.O. *Tetrahedron* **2004**, *60*, 8633.
62 Wilson, N.S.; Sarko, C.R.; Roth, G.P. *Tetrahedron Lett.* **2002**, *43*, 581.
63 Tu, S.; Miao, C.; Gao, Y.; Fang, F.; Zhuang, Q.; Feng, Y.; Shi, D. *Synlett* **2004**, 255.
64 Larhed, M.; Hoshino, M.; Hadida, S.; Curran, D.P.; Hallberg, A. *J. Org. Chem.* **1997**, *62*, 5583.
65 Bentz, E.; Moloney, M.G.; Westaway, S.M. *Tetrahedron Lett.* **2004**, *45*, 7395.
66 Stanetty, P.; Schnürch, M.; Mihovilovic, M.D. *Synlett* **2003**, 1862.
67 Walla, P.; Kappe, C.O. *Chem. Commun.* **2004**, 564.
68 He, H.; Wu, Y.J. *Tetrahedron Lett.* **2004**, *45*, 3237.
69 Wang, T.; Magnin, D.R.; Hamann, L.G. *Org. Lett.* **2003**, *5*, 897.
70 Maes, B.U.W.; Loones, K.T.J.; Lemière, G.L.F.; Dommisse, R.A. *Synlett* **2003**, 1822.
71 Burton, G.; Cao, P.; Li, G.; Rivero, R. *Org. Lett.* **2003**, *5*, 4373.
72 Harmata, M.; Hong, X.; Ghosh, S.K. *Tetrahedron Lett.* **2004**, *45*, 5233.
73 Cherng, Y.J. *Tetrahedron* **2002**, *58*, 4931.
74 Cherng, Y.J. *Tetrahedron* **2002**, *58*, 1125.
75 Cherng, Y.J. *Tetrahedron* **2002**, *58*, 887.
76 Srivastava, R.R.; Collibee, S.E. *Tetrahedron Lett.* **2004**, *45*, 8895.
77 Cid, J.M.; Romera, J.L.; Trabanco, A.A. *Tetrahedron Lett.* **2004**, *45*, 1133.
78 Pemberton, N.; Aberg, V.; Almstedt, H.; Westermark, A.; Almqvist, F. *J. Org. Chem.* **2004**, *69*, 7830.
79 Zhang, W.; Nagashima, T.; Lu, Y.; Chen, C.H.T. *Tetrahedron Lett.* **2004**, *45*, 4611.
80 Alajarín, R.; Vaquero, J.J.; García-Navío, J.L.; Alvarez-Builla, J. *Synlett* **1992**, 297.
81 Khadilkar, B.M.; Madyar, V.R. *Org. Proc. Res. Dev.* **2001**, *5*, 452.
82 Vanden Eynde, J.J.; Rutot, D. *Tetrahedron* **1999**, *55*, 2687.
83 Fawcett, J.; Griffith, G.A.; Percy, J.M.; Uneyama, E. *Org. Lett.* **2004**, *6*, 1277.
84 (a) Moghaddam, F.M.; Sharifi, A.; Saidi, M.R. *J. Chem. Res. (S)* **1996**, 338; (b) Saidi, M.R.; Bigdeli, M.R. *J. Chem. Res. (S)* **1998**, 800. (c) Saidi, M.R.; Rajabi, F. *Heterocycles* **2001**, *55*, 1805.
85 Shanmugasundaram, M.; Manikandan, S.; Raghunathan, R. *Tetrahedron* **2002**, *58*, 997.
86 Han, J.W.; Castro, J.C.; Burgess, K. *Tetrahedron Lett.* **2003**, *44*, 9359.
87 Santagada, V.; Fiorino, F.; Perissutti, E.; Severino, B.; Terracciano, S.; Cirino, G.; Caliendo, G. *Tetrahedron Lett.* **2003**, *44*, 1145.
88 Lange, J.H.M.; Hofmeyer, L.J.F.; Hout, F.A.S.; Osnabrug, S.J.M.; Verveer, P.C.; Kruse, C.G.; Feenstra, R.W. *Tetrahedron Lett.* **2002**, *43*, 1101.
89 Sandin, H.; Swanstein, M.L.; Wellner, E. *J. Org. Chem.* **2004**, *69*, 1571.
90 Yeh, W.-B.; Sun, C.-M. *J. Comb. Chem.* **2004**, *6*, 279.
91 Bagley, M.C.; Hughes, D.D.; Taylor, P.H. *Synlett* **2003**, 259.
92 Abdel-Rahman, A.A.H.; El Ashry, E.S.H. *Synlett* **2002**, 2043.

93 Paolini, L.; Petricci, E.; Corelli, F.; Botta, M. *Synthesis* **2003**, 1039.
94 Villemin, D.; Gomez-Escolinilla, M.J.; Saint-Clair, J.F. *Tetrahedron Lett.* **2001**, *42*, 635.
95 Yoon, D.S.; Han, Y.; Stark, T.M.; Haber, J.C.; Gregg, B.T.; Stankovich, S.B. *Org. Lett.* **2004**, *6*, 4775.
96 Alexandre, F.-R.; Berecibar, A.; Besson, T. *Tetrahedron Lett.* **2002**, *43*, 3911.
97 (a) Wu, T.Y.H.; Schultz, P.G.; Ding, S. *Org. Lett.* **2003**, *5*, 3587; (b) Luo, G.; Chen, L.; Poindexter, G.S. *Tetrahedron Lett.* **2002**, *43*, 5739; (c) Takvorian, A.G.; Combs, A.P. *J. Comb. Chem.* **2004**, *6*, 171.
98 Dandia, A.; Saha, M.; Taneja, H. *J. Fluorine Chem.* **1998**, *90*, 17.
99 Hazarkhani, H.; Karimi, B. *Synthesis* **2004**, 1239.
100 Kappe, C.O.; Kumar, D.; Varma, R.S. *Synthesis* **1999**, 1799.
101 Bowman, M.D.; Jeske, R.C.; Blackwell, H.E. *Org. Lett.* **2004**, *6*, 2019.
102 Zhang, W.; Tempest, P. *Tetrahedron Lett.* **2004**, *45*, 6757.
103 Goswami, S.; Adak, A.K. *Tetrahedron Lett.* **2002**, *43*, 8371.
104 Zhao, Z.; Wisnoski, D.D.; Wolkenberg, S.E.; Leister, W.H.; Wang, Y.; Lindsley, C.W. *Tetrahedron Lett.* **2004**, *45*, 4873.
105 Efskind, J.; Undheim, K. *Tetrahedron Lett.* **2003**, *44*, 2837.
106 Tung, C.-L.; Sun, C.M. *Tetrahedron Lett.* **2004**, *45*, 1159.
107 Zhao, Z.; Leister, W.H.; Strauss, K.A.; Wisnoski, D.D.; Lindsley, C.W. *Tetrahedron Lett.* **2003**, *44*, 1123.
108 Lee, H.-K.; Rana, T.M. *J. Comb. Chem.* **2004**, *6*, 504.
109 Besson, T.; Guillard, J.; Rees, C.W. *Tetrahedron Lett.* **2000**, *41*, 1027.
110 Soukri, M.; Guillaumet, G.; Besson, T.; Aziane, D.; Aadil, M.; Essassi, El M.; Akssira, M. *Tetrahedron Lett.* **2000**, *41*, 5857.
111 Santagada, V.; Persutti, E.; Fiorino, F.; Vivenzio, B.; Caliendo, G. *Tetrahedron Lett.* **2001**, *42*, 2397.
112 Fabis, F.; Jolivet-Fouchet, S.; Robba, M.; Landelle, M.; Rault, S. *Tetrahedron* **1998**, *54*, 10789.
113 Rault, S.; Derobert, M. *Eur. Pat. Appl.* 059 5084 A1, 1993.
114 Hinschberger, A.; Gillard, A.C.; Bureau, I.; Rault, S. *Tetrahedron* **2000**, *56*, 1361.
115 Bagley, M.C.; Singh, N. *Synlett* **2002**, 1718.
116 El Latif, F.M.A.; Barsy, M.A.; Aref, A.M.; Sadek, K.U. *Green Chem.* **2002**, *4*, 196.
117 Petricci, E.; Radi, M.; Corelli, F.; Botta, M. *Tetrahedron Lett.* **2003**, *44*, 9181.
118 Hopkins, C.R.; Collar, N. *Tetrahedron Lett.* **2004**, *45*, 8631.
119 Ireland, S.M.; Tye, H.; Whittaker, M. *Tetrahedron Lett.* **2003**, *44*, 4369.
120 Raboisson, P.; Norberg, B.; Casimir, J.R.; Bourguignon, J.J. *Synlett* **2002**, 519.
121 Perzyna, A.; Houssin, R.; Barbry, D.; Hénichart, J.P. *Synlett* **2002**, 2077.
122 Lindsley, C.W.; Wisnoski, D.D.; Wang, Y.; Leister, W.H.; Zhao, Z. *Tetrahedron Lett.* **2003**, *44*, 4495.
123 Wu, C.Y.; Sun, C.M. *Synlett* **2002**, 1709.
124 Kuo, F.-M.; Tseng, M.C.; Yen, Y.-H.; Chu, Y.H. *Tetrahedron* **2004**, *60*, 12075.
125 El Ashry, E.S.H.; Ramadan, E.S.; Hamid, H.M.A.; Hagar, M. *Synlett* **2004**, 723.
126 (a) Kaval, N.; Dehaen, W.; Mátyus, P.; Van der Eycken, E. *Green Chem.* **2004**, *6*, 125; (b) Kaval, N.; Halasz-Dajka, B.; Vo-Thanh, G.; Dehaen, W.; Van der Eycken, J.; Matyus, P.; Loupy, A.; Van der Eycken, E. *Tetrahedron*, **2005**, *61*, 9052.
127 Yadav, L.D.S.; Kapoor, R. *Synthesis* **2002**, 2344.
128 Olah, G.A.; Keumi, T.; Fung, A.P. *Synthesis* **1979**, 537.
129 Laurent, A.; Jacquault, P.; Di Martino, J.L.; Hamelin, J. *J. Chem. Soc. Chem. Commun.* **1995**, 1101.
130 Texier-Boullet, F.; Latouche, R.; Hamelin, J. *Tetrahedron Lett.* **1993**, *34*, 2123.

131 Pilard, J.F.; Klein, B.; Texier-Boullet, F.; Hamelin, J. *Synlett* **1992**, 219.
132 Vidal, T.; Petit, A.; Loupy, A.; Gedye, R.N. *Tetrahedron* **2000**, *56*, 5473.
133 Seijas, J.A.; Vázquez-Tato, M.P.; González-Bande, C.; Martínez, M.M.; Pacios-López, B. *Synthesis* **2001**, 999.
134 Hajra, A.; Ranu, B.C. *Tetrahedron Lett.* **2001**, *57*, 4767.
135 de la Hoz, A.; Díaz-Ortiz, A.; Gómez, M.V.; Mayoral, J.A.; Moreno, A.; Sánchez-Migallón, A.M.; Vázquez, E. *Tetrahedron Lett.* **2001**, *57*, 5421.
136 Carillo, J.R.; Diaz-Ortiz, A.; Gomez-Escalonilla, M.J.; de la Hoz, A.; Moreno, A.; Prieto, P. *Tetrahedron* **1999**, *55*, 9623.
137 Bora, U.; Saikia, A.; Boruah, R.C. *Org. Lett.* **2003**, *5*, 435.
138 Barchín, B.M.; Cuadro, A.M.; Alvarez-Builla, J. *Synlett* **2002**, 343.
139 Mortoni, A.; Martinelli, M.; Piarulli, U.; Regalia, N.; Gagliardi, S. *Tetrahedron Lett.* **2004**, *45*, 6623.
140 Collina, S.; Loddo, G.; Barbieri, A.; Linati, L.; Alcaro, S.; Chimenti, P.; Azzolina, O. *Tetrahedron: Asymmetry* **2004**, *15*, 3601.
141 (a) Lipinska, T.; Guibé-Jampel, E.; Petit, A.; Loupy, A. *Synth. Commun.* **1999**, *29*, 1349; (b) Lipinska, T. *Tetrahedron Lett.* **2004**, *45*, 8831.
142 Chakrabarty, M.; Basak, R.; Ghosh, N. *Tetrahedron Lett.* **2001**, *42*, 3913.
143 Petit, A.; Loupy, A.; Maillard, P.; Momenteau, M. *Synth. Commun.* **1992**, *22*, 1137.
144 Collman, J.P.; Decréau, R.A. *Tetrahedron Lett.* **2003**, *44*, 1207.
145 Lu, M.-W.; Hu, W.-X.; Yun, L.H. Proceedings of the first national meeting on Microwaves, Beijing (China), **1996**, 54.
146 Abenhaïm, D.; Loupy, A.; Mahieu, C.; Séméria, D. *Synth. Commun.* **1994**, *24*, 1809.
147 Chatti, S.; Bortolussi, M.; Loupy, A. *Tetrahedron Lett.* **2000**, *41*, 3367.
148 Chatti, S.; Bortolussi, M.; Loupy, A. *Tetrahedron* **2000**, *56*, 5877.
149 Kotha, S.; Mandal, K.; Deb, A.C.; Banerjee, S. *Tetrahedron Lett.* **2004**, *45*, 9603.
150 Filippi, J.-J.; Fernandez, X.; Lizzani-Cuvelier, L.; Loiseau, A.M. *Tetrahedron Lett.* **2003**, *44*, 6647.
151 Chowdhury, M.A.; Senboku, H.; Tokuda, M. *Synlett* **2004**, 1933.
152 Bogdal, D.; Warzala, M. *Tetrahedron Lett.* **2000**, *56*, 8769.
153 Vo-Thanh, G.; Loupy, A. *Tetrahedron Lett.* **2003**, *44*, 9091.
154 Majdoub, M.; Loupy, A.; Petit, A.; Roudesli, S. *Tetrahedron* **1996**, *52*, 617.
155 Yadav, J.S.; Anuradha, K.; Reddy, B.V.S.; Eeshwaraiah, B. *Tetrahedron Lett.* **2003**, *44*, 8959.
156 Melucci, M.; Barbarella, G.; Sotgiu, G. *J. Org. Chem.* **2002**, *67*, 8877.
157 Lee, J.C.; Park, J.Y.; Yoon, S.Y.; Bae, Y.H.; Lee, S.J. *Tetrahedron Lett.* **2004**, *45*, 191.
158 Shaabani, A.; Bazgir, A. *Tetrahedron Lett.* **2004**, *45*, 2575.
159 Paul, S.; Gupta, M.; Gupta, R.; Loupy, A. *Synthesis* **2002**, 75.
160 Pleier, K.; Glas, H.; Grosche, M.; Sirsch, P.; Thiel, W.R. *Synthesis* **2001**, 55.
161 Azarifar, D.; Zolfigol, M.A.; Maleki, B. *Synthesis* **2004**, 1744.
162 Villemin, D.; Hammadi, M.; Martin, B. *Synth. Commun.* **1996**, *26*, 2895.
163 Bougrin, K.; Loupy, A.; Petit, A.; Daou, B.; Soufiaoui, M. *Tetrahedron* **2001**, *57*, 63.
164 Martín-Aranda, R.M.; Vicente-Rodríguez, M.A.; López-Pestaña, J.M.; Banares-Muñoz, M.A. *J. Mol. Catal. A* **1997**, *124*, 115.
165 Kidwai, M.; Sapra, P.; Bushan, K.R.; Misra, P. *Synthesis* **2001**, 1509.
166 Chérouvrier, J.R.; Boissel, J.; Carreaux, F.; Bazureau, J.P. *Green Chem.* **2001**, *3*, 165.
167 Chérouvrier, J.-R.; Carreaux, F.; Bazureau, J.P. *Tetrahedron Lett.* **2002**, *13*, 3581.
168 Meddad, N.; Rhamouni, M.; Derdour, A.; Bazureau, J.P.; Hamelin, J. *Synthesis* **2001**, 581.

169 Balalaie, S.; Hashemi, M.M.; Akhbari, M. *Tetrahedron Lett.* **2003**, *44*, 1709.

170 Martínez-Palou, R.; de Paz, G.; Marín-Cruz, J.; Zepeda, L.G. *Synlett* **2003**, 1847.

171 Uchida, H.; Tanikoshi, H.; Nakamura, S.; Reddy, P.Y.; Toru, T. *Synlett* **2003**, 1117.

172 Lee, J.C.; Choi, H.J.; Lee, Y.C. *Tetrahedron Lett.* **2003**, *44*, 123.

173 Csiba, M.; Cléophax, J.; Loupy, A.; Malthête, J.; Géro, S.D. *Tetrahedron Lett.* **1993**, *34*, 1787.

174 Pério, B.; Dozias, M.J.; Jacquault, P.; Hamelin, J. *Tetrahedron Lett.* **1997**, *38*, 7867.

175 Matloubi-Moghaddam, F.; Sharifi, A. *Synth. Commun.* **1995**, *25*, 2457.

176 Rotstein, D.M.; Kertesz, D.J.; Walker, K.A.; Swinney, D.C. *J. Med. Chem.* **1992**, *35*, 2818.

177 Pério, B.; Dozias, M.J.; Hamelin, J. *Org. Proc. Res. Dev.* **1998**, *2*, 428.

178 Pério, B.; Hamelin, J. *Green Chem.* **2000**, *2*, 252.

179 Díaz-Ortiz, A.; Prieto, P.; Loupy, A.; Abenhaïm, D. *Tetrahedron Lett.* **1996**, *37*, 1695.

180 Ferrett, R.R.; Hyde, M.J.; Lahti, K.A.; Friebe, T.L. *Tetrahedron Lett.* **2003**, *44*, 2573.

181 Chakraborti, A.K.; Selvam, C.; Kaur, G.; Bhagat, S. *Synlett* **2004**, 851.

182 Paul, S.; Nanda, P.; Gupta, R.; Loupy, A. *Tetrahedron Lett.* **2002**, *43*, 4261.

183 (a) Cléophax, J.; Liagre, M.; Loupy, A.; Petit, A. *Org. Proc. Res. Dev.* **2000**, *6*, 498; (b) Loupy, A.; Perreux, L.; Liagre, M.; Burle, K.; Moneuse, M. *Pure Appl. Chem.* **2001**, *73*, 161; (c) Perez, E.R.; Loupy, A.; Liagre, M.; de Guzzi Plepis, A.M.; Cordeiro, P. *Tetrahedron* **2003**, *59*, 865.

184 Abenhaïm, D.; Diaz-Barra, E.; de la Hoz, A.; Loupy, A.; Sanchez-Migallon, A. *Heterocycles* **1994**, *38*, 793.

185 Sharma, U.; Ahmed, S.; Boruah, R.C. *Tetrahedron Lett.* **2000**, *41*, 3493.

186 Chetia, A.; Longchar, M.; Lekhok, K.C.; Boruah, R.C. *Synlett* **2004**, 1309.

187 Balalaie, S.; Kowsari, E. *Monatsh. Chem.* **2001**, *132*, 1551.

188 Rodríguez, H.; Suarez, M.; Pérez, R.; Petit, A.; Loupy, A. *Tetrahedron Lett.* **2003**, *44*, 3709.

189 Ranu, R.C.; Hajra, A.; Jana, U. *Tetrahedron Lett.* **2000**, *41*, 531.

190 Song, S.J.; Cho, S.J.; Park, D.K.; Kwon, T.W.; Jenekhe, S.A. *Tetrahedron Lett.* **2003**, *44*, 255.

191 Yadav, J.S.; Reddy, B.V.S.; Rao, R.S.; Naveenkumar, V.; Nagaiah, K. *Synthesis* **2003**, 1610.

192 Zhang, J.M.; Yang, W.; Song, L.P.; Cai, X.; Zhu, S.Z. *Tetrahedron Lett.* **2004**, *45*, 5771.

193 Seijas, J.A.; Vázquez-Tato, M.P.; Martínez, M.M.; Rodríguez-Parga, J. *Green Chem.* **2002**, *4*, 390.

194 Silveira, C.C.; Bernardi, C.R.; Braga, A.L.; Kaufman, T.S. *Synlett* **2002**, 907.

195 Vanden Eynde, J.J.; Mayence, A. *Molecules* **2003**, *8*, 381.

196 Zhang, Y.W.; Shen, Z.X.; Pan, B.; Lu, X.H.; Chen, M.H. *Synth. Commun.* **1995**, *85*, 857.

197 Almena, I.; Díaz-Ortiz, A.; Díez-Barra, E.; de la Hoz, A.; Loupy, A. *Chem. Lett.* **1996**, 333.

198 Stoyanov, E.V.; Heber, D. *Synlett.* **1999**, *11*, 1747.

199 Ohberg, L.; Westman, J. *Synlett* **2001**, 1296.

200 Lange, Jos.H.M.; Verveer, P.C.; Osnabrug, S.J.M.; Visser, G.M. *Tetrahedron Lett.* **2001**, *42*, 1367.

201 Basu, B.; Das, P.; Bhuiyan, M.M.H.; Jha, S. *Tetrahedron Lett.* **2003**, *44*, 3817.

202 Narayan, S.; Seelhammer, T.; Gawley, R.E. *Tetrahedron Lett.* **2004**, *45*, 757.

203 Paul, S.; Nanda, P.; Gupta, R.; Loupy, A. *Synthesis* **2003**, 2877.

204 Khalafi-Nezhad, A.; Mokhtari, B.; Rad, M.N.S. *Tetrahedron Lett.* **2003**, *44*, 7325.

205 (a) Mariani, E.; Genta, M.T.; Bargagna, A.; Neuhoff, C.; Loupy, A. in "Application of the microwave technology to synthesis and materials processing" Edited by Acierno, D.; Leonelli, C.; Pellacini, G.C. Mucchi

Editore **2000**, 157; (b) Villa, C.; Mariani, E.; Loupy, A.; Grippo, C.; Grossi, G.C.; Bergagna, A. *Green Chemistry* **2003**, *5*, 623.

206 Besson, T.; Thiéry, V.; Frère, S. *Tetrahedron Lett.* **2001**, *42*, 2791.

207 de la Hoz, A.; Moreno, A.; Vasquez, E. *Synlett* **1999**, 608.

208 Jeselnik, M.; Varma, R.S.; Polanc, S.; Kocevar, M. *Green Chem.* **2002**, *4*, 35.

209 Devi, I.; Bhuyan, P.J. *Tetrahedron Lett.* **2004**, *45*, 8625.

210 Kappe, C.O.; Stadler, A. *J. Chem. Soc., Perkin Trans. 1* **2000**, *2*, 1363.

211 Vanden Eynde, J.J.; Hecq, N.; Kataeva, O.; Kappe, C.O. *Tetrahedron* **2001**, *57*, 1785.

212 Gohain, M.; Prajapati, D.; Sandhu, J.S. *Synlett* **2004**, 235.

213 Kamal, A.; Reddy, K.S.; Prasad, B.R.; Babu, A.H. *Tetrahedron Lett.* **2004**, *45*, 6517.

214 Vázquez, E.; de la Hoz, A.; Elander, N.; Moreno, A.; Stone-Elander, S. *Heterocycles* **2001**, *55*, 109.

215 Paul, S.; Gupta, M.; Gupta, R.; Loupy, A. *Tetrahedron Lett.* **2001**, *42*, 3827.

216 (a) Jolivet, S.; Texier-Boullet, F.; Hamelin, J.; Jacquault, P. *Heteroatom Chem.* **1995**, *6*, 469; (b) Jolivet, S.; Toupet, L.; Texier-Boullet, F.; Hamelin, J. *Tetrahedron* **1996**, *52*, 5819.

217 Benhaliliba, H.; Derdour, A.; Bazureau, J.P.; Texier-Boullet, F.; Hamelin, J. *Tetrahedron Lett.* **1998**, *39*, 541.

218 Yadav, L.D.S.; Singh, S.; Singh, A. *Tetrahedron Lett.* **2002**, *43*, 8551.

219 Yadav, L.D.S.; Yadav, B.S.; Dubey, S. *Tetrahedron* **2004**, *60*, 131.

220 Yadav, L.D.S.; Singh, A. *Tetrahedron Lett.* **2003**, *44*, 5637.

221 Dandia, A.; Arya, K.; Sati, M.; Gautam, S. *Tetrahedron* **2004**, *60*, 5253.

222 Kurteva, V.B.; Afonso, C.A.M. *Green Chem.* **2004**, *6*, 183.

223 Varma, R.S.; Kumar, D. *Tetrahedron Lett.* **1999**, *40*, 7665.

224 Paudrey, S.K.; Awasthi, K.K.; Saxena, A.K. *Tetrahedron* **2001**, *57*, 4437.

225 Rahmouni, M.; Derdour, A.; Bazureau, J.P.; Hamelin, J. *Synth. Commun.* **1996**, *26*, 453.

226 Dandia, A.; Sati, M.; Loupy, A. *Green Chem.* **2002**, *4*, 599.

227 Nemes, P.; Vincze, Z.; Balázs, B.; Tóth, G.; Scheiber, P. *Synlett* **2003**, 250.

228 Pessoa-Mahana, H.; Pessoa-Mahana, C.D.; Salazar, R.; Valderrama, J.A.; Saez, E.; Araya-Maturana, R. *Synthesis* **2004**, 436.

229 Devi, I.; Kumar, B.S.D.; Bhuyan, P.J. *Tetrahedron Lett.* **2003**, *44*, 8307.

230 Devi, I.; Bhuyan, P.J. *Synlett* **2004**, 283.

231 Devi, I.; Borah, H.N.; Bhuyan, P.J. *Tetrahedron Lett.* **2004**, *45*, 2405.

232 Cledera, P.; Sánchez, J.D.; Caballero, E.; Avendaño, C.; Ramos, M.T.; Menéndez, J.C. *Synlett* **2004**, 803.

Microwaves in Organic Synthesis
Volume 2

Edited by
André Loupy

Related Titles

Wasserscheid, P., Welton, T. (eds.)

Ionic Liquids in Synthesis

2006
ISBN 3-527-31239-0

Kappe, C. O., Stadler, A.

Microwaves in Organic and Medicinal Chemistry

2005
ISBN 3-527-31210-2

de Meijere, A., Diederich, F. (eds.)

Metal-Catalyzed Cross-Coupling Reactions

2 volumes

2004
ISBN 3-527-30518-1

Yamamoto, H., Oshima, K. (eds.)

Main Group Metals in Organic Synthesis

2 volumes

2004
ISBN 3-527-30508-4

Tanaka, K.

Solvent-free Organic Synthesis

2003
ISBN 3-527-30612-9

Microwaves in Organic Synthesis

Volume 2

Edited by André Loupy

Second, Completely Revised and Enlarged Edition

WILEY-VCH Verlag GmbH & Co. KGaA

The Editor

André Loupy
Laboratoire des Réactions Sélectives sur Supports
Université Paris-Sud
Batiment 410
91405 Orsay cedex
France

■ All books published by Wiley-VCH are carefully produced. Nevertheless, authors, editors, and publisher do not warrant the information contained in these books, including this book, to be free of errors. Readers are advised to keep in mind that statements, data, illustrations, procedural details or other items may inadvertently be inaccurate.

Library of Congress Card No.: applied for
British Library Cataloguing-in-Publication Data
A catalogue record for this book is available from the British Library.

Bibliographic information published by the Deutsche Nationalbibliothek
Die Deutsche Nationalbibliothek lists this publication in the Deutsche Nationalbibliografie; detailed bibliographic data are available in the Internet at ⟨http://dnb.d-nb.de⟩.

© 2006 WILEY-VCH Verlag GmbH & Co. KGaA, Weinheim

All rights reserved (including those of translation into other languages). No part of this book may be reproduced in any form – by photoprinting, microfilm, or any other means – nor transmitted or translated into a machine language without written permission from the publishers. Registered names, trademarks, etc. used in this book, even when not specifically marked as such, are not to be considered unprotected by law.

Typesetting Asco Typesetters, Hong Kong
Printing betz-druck GmbH, Darmstadt
Binding Litges & Dopf GmbH, Heppenheim
Cover Design
Grafik-Design Schulz, Fußgönheim

Printed in the Federal Republic of Germany
Printed on acid-free paper

ISBN-13: 978-3-527-31452-2
ISBN-10: 3-527-31452-0

Contents

Volume 1

Preface *XVII*

About European Cooperation in COST Chemistry Programs *XIX*

List of Authors *XXIII*

1	**Microwave–Material Interactions and Dielectric Properties, Key Ingredients for Mastery of Chemical Microwave Processes** *1*	
	Didier Stuerga	
1.1	Fundamentals of Microwave–Matter Interactions *1*	
1.1.1	Introduction *2*	
1.1.2	The Complex Dielectric Permittivity *9*	
1.1.3	Dielectric Properties and Molecular Behavior *29*	
1.2	Key Ingredients for Mastery of Chemical Microwave Processes *43*	
1.2.1	Systemic Approach *43*	
1.2.2	The Thermal Dependence of Dielectric Loss *45*	
1.2.3	The Electric Field Effects *46*	
1.2.4	Hydrodynamic Aspects *49*	
1.2.5	Thermodynamic and Other Effects of Electric Fields *51*	
1.2.6	The Athermal and Specific Effects of Electric Fields *52*	
1.2.7	The Thermal Path Effect: Anisothermal Conditions *54*	
1.2.8	Hot Spots and Heterogeneous Kinetics *56*	
	References *57*	
2	**Development and Design of Laboratory and Pilot Scale Reactors for Microwave-assisted Chemistry** *62*	
	Bernd Ondruschka, Werner Bonrath, and Didier Stuerga	
2.1	Introduction *62*	
2.2	Basic Concepts for Reactions and Reactors in Organic Synthesis *63*	
2.3	Methods for Enhancing the Rates of Organic Reactions *64*	
2.4	Microwave-assisted Organic Syntheses (MAOS) *66*	

Microwaves in Organic Synthesis, Second edition. Edited by A. Loupy
Copyright © 2006 WILEY-VCH Verlag GmbH & Co. KGaA, Weinheim
ISBN: 3-527-31452-0

2.4.1	Microwave Ovens and Reactors – Background 67
2.4.2	Scale-up of Microwave Cavities 72
2.4.3	Efficiency of Energy and Power 73
2.4.4	Field Homogeneity and Penetration Depth 73
2.4.5	Continuous Tube Reactors 74
2.4.6	MAOS – An Interdisciplinary Field 74
2.5	Commercial Microwave Reactors – Market Overview 75
2.5.1	Prolabo's Products 77
2.5.2	CEM's products 78
2.5.3	Milestone's Products 80
2.5.4	Biotage's Products 84
2.6	Selected Equipment and Applications 85
2.6.1	Heterogeneous Catalysis 89
2.6.2	Hyphenated Techniques in Combination with Microwaves 90
2.6.3	Combination of Microwave Irradiation with Pressure Setup 92
2.6.4	Synthesis of Laurydone® 97
2.6.5	Industrial Equipment: Batch or Continuous Flow? 98
2.7	Qualification and Validation of Reactors and Results 101
2.8	Conclusion and Future 102
	References 103

3 Roles of Pressurized Microwave Reactors in the Development of Microwave-assisted Organic Chemistry 108

Thach Le Ngoc, Brett A. Roberts, and Christopher R. Strauss

3.1	Introduction 108
3.2	Toward Dedicated Microwave Reactors 109
3.2.1	The Continuous Microwave Reactor (CMR) 110
3.2.2	Microwave Batch Reactors 111
3.3	Applications of the New Reactors 112
3.3.1	With Low-boiling Solvents 112
3.3.2	Kinetics Measurements 113
3.3.3	Core Enabling Technology 114
3.4	Commercial Release of MBRs and CMRs 114
3.5	Advantages of Pressurized Microwave Reactors 116
3.6	Applications 117
3.6.1	General Synthesis 117
3.6.2	Kinetic Products 121
3.6.3	Selective Synthesis of O-Glycofuranosides and Pyranosides from Unprotected Sugars 121
3.6.4	Synthesis in High-temperature Water 123
3.6.5	Biomimetic Reactions 125
3.6.6	Indoles 126
3.7	Effect of the Properties of Microwave Heating on the Scale-up of Methods in Pressurized Reactors 127
3.8	Software Technology for Translation of Reaction Conditions 129

3.9	Conclusion 131
	Acknowledgments 132
	References 132

4	**Nonthermal Effects of Microwaves in Organic Synthesis** 134
	Laurence Perreux and André Loupy
4.1	Introduction 134
4.2	Origin of Microwave Effects 135
4.3	Specific Nonthermal Microwave Effects 137
4.4	Effects of the Medium 141
4.4.1	Polar Solvents 141
4.4.2	Nonpolar Solvents 143
4.4.3	Solvent-free Reactions 146
4.5	Effects Depending on Reaction Mechanisms 147
4.5.1	Isopolar Transition-state Reactions 148
4.5.2	Bimolecular Reactions Between Neutral Reactants Leading to Charged Products 152
4.5.3	Anionic Bimolecular Reactions Involving Neutral Electrophiles 153
4.5.4	Unimolecular Reactions 154
4.6	Effects Depending on the Position of the Transition State Along the Reaction Coordinates 155
4.7	Effects on Selectivity 156
4.8	Some Illustrative Examples 157
4.8.1	Bimolecular Reactions Between Neutral Reactants 157
4.8.2	Bimolecular Reactions with One Charged Reactant 184
4.8.3	Unimolecular Reactions 198
4.8.4	Some Illustrative Examples of Effects on Selectivity 204
4.9	Concerning the Absence of Microwave Effects 209
4.10	Conclusions: Suitable Conditions for Observation of Specific MW Effects 210
	References 212

5	**Selectivity Under the Action of Microwave Irradiation** 219
	Antonio de la Hoz, Angel Díaz-Ortiz, and Andrés Moreno
5.1	Introduction 219
5.2	Selective Heating 220
5.2.1	Solvents 220
5.2.2	Catalysts 223
5.2.3	Reagents; Molecular Radiators 230
5.2.4	Susceptors 232
5.3	Modification of Chemoselectivity and Regioselectivity 233
5.3.1	Protection and Deprotection of Alcohols 233
5.3.2	Electrophilic Aromatic Substitution 236
5.3.3	Synthesis and Reactivity of Heterocyclic Compounds 241
5.3.4	Cycloaddition Reactions 247

5.3.5	Polymerization	*252*
5.3.6	Miscellaneous	*257*
5.4	Modification of Stereo and Enantioselectivity	*264*
5.5	Conclusions	*272*
	Acknowledgments	*273*
	References	*273*

6   Microwaves and Phase-transfer Catalysis   *278*
André Loupy, Alain Petit, and Dariusz Bogdal

6.1	Phase-transfer Catalysis	*278*
6.2	Synthetic Applications of Phase-transfer Processes	*281*
6.2.1	O-Alkylations	*281*
6.2.2	N-Alkylations	*293*
6.2.3	C-Alkylations of Active Methylene Groups	*297*
6.2.4	Alkylations with Dihalogenoalkanes	*300*
6.2.5	Nucleophilic Addition to Carbonyl Compounds	*302*
6.2.6	Deprotonations	*307*
6.2.7	Miscellaneous Reactions	*309*
6.3	Conclusion	*322*
	References	*322*

7   Microwaves and Ionic Liquids   *327*
Nicholas E. Leadbeater and Hanna M. Torenius

7.1	Introduction	*327*
7.2	Ionic Liquids in Conjunction with Microwave Activation	*330*
7.2.1	Synthesis of Ionic Liquids Using Microwave Heating	*330*
7.2.2	Reactions Using Microwave Irradiation and Ionic Liquids as Solvents and Reagents	*333*
7.2.3	Use of Ionic Liquids and Microwaves in Multicomponent Reactions	*349*
7.2.4	Use of Ionic Liquids as Heating Aids	*354*
7.3	Conclusions	*357*
	Abbreviations	*357*
	References	*358*

8   Organic Synthesis Using Microwaves and Supported Reagents   *362*
Rajender S. Varma and Yuhong Ju

8.1	Introduction	*362*
8.2	Microwave-accelerated Solvent-free Organic Reactions	*363*
8.2.1	Protection–Deprotection Reactions	*364*
8.2.2	Condensation Reactions	*372*
8.2.3	Isomerization and Rearrangement Reactions	*379*
8.2.4	Oxidation Reactions – Oxidation of Alcohols and Sulfides	*382*
8.2.5	Reduction Reactions	*388*
8.2.6	Synthesis of Heterocyclic Compounds	*391*
8.2.7	Miscellaneous Reactions	*401*

8.3	Conclusions 406
	References 407

9	**Microwave-assisted Reactions on Graphite** 416
	Thierry Besson, Valérie Thiery, and Jacques Dubac
9.1	Introduction 416
9.2	Graphite as a Sensitizer 418
9.2.1	Diels–Alder Reactions 418
9.2.2	Ene Reactions 423
9.2.3	Oxidation of Propan-2-ol 425
9.2.4	Thermolysis of Esters 425
9.2.5	Thermal Reactions in Heterocyclic Syntheses 427
9.2.6	Decomplexation of Metal Complexes 433
9.2.7	Redistribution Reactions between Tetraalkyl or Tetraarylgermanes, and Germanium Tetrahalides 434
9.2.8	Pyrolysis of Urea 436
9.2.9	Esterification of Stearic Acid with *n*-Butanol 437
9.3	Graphite as Sensitizer and Catalyst 437
9.3.1	Analysis of Two Synthetic Commercial Graphites 438
9.3.2	Acylation of Aromatic Compounds 438
9.3.3	Acylative Cleavage of Ethers 443
9.3.4	Ketodecarboxylation of Carboxylic Diacids 444
9.4	Notes 447
9.4.1	MW Apparatus, Typical Procedures, and Safety Measures 447
9.4.2	Temperature Measurement 448
9.4.3	Mechanism of Retention of Reactants on Graphite 449
9.4.4	Graphite or Amorphous Carbon for C/MW Coupling 449
9.5	Conclusion 450
	References 451

10	**Microwaves in Heterocyclic Chemistry** 456
	Jean Pierre Bazureau, Jack Hamelin, Florence Mongin, and Françoise Texier-Boullet
10.1	Introduction 456
10.2	Microwave-assisted Reactions in Solvents 456
10.2.1	Three, Four, and Five-membered Systems with One Heteroatom 456
10.2.2	Five-membered Systems with Two Heteroatoms 463
10.2.3	Five-membered Systems with More than Two Heteroatoms 468
10.2.4	Six-membered Systems with One Heteroatom 469
10.2.5	Six-membered Systems with More than One Heteroatom 479
10.2.6	Synthesis of More Complex or Polyheterocyclic Systems 485
10.3	Solvent-free Synthesis 492
10.3.1	Three, Four and Five-membered Systems with One Heteroatom 492
10.3.2	Five-membered Systems with Two Heteroatoms 499
10.3.3	Five-membered Systems with More than Two Heteroatoms – Synthesis of Triazoles and Related Compounds 506

10.3.4 Six-membered Systems with One Heteroatom *506*
10.3.5 Six-membered Systems with More than One Heteroatom *511*
10.3.6 Synthesis of More Complex or Polyheterocyclic Systems *515*
10.4 Conclusion *517*
References and Notes *517*

Volume 2

11 Microwaves in Cycloadditions *524*
Khalid Bougrin, Mohamed Soufiaoui, and George Bashiardes
11.1 Cycloaddition Reactions *524*
11.2 Reactions with Solvent *525*
11.3 Reactions under Solvent-free Conditions *526*
11.3.1 Reaction on Mineral Supports *527*
11.3.2 Reaction with Neat Reactants *528*
11.4 [4+2] Cycloadditions *532*
11.4.1 Intramolecular Hetero and Diels–Alder Reactions *533*
11.4.2 Intermolecular Hetero and Diels–Alder Reactions *538*
11.5 [3+2] Cycloadditions *546*
11.5.1 Nitrile Oxides, Nitrile Sulfides, and Nitrones *547*
11.5.2 Azomethine Ylides, and Nitrile Imines *556*
11.5.3 Azides *562*
11.6 [2+2] Cycloadditions *567*
11.6.1 Cycloadditions of Ketenes with Imines *567*
11.6.2 Cycloadditions of β-Formyl Enamides with Alkynes *569*
11.7 Other Cycloadditions *570*
11.8 Conclusions *571*
Acknowledgments *572*
References *573*

12 Microwave-assisted Chemistry of Carbohydrates *579*
Antonino Corsaro, Ugo Chiacchio, Venerando Pistarà, and Giovanni Romeo
12.1 Introduction *579*
12.2 Protection *580*
12.2.1 Acylation *580*
12.2.2 Acetalation *583*
12.2.3 Silylation *585*
12.2.4 Methylation of Carbohydrate Carboxylic Acids *586*
12.2.5 Synthesis of 1,6-Anhydro-β-D-hexopyranoses *586*
12.3 Deprotection *586*
12.3.1 Deacylation *587*
12.3.2 Deacetalation *588*
12.3.3 Desilylation *588*
12.4 Glycosylation *589*
12.4.1 *O*-Glycosylation *589*
12.4.2 *C*-Glycosylation Reactions *594*

12.4.3	N-Glycosylation	594
12.5	Hydrogenation (Catalytic Transfer Hydrogenation)	594
12.6	Oxidation	595
12.7	Halogenation	595
12.7.1	Chlorination and Bromination	595
12.7.2	Fluorination	596
12.7.3	Iodination	597
12.8	Stereospecific C–H Bond Activation for Rapid Deuterium Labeling	597
12.9	Reaction of Carbohydrates with Amino-derivatized Labels	598
12.10	Ferrier (II) Rearrangement to Carbasugars	599
12.11	Synthesis of Unsaturated Monosaccharides	599
12.12	Synthesis of Dimers and Polysaccharides, and their Derivatives	601
12.13	Synthesis of Heterocycles and Amino Acids	602
12.13.1	Spiroisoxazoli(di)ne Derivatives	602
12.13.2	Triazole-linked Glycodendrimers	603
12.13.3	5,6-Dihydro-1,2,3,4-tetrazenes	604
12.13.4	C-Nucleosides	605
12.13.5	Pterins	605
12.13.6	Epoxides	606
12.13.7	Azetidinones	607
12.13.8	Substituted Pyrazoles	608
12.13.9	3H-Pyrido[3,4-b]indole Derivatives	608
12.14	Enzymatic Reactions	608
12.15	Conclusion	611
	References	611
13	**Microwave Catalysis in Organic Synthesis**	**615**
	Milan Hájek	
13.1	Introduction	615
13.1.1	Definitions	616
13.2	Preparation of Heterogeneous Catalysts	617
13.2.1	Drying and Calcination	617
13.2.2	Catalyst Activation and Reactivation (Regeneration)	620
13.3	Microwave Activation of Catalytic Reactions	621
13.3.1	Reactions in the Liquid Phase	622
13.3.2	Reactions in the Gas Phase	628
13.3.3	Microwave Effects	634
13.3.4	Microwave Catalytic Reactors	641
13.4	Industrial Applications	645
	References	647
14	**Polymer Chemistry Under the Action of Microwave Irradiation**	**653**
	Dariusz Bogdal and Katarzyna Matras	
14.1	Introduction	653
14.2	Synthesis of Polymers Under the Action of Microwave Irradiation	653
14.2.1	Chain Polymerizations	654

14.2.2	Step-growth Polymerization 663
14.2.3	Miscellaneous Polymers 675
14.2.4	Polymer Modification 679
14.3	Conclusion 681
	References 682

15 Microwave-assisted Transition Metal-catalyzed Coupling Reactions 685
Kristofer Olofsson, Peter Nilsson, and Mats Larhed

15.1	Introduction 685
15.2	Cross-coupling Reactions 686
15.2.1	The Suzuki–Miyaura Reaction 686
15.2.2	The Stille Reaction 700
15.2.3	The Negishi Reaction 703
15.2.4	The Kumada Reaction 704
15.2.5	The Hiyama Reaction 705
15.3	Arylation of C, N, O, S, P and Halogen Nucleophiles 706
15.3.1	The Sonogashira Coupling Reaction 706
15.3.2	The Nitrile Coupling 707
15.3.3	Aryl–Nitrogen Coupling 708
15.3.4	Aryl–Oxygen Bond Formation 713
15.3.5	Aryl–Phosphorus Coupling 713
15.3.6	Aryl–Sulfur Bond Formation 715
15.3.7	Aryl Halide Exchange Reactions 716
15.4	The Heck Reaction 717
15.5	Carbonylative Coupling Reactions 719
15.6	Summary 721
	Acknowledgment 721
	References 722

16 Microwave-assisted Combinatorial and High-throughput Synthesis 726
Alexander Stadler and C. Oliver Kappe

16.1	Solid-phase Organic Synthesis 726
16.1.1	Introduction 726
16.1.2	Microwave Chemistry and Solid-phase Organic Synthesis 727
16.1.3	Peptide Synthesis and Related Examples 728
16.1.4	Resin Functionalization 729
16.1.5	Transition-metal Catalysis 734
16.1.6	Substitution Reactions 740
16.1.7	Multicomponent Chemistry 745
16.1.8	Microwave-assisted Condensation Reactions 747
16.1.9	Rearrangements 748
16.1.10	Cleavage Reactions 749
16.1.11	Miscellaneous 752
16.2	Soluble Polymer-supported Synthesis 756
16.3	Fluorous-phase Organic Synthesis 762

16.4	Polymer-supported Reagents	769
16.5	Polymer-supported Catalysts	778
16.6	Polymer-supported Scavengers	781
16.7	Conclusion	783
	References	784

17 Multicomponent Reactions Under Microwave Irradiation Conditions   788
Tijmen de Boer, Alessia Amore, and Romano V.A. Orru

17.1	Introduction	788
17.1.1	General	788
17.1.2	Tandem Reactions and Multicomponent Reactions	789
17.1.3	Microwaves and Multicomponent Reactions	791
17.2	Nitrogen-containing Heterocycles	793
17.2.1	Dihydropyridines	793
17.2.2	Pyridones	797
17.2.3	Pyridines	798
17.2.4	Dihydropyrimidinones	800
17.2.5	Imidazoles	802
17.2.6	Pyrroles	805
17.2.7	Other N-containing Heterocycles	807
17.3	Oxygen-containing Heterocycles	809
17.4	Other Ring Systems	812
17.5	Linear Structures	814
17.6	Conclusions and Outlook	816
	References	816

18 Microwave-enhanced Radiochemistry   820
John R. Jones and Shui-Yu Lu

18.1	Introduction	820
18.1.1	Methods for Incorporating Tritium into Organic Compounds	821
18.1.2	Problems and Possible Solutions	822
18.1.3	Use of Microwaves	824
18.1.4	Instrumentation	826
18.2	Microwave-enhanced Tritiation Reactions	827
18.2.1	Hydrogen Isotope Exchange	827
18.2.2	Hydrogenation	832
18.2.3	Aromatic Dehalogenation	833
18.2.4	Borohydride Reductions	834
18.2.5	Methylation Reactions	835
18.2.6	Aromatic Decarboxylation	836
18.2.7	The Development of Parallel Procedures	838
18.2.8	Combined Methodology	839
18.3	Microwave-enhanced Detritiation Reactions	841
18.4	Microwave-enhanced PET Radiochemistry	842
18.4.1	^{11}C-labeled Compounds	843

18.4.2	^{18}F-labeled Compounds	846
18.4.3	Compounds Labeled with Other Positron Emitters	852
18.5	Conclusion	854
	Acknowledgments	854
	References	855

19 Microwaves in Photochemistry 860

Petr Klán and Vladimír Církva

19.1	Introduction	860
19.2	Ultraviolet Discharge in Electrodeless Lamps	861
19.2.1	Theoretical Aspects of the Discharge in EDL	862
19.2.2	The Fundamentals of EDL Construction and Performance	863
19.2.3	EDL Manufacture and Performance Testing	865
19.2.4	Spectral Characteristics of EDL	866
19.3	Photochemical Reactor and Microwaves	869
19.4	Interactions of Ultraviolet and Microwave Radiation with Matter	877
19.5	Photochemical Reactions in the Microwave Field	878
19.5.1	Thermal Effects	878
19.5.2	Microwaves and Catalyzed Photoreactions	883
19.5.3	Intersystem Crossing in Radical Recombination Reactions in the Microwave Field – Nonthermal Microwave Effects	885
19.6	Applications	888
19.6.1	Analytical Applications	888
19.6.2	Environmental Applications	888
19.6.3	Other Applications	891
19.7	Concluding Remarks	891
	Acknowledgments	892
	References	892

20 Microwave-enhanced Solid-phase Peptide Synthesis 898

Jonathan M. Collins and Michael J. Collins

20.1	Introduction	898
20.2	Solid-phase Peptide Synthesis	899
20.2.1	Boc Chemistry	900
20.2.2	Fmoc Chemistry	901
20.2.3	Microwave Synthesis	905
20.2.4	Tools for Microwave SPPS	906
20.2.5	Microwave Enhanced N-Fmoc Deprotection	907
20.2.6	Microwave-enhanced Coupling	912
20.2.7	Longer Peptides	922
20.3	Conclusion	925
20.4	Future Trends	926
	Abbreviations	927
	References	928

21	**Application of Microwave Irradiation in Fullerene and Carbon Nanotube Chemistry** *931*	

Fernando Langa and Pilar de la Cruz

21.1	Fullerenes Under the Action of Microwave Irradiation	*931*
21.1.1	Introduction *931*	
21.1.2	Functionalization of Fullerenes *932*	
21.2	Microwave Irradiation in Carbon Nanotube Chemistry	*949*
21.2.1	Synthesis and Purification *949*	
21.2.2	Functionalization of Carbon Nanotubes *950*	
21.3	Conclusions *953*	
	References *954*	

22	**Microwave-assisted Extraction of Essential Oils** *959*	

Farid Chemat and Marie-Elisabeth Lucchesi

22.1	Introduction *959*
22.2	Essential Oils: Composition, Properties, and Applications *960*
22.3	Essential Oils: Conventional Recovery Methods *963*
22.4	Microwave Extraction Techniques *965*
22.4.1	Microwave-assisted Solvent Extraction (MASE) *965*
22.4.2	Compressed Air Microwave Distillation (CAMD) *968*
22.4.3	Vacuum Microwave Hydrodistillation (VMHD) *968*
22.4.4	Microwave Headspace (MHS) *969*
22.4.5	Microwave Hydrodistillation (MWHD) *969*
22.4.6	Solvent-free Microwave Hydrodistillation (SFME) *969*
22.5	Importance of the Extraction Step *970*
22.6	Solvent-free Microwave Extraction: Concept, Application, and Future *970*
22.6.1	Concept and Design *971*
22.6.2	Applications *974*
22.6.3	Cost, Cleanliness, Scale-up, and Safety Considerations *976*
22.7	Solvent-free Microwave Extraction: Specific Effects and Proposed Mechanisms *976*
22.7.1	Effect of Operating Conditions *976*
22.7.2	Effect of the Nature of the Matrix *977*
22.7.3	Effect of Temperature *979*
22.7.4	Effect of Solubility *980*
22.7.5	Effect of Molecular Polarity *981*
22.7.6	Proposed Mechanisms *982*
22.8	Conclusions *983*
	References *983*

Index *986*

Preface

The domestic microwave oven was a serendipitous invention. Percy Spencer was working for Raytheon, a company very involved with radar during World War II, when he noticed the heat generated by a radar antenna. In 1947 an appliance called a Radarange appeared on the market for food processing. The first kitchen microwave oven was introduced by Tappan in 1955. Sales of inexpensive domestic ovens now total many billions of dollars (euros) annually.

Numerous other uses of microwaves have been discovered in the recent years – essentially drying of different types of material (paper, rubber, tobacco, leather ...), treatment of elastomers and vulcanization, extraction, polymerization, and many applications in the food-processing industry.

The field of microwave-assisted organic chemistry is therefore quite young. The first pioneering publication was in 1981, the not very well known patent of Bhargava Naresh, from BASF Canada Inc., dealing with the use of microwave energy for production of plasticizer esters. Next, two famous papers from the groups of R. Gedye and R.J. Giguere appeared in 1986. These authors described several reactions completed within a few minutes when conducted in sealed vessels (glass or Teflon) in domestic microwave ovens. Although the feasibility of the procedure was thus apparent, a few explosions caused by the rapid development of high pressure in closed systems were also reported. To prevent such drawbacks, safer techniques were developed – reactions in open beakers or flasks, or solvent-free reactions, as developed since 1987 in France – in Orsay (G. Bram and A. Loupy), Caen (D. Villemin), and Rennes (J. Hamelin and F. Texier-Boullet). Combination of solvent-free procedures with microwave irradiation is an interesting and well-accepted approach within the concepts of green chemistry. This combination takes advantage both of the absence of solvent and of microwave technology under economical, efficient, and safe conditions with minimization of waste and pollution.

Approximately 2500 publications on microwave activation in organic synthesis have appeared in the literature. The spectacular growth of this technique is undoubtedly connected with the development of new, adapted, reactors enabling accurate control and reproducibility of the microwave-assisted procedures but also to the increasing involvement of industrial and pharmaceutical laboratories. This interest has been also been manifested by several reviews and books in the field including, very recently, the Wiley–VCH book by Oliver Kappe and Alexander Stadler

Microwaves in Organic Synthesis, Second edition. Edited by A. Loupy
Copyright © 2006 WILEY-VCH Verlag GmbH & Co. KGaA, Weinheim
ISBN: 3-527-31452-0

on Methods and Principles in Medicinal Chemistry. Many international congresses, all over the world, have been devoted to microwaves.

The objective of this book is to focus on the different fields of application of this technology in several aspects of organic synthesis. This second edition is a revised and enlarged version of the first. Eight new fields of application have been included. The chapters, which complement each other, are written by the most eminent scientists, all well-recognized in their fields.

After essential revision, and description of wave–material interactions, microwave technology, and equipment (Chapter 1), the development and design of reactors including scaling-up (Chapter 2) and the concepts of microwave-assisted organic chemistry in pressurized reactors (Chapter 3) are described. Special emphasis on the possible intervention of specific (not purely thermal) microwave effects is discussed in Chapter 4 and this is followed by up-to-date reviews of their effects on selectivity (Chapter 5). The next topics to be treated are coupling of microwave activation with phase-transfer catalysis (Chapter 6), ionic liquids (Chapter 7), mineral solid supports under "dry media" conditions (Chapter 8) and, more specifically, on graphite (Chapter 9). Applications in which microwave-assisted techniques have afforded spectacular results and applications are discussed extensively Chapters 10, 11, and 12 for heterocyclic chemistry, cycloadditions, and carbohydrate chemistry, respectively. Finally, the techniques have led to fruitful advances in microwave catalysis (Chapter 13) and have been applied to polymer chemistry (Chapter 14) and to organometallic chemistry using transition metal complexes (Chapter 15). Very promising techniques are now under intense development as a result of application of microwave irradiation to combinatorial chemistry (Chapter 16), multi-component reactions (Chapter 17), radiochemistry (Chapter 18), and photochemistry (Chapter 19). Other promising fields are solid-phase peptide synthesis (Chapter 20), fullerene and carbon nanotube chemistry (Chapter 21), and extraction of essential oils (Chapter 22).

I wish to thank sincerely all my colleagues and, nevertheless, (essentially), friends involved in the realization of this book. I hope to express to them, all eminent specialists, my friendly and scientific gratitude for agreeing to devote their competence and time to submitting and reviewing papers to ensure the success of this book.

Gif sur Yvette, February 22nd, 2006 *André Loupy*

About European Cooperation in COST Chemistry Programs

Microwave irradiation is a means of rapidly introducing energy into a chemical system in a manner different from the traditional methods of thermal heating. Other techniques can also be used to introduce energy into a system and, in this way, by accelerating reactions and also causing other more specific individual effects, both microwaves and ultrasonic irradiation provide means to achieve "Chemistry in High-Energy Microenvironments". Comparison of microwaves with ultrasound is the subject of COST Chemistry Action D32, which seeks to explore further uses of these methods individually, and also in combination. COST is the acronym for, in French "Co-opération Européenne dans le domaine de la Recherche Scientifique et Technique" or, in English, "European Co-operation in the Field of Scientific and Technical Research". COST started in 1992 and earlier initiatives D6 "Chemistry under Extreme and Non-Classical Conditions", and D10 "Innovative Transformations in Chemistry" involved pioneering studies of aspects of ultrasound and, more recently, microwaves in chemistry. Now D32, which started in 2004, has, as one of its objectives to take these techniques further.

Cost Actions are interdisciplinary and the objectives encourage the collaboration of chemists, biochemists and material scientists, and enable the transfer of experience between fundamentally orientated and applied research in these areas. COST Actions are split into working groups which have a cohesive theme within the overall objectives of the action. In D32 there are four microwave-based working groups involving collaboration between scientists with different expertise in modern technology:

- "Diversity oriented synthesis under (highly efficient) microwave conditions". The objectives of this working group are:
  - to apply solvent-less reaction systems and alternative reagents (those which are optimum for microwave activation) to already known reactions;
  - to perform organic reactions in aqueous media, and in this way substitute VOCs by water as the reaction media;
  - to evaluate multicomponent reactions under microwave and/or microreactor conditions to minimize waste and optimize atom-efficiency in comparison with conventional stepwise reactions under traditional heating conditions; and
  - to establish, in a combinatorial approach, novel atom economic multi-

Microwaves in Organic Synthesis, Second edition. Edited by A. Loupy

ISBN: 3-527-31452-0

component reactions assisted by microwave and/or microreactor technology to efficiently access pharmaceutically relevant N-heterocyclic compounds.

In such ways, microwaves can play an important role in the development of green chemistry. The use of microwaves for "green" reactions for industrial purposes can be accomplished by minimization of waste, replacement of hazardous and polluting materials, recycling of materials, and energy saving.

- "Microwave and High-intensity Ultrasound in the Synthesis of Fine Chemicals" The concern of this group is the application of high-intensity ultrasound (HIU) and microwave (MW), in combination, to the synthesis of fine chemicals, because these forms of energy can induce reactions that would be otherwise very laborious and bring about peculiar chemoselectivity. The synergic effect of these techniques can be exploited in such diverse fields as organic synthesis and sample preparation for analytical procedures. This results in challenging practicalities and technical hurdles, however, and the potential of this promising combination has yet to be investigated systematically. One objective of COST D32 is to bring experts in microwave-promoted reactions and in sonochemistry together for a joint effort. This is offering new opportunities of growth for both techniques, opening up new applications. Another important objective is to have, in each working group, as many scientists active in both fundamental and industrial research as possible, with collaboration that will lead to stimulating exchanges and innovative projects. Many important reactions shall be studied in unusual reaction media (water, ionic liquids, supercritical fluids) and under unusual conditions (sonication under pressure and HIU/MW combined).

- Finally, a significant issue that COST D32 intends to address is the challenge of scaling up results obtained in the laboratory. This is the theme of the third working group "Development and design of reactors for microwave-assisted chemistry in the laboratory and on the pilot scale". The aim of this working group is interdisciplinary comparative study of different fields of chemical technology and reactor design, to demonstrate the real advantages of microwave technology as a method for activation of chemical reactions and processes with a high radiation flow density (or high power density). The different concepts of technical microwave systems (monomode and/or multimode), different methods of temperature measurement (IR, fiber optics, metal sensors), and of energy entry (pulsed or unpulsed) will be compared and the advantages of the different options determined for some typical reactions. Safety issues (dielectric properties, temperature and power measurement, materials) will be addressed as also will risk management for the new technology. Scale-up for technical usefulness will be tested for different reactors and different types of reaction (heterogeneous catalysis, two liquid phases) and combination of microwaves with other forms of energy (UV–visible, ultrasound, plasma) will be examined, together with transfer from batch to continuous process, which is important for the economic use of microwave-assisted reactions. The energy efficiency of microwave-assisted reactions and processes will be compared with that of conventional reactions and processes.

- "Ultrasonic and Microwave-assisted Synthesis of Nanometric Particles". The objectives of this fourth working group are to use the techniques singly or in tandem to produce technologically-useful ceramics, metals, alloys, and other nanoparticulate systems for the wide range of applications for which these materials are being studied.

In summary, one set of objectives of COST D32 is to establish a firm EU base in microwave chemistry and to exploit the new opportunities provided by microwave techniques, singly or in appropriate combination, for the widest range of applications in modern chemistry.

Coventry (UK), February 22nd, 2006 *David Walton*

List of Authors

Alessia Amore
Vrije Universiteit
Department of Chemistry
Synthetic Bio-organic Chemistry
De Boelelaan 1083
1081 HV Amsterdam
The Netherlands

George Bashiardes
Université de Poitiers
40 avenue du Recteur Pineau
86022 Poitiers
France

Jean Pierre Bazureau
Université de Rennes 1
Sciences Chimiques de Rennes
UMR CNRS 6226
Bât. 10A
Campus de Beaulieu
Avenue du Général Leclerc
CS 74205
35042 Rennes cedex 1
France

Thierry Besson
Université de Rouen
UFR Médicine-Pharmacie
UMR CNRS 6014
22, bld Gambetta
76183 Rouen cedex 1
France

Tijmen de Boer
Vrije Universiteit
Department of Chemistry
Synthetic Bio-organic Chemistry
De Boelelaan 1083
1081 HV Amsterdam
The Netherlands

Dariusz Bogdal
Politechnika Krakowska
Department of Chemistry
ul. Warszawska 24
31-155 Krakow
Poland

Werner Bonrath
DSM Nutritional Products
R&D Chemical Process Technology
PO Box 3255
4002 Basel
Switzerland

Khalid Bougrin
Université Mohammed V
Faculté des Sciences
Laboratoire de Chimie des Plantes et de
Synthèse Organique et Bio-organique
Avenue Ibn Batouta
BP 1014
Rabat
Morocco

Farid Chemat
Université d'Avignon
Sécurité et Qualité
des Produits d'Origine
Végétale
33, rue Louis Pasteur
84029 Avignon cedex 1
France

Ugo Chiacchio
Università di Catania
Dipartimento di Scienze Chimiche
viale A. Doria 6
95125 Catania
Italy

List of Authors

Vladimír Církva
Academy of Sciences of the Czech Republic
Institute of Chemical Process Fundamentals
Rozvojova 135
165 02 Prague
Czech Republic

Jonathan M. Collins
CEM Corporation
PO Box 200
3100 Smith Farm Road
Matthews
NC 28106-0200
USA

Michael J. Collins
CEM Corporation
PO Box 200
3100 Smith Farm Road
Matthews
NC 28106-0200
USA

Antonino Corsaro
Università di Catania
Dipartimento di Scienze Chimiche
viale A. Doria 6
95125 Catania
Italy

Pilar de la Cruz
Universidad de Castilla-La Mancha
Departamento de Química Orgánica
Facultad de Ciencias del Medio Ambiente
45071 Toledo
Spain

Angel Díaz-Ortiz
Universidad de Castilla-La Mancha
Departamento de Química Orgánica
Facultad de Química
13071 Ciudad Real
Spain

Jacques Dubac
2 Chemin du Taur
31320 Pechbusque
France

Milan Hájek
Academy of Sciences of the Czech Republic
Institute of Chemical Process Fundamentals
Rozvojová 135
165 02 Prague 6
Suchdol
Czech Republic

Jack Hamelin
Université de Rennes 1
Sciences Chimiques de Rennes
UMR CNRS 6226
Bât. 10A
Campus de Beaulieu
Avenue du Général Leclerc
CS 74205
35042 Rennes cedex 1
France

Antonio de la Hoz
Universidad de Castilla-La Mancha
Departamento de Química Orgánica
13071 Ciudad Real
Spain

John R. Jones
University of Surrey
School of Biomedical and Molecular Sciences
Department of Chemistry
Guildford
Surrey GU2 7XH
UK

Yuhong Ju
US Environmental Protection Agency
National Risk Management Research
Laboratory
Clean Processes Branch
26 West Martin Luther King Drive
MS 443
Cincinnati, OH 45268
USA

C. Oliver Kappe
Karl-Franzens-University
Institute of Chemistry
Heinrichstrasse 28
8010 Graz
Austria

Petr Klán
Masaryk University
Faculty of Science
Department of Organic Chemistry
Kotlářská 2
611 37 Brno
Czech Republic

Fernando Langa
Universidad de Castilla-La Mancha
Departamento de Química Orgánica
Facultad de Ciencias del Medio Ambiente
45071 Toledo
Spain

List of Authors | XXV

Mats Larhed
Uppsala University
Department of Medicinal Chemistry
Organic Pharmaceutical Chemistry
BMC
Box 574
75123 Uppsala
Sweden

Nicholas E. Leadbeater
University of Connecticut
Department of Chemistry
55 North Eagleville Road
Storrs, CT 06269-3060
USA

André Loupy
29 Allée de la Gambeauderie
91190 Gif sur Yvette
France

Shui-Yu Lu
National Institutes of Health
National Institute of Mental Health
Molecular Imaging Branch
Building 10. Room B3 C346
10 Center Drive, MSC 1003
Bethesda, MD 20892-1003
USA

Marie-Elisabeth Lucchesi
Université de la Réunion
Naturelles et de Science des Aliments
Laboratoire de Chimie des Substances
BP 7151
15 rue René Cassin
97715 Saint Denis Messageries cedex 9
France

Katarzyna Matras
Politechnika Krakowska
Department of Chemistry
ul. Warszawska 24
31-155 Krakow
Poland

Florence Mongin
Université de Rennes 1
Institut de Chimie, Synthèse and
Electrosynthèse Organiques 3
UMR 6510
Bât. 10A
Campus de Beaulieu
Avenue du Général Leclerc
CS 74205
35042 Rennes cedex 1
France

Andrés Moreno
Universidad de Castilla-La Mancha
Facultad de Química
Departamento de Química Orgánica
14071 Ciudad Real
Spain

Thach Le Ngoc
National University of Hochiminh City
University of Natural Sciences
Department of Organic Chemistry
227 Nguyen Van Cu Str., Dist. 5
Hochiminh City
Vietnam

Peter Nilsson
Biolipox
Medicinal Chemistry
Berzelius väg 3, plan 5
17165 Solna
Sweden

Kristofer Olofsson
Biolipox
Medicinal Chemistry
Berzelius väg 3, plan 5
17165 Solna
Sweden

Bernd Ondruschka
Friedrich Schiller University of Jena
Department of Technical Chemistry and
Environmental Chemistry
Lessingstr. 12
07743 Jena
Germany

Romano V.A. Orru
Vrije Universiteit
Department of Chemistry
Synthetic Bio-organic Chemistry
De Boelelaan 1083
1081 HV Amsterdam
The Netherlands

Laurence Perreux
Université Paris-Sud
ICMMO
Laboratoire des Réactions Sélectives sur
Supports
bât 410
91405 Orsay cedex
France

List of Authors

Alain Petit
Université Paris-Sud
ICMMO
Laboratoire des Réactions Sélectives sur Supports
bâtiment 410,
91405 Orsay cedex
France

Venerando Pistarà
Università di Catania
Dipartimento di Scienze Chimiche
viale A. Doria 6
95125 Catania
Italy

Brett A. Roberts
Monash University
ARC Special Research Centre for Green Chemistry
Clayton Campus
Victoria 3800
Australia

Giovanni Romeo
Università di Messina
Dipartimento Farmaco-Chimico
viale SS. Annunziata 6
98168 Messina
Italy

Mohamed Soufiaoui
Université Mohammed V
Faculté des Sciences
Laboratoire de Chimie des Plantes et de Synthèse Organique et Bio-organique
Avenue Ibn Batouta
BP 1014
Rabat
Morocco

Alexander Stadler
Anton Paar GmbH
Anton-Paar Strasse 20
8054 Graz
Austria

Christopher R. Strauss
CSIRO Molecular and Health Technologies
Private Bag 10
Clayton South
Victoria 3169
Australia

Didier Stuerga
University of Burgundy
GERM/LRRS and Naxagoras Technologie
BP 47870
21078 Dijon cedex
France

Françoise Texier-Boullet
Université de Rennes 1
Institut de Chimie, Synthèse and Electrosynthèse Organiques 3
UMR 6510
Bât. 10A
Campus de Beaulieu
Avenue du Général Leclerc
CS 74205
35042 Rennes cedex 1
France

Valérie Thiéry
Université de La Rochelle
UFR Sciences Fondamentales et Sciences pour l'Ingénieur, LBCB, FRE CNRS 2766
Avenue Marillac, Bâtiment Marie Curie
17042 La Rochelle cedex 1
France

Hanna M. Torenius
King's College London
Strand
London WC2R 2LS
UK

Rajender S. Varma
US Environmental Protection Agency
National Risk Management Research Laboratory
Clean Processes Branch
26 West Martin Luther King Drive
MS 443
Cincinnati, OH 45268
USA

11
Microwaves in Cycloadditions

Khalid Bougrin, Mohamed Soufiaoui, and George Bashiardes

11.1
Cycloaddition Reactions

Cycloadditions are particularly useful reactions and are the most effective methods for generating cyclic and heterocyclic compounds from unsaturated precursors by forming two bonds in a single operation [1]. As such, they are extremely useful for construction of many natural or biologically active molecules.

The introduction of microwave-assisted chemistry has had a significant impact on synthetic chemistry. Reductions in reaction time, increases in yield, and suppression of side-product formation have been reported for microwave conditions compared with conventional thermal heating [2]. Although the basis of these practical benefits remains speculative, the preparative advantages are evident and have motivated a large and continuing survey of nearly all classes of thermal reaction for improvement by use of microwave activation. This exploration has been extended to a variety of topics in synthetic organic chemistry [3] including cycloaddition reactions, and numerous examples have been summarized in previous review articles and book chapters [3]. Conventional cycloaddition reactions often need the use of harsh conditions, for example high temperatures, toxic solvents, and long reaction times; such conditions often lead to decomposition of reagents and products and polymerization of the diene or dienophile in Diels–Alder cycloadditions. In a great majority of examples, all these problems have been readily solved successfully by use of microwave irradiation [4]. In this chapter, we review the field of [4+2] and [3+2] cycloadditions, the major topics of interest in the study of the effect of microwave activation, and we will discuss some of the underlying phenomena and questions involved. A few [2+2] and other cycloadditions, for example [2+2+1], [6+4], and tandem [6+4]–[4+2] cycloadditions will also be described herein. Few examples of cycloaddition reactions have been described in which changes in selectivity have been observed as a result of the use of microwave irradiation. Indeed, in concerted processes, the regio and stereoselectivity of the reaction is governed by frontier orbital interactions, so microwaves are not expected to affect the selectivity unless a change in the reaction mechanism occurs. When

Microwaves in Organic Synthesis, Second edition. Edited by A. Loupy
Copyright © 2006 WILEY-VCH Verlag GmbH & Co. KGaA, Weinheim
ISBN: 3-527-31452-0

chemo, regio, or stereoselectivity is modified as a result of the use of microwaves, possible explanations of this behavior will be given.

The objective of this chapter is to emphasize some of the most recent applications and trends in microwave cycloaddition reactions, and to discuss the impact and future potential of this technology.

11.2
Reactions with Solvent

The first experimental microwave-induced reactions were cycloadditions performed with solvent under pressure [5]. Reactions were performed in sealed, thick-walled glass tubes or in Teflon acid-digestion vessels, in domestic microwave ovens [6]. Elevated temperatures are developed and the solvents rapidly reach their boiling points, because of energy transfer between the polar molecules (or polar solvent) and the microwave radiation [2b, 7]. In the absence of temperature and pressure controls in these systems, however, safety problems become a major issue, because of overpressure resulting from the rate of heating caused by microwaves.

Such problems have been reduced, first by working under atmospheric pressure with nonpolar solvents in open vessels [8] and, subsequently, under polar solvent reflux in modified commercial microwave ovens or by monomode reactors specially designed for chemical synthesis. Several such reactors are available commercially from manufacturers such as Prolabo [3f], Milestone [9], Biotage AB [10], CEM [11], and Anton–Paar [12]. These products are equipped with temperature control within the reaction media, by means of optical fibers, or on the surface of the mixture, by means of infrared detection, enabling good temperature control by regulation of microwave power output (15–300 W). Most often, monitoring of the reaction is performed by software, by means of which it is possible to program power and temperature conditions. Monomode reactors lead to considerable improvements in yields of organic synthesis by preserving thermal stabilities of products with low emitted power and good temperature homogeneity [3f].

These problems can also be reduced by using solvent-free methodology, which also enables the use of larger quantities of reagents. Yields have been greatly improved, and reaction times reduced, compared with conventional procedures in solvents under reflux [13]. One advantage of reactions using solvent could be that the reaction temperature is controlled by the reflux temperature [14]. Depending on the solvent employed, when heating by microwave irradiation at atmospheric pressure in an open vessel the reaction temperature is typically limited to the boiling point of the solvent. In the absence of any specific or nonthermal microwave effects (for example the superheating effect at atmospheric pressure, when the boiling temperature can be up to 40 °C above the normal boiling point of the polar solvent [15]) the expected rate enhancements would be comparatively small. To achieve high reaction rates, however, high-boiling, microwave-absorbing (polar) solvents such as DMSO, 1,2-dichlorobenzene (DCB), N-methyl-2-pyrrolidine

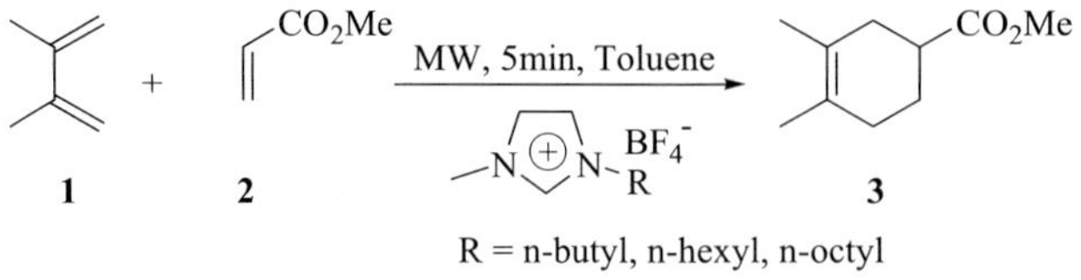

Scheme 11.1

R = n-butyl, n-hexyl, n-octyl

(NMP), or ethylene glycol have frequently been used in open-vessel microwave synthesis [16, 17].

A significant application of this procedure has been described for [60]fullerene cycloaddition under microwave conditions by Langa et al. [17]. More recently, Chi et al. have applied the concept to the Diels–Alder reaction of [60]fullerene with o-quinodimethane derivatives, generated *in situ* from 4,5-benzo-3,6-dihydro-1,2-oxathiine-2-oxide derivatives (thienosultines) under reflux in 1,2-dichlorobenzene solution; the reaction was highly accelerated by microwave irradiation giving comparable yields of the mono and bis cycloadducts [18].

A very recent addition to the already powerful range of microwave cycloaddition chemistry is the development of a general procedure applying a catalyst/ionic liquid system [19]. Several studies in this area have used ionic liquids, or mixtures of ionic liquids and other solvents, as reaction media in several important microwave-heated organic syntheses [20], including Diels–Alder reactions [21, 22] and 1,3-dipolar cycloaddition reactions [23].

Some ionic liquids are soluble in nonpolar organic solvents and can therefore be used as microwave coupling agents when microwave-transparent solvents are employed. For example, in Diels–Alder reactions, when adding ionic liquids to toluene, the temperature can reach 195 °C within 150 s of irradiation in contrast to 109 °C without ionic liquids [24]. Leadbeater et al. used this method to increase the rate of the Diels–Alder reaction (Scheme 11.1) (see Chapter 7 of this book).

A key feature of this catalyst/ionic liquid system is its recyclability [21c, 25]. Because ionic liquids can be very costly to use as solvents, several research groups use them instead as "doping agents" for microwave heating of otherwise nonpolar solvents, for example hexane, toluene, THF, or dioxane. This technique, first introduced by Ley et al. in 2001 [26] is becoming increasingly popular, as demonstrated by many recently published examples [21b, 24, 27].

11.3
Reactions under Solvent-free Conditions

To avoid pollution and high cost, solvent-free methods are of great interest to improve on conventional procedures and render them cleaner, safer, and easier to perform. Several advantages of this approach in cycloaddition reactions are described in reviews by de la Hoz [3j] and Bougrin [13c]. This interest in solvent-free processes is increased by noticeable increases in reactivity and selectivity. Comprehensive reviews of these techniques are presented in the literature [13c, 28]. They can be further improved by taking advantage of microwave activation as an advantageous alternative to conventional heating under safe and efficient conditions, again resulting in enhanced yields and reaction times.

Solvent-free techniques hold a strategic position in environmentally improved processes, because solvents are very often toxic, costly, and difficult to use and remove. These approaches also enable experiments to be run without strong mineral acids by replacing them with montmorillonites (acidic clays) thus avoiding causes of corrosion, unsafe manipulations, and pollution by waste. Indeed, such improvements can enable safe use of domestic microwave ovens and standard open-vessel technology. Recent processing techniques have employed microwave-assisted solvent-free (or "dry-media") procedures in which reactions are performed between the neat reagents or preadsorbed either on variably microwave-transparent supports (silica, alumina, zeolite, or clay) [3e–f, 29] or on strongly absorbing (e.g. graphite) [30] inorganic supports which can also be doped with a catalyst or reagent (see Chapter 9 of this book).

Although many interesting transformations with "dry-media" reactions, including cycloadditions, have been published in the literature [13c, 31], technical difficulties relating to nonuniform heating, mixing, and the precise determination of the reaction temperature remain unsolved, in particular when scale-up issues need to be addressed. In addition, phase-transfer catalysis (PTC) solvent-free conditions have rarely been applied to cycloaddition reactions [4c].

Nonetheless, microwave-assisted solvent-free conditions have wide applications in industrial processes and enable classification of microwave chemistry as an environmentally benign method, or "green chemistry" [3e–g, 28, 32].

11.3.1
Reaction on Mineral Supports

Solid mineral-supported organic synthesis, coupled with solvent-free microwave irradiation has several advantages compared with classical procedures in solution. The reactants are impregnated as neat liquids on solid supports such as aluminas, silicas, zeolites, and clays, or, for solids, as solutions in appropriate organic solvent, which is subsequently removed. Reaction in "dry media" is performed between individually impregnated reactants, possibly followed by heating. At the end of the reaction, organic products are simply removed by extraction with an appropriate solvent. Considering the small quantities of solid support used, microwave absorption is considered insignificant, thus all the energy is absorbed by the reagents. This results in efficient thermal activation, thus dramatically reducing reaction times (from days and hours to minutes and seconds), and higher yields and purity of the reaction products. Occasionally, however, changes in the selectivity of the cycloaddition reactions have been observed (see Chapter 5 of this book).

Alumina, silica, clay, and zeolites are increasingly bring used as acidic or basic supports in organic synthesis [3f, 13c, 33]. Several groups have reported improve-

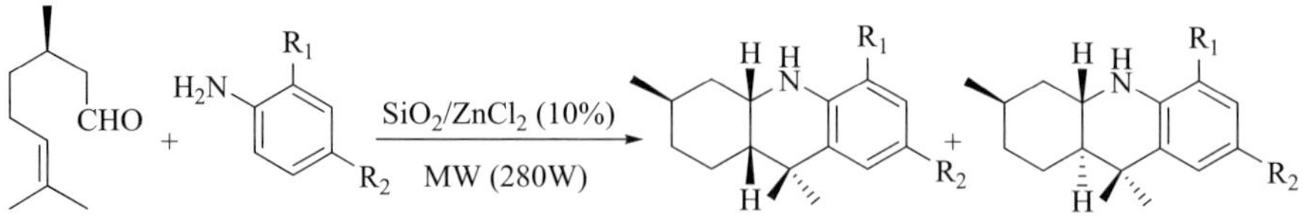

Scheme 11.2

	Microwave	Conventional heating (40–45°C)
Time	3 min	80 min
Yields	78%	79%
cis:trans	1:1	1:1

($R_1=R_2=H$)

ments in [4+2] cycloadditions when using clay [13c, 34], silica [13c, 35, 36], or alumina [13c, 37] under solvent-free microwave-assisted conditions. In one recent example Jacob et al. reported a microwave-induced one-pot intramolecular N-arylimino Diels–Alder reaction without solvent, catalyzed by silica gel impregnated with $ZnCl_2$. This economical ("green") method was used in the direct synthesis of the cis:trans octahydroacridines **6** in good yields from (+)-citronellal and arylamines, with short reaction times (Scheme 11.2) [36].

Several research groups have used solid supports such as alumina [38] and silica gel [39] to generate 1,3-dipoles, which undergo cycloadditions with alkene or alkyne dipolarophiles to furnish five-membered heterocyclic compounds in a one-pot procedure. Use of microwaves has occasionally affected selectivity, changing the chemo or regioselectivity of the reactions compared with conventional heating. Several reports on [2+2] cycloadditions in "dry media" using solid supports have been published in the literature [40].

11.3.2
Reaction with Neat Reactants

Organic reactions under microwave conditions, including cycloadditions, frequently involve solvent-free "dry media" procedures. The reaction is performed between neat reagents, at least one of which must be a liquid and polar substance. These are liquid–liquid or liquid–solid systems; the latter implies the solid is soluble in the liquid phase or at least that the liquid counterpart is adsorbed on the solid, so reaction occurs at the interface [3e, 32a]. Because of the absence of solvent, the radiation is absorbed directly by the reagents, so the effect of the microwaves is more marked. As a result, high overall yields and purity of the desired products are usually obtained more quickly.

De la Hoz et al. describe several of the advantages of microwave irradiation by this method in Diels–Alder and 1,3-dipolar reactions of ketene acetals [41]. A specific improvement is the absence of polymerization of the ketene acetals. The same

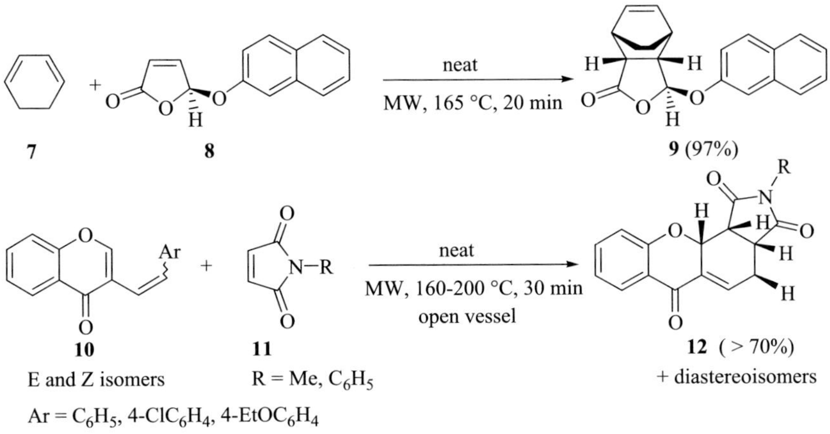

Scheme 11.3

research group has recently reported numerous examples of Diels–Alder reactions and 1,3-dipolar cycloadditions [3j].

Using these procedures, the most important advantages are simplification of reaction work-up, because the products are isolated directly from the crude reaction mixture. In addition, rates of neat reactions are markedly accelerated compared with microwave activation with solvents or classical reflux conditions with the same solvents. The reduced reactivity observed using solvent can be associated to the limitation of the reaction temperature reaction (at maximum, the boiling point of the solvent) and the dilution.

Two recent examples of Diels–Alder cycloadditions performed by microwave dielectric heating are described in Scheme 11.3. In both reactions diene and dienophile were employed neat, i.e. without the addition of solvent. The reaction described by Trost and Crawley between compounds **7** and **8** (irradiation for 20 min at 165 °C or for 60 min at 150 °C) produced the cycloadduct **9** in nearly quantitative yield [42]. In the process reported by de la Hoz et al., open-vessel irradiation of 3-(2-arylethenyl)chromones **10** with maleimides **11** at 160–200 °C for 30 min furnished tetracyclic adducts of type **12** with minor amounts of other diastereoisomers [43].

11.3.2.1 Catalyst Addition

Homogeneous or heterogeneous, solvent-free, catalyzed reactions are some of the most important examples of microwave organic synthesis [44]. Use of catalysts in conjunction with microwaves may have significant advantages over conventional heating methods. The reagents and/or catalysts are likely to be polar, and the overall dielectric properties of the reaction medium will usually enable sufficient heating by microwaves. Phase-transfer catalysis (PTC) under solvent-free conditions, in particular, have also been used as a processing technique in microwave-assisted organic synthesis (MAOS) [3g, 45].

Catalyzed reactions under microwave, solvent-free conditions are usually conducted with at least one liquid reactant or catalyst, although few applications of this technique to cycloaddition reactions have appeared in the literature [4c, 46]. Nonetheless, two examples of 1,3-dipolar cycloaddition reactions successfully performed under microwave-assisted solvent-free catalyzed conditions are illustrated in Scheme 11.4 [4c, 47]. In the first example, Bougrin achieved the first practical [3+2] cycloaddition of diphenylnitrilimine (DPNI) to dipolarophiles **14** in dry media under microwave irradiation conditions. The cycloaddition reaction between **13** and **14** was performed by microwave irradiation (power = 30 W, Synthewave 402 monomode reactor) under solvent-free, solid–liquid phase-transfer catalysis conditions. The optimized conditions involved the use of 0.15 equiv. Aliquat 336 as phase-transfer catalyst and 1.5 equiv. KF as base. DPNI is formed *in situ* and reacts in a one-pot procedure (Scheme 11.4). The best yields (56–68%) for adducts **15** were achieved in very short reaction times (3–5 min) using controlled microwave exposure at 120–161 °C. Importantly, similar yields were also obtained when the cycloaddition reactions were performed in a preheated oil bath at temperatures identical with those used under microwave irradiation, but the reaction times were much longer (20 h). With conventional heating at the same high temperature and short reaction time conditions as with the microwaves, no reaction was observed, clearly indicating the involvement of specific microwave effects [3j]. In addition, the reaction mixtures obtained from the microwave reactions towards **15** were clean. The regioselectivity for **I** and **II** was unchanged.

In the second example, Hamelin et al. [47] reported the first microwave-assisted solvent-free catalyzed conditions for generation of nitrile oxide intermediates **17**. In this work, a mixture of methyl nitroacetate **16** (as 1,3-dipole precursor), dimethyl acetylene dicarboxylate (DMAD) **18** (as dipolarophile), and *p*-toluene sulfonic acid (PTSA) as catalyst (10% w/w) was irradiated neat for 30 min. The reaction was performed in an open vessel from which water was continuously removed [48]. In this manner, the desired isoxazole adduct **19** was obtained in excellent yield (91%).

11.3.2.2 Heat Captor Addition

Graphitized carbon is among the solids most efficiently heated by microwaves and is also known for its good absorption of organic molecules [30]. The excellent capacity of graphite to absorb microwave radiation [49] can be explained by the partly delocalized electron system, the high electrical internal resistance resulting from the pronounced pore structure, and, indeed, by the dielectricity of the substance. Thus, graphitized carbon is a chemically inert support that couples strongly with microwaves by a conduction process [49a]. Numerous microwave-assisted graphite-supported reactions have emerged as useful methods for the construction of small, medium, and polycyclic ring systems, including cycloaddition reactions [49b]. In this context, several groups, in particular those of Dubac and Besson, have described numerous examples which are summarized in several articles [50] and a book chapter (Ref. [30] and Chapter 9 in this book).

In the past few years, Garrigues et al. have reported two papers dealing with graphite-supported, solvent-free "dry media" conditions as a powerful technique

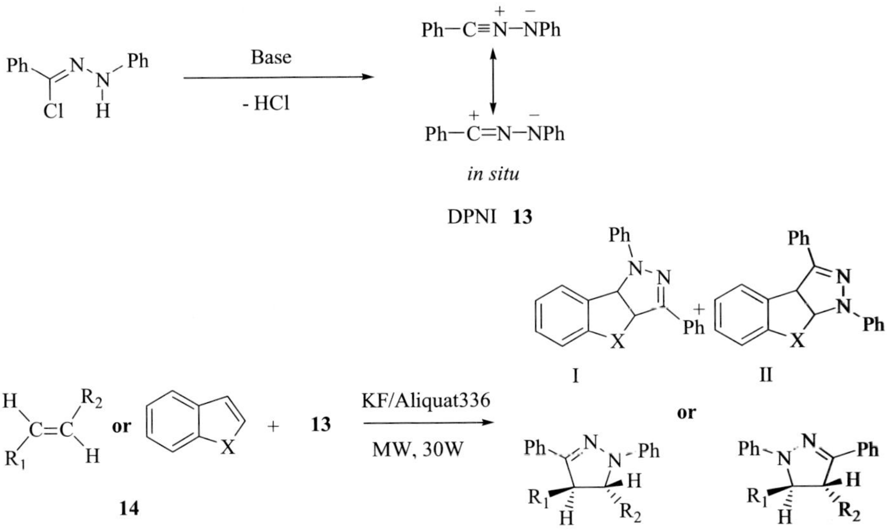

11.3 Reactions under Solvent-free Conditions | 531

DPNI 13

14
14a; $R_1 = C_6H_5CO$, $R_2 = C_6H_5$
14b; $R_1 = 4\text{-}ClC_6H_4CO$, $R_2 = C_6H_5$
14c; $R_1 = n\text{-}C_4H_9CO_2$, $R_2 = H$
14d; $X = CH_2$
14e; $X = SO_2$

15 (mixture of regioisomers : I and II)

Adducts	Reaction time (min)	Final Temperature T (°C)	Yield (%)	Ratio I/II
15a	5	160	61	91/9
15b	4	161	59	93/7
15c	4	120	56	100/0
15d	3	126	68	93/7
15e	4	153	67	100/0

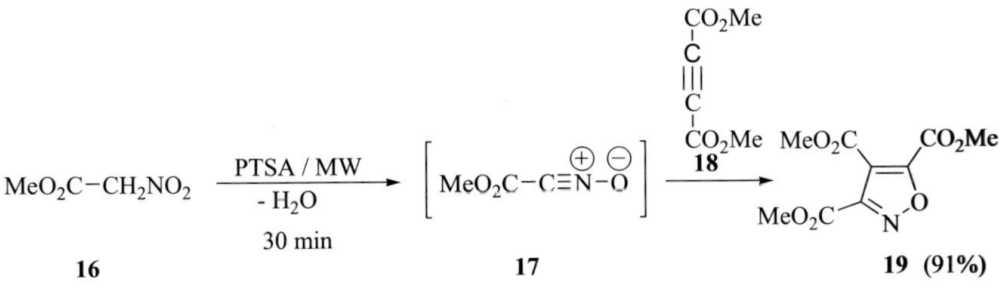

Scheme 11.4. 1,3-Dipolar cycloadditions for synthesis of pyrazolines **15** [4c] or isoxazoles **19** [47].

in hetero and retro Diels–Alder reactions [51]. In the first, Diels–Alder reactions between dienophile and diene supported on graphite were reported as dramatically accelerated under the action of microwave irradiation [52]. For example, reaction of **20** and **21** under microwave irradiation conditions (power = 30 W) for 10 × 1 min

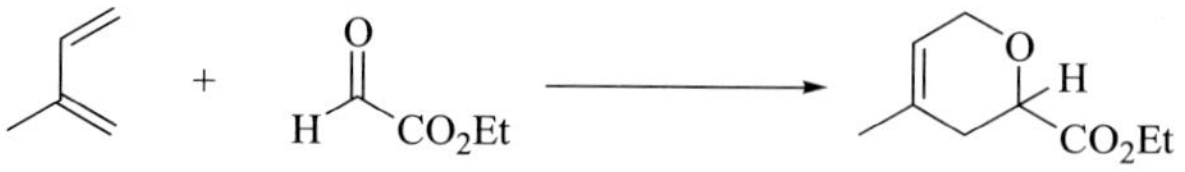

Scheme 11.5

MW	Graphite / 5% $ZnCl_2$	P=30W/ 1min x 10/ 146°C	73% (22)
Classical conditions	CH_2Cl_2 / 5% $SnCl_4$	18h / 20°C	10% (22)

Scheme 11.5

at 146 °C in the presence of 5% $ZnCl_2$ at atmospheric pressure in an open vessel, provided cycloadduct 22 in 73% yield (Scheme 11.5). Under conventional conditions the same cycloaddition reaction had to be conducted at 20 °C in dichloromethane for 18 h in the presence of 5% $SnCl_4$ to afford 22 in poor yield (10%) [52]. Importantly, the authors also described changes in regio and chemoselectivity under microwave conditions compared with classical conditions [52].

In the second reference, under similar conditions, the combination of graphite and microwave irradiation between "dry media" conditions was successfully used for cycloaddition of anthracene to several C–C dienophiles and carbonyl compounds in hetero Diels–Alder reactions [53]. This method led to a shortening of reaction times (3 × 1 min) and enabled work under nitrogen in open vessels. After sequential irradiation at low power (30 W) and at reaction temperatures in the range of 130–270 °C the adducts were obtained in good to excellent yields (75–97%). Evaporation of volatile reagents was, moreover, avoided, probably because of their retention or confinement on the graphite. Importantly, when using more powerful irradiation (90 or 120 W) the adducts decomposed in a retro Diels–Alder reaction at higher temperatures (301–421 °C) [53].

11.4
[4+2] Cycloadditions

The [4+2] cycloaddition is certainly one of the most important reactions in organic chemistry, and many books and reviews are dedicated to this topic [54]. In particular, Diels–Alder reaction of (hetero)dienes with (hetero)dienophiles is extensively used at the early stages of numerous syntheses to establish a structural scaffold which is then usually further elaborated toward more complex target structures. The [4+2] cycloaddition is usually an efficient method with predictably high regio and stereoselectivity; as such it can enable the synthesis of highly functionalized polycyclic systems.

According to the literature some Diels–Alder reactions occur spontaneously whereas others need catalysis, pressure, and/or heat. Because of its usefulness, investigators are encouraged to develop further improvements, often by using new activation methods [55]. One such approach is the use of microwave irradiation as an unconventional activation technique to improve reactions involving (hetero)dienophiles and (hetero)dienes of low reactivity. Thus, many classical Diels–Alder reactions have been greatly improved upon by use of microwaves, with reduced reaction times and increased yields. Examples of Diels–Alder reactions were most recently reviewed by de la Hoz et al. [3j].

In Section 11.4.1 some characteristic examples are discussed, illustrating the two major series of [4+2] cycloadditions, namely inter and intramolecular Diels–Alder reactions, that have been performed successfully under microwave conditions.

11.4.1
Intramolecular Hetero and Diels–Alder Reactions

The intramolecular Diels–Alder (IMDA) or hetero Diels–Alder (IMHDA) cycloaddition reactions are often the key step in the synthesis of functionalized bicyclic bridged and polycyclic systems with high regio and stereoselectivity. The application of microwave irradiation to IMDA and IMHDA reactions may have significant advantages compared with conventional heating methods. Early examples of microwave activation in these reactions for synthesis of polycyclic frameworks, have been extensively reviewed by de la Hoz and will not be discussed herein [3j].

In many of the more recent reports of IMDA and IMHDA cycloadditions, the reactions are performed at high temperature under microwave activation. Different conditions have been used, including closed-vessel processes [56–60] with highly polar solvents or ionic liquids as "doping agents" in nonpolar solvents, or indeed, microwave irradiation under open-vessel reflux conditions [61–63, 36].

Intramolecular cycloaddition of furan has been performed successfully on a solid support in the presence of solvent under open-vessel or sealed-vessel microwave irradiation conditions. Whereas intramolecular reaction of furan **23** does not occur with classical heating [61], the reaction was performed susessfully in 64% yield by using microwave activation (Scheme 11.6).

Mance and Jakopčić have recently described the intramolecular Diels–Alder reaction (IMDA) of the furanic tertiary amine **25** in benzene under microwave irradiation conditions in sealed vessels [57]. Irradiation for 5 min at 140 °C afforded IMDA product **26** in 46% yield with little decomposition. Under classical conditions, heating a benzene solution of the amine **25** at 80 °C for 72 h under a nitrogen atmosphere, produced the IMDA product **26** in lower yield (12%) (Scheme 11.7).

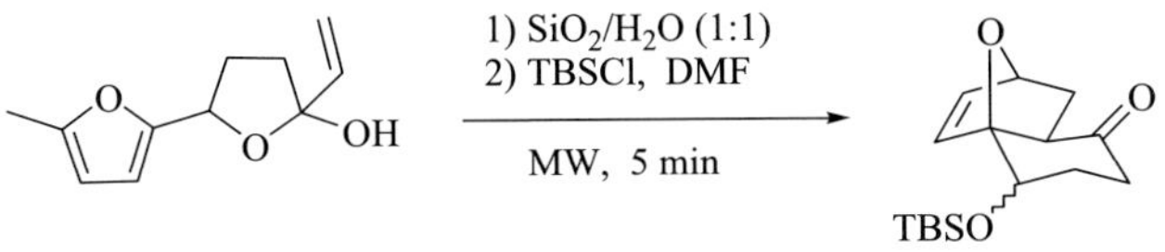

Scheme 11.6

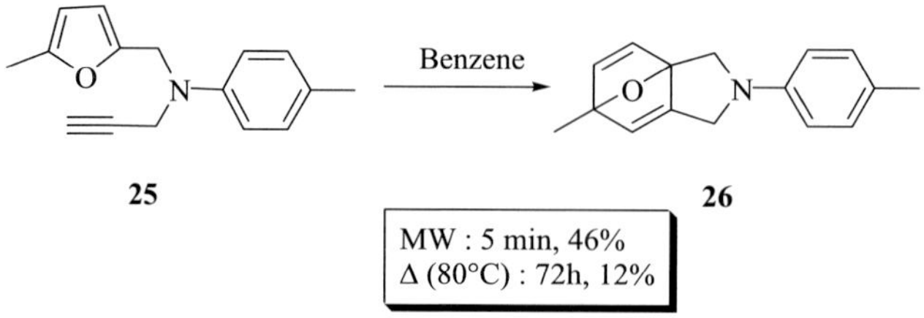

Scheme 11.7

[Structure 25] → Benzene → [Structure 26]

MW : 5 min, 46%
Δ (80°C) : 72h, 12%

The IMHDA reaction can be performed efficiently by employing ionic liquids either as solvents or as additives in conjunction with nonpolar solvents under microwave irradiation conditions. Ionic liquids can provide an ideal medium for IMHDA reactions, which involve reactive ionic intermediates. Because of the stabilization of ionic or polar reaction intermediates, ionic liquids can afford enhanced selectivities and reaction rates [3k, 64].

The hetero Diels–Alder reaction of a series of functionalized 2(1H)-pyrazinones was studied in detail by Van der Eycken et al. [58, 65]. For example, in a series of intramolecular cycloadditions of alkenyl-tethered 2(1H)-pyrazinones **27** the reaction required 1–2 days under conventional thermal conditions (chlorobenzene, reflux, 132 °C) whereas use of 1,2-dichloroethane doped with the ionic liquid 1-butyl-3-methylimidazolium hexafluorophosphate (bmimPF$_6$) and use of microwaves up to a temperature of 190 °C (sealed vessels) enabled the same transformations to be completed within 8–15 min. The primary imidoyl chloride cycloadducts were not isolated, but were rapidly hydrolyzed under the action of microwaves by addition of small amounts of water (130 °C, 5 min). The overall yields of **28** were in the same range as reported for the conventional thermal procedures (Scheme 11.8) [58].

Snyder et al. reported the use of NH$_4$OAc acting as an energy-transfer agent, surprisingly, to perform the intramolecular Diels–Alder cycloaddition depicted in Scheme 11.9, in o-dichlorobenzene under pressure. Cycloaddition of the imidazole/triazine-tethered partners **29** was complete within 20 to 30 min under the action of microwave irradiation at 210–225 °C, producing the pure cycloadducts **30** in 56–85% yields. By the conventionally heated process (under reflux in diphenyl ether at 259 °C) the yields were comparable but required 1.5–5 h to

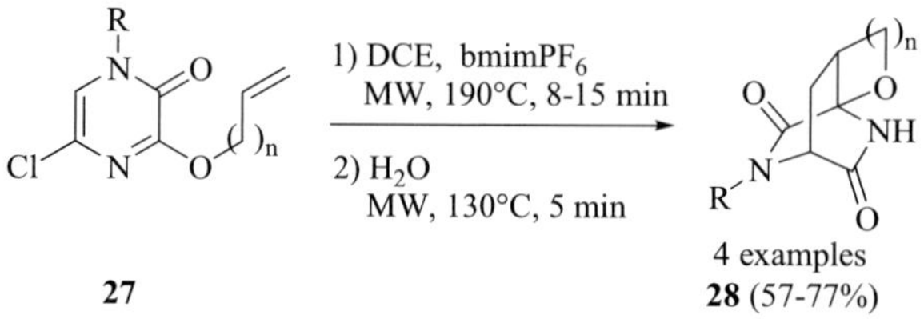

[Structure 27]
1) DCE, bmimPF$_6$
   MW, 190°C, 8-15 min
2) H$_2$O
   MW, 130°C, 5 min
→ [Structure 28]

4 examples
28 (57-77%)

Scheme 11.8

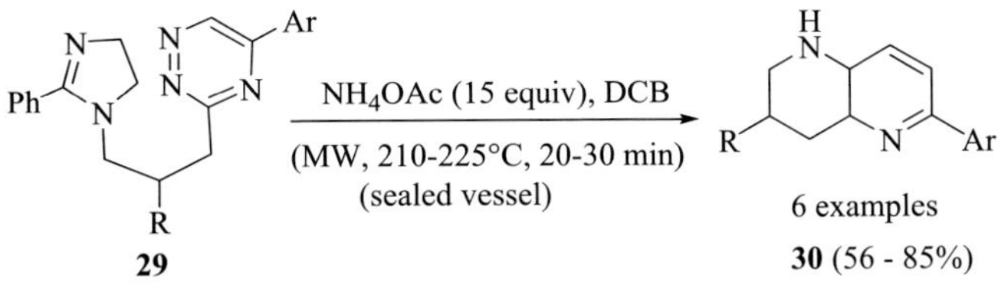

Scheme 11.9

reach completion [59]. In the same paper the authors report that use of other salts, for example $(n\text{-Bu})_4\text{NPF}_6$ or bmimPF$_6$, was considerably less effective.

Two other interesting reports on IMHDA reactions describe the application of solvent-free, solid-support catalyzed microwave technology. Indeed, as mentioned previously in Scheme 11.2, the IMHDA reaction for synthesis of octahydroacridine 6 was efficiently catalyzed by SiO$_2$/ZnCl$_2$ under microwave irradiation conditions [36]. In another example (Scheme 11.10), sesamol 31 and 3-methyl-2-butenal 32 reacted to provide 35, under basic K10–K$^+$ clay-catalyzed solvent-free microwave conditions. The IMHDA reaction of o-quinone methine 34 produced the expected chromene 35 in 84% yield within 8 min. Conventional heating conditions at 110 °C (K10-K$^+$, 60 min) were longer and provided the desired product 35 in an equivalent 91% yield but with 9% of another, unidentified product (Scheme 11.10) [34b].

Shao et al. successfully exploited a combination of microwave irradiation and a polar solvent in closed-vessel for the IMHDA cycloaddition of ω-acetylenic pyrimidines 36 (Scheme 11.11) [60]. Using a Microsynth reactor from Milestone at a max-

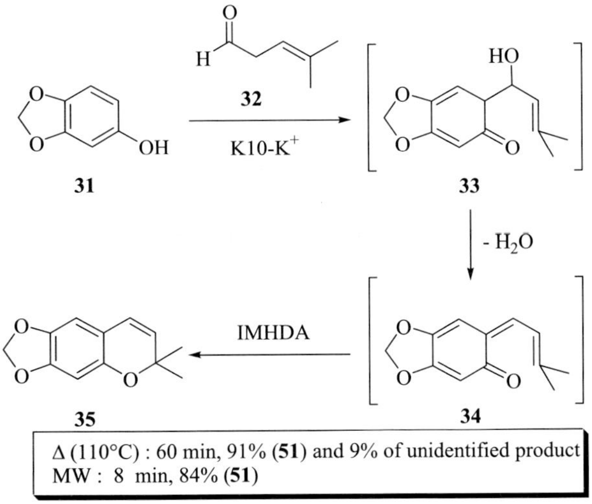

Scheme 11.10

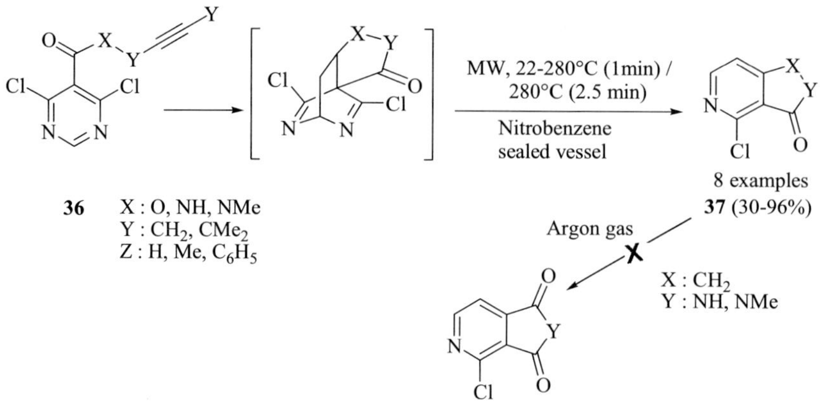

Scheme 11.11

imum power output of 1000 W, in nitrobenzene as polar solvent, the acetylenic pyrimidines **36** were irradiated with temperature profiles of 22 to 280 °C in 1 min, then 280 °C for 2.5 min, then cooling to room temperature under sealed vessel conditions. The desired fused lactones and lactams **37** were obtained in moderate to excellent yields under these optimized conditions (Scheme 11.11). An important aspect of these reactions was that the nitrobenzene solution containing the reactants had to be deoxygenated before heating, by bubbling an inert gas (argon) through the solution. Under classical heating conditions (150–190 °C), the same IMHDA reactions were complete only after 15–144 h with significantly lower yields (0–45%), with one exception (94%, X: CMe$_2$, Y: O) [61].

Raghunathan's group has recently published two reports on the synthesis of polycyclic compounds by IMHDA [62]. In the first example the authors reported the chemoselective synthesis of novel pyrano[3-2c]coumarin derivatives via domino Knoevenagel-IMHDA reactions of 4-hydroxycoumarin and other analogs **38** with O-prenylated salicylaldehyde **39**, using a one-pot procedure. The polycyclic cycloadducts **40** and **41** were obtained with high chemoselectivity in favor of **40** (ratio 93:7 **40/41**) in good yields (82%) by microwave irradiation in absolute ethanol under reflux for 15 s. Conventional reflux conditions, again in EtOH for 4 h, resulted in somewhat lower yields and poorer chemoselectivity (57% yield, ratio 68:32 **40/41**) (Scheme 11.12).

In the second example, by a similar approach, shown in Scheme 11.13, O-prenylated aromatic aldehydes **43** reacted with 4-hydroxyquiniolinone **42** and its benzo analogues to afford pyrano quinolinones **44** and **45** [63]. The polycyclic products **44** and **45** were obtained in 68% yield with good chemoselectivity (**44/45**, 85:15) after 3 min under the action of microwaves in absolute EtOH solvent and with triethylamine as base. Under conventional reflux conditions with the same solvent and base a 41% yield of products **44** and **45** was obtained after 16 h, with a chemostereomeric ratio of 46:54 (Scheme 11.13).

Scheme 11.14 shows an interesting example of an IMDA reaction used to

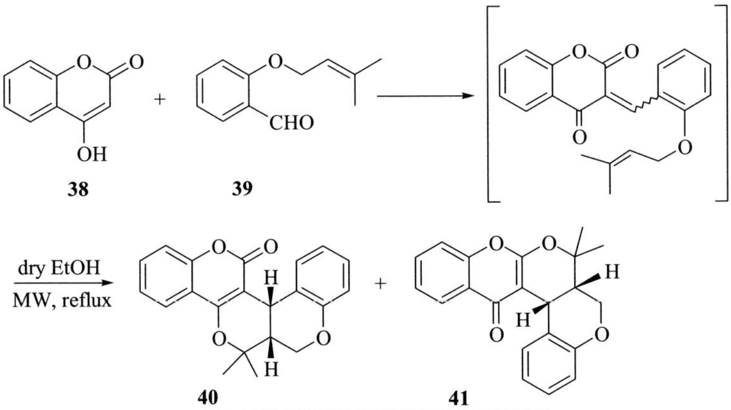

MW (reflux): 82%, 15s, 93:7 (**40**/**41**)
Δ (reflux): 57%, 4h, 68:32 (**40**/**41**)

Scheme 11.12

construct highly functionalized bicyclic bridged compounds **47**, **49**, and **51** from different ω-alkenic cyclohexadienes amide **46**, allyl ether **48**, and ester **50**. Optimized reaction conditions with conventional heating (toluene, under reflux) were compared with microwave irradiation (toluene 135–210 °C) [66]. Microwave irradiation generally had the advantages of significantly increased reaction rates accompanied by an important decrease of side reactions (Scheme 11.14).

Δ (reflux): EDDA, 51%, 14 h, 54:46 (**44**/**45**)	MW (reflux): EDDA, 70%, 5 min, 84:16 (**44**/**45**)
Δ (reflux): Et_3N, 41%, 16 h, 46:54 (**44**/**45**)	MW (reflux): Et_3N, 68%, 3 min, 85:15 (**44**/**45**)
Δ (reflux): Piperidine, 43%, 12 h, 53:47 (**44**/**45**)	MW (reflux): Piperidine, 70%, 5 min, 91:9 (**44**/**45**)

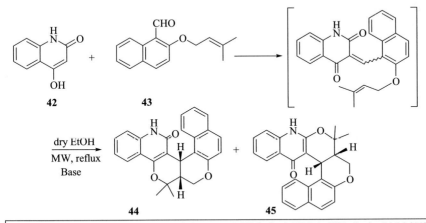

Scheme 11.13

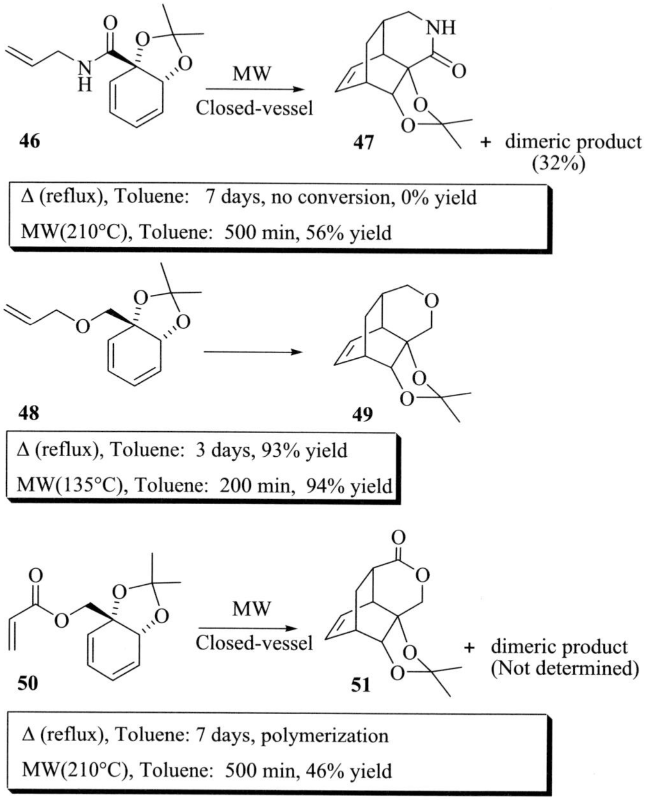

Scheme 11.14

The examples provided above from the recent literature on IMDA and IMHDA reactions, typically illustrate the significant advantages of the use of solvent and ionic liquids under microwave conditions in closed or open vessels compared with classical heating procedures.

11.4.2
Intermolecular Hetero and Diels–Alder Reactions

Intermolecular Diels–Alder or hetero Diels–Alder reactions have been greatly improved by using microwave technology – again with higher reaction rates and improved yields [3j]. Remarkable improvements in rate acceleration and selectivity enhancement for a variety of intermolecular Diels–Alder reactions have also been accomplished in the past two decades by application of catalysts such as Lewis acids. Recently, many such examples have been reported under microwave conditions in polar solvents or ionic liquids as energy-transfer medium. These reactions have also been developed in open vessels by adsorption of the reactants on mineral solid supports or using neat reactants.

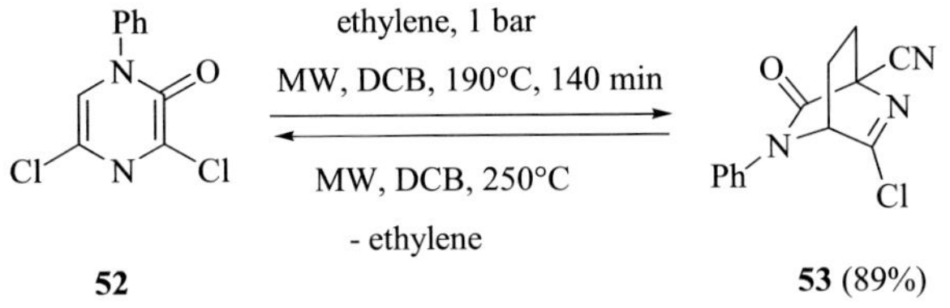

Scheme 11.15

In a study described by Kappe et al. (Section 11.4.1) [58], the intermolecular Diels–Alder cycloaddition reaction of the pyrazinone heterodiene **52** with ethylene led to the bicyclic cycloadduct **53** (Scheme 11.15). Under conventional conditions these cycloaddition reactions must be conducted in an autoclave at an ethylene pressure of 25 bar at 110 °C for 12 h. In contrast, under the action of microwaves, he Diels–Alder addition of pyrazinone precursor **52** to ethylene in a sealed vessel flushed with ethylene before sealing was complete after 140 min at 190 °C. It was not, however, possible to further increase the reaction rate by increasing the temperature. At temperatures above 200 °C an equilibrium between the cycloaddition **52** → **53** and the competing retro Diels–Alder fragmentation process was observed (Scheme 11.15) [58]. By use of a microwave reactor enabling pre-pressurization of the reaction vessel with 10 bar ethylene, however, the Diels–Alder addition **52** → **53** was definitely more efficient; at 190 °C 85% yield of adduct **53** was obtained within 20 min [65b].

Under solvent-free conditions with microwave irradiation, Moody et al. have improved the hetero Diels–Alder reaction of eneacylamines with 1-alkoxy-2-aza-1,3-dienes [67]. A mixture of 1-(2-thiazolyl)-1-acetylaminoethylene **54** and the starting 2-aza-1,3-diene **55**, irradiated in a CEM Focused Synthesizer at 180 °C for 15 min produced a 64% yield of adduct **56** (Scheme 11.16). Carboxylic acid esters and thiazole rings are tolerated under these reaction conditions (Scheme 11.16). Without microwave irradiation yields were in the range 25 to 42% [67].

Another technique used for intermolecular Diels–Alder (DA) reaction with microwave irradiation takes advantage of the intrinsic properties of the solvent [3j]. In this context, an interesting example is the microwave-induced Diels–

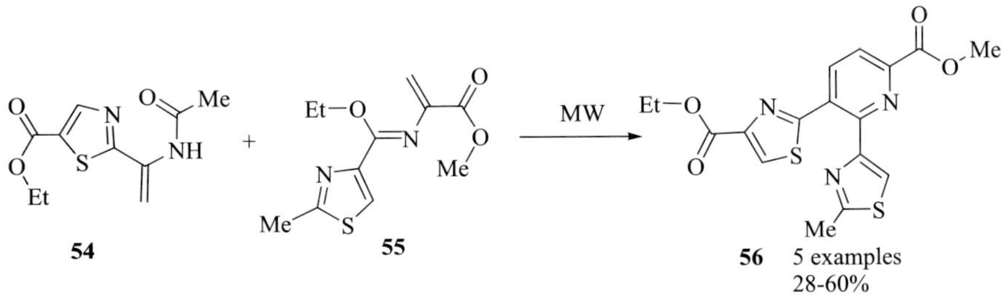

Scheme 11.16

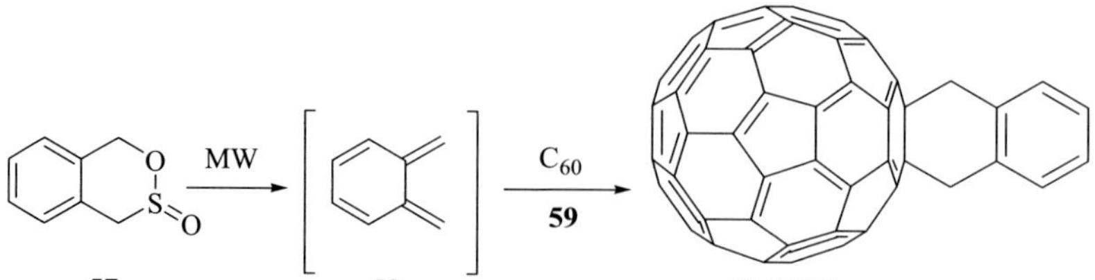

Scheme 11.17

Alder reaction of *ortho*-quinodimethane (*o*-QDM) **58**, generated *in situ* from 1,4-dihydrobenzo-l,2-oxathiin-2-oxide (sultine) **57**, with 60[fullerene] (C_{60}) **59** as dienophile, to produce cycloadduct **60** (Scheme 11.17). This reaction, reported by Langa et al. [68], was the first use of microwave irradiation to prepare a functionalized C_{60} [68].

Another interesting example of the use of *o*-QDM as diene was reported by Delgado et al. [69]. The first DA cycloaddition of *o*-QDM **58** to ester-functionalized single-wall carbon nanotubes (SWNT) **61** was performed in accordance with to Scheme 11.18. The authors showed that such DA reactions could be performed very efficiently in *o*-dichlorobenzene (*o*-DCB) under the action of microwave irradiation at 150 W for 45 min (Scheme 11.18). Conventional heating in *o*-DCB under reflux required 72 h and resulted in less conversion.

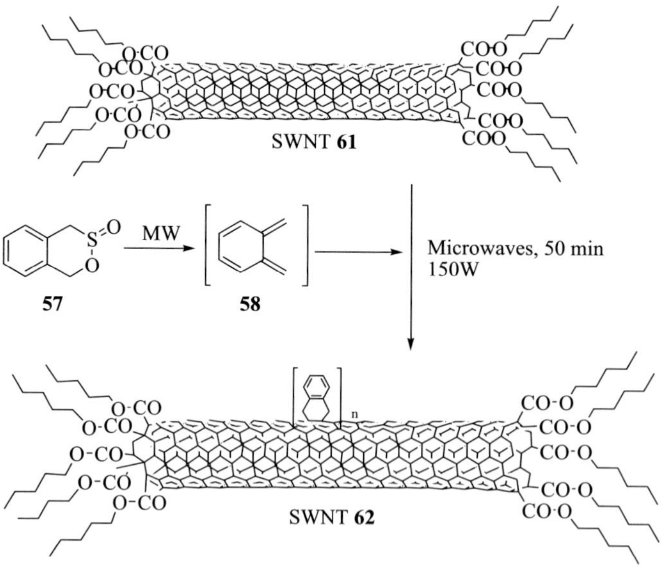

Scheme 11.18

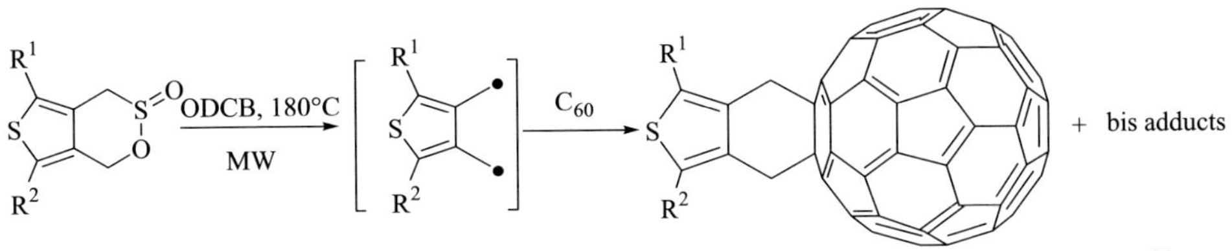

11.4 [4+2] Cycloadditions

Scheme 11.19

R¹, R²: CH_3, Cl, Ph, SCH_3, SPh

Chung's research group recently reported a different type of sultine-type precursor [70]. The synthesis of heterocyclic sultines **63** has been reported elsewhere [71]. Conventional heating of 2,5-disubstituted thienosultines **63** with [60]fullerene **59** in o-DCB under reflux for 2–24 h produces the biradical intermediates **64** which are subsequently trapped by [60]fullerene **59** to produce the cycloadducts **65** and bis adducts **66** in 37–79% yields in 2:1 to 3:1 ratios (Scheme 11.19). The reaction was highly accelerated under microwave conditions (900 W, ≤180 °C) for only 4 min. The ratio of mono adducts **65** to bis adducts **66** was also increased from 2–3:1 to 3.5–6:1 when microwaves were applied [18].

Another series of ring systems has been successfully synthesized by microwave-assisted Lewis acid-catalyzed DA reactions. Examples of these reactions are shown in Scheme 11.20 [72]. The authors describe the [4+2] cycloaddition reactions between dimethylacetylenedicarboxylate **67** or diethylacetylenedicarboxylate **68** dienophiles and dienic substrates, for example furan **69**, 2,5-dimethylfuran **70**, 1,3-

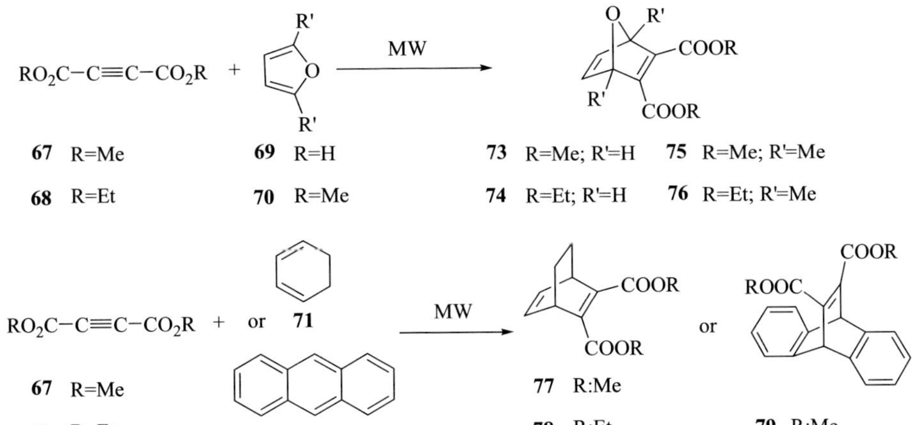

| 67 R=Me | 69 R=H | 73 R=Me; R'=H | 75 R=Me; R'=Me |
| 68 R=Et | 70 R=Me | 74 R=Et; R'=H | 76 R=Et; R'=Me |

| 67 R=Me | | 77 R:Me | 79 R:Me |
| 68 R=Et | | 78 R:Et | |

Scheme 11.20

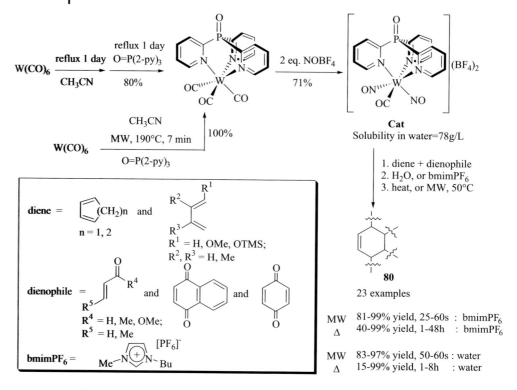

Scheme 11.21

cyclohexadiene **71**, or anthracene **72**. The classical method for these Diels–Alder reactions requires long heating times (1–48 h) at high temperatures (100–200 °C) in the absence of solvent to produce adducts **73–79** in excellent yields (81–95%). The reactions can be accelerated and driven to completion by using a Lewis acid ($AlCl_3$) in the presence of a minimum amount of solvent (dichloromethane, CH_2Cl_2) with microwave activation in closed vessels. Continuous MW irradiation with incident 650 W power for 80–120 s gave the adducts **73–79** in excellent yields (90–95%).

A significant advance in DA reactions is use of a recyclable organotungsten Lewis acid, readily employed in water or ionic liquid ([bmim]PF_6) as solvent, in conjunction with microwave technology [21c]. In the DA reactions shown in Scheme 11.21, 3 mol% organotungsten Lewis acid catalyst [O=P(2-py)$_3$W(CO)(NO)$_2$](BF$_4$)$_2$ was used. The authors showed that such DA reactions could be performed very efficiently with the combined effects of the Lewis acid catalyst in water (or in bmimPF$_6$) and controlled microwave irradiation. Full conversion was achieved under such conditions at 50 °C within 50–60 s in water (25–60 s in bmimPF$_6$) compared with thermal heating which required 1–8 h in water or 1–48 h in bmimPF$_6$.

Water and the ionic liquid bmimPF$_6$ act as powerful reaction media not only for rate acceleration (for adduct **80**, in water, conversion = 92–99%, yield = 83–97%, and in bmimPF$_6$, conversion = 81–99%, yield = 71–96%) and chemoelectivity enhancement but also for facilitating catalyst recycling in the [O=P(2-py)$_3$W(CO)(NO)$_2$](BF$_4$)$_2$-catalyzed Diels–Alder reaction systems. A key feature of this catalyst–water or catalyst–ionic liquid system is that the catalyst was recycled many times. In addition, the authors illustrated the development of the catalyst by conventional heating or under the action of microwave irradiation, the results of which are summarized in Scheme 11.21.

The intermolecular Diels–Alder or hetero Diels–Alder reaction has also been performed under solvent-free microwave conditions in open vessels (Section 11.3). In the past few years solvent-free DA reactions have been regularly performed successfully under microwave irradiation conditions, reducing reaction times to a few minutes compared with several hours under conventional reflux conditions [3j]. Scheme 11.3, mentioned above, shows two recent examples of Diels–Alder cycloadditions performed by microwave dielectric heating [42, 43]. In both examples the diene and dienophile were reacted neat under solvent-free conditions.

Loupy et al. studied possible specific MW effects in irreversible Diels–Alder reactions with acetylenic dienophiles under solvent-free conditions [73]. Strict comparisons of microwave irradiation and conventional heating were conducted and substantial specific kinetic MW effects were observed for the reaction of 3-carbomethoxy-2-pyrone with acetylenic compounds (Scheme 11.22). This were in

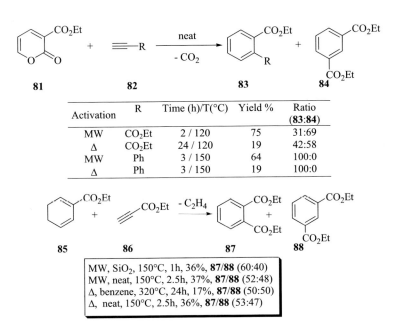

Scheme 11.22

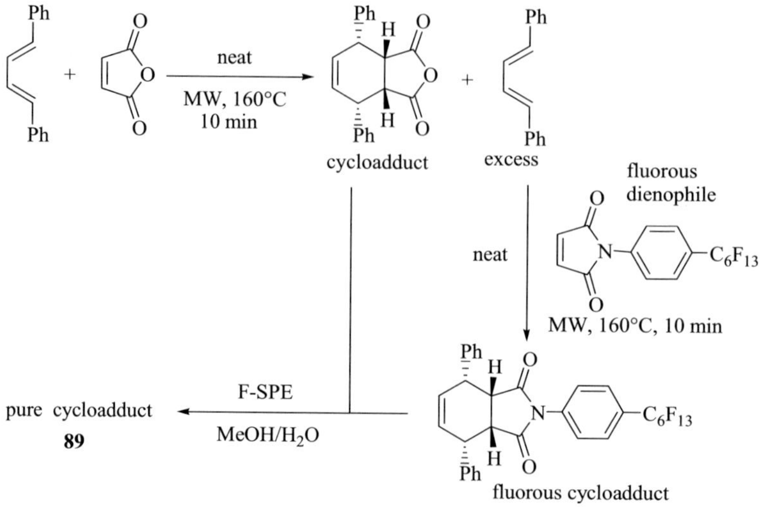

Scheme 11.23

accordance with results from *ab initio* calculations indicative of an asynchronous mechanism with increasing polarity from ground state to transition state (Scheme 11.22).

In the same work, the authors performed reactions in open vessel systems, either in the absence of solvent or on silica gel support in "dry media". The cycloaddition of ethyl 1,3-cyclohexadiene carboxylate **81** to ethyl propiolate **82** was improved compared with the harsh classical conditions (benzene at 320 °C for 24 h) (Scheme 11.22). No specific microwave effects were observed in this reaction; this was in accordance with results from *ab initio* calculations which indicated a synchronous mechanism with similar polarity (dipole moment) in both ground and transition states.

Another interesting development is the use of fluorous-based scavengers in conjunction with microwave synthesis and fluorous solid-phase extraction (F-SPE) for purification. This was recently illustrated by Werner and Curran [74] in their investigation of the Diels–Alder cycloaddition of maleic anhydride to diphenylbutadiene (Scheme 11.23). After performing microwave-assisted cycloaddition (160 °C, 10 min) with a 50% excess of the diene, the excess diene reagent was scavenged by a structurally related maleimide fluorous dienophile under the same reaction conditions. Elution of the product mixture from an F-SPE column with MeOH–H$_2$O provided the desired cycloadduct **89** in 79% yield and 90% purity. Subsequent elution with diethyl ether furnished the fluorous Diels–Alder cycloadduct.

A range of other six-membered cycloadducts synthesized by different research groups [75–78] using microwave-assisted intermolecular [4+2] cycloaddition procedures are illustrated in Scheme 11.24.

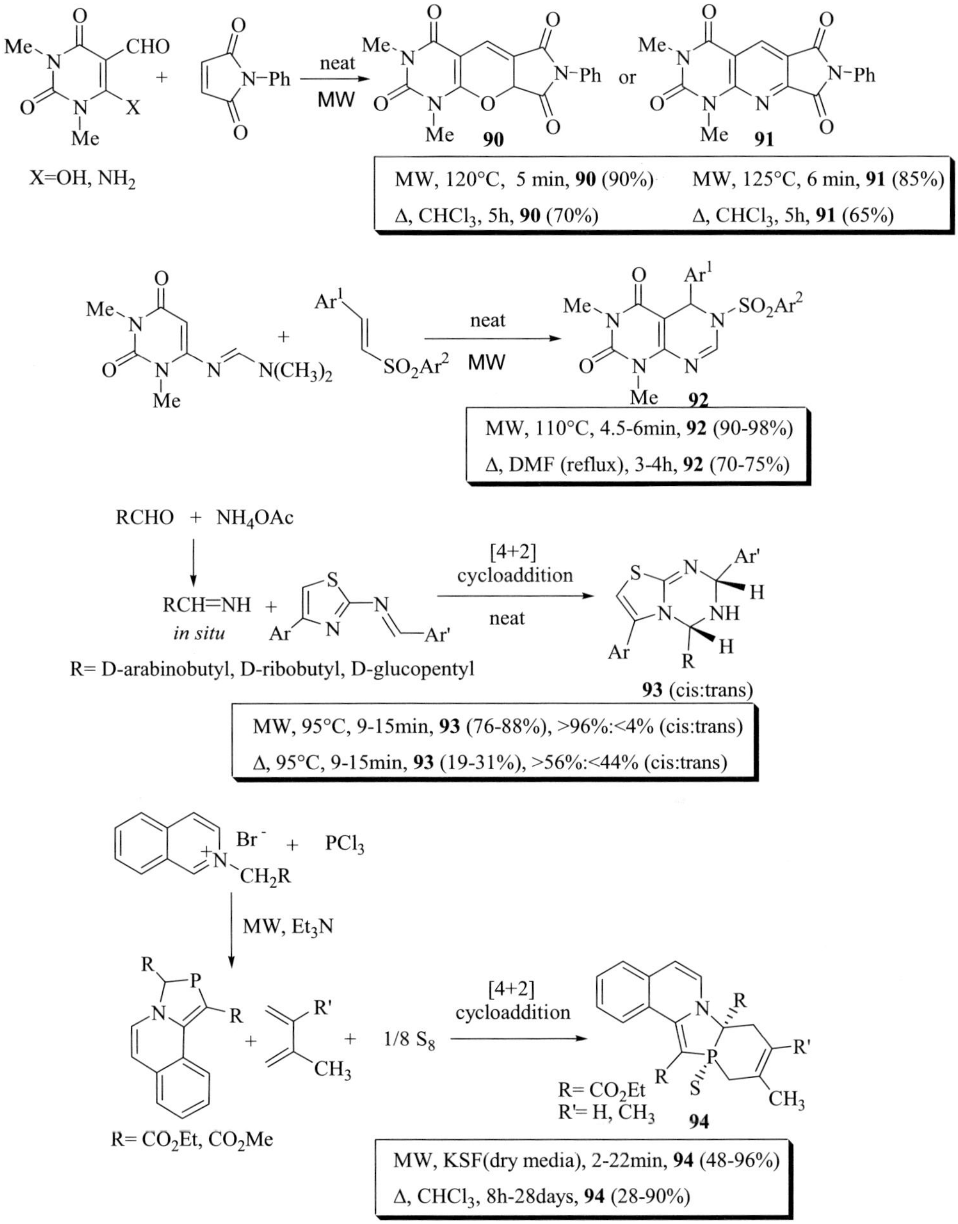

Scheme 11.24. One-pot synthesis of pyrimidine systems [75, 76], thiazolo-s-triazine C-nucleosides [77], and 1,3-azaphospholes [78].

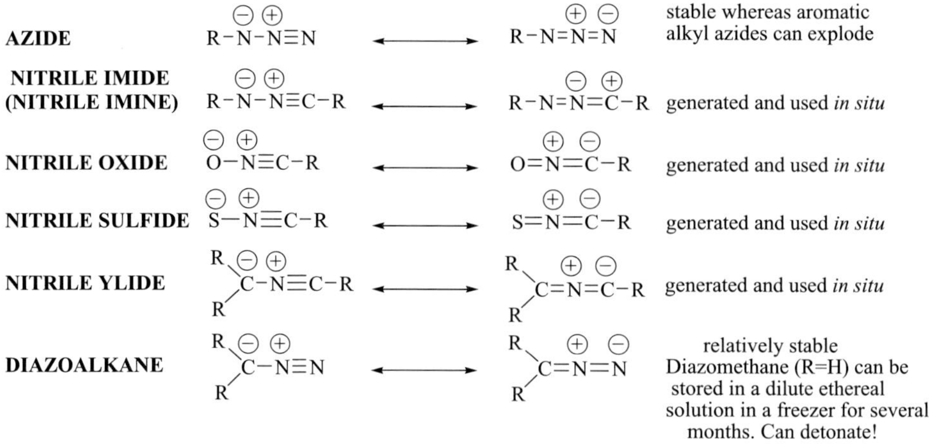

AZIDE	R−N⁻−N⁺≡N	⟷ R−N⁺=N=N⁻	stable whereas aromatic alkyl azides can explode
NITRILE IMIDE (NITRILE IMINE)	R−N⁻−N⁺≡C−R	⟷ R−N⁻=N⁺=C−R	generated and used *in situ*
NITRILE OXIDE	O⁻−N⁺≡C−R	⟷ O=N⁺=C⁻−R	generated and used *in situ*
NITRILE SULFIDE	S⁻−N⁺≡C−R	⟷ S=N⁺=C⁻−R	generated and used *in situ*
NITRILE YLIDE	R₂C⁻−N⁺≡C−R	⟷ R₂C=N⁺=C⁻−R	generated and used *in situ*
DIAZOALKANE	R₂C⁻−N⁺≡N	⟷ R₂C=N⁺=N⁻	relatively stable Diazomethane (R=H) can be stored in a dilute ethereal solution in a freezer for several months. Can detonate!

Fig. 11.1. sp hybridized (linear dipoles like the propargyl anion).

11.5
[3+2] Cycloadditions

Intermolecular and intramolecular [3+2] cycloaddition reactions are among the most efficient and widely used procedures for synthesis of five-membered heterocycles. The reactive partners in these reactions are 1,3-dipoles and dipolarophiles such as alkenes and alkynes. 1,3-Dipoles vary in stability: some can be isolated and stored, others are relatively stable, but they are usually employed immediately. Others are so unstable that they have to be generated and reacted *in situ*. There are two general classes of dipole, often referred to as sp (Fig. 11.1) and sp²-hybridized dipoles (Fig. 11.2).

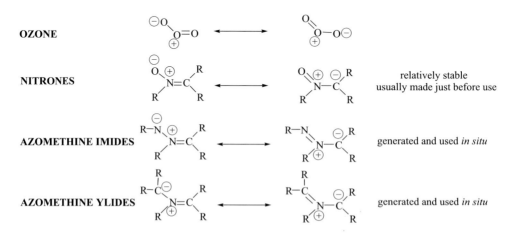

Fig. 11.2. sp2 hybridized (bent dipoles like the allyl anion).

1,3-Dipoles are usually very unstable and their formation requires high temperatures; the subsequent cycloadditions often require long reaction times. In addition, the use of harsh conditions under solvent reflux at high temperatures lead to the formation of undesirable byproducts, thus reducing the purity and yields of products. The rapid and efficient heating induced by microwave irradiation facilitates 1,3-dipolar cycloadditions, even those reputedly difficult (or impossible) to achieve otherwise. The [3+2] cycloaddition reactions were among the first transformations to be studied using microwave technology, and numerous examples have been summarized in previous review articles [3a–3i, 3k, 13c] and a book chapter [3j]. Product yields and/or reaction times were usually improved and the method has been used to prepare valuable compounds that could not be obtained by classical heating. Sometimes the regioselectivity of the reaction can be modified. Most of these processes have been performed under solvent-free conditions.

In the sections below the more representative 1,3-dipolar cycloadditions performed under microwave conditions will be reviewed according to the nature of the dipole.

11.5.1
Nitrile Oxides, Nitrile Sulfides, and Nitrones

11.5.1.1 Nitrile Oxides

Classical methods for use of nitrile oxides in [3+2] cycloadditions have been studied extensively [79], and several procedures have been reported for generation of nitrile oxides. These dipoles undergo cycloadditions to alkenes and alkynes to furnish isoxazolines and isoxazoles, respectively.

The dehydrogenation of aldoximes via the formation of halogenated derivatives such as hydroximoyl chlorides and bromides (see below) has been frequently employed and is an effective procedure [80]. The reaction of primary nitroalkanes with a dehydrating agent and aromatic isocyanates [81] or ethyl chloroformate [82] has also been employed. Both of these methods suffer from limitations such as low tolerability of some functional groups, in the first instance, and the relatively high reaction temperatures in the other instance, which may cause polymerization of the nitrile oxides. In the last decade reports have appeared describing new strategies for improving nitrile oxide-1,3-dipolar cycloaddition reactions by use of microwave activation; these have, in general, overcome the problem of polymerization and dimerization of nitrile oxides [4d]. Thus Hamelin et al. [47] and Bougrin et al. [4d] have shown that the rates of many of these transformations can be enhanced substantially by microwave activation under solvent-free conditions and, indeed, for most of the reactions the authors specifically note that no dimerization or polymerization of the nitrile oxides is observed.

In 1994, Hamelin et al. [47] published the first report of the 1,3-dipolar cycloaddition of nitrile oxide under microwave conditions (*vide supra*, Scheme 11.4). They also reported successful 1,3-dipolar cycloaddition of nitrile oxide to an alkyne under two different conditions – on alumina in "dry media" or in xylene under the action of focused microwaves (Scheme 11.25). The products **96** were obtained in

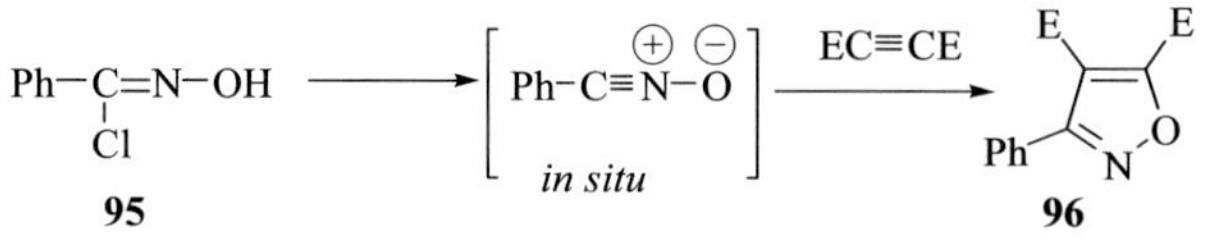

Scheme 11.25

alumina / neat/ MW/ 30 min	60%
NMM/ xylene/ MW/ 130°C, 30 min	65%
classical heating	30%

slightly better yields in solvent compared with reactions in dry media. Use of classical heating always resulted in lower yields (Scheme 11.25).

Bougrin et al. [4d] reported two approaches for 1,3-dipolar cycloaddition of nitrile oxides in open vessels under the action of microwave irradiation. The arylnitrile oxides **99** can be generated *in situ* under the action of microwaves by two different methods – from aryloximes for example **97**, by addition of a chlorinating agent such as NCS (*N*-chlorosuccinimide) supported on alumina (Method A) and from hydroxamoic chloride, for example **98**, under the action of alumina (Method B) (Scheme 11.26).

Both methods afforded identical cycloadducts in similar yields. With dipolarophiles **100** the isolated adducts, **101**, were obtained as two regioisomers, in good yields (65–76%) after activation at 130–162 °C for 10 min. Conventional heating

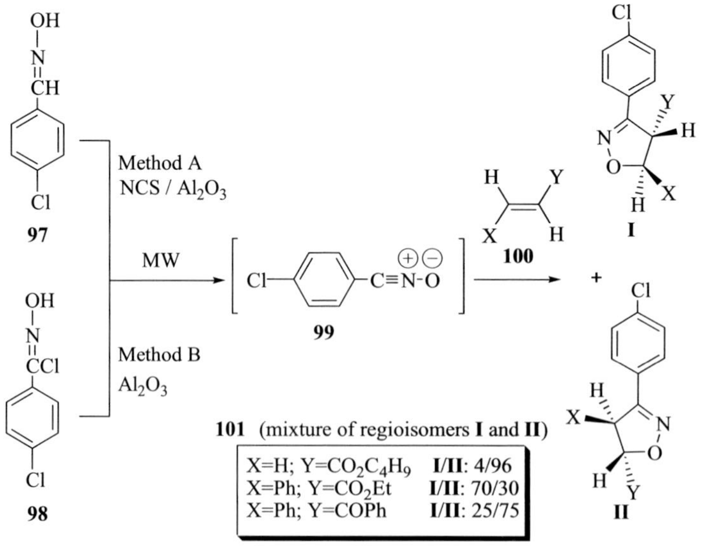

101 (mixture of regioisomers **I** and **II**)

X=H; Y=CO$_2$C$_4$H$_9$	I/II: 4/96
X=Ph; Y=CO$_2$Et	I/II: 70/30
X=Ph; Y=COPh	I/II: 25/75

Scheme 11.26

at the same temperature for 10 min, on the other hand, furnished products in poor yields (10–24%), but with similar regioselectivity (Scheme 11.26).

De la Hoz recently described, in a book chapter, selected reports highlighting particularly interesting [3+2] cycloaddition reactions of nitrile oxides, and their applications [3j]. Earlier examples of controlled MAOS are limited and can be found in the most recent review article by Bougrin [13c]. From these, and for the sake of completeness, some of the most significant examples can also be found in this review. The application of microwaves to the cycloaddition reactions of allyl alcohols with nitrile oxides not only achieved a substantial reduction in reaction time and improved yields, but also altered the regioselectivity of the cycloaddition in favor of the non hydrogen bond-directed cycloadduct [83].

Diaz-Ortiz described the microwave-induced 1,3-dipolar cycloadditions of mesitonitrile oxide to aliphatic and aromatic nitriles under solvent-free conditions. The adducts were obtained in shorter reaction times and better yields than with classical methods [84]. By the same approach, Corsaro reported the successful cycloaddition of arylnitrile oxides to naphthalene or aromatic polycyclic dipolarophiles under solvent-free microwave activation [85]. In the same report, the yields of bis-cycloadducts were also improved by microwave exposure [85].

Occasionally, [3+2] cycloaddition to nitrile oxide dipoles has also been significantly improved by the presence of solvent. Thus, Fišera et al. [86] performed cycloadditions between mesitonitrile oxide and Baylis–Hillman adducts in toluene or chlorobenzene under the action of microwave irradiation. In these reactions the stereoselectivity could be inverted by using methylmagnesium bromide as Grignard reagent. Another important example of microwave nitrile oxide [3+2] cycloaddition in solvent was reported by de la Hoz [87]. In benzene solution, using NCS or NBS and triethylamine as base, the nitrile oxides generated *in situ* from the corresponding oximes reacted with [60]fullerene as dipolarophile to form adducts in good yields after 4 min, compared to 24 h under thermal conditions [87].

A literature survey of nitrile oxide [3+2] cycloaddition reactions with MW activation for the period 2002–2005 reveals that only a limited number of examples have been reported. Among these examples, nitroalkenes are converted *in situ* into nitrile oxides using 4-(4,6-dimethoxy[1,3,5]triazin-2-yl)-4-methylmorpholinium chloride (DMTMM) and 4-N,N-dimethylaminopyridine (DMAP) (Scheme 11.27) [38]. The generated 1,3-dipoles undergo cycloaddition to alkene **103** or alkyne **106** dipolarophiles (5 equiv.), to furnish 4,5-dihydroisoxazoles **104** or isoxazoles **107**, respectively. Open-vessel conditions were used and full conversions with very high yields of products were achieved within 3 min at 80 °C.

By using a multicomponent cascade reaction, Parsons et al. [88] achieved one-pot sequential [1+4] and [3+2] cycloadditions to synthesize highly substituted isoxazolines via nitrile oxides (Scheme 11.28). These five-component reactions proceed by initial formation of isonitriles **109** that react with nitroalkenes **110** to form unstable N-(isoxazolylidene)alkylamines, which in turn fragment to generate the nitrile oxides **111**. Cycloaddition then occurs with methyl acrylate, chosen for its expected reactivity with nitrile oxide dipoles, to generate the isoxazolines **112**. Reactions using standard thermal conditions and microwave irradiation were com-

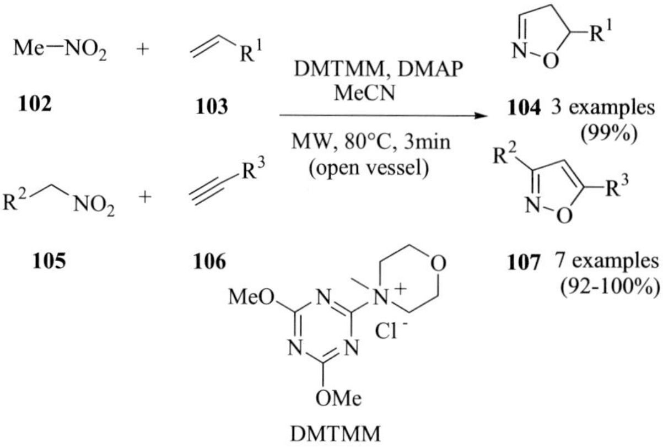

Scheme 11.27

pared; under the latter conditions reaction times were reduced substantially from several hours to 15 min whereas yields remained approximately the same.

Bravo et al. have demonstrated a procedure employing microwave irradiation without solvent that facilitated the nitrile oxide-[3+2] cycloaddition reaction. The authors applied this concept in their approach toward the synthesis of bioactive molecules such as psammaplysins and ceratinamides (metabolites isolated from marine sponges) (Scheme 11.29) [89]. For example, synthesis of the spiro-isoxazoline was successfully achieved by 1,3-dipolar cycloaddition of the E and Z olefinated sugars **113** as dipolarophiles to in situ-generated ethoxycarbonylnitrile oxide. A good yield of adduct **115**, with high stereoselectivity (d.r. > 95:5), was achieved by using Al_2O_3 as support under microwave solvent-free conditions for 15–30 min (Scheme 11.29). The classical method using CH_2Cl_2 as solvent and tri-

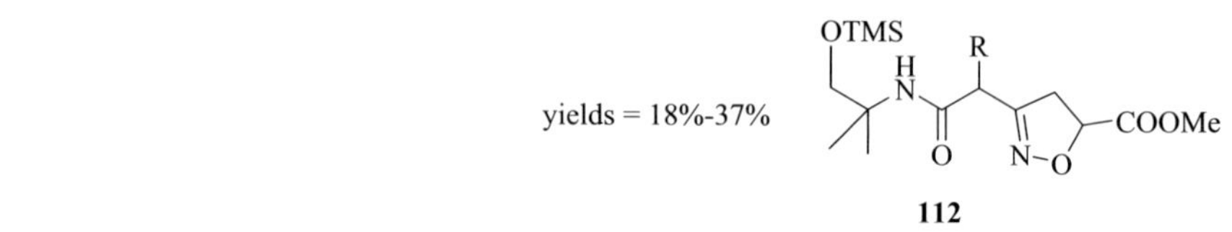

Scheme 11.28

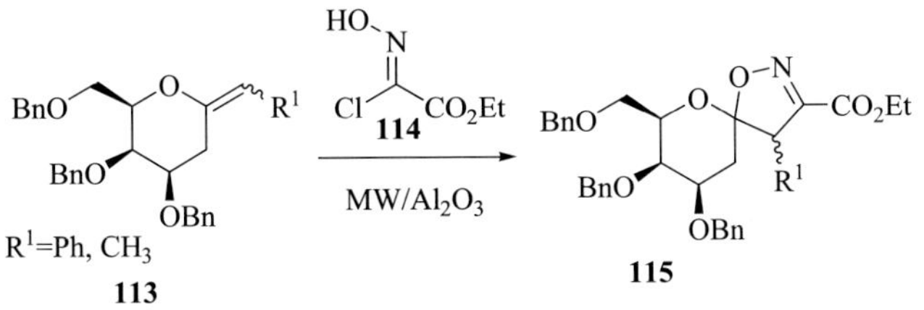

11.5 [3+2] Cycloadditions

Scheme 11.29

ethylamine (Et$_3$N) as base at room temperature for 18 h usually produced 30–78% yields of a furoxan byproduct derived from dimerization of the dipole.

11.5.1.2 Nitrile Sulfides

The chemistry of nitrile sulfides dates back to the 1970s. Franz et al. studied the thermal decomposition of 5-phenyl-1,3,4-oxathiazol-2-one **116**, which produced benzonitrile, sulfur, and CO$_2$ [90]. The benzonitrile sulfide **117** generated from 1,3,4-oxathiazol-2-one precursors, undergoes cycloaddition to nitriles or alkynes to furnish 1,2,4-thiadiazoles or isothiazoles, respectively. Nitrile sulfides are short-lived intermediates that are prone to decompose to sulfur and the corresponding nitrile (Scheme 11.30) [90]. Indeed, typical experimental procedures involve heating under reflux in chlorobenzene or p-xylene (130–135 °C) for prolonged periods (5–170 h) an lead to significant decomposition. These harsh conditions can be avoided by use of microwaves.

To the best of our knowledge only one example of nitrile sulfide [3+2] cycloaddi-

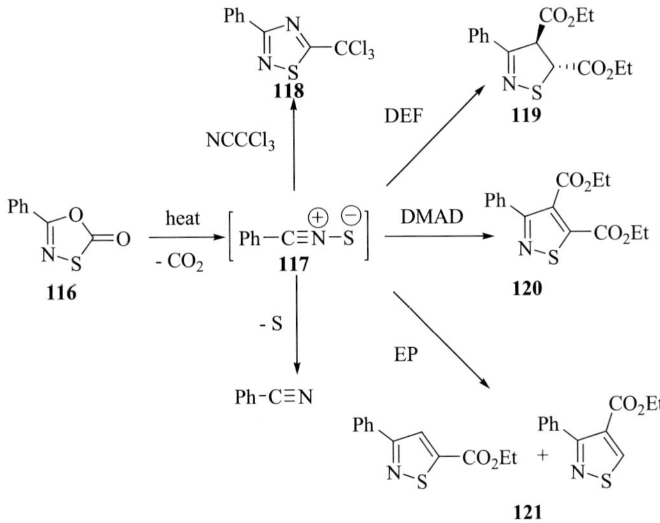

Scheme 11.30

tion reactions with microwave activation has been described in the literature [91], in 2005. Morrison et al. applied the first nitrile sulfide-[3+2] cycloaddition reaction in *p*-xylene as solvent under microwave conditions. The dipolarophiles used were DMAD, ethyl propiolate (EP), diethyl fumarate (DEF), or trichloroacetonitrile (Scheme 11.30). Reaction times were reduced from hours or days at 130–135 °C in *p*-xylene under reflux, with conventional heating, to 10–15 min under microwave irradiation conditions at 160 °C in acetonitrile or *p*-xylene. The yields (45–65%) of cycloadducts **118–121** were comparable with or better than those obtained under traditional conditions.

11.5.1.3 Nitrones

The 1,3-dipolar cycloaddition reaction between nitrones and unsaturated systems under the action of microwave irradiation has been shown to be a powerful method for the synthesis of a wide variety of novel five-membered heterocycles. In 2002, several nitrone-mediated 1,3-dipolar cycloaddition reactions were reported by de la Hoz in the original edition of this book [3j]. Most of the examples provided had been described between 1995 and 2001 [9a, 13b, 84, 92].

More recently, Parmar et al. reported the cycloaddition reaction between ethyl (3-indolylidene)acetate **122** and a variety of substituted α,*N*-diphenylnitrones **123** [93]. The regioisomeric adducts **124** and **125** were obtained without solvent within 4–5 min at 132–140 °C in 35–36% combined yields. Conventional heating in benzene at 60 °C for 150–180 h provided products in only 10–15% combined yields (Scheme 11.31).

Bashiardes et al. [94] described an intramolecular cycloaddition reaction of unprotected carbohydrates **126**, involving a nitrone ylide dipole **127** derived from the 1-aldehydic position, and an ω-olefinic moiety constructed from the 6-hydroxyl function (Scheme 11.32). In this enantiomeric synthesis of novel bicyclic oxazolidines bearing a quaternary bridgehead, **128**, a comparative study was performed of classical heating conditions and microwave-assisted cycloaddition, both in the same reaction medium, aqueous ethanol. All the examples provided products in yields which were improved from approximately 60% to 90%, basically because of the cleaner reactions. The reaction times were reduced from 48 h to just 1 h.

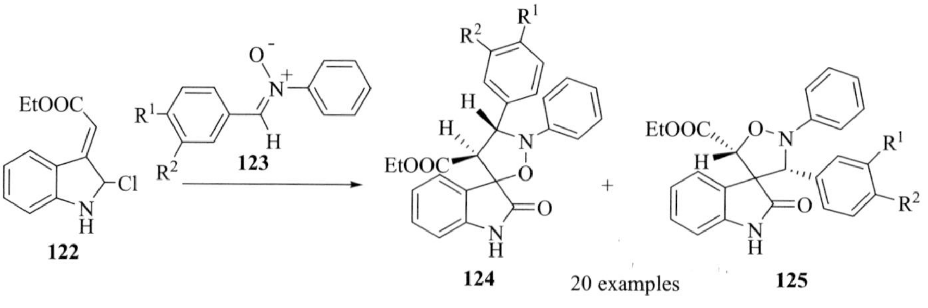

Scheme 11.31

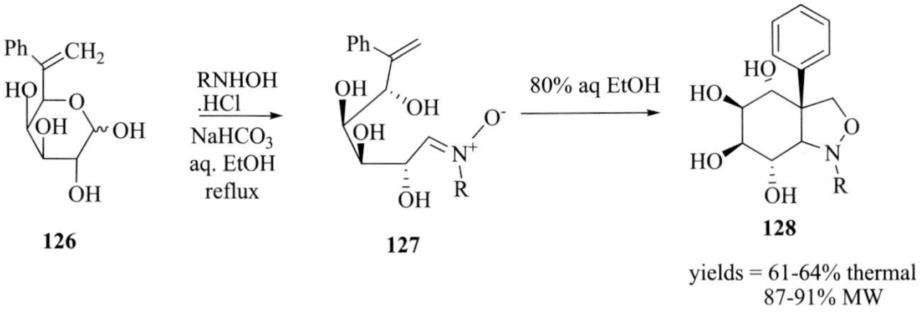

11.5 [3+2] Cycloadditions

[Scheme 11.32 structural diagram showing compounds 126, 127, and 128 with reagents RNHOH·HCl, NaHCO₃, aq. EtOH reflux, then 80% aq EtOH; yields = 61–64% thermal, 87–91% MW]

Scheme 11.32

The [3+2] cycloaddition of nitrones to organonitriles bound to Pt(II), Pt(IV), or Pd(II), by use of a combination of the Lewis acidity of the transition metals and microwave irradiation, has been described in three independent reports. In the first example, Wagner et al. [95] reported that the [3+2] cycloaddition of N-methyl-C-phenylnitrone **130** to transition metal coordinated (E)-cinnamonitrile **129** occurred exclusively at the nitrile C≡N bond, leading to bis-oxadiazoline complexes **132** in high yield (Scheme 11.33). In contrast, reaction of the nitrone **130**

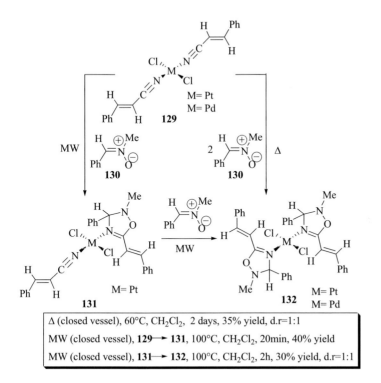

Δ (closed vessel), 60°C, CH₂Cl₂, 2 days, 35% yield, d.r=1:1
MW (closed vessel), **129** → **131**, 100°C, CH₂Cl₂, 20min, 40% yield
MW (closed vessel), **131** → **132**, 100°C, CH₂Cl₂, 2h, 30% yield, d.r=1:1

Scheme 11.33

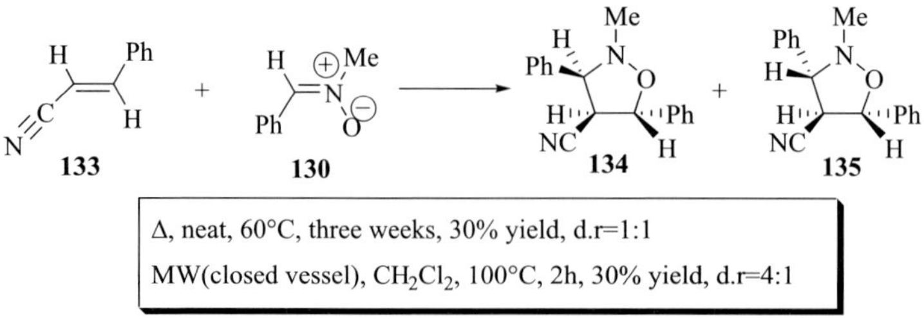

Scheme 11.34

Δ, neat, 60 °C, three weeks, 30% yield, d.r=1:1
MW(closed vessel), CH$_2$Cl$_2$, 100 °C, 2h, 30% yield, d.r=4:1

with free cinnamonitrile **133** involves the C=C bond only, yielding a diastereomeric mixture of isoxazolidine-4-carbonitriles **134** and **135** (Scheme 11.34). Microwave irradiation substantially enhanced the rates of reaction in both transformations, without changing their regioselectivity compared with that from conventional thermal conditions. The two nitrile ligands in the complexes of the type [MCl$_2$(cinnamonitrile)$_2$] (M = Pt or Pd) are significantly different in reactivity. Short microwave irradiation enables the selective synthesis of the mono-cycloaddition product **131** [PtCl$_2$(cinnamonitrile)(oxadiazoline)], even in the presence of excess nitrone. By extending irradiation times, this complex is further transformed to the bis-cycloaddition product [PtCl$_2$(oxadiazoline)$_2$]. The latter compound is also produced when thermal heating is used. Formation of the mono-cycloaddition product is not selective under thermal conditions, however.

The same authors have successfully used this technology in the [3+2] cycloaddition of N-methyl-C-arylnitrone **137** or **139** to cis or trans-[PtCl$_2$(PhCN)$_2$] **136** or **141** [96]. The reaction rates of the cycloadditions were substantially enhanced by use of microwave irradiation, with high selectivity for the mono-cycloadducts **138** or **142**, because the first cycloaddition is accelerated to a greater extent than the second. The cycloaddition reaction of the trans-[PtCl$_2$(PhCN)-(oxadiazoline)] **138** with nitrone **139** afforded bis-oxadiazoline compounds **140**. The corresponding cis-configured complexes **142** did not undergo further cycloaddition (Scheme 11.35).

Using a similar strategy, Pombeiro et al. described an efficient procedure for metal-catalyzed [3+2] cycloaddition of nitrones to Pt-liganded ethanonitrile (Scheme 11.36) [97]. The [3+2] cycloadditions are typically conducted with anionic platinum(II) [Ph$_3$PCH$_2$Ph]PtCl$_3$(EtCN)] **144**, neutral platinum(II) [PtCl$_2$(EtCN)$_2$] **145**, or neutral platinum(IV) [PtCl$_4$(EtCN)$_2$] **146** complexes as dipolarophiles and cyclic nonaromatic nitrone dipoles **143**. These reactions are greatly accelerated by microwave irradiation to provide bicyclic fused oxadiazolines **147**–**149**, with high stereoselectivity and 60–90% yields. Under similar experimental conditions, no reaction was observed between the free nitrile and the nitrone, indicating that the cycloaddition is metal-mediated. Under controlled microwave irradiation conditions, without solvent, on silica gel as support, adduct **149** was obtained in 90% yield.

In another study, Fišera et al. studied the same nitrone, **130**, in 1,3-dipolar cycloadditions to β-hydroxy-α-methylene esters **150** under thermal heating or microwave

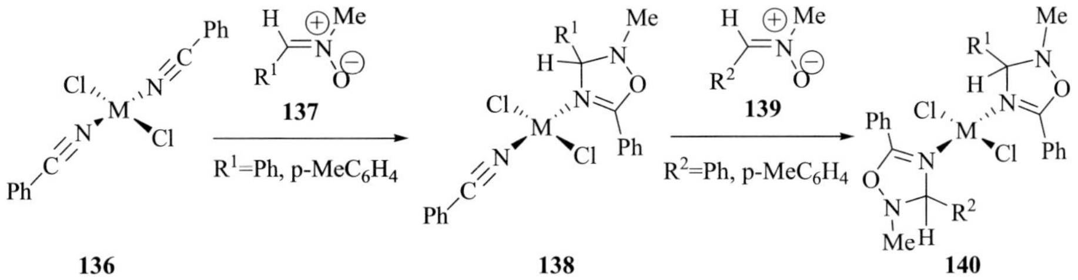

11.5 [3+2] Cycloadditions

Classical method, **136** → **138**, RT, CH$_2$Cl$_2$, overnight, 68-72% yields.
MW(closed vessel), **136** → **1386**, 60°C, CH$_2$Cl$_2$, 20min, 68-72% yields.
Classical method, **138** → **140**, RT, CH$_2$Cl$_2$, overnight, 76% yield.
MW(closed vessel), **138** → **140**, 60°C, CH$_2$Cl$_2$, 2.5h, 76% yield.

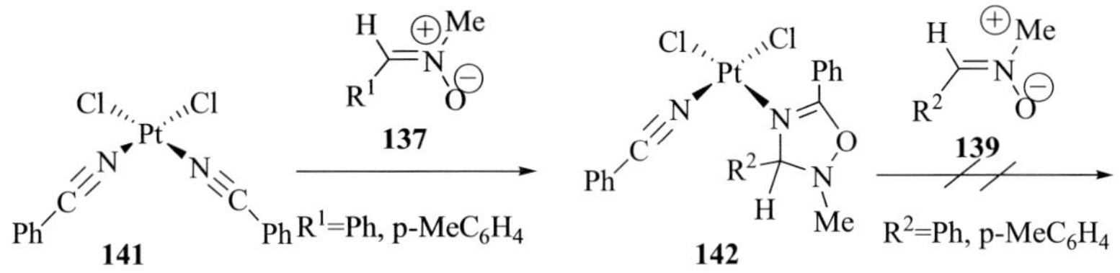

Classical method, **141** → **142**, RT, CH$_2$Cl$_2$, overnight, 43-48% yields.
MW(closed vessel), **141** → **142**, 60°C, CH$_2$Cl$_2$, 20min, 43-48% yields.

Scheme 11.35

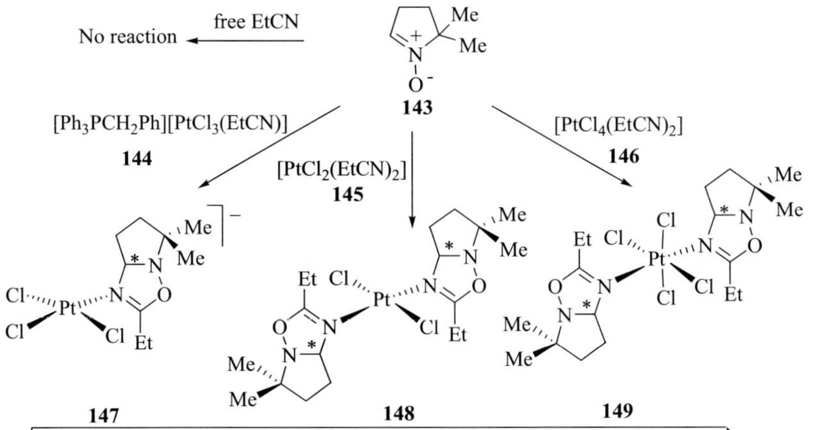

Classical method, CH$_2$Cl$_2$, 25°C, 15min-48h, 60-85% yields (**147**, **148** and **149**).
MW, CH$_2$Cl$_2$, 25-30°C, 10min-3h, 62-90% yields (**147**, **148** and **1497**).
MW, CH$_2$Cl$_2$, SiO$_2$, 25°C, 10min, 90% yield (**149**).

Scheme 11.36

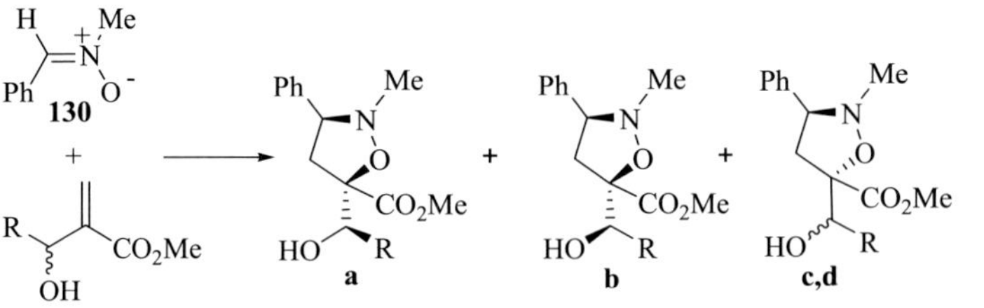

Scheme 11.37

R= iPr, Ph	Δ(reflux), R=iPr, CCl$_4$/MgI$_2$-I$_2$, 7days, 81% yield, d r: 87:9:3:1
	MW, R=iPr, CCl$_4$/MgI$_2$-I$_2$, 1h, 68% yield, d.r: 89:8:3:0
	Δ(reflux), R=Ph, CCl$_4$/MeMgBr, 7days, 84% yield, d r: 78:15:5:2
	MW, R=Ph, CCl$_4$/MeMgBr, 1h, 74% yield, d.r: 79:15:6:0

irradiation conditions [98]. When combination of microwave irradiation and a Lewis acid [Mg(II)] was used the reactions were extremely clean and rapid, affording higher yields. Little effect on the diastereoisomeric ratio of products **151** (Scheme 11.37) was observed.

Some excellent results for cycloadditions involving nitrones were observed by Merino et al. [99] using a solvent-free [3+2] cycloaddition of N-benzyl-C-glycosyl nitrones to methyl acrylate. Microwave irradiation for 0.1 h was sufficient to afford a 100% yield of the isolated isoxazolidine whereas thermal heating under reflux conditions (80 °C) required 0.5 h and produced the same yield and the same stereoisomeric ratio.

11.5.2
Azomethine Ylides, and Nitrile Imines

11.5.2.1 Azomethine Ylides

Microwave-induced 1,3-dipolar cycloaddition reactions involving azomethine ylides have been widely reported in the literature. In 2002; many examples were described in a book chapter by de la Hoz [3j], which provides extensive coverage of the subject. The objective of this section is to highlight some of the most recent applications and trends in microwave synthesis, and to discuss the impact of this technology. Highly stereoselective intramolecular cycloadditions of azomethine ylides have been performed under solvent-free microwave conditions.

Bashiardes et al. reported a study involving cycloaddition of O-allylic or O-propargylic salicylaldehydes **152** or **153** to α-amino esters **154** treated under different conditions: toluene at reflux (A); conventional heating without solvent (B); or microwave irradiation under solvent-free conditions (C) (Scheme 11.38) [100]. The cycloaddition reaction involves polar 1,3-dipolar azomethine ylide intermediates **155** and **156**. The best results were obtained in a one-pot procedure under microwave conditions providing pyrrolidines **157** or pyrroles **158** in excellent yields ranging from 70 to 98% at 130 °C for 5–30 min.

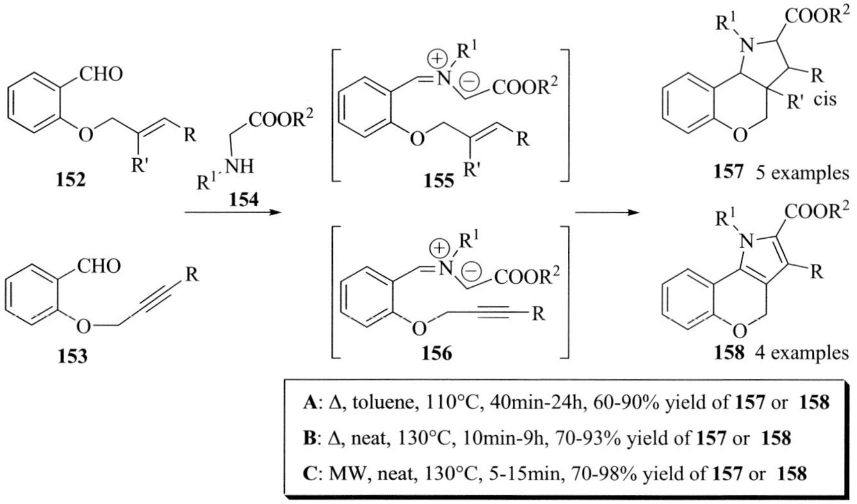
CHO
152
R'
R
COOR2
R1 NH
154
155
156
153
157 5 examples
158 4 examples
COOR2
A: Δ, toluene, 110°C, 40min-24h, 60-90% yield of 157 or 158
B: Δ, neat, 130°C, 10min-9h, 70-93% yield of 157 or 158
C: MW, neat, 130°C, 5-15min, 70-98% yield of 157 or 158

Scheme 11.38

Díaz-Ortiz et al. reported the intermolecular cycloaddition of azomethine ylides to substituted *β*-nitrostyrenes **160** under solvent-free, microwave-assisted, open vessel conditions at 110–120 °C for 10–15 min to afford the three stereoisomeric pyrrolidines **161a**, **b**, and **c** in 81–86% yield. Under classical heating conditions in toluene under reflux for 24 h these cycloadditions afforded yields below 50% and only stereoisomers **a** and **b** were obtained (Scheme 11.39) [101].

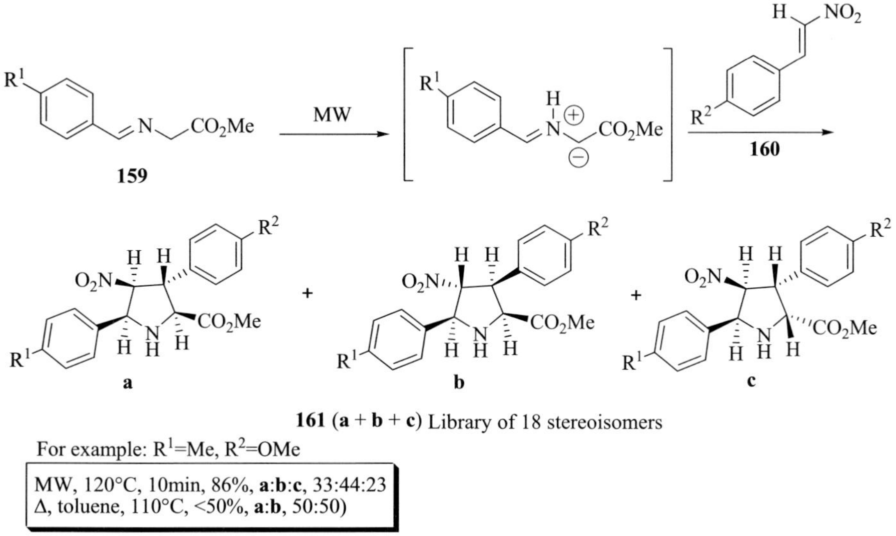
R1
CO2Me
159
MW
160
R2
a
b
c
161 (a + b + c) Library of 18 stereoisomers
For example: R1=Me, R2=OMe
MW, 120°C, 10min, 86%, a:b:c, 33:44:23
Δ, toluene, 110°C, <50%, a:b, 50:50)

Scheme 11.39

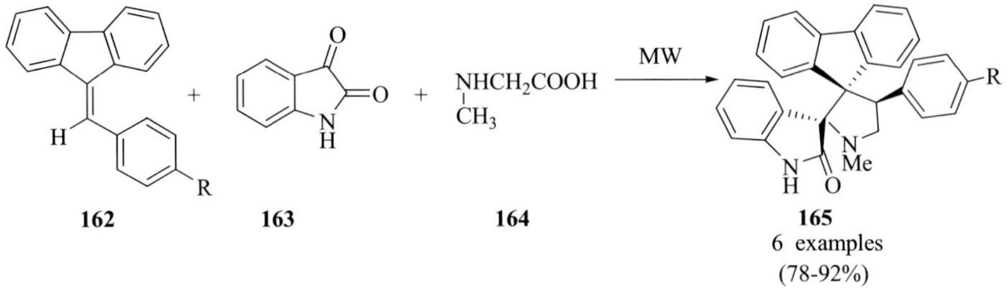

Scheme 11.40

Azomethine ylides have been successfully used for rapid synthesis of dispiroheterocycles by cycloaddition to 9-azylidenefluorene **162** dipolarophile in a 3-component solvent-free microwave reaction either neat or in the presence K-10 montmorillonite (Scheme 11.40) [3k, 102]. Although extremely sterically demanding, these reactions were performed successfully, reducing reaction times from 5–8 h under conventional conditions (methanol reflux) to only few minutes under the action of microwaves. The reactions were clean and products **165** were obtained in high yields (78–92%) compared with the conventional conditions (3–60%).

Azizian et al. [103] described a novel four-component reaction for the diastereoselective synthesis of spiro-pyrrolizidines **170** using a 1,3-dipolar cycloaddition of azomethine ylides to N-aryl maleimides. Reaction of ninhydrin **166** with 1,2-phenylenediamine **167** in DMSO, and addition of L-proline **168** and N-aryl maleimides **169** in a one-pot four-component reaction (Scheme 11.41) usually required extended reaction times of up to 3 h at 100 °C to obtain the desired spiropyrrolizidines **170** in 76–86% yield as single diastereoisomers. This process can be accelerated substantially (3–5 min) with good yields (87–95%) by performing the reaction under microwave irradiation conditions.

Zhu et al. described two independent one-pot three-component routes to C_{60} pyrrolidines **174** and **175** involving microwave activation (Scheme 11.42) [104]. In the first approach, a solution of C_{60}, **171**, amino acid **172** and fluorinated benzaldehyde **173** was heated at reflux in o-dichlorobenzene under the action of microwave irradiation. The in situ-generated fluorinated azomethine ylides undergo cycloaddition

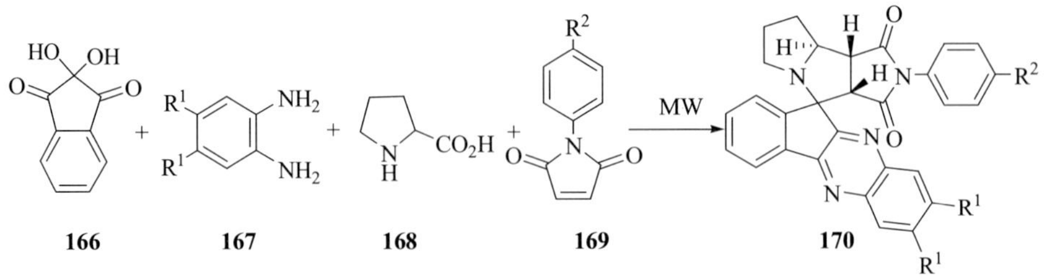

Scheme 11.41

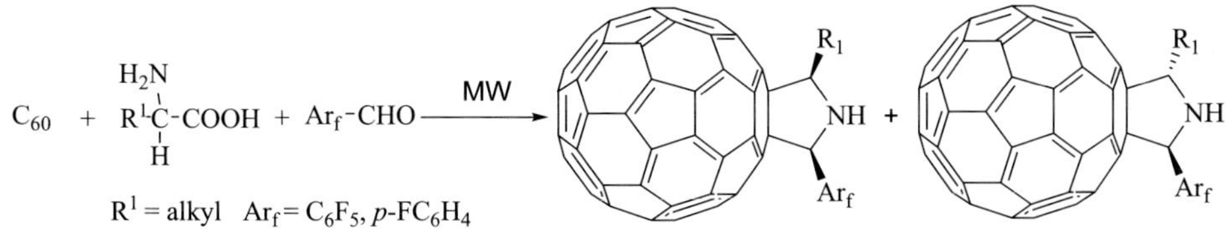

11.5 [3+2] Cycloadditions

Scheme 11.42

C_{60} + R^{1}C(H)(NH$_2$)-COOH + Ar$_f$-CHO $\xrightarrow{MW}$ pyrrolidines **174** + **175**

R^1 = alkyl   Ar$_f$ = C$_6$F$_5$, p-FC$_6$H$_4$

171   **172**   **173**   **174**   **175**

to C_{60} to furnish cis and trans-isomeric C_{60} pyrrolidines **174** and **175** within 50–90 min with 11–29% yields of **174** and 8–26% yields of the trans isomer **175**. Conventional reflux in toluene afforded 5–23% **174** and 8–20% **175** after 10 h. With the second method the authors reported another microwave-promoted one-pot three-component reaction of C_{60}, **171**, amino acids, **172**, and aldehydes, **176**, under solvent-free conditions, to furnish products **177** in moderate to good yields (55–71%) within 8–10 min. Traditional thermal conditions (toluene, 110 °C) afforded 40–55% yields of **177** after 20–22 h (Scheme 11.43).

In the same context, Zhang et al. described use of an attractive microwave-assisted three-component 1,3-dipolar cycloaddition to access a library of 15 biaryl-substituted proline compounds using fluorous benzaldehydes as starting materials [105]. By irradiating (250 W, 150 °C, 15 min) mixtures of aminoesters, **178**, benzaldehydes, **179**, and maleimides, **180**, in the presence of Et$_3$N in DMF, they obtained the desired bicyclic proline **181** as a single diastereoisomer in moderate to excellent yields (40–75%) (Scheme 11.44). Figure 11.3 Shows the ^{1}H NMR spectrum of the fluorous proline analog (**181a**: R^1 = Me, R^2 = H) purified by solid-phase extraction (SPE).

11.5.2.2 Nitrile Imines

1,3-Dipolar cycloadditions of nitrile imines under the action of microwave irradiation have rarely been described. Under classical conditions these reactions are nor-

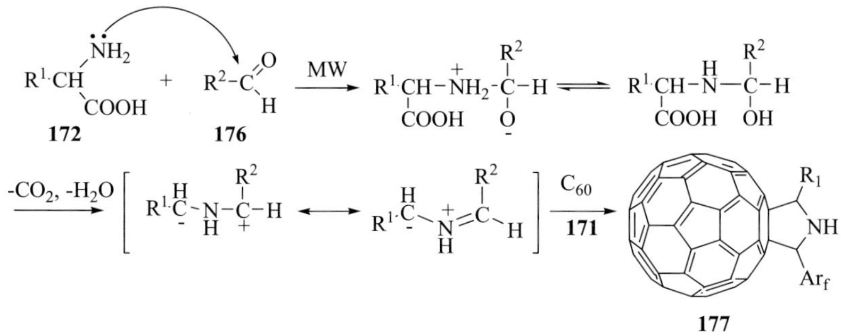

Scheme 11.43

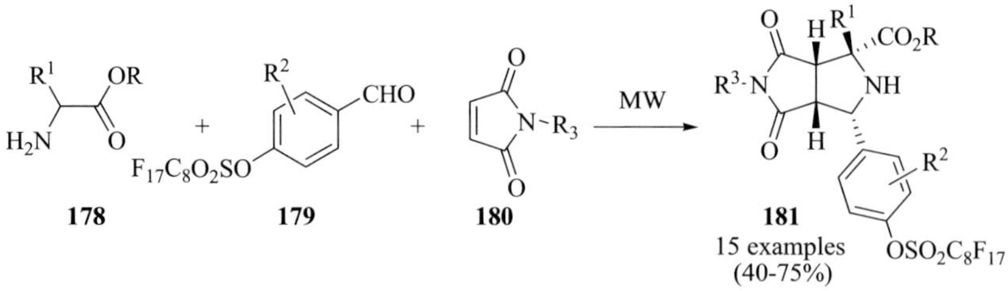

Scheme 11.44

mally performed in potentially toxic solvents (benzene, toluene,...) for long periods, ranging from hours to days and necessitating harsh thermal conditions (≥ 100 °C). The reactions are often accompanied by dimerization of the dipole and decomposition of products. Combination of microwave irradiation with solvent-free techniques and basic conditions, however, usually enables efficient cycloaddition reactions.

In 1995 Bougrin et al. [4c] reported the first practical use of microwave irradiation with nitrile imine **13** dipole using solvent-free conditions. In this work the reactivity of diphenylnitrilimine (DPNI) with different dipolarophiles **14a**, **b**, and **c**, in dry media was compared. Diphenyhydrazonoyl chloride and **14a**, **b**, or **c**, were either applied to KF/alumina as basic support (Schemes 11.4 and 11.45) or reacted under phase-transfer catalysis (PTC) conditions using KF-Aliquat as base (Scheme 11.4) to generate reactive intermediate DPNI **13**. The results showed that the best yields of adducts (87–93%) were achieved on KF/Al$_2$O$_3$ under controlled microwave conditions (Scheme 11.45). The synthesized adducts were obtained in high purity at 123–130 °C after 12 min, with unchanged regioselectivity of **15a**, **b**, and **c**. When classical heating under comparable conditions (time, temperature) was used, very low yields (0–6%) of products were always obtained.

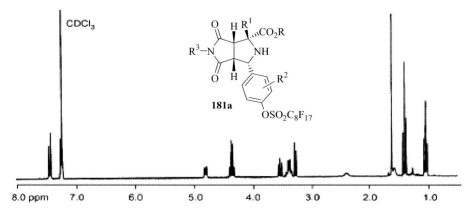

Fig. 11.3. ^{1}H NMR (270 MHz) spectrum of fluorous bicyclic proline.

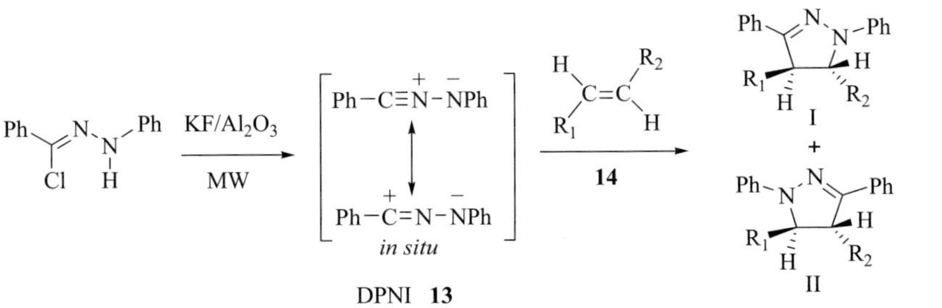

11.5 [3+2] Cycloadditions

Scheme 11.45

15 (mixture of regioisomers : I and II)
15a; $R_1 = C_6H_5CO$, $R_2 = C_6H_5$
15b; $R_1 = 4\text{-}ClC_6H_4CO$, $R_2 = C_6H_5$
15c; $R_1 = n\text{-}C_4H_9CO_2$, $R_2 = H$

Ben Alloum et al. [31a] described new spiro-rhodanine-pyrazolines **183** prepared by 1,3-dipolar addition of DPNI **13** and 5-arylidenerhodanines **182** on alumina or KF/alumina in "dry media" under microwave irradiation conditions (Scheme 11.46). The adducts were always obtained in excellent yields after 10 min whereas no reaction occurred when conventional heating was used under similar conditions (time and temperature).

In 2005, Bougrin et al. [106] reported simple modifications of this general method – use of porous calcium hydroxyapatite, $[Ca_{10}(PO_4)_6(OH)_2]$ (p-HAP), as base – enabling the synthesis of the corresponding pyrazolines **185** (Scheme 11.47). Solvent-free 1,3-dipolar cycloaddition of diphenylnitrilimine **13**, generated *in situ* from hydrazonoyl chloride, to olefins **184** was efficiently catalyzed by a combination of porous calcium hydroxyapatite, serving both as base and as solid support, and microwaves. A key feature of this catalyst/solvent-free system is the recy-

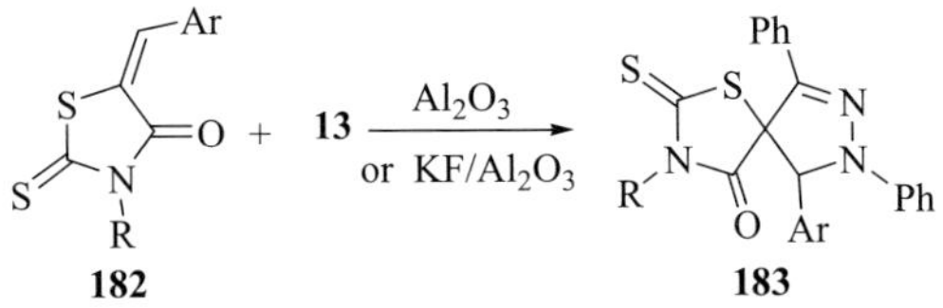

Entry	R, Ar	Reaction time (min)	Technique	Yield (%)
1	H, C_6H_5	10	MW (multimode)	80
2		10	Δ, 117°C	traces
3	CH_3, 4-$NO_2C_6H_4$	10	MW (multimode)	94
4		10	Δ, 117°C	traces
5	CH_3, 4-ClC_6H_4	10	MW (multimode)	90
6		10	Δ, 125°C	traces

Scheme 11.46

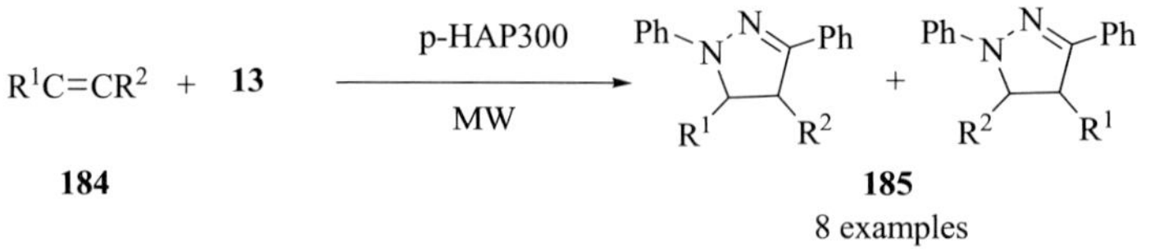

Scheme 11.47

clability of p-HAP, which, after heating at 300 °C under reduced pressure for 20 min was recycled at least ten times. After each cycle, the adduct was directly isolated in high yield (87–95%) by rapid elution from the support with CH_2Cl_2 or THF under reduced pressure. The pyrazolines **185** were obtained after 3 min at 104–181 °C. Several effects, for example the specific surface of porous calcium hydroxyapatite and nonthermal microwave effects, are discussed in this work [106].

11.5.3
Azides

The Huisgen 1,3-dipolar cycloaddition of azides to alkynes or nitriles as dipolarophiles, resulting in 1,2,3-triazoles or tetrazoles, is one of the most powerful "click reactions". A limitation of this approach, however, is the absence of regiospecificity normally found in thermal 1,3-cycloaddition of nonsymmetrical alkynes; this leads to mixtures of the different possible regioisomers. In other reports, classical 1,3-dipolar cycloadditions of azides to metal acetylides, alkynic Grignard reagents, phosphonium salts and acetylenic amides have been described. Extended reaction times and high temperatures are required in most of the reactions, but they can also be performed more effectively with the aid of microwave irradiation. The main results reported are reviewed in this section.

An interesting report from Loupy et al. described the use of solvent-free conditions and microwaves for the synthesis of regioisomeric 1,2,3-triazoles **188** and **189** with substantial reduction for reaction times – down to 5–30 min compared with 30–40 h under reflux in toluene under thermal conditions for the 1,3-dipolar cycloaddition between phosphonate azides **186** and acetylenic esters **187** (Scheme 11.48) [31c].

More recently, microwave-assisted cycloaddition employing azide dipoles under solvent-free microwave conditions has been reported by the research group of Katritzky [31h]. A variety of two regioisomeric 1,2,3-triazoles **194** were synthesized by cycloaddition between organic azides **190, 191**, or **192** and dipolarophile acetylenic amides **193** under solvent-free microwave conditions (Scheme 11.49). The reactions were performed in sealed vessels at fixed temperature (55 or 85 °C) for 30 min with the power set at 120 W or 170 W. The resulting 1,3-dipolar cycloadditions to different acetylenic amides **193** produced the corresponding N-substituted C-carbamoyl-1,2,3-triazoles **194** in good yields (65–84%). In the same paper the authors described the preparation of azidotriazole **196** as the major regioisomer

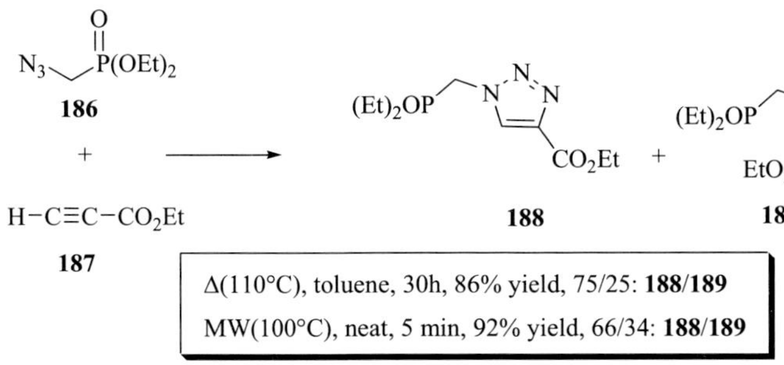

Scheme 11.48

in moderate yield (56%) from bis-azide **195** and *N*-benzyl-2-propynamide (**193a**: $R^1 = R^2 = \text{CONHCH}_2\text{Ph}$) under solvent-free microwave conditions. Further reaction of **196** with the second amide (**189b**: $R^1 = R^2 = \text{COpiperidinyl}$) failed to give the bis-triazole **197**, however (Scheme 11.50).

In 2003 the same authors introduced a technique for convenient and general preparation of substituted *C*-carbamoyl mono and bis-triazoles in good yields by 1,3-dipolar cycloaddition of a variety of organic azides **190**, **191**, or **195** to ester or benzotriazolylcarbonyl-activated acetylenic amides under microwave-assisted reaction conditions [107]. For example, reaction of 1,4-bis(azidomethyl)benzene **195** with 2 equiv. ethyl 4-anilino-4-oxo-2-butynoate (**198a**: R = Ph) in toluene under continuous microwave irradiation (120 W) with simultaneous cooling at 75 °C for 1 h furnished a mixture of regioisomeric bis-triazoles **199a** and **199b**. The major regioisomer **199a** was isolated by column chromatography in 54% yield. (Scheme 11.51).

Regiospecific catalytic cycloaddition of alkynes and azides using copper complexes as soluble organic catalysts enables the preparation of "divalent" and "mul-

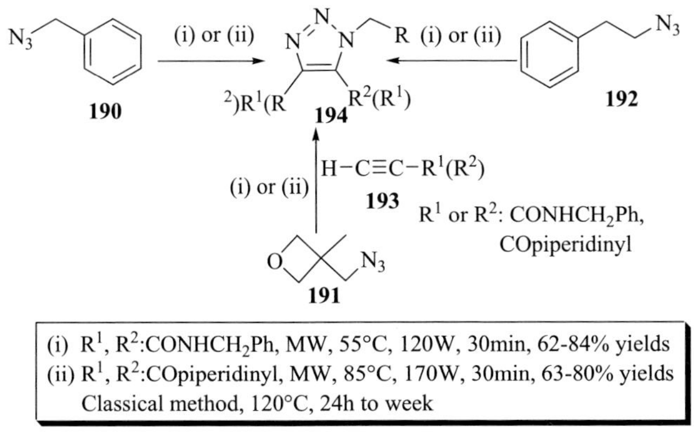

Scheme 11.49

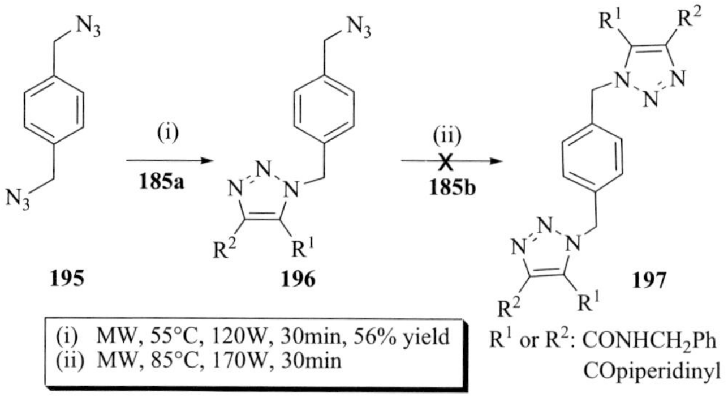

Scheme 11.50

(i) MW, 55°C, 120W, 30min, 56% yield
(ii) MW, 85°C, 170W, 30min

R^1 or R^2: CONHCH$_2$Ph
COpiperidinyl

tivalent neoglycoconjugates" (1,4-disubstituted 1,2,3-triazoles) **202**, **204**, and **207** in toluene under microwave conditions [108]. Thus, the microwave irradiation of alkynes **200** and **205** with azide derivatives **201**, **203**, and **206** and diisopropylethylamine, DIPEA, or diazabicycloundecane, DBU, in the presence of 10–20% copper catalyst [(Ph$_3$P)$_3$, CuBr, or (EtO)$_3$P, CuI] in toluene afforded adducts **202**, **204**, and **207**, respectively, in moderate to high yields (51–99%) within 6 to 45 min irradiation using 2-min bursts (Scheme 11.52). The cycloaddition reactions were also performed at room temperature without the assistance of microwave irradiation. The catalyzed reactions proved to be equally efficient and regiospecific, with similar isolated yields, although reaction times were longer (24–72 h) than those observed in the microwave-irradiated reactions.

Fokin and Van der Eycken, et al. also described the microwave-assisted three-component one-pot synthesis of 1,4-disubstituted-1,2,3-triazoles **211** (Scheme 11.53) [109]. A suspension of alkyl halides **208**, sodium azide **209**, and alkynes **210** in a 1:1 mixture of t-BuOH and water, together with a copper catalyst, in a

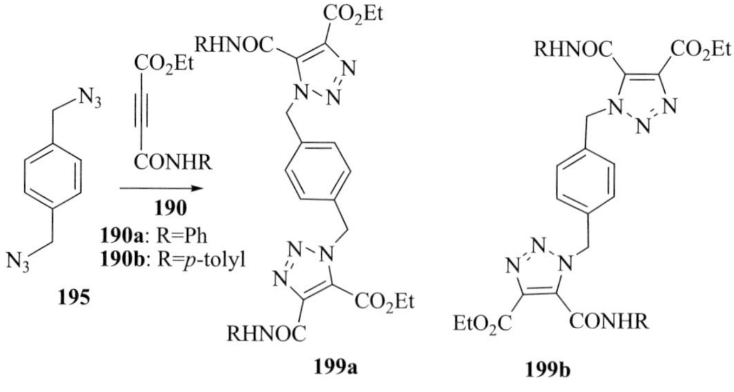

Scheme 11.51

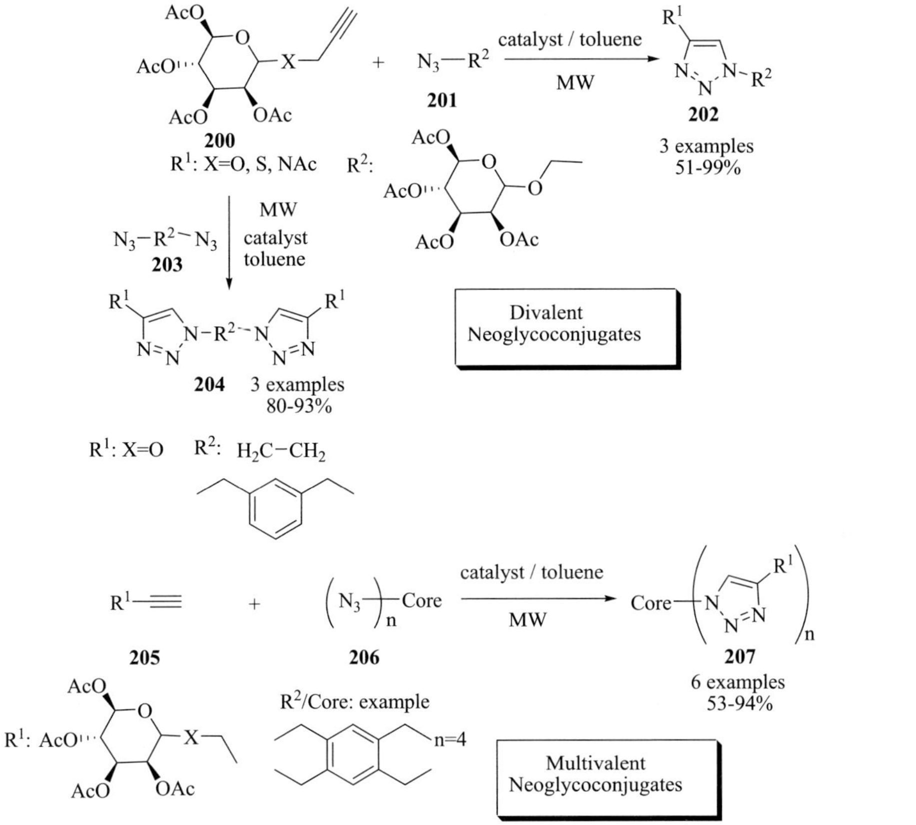

Scheme 11.52

sealed vessel were treated with microwaves. The reactions were complete within 10 to 15 min of irradiation at 125 °C, and the products were isolated as single regioisomers **211** in excellent yields (81–93%) and high purity. The Cu(I) catalyst was prepared *in situ* by the reaction of Cu(0) and Cu(II).

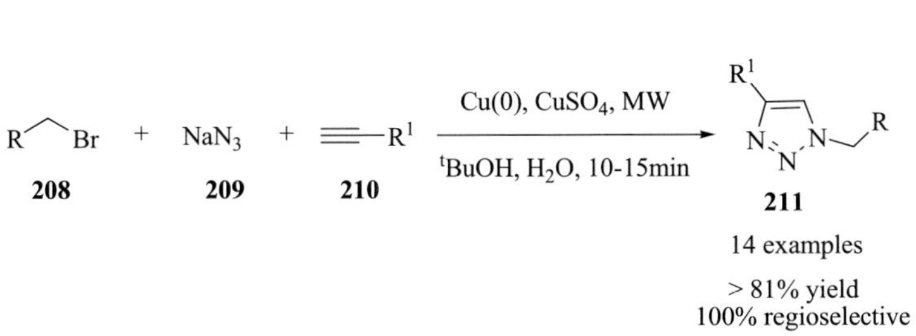

Scheme 11.53

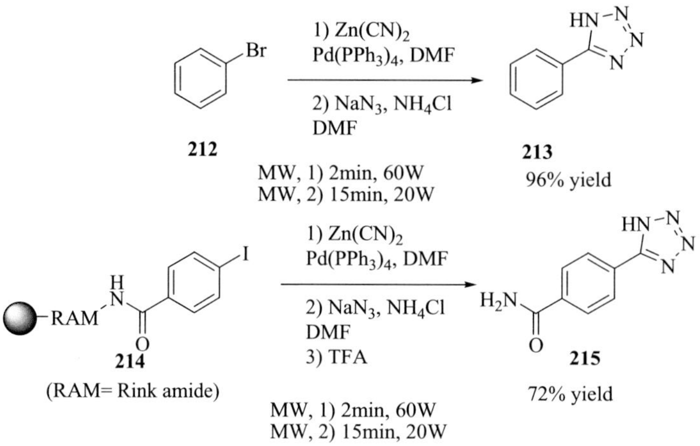

Scheme 11.54

Another interesting type of 1,3-dipolar cycloaddition with azides involves condensation with nitriles as dipolarophiles to form tetrazoles. These products are of particular interest to the medicinal chemist, because they probably constitute the most commonly used bioisostere of the carboxyl group. Reaction times of many hours are typically required for the palladium-catalyzed cyanation of aryl bromides under the action of conventional heating. The subsequent conversion of nitriles to tetrazoles requires even longer reaction times of up to 10 days to achieve completion. Under microwave irradiation conditions, however, the nitriles are rapidly and smoothly converted to tetrazoles in high yields. An example of a one-pot reaction is shown in Scheme 11.54 [110], in which the second step, i.e. the cycloaddition, was achieved successfully under the action of careful microwave irradiation. The flash heating method is also suitable for conversion of **212** and **214** to tetrazoles **213** and **215**, respectively, on a solid support, as shown in Scheme 11.54.

New sterically hindered 3-(5-tetrazolyl)pyridines **217** and **219** have been successfully prepared by cycloaddition of trimethylsilylazide (TMSN$_3$/Bu$_2$SnO reagent) to diversely substituted nicotinonitriles **216** under microwave irradiation conditions [111] (Scheme 11.55). The reaction had previously been explored under thermal conditions and was usually unsuccessful. It should be noted that the optimum experimental conditions had been determined using model hindered gem-substituted carbocyclic nitriles **216**. Indeed, after this exploration the use of potentially toxic and explosive sodium azide could be avoided and advantageously replaced with the TMSN$_3$/Bu$_2$SnO system. Thus, diverse nicotinonitriles **216** were added to the azide dipole under microwave irradiation conditions to produce several 3-(5-tetrazolyl)pyridines **217** in good yields. In an intramolecular version of the reaction, novel fused heterocycles, (tetrazolo)azaisoindole systems **219**, were obtained in 80–99% yields after 2–4 h under the action of microwave activation at 140 °C, compared with 4–5 days thermally, in toluene under reflux.

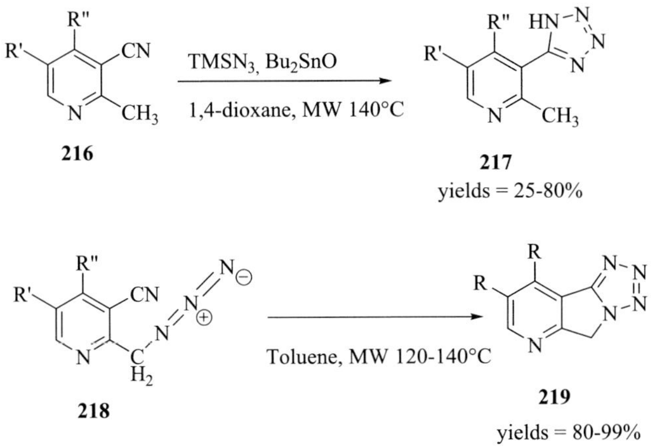

11.6 [2+2] Cycloadditions

[Scheme 11.55 showing conversion of 216 to 217 with TMSN$_3$, Bu$_2$SnO in 1,4-dioxane, MW 140°C, yields = 25–80%; and conversion of 218 to 219 in Toluene, MW 120–140°C, yields = 80–99%]

Scheme 11.55

11.6
[2+2] Cycloadditions

[2+2] Cycloaddition reactions enable easy access to four-membered rings such as β-lactams, oxetenes, cyclobutanones, and heterocyclic compounds. Only two examples, reaction of ketenes or β-formyl enamides with imines or alkynes, respectively, have been investigated under the action of microwaves [3j], however. These limitations could be because of the facile polymerization of ketenes.

11.6.1
Cycloadditions of Ketenes with Imines

The three most important reactions for generating ketenes are the base-mediated dehydrochlorination of acyl chlorides, the dehalogenation of α-chloroacyl chlorides, and the transformation of α-diazoketones. These three methods have been explored under microwave conditions.

The Bose research group has developed a large variety of [2+2] cycloaddition reactions of ketenes with imines to form β-lactams [16, 112, 113]. In 1991, Bose et al. described the synthesis of α-vinyl β-lactams by [2+2] cycloaddition of α,β-unsaturated acyl chlorides with a Schiff base in chlorobenzene. Reaction times were reduced from several hours with conventional methods to 5 min under the action of microwaves [112], and yields of the desired adducts were improved from very low to approximately 70% when compared with classical methods.

An efficient and rapid method for synthesis of β-lactams under solvent-free conditions in the presence of a phase-transfer agent (18-crown-6) in a closed Teflon vessel has also been developed by Hamelin et al. [46a]. The same solvent-free approach has also been successfully used by Kidwai et al. for synthesis of N-(4-hydroxycyclohexyl)-3-mercapto-3-cyano-4-arylazetidine-2-ones [40]. The adducts

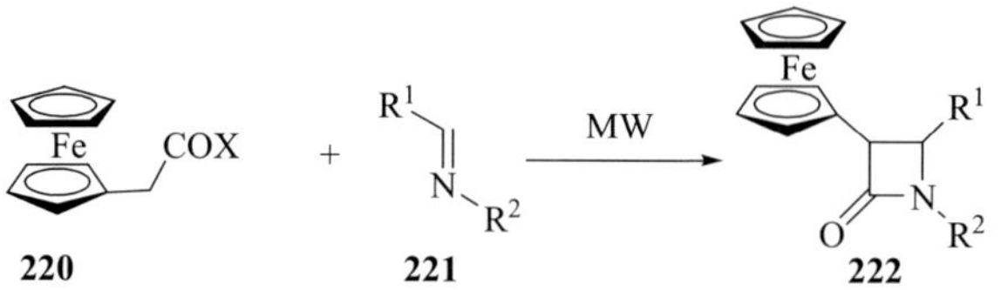

Scheme 11.56

were synthesized by [2+2] cycloaddition reactions of N-(4-hydroxycyclohexyl)-azyldiimine with ethyl α-mercapto-α-cyanoacetate on basic alumina under the action of microwaves. The reaction time was reduced to minutes from hours, in comparison with conventional heating and, moreover, the yields were improved [40].

In an interesting example, Bonini et al. [114] have shown that the one-pot synthesis of β-lactams **222** containing the ferrocene moiety can be performed by microwave-assisted [2+2] cycloaddition (Scheme 11.56).

Podlech et al. [115] recently reported the transformation of diazoketones **223** to ketenes, which react further with imines **224** to form β-lactams **225a** and **225b** in moderate to excellent yields (40–85%) under microwave irradiation conditions or photochemical conditions (Scheme 11.57).

In 2004 Xu et al. investigated the preparation of β-lactams by reaction of ketenes generated from α-diazoketones with a series of acyclic and cyclic imines under the action of either microwaves in dichlorobenzene (DCB) or photoirradiation in CH_2Cl_2 by (Scheme 11.58) [116].

These results indicate that the zwitterionic azabutadiene intermediates generated from imines **227** or **228** and ketenes undergo a *conrotatory* ring closure only to produce β-lactams **229** or **230**. This reveals the inapplicability of the Woodward–Hoffmann rules to the photoirradiation-induced Staudinger reaction. Second, when acyclic imines **227** were employed, E to Z isomerization of the imine moiety in the zwitterionic intermediates occurs, induced by microwave or ultraviolet irradiation. The results also provide direct experimental evidence for the proposed stereochemical process of the microwave and photoirradiation-induced Staudinger reactions, which, in conjunction with the results of Podlech

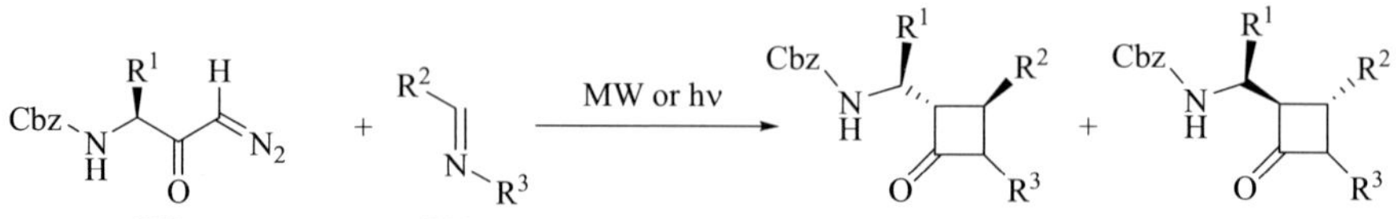

Scheme 11.57

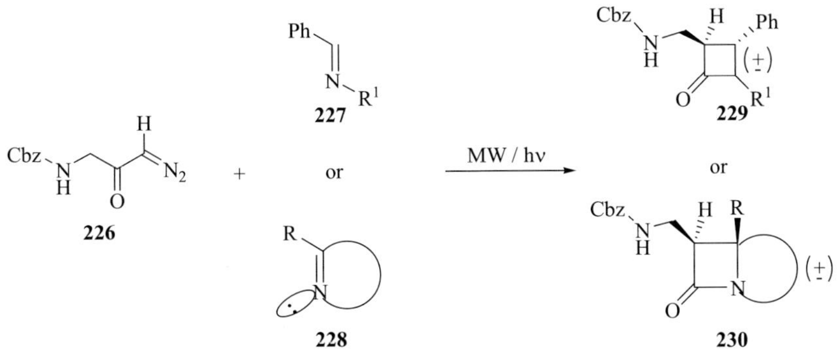

Scheme 11.58

et al. [115], indicate that the microwave-assisted reaction produces β-lactams with the same stereochemistry as the photoirradiation-induced reaction. Whereas the photochemical reaction often generates several byproducts and is sometimes inapplicable to the synthesis of some bicyclic β-lactams, the microwave-assisted reaction is more efficient, convenient, and practical, making this technique a particularly suitable method for the synthesis of bicyclic β-lactams.

11.6.2
Cycloadditions of β-Formyl Enamides with Alkynes

Systematic examination of the [2+2] cycloaddition reactions of β-formyl enamides with alkynes with use of microwave technology has been performed only recently. Novel utility of β-formyl enamides **231** in cycloaddition with acetylenic dienophiles **232**, solvent-free on basic alumina, under the action of microwave irradiation was reported by Boruah et al. (Scheme 11.59) [117]. Indeed, the desired steroidal oxetenes **233** were obtained in 70–80% yields under the optimized conditions (1 equiv. **231**, 1.5 equiv. **232**, and 2 equiv. triphenylphosphine for 5 min). Conventional heating in a protic solvent for 72 h led to very poor yields (15–23%) of oxetenes **233**.

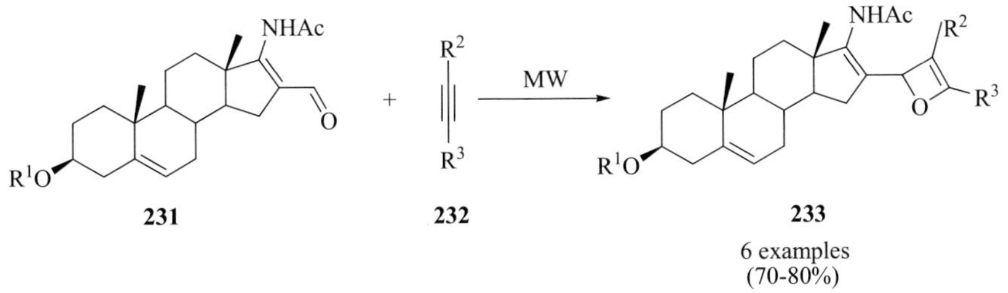

Scheme 11.59

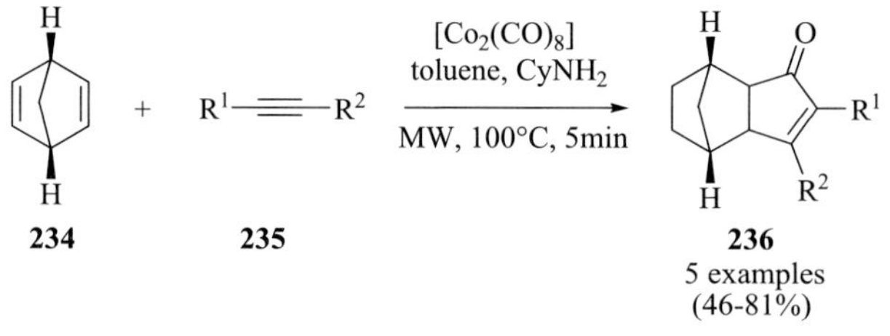

Scheme 11.60

11.7
Other Cycloadditions

Few other cycloaddition reactions with microwave irradiation have appeared in the literature since the first applications were published. In particular, the [2+2+1] reaction, known as the Pauson–Khand reaction, [6+4], and tandem [6+4]–[4+2] cycloadditions have been reported as being performed successfully under microwave conditions. A range of examples is described below.

The [2+2+1] cycloaddition of an alkene, an alkyne, and carbon monoxide is often the method of choice for preparation of cyclopentenones [118]. Groth et al. have demonstrated that such Pauson–Khand reactions can be performed very efficiently under the action of microwaves (Scheme 11.60) [119]. A small quantity, 20 mol%, of $[Co_2(CO)_8]$ was sufficient to drive the reactions to completion under sealed vessel conditions, without the need for additional carbon monoxide. Under carefully optimized reaction conditions with 1.2 equiv. cyclohexylamine as an additive in toluene, microwave exposure for 5 min at 100 °C provided good yields of the desired cycloadducts. Similar results were published independently by Evans et al. (Scheme 11.61) [120].

Hong et al. reported a [6+4] cycloaddition of fulvenes **239** with α-pyrones **240** to provide azulene-indoles **241** in good yields [121] (Scheme 11.62). Reactions were performed in DMF solution using a monomode reactor (30 W for 60 min). It is worthy of note that reactions performed using conventional techniques required

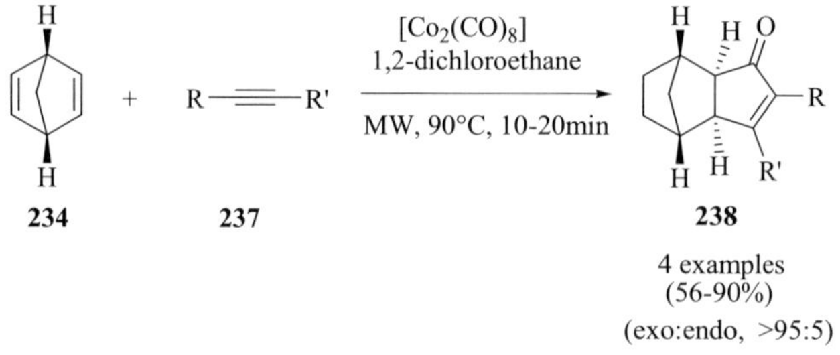

Scheme 11.61

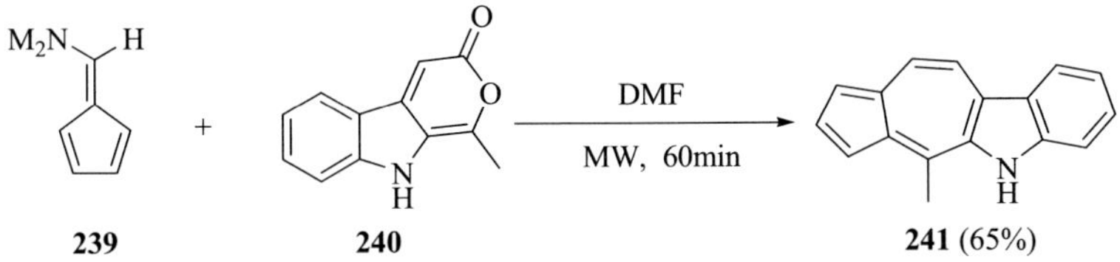

Scheme 11.62

heating for 5 days and provided products in lower yields, with recovered starting materials and decomposition products.

In a different study, the same authors showed that microwave irradiation can alter the reaction pathway. With conventional heating (methods **B** and **C**) the Diels–Alder adducts are favored, whereas under the action of microwaves (method **A**), the tandem [6+4]–[4+2] cycloaddition products were obtained exclusively (Scheme 11.63) [122].

11.8
Conclusions

This chapter emphasizes recent applications of microwave technology in the synthesis of carbocycles and heterocycles by cycloaddition reactions. In these applications, it is clearly demonstrated that it is possible to use two different types of

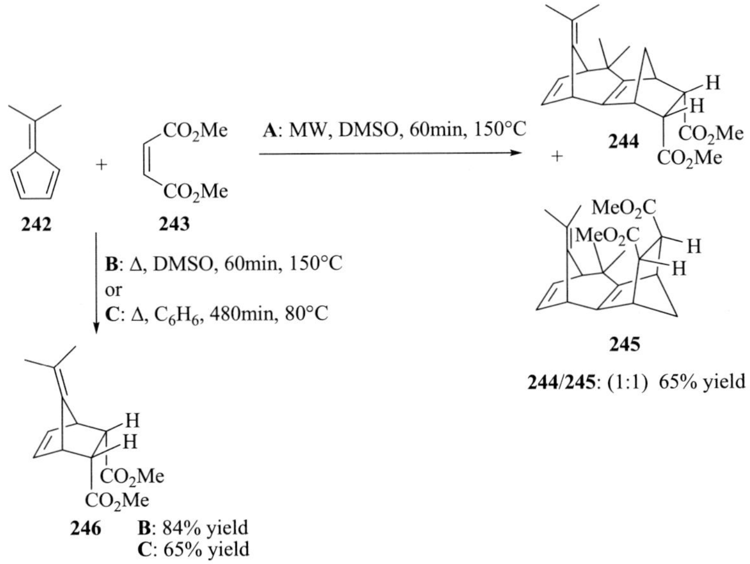

Scheme 11.63

microwave reactor technology – multimode (for example domestic ovens) and monomode (also referred to as single-mode) reactors. Several procedures has been designed to work with a combination of open or sealed vessel technology in solution or under solvent-free conditions. The choice of the method depends on the nature of the reagents and the type of cycloaddition reaction, although solvent-free, open-vessel conditions using solid supports are especially appropriate to microwave conditions. There are, furthermore, other advantages of these solvent-free open-vessel procedures in *"green chemistry"* because the absence of solvent renders the reactions environmentally friendly.

The benefits of microwave heating, in conjunction with polar solvents and sealed-vessel systems, are diverse:

- Significantly, microwave processing frequently leads to noticeable reduction in reaction times and higher yields. The rate improvements observed may simply be a consequence of the high reaction temperatures that can be achieved rapidly by use of this nonclassical heating method, or may result from the involvement of so-called specific or nonthermal microwave effects (cf. Chapter 4 in this book).
- The choice of solvent for a given reaction is not governed by the boiling point (as in conventional reflux) but rather by the dielectric properties of the reaction medium which can be easily tuned by, for example, addition of highly polar materials such as ionic liquids.
- The overall process is more energy-efficient than classical oil-bath heating, because direct "in-core" heating of the medium achieves uniform and stable heating of the reaction medium.

It is noticeable that nonthermal effects are often important in determining the rates and chemo, regio, or stereoselectivity of the cycloadditions. There are, moreover, many cycloaddition reactions with great potential for automated synthesis of small compound libraries, traditionally performed in lengthy processes, that could be dramatically accelerated by solvent-free microwave irradiation. New therapeutic hetero-macromolecular targets are constantly being identified as drug targets, and lead-generation and optimization processes must be accelerated. Traditional methods of organic heterocyclic synthesis can be too slow to satisfy the future demand for compounds. Microwave-assisted cycloaddition reaction is a technique that can complement traditional methods and has the potential of accelerating the generation of organic molecules.

Acknowledgments

The authors work in microwave chemistry has been generously supported by the CNRST-Morocco (*convention France-Morocco, CHIMIE 04/05*) Fund, PROTARS III financial support (*Programme Thématique d'Appui à la Recherche Scientifique, ref. number: D13/57*) (KB and MS), by the CNRS, and by the French Ministry of Education (GB). We thank all our former and present coworkers for their dedication, enthusiasm, and for their essential contributions.

References

1. (a) S. Kobayashi, K. A. Jorgensen, *Cycloaddition in Organic Synthesis*, Hardcover, **2002**; (b) B. E. Hanson, *Comm. Inorg. Chem.* **2002**, *23*, 289–318; (c) H. Suga, *Angew. Chem. Int. Ed.* **2003**, *42*, 4569–4570; (d) H. W. Carruthers, I. Coldham, *Modern Methods of Organic Synthesis*, 4th ed. Cambridge, **2004**, 159–222.
2. (a) S. Caddick, *Tetrahedron* **1995**, *51*, 10403–10432; (b) A. K. Bose, B. K. Banik, N. Lavlinskaia, M. Jayaraman, M. S. Manhas, *Chemtech* **1997**, *27*, 18–24; (c) C. R. Strauss, R. W. Trainor, *Aust. J. Chem.* **1995**, *48*, 1665–1692.
3. (a) A. de la Hoz, A. Diaz-Ortis, A. Moreno, F. Langa, *Eur. J. Org. Chem.* **2000**, 3659–3673; (b) R. S. Varma, *J. Heterocycl. Chem.* **1999**, *36*, 1565–1571; (c) N. Elander, J. R. Jones, S.-Y. Lu, S. Stone-Elander, *Chem. Soc. Rev.* **2000**, 239–250; (d) N.-F. K. Kaiser, U. Bremberg, M. Larhed, C. Moberg, A. Hallberg, *Angew. Chem.* **2000**, *112*, 3742–3744; (e) R. S. Varma, *Green Chem.* **1999**, 43–55; (f) A. Loupy, A. Petit, J. Hamelin, F. Texier-Boullet, P. Jacquault, D. Mathé, *Synthesis* **1998**, 1213–1234; (g) S. Deshayes, M. Liagre, A. Loupy, J.-L. Luche, A. Petit, *Tetrahedron* **1999**, *55*, 10851–10870; (h) M. Larhed, A. Hallberg, *Drug Discovery Today* **2001**, *6*, 406–416; (i) J. L. Krstenansky, I. Cotterill, *Curr. Opin. Drug Discovery Dev.* **2000**, *4*, 454–461; (j) A. de la Hoz, A. Diaz-Ortis, F. Langa in *Microwaves in Organic Synthesis* (Ed.: A. Loupy), Wiley–VCH, Weinheim, **2002**, pp. 295–343; (k) C. O. Kappe, *Angew. Chem. Int. Ed.* **2004**, *43*, 6250–6284.
4. (a) C. J. Dell, *J. Chem. Soc. Perkin Trans. 1* **1998**, 3873–3905; (b) L. F. Tietze, G. Kettschau, *Hetero Diels–Alder Reactions in Organic Chemistry in Topics in Current Chemistry*, Vol. 190, **1998**, 1–121; (c) K. Bougrin, M. Soufiaoui, A. Loupy, P. Jacquault, *New J. Chem.* **1995**, *19*, 213–219; (d) B. Syassi, K. Bougrin, M. Soufiaoui, *Tetrahedron Lett.* **1997**, *38*, 8855–8858.
5. (a) R. Gedye, F. Smith, K. Westaway, A. Humera, L. Baldisera, L. Laberge, L. Rousell, *Tetrahedron Lett.* **1986**, *26*, 279–282; (b) R J. Giguere, T. L. Bray, S. M. Duncan, G. Majetich, *Tetrahedron Lett.* **1986**, 27, 4945–4948.
6. (a) G. Majetich, R. Hicks, *Radiat. Rhys. Chem.* **1995**, *45*, 567–579; (b) C. R. Strauss in *Microwaves in Organic Synthesis* (Ed.: A. Loupy), Wiley–VCH, Weinheim, **2002**, 35–60.
7. A. K. Bose, M. S. Manhas, S. N. Ganguly, A. H. Sharma, B. K. Banik, *Synthesis* **2002**, 1578–1591.
8. (a) E. Guttierez, A. Loupy, G. Bram, E. Ruiz-Hitzky, *Tetrahedron Lett.* **1989**, *30*, 945–948; (b) A. Ben Alloum, B. Labiad, D. Villemin, *Chem. Commun.* **1989**, 386–387; (c) A. Stambouli, M. Chastrette, M. Soufiaoui, *Tetrahedron Lett.* **1991**, *32*, 1723–1726.
9. Milestone, www.milestonesci.com
10. Biotage AB, www.biotage.com
11. CEM Discover, www.cemsynthesis.com
12. Synthos 3000 from Anton-Paar, www.anton-paar.com
13. (a) K. D. Raner, C. R. Strauss, F. Vyskov, L. Mokbel, *J. Org. Chem.* **1993**, *58*, 950–953; (b) Q. Cheng, W. Zhang, Y. Tagami, T. Oritani, *J. Chem. Soc. Perkin Trans. 1* **2001**, 452–456; (c) K. Bougrin, A. Loupy, M. Soufiaoui, *J. Photochem. Photobiol. C: Photochemistry Reviews* **2005**, *6*, 139–167.
14. D. M. P. Mingos, A. G. Whittaker, *Microwave Dielectric Heating Effects in Chemical Synthesis. In Chemistry under Extreme or non-Classical Conditions*, R. Van Eldik, C. D. Hubbard (Eds.), John Wiley and Sons, New York, **1997**, 479–545.
15. (a) D. R. Baghurst, D. M. P. Mingos, *J. Chem. Soc., Chem. Commun.* **1992**, 674–675; (b) R. Saillard, M. Poux, J. Berlan, M. Audhuy-Peaudecerf, *Tetrahedron* **1995**, *51*, 4033–4042;

(c) F. Chemat, E. Esveld, *Chem. Eng. Technol.* **2001**, *24*, 735–744.

16 A. K. Bose, M. S. Manhas, B. K. Banik, E. W. Robb, *Res. Chem. Intermediat.* **1994**, *20*, 1–11.

17 P. de La Cruz, A. de la Hoz, F. Langa, B. Illescas, N. Martin, C. Seoane, *Synth. Metals* **1997**, *87*, 2283–2284.

18 C.-C. Chi, I-F. Pai, W.-S. Chung, *Tetrahedron* **2004**, *60*, 10869–10876.

19 P. Wasserscheid, T. Welton (Eds.), *Ionic liquids in Synthesis*, Wiley–VCH, Weinheim, **2002**.

20 (a) R. S. Varma, V. V. Namboodiri, *Chem. Commun.* **2001**, 643–644; (b) K. G. Mayo, E. H. Nearhoof, J. J. Kiddle, *Org. Lett.* **2002**, *4*, 1567–1570; (c) K. S. A. Vallin, P. Emilsson, M. Larhed, A. Hallberg, *J. Org. Chem.* **2002**, *67*, 6243–6246; (d) G. Vo-Thanh, B. Pégot, A. Loupy, *Eur. J. Org. Chem.* **2004**, 1112–1116; (e) F. Mavandadi, P. Lidström, *Curr. Top. Med. Chem.* **2004**, *4*, 773–792.

21 M. Avalos, R. Babiano, P. Cintas, F. R. Clemente, J. L. Jiménez, J. C. Palacios, J. B. Sánchez, *J. Org. Chem.* **1999**, *64*, 6297–6305. See also refs 58 and 65a.

22 I.-H. Chen, J.-N. Young, S. J. Yu, *Tetrahedron* **2004**, *60*, 11903–11909.

23 J. Fraga Dubreuil, J. P. Bazureau, *Tetrahedron Lett.* **2000**, *41*, 7351–7355.

24 N. E. Leadbeater, H. M. Torenius, *J. Org. Chem.* **2002**, *67*, 3145–3148.

25 K. S. A. Vallin, P. Emilsson, M. Larhed, A. Hallberg, *J. Org. Chem.* **2002**, *67*, 6243–6246.

26 S. V. Ley, A. G. Leach, R. I. Storer, *J. Chem. Soc.* **2001**, 358–361.

27 (a) J. Hoffmann, M. Nüchter, B. Ondruschka, P. Wasserscheid, *Green Chem.* **2003**, *5*, 296–299; (b) S. Garbacia, B. Desai, O. Lavastre, C. O. Kappe, *J. Org. Chem.* **2003**, *68*, 9136–9139; (c) G. K. Datta, K. S. A. Vallin, M. Larhed, *Mol. Diversity* **2003**, *7*, 107–114; (d) N. Srinivasan, A. Ganesan, *Chem. Commun.* **2003**, 916–917; (e) N. E. Leadbeater, H. M. Torenius, H. Tye, *Mol. Diversity* **2003**, *7*, 135–144; (f) W.-C. Shieh, M. Lozanov, O. Repic, *Tetrahedron Lett.*

2003, *44*, 6943–6945; (g) M. J. Gaunt, A. S. Jessiman, P. Orsini, H. R. Tanner, D. F. Hook, S. V. Ley, *Org. Lett.* **2003**, *5*, 4819–4822.

28 A. Loupy (Ed.), *Microwaves in Organic Synthesis*, Wiley–VCH, Weinheim, **2002**.

29 (a) M. Kidwai, *Pure Appl. Chem.*, **2001**, *73*, 147–151; (b) R. S. Varma, *Pure Appl. Chem.*, **2001**, *73*, 193–198; (c) R. S. Varma, *Tetrahedron* **2002**, *58*, 1235–1255.

30 A. Laporterie, J. Marquié, J. Dubac in *Microwaves in Organic Synthesis* (Ed.: A. Loupy), Wiley–VCH, Weinheim, **2002**, 219–252.

31 (a) A. Ben Alloum, S. Bakkas, K. Bougrin, M. Soufiaoui, *New J. Chem.* **1998**, 809–812; (b) B. Syassi, K. Bougrin, M. Lamiri, M. Soufiaoui, *New J. Chem.* **1998**, 1545–1548; (c) F. Louërat, K. Bougrin, A. Loupy, A. M. Ochoa de Retana, J. Pagalday, F. Palacios, *Heterocycles* **1998**, *48*, 161–168; (d) H. Kaddar, J. Hamelin, H. Benhaoua, *J. Chem. Res. (S)* **1999**, 718–719; (e) R. M. Dinica, I. I. Druta, C. Pettinari, *Synlett* **2000**, 1013–1015; (f) Q. Cheng, W. Zhang, Y. Tagami, T. Oritani, *J. Chem. Soc. Perkin Trans. 1* **2001**, 452–456; (g) J. Azizian, A. Asadi, K. Jadidi, *Synth. Commun.* **2001**, *31*, 2727–2733; (h) A. R. Katritzky, S. K. Singh, *J. Org. Chem.* **2002**, *67*, 9077–9079; (i) A. Krishnaiah, B. Narsaiah, *J. Fluorine Chem.* **2002**, *113*, 133–137; (j) J. Jayashankaran, R. D. R. S. Manian, R. Raghunathan, *Tetrahedron Lett.* **2004**, *45*, 7303–7305.

32 (a) A. Loupy, *Solvent-free Reactions in Modern Solvent in Organic Chemistry. Topic. Curr. Chem.* **1999**, Vol. 206, 155–207; (b) G. W. V. Cave, C. L. Raston, J. L. Scott, *Chem. Commun.* **2001**, 2159–2169; (c) J. Freitag, M. Nüchter, B. Ondruschka, *Green Chem.* **2003**, *5*, 291–295.

33 K. Smith (Ed.), *Solid Supports and Catalysts in Organic Synthesis*, Ellis Horwood, P.T.R. Prentice Hall, Chichester, **1992**.

34 (a) M. Avalos, R. Babiano, J. L. Bravo, P. Cintas, J. L. Jiménez, J. C.

Palacios, *Tetrahedron Lett.* **1998**, *39*, 9301–9304; (b) M. R. Dintzner, T. W. Lyons, M. H. Akroush, P. Wucka, A. T. Rzepka, *Synlett* **2005**, 785–788.
35 (a) A. de la Hoz, A. Diaz-Ortis, J. M. Fraile, M. V. Gómez, J. A. Mayoral, A. Moreno, A. Saiz, E. Vázquez, *Synlett* **2001**, 236–237.
36 R. G. Jacob, G. Perin, G. V. Botteselle, E. J. Lenardão, *Tetrahedron Lett.* **2003**, *44*, 6809–6812.
37 A. Krishnaiah, B. Narsaiah, *J. Fluorine Chem.* **2002**, *113*, 133–137.
38 G. Giacomelli, L. De Luca, A. Porcheddu, *Tetrahedron* **2003**, *59*, 5437–5440.
39 M. A. J. Charmier, V. Y. Kukushkin, A. J. L. Pombeiro, *J. Chem. Soc., Dalton Trans.* **2003**, 2540–2543.
40 M. Kidwai, R. Venkataramanan, S. Kohli, *Synth. Commun.* **2000**, *30*, 989–1002.
41 (a) A. Diaz-Ortiz, E. Diez-Barra, A. de la Hoz, P. Prieto, A. Moreno, *J. Chem. Soc. Perkin Trans. 1* **1994**, 3595–3598; (b) A. Diaz-Ortiz, E. Diez-Barra, A. de la Hoz, P. Prieto, A. Moreno, F. Langa, T. Prangé, A. Neuman, *J. Org. Chem.* **1995**, *60*, 4160–4166.
42 B. M. Trost, M. L. Crawley, *J. Am. Chem. Soc.* **2002**, *124*, 9328–9329.
43 D. C. G. A. Pinto, A. M. S. Silva, L. M. P. M. Almeida, J. R. Carrillo, A. Diaz-Ortiz, A. De la Hoz, J. A. S. Cavaleiro, *Synlett* **2003**, 1415–1418.
44 (a) R. S. Varma, D. Kumar, *Tetrahedron Lett.* **1999**, *40*, 7665–7669; (b) C. O. Kappe, D. Kumar, R. S. Varma, *Synthesis* **1999**, 1799–1803; (d) S. J. Song, S. J. Cho, D. K. Park, T. W. Kwon, S. A. Jenekhe, *Tetrahedron Lett.* **2003**, *44*, 255–257.
45 (a) G. Bram. A. Loupy, J. Sansoulet, *Isr. J. Chem.* **1985**, *26*, 291–298; (b) A. Loupy, J. L. Luche, *Sonochemical and Microwave Activation in Phase Tranfer Catalysis*, in *Handbook of Phase-transfer catalysis*, Sasson, Y.; Neumann, R., Eds.; Blackie Academic & Professional, Chapman & Hall, London, **1996**, 369–404.
46 (a) F. Texier-Boullet, R. Latouche, J. Hamelin, *Tetrahedron Lett.* **1993**, *34*, 2123–2126; (b) F. Pérez-Balderas, M. Ortega-Munoz, J. Morales-Sanfrutos, F. Hernández-Mateo, F. G. Calvo-Flores, J. A. Calvo-Asin, J. Isac-Garcia, F. Santoyo-González, *Org. Lett.* **2003**, *5*, 1951–1954.
47 B. Touaux, B. Klein, F. Texier-Boullet, J. Hamelin, *J. Chem. Res. (S)* **1994**, 116–117.
48 B. Touaux, F. Texier-Boullet, J. Hamelin, *Heteroatom Chem.* **1998**, *9*, 351–354.
49 (a) J. W. Walkiewicz, G. Kazonicii, S. L. McGill, *Miner. Metal. Proc.* **1988**, *5*, 39–41; (b) M. De Fatima Pereira, L. Picot, J. Guillon, J. M. Léger, C. Jarry, V. Thiéry, C. Besson, *Tetrahedron Lett.* **2005**, *46*, 3445–3447.
50 (a) J. J. Vanden Eynde, N. Labuche, Y. Van Haverbeke, *Synth. Commun.* **1997**, *27*, 3683–3690; (b) S. Frère, V. Thiéry, T. Besson, *Tetrahedron Lett.* **2001**, *42*, 2791–2794; (c) F. R. Alexandre, A. Berecibar, R. Wrigglesworth, T. Besson, *Tetrahedron* **2003**, *59*, 1413–1419.
51 (a) B. Garrigues, C. Laporte, R. Laurent, A. Laporterie, J. Dubac, *Liebigs Ann.* **1996**, 739–741; (b) P. Garrigues, B. Garrigues, *C.R. Acad. Sci. Paris Ser. II* **1998**, 545–550; (c) B. Garrigues, R. Laurent, C. Laporte, A. Laporterie, J. Dubac, *Liebigs Ann.* **1996**, 743–744.
52 C. Laporte, A. Oussaid, B. Garrigues, *C. R. Acad. Sci. Paris, série llc* **2000**, *3*, 321–325.
53 C. Laporte, A. Oussaid, B. Garrigues, *J. Nat.* **2001**, *13*, 16–18.
54 (a) D. Boger, S. Weinreb (Eds.), *Hetero Diels–Alder Methodology in Organic Synthesis*, Hardbound, **1987**; (b) R. J. Hamers, S. K. Coulter, M. D. Ellison, J. S. Hovis, D. F. Padowitz, M. P. Schwartz, C. M. Greenlief, J. N. J. Russell, *Acc. Chem. Res.* **2000**, *33*, 617–624; (c) F. Francesco, T. Aldo (Eds.), *The Diels–Alder Reaction; Selected Practical Methods*, John Wiley & Sons, **2002**; (d) H. Oikawa, T. Tokiwano, *Nat. Prod. Rep.* **2004**, *21*, 321–352.
55 (a) H. Fillion, J. L. Luche, *Synthetic Organic Sonochemistry: Cycloadditions,*

J. L. Luche (Ed.), Plenum, New York, **1998**, 91–106; (b) M. Avalos, R. Babiano, N. Cabello, P. Cintas, M. B. Hursthouse, J. L. Jiménez, M. E. Light, J. C. Palacios, *J. Org. Chem.* **2003**, *68*, 7193–7203; (c) W. Bonrath, R. A. P. Schmidt, *Advances in Organic Synthesis: Ultrasound in Synthetic Organic Chemistry*, G. Jenner (Ed.), **2005**, Vol. 1, 81–117; (d) A. de la Hoz, A. Diaz-Ortiz, A. Moreno, *Advances in Organic Synthesis: Activation of Organic Reactions by Microwaves*, G. Jenner (Ed.), **2005**, Vol. 1, 119–171.

56 B. Alcaide, P. Almendros, C. Aragoncillo, M. C. Redondo, *Chem. Commun.* **2002**, *14*, 1472–1473.

57 A. D. Mance, K. Jakopčić, *Mol. Diversity* **2005**, *9*, 229–232.

58 E. Van der Eycken, P. Appukkuttan, W. De Borggraeve, W. Dehaen, D. Dallinger, C. O. Kappe, *J. Org. Chem.* **2002**, *67*, 7904–7909.

59 B. R. Lahue; S.-M. Lo, Z.-K. Wan, G. H. C. Woo, J. K. Snyder, *J. Org. Chem.* **2004**, *69*, 7171–7182.

60 B. Shao, *Tetrahedron Lett.* **2005**, *46*, 3423–3427.

61 W. B. Wang, E. J. Roskamp, *Tetrahedron Lett.* **1992**, *33*, 7631–7633.

62 M. Shanmugasundaram, S. Manikandan, R. Raghunathan, *Tetrahedron* **2002**, *58*, 997–1003.

63 S. Manikandan, M. Shanmugasundaram, R. Raghunathan, *Tetrahedron* **2002**, *58*, 8957–8962.

64 C. M. Gordon, *Appl. Catal. A* **2001**, *222*, 101–117.

65 (a) N. Kaval, J. Van der Eycken, J. Caroen, W. Dehaen, G. A. Strohmeier, C. O. Kappe, E. Van der Eycken, *J. Comb. Chem.* **2003**, *5*, 560–568; (b) N. Kaval, W. Dehaen, C. O. Kappe, E. Van der Eycken, *Org. Biomol. Chem.* **2004**, *2*, 154–156.

66 M. D. Mihovilovic, H. G. Leisch, K. Mereiter, *Tetrahedron Lett.* **2004**, *45*, 7087–7090.

67 C. J. Moody, R. A. Hughes, S. P. Thompson, L. Alcaraz, *Chem. Commun.* **2002**, *16*, 1760–1761.

68 B. Illescas, N. Martín, C. Seoane, P. de la Cruz, F. Langa, F. Wudl, *Tetrahedron Lett.* **1995**, *36*, 8307–8310.

69 J. L. Delgado, P. de la Cruz, F. Langa, A. Urbina, J. Casado, J. T. L. Navarrete, *Chem. Commun.* **2004**, 1734–1735.

70 C.-C. Chi, I. F. Pai, W.-S. Chung, *Tetrahedron* **2004**, *60*, 10869–10876.

71 (a) W.-S. Chung, W.-J. Lin, W.-D. Liu, L.-G. Chen, *J. Chem. Soc. Chem. Commun.* **1995**, 2537–2539; (b) W.-D. Liu; C.-C. Chi; I.-F. Pai; A.-T. Wu; W.-S. Chung, *J. Org. Chem.* **2002**, *67*, 9267–9275.

72 J. Safaei-Ghomi, M. Tajbakhsh, Z. Kazemi-Kani, *Acta Chim. Slov.* **2004**, *51*, 545–550.

73 A. Loupy, F. Maurel, A. Sabatié-Gogová, *Tetrahedron* **2004**, *60*, 1683–1691.

74 S. Werner, D. P. Curran, *Org. Lett.* **2003**, *5*, 3293–3296.

75 I. Devi, H. N. Borah, P. J. Bhuyan, *Tetrahedron Lett.* **2004**, *45*, 2405–2408.

76 M. Gohain, D. Prajapati, B. J. Gogoi, J. S. Sandhu, *Synlett* **2004**, 1179–1182.

77 L. D. S. Yadav, R. Kapoor, *Tetrahedron Lett.* **2003**, *40*, 8951–8954.

78 R. K. Bansal, A. Dandia, N. Gupta, D. Jain, *Heteroatom Chem.* **2003**, *14*, 560–563.

79 (a) T. V. Rajanbabu, G. S. Reddy, *J. Org. Chem.* **1986**, *51*, 5458–5461; (b) K. B. G. Torssell, *Nitrile Oxides, Nitrones and Nitronates in Organic Synthesis*; VCH: New York, **1988**.

80 (a) G. Just, L. Dahl, *Tetrahedron* **1968**, *24*, 5251–5269; (b) C. Grundmann, J. M. Dean, *J. Org. Chem.* **1965**, *30*, 2809–2812; (c) H. Hassner, K. M. L. Rai, *Synthesis* **1989**, 57–59.

81 (a) T. Mukaiyama, T. Hoshino, *J. Am. Chem. Soc.* **1960**, *82*, 5339–9342; (b) M. Falorni, G. Giacomelli, E. Spanu, *Tetrahedron Lett.* **1998**, *39*, 9241–9244.

82 T. Shimizu, Y. Hayashi, H. Shibafuchi, K. Teramura, *Bull. Chem. Soc. Jpn* **1986**, *59*, 2827.

83 T. J. Lu, G. M. Tzeng, *J. Chin. Chem. Soc.* **2000**, *47*, 189–196.

84 A. Díaz-Ortiz, E. Díez-Barra, A. de la Hoz, A. Moreno, M. J. Gómez-

Escalonilla, A. Loupy, *Heterocycles* **1996**, *43*, 1021–1030.

85 A. Corsaro, U. Chiacchio, V. Librando, S. Fisichella, V. Pistarà, *Heterocycles* **1997**, *45*, 1567–1572.

86 (a) P. Mičúch, L. Fišera, M. K. Cyrañski, T. M. Krygowski, *Tetrahedron Lett.* **1999**, *40*, 167–170; (b) P. Mičúch, L. Fišera, M. K. Cyračski, T. M. Krygowski, J. Krajčík, *Tetrahedron* **2000**, *56*, 5465–5472.

87 F. Langa, P. de la Cruz, E. Espíldora, A. González-Cortés, A. de la Hoz, V. López-Arza, *J. Org. Chem.* **2000**, *65*, 8675–8684.

88 N. M. Fédou, P. J. Parsons, E. Viseux, A. J. Whittle, *Org. Lett.* **2005**, *15*, 3179–3182.

89 P. A. Colinas, V. Jäger, A. Lieberknecht, R. D. Bravo, *Tetrahedron Lett.* **2003**, *44*, 1071–1074.

90 R. K. Howe, T. A. Gruner, L. G. Carter, L. L. Black, J. E. Franz, *J. Org. Chem.* **1978**, *43*, 3736–3742.

91 A. J. Morrison, R. M. Paton, R. D. Sharp, *Synth. Commun.* **2005**, *35*, 813–813.

92 (a) B. Baruah, D. Prajapati, A. Boruah, J. S. Sandhu, *Synth. Commun.* **1997**, *27*, 2563–2567; (b) S. Rigolet, P. Goncalo, J. M. Mélot, J. Vébrel, *J. Chem. Res. (S)* **1998**, 686–687; (c) V. Ondrus, M. Orság, L. Fišera, N. Prónayosá, *Tetrahedron* **1999**, *55*, 10425–10436; (d) R. S. Kusurcar, V D. Kannadkar, *Synth. Commun.* **2001**, *31*, 2235–2239.

93 G. Raunak, V. Kumar, S. Mukherjee, G. Poonam, A. K. Prasad, C. E. Olsen, S. J. C. Schäffer, S. K. Sharma, A. C. Watterson, W. Errington, V. S. Parmar, *Tetrahedron* **2005**, *61*, 5687–5697.

94 G. Bashiardes, C. Cano, B. Mauzé, *Synth. Commun.* **2005**, *35*, 587–590.

95 B. Desai, T. N. Danks, G. Wagner, *J. Chem. Soc., Dalton Trans.* **2003**, 2544–2549.

96 B. Desai, T. N. Danks, G. Wagner, *J. Chem. Soc., Dalton Trans.* **2004**, 166–171.

97 M. A. J. Charmier, V. Y. Kukushkin, A. J. L. Pombeiro, *J. Chem. Soc., Dalton Trans.* **2003**, 2540–2543.

98 B. Dugovič, L. Fišera, C. Hametner, N. Prónayová, *ARKIVOC* **2003**, 162–169.

99 P. Merino, S. Franco, F. L. Merchan, P. Romero, T. Tejero, S. Uriel, *Tetrahedron Asymmetry* **2003**, *14*, 3731–3743.

100 G. Bashiardes, I. Safir, A. S. Mohamed, F. Barbot, J. Laduranty, *Org. Lett.* **2003**, *5*, 4915–4918.

101 A. Díaz-Ortiz, A. de la Hoz, A. Herrero, P. Prieto, A. Sánchez-Migallón, F. P. Cossío, A. Arrieta, S. Vivanco, C. Foces-Foces, *Mol. Diversity* **2003**, *7*, 175–180.

102 J. Jayashankaran, R. D. R. S. Manian, R. Venkatesan, R. Raghunathan, *Tetrahedron* **2005**, *61*, 5595–5598.

103 J. Azizian, A. R. Karimi, A. A. Mohammadi, M. R. Mohammadizadeh, *Synthesis* **2004**, 2263–2265.

104 (a) S. Wang, J.-M. Zhang, L.-P. Song, H. Jiang, S.-Z. Zhu, *J. Fluorine Chem.* **2005**, *126*, 349–353; (b) J.-M. Zhang, W. Yang, S. Wang, P. He, S. Zhu, *Synth. Commun.* **2005**, *35*, 89–96.

105 W. Zhang, C. H.-T. Chen, *Tetrahedron Lett.* **2005**, *46*, 1807–1810.

106 R. Atir, S. Mallouk, K. Bougrin, A. Laghzizil, M. Soufiaoui, *Synth. Commun.* **2005**, *36*, 111–120.

107 A. R. Katritzky, Y. Zhang, S. K. Singh, P. J. Steelb, *ARKIVOC* **2003**, *(xv)*, 47–64.

108 F. Pérez-Balderas, M. Ortega-Muñoz, J. Morales-Sanfrutos, F. Hernández-Mateo, F. G. Calvo-Flores, J. A. Calvo-Asin, J. Isac-Garcia, F. Santoyo-González, *Org. Lett.* **2003**, *5*, 1951–1954.

109 P. Appukkuttan, W. Dehaen, V. V. Fokin, E. Van der Eycken, *Org. Lett.* **2004**, *6*, 4223–4225.

110 M. Alterman, A. Hallberg, *J. Org. Chem.* **2000**, *65*, 7984–7989.

111 S. M. Lukyanov, I. V. Bliznets, S. V. Shorshnev, G. G. Aleksandrov, A. E. Stepanov, A. A Vasil'ev, *Tetrahedron*, **2006**, *62*, in press.

112 A. K. Bose, M. S. Manhas, M. Ghosh, M. Shah, V S. Raju, S. S. Bari, S. N. Newaz, B. K. Banik, A. G. Chaudhary, K. J. Barakat, *J. Org. Chem.* **1991**, *56*, 6968–6970.

113 (a) A. K. Bose, M. S. Manhas, M. Ghosh, M. Shah, V. S. Raju, S. S. K. Tabey, Z. Urbanczyck-Lipkowska, *Heterocycles* **1990**, *30*, 741–744; (b) A. K. Bose, B. K. Banik, S. N. Newaz, M. S. Manhas, *Synlett* **1993**, 897–899; (c) A. K. Bose, B. K. Banik, M. S. Manhas, *Tetrahedron Lett.* **1995**, *36*, 213–216; (d) A. K. Bose, M. Jayaraman, A. Okawa, S. S. Bari, E. W. Robb, M. S. Manhas, *Tetrahedron Lett.* **1996**, *37*, 6989–6992; (e) M. S. Manhas, B. K. Banik, A. Mathur, J. E. Vincent, A. K. Bose, *Tetrahedron* **2000**, *56*, 5587–5601; (f) A. K. Bose, B. K. Banik, A. Mathur, D. R. Wagle, M. S. Manhas, *Tetrahedron* **2000**, *56*, 5603–5619.

114 B. F. Bonini, C. Femoni, M. Comes-Franchini, M. Fochi, G. Mazanti, *Synlett* **2001**, 1092–1096.

115 M. R. Linder, J. Podlech, *Org. Lett.* **2001**, *3*, 1849–1851.

116 Y. Liang, L. Jiao, S. Zhang, J. Xu, *J. Org. Chem.* **2004**, *70*, 334–337.

117 M. Longchar, U. Bora, R. C. Boruah, J. S. Sandhu, *Synth. Commun.* **2002**, *32*, 3611–3616.

118 P. L. Pauson, *Tetrahedron* **1985**, *41*, 5855–5860.

119 S. Fischer, U. Groth, M. Jung, A. Schneider, *Synlett* **2002**, 2023–2026.

120 M. Iqbal, N. Vyse, J. Dauvergne, P. Evans, *Tetrahedron Lett.* **2002**, *43*, 7859–7862.

121 B.-C. Hong, Y.-F. Jiang, E. S. Kumar, *Biorg. Med. Chem. Lett.* **2001**, *11*, 1981–1984.

122 B.-C. Hong, Y.-J. Shr, J.-H. Liao, *Org. Lett.* **2002**, *4*, 663–666.

12
Microwave-assisted Chemistry of Carbohydrates

Antonino Corsaro, Ugo Chiacchio, Venerando Pistarà, and Giovanni Romeo

12.1
Introduction

Activation by using microwave irradiation has recently become very common covering wide fields of applications, for example alimentary, analytical, biomedical and other strictly chemical fields. Initially, the extension of this method of heating to the organic chemistry has suffered a certain delay, both in the academic research, mainly owing to the control of the temperature and pressure (absent in the domestic microwave oven) of several processes in which microwave irradiation was used, and in the industrial research, for the design of reactors necessary for its use in large scale, and the control and containment of microwave energy.

Because of new technology, microwave energy has been widely adopted as an alternative or complementary source to the thermal energy in all areas of chemistry and, particularly, in synthetic chemistry because its use provides higher yields and dramatically shorter reaction times, and sometimes it enables reactions to be conducted under solvent-free conditions or with much reduction of solvent use.

In this chapter, we intend to provide insight into the variety of applications of microwave irradiation in carbohydrate chemistry and to discuss the benefits and limitations of methods connected with microwave activation. Carbohydrates, which have been developed to a limited extent in the recent past, because of their notorious sensitivity to heat, are ideal candidates for microwave-assisted processes which are less destructive than thermal heating because their reactions are favored by interactions between their polar molecules and the electromagnetic field. Although there are several excellent reviews [1] and books [2] on the general utilization of microwave techniques in organic chemistry, some include only a few examples involving carbohydrates and their derivatives. To the best of our knowledge, there are only our review [3] and that by S.K. Das [4] that cover several aspects of carbohydrate chemistry activated by microwave energy.

Most arguments already treated in our previous review [3] have again been discussed in brief, but expanded with the most recent work published until July 2005. Topics include selective and nonselective protection (for example acylation, acetala-

Microwaves in Organic Synthesis, Second edition. Edited by A. Loupy
Copyright © 2006 WILEY-VCH Verlag GmbH & Co. KGaA, Weinheim
ISBN: 3-527-31452-0

tion and, silylation) and deprotection of hydroxyl functionality, O-glycosylation performed by Fischer alcoholysis, by Ferrier's rearrangement, and by formation of 1,6-anhydro-β-D-hexopyranoses, N- and C-glycosylation, catalytic transfer hydrogenation, oxidation, and halogenation. Other subjects treated are stereospecific C–H bond activation for rapid deuterium labeling, reactions with amino-derivatized labels, Ferrier's rearrangement to carbasugars, synthesis of unsaturated monosaccharides, synthesis of monosaccharide dimers, polysaccharides and their derivatives, synthesis of heterocycles and amino acids, and, finally, enzymatic reactions.

12.2
Protection

12.2.1
Acylation

Several methods have been described for rapid and high-yielding microwave-assisted protection of carbohydrate hydroxyl functionalities with acetic anhydride, acetyl, chloroacetyl, pivaloyl, dodecanoyl, and benzoyl chlorides, and with a conventional base or a resin-linked amine. The use of solid supported reagent in these reactions resulted in reactions which are slightly slower but still comparable with conventional methods.

12.2.1.1 Acetylation

In a recent paper, the protected α-D-glucofuranose (**1**) was selected as a carbohydrate model compound by Oscarson et al. [5] for acetylation, under the action of microwave irradiation, with acetyl chloride (2 equiv.) and pyridine, N,N-(diisopropyl)aminoethylpolystyrene (PS-DIEA), or N-(methylpolystyrene)-4-(methylamino)pyridine (PDS-DMAP) as conventional bases. The acetyl derivative **2** was produced in good to excellent yields in a very short time (Scheme 12.1).

12.2.1.2 Peracetylation

Microwave-assisted peracetylation of D-glucose (**3**) with a small excess of acetic anhydride and catalysis by anhydrous salts such as potassium acetate or zinc chloride has been found by Limousin et al. [6] to give the corresponding acetyl derivative **4** almost quantitatively in less than 15 min (Scheme 12.2).

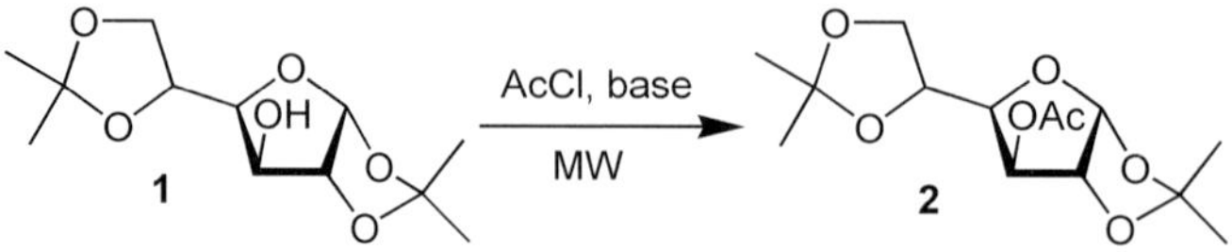

Scheme 12.1

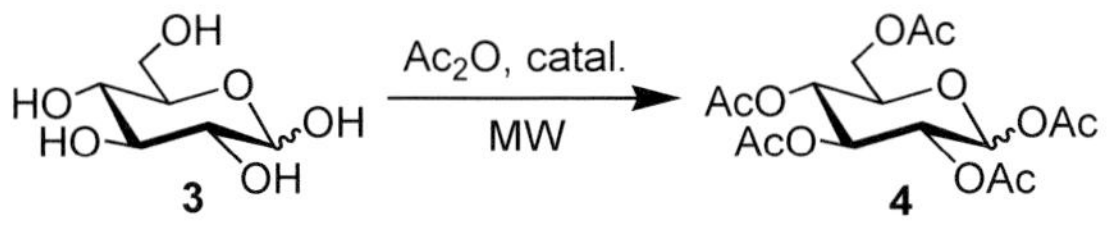

12.2 Protection

Scheme 12.2

Attempts with D-galactose, D-mannose or *N*-troc-D-glucosamine and sodium acetate as catalyst gave also excellent yields of the corresponding acetyl derivates within 11 min. Loupy et al. [7] performed peracetylation of **3** with a slight excess of acetic anhydride and catalytic amounts of zinc dichloride either with classical heating or under microwave activation, but they obtained rather similar yields irrespective of the conditions used.

Das et al. [8] described an operationally simple procedure for microwave-activated acetylation of carbohydrate alcohols by acetic anhydride, which uses only 0.1 equiv. of the powerful indium chloride catalyst. Peracetates were produced in good to excellent yields. Comparison of this method with others which use well established catalysts (pyridine, I_2, $Cu(OTf)_2$, $CoCl_2$), either with an appropriate solvent or neat (following standard conditions), shows that under microwave irradiation conditions indium chloride is the best choice providing higher yields than those without microwave energy.

12.2.1.3 Benzoylation

In connection with their current endeavors toward the synthesis of polyoxygenated natural products and oligosaccharides [9], Herradón et al. [10] examined the effect of microwave activation and the influences of both the solvent and oven electric power [11] on regioselective benzoylation of different carbohydrate-derived polyols. The interesting feature of this conversion is the intermediacy of not isolable dibutylstannylene acetals, which are obtained by irradiating a mixture of the polyol and dibutyltin oxide in a commercial oven for few minutes, compared with several hours for the standard method. The benzoylation of methyl β-D-pyranoside **5**, as previously documented in the literature [12], is not selective, giving a mixture of tribenzoate **6** and dibenzoates **7** and **8** (Scheme 12.3), while that of derivatives **9**, **11**, and **13** gave the corresponding selectively protected derivatives **10**, **12**, and **14** (Scheme 12.4).

Oscarson et al. [5] found that benzoylation of **1** is often very slow when conventional methods are used, but it is dramatically accelerated under microwave activation. Benzoylation of selected substrates, for example acetonated α-D-galactopyranose and benzylidene 4,6-protected α-D-glucopyranose, which are not stable in acidic conditions, was performed by Cléophax et al. [13] under the action of microwave irradiation in a monomode system with methyl benzoate for 4–15 min.

Good yields of **16** were isolated by *trans*-esterification of **15** under basic conditions, using tetrabutylammonium bromide as phase-transfer agent, either in the absence of solvent (66%) or in the presence of small amounts of different solvents (up 76%) (Scheme 12.5, a representative example).

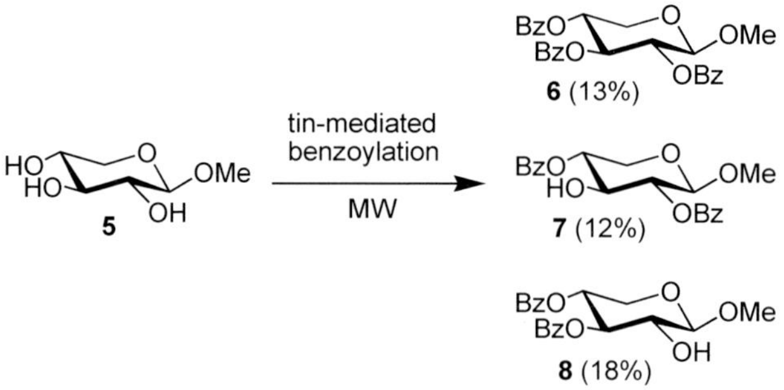

Scheme 12.3

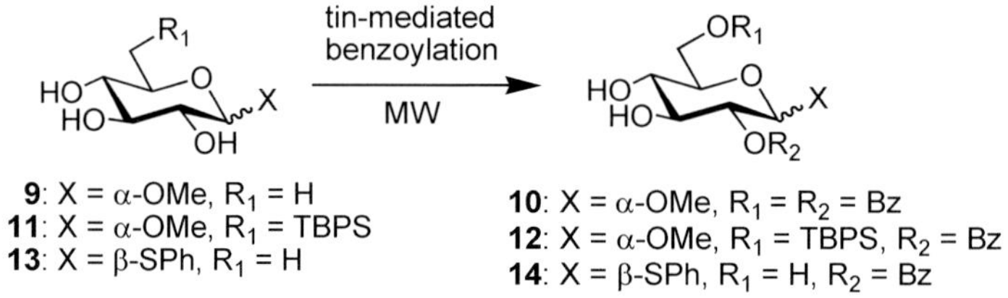

Scheme 12.4

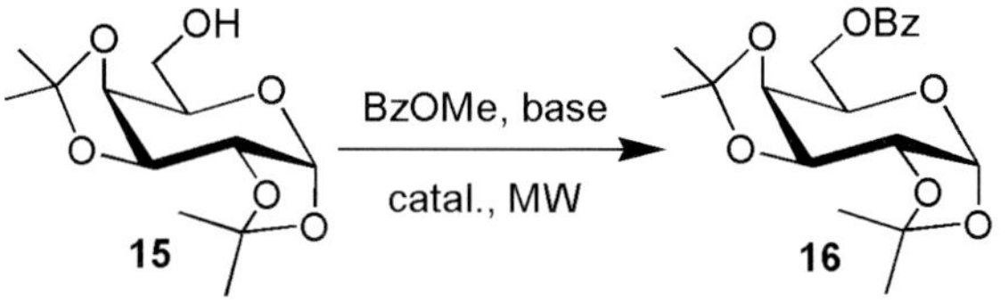

9: X = α-OMe, R₁ = H
11: X = α-OMe, R₁ = TBPS
13: X = β-SPh, R₁ = H

10: X = α-OMe, R₁ = R₂ = Bz
12: X = α-OMe, R₁ = TBPS, R₂ = Bz
14: X = β-SPh, R₁ = H, R₂ = Bz

Scheme 12.5

Several attempts at selective monobenzoylation were performed with 4,6-dibenzylidene-α-D-glucopyranose, but without success.

12.2.1.4 Pivaloylation

Of special interest is the formation of pivaloyl esters, because of the steric bulk of the pivaloyl group. Oscarson et al. [5] obtained the same results as for benzoylation when they subjected **1** to pivaloylation under microwave irradiation conditions, but with a α:β ratio of 1:2 (Scheme 12.6).

It is worthy of note that, during pivaloylation of **1** with pyridine, the authors [5] isolated, with **17**, the 6-O-pivaloyl derivative **18** (ca. 25%) also; it was shown the latter was produced by 5,6- to 3,5-acetal migration before the acylation. The acetal migration was not observed for acetylation and chloroacetylation reactions, indicating that the acetylation is too rapid to enable acetal rearrangement. Pivaloylation of

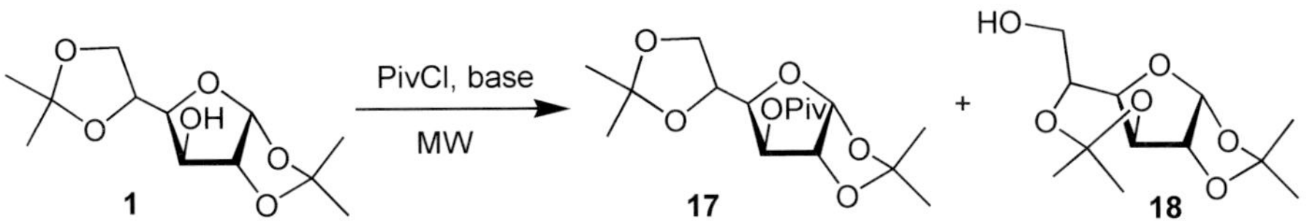

12.2 Protection

Scheme 12.6

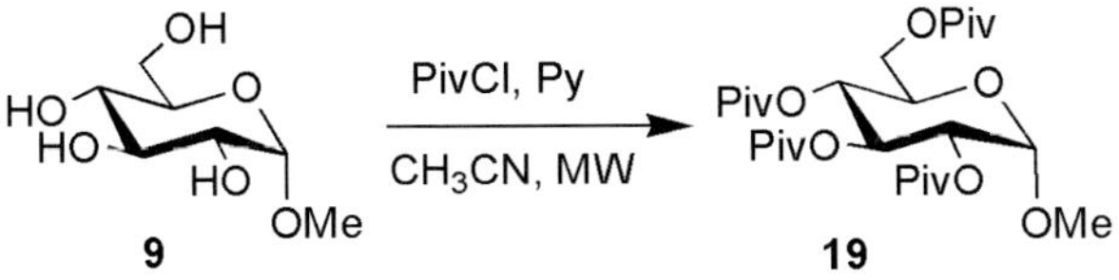

Scheme 12.7

methyl α-D-glucopyranoside **9** in pyridine gave 68% of the pivaloyl derivative **19** in 10 min (Scheme 12.7).

12.2.1.5 Dodecanoylation

Cléophax et al. [13] achieved *trans*-esterification of **1** by treatment with methyl laurate, potassium carbonate as the base, and tetrabutylammonium bromide as a phase-transfer catalyst. The best yield (88%) was obtained in the presence of dimethylformamide with a very important specific effect of irradiation. With **20**, the product of diesterification **21** and products of monoesterification **22** and **23** were obtained (Scheme 12.8).

12.2.2 Acetalation

The method developed by Garegg et al. [14] to form benzylidene acetals was modified by Oscarson et al. [5], who treated compound **9** with benzal bromide and N-(methylpolystyrene)-4-(methylamino)pyridine (PS-DMAP) in acetonitrile under microwave irradiation conditions to form the 4,6-*O*-benzylidene derivative **20** (83%) after 5 min (Scheme 12.9).

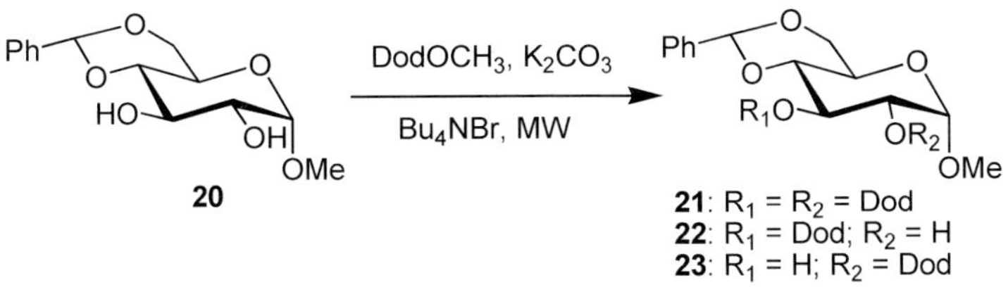

21: $R_1 = R_2 = $ Dod
22: $R_1 = $ Dod; $R_2 = $ H
23: $R_1 = $ H; $R_2 = $ Dod

Scheme 12.8

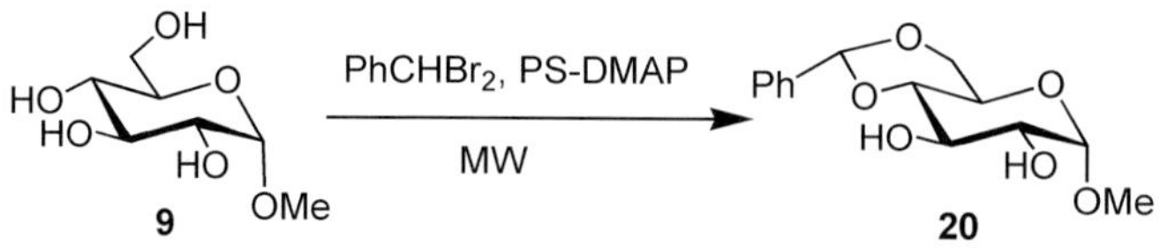

Scheme 12.9

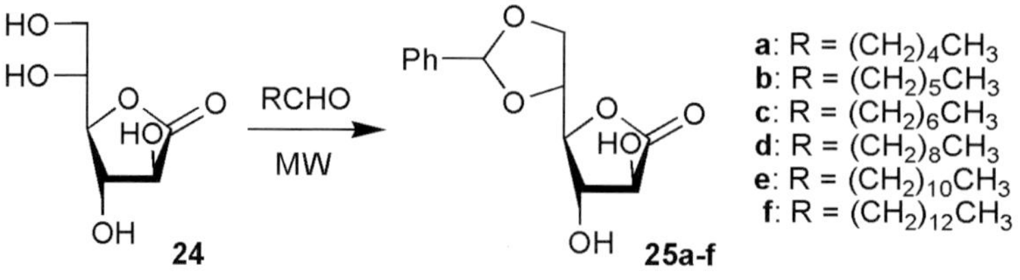

Scheme 12.10

Because of their interest in products with liquid crystalline and surfactant properties, Csiba et al. [15] synthesized some amphiphilic derivatives of L-galactono-1,4-lactone 24, a by-product of the sugar beet industry available in large quantities. Reactions between 24 and hexyl, heptyl, octyl, decyl, dodecyl, and myristyl aldehydes on montmorillonite KSF or K10 were performed in a focused open-vessel microwave system for 10 min in the absence of solvent (Scheme 12.10). Protected derivatives 25a–f were afforded with yields (60–89%) considerably better than those (22–38%) obtained by conventional heating for 24 h in the presence of sulfuric acid.

In a research directed toward monoacetalation of sucrose (26) by treatment with citral 27 (geranial and neral), widely used as chemical intermediates in the perfume industry [16], Queneau et al. [17] determined conditions for optimized acidic catalysis in dimethylformamide (DMF) which afforded good yields (>80%) of acetals 28 directly from the unprotected 26 in comparison with those obtained in a oil bath. With prolonged reaction times, cleavage to acetal 29 occurred (Scheme 12.11).

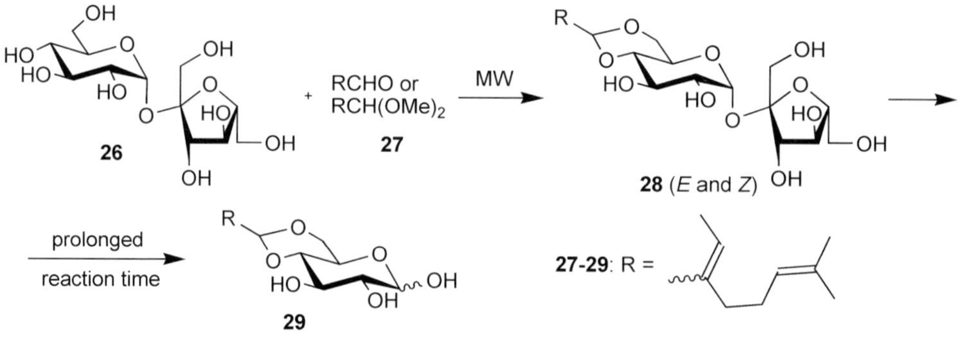

Scheme 12.11

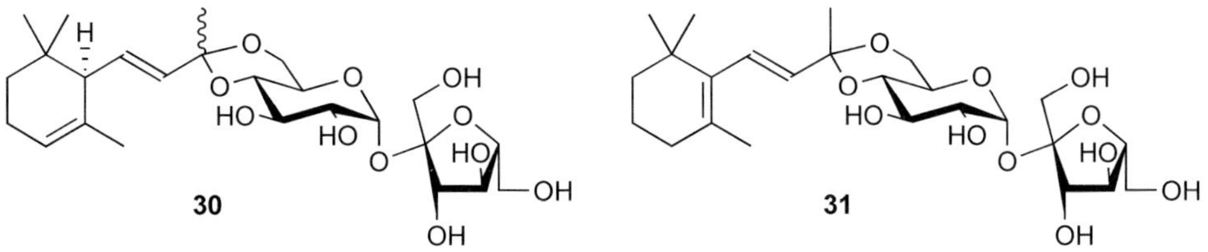

Fig. 12.1. Structures of **30** (1:1 mixture of epimers) and **31**.

By applying reaction conditions optimized for conversion of citral to the dimethyl acetals of α- and β-ionone, *trans*-esterification of **26** was performed to furnish the corresponding acetals **30** (66%) and **31** (76%) (Fig. 12.1), which were peracetylated for characterization purposes.

12.2.3
Silylation

Silylation of the 3-hydroxyl group of **1** was achieved by Oscarson [5], who used *tert*-butyldimethylsilyl chloride (TBDMSCl) in the presence of pyridine or N-(methylpolystyrene)-4-(methylamino)pyridine (PDS-DMAP), in a Smith Synthesizer (Scheme 12.12). In addition to the silylated product **32a**, the secondary product **33a**, formed by acetal migration, was also observed.

When N,O-bis-(trimethylsilyl)acetamide (BSA), a reagent which does not require any additional base, was used for the total trimethyl silylation of **9** into **34**, short reaction times and high yields were observed (Scheme 12.13).

Khalafi-Nezhad et al. [18] achieved silylation of the 5′-hydroxyl function of uridine by use of imidazole and triisopropylsilyl chloride in a microwave oven at 200 W for 60 s, obtaining 80% yield of silylated product. The same reaction in dimethylformamide with conventional heating [19] gave a 76% yield of silylated product after 24 h.

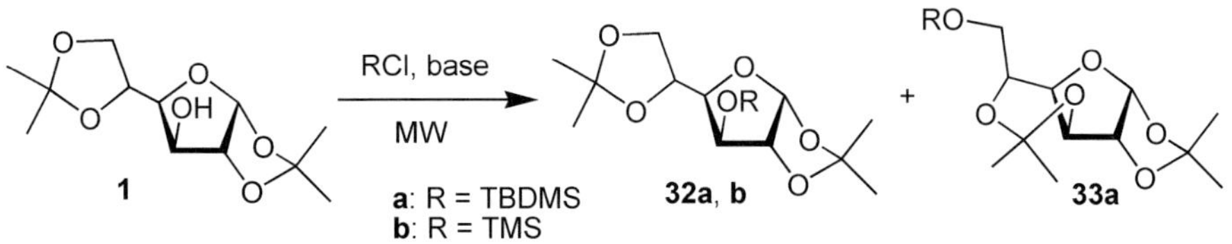

Scheme 12.12

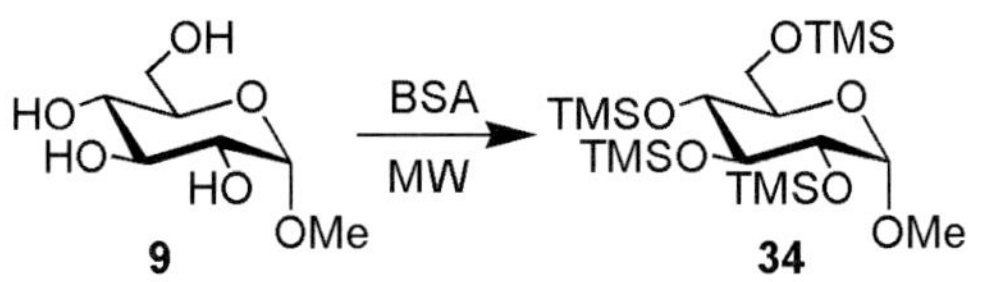

Scheme 12.13

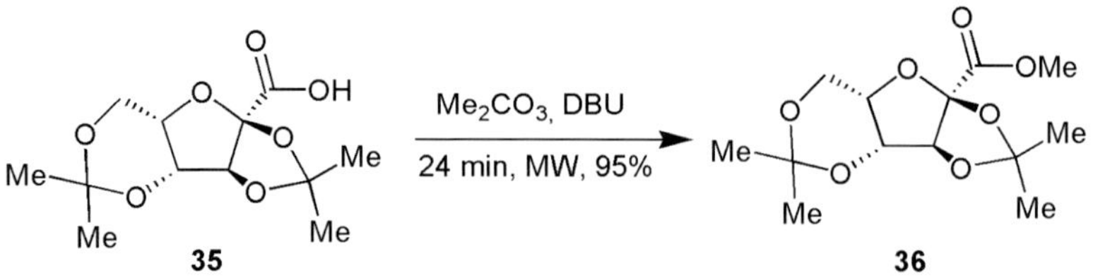

Scheme 12.14

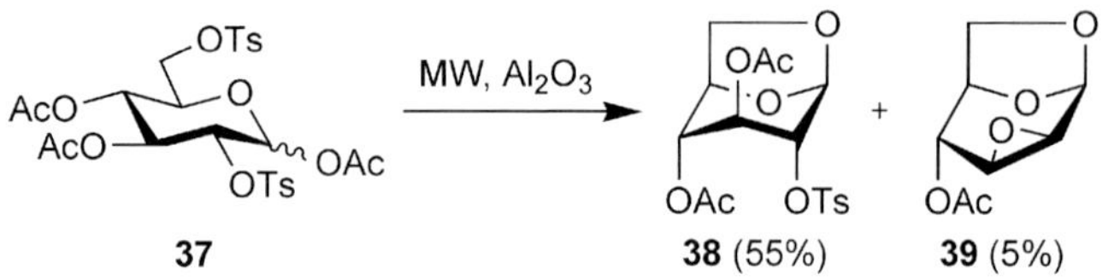

Scheme 12.15

12.2.4
Methylation of Carbohydrate Carboxylic Acids

Shieh et al. [20] have reported esterification of carbohydrate carboxylic acid **35** to form the corresponding ester **36**. Dimethyl carbonate was used both as solvent and reagent, in the presence of 1,8-diazabicyclo[5,4,0]undec-7-ene (DBU), under the action of microwave irradiation for 24 min (Scheme 12.14).

12.2.5
Synthesis of 1,6-Anhydro-β-D-hexopyranoses

Cléophax et al. [21] used intramolecular protection in the large-scale preparation of 1,6-anhydro-D-hexopyranoses (45–86%) from the corresponding monotosylates. For ditosylate **37**, a mild, rapid, and solvent-free microwave procedure was used. Interestingly, **38** (55%) and 1,6:2,3-di-anhydro-4-O-acetyl-β-D-glucopyranose (**39**) (approx. 5%) were obtained after re-acetylation (Scheme 12.15).

12.3
Deprotection

Mild conditions have been used in deprotection studies to determine whether the extended reaction times often needed with conventional heating could be shortened by use of microwave irradiation. Some solid-supported reagents have also been tested.

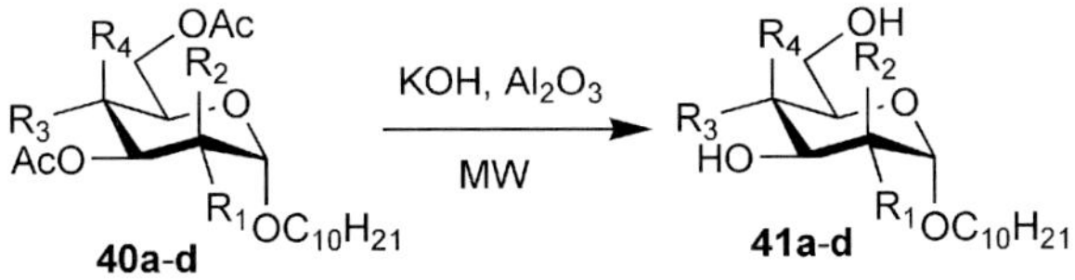

Scheme 12.16

40, 41	R$_1$	R$_2$	R$_3$	R$_4$
a	OAc/OH	H	OAc/OH	H
b	OAc/OH	H	H	OAc/OH
c	H	OAc/OH	OAc/OH	H
d	NHTroc	H	OAc/OH	H

12.3.1
Deacylation

12.3.1.1 Deacetylation

Whereas no reaction was observed on microwave irradiation of decyl 2,3,4,6-tetra-O-acetyl-α-D-glucopyranoside 40 adsorbed on alumina (Varma's conditions) [22], saponification of the protected glucoside 40a to decyl glucoside 41a was achieved by Limousin et al. [6] by use of alumina impregnated with a small excess of potassium hydroxide under dry conditions (Scheme 12.16). The same procedures were applied to the acetates 40b–40d of decyl D-galactoside 41b, D-mannoside 41c, and D-glucoside 41d.

12.3.1.2 Debenzoylation

By following methods described by Tsuzuki et al. [23] and Mori et al. [24], except with microwave activation, Oscarson et al. [5] debenzoylated derivative 42 into 1 (71%) by irradiation at 150 °C for 15 min in 1:5:1 triethylamine–methanol–water (Scheme 12.17). Under the same conditions, ethyl 2,3-di-O-benzoyl-4,6-O-benzylidene-1-thio-β-D-glucopyranoside gave the corresponding 2,3-debenzoylated monosaccharide (78%).

12.3.1.3 Depivaloylation

In a effort to devise a mild, high-yielding deprotection strategy for hindered esters, Ley and Mynett [25] investigated the possibility of microwave-assisted hydrolysis of

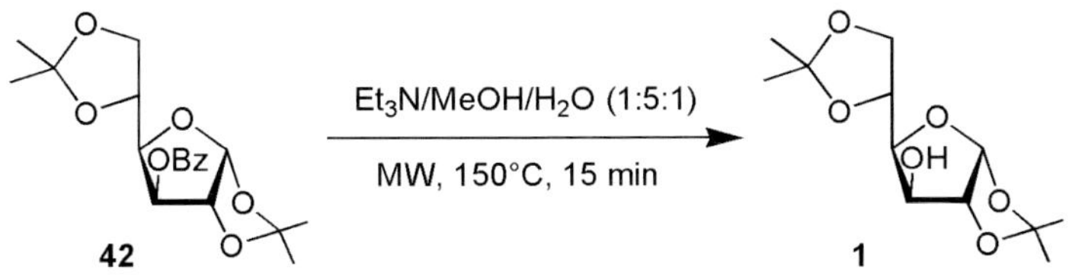

Scheme 12.17

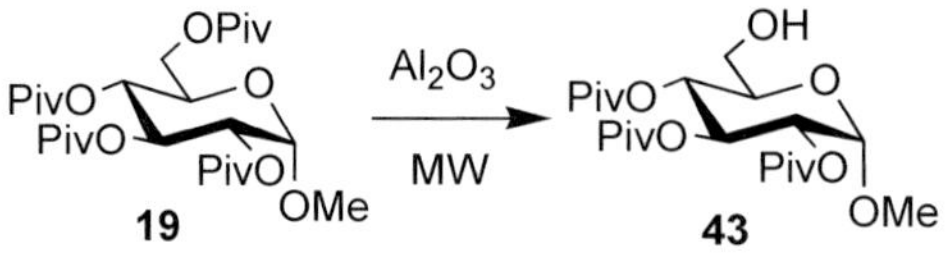

Scheme 12.18

a series of pivaloyl-protected alcohols on neutral alumina. The authors showed that the deprotection of position 6 of compound **19** occurs selectively, cleanly, and efficiently without group migration or isomerization of the anomeric center to give **43** (Scheme 12.18, a representative example). Deprotection could not be achieved by using conventional heating (oil bath at 75 °C).

12.3.2
Deacetalation

Oscarson et al. [5] described the hydrolysis of compound **20** to **9** (89%) by treatment with K10 clay in methanol–water for 2 min under the action of microwave irradiation at 150 °C. Because polymer-supported reagents worked out well in acylation reactions, they also submitted **20** to microwave irradiation with polymer supported poly(4-vinylpyridinium-p-toluenesulfonate) (PPTS) in ethanol–water at 150 °C for 2 min and obtained **9** in 94% yield (Scheme 12.19).

Simple and rapid deacetalation was accomplished by Couri et al. [26] by irradiation in a domestic oven under solvent-free conditions with silica gel-supported reagents.

12.3.3
Desilylation

Several methods can be used for deprotection of silyl groups under acidic or basic conditions [27]. Under basic conditions, Oscarson et al. [5] converted **37a** into **1** (61%) with tetrabutylammonium bromide and potassium fluoride under the action of microwave irradiation at 180 °C for 2 min. Under slightly acidic conditions, they used montmorillonite K10 clay in methanol–water (1:1) at 150 °C for 4 min and 3 min, to obtain **3** in high yields (85% and 91%, respectively) (Scheme 12.20). Under

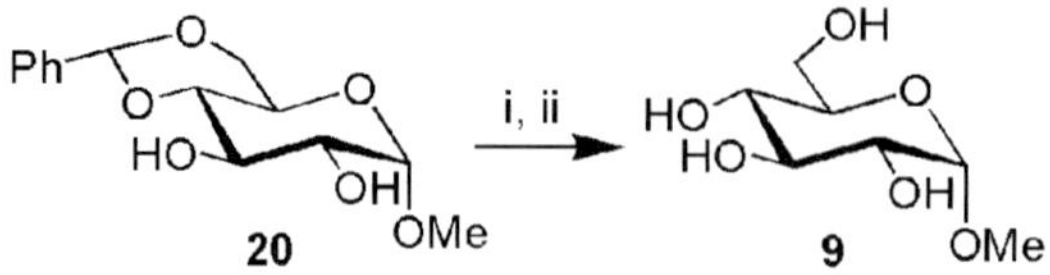

i) K 10 Clay, MeOH/H$_2$O, 150 °C, MW, 2 min;
ii) PPTS, EtOH/H$_2$O, 150 °C, MW, 2 min.

Scheme 12.19

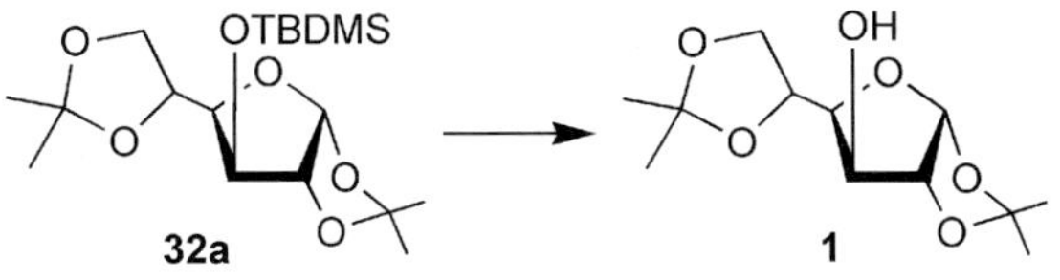

Scheme 12.20

the original conditions (75 °C for 72 h), the desilylated derivative **1** was obtained in a yield of 77% [28].

12.4 Glycosylation

Interest in the glycosylation is connected with the physical properties of glycosides, which are used as liquid crystals and as non toxic and biodegradable surfactants [29]. In particular, alkyl glycosides with long-chain alkyl groups derived from natural raw materials [30] are used for the production of tenside formulations, because of their technological degradability [31]. They are also important chemical intermediates in biotechnology [32].

12.4.1 O-Glycosylation

12.4.1.1 *O-Glycosylation by Fischer Alcoholysis*

Fischer's classical O-glycosylation with alcohols and an acid catalyst was applied by Limousin et al. [6], who described the reaction of peracetylated D-glucopyranose **4** with 1-decanol in the presence of zinc chloride to afford the corresponding α- (**44a**) and β-glucoside (**44b**) in 74% yield (**44a:44b** = 7:1), with a small amount of the 2-de-O-acetyl α-derivatives **45a** in 3 min (Scheme 12.21).

The same glycosylation was extended to D-galactose, D-mannose, and N-troc-D-glucosamine with similar results. The same authors also achieved glycosylation of

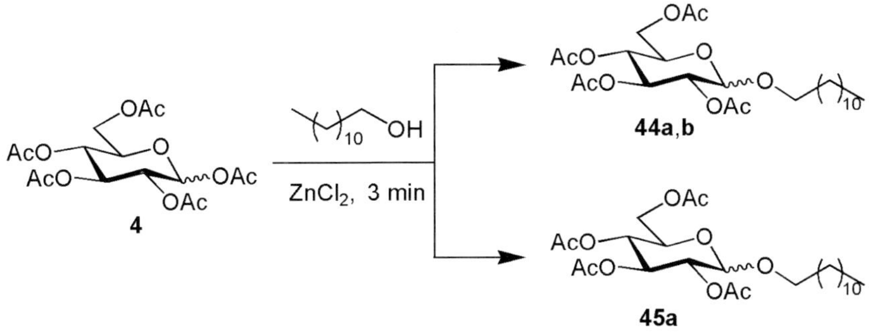

Scheme 12.21

unprotected glucose, on the solid support Hyflo Super Cel, by use of 1-decanol in the presence of *p*-toluenesulfonic acid, although with very poor yield. Scale-up of this procedure was found to be fairly interesting by Loupy et al. [7], who obtained yields similar to those obtained under very similar conditions from glycosylation of decanol with peracetyl derivatives of **4** at 110 °C in Prolabo Synthewave systems. An important not purely thermal effect is involved if yields of 72–74% are compared with those obtained under classical heating conditions (less than 25%).

By means of an ETHOS MR oven, Nuchter et al. [33] accomplished scaling-up of a microwave-assisted Fischer glycosylation to the kilogram scale with improved economic efficiency. In batch reactions, carbohydrates (D-glucose, D-mannose, D-galactose, butyl D-galactose, starch) were converted on the 50-g scale (95–100% yield, $\alpha{:}\beta$ from 95:5 to 100%) with 3–30-fold molar excesses of an alcohol (methanol, ethanol, butanol, octanol) in the presence of a catalytic amount of acetyl chloride under pressure (microwave flow reactor, 120–140 °C, 12–16 bar, 5–12 min) or without applying pressure (120–140 °C or reflux temperature, 20–60 min). Furanosides are not stable under these reaction conditions.

Fraser-Reid et al. [34] found that the *n*-pentenyl family of armed and disarmed glycosyl donors and ortho esters undergo ready coupling with hindered or unhindered acceptors under microwave activation conditions to give the corresponding disaccharides in modest to good yields. Reactions performed in acetonitrile under neutral conditions with *N*-iodosuccinimide as promoter were always rapid, clean, and ideal for use with reactants bearing acid and labile protecting groups.

Cléophax et al. [35] performed a microwave-assisted synthesis of dichlorovinyl glycopyranosides **47** from per-*O*-acetylated glycopyranoses **46** via Wittig-type reaction with triphenylphosphine–tetrachloromethane either in toluene and pyridine or dichloromethane on the anomeric acetate group (Scheme 12.22).

Seibel et al. [36] reported, for the first time, the use of microwave heating in the glycosylation of amino acids. They performed glycosyl transfer reactions of peracetylated monosaccharides (glucose and galactose) and disaccharides (maltose and lactose) with *N*-9-fluorenylmethoxycarbonyl-L-serine benzyl ester in the presence of iron trichloride with short reaction times (4 min, compared with 5–10 h by conventional heating) and with improved yields (52–61% compared with 10–31% by conventional heating). It is worthy of note that in these reactions heavy metal compounds, for example silver trifluoromethanesulfonate, mercury dibromide and dicyanide, or boron trifluoride–diethyl etherate, can be replaced by the environmentally safe promoter iron trichloride.

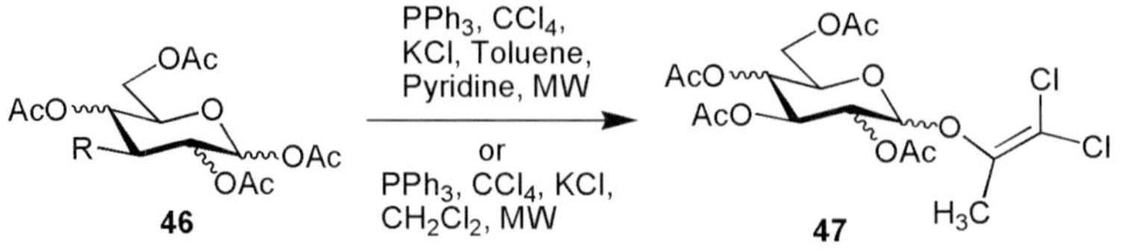

Scheme 12.22

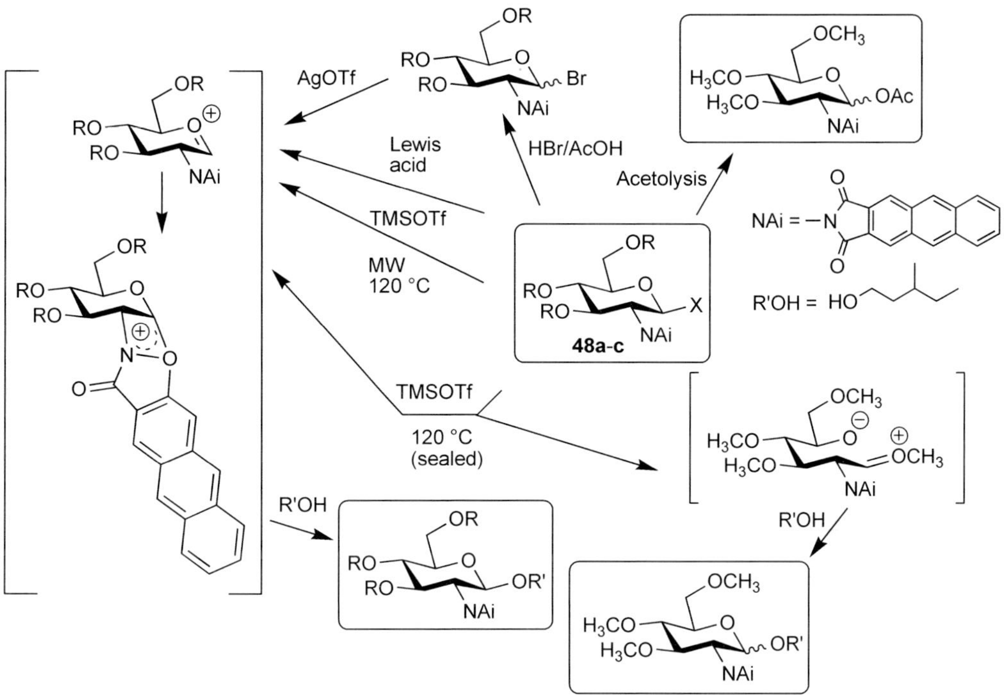

Scheme 12.23

Ohrui et al. [37] succeeded in synthesizing new chiral label reagents **48a–c** with a 2,3-anthracenedicarboxamide group, starting from D-glucosamine and introducing target alcohols (methyl-1-pentanol, 4,8,12,16-tetramethylheptadecanol) at the anomeric position with β-selectivity. They used methylglycoside reagents as glycosyl donors with a Lewis acid (silver and trimethylsilyl trifluoromethanesulfonate, zinc dichloride), and microwave irradiation. In the absence of microwave irradiation, α- and β-mixed glycosides were afforded (Scheme 12.23).

Bornaghi and Poulsen [38] have demonstrated that microwave irradiation is potentially one of the easiest and quickest routes to simple glycosides. This was apparent in Fischer glycosylation reactions of *N*-acetyl-D-glucosamine, *N*-acetyl-D-galactosamine, D-glucose, D-galactose, and D-mannose with a variety of alcohols (methanol, ethanol, benzyl alcohol, and allyl alcohol) from the impressive acceleration of the reaction time (minutes compared with hours) and good α-glycoside product selectivity.

12.4.1.2 *O*-Glycosylation via Ferrier's Rearrangement

O-Glycosylation via Ferrier's rearrangement [39], an allylic rearrangement of glycals (1,2-unsaturated pyranosides) in the presence of nucleophiles, has acquired great significance in carbohydrate chemistry, because of the further transformation of the products into other interesting carbohydrates [40]. In addition to affording

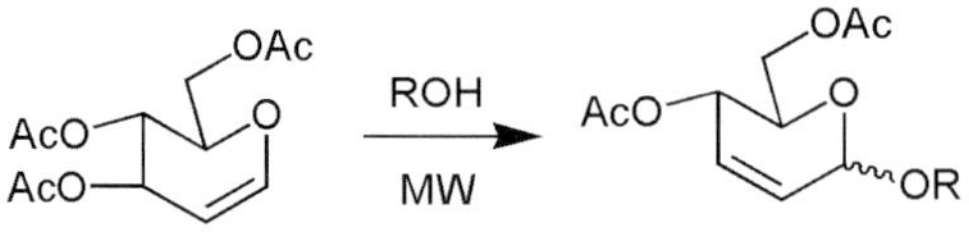

50: R = aryl; **51**: R = ethyl, cyclohexyl, allyl; **52**: R = N-hydroxymethyl-, N-hydroxyethyl-, (S)-(+)-N-hydroxy-isopropyl-phthalimide, (R,S)-2-phenyl-, (R,S)-(4-tolyl)-1,2,4-oxadiazol-5-yl-2-butanol; **53**: R = allyl, propargyl, heptyl, benzyl, pent-4-enyl, phenethyl.

Scheme 12.24

clean products, higher yields, and shorter times, use of microwaves in this reaction has enabled avoidance of limitations frequently encountered under conventional conditions, for example strong oxidizing conditions, high acidic media, low yields, high temperatures, longer reaction times, incompatibilities with functional groups, and the types of catalyst and reagent used, particularly, their amounts and, especially, their cost.

By using the microwave-induced organic reaction enhancement (MORE) method Balasubramanian et al. [41] achieved solvent-free glycosylations by treating tri-O-acetyl-D-glucal **49** with substituted phenols to afford 2,3-unsaturated O-aryl glycosides **50**; reaction times reduced several fold compared with those required under thermal conditions (Scheme 12.24).

Balasubramanian et al. [42] accomplished microwave-assisted glycosylations in solution by irradiation of a mixture of **49a** and tri-O-acetyl-D-galactal **49b** with an appropriate alcohol or substituted phenol in the presence of montmorillonite K-10 as catalyst; they obtained solely the corresponding α-anomers of unsaturated glycosides **50** and **51** in much shorter times and with better yields than those obtained by conventional heating (Scheme 12.25). Silica gel and ferric chloride were also used as catalysts, but either no reaction occurred or extensive decomposition occurred when these acids were used.

Srivastava et al. [43] synthesized 2,3-unsaturated α-glucosides **52** in good yields by using montmorillonite K10 catalyst (71–87%) (Scheme 12.24). It is worthy of

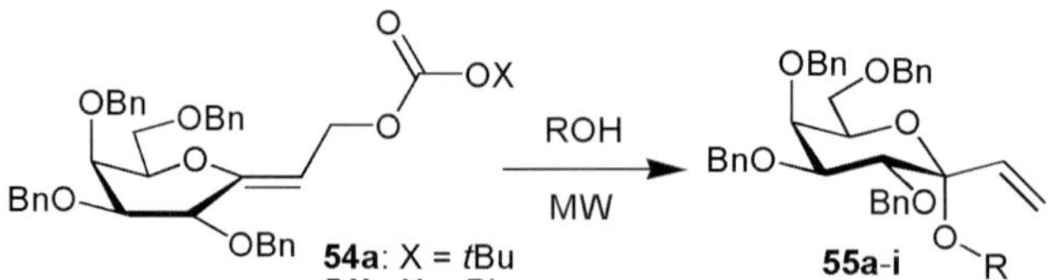

ROH = hexanol, cyclohexanol, isopropanol, 2-methoxyethanol, 1,3-propandiol, benzyl and allylic alcohol, acetonated D-glucopyranose.

Scheme 12.25

note that no reaction occurred when cyclohexanol was used or the catalyst was changed for silica gel and iron trichloride. The latter change also resulted in extensive decomposition.

By his efficient, mild, rapid eco-friendly method, Das et al. [44] treated different per-O-acetylglycals (**49a,b** and L-rhamnal, **49c**, and L-arabinal, **49d**) with aliphatic primary alcohols, in the presence of indium(III) chloride in acetonitrile solution in an open vessel, to afford the corresponding 2,3-unsaturated products **53** in good to excellent yields in a few seconds (Scheme 12.24).

Hotha and Tripathi [45] reported a practical method based on microwave irradiation for Ferrier reaction of per-O-acetylated glycals **49a,b** with primary, secondary, allylic, benzylic, and monosaccharide alcohols; they used a catalytic amount of niobium pentachloride to give the corresponding 2,3-unsaturated O-glycosides **51** and **53** with α stereoselectivity in high yields and with short reaction times (Scheme 12.24).

exo-Glycals **54a,b**, **56**, and **58**, which have shown great promise in the synthesis of C- and N-glycosides, ketoses and ketosides, and carbasugars [46], were reported by Lin et al. to be efficient glycosyl donors in stereoselective glycosylations [47]. As carbonates, their reactivity is further increased by decarboxylation, which results in an extra driving force during the heating process. α-Glycosyl additions occur affording **55a–i** (75–92%), **57a–d** (50–74%, α:β = 5–7:1), and **59a,b** (90–93%) under solvent-free conditions; yields can, however, be improved by catalysis of a Lewis acid, for example indium trichloride. So, for example, the improvement was from 50% (α:β = 5:1) to 91% (α:β = 7:1) with **56** and hexanol and from 25% (α:β = 7:1) to 85% (α:β = 11:1) with **56** and cyclohexanol. Interestingly, *exo*-glycals were found to be more active than *endo*-glycals (Schemes 12.25, 12.26, and 12.27).

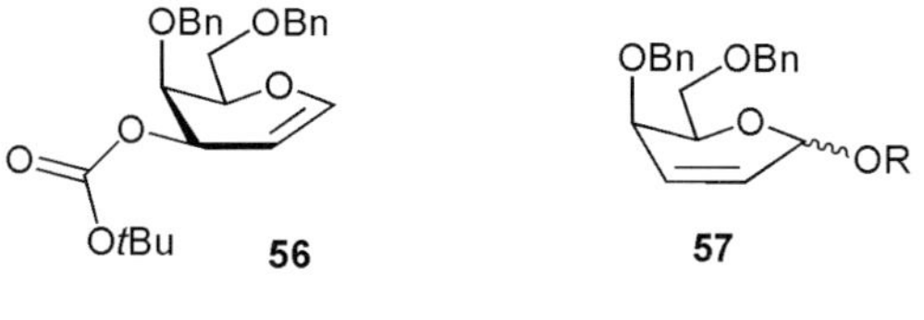

ROH = hexanol, 2-methoxyethanol, benzyl alcohol, cyclohexanol.

Scheme 12.26

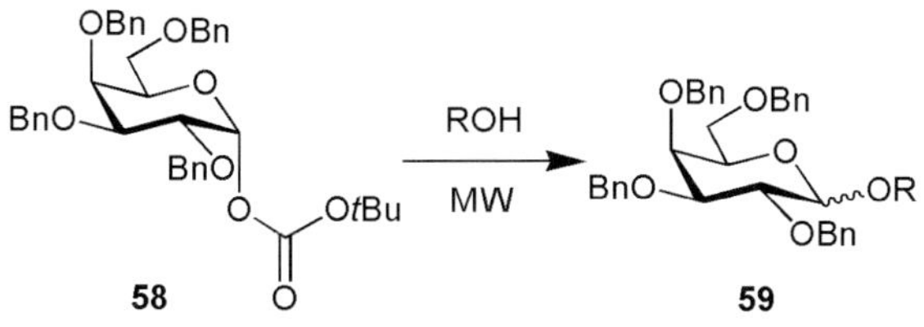

ROH = 2-methoxyethanol, 1,3-propanediol.

Scheme 12.27

12.4.2
C-Glycosylation Reactions

C-Glycosides are versatile chiral building blocks in the synthesis of many biologically interesting and potent natural products [48]. Their synthesis in solution using microwave activation is rare and so far there is no report of use of solvent-free conditions. To the best of our knowledge, the literature reports only one example – reduction of C-glycosylidene nitriles with hydrogen in the presence of Pd/C with complete stereocontrol at the anomeric center [49].

12.4.3
N-Glycosylation

N-Glycosides occur widely in the nature – they are, for example, components of nucleic acids and constituents of glycoproteins. New methods for synthesis of these derivatives are important in the search for compounds that may serve as glycomimetics or as novel carbohydrate-containing materials; synthesis is, however, often complicated by anomerization in intermediate glycosyl amines. Glycosyl 1,2,3-triazoles and glycosyl amides are potentially useful linkers for attaching carbohydrates in novel oligomers. Norris et al. [50] have developed new methods of avoiding such anomerization reactions, including direct conversion of glycosylazides to glycosylamides by use of the Staudinger reaction, and dipolar cycloaddition chemistry and rapid purification of the N-glycosyl products.

12.5
Hydrogenation (Catalytic Transfer Hydrogenation)

In recent years a few laboratories have started to use catalytic transfer hydrogenation (CTH) [51] consisting of a safe and simple operation in which a catalyst and hydrogen are replaced with a catalyst and a hydrogen donor. Bose et al. [52] recently demonstrated that CTH can be conducted very rapidly, and essentially in quantitative yield, inside an unmodified domestic microwave oven. He gave several examples, including a microwave-assisted CTH reaction (10% Pd/C, HCO_2NH_4, $HOCH_2CH_2OH$, 90 s) on the 2,3-unsaturated Ferrier's rearranged O-glycoside of rhamnal **60** with hydroxy-β-lactam (Scheme 12.28) resulting in the formation of a reduced and deacetylated product **61**.

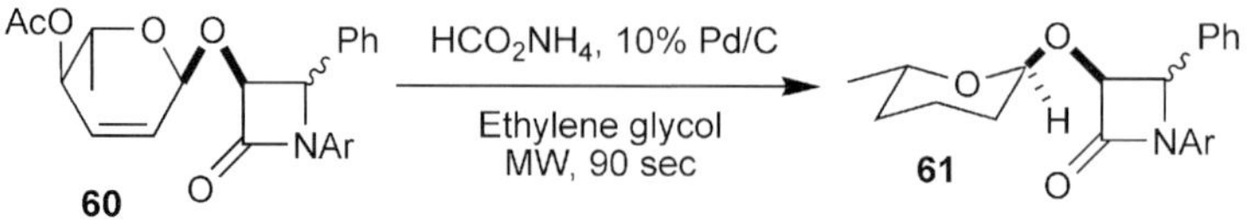

Scheme 12.28

12.6
Oxidation

New oxidation reactions of organic substances and the reasons for their acceleration under microwave irradiation have recently been investigated. In particular, in a search for highly efficient oxidation procedures, Chakraborty and Bordoloi [53] used pyridinium chlorochromate (PCC) under the action of microwave irradiation for oxidation of protected α-glucofuranose **62** to the corresponding ketone **63** (99%) much more quickly (10 min) than using the conventional technique (4 h under reflux) (Scheme 12.29) and with an easier work-up procedure. They found the oxidation can also be performed with moist PCC under solvent-free conditions and with the same yield.

Kubrakova et al. [54] studied the mechanism of oxidation of several organic compounds, including galactose and sucrose, with nitric acid. They analyzed activation energy values and observed changes of the kinetic characteristics which were attributed to an effect directly associated with the action of microwave radiation.

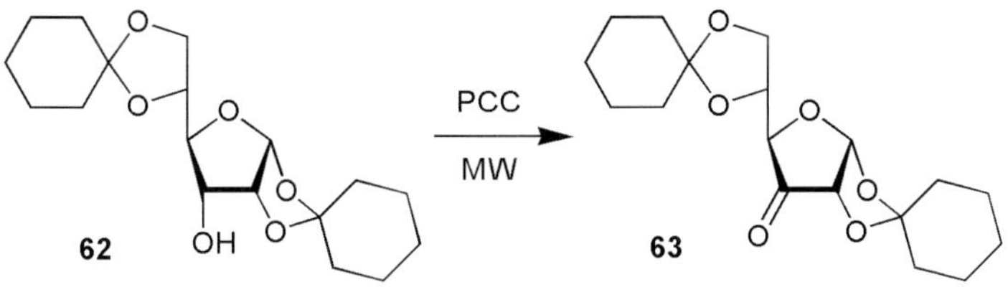

Scheme 12.29

12.7
Halogenation

12.7.1
Chlorination and Bromination

In the last decade of the twentieth century much effort was devoted to the microwave-assisted chlorination and bromination of several sugars, which were analyzed from the standpoints of the halogen donor, solvent effects (presence and nature), added salts, and halogenated products such as 6-halogenated and/or 4- and 4′ and 6-halogenated derivatives. Cléophax et al. [55] studied the halogenation of **10** and its 2,3-di-O-benzyl, 2,3,4-tri-O-benzyl, 2,3,6-tri-O-benzyl, and 6-acetyl-2,3-di-O-benzyl derivatives, and methyl 2,3-di-O-benzyl-α-D-galactopyranoside with triphenylphosphine and chlorine or bromine donors. Carbon tetrachloride, hexachloroethane, and 1,2-dibromotetrachloroethane were shown to be good halogen donors whereas N-chloro and N-bromosuccinimide and tetrabromomethane decomposed almost instantaneously under the action of microwave irradiation. It also seemed that in highly concentrated solutions of nonpolar solvents such as toluene or 1,2-

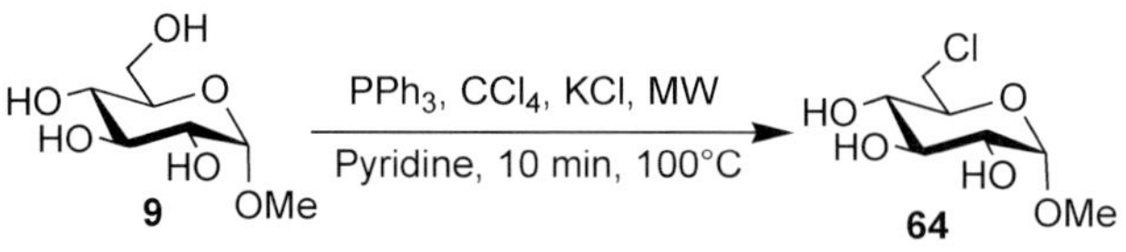

Scheme 12.30

dichloroethane with, occasionally, addition of potassium chloride, potassium bromide, and/or pyridine, it was possible to halogenate primary and secondary alcohol groups in good yields and with short reaction times. Pyridine seems to inhibit extensive decomposition and enables the reaction temperature to be increased. Furthermore, in contrast with reactions in dilute solution, dihalogenation was possible. The reactions were observed to be very fast (2 to 30 min) under the action of microwave irradiation and with classical heating at 80–120 °C (2 to 45 min). Most often, yields were better under microwave irradiation conditions and the reaction product distribution was sometimes different from that obtained by use of classical methods with, in all experiments, much reduced use of solvent.

Scale-up of chlorination of **9**, conducted by Loupy et al. [7] with triphenylphosphine and carbon tetrachloride in the presence of pyridine, after addition of excess potassium chloride, led to improvements in yields of **64** irrespective of whether microwave activation in Synthewave equipment or classical heating (Δ, oil bath) were used (Scheme 12.30). A specific microwave effect led only to a 10% difference in yield.

12.7.2
Fluorination

Fluorinated carbohydrates have recently received much attention because of their importance in the study of the enzyme–carbohydrate interactions and their interesting biological activity [56]. Hara et al. [57] achieved the deoxyfluorination of primary and anomeric hydroxyl groups of a series of protected carbohydrates, for example **65** into **66** (70%) (Scheme 12.31, a representative example) and **67** into **68** (90%) (Scheme 12.32, a representative example), with good yields, by use of N,N-diethyl-α,α-difluoro-(m-methylbenzyl)amine (DFMBA), which has previously proved to be a selective reagent for the synthesis of fluorinated alcohols and carboxylic acids [58]. Chemoselective deoxyfluorination at the anomeric position

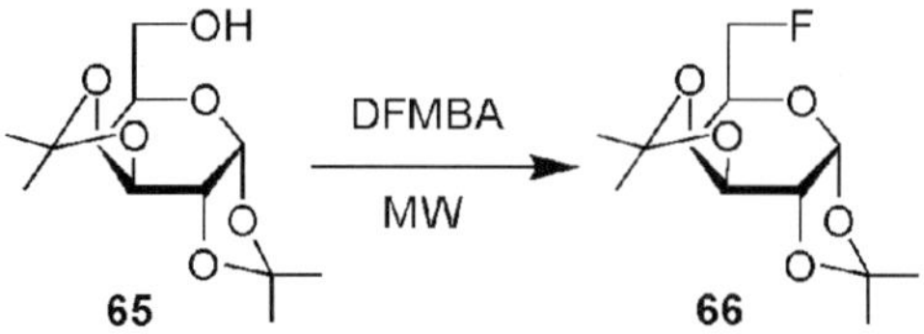

Scheme 12.31

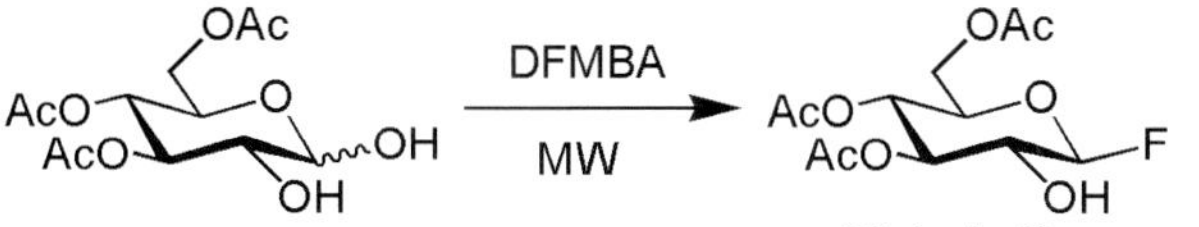

67 (α:β = 60:40) **68** (only β)

Scheme 12.32

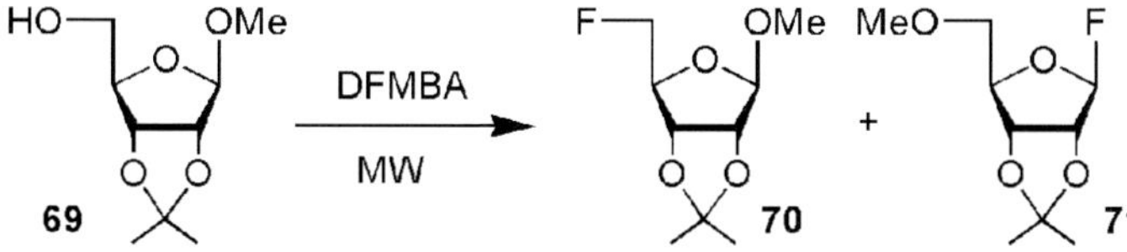

69 **70** **71**

Scheme 12.33

proceeded at below room temperature, conditions under which no other hydroxyl group is converted and most protecting groups, for example acetonide, benzyl ether, acetate, and silyl ether, can survive.

For methyl 2,3-*O*-isopropylidene-*β*-D-ribofuranose **69** migration of the methoxyl group from the 1- to 5-position, to give **70** and **71**, was observed. This, however, could be prevented by performing the reaction in dioxane at 100 °C in the presence of potassium fluoride (Scheme 12.33).

12.7.3
Iodination

In the synthesis of the carbocyclic version of the most common naturally occurring glucose-1-phosphate, iodination of the 6-hydroxy group in protected glucose **72** was achieved with 98% yield by Pohl et al. [59], who used iodine in the presence of triphenylphosphine and imidazole, and only 1 min microwave irradiation (Scheme 12.34).

12.8
Stereospecific C–H Bond Activation for Rapid Deuterium Labeling

Deuterium labeling of carbohydrates and glycoconjugates has wide applicability in biochemistry and biophysics [60], but its execution has required many chemical

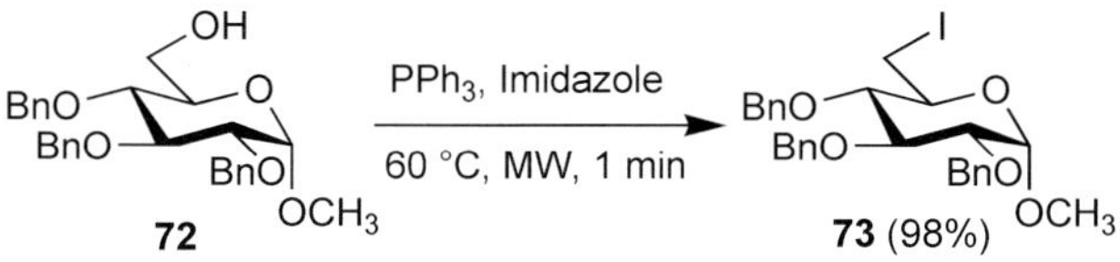

72 **73** (98%)

Scheme 12.34

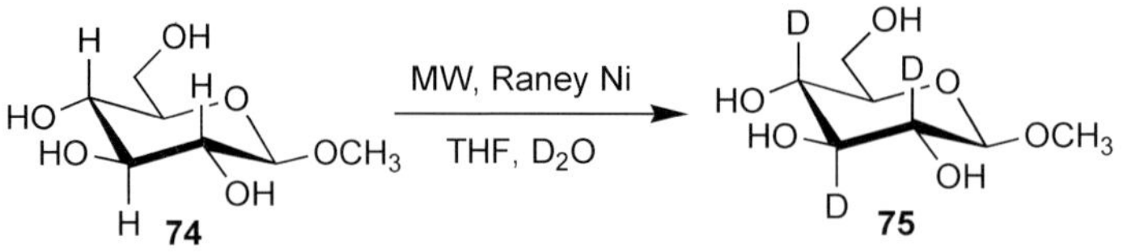

Scheme 12.35. Deuterium exchange at C_2, C_3, and C_4 positions in 1-O-methyl-β-D-galactopyranoside.

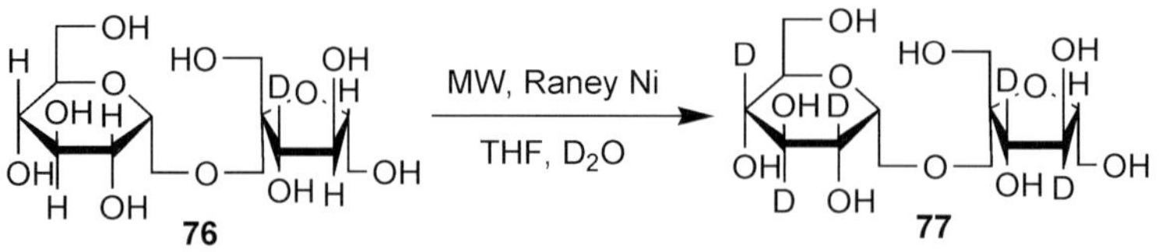

Scheme 12.36. Deuterium exchange at the $C_{2'}$, $C_{3'}$, $C_{4'}$, C_2, and C_3 positions in sucrose.

and enzymatic synthetic pathways [61]. Among these, Cioffi et al. [62] found that use of a simple multimode domestic microwave oven for irradiation of pre-activated Raney Nickel catalyst and a deuterium isotope source readily promoted stereospecific C–H → C–D exchange in nonreducing carbohydrates within a short reaction time [63]. In particular, 1-O-methyl-β-D-galactopyranoside **74** and sucrose **76** (Schemes 12.35 and 12.36), selected as monosaccharide and disaccharide models, respectively, resulted in significant percentages of ^{2}H incorporation at one C–H site in the two sugars, in less than 10 min total irradiation. If ice bath cooling is used between irradiation intervals, irradiation times can be extended to afford very high levels of ^{2}H incorporation without decomposition or epimerization.

12.9
Reaction of Carbohydrates with Amino-derivatized Labels

The condensation reaction between carbohydrates and amino derivatives to give Schiff's bases is a well known reaction [64] which has recently been widely exploited for sugar labeling and saccharide sequencing [65]. Its use for attachment of sugars to glass surfaces has became an attractive prospect, particularly because of the wide-ranging biological activity and potential medical applications of carbohydrates [66]. Yates et al. [67] improved the rate of this reaction by using microwave irradiation in the attachment of bioactive and medically important carbohydrates to an aminosilane-derivatized glass surface suitable for construction of functional carbohydrate microarray systems. Compared with conventional heating, which is slow (hours), microwave activation is quicker (some minutes) and convenient. The reaction rates were measured for hexoses, pentoses, aminosugars, and disaccharides and rate improvements for amination of model sugars with 7-aminonaphthalene-1,3-disulfonic acid in formamide were, except for sucrose, approximately 10^3. The authors showed that the substantial improvement in reaction rates was mainly because of solvent-mediated heating.

Another application of this process is the rapid and facile attachment of saccharides of the heparin sulfate family, glycosaminoglycans with many important biological and pharmacological properties, to glass slides derivatized with γ-aminopropylsilane [68].

12.10
Ferrier (II) Rearrangement to Carbasugars

The so-called Ferrier (II) rearrangement is the most common approach to carbasugars, although it requires use of mercury and a reaction time (several hours). Pohl et al. [59], found that the Ferrier (II) rearrangement of **78** into **79** in the presence of palladium dichloride can be performed in less time and with higher yields by use of microwave irradiation rather than conventional heating (Scheme 12.37).

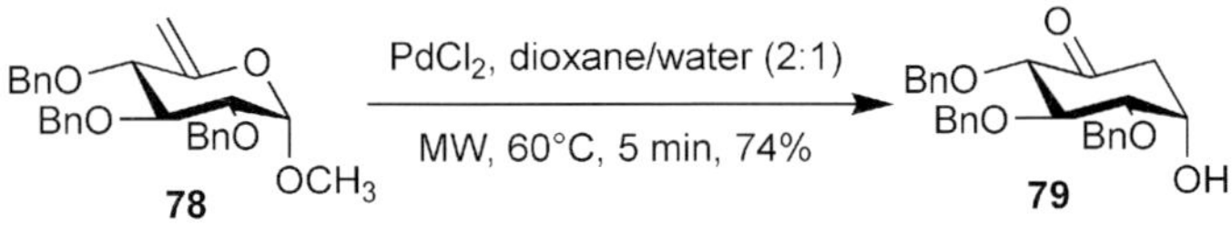

Scheme 12.37

12.11
Synthesis of Unsaturated Monosaccharides

Introduction of a double bond into a sugar is a very important reaction because the double bond can easily be further functionalized. The olefination of sugars can give rise to two types of unsaturated derivative which are categorized as *endo-* and *exo-*glycals according to whether the double bond is located inside or outside the sugar ring. The several literature methods which use conventional heating for formation of these unsaturated sugars [69] require prolonged reaction times and afford low yields of unsaturated glycosides, because of decomposition of the reagents. In contrast, microwave procedures have resulted in faster reactions and improved product yields. The best method of obtaining *endo-*glycals, in which the double bond is located in the 2,3-positions (2,3-unsaturated glycosides), remains Ferrier's rearrangement, discussed in Section 12.4.1.2.

By applying Ferrier's rearrangement to per-*O*-acetylglycals **80** (**a**, 3,4,6-tri-*O*-acetyl-1,5-anhydro-2-deoxy-D-*arabino*-hex-1-enitol, and **b**, -D-*lyxo*-hex-1-enitol, **c**, 3,4-

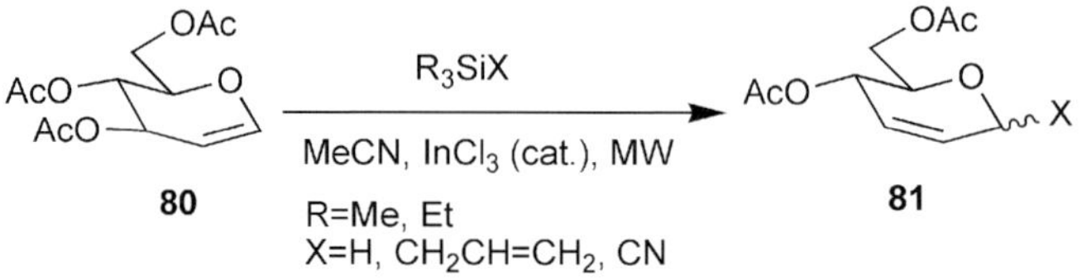

Scheme 12.38

di-*O*-acetyl-1,5-anhydro-2,6-deoxy-L-*arabino*-hex-1-enitol, and **d**, -2-deoxy-L-*arabino*-pent-1-enitol), in the presence of alkyl (trimethylallyl, triethyl, trimethylcyano) silanes and catalytic amounts of indium trichloride, Das et al. [70] obtained good to excellent yields of the corresponding 2,3-unsaturated *C*-glycosides **81** (Scheme 12.38, as a representative example). The microwave-assisted reaction was conducted in acetonitrile solution in an open vessel and resulted in a large reduction in reaction time compared with the conventional reflux method [71].

Methods for preparation of *endo*-glycals include that of Baptistella et al. [72], who used microwave irradiation to obtain *endo*-glycals from gluco- and galacto-di-*O*-mesyl or di-*O*-tosyl derivatives, by means of Tipson–Cohen elimination reactions [73] (Scheme 12.39).

The results obtained clearly show the benefits of microwave activation compared with conventional heating.

Current preparations of *exo*-glycals with microwave activation include those of 5,6-unsaturated pyranosides [59] and *C*-glycosylidene derivatives [49]. *exo*-Glycals with the double bond located at the 5,6-positions, of which **78** is an example, were obtained in 98% yield from the 6-iodo selectively protected glucose derivative **82** by treatment with DBU in DMF for 30 min under the action of microwave irradiation (Scheme 12.40).

New *C*-glycosylidene nitriles of special interest as carbohydrate mimics [74] have

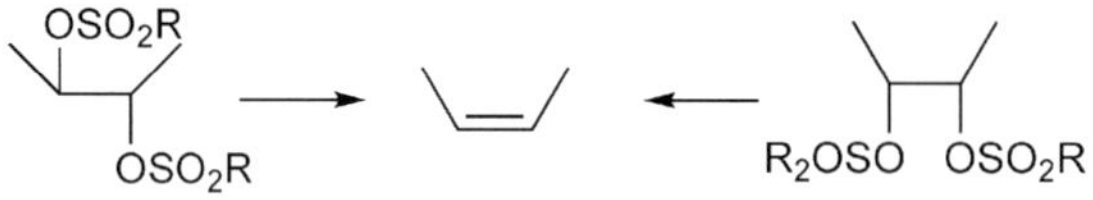

Scheme 12.39

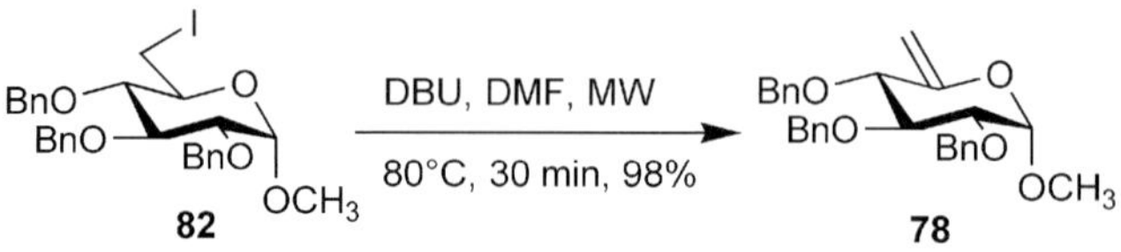

Scheme 12.40

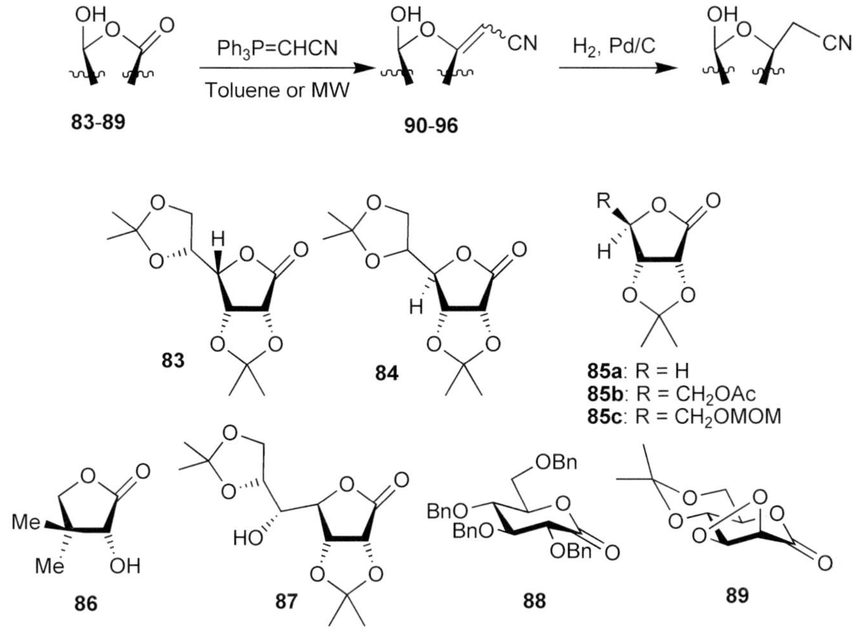

12.12 Synthesis of Dimers and Polysaccharides, and their Derivatives

Scheme 12.41

been prepared by Chapleur et al. [49] by microwave activated reaction of cyanomethyltriphenylphosphorane with different sugar lactones **83–89**. E–Z mixtures of **90–96** were produced in almost equal amounts, in good to excellent yields, and with dramatic reduction of the reaction time (Scheme 12.41).

These C-glycosylidene nitriles were reduced to the corresponding C-glycosyl compounds with complete stereocontrol at the anomeric center.

12.12
Synthesis of Dimers and Polysaccharides, and their Derivatives

Glycosylation using oxazoline donors is rather slow and requires extended reflux heating, because of their low reactivity. Microwave activation seemed likely to be an appropriate technique because reaction time should be reduced quite drastically, hopefully with increased yield, because of the formation of fewer byproducts.

Oscarson et al. [75] performed easy and efficient synthesis of spacer-linked dimers of N-acetyllactosamine from the corresponding oxazoline donor **97**, by applying pyridinium triflate as promoter combined with microwave heating. Compared with the promoter pyridinium p-toluenesulfonate, previously proven to be more effective than other sulfonates [76], the triflate salt increased the yield of dimers **100** by more than 30%. Microwave irradiation reduced the reaction time

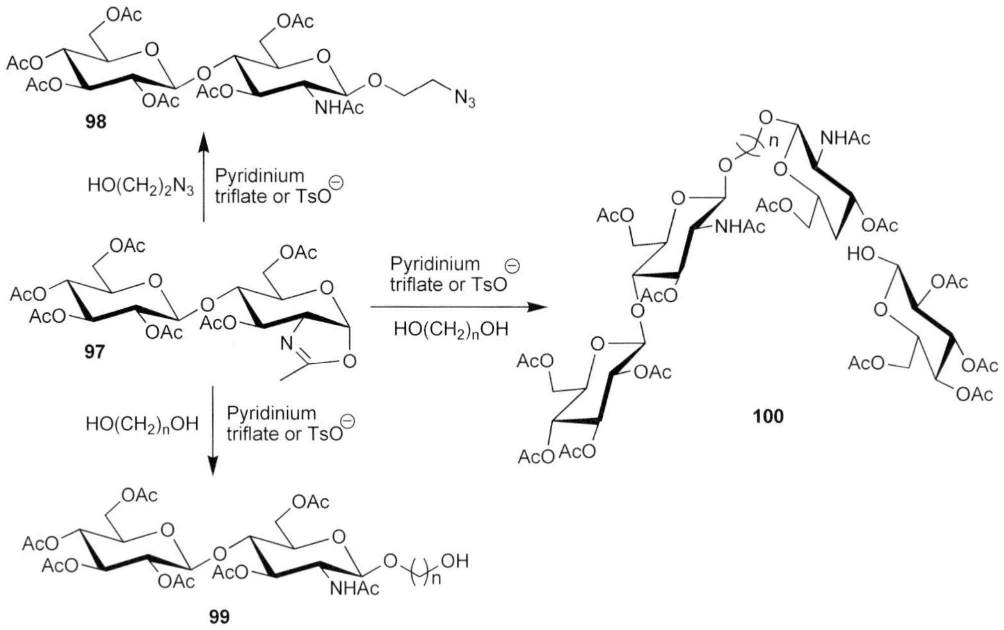

Scheme 12.42

from approximately 3 h to less than 20 min and improved yields by a further 12–15% (Scheme 12.42).

12.13
Synthesis of Heterocycles and Amino Acids

12.13.1
Spiroisoxazoli(di)ne Derivatives

Cycloaddition reactions of the activated exo glycal **101**, in the *E*-form, which can be regarded as a capto-dative olefin, with nitrones **102a–d** have been reported to occur only under the action of microwave activation to give spiroisoxazoli(di)nes **103a–c**, with good stereoselectivity and yields (Scheme 12.43) [77]. These cycloadditions, and those of the *Z*-isomer **101** and other ribo derivatives, were characterized by the endo mode of the *Z*-nitrone, as already postulated in most NOC reactions, and excellent facial selectivity of the furanoglycosylidenes, because of the presence of the dioxolane-protecting group near the anomeric center.

Pyranose *exo*-glycal **104** also gave good yields in NOC reactions with nitrones **102a–c**. Because of steric hindrance of both faces of the sugar ring, however, facial selectivity was incomplete and cycloaddition reactions gave α/β mixtures. Here again, the endo mode of cycloaddition is preferred with the *Z*-nitrone (Scheme

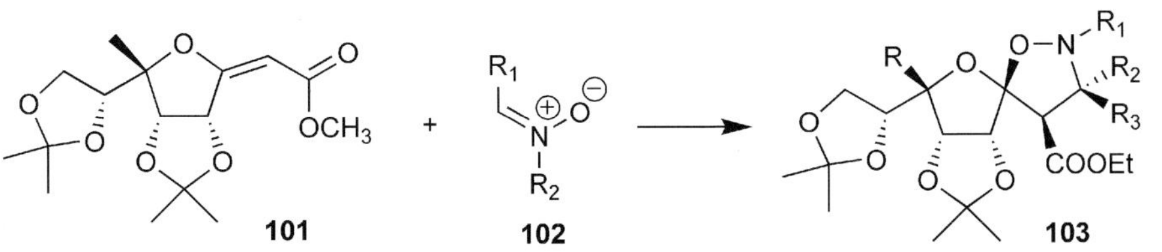

102: a) R_1 = nPr, R_2 = Me; b) R_1 = Ph, R_2 = Me; c) R_1 = nPr, R_2 = CH_2Ph; d) R_1 = H, R_2 = CH_2Ph
103: a) R_1 = Me, R_2 = nPr, R_3 = H; b) R_1 = Me, R_2 = H, R_3 = nPr; c) R_1 = Me, R_2 = Ph, R_3 = H; d) R_1 = Me, R_2 = H, R_3 = Ph; e) R_1 = CH_2Ph, R_2 = R_3 = H.

Scheme 12.43

12.44). Interestingly, only one cycloadduct, **105d**, was obtained in 68% yield with complete α selectivity.

The reactivity was similar in the reactions of benzonitrile oxide (**107**) with furan and pyran *exo*-glycals **101** and **104** and their Z-isomers. These cycloaddition reactions proceed at room temperature and give open-chain isoxazoles, for example **108**, because of facile β-elimination of the sugar ring oxygen on the intermediate isoxazoline ring system (Scheme 12.45, as a representative example of cycloadditions to furan *exo*-glycals).

12.13.2
Triazole-linked Glycodendrimers

Pieters et al. [78] described straightforward microwave-assisted procedures for rapid preparation of azido carbohydrates **109** which were applied in a general microwave-enhanced regioselective Cu^I-catalyzed [3+2] cycloaddition reaction with different

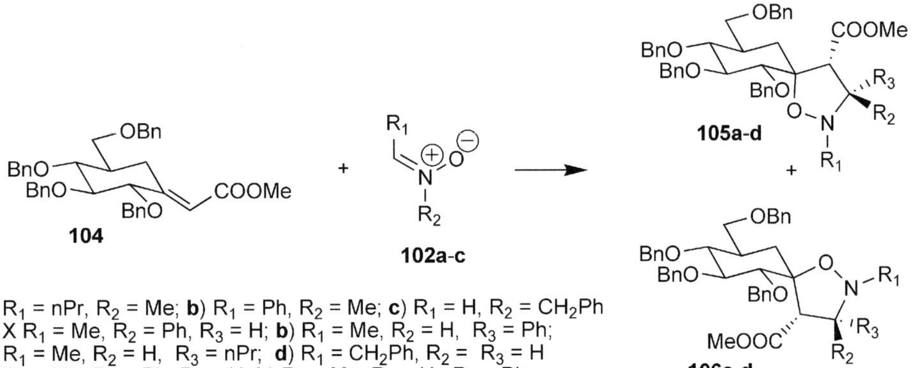

102: a) R_1 = nPr, R_2 = Me; b) R_1 = Ph, R_2 = Me; c) R_1 = H, R_2 = CH_2Ph
105: a) X R_1 = Me, R_2 = Ph, R_3 = H; b) R_1 = Me, R_2 = H, R_3 = Ph; c) R_1 = Me, R_2 = H, R_3 = nPr; d) R_1 = CH_2Ph, R_2 = R_3 = H
106: a) R_1 = Me, R_2 = Ph, R_3 = H; b) R_1 = Me, R_2 = H, R_3 = Ph; c) R_1 = Me, R_2 = nPr, R_3 = H; d) R_1 = Me, R_2 = H, R_3 = nPr

Scheme 12.44

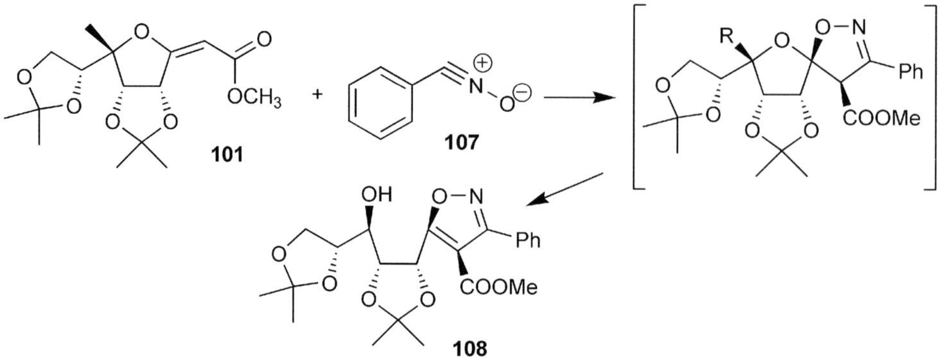

Scheme 12.45

types of alkyne-bearing dendrimer **110**. Triazole glycodendrimers **111**, up to the nonavalent level, were produced in high yields. Two biologically important glycoconjugates containing a fluorescent label were also prepared (Scheme 12.46).

12.13.3 5,6-Dihydro-1,2,3,4-tetrazenes

The stereoselective normal electron-demand Diels–Alder reactions of chiral 1,2-diaza-1,3-butadienes, derived from acyclic carbohydrates of different configuration, with diethyl azodicarboxylate (DEAD) are impractical at room temperature. In contrast, these are completed within a few hours by use of a focused microwave reactor.

Jimenez et al. [79] obtained a mixture of (6S) and (6R) configured 1,2,3,6-tetrahydro-1,2,3,4-tetrazines **113** and **114** in good yields (up 90%) and diastereoisomeric ratios (about 85:15 = α:β) (Scheme 12.47). The stereoselectivity of reactions was explained by theoretical calculations at semi-empirical level.

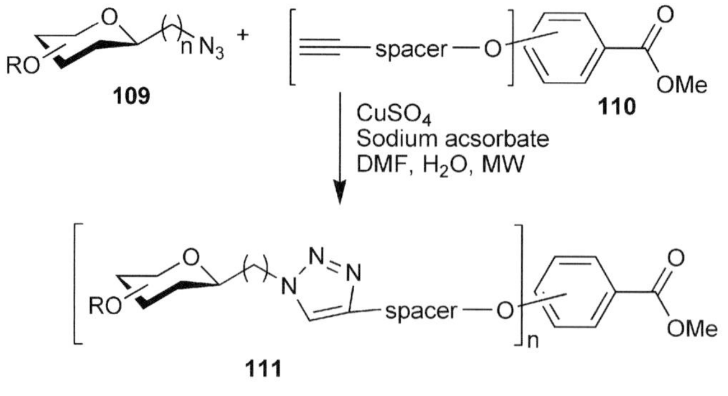

Scheme 12.46

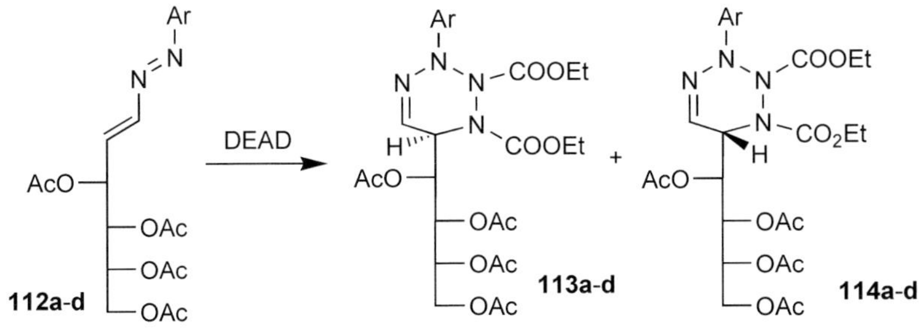

12.13 Synthesis of Heterocycles and Amino Acids | 605

Ar: **a** = C$_6$H$_5$; **b** = 4-ClC$_6$H$_4$; **c** = 4-CH$_3$C$_6$H$_4$; **d** = 4-CH$_3$OC$_6$H$_4$

Scheme 12.47

12.13.4 C-Nucleosides

Based on the pyrolysis of the neat hydrochloride salts of imidazolyl tetritols, e.g. **115a–c**, which leads to the 4-glycofuranosyl-1H-imidazoles, e.g. **116a–c** and their anomer mixtures, Tschamber et al. [80] obtained the same pairs of C-nucleosides in one-pot procedures by microwave irradiation with a domestic Whirlpool MO 100 oven for 1.7–3 min. Mixtures containing formamidine acetate, a few drops of water, and the appropriate hexose or hexulose, i.e. D-fructose (or D-glucose), D-galactose, or L-sorbose, resulted in overall yields of 19–28% (Scheme 12.48). Irradiation for longer periods led to caramelization.

12.13.5
Pterins

By Isay type condensation of a pyrimidinediamine **117** with aldohexoses, under microwave irradiation conditions (300 W for 270 s), Goswami and Adak [81] achieved construction of a pyrazine ring to afford pterins **118a–d** with a sugar substituent (Scheme 12.49a). Interestingly, the desired isomer-free 6-substituted sugar derivatives were synthesized in moderate to good yields whereas mixtures of both 6- and 7-isomers (major) were usually obtained from conventional Isay type conden-

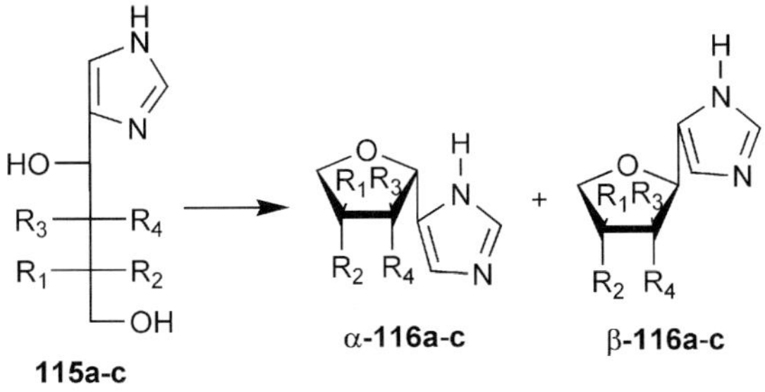

	R$_1$	R$_2$	R$_3$	R$_4$
115a, α,β-**116**	H	OH	H	OH
115b, α,β-**116**	H	OH	OH	H
115c, α,β-**116**	OH	H	H	OH

Scheme 12.48

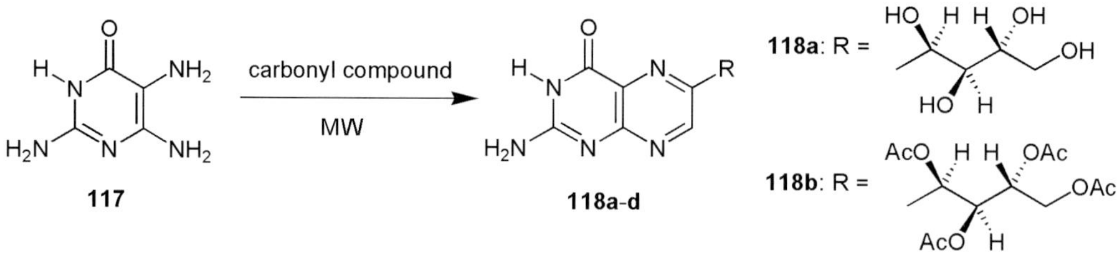

Scheme 12.49a

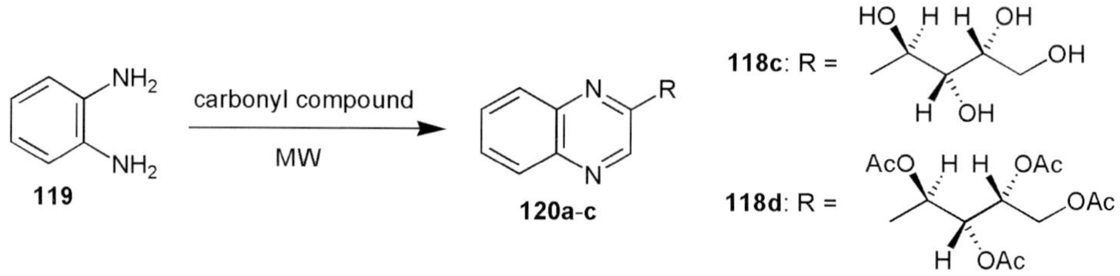

Scheme 12.49b

sations. By using benzenediamine **119**, quinoxaline systems **120** were obtained (Scheme 12.49b).

12.13.6
Epoxides

By treatment of the protected 2-O-tosyl-α-D-glucopyranoside **121** under microwave irradiation conditions at 100 °C for 6 min in the presence of potassium hydroxide–alumina, Cléophax et al. [82] obtained the epoxide **122** in nearly quantitative yield. Under classical heating conditions (same conditions, oil bath), the required epoxide was obtained in 25% yield only (Scheme 12.50).

Loupy et al. [7] scaled up the same reactions, using the conditions listed in Table 12.1.

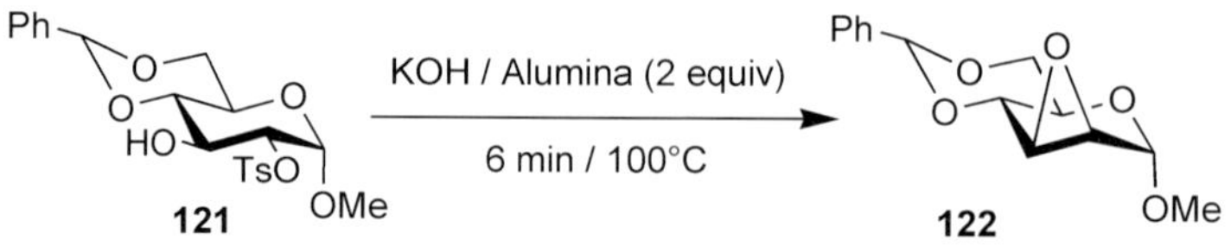

Scheme 12.50

Tab. 12.1. Epoxidation of **121** into **122** in 6 min at 100 °C [7].

Activation method	Yield[a] of 122
MW, S402	99
Δ, oil bath	25
MW, S1000 (on a larger scale of twenty times)	95

[a] Yields in isolated products

12.13.7
Azetidinones

During the course of their studies on the synthesis of β-lactams [83], Bose et al. [84] conveniently conducted several types of synthetic step under the action of microwave irradiation. Enantiospecific synthesis was developed for α-hydroxy-α-lactams of predictable absolute configuration starting from readily available carbohydrates. Stereospecific approaches and microwave-assisted chemical reactions have been used for preparation of these 3-hydroxy-2-azetidinones and their conversion to natural or non-natural enantiomers of intermediates in the preparation of gentosamine, 6-epilincosamine, γ-hydroxythreonine, and polyoxamic acid. They showed that reaction of acyl chloride **123** and triethylamine with a Schiff base such as **124**, with low levels of microwave irradiation, gives mostly cis-β-lactams **125**. If higher energy levels are used, producing higher temperatures, more than 90% of the β-lactam formed may be the trans isomer **126** (Scheme 12.51). Interest-

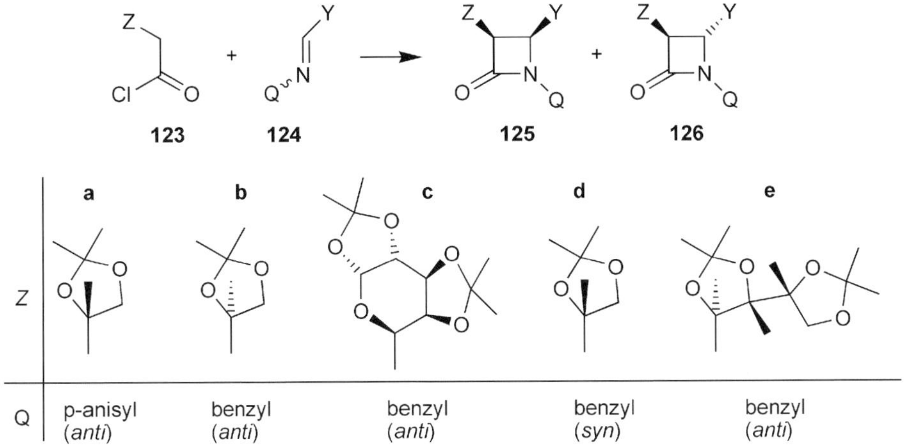

Scheme 12.51

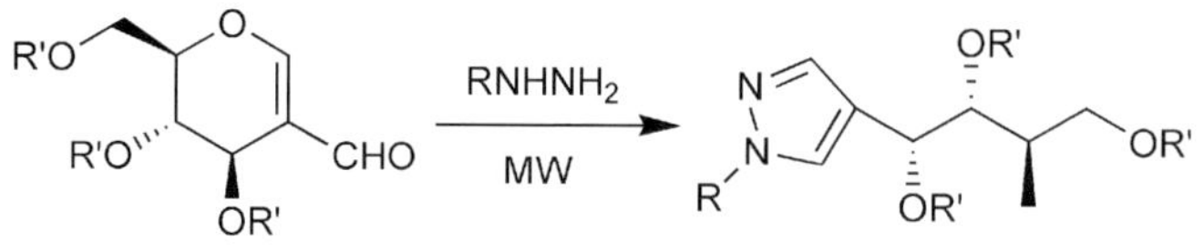

Scheme 12.52

ingly with Schiff bases of the type **181a–e**, the *cis-β*-lactams are formed irrespective of the level of microwave irradiation. On the few grams scale, optically active *cis-β*-lactams derived from **124d,e** were obtained in high yield after approximately 3 min irradiation in an 800 W commercial microwave oven.

12.13.8
Substituted Pyrazoles

In continuing efforts to develop more general and versatile methods for the synthesis of pyrazoles contained in natural products, and at the same time to use microwave irradiation, instead of conventional heating, particularly the solvent-free technique, because it provides the opportunity to work with open vessels, Yadav et al. [85] prepared a new class of optically pure 4-substituted pyrazoles **128** from 2-formyl glycols **127** and aryl hydrazines (Scheme 12.52).

Comparison of reactions conducted under both thermal and microwave conditions (the lowest observed temperature being 80 °C after irradiation for 1 min at 450 W and the highest temperature being 110 °C after irradiation for 3 min at the same power) showed that reaction rates and yields were dramatically enhanced by use of microwave conditions.

12.13.9
3*H*-Pyrido[3,4-*b*]indole Derivatives

Pal et al. [86] prepared 3*H*-pyrido[3,4-*b*]indole derivatives **130** by means of the Pictet–Spengler and Bischler–Napieralski reactions starting from α-D-*xylo*-pentodialdo-1,4-furanose derivatives **129** and tryptamines. Microwave-assisted reactions conducted on silica gel support under solvent-free conditions resulted in better yields than those performed under conventional conditions (Scheme 12.53, a representative example).

12.14
Enzymatic Reactions

Enzymatic methods of synthesis are promising alternatives to chemical methods, because the latter are usually complicated by the formation of unwanted enan-

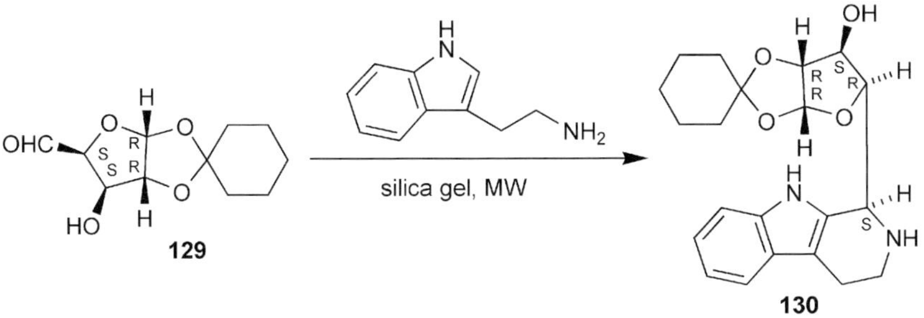

Scheme 12.53

tiomers, require long protection–deprotection steps for selectivity control, and result in low final yields. Use of microwave irradiation in enzymatic reactions has recently been proven to be efficient, when the temperature does not harm the enzyme properties. In comparison with reactions performed with conventional heating, use of microwaves affords increased selectivity and yields.

During research on enzymatic acylation, Gelo-Pujic et al. [87] performed Novozym immobilized lipase-catalyzed esterification of **131a,b** and **7** with dodecanoic acid in dry media both with microwave irradiation and with classical heating, using the same conditions of time and temperature (Scheme 12.54). Esterification of **131a** by use of a focused Synthewave reactor and oil bath, both at 95 °C, afforded **132** in 95% and 55% yield, respectively.

Starting from α,α-trehalose (**133**), the main product obtained was the 6,6′-di-ester **134** (88%) accompanied by a minor amount of 6-mono-ester **135** (4%) (Scheme 12.55).

Later, the same authors [88] performed some *trans*-glycosylations in a Synthewave 402 monomode reactor, using almond-β-glucosidase in an open system, except for one reaction for which a closed system was required. They performed the synthesis by reversed hydrolysis using glucose **131a** and hexane-1,6-diol **136**; *trans*-glycosylation was studied with phenyl β-D-glucoside **137** and cellobiose **138** as donors and propane-1,2-diol **134** as acceptor (Scheme 12.56).

The advantage of *trans*-glycosylation under microwave irradiation conditions rather compared with classical heating is complete conversion within 2–3 h, hydrolysis reduced to 10%, and only 2 equiv. excess of acceptor.

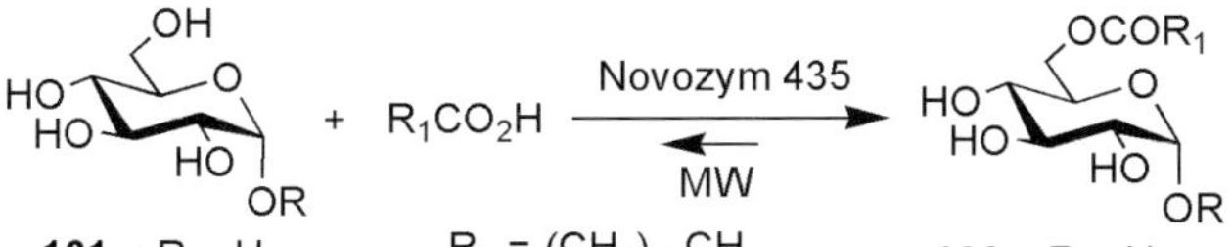

Scheme 12.54

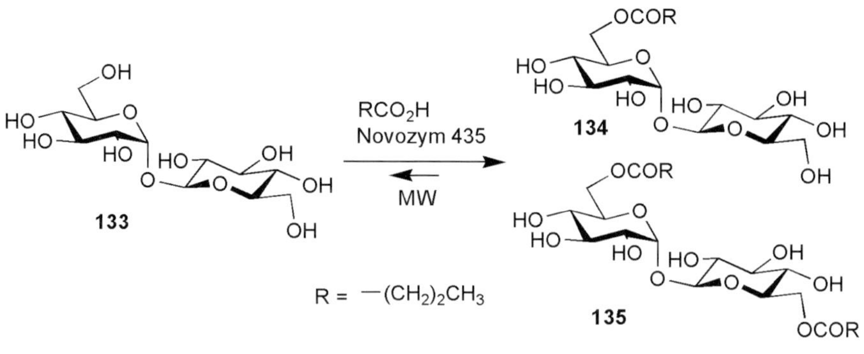

Scheme 12.55

Enhancement of the rates of enzyme reactions by microwave irradiation has also been reported by Chen et al. [89] who studied enzyme activity in the regioselective acylation of sugar derivatives in nonaqueous solvents.

In a screening procedure for characterization of lipase selectivity, Bradoo et al. [90] esterified sucrose and ascorbic acid with different fatty acids in a microwave oven using porcine pancreas, *B. stearothermophilus* SB-1 and *B. cepacia* RGP-10 lipases. Microwave-assisted enzyme catalysis was found to be an attractive procedure for rapid characterization of a large number of enzyme samples and substrates; this would otherwise have been a cumbersome and time-consuming exercise.

Maugard et al. [91] showed for the first time that it is possible to produce oligosaccharides by use of β-glycosidases under the action of focused microwave irradiation. As model reaction, they studied the synthesis of galacto-oligosaccharides (GOS) from lactose using β-galactosidase from *Kluyveromyces lactis* (free and immobilized on Duolite A-568) under the action of focused microwave irradiation and under conventional heating conditions. By reducing the water activity of media and substantially increasing the initial lactose concentration the GOS selectivity (GOS synthesis/lactose hydrolysis ratio) was increased 217-fold by exposing immobilized enzyme to microwave irradiation and by the addition of cosolvents such as hexanol, in comparison with the reaction performed with conventional heating (in water, with free enzyme).

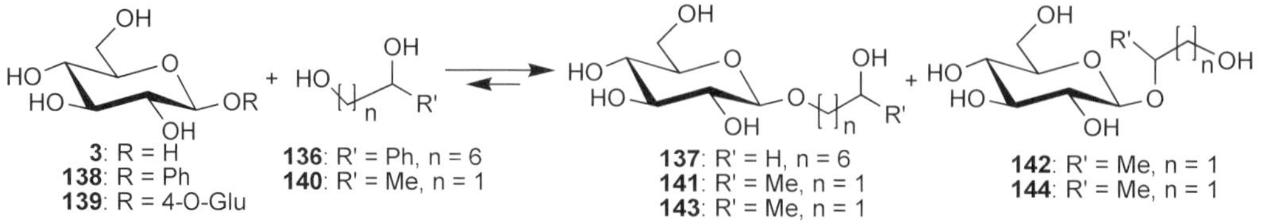

Scheme 12.56

12.15
Conclusion

Microwave activation is emerging as an alternative and (or) a complement to thermal heating which enables chemical transformations to be accomplished with very short reaction times, and higher yields of cleaner products. Its applications are rapidly expanding to any type of chemistry, including carbohydrate chemistry, in which energy transfer directly to the reactive species can create many new possibilities of performing chemical reactions currently not practicable with conventional heating.

Today, these new possibilities assume an important significance because of the growing development of carbohydrate chemistry as a consequence of the change from petroleum as raw material to natural feedstocks. Growth of microwave-assisted chemistry in industry is also likely, because of the demonstrable possibility of moving from small-scale (g) to the multigram (kg) synthesis of carbohydrate derivatives known to be valuable intermediates in the synthesis of diverse natural products and their analogs.

The short reaction times, which prevent decomposition of the sugars, in combination with easy performance and work-up, especially when solid-supported reagents and environmentally benign solvent-free conditions are used, make these microwave methods most attractive, and also suitable for automation.

References

1 (a) ABRAMOVICH, R.A. *Org. Prep. Proc. Int.* **1991**, *23*, 685; (b) MINGOS, D.M.P.; BAGHURST, D.R. *Chem. Soc. Rev.* **1991**, *20*, 1; (c) CADDICK, S. *Tetrahedron*, **1995**, *51*, 10403; (d) STRAUSS, C.R.; TRAINOR, R.W. *Aust. J. Chem.* **1995**, *48*, 1665; (e) LANGA, F.; DE LA CRUZ, P.; DE LA HOZ, A.; DÌAZ-ORTIZ, A.; DÌEZ-BARRA, E. *Contemp. Org. Synth.* **1997**, *4*, 373; (f) GEDYE, R.N.; WEI, J.B. *Can. J. Chem.* **1998**, *76*, 525; (g) GABRIEL, C.; GABRIEL, S.; GRANT, E.H.; HALSTEAD, B.S.J.; MINGOS, D.M.P. *Chem. Soc. Rev.* **1998**, *27*, 213; (h) LOUPY, A.; PETIT, A.; HAMELIN, J.; TEXIER-BOULLET, F.; JACQUAULT, P.; MATHÉ, D. *Synthesis*, **1998**, 1213; (i) STRAUSS, C.R. *Aust. J. Chem.* **1999**, *52*, 83; (j) VARMA, R.S. *Green Chemistry* **1999**, 43; (k) VARMA, R.S. *Clean Products and Processes*, **1999**, *1*, 132; (l) PERREUX, L.; LOUPY, A. *Tetrahedron*, **2001**, *57*, 9199; (m) LIDSTRÖM, P.; TIERNEY, J.; WATHEY, B.; WESTMAN, J. *Tetrahedron*, **2001**, *57*, 9225; (n) LARHED, M.; MOBERG, C.; HALLBERG, A. *Acc. Chem. Res.* **2002**, *35*, 717; (o) HAYES, B.L. *Aldrichimica Acta*, **2004**, *37*, 66; (p) KAPPE, C.O. *Angew. Chem. Int. Ed.* **2004**, *43*, 6250.

2 (a) KINGSTON, H.M., HASWELL, S.J. (Eds.), *Microwave-Enhanced Chemistry: Fundamental, Sample Preparation, and Applications*, American Chemical Society, 1997; (b) LOUPY, A. (Ed.), *Microwaves in Organic Synthesis*, Wiley–VCH, 2003.

3 CORSARO, A.; CHIACCHIO, U.; PISTARÀ, V.; ROMEO, G. *Current Organic Chemistry*, **2004**, *8*, 511.

4 DAS, S.K. *SynLett*, **2004**, 915.

5 SODERBERG, E.; WESTMAN, J.; OSCARSON, S. *J. Carbohydr. Chem.* **2001**, *20*, 397.

6 LIMOUSIN, C.; CLÉOPHAX, J.; PETIT, A.; LOUPY, A.; LUKACS, G. *J. Carbohydr. Chem.* **1997**, *16*, 327.

7 CLÉOPHAX, J.; LIAGRE, M.; LOUPY, A.;

Petit, A. *Organic Process Research & Development*, **2000**, *4*, 498.
8 Das, S.K.; Reddy, K.A.; Krovvidi, V.L.N.R.; Mukkanti, K. *Carbohydr. Res.* **2005**, *340*, 1387.
9 Valverde, S.; Hernandez, A.; Herradón, B.; Rabanal, R.M.; Martin-Lomas, M. *Tetrahedron*, **1987**, *43*, 3499.
10 Morcuende, A.; Valverde, S.; Herradón, B. *Synlett*, **1994**, 89.
11 Herradón, B.; Morcuende, A.; Valverde, S. *Synlett*, **1995**, 455.
12 Helm, R.F.; Ralph, J.; Anderson, L. *J. Org. Chem.* **1991**, *56*, 7015.
13 Limousin, C.; Cléophax, J.; Loupy, A.; Petit, A. *Tetrahedron*, **1998**, *54*, 13567.
14 Garegg, P.J.; Swahn, C.-G. *Acta Chem. Scand.* **1972**, *26*, 3895.
15 Csiba, M.; Cléophax, J.; Loupy, A.; Malthête, J.; Gero, S.D. *Tetrahedron Lett.* **1993**, *34*, 1787.
16 Kula, J.; Bragiel, B.; Gora, J. *Parf. Kosmetik*, **1995**, *76*, 368.
17 Salanski, P.; Descotes, G.; Bouchu, A.; Queneau, Y. *J. Carbohydr. Chem.* **1998**, *17*, 129.
18 Khalafi-Nezhad, A.; Alamdari, R.F.; Zekri, N. *Tetrahedron*, **2000**, *56*, 7503.
19 (a) Hanessian, S.; Lavallée, P. *Can. J. Chem.* **1975**, *53*, 2975; (b) Hanessian, S.; Lavallée, P. *Can. J. Chem.* **1977**, *55*, 562.
20 Shieh, W.-C.; Dell, S.; Repič, O. *Tetrahedron Lett.* **2002**, *43*, 5067.
21 Bailliez, V.; de Figueiredo, R.M.; Olesker, A.; Cléophax, J. *Synthesis*, **2003**, 1015.
22 Varma, R.S.; Varma, M.; Chatterjee, A.K. *J. Chem. Soc., Perkin Trans. 1*, **1993**, 999.
23 Tsuzuki, K.; Nakajima, I.; Watanabe, T.; Yanagiya, M.; Matsumoto, T. *Tetrahedron Lett.* **1978**, 989.
24 Mori, K.; Tominaga, M.; Takigawa, T.; Matsui, M. *Synthesis*, **1973**, 790.
25 Ley, S.V.; Mynett, D.M. *Synlett.*, **1993**, 793.
26 Couri, M.R.C.; Evangelista, E.A.; Alves, R.B.; Prado, M.A.F.; Gil, R.P.F.; De Almeida, M.V.; Raslan, D.S. *Synth. Commun.* **2005**, *35*, 2025.
27 Vaino, A.R.; Szarek, W.A. *Chem. Commun.* **1996**, 2351, and references cited therein.
28 Szarek, W.A.; Zamojski, A.; Tiwari, K.N.; Ison, E.R. *Tetrahedron Lett.* **1986**, *27*, 3827.
29 (a) Ames, G.R. *Chem. Rev.*, **1960**, *60*, 541; (b) Havlinova, B.; Kosik, M.; Kovak, P.; Blazej, A. *Tenside Detergents*, **1978**, *14*, 72; (c) Jones, R.F.D.; Camilleri, P.; Kirby, A.J.; Okafo, G.N. *J. Chem. Soc., Chem. Commun.* **1994**, 1311.
30 Hill, K.; Rybinski, W.V.; Stoll, G. *Alkyl Polyglycosides*, VCH, Weinheim, **1997**.
31 Stubbs, G.W. *Biochem. Biophys. Acta*, **1976**, *426*, 46.
32 (a) Fischer, E. *Ber. Dtsch. Chem. Ges.*, **1893**, *26*, 40; (b) Fischer, E. *Ber. Dtsch. Chem. Ges.*, **1895**, *28*, 1145.
33 Nuchter, M.; Ondruschka, B.; Lautenschlarger, W. *Synth. Commun.* **2001**, *31*, 1277.
34 Mathew, F.; Jayaprakash, K.N.; Fraser-Reid, B.; Mathew, J.; Scicinski, J. *Tetrahedron Lett.* **2003**, *44*, 9051.
35 De Figueiredo, R.M.; Bailliez, V.; Dubreuil, D.; Olesker, A.; Cléophax, J. *Synthesis*, **2003**, 2831.
36 Seibel, J.; Hillringhaus, L.; Moraru, R. *Carbohydr. Res.* **2005**, *340*, 507.
37 Ohrui, H.; Kato, R.; Kodara, T.; Shimizu, H.; Akasaka, K.; Kitara, T. *Biosci. Biotechnol. Biochem.* **2005**, *69*, 1054.
38 Bornaghi, L.F.; Poulsen, S.-A. *Tetrahedron Lett.* **2005**, *46*, 3485.
39 Ferrier, R.J.; Prasad, N. *J. Chem. Soc. (C)*, **1969**, 570.
40 (a) Holder, N. *Chem. Rev.* **1982**, *82*, 287; (b) Liu, Z.J.; Zhou, M.; Min, J.M.; Zhang, L.H. *Tetrahedron: Asymmetry*, **1999**, *10*, 2119; (c) Brito, T.M.B.; Silva, L.P.; Siqueira, V.L.; Srivastava, R.M. *J. Carbohydr. Chem.* **1999**, *18*, 609; (d) Srivastava, R.M.; Oliveira, F.J.S.; da Silva, L.P.; de Freitas Filho, J.R.; Oliveira, S.P.; Lima, V.L.M. *Carbohydr. Res.* **2001**, *332*, 335.
41 Sowmya, S.; Balasubramanian, K.K. *Synth. Commun.* **1994**, *24*, 2097.

42 Shanmugasundaram, B.; Bose, A.K.; Balasubramanian, K.K. *Tetrahedron Lett.* **2002**, *43*, 6795.
43 de Oliveira, R.N.; de Freitas Filho, J.R.; Srivastava, R.M. *Tetrahedron Lett.* **2002**, *43*, 2141.
44 Das, S.K.; Reddy, K.A.; Roy, J. *Synlett*, **2003**, *11*, 1607.
45 Hotha, S.; Tripathi, A. *Tetrahedron Lett.* **2005**, *46*, 4555.
46 Taillefumier, C.; Chapleur, Y. *Chem. Rev.* **2004**, *104*, 263.
47 Lin, H.-C.; Chang, C.-C.; Chen, J.-Y.; Lin, C.-H. *Tetrahedron: Asymmetry*, **2005**, *16*, 297.
48 (a) Lewis, M.D.; Cha, J.K.; Kishi, Y. *J. Am. Chem. Soc.* **1982**, *104*, 4976; (b) Paterson, L.; Keown, L.E. *Tetrahedron Lett.* **1997**, *38*, 5727; (c) Horita, K.; Sakurai, Y.; Nagasawa, M; Hachiya, S.; Yonemitsu, O. *Synlett* **1994**, 43; (d) Suhadolnik, R.J. *Nucleoside Antibiotics*, Wiley Interscience, New York, **1970**.
49 Lakhrissi, Y.; Taillefumier, C.; Lakhrissi, M.; Chapleur, Y. *Tetrahedron: Asymmetry*, **2000**, *11*, 417.
50 (a) Norris, P.; Myers, T.; Hurt, R.; Duncan, S., Sacui, J.; McKee, S. *35th ACS Central Regional Meeting*, Pittsburg (PA), U.S.A., October 19–22, **2003**; (b) Myers, T.; Hurt, R.; Duncan, S.; McKee, S.; Dobosh, B.; Norris, P. *227th ACS National Meeting*, Anaheim (CA), U.S.A., March 28–April 1, **2004**.
51 (a) Rao, H.S.; Reedy, K.S. *Tetrahedron Lett.* **1994**, *35*, 171; (b) Ram, S.; Ehrenkaufer, R.E. *Synthesis*, **1988**, 91.
52 Banik, B.K.; Barakat, K.J.; Wagle, D.R.; Manhas, M.S.; Bose, A.K. *J. Org. Chem.* **1999**, *64*, 5746.
53 Chakraborty, V.; Bordoloi, M. *J. Chem. Res. (S)*, **1999**, 118.
54 Kubrakova, I.V.; Formanovskii, A.A.; Kudinova, T.F.; Kuz'min, N.M. *J. Anal. Chem. (Zhurnal Analiticheskoi Khimii)*, **1999**, *54*, 460.
55 Limousin, C.; Olesker, A.; Cléophax, J.; Petit, A.; Loupy, A.; Lukacs, G. *Carbohydr. Res.* **1998**, *312*, 23.
56 Dax, K.; Albert, M.; Ortner, J.; Paul, B.J. *Carbohydr. Res.* **2000**, *327*, 47.

57 (a) Kobayashi, S.; Yoneda, A.; Fukuhara, T.; Hara, S. *Tetrahedron Lett.* **2004**, *45*, 1287; (b) Kobayashi, S.; Yoneda, A.; Fukuhara, T.; Hara, S. *Tetrahedron*, **2004**, *60*, 6923; (c) Hara, S.; Fukuhara, T. WO 2004050676, **2004**.
58 Dmowski, W.; Kaminski, M.J. *J. Fluorine Chem.* **1983**, *23*, 219.
59 Ko, K.-S.; Zea, C.J.; Pohl, N. *J. Am. Chem. Soc.* **2004**, *126*, 13188.
60 Cameron, D.G.; Martin, A.I.; Mantsch, H.H. *Science*, **1983**, *219*, 180.
61 Barnett, J.E.G. *Adv. Carbohydr. Chem. Biochem.* **1972**, *27*, 127.
62 Cioffi E.A. in *Synthesis and Applications of Isotopically Labeled Compounds*, Vol. 7, pp 89–92, Pleiss, U., Voges, R. (Eds.), John Wiley & Sons LTD, **2001**.
63 (a) Maier, M.L.; Cioffi, E.A. 225th ACS National Meeting, New Orleans (LA), U.S.A., March 23–27, **2003**; (b) Cioffi, E.A.; Cook, M.L. 226th ACS National Meeting, New York (NY), U.S.A., September 7–11, **2003**; (c) Cioffi, E.A.; Bell, R.H.; Le, B. *Tetrahedron: Asymmetry*, **2005**, *16*, 471.
64 (a) Likhosherstov, L.M.; Novikova, O.S.; Derevitskaja, V.A.; Kochetkov, N.K. *Carbohydr. Res.* **1986**, *146*, C1–C5; (b) Cohen-Anisfeld, S.T.; Lansbury, P.T. Jr. *J. Am. Chem. Soc.* **1993**, *115*, 10531.
65 (a) Turnbull, J.E.; Hopwood, J.J.; Gallagher, J.T. *Proc. Natl. Acad. Sci. USA*, **1999**, *96*, 2698; (b) Drummond, K.J.; Yates, E.A.; Turnbull, J.E. *Proteomics*, **2000**, *1*, 304; (c) Lee, K.-B.; Al-Hakim, A.; Loganathan, D.; Linhardt, R. *Carbohydr. Res.* **1991**, *214*, 155.
66 Dove, A. *Nature Biotechnol.* **2001**, *19*, 913.
67 Yates, E.A.; Jones, M.O.; Clarke, C.E.; Powell, A.K.; Johnson, S.R.; Porch, A.; Edwards, P.P.; Turnbull, J.E. *J. Mat. Chem.* **2003**, *13*, 2061.
68 Lever, R.; Page, C.P. *Nat. Rev. Drug Discov.* **2002**, *1*, 140.
69 For a review of these methods see unsaturated derivatives in *Carbo-*

hydrate Chemistry, Royal Society of Chemistry.
70 Das, S.K.; Reddy, K.A.; Abbineni, C.; Roy, J.; Rao, K.V.L.N.; Sachwani, R.H.; Iqbal, J. Tetrahedron Lett. **2003**, 44, 4507.
71 Ghosh, R.; De, D.; Shown, B.; Maiti, S.B. Carbohydr. Res. **1999**, 321, 1, and references cited therein.
72 Baptistella, L.H.B.; Neto, A.Z.; Onaga, H.; Godoi, E.A.M. Tetrahedron Lett. **1993**, 34, 8407.
73 (a) Umezawa, S; Okazaki, Y.; Tsuchiya, T. Bull. Chem. Soc. Jpn. **1972**, 45, 619; (b) Tipson, R.S.; Cohen, A. Carbohydr. Res. **1965**, 1, 338; (c) Albano, E.; Horton, D.; Tsuchiya, T. Carbohydr. Res. **1966**, 2, 349.
74 (a) Chapleur, Y. Carbohydrate Mimics, Wiley–VCH: Weinheim, **1998**; 58; (b) Lakhrissi, M.; Bandzouri, A.; Chapleur, Y. Carbohydr. Lett., **1995**, 307, and references cited therein.
75 Mohan, H.; Gemma, E.; Ruda, K.; Oscarson, S. Synlett, **2003**, 9, 1255.
76 Yohino, T.; Sato, K.-I.; Wanme, F.; Takai, I.; Ishido, Y. Glycoconj. J. **1992**, 9, 287.
77 (a) Taillefumier, C.; Enderlin, G.; Chapleur, Y. Lett. Org. Chem. **2005**, 2, 226; (b) Enderlin, G.; Taillefumier, C.; Didierjean, C.; Chapleur, Y. Tetrahedron: Asymmetry, **2005**, 16, 2459.
78 Joosten, J.A.F.; Tholen, N.T.H.; El Maate, F.A.; Brouwer, A.J.; van Esse, G.W.; Rijkers, D.T.S.; Liskamp, R.M.J.; Pieters, R.J. Eur. J. Org. Chem. **2005**, 3182.
79 Avalos, M.; Babiano, R.; Cintas, P.; Clemente, F.R.; Jimenez, J.L.; Palacios, J.C.; Sanchez, J.B. J. Org. Chem. **1999**, 64, 6297.
80 Tschamber, T.; Rudyk, H.; Le Nouen, D. Helv. Chim. Acta, **1999**, 82, 2015.
81 Goswami, S.; Adak, A.K. Tetrahedron Lett. **2002**, 43, 8371.
82 Hladezuk, I.; Olesker, A.; Cléophax, J.; Lukacs, G. J. Carbohydr. Chem. **1998**, 17, 869.
83 Manhas, M.S.; Banik, B.K.; Mathur, A.; Vincent, J.E.; Bose, A.K. Tetrahedron, **2000**, 56, 5587.
84 Bose, A.K.; Banik, B.K.; Mathur, C.; Wagle, D.R.; Manhas, M.S. Tetrahedron, **2000**, 56, 5603.
85 Yadav, J.S.; Reddy, B.V.S.; Satheesh, G.; Naga Lakshmi, P.; Kiran Kumar, S.; Kunwar, A.C. Tetrahedron Lett. **2004**, 45, 8587.
86 Pal, B.; Jaisankar, P.; Giri, V.S. Synth. Commun. **2003**, 33, 2339.
87 Gelo-Pujic, M.; Guibé-Jampel, E.; Loupy, A.; Galema, S.A.; Mathé, D. J. Chem. Soc., Perkin Trans. 1, **1996**, 2777.
88 Gelo-Pujic, M.; Guibé-Jampel, E.; Loupy, A.; Trincone, A. J. Chem. Soc., Perkin Trans. 1, **1997**, 1001.
89 Chen, S.-T.; Sookkhedo, B.; Phutrahul, S.; Wang, K.-T. Methods in Biotechnology, **2001**, 15 (Enzymes in non-aqueous Solvents), 373.
90 Bradoo, S.; Rathhi, P.; Saxena, R.K.; Gupta, R. J. of Biochem. and Biophys. Methods, **2002**, 51, 115.
91 Maugard, T.; Gaunt, D.; Legoy, M.D.; Besson, T. Biotechnology Lett., **2003**, 25, 623.

13
Microwave Catalysis in Organic Synthesis

Milan Hájek

13.1
Introduction

The objective of this chapter is to draw the attention of experimental organic and catalytic chemists to a new field of catalysis, especially to catalytic methods which use microwave irradiation as a new means of activation of chemical reactions, called "microwave catalysis". It is intended to advise synthetic organic chemists about the choice of catalytic steps which might be more efficient than conventional synthetic methods.

This chapter focuses exclusively on *microwave heterogeneous catalysis*. Microwave homogeneous catalysis by transition metal complexes is treated in Chapter 15, phase-transfer catalysis in Chapter 6 and photocatalytic reactions in Chapter 19.

The development of microwave heterogeneous catalysis is, however, impeded because most synthetic chemists are not well acquainted with factors affecting both heterogeneous catalysis and microwaves. Although some attempts have been made to accelerate reaction rate or to improve yield and selectivity, they were based on the idea "let's try microwaves and see what happens".

This chapter is written to help the synthetic chemist to understand the effect of microwaves on heterogeneously catalyzed reactions. Not only metal catalysts are discussed, but also metal oxides, zeolites, clays, and similar materials that act either as catalysts or as potential supports for the catalytically active species. The factors involved in the preparation of catalysts under microwave conditions are summarized and discussed in terms of their effects on catalyst activity and selectivity. Basic mechanistic understanding can frequently be used to modify reaction conditions to achieve product formation in high yield. Such an approach cannot be readily applied to heterogeneously catalyzed processes, because the mechanistic understanding needed by the synthetic chemist is not yet commonly available for this type of reaction. This chapter is intended to call the attention of synthetic chemists to microwave catalysis, because it has advantages over conventional heterogeneous catalysis (there are also some problems). Hopefully, this may encourage synthetic and catalytic chemists to use such process more frequently and thus extend the application of microwaves to the preparation of a variety of materials.

Microwaves in Organic Synthesis, Second edition. Edited by A. Loupy
Copyright © 2006 WILEY-VCH Verlag GmbH & Co. KGaA, Weinheim
ISBN: 3-527-31452-0

13.1.1
Definitions

A *catalyst* is a substance that increases the rate of a chemical reaction without being changed in the process. During the reaction it can become a different entity, but after the catalytic cycle is complete, the catalyst is the same as at the start. The function of the catalyst is to reduce the activation energy of the reaction pathway. Catalysts can be heterogeneous, homogeneous or biological. Typical types of heterogeneous catalyst are bulk metals, supported inorganic metallic compounds, and supported organometallic complexes. Catalyst interaction with reactants can occur homogeneously, i.e. with the reactants and the catalyst in the same phase (usually liquid), or heterogeneously at the interface between two phases. The latter type of catalyzed reaction utilizes a solid catalyst, and the interaction occurs at either the liquid–solid or gas–solid interface.

Microwave catalysis is a catalytic process performed in the presence of a microwave (electromagnetic) field in which the catalyst acts as an energy "convertor". It uses microwave irradiation to stimulate catalytic reactions. It is necessary to stress that any sort of electromagnetic or microwave radiation is not itself a catalyst, as has sometimes been erroneously stated [1]. Similarly, it is not correct to say that microwave irradiation catalyzes chemical reactions [1]. The principles of microwave catalysis will be described below.

A *catalytic reaction* is one in which more than one turnover or event occurs per reaction center or catalytically active site. If, however, less than one turnover occurs per active site, it cannot be a true catalytic reaction. This occurs for reactions on supported reagents in the absence of solvent, when support or catalyst is used in excess (called also solvent-free, solvent-less, dry media etc., see Chapter 8). Catalytic reactors usually contain a small amount of solid catalyst compared with the amounts of reactants that pass over the catalyst as liquids or gases. For microwave heating to be successfully applied in catalytic systems the catalyst itself must absorb microwave energy. Many supported catalysts (metal, metal oxides, etc.) are known to absorb microwave energy readily to different extents, whereas the support (silica, alumina) does not. Thus, after conversion, a liquid or gas stream passing over the catalyst reaches the cooler part of the reactor more quickly, thus preserving the reactive products otherwise destroyed in a conventional system. Microwave catalyzed reactions thus occur at lower temperatures with substantial energy savings and often with higher yields of the desired products.

Heterogeneously catalyzed reactions are rather complex processes. Considering a two-phase system, either liquid–solid or gas–solid, several steps are needed to complete the catalytic cycle:

- transport of the reactants to the catalyst;
- interaction of the reactants with the catalyst (adsorption);
- reaction of adsorbed species to give the products (surface reaction);
- desorption of the products; and
- transport of the products away from the catalyst.

Microwave radiation can be used to prepare new catalysts, enhance the rates of chemical reactions, by microwave activation, and improve their selectivity, by selective heating. The heating of the catalytic material usually depends on several factors including the size and shape of the material and the exact location of the material in the microwave field. Its location depends on the type of the microwave cavity used [2].

The objective of this chapter is to describe some advances in catalysis achieved by use of microwave irradiation. The aspects of microwave catalytic reactions which differ from traditional thermal methods are emphasized. The input of microwave energy into a reaction mixture is quite different from conventional (thermal) heating and it is the task of the synthetic chemists to exploit this special situation as fully as possible.

13.2
Preparation of Heterogeneous Catalysts

The interaction of microwaves (MW) with solid materials has proven attractive for the preparation and activation of heterogeneous catalysts. It has been suggested that microwave irradiation modifies the catalytic properties of solid catalysts, resulting in increasing rates of chemical reactions. It is evident that microwave irradiation creates catalysts with different structure, activity and/or selectivity. Current studies document a growing interest in microwave-assisted catalyst preparation and in the favorable effects of microwaves on catalytic reactions.

Preparation of catalysts usually involves impregnation of a support with a solution of active metal salts. The impregnated support is then dried, calcined to decompose the metal salt, and then reduced (activated) to produce the catalyst in its active form. Microwaves have been used in all stages of catalyst preparation. Beneficial effects of microwave heating, compared with conventional methods, have been observed especially in the *drying, calcination*, and *activation* steps.

13.2.1
Drying and Calcination

It is well known that microwave (MW) drying of many solid materials is a very efficient and widely used process even on an industrial scale [3]; it is also an attractive means of drying of heterogeneous catalysts. Microwave drying of catalysts and supported adsorbents has several advantages:

- reduction of drying time;
- higher surface area and catalytic activity;
- thermal dispersion of active species facilitated by microwave energy, providing more uniform metal distribution; and
- higher mechanical strength of catalyst pellets.

Microwave heating has been reported to produce materials with particular physical and chemical properties [4]. Stable solid structures are formed at low reaction temperatures with unusually high surface areas, making them very useful as catalysts or catalyst supports. Calcination of solid precursors in a microwave field has significant advantages over conventional heating. The effective synthesis of the catalysts and supporting adsorbents has been reported for the examples below.

Microwave drying of an alumina-supported nickel catalyst reduced the drying time by a factor of 2 to 3 [4]. The results indicated that the better dispersion of nickel was achieved by microwave drying, presumably because of minimization of moisture gradients during drying. Analysis of the results confirmed, moreover, that the microwave-dried samples were significantly stronger than those dried conventionally. As a consequence of minimum moisture gradients, the metal ions are not redistributed to the same extent during the microwave-drying process as they are when the samples are dried conventionally, because of moisture leveling [2, 3]. This causes the resulting dry pellets to be stronger, as shown by analysis of crushing strengths [4]. Because the heating effect is approximately proportional to moisture content, microwaves are ideal for equalizing moisture within the product in which the moisture distribution is initially nonuniform. Microwave drying also proceeds at relatively low temperatures, and no part of the product needs to be hotter than the evaporating temperature.

The microwave technique has also been found to be a potential method for the preparation of the catalysts containing highly dispersed metal compounds on high-porosity materials. The process is based on thermal dispersion of active species, facilitated by microwave energy, into the internal pore surface of a microporous support. Dealuminated Y zeolite-supported CuO and CuCl adsorbents were prepared by this method and used for SO_2 removal and industrial gas separation, respectively [5]. The results demonstrated the effective preparation of supported adsorbents by microwave heating. The method was simple, rapid, and energy-efficient, because synthesis of both adsorbents required much lower temperatures and much less time compared with conventional thermal dispersion. Similarly, MW drying of Ni–Al_2O_3–cordierite catalyst yielded more uniform active phase distribution compared with conventional freeze-drying [6].

The V_2O_5–SiO_2 catalyst for o-xylene oxidation prepared by wet impregnation under the action of microwave irradiation had several advantages [7] compared with that prepared by the conventional thermal method:

- dispersion of V_2O_5 on the surface of SiO_2 was more homogeneous;
- the nonisothermal process was minimized;
- dispersion of active phase (V_2O_5) was highly uniform; and
- in the catalytic microwave process the optimum reaction temperature of o-xylene oxidation was reduced by 100 K (from 653 to 553 K).

The more active cobalt catalyst for pyrolytic reactions was prepared by MW calcination of cobalt nitrate, which was converted to cobalt oxide by rapid microwave heating [8].

The high dispersion of inorganic salts ($CuCl_2$, $NiCl_2$, $AuCl_3$, $RuCl_3$, etc.) on the surface of zeolites (NaZSM-5, NaY, NaBeta) and alumina, with high loading of the active components, has recently been achieved by microwave techniques [9–11]. The catalysts were very active in NO_x decomposition, even at room temperature, CO and NO were partially converted to CO_2 and N_2. It was concluded that microwave treatment is a new route for dispersal and high loading of inorganic compounds on to the surface of supports to form highly active catalysts.

The microwave technique has been also found to be the best method for preparing strongly basic zeolites (ZSM-5, L, Beta, etc.) by direct dispersion of MgO and KF. This novel procedure enabled the preparation of shape-selective, solid, strongly basic catalysts by a simple, cost-effective, and environmentally friendly process [12, 13]. The new solid bases formed were efficient catalysts for dehydrogenation of 2-propanol and isomerization of cis-2-butene.

In the microwave synthesis of zeolites, a mixture of a precursor and a zeolite support is heated in a microwave oven. The sample is then tested for its catalytic activity and the results compared with those from the sample obtained by the conventional method. Microwave irradiation at the calcination stage led to samples with more uniform particle-size distribution and microstructure and to bimetallic catalysts with different morphology.

Microwave calcination of magnesia, alumina, and silica-supported Pd and Pd–Fe catalysts resulted in their having enhanced catalytic activity in test reactions – hydrogenation of benzene and hydrodechlorination of chlorobenzene – compared with conventionally prepared catalysts [14–16]. The greater catalytic activity was attributed to prevention of the formation of a Pd–Fe alloy of low activity, which occurs at the high reduction temperature used in conventional heating.

The microwave technique for drying then calcination is an excellent way of obtaining highly porous silica gel with a high surface area (as high as 635 $m^2\ g^{-1}$) for use as a catalyst and as a catalyst support [17].

The dispersion and solid-state ion exchange of $ZnCl_2$ on the surface of NaY zeolite by use of microwave irradiation [18] and modification of the surface of active carbon as catalyst support by means of microwave-induced treatment have also been reported [19]. The ion-exchange reactions of both cationic (montmorillonites) and anionic clays (layered double hydroxides) were greatly accelerated under conditions of microwave heating compared with other techniques currently available [20].

Microwave irradiation has also been applied to the preparation of Fe_2O_3–SO_4^{2-} superacid [21, 22] and high-surface aluminum pillared montmorillonites [21].

The most successful application of microwave energy in the preparation of heterogeneous solid catalysts has been the microwave synthesis and modification of zeolites [22, 23]. For example, cracking catalysts in the form of uniformly sized Y zeolite crystallites were prepared in 10 min by microwave irradiation, whereas 10–50 h were required by conventional heating techniques. Similarly, ZSM-5 was synthesized in 30 min by use of this technique. The rapid internal heating induced by microwaves not only led to shorter synthesis time and high crystallinity, but also enhanced substitution and ion exchange [23, 24].

Microwave processing of zeolites and their application in the catalysis of synthetic organic reactions has recently been excellently reviewed by Cundy [25] and other authors [26]. The microwave synthesis of zeolites and mesoporous materials was surveyed, with emphasis on those aspects which differ from conventional thermal methods. The observed rate enhancement of microwave-mediated organic synthesis achieved by use of these catalysts was caused by a variety of thermal effects, including very high rates of temperature increase, bulk superheating, and differential heating. Examples of microwave activation of chemical reactions catalyzed by zeolites will be presented in Section 13.3.

An efficient oxidation catalyst OMS-1 (octahedral molecular sieve) has been prepared by microwave irradiation of a family of layered and tunnel-structured manganese oxide materials. These materials are known to interact strongly with microwave radiation, and thus pronounced effects on the microstructure were expected. Their catalytic activity was tested in the oxidative dehydrogenation of ethylbenzene to styrene [27].

In the preparation of microporous manganese oxide materials, different chemical properties were observed after microwave and thermal preparation. In the conversion of ethylbenzene to styrene the activity and selectivity of the materials was different [28].

Greater catalyst activity when prepared with MW irradiation was observed in the hydrodechlorination of chlorobenzene over Pd–Fe catalyst [29] and in hydrocracking of liquid fuels over pillared clays as catalyst [30]. Similar catalytic activity, however, was observed in catalytic transfer hydrogenation of aromatic nitro compounds in the presence of perovskites as catalysts [31] or in phenol oxidation with titanium silicate catalyst [32]. Lower catalytic activity of Au–Fe_2O_3 catalyst in the water-gas shift reaction was explained by the larger crystallinites formed by MW irradiation, which led to reduction of surface area and, therefore, activity [33], similarly to methane combustion over PdO–$NaZr_2P_2O_3$ and Fe–SiO_2 catalysts [34]. Lower overall catalytic activity, however, can have a favorable effect on selectivity as was recorded in the aromatization of propane over Zn-ZSM-5 and Fe–SiO_2 catalysts [35].

An alternative approach for the preparation of a supported metal catalyst is based on the use of a microwave-generated plasma [36]. Several new materials prepared by this method are unlikely to be obtained by other methods. It is accepted that use of a microwave plasma results in a unique mechanism, because of the generation of a nonthermodynamic equilibrium in discharges during catalytic reactions. This can lead to significant changes in the activity and selectivity of the catalyst.

13.2.2
Catalyst Activation and Reactivation (Regeneration)

Microwave irradiation of catalysts before their use in chemical reactions has been found to be a promising new tool for catalyst activation. Microwave irradiation has been found to modify not only the size and distribution of metal particles but probably also their shape and, consequently, the nature of their active sites. These phenomena may have a significant effect on the activity and selectivity of catalysts, as found in the isomerization of 2-methylpentene on a Pt catalyst [2].

As an example, microwave activation of the platinum catalyst under conditions when it was highly sensitive to thermal treatment resulted in an increase of its catalytic activity and selectivity (from 40% to 80%) [37].

Durable changes of the catalytic properties of supported platinum induced by microwave irradiation have been also recorded [38]. A substantial reduction of the time taken for activation (from 9 h to 10 min) was observed in the activation of NaY zeolite catalyst by microwave dehydration in comparison with conventional thermal activation [39]. The very efficient activation and regeneration of zeolites by microwave heating can be explained by the direct desorption of water molecules from zeolite by the electromagnetic field; this process does not depend on the temperature of the solid [40]. Interaction of the adsorbed molecules with the microwave field does not result simply in heating of the system. Desorption is much faster than in the conventional thermal process, because transport of water molecules from the inside of the zeolite pores is much faster than the usual diffusion process.

Very little is known about the reactivation (regeneration) of used catalysts by microwave irradiation. Catalyst activity has been shown to decay with increasing carbon deposition, and several patents discuss the decarbonization of cracking zeolite catalysts by use of microwaves [41]. Because carbon is a very lossy material (it absorbs microwaves very efficiently), any carbon deposited on the surface of the catalyst is strongly heated. In the presence of air or hydrogen, the carbon is removed in the form of carbon dioxide or methane, respectively. When carbon deposition reaches a certain level it starts to absorb microwave energy strongly and is, therefore, subsequently removed, leading to an increase in the activity of the catalyst.

Alumina spheres polluted by carbon residues have been also reactivated by use of microwaves [42]. Their regeneration has been performed in a stream of air and in the presence of silicon carbide as an auxiliary microwave absorber. Microwave heat treatment led to full recovery of the catalyst in times varying from a half to a quarter of those required by conventional treatment. Regeneration of a commercial Ni catalyst ($Ni-Al_2O_3$) deactivated, presumably, by coke formation, by means of a flow of hydrogen or oxygen and water vapor under the action of microwave irradiation was, however, unsuccessful [43].

Microwaves are frequently used in the laboratory by synthetic organic chemists for regeneration and activation of solids such as molecular sieves, silica gel, or alumina when fast and complete drying is required.

13.3
Microwave Activation of Catalytic Reactions

Heterogeneous catalysts have been used in a number of organic reactions in which microwave heating was used. There is, unfortunately, a limited number of control data, which makes comparison with conventional heating difficult. Nevertheless, from the data so far reported, one can conclude that use of microwave radiation has yielded some remarkable results relating to rate enhancement and selectivity improvement. Activation can be achieved by superheating of the catalyst or by selective heating of active sites (Section 13.3.3), which cannot be achieved by conventional heating. Numerous reactions performed using heterogeneous catalysts can be divided into two groups – liquid-phase and gas-phase reactions.

13.3.1
Reactions in the Liquid Phase

Heterogeneously catalyzed reactions in the liquid phase have been described when both catalyst and reactants or solvents absorb microwaves. Reactions in nonpolar media, where only solid catalyst absorbs microwaves, providing interesting mechanistic information, are under examination [44].

13.3.1.1 Esterification, Transesterification

Chemat et al. [45] examined the rates of esterification of stearic acid by butanols, as a model reaction, to compare differences between homogeneously and heterogeneously catalyzed reactions, Scheme 13.1. Particular attention was paid to whether the effect of MW can be advantageously used to improve yields and to accelerate the rate of esterification.

$$CH_3(CH_2)_{14}COOH + CH_3(CH_2)_2CH_2OH \underset{}{\overset{Cat.}{\rightleftharpoons}} CH_3(CH_2)_{14}COOCH_2(CH_2)_2CH_3 + H_2O$$

Scheme 13.1. Esterification of stearic acid with 1-butanol.

It was found that the reaction under microwave conditions was faster than that heated conventionally. Yields were greatly improved, from 50–82% for conventional heating to 71–95% for microwave heating, when heterogeneous catalysts were used. The catalysts [$Fe_2(SO_4)_3$, $TiBu_4$, KF, KSF and PTSA] were used in a continuous-flow reactor. The results were further improved by use of a heat captor (graphite) and simultaneous use of ultrasound. Better results (yield and rate) under MW conditions were also achieved in other esterification reactions of stearic and acetic acid in the presence of a heterogeneous catalyst ($Fe_2(SO_4)_3$) adsorbed on montmorillonite clay pellets 4–5 mm in diameter [46, 47]. The rate of the reaction was 50–150% faster than with conventional heating. This increase was most probably because of superheating of voluminous pellets (5 mm), the temperature of which was calculated to be 9–18 K above the bulk temperature. When the esterification was homogeneously catalyzed by sulfuric acid using the two modes of heating, no differences between yield and reaction rates were observed.

The effect of the mode of heating was also studied for heterogeneously catalyzed esterification of acetic acid by isopentyl alcohol in the presence of Amberlyst-15 cation-exchange resin catalyst [48], Scheme 13.2.

$$\text{CH}_3\text{COOH} + (\text{CH}_3)_2\text{CHCH}_2\text{CH}_2\text{OH} \xrightleftharpoons{\text{Amberlyst}} \text{CH}_3\text{COOCH}_2\text{CH}_2\text{CH}(\text{CH}_3)_2 + \text{H}_2\text{O}$$

Scheme 13.2. Esterification of acetic acid with isopentyl alcohol.

Because the reaction is driven by protonation of the carbonyl functionality, reacting species were expected to be localized on the bed of the acid catalyst subjected to MW irradiation. Hexane was used as a nonpolar solvent to minimize solvent absorption and superheating. Elimination of catalyst superheating in a continuous-flow reactor was most probably the reason why no significant differences were observed between the reaction rates under the action of microwave and conventional heating.

Similar results were obtained in the esterification of acetic acid with 1-propanol performed in the presence of a heterogeneous silica catalyst [49]. The results showed that for this reaction MW irradiation and conventional heating had similar effects on the reaction rate.

Substantially reduced reaction times, from hours to minutes, and improved selectivity in the esterification of acetic acid and benzoic acid by different alcohols, using heteropolyacids as catalysts providing high yields of esters (98.0–99.7%) has been reported by Turkish authors [50].

The rates of transesterification of triglycerides to methyl esters (Bio-Diesel), efficiently catalyzed by boron carbide (B_4C), were faster under MW conditions, probably because of superheating of boron carbide catalyst, which is known to be a very strong absorber of microwaves (Scheme 13.3) [51]. Yields of the methyl ester of up to 98% were achieved. $Ba(OH)_2$ and other alkaline catalysts were also used to advantage [52].

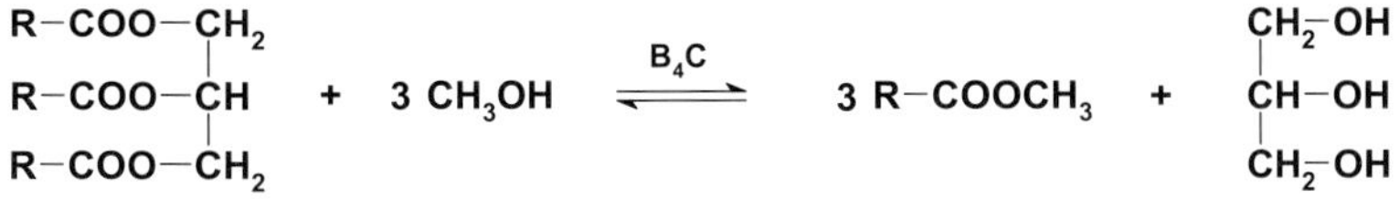

$$\begin{array}{l} \text{R–COO–CH}_2 \\ |\\ \text{R–COO–CH} \\ | \\ \text{R–COO–CH}_2 \end{array} + 3\,\text{CH}_3\text{OH} \xrightleftharpoons{B_4C} 3\,\text{R–COOCH}_3 + \begin{array}{l} \text{CH}_2\text{–OH} \\ | \\ \text{CH–OH} \\ | \\ \text{CH}_2\text{–OH} \end{array}$$

Scheme 13.3. Transesterification of triglycerides with methanol.

13.3.1.2 Hydrogenation, Hydrogenolysis

Another attractive heterogeneous catalytic reaction of much interest in organic synthesis is hydrogenation. Catalytic transfer hydrogenation of soybean oil in the presence of palladium catalyst (10% Pd–C) has been investigated using microwave and conventional heating [53]. Sodium formate was used as hydrogen donor. Kinetic results revealed the reaction rate increased significantly when MW heating was used. The rate enhancement was attributed to MW heating assistance of transport processes at the catalyst and oil–water interface.

Hydrogenation of nitrobenzene over Pd–Al_2O_3 catalyst was successfully used as a model reaction for testing a continuous-flow MW reactor under pressure for both heterogeneous and homogeneous reactions in gas–liquid media. From thermal modeling it was concluded that temperature gradients were present in both reactions [54].

Microwave-assisted catalytic hydrogenation of steroid compounds, e.g. cholesterol, campesterol, sitosterol, etc., in the presence of Pd–C catalyst and ammonium formate, in glycol as solvent, was rapid and afforded the corresponding products in high yield (80–95%) and purity [55].

A simplified and rapid hydrogenation of β-lactams under the action of MW irradiation has been described by Bose et al. (Scheme 13.4) [56, 57]. Pd–C and Raney nickel were used as catalysts and a high-boiling solvent, for example ethylene glycol, was used as the microwave energy-transfer agent in the presence of ammonium formate as the hydrogen donor. The yields of the corresponding amides were 80–90%.

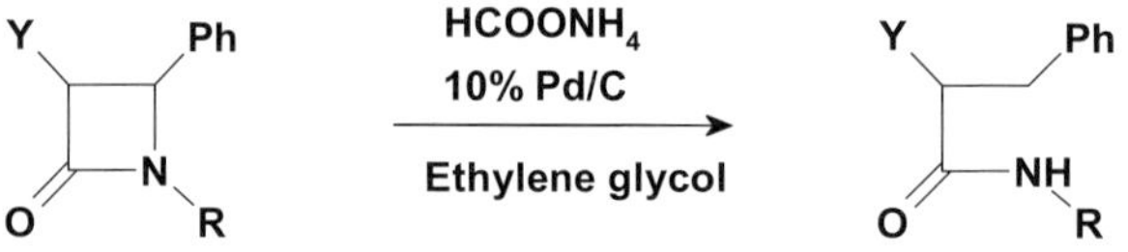

Scheme 13.4. Selective transformation of β-lactams.

Hydrogenation of C–C double bonds and hydrogenolysis of several functional groups have been described as safe, rapid, and efficient methods resulting in high yields (80–90%) of products. The technique described have been recommended as suitable for research and for undergraduate and high school exercises [56, 57]. Hydrogenolysis and dehalogenation of aromatic compounds with the same catalytic system was also successful. Thus, several β-lactams and isoquinoline derivatives were smoothly dehalogenated in a few minutes [57].

Asymmetric heterogeneous catalysis has been demonstrated by the enantioselective and racemic hydrogenation of ethyl pyruvate over Pt–Al$_2$O$_3$ catalyst. Higher reaction rate and enantioselectivity (75%) were observed in nonpolar solvent (toluene) whereas in a polar solvent (ethanol) they decreased substantially (Scheme 13.5) [58].

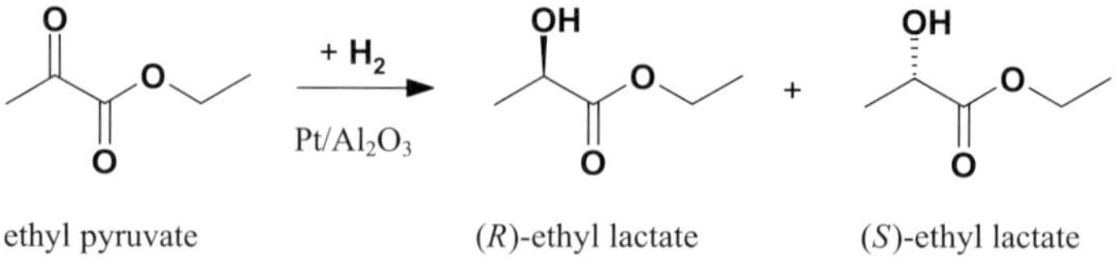

Scheme 13.5. Enantioselective hydrogenation of ethyl pyruvate.

13.3.1.3 Miscellaneous Reactions

The hydrolysis of sucrose catalyzed by the strongly acidic cation-exchange resin Amberlite 200C in the RH form was chosen as a model reaction to compare the use of stirred tank reactor and continuous-flow reactors (Scheme 13.6) [59–61].

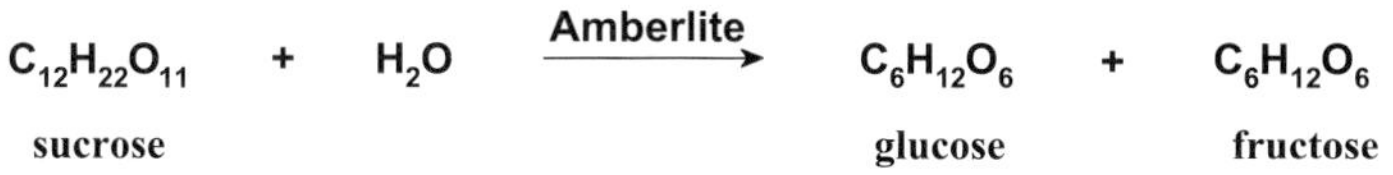

$C_{12}H_{22}O_{11}$ + H_2O $\xrightarrow{\text{Amberlite}}$ $C_6H_{12}O_6$ + $C_6H_{12}O_6$

sucrose   glucose   fructose

Scheme 13.6. Hydrolysis of sucrose.

No rate enhancement was observed when the reaction was performed under MW irradiation conditions at the same temperature as in conventional heating [59]. Similar reaction kinetics were found in both experiments, presumably because mass and heat effects were eliminated by vigorous stirring [59]. The model developed enabled accurate description of microwave heating in a continuous-flow reactor equipped with specific regulation of MW power [59, 60]. Calculated conversions and yields of sucrose based on predicted temperature profiles agreed with experimental data.

A microwave continuous-flow capillary reactor was capable of local heating of Pd–SiO$_2$ and Pd–Al$_2$O$_3$ catalysts enhancing the rate of the Suzuki reaction to give 70% product yield in 60 s contact time [62]. A thin layer of gold metal on the outside surface enabled effective heating of the reaction mixture.

Several other miscellaneous heterogeneously catalyzed reactions have been performed in the liquid phase. For instance, MW irradiation in the acylation of anisole over a range of solid acid catalysts led to reaction rates up to eight times higher that depended on the nature of the catalyst, even though absorption of microwaves was relatively low [63]. Hexane was successfully oxyfunctionalized with aqueous hydrogen peroxide using the zeolite TS-1 catalyst [64]. MW-promoted acetalization of several aldehydes and ketones with ethylene glycol proceeded readily (2 min) in the presence both of heterogeneous (acidic alumina) and homogeneous (PTSA, Lewis acids) catalysts (Scheme 13.7) [65].

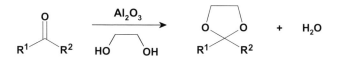

Scheme 13.7. Acetalization of ketones with ethylene glycol.

Yields were high (up to 97%) and comparable with those of homogeneously catalyzed reactions. A much higher catalyst-to-substrate ratio had to be used with heterogeneous alumina (10–15) than with the homogeneous catalysts (0.015–0.050), however. It was concluded that the MW method led to substantial improvement of acetalization reactions compared with conventional methods.

Friedel–Crafts acylation of aromatic ethers has been performed in the presence of a variety of metal chlorides and oxides (FeCl$_3$, ZnCl$_2$, AlCl$_3$, Fe$_2$O$_3$, Fe$_3$O$_4$, etc.) but without temperature control (Scheme 13.8) [66].

The short reaction time (1 min, 160 °C) in the benzoylation of anisole was probably a result of large temperature gradients rather than a nonthermal MW effect.

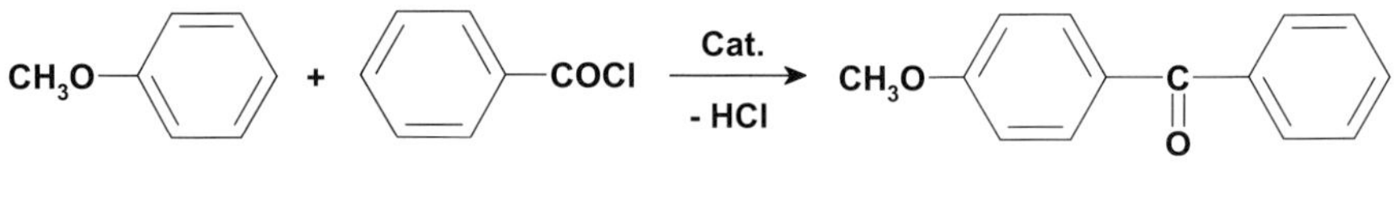

$$CH_3O-\langle\text{Ar}\rangle + \langle\text{Ar}\rangle-COCl \xrightarrow[-HCl]{Cat.} CH_3O-\langle\text{Ar}\rangle-C(=O)-\langle\text{Ar}\rangle$$

Scheme 13.8. Benzoylation of anisole.

MW irradiation also greatly shortened the reaction time for wet hydrogen peroxide oxidation of phenol catalyzed by Fe–Ti pillared montmorillonite [67].

Some reactions have been found to proceed with better results in the absence of solvent, probably because of the creation of temperature gradients, which are eliminated in the presence of a stirred solvent. This was observed for Diels–Alder reaction of α-amino acid precursors with cyclopentadiene catalyzed by heterogeneous catalysts (SiO_2–Al, SiO_2–Ti), when the reaction was performed in toluene or in the absence of solvent [68]. Microwave activation increased the rate of reaction without reducing its selectivity.

The MW activation of Michael additions in the preparation of N-substituted imidazoles afforded excellent yields in very short reaction times under mild reaction conditions, Scheme 13.9. Basic clays (Li^+, Cs^+) exchanged montmorillonites were found to be very active and selective catalysts for the Michael addition of imidazole and ethyl acrylate [69].

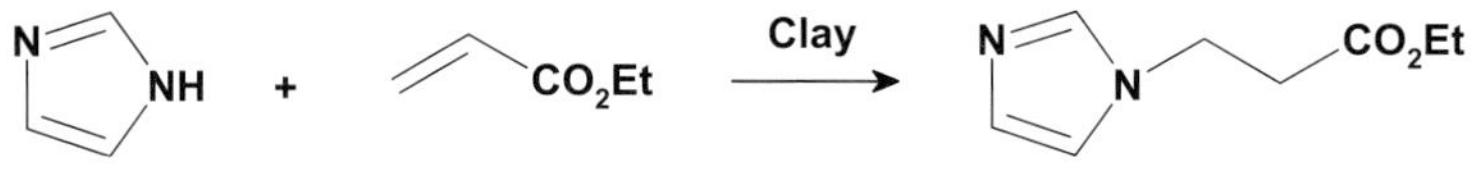

$$\text{Imidazole-NH} + \text{CH}_2=\text{CH-CO}_2\text{Et} \xrightarrow{\text{Clay}} \text{Imidazole-N-CH}_2\text{CH}_2\text{-CO}_2\text{Et}$$

Scheme 13.9. Michael addition of imidazole with ethyl acrylate.

A total of 75% conversion with 100% of selectivity was obtained in only 5 min in the absence of solvents.

An efficient method for conversion of a variety of acids into their corresponding amides in the presence of zeolite-HY under the action of MW irradiation has recently been described (Scheme 13.10) [70].

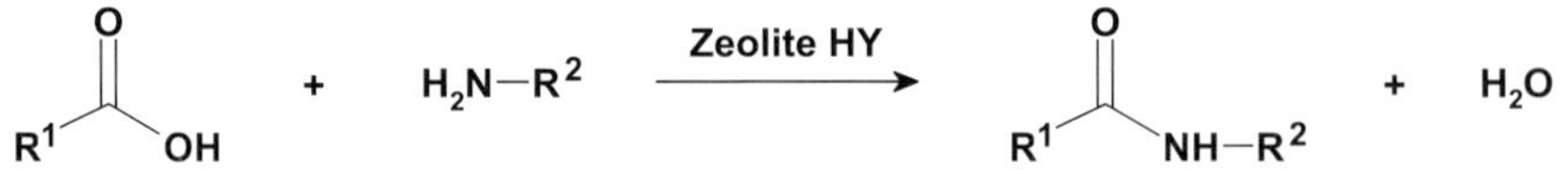

$$R^1-C(=O)-OH + H_2N-R^2 \xrightarrow{\text{Zeolite HY}} R^1-C(=O)-NH-R^2 + H_2O$$

Scheme 13.10. Preparation of amides.

The reactions proceeded at atmospheric pressure with high yields (80–97%) in the absence of solvent.

Facile N-alkylation of anilines with alcohols in the presence of Raney nickel proved the high efficiency of the catalyst, which is a highly absorptive material for microwaves [71], Scheme 13.11.

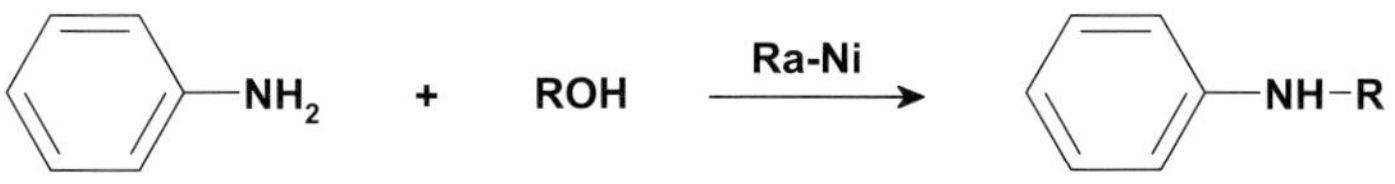

13.3 Microwave Activation of Catalytic Reactions

Scheme 13.11. Alkylation of aniline with alcohols.

A 6 to 48-fold rate enhancement was observed for this reaction. The authors suggested that a nonthermal effect might explain this [71].

Hydroxylation of benzene with hydrogen peroxide in the presence of solid zeolite catalyst (titan–silicate–zeolite) was chosen by Radoiu et al. [72] as a model heterogeneous liquid-phase reaction for study of microwave effects, Scheme 13.12.

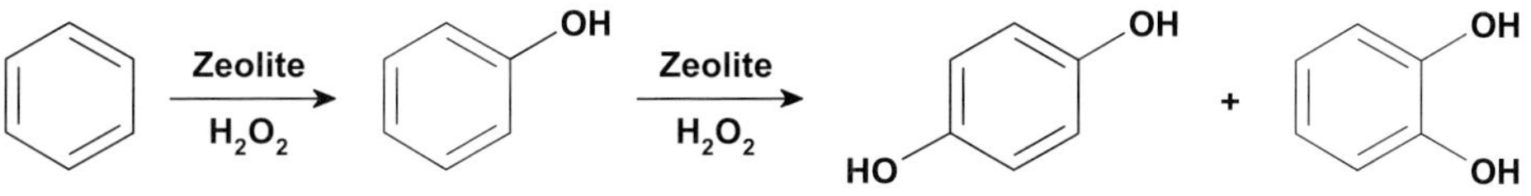

Scheme 13.12. Hydrogen peroxide oxidation of benzene.

Titanium-containing zeolite is an efficient catalyst for oxidation of benzene with hydrogen peroxide in a MW field, affording phenol with high selectivity. It was reported that microwaves had a strong effect on the selectivity of the reaction [72].

Several different organic transformations in the presence of homogeneous and heterogeneous catalysts have been described by Kappe et al. [73]. A review highlighting the immobilization of different catalysts used in organic synthesis induced by microwave heating has been published by the same research team [74].

A detailed study of microwave activation of catalytic reactions in the liquid phase has recently been performed by Hájek et al. (Scheme 13.13) [75–77]. It was found that in the catalytic transformation of 2- and 4-t-butylphenol in the liquid phase on heterogeneous KSF and K10 montmorillonite catalysts under MW and conventional conditions the microwaves affected both the rate and selectivity of the reaction.

The rate enhancement was up to 2.6-fold for reaction of the 2-isomer and up to 14-fold for the 4-isomer. Product distribution in the final reaction mixtures was

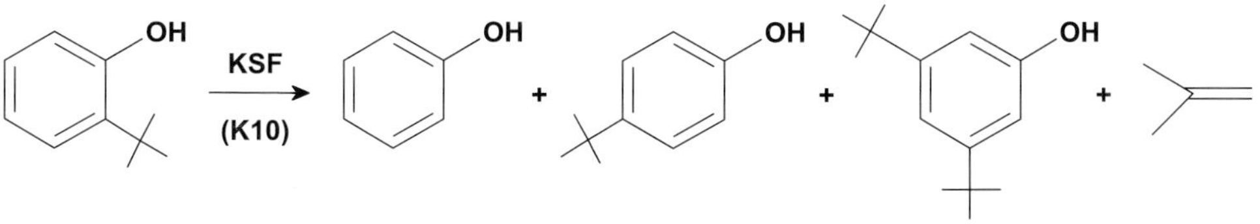

Scheme 13.13. Catalytic transformation of t-butylphenols.

always somewhat different when microwave heating was used. The results were explained in terms of efficient interaction of microwaves with a highly polarized reagent molecule adsorbed on the acidic active site. Possible superheating of the active sites was difficult to detect (Section 13.3.3).

To elucidate the cause of the microwave-induced enhancement of the rate of this reaction in more detail, transformation of 2-t-butylphenol was performed at low temperatures (up to −176 °C). This method, which is called "*Simultaneous cooling*" whilst microwave heating, was used for the first time in heterogeneous catalytic reactions [75]. At temperatures below zero the reaction did not proceed under conventional conditions. When the reaction was performed under microwave conditions in this low-temperature region, however, product formation was always detected (conversion ranged from 0.5 to 31.4%). It was assumed that the catalyst was superheated or selectively heated by microwaves to a temperature calculated to be more than 80–115 °C above the low bulk temperature. Limited heat transfer in the solidified reaction mixture caused superheating of the catalyst particles and this was responsible for initiation of the reaction even at very low temperatures. If superheating of the catalyst was eliminated by the use of a nonpolar solvent, no reaction products were detected at temperatures below zero (Section 13.3.3).

Other new methods used in heterogeneous catalytic reactions include the use of *ionic liquids* as catalysts and simultaneous applications of *ultrasound and microwave irradiation*. An ionic liquid as an acidic catalyst has been used for synthesis of coumarins [78], in esterification of acetic acid with benzyl alcohol [79], in catalytic esterification at room temperature [80], in alkylation of α-methylnaphthalene with long-chain alkenes [81], and in hydrogenation of benzene [82].

Use of electromagnetic and acoustic irradiation to enhance heterogeneous catalytic reactions has been recently reviewed [83]. Alternative use of ultrasound and microwave irradiation for hydrocarbon oxidation and catalyst preparation, to improve selectivity, was also recently reported [84].

13.3.2
Reactions in the Gas Phase

Heterogeneous catalytic gas-phase reactions are most important in industrial processes, especially in petrochemistry and related fields; most petrochemical and chemical products are manufactured by this method. These reactions are currently being studied in many laboratories, and the results of this research can be also used for synthetic purposes. The reactions are usually performed [85] in a continuous system on a fixed catalyst bed (exceptionally a fluidized bed).

In catalytic reactions, sufficient heat is usually required to overcome the activation energy barrier. In kinetic terms, the activation energy is the minimum energy required to form an activated complex undergoing transformation to the reaction products. Microwaves can be used as the source of thermal energy to induce catalytic reactions. The advantage of microwaves is that they can heat microwave-absorptive catalysts selectively to a temperature well above the bulk temperature of the reactants. The ability to heat the catalyst in a microwave field enables heating of the bed from the interior and not from a hot exterior, as by conventional thermal means. Thus, the gas stream passing over the catalyst enters the cooler part of the reactor more quickly, so preserving the reactive products which would otherwise be destroyed in a conventional system. During microwave irradiation the catalyst active sites are, ideally, heated to the required temperature well before the adsorbed reactants have time to desorb. In general, because the bond-breaking step occurs on the catalyst surface (the surface reaction is the rate-determining step), it is understandable that the reaction is enhanced. Because the temperature of the bulk reactants (and nonabsorptive support) is lower, the back, side, or consecutive reactions may occur to a lesser extent, i.e. the selectivity of the reaction can be significantly improved. For example, in the formation of ethane and ethylene from methane in the presence of doped catalysts, oxygen, and microwave irradiation, the undesirable reactions producing gaseous carbon dioxide and carbon monoxide were slower by one to three orders of magnitude, because the temperature of the bulk gas was lower than when conventional heating was used [86].

13.3.2.1 Reactions of Methane

The most studied catalytic gas-phase reaction has been the transformation of methane to higher hydrocarbons or oxygenated products. This reflects the large effort being made by catalytic chemists to find a simple process by which world's large resources of natural gas can be utilized. The existence of vast worldwide reserves of the natural gas, of which methane is the main component, has focused attention on the possibility of converting it directly to more valuable chemicals. A single-step conversion of methane to more valuable products is of immense industrial significance and economic importance.

The activation of methane by microwaves has long been a goal of scientists in attempts to convert this natural gas component into higher hydrocarbons valuable in the petrochemistry and chemical industry. Two pathways are being extensively investigated by research groups all over the world:

- oxidative coupling of methane to yield C_2 and higher hydrocarbons; and
- direct partial oxidation of methane to produce methanol and other oxygenates.

Oxidative coupling of methane to yield C_2 and higher hydrocarbons The oxidative coupling of methane has been studied by several authors. The most elusive transformation has been the oxidative coupling of methane into C_2 hydrocarbons (ethene, ethane), because the reaction is more endothermic than other transformations [2]. The application of rapid and efficient MW heating to endothermic reactions is particularly interesting.

Many attempts had been made to solve this problem. One is partial oxidation of methane over a special catalyst capable of inhibiting further oxidation into CO and CO_2 during the transformation. The reactions in question are shown in Scheme 13.14.

The most efficient catalysts for the desired transformation are metal oxides such

$$2\ CH_4 + O_2 \longrightarrow C_2H_4 + 2\ H_2O$$

$$2\ CH_4 + 1/2\ O_2 \longrightarrow C_2H_6 + H_2O$$

$$CH_4 + 3/2\ O_2 \longrightarrow CO + 2\ H_2O$$

$$CH_4 + 2\ O_2 \longrightarrow CO_2 + 2\ H_2O$$

or

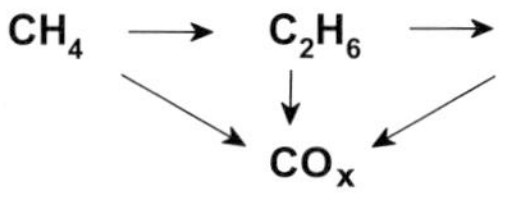

$$CH_4 \longrightarrow C_2H_6 \longrightarrow C_2H_4$$
$$\searrow CO_x \nearrow$$

Scheme 13.14. Oxidative coupling of methane.

as MgO, CaO, La_2O_3, Sm_2O_3 and LiO_2. The problem is, however, far from being solved. The best C_2 selectivity obtained so far is only 60% [2].

In conventional experiments, the gas and catalyst are maintained at the same temperature. In microwave experiments the power is deposited within the catalyst, which is cooled by the gas flow and thermal conduction to the surroundings. If the catalyst bed is not thick, the gas is always at a lower temperature than the solid catalyst. The increased loss factor of the catalyst favors the formation of CH_3 radicals because they are produced at active "O_2" sites and these *specific sites* are preferentially excited by the microwave field. Hence, the observed enhancement of C_2 selectivity is, in fact, a "thermal" phenomenon in that it can be completely explained by temperature gradients within catalyst particles, i.e. by locally excited catalytic sites. This observed effect can be explained by assuming that under the action of microwave irradiation the temperature of the reaction sites is higher than the mean temperature of the catalyst bed (Section 13.3.3).

Similarly, Bond et al. [4] confirmed that MW stimulation of methane-transformation reactions in the presence of several rare earth basic oxides to form C_2 hydrocarbons (ethene, ethane) was achieved at a lower temperature and with increased selectivity. Microwave irradiation resulted in an increase of ethene-to-ethane ratio, which was desirable. The results obtained were explained by the formation of hot spots (Section 13.3.3) of higher temperature than the bulk catalyst. This means that methane is activated at these hot spots.

The conversion of methane into C_2 hydrocarbons over an alumina supported La_2O_3–CeO_2 catalyst with and without oxygen has been investigated using both microwave and conventional heating [87]. For oxidative coupling reactions occurring in the presence of oxygen, microwave heating had no significant effect that could make it an advantageous method of heating; neither were there any selectivity differences. When methane was converted into hydrocarbons in the absence of oxygen, however, microwave heating was found to have a dramatic effect on the reaction. The products were formed at a temperature 250 °C lower than when conventional heating was used [87]. This was explained by CH_4 plasma formation and arcing between catalyst particles leading to the establishment of a high-temperature gradient within the catalyst bed.

Similar results were obtained by Radoiu et al. [88] in the application of a pulsed microwave discharge as potential technology in the decomposition of methane to acetylene. Discharges occurred on the metal surface (Ni, Cu, and W), where methane is absorbed, and therefore initiated chemical reaction.

Wan [89, 90] has mainly investigated microwave-induced catalytic gas-phase reactions. Wan et al. [91] used pulsed microwave radiation (millisecond high-energy pulses) to study the reaction of methane in the absence of oxygen. The reaction was performed by use of a series of nickel catalysts. The structure of the products seemed to depend both on the catalyst and on the power and frequency of microwave pulses. An Ni–SiO_2 catalyst has been reported to produce 93% ethyne, whereas under the same irradiation conditions a Ni powder catalyst produced 83% ethene and 8.5% ethane, but no ethyne.

Wan et al. [85] also reported the highly effective conversion of methane into aromatic hydrocarbons over Cu, Ni, Fe and Al catalysts. The effects of the type of catalyst, its configuration, and the microwave irradiation conditions on reaction path and product selectivity were examined under both batch and continuous-flow conditions.

Catalytic microwave heating has been used to oligomerize methane to higher hydrocarbons in the presence of Ni and Fe powder and activated carbon as the most active and efficient catalysts [92]. Oligomers ranging from C_2 to C_6 hydrocarbons, including benzene, have been prepared with good selectivity, depending on the nature of the diluent (helium) which favors the oligomerization of methane by microwave heating.

Mechanistic details of microwave-induced oligomerization of methane on a microporous MnO_2 catalyst were studied by Suib et al. [93], with emphasis on fundamental aspects such as reactor configuration, additives (chain propagators, dielectrics), temperature measurements, magnetic field effect, and other reaction conditions.

Oxidative coupling of methane has also been examined by other authors [2, 94–96], who have used different catalysts.

Pyrolysis of methane under the action of pulsed microwave radiation in the presence of solid catalysts has been reported by Russian authors [97, 98]. The application of pulsed microwave power was shown to be a promising means of production of hydrogen syngas, ethyne, and filamental carbon.

Partial oxidation of methane to syngas over Ni and Co catalysts was effected by use of microwave irradiation and compared with conventional heating [99]. Although the same conversions and H_2-to-CO ratio (2.0 ± 0.2) were observed, the temperature of the catalyst bed was much lower (200 K) when microwave irradiation was used than with conventional heating. Under both activation modes Ni-based catalysts (Ni–ZrO_2, Ni–La_2O_3) were more active and selective than Co catalysts (Co–ZrO_2, Co–La_2O_3). It was proposed that nonuniform distribution of the microwave energy on the catalyst surface created "hot spots" (Section 13.3.3) as active centers for the catalytic reaction.

Microwaves have been used to generate plasma in methane at 5–50 Torr. The radicals produced in these systems were then reacted over a nickel catalyst, affording a mixture of ethane, ethene, and ethyne [100].

Methane has also been used as the reducing agent in the catalytic conversion of NO to N_2 over Co-ZSM-5 zeolites [101] in the presence of oxygen. High NO conversions (>70%) were achieved by microwave irradiation at 250–400 °C whereas under similar conditions thermal runs failed to convert either NO or methane in significant amounts. The high activity and selectivity of the reduction of NO by methane achieved with microwave irradiation was probably because of the activation of methane to form methyl radicals at relatively low reaction temperatures.

Direct partial oxidation of methane to produce methanol and other oxygenates The formation of oxygenates by reaction of methane with oxygen and vapor under the action of conventional heating has received much less attention than the methane coupling reaction. The problem was poor selectivity and the formation of large amounts of CO_x. The use of microwave radiation for formation of the oxygenates from methane has, nevertheless, yielded some encouraging results. Wan et al. [102] used the vapor as the oxidant in preference to oxygen. They showed that acetone, 2-propanol, methanol and dimethyl ether can be produced by this method.

Suib et al. [103] used microwaves to generate a plasma in an atmosphere containing methane and oxygen. The plasma passing over a metal or metal oxide catalyst led to formation of C_2 hydrocarbons and some oxygenates.

Wan and Koch [104] also developed a method for producing HCN on a small scale by reaction of methane with ammonia, Scheme 13.15.

$$CH_4 + NH_3 \xrightarrow{M / Al_2O_3} HCN + 3 H_2$$

Scheme 13.15. Synthesis of hydrogen cyanide.

The reaction was performed over a series of $Pt–Al_2O_3$, $Ru–Al_2O_3$, and carbon-supported catalysts under the action of pulsed microwave radiation; conversions exceeded 90% and acetonitrile was formed as the by-product.

13.3.2.2 Reactions of Higher Hydrocarbons

Microwave activation of alkane transformations was studied in detail by Roussy et al., who summarized their results in several papers [2, 37, 38, 105]. Isomerization of hexane, 2-methylpentane, and 2-methyl-2-pentene, and hydrogenolysis of methylcyclopentane have been investigated, and the diversity of possible effects was specified [2]. The course of 2-methylpentane isomerization on 0.3% $Pt–Al_2O_3$ as catalyst depended on the mode of heating – the distribution of hexane products was different under the action of conventional and microwave heating [37, 105]. The isomer selectivity of the transformation of 2-methylpentane to 3-methylpentane, methylcyclopentane and hexane on this catalyst increased from 40 to 80% when the catalyst was pretreated with microwave energy [105]. The beneficial change in selectivity was found to be permanent. Microwave irradiation has been found to be a new, original means of activation of supported metal catalysts, in particular reforming catalysts [37, 38]. Hydrogenolysis of methylcyclopentane was studied mechanistically, because methylcyclopentane is one of the four molecules in equilibrium with intermediate cyclopentane in the cyclic mechanism [2].

Four possible effects of the electromagnetic field on the catalytic reactions were identified:

1. Microwave treatment can permanently modify the catalyst.
2. Heterogeneous phases (solid and gas) may give rise to inter-phase temperature gradients.
3. The electromagnetic field can act directly on the reaction sites to activate adsorbed organic molecules.
4. The electromagnetic field can act on polar intermediate species, e.g. carbenium ions.

Hydrogenolysis of 2-methylpentane, hexane, and methylcyclopentane on tungsten carbide, WC, a highly absorptive catalyst, at 150–350 °C in a flow reactor, has also been studied [106]. These reforming reactions were mainly cracking reactions leading to lower-molar-mass hydrocarbons. At the highest temperature (350 °C), all the carbon–carbon bonds were broken, and only methane was formed. At lower temperatures (150–200 °C) product molecules contained several carbon atoms.

The effect of microwave irradiation on the catalytic hydrogenation, dehydrogenation, and hydrogenolysis of cyclohexene was studied by Wolf et al. [107]. Optimum conditions for benzene formation were hydrogen flow, N-CaNi$_5$-catalyst, atmospheric pressure, and 70 s irradiation time. Cyclohexane was the main product when the irradiation time was 20 s, or in a batch/static system.

Oxidation of toluene to benzaldehyde and benzoic acid over V_2O_5–TiO_2 assisted by microwaves was studied by Liu et al. [108]. The authors concluded that microwave energy can greatly improve the process of selective toluene oxidation. The highest yields of benzoic acid were, however, only 38–41% and the highest selectivity for benzoic acid was 51% at 80% conversion to the acid.

In the cracking of benzene to acetylene over alumina and silica-supported nickel catalysts it was observed that the selectivity of the reaction, expressed as the ethyne-to-ethene ratio, was dramatically affected (from 1:9 to 9:1) by controlling the microwave energy input (i.e. 90% selectivity could be achieved) [109].

Other catalytic hydrocarbon reactions include decomposition of olefins over a powdered nickel catalyst [110], hydrogenation of alkenes, hydrocracking of cycloalkenes, and water-gas shift reactions [90].

13.3.2.3 Miscellaneous Reactions

Several reactions have already been mentioned in Section 13.2 – NO_x decomposition over modified zeolites [9–11], dehydrogenation of 2-propanol, isomerization of *cis*-butene over basic zeolites [12, 13], hydrogenation of benzene and hydrodechlorination of chlorobenzene [14–16], and oxidative dehydrogenation of ethylbenzene to styrene [28, 36]. Microwave-assisted decomposition of NO_x or its mixture with SO_2 has been reported by Tse [102] and other authors [101, 111–113]. Oxygen, nitrogen, and solid sulfur were the main products of the decomposition performed over a range of nickel and copper catalysts [102]. The reaction of nitrogen oxides under an oxidizing atmosphere is a very extensively studied reaction, but the true reaction pathways are still very difficult to understand. The high activity and selectivity of Co-ZSM-5 zeolite catalysts for NO reduction by methane is probably because of activation of methane to form methyl radicals at relatively low temperatures [101] as already mentioned. Wan et al. [114] have also shown that carbon dioxide can be reacted over a supported metal catalyst in the presence of water vapor to yield alcohols and other oxygenates if the catalyst is irradiated with MW. The reaction proceeded at relatively low temperatures (220–350 °C). It seems likely that the surface temperature of the metal is several hundred degrees higher than the bulk temperature of the catalyst. Decomposition of chloromethane over a metal catalyst irradiated with pulsed MW radiation afforded methane, although, the metal catalyst was inhibited by partial formation of the metal chloride by reaction with hydrogen chloride [115].

Microwave irradiation has recently been used in hydrocarbon oxidation catalyzed by doped perovskites (with the objective of cleaner diesel power) via the creation of hot spots [116].

The reaction of 2-propanol to propanone and propene over a series of alkali-metal-doped catalysts with use of microwave irradiation has been studied by Bond et al. [117]. The nature of the carbon support was shown to affect the selectivity of the catalyst. Under the action of microwave irradiation the threshold reaction temperature (i.e. the lowest temperature at which the reaction proceeded) was substantially reduced; this was explained by "hot spots" (Section 13.3.3) formed within the catalyst bed.

The effect of microwave irradiation on the catalytic properties of a silver catalyst ($Ag–Al_2O_3$) in ethene epoxidation was studied by Klimov et al. [118]. It was found that on the catalyst previously reduced with hydrogen the reaction rates of both epoxidation and carbon dioxide formation increased substantially on exposure to a microwave field. This effect gradually decreased or even disappeared as the catalyst achieved the steady state. It was suggested that this was very probably because of modification of the electronic properties of the catalyst exposed to microwave irradiation.

Catalytic transformation of 1-butene over acidic (low polarity) zeolites resulted in somewhat higher yield and selectivity toward isobutene. A single mode continuous-flow reactor was used and the results were explained by formation of "hot spots" [119].

13.3.3
Microwave Effects

Although microwave activation of catalytic reactions has been the subject of many studies (Sections 13.3.1 and 13.3.2), the mechanism of these reactions is not yet fully understood. In heterogeneous catalytic liquid–solid and gas–solid systems, many results have revealed significant differences between the rates of conventionally and microwave-heated reactions. At the same temperature, microwave-heated reactions were usually faster than conventionally heated and their rate enhancement was over one order of magnitude. Such rate differences have not been reported for other catalytic reactions, so how should they be interpreted?

Several reasons have been proposed to account for the effect of microwaves on chemical reactions and catalytic systems. The results obtained by different authors show that under specific conditions MW irradiation favorably affects reaction rates of both the liquid and gas-phase processes. This phenomenon has been explained in terms of *microwave effects*.

What are *microwave effects*? Microwave effects are usually effects which cannot be achieved by conventional heating. These microwave effects can be regarded as *thermal* or *nonthermal*. *Thermal effects* result from microwave heating, which may result in a different temperature regime, whereas *nonthermal effects* are *specific effects* resulting from nonthermal interaction between the substrate and the microwaves.

The proposal of some authors on the operation of *nonthermal effects* is still controversial [120–122]. In the literature microwave effects are the subject of some misunderstanding. Most mistakes was recorded when authors considered that "microwave effect" means "specific effect", i.e. a nonthermal effect. For that reason, let us discuss the matter in more detail. In heterogeneous catalytic reactions, differences between reaction rates or selectivity under microwave and conventional heating conditions have been explained by thermal effects.

Microwave thermal effects can be summarized as:

- rapid heating,
- volumetric heating,
- superheating,
- hot spots,
- selective heating,
- simultaneous cooling.

Rapid heating is a heating of a substrate or catalyst achieved by high absorption of microwaves by the whole volume irrespective of low thermal conductivity. It accelerates the rate of the reaction, because the rate of conventional heating is limited by lower heat transfer.

Volumetric heating means heating of the whole volume of the sample from the center, i.e. in the opposite direction to conventional heating. It results in the opposite temperature profile and contributes to rapid heating of the sample. For a highly absorptive mixture or solid material one must be careful with the depth of penetration of the microwaves, which is very limited, especially for highly absorptive material.

Superheating causes heating to higher temperature than is expected by conventional heating. It can occur as superheating of a liquid reaction mixture or as localized superheating of solids, e.g. a catalyst. Superheating of a liquid reaction mixture above the boiling point [123, 124] and localized superheating of solid samples generating temperature gradients [125] is very often responsible for enhancement of reaction rates in both homogeneous and, especially, heterogeneous reactions. It is generally known that irradiation of a heterogeneous system (e.g. a suspension) by means of an electromagnetic field leads to nonuniform temperature distribution. If the reaction is performed under reflux, the reaction temperature can be significantly higher under the action of microwaves than under conventional conditions, because of the superheating effect of microwaves on the polar unstirred reaction mixture. Superheating occurs because the sample is heated so quickly that convection to the surface of the liquid and vaporization cannot adequately dissipate the excess energy. Organic chemists can use the superheating effect to accelerate both heterogeneous and homogeneous liquid-phase reactions. Under these conditions, the reactions proceed at temperatures higher than the boiling point of the liquid mixture, without the need to work under pressure. The superheating of liquids can be reduced or suppressed by addition of boiling stone or by efficient stirring. Localized superheating accompanied by the creation of hot zones is more significant for solids than for liquids.

Enhancement of the rate of gas–solid phase reactions by microwaves can be a result of localized superheating of a solid catalyst bed to much greater extent than the liquid phase especially in scaling up of the catalyst bed. Both the differential coupling properties of materials and the distribution of electromagnetic fields can result in localized temperature distribution in the catalytic bed, although the contribution of these effects is difficult to quantify. Localized superheating creates hot zones, and consequently, leads to temperature gradients in solid materials. It also increases the frequency of collision of reactants. The preferential alignment of dipolar or ionic functional groups because of the microwave field may create a favorable situation for collisions to occur at the active sites of catalyst, thus increasing collision efficiency. The large difference between temperature profiles of the catalyst bed under the action of microwave and conventional heating can result in a different product distribution for catalytic reactions. Stuerga et al. [125] have suggested that the localized rate enhancements might be responsible for nonisothermal and heterogeneous kinetic phenomena.

Measurement and estimation of temperature distributions induced by microwave heating in solid materials is very difficult, however. Consequently, most local temperature fluctuations are greater than those measured. Under stronger microwave irradiation conditions it is, therefore, very easy to obtain local temperature gradients. Temperature measurements usually yield an average temperature, because temperature gradients induce convective motion. Despite these difficulties, some methods, e.g. IR thermography, can reveal the surface temperature distribution without any contact with the sample studied. By this method, temperature gradients between 7 and 70° cm^{-1} in an alumina layer [125] and 50° mm^{-1} in a magnetite sample [126] have been monitored. Similar temperature gradients, 30–50° cm^{-1}, have been found by Hajek et al. [127] for a cylindrical sample of alumina.

A systematic theoretical study of temperature profiles in MW-heated solids has been performed by Dolande et al. [128]. According to their results, differences between reaction rates and/or yields observed for conventional and MW heating can be explained in terms of localized superheating. The main interest in the use of MW heating is its ability to produce very steep thermal gradients and very fast heating rates. One can also induce localized superheating, which leads to localized rate enhancement [125–127]. Localized superheating can be advantageously used by synthetic chemists, especially for reactions performed on inorganic supports under solvent-free conditions (Chapter 8). In conclusion, it should be stressed that localized superheating can have a beneficial effect on heterogeneous catalytic reactions because of the possibility of both rate enhancement and selectivity improvement.

Localized superheating can create "hot spots", which are also frequently a cause of rate enhancement of heterogeneous catalytic reactions.

Hot spots are created by a nonlinear dependence on temperature of the thermal and electromagnetic properties of the material being heated. If the rate at which microwave energy is absorbed by the material increases more rapidly than linearly with temperature, heating does not take place uniformly and the regions of very high temperature can create hot spots. Inhomogeneity of the electromagnetic field contributes substantially to the creation of hot zones. Hot spots can be created by localized superheating, selective heating, and as a result of inhomogeneity of the electromagnetic field.

In a microwave-heated packed catalyst bed, two different forms of hot spot can be created. Hot spots are either *macroscopic hot spots* (measurable) or *microscopic hot spots* (on molecular level, immeasurable and similar to those in sonochemistry).

Macroscopic hot spots as large-scale nonisothermalities which can be detected and measured by use of optical pyrometers (fiber optic or IR pyrometers), i.e. these are macro scale hot zones.

Microscopic hot spots, for example temperature gradients between the metal particles and the support, which cannot be detected and measured because they are close to micro scale, i.e. have molecular dimensions, are close to selective heating of active sites. Unfortunately, microwave radiation effects at the molecular level are not well understood.

Zhang et al. [119] presented evidence of the formation of hot spots in the microwave experiments and demonstrated that these hot spots need not be exclusively localized on the active sites but may also involve the support material (γ-Al$_2$O$_3$). They also estimated the dimensions of these hot spots to be in the region of 90–1000 µm. Development of hot spots in the catalyst bed during the course of the gas phase decomposition of H$_2$S catalyzed by MoS$_2$–Al$_2$O$_3$ was probably the reason for a significant apparent shift in the equilibrium constant. The temperature of these hot spots was probably 100–200 K above the bulk temperature. The formation of hot spots in the support could be because of the absorptive properties of γ-Al$_2$O$_3$ (compared with the low absorptivity of α-Al$_2$O$_3$). It was concluded that these hot spots also induce considerable reorganization of the catalyst under microwave conditions. Such selective or localized superheating to create hot spots, as is effected with microwave heating, cannot be achieved by conventional heating methods.

If MW heating leads to enhanced reactions rates, it is plausible to assume the active sites on the surface of the catalyst (microscopic hot spots) are exposed to selective heating which causes some pathways to predominate. With metal supported catalysts the metal can be heated without heating of the support, because of the different dielectric properties of both catalyst components. The nonisothermal nature of the MW-heated catalyst and the lower reaction temperature favorably affects not only reaction rate but also selectivity of such reactions.

Chen et al. [95] suggested that temperature gradients may have been responsible for more than 90% selectivity in the formation of acetylene from methane in a microwave-heated activated carbon bed. The authors believed that the highly nonisothermal nature of the packed bed might enable reaction intermediates formed on the surface to desorb into a relatively cool gas stream where they are transformed via a reaction pathway different from that in a conventional isothermal reactor. The results indicated that temperature gradients were approximately 20 K. The nonisothermal nature of this packed bed resulted in an apparent rate enhancement and altered the activation energy and pre-exponential factor [129]. Formation of hot spots has been modeled by calculation and, for solid materials, studied by several authors [130–133].

It is obvious that nonisothermal conditions induced by microwave heating lead to very different results from those obtained under conventional heating conditions. In summary, microwave effects such as superheating, selective heating, and hot spots, can all be characterized by temperature gradients ranging from macroscopic to molecular scale dimensions.

Let us now return to the question "How to interpret or explain the fact that some reactions are affected by microwaves and some reactions are not". A detailed study of this subject has been performed by Hájek et al. [75–77] for heterogeneous catalytic liquid-phase reactions. Transformation of 2-t-butylphenol into phenol, 4-t-butylphenol, 2,4-di-t-butylphenol, and isobutene on montmorillonites as catalysts (KSF, K10) was chosen as model reaction, Scheme 13.13. Both reactant and catalysts coupled very well with microwaves. KSF and K10 catalysts in the form of a fine powder (10–15 μm) were used to avoid creation of macroscopic hot spots (as in the presence of voluminous catalyst pellets, e.g. 5 mm [46, 47]). The results are summarized below.

13.3.3.1 Effect of Microwaves on Selectivity

The product distribution determined for reactions performed over a broad temperature range (from −176 to 199 °C) under the action of microwave heating was always different from that obtained by use of a conventional method. Thus, vigorous formation of isobutene under reflux using MW heating indicates superheating of the catalyst particles to a higher temperature. This facilitates the dealkylation reaction, which is promoted by elevated temperatures. A strong effect of simultaneous cooling (*vide infra*) on selectivity was observed, indicating creation of microscopic hot spots on catalyst particles or on active acidic sites.

13.3.3.2 Effect of Microwaves on Rate Enhancement

Temperature and solvent effects were also examined. When the reaction temperature was gradually reduced from reflux temperature (199 °C), the rate enhancement factor increased from 1.0 at 199 °C to 1.4 at 105 °C, to 2.6 at 75 °C, and to 4.1 at 24 °C. These results may indicate that superheating of the catalyst is more pronounced at lower temperatures. In the presence of nonpolar solvents, for example hexane and heptane, more efficient heat transfer from the superheated catalyst particles can explain the lower values of the rate enhancement 1.2 at 105 °C (heptane) and 1.6 at 75 °C (hexane). To suppress the reaction under conventional heating conditions, the reaction was performed at very low temperatures ranging from −176 to 0 °C, i.e. in the temperature region in which conventional conditions [77] do not lead to any reaction. Efficient cooling of the reaction mixture by use of liquid nitrogen or dry ice enabled the reaction to be performed at the low temperatures mentioned above. The reaction was taking place in the solid state, because 2-t-butylphenol (m.p. −7 °C) solidified at these temperatures. It seems likely that, under these conditions, heat transfer from the superheated catalyst particles in the solid reaction mixture at −176 to 0 °C was not so efficient, and rather significant conversions (0.5–31.4%) were recorded. Superheating of catalyst particles (or acidic active sites?) was calculated to be 80–115 °C throughout the bulk. When the reaction mixture was diluted with a nonpolar solvent (hexane + tetrachloromethane) under the action of efficient stirring, no reaction was observed to occur, similarly to under conventional conditions [77]. This finding can be accounted for by efficient heat transfer from the catalyst particles, because of efficient stirring of the liquid reaction mixture. The question is the extent of catalyst superheating. Are catalyst particles of 10–15 μm or polar acidic sites superheated and then participating in the interaction of MW energy with highly polarized reagent molecules via adsorption of these molecules on their active sites? Although this suggestion, i.e. selective heating of active acidic sites, has not been excluded, no direct evidence has been found for the assertion.

For both the liquid and gaseous phases, one can conclude that if reactions are performed under conditions which completely eliminate temperature gradients, e.g. by use of efficient stirring, nonpolar solvents, a high flow of reactants in continuous flow system, or a fluidized bed, any differences between microwave and conventional heating conditions may disappear. It is, however, synthetically more attractive to perform reactions using microwaves under conditions favorable to the production of temperature gradients, because higher reaction rates and improved selectivity can be obtained more easily compared with conventional heating methods. In conclusion, the effect of hot spots on catalytic reactions can be as effective as selective heating.

Selective heating usually means that in a sample containing more than one component, only that component which couples with microwaves is selectively heated. This is a very important effect in catalytic reactions in which the catalyst can be selectively heated. The nonabsorbing components are, therefore, not heated directly but only by heat transfer from the heated component. Examples include:

- Liquid–liquid system – polar reactants in nonpolar solvents (e.g. mixture of H_2O with CCl_4, only H_2O is heated).
- Solid–liquid system – solid reactants or solid catalysts in nonpolar reaction mixtures (e.g. KSF in hydrocarbons, only the catalyst is heated).
- Solid–solid system – mixtures of strong microwave absorbers (e.g. C, SiC, CuO,

etc.) with transparent materials (e.g. SiO_2, MgO etc.) or supported catalysts (e.g. Pt–SiO_2, Pd–Al_2O_3 etc., only metallic active sites are heated).

The selective heating is more enhanced when the catalyst contains a poorly microwave-absorbing support (e.g. α-Al_2O_3 rather than γ-Al_2O_3). Metal or metal oxides supported on a transparent support may therefore be heated selectively. Thus, for example, Pt sites in a Pt–SiO_2 catalyst can be selectively heated, in contrast with the transparent, unheated SiO_2 support. Hence, selective heating of the active sites to a temperature higher than that of the support may occur if heat loss from the metal particles is not too fast. Although the microwave field does not couple with transparent ceramic support, it may couple strongly with metal particles, because of their high electrical conductivity. Unfortunately, obtaining direct experimental proof of this concept seems very unlikely, because measurement of the temperature of individual active sites is beyond current experimental possibilities. So far, the possibility of selective heating of the active sites without increasing the temperature of the catalyst bed has only been modeled [134] and found to be strongly dependent on the size of catalyst particles and microwave frequency [135]. Several experimental studies on catalytic reactions under microwave conditions have been concerned with kinetic aspects and showed that the reaction rates and product distributions correspond to a higher reaction temperature than was measured for the bulk of catalyst bed. Both reaction characteristics were often explained in terms of the higher local temperature of certain active centers within the catalyst bulk. It is worth mentioning that the results of theoretical calculations reported by Thomas [134] did not substantiate the possibility of achieving selective heating of supported metal catalysts by use of microwaves. Recent detailed modeling of small-scale, microwave-heated, packed bed, and fluidized bed catalytic reactors has, however, indicated that under specific conditions the active sites may be selectively heated in both types of catalyst bed [135, 136]. Selective heating of active sites (Na atoms in NaY zeolite) was recently supported by Auerbach's results [137] from use of equilibrium molecular dynamics in zeolite–guest systems after experimental work by Conner [138]. At equilibrium, all the atoms in the system are at the same temperature. In contrast, when NaY zeolite is exposed to microwave energy, the effective steady-state temperature of Na atoms is considerably higher (20 K) than that of the rest of the framework. This is not an effect of athermal energy distribution as was suggested [137, 139], but can be explained by simple selective heating of Na atoms.

Simultaneous cooling whilst microwave heating is a new method in synthetic organic chemistry. It is most recently discovered microwave effect in heterogeneous catalysis and can substantially improve yields and selectivity of catalytic reactions [77].

It is obvious that nonisothermal conditions induced in catalysts by microwave heating can lead to very different results from those obtained under conventional heating conditions. When intensive cooling is used, e.g. with liquid nitrogen, these differences are more profound [77]. Because the temperature of the reaction mixture is lower than the temperature of the catalyst, reverse, side, or consecutive reaction can occur to a lesser extent, i.e. the selectivity of the reaction can be significantly improved.

Simultaneous external cooling enables more microwave power to be distributed directly to the reaction mixture. Published applications of the simultaneous cooling technique are very rare so far, probably because of the lack of availability of a convenient instrument. This problem was recently overcome by release of the CEM Discover CoolMate monomode instrument for small-scale experiments which can maintain low temperatures in the region -80 to $+35$ °C [140]. Scaling up of this method has been solved by use of a cryostat in combination with multimode equipment (Milestone, Italy; Shikoku Instrumentation, Japan).

13.3.4
Microwave Catalytic Reactors

13.3.4.1 Batch Reactors

The instrumentation which has been used for microwave catalytic reactions varies from domestic multi-mode ovens to continuous-flow single-mode reactors. In domestic microwave ovens the microwave output is changed by varying the patterns of on–off cycles. Thus, for example, one-half microwave output does not really mean half output power but only that full power is switched on and off for certain times (e.g. half power corresponds to full power for 10 s then no power for 10 s). Under these conditions the catalyst and/or reaction mixture suffer thermal shocks, which is not desirable. In domestic microwave ovens, moreover, the microwaves are randomly distributed throughout the oven space, which results in ill-defined regions of high and low intensity inside the oven. A second disadvantage of domestic ovens is the problem of temperature measurement, because, in the absence of a correct temperature profile in the reaction mixture, the reaction is not comparable with conventional heating and may not be reproducible when performed in two different microwave systems.

The problem of the switch on–off system of most domestic ovens has been overcome by use of an inverter circuit that enables power levels to be adjusted in increments (e.g. 10% of output) [141]. The desired power is continuous at different levels, compared with the long pulsed operations of the magnetron in most domestic ovens. The advantage of this system for laboratory applications, in which small loads are normally used, is that lower power levels can be applied, which minimizes the amount of reflected power reaching the magnetron. A simple unmodified domestic microwave oven operating on a multi-mode system cannot, therefore, be recommended for catalytic reactions (or, in general, for organic synthesis) induced by microwaves. The disadvantages are quite evident:

- no possibility of temperature measurement and no temperature control;
- reaction mixtures subjected to thermal shocks by switch on–off cycles; and
- inhomogeneity of microwave field.

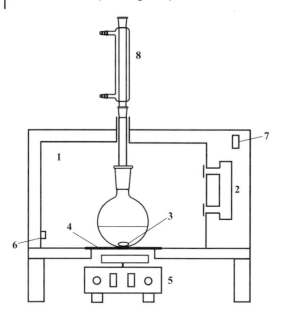

Fig. 13.1. Microwave batch reactor: 1, microwave cavity; 2, magnetron; 3, stirring bar; 4, aluminum plate; 5, magnetic stirrer; 6, IR pyrometer; 7, on–off switch; 8, water cooler.

Domestic ovens can be inexpensively and safely modified, however; this almost eliminates these disadvantages and enables independent temperature measurement and reasonable temperature control. For temperature measurement an IR thermometer or, better, a fiber-optic thermometer [75–77] has been recommended. Such a batch microwave reactor made by modification of a domestic microwave oven is depicted in Fig. 13.1 and has been described elsewhere (Refs. [51, 75–77, 141–144] and references cited therein).

A complementary, more advanced, laboratory-scale microwave batch reactor for synthetic and kinetic studies has been developed by Strauss et al. (Fig. 13.2) [145].

The reactor is equipped with magnetic stirrer, microwave power, and temperature control by computer and can operate under pressure. Although it was developed for homogeneous organic synthetic reactions it can also be used for heterogeneous catalytic reactions in the liquid phase.

13.3.4.2 Continuous-flow Reactors

The first continuous-flow reactor was developed by Strauss [144–148] and has recently been commercialized. It consists of microwave cavity fitted with a tubular coil (3 m × 3 mm) of microwave-transparent, inert material (Fig. 13.3). The coil is attached to a metering pump and pressure gauge at the inlet end and to a heat exchanger and pressure-regulating valve at the effluent end. Temperature is monitored outside the cavity at the inlet and the outlet.

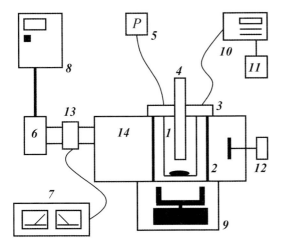

Fig. 13.2. Schematic diagram of a microwave batch reactor: 1, reaction vessel; 2, retaining cylinder; 3, top flange; 4, cold finger; 5, pressure meter; 6, magnetron; 7, power meters; 8, power supply; 9, stirrer; 10, fiber-optic thermometer; 11, computer; 12, load-matching device; 13, waveguide; 14, microwave cavity. (Reproduced from Ref. 146 by permission of CSIRO Publishing.)

The reactor has facilitated a diverse range of synthetic reactions at temperatures up to 200 °C and pressures up to 1.4 MPa. Temperature measurement at the microwave zone exit indicates the maximum temperature is attained but gives insufficient information about thermal gradients within the coil. Accurate kinetic data for the reactions studied are, therefore, difficult to obtain. This problem

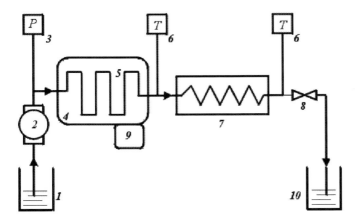

Fig. 13.3. Schematic diagram of a continuous-flow microwave reactor: 1, reactants for processing; 2, metering pump; 3, pressure transducer; 4, microwave cavity; 5, reaction coil; 6, temperature sensor; 7, heat exchanger; 8, pressure regulator; 9, microprocessor regulator; 10, product vessel. (Reproduced from Ref. 146 by permission of CSIRO Publishing.)

has recently been avoided by using a fiber-optic thermometer. An advantage of continuous-flow reactors is the possibility of processing large amounts of starting material in a small-volume reactor (50 mL, flow rate 1 L h^{-1}). A similar reactor, but of smaller volume (10 mL), has been described by Chen et al. [149].

It has been observed that hydrolysis of sucrose using a strongly acidic cation-exchange resin as heterogeneous catalyst gives better results than with soluble mineral acids [150]. Hydrolysis of sucrose catalyzed by the strongly acidic cation-exchange resin Amberlite 200C has been used by Plazl et al. [60, 61] for successful testing of a continuous-flow catalytic reactor with a packed catalyst bed made by modification of a commercial microwave oven (Panasonic NE-1780). Other authors have also built continuous-flow catalytic reactors by modification of domestic ovens (Panasonic, General Electric) [48, 51]. Continuous-flow microwave reactors based on a single-mode system have been described by Chemat [47] and Marquié [150]. Better isothermal conditions, the possibility of selective heating of the catalyst bed, and the possibility of scaling-up are the main advantages of single-mode continuous-flow reactors compared with multi-mode systems for catalytic reactions in the liquid phase.

For gas-phase heterogeneous catalytic reactions continuous-flow integral catalytic reactors with packed catalyst beds have been exclusively used [85–118]. Continuous or short pulsed radiation (milliseconds) has been used in catalytic studies (Section 13.3.2). To avoid the creation of temperature gradients in the catalyst bed, a single-mode radiation system can be recommended. A typical example of the most advanced laboratory-scale microwave, continuous single-mode catalytic reactor has been described by Roussy et al. [105] and is shown in Figs 13.4 and 13.5.

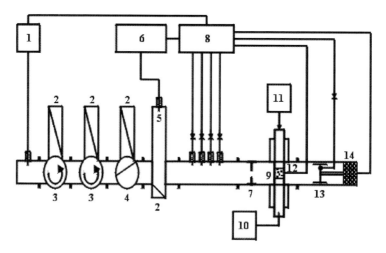

Fig. 13.4. Schematic diagram of a microwave system for catalyst studies: 1, generator; 2, water load; 3, circulator; 4, switch; 5, input power; 6, power meter; 7, coupling iris; 8, microcomputer; 9, catalyst; 10, chromatograph; 11, gas production set-up; 12, thermocouple; 13, short-circuit; 14, stepping motor. (Reprinted from Ref. 151 with permission from Elsevier Science.)

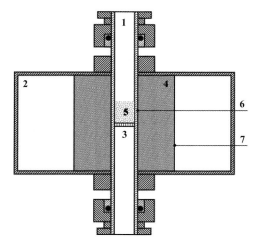

Fig. 13.5. Schematic diagram of the microwave applicator: 1, reactor; 2, wave guide; 3, fritted silica disc; 4, thermal insulation; 5, catalyst; 6, 7, thermocouples. (Reprinted from Ref. 151 with permission from Elsevier Science.)

The principles of this microwave irradiation system are, briefly:

- continuously variable microwave power from 20 to 200 W;
- temperature of the catalyst bed measured by use of two thermocouples; and
- homogeneous distribution of the electromagnetic field.

In addition to a packed catalyst bed, a *fluidized* bed irradiated by single and multi-mode microwave fields was also modeled by Roussy et al. [152]. It was proved that equality of solid and gas temperatures could be achieved in the stationary state and during cooling in a single mode system. The single-mode cavity eliminates the effect of particle movements on the electric field distribution. When the bed was irradiated in the multimode cavity, the model failed. Nevertheless, combination of the fluidization technique with microwave heating provides significant benefits, for example reduced energy costs, appreciable reduction in processing time, and higher product quality. A fluidized bed reactor has been successfully used to destroy hazardous/toxic organic compounds, for example trichloroethylene, *p*-xylene, naphthalene, and some gasoline-like hydrocarbon materials [153]. Granulated activation charcoal was used as the catalyst. All these organic compounds could be completely decomposed by combining these two technologies.

13.4
Industrial Applications

Since the early 1980s there has been much interest in the development of industrial applications of microwave irradiation in catalytic processes. This interest has been driven by the unique attributes of the selective interactions of microwave radiation with different materials as the basis for the initiation and control of catalytic chemical reactions. The microwave mode of energy conversion has many attractive features for the chemist, because some liquids and solids can absorb in situ and transform electromagnetic energy into heat very effectively [2]. Microwave technology has so far only rarely attracted industrial attention, however. Several potential chemical processes using this technique have been established in research laboratories, but the large scale-up to commercial practice remains elusive. Microwave technology is not new to physicists and electrical engineers, but the concept of using microwaves as an energy source for chemical reactions has only recently been appreciated by chemists. In the subsequent stage of the scale-up of the process some of the potential advantages of the microwave-induced catalysis proved in laboratory experiments might not be easy to realize in commercial scale production. This is because of the need to identify many difficulties and modifications, technical and economic, before the final choice of a proper system can be designed and installed [154].

Wan et al. [154] demonstrated problems of scale-up in the catalytic conversion ethane into ethyne by finding that optimization of the process for the large-scale reaction system was quite different from that in the laboratory experiment. It is not practical to construct a huge quartz reactor enclosed by a huge microwave oven. Rather, the microwave radiation can be introduced into a stainless steel reactor. There are obviously many other engineering problems to be encountered, for example penetration depth, uniformity of temperature profile of the catalyst bed, etc., which are specific to a microwave reactor. In large-scale processes, spatial uniformity may often be critical. Engineering and scale-up aspects of microwave-induced catalytic reactions were summarized by Mehdizadeh [155] of DuPont de Nemours. It was recommended that the choice of catalyst should be made not only on the basis of its chemical merits, but also because of its microwave properties. For example, a catalyst with too low a dielectric loss would be inefficient, and could cause control difficulties. On the other hand, a catalyst with too high a dielectric loss can cause low penetration and uniformity problems.

Despite the scaling-up problems, several industrial processes have been proposed and disclosed, mainly in the patent literature:

- microwave-induced catalytic conversion of methane to ethene and hydrogen [89];
- microwave-induced catalytic conversion of methane to C_3 oxygenates [156];
- oxidative coupling of methane to C_2–C_4 oxygenates [157];
- oxidative coupling of methane to ethane and ethene [158];
- cracking of heavy hydrocarbon feeds [159];
- reforming of hydrocarbons [160];
- production of terminal alkenes from long-chain alkanes [161–163];
- formation of shorter-chain hydrocarbons from higher hydrocarbons [164];
- preparation of vinyl acetate [165];
- production of organic nitrogen compounds by direct conversion of elemental nitrogen [166];
- preparation of lower alpha-alkenes [167];

- Heck arylation [168];
- synthesis of a porphyrin compound [169];
- preparation of cellulosic or starchy material [170]; and
- microcomposit catalyst or support with highly dispersed particles [171].

In conclusion, one can ask the question "Which microwave catalytic process is operated on an industrial scale"? The answer has frequently been sought at Ampere meetings, in particular at conferences in Prague (1998), Valencia (1999), the Antibes (2000), Bayreuth (2001), Loughborough (2003), and Modena (2005) and also at world congresses in Orlando 1998, 2000, Sydney 2002, and Austin 2004. It has been stated that relevant information concerning this matter was missing, because an industrial microwave catalytic process put into operation had not been disclosed. The reason could be either that such a process has not yet been successfully realized or that the disclosure of new technology has been strictly protected by industry. Nevertheless, the potential of the microwave catalysis for industrial application is substantial, both for gas-phase processes [172–174] and continuous-flow liquid-phase reactors [175, 176], for which even small equipment can produce significant amounts of products. Another potential application in environmental catalysis has resulted from recent legislative measures which have stimulated the development of new technology to reduce toxic emissions or to remove and dispose of hazardous wastes.

References

1 V. Sridar, *Curr. Sci.* **1998**, *74*, 446–450.
2 G. Roussy, J.A. Pearce, *Foundations and Industrial Applications of Microwave and Radiofrequency Fields*, John Wiley & Sons, New York, **1995**.
3 A.C. Metaxas, R.J. Meredith, *Industrial Microwave Heating*, Peter Peregrinns Ltd., **1993**.
4 G. Bond, R.B. Moyes, D.A. Whan, *Catal. Today* **1993**, *17*, 427–437.
5 S.G. Deng, Y.S. Lin, *Chem. Eng. Sci.* **1997**, *52*, 1563–1575.
6 T. Vergunst, F. Kapteijn, J.A. Moulijn, *Appl. Catal. A: General* **2001**, *213*, 179–187.
7 Y. Liu, Y. Lu, S. Liu, Y. Yin, *Catal. Today* **1999**, *51*, 147–151.
8 J.L. Kuester, *Res. Chem. Intermed.* **1994**, *20*, 51–59.
9 F.-S. Xiao, W. Xu, S. Qiu, R. Xu, *J. Mater. Sci. Lett.* **1995**, *14*, 598–599.
10 F.-S. Xiao, W. Xu, S. Qiu, R. Xu, *Catal. Lett.* **1994**, *26*, 209–215.
11 F.-S. Xiao, W. Zhang, M. Jia, Y. Yu, C. Fang, G. Tu, S. Zheng, S. Qiu, R. Xu, *Catal. Today* **1999**, *50*, 117–123.
12 Y. Wang, J.H. Zhu, J.M. Cao, Y. Chun, Q.H. Xu, *Micropor. Mesopor. Mater.* **1998**, *26*, 175–184.
13 J. Zhu, Y. Chun, Y. Wang, Q. Xu, *Catal. Today* **1999**, *51*, 103–111.
14 N. Lingaiah, P.S.S. Prasad, P.K. Rao, L.E. Smart, F.J. Berry, *Appl. Catal. A: General* **2001**, *213*, 189–196.
15 F.J. Berry, L.E. Smart, P.S.S. Prasad, N. Lingaiah, P.K. Rao, *Appl. Catal. A: General* **2000**, *204*, 191–201.
16 P.S.S. Prasad, N. Lingaiah, P.K. Rao, F.J. Berry, L.E. Smart, *Catal. Letters* **1995**, *35*, 345–351.
17 S. Rajeshkumar, G.M. Anilkumar, S. Ananthakumar, K.G.K. Warrier, *J. Porous. Mat.* **1998**, *5*, 59–63.
18 D.H. Yin, D.L. Yin, *Micropor. Mesopor. Mat.* **1998**, *24*, 123–126.
19 J.A. Menendez, E.M. Menendez, M.J.

Iglesias, A. Garcia, J.J. Pis, *Carbon* **1999**, *37*, 1115–1121.
20. P. Monsef-Mirzai, D.M. Kavanagh, S. Bodman, S. Lange, W.R. McWhinie, *J. Microwave Power Electromagn. Energy* **1999**, *34*, 216–220.
21. Y. Ning, J. Wang, P. Hong, *J. Mater. Sci. Technol.* **1996**, *12*, 307–311.
22. G. Fetter, G. Heredia, L.A. Velazquez, A.M. Maubert, P. Bosch, *Appl. Catal. A: General* **1997**, *162*, 41–45.
23. A. Arafat, J.C. Jansen, A.R. Ebaid, H. van Bekkum, *Zeolites* **1993**, *13*, 162–165.
24. M.D. Romero, G. Ovejero, M.A. Uguina, A. Rodrigues, J.M. Gomez, *Catal. Commun.* **2004**, *5*, 157–160.
25. C.S. Cundy, *Collect. Czech. Chem. Commun.* **1998**, *63*, 1699–1723 (118 references).
26. H. Kosslick, H.-L. Zubova, U. Lohse, H. Landmesser, *Ceram. Trans.* **1997**, *80*, 523–536.
27. E. Vileno, H. Zhou, Q.H. Zhang, S.L. Suib, D.R. Corbin, T.A. Koch, *J. Catal.* **1999**, *187*, 285–297.
28. D.R. Baghurst, D.M.P. Mingos, *J. Chem. Soc., Chem. Commun.* **1988**, 829–830.
29. L. Lingaiah, P.S. Sai Prasad, P.K. Rao, F.J. Berry, L.E. Smart, *Catal. Commun.* **2002**, *3*, 391–397.
30. M.E. Gyftopoulou, M. Millan, A.V. Bridgwater, D. Dugwell, R. Kandiyoti, J.A. Hriljac, *Appl. Catal. A: Gen.* **2005**, *282*, 205–214.
31. A.S. Kulkarni, R.V. Jayaram, *Appl. Catal. A: Gen.* **2003**, *252*, 225–230.
32. M.R. Prasad, G. Kamalakar, S.J. Kulkarni, K.V. Raghavan, K.N. Rao, P.S.S. Prasad, S.S. Madhavendra, *Catal. Commun.* **2002**, *3*, 399–404.
33. A. Venugopal, M.S. Scurrell, *Appl. Catal. A: Gen.* **2003**, *258*, 241–249.
34. A. Kaddouri, Proc. 10th International Conference on Microwave and RF Heating, Modena, Italy, 12–15 Sept., **2005**, 204–207.
35. N. Janjic, M.S. Scurrel, *Catal. Commun.* **2002**, *3*, 253–256.
36. S.L. Suib, *CATTECH* **1998**, *2*, 75–84.
37. L. Seyfried, F. Garin, G. Maire, J.-M. Thiébaut, G. Roussy, *J. Catal.* **1994**, *148*, 281–287.
38. J.-M. Thiébaut, G. Roussy, M.-S. Medjram, F. Garin, L. Seyfried, G. Maire, *Catal. Lett.* **1993**, *21*, 133–138.
39. M. Hájek, *Coll. Czech. Chem. Commun.* **1997**, *62*, 347–354.
40. G. Roussy, A. Zoulalian, M. Charreyre, J.-M. Thiébaut, *J. Phys. Chem.* **1984**, *88*, 5702–5708.
41. J.A. Herbst, C.L. Markham, A.V. Sapre, G.J. Teitman, **1990**, US Patent 4968403; W.J. Murphy, **1993**, US Patent 5205912.
42. P. Veronesi, C. Leonelli, A.B. Corradi, *Ceram. Trans.* **1997**, *80*, 249–256.
43. T.A. Treado, Proc. 30th Intl. Microwave Power Symp. Colorado, USA, 9–12.07.**1995**, 4–7.
44. M. Hájek, J. Kurfürstová, Proc. 10th International Conference on Microwave and RF Heating, Modena, Italy, 12–15 Sept., **2005**, 150–153.
45. F. Chemat, M. Poux, S.A. Galema, *J. Chem. Soc., Perkin Trans. 2* **1997**, 2371–2374.
46. F. Chemat, M. Poux, J.-L. di Martino, J. Berlan, *Chem. Eng. Technol.* **1996**, *19*, 420–424.
47. F. Chemat, D.C. Esveld, M. Poux, J.L. di Martino, *J. Microwave Power Electromagn. Energy* **1998**, *33*, 88–94.
48. K.G. Kabza, B.R. Chapados, J.E. Gestwicki, J.L. McGrath, *J. Org. Chem.* **2000**, *65*, 1210–1214.
49. S.D. Pollington, G. Bond, R.B. Moyes, D.A. Whan, J.P. Candlin, J.R. Jennings, *J. Org. Chem.* **1991**, *56*, 1313–1314.
50. G. Öztűrk, B. Gűmgűm, Proc. International Symposium on Microwave Science and Its Application to Related Fields, Takamatsu, Japan, 27–30 July, **2004**, 398–399.
51. A. Breccia, B. Esposito, G.B. Fratadocchi, A. Fini, *J. Microwave Power Electromagn. Energy* **1999**, *34*, 3–8.
52. C. Mazzocchia, A. Kaddouri, G. Modica, R. Nannicini, F. Martini, S. Marengo, Proc. 10th International Conference on Microwave and RF

Heating, Modena, Italy, 12–15 Sept., **2005**, 136–139.
53 S. Leskovšek, A. Šmidovnik, T. Koloini, *J. Org. Chem.* **1994**, *59*, 7433–7436.
54 C. Bonnet, L. Estel, A. Ledoux, B. Mazari, A. Louis, *Chem. Eng. Proc.* **2004**, *43*, 1435–1440.
55 B. Dayal, N.H. Ertel, K.R. Rapole, A. Asgaonkar, G. Salen, *Steroids* **1997**, *62*, 451–454.
56 A.K. Bose, B.K. Banik, K.J. Barakat, M.S. Manhas, *SYNLETT* **1993**, 575–576.
57 B.K. Banik, K.J. Barakat, D.R. Wagle, M.S. Manhas, A.K. Bose, *J. Org. Chem.* **1999**, *64*, 5746–5753.
58 B. Toukoniitty, J.P. Mikkola, T. Salmi, D. Yu. Murzin, *Proc. 10th International Conference on Microwave and RF Heating, Modena, Italy, 12–15 Sept.,* **2005**, 146–149.
59 I. Plazl, S. Leskovšek, T. Koloini, *Chem. Eng. J.* **1995**, *59*, 253–257.
60 S. Leskovšek, I. Plazl, T. Koloini, *Chem. Biochem. Eng. Q.* **1996**, *10*, 21–26.
61 I. Plazl, G. Pipus, T. Koloini, *AIChE J.* **1997**, *43*, 754–760.
62 P. He, S.J. Haswell, P.D.I. Fletcher, *Appl. Catal. A: General* **2004**, *274*, 111–114.
63 J.A. Gardner, G. Bond, *Proc. 9th International Conference on Microwave and RF Heating, Loughborough, UK, 1–5 Sept.,* **2005**, 125–128.
64 P.J. Kooyman, G.C.A. Luijkx, A. Arafat, H. van Bekkum, *J. Mol. Cat. A: Chem.* **1996**, *111*, 167–174.
65 F. Matloubi Moghaddam, A. Shariff, *Syn. Commun.* **1995**, *25*, 2457–2461.
66 C. Laporte, J. Marquié, A. Laporterie, J.-R. Desmurs, J. Dubac, *C. R. Acad. Sci. Paris* **1999**, *2*, 455–465.
67 J.G. Mei, S.M. Yu, J. Cheng, *Catal. Commun.* **2004**, *5*, 437–440.
68 C. Cativiela, J.I. García, J.A. Mayoral, E. Pires, A.J. Royo, F. Figueras, *Appl. Catal. A: General* **1995**, *131*, 159–166.
69 R.M. Martín-Aranda, M.A. Vicente-Rodrígues, J.M. López-Pestana, A.J. López-Peinado, A. Jerez, J.D. López-Gonzáles, M.A. Bañarez-Muñoz, *J. Mol. Cat. A: Chem.* **1997**, *124*, 115–121.
70 S. Gadhwal, M.P. Dutta, A. Boruah, D. Prajapati, J.S. Sandhu, *Indian J. Chem.* **1998**, *37B*, 725–727.
71 Y.-L. Jiang, Y.-Q. Hu, S.-Q. Feng, J.-S. Wu, Z.-W. Wu, Y.-C. Yuan, *Synth. Commun.* **1996**, *26*, 161–164.
72 M.T. Radoiu, I. Calinescu, P. Chipurici, D.I. Martin, *J. Microwave Power Electromagn. Energy* **2000**, *35*, 86–91.
73 A. Stadler, B.H. Yousefi, D. Dalinger, P. Walla, E. van Eycken, N. Kaval, C.O. Kappe, *Org. Proc. Res. Dev.* **2003**, *7*, 707–716.
74 B. Desai, C.O. Kappe, *Topics in Curr. Chem.* **2004**, *242*, 177–208.
75 M. Hájek, M.T. Radoiu, *J. Mol. Catal. A: Chem.* **2000**, *160*, 383–392.
76 M. Hájek, M.T. Radoiu, *Ceram. Trans.* **2001**, *111*, 257–264.
77 J. Kurfürstová, M. Hájek, *Res. Chem. Intermed.* **2004**, *30*, 673–681.
78 V. Singh, S. Kaur, V. Sapehiyia, J. Singh, G.L. Kad, *Catal. Commun.* **2005**, *6*, 57–60.
79 T. Joseph, S. Sahoo, S.B. Halligudi, *J. Mol. Catal. A: Chem.* **2005**, *234*, 107–110.
80 J. Fraga-Dubreuil, K. Bourahla, M. Rahmouni, J.P. Bazureau, J. Hamelin, *Catal. Commun.* **2002**, *3*, 185–190.
81 Z. Zhao, B. Yuan, W. Qiao, Z. Li, G. Wang, L. Cheng, *J. Mol. Catal. A: Chem.* **2005**, *235*, 74–80.
82 P.J. Dyson, T. Geldbach, F. Moro, C. Taeschler, D.B. Zhao, *ACS Symposium series* **2005**, *902*, 322–333.
83 B. Toukonitty, J.P. Mikkola, D.Y. Murzin, T. Salmi, *Appl. Catal. A: Gen.* **2005**, *279*, 1–22.
84 U.R. Pillai, E.S. Demessie, R. Varma, *Appl. Catal. A: Gen.* **2003**, *252*, 1–8.
85 J.K.S. Wan, Y.G. Chen, Y.J. Lee, M.C. Depew, *Res. Chem. Intermed.* **2000**, *26*, 599–619.
86 G. Roussy, J.M. Thiébaut, M. Souiri, A. Kinnenmann, G. Maire, *Proc. 28th Intl. Microwave Power Symp,*

Montreal, Canada, 11–14.07.1993, pp. 79–82.
87 X. Zhang, C.S.M. Lee, D.M.P. Mingos, D.O. Hayward, *Appl. Catal. A: Gen.* **2003**, *249*, 151–164.
88 M.T. Radoiu, Y. Chen, M.C. Depew, *Appl. Catal. B: Environ.* **2003**, *43*, 187–193.
89 J.K.S. Wan, *US Patent 4,574,038*, **1986**.
90 J.K.S. Wan, K. Wolf, R.D. Heyding, *Stud. Surf. Sci. Catal.* **1984**, *19*, 561–568.
91 J.K.S. Wan, M. Tse, T.H. Husby, M.C. Depew, *J. Microwave Power Electromagn. Energy* **1990**, *25*, 32–38.
92 C. Marun, L.D. Conde, S.L. Suib, *J. Phys. Chem. A* **1999**, *103*, 4332–4340.
93 S.L. Suib, E. Vileno, Q. Zhang, C. Marun, L.D. Conde, *Ceram. Trans.* **1997**, *80*, 331–339.
94 G. Roussy, J.M. Thiebaut, M. Souiri, E. Marchal, A. Kiennemann, G. Maire, *Catal. Today* **1994**, *21*, 349–355.
95 C.-L. Chen, P.-J. Hong, S.-S. Dai, C.-C. Zhang, *React. Kinet. Catal. Lett.* **1997**, *61*, 181–185.
96 C.-G. Chen, P.-J. Hong, S.-H. Dai, J.-D. Kan, *J. Chem. Soc., Faraday Trans.* **1995**, *91*, 1179–1180.
97 Y.Y. Tanashev, V.I. Fedoseev, Y.I. Aristov, V.V. Pushkarev, L.B. Avdeeva, V.I. Zaikovskii, V.N. Parmon, *Catal. Today* **1998**, *42*, 333–336.
98 V.I. Fedoseev, Y.I. Aristov, Y.Y. Tanashev, V.N. Parmon, *Kinet. Catal.* **1996**, *37*, 808–811.
99 X.-J. Bi, P.-J. Hong, X.-G. Xie, S.-S. Dai, *React. Kinet. Catal. Lett.* **1999**, *66*, 381–386.
100 S.L. Suib, R.P. Zerger, *J. Catal.* **1993**, *139*, 383–391.
101 Y.-F. Chang, A. Sanjurjo, J.G. McCarty, G. Krishnan, B. Woods, E. Wachsman, *Catal. Lett.* **1999**, *57*, 187–191.
102 M.Y. Tse, M.C. Depew, J.K.S. Wan, *Res. Chem. Intermed.* **1990**, *13*, 221–236.
103 S.L. Suib, R.P. Zerger, Z.C. Zhang, *Abstr. Pap. Am. Chem. Soc.* **1992**, *203*, 133.
104 J.K.S. Wan, T.A. Koch, *Res. Chem. Intermed.* **1994**, *20*, 29–37.
105 G. Roussy, S. Hilaire, J.-M. Thiébaut, G. Maire, F. Garin, S. Ringler, *Appl. Catal. A: General* **1997**, *156*, 167–180.
106 V. Keller, P. Wehrer, F. Garin, R. Ducros, G. Maire, *J. Catal.* **1995**, *153*, 9–16.
107 K. Wolf, H.K.J. Choi, J.K.S. Wan, *AOSTRA J. Res.* **1986**, *3*, 53–59.
108 Y. Liu, P. Liu, R.X. Gao, Y. Yin, *Appl. Catal. A: General* **1998**, *170*, 207–214.
109 G. Bamwenda, M.C. Depew, J.K.S. Wan, *Res. Chem. Intermed.* **1993**, *19*, 533–564.
110 K.L. Cameron, M.C. Depew, J.K.S. Wan, *Res. Chem. Intermed.* **1991**, *16*, 57–70.
111 J.K.S. Wan, *Res. Chem. Intermed.* **1993**, *19*, 147–158.
112 J.W. Tang, T. Zhang, D.B. Liang, C.H. Xu, X.Y. Sun, L.W. Lin, *Chem. Commun.* **2000**, 1861–1862.
113 S. Ringler, P. Girard, G. Maire, S. Hilaire, G. Roussy, F. Garin, *Appl. Catal. B: Environmental* **1999**, *20*, 219–233.
114 J.K.S. Wan, G. Bamwenda, M.C. Depew, *Res. Chem. Intermed.* **1991**, *16*, 241–255.
115 T.R.J. Dinesen, M.Y. Tse, M.C. Depew, J.K.S. Wan, *Res. Chem. Intermed.* **1991**, *15*, 113–127.
116 J. Beckers, G. Rothenberg, *Chem. Phys. Chem.* **2005**, *6*, 223–225.
117 G. Bond, R.B. Moyes, I. Theaker, D.A. Whan, *Top. Catal.* **1994**, *1*, 177–182.
118 A.Y. Klimov, B.S. Balzhinimaev, L.L. Makarshin, V.I. Zaikovskii, V.N. Parmon, *Kinet. Catal.* **1998**, *39*, 511–515.
119 X.L. Zhang, D.O. Hayward, D.M.P. Mingos, *Catal. Lett.* **2003**, *88*, 33–38.
120 N. Kuhnert, *Angew. Chem. Int. Ed.* **2002**, *41*, 1863–1866.
121 C.R. Strauss, *Angew. Chem. Int. Ed.* **2002**, *41*, 3589–3590.
122 L. Perreux, A. Loupy, *Tetrahedron* **2001**, *57*, 9199–9223.
123 D.R. Baghurst, D.M.P. Mingos, *J. Chem. Soc., Chem. Commun.* **1992**, 674–677.

124 D.M.P. Mingos, *Res. Chem. Intermed.* **1994**, *20*, 85–91.
125 D. Stuerga, P. Gaillard, *Tetrahedron* **1996**, *52*, 5505–5510.
126 P. Gaillard, N.R. Roudergue, D. Stuerga, *Proc. 6th International Conference on Microwave and High Frequency Heating, Fermo, Italy,* 9–13.09.1997, p. 82–83.
127 M. Hájek, H. Richterová, *Proc. 6th International Conference on Microwave and High Frequency Heating, Fermo, Italy,* 9–13.09.**1997**, p. 79–81.
128 J. Dolande, A. Datta, *J. Microwave Power Electromagn. Energy* **1993**, *28*, 58–67.
129 W.L. Perry, J.D. Katz, D. Rees, M.T. Paffer, A.K. Datye, *J. Catal.* **1997**, *171*, 431–438.
130 J.M. Hill, N.F. Smith, *Math. Eng. Ind.* **1990**, *2*, 267–278.
131 C.J. Coleman, *J. Aust. Math. Soc. B* **1991**, *33*, 1–8.
132 G.A. Kriegsman, *Ceram. Trans.* **1991**, *21*, 177–183.
133 G.A. Kriegsman, P. Varatharajah, *J. Appl. Phys.* **1993**, *36*, 221–228.
134 J.R. Thomas, *Ceram. Trans.* **1997**, *80*, 397–406.
135 J.R. Thomas, *Catal. Letters* **1997**, *49*, 137–141.
136 J.R. Thomas, J.F. Faucher, *J. Microwave Power Electromag. Energy* **2000**, *35*, 165–174.
137 C. Blanco, S.M. Auerbach, *J. Am. Chem. Soc.* **2002**, *124*, 6250–6251.
138 M.D. Turner, R.L. Laurence, W.C. Conner, K.S. Yngvesson, *AIChE J.* **2000**, *46*, 758–768.
139 A. de la Hoz, A. Diaz-Ortiz, A. Moreno, *Chem. Soc. Rev.* **2005**, *34*, 164–178.
140 C.O. Kappe, A. Stadler, *Microwaves in Organic and Medicinal Chemistry*, Wiley–VCH, Ed. R. Mannhold, H. Kubinyi, G. Folkers, **2005**, p. 54.
141 M.A.B. Pougnet, *Rev. Sci. Instrum.* **1993**, *64*, 529–531.
142 M. Pagnotta, A. Nolan, L. Kim, *J. Chem. Educ.* **1992**, *69*, 599–600.
143 D.M.P. Mingos, *Chem. Ind.* **1994**, 596–599.
144 K.D. Raner, C.R. Strauss, R.W. Trainor, J.S. Thorn, *J. Org. Chem.* **1995**, *60*, 2456–2460.
145 C.R. Strauss, R.W. Trainor, *Aust. J. Chem.* **1995**, *48*, 1665–1692.
146 C.R. Strauss, R.W. Trainor, *Aust. J. Chem.* **1995**, *48*, 1674, 1675.
147 S. Stinson, *Chem. Eng. News* **1990**, *68*, 33–34.
148 T. Cablewski, A.F. Faux, C.R. Strauss, *J. Org. Chem.* **1994**, *59*, 3408–3412.
149 S.-T. Chen, S.-H. Chiou, K.-T. Wang, *J. Chem. Soc., Chem. Commun* **1990**, 807–809.
150 J. Marquié, G. Salmoria, M. Poux, A. Laporterie, J. Dubac, N. Roques, *Ind. Eng. Chem. Res.* **2001**, *40*, 4485–4490.
151 G. Roussy, S. Hillaire, J.-M. Thiébaut, G. Maire, F. Garin, S. Ringler, *Appl. Catal. A: Gen.* **1997**, *156*, 169, 170.
152 G. Roussy, S. Jassm, J.-M. Thiébaut, *J. Microwave Power Electromagn. Energy* **1995**, *30*, 178–187.
153 G.C.-J. Jou, *Carbon* **1998**, *36*, 1643–1648.
154 J.K.S. Wan, M.C. Depew, *Ceram. Trans.* **2001**, *111*, 241–247.
155 M. Mehdizadeh, *Res. Chem. Intermed.* **1994**, *20*, 79–84.
156 J.K.S. Wan, *Canadian Patent* CA 2,031,608, **1995**.
157 C.L. O'Young, *Eur. Pat. Appl.* EP 634,211, **1995**.
158 G. Roussy, J.-M. Thiébaut, O.M. Souiri, A. Kiennemann, K.C. Petit, G. Maire, *Fr. Pat. Appl.* FR 92 11676, **1994**.
159 F.G. Dwyer, J.A. Herbst, Y.Y. Huang, H. Owen, P.H. Scipper, A.B. Schwartz, *Int. Pat. Appl.* PCT 87/02227, **1988**.
160 G. Roussy, G. Maire, *Eur. Pat. Appl.* EP 519 824, **1992**.
161 D.D. Tanner, Q.Z. Ding, P. Kandanarachchi, J.A. Franz, *Prep. Symp. Amer. Chem. Soc., Div. Fuel Chem.* **1999**, *44*, 133–139.
162 D.D. Tanner, Q.Z. Ding, P. Kandanarachchi, J.A. Franz, *Int. Pat. Appl.* PCT 99/16851, **1999**.
163 W. Cho, Y. Baek, D. Park, Y.C. Kim,

M. Anpo, *Res. Chem. Intermed.* **1998**, 24, 55–66.
164 T. Hill, D.R. Weaver, *Ger. Pat. Appl.* DE 2535119, **1976**.
165 B. Herzog, *Eur. Pat. Appl.* EP 922,491, **1999**.
166 D.J. Harper, D.J. Wheeler, R.M. Henson, *Eur. Pat. Appl.* EP 742,189, **1996**.
167 S. Bhaduri, V. Gupta, *World Pat. Appl.* WO2004089998, **2004**.
168 S.M. Pillai, A. Vali, *US Pat. Appl.* US2003100625, **2003**.
169 S.J. Kulkarni, K.V. Raghavan, K.V. *US Pat. Appl.* US2002143175, **2002**.
170 G.C. Vaca, S. Girardeau, *World Pat. Appl.* WO0050492, **2000**.
171 A. Heidekum, *DE Pat.* DE19954827, **2001**.
172 H. Will, P. Scholz, B. Ondruschka, *Chem. Eng. Technol.* **2004**, 27.2, 113–122.
173 H. Will, B. Ondruschka, *Topics in Catalysis*, **2004**, 29, 175–182.
174 H. Will, P. Scholz, B. Ondruschka, W. Burckhardt, *Chem. Eng. Technol.*, **2003**, 26, 1146–1149.
175 W. Bonrath, K. Reinhard, M. Nuechter, B. Ondruschka, *World Pat. Appl.* WO03091235.
176 G. Bond, J.A. Gardner, J.G. McGuire, *Proc. 4th World Congress on Microwave and Radio Frequency Applications, Austin, Texas, November 7–12.* **2004**, 131–138.

14
Polymer Chemistry Under the Action of Microwave Irradiation

Dariusz Bogdal and Katarzyna Matras

14.1
Introduction

Polymer processing is probably one of the most developed and well established applications of microwave technology. The vulcanization of rubber in the tire industry was the first industrial application of microwave irradiation for processing of polymeric materials. Preparation and processing of polymeric dental restorative materials is another discipline in which microwaves have found commercial applications. In recent years it has been shown that polymer synthesis can, in the same way as processing, greatly benefit from the unique features of modern microwave technology, for example shorter reaction time, increased yield, and temperature uniformity during polymerization and crosslinking, which have been demonstrated in many successful laboratory-scale applications [1–3].

Since very recently the number of papers on microwave-assisted polymerization reactions has been growing almost exponentially [3], the purpose of this chapter is to provide useful details about the application of microwave irradiation to polymer chemistry during the last few years. A survey of past achievements in polymer synthesis and polymer composites can be found in review papers [1–8] whereas fundamentals of electromagnetic heating and processing of polymers, resins and related composites have been summarized by Parodi [9].

14.2
Synthesis of Polymers Under the Action of Microwave Irradiation

The effect of microwave irradiation on chemical reactions is usually described by comparing time needed to obtain a desired yield of final products compared with conventional thermal heating. Research in the area of chemical synthesis has shown potential advantages in the ability not only to drive chemical reactions but to perform them more quickly. In polymer synthesis other factors can be considered, for example molecular weight, polydispersity index, crystallinity, mechanical properties (i.e. strength, elongation, modulus, toughness), and thermal properties

Microwaves in Organic Synthesis, Second edition. Edited by A. Loupy
Copyright © 2006 WILEY-VCH Verlag GmbH & Co. KGaA, Weinheim
ISBN: 3-527-31452-0

(i.e. glass transition and melting temperatures). Occasionally the products obtained have properties that may not be possible after use of conventional thermal treatment. Such features can be both advantages and disadvantages of microwave technology. In turn, changing of material properties enables modeling of materials for different applications.

This paper reports and discusses microwave-assisted polymer syntheses, cross-linking, and processing with the stress on the chemistry of those processes. For this purpose, syntheses under microwave conditions have been compared with those under the action of conventional heating methods. In the most examples reported in the literature, amounts of reagents vary from a few milligrams to a few grams. The shape and size of the reaction vessel are also important in the processing of material under microwave irradiation, as is the microwave system used (i.e. applicator and temperature detection). In each example we have, therefore, tried to briefly describe the microwave systems used by the different research groups and the amounts of reagents they used, to give to readers deeper insight into these microwave experiments.

14.2.1
Chain Polymerizations

Chain polymerization reactions under microwave irradiation conditions have been investigated for both free-radical and controlled "living" polymerization and ring-opening polymerization.

14.2.1.1 Free-radical Polymerization Reactions

Free-radical polymerization reactions have recently been studied for different monomers, for example mono and disubstituted vinyl monomers and dienes. The bulk polymerization of vinyl monomers (e.g. vinyl acetate, styrene, methyl methacrylate, and acrylonitrile) has been investigated by Amorim et al. [10]. The reactions were conducted in the presence of catalytic amounts of AIBN (or benzoyl peroxide). It was found that the rate of polymerization depends on the structure of the monomers and the power and time of microwave irradiation. In a typical experiment 10.0 mL of each monomer and 50 mg AIBN was irradiated in a domestic microwave oven for 1 to 20 min to afford the polymers polystyrene, poly(vinyl acetate), and poly(methyl methacrylate) with weight-average molecular weights 48 400, 150 200, 176 700 g mol^{-1}, respectively (Scheme 14.1). The experiments were performed without temperature control.

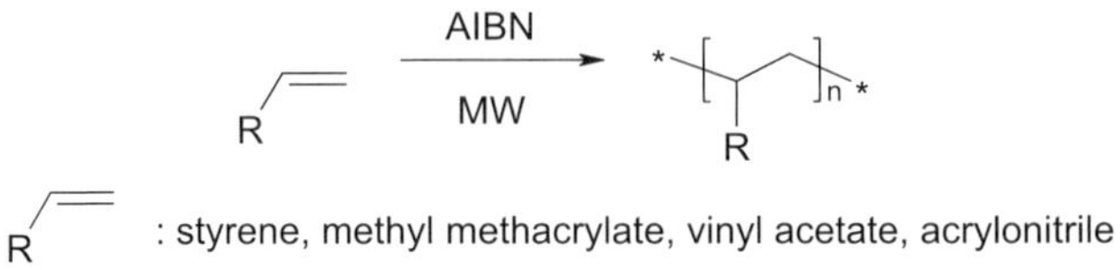

R⟨= : styrene, methyl methacrylate, vinyl acetate, acrylonitrile

Scheme 14.1

Emulsion polymerization of methyl methacrylate under the action of pulsed microwave irradiation was studied by Zhu et al. [11]. The reactions were conducted in a self-designed single-mode microwave reaction apparatus with a frequency of 1250 MHz and a pulse width of 1.5 or 3.5 μs. The output peak pulse power, duty cycles, and mean output power were continuously adjustable within the ranges 20–350 kW, 0.1–0.2%, and 2–350 W, respectively. Temperature during microwave experiments was maintained by immersing the reaction flask in a thermostatted jacket with a thermostatic medium with little microwave absorption (for example tetrachloroethylene). In a typical experiment, 8.0 mL methyl methacrylate, 20 mL deionized water, and 0.2 g sodium dodecylsulfonate were transferred to a 100-mL reaction flask which was placed in the microwave cavity. When the temperature reached a preset temperature, 10 mL of an aqueous solution of the initiator (potassium persulfate) was added and the flask was exposed to microwave irradiation. The results obtained under microwave irradiation were compared with those from conventional experiments. Under the action of microwaves the amount of initiator needed to achieve constant conversion was reduced by 50% at the same polymerization rate; if the same initiator concentration was used (0.15 and 0.20% w/w) the rate of polymerization increased by factors of 131 and 163%, respectively. The molecular weights of the polymers were 1.1 to 2.0 times higher than those obtained by use of conventional conditions. The glass transition temperatures (T_g), polydispersity index, and regularity of the polymers obtained by use of the two processes (microwave and conventional) were similar, indicating an analogous mechanism of polymerization.

Polymerization of isoprene in the presence of organolanthanide catalysts under the action of microwave irradiation has been performed by Barbier-Baudry et al. [12]. The reactions were conducted in a single-mode microwave reactor with calibrated IR temperature detection. The main power values necessary to reach and maintain temperature were 15, 32, 55, and 95 W for 60, 80, 100, and 120 °C, respectively. The polymerization of isoprene was performed in toluene solution, in which isoprene was mixed with the catalyst [Nd(BH$_4$)$_3$(THF)$_3$] and co-catalyst, either Mg(Bu)$_2$ and Al(Et)$_3$, in 10-mL vessels that were sealed for both microwave and conventional experiments. The reaction time was in the range 15 to 120 min. In some experiments, the reaction mixture was kept for 2 h at room temperature before the reaction. The study revealed enhancement of reactivity under microwave conditions compared with conventional conditions; the selectivity was only slightly modified. The highest yields (85–94%) of polyisoprene were obtained within a reaction time of 2 h at 80 °C to afford polymer with a number-average molecular weight in the range 17 000 to 27 000 g mol^{-1} and polydispersity index 1.6 to 2.5. Interestingly, reaction at 120 °C afforded polyisoprene with higher yield under conventional conditions; this was explained by a depolymerization reaction under microwave irradiation at high temperature (Scheme 14.2).

Chain-transfer polymerizations (telomerizations), which were developed to produce polymers of narrower molecular weight distributions than in conventional free-radical polymerization, of poly-N-isopropylacrylamide (PNIPAM), poly-N,N-dimethylacrylamide (PNDMAM), and poly-N-{3-(dimethylamino)propyl}acrylamide

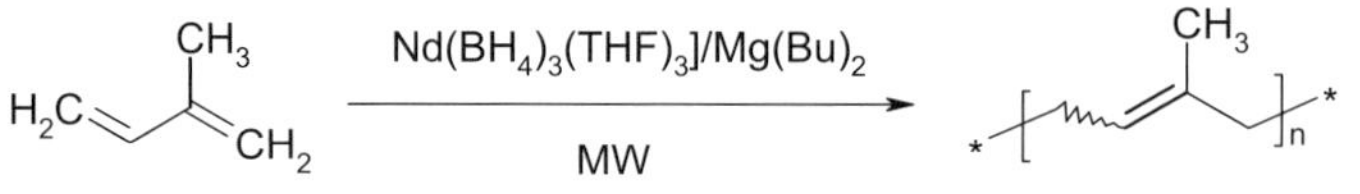

Scheme 14.2

$H_2C=C(CH_3)-CH_2CH_3 \xrightarrow[\text{MW}]{Nd(BH_4)_3(THF)_3]/Mg(Bu)_2}$ polymer with CH$_3$ branches

(PN3DMAPAM), and copolymers of PNIPAM and PNDMAM, under microwave irradiation conditions were studied by Fisher et al. (Scheme 14.3) [13]. Under conventional conditions the telomerization reactions were performed in superheated (80–170 °C) methanol solution in an autoclave in which 100 mmol N-alkylacrylamide was mixed with 4 mmol 3-mercaptopropionic acid, 0.5 mmol AIBN, and 15 mL methanol. In the microwave experiments 50 mmol N-alkylacrylamide was mixed with 2 mmol 3-mercaptopropionic acid and 0.25 mmol AIBN in an open flask which was irradiated in a domestic microwave oven. It was found that in superheated methanol the reaction time was reduced by 66% and the average molecular weight and yield remained unchanged in comparison with the standard reflux conditions. In microwave experiments reduction of the reaction time was even greater, although the average molecular weight dropped by 30% in these experiments. In our opinion, it is difficult to compare bulk polymerization under microwave conditions without temperature control with solution polymerization under pressurized and nonpressurized conditions.

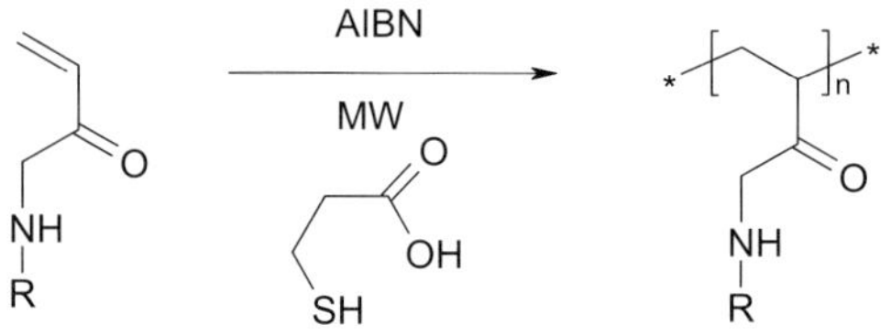

Scheme 14.3

Ritter et al. have synthesized a variety of (meth)acrylamides in good yields from (meth)acrylic acid and aliphatic and aromatic amines under solvent-free microwave irradiation conditions [14]. It was found that addition of a polymerization initiator (AIBN) to the reaction mixture led directly to poly(meth)acrylamides in a single step. In these polymerization procedures, 11.6 mmol methacrylic acid was mixed with 11.6 mmol amine and 0.58 mmol AIBN in a pressure-resistant test tube. The tubes were sealed and irradiated in a single-mode microwave reactor for 30 min at 140 W (Scheme 14.4). The experiments were performed without temperature control and were not compared with conventional conditions.

It was subsequently demonstrated that under microwave conditions it was possible to obtain chiral (R)-N-(1-phenylethyl) methacrylamide directly from methacrylic acid and (R)-1-phenylethylamine under solvent-free conditions. Addition of free-radical initiator (AIBN) again led, in a single-step reaction, to optically active poly-

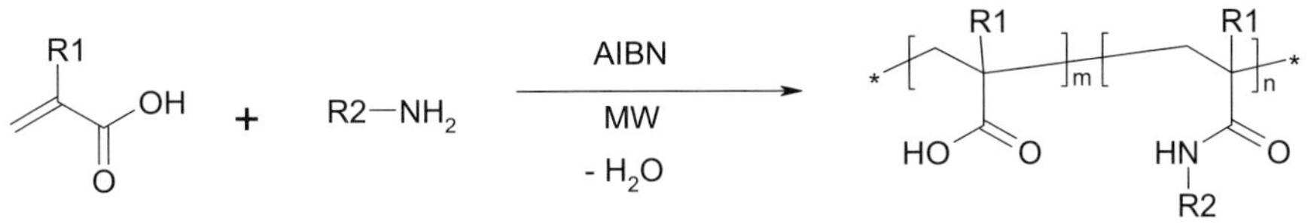

14.2 Synthesis of Polymers Under the Action of Microwave Irradiation | 657

Scheme 14.4

mers containing both methacrylamide and imide moieties (Scheme 14.5) [15, 16]. In a typical polymerization reaction, 16.5 mmol methacrylic acid were mixed with 16.5 mmol (R)-1-phenylethylamine and 0.83 mmol AIBN in a pressure-resistant test tube. The tubes were sealed and irradiated in a single-mode microwave reactor for 30 min at 140 W at constant temperature (120 °C). It was found that microwave irradiation substantially accelerates the process of condensation between the acid and the amine, which is also more selective under the action of microwaves than with thermal heating. The one-pot polymerization under microwave conditions afforded the polymers with relatively high yields (80%) which depended on the applied power. The yield under the action of classical heating in an oil bath was only 40%.

The bulk polymerization of N-phenylmaleimide, prepared from maleic anhydride and aniline before the reaction, under microwave irradiation conditions, has been reported (Scheme 14.6) [17]. For this purpose 5.9 mmol N-phenylmaleimide

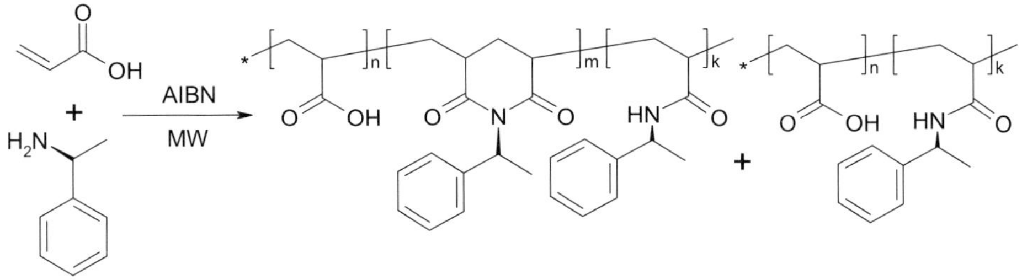

Scheme 14.5

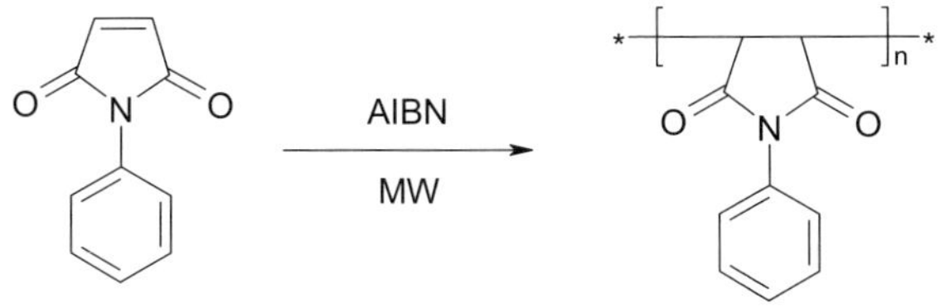

Scheme 14.6

was mixed with 2.8 mmol AIBN in a pressure-resistant test tube. The tube was sealed with a septum, flushed with nitrogen and irradiated in a single-mode microwave reactor for 15 min at 90 °C (IR–pyrometry). The final yield of the polymer was 57%. Bulk homopolymerization under conventional conditions (oil bath preheated to 95 °C) afforded the polymer with a relatively low yield, ca. 19%.

Solution free-radical polymerization of carbazole-containing monomers such as N-vinylcarbazole and 2-(9-carbazolyl)ethyl methacrylate under microwave irradiation conditions has been investigated by Bogdal et al. (Scheme 14.7) [18]. The reactions were conducted in pressure-resistant tubes, in different solvents, for example toluene, hexane, nitromethane, and diethylene glycol. In a standard procedure, 1 g monomer was mixed with 1 mL solvent and 0.060 mmol AIBN. The vessel was purged with argon, sealed, and irradiated in a single-mode microwave reactor for 10 min at 65 °C. After precipitation, polymers characterized by weight-average molecular weights of 20 000 to 50 000 g mol^{-1} were afforded in high yields (80 to 99%). Interestingly, in experiments under conventional conditions (a preheated oil bath) the polymers were obtained in very low yields ca. 1%.

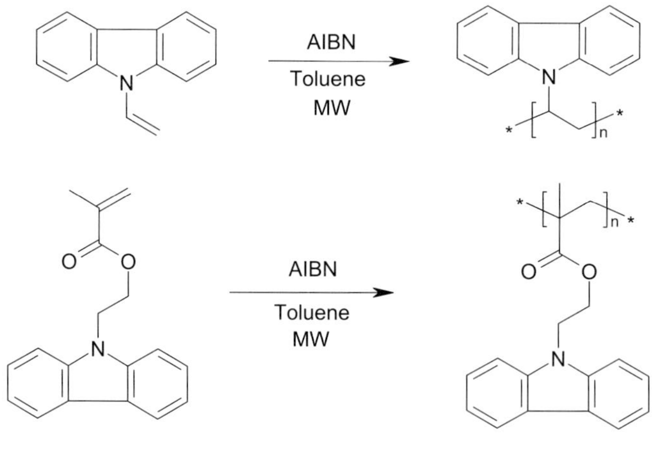

Scheme 14.7

The copolymerization of methacrylic acid, 2-(dimethylamino)ethyl ester, and thiourea under microwave irradiation was studied by Lu et al. (Scheme 14.8) [19]. It was shown that the copolymers can be used to coordinate Cu(II) to afford coordinated copolymers which can, in turn, be used as heterogeneous catalysts in the polymerization of methyl methacrylate. The reactions were performed in a modified domestic microwave oven with a continuous power regulation.

14.2.1.2 Controlled "Living" Radical Polymerization

Controlled "living" radical polymerization methods were developed to produce polymers with predetermined molecular weights, low polydispersity index, specific

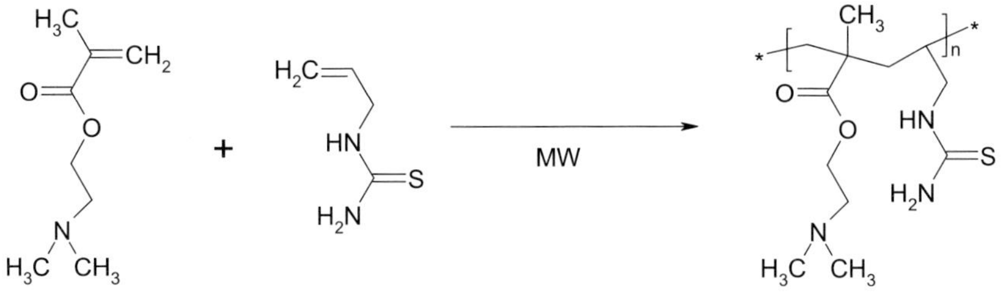

Scheme 14.8

functionality, and more diverse architecture than in conventional free-radical polymerization [37].

Bulk polymerization of methyl methacrylate by atom-transfer radical polymerization (ATRP) under microwave irradiation conditions has been investigated by Zhu et al. (Scheme 14.9) [20–22]. The reactions were conducted in sealed tubes which were placed inside two-necked flasks filled with a solvent transparent to microwaves (hexane, CCl_4). The flask was irradiated in a modified microwave oven so that temperature was controlled by the solvent boiling point. The reactions were performed with different activator–initiator systems including benzyl chloride and bromide–CuCl–2,2′-bipyridine [20], AIBN–CuBr$_2$–2,2′-bipyridine [21], and α,α′-dichloroxylene–CuCl–N,N,N',N'',N''-pentamethyldiethylenetriamine [22]. It was found that microwave irradiation always enhanced the rate of polymerization and gave polymers with narrower molecular weight distributions (polydispersity index). Linear first-order rate plots, a linear increase of number-average molecular weight with conversion, and low polydispersity were also observed; this indicated that ATRP of methyl methacrylate was controlled under microwave conditions.

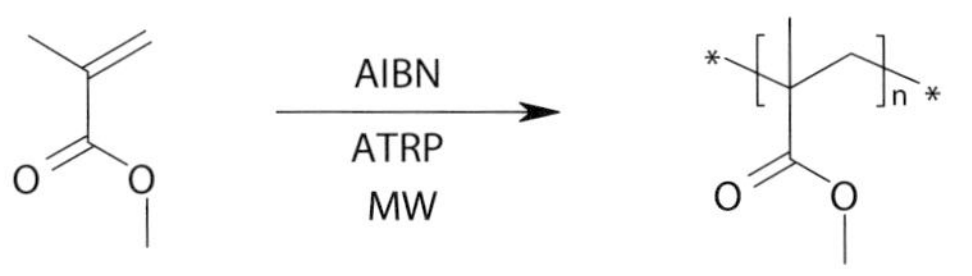

Scheme 14.9

Homogenous atom-transfer radical polymerization of methyl methacrylate under microwave irradiation conditions has also been studied by Zhu et al. [23, 24]. In a typical run, a small amount of CuCl, N,N,N',N'',N''-pentamethyldiethylenetriamine, with ethyl 2-bromobutyrate as activator–initiator system, were placed in a 10-mL glass tube with 1 mL DMF and 5 mL methyl methacrylate. The tube was sealed and placed in a two-necked reaction flask filled with hexane, so that temperature was controlled by the boiling point of the solvent during reflux in a modified domestic microwave oven. Linear first-order rate plots, linear increases of the number-average molecular weight with conversion, and low polydispersities were observed. It was found that microwave irradiation enhanced the rate of polymerization. For example, after 2.5 h of microwave irradiation monomer conversion reached 27% and the polymers were afforded with a number-average molecular weight of 57 300 g mol^{-1} and polydispersity index of 1.19. Under conventional conditions, similar conversion was achieved after 16 h and the polymers were characterized by number-average molecular weight of 64 000 g mol^{-1} and a polydispersity index of 1.19. Similar results were obtained for the atom-transfer radical polymerization of n-octyl acrylate in acetonitrile solution in the presence of 2-bromobutyrate, CuBr, and 2,2′-bipyridine under microwave conditions [25].

In contrast with this investigation, Schubert et al. [26] reported that atom-transfer radical polymerization of methyl methacrylate in p-xylene solution under microwave conditions did not result in rate enhancement in comparison with conventional conditions. In a typical experiment, a stock solution of 70 mmol methyl methacrylate, 15 mL p-xylene, 0.47 mmol ethyl 2-bromoisobutyrate, and N-hexyl-2-pyridylmethanimine was divided into six vials containing 0.078 mmol CuBr. The vials were sealed, purged with argon, and irradiated in a single-mode microwave reactor for different periods up to 6 h. Good control of the polymerization reaction was observed – linear first-order rate plots, a linear increase of the number-average molecular weight with conversion, and low polydispersity. Results were, however, almost the same as those obtained under conventional conditions (Scheme 14.10).

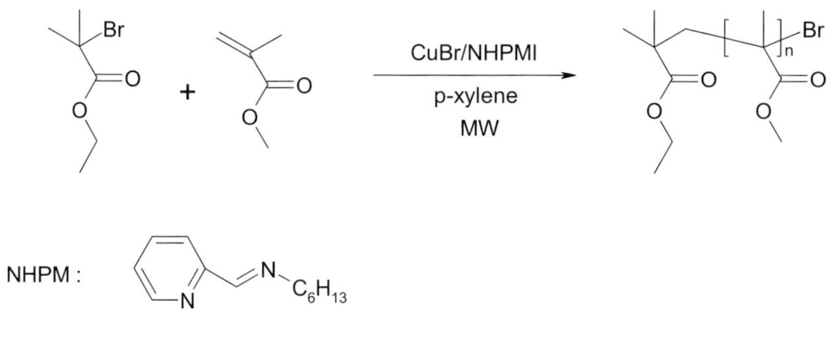

Scheme 14.10

Solid-supported TEMPO-mediated controlled "living" polymerization in the preparation of novel high-loading functionalized styrenyl resins has been reported by Wisnoski et al. (Scheme 14.11) [27]. The resin was prepared in a neat reaction of TEMPO-methyl resin with styrene derivatives. For example, 200 mg TEMPO-methyl resin suspended in 16.8 mmol p-bromostyrene was transferred to a 5 mL vial which was sealed and placed in a single-mode microwave reactor. The mixture was irradiated for 10 min at 185 °C to afford resin with a 7.25-fold increase in mass. By applying this procedure it was possible to obtain high-loading Rasta Merrifield resin (5.8 mmol g^{-1}). It was stressed that the microwave procedure was 150 times than that described in literature performed under conventional conditions.

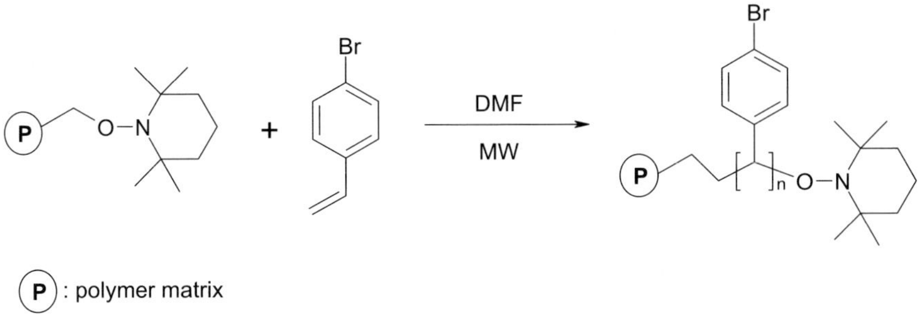

14.2 Synthesis of Polymers Under the Action of Microwave Irradiation

Scheme 14.11

Ⓟ : polymer matrix

14.2.1.3 Ring-opening Polymerization

Ring-opening polymerization of ε-caprolactone under microwave irradiation conditions has been reported by Liu et al. [28]. The reactions were performed in the presence of Sn(Oct)$_2$ and zinc powder as catalysts. Typically, a reaction mixture consisting of ε-caprolactone and the catalyst in a vacuum-sealed ampoule was irradiated in a multimode microwave oven at different temperatures ranging from 80 to 210 °C. For example, poly(ε-caprolactone) with a weight-average molecular weight of 124 000 g mol^{-1} was obtained in 90% yield after 30 min irradiation at 680 W using 0.1% mol/mol Sn(Oct)$_2$ whereas polymerization catalyzed by 1% mol/mol zinc powder afforded poly(ε-caprolactone) with a weight-average molecular weight of 92 300 g mol^{-1} after 30 min irradiation at 680 W. Without microwave irradiation the rate of polymerization was much slower – at 120 °C with Sn(Oct)$_2$ as catalyst poly(ε-caprolactone) was afforded with a weight-average molecular weight of 60 000 g mol^{-1} after 24 h and with zinc powder 27 000 g mol^{-1} poly(ε-caprolactone) was obtained after 48 h (Scheme 14.12).

A similar procedure has been used for the metal-free synthesis of poly(ε-caprolactone) from ε-caprolactone in the presence of benzoic acid [29]. The molar ratios of ε-caprolactone to benzoic acid were in the range 5 to 25, and the reaction mixtures were heated in the so called "self-regulated" temperature range of 204 to 240 °C in a multi-mode microwave reactor. The advantage of the microwave procedure is enhancement of propagation rate; above 240 °C, however, degradation of poly(ε-caprolactone) became significant. With the metal-free method the weight-average molecular weight was ca. 40 000 g mol^{-1}.

In their next paper [30] Lu et al. reported the ring-opening polymerization of D,L-lactide in the presence of Sn(Oct)$_2$ under the action of microwave irradiation

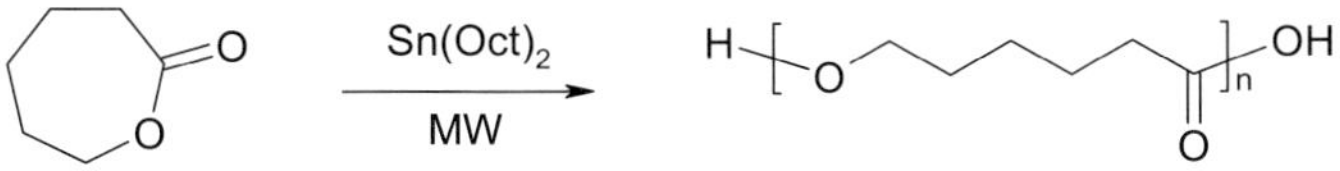

Scheme 14.12

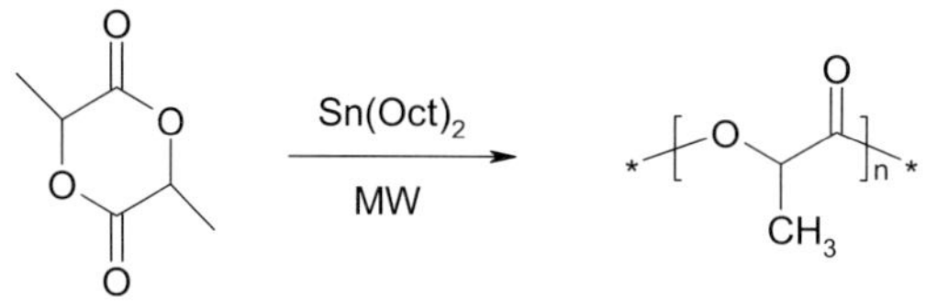

Scheme 14.13

(Scheme 14.13). The procedure was similar to that in their two previous papers. It was stated that polymerization of D,L-lactide proceeded quickly, but no comparison with a conventional procedure was performed. Under optimum conditions, poly(D,L-lactide) (weight-average molecular weight 400 000 g mol^{-1}) was obtained with 90% yield after 10 min. It was stated, however, that degradation of the product was much affected by microwave irradiation.

Ring-opening polymerization of ε-caprolactone in the presence of lanthanide halide catalysts under microwave irradiation conditions was also recently described by Barbier-Baudry et al. [31]. In contrast with the previous report the reactions were performed in open vessels in a single-mode microwave reactor. For this purpose, 1 mL ε-caprolactone was mixed with 2 to 50 mg catalyst and irradiated in a Pyrex tube at 200–230 °C. The highest number-average molecular weight polymers were obtained when the mixture of the monomer and catalyst was intensely irradiated in a microwave reactor so that the boiling point of ε-caprolactone was reached in ca. 1 min. The number-average molecular weights of the polymers obtained were between 3000 and 16 000 g mol^{-1}. Compared with conventional thermal processes, under microwave conditions the polymers had higher molecular weights and lower polydispersity indexes. For completion of the reactions under thermal heating conditions longer polymerization times were necessary.

More recently, oxazoline derivatives became the next group of cyclic monomers to be studied by use of ring-opening polymerization reactions under microwave irradiation conditions. Ritter et al. [32] investigated the ring-opening polymerization of 2-phenyl-2-oxazoline (Scheme 14.14). For this purpose, 7.1 mL 2-phenyl-2-oxazoline was mixed with 7.1 mL acetonitrile and 0.82 mmol methyl tosylate. This mixture (2.5 mL) was transferred to a 10-mL vial sealed with septum and irradiated in a single-mode microwave reactor for 30 to 150 min at 125 °C, with monitoring by use of a fiber-optic sensor. Comparison with thermal heating experiments revealed great enhancement of reaction rates while conserving the living

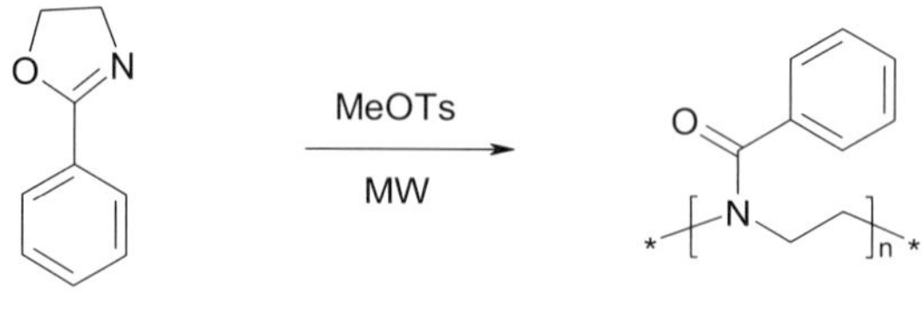

Scheme 14.14

character of the polymerization. For example, under the action of microwave irradiation for 90 min conversion of the monomer was nearly quantitative. In contrast, polymerization under conventional conditions resulted in 71% conversion only after 90 min. Interestingly, the reaction rate coefficient under conventional conditions was the same for reactions performed in open and closed vessels (1.1×10^{-2} min^{-1}) whereas for the microwave experiments the reaction rate coefficient was different for the reaction in open and closed reaction vessels (3.6×10^{-2} and 4.2×10^{-2} min^{-1}, respectively).

Ring-opening polymerization reactions of several 2-substituted-2-oxazolines (i.e. 2-methyl, 2-ethyl, 2-nonyl, and 2-phenyl) in the presence of methyl tosylate as catalyst have been described by Schubert et al. (Scheme 14.14) [33–35]. The reactions were performed in the temperature range 80 to 200 °C inside a single-mode microwave reactor. In a typical run, 25 mL stock solutions of monomer–initiator–solvent were prepared before the polymerization. These stock solutions were divided among different reaction vials so each experiment was performed on a 1-mL scale. It was found that when the reaction rate was enhanced by factors of up to 400, by changing from 80 to 200 °C, activation energies for the polymerization (E_A, 73 to 84 kJ mol^{-1}) were within the range of values obtained when conventional heating was used. The first-order kinetics of monomer conversion and living character of the polymerization were maintained. The polymerization can also be conducted in concentrated solutions or even under bulk conditions to afford well-defined monomers ($PDI < 1.20$). Under such conditions a maximum of 300 monomers can be incorporated into the polymer chains.

The same technique of the ring-opening polymerization under microwave irradiation conditions was subsequently applied to the synthesis of a library of diblock copoly(2-oxazoline)s in which a total number of 100 (50 + 50) monomer units were incorporated into the polymer chains [36]. As a result, 16 polymers were obtained with narrow polydispersity indexes ($PDI < 1.30$). The reactions were initiated by methyl tosylate and conducted in acetonitrile solution at 140 °C. After polymerization of the first monomer the reaction vessels were re-transferred to an inert atmosphere of argon, the second monomer was added, and the reaction mixture was again irradiated in a microwave reactor.

14.2.2
Step-growth Polymerization

14.2.2.1 Epoxy Resins

Microwave-assisted curing of epoxy resin systems was one of the first applications of MW in polymer chemistry and is the most widely studied area in polymer chemistry under both continuous and pulse microwave conditions. The structure, dielectric properties, toughness, mechanical strength, percentage of cure, and glass transition temperature of the epoxy formulations have been investigated [1].

It was found that the pulse method led to the fastest heating of the resins [38] and improved their mechanical properties [39]. It was, for example, shown that computer-controlled pulsed microwave processing of epoxy systems consisting of

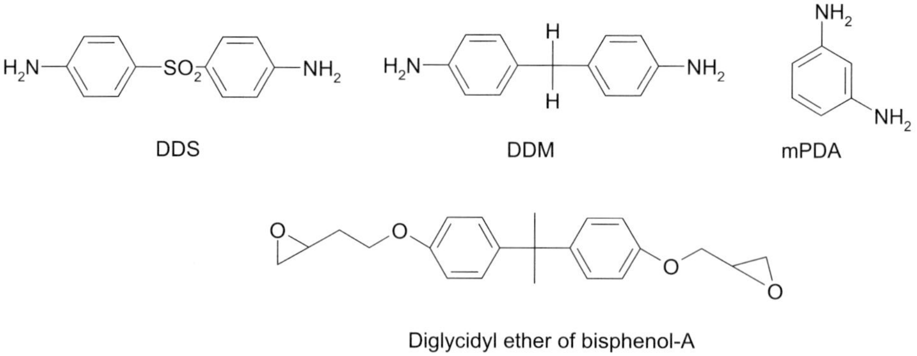

Scheme 14.15

diglycidyl ether of bisphenol-A (DER 332) and 4,4'-diaminodiphenyl sulfone (DDS) (Scheme 14.15) in a cavity operating in TM_{012} mode could be successfully used to eliminate the exothermic temperature peak and maintain the same cure temperature at the end of the reaction [40]. The epoxy systems were cured more rapidly under pulsed microwave irradiation and it was possible to cure them at higher temperatures than when continuous microwaves or conventional thermal processing were used.

Microwave curing processes of epoxy–amine systems were recently studied by Boey et al. [41], who investigated the effect of using different curing agents on the final cured glass-transition temperature (T_g). Microwave irradiation and thermal heating were performed on a diglycidyl ether of bisphenol-A (DGEBA) with three curing agents – 4,4'-diaminodiphenyl sulfone (DDS), 4,4'-diaminodiphenylmethane (DDM), and m-phenylenediamine (mPDA) (Scheme 14.15). Microwave curing was conducted in a multimode cavity powered by a variable-power generator up to 1.26 kW at 2.45 GHz.

For all three systems (DDS, DDM, mPDA) a shorter curing time was needed to reach the maximum percentage cure and T_g was substantially lower than for thermal curing. It was found that during microwave curing, although the rate was faster, the presence of highly electron-attracting groups (for example SO_2 in DDS) seemed to induce a delay in the reactivity of the amine functions and effectively inhibit further curing. In contrast, maximum percentage cure was achieved for both DDM and mPDA systems and T_g values were equal to those for thermal curing after substantially shorter curing times, because of greater reduction in the effective cure time than in the lag time. Microwave curing was more effective at reducing the overall cure time for the mPDA system, which suggested that microwave curing was more effective in enhancing the reaction rates during crosslinking rather than shortening the lag time.

A similar investigation by Zhou et al. [42] on the microwave-assisted curing of the diglycidyl ether of bisphenol-A (DGEBA) (Scheme 14.15) with maleic anhydride showed that in comparison with conventional thermal cure microwave cure can reduce curing temperature by 15–20 °C, increase the compressive strength and bending strength of the epoxy resin, and reduce the amount of maleic anhydride needed by 5%.

Microwave-assisted synthesis of elevated molecular weight epoxy resins (solid epoxy resins) has been described by Bogdal et al. [43, 44]. The method is based on the polyaddition reaction of bisphenol-A to a low-molecular-weight epoxy resin or the diglycidyl ether of bisphenol-A (DGEBA) in the presence of 2-methylimidazole (2-MI) as catalyst (Scheme 14.16). The syntheses were also performed using conventional thermal heating for comparison of the properties of the epoxy resins obtained under both conditions. The microwave reactions were performed in a 250-mL round-bottomed flask charged with 11.6 g bisphenol-A, 25 g low-molecular-weight epoxy resin (epoxy value 0.57 mol/100 g), and a small amount of a catalyst. For microwave experiments the flask was irradiated in a multimode microwave reactor at a frequency of 2.45 GHz and maximum power of 600 W. The main advantage of the microwave process is twofold reduction of reaction time in comparison with conventional conditions. Number-average molecular weight, weight-average molecular weight, polydispersity index, epoxy value, and degree of branching of the resins were determined for both microwave and conventional conditions. It was found that the molecular weight distribution and degree of branching of the epoxy resins synthesized under microwave irradiation were comparable with those obtained under conventional heating conditions and were not affected by reduction of the reaction time.

The same approach was used by Brzozowski et al. [45] for synthesis of elevated-molecular-weight (E-M) epoxy resins (solid epoxy resins) with reduced flammabil-

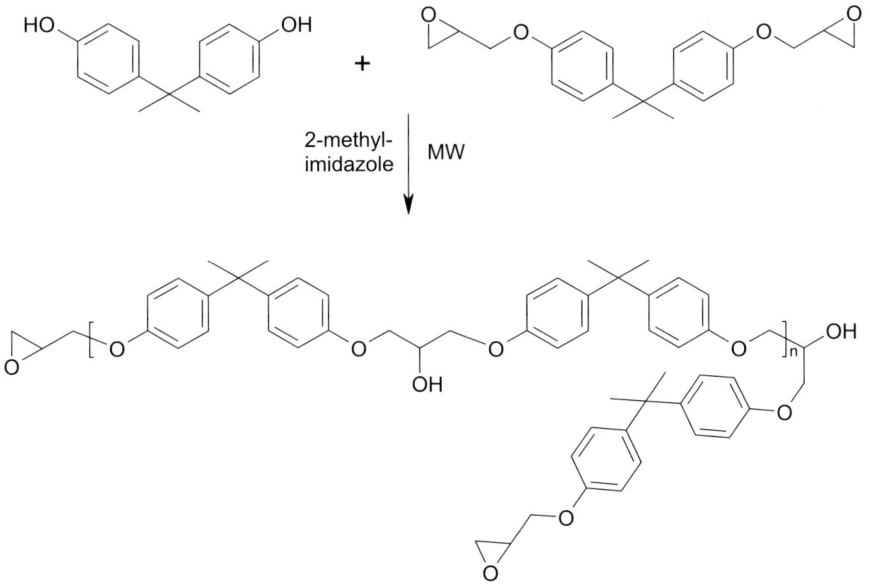

Scheme 14.16

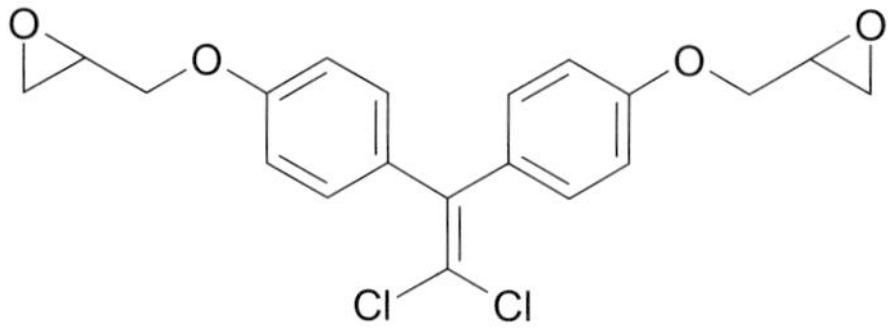

Scheme 14.17

ity. For this purpose, bisphenol-A was either substituted or partially substituted with 1,1-dichloro-2,2-bis(4-hydroxyphenyl)ethylene (Scheme 14.17).

Polymerization of epoxides (3,4-epoxycyclohexylmethyl, 3,4-epoxycyclohexylcarboxylate) initiated by diaryliodonium or triarylsulfonium salts under microwave conditions was investigated by Stoffer et al. (Scheme 14.18) [46]. The reactions were conducted in 20 mL vials in which 4 g samples were placed and irradiated in a multimode microwave reactor. Temperature during microwave polymerizations was measured with a thermocouple immediately after the reaction vessel was removed from the microwave reactor. The extent of polymerization was determined by means of DSC and FTIR and compared with that for samples cured under conventional conditions.

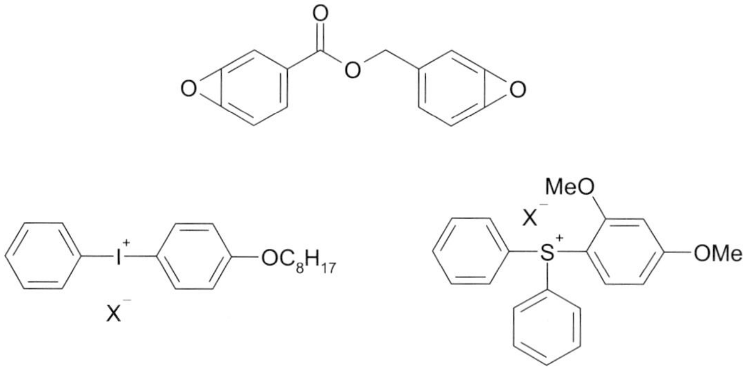

Scheme 14.18

This study revealed that under microwave conditions polymerization phenomena such as polymerization selectivity, polymerization temperature shift, and polymerization temperature shift as a result of the microwave power setting, can be observed when products are compared with those obtained under conventional conditions. To explain these phenomena it was proposed that a new dipole partition function is present in the microwave field, so values of thermodynamic properties such as internal energy and Gibbs free energy of materials with permanent dipole moments change under microwave conditions, which in turn leads to shifts in the reaction equilibrium and kinetics compared with conventional conditions at the same temperature [46].

14.2.2.2 Polyethers and Polyesters

The synthesis of linear polyethers from either isosorbide or isoidide and alkyl dibromides or dimethanesulfonates using microwave irradiation under solid–liquid PTC conditions was described by Loupy et al. (Scheme 14.19) [47, 48].

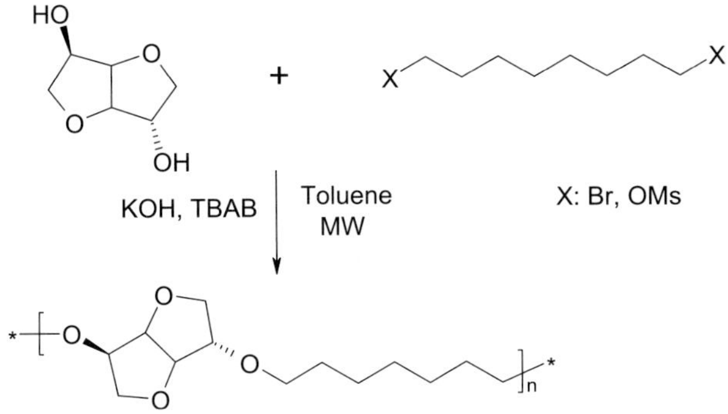

Scheme 14.19

The reactions were performed in a single-mode microwave reactor with an infrared temperature detector previously calibrated by use of an optical fiber detector introduced into a reaction mixture. Reaction mixtures consisting of 5 mmol isosorbide or isoidide, 5 mmol alkyl dibromide–dimesylate, 1.25 mmol tetrabutylammonium bromide (TBAB), and 12.5 mmol powdered KOH were irradiated for 30 min to afford the polyethers in 70–90% yield. It was found that use of a small amount of solvent was necessary to ensure good temperature control and to reduce the viscosity of the reaction medium. With isosorbide the microwave-assisted synthesis proceeded more rapidly compared with conventional heating; reaction time was reduced to 30 min and yields of 69–78% were obtained. Under conventional conditions, the polyethers were afforded in 28–30% yield within 30 min. Similar yields of the polyethers were obtained while the reaction time was extended to 24 h. They remained practically unchanged even though the synthesis was conducted for another 7 days. Analysis of properties of the synthesized polyethers revealed that the structure of the products was strictly dependent on the mode of activation (microwave or conventional activation). Under microwave conditions the polyethers were characterized by higher molecular weights and better homogeneity. For example, after 30 min reaction under conventional heating conditions polyesters with higher molecular weight were not observed. It was also found that the chain terminations were different under microwave and conventional conditions. Polyesters prepared by use of conventional heating had shorter chains with terminal hydroxyl ends whereas under microwave irradiation the polymer chains were longer with terminal ethylenic ends. Under the action of microwave irradiation terminal ethylenic ends were formed rapidly and hindered further polymer growth. In contrast, under conventional conditions terminations were essentially by hydroxyl functions; again, further polymerization was terminated.

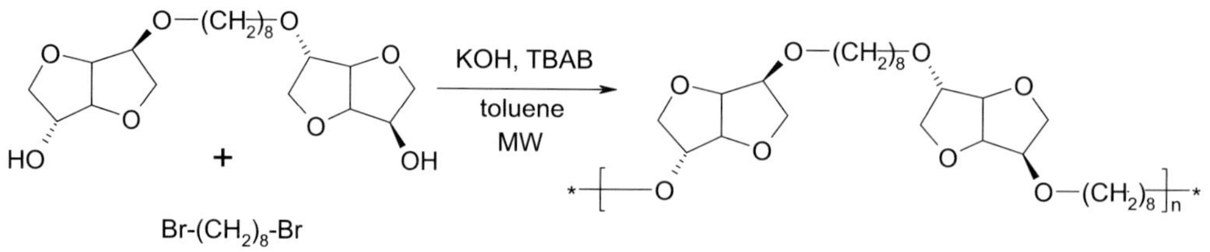

Scheme 14.20

Later, the same procedure was applied to the polycondensation of aliphatic diols of isosorbide with 1,8-dimesyloctane and other dialkylating agents (Scheme 14.20) [49]. It was always found that microwave-assisted polycondensations proceeded more efficiently compared with conventional heating (the reaction time was reduced from 24 h to 30 min – ratio 1:50). Polycondensation under microwave conditions yielded 63% polyethers with relatively high average-weight molecular weights (weight-average molecular weight up to 7000 g mol^{-1}). The polyethers were characterized by ^{1}H and ^{13}C NMR and FTIR spectroscopy, SEC measurement, and MALDI-TOF mass spectrometry.

It has been shown that application of previously synthesized ethers of isosorbide was beneficial and enabled the preparation of polyethers in better yields than the polyethers obtained in direct reactions of isosorbide and dibromo or dimesyl alkanes [47, 48]. Moreover, the molecular weights of the polyethers were higher than of those prepared in the earlier work [47] and the molecular weight distributions of new polyethers were similar or lower. Such a microwave-assisted procedure can contribute to the synthesis of alternating polyethers and further modification of their properties.

The synthesis of poly(ether imide)s by condensation of the disodium salt of bisphenol-A with bis(chlorophthalimide)s under microwave irradiation conditions has been described by Zhang et al. (Scheme 14.21) [50]. The polymerization reactions were performed under phase-transfer catalysis (PTC) conditions in o-dichlorobenzene solution. For this purpose a mixture of 16.12 mmol bis(chlorophthalimide)s and 16.12 mmol disodium salt of bisphenol-A in 60 mL o-dichlorobenzene with 0.56 mmol hexaethylguanidinium bromide was irradiated in a domestic microwave oven for 25 min and the product was precipitated by addition of methanol. The polymerization reactions, in comparison with those under the action of conventional heating, proceeded rapidly (25 min compared with 4 h at 200 °C) and polymers with inherent viscosities in the range 0.55 to 0.90 dL g^{-1} were obtained.

Bogdal et al. have obtained unsaturated polyesters in polyaddition reactions of alkylene oxides such as epichlorohydrin and acid anhydrides (maleic and phthalic anhydrides) in the presence of lithium chloride as catalyst under microwave irradiation conditions (Scheme 14.22) [51, 52]. In the standard procedure, a mixture of 0.10 mol phthalic anhydride and 0.10 mol maleic anhydride with 0.010 mol ethylene glycol, 0.20 mol epichlorohydrin, lithium chloride (0.1% w/w) was placed in a

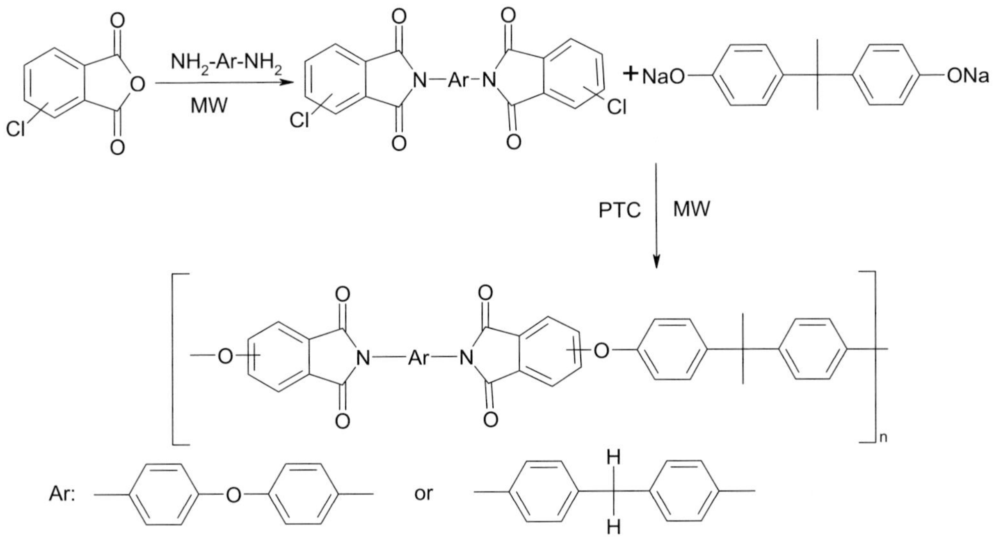

14.2 Synthesis of Polymers Under the Action of Microwave Irradiation

Scheme 14.21

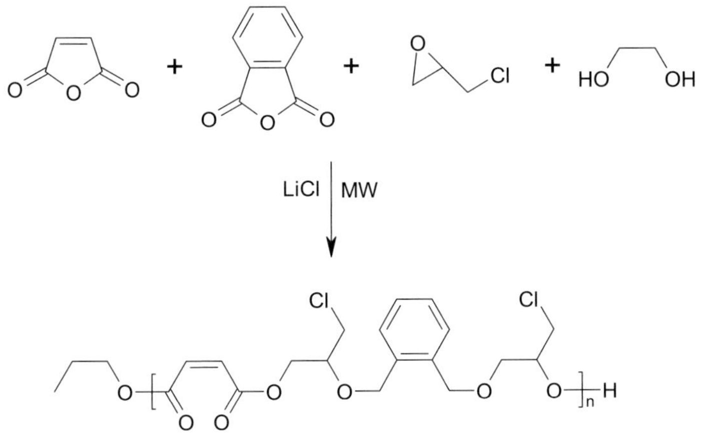

Scheme 14.22

three-necked round-bottomed flask and irradiated in a multimode microwave reactor equipped with an IR temperature sensor, a magnetic stirrer, and an upright condenser. The reaction temperature was maintained in the range 120–140 °C under an inert atmosphere. Polymerization was continued until the acid number value of polyesters dropped below 50 mg KOH g^{-1}. Compared with polycondensation reactions of acid anhydrides with diols, these reactions proceeded without release of by-products. At the same time the polyaddition reactions were performed under conventional thermal conditions, applying a similar set of reaction conditions. In comparison with these experiments, twofold reduction of reaction times was observed under microwave conditions whereas other properties, for example number-average molecular weight and polydispersity index, remained comparable.

Polycondensation of acid anhydrides (maleic and phthalic anhydrides) with diols (e.g. ethylene glycol) under microwave irradiation conditions has also been described for synthesis of unsaturated polyesters [52]. In addition to the previous procedure, the reaction temperature was increased to 200 °C and a Dean–Stark trap was used to remove water from the reaction mixture (Scheme 14.23). It was found that reaction times for the microwave and conventional procedures were comparable and depended on the rate of removal of water from the reaction system.

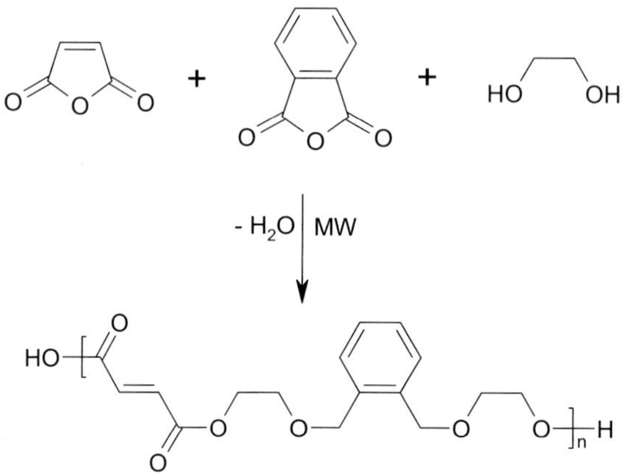

Scheme 14.23

Poly(ester imide)s have been synthesized by Mallakpour et al. [53] via a route involving reaction of pyromellitic anhydride with *l*-leucine, then conversion of the resulting diacid into its diacid chloride, which in turn reacted with several diols (for example phenolphthalein, bisphenol-A, and 4,4′-hydroquinone) under microwave irradiation conditions (Scheme 14.24). The polymerization reactions were conducted in 10 min in a domestic microwave oven in a porcelain dish in which

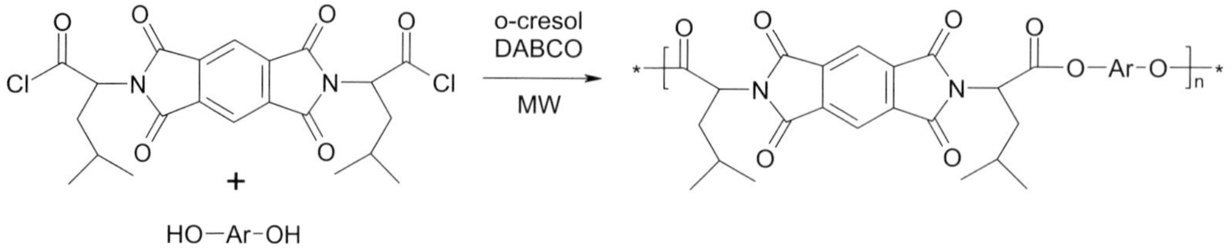

Scheme 14.24

0.10 g diacid chloride was mixed with an equimolar amount of diol in the presence of small amounts of o-cresol and DABCO as catalysts. A series of optically active poly(ester imide)s were obtained in good yields and with moderate inherent viscosities of 0.10–0.27 dL g^{-1}.

Mallapragada et al. have prepared polyanhydrides, biodegradable polymers, from aliphatic and aromatic diacids under the action of microwave irradiation (Scheme 14.25) [54]. The reactions were performed in a household microwave oven. For this purpose, 0.49 mmol aliphatic or aromatic acid were mixed with 2.95 mmol acetic anhydride and irradiated at full power in a sealed borosilicate vial for 2 min. The anhydride was then removed by evaporation and the vial was irradiated for another 5 to 25 min. It was found that by use of this method it was possible to obtain polymers with number-average molecular weights (1700 to 11 300 g mol^{-1}) rather similar to those obtained under conventional conditions while reducing the reaction time from hours to 6–20 min. It was also possible to prepare copolymers of sebacinic acid prepolymer and 1,6-bis-(p-carboxyphenoxy)hexane.

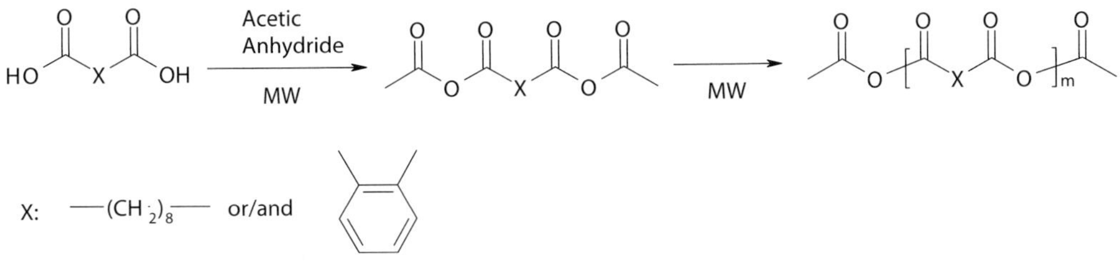

Scheme 14.25

14.2.2.3 Polyamides and Polyimides

Polyamides have been synthesized under microwave conditions from both ω-amino acids and Nylon-salt-type monomers, and polyimides have been obtained from salt monomers comprising aliphatic diamines and pyromellitic acid or its diethyl ester or derivatives of pyromellitic acid chlorides and aromatic diamines in the presence of a small amount of an organic solvent [1].

Synthesis, under microwave irradiation conditions, of polyamides containing azobenzene units and hydantoin derivatives in the main chains has recently been proposed by Faghihi et al. [55]. Polycondensation of 4,4'-azodibenzoyl chloride with eight 5,5-disubstituted hydantoin moieties has been achieved in the presence of a small amount of o-cresol (Scheme 14.26). The polycondensations were performed in 8 min, in a domestic microwave oven, in a porcelain dish in which 1.0 mmol diacid chloride was mixed with an equimolar amount of diol in the presence of small amounts of o-cresol. The polymerization proceeded rapidly, compared with the bulk reactions under conventional conditions (8 min compared with 1 h), producing a series of polyamides in high yield and inherent viscosity between 0.35 to 0.60 dL g^{-1}.

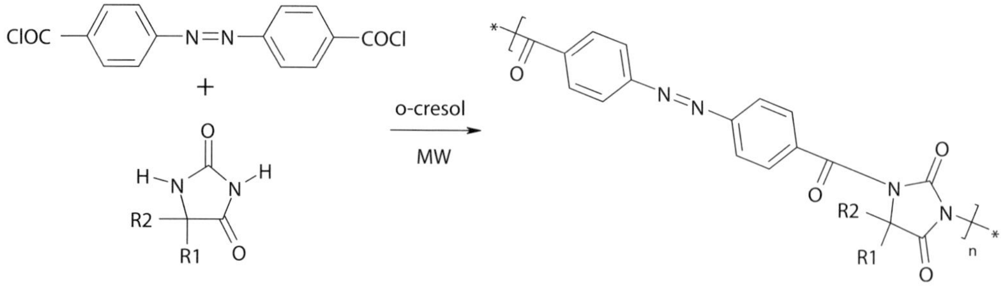

Scheme 14.26

Polyimides with third-order NLO properties have been prepared by Lu et al. [56] by polycondensation pyromellitic dianhydride with either benzoguanamine or 3,3'-diaminobenzophenone under microwave irradiation conditions (Scheme 14.27). The polymers obtained under microwave conditions were characterized by large third-order nonlinearities and time response.

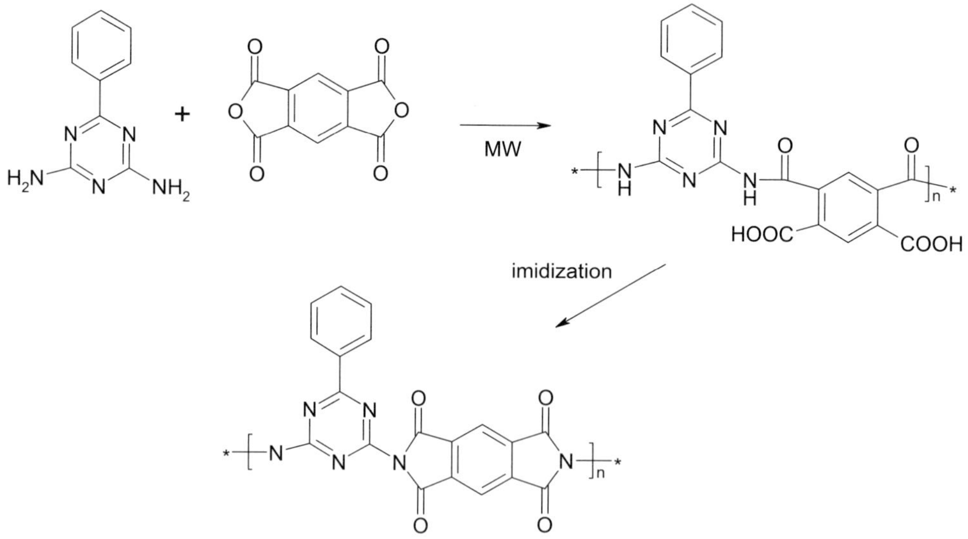

Scheme 14.27

A series of poly(amide–imide)s have been obtained by polycondensation of hydantoin and thiohydantoin derivatives of pyromellitic acid chlorides with [N,N'-(4,4'-carbonyldiphthaloyl)] bisalanine diacid chloride [57], N,N'-(pyromellitoyl)-bis-l-phenylalanine diacid chloride [58], and N,N'-(4,4'-diphenyl ether) bistrimellitide

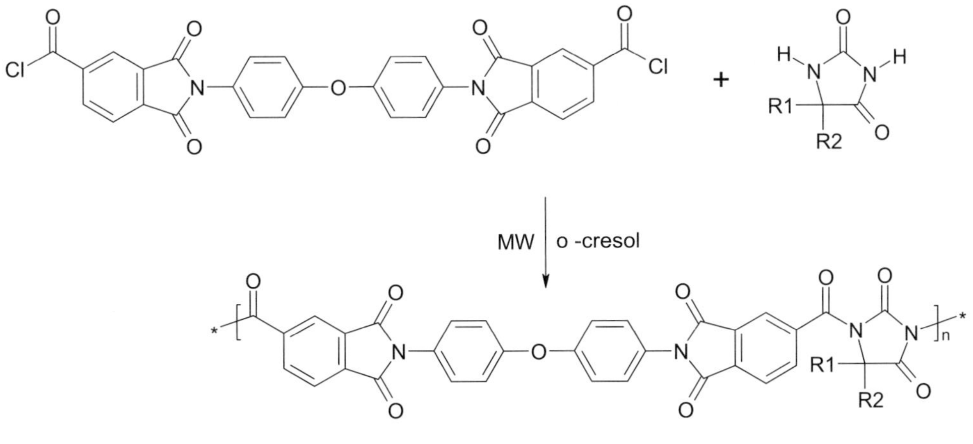

Scheme 14.28

diacid chloride [59] under the action of microwave irradiation (Scheme 14.28). Typically, 1.0 mmol diacid chloride was mixed with an equimolar amount of the hydantoin derivative in the presence of 1 mL o-cresol and the reactions were performed in a domestic microwave oven for 10 min without temperature control. The resulting poly(amide–imide)s were obtained in good yields with inherent viscosities ca. 0.28 to 0.66 dL g^{-1}.

A similar series of poly(amide-imide)s has been obtained by Mallakpour et al. by polycondensation of several diacid chlorides, for example [N,N'-(4,4'-carbonyldiphthaloyl)]-bis-isoleucine diacid chloride [60], N,N'-(4,4'-carbonyldiphthaloyl)-bis-l-phenylalanine)diacid chloride [61], and N,N'-(4,4'-sulfonediphthaloyl)-bis-l-phenylalanine) diacid chloride [62] with aromatic amines (Scheme 14.29) and with diacid chlorides obtained by reaction of Epiclon B-4400 and phenylalanine [63] or reaction of l-leucine [64] with aromatic amines (Scheme 14.30).

The reactions were performed in a domestic microwave oven that was used without any modification and temperature control. Prior microwave irradiation, 0.1 g diacid chloride was ground with an equimolar amount of an aromatic amine or diphenol and a small amount of a polar high-boiling solvent (o-cresol or NMP, 0.05

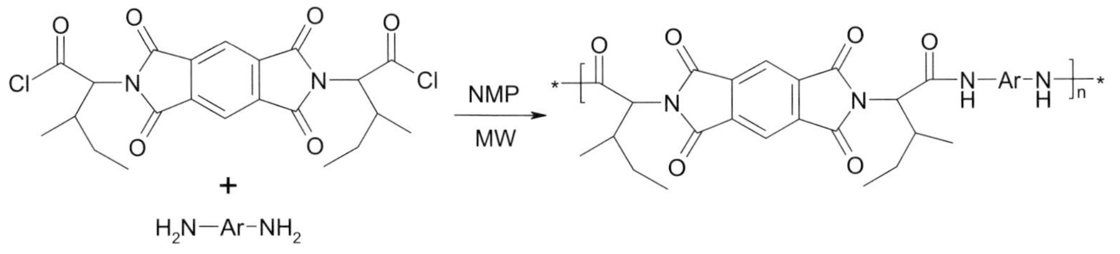

Scheme 14.29

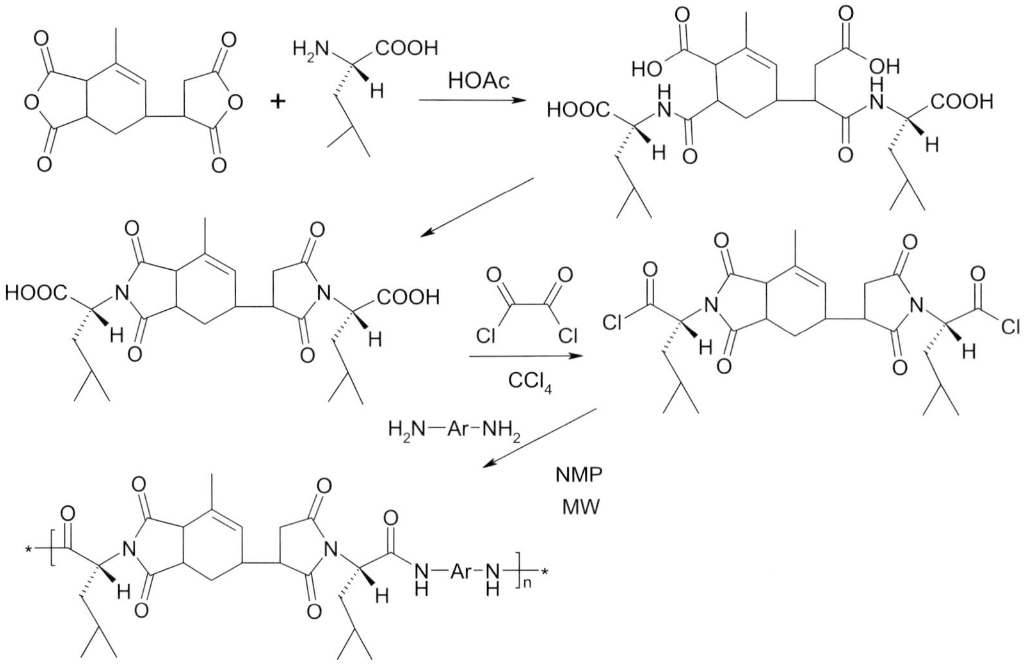

Scheme 14.30

to 0.45 mL), which acted as a primary microwave absorber. Under microwave irradiation conditions the polycondensation reactions proceeded rapidly (6–12 min) compared with conventional solution polymerization (reflux for 5–10 h in chloroform or NMP) to give a series of optically active polymers with inherent viscosities in the range 0.12–0.52 dL g^{-1}.

Lu et al. have obtained poly(amic acid) side-chain polymers by polycondensation of benzoguanamine and pyromellitic dianhydride under microwave irradiation conditions [65–67]. The reactions were performed in a household microwave oven in which 100 mL DMF solution of 33 mmol benzoguanamine and an equimolar amount of pyromellitic dianhydride were stirred and irradiated for 1 h at 60 °C (Scheme 14.31). The resulting poly(amic acid) was precipitated from the solution and then modified to obtain side-chain polymers with fluorescent and third-order NLO properties.

Polyureas and polythioureas have been synthesized by Banihashemi et al. [68] by reaction of aromatic and aliphatic amines with urea and thiourea, respectively (Scheme 14.32). In a typical procedure, a solution of 10 mmol amine, 10 mmol urea, and a small amount of p-toluenesulfonic acid (1 mmol) in 5 mL N,N-dimethylacetamide was irradiated for 7 min at 220 W and then for 8 min at 400 W in a tall beaker placed in a household microwave oven. As the result, a series of polyureas and polythioureas was obtained in good yields and with moderate inherent viscosities of 0.13–0.25 dL g^{-1}.

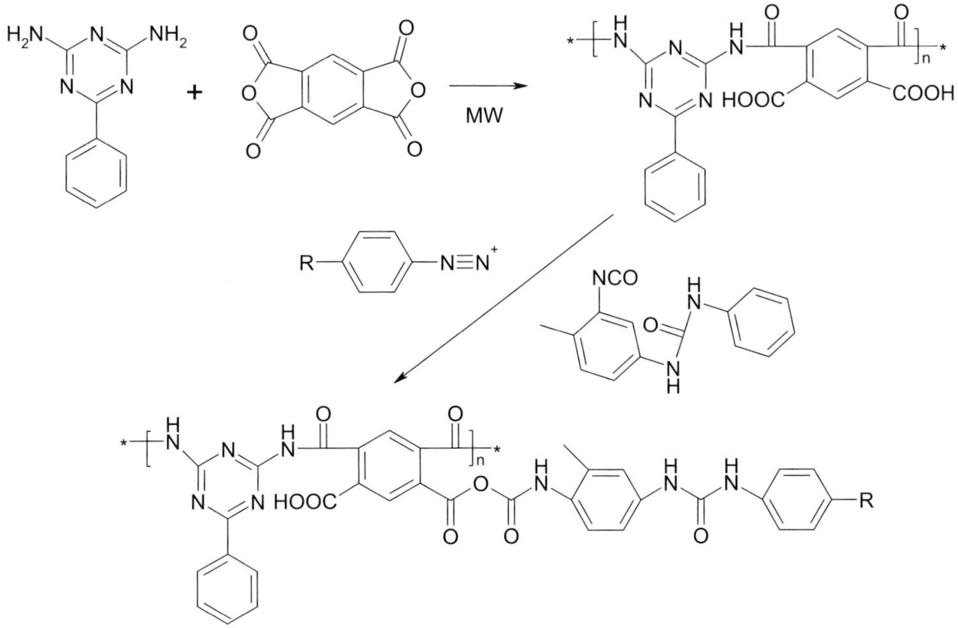

Scheme 14.31

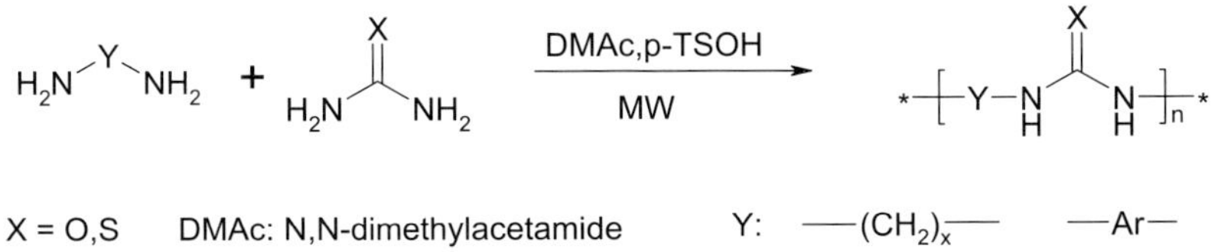

Scheme 14.32

X = O,S   DMAc: N,N-dimethylacetamide   Y: —(CH$_2$)$_x$—   —Ar—

Synthesis of poly(aspartic acid) from maleic acid derivatives and from aspartic acid under microwave irradiation conditions has been described by Pielichowski et al. (Scheme 14.33) [69, 70]. The reactions were performed in propylene glycol solutions in the temperature range 160 to 230 °C in a multimode microwave reactor. Poly(aspartic acid) with number-average molecular weights of 6150 to 18 500 g mol^{-1} was obtained in good yield (50 to 85%).

14.2.3
Miscellaneous Polymers

Carter prepared a polyarylene-type polymer by polymerization of 2,7-dibromo-9,9-dihexylfluorene under the action of microwave irradiation (Scheme 14.34) [71]. In a typical procedure, a catalyst stock solution was prepared consisting of 704 mg

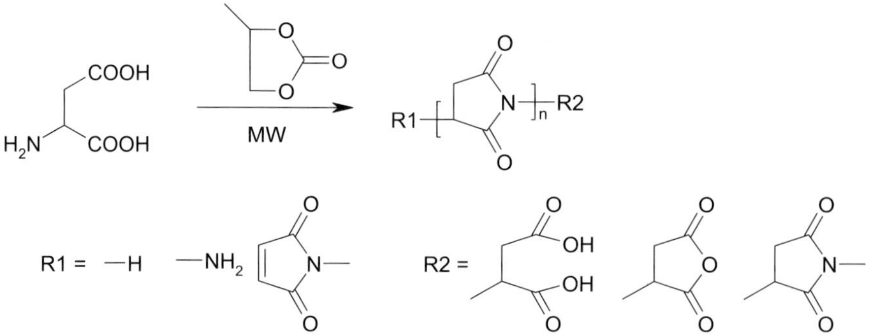

Scheme 14.33

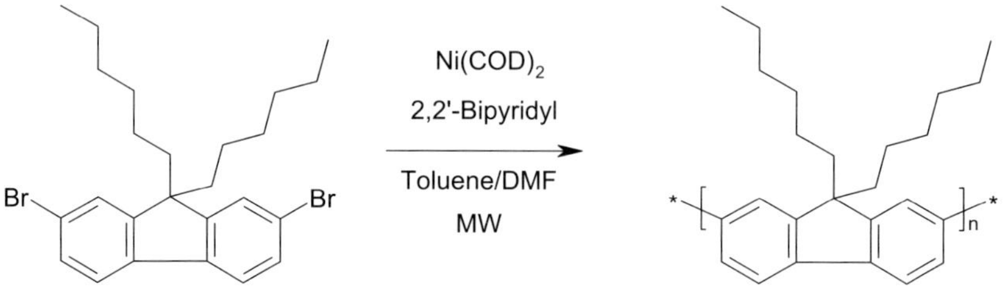

Scheme 14.34

bis(1,5-cyclooctadiene)nickel(0), 410 mg 2,2′-bipyridine, and 281 mg cyclooctadiene in 17.5 mL toluene–DMF (1.15:1). A monomer solution containing 600 mg 2,7-dibromo-9,9-dihexylfluorene in 17.3 mL toluene was then prepared. Polymerization was accomplished by charging a 10-mL reaction vial with 2.2 g the catalyst solution and 1.77 g monomer solution. The vial was sealed and irradiated in single-mode microwave up to 250 °C for 10 min. Eventually, polymers of number-average molecular weight in the range 5000 to 10 000 g mol^{-1} with polydispersity index between 1.65 to 2.22 were obtained.

Synthesis of conjugated p-phenylene ladder polymers by means of a microwave-assisted reaction has been achieved by Scherf et al. (Scheme 14.35) [72]. The polymerization reactions were performed in THF solution at 130 °C in the presence of palladium catalyst with phosphine ligands with irradiation in a single-mode microwave reactor for 11 min. Compared with conventional thermal procedures, the reaction time was reduced from days to a couple of minutes and molecular weight distributions (*PDI* ca 1.8) of the polymers were changed substantially.

Thiophene oligomers (up to six units) have been obtained by Barbarella et al. [73] under microwave conditions from 2-thiophene boronic acid and dibromo precursors with three thiophene units in the presence of a palladium catalyst and KF with KOH (Scheme 14.36). The reactions were run in a single-mode microwave reactor at 70 °C for 10 min.

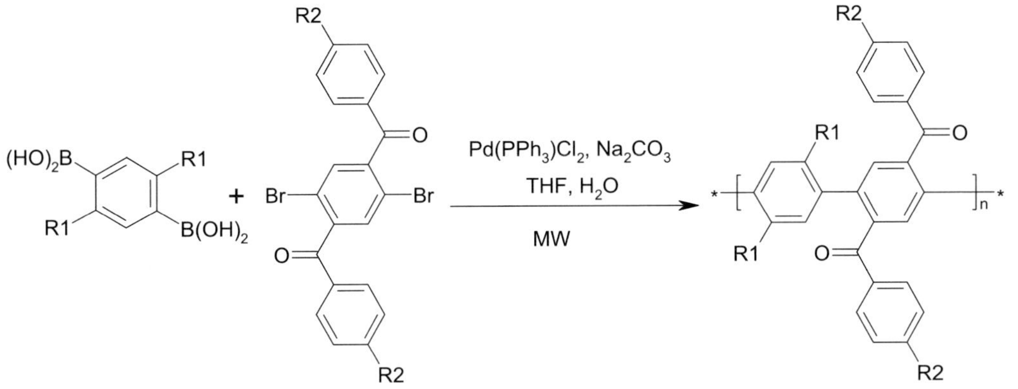

14.2 Synthesis of Polymers Under the Action of Microwave Irradiation | 677

Scheme 14.35

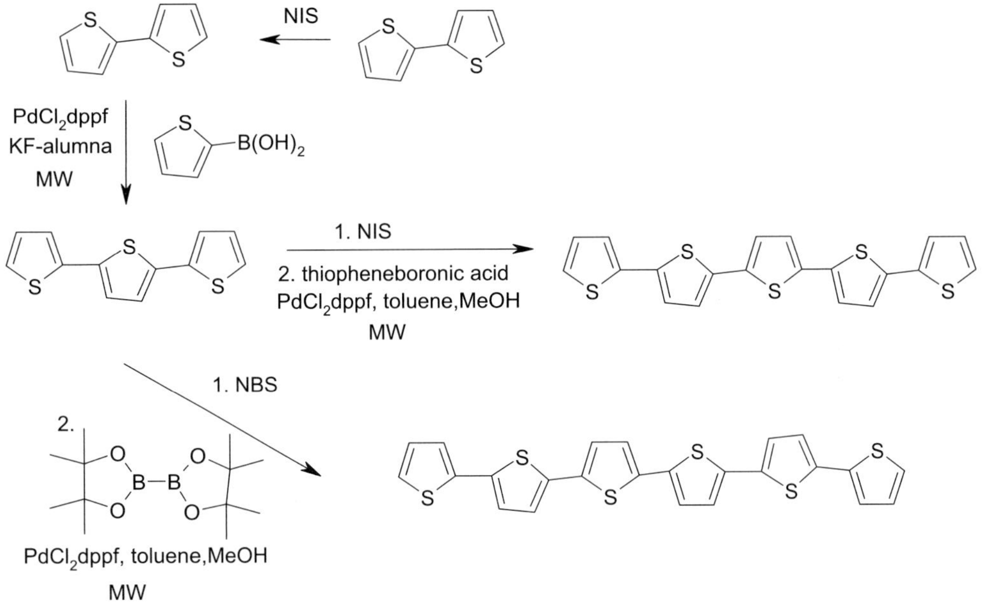

Scheme 14.36

Poly(pyrazine-2,5-diyl) has been prepared by Yamamoto et al. [74] by organometallic dehalogenative polycondensation of 2,5-dibromopyrazine (Scheme 14.37). The reaction was performed by irradiating a mixture of 2.62 mmol 2,5-dibromopyrazine, 5.23 mmol bis(1,5-cyclopentadiene)nickel(0), and 5.23 mmol 2,2′-bipyridyl in a single-mode microwave reactor for 10 min, either in toluene or DMF solution. Under microwave conditions, the polymer was afforded in 83–95%

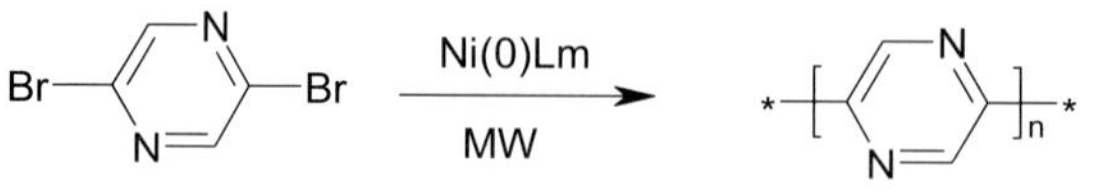

Scheme 14.37

yield; under conventional conditions at 60 °C a similar yield was obtained after 2 days.

Poly(dichlorophenylene oxide) and a conducting polymer have been simultaneously obtained from 2,4,6-trichlorophenol by Cakmak et al. (Scheme 14.38) [75]. Microwave-initiated polymerization was performed in a Pyrex vessel in which 2.5 g 2,4,6-trichlorophenol was mixed with 0.5 g NaOH and 1–2 mL triple-distilled water. The reaction mixtures were irradiated in a domestic microwave oven for times from 1 to 7 min. The resulting polymers, i.e. poly(dichlorophenylene oxides) and the conducting polymer (0.3 S cm^{-2}), were separated by precipitation from toluene. The optimum conditions for poly(dichlorophenylene oxide) and the conducting polymer were 70 W for 5 min and 100 W for 1 min, respectively.

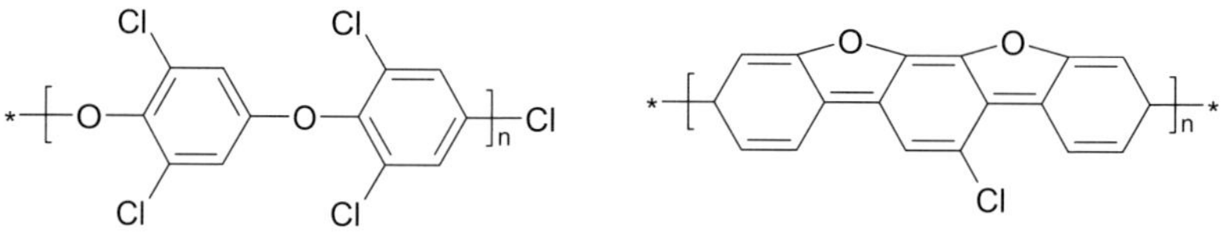

Scheme 14.38

Poly(alkylene hydrogen phosphonate)s have been obtained by Ritter et al. by transesterification of dimethyl hydrogen phosphonate and poly(ethylene glycol) (PEG 400) under microwave irradiation conditions (Scheme 14.39) [76]. The reaction was performed in a round-bottomed vessel, equipped with an upright condenser. A mixture of 52.5 mmol dimethyl hydrogen peroxide and 50 mmol PEG 400 was irradiated in a single-mode microwave reactor for 55 min at 140–190 °C. The temperature was monitored by use of an IR sensor. It was observed that microwave conditions avoid the undesirable thermal degradation of dimethyl hydrogen phosphonate, because of the short reaction times. Microwave (55 min) and conventional (9 h) conditions gave poly(alkene hydrogen phosphonate)s with number-average molecular weights of 3100 and 4900 g mol^{-1}, respectively.

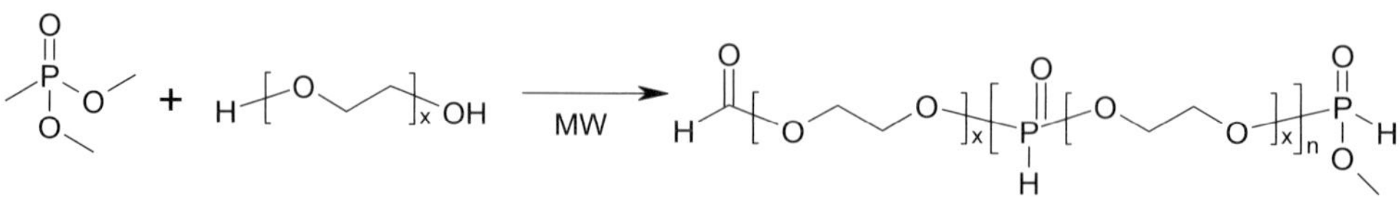

Scheme 14.39

14.2.4
Polymer Modification

Graft polymerization of ε-caprolactone on to chitosan has been achieved by Fang et al. [77]. The reactions run smoothly under the action of microwave irradiation, via a protection–graft–deprotection procedure with phthaloylchitosan as precursor and stannous octoate as catalyst. The copolymerization reactions were performed in a household microwave oven, and chitosan-g-polycaprolactone with high grafting percentage above 100% was achieved. After deprotection, the phthaloyl group was removed and the amino group was regenerated. Thus, the chitosan-g-polycaprolactone copolymer was an amphoteric hybrid with both a large amount of free amino groups and hydrophobic polycaprolactone side chains (Scheme 14.40).

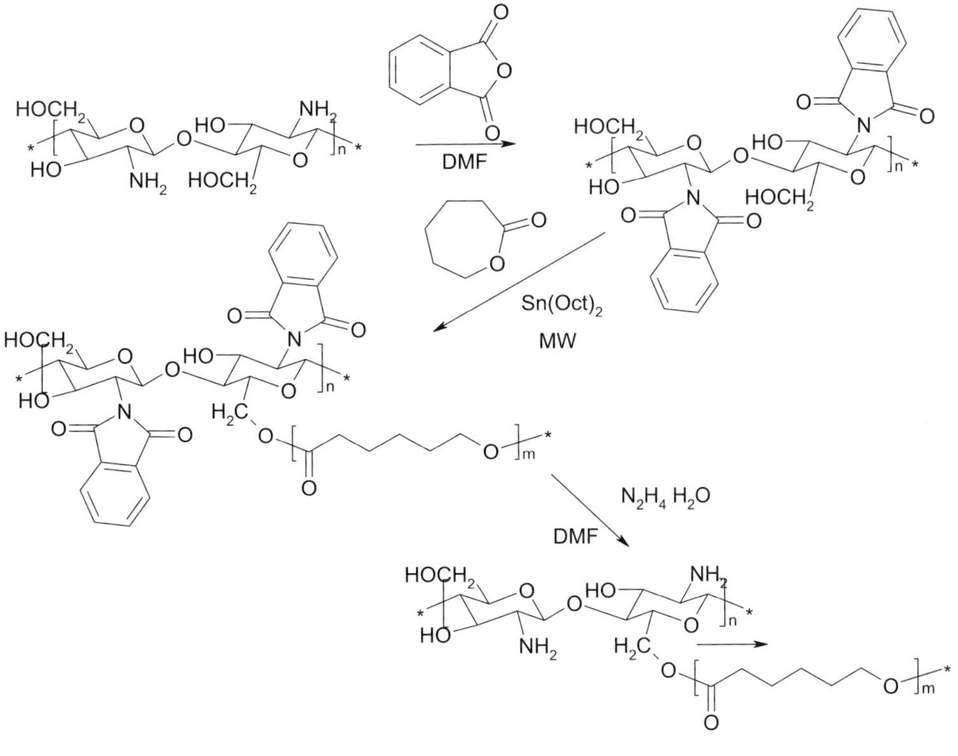

Scheme 14.40

The effect of inorganic salts, for example sodium chloride, on the hydrolysis of chitosan in a microwave field was investigated by Li et al. [78]. The reactions were conducted in a domestic microwave oven. It was found that the molecular weight of the degraded chitosan obtained by microwave irradiation was considerably lower than that obtained by convectional heating.

Using microwave (MW) irradiation, Singh et al. [79] grafted polyacrylonitrile on to chitosan with 170% grafting yield under homogeneous conditions in 1.5 min in the absence of any radical initiator or catalyst. The reactions were conducted in a domestic microwave oven. It was stated that under similar conditions maximum grafting of 105% could be achieved when the $K_2S_2O_8$–ascorbic acid redox system was used as radical initiator in a thermostatic water bath at 35 °C. Unfortunately, the temperature was not recorded in the microwave experiments. Grafting was found to be increased with enhancement in the initial concentration of the monomer in the range $10-28 \times 10^{-2}$ mol L^{-1}. Grafting was also found to increase for MW power up to 80% and then decrease; this may be because of either more homopolymerization or decomposition of grafted copolymer occurring at MW power higher than 80%.

Microwave-assisted synthesis of a guar-g-polyacrylamide (G-g-PAA) has also been reported [80]. The reactions were performed in a domestic microwave oven. Graft copolymerization of the guar gum (GG) with acrylamide (AA) under the action of microwave irradiation in the absence of any radical initiators and catalyst resulted in grafting yields comparable with redox (potassium persulfate–ascorbic acid) initiated by conventional heating but in a very short reaction time. Grafting efficiency up to 20% was further increased when initiators and catalyst were used under microwave irradiation conditions. Maximum grafting efficiency achieved under MW conditions was 66.66% in 0.22 min, compared with 49.12% in 90 min by the conventional method.

Sodium acrylate has been grafted on to corn starch by Tong et al. [81] to furnish a superabsorbent. Potassium persulfate (PPS) was used as the initiator and poly(ethylene glycol) diacrylate as the crosslinker, which in turn was obtained from poly(ethylene glycol) esterified with acrylate. It was found that microwave irradiation substantially accelerated the synthesis without the need to remove O_2 or inhibitor. Microwave power was also believed to be the most significant factor affecting the swelling ratio and solubility of the product. Optimized experimental results showed that microwave irradiation for 10 min at 85–90 W could produce a corn starch-based superabsorbent with a swelling ratio of 520–620 g g^{-1} in distilled water and solubility of 8.5–9.5% (w/w).

Krausz et al. obtained cellulosic plastic films under homogeneous conditions by microwave-induced acylation of commercial or chestnut tree sawdust cellulose by fatty acids (Scheme 14.41) [82]. They studied the effect on the acylation reaction of the amount of N,N-dimethyl-4-aminopyridine (DMAP), which simultaneously

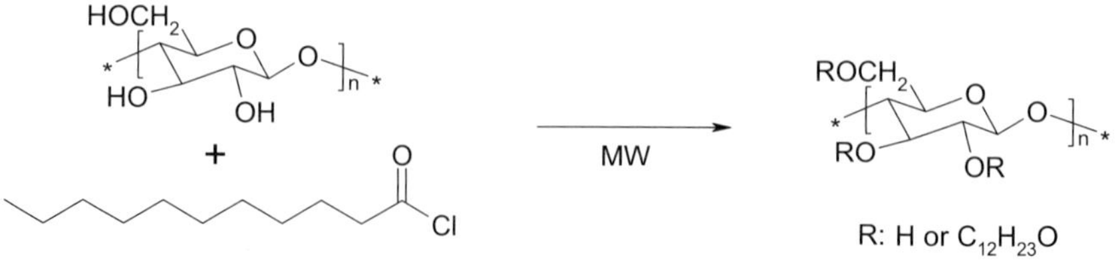

Scheme 14.41

acts as a catalyst and a proton-trapping base. Plastic films synthesized in the absence of DMAP had inferior mechanical behavior. Organic (tributylamine) or inorganic bases ($CaCO_3$, Na_2CO_3) were then added to replace DMAP basic activity, but no changes were observed. The thermal and mechanical properties of plastics obtained by use of different bases, glass transition temperatures (T_g), and degradation temperatures (T_d) were constant, irrespective of the base. The best mechanical properties were obtained for films synthesized in the presence of $CaCO_3$. The same remarks were made about the valorization of chestnut tree sawdust cellulose.

Phosphorylation of microcrystalline cellulose under the action of microwave irradiation was achieved by Gospodinova et al. [83]. The reactions were performed in a single-mode microwave reactor under an argon atmosphere. Mixtures of 29.0 mmol urea, 17.6 mmol phosphorous acid, and 1.8 mmol cellulose were irradiated for 60 to 120 min at temperatures from 75 to 150 °C (Scheme 14.42). The process led to monosubstituted phosphorous acid esters of cellulose with different degrees of substitution of hydroxy functions (0.2 to 2.8) without pretreatment with solvents. The best results (degree of substitution) were obtained at 105 °C after irradiation for 2 h.

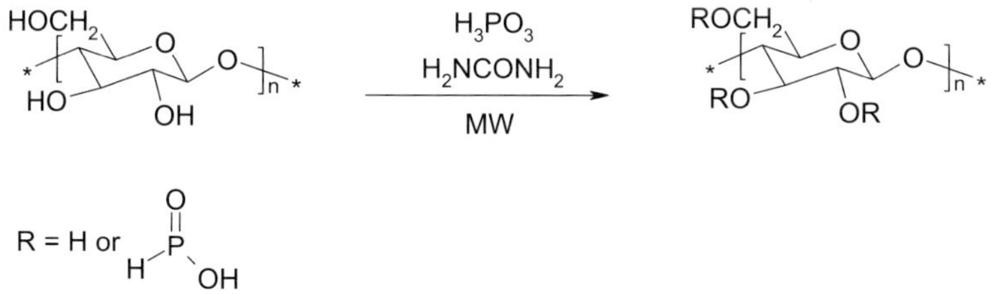

Scheme 14.42

The kinetics of the thermal and microwave-assisted oxidative degradation of poly(ethylene oxide), with potassium persulfate as the oxidizing agent, were determined by Madras [84]. The degradation was studied as a function of temperature and persulfate concentration and it was found that the rate of degradation increased with increasing temperature and persulfate concentration. Continuous distribution kinetics were used to determine the rate coefficients for the degradation process, and the activation energies were obtained. The results indicated that the microwave-assisted process had a lower activation energy (10.3 kcal mol^{-1}) than thermal degradation (25.2 kcal mol^{-1}). Similar investigations have been conducted for the degradation of polystyrene [85].

14.3
Conclusion

In conclusion, polymer synthesis can benefit greatly from the unique features of modern microwave technology recently demonstrated in the large number of successful laboratory-scale applications presented in this chapter. These can include such issues as shorter processing times, increased process yields, and temperature uniformity during polymerization and crosslinking.

The mode of action of microwave irradiation on chemical reactions is still under debate, and some research groups have proposed the existence of so-called nonthermal microwave effects, i.e. sudden acceleration of reaction rates which cannot be explained by the reaction temperatures observed. Recent critical reviews of both groups of theories have been published by Loupy et al. [86, 87], Nuchter et al. [88], and de la Hoz et al. [89].

Irrespective of the type of activation (thermal) or the type of microwave effect (nonthermal), microwave energy has advantages which are still waiting to be fully understood and applied to chemical processes.

References

1 D. BOGDAL, P. PENCZEK, J. PIELICHOWSKI, A. PROCIAK, *Adv. Polym. Sci.* **2003**, *163*, 193.
2 D. BOGDAL, J. PIELICHOWSKI, *A Review of Microwave Assisted Synthesis and Crosslinking of Polymeric Materials.* In: Microwave and Radio Frequency Applications, R. L. SCHULZ, D. FOLZ (eds.), *Proceedings of Fourth World Congress on Microwave and Radio Frequency Applications*, **2004**, 211.
3 F. WIESBROCK, R. HOOGENBOOM, U. S. SCHUBERT, *Macromol. Rapid Commun.* **2004**, *25*, 1739.
4 J. B. WEI, T. SHIDAKER, M. C. HAWLEY, *TRIP*, **1996**, *4*, 18.
5 F. PARODI, *Microwave Heating and the Acceleration of Polymerization Processes.* In: Polymers and Liquid Crystals, WŁOCHOWICZ (Ed.) *Proceedings of SPIE – The International Society for Optical Engineering*, **1999**, *4017*, 2.
6 E. T. THOSTENSON, T. W. CHOU, *Composites, Part A* **1999**, *30*, 1055.
7 J. MIJOVIC, J. WIJAYA, *Polym. Composites* **1990**, *11*, 191.
8 J. JACOB, L. H. L. CHIA, F. Y. C. BOEY, *J. Mater. Sci.* **1995**, *30*, 5321.
9 F. PARODI, *Physics and Chemistry of Microwave Processing.* In: S. L. AGARWAL, S. RUSSO (Eds), Comprehensive Polymer Science, 2nd Supplement Volume, Pergamon–Elsevier, Oxford, **1996**.
10 A. F. PORTO, B. L. SADICOFF, M. C. V. AMORIM, M. C. S. DE MATTOS, *Polym. Test.* **2002**, *21*, 145.
11 X. ZHU, J. CHEN, N. ZHOU, Z. CHENG, J. LU, *Eur. Polym. J.* **2003**, *39*, 1187.
12 P. ZINCK, D. BARBIER-BAUDRY, A. LOUPY, *Macromol. Rapid Commun.* **2005**, *26*, 46.
13 F. FISCHER, R. TABIB, R. FREITAG, *Eur. Polym. J.* **2005**, *41*, 403.
14 Ch. GORETZKI, A. KRLEJ, Ch. STEFFENS, H. RITTER, *Macromol. Rapid Commun.* **2004**, *25*, 513.
15 M. IANNELLI, V. ALUPEI, H. RITTER, *Tetrahedron* **2005**, *61*, 1509.
16 M. IANNELLI, H. RITTER, *Macromol. Rapid Commun.* **2005**, *206*, 349.
17 E. BEZDUSHNA, H. RITTER, *Macromol. Rapid Commun.* **2005**, *26*, 1087.
18 M. PAJDA, D. BOGDAL, R. ORRU, Modern Polymeric Materials for Environmental Applications, J. PIELICHOWSKI (Ed.), **2004**, vol. 1, 113.
19 J. LU, J. WU, L. WANG, S. YAO, *J. Appl. Polym. Sci.* **2004**, *97*, 2072.
20 X. ZHU, N. ZHOU, X. HE, Z. CHENG, J. LU, *J. Appl. Polym. Sci.* **2003**, *88*, 1787.
21 G. CHEN, X. ZHU, Z. CHENG, W. XU, J. LU, *Radiat. Phys. Chem.* **2004**, *69*, 129.
22 X. LU, X. ZHU, Z. CHENG, W. XU, G. CHEN, *J. Appl. Polym. Sci.* **2004**, *92*, 2189.

23 Z. Cheng, X. Zhu, L. Zhang, N. Zhou, X. Xue, *Polym. Bull.* **2003**, *49*, 363.

24 Z. Cheng, X. Zhu, M. Chen, J. Chen, L. Zhang, *Polymer* **2003**, *44*, 2243.

25 W. Xu, X. Zhu, Z. Cheng, G. Chen, J. Lu, *Eur. Polym. J.* **2003**, *39*, 1349.

26 H. Zhang, U. S. Schubert, *Macromol. Rapid Commun.* **2004**, *25*, 1225.

27 D. D. Wisnoski, W. H. Leister, K. A. Strauss, Z. Zhao, C. W. Lindsley, *Tetrahedron Lett.* **2003**, *44*, 4321.

28 L. Q. Liao, L. J. Liu, C. Zhang, F. He, R. X. Zhuo, K. Wan, *J. Polym. Sci., Part A: Polym. Chem.* **2002**, *40*, 1749.

29 Z. J. Yu, L. J. Liu, *Eur. Polym. J.* **2004**, *40*, 2213.

30 Ch. Zhang, L. Liao, L. Liu, *Macromol. Rapid Commun.* **2004**, *25*, 1402.

31 D. Barbier-Baudry, M. H. Brachais, A. Cretu, A. Loupy, D. Stuerga, *Macromol. Rapid Commun.* **2002**, *23*, 200.

32 S. Sinnwell, H. Ritter, *Macromol. Rapid Commun.* **2005**, *26*, 160.

33 F. Wiesbrock, R. Hoogenboom, C. H. Abeln, U. S. Schubert, *Macromol. Rapid Commun.* **2004**, *25*, 1895.

34 R. Hoogenboom, F. Wiesbrock, M. A. M. Leenen, M. A. R. Meier, U. S. Schubert, *J. Comb. Chem.* **2005**, *7*, 10.

35 F. Wiesbrock, R. Hoogenboom, M. A. M. Leenen, M. A. R. Meier, U. S. Schubert, *Macromolecules* **2005**, *38*, 5025.

36 F. Wiesbrock, R. Hoogenboom, S. F. G. M. Van Nispen, M. Van Nispen, M. Van der Loop, C. H. Abeln, A. M. J. Van den Berg, U. S. Schubert, *Macromolecules* **2005**, *38*, 7957.

37 K. Matyjaszewski, J. Xia, *Chem. Rev.* **2001**, *101*, 2921.

38 N. Beldjoudi, A. Gourdenne, *Eur. Polym. J.* **1988**, *24*, 265.

39 F. M. Thuillier, H. Jullien, *Makromol. Chem., Macromol. Sym.*, **1989**, *25*, 63.

40 J. Jow, J. D. DeLong, M. C. Hawley, *SAMPE Quart.*, **1989**, *20*, 46.

41 F. Y. C. Boey, B. H. Yap, *Polym. Test.* **2001**, *20*, 837.

42 J. Zhou, Ch. Shi, B. Mei, R. Yuan, Z. Fu, *J. Mater. Process. Technol.* **2003**, *137*, 156.

43 D. Bogdal, J. Gorczyk, *J. Appl. Polym. Sci.* **2004**, *94*, 1969.

44 D. Bogdal, J. Gorczyk, *Polymer* **2003**, *44*, 7795.

45 Z. K. Brzozowski, S. K. Staszczak, L. K. Hadam, S. Rupinski, D. Bogdal, J. Gorczyk, *J. Appl. Polym. Sci.* **2005**, in the press.

46 D. Zhang, J. V. Crivello, J. O. Stoffer, *J. Polym. Sci., Part B: Polym. Phys.* **2004**, *42*, 4230.

47 S. Chatti, M. Bortolussi, A. Loupy, J. C. Blais, D. Bogdal, M. Majdoub, *Eur. Polym. J.* **2002**, *38*, 1851.

48 S. Chatti, M. Bortolussi, A. Loupy, J. C. Blais, D. Bogdal, P. Roger, *J. Appl. Polym. Sci.* **2003**, *90*, 1255.

49 S. Chatti, M. Bortolussi, D. Bogdal, J. C. Blais, A. Loupy, *Eur. Polym. J.* **2004**, *40*, 561.

50 Ch. Gao, S. Zhang, L. Gao, M. Ding, *J. Appl. Polym. Sci.* **2004**, *92*, 2414.

51 J. Pielichowski, P. Penczek, D. Bogdal, E. Wolff, J. Gorczyk, *Polimery*, **2004**, *49*, 763.

52 J. Pielichowski, D. Bogdal, E. Wolff, *Przem. Chem.* **2003**, *82*, 8.

53 S. Mallakpour, S. Habibi, *Eur. Polym. J.* **2003**, *39*, 1823.

54 B. M. Vogel, S. K. Mallapragada, B. Narasimhan, *Macromol. Rapid Commun.* **2004**, *25*, 330.

55 K. Faghihi, M. Hagibeygi, *Eur. Polym. J.* **2003**, *39*, 2307.

56 J. M. Lu, S. J. Ji, N. Y. Chen, Z. B. Zhang, Z. R. Sun, X. L. Zhu, W. P. Shi, *J. Appl. Polym. Sci.* **2003**, *87*, 1739.

57 K. Faghihi, K. Zamani, A. Mirsaamie, M. R. Sangi, *Eur. Polym. J.* **2003**, *39*, 247.

58 K. Faghihi, K. Zamani, A. Mirsaamie, S. Mallakpour, *J. Appl. Polym. Sci.* **2004**, *91*, 516.

59 K. Faghihi, M. Hajibeygi, *J. Appl. Polym. Sci.* **2004**, *92*, 3447.

60 S. Mallakpour, M. H. Shahmohammadi, *J. Appl. Polym. Sci.* **2004**, *92*, 951.

61 S. Mallakpour, A. R. Hajipour, M. R. Zamanlou, *Eur. Polym. J.* **2002**, *38*, 475.
62 S. Mallakpour, E. Kowsari, *J. Polym. Sci., Part A: Polym. Chem.* **2003**, *41*, 3974.
63 S. Mallakpour, M. R. Zamanlou, *J. Appl. Polym. Sci.* **2004**, *91*, 3281.
64 S. Mallakpour, A. R. Hajipour, M. R. Zamanlou, *J. Polym. Sci. Part A: Polym. Sci.* **2003**, *41*, 1077.
65 N. Li, J. Lu, S. Yao, *Macromol. Chem. Phys.* **2005**, *206*, 559.
66 N. Li, J. Lu, S. Yao, X. Xia, X. Zhu, *Mater. Lett.* **2004**, *58*, 3115.
67 J. Lu, S. Yao, X. Tang, M. Sun, X. Zhu, *Opt. Mat.* **2003**, *25*, 359.
68 A. Banihashemi, H. Hazarkhani, A. Abidolmaleki, *Polym. Sci. Part A: Polym. Sci.* **2004**, *42*, 2106.
69 J. Pielichowski, E. Dziki, J. Polaczek, *Pol. J. Chem. Technol.* **2003**, *5*, 3.
70 J. Polaczek, E. Dziki, J. Pielichowski, *Polimery* **2003**, *23*, 61.
71 K. R. Carter, *Macromolecules* **2002**, *35*, 6757.
72 B. S. Nehls, U. Asawapirom, S. Fuldner, E. Preis, T. Farrell, U. Scherf, *Adv. Funct. Mater.* **2004**, *14*, 352.
73 M. Melucci, G. Barbarella, M. Zambianchi, P. Di Pietro, A. Bongini, *J. Org. Chem.* **2004**, *69*, 4821.
74 T. Yamamoto, Y. Fujiwara, H. Fukumoto, Y. Nalamura, S. Koshihara, T. Ishikawa, *Polymer* **2003**, *44*, 4487.
75 O. Cakmak, M. Bastukmen, D. Kisakerek, *Polymer* **2004**, *45*, 5451.
76 E. Beydushna, H. Ritter, K. Troev, *Macromol. Rapid Commun.* **2005**, *26*, 471.
77 L. Liu, Z. Fang, L. Chen, *Carbohydr. Polym.* **2005**, *60*, 351.
78 R. Xing, S. Liu, H. Zu, Y. Guo, P. Wang, C. Li, Y. Li, P. Li, *Carbohydr. Res.* **2005**, *340*, 2150.
79 V. Singh, D. N. Tripathi, A. Tiwari, R. Sanghi, *J. Appl. Polym. Sci.* **2005**, *95*, 820.
80 V. Singh, D. N. Tripathi, A. Tiwari, R. Sanghi, *Carbohydr. Polym.* **2004**, *58*, 1.
81 Z. Tong, W. Peng, Z. Zhiqian, Z. Baoxiu, *J. Appl. Polym. Sci.* **2005**, *95*, 264.
82 N. Joly, R. Granet, P. Branland, B. Verneuil, P. Krausz, *J. Appl. Polym. Sci.* **2005**, *99*, 1266.
83 N. Gospodinova, A. Grelard, M. Jeannin, G. C. Chitanu, A. Carpov, V. Thiery, T. Besson, *Green Chem.* **2002**, *4*, 220.
84 S. P. Vijayalakshmi, J. Chakraborty, G. Madras, *J. Appl. Polym. Sci.* **2005**, *96*, 2090.
85 G. Sivalingam, N. Agarwal, G. Madras, *AIChe Journal* **2003**, *49*, 1821.
86 L. Perreux, A. Loupy, *Tetrahedron* **2001**, *57*, 9199.
87 A. Loupy, F. Maurel, A. Sabatié-Gogova, *Tetrahedron* **2004**, *60*, 1683.
88 M. Nuchter, B. Ondruschka, W. Bonrath, A. Gum, *Green Chem.* **2004**, *6*, 128.
89 A. de la Hoz, A. Diaz-Ortiz, A. Moreno, *Chem. Soc. Rev.* **2005**, *34*, 164.

15
Microwave-assisted Transition Metal-catalyzed Coupling Reactions

Kristofer Olofsson, Peter Nilsson, and Mats Larhed

15.1
Introduction

The development of microwave-assisted chemistry has been remarkable in many ways, which also has implications for the preparation of this volume. When the first edition of Microwaves in Organic Synthesis was compiled in 2001, it was possible to nurse an ambition to cover all aspects of the literature on microwave-assisted homogeneous transition metal-catalyzed reactions within the scope of a book chapter [1]. Today, this is not easily done, because the number of publications has increased substantially together with the range of transformations investigated [2, 3]. The safety and reproducibility of microwave (MW)-assisted chemistry is also better, because most chemists nowadays use dedicated single or modified multi-mode equipment that is safe to use and generally shows good inter-lab reproducibility [4]. Several articles have also reported successful scale-up reactions with little or no change needed in heating procedure when applying reaction conditions optimized for the small-scale reaction [5, 6]. Another development is the source of the publications. Whereas four years ago most papers on microwave chemistry tended to come from a limited number of academic groups, today the trend is for an increasing number of papers to come from the pharmaceutical industry and research groups worldwide [4]. This mirrors the growing acceptance of microwave technology in modern chemical applications such as medicinal chemistry [7, 8], high-throughput chemistry using polymer-supported reagents or scavengers [9], and fluorous chemistry [10, 11]. Microwave activation is no longer viewed as the last resource but as the first-choice heating method.

In recent decades much effort has been devoted to extending the scope of palladium, copper, and nickel-catalyzed reactions proceeding via aryl or vinyl metal intermediates [12]. These coupling reactions have enabled the formation of many kinds of carbon–carbon and carbon–heteroatom connections that were previously very difficult to realize. Metal-mediated transformations have proven especially valuable for introduction of substituents to aromatic core structures. They allow the presence of a wide variety of functional groups and perform equally well in both inter and intramolecular applications. Furthermore, in homogeneous catalysis,

Microwaves in Organic Synthesis, Second edition. Edited by A. Loupy
Copyright © 2006 WILEY-VCH Verlag GmbH & Co. KGaA, Weinheim
ISBN: 3-527-31452-0

many different ligands can now be used to fine-tune the activity of catalytic complexes and the selectivity of reactions.

However, the long reaction times frequently required with classical heating (ranging from hours to days) have previously limited the exploitation of these transformations in laboratory-scale medicinal chemistry, fine chemical synthesis, and high-throughput processing. Rapid and reliable microwave applications are therefore superior, not only for rapid production of new chemical entities in drug-discovery efforts [13] but also for efficient optimization of metal-catalyzed methods in general [14, 15]. Although microwave-assisted organic reactions can sometimes be smoothly conducted in open vessels, it is often of interest to work with closed systems, especially if superheating with its associated time reductions is desired [4]. The use of disposable septum-sealed vessels designed for straightforward pressurized processing and automation is here essential for both safety and productivity. Importantly, when applying pressurized conditions it is strongly recommended to use purpose-built reactors equipped with accurate temperature and pressure feedback systems coupled to the power control to avoid vessel rupture.

It has been our ambition in writing this chapter not to give a complete overview of microwave-assisted metal-catalyzed transformations of aryl and vinyl halides (or pseudo halides) – which is indeed impossible within the scope of this edition – but rather to focus selectively both on medicinal chemistry and papers with a recent publication date. Especially in sections in which the number of publications is large we have opted for a deeper discussion of selected topics and a few topics covered in the last edition have been removed.

15.2
Cross-coupling Reactions

15.2.1
The Suzuki–Miyaura Reaction

Organoboron compounds were at first thought to be poor coupling partners in cross-coupling reactions because the organic groups on boron are only weakly nucleophilic. However, in 1979, Suzuki discovered that coupling reactions of organoboron compounds proceeded in the presence of ordinary bases, for example hydroxide or alkoxide ions [16]. This modification proved to be generally applicable and the Suzuki reaction is today arguably the most versatile cross-coupling reaction. For example, the reaction has attracted the interest of several research teams involved in high-throughput chemistry, as a large variety of boronic acids are commercially available [7]. In addition, high-speed synthesis of aryl boronates (Suzuki coupling reactants) has been performed under single-mode irradiation conditions with a palladium carbene catalyst generated in situ [17].

The first MW-promoted Suzuki couplings were published in 1996 (Scheme 15.1). Phenyl boronic acid was coupled with 4-methylphenyl bromide to give a fair yield of product after a reaction time of less than 4 min under the action of single-mode

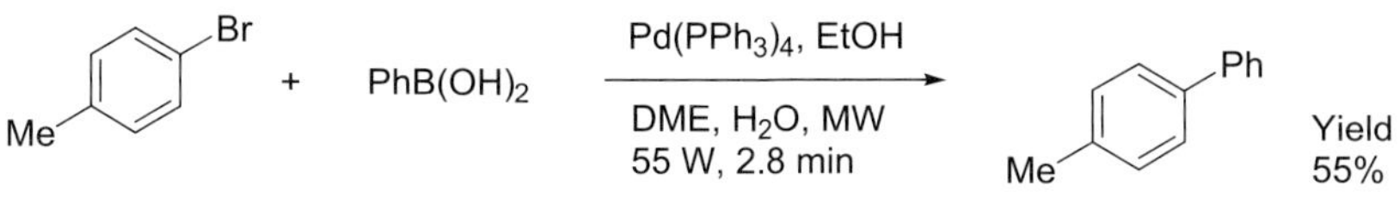

Scheme 15.1. Suzuki coupling of phenyl boronic acid with 4-methylphenyl bromide.

MW irradiation. The same reaction had previously been conducted with conventional heating with a reported reaction time of 4 h [18].

In 2005, very similar reactions were shown to proceed smoothly with continuous flow reactors (Scheme 15.2). The yield of couplings with aryl bromides and iodides were high, although the authors noted it was unclear exactly how long a sample was irradiated, because of uncertainty about the focus of the irradiation over the capillary column [19].

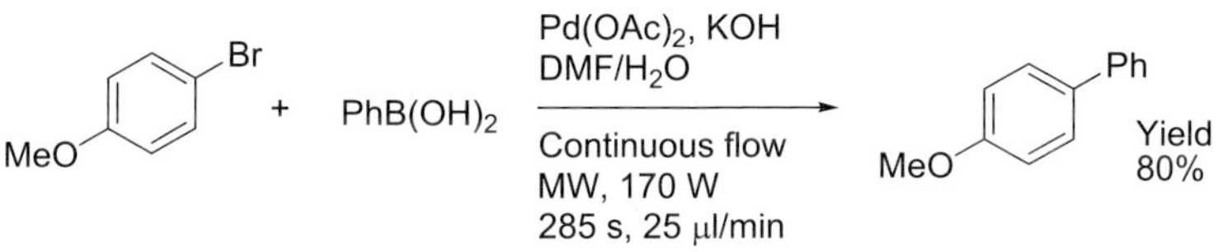

Scheme 15.2. Microwave-assisted continuous flow reaction.

Another topic often encountered in the literature on microwave-promoted reactions is the lower consumption of energy associated with the use of MW technology in small-scale chemistry. For the palladium-catalyzed Suzuki reaction there have been attempts to investigate this matter in more detail. Clark et al. have performed a comparative study of the energy efficiency of the different reaction techniques. The Suzuki reaction was analyzed and under the reaction conditions used the MW-assisted reaction was 85 times more energy-efficient than the corresponding oil-bath-heated reaction. As there are a multitude of reaction conditions for the Suzuki coupling, this value should be seen as an example, rather than a definite value [20].

The development of MW-assisted reactions using aryl chlorides has attracted the interest of several research groups. Transition metal-catalyzed reactions with aryl chlorides were elusive for a long time and were generally only successful with very high reaction temperatures and special reaction conditions. Lately, new catalytic systems, most notably those presented by Fu [21], have spurred the development of several schemes for MW-assisted activation of aryl chlorides.

Efficient reactions using electron-rich aryl chlorides have been reported with the $Pd(OAc)_2$–PCy_3 catalytic combination. The reaction conditions enabled the use of several bases but the authors chose the inexpensive potassium phosphate (Scheme 15.3) [22].

N-Heterocyclic carbene ligands were used with good results in the sluggish coupling of electron-rich aryl chlorides with phenylboronic acid. These ligands enabled

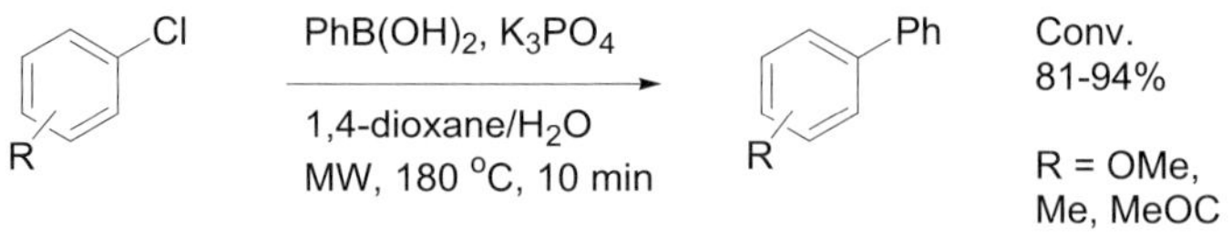

Scheme 15.3. Suzuki–Miyaura reaction with an aryl chloride.

IPr = (N,N-bis(2,6-diisopropylphenyl)imidazol)-2-ylidene

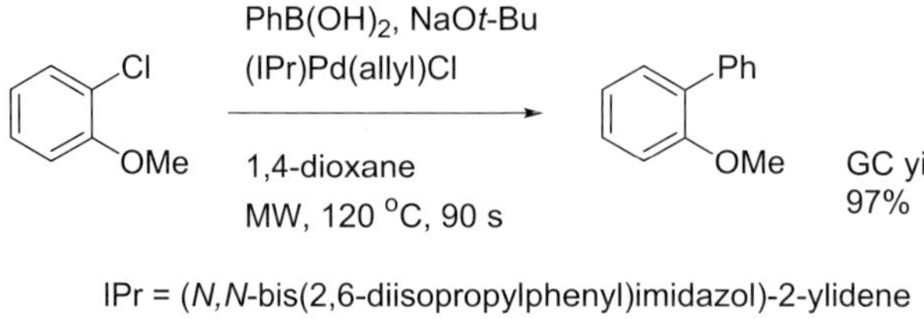

Scheme 15.4. Reaction with an aryl chloride using a carbene ligand.

a reaction with a comparatively low reaction temperature and a short reaction time (Scheme 15.4) [23].

A recent paper has described Suzuki–Miyaura couplings of aryl chlorides in DMF and water, using the air and moisture-stable dihydrogen di-μ-chlorodichlorobis(di-*tert*-butylphosphinito-κP)dipalladate (POPd2) catalyst (Scheme 15.5) [24].

Development of catalytic systems with water as solvent is very important for industrial and environmentally friendly applications. Water is, in this respect, perhaps the ultimate solvent, because of its nontoxicity and ready availability. Leadbeater has published several papers reporting optimization of the Suzuki–Miyaura reaction for aqueous conditions [25, 26]. Aryl bromides and iodides were coupled and isolated in good yields with an attractive ligandless procedure (Scheme 15.6).

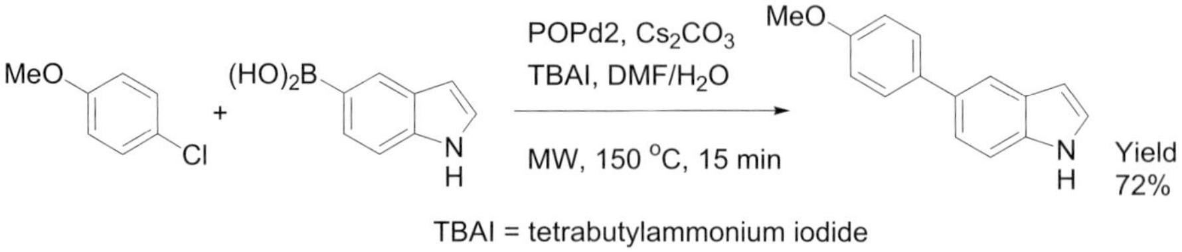

TBAI = tetrabutylammonium iodide

Scheme 15.5. Suzuki–Miyaura coupling with the POPd2 catalyst.

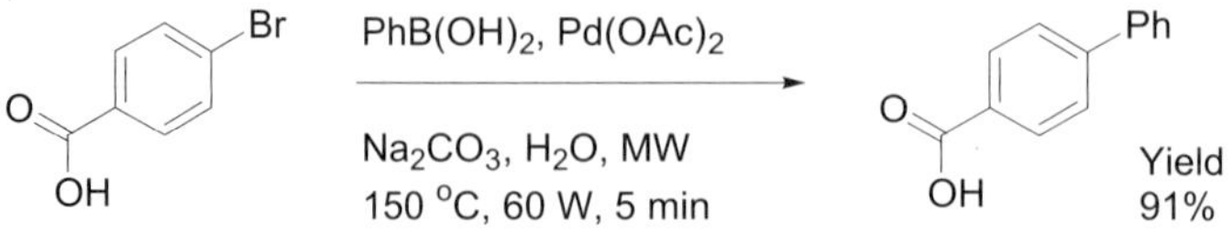

Scheme 15.6. Ligandless Suzuki–Miyaura coupling in water.

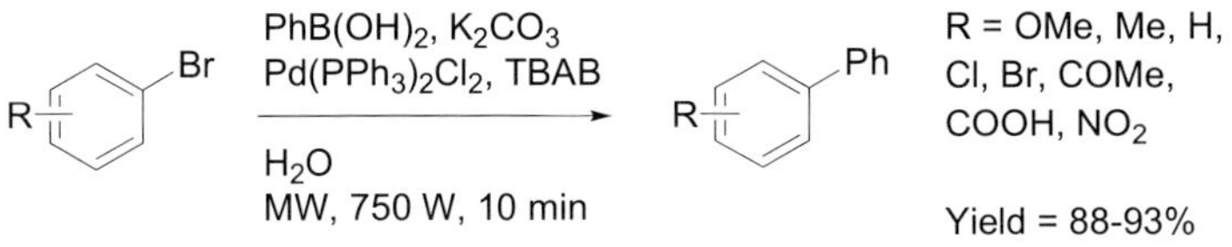

15.2 Cross-coupling Reactions

Scheme 15.7. Suzuki–Miyaura couplings in water.

Reaction conditions: $PhB(OH)_2$, K_2CO_3, $Pd(PPh_3)_2Cl_2$, TBAB, H_2O, MW, 750 W, 10 min. R = OMe, Me, H, Cl, Br, COMe, COOH, NO_2. Yield = 88–93%.

Some reactions gave increased yields on addition of tetrabutylammonium bromide (TBAB) [27].

Aryl bromides have been used by another team in the coupling of phenylboronic acid in water, using easily available starting materials and an improved commercial microwave oven (Scheme 15.7) [28].

Water has also been used in the coupling of aryl chlorides with the electron-neutral phenylboronic acid. Here Leadbeater and Arvela took advantage of simultaneous cooling of the reaction vessel by compressed air while heating the bulk of the reaction mixture with microwaves. This enabled the successful coupling of both electron-rich and electron-poor aryl chlorides in moderate to excellent yields. The rationale was that a careful control of the reaction temperature is needed to prevent the destruction of the aryl chloride in the reaction mixture. A significant advantage of the simultaneous cooling technique, compared to reactions without cooling, was especially apparent for electron-rich and neutral aryl chlorides [29]. Suzuki–Miyaura couplings in water using di(2-pyridyl)methylamine–palladium dichloride complexes [30] and $Pd(PPh_3)_2Cl_2$ [31] have also been reported.

An early report by König deals with rapid parallel Suzuki reactions in water with phase-transfer catalysts. The solid support used in these reactions was PEG, and a variety of aryl palladium precursors were evaluated; aryl halides as well as aryl triflates and nonaflates (Scheme 15.8). The inclusion of PEG is appealing, because it not only helps to solubilize the reagents but is also suggested to stabilize the palladium catalyst in the absence of phosphine ligands. Both the polymers and the esters were reported to withstand 10 min of 900 W multimode MW irradiation whereas the thermal conditions induced substantial ester cleavage (up to 45%). Nonaflates were found to be associated with lower yields and diminished product purity [32].

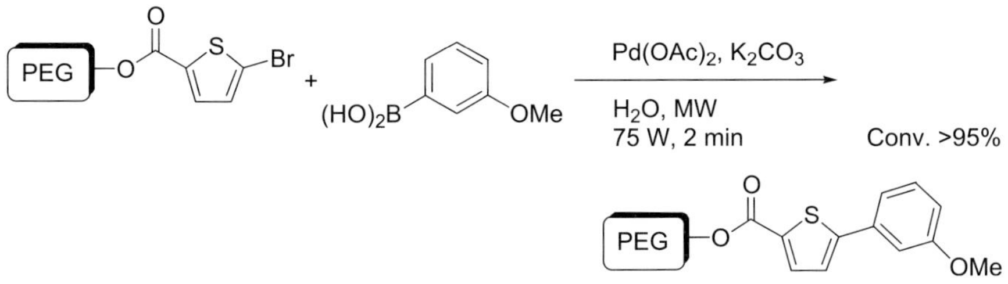

Scheme 15.8. A PEG-supported, aqueous Suzuki coupling.

Reaction conditions: $Pd(OAc)_2$, K_2CO_3, H_2O, MW, 75 W, 2 min. Conv. >95%.

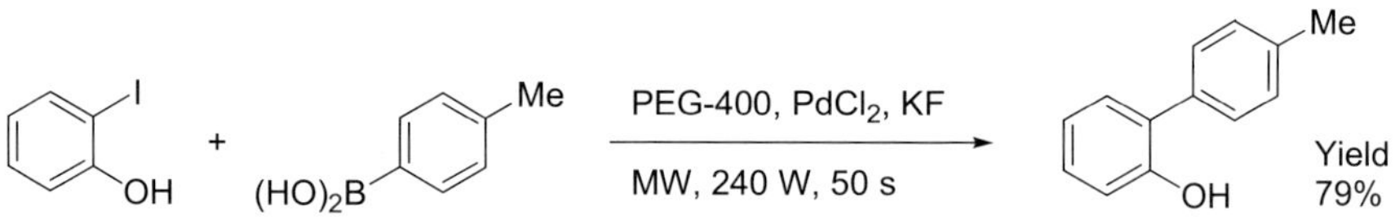

Scheme 15.9. Suzuki coupling in PEG as a nontoxic reaction medium.

Similar Suzuki couplings performed with PEG as a nontoxic reaction medium have been reported by Varma (Scheme 15.9) [33].

Villemin has reported the use of sodium tetraphenylborate as a stable reactant for Suzuki couplings performed with water or monomethylformamide (MMF) as solvent. The high dielectric constant of MMF resulted in very efficient MW heating (Scheme 15.10) [34].

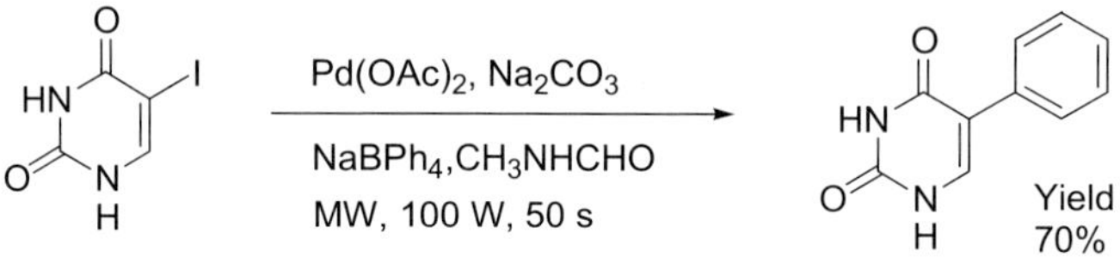

Scheme 15.10. Suzuki coupling with sodium tetraphenylborate.

A few interesting papers have appeared with ligandless and solvent-free Suzuki–Miyaura reactions using cheap palladium powder and potassium fluoride on alumina. The catalysts have been recycled and used through several reaction cycles and the products were collected by a simple filtration, adding to the preparative ease of the method [35]. Potassium fluoride on alumina has also been used in the solvent-free synthesis of unsymmetrical ketones, with good results [36, 37].

Interestingly, the Suzuki reaction was shown to proceed smoothly on polymeric supports as long as ten years ago and high yields of a variety of products were reported under these reaction conditions (Scheme 15.11) [38]. 4-Bromo and 4-iodobenzoic acids linked to Rink-amide TentaGel resulted in a conversion of more than 99% within 4 min. The yields suggested high potential for use of microwave-assisted reactions on polymeric resins [39].

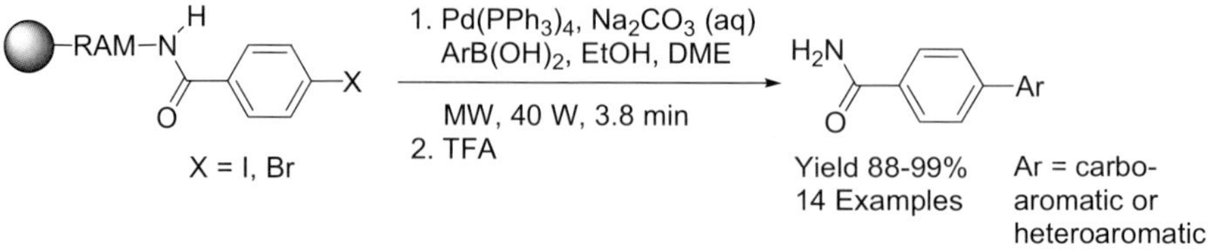

Scheme 15.11. Suzuki couplings on polymer supports.

For applications in high-speed synthesis, one interesting paper looked at different polyethylene-supported palladium catalysts (FiberCat) and their efficiency in the Suzuki–Miyaura reaction. The supported catalysts have attracted some interest, because of the possibility of recycling and their convenience at the work-up stage, where the catalyst can easily be removed by filtration. The more electron-rich and more reactive systems usually resulted in higher conversions and shorter reaction times. The reactions could all be conducted under ambient atmosphere and performed better than standard homogeneous systems, as measured by the purity of the products. Reactions performed with supported palladium were, when the conversion was quantitative, pure enough to be collected simply by solid-phase extraction over Si-Carbonate. Aryl iodides, bromides, triflates, and electron-poor chlorides all gave excellent yields and electron-rich aryl chlorides gave moderate yields (Scheme 15.12) [40].

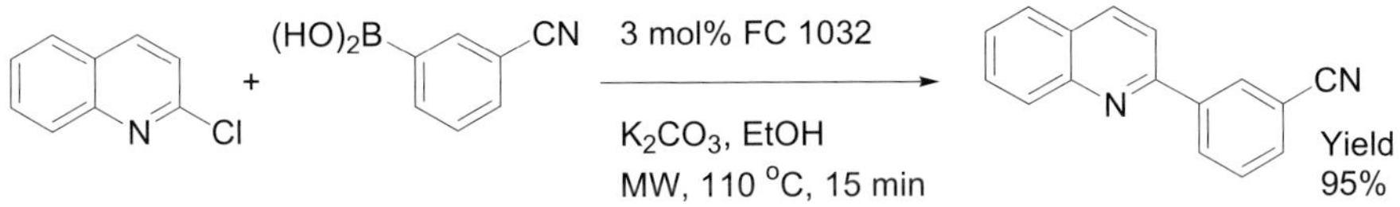

Scheme 15.12. Suzuki–Miyaura coupling with an alkyl phosphine-supported catalyst (FC 1032).

Clean reactions and high yields were reported in the Suzuki–Miyaura reaction with polystyrene-supported palladium using an improved commercial MW oven and reflux conditions. The polystyrene-based catalysts are described as a borderline class of catalysts that retain the advantages of homogeneous catalysts while securing the ease of recovery and workup of heterogeneous catalysts [41]. In this method polystyrene-supported Pd(II) was prepared rapidly by ultrasound treatment and subsequently used in the reactions. The catalysts were stable under heat and there was no need for an inert atmosphere. Initial attempts to conduct the reaction in pure water failed and a mixture of toluene and water was used throughout. Electron-rich and electron-poor aryl bromides were all reactive and comparison with conventional oil-bath heating revealed yields that were comparable, although with reaction times that were longer (Scheme 15.13) [41].

Alternatives to solid-supported catalysts are catalysts that are themselves insoluble [42]. A pyridine-aldoxime ligand has been evaluated in the Suzuki–Miyaura reaction using water as solvent. When an Irori Kan was used to contain the polymeric catalyst, the reaction could be repeated 14 times without noticeable reduc-

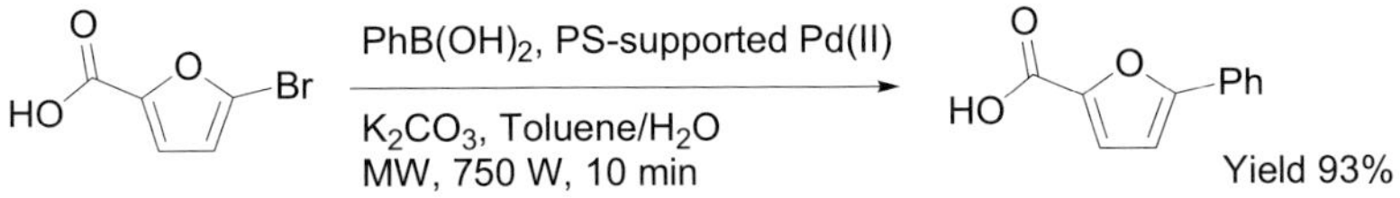

Scheme 15.13. Suzuki–Miyaura coupling with polystyrene-supported catalyst.

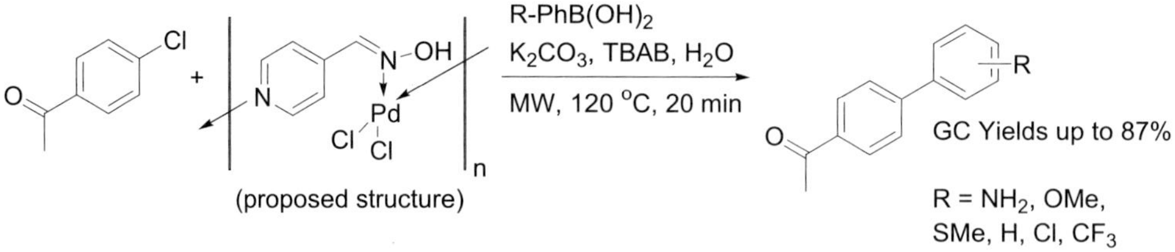

Scheme 15.14. Suzuki–Miyaura couplings with an insoluble pyridine–aldoxime catalyst.

tion of efficiency. The optimized reaction conditions were then used to create a small library of approximately 30 biaryl compounds using aryl iodides, bromides, triflates, and an activated chloride (Scheme 15.14) [42].

A modern development and variation of solid support is fluorous chemistry [43, 44], an emerging technique that takes advantage of the unique physical and solubility properties of perfluorinated organic compounds. Many publications have recently drawn attention to the special properties of fluorous chemistry in which the attractive features of solution-phase reactions are combined with the convenient workup of solid-phase reactions, without the disadvantages of the latter. Zhang used perfluorooctylsulfonates as a coupling partner in a Suzuki–Miyaura reaction in which the perfluorooctylsulfonate group filled the function of leaving group (a pseudo triflate) in the coupling while also having sufficiently high fluoricity to function as a fluorous tag in fluorous separations. The perfluorooctylsulfonate group was highly soluble in organic solvents and was thermostable under the reaction conditions used. The example illustrated was a quite challenging coupling in which the product was isolated in a useful yield (Scheme 15.15). The use of the fluorous sulfonyl group was further demonstrated in a multistep synthesis of a biaryl-substituted hydantoin [45].

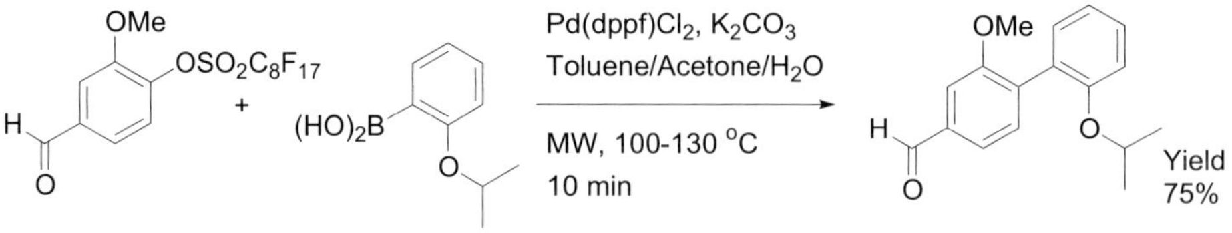

Scheme 15.15. Fluorous Suzuki–Miyaura reaction.

The same group reported the synthesis of a library of 3-aminoimidazo[1,2-*a*] pyridines/pyrazines by fluorous multicomponent reactions. Here the overall yields, and the yields for the separate Suzuki–Miyaura reactions that were a part of the synthesis were relatively low, because of competing reactions and poor reactivity of the substrates, but the speed of the microwave-mediated syntheses and ease of separation emphasized the usefulness of fluorous reagents [46]. A recent paper further illustrated the use of Suzuki–Miyaura couplings of aryl perfluorooctylsulfonates in the decoration of products derived from 1,3-dipolar cycloadditions [47].

Several publications have appeared dealing with Suzuki–Miyaura reactions producing heterocyclic products. These are interesting not only technically, because not all heterocyclic compounds are readily compatible with transition metal catalysis, but also because of the many applications of heterocyclic molecules in drug discovery. Indeed, several pharmaceutical companies have published high-speed or combinatorial synthesis-related reports describing the production of a variety of heterocyclic compounds [48].

Unprotected 4-heteroaryl phenylalanines have been prepared by microwave-assisted Suzuki–Miyaura reactions. Amino acids containing the biaryl motif have several interesting applications in medicinal chemistry and this method enabled their synthesis without protection of the amino acid. Optically pure boronic acids could be used without racemization (Scheme 15.16) [49].

Aryl-substituted aminopyrimidines have also been prepared under microwave irradiation conditions with very attractive reaction times compared with the oil-bath heated reactions (Scheme 15.17) [50].

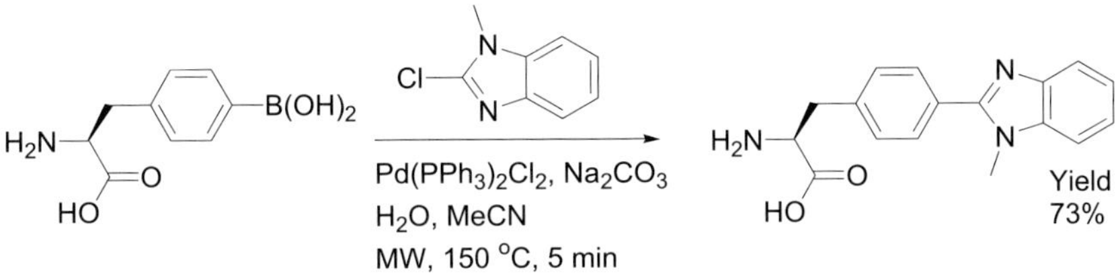

Scheme 15.16. Synthesis of a 4-heteroaryl phenylalanine analog.

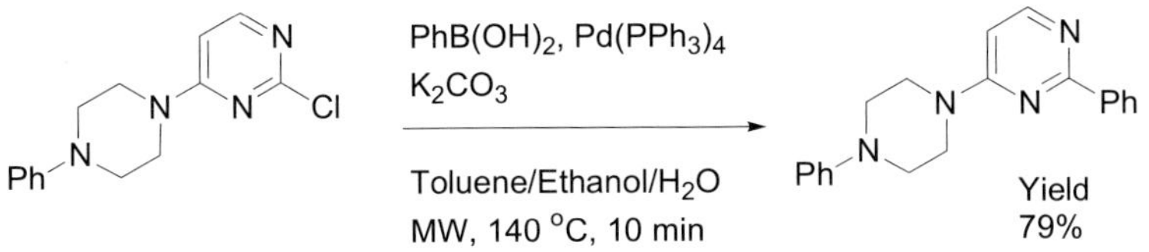

Scheme 15.17. Suzuki phenylation of a chloropyrimidine.

The first example of a Suzuki–Miyaura reaction on a pyridopyrimidine skeleton was used for decoration of a 4-chloro[2,3-d]pyrimidin-7(8H)one scaffold. Phenylboronic acid was used as a coupling partner with a good isolated yield (Scheme 15.18) [51].

Another interesting MW-mediated synthesis is the one-pot stepwise construction of 4,5-disubstituted pyrazolopyrimidines in which the 4-position was functionalized via an S_NAr reaction and the 5-position via a Suzuki–Miyaura coupling. The strength of this synthesis is the possibility of choosing different 4 and

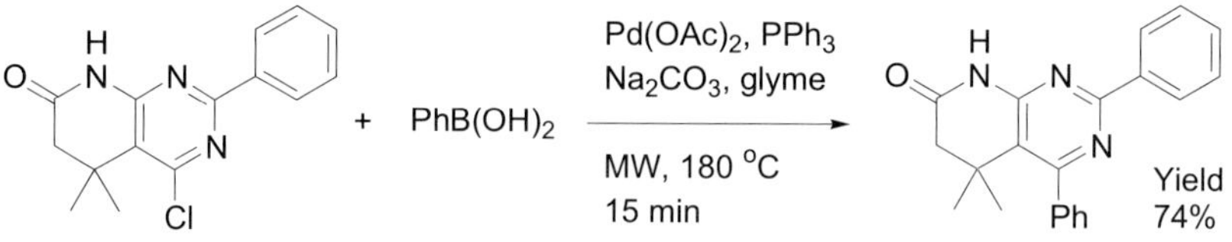

Scheme 15.18. Suzuki–Miyaura reaction on pyridopyrimidinone.

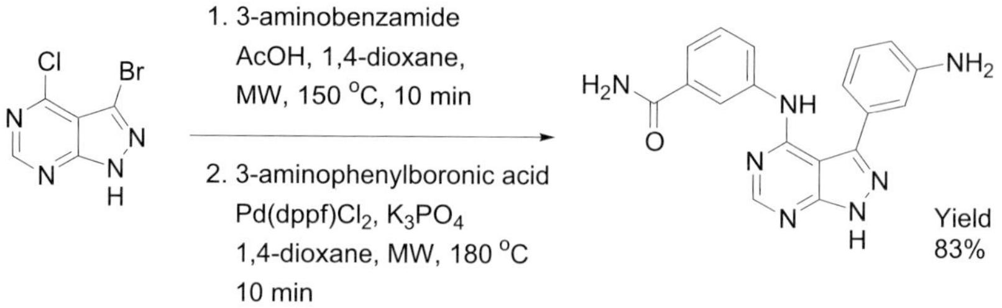

Scheme 15.19. Suzuki–Miyaura reaction of pyrazolopyrimidines.

5-substituents at a late stage in the synthesis [52]. The MW procedure was also more tolerant of functional groups than previously reported oil-bath-heated reaction routes (Scheme 15.19). This method was also revealed to be compatible with the similar pyrrolopyrimidine scaffold [52].

Several papers on Suzuki–Miyaura mediated syntheses of heterocyclic compounds have been published by Erik Van der Eycken's research group. 2-Substituted carbazoles are present in several naturally occurring and biologically active molecules. Most of the compounds reported in this class have been made by the Cadogan synthesis. This method, however, often requires drastic conditions and long reaction times. In this approach Suzuki–Miyaura couplings between an *ortho*-nitro-substituted boronic acid and sixteen aryl bromides furnished the necessary substrates for the Cadogan reductive cyclization. The coupling of the *ortho*-nitro-substituted boronic acid is known to be troublesome, because of competing proto-deboronation, but this side reaction could be minimized by use of rapid microwave methods (Scheme 15.20). The ensuing cyclization was greatly facilitated by MW irradiation and was conducted at 210 °C for 20 min with a maximum

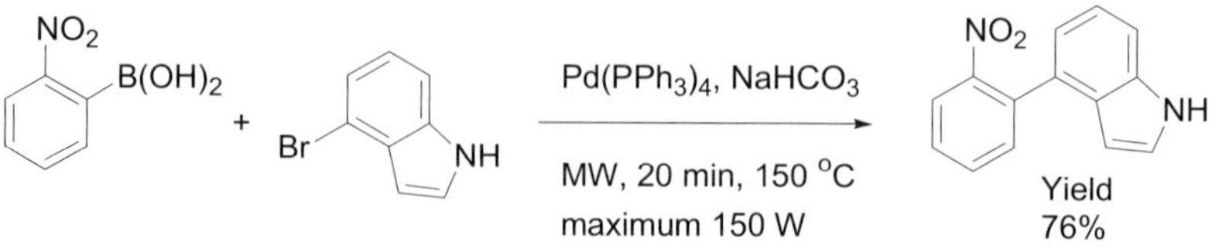

Scheme 15.20. Suzuki–Miyaura coupling of an *ortho*-nitro-substituted boronic acid.

irradiation power of 300 W [53]. Difficult Suzuki–Miyaura couplings were also performed in high yield by the same group in the synthesis of buflavine [54] and apogalanthamine [55] analogs. In the latter paper the corresponding reaction performed in oil-baths was associated with distinctly lower yields.

Several different transition metal-catalyzed reactions with the 2(1H)pyrazinone template have been evaluated. The Suzuki–Miyaura coupling was efficient in introducing aryl groups to both the 3 and the 5-positions of the heterocycle. The 3-arylated product could be isolated in 75% yield by using 1.1 equivalents of boronic acid and sodium carbonate as base whereas use of 2.2 equivalents of boronic acid with cesium carbonate yielded the 3,5-disubstituted compound in 52% yield (Scheme 15.21) [56]. Efforts to widen the utility of this Suzuki–Miyaura reaction to include solid-phase reactions met with difficulties, because the reaction was problematic to drive to completion [57]. Other teams have also reported problems with Suzuki–Miyaura couplings on polymeric supports [44, 58].

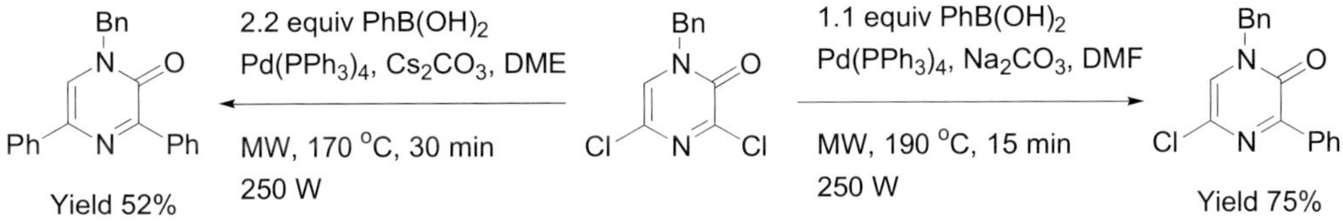

Scheme 15.21. Synthesis of mono and diarylated 2(1H)pyrazinones.

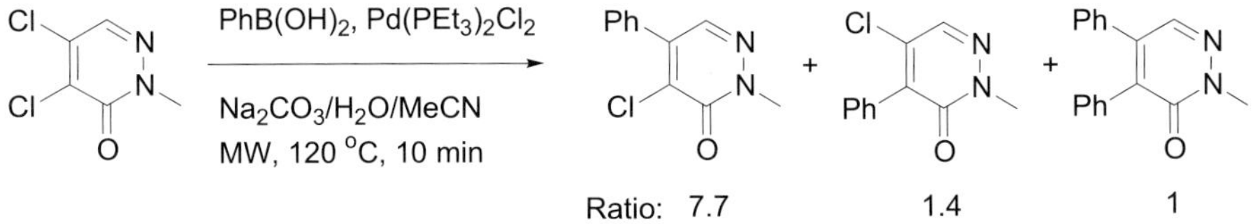

Scheme 15.22. Suzuki–Miyaura reaction with 4,5-dichloropyridazinones.

An MW-assisted coupling of 4,5-disubstituted pyridazinones has been reported. Many ligands and catalytic systems were evaluated but, as indicated in Scheme 15.22, selectivity between the 3-, 4-, and 3,4-disubstituted products was usually low. One of the best ligands was, somewhat surprisingly, the sensitive PEt$_3$ alkyl phosphine ligand [59].

para-Biaryl-substituted dihydropyrimidones have recently been synthesized using Pd–C under microwave conditions. In this reaction the inexpensive Pd–C was superior to palladium acetate (Scheme 15.23) [60].

A microwave mediated Suzuki–Miyaura reaction with an unprotected tetrazole moiety has also been evaluated. This was reported as the first Suzuki–Miyaura coupling known to proceed without protection of the N-2 position of the tetrazole (Scheme 15.24) [61].

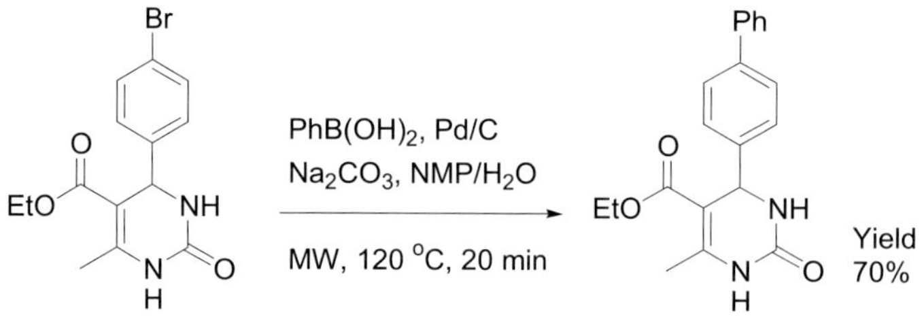

Scheme 15.23. Suzuki–Miyaura reaction with 4,5-disubstituted pyridazinones.

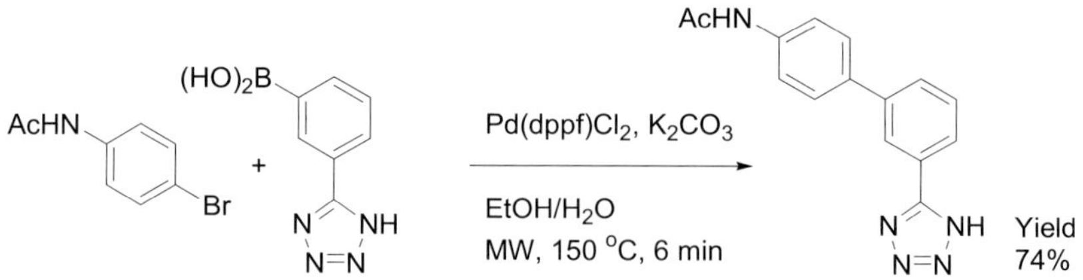

Scheme 15.24. Suzuki–Miyaura reaction with an unprotected tetrazole.

Several Suzuki–Miyaura couplings of more complex natural products and other large compounds have been reported. Fluorescein and rhodamine derivatives were successfully synthesized by Burgess using a water-soluble phosphine ligand [62], and an intramolecular cyclization using a Suzuki-reaction in the total synthesis of Biphenomycin B has been reported [63]. This macrocyclization was not very effective using oil-bath heating but under controlled microwave irradiation the yield could be more than doubled to 50%.

Isoflavones could be prepared in a water, DME, and ethanol solution at a relatively low reaction temperature, as depicted in Scheme 15.25 [64].

A Suzuki–Miyaura reaction for functionalization of quinolin-2(1H)ones at the 4-position was recently published by Kappe. This biologically active structural class has attracted interest in the treatment of several diseases. The suggested synthesis enabled a late introduction of the moiety at the 4-position and this was found to be an advantage over previously reported procedures. The reaction could be run with a

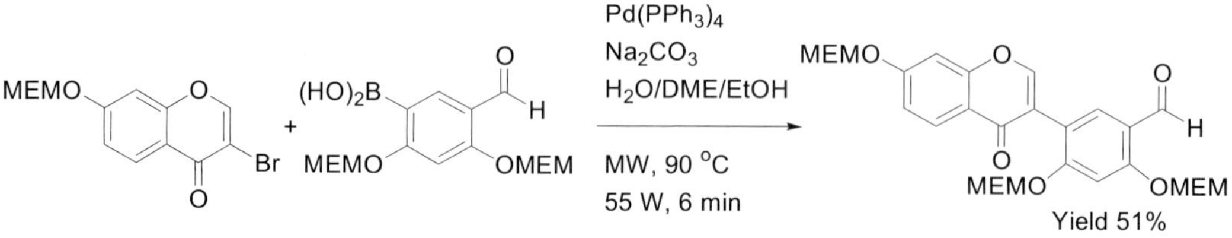

Scheme 15.25. Reaction of isoflavones under Suzuki–Miyaura reaction conditions.

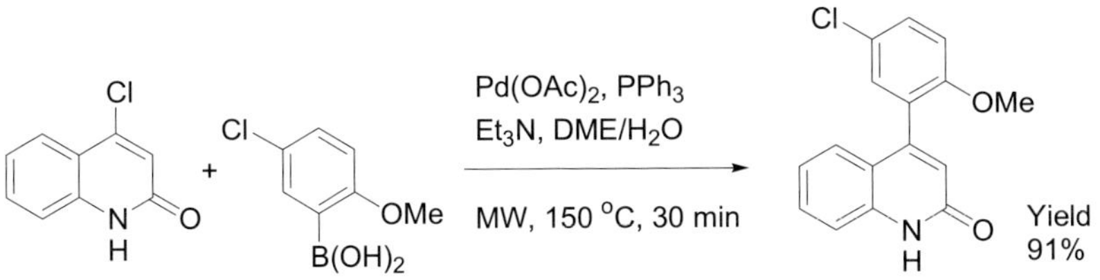

Scheme 15.26. Chemoselective arylation of quinolin-2(1H)ones.

comparatively low loading of catalyst (0.5 mol%). The precise amount of water was also found to be important when optimizing the yield (Scheme 15.26) [65]. A note of interest is that many of the reactions described in this paper, including the Suzuki–Miyaura reactions, were easily scaled up using a multimode batch reactor, without re-optimization of the reaction conditions [6c].

Coats reported a parallel synthesis of delta/mu agonists as depicted in Scheme 15.27. Both solid and solution-phase techniques were evaluated with regard to the reactivity of the vinyl bromide template but solution-phase couplings resulted in more rapid reactions. In the latter case it was found that the Suzuki–Miyaura reaction could be applied directly to the reaction mixture of the preceding reductive amination, thus ensuring a relatively rapid and easy synthetic route to a library of 192 compounds [66].

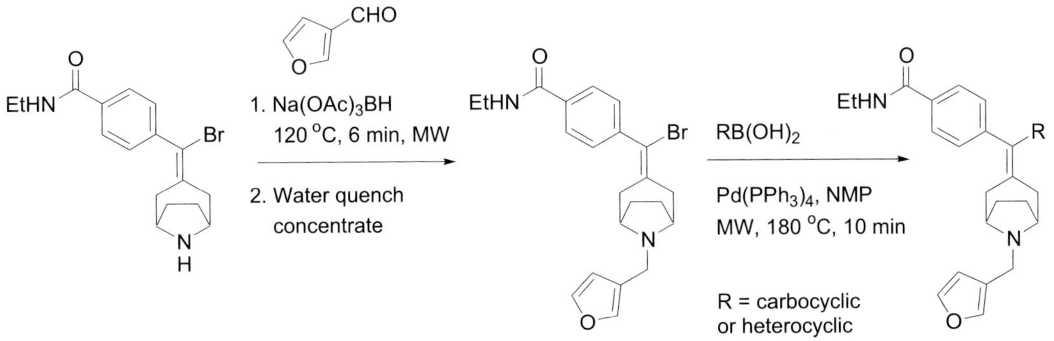

Scheme 15.27. Parallel synthesis of delta/mu agonists.

Antifungal 3-aryl-5-methyl-2,5-dihydrofuran-2-ones have been reported in which the 3-aryl group was introduced by palladium chemistry. The yields were usually moderate, possibly because of instability of the core structure at high temperature (Scheme 15.28) [67].

Antimicrobial oxazolidinones have been successfully synthesized by means of single-mode MW irradiation on a polystyrene resin. In this reaction the use of domestic multimode ovens was associated with inconsistent yields and purity, presumably because of the inhomogeneity of the MW field and lack of sufficient tem-

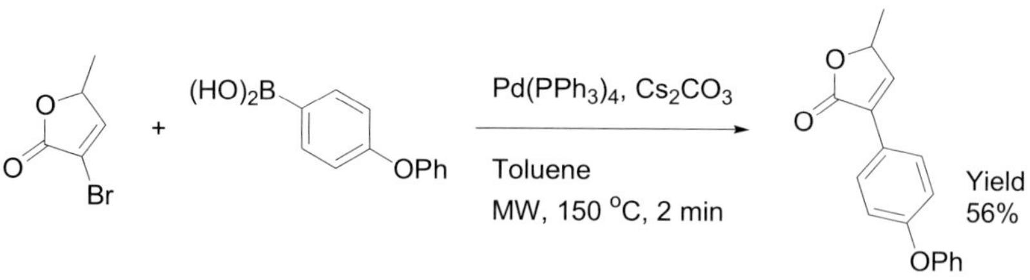

Scheme 15.28. Preparation of 3-aryl-5-methyl-2,5-dihydrofuran-2-ones.

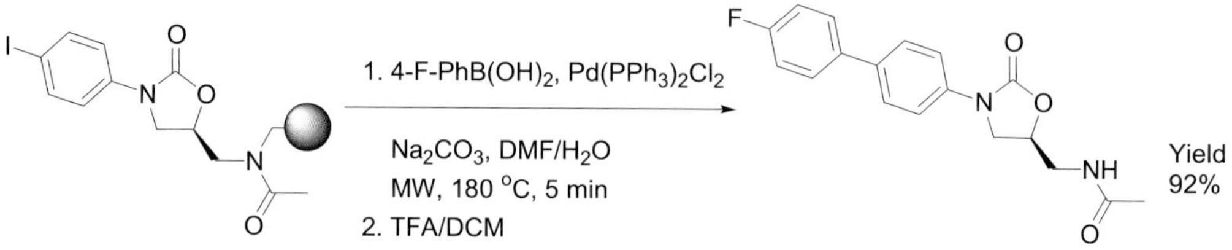

Scheme 15.29. Polymer-supported synthesis of antimicrobial oxazolidinones.

perature and pressure control. A representative reaction is presented in Scheme 15.29. These solid-supported reactions proceeded smoothly in 5–10 min on addition of 6 equiv. boronic acid, and a small library with variations at both the N-acyl and the biaryl functionalities was created [68].

Pyrazole-based COX-inhibitors have been synthesized using Pd–C as a heterogeneous and ready filterable palladium source. Electron-deficient boronic acids coupled well whereas ortho-substituted and electron-rich boronic acids were less reactive (Scheme 15.30) [69]. The same team also developed a two-step, one-pot procedure for synthesis of styrene-based nicotinic acetylcholine receptor antagonists.

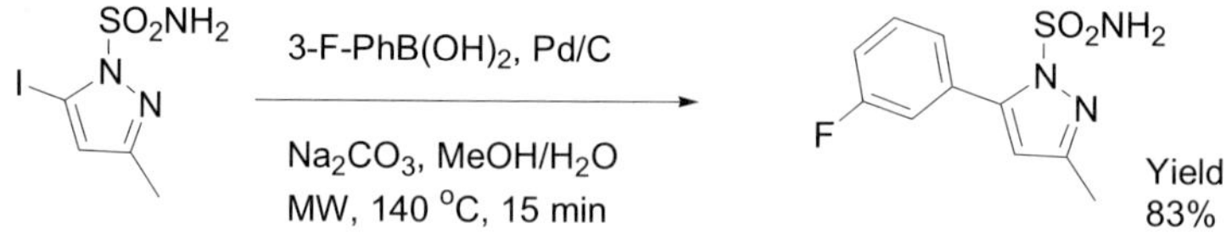

Scheme 15.30. Palladium-on-carbon-catalyzed synthesis of COX-inhibitors.

For several years Hallberg has used MW-promoted reactions for optimization of different types of aspartyl protease inhibitors [7]. The Suzuki–Miyaura coupling was recently used to introduce biaryl moieties in cyclic sulfonamide HIV-1 protease inhibitors. A series of sixteen reactions were performed with fair to moderate yields and the reaction times were, in all examples except two, limited to only 5 min

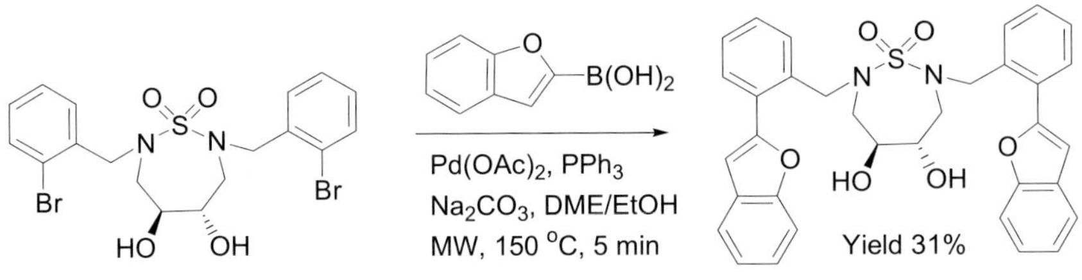

Scheme 15.31. Synthesis of a cyclic sulfonamide HIV-1 protease inhibitor.

(Scheme 15.31) [70]. Other Suzuki–Miyaura couplings with similar structures have previously been reported, with products having K_i values in the nanomolar range [71].

Similarly, eight relatively complex C2-symmetric plasmepsin I and II inhibitors against the malaria-causing protozoon *Plasmodium falciparum* were effectively synthesized with an MW method by substitution of two vinyl bromide groups. Here, the reaction temperatures were held low in the Suzuki–Miyaura couplings, presumably to minimize decomposition of the peptide mimetic. Heck couplings in the same series could, however, be executed at 150–170 °C by use of organic bases (Scheme 15.32) [72].

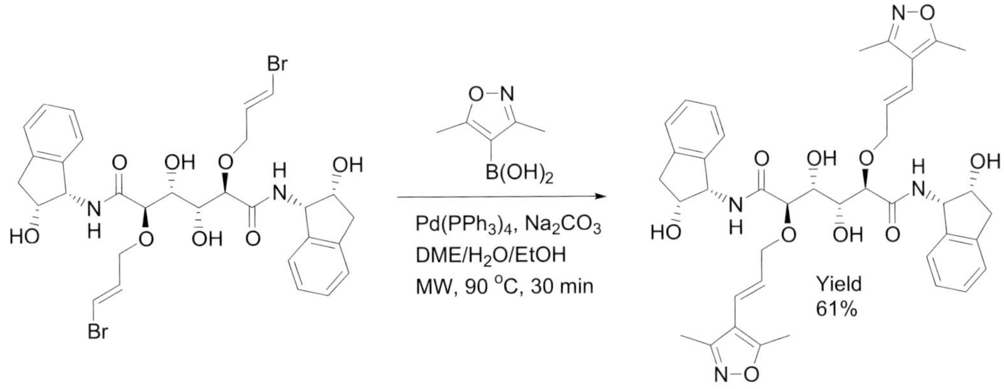

Scheme 15.32. Synthesis of a plasmepsin I and II inhibitor.

Suzuki–Miyaura reactions with aryl bromides and triflates have been reported in the synthesis of plasmepsin I and II inhibitors using a hydroxyethylamine transition-state-mimicking scaffold [73]. Four libraries of similar compounds were prepared where the Suzuki–Miyaura reaction was used for direct derivatization of the P1′-position without protection of the hydroxyethylamine center. It was noted in this context that no epimerization occurred during the reaction and that exchange of cesium carbonate for sodium carbonate resulted in better yields (Scheme 15.33) [74].

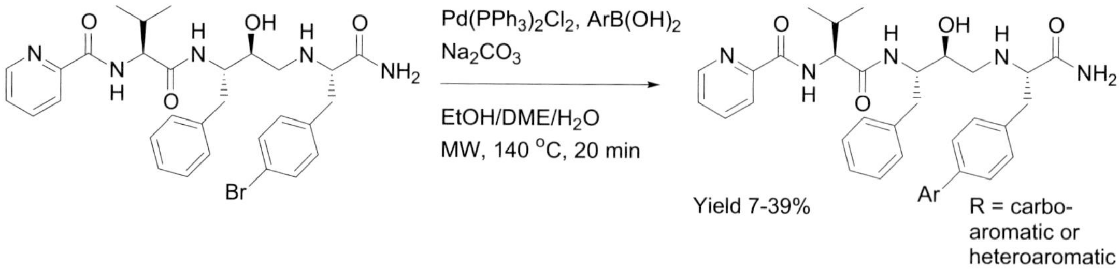

Scheme 15.33. Library production of plasmepsin I and II inhibitors.

15.2.2
The Stille Reaction

The defining feature of the Stille cross-coupling reaction (or the Migita–Kosugi–Stille coupling) is the use of organotin moieties in combination with palladium catalysts [75]. This base-free reaction is, just as the Suzuki reaction, very reliable, high yielding, and tolerant of many functionalities. The main drawback is the modest reactivity of the organotin reactants, but this limitation can often be overcome by a judicious choice of experimental conditions. The nonreacting ligands are usually methyl or butyl, although newer dummy ligands have been proposed. Typically, the transferable fourth ligand on tin is an unsaturated moiety. The group migration order is believed to be alkynyl > vinyl > aryl > alkyl.

The Stille reaction was one of the earliest transition metal-catalyzed reactions to be accelerated with MW assistance. Single-mode irradiation with very short reaction times was easily applied during Stille reactions in solution [38] (Scheme 15.34) and on a resin support [38] (Scheme 15.35).

Different substrates for the Stille reaction have been used in two one-pot microwave-assisted hydrostannylation Stille-coupling sequences. High isolated

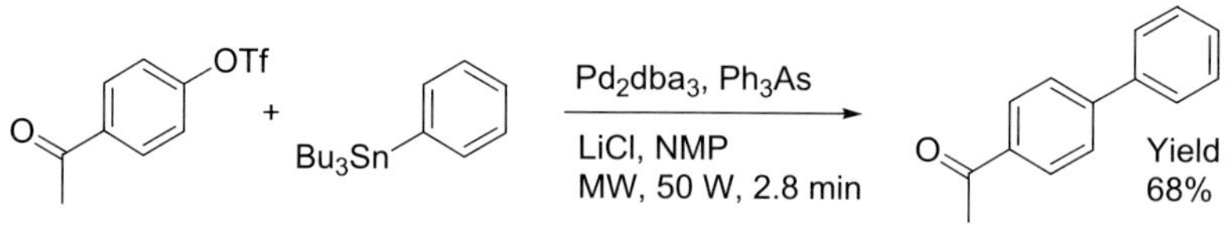

Scheme 15.34. Stille coupling in solution with 4-acetylphenyl triflate.

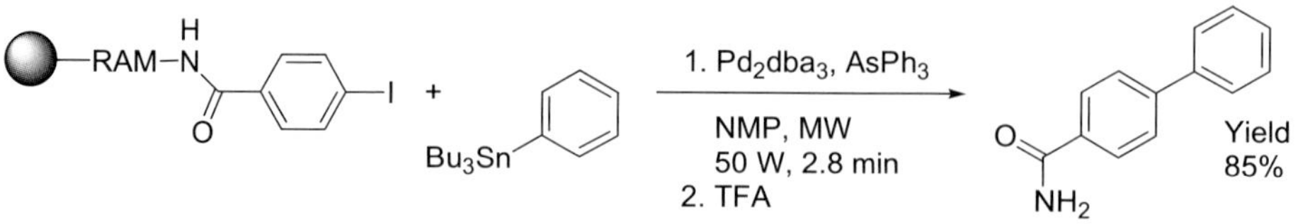

Scheme 15.35. Stille coupling using a RAM-linker on a polymer support.

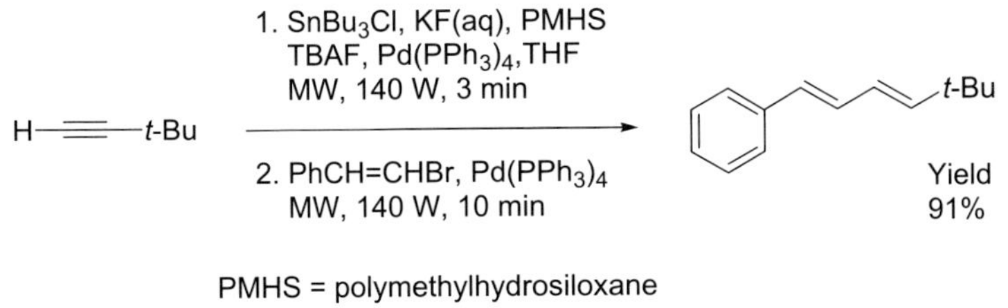

Scheme 15.36. One-pot hydrostannylation and Stille coupling.

Reagents/conditions shown:
1. SnBu₃Cl, KF(aq), PMHS, TBAF, Pd(PPh₃)₄, THF, MW, 140 W, 3 min
2. PhCH=CHBr, Pd(PPh₃)₄, MW, 140 W, 10 min

H—≡—t-Bu → Ph-CH=CH-CH=CH-t-Bu, Yield 91%

PMHS = polymethylhydrosiloxane

yields were reported in both of these papers (Scheme 15.36; see also Scheme 15.39) [76, 77].

Fluorous chemistry has been applied to Stille couplings as well as to Suzuki reactions as described previously. One of the many applications reported is the Stille coupling of tin reagents with fluorinated tags, in which the products and excess of the toxic tin-containing reagents can be easily separated from the reaction mixture and, in the case of the reagents, be recycled. One example of the use of the $-CH_2CH_2C_6F_{13}$ (shortened F-13) tagged organostannanes is presented in Scheme 15.37 [78].

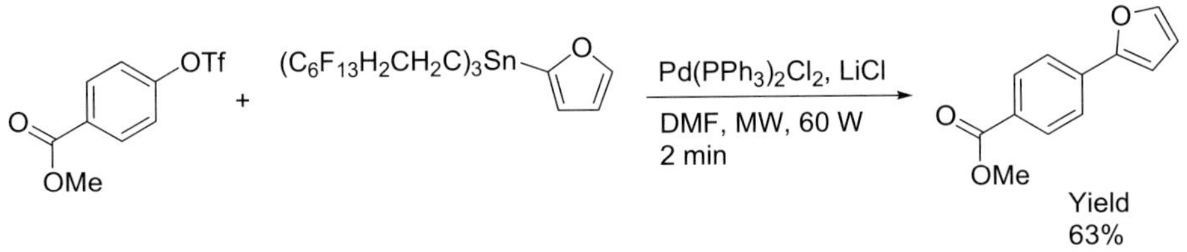

ArOTf + $(C_6F_{13}H_2CH_2C)_3$Sn-furan → Pd(PPh₃)₂Cl₂, LiCl, DMF, MW, 60 W, 2 min → Ar-furan, Yield 63%

Scheme 15.37. Stille reaction with the F-13-tagged furan stannane reagent.

It was sometimes apparent that the fluoricity of the F-13 tags was not enough to enable full partitioning of the products into the liquid fluorous phase. The concept of using more heavily fluorinated tags, for example the $-CH_2CH_2C_{10}F_{21}$ (F-21) tag, was easy to suggest but proved to be preparatively elusive, because the solubility of these compounds is very poor. Heating the reactions to 80 °C in fluorinated solvents resulted in very sluggish and irreproducible reactions. However, application of single-mode heating enabled rapid and efficient reactions in standard DMF (Scheme 15.38) [76]. The insolubility of the F-21-tagged compounds at room tem-

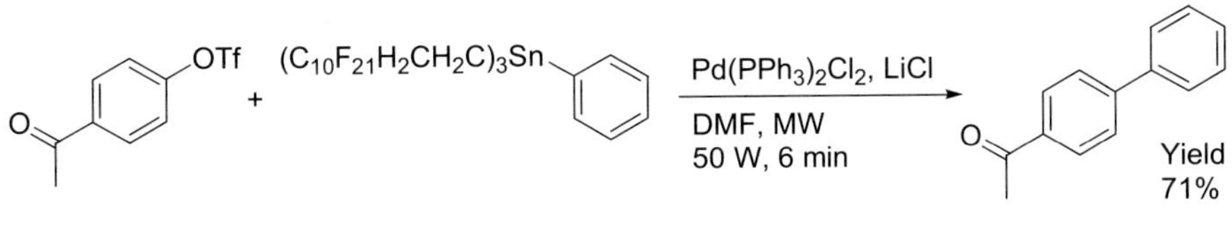

ArOTf + $(C_{10}F_{21}H_2CH_2C)_3$Sn-Ph → Pd(PPh₃)₂Cl₂, LiCl, DMF, MW, 50 W, 6 min → Ar-Ph, Yield 71%

Scheme 15.38. Stille reaction with the F-21-tagged phenyl stannane reagent.

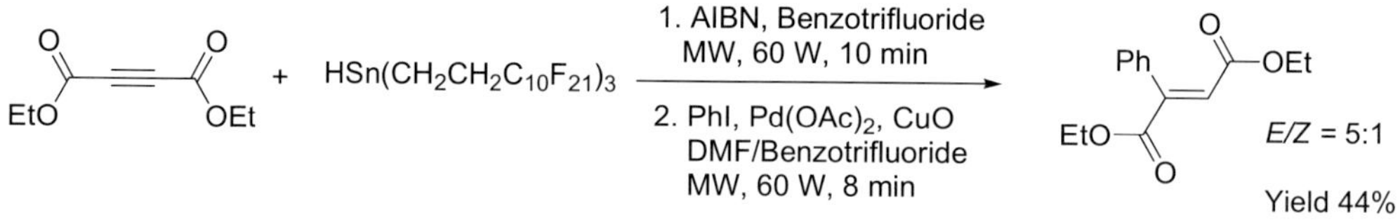

Scheme 15.39. One-pot hydrostannylation and Stille reaction with F-21-tagged reagents.

perature resulted in a very convenient method for removing the fluorous tin compound by filtration.

In the same paper the fluorous Stille procedure was applied to a one-pot hydrostannylation of an acetylene in the hybrid fluorous–organic solvent benzotrifluoride (BTF) with subsequent cross-coupling of the product in BTF–DMF, as shown in Scheme 15.39 [76].

A few important papers have appeared describing solvent-free Stille reactions on palladium doped Al_2O_3. Villemin performed several different reactions under MW irradiation conditions, including Stille couplings, with potassium fluoride on alumina as base. These reactions were attractive because the unpleasant and toxic stannous reagents and side-products remained absorbed on the solid Al_2O_3 support, thus enabling simplified work-up compared with classic Stille reactions. To ensure reproducibility, the use of a dedicated single-mode cavity was important [79].

2(1H)-Pyrazinones on polystyrene resins were more reactive in the Stille reaction than in the Suzuki–Miyaura coupling (Scheme 15.40) [57].

The Stille reaction has also been used for synthesis of melatonin derivatives. Two heating cycles were used to achieve a yield comparable with that of oil-bath heating. The reaction time, two irradiation cycles of 20 min, was notably shorter than the 24-h reaction with standard heating (Scheme 15.41) [58].

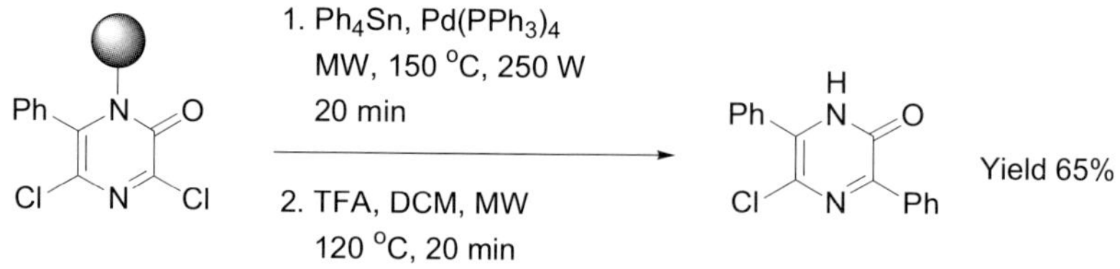

Scheme 15.40. Solid-phase reaction of 2(1H)pyrazinones.

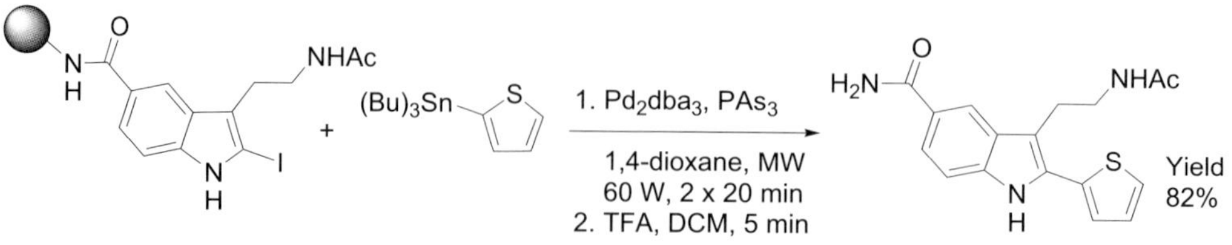

Scheme 15.41. Solid-phase synthesis of melatonin derivatives.

A Stille-related paper reported different regioselectivities in a cyclocarbopalladation in which the multistep reaction described terminated in a Stille cross-coupling. Mechanistic studies with deuterium provided a background for a discussion of the different outcomes when alkynyl or vinyl stannanes were used. Microwave heating was found to increase the rate of the reaction with two vinyl stannanes and one heteroaromatic stannane as substrates [80].

15.2.3
The Negishi Reaction

The first examples of MW-assisted cross-couplings with organozinc compounds (Scheme 15.42) were reported in 2001. Aryl and alkylzinc bromides were effectively coupled with short reaction times [81].

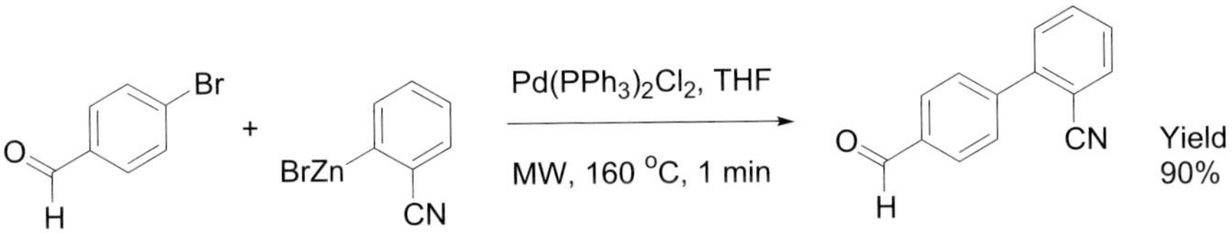

Scheme 15.42. Negishi coupling of an unprotected aryl bromide.

Kappe has reported a general method for microwave-assisted Negishi couplings. The organozinc reagents were prepared from activated Rieke zinc and aryl bromides or iodides. Nickel-catalyzed reactions were reported to result in high degrees of homo-coupling and could not be driven to completion with electron-rich, deactivated aryl chlorides. However, palladium in combination with electron-rich phosphines was found to be more effective with both electron-rich and poor aryl chlorides. (Scheme 15.43). n-Butylzinc chloride and resin bound aryl chlorides could also be coupled and the products could be isolated in good yield [82].

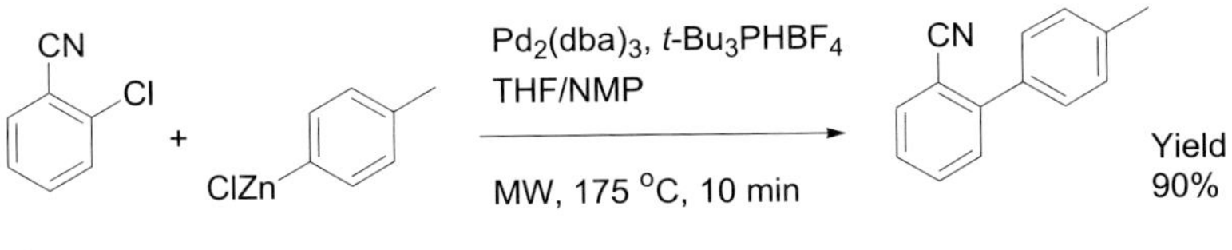

Scheme 15.43. Negishi coupling of an aryl chloride.

Enantiomerically pure 1,1'-binaphthyl derivatives have been prepared from binaphthyl iodides or triflates without loss of enantiomeric purity. The same reaction performed with oil-bath heating was associated with slower reactions and lower yields [83].

Different pyridinyl pyrimidines have been prepared by Negishi couplings (Scheme 15.44). The reported procedure took advantage of the lower hygroscopicity

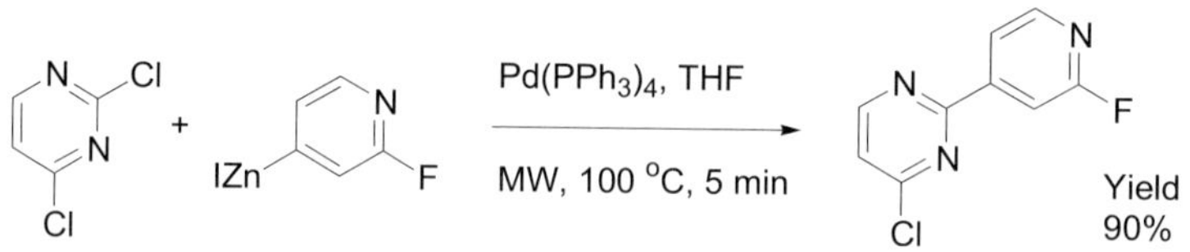

Scheme 15.44. Synthesis of pyridinyl pyrimidines by the Negishi reaction.

and higher solubility in ethereal solvents of zinc iodide compared with zinc chloride when the organozinc substrates were first prepared by conventional lithiation and subsequent transmetallation. The latter reactant usually resulted in a higher amount of the undesired homo-coupling products [84].

A different method of preparing the arylzinc reagents is the reaction between activated zinc dust and aryl iodides. The generated aryl-zinc reagent was then used in Negishi cross-couplings to generate thirteen biaryl formaldehydes in good to excellent yields (Scheme 15.45). Both nickel and palladium catalysts could be used, but palladium was chosen because of its superior performance in DMF – a solvent popular in microwave promoted reactions because of its high tan δ value [85]. It should be noted that DMF might decompose into carbon monoxide and dimethylamine at high temperatures [86].

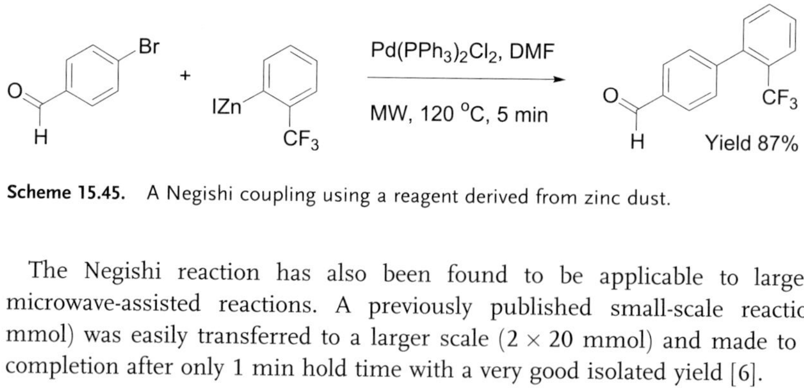

Scheme 15.45. A Negishi coupling using a reagent derived from zinc dust.

The Negishi reaction has also been found to be applicable to large-scale microwave-assisted reactions. A previously published small-scale reaction (1 mmol) was easily transferred to a larger scale (2 × 20 mmol) and made to go to completion after only 1 min hold time with a very good isolated yield [6].

15.2.4
The Kumada Reaction

Microwaves have been used both in the preparation of the Grignard reagent and in the Kumada coupling [87] with aryl chlorides. It was noted in this reaction that more homo-coupling side-products were typically formed when microwaves were used as an activation source than when the reaction was performed using ultrasound at ambient temperature (Scheme 15.46) [82].

Grignard reagents have been generated from sluggish aryl chlorides and bromides by use of controlled MW irradiation in a safe, productive, and reproducible

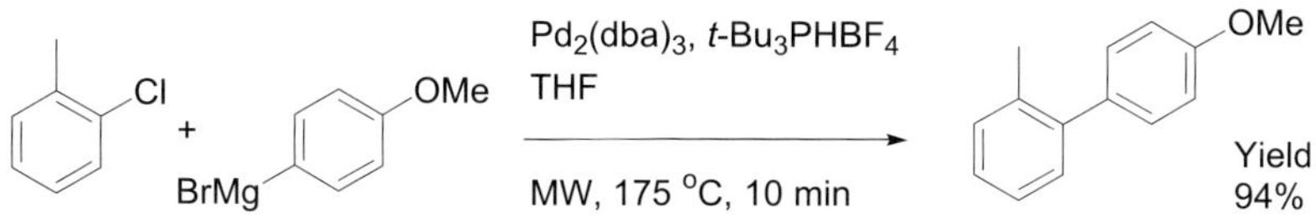

Scheme 15.46. Kumada reaction with aryl chlorides.

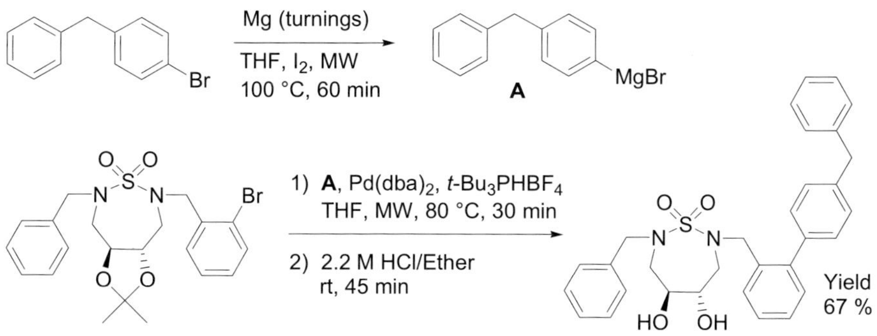

Scheme 15.47. Preparation of a cyclic HIV-1 protease inhibitor by Grignard and Kumada chemistry.

method. In the synthesis of a novel HIV-1 protease inhibitor, microwave irradiation was used both to generate the starting arylmagnesium halide and to promote the subsequent Kumada coupling [88] (Scheme 15.47).

15.2.5
The Hiyama Reaction

Clarke recently reported the first MW-accelerated Hiyama coupling [89, 90]. It was noted that the availability and nontoxic attributes of the organosilicon reactants make them very attractive in synthesis, but their low nucleophilicity limits their potential. Use of microwaves enabled aryl bromides and activated aryl chlorides to react under palladium catalysis with an electron-rich N-methyl piperazine/cyclohexyl phosphine ligand (Scheme 15.48). A vinylation reaction with vinyl trimethoxysilane was also reported [90].

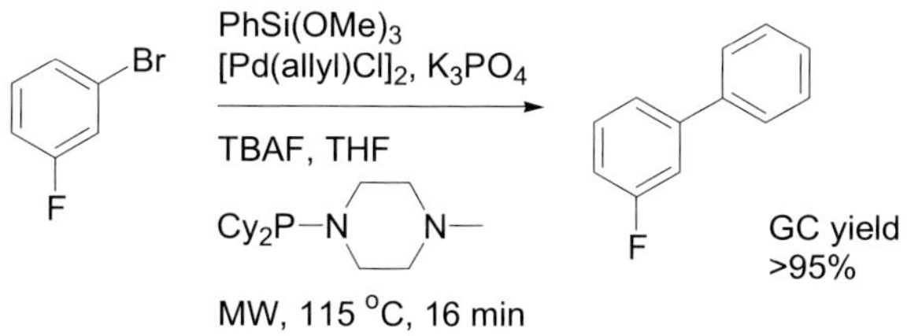

Scheme 15.48. Hiyama reaction with an aryl bromide.

15.3
Arylation of C, N, O, S, P and Halogen Nucleophiles

15.3.1
The Sonogashira Coupling Reaction

The copper or palladium-catalyzed cross-coupling between terminal alkynes and aryl/vinyl halides, the Sonogashira reaction, is a general and robust procedure enabling straightforward formation of unsymmetrical aryl alkynes [91]. In 2000, Kabalka and his group reported a solvent-free procedure employing an alumina-supported palladium catalyst for coupling of aryl iodides with alkyl and phenyl acetylenes in a domestic oven [92]. In 2001, Erdélyi and Gogoll published a pivotal work on the effect of directed microwave activation on efficiency and productivity in the Sonogashira coupling employing several different aryl precursors (see Scheme 15.49) [93].

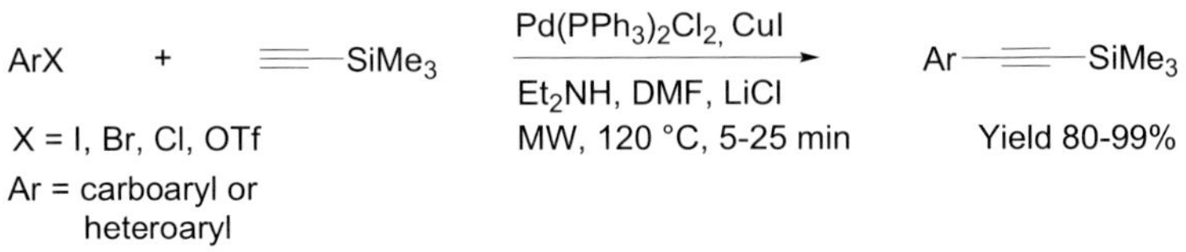

ArX + ≡—SiMe$_3$ → Ar—≡—SiMe$_3$

X = I, Br, Cl, OTf
Ar = carboaryl or heteroaryl

Pd(PPh$_3$)$_2$Cl$_2$, CuI
Et$_2$NH, DMF, LiCl
MW, 120 °C, 5-25 min

Yield 80-99%

Scheme 15.49. Palladium-catalyzed Sonogashira reaction with trimethylsilylacetylene.

Recent reports of successful MW-assisted Sonogashira reactions involve attachment of the aryl halide on a solid support [58, 94–96] (polystyrene and PEG 4000) and use of solvent-free conditions [97]. Interestingly, new nickel [98] and copper [99] catalyst systems have been introduced, and even "transition metal-free" reactions have been reported [100]. The example by Wang and coworkers using nickel catalysis deserves attention because it involves the use of 1,1-dibromostyrene precursors forming aryl acetylenes *in situ* (Scheme 15.50). The authors state that the reaction goes to completion without copper but that addition of copper greatly accelerates the reaction resulting in full conversion after irradiation for only 3 min in a domestic oven.

The possibility of performing Sonogashira reactions without use of transition metals is undoubtedly very appealing. By using microwaves Erik Van der Eycken and his group were able to couple phenyl acetylene with different aryl halides at 175 °C using sodium carbonate and tetrabutylammonium bromide in water [100]. Reaction times varied from 5 to 25 min usually with good to excellent yields. The

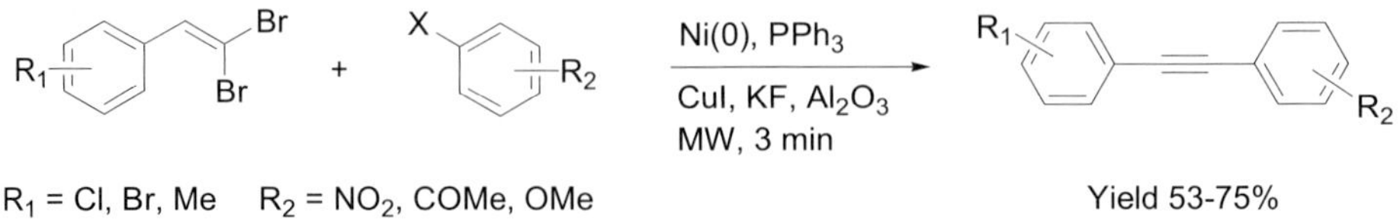

R$_1$ = Cl, Br, Me   R$_2$ = NO$_2$, COMe, OMe

Ni(0), PPh$_3$
CuI, KF, Al$_2$O$_3$
MW, 3 min

Yield 53-75%

Scheme 15.50. A solvent-free Sonogashira coupling employing nickel catalysis.

crude reaction mixture was studied by atomic absorption spectrophotometry to exclude the possibility of contamination with levels of significant transition metals exceeding 1 ppm. Leadbeater has also reported a similar transition metal-free water-based procedure using PEG instead of TBAB as the phase-transfer reagent [101]. However, in the light of Leadbeater's recent finding that Suzuki couplings can proceed with palladium concentrations as low as 0.5 ppb it is still debatable whether the term "transition metal-free" can be used mechanistically [102].

15.3.2
The Nitrile Coupling

The synthesis of aryl or alkyl nitriles from halides is valued in medicinal chemistry, because the nitriles themselves constitute a flexible building block that easily can be converted into carboxylic acids, amides, amines, or a variety of heterocyclic compounds [103], for example thiazoles, oxazolidones, triazoles, and tetrazoles [104]. The importance of the tetrazole group in medicinal chemistry is easily understood if one remembers it is the most commonly used bioisostere of the carboxyl group.

An improvement of the palladium-catalyzed cyanation of aryl bromides, in which zinc cyanide was used as the cyanide source, was reported in the middle of the nineties [105]. Typically, conversion from halide to nitrile required at least 5 h by this method and the subsequent cycloaddition to the tetrazole is known to require even longer reaction times. The Hallberg group has described a single-mode MW procedure, using zinc cyanide, for palladium-catalyzed preparation of both aryl and vinyl nitriles from the corresponding bromides [106]. The reaction times were short and full conversion was achieved in just a few minutes (Scheme 15.51).

In the same publication, a nitrile coupling followed by a subsequent cycloaddition, forming a tetrazole, was executed as a one-pot procedure on a TentaGel-support, as depicted in Scheme 15.52. Negligible decomposition of the solid support was reported.

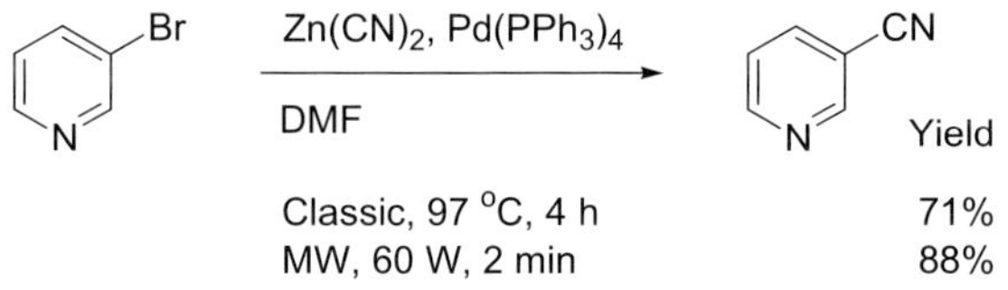

Scheme 15.51. Palladium-catalyzed conversion of aryl bromides into aryl nitriles.

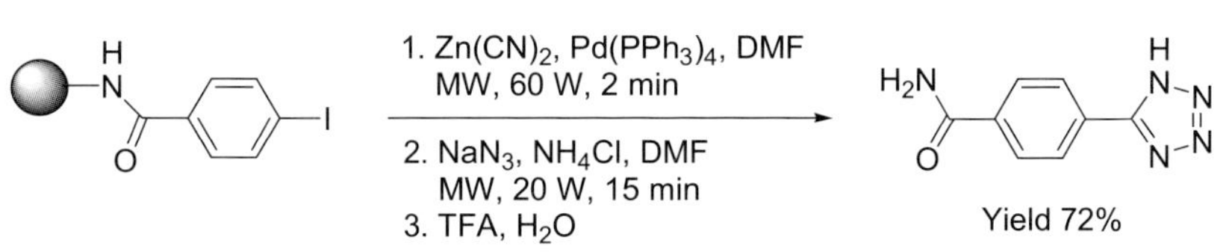

Scheme 15.52. One-pot procedure for tetrazole synthesis on polymer support.

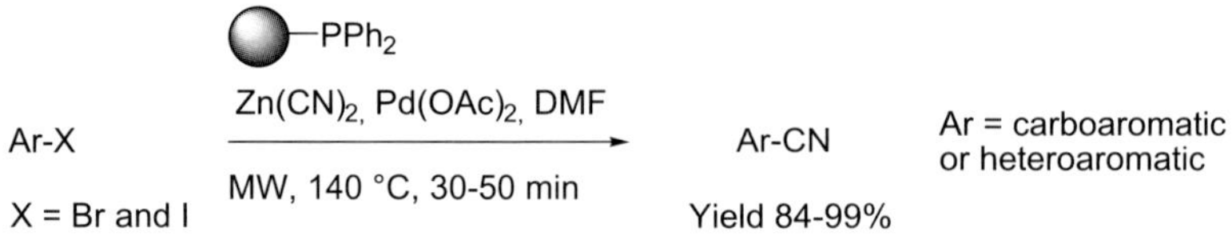

15 Microwave-assisted Transition Metal-catalyzed Coupling Reactions

Ar–X →[Zn(CN)₂, Pd(OAc)₂, DMF / MW, 140 °C, 30–50 min] Ar–CN

X = Br and I

Ar = carboaromatic or heteroaromatic

Yield 84–99%

Scheme 15.53. Aryl cyanation using palladium and polymer-supported triphenyl phosphine.

In the last few years several articles have been published on MW-enhanced aryl halide cyanation employing nickel [107], palladium [108, 109], and copper [110, 111] catalysts. Both water [111] and ionic liquids [112] have proven useful as solvents for these transformations. Srivastava and Collibee have reported a rapid and high-yielding procedure in which polymer-supported triphenylphosphine enables easy purification [108]. As shown in Scheme 15.53, both bromides and iodides can be activated using palladium catalysis in DMF. Although reaction times have not been optimized, the overall process time involving simple filtration and extraction for compound isolation seems to be short, thus rendering the protocol well adapted to high-throughput synthesis.

Aryl triflates, which are readily prepared by use of microwaves [113] from the corresponding aryl alcohol using triflic anhydride or triflic imide, are important halide alternatives but are, occasionally, poorly reactive. Nevertheless, Zhang and Neumeyer have reported palladium-catalyzed activation of triflates in nitrile couplings for preparation of different κ-opioid receptor ligands [109]. As shown in Scheme 15.54, a reaction time of 15 min at 200 °C, in sealed reaction vessels, was sufficient for complete displacement, generating yields between 86 and 92%.

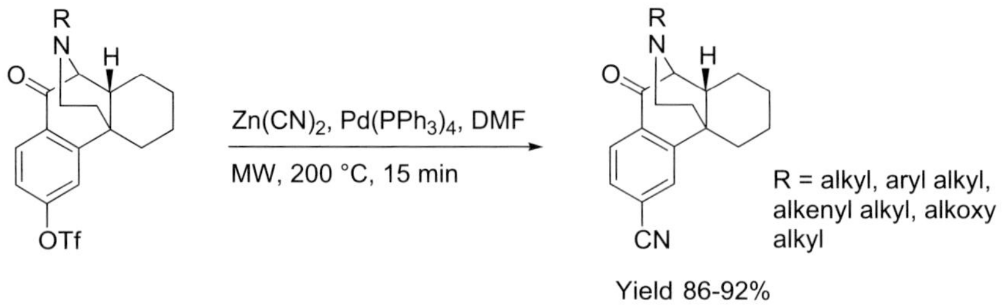

Zn(CN)₂, Pd(PPh₃)₄, DMF
MW, 200 °C, 15 min

R = alkyl, aryl alkyl, alkenyl alkyl, alkoxy alkyl

Yield 86–92%

Scheme 15.54. Smooth cyanation using aryl triflates and palladium catalysis.

15.3.3
Aryl–Nitrogen Coupling

The seminal work by the groups of Hartwig and Buchwald in 1994 on aryl amination chemistry has spurred substantial research on C–N bond formation in general and aryl–nitrogen bond formation in particular [114]. Catalytic aryl amine couplings are usually slow processes, especially when copper catalysis is used, often

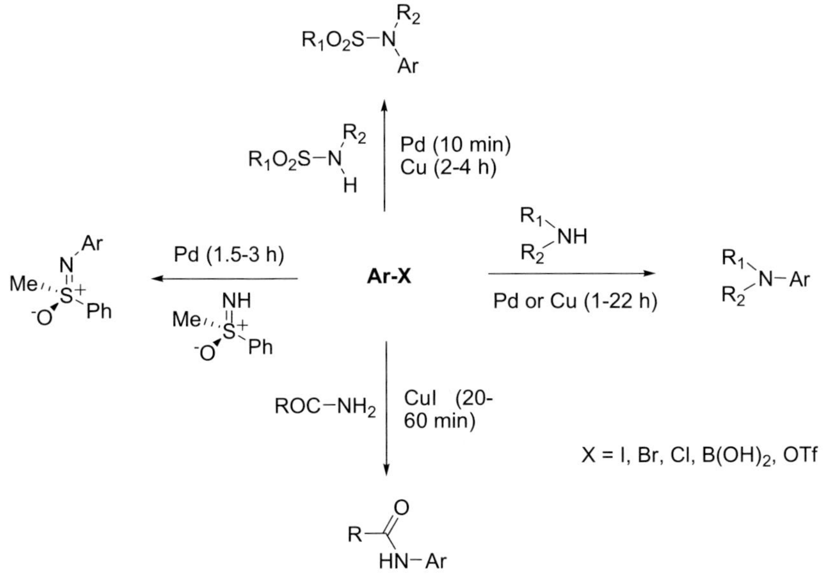

Scheme 15.55. Microwave-assisted copper and palladium-catalyzed N-arylations.

demanding days for completion. Thus several MW-enhanced methods have been developed in the wake of all newly discovered catalytic procedures in this area. Some of the transformations that will be discussed are summarized in Scheme 15.55.

In work published by Sharifi using a domestic oven and nonpressurized vessels, couplings between aryl bromides and alkyl amines were performed with varying results (yields 32–86%) [115]. The palladium precatalyst $Pd[P(o\text{-tolyl})_3]_2Cl_2$ was found to be most efficient when toluene was used as solvent and sodium tert-butoxide as base. Almost simultaneously a similar method for arylation of alkyl and aryl amines was reported by Hallberg and coworkers [116]. Directed MW irradiation of the closed reaction vessel, following the original procedure introduced by Hartwig and Buchwald, caused the transformations to be complete in only 4 min. In a medicinal chemistry project executed by Skjaerbaeck's group microwaves were integrated into a procedure to synthesize p38 MAP kinase inhibitors [117]. Systematic optimization of the catalyst, solvent, base, and reaction time/temperature resulted in a general procedure for high-speed production of the desired aryl aminobenzophenones within 3–15 min, as illustrated in Scheme 15.56.

Amination of azaheteroaryl bromides and chlorides has also been reported to be smoothly executed within 10 min by use of standard reaction conditions and microwave irradiation [118]. Benzimidazoles have been prepared by Brain and Steer via intramolecular cyclization using an amidine moiety as the N-nucleophile [119]. The reaction was rapid and high yielding, and in combination with a "catch and release" strategy featuring capture of the benzimidazole on an acidic resin in the

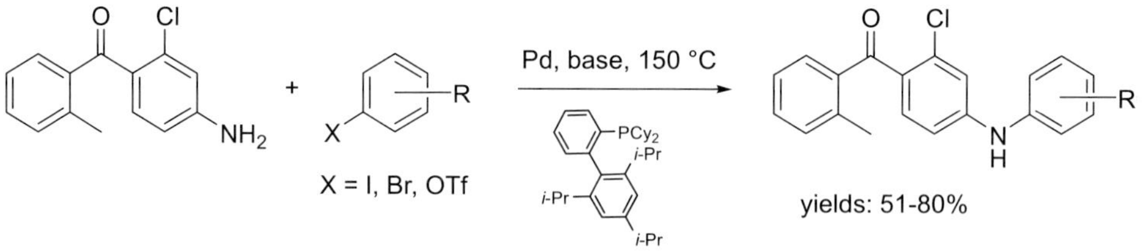

Scheme 15.56. Synthesis of p38 MAP kinase inhibitors by use of rapid microwave chemistry.

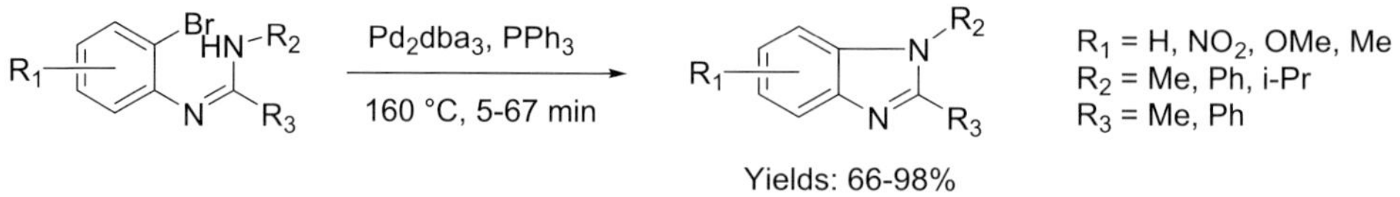

Scheme 15.57. Palladium-catalyzed intramolecular amidination producing different benzimidazoles.

purification step, a very appealing procedure for rapid compound production was obtained (Scheme 15.57).

Rapid aminations of aryl chlorides and bromides using amine resins as the nitrogen nucleophile have been developed by Weigand and Pelka [120]. The normally very sluggish reaction (18 h, reflux) between the polystyrene Rink resin and electron-poor chlorides and bromides could be performed within 15 min under the action of microwave irradiation in a closed vessel (solvent DME–t-BuOH, 1:1) at 130 °C. This high-speed reaction was equally high yielding as the classic method.

Aryl chlorides are more reluctant to undergo amination than most other aryl halides/pseudohalides. To address this problem, Caddick and coworkers investigated the outcome of palladium–N-heterocyclic carbenes as catalysts in rapid microwave promoted reactions [121]. para-Tolyl and para-anisyl chloride were coupled with aromatic and aliphatic amines in generally good yields within 6 min at 160 °C. Reactions between tolyl/anisyl/phenyl chlorides and aliphatic amines have also been reported by Maes et al. using a more classic reaction system with a phosphine ligand and a strong base; the reaction afforded the desired products after 10 min irradiation at 110–200 °C [122].

Copper-catalyzed N-arylation, commonly referred to as the Ullmann coupling, is recognized to be more sluggish than the corresponding palladium-catalyzed transformation. Nevertheless, Wu and coworkers have managed to accelerate reactions to only 1 h reaction time with retained chemoselectivity (Scheme 15.58) [123]. For a set of aromatic aza-heterocycles, yields were between 49 and 91% and processing times between 1 and 22 h.

In 1999 Combs and colleagues published a procedure for N-arylation of imidazoles, pyrazole, and 1,2,3-triazole attached to a solid support [124]. Interestingly, they used p-tolyl boronic acid as the arylating agent when employing Cu(II) cataly-

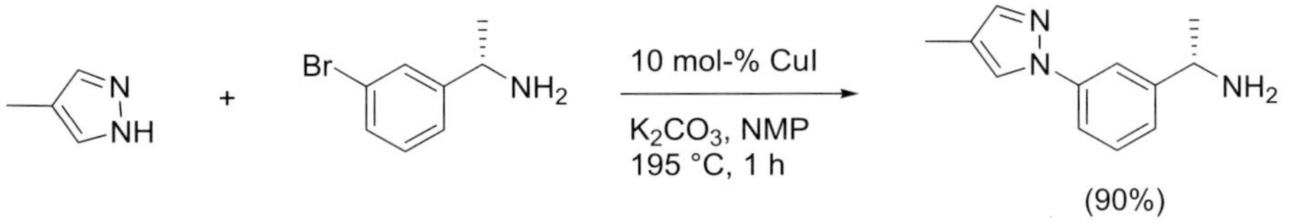

Scheme 15.58. Microwave-promoted Ullman coupling.

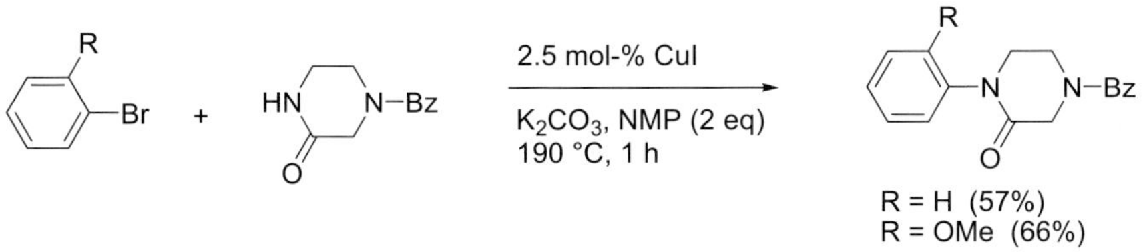

Scheme 15.59. Copper-catalyzed Goldberg amidation of phenyl and anisyl bromide.

sis. The reaction mixture was irradiated in a domestic oven for three ten-second periods with manual agitation in between, producing 55–64% yield of the cleaved product. According to Lange the closely related Goldberg reaction at high concentrations in the production of N-aryl-2-piperazinones also benefits from use of MW (Scheme 15.59) [125].

Dihydropyrimidones are valuable templates for development of pharmaceuticals and, consequently, a method for efficient N-arylation of the urea moiety, outlined in Scheme 15.60, has been reported by the groups of Larhed and Kappe [60]. As the authors remark in the paper, N3-arylated dihydropyrimidone derivatives cannot be synthesized by common Biginelli condensations, thus increasing the utility of this rapid and convenient procedure for introduction of chemical diversity to the heterocyclic backbone.

An efficient intramolecular, one-pot, two-step, Goldberg aryl amidation producing useful N-substituted oxindoles was recently described by Turner and Poondra (Scheme 15.61) [126].

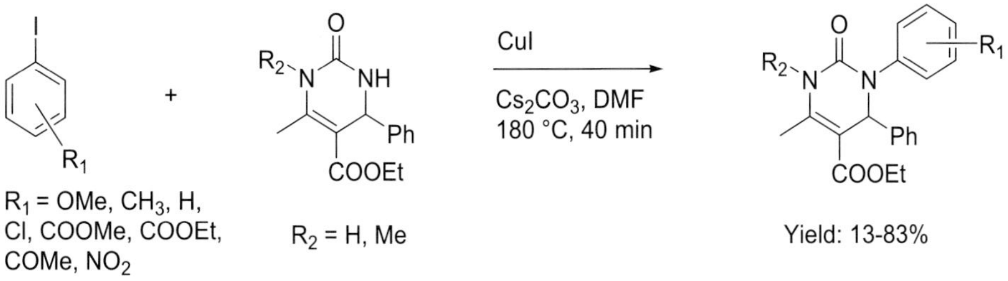

Scheme 15.60. Copper-catalyzed arylation of dihydropyrimidones.

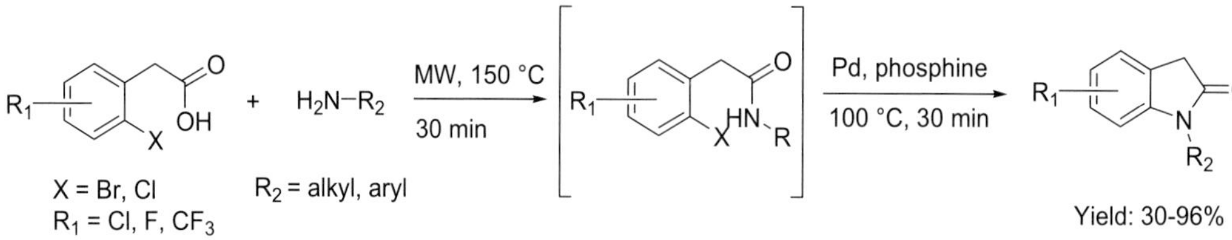

Scheme 15.61. Two-step construction of substituted oxindoles.

The N-arylsulfonamide moiety is ubiquitous in medicinal chemistry, because of its varied pharmacological profile. The possibility of connecting aryl groups directly to the sulfonamide, via an N-aryl coupling, is an attractive alternative to the reaction between sulfonyl chloride and arylamines, especially because arylamines are often poor nucleophiles. Wu and He recently reported a simple and straightforward MW procedure, employing aryl iodides and bromides, for copper-catalyzed N-arylations [127]. Under sealed vessel conditions at 195 °C, moderate to good yields (54–90%) were obtained within 2–4 h using NMP as the solvent and potassium carbonate as the base. A possibly even more productive method has been published by Cao and his colleagues, who used palladium catalysis and aryl chlorides to arylate several aryl and alkyl sulfonamides (Scheme 15.62) [128].

Aryl chlorides were found to successfully couple with methyl phenyl sulfoximines in a series of experiments reported by Harmata's group [129]. Using palladium acetate and binap with a large excess of aryl chlorides as coupling partners and cesium carbonate as the base, yields between 10–94% were obtained after one or two 1.5-h irradiation periods at 135 °C. Switching to an aryl triflate and use of an excess of the sulfoximines (5 equiv.) furnished an impressive 94% yield (Scheme 15.63).

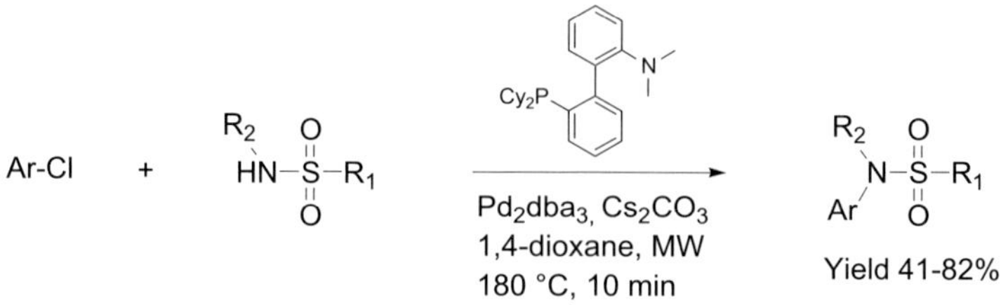

Scheme 15.62. Palladium-catalyzed N-arylation of sulfonamides.

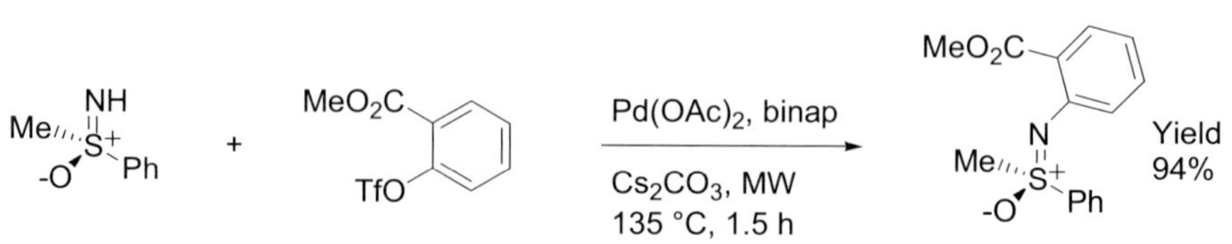

Scheme 15.63. Palladium-catalyzed N-arylation of a sulfoximine using an aryl triflate.

15.3.4
Aryl–Oxygen Bond Formation

The coupling of organohydroxy compounds with aryl halides using copper or palladium species as catalysts is a commonplace method for producing aryl ethers. The copper-catalyzed Ullmann aryl ether synthesis is attractive because the metal is cheap and functional group tolerance is very good. However, compared with the palladium-catalyzed counterparts Ullmann reactions require larger amounts of catalyst and the reactions are generally slower. To deal with these problems, Stockland Jr and coworkers have developed a method using organosoluble copper clusters [130]. A series of alkyl aryl ethers was prepared using only 0.4 mol% copper cluster to evaluate conventional and microwave heating (Scheme 15.64). The yields obtained after classic heating for 11 h at 110 °C were comparable with the MW results. When the reactions were performed under an air atmosphere, however, yields dropped for both methods of activation.

He and Wu have developed a method for arylation of phenols using aryl iodides and bromides [131]. It is notable that coupling of 1-iodo-4-t-butylbenzene with phenol was conducted both thermally and with directed MW heating, producing 74% and 90% yields, respectively, at the same reaction temperature (195 °C). Unfortunately, the authors report the procedure to be incompatible with the less expensive aryl chlorides (Scheme 15.65).

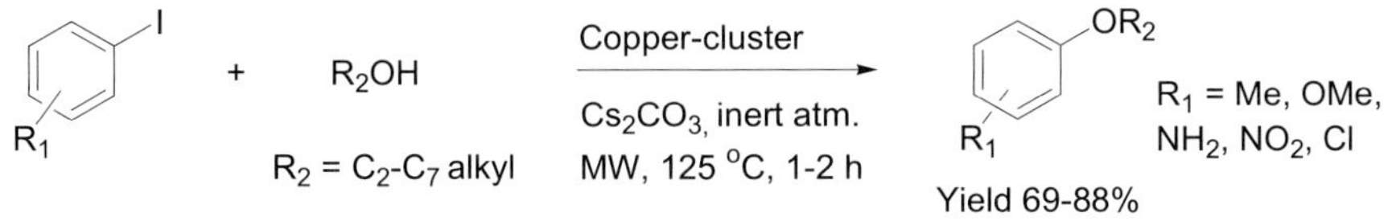

Scheme 15.64. Copper-catalyzed generation of alkyl aryl ethers.

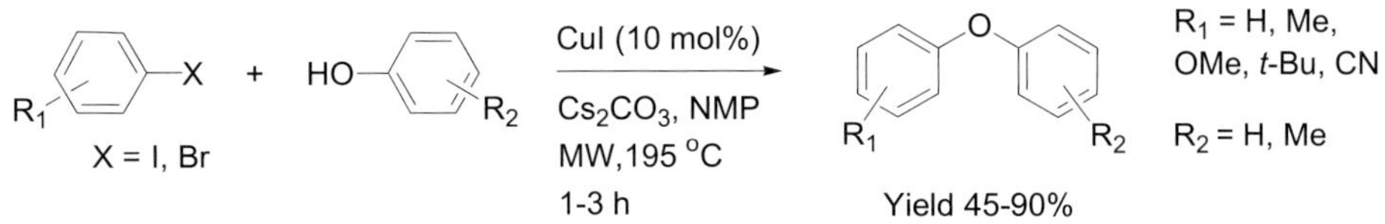

Reaction conditions: ArI + R_2OH, $R_2 = C_2$–C_7 alkyl, Copper-cluster, Cs_2CO_3, inert atm., MW, 125 °C, 1-2 h. Product: aryl ether with R_1 = Me, OMe, NH_2, NO_2, Cl. Yield 69-88%.

Scheme 15.65. Copper(I)-catalyzed synthesis of diaryl ethers from aryl iodides and bromides.

Reaction conditions: Ar-X + HO-Ar', X = I, Br, CuI (10 mol%), Cs_2CO_3, NMP, MW, 195 °C, 1-3 h. Product: diaryl ether with R_1 = H, Me, OMe, t-Bu, CN; R_2 = H, Me. Yield 45-90%.

15.3.5
Aryl–Phosphorus Coupling

The conversion of aryl iodides to aryl phosphonates, useful precursors to aryl phosphonic acids, has been conducted in a Teflon autoclave by Villemin and colleagues [132]. A domestic MW oven was used for these experiments and the reaction times using classic heating were effectively reduced from 10 h to 4–22 min. The reactiv-

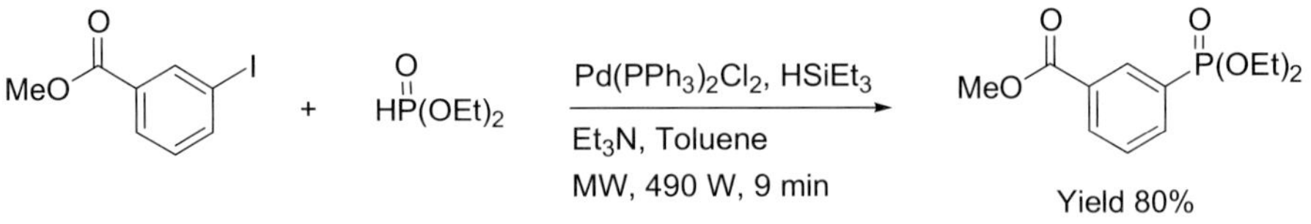

Scheme 15.66. Rapid phosphonylation of methyl 3-iodobenzoate.

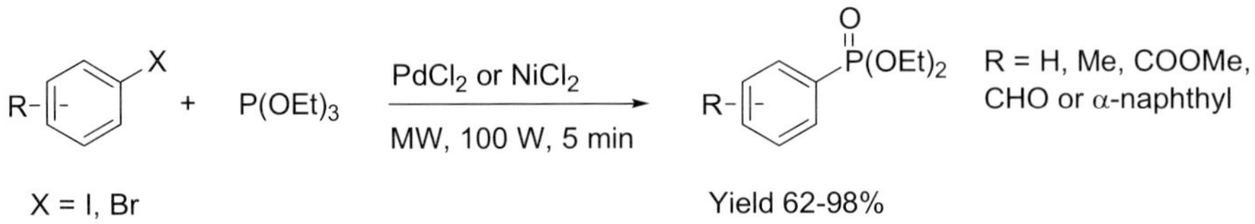

Scheme 15.67. Palladium and nickel-catalyzed preparation of aryl phosphonates.

ity of aryl iodides was good whereas use of bromides resulted in lower yields and triflates in very slow reactions (Scheme 15.66). It is interesting that the reactions were brought to completion with short reaction times in the nonpolar solvent toluene.

In a separate study, Villemin achieved coupling between triethyl phosphite and aryl halides. The process was successfully catalyzed by nickel and palladium among the transition metals investigated (Ni, Pd, Co, Fe, Cu) [133]. Using sealed vessels and an inert atmosphere, reactions could be accomplished within 5 min reaching a final temperature of approximately 200 °C (Scheme 15.67).

Kappe and Stadler have developed an MW procedure for rapid production of triaryl phosphines by coupling diphenylphospine with aryl halides and triflates [134]. Taking into account the importance of phosphine ligands in a variety of transition metal-catalyzed reactions, convenient procedures for their production is valuable. Both homogeneous Pd–Ni and heterogeneous Pd catalysts were explored and the more unusual substrate phenyl triflate could also be coupled swiftly by use of nickel catalysis (Scheme 15.68). Couplings with other aryl halides proceeded in 26–85% yield after 3–30 min microwave irradiation at 180–200 °C.

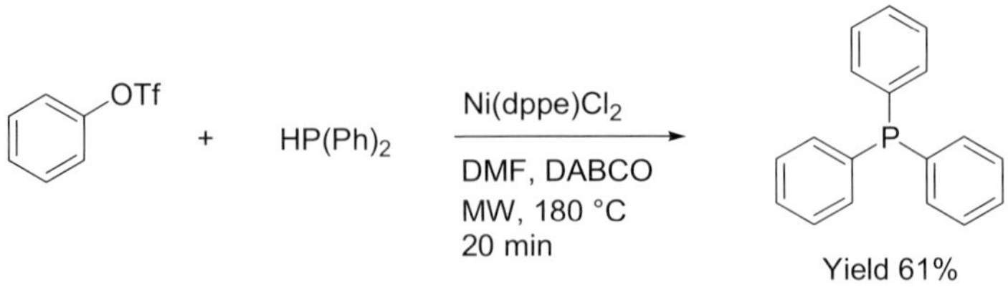

Scheme 15.68. Nickel-catalyzed synthesis of triphenylphosphine.

15.3.6
Aryl–Sulfur Bond Formation

Thiation of arenes is used both for preparation of aryl sulfides and for generation of heterocyclic sulfur aromatic compounds. The diaryl thioether group is found in the structure of several approved pharmaceuticals, for example antihistamines. In a recent letter Wu and He disclosed that Cu(I) catalysis is very efficient for coupling of aryl thiols with aryl iodides (Scheme 15.69) [135]. The procedure is very similar to that reported by the same authors for diaryl ether synthesis under the action of microwave irradiation (Scheme 15.65).

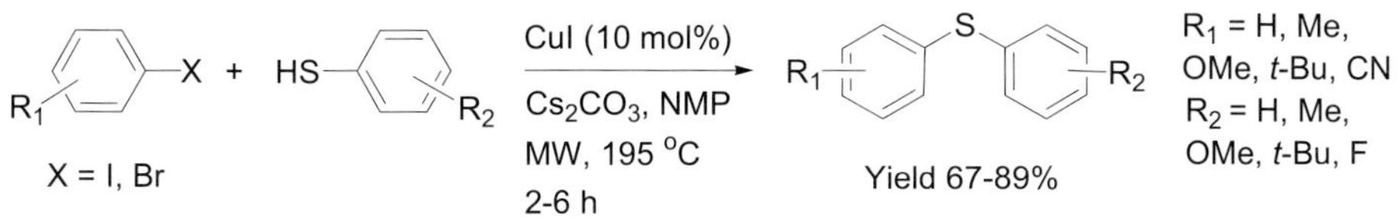

Scheme 15.69. Formation of a diaryl sulfide linkage with copper catalysis.

Sulfur-containing aromatic heterocyclic compounds are very common in biologically active compounds and, accordingly, Besson's group has developed an intramolecular aryl sulfur coupling to establish a benzothiazol substructure during a multistep synthesis [136]. The cyclization–elimination process was conveniently performed under the action of microwave irradiation for 15 min at 115 °C (75% yield). In a previous report the same group investigated the scope and limitation of this key-step transformation as presented below in Scheme 15.70 [137]. All reactions were duplicated using conventional heating (oil-bath) at reflux temperature and produced similar yields after 45–60 min.

Sulfonylation of arenes is normally performed using sulfonyl chloride and a stoichiometric amount of a Lewis or Brönstedt acid as the catalyst. Dubac and coworkers found a practical method using MW high-temperature conditions in which only 5–10 mol% $FeCl_3$ (relative to the sulfonyl chloride) sufficed for complete reactions to occur [138]. A number of arenes encompassing alkylbenzenes, anisole, and halobenzenes were sulfonylated by use of several different arylsulfonyl

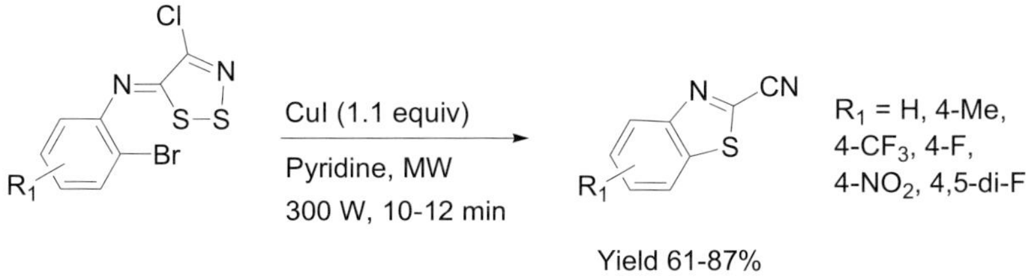

Scheme 15.70. Benzothiazole formation via copper-catalyzed cyclization–elimination.

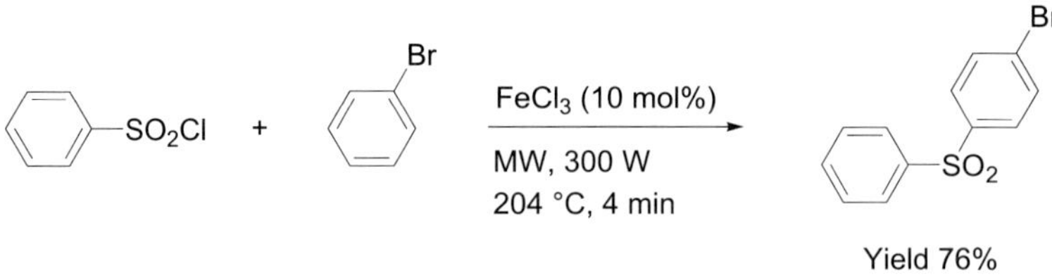

Scheme 15.71. Regio and chemoselective sulfonylation catalyzed by iron(III) chloride.

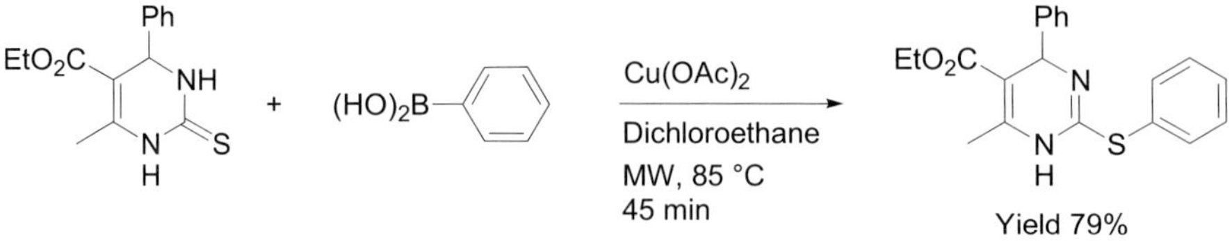

Scheme 15.72. Swift phenylation of a dihydropyrimidin-2-thione using phenyl boronic acid under stoichiometric Cu(II) conditions.

chlorides. A representative example is depicted in Scheme 15.71. The sulfonylations usually occurred with good para regioselectivity, with the exception of the electron-rich bromoanisole substrate.

Kappe and Lengar have shown the versatility of microwaves in a sulfur phenylation of a thiourea [139]. Under pressurized conditions and rapid microwave irradiation the reactions could be completed within an hour, as presented in Scheme 15.72. The corresponding standard reaction performed at room temperature with dichloromethane takes four days to reach completion, delivering a similar yield (72%).

15.3.7
Aryl Halide Exchange Reactions

Halogen exchanges in aryl halides are important for several reasons:

1. Several useful transition metal-catalyzed processes exploit the halide as a leaving group to form the essential aryl metal complex in the catalytic cycle. Because reactivity differs substantially among the halides, depending on the chosen reaction conditions, a halogen exchange reaction can obviate problems at a later stage.
2. In drug optimization endeavors, a common practice is to introduce fluorine or chlorine in metabolically sensitive positions, because of their bioisosteric properties with hydrogen, thus blocking, for instance, hydroxylation by cytochrome P-450.

3. Incorporation of ^{124}I, ^{76}Br, and ^{18}F nuclides is useful for radiotracer synthesis and subsequent positron emission tomography (PET) imaging studies [140]. Because radionuclides are extraordinarily expensive and often have short half-lives it is necessary to introduce the radioactive nucleus at a very late stage of the synthesis to achieve good radiochemical yields [141]. The time to complete the reaction involving the radioactive species is also highly critical when using short-lived nuclides.

The benefit of using high-speed microwave-promoted reactions is, in this sense, indisputable. Illustrated below is the important contribution to this topic made by Leadbeater and his research group by development of MW-enhanced nickel-catalyzed halogen-exchange procedures (Scheme 15.73) [142]. The yields obtained were good to excellent except for 4-iodophenol (only 3% when reacted with $NiCl_2$). A procedure for oil-bath heating was also disclosed using a closed vessel at 170 °C for 4 h, furnishing equivalent yields. Unfortunately, no procedure for fluorine incorporation was reported and, moreover, activation of heteroaryl substrates would significantly expand the scope and utility of the procedure.

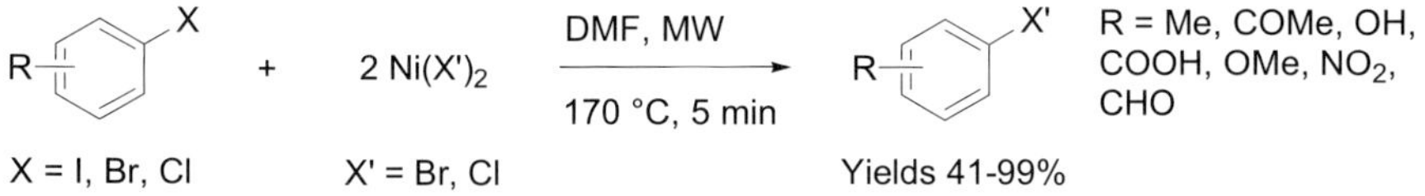

Scheme 15.73. Microwave-promoted nickel-catalyzed halogen-exchange reactions.

15.4
The Heck Reaction

The Heck reaction, a palladium-catalyzed vinylic substitution, is conducted with olefins; organohalides or pseudohalides are frequently used as organopalladium precursors [143]. One of the strengths of the method is that it enables direct monofunctionalization of a vinylic carbon, which is difficult to achieve by other means. The Heck arylation in Scheme 15.74, reported in 1996, was the first example of a microwave-promoted, palladium-catalyzed C–C bond formation [18]. The power

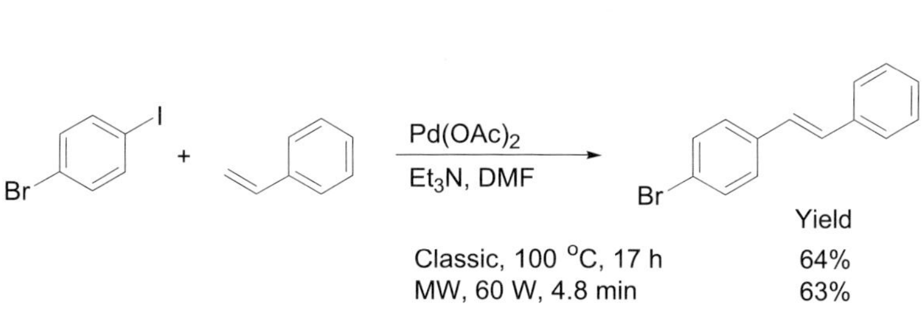

Scheme 15.74. Chemoselective Heck coupling of 4-bromoiodobenzene and styrene.

of the flash-heating method is amply manifested by the short reaction times and good yields of these couplings. The reactions were conducted in a single-mode cavity in septum-sealed Pyrex vessels without temperature control. The reaction in Scheme 15.74 (and in seven additional Heck coupling examples) was originally conducted with classic heating in the absence of solvent. To enhance yields and reduce reaction times, 0.5 mL DMF was added to increase the polarity and dielectric loss tangent of the reaction mixture. This small modification of the original reaction conditions enabled isolation of the products in high purity after very short reaction times (2.8–4.8 min). The same high chemo and regioselectivity as experienced with classical, oil-bath heating was found to apply to these MW promoted reactions [18].

Compared with the dramatic development of MW techniques in cross-coupling and N-arylation chemistry, the number of recent MW-assisted Heck reactions remains limited. Thus, only five selected new examples emphasizing different concepts will be presented in this section.

The use of ionic liquids in combination with MW irradiation has great benefits, because the high boiling point and low vapor pressure of ionic liquids is combined with a propensity to interact strongly with microwave fields (Chapter 7). 1-Butyl-3-methylimidazolium hexafluorophosphate (bmimPF$_6$) was therefore recently evaluated as a solvent for the Heck reaction. Terminal arylations of electron-poor butyl acrylate, using palladium chloride as the precatalyst, were conducted under the action of high-density irradiation and afforded good to excellent yields (Scheme 15.75) [144]. The authors also showed that the catalyst was immobilized in the ionic liquid, enabling recycling of the "ionic catalyst phase" in five consecutive Heck reactions.

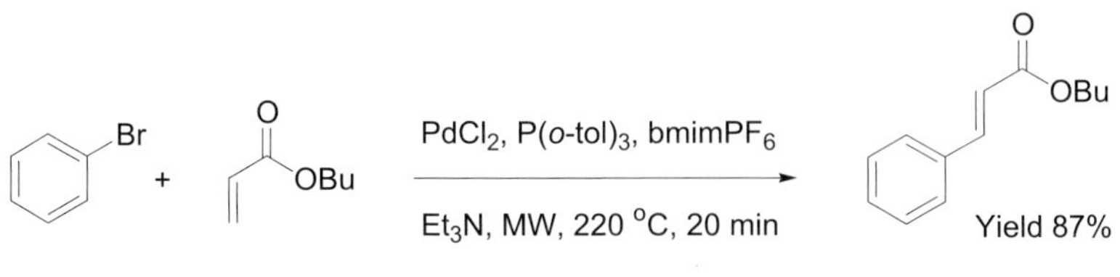

Scheme 15.75. Heck arylation with an ionic liquid as solvent.

The recent advances using the relatively cheap and readily available chloroarenes in organometallic chemistry, instead of bromo or iodo-arenes, is arguably one of the most exciting developments in chemistry today [21]. A paper published in 2003 dealt with Heck couplings performed with both activated and deactivated chloroarenes in ionic liquid-doped 1,4-dioxane [145]. The coupling of butyl acrylate and 2-chloro-m-xylene took 1 h at 180 °C when microwaves were used whereas standard heating at the same temperature required 1.5 h and resulted in a reduced yield (Scheme 15.76).

Microwave-promoted Heck reactions in water using ultra-low concentrations of palladium catalyst have also been performed. Different catalyst concentrations

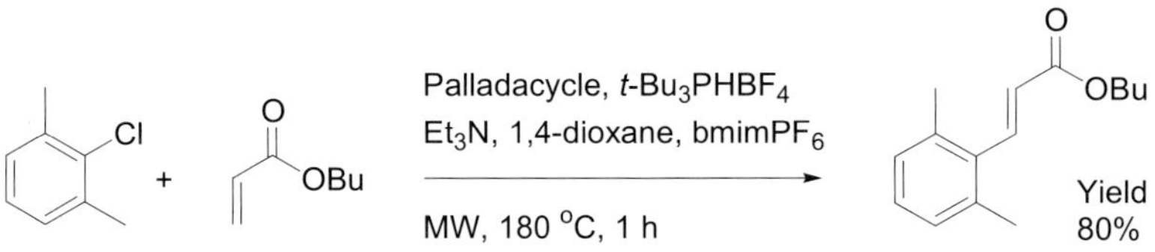

Scheme 15.76. Heck coupling of aryl chlorides.

were investigated, using a commercially available 1000 ppm palladium solution as the catalyst source [146]. Impressively, useful Heck arylations were performed with palladium concentrations as low as 500 ppb.

Heck vinylation of electron-rich olefins usually affords the branched 1,3-butadiene product [143]. However, by incorporating a palladium(II)-coordinating tertiary amino group into a vinyl ether efficient substrate presentation and full terminal selectivity were realized. These highly regioselective vinylations were complete in less than 30 min under single-mode MW irradiation conditions, compared with overnight reactions with conventional heating [147]. Slightly lower E/Z stereoselectivity and chemical yields were often obtained in the high-temperature microwave-mediated couplings compared with the corresponding traditional reactions (Scheme 15.77).

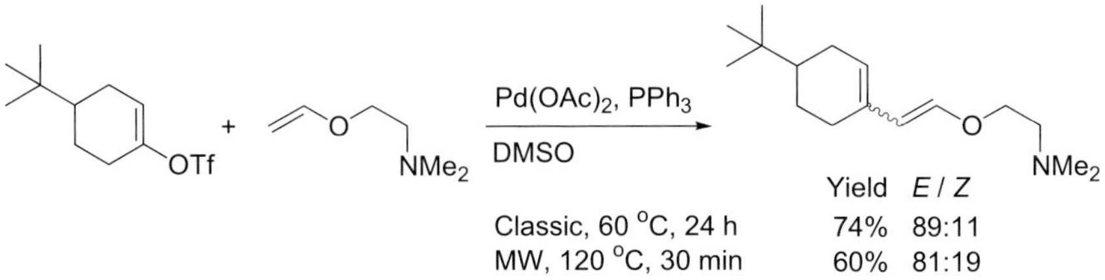

Scheme 15.77. Terminal Heck vinylation of chelating vinyl ether.

2,3-Epoxycyclohexanone is an unusual substrate for the Heck reaction. The reactivity of this molecule under Heck coupling conditions is most likely attributed to its *in-situ* isomerization to 1,2-cyclohexanedione. The 1,2-diketone is subsequently reacting as an olefin via the enol tautomer. Thus, within 5 to 30 min of directed microwave irradiation of the aqueous PEG mixture, with less than 0.05% palladium acetate and no phosphine ligand, up to 72% C3-arylated product was isolated (Scheme 15.78) [148].

15.5
Carbonylative Coupling Reactions

The palladium-catalyzed carbonylation reaction with aryl halides is a powerful method for generating aromatic amides, hydrazides, esters, and carboxylic acids

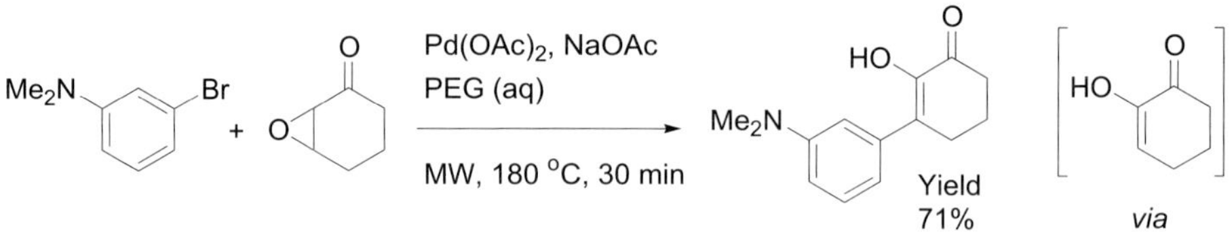

Scheme 15.78. One-pot isomerization–arylation of 2,3-epoxycyclohexanone.

[149]. The development of rapid, reliable and convenient procedures for introduction of carbonyl groups is important for high-throughput chemistry in general and high-speed microwave-mediated chemistry in particular. Unfortunately, the traditional method of introducing carbon monoxide to a reaction mixture via a balloon or gas tube is not practical, because of the special requirements of MW synthesis.

The molybdenum hexacarbonyl complex has recently been introduced as a condensed source of carbon monoxide for small-scale carbonylation chemistry [150]. This easily handled and inexpensive solid delivers a fixed amount of carbon monoxide on heating or on addition of a competing molybdenum ligand (for example DBU). This enables direct liberation of carbon monoxide in the reaction mixture without the need for external devices.

In the presence of molybdenum hexacarbonyl a variety of acyl sulfonamides have been formed in high to excellent yields with aryl bromides or iodides as aryl precursors using controlled MW irradiation for 15 min at 110–140 °C (Scheme 15.79) [151]. Under these conditions, primary sulfonamides reacted readily, whereas N-methylated sulfonamides afforded lower yields and incomplete conversions. The use of this carbonylation method in a key transformation step enabled an efficient synthesis of a novel hepatitis C virus NS3 protease inhibitor ($K_i = 85$ nM).

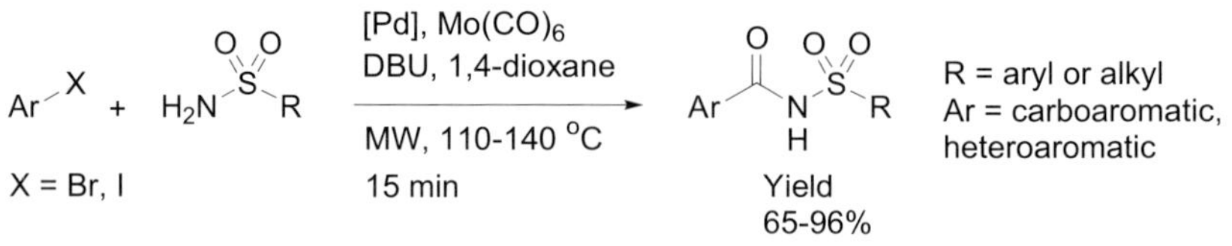

Scheme 15.79. Microwave-assisted synthesis of acyl sulfonamides using $Mo(CO)_6$ as a solid source of carbon monoxide.

$Mo(CO)_6$ has recently been used as the source of CO in the palladium-catalyzed generation of 3-acylaminoindanones [152]. These target structures were prepared from o-bromoaryl-substituted enamides after microwave irradiation for 30 min (Scheme 15.80).

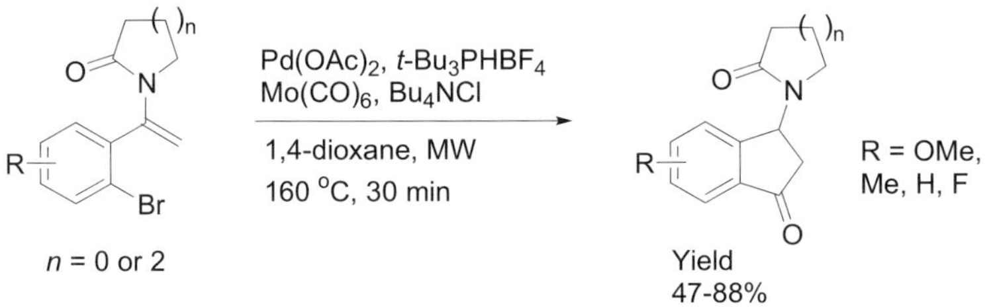

Scheme 15.80. Preparation of 3-acylaminoindanones by in-situ carboannulation.

15.6
Summary

The development of microwave equipment and microwave chemistry has been remarkable during the last two decades – from the first reports, in which standard synthetic transformations and domestic ovens were used, to modern multi-step applications where state-of-the-art single-mode cavities for small-scale synthesis and batch or continuous-flow reactors for increased scale are exploited. Indeed, it is now possible to perform direct up-scaling of almost any microwave-assisted procedure from the milligram to the kilogram scale, although safe scale-up using high-temperature conditions always requires purpose-built equipment.

The examples presented indicate that the combined approach of microwave irradiation and homogeneous catalysis can offer a nearly synergistic strategy, in the sense that the combination has greater potential than its two separate parts in isolation. The synthetic chemist can now take advantage of unique carbon–carbon and carbon–heteroatom bond formation reactions enabled by organometallic activation and make the reaction occur in seconds or minutes by microwave flash heating, an important feat because many transition metal-catalyzed reactions are known to be time-consuming. Furthermore, there are still many other catalytic reactions with great potential for microwave heating. For example, it might be expected that an increasing number of reactions in water or brine will be performed at elevated temperatures using microwave irradiation. We believe that within five to ten years microwave reactors will be the most common energy source in the organic chemistry laboratory. It is already clear that this technology is now changing and improving the way laboratory-scale organic chemistry is being performed worldwide.

Acknowledgment

We thank the Swedish Natural Science Research Council, the Swedish Foundation for Strategic Research, Knut and Alice Wallenberg's Foundation, Biolipox AB, Biotage AB, Medivir AB, and Gunnar Wikman.

References

1 K. Olofsson, A. Hallberg, M. Larhed, In *Microwaves in Organic Synthesis*, A. Loupy, Ed., Wiley: Weinheim, Germany, **2002**, pp 379–403.
2 (a) L. Perreux, A. Loupy, *Tetrahedron* **2001**, *57*, 9199–9223; (b) P. Lidström, J. Tierney, B. Wathey, J. Westman, *Tetrahedron* **2001**, *57*, 9225–9283.
3 C. O. Kappe, *Angew. Chem. Int. Ed.* **2004**, *43*, 6250–6284.
4 C. O. Kappe, A. Stadler, *Microwaves in Organic and Medicinal Chemistry*, Wiley–VCH: Weinheim, Germany, **2005**.
5 B. A. Roberts, C. R. Strauss, In *Microwave Assisted Organic Synthesis*, J. P. Tierney, P. Lidström, Eds., Blackwell: Oxford, UK, **2005**, pp 237–271.
6 (a) B. Pério, M. J. Dozias, J. Hamelin, *Org. Proc. Res. Dev.* **1998**, *2*, 428–435; (b) J. Cléophax, M. Liagre, A. Loupy, A. Petit, *Org. Proc. Res Dev.* **2000**, *4*, 498–508; (c) A. Stadler, B. H. Yousefi, D. Dallinger, P. Walla, E. Van der Eycken, N. Kaval, C. O. Kappe, *Org. Proc. Res Dev.* **2003**, *7*, 701–716.
7 K. Ersmark, M. Larhed, J. Wannberg, *Curr. Opin. Drug Discov. Devel.* **2004**, *7*, 417–427.
8 C. R. Sarko, In *Microwave Assisted Organic Synthesis*, J. P. Tierney, P. Lidström, Eds., Blackwell: Oxford, UK, **2005**, pp 222–236.
9 I. R. Baxendale, A. L. Lee, S. V. Ley, In *Microwave Assisted Organic Synthesis*, J. P. Tierney, P. Lidström, Eds., Blackwell: Oxford, UK, **2005**, pp 133–176.
10 W. Zhang, *Tetrahedron* **2003**, *59*, 4475–4489.
11 K. Olofsson, M. Larhed, In *Handbook of Fluorous Chemistry*, J. A. Gladysz, D. P. Curran, I. T. Horváth, Eds., Wiley: Weinheim, Germany, **2004**, pp 359–365.
12 A. de Meijere, F. Diederich, Eds., *Metal-Catalyzed Cross-Coupling Reactions*, 2nd Edn, Wiley–VCH: Weinheim, Germany, **2004**.
13 M. Larhed, A. Hallberg, *Drug Discov. Today* **2001**, *6*, 406–416.
14 M. Larhed, C. Moberg, A. Hallberg, *Acc. Chem. Res.* **2002**, *35*, 717–727.
15 K. Olofsson, M. Larhed, In *Microwave Assisted Organic Synthesis*, J. P. Tierney, P. Lidström, Eds., Blackwell: Oxford, UK, **2005**, pp 23–43.
16 N. Miyaura, A. Suzuki, *Chem. Rev.* **1995**, *95*, 2457–2483.
17 A. Furstner, G. Seidel, *Org. Lett.* **2002**, *4*, 541–543.
18 M. Larhed, A. Hallberg, *J. Org. Chem.* **1996**, *61*, 9582–9584.
19 E. Comer, M. G. Organ, *J. Am. Chem. Soc.* **2005**, *127*, 8160–8167.
20 M. J. Gronnow, R. J. White, J. H. Clark, D. J. Macquarrie, *Org. Process Res. Dev.* **2005**, *9*, 516–518.
21 A. F. Littke, G. C. Fu, *Angew. Chem. Int. Ed.* **2002**, *41*, 4176–4211.
22 R. B. Bedford, C. P. Butts, T. E. Hurst, P. Lidström, *Adv. Synth. Catal.* **2004**, *346*, 1627–1630.
23 O. Navarro, H. Kaur, P. Mahjoor, S. P. Nolan, *J. Org. Chem.* **2004**, *69*, 3173–3180.
24 G. B. Miao, P. Ye, L. B. Yu, C. M. Baldino, *J. Org. Chem.* **2005**, *70*, 2332–2334.
25 N. E. Leadbeater, M. Marco, *J. Org. Chem.* **2003**, *68*, 888–892.
26 N. E. Leadbeater, *Chem. Commun.* **2005**, 2881–2902.
27 N. E. Leadbeater, M. Marco, *Org. Lett.* **2002**, *4*, 2973–2976.
28 L. Bai, J. X. Wang, Y. M. Zhang, *Green Chem.* **2003**, *5*, 615–617.
29 R. K. Arvela, N. E. Leadbeater, *Org. Lett.* **2005**, *7*, 2101–2104.
30 C. Najera, J. Gil-Molto, S. Karlstrom, *Adv. Synth. Catal.* **2004**, *346*, 1798–1811.
31 L. Bai, J. X. Wang, *Chin. Chem. Lett.* **2004**, *15*, 286–287.
32 C. G. Blettner, W. A. König, W. Stenzel, T. Schotten, *J. Org. Chem.* **1999**, *64*, 3885–3890.
33 V. V. Namboodiri, R. S. Varma, *Green Chem.* **2001**, *3*, 146–148.
34 D. Villemin, M. J. Gómez-

Escalonilla, J. F. Saint-Clair, *Tetrahedron Lett.* **2001**, *42*, 635–637.
35 G. W. Kabalka, L. Wang, R. M. Pagni, C. M. Hair, V. Namboodiri, *Synthesis* **2003**, 217–222.
36 J. X. Wang, Y. H. Yang, B. G. Wei, Y. L. Hu, Y. Fu, *Bull. Chem. Soc. Jpn.* **2002**, *75*, 1381–1382.
37 J. X. Wang, B. G. Wei, Y. L. Hu, Z. X. Liu, Y. H. Yang, *Synth. Commun.* **2001**, *31*, 3885–3890.
38 M. Larhed, G. Lindeberg, A. Hallberg, *Tetrahedron Lett.* **1996**, *37*, 8219–8222.
39 A. Stadler, C. O. Kappe, *Eur. J. Org. Chem.* **2001**, *2001*, 919–925.
40 Y. Wang, D. R. Sauer, *Org. Lett.* **2004**, *6*, 2793–2796.
41 L. Bai, Y. M. Zhang, J. X. Wang, *QSAR Comb. Sci.* **2004**, *23*, 875–882.
42 W. Solodenko, U. Schon, J. Messinger, A. Glinschert, A. Kirschning, *Synlett* **2004**, 1699–1702.
43 J. A. Gladysz, D. P. Curran, I. T. Horváth, Eds., *Handbook of Fluorous Chemistry*, Wiley: Weinheim, Germany, **2004**.
44 W. Zhang, *Chem. Rev.* **2004**, *104*, 2531–2556.
45 W. Zhang, C. H. T. Chen, Y. M. Lu, T. Nagashima, *Org. Lett.* **2004**, *6*, 1473–1476.
46 Y. M. Lu, W. Zhang, *QSAR Comb. Sci.* **2004**, *23*, 827–835.
47 W. Zhang, C. H. T. Chen, *Tetrahedron Lett.* **2005**, *46*, 1807–1810.
48 T. Besson, C. T. Brain, In *Microwave Assisted Organic Synthesis*, J. P. Tierney, P. Lidström, Eds., Blackwell: Oxford, UK, **2005**, pp 44–74.
49 Y. Gong, W. He, *Org. Lett.* **2002**, *4*, 3803–3805.
50 G. L. Luo, L. Chen, G. S. Poindexter, *Tetrahedron Lett.* **2002**, *43*, 5739–5742.
51 N. Mont, L. Fernandez-Megido, J. Teixido, C. O. Kappe, J. I. Borrell, *QSAR Comb. Sci.* **2004**, *23*, 836–849.
52 T. Y. H. Wu, P. G. Schultz, S. Ding, *Org. Lett.* **2003**, *5*, 3587–3590.
53 P. Appukkuttan, E. Van der Eycken, W. Dehaen, *Synlett* **2005**, 127–133.
54 P. Appukkuttan, W. Dehaen, E. Van der Eycken, *Org. Lett.* **2005**, *7*, 2723–2726.
55 P. Appukkuttan, A. B. Orts, R. P. Chandran, J. L. Goeman, J. Van der Eycken, W. Dehaen, E. Van der Eycken, *Eur. J. Org. Chem.* **2004**, 3277–3285.
56 N. Kaval, K. Bisztray, W. Dehaen, C. O. Kappe, E. Van der Eycken, *Mol. Divers.* **2003**, *7*, 125–133.
57 N. Kaval, W. Dehaen, E. Van der Eycken, *J. Comb. Chem.* **2005**, *7*, 90–95.
58 A. Berthault, S. Berteina-Raboin, A. Finaru, G. Guillaumet, *QSAR Comb. Sci.* **2004**, *23*, 850–853.
59 Y. Gong, W. He, *Heterocycles* **2004**, *62*, 851–856.
60 J. Wannberg, D. Dallinger, C. O. Kappe, M. Larhed, *J. Comb. Chem.* **2005**, *7*, 574–583.
61 M. J. Schulz, S. J. Coats, D. J. Hlasta, *Org. Lett.* **2004**, *6*, 3265–3268.
62 J. W. Han, J. C. Castro, K. Burgess, *Tetrahedron Lett.* **2003**, *44*, 9359–9362.
63 R. Lepine, J. P. Zhu, *Org. Lett.* **2005**, *7*, 2981–2984.
64 F. Ito, M. Iwasaki, T. Watanabe, T. Ishikawa, Y. Higuchi, *Org. Biomol. Chem.* **2005**, *3*, 674–681.
65 T. N. Glasnov, W. Stadlbauer, C. O. Kappe, *J. Org. Chem.* **2005**, *70*, 3864–3870.
66 S. J. Coats, M. J. Schulz, J. R. Carson, E. E. Codd, D. J. Hlasta, P. M. Pitis, D. J. Stone, Jr., S.-P. Zhang, R. W. Colburn, S. L. Dax, *Bioorg. Med. Chem. Lett.* **2004**, *14*, 5493–5498.
67 C. J. Mathews, J. Taylor, M. J. Tyte, P. A. Worthington, *Synlett* **2005**, 538–540.
68 A. P. Combs, B. M. Glass, S. A. Jackson, *Meth. Enzymol.* **1999**, *369*, 223–231.
69 M. G. Organ, S. Mayer, F. Lepifre, B. N'Zemba, J. Khatri, *Mol. Divers.* **2003**, *7*, 211–227.
70 A. Ax, W. Schaal, L. Vrang, B. Samuelsson, A. Hallberg, A. Karlen, *Bioorg. Med. Chem.* **2005**, *13*, 755–764.
71 W. Schaal, A. Karlsson, G. Ahlsen, J. Lindberg, H. O. Andersson, U. H. Danielson, B. Classon, T. Unge, B. Samuelsson, J. Hulten, A. Hallberg, A. Karlen, *J. Med. Chem.* **2001**, *44*, 155–169.

72 K. Ersmark, I. Feierberg, S. Bjelic, E. Hamelink, F. Hackett, M. J. Blackman, J. Hulten, B. Samuelsson, J. Åqvist, A. Hallberg, *J. Med. Chem.* **2004**, *47*, 110–122.

73 D. Nöteberg, E. Hamelink, J. Hultén, M. Wahlgren, L. Vrang, B. Samuelsson, A. Hallberg, *J. Med. Chem.* **2003**, *46*, 734–746.

74 D. Nöteberg, W. Schaal, E. Hamelink, L. Vrang, M. Larhed, *J. Comb. Chem.* **2003**, *5*, 456–464.

75 P. Espinet, A. M. Echavarren, *Angew. Chem. Int. Ed.* **2004**, *43*, 4704–4734.

76 K. Olofsson, S. Y. Kim, M. Larhed, D. P. Curran, A. Hallberg, *J. Org. Chem.* **1999**, *64*, 4539–4541.

77 R. E. Maleczka, Jr., J. M. Lavis, D. H. Clark, W. P. Gallagher, *Org. Lett.* **2000**, *2*, 3655–3658.

78 M. Larhed, M. Hoshino, S. Hadida, D. P. Curran, A. Hallberg, *J. Org. Chem.* **1997**, *62*, 5583–5587.

79 D. Villemin, F. Caillot, *Tetrahedron Lett.* **2001**, *42*, 639–642.

80 C. Bour, J. Suffert, *Org. Lett.* **2005**, *7*, 653–656.

81 L. Öhberg, J. Westman, *Synlett* **2001**, 1893–1896.

82 P. Walla, C. O. Kappe, *Chem. Commun.* **2004**, 564–565.

83 K. Krascsenicsova, P. Walla, P. Kasak, G. Uray, C. O. Kappe, M. Putala, *Chem. Commun.* **2004**, 2606–2607.

84 P. Stanetty, M. Schnurch, M. D. Mihovilovic, *Synlett* **2003**, 1862–1864.

85 I. Mutule, E. Suna, *Tetrahedron Lett.* **2004**, *45*, 3909–3912.

86 Y. Q. Wan, M. Alterman, M. Larhed, A. Hallberg, *J. Org. Chem.* **2002**, *67*, 6232–6235.

87 J. Hassan, M. Sevignon, C. Gozzi, E. Schulz, M. Lemaire, *Chem. Rev.* **2002**, *102*, 1359–1469.

88 H. Gold, M. Larhed, P. Nilsson, *Synlett* **2005**, 1596–1600.

89 T. Hiyama, E. Shirakawa, **2002**, *1*, 285–309.

90 M. L. Clarke, *Adv. Synth. Catal.* **2005**, *347*, 303–307.

91 K. Sonogashira, In Negishi, E. I., editor, *Handbook of Organopalladium Chemistry for Organic Synthesis*, **2002**, *1*, 493–529.

92 G. W. Kabalka, L. Wang, V. Namboodiri, R. M. Pagni, *Tetrahedron Lett.* **2000**, *41*, 5151–5154.

93 M. Erdelyi, A. Gogoll, *J. Org. Chem.* **2001**, *66*, 4165–4169.

94 M. Erdelyi, A. Gogoll, *J. Org. Chem.* **2003**, *68*, 6431–6434.

95 M. Xia, Y.-G. Wang, *J. Chem. Res.* **2002**, 173–175.

96 M. Xia, Y. G. Wang, *Chin. Chem. Lett.* **2002**, *13*, 1–2.

97 G. W. Kabalka, L. Wang, R. M. Pagni, *Tetrahedron* **2001**, *57*, 8017–8028.

98 J. Yan, Z. Wang, L. Wang, *J. Chem. Res.(S)* **2004**, 71–73.

99 H. He, Y. J. Wu, *Tetrahedron Lett.* **2004**, *45*, 3237–3239.

100 P. Appukkuttan, W. Dehaen, E. Van der Eycken, *Eur. J. Org. Chem.* **2003**, 4713–4716.

101 N. E. Leadbeater, M. Marco, B. J. Tominack, *Org. Lett.* **2003**, *5*, 3919–3922.

102 R. K. Arvela, N. E. Leadbeater, M. S. Sangi, V. A. Williams, P. Granados, R. D. Singer, *J. Org. Chem.* **2005**, *70*, 161–168.

103 Y. Kohara, K. Kubo, E. Imamiya, T. Wada, Y. Inada, T. Naka, *J. Med. Chem.* **1996**, *39*, 5228–5235.

104 C. Liljebris, S. D. Larsen, D. Ogg, B. J. Palazuk, J. E. Bleasdale, *J. Med. Chem.* **2002**, *45*, 1785–1798.

105 D. M. Tschaen, R. Desmond, A. O. King, M. C. Fortin, B. Pipik, S. King, T. R. Verhoeven, *Synth. Commun.* **1994**, *24*, 887–890.

106 M. Alterman, A. Hallberg, *J. Org. Chem.* **2000**, *65*, 7984–7989.

107 R. K. Arvela, N. E. Leadbeater, *J. Org. Chem.* **2003**, *68*, 9122–9125.

108 R. R. Srivastava, S. E. Collibee, *Tetrahedron Lett.* **2004**, *45*, 8895–8897.

109 A. Zhang, J. L. Neumeyer, *Org. Lett.* **2003**, *5*, 201–203.

110 L. Cai, X. Liu, X. Tao, D. Shen, *Synth. Commun.* **2004**, *34*, 1215–1221.

111 R. K. Arvela, N. E. Leadbeater, H. M. Torenius, H. Tye, *Org. Biomol. Chem.* **2003**, *1*, 1119–1121.

112 N. E. Leadbeater, H. M. Torenius, H. Tye, *Tetrahedron* **2003**, *59*, 2253–2258.

113 A. BENGTSON, A. HALLBERG, M. LARHED, *Org. Lett.* **2002**, *4*, 1231–1233.
114 J. F. HARTWIG, *Angew. Chem. Int. Ed.* **1998**, *37*, 2046–2067.
115 A. SHARIFI, R. HOSSEINZADEH, M. MIRZAEI, *Monatsh Chem* 2002; Vol. 133, p 329–332.
116 Y. WAN, M. ALTERMAN, A. HALLBERG, *Synthesis* **2002**, 1597–1600.
117 T. A. JENSEN, X. LIANG, D. TANNER, N. SKJAERBAEK, **2004**, *69*, 4936–4947.
118 B. U. W. MAES, K. T. J. LOONES, G. L. F. LEMIERE, R. A. DOMMISSE, *Synlett* **2003**, 1822–1825.
119 C. T. BRAIN, J. T. STEER, *J. Org. Chem.* **2003**, *68*, 6814–6816.
120 K. WEIGAND, S. PELKA, *Mol. Divers.* **2003**, *7*, 181–184.
121 A. J. MCCARROLL, D. A. SANDHAM, L. R. TITCOMB, A. K. d. K. LEWIS, F. G. N. CLOKE, B. P. DAVIES, A. PEREZ DE SANTANA, W. HILLER, S. CADDICK, *Mol. Divers.* **2003**, *7*, 115–123.
122 B. U. W. MAES, K. T. J. LOONES, S. HOSTYN, G. DIELS, G. ROMBOUTS, *Tetrahedron* **2004**, *60*, 11559–11564.
123 Y. J. WU, H. HE, A. L'HEUREUX, *Tetrahedron Lett.* **2003**, *44*, 4217–4218.
124 A. P. COMBS, S. SAUBERN, M. RAFALSKI, P. Y. S. LAM, *Tetrahedron Lett.* **1999**, *40*, 1623–1626.
125 J. H. M. LANGE, L. J. F. HOFMEYER, F. A. S. HOUT, S. J. M. OSNABRUG, P. C. VERVEER, C. G. KRUSE, R. W. FEENSTRA, *Tetrahedron Lett.* **2002**, *43*, 1101–1104.
126 R. R. POONDRA, N. J. TURNER, *Org. Lett.* **2005**, *7*, 863–866.
127 H. HE, Y. J. WU, *Tetrahedron Lett.* **2003**, *44*, 3385–3386.
128 G. BURTON, P. CAO, G. LI, R. RIVERO, *Org. Lett.* **2003**, *5*, 4373–4376.
129 M. HARMATA, X. C. HONG, S. K. GHOSH, *Tetrahedron Lett.* **2004**, *45*, 5233–5236.
130 G. F. MANBECK, A. J. LIPMAN, R. A. STOCKLAND, A. L. FREIDL, A. F. HASLER, J. J. STONE, I. A. GUZEI, *J. Org. Chem.* **2005**, *70*, 244–250.
131 H. HE, Y. J. WU, *Tetrahedron Lett.* **2003**, *44*, 3445–3446.
132 D. VILLEMIN, P. A. JAFFRES, F. SIMEON, *Phosphorus, Sulfur Silicon Relat. Elem.* **1997**, *130*, 59–63.
133 D. VILLEMIN, A. ELBILALI, F. SIMEON, P. A. JAFFRES, G. MAHEUT, M. MOSADDAK, A. HAKIKI, *J. Chem. Res.(S)* **2003**, 436–437.
134 A. STADLER, C. O. KAPPE, *Org. Lett.* **2002**, *4*, 3541–3543.
135 Y. J. WU, H. HE, *Synlett* **2003**, 1789–1790.
136 F. R. ALEXANDRE, A. BERECIBAR, R. WRIGGLESWORTH, T. BESSON, *Tetrahedron Lett.* **2003**, *44*, 4455–4458.
137 T. BESSON, M. J. DOZIAS, J. GUILLARD, C. W. REES, *J. Chem. Soc., Perkin Trans. 1* **1998**, 3925–3926.
138 J. MARQUIÉ, A. LAPORTERIE, J. DUBAC, N. ROQUES, J. R. DESMURS, *J. Org. Chem.* **2001**, *66*, 421–425.
139 A. LENGAR, C. O. KAPPE, *Org. Lett.* **2004**, *6*, 771–774.
140 G. W. KABALKA, R. S. VARMA, *Tetrahedron* **1989**, *45*, 6601–6621.
141 N. ELANDER, J. R. JONES, S. Y. LU, S. STONE-ELANDER, *Chem. Soc. Rev.* **2000**, *29*, 239–249.
142 R. K. ARVELA, N. E. LEADBEATER, *Synlett* **2003**, 1145–1148.
143 M. LARHED, A. HALLBERG, In *Handbook of Organopalladium Chemistry*, E. NEGISHI (ed.), vol 1, **2002**; Wiley-Interscience: New York, USA, 1133–1178.
144 K. S. A. VALLIN, P. EMILSSON, M. LARHED, A. HALLBERG, *J. Org. Chem.* **2002**, *67*, 6243–6246.
145 G. K. DATTA, K. S. A. VALLIN, M. LARHED, *Mol. Divers.* **2003**, *7*, 107–114.
146 R. K. ARVELA, N. E. LEADBEATER, *J. Org. Chem.* **2005**, *70*, 1786–1790.
147 A. STADLER, H. VON SCHENCK, K. S. A. VALLIN, M. LARHED, A. HALLBERG, *Adv. Synth. Catal.* **2004**, *346*, 1773–1781.
148 A. SVENNEBRING, N. GARG, P. NILSSON, A. HALLBERG, M. LARHED, *J. Org. Chem.* **2005**, *70*, 4720–4725.
149 R. SKODA-FOLDES, L. KOLLAR, *Curr. Org. Chem.* **2002**, *6*, 1097–1119.
150 J. WANNBERG, M. LARHED, *J. Org. Chem.*, *68*, 5750–5753.
151 X. WU, R. RÖNN, T. GOSSAS, M. LARHED, *J. Org. Chem.* **2005**, *70*, 3094–3098.
152 X. WU, P. NILSSON, M. LARHED, *J. Org. Chem.*, *70*, 346–349.

16
Microwave-assisted Combinatorial and High-throughput Synthesis

Alexander Stadler and C. Oliver Kappe

16.1
Solid-phase Organic Synthesis

16.1.1
Introduction

Combinatorial chemistry, the art and science of rapidly synthesizing and testing potential lead compounds for any desired property, has been found to be one of the most promising approaches in drug discovery [1]. This technique is, therefore, used to generate libraries of potential lead compounds which can immediately be screened for biological efficiency. Whereas chemistry has, in the past, been characterized by slow, steady, and painstaking work, combinatorial chemistry (also referred to as high-throughput synthesis) has changed the characteristics of chemical research and enabled a level of productivity thought impossible a few years ago.

One of the cornerstones of combinatorial chemistry is solid-phase organic synthesis (SPOS) [2], originally developed by Merrifield in 1963 for the synthesis of peptides [3]. In SPOS, a molecule (scaffold) is attached to a solid support, for example a "polymer resin" (Fig. 16.1). In general, resins are insoluble base polymers with a "linker" molecule attached. Spacers are probably included to reduce steric hindrance by the bulk of the resin. Linkers on the other hand are functional moieties, which enable attachment and cleavage of scaffolds under controlled conditions. Subsequent chemistry is then performed on the molecule attached to the support until, at the end of the often multistep synthesis, the desired molecule is released from the support.

To accelerate reactions and to drive them to completion, a large excess of reagents can be used, because this can easily be removed by filtration and subsequent washing of the solid support. Thus, even final purification of the desired products is simplified, as byproducts formed in solution do not affect the outcome of the target. Nowadays, many linkers and various classic or microwave-assisted cleavage conditions are available, enabling efficient processing of a variety of chemical transformations.

SPOS can also easily be automated by using appropriate robotics for filtration and evaporation of the reaction mixture to isolate the cleaved product. SPOS can,

Microwaves in Organic Synthesis, Second edition. Edited by A. Loupy
Copyright © 2006 WILEY-VCH Verlag GmbH & Co. KGaA, Weinheim
ISBN: 3-527-31452-0

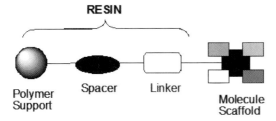

Fig. 16.1. The concept of solid-phase organic synthesis.

furthermore, be applied to the powerful "split-and-mix" strategy, which has been found to be an important tool for combinatorial chemistry since it was introduced in the late 1980s [4].

Since the 1990s an ever-growing number of publications discussing combinatorial approaches in organic synthesis have been published, including reactions on soluble polymers, performed in homogeneous solution, or on a well-defined fluorous support. Many common solution-phase reactions have been performed equally well on the solid phase, and a great variety of reagents, catalysts, or scavengers have been attached to polymer supports, enabling synthesis of a variety of desired target molecules with increased efficiency and productivity.

16.1.2
Microwave Chemistry and Solid-phase Organic Synthesis

Parallel to these developments in solid-phase synthesis and combinatorial chemistry, microwave-enhanced organic synthesis has attracted much attention in recent years. As is evident from the other chapters in this book and the comprehensive reviews available on this subject [5, 6], high-speed microwave-assisted synthesis has been applied successfully in many fields of synthetic organic chemistry. Any technique which can speed the process of rather time-consuming solid-phase synthesis is of substantial interest, particularly in research laboratories involved in high-throughput synthesis.

Not surprisingly, the benefits of microwave-assisted organic synthesis (MAOS) have also attracted interest from the combinatorial/medicinal chemistry community, to whom reaction speed is of great importance [7–10], and the number of publications reporting rate enhancements in solid-phase organic synthesis utilizing microwaves is steadily growing. This attractive link between combinatorial/high-throughput processing and microwave heating is a logical consequence of the increased speed and effectiveness offered by the microwave approach.

Although the examples of microwave-assisted solid-phase reactions presented in this chapter reveal that rapid synthetic transformations can often be achieved by using microwave irradiation, the possibility of high-speed synthesis does not necessarily mean that these processes can also be adapted to a truly high-throughput format. In the past few years all commercial suppliers of microwave instrumentation [11–14] for organic synthesis have moved toward combinatorial/high-throughput

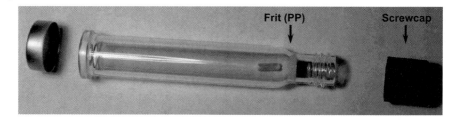

Fig. 16.2. Modified microwave vial for use in the Emrys series (reproduced with permission from Ref. 21).

platforms for conventional solution-phase synthesis [15], and special equipment for solid-phase synthesis under microwave conditions is now available.

A recent development in this context is the Liberty System introduced by CEM in 2004. This instrument is an automated microwave peptide synthesizer, equipped with special vessels, suitable for unattended synthesis of up to 12 peptides employing 25 different amino acids. For details on solid-phase peptide synthesis refer to Chapter 19.

In addition, several articles on microwave-assisted parallel synthesis have described irradiation of 96-well filter-bottom polypropylene plates in conventional household microwave ovens for high-throughput synthesis [16–19]. One interesting article described the construction and use of a parallel polypropylene reactor comprising cylindrical, expandable reaction vessels with porous frits at the bottom [20]. This work presented the very first description of reaction vessels for microwave-assisted synthesis that may be useful for performing solid-phase synthesis using bottom-filtration techniques in conjunction with microwave heating.

A prototype microwave reaction vessel that takes advantage of bottom-filtration techniques was presented in a more recent publication. The authors described the use of a modified reaction vessel (Fig. 16.2) for a Biotage monomode instrument, with a polypropylene frit, suitable for the filtration/cleavage steps in their microwave-mediated solid-phase Sonogashira coupling (Scheme 16.15) [21].

For general solid-phase reactions in a dedicated multimode instrument, an adaptable filtration unit is available from Anton Paar (Fig. 16.3). This tool is connected to the appropriate reaction vessel by a simple screw cap and, after turning over the vessel, the resin is filtered by applying a slight pressure up to 5 bar. The resin can then be used for further reaction sequences or cleavage steps in the same reaction vessel without loss of material. At the time of writing, however, no applications of this system for solid-phase synthesis had been reported.

16.1.3
Peptide Synthesis and Related Examples

One of the first dedicated applications of microwaves in solid-phase chemistry was the synthesis of small peptide molecules [22]. These will be dealt with in detail in Chapter 19.

Fig. 16.3. Filtration unit for solid-phase syntheses utilizing the Synthos 3000 (Anton Paar).

16.1.4
Resin Functionalization

The functionalization of commercially available standard solid supports is of particular interest for combinatorial purposes to enable a broad range of reactions to be studied. Because these transformations usually require long reaction times under conventional thermal conditions, it was obvious to combine microwave chemistry with the art of resin functionalization.

As a suitable model reaction, coupling of a variety of substituted carboxylic acids to polymer resins has been investigated [23]. The resulting polymer-bound esters served as useful building blocks in a variety of further solid-phase transformations. This functionalization was also used to determine the effect of microwave irradiation on the cleavage of substrates from polymer supports (Section 16.1.10). Thirty-four substituted carboxylic acids were coupled to chlorinated Wang resin by use of an identical reaction procedure (Scheme 16.1). For most of these the yield from microwave-mediated conversion was at least 85% after 3–15 min at 200 °C in standard glassware under reflux at atmospheric pressure. Importantly, in all these examples the loadings accomplished after microwave irradiation for 15 min were actually higher than those achieved using the thermally heated procedures.

In a related study the same authors investigated the effect of microwave irradiation on carbodiimide-mediated esterification of benzoic acid on a solid support [24]. The carboxylic acid was activated with N,N'-diisopropylcarbodiimide (DIC) via the O-acyl isourea or the symmetrical anhydride procedure. The isourea proce-

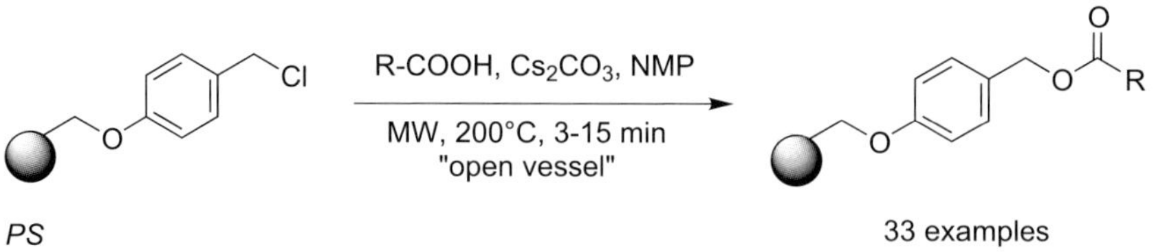

Scheme 16.1. Resin functionalization with carboxylic acids using a polystyrene-based polymer support.

dure had some deficiencies – complete conversion could not be achieved, because of unexpected side-reactions at higher temperatures. The anhydride procedure was superior to this method, because it could be performed quantitatively at 200 °C within 10 min under "open vessel" conditions, i.e. without the need for high-pressure vessels.

A more detailed approach to the synthesis of a series of functionalized Merrifield resins has been reported [25]. In a modified domestic microwave oven, using a reflux condenser, reaction rates were dramatically enhanced compared with conventional methods – as high conversions were achieved within 25 min (Scheme 16.2). These microwave-mediated pathways are convenient methods for rapid and efficient solid-phase synthesis, using PS-Merrifield resin as either a support or a scavenger (see Section 16.6).

An interesting attempt at resin functionalization, microwave-assisted PEGylation of Merrifield resin, has been described [26]. Treating commercially available polystyrene Merrifield resin with poly(ethylene glycol) (PEG 200) at 170 °C for only 2 min afforded the corresponding hybrid polymer combining the advantages of

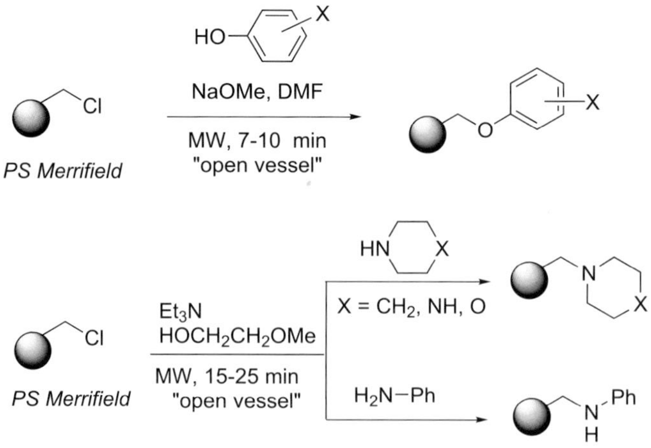

Scheme 16.2. Efficient preparation of functionalized resins.

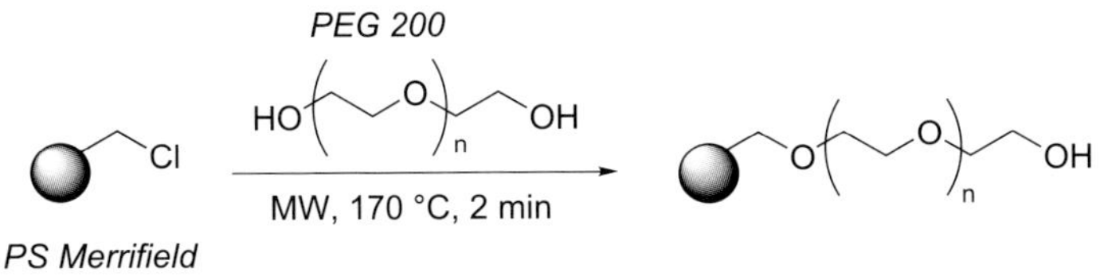

Scheme 16.3. PEGylation of Merrifield resin.

both insoluble and soluble polymers (Scheme 16.3). The Merrifield resin was suspended in excess PEG, which acts as the solvent and at the same time prevents cross-linking of the Merrifield resin. The product was simply purified by successive washing with water, 10% hydrochloric acid, and methanol.

In more recent work the same group immobilized β-cyclodextrin (CD) on solid support [27]. Initially 1,6-hexamethylene diisocyanate (HMDI) was quantitatively introduced as a linker to conventional PEGylated Merrifield resin (PS-PEG). β-CD was subsequently attached to the prepared PS-PEG-HMDI polymer by microwave irradiation at 70 °C for 30 min (Scheme 16.4a). The resulting β-CD resins are insoluble in water and can be used to trap volatile organic compounds from water.

In closely related work the same group also attached 2,6-dichloroindophenol (DCIP) to PEGylated Merrifield resin [28]. The resulting blue-pigmented resin can be reversibly reduced by ascorbic acid and can therefore be used as immobilized redox indicator. After initial activation of commercially available PEGylated Merrifield resin with excess thionyl chloride the organic dye was coupled by microwave-mediated nucleophilic substitution (Scheme 16.4b). The chlorinated resin was suspended in acetonitrile, treated with DCIP sodium salt, and irradiated at atmospheric pressure for 5 min at 75 °C. Filtration and washing with methanol and water furnished the desired polymer in good yields.

In an approach toward heterocycle synthesis two different three-step methods for solid-phase preparation of aminopropenones and aminopropenoates have been developed [29]. The first involved formation of the respective ester from N-protected glycine derivatives and Merrifield resin (Scheme 16.5) whereas the second used a different approach, use of simple aqueous methylamine solution for functionalization of the solid support [29]. The desired heterocycles were obtained by treatment of the generated polymer-bound benzylamine with the corresponding acetophenones, using N,N-dimethylformamide diethylacetal (DMFDEA) as reagent. The final step in the synthesis of the pyridopyrimidinones involved release of the products from the solid support by intramolecular cyclization.

A dedicated combinatorial approach for rapid parallel synthesis of polymer-bound enones has been reported [30]. The two-step procedure combines initial high-speed acetoacetylation of Wang resin with a selection of common β-ketoesters (Scheme 16.6) with subsequent microwave-mediated Knoevenagel condensations with a set of 13 different aldehydes (Section 16.1.8). Acetoacetylations were performed successfully in parallel within 10 min in open PFA vessels, in 1,2 dichlorobenzene at 170 °C, using a multi-vessel rotor system.

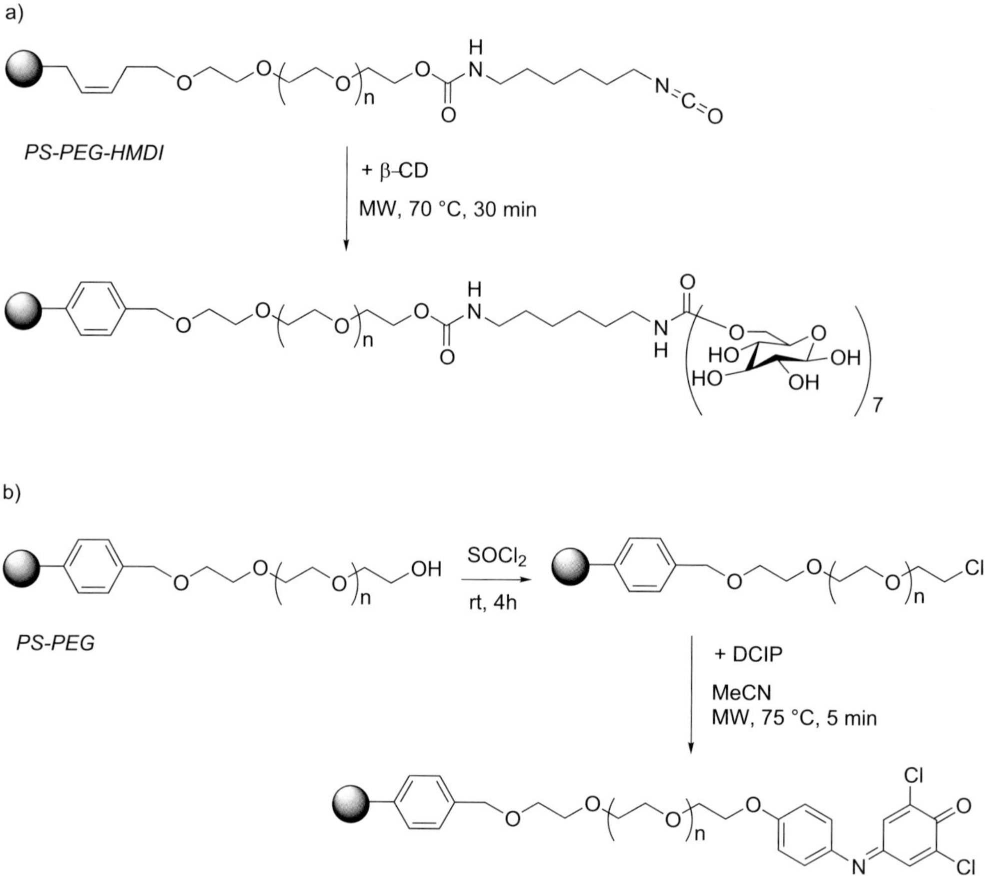

Scheme 16.4. Immobilization of β-cyclodextrin (a) and DCIP (b) on PEGylated Merrifield resin.

Very recently a rapid method for the preparation of effective polymer-supported isocyanate resins has been reported [31]. Gel-type isocyanate resins were generated from aminomethyl resins and inexpensive substrates as alternatives to the commercially available, expensive macroporous polystyrene isocyanate supports. Several isocyanates have been investigated; phenyl diisocyanate (PDI) was found to be the most efficient. Aminomethyl resin was pre-swollen in NMP, mixed with 2 equiv. PDI, and irradiated at 100 °C for 5 min (Scheme 16.7). Filtration, washing with NMP and DCM and drying under vacuum furnished the corresponding isocyanate resin. The reactivity of this novel gel-type resin was better than that of commercially available methyl isocyanate resins and it was successfully used for purification of a small amide library [31].

Interesting work on the synthesis of resin-bound naphthoquinones has recently been reported [32]. Resin-bound amines form intensely colored beads when treated with 2,3-dichloro-5-nitro-1,4-naphthoquinone; this can be used as a qualitative test

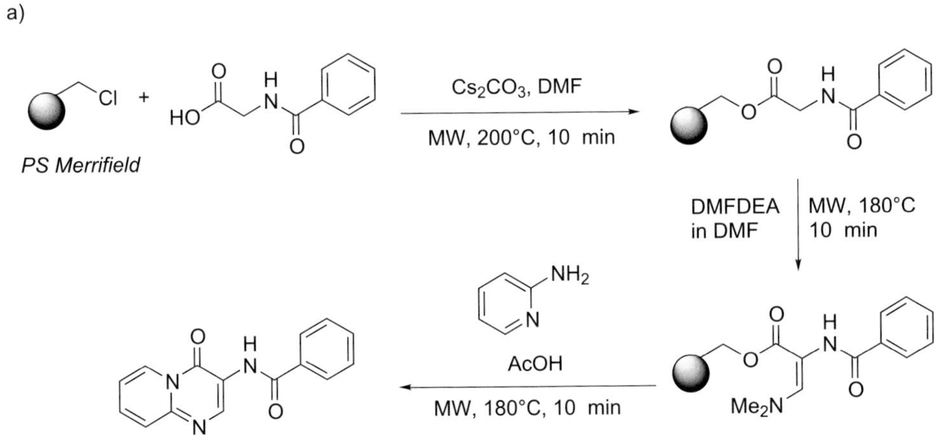

Scheme 16.5. Polymer-supported synthesis of dialkylaminopropenones.

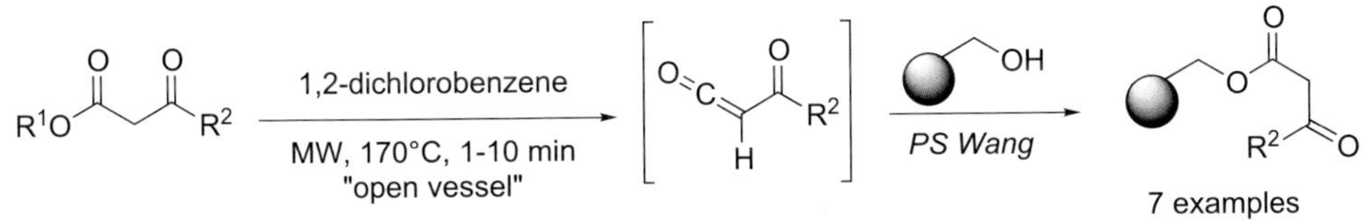

Scheme 16.6. Microwave-mediated acetoacetylation reactions.

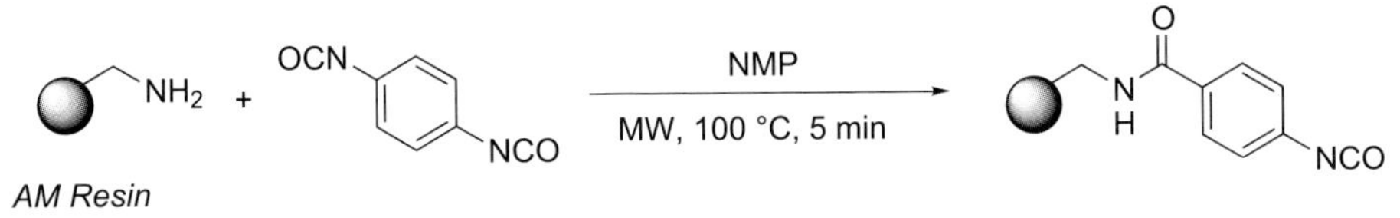

Scheme 16.7. Preparation of gel-type isocyanate resin from aminomethyl resin.

for amino groups on a resin. In a general procedure, several primary amines were anchored to commercially available FDMP resin under room-temperature conditions. The resulting polymers react readily with quinones in the presence of 2,6-di-*tert*-butylpyridine to form red polymer-bound quinones. In the presence of a phenyl ring on the resin-bound amine, however, the reactions were significantly slower. Thus, microwave irradiation of the mixture in dichloroethane at 80 °C for 10 min furnished the desired red compounds (Scheme 16.8). The reaction could also be performed employing H-Ala-RAM resin to introduce a steric center into the scaffold [32]. Conventional cleavage with TFA–DCM furnished the corresponding 3-amino naphthoquinones in good yields but varying purity.

Very recently an efficient synthesis of novel photocleavable linkers has been described [33]. Commercially available PEG 4000 was modified by standard esterifica-

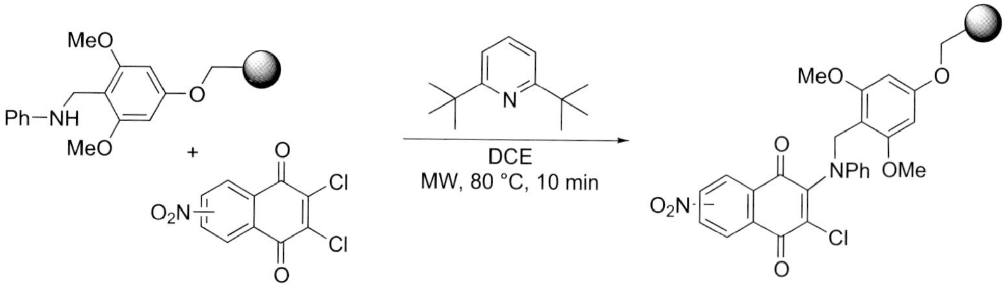

Scheme 16.8. Coupling of 2,3-dichloro-5-nitro-1,4-naphthoquinone to FDMP resin.

tion with 4-methyl-3-chloroacetylphenylacetic acid. In the presence of Cs_2CO_3 a variety of carboxylic acids and protected amino acids can be quantitatively attached to the resulting polymer by microwave-mediated reflux in DMF for 10 min. The model linkers have been subjected to photolysis in benzene (20 mL per 100 mg polymer) with a mercury lamp (280–366 nm) to release the corresponding acids from the PEG-bound indanone (Scheme 16.9). After 5–8 h all products could be obtained in good yields and purity > 90% (GC–MS).

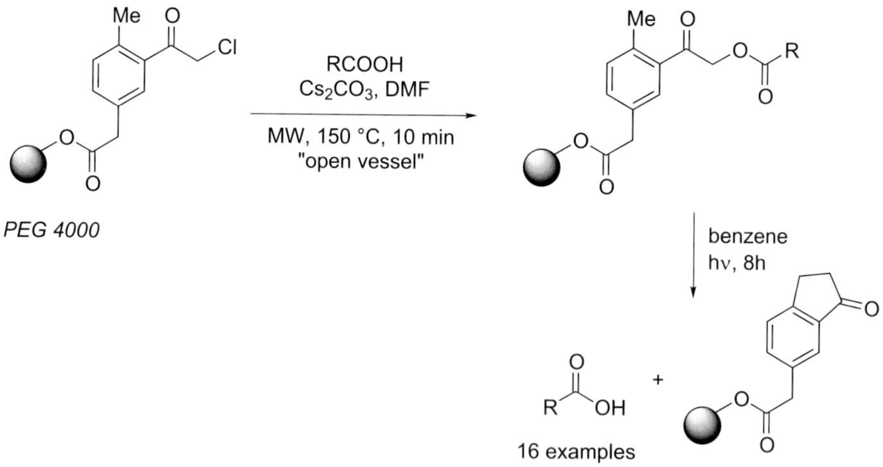

Scheme 16.9. Modification and photolytic cleavage of a phenacyl ester-based linker.

16.1.5
Transition-metal Catalysis

Palladium-catalyzed cross-coupling reactions are a cornerstone of modern organic synthesis. Combining microwave irradiation and solid-phase synthesis with these chemical transformations is therefore of great interest. One of the first publica-

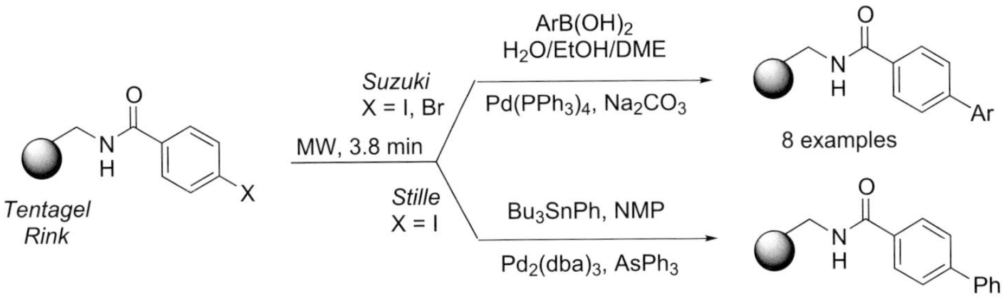

Scheme 16.10. Palladium-catalyzed cross-coupling reactions.

tions dealing with such reactions was presented in 1996 [34] and reported the effect of microwave irradiation on Suzuki and Stille-type cross-coupling reactions on solid phases. The reactions, utilizing TentaGel resin, were performed in sealed vials, using a prototype single-mode microwave cavity. Rather short reaction times resulted in almost quantitative conversions and minimum degradation of the solid support (Scheme 16.10).

In a related study the same group performed molybdenum-catalyzed alkylations in solution and on solid phases [35] and demonstrated that microwave irradiation could also be applied to highly enantioselective reactions (Scheme 16.11). In these examples, commercially available and stable molybdenum hexacarbonyl [Mo(CO)$_6$] was used to generate the catalytic system *in situ*. Although the conversion rates for the solid-phase examples were rather poor, enantioselectivity was excellent (>99% ee).

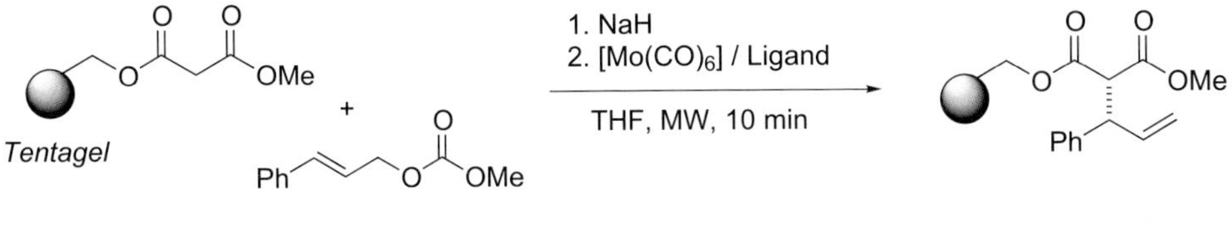

Scheme 16.11. Solid-phase molybdenum-catalyzed allylic alkylation.

Microwave-promoted preparation of tetrazoles from organonitriles has also been described [36]. Full conversion to the corresponding nitriles was achieved after very short reaction times at a maximum temperature of 175 °C (Scheme 16.12). The nitriles were subsequently treated with sodium azide to form the desired tetrazoles at 220 °C within 16 min. It is worthy of note that tetrazoles were readily formed by a one-pot reaction in good yields, eliminating the need for tetrakis(triphenylphosphine)palladium catalyst in the reaction mixture.

In a more recent study, solid-supported aminocarbonylations were achieved by using molybdenum hexacarbonyl as solid source of carbon monoxide [37]. The carbon monoxide is liberated smoothly at the reaction temperature by addition of a

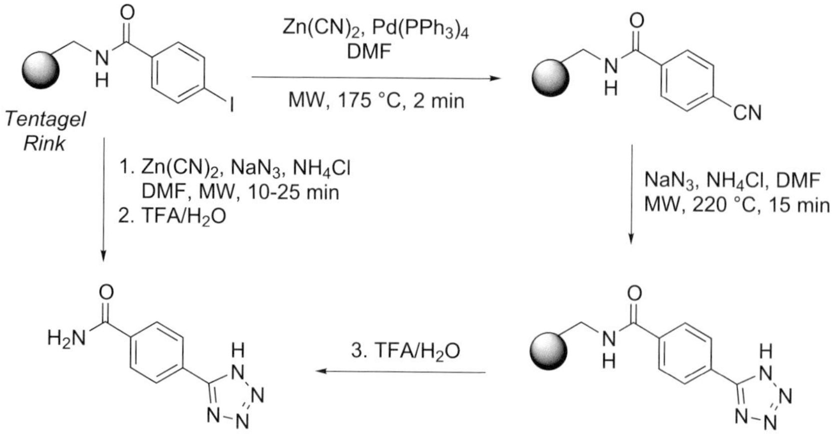

Scheme 16.12. Rapid synthesis of tetrazoles on solid-phases.

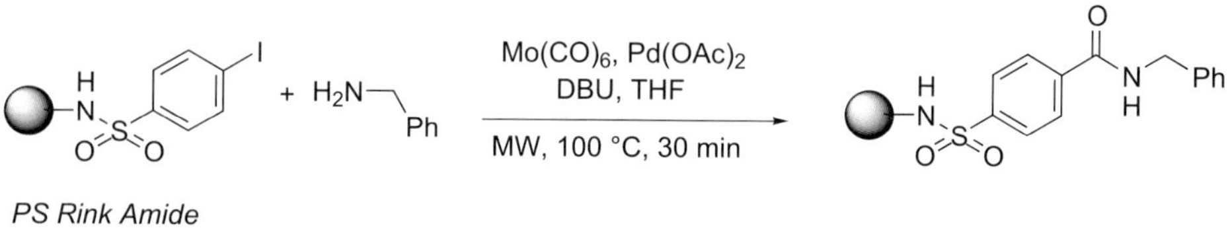

Scheme 16.13. Aminocarbonylation on solid phases.

large excess of the strong base 1,8-diazabicyclo[2.2.2]octane (DBU). Cleavage with a conventional mixture of trifluoroacetic acid and dichloromethane furnished the desired sulfamoylbenzamide in good yields (Scheme 16.13).

Another investigation of microwave-mediated Suzuki couplings has led to the solid phase synthesis of oxazolidinone antimicrobials [38]. A valuable oxazolidinone scaffold was coupled to Bal-resin (PS-PEG resin with a 4-formyl-3,5-dimethoxyphenoxy linker) to afford the corresponding resin-bound secondary amine. After subsequent acylation the resulting intermediate was transformed, by microwave-assisted Suzuki coupling, into the corresponding biaryl compound (Scheme 16.14). Cleavage with trifluoroacetic acid–dichloromethane yielded the target oxazolidinones.

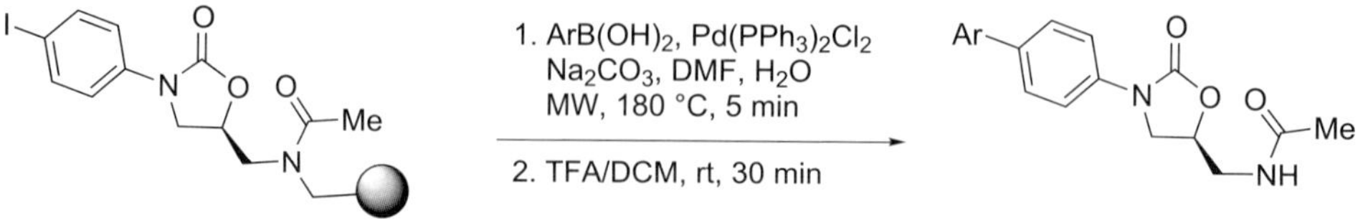

Scheme 16.14. Synthesis of biaryl oxazolidinones.

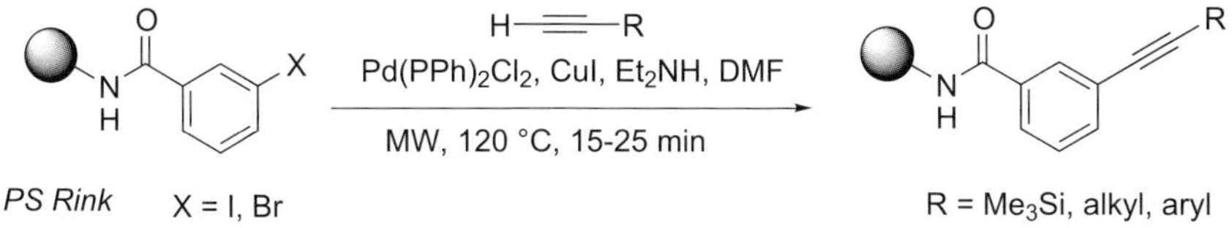

Scheme 16.15. Sonogashira couplings on solid phases.

Another palladium-catalyzed carbon–carbon coupling which could be accelerated efficiently by microwave heating is the Sonogashira reaction [21]. Aryl bromides and iodides have been coupled successfully with a variety of terminal acetylene derivatives (Scheme 16.15). It has been found that 5 mol% palladium catalyst and 10 mol% copper(I) iodide as cocatalyst in a mixture of diethylamine and N,N-dimethylformamide gave the best results. Depending on the substrates used, different irradiation times were required to achieve full conversion. The reactions were performed in a dedicated single-mode instrument using a modified reaction vessel (Fig. 16.2) for simplified resin handling.

Research on a solid-phase Sonogashira–Nicholas reaction sequence for synthesis of substituted alkynes was recently described [39]. Microwave irradiation was used to accelerate attachment of alkynols to Merrifield resin. Initial modification of the polymer support was achieved within 10 min by treating a mixture of 3-butynol and 15-crown-5 in THF in the presence of sodium hydride with microwave irradiation at 60 °C. Subsequent Sonogashira coupling was performed under the action of microwave irradiation at 120 °C to achieve full conversion within 15 min (Scheme 16.16). After filtration and washing the polymer was immediately used for complexation and subsequent Nicholas reaction to introduce a corresponding nucleophile. Although these last steps are conducted at low temperatures the five-step reaction can be performed within one day because the time required for the two initial time-consuming steps could be reduced to 10 and 15 min, respectively, by microwave assistance instead of 48 and 18 h under conventional thermal heating conditions.

A comprehensive study of the decoration of polymer-bound 2(1H)-pyrazinone scaffolds, by means of a variety of transition metal-catalyzed transformations, has been described [40]. The readily prepared pyrazinone was specifically decorated at the C3 position by use of microwave-mediated Suzuki, Stille, Sonogashira, and Ullmann procedures (Scheme 16.17) to introduce additional diversity. In all these

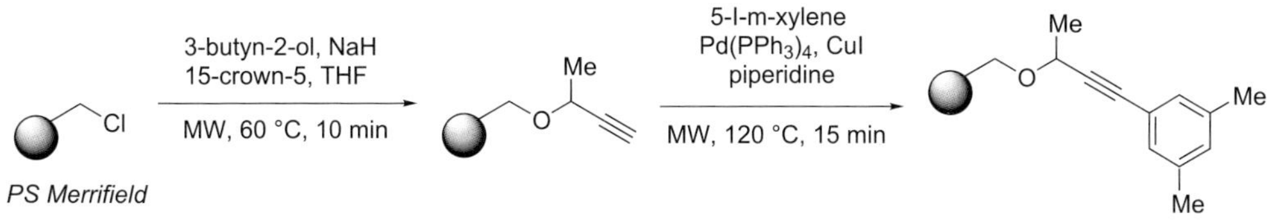

Scheme 16.16. Optimized Sonogashira–Nicholas reaction sequence.

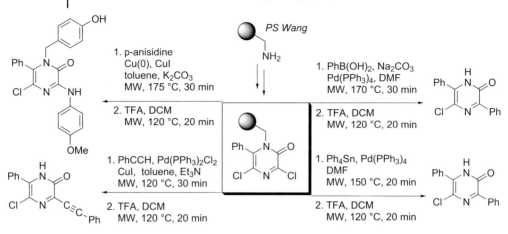

Scheme 16.17. Scaffold-decoration of polymer-bound pyrazinone scaffolds.

reactions smooth cleavage of the desired products was achieved by treatment of the resin with a 1:2 mixture of trifluoroacetic acid (TFA) and dichloromethane on microwave irradiation for 20 min at 120 °C.

The first polymer-supported cross-coupling reaction involving aryl chlorides was described in 2004 [41]. Rapid Negishi couplings were performed in which organozinc reagents were used to prepare biaryl compounds from a variety of aryl halides. The highly reactive but air-stable tri-*tert*-butylphosphonium tetrafluoroborate and *tris*(dibenzylideneacetone)dipalladium(0) were used as catalytic system (Scheme 16.18). Subsequent rapid cleavage under microwave conditions furnished the desired biaryl carboxylic acids in high yield and excellent purity.

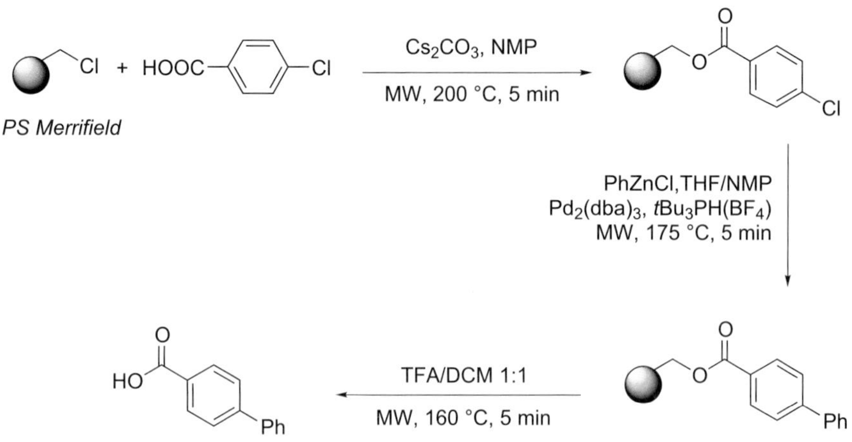

Scheme 16.18. Negishi coupling on a polymer support.

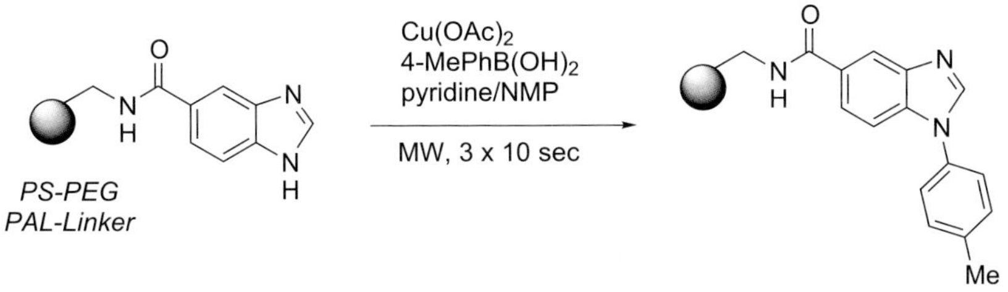

Scheme 16.19. Copper(II)-mediated N-arylation of polymer-bound benzimidazole.

The first examples of microwave-mediated solid-phase carbon–nitrogen cross-coupling reactions, using a boronic acid and a copper(II) catalyst, were reported in 1999 [16]. The reactions were performed in a domestic microwave oven at full power for 3 × 10 s (Scheme 16.19). After five cycles of heating with addition of fresh reagents none of the remaining starting benzimidazole amide could be detected after cleavage (TFA–DCM) from the solid support. This reduced the reaction time from 48 h under conventional heating conditions at 80 °C to less than 5 min by use of microwave heating. It should, however, be noted that both possible N-arylated regioisomeric products were obtained with this microwave-heated procedure.

A more recent publication disclosed a polymer-bound Buchwald–Hartwig amination employing activated, electron-deficient aryl halides under the action of microwave irradiation [42]. Subsequent acidic cleavage afforded the desired aryl amines in moderate to good yields (Scheme 16.20). Commercially available Fmoc-protected Rink amide resin was placed in a silylated microwave vessel and suspended under an argon atmosphere in dimethoxyethane (DME)–tert-butanol. After approximately 10 min the aryl halide was added and the mixture was stirred under argon for an additional 10 min. The other reagents were subsequently added and the sealed vessel submitted to microwave irradiation at 130 °C for 15 min. After conventional cleavage with trifluoroacetic acid in dichloromethane and purification the desired anilines were obtained in good yields and excellent purity.

Transition-metal catalysis on solid supports can also be used for indole formation [43]. A palladium or copper-catalyzed procedure has been described for generation of a small indole library (Scheme 16.21); this is the first example of solid-phase synthesis of 5-arylsulfamoyl substituted indole derivatives. Whereas the best

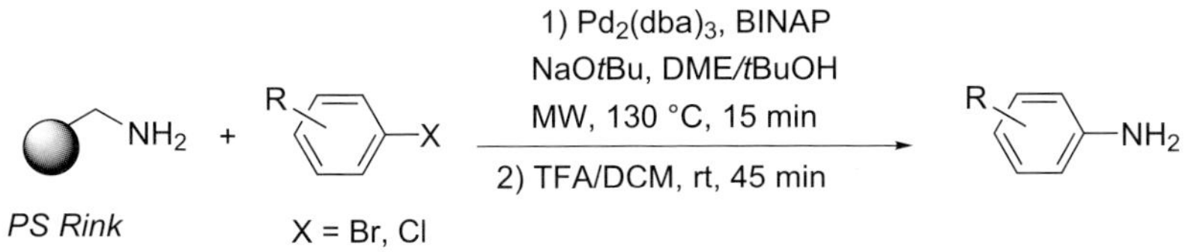

Scheme 16.20. Solid-supported Buchwald–Hartwig amination.

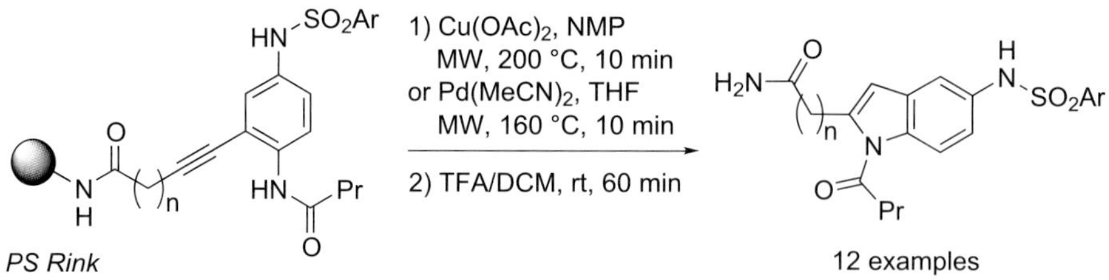

Scheme 16.21. Transition-metal catalyzed indole-formation on solid phases.

results for the copper-mediated cyclization of the polymer-bound precursor were achieved using 1-methyl-2-pyrrolidinone (NMP) as solvent, the palladium-catalyzed variation required tetrahydrofuran to achieve comparable results. Both procedures afforded the desired indoles in good yields and excellent purity [43].

16.1.6
Substitution Reactions

Another interesting topic is microwave-mediated substitution reactions on solid phases. In this context, an innovative study has achieved the synthesis of trisamino and aminooxy-1,3,5-triazines on cellulose and polypropylene membranes by applying SPOT-synthetic techniques [44]. The development of the microwave-mediated SPOT-synthetic procedures has enabled the rapid generation of highly diverse spatially addressed single compounds under mild conditions (Scheme 16.22). This SPOT-synthetic procedure required investigation of suitable planar polymeric supports, bearing an orthogonal ester-free linker system, which were cleavable under dry conditions.

Similar to the work described above, several publications on substitution reactions have dealt with solid-phase peptide synthesis. As already mentioned, these experiments are not addressed within this section and are summarized in Chapter 19.

A more recent study applied SPOT synthesis for the preparation of pyrimidines after appropriate modification of commercially available cellulose sheets as supports (Scheme 16.23) [45]. Initial introduction of the amine spacer was achieved within 15 min by use of microwave irradiation compared with 6 h by conventional heating. Attachment of the acid-cleavable Wang-type linker was performed by classical methods at ambient temperature.

The readily prepared support was then used for dihydropyrimidine and chalcone synthesis (Scheme 16.24). Thus, the modified support was activated before reaction by treatment with tosyl chloride. Solutions of the corresponding acetophenones were then spotted on to the membrane and the support was submitted to microwave irradiation for 10 min [45]. In the next step, several aryl aldehydes were attached under microwave irradiation conditions to form a set of the corresponding chalcones in a Claisen–Schmidt condensation.

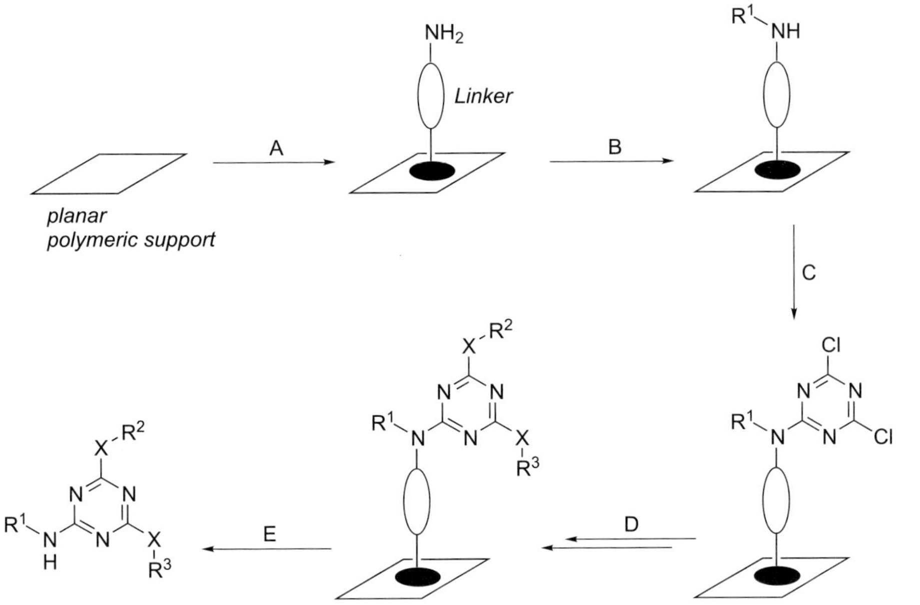

Scheme 16.22. General strategy for triazine synthesis on planar surfaces: (A) functionalization; (B) introduction of the first building block; (C) attachment of cyanuric chloride; (D) stepwise chlorine substitution; (E) cleavage.

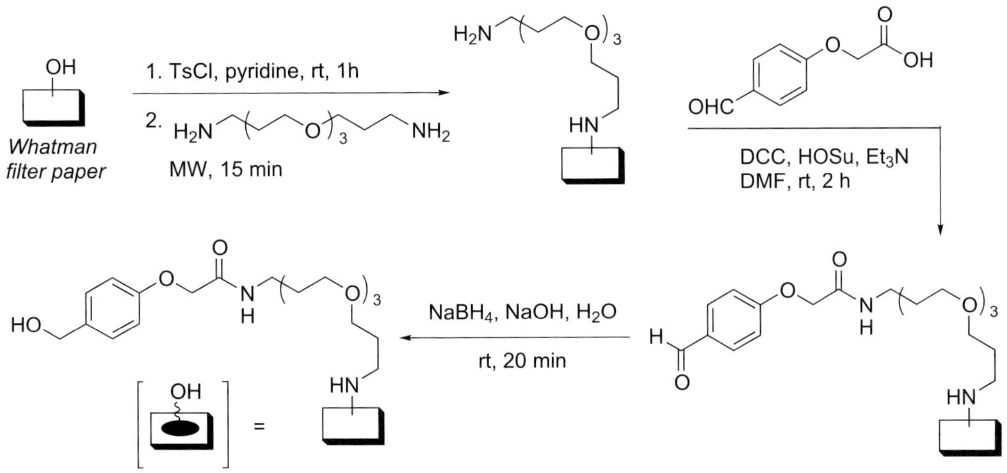

Scheme 16.23. Support modification for SPOT synthesis of pyrimidines.

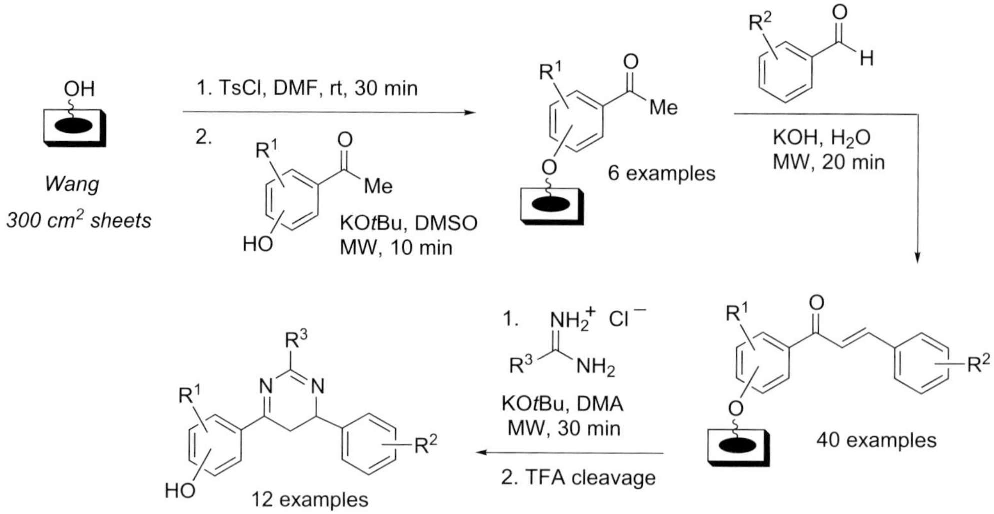

Scheme 16.24. Microwave-assisted SPOT synthesis of pyrimidines.

The successfully generated chalcones could be cleaved by treatment with trifluoroacetic acid or used for subsequent synthesis of pyrimidines [45]. Condensation of the polymer-bound chalcones with benzamidine hydrochloride under the action of microwave irradiation for 30 min furnished the corresponding pyrimidines in good yields after TFA-induced cleavage.

A SPOS approach to DNA synthesis was reported very recently. It involves a ten-step synthesis of a phosphoramidite building block of 1′-aminomethylthymidine starting from 2-deoxyribose [46]. Microwave-mediated deprotection of a corresponding N-allyloxycarbonyl (alloc-) protected nucleoside and acylation with the residue of pyrene-1-ylbutanolic acid (pyBA) were described (Scheme 16.25). Removal of the alloc protecting group was achieved on a support (controlled pore glass, cpg) under microwave conditions (80 °C, 10 min) to ensure full conversion.

After the reaction, the solid supported product was washed with N,N-dimethylformamide and dichloromethane and dried before being subjected to acylation. Coupling of the pyrene butanoic acid (PyBA) was again performed under the action of microwave irradiation at 80 °C for 10 min.

A related study led to a reactivity test for HBTU-activated carboxylic acids involving N-methylpyrrole derivatives as building blocks for oligopyrrolamides to bind DNA duplexes [47]. Polymer-bound DNA was decorated with Fmoc-protected N-methylpyrrole-2-carboxylic acid (Scheme 16.26) under the action of microwave irradiation for 10 min at 80 °C; subsequent deprotection furnished the desired modified oligonucleotide in moderate yields.

In a comparable study glass was used as solid support – carbohydrates were attached to amino-derived glass slides. Significant rate enhancement was observed when this step was conducted under the action of microwave irradiation rather

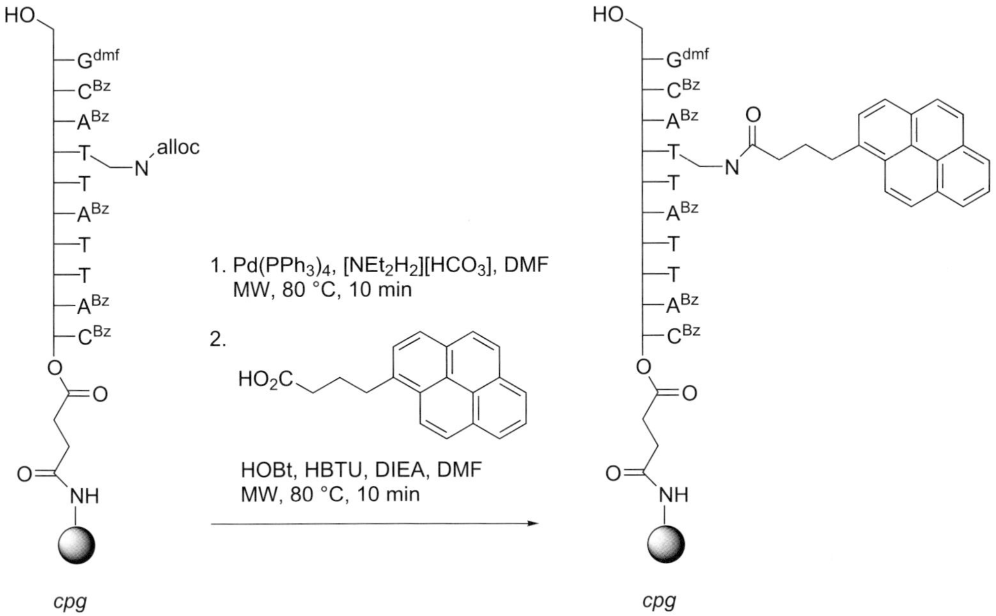

16.1 Solid-phase Organic Synthesis | 743

Scheme 16.25. Deprotection of oligonucleotides on controlled pore glass (cpg).

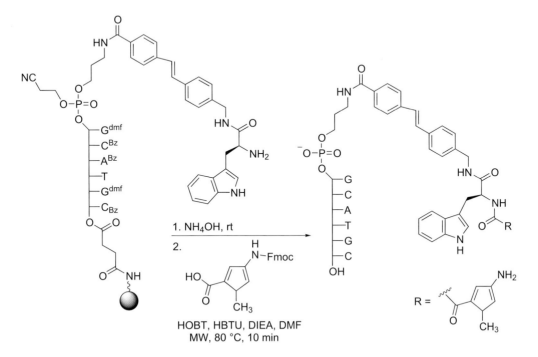

Scheme 16.26. Coupling of pyrrole carboxylic acids with oligonucleotides.

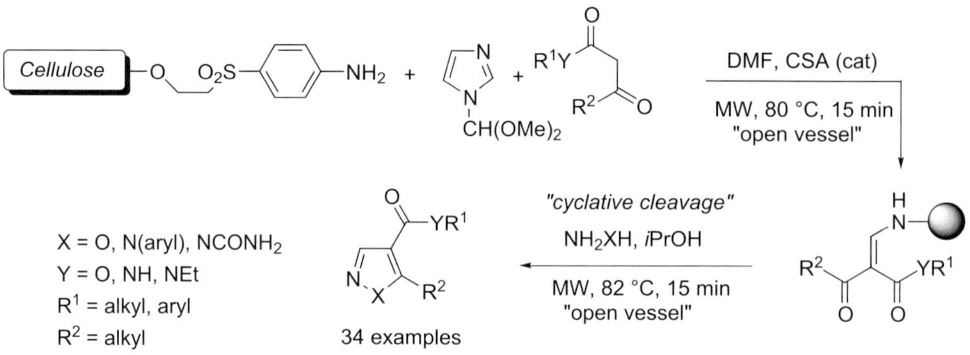

Scheme 16.27. Generation of pyrazole and oxazole libraries on cellulose beads.

than classical heating [48]. This method should be an efficient aid in the construction of functional carbohydrate array systems.

The first report of the use of cellulose beads as support for microwave-assisted SPOS was published in 2003 and described the generation of a library of pyrazoles and isoxazoles [49]. The synthesis was performed using commercially available amino cellulose (Perloza VT-100) containing aminoaryl ethyl sulfone groups in flexible chains (Scheme 16.27). Initially, the solid support was treated with excess formyl imidazole and the corresponding β-keto compounds to generate cellulose-bound enaminones in a one-pot Bredereck-type condensation. The reaction was catalyzed by (±)-camphor-10-sulfonic acid (CSA) and performed under microwave-irradiation conditions in an open vessel to enable the methanol formed to be removed from the reaction equilibrium [49].

Subsequent cyclization and cleavage from the support were achieved by microwave heating with several hydrazines or hydroxylamines to afford the desired heterocyclic targets. The cellulose-bound aniline could be recycled by simple washing and re-used up to ten times without loss of efficiency or reduced purity of the resulting compounds.

Microwave-assisted solid-phase synthesis of purines on an acid-sensitive methoxybenzaldehyde (AMEBA)-linked polystyrene has been reported [50]. The heterocyclic scaffold was first attached to the polymer support via an aromatic nucleophilic substitution reaction by conventional heating in 1-methyl-2-pyrrolidinone (NMP) in the presence of N,N-diisopropylethylamine. The key aromatic nucleophilic substitution of the iodine with primary and secondary amines was conducted by microwave heating for 30 min at 200 °C in 1-methyl-2-pyrrolidone (Scheme 16.28). After reaction the products were cleaved from the solid support by use of trifluoroacetic acid–water at 60 °C.

In related work the same group achieved efficient displacement of the substituent at C-2 from polymer-bound 2,6-dihalopurines [51]. Whereas substitution of Cl at C-6 was readily achieved with primary and secondary amines, displacement of the C-2 substituent proved difficult. Under the action of microwave irradiation at 150 °C for 30 min, however, a variety of amines could be used for further decora-

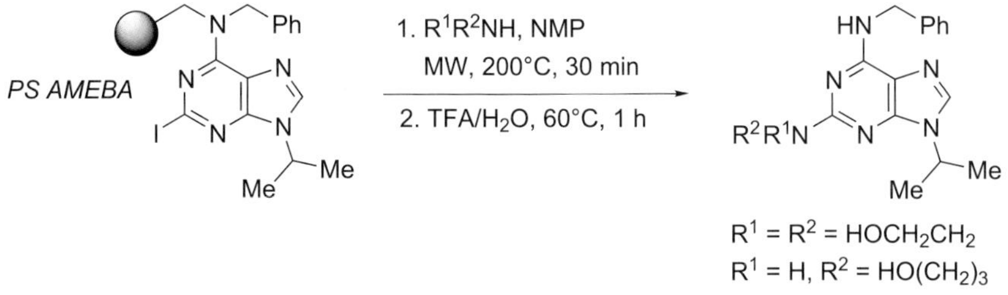

Scheme 16.28. Purine synthesis on AMEBA-linked solid support.

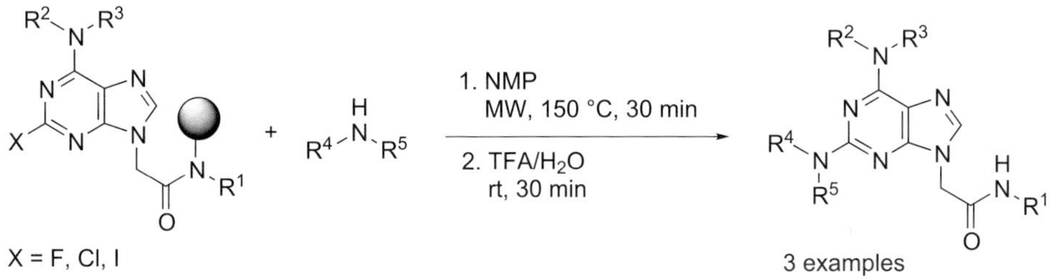

Scheme 16.29. Polymer-supported synthesis of trisubstituted purin-9-yl acetamides.

tion of the purine scaffold (Scheme 16.29). Standard cleavage with TFA–DCM furnished the trisubstituted 2-(2,6-purin-9-yl)acetamides in varying yields.

16.1.7
Multicomponent Chemistry

Multicomponent reactions in which three or more components build a single product have attracted much interest for several years. Because most of these reactions tolerate a wide range of building block combinations they are frequently used for combinatorial purposes.

The first solid-phase application toward the Ugi four-component condensation generating an acylamino amide library was reported in 1999. Amino-functionalized PEG-polystyrene (TentaGel S RAM) was used as the solid support [52]. A set of three aldehydes, three carboxylic acids, and two isonitriles was used for generation of the 18-member library.

Very recently, a polymer-supported construct for quantifying the reactivity of monomers involving multicomponent Ugi strategies was reported [53]. The authors investigated the Ugi four-component condensation of a set of aldehydes, acids, and isonitriles. The amine compound used was bound to a specially modified Rink amide polystyrene resin bearing an ionization leveler and bromine functionality as a mass splitter for accurate analysis by positive electrospray ionization mass spectrometry. The microwave-assisted experiment was performed with a set of ten carboxylic acids, hydrocinnamaldehyde, and cyclohexyl isonitrile in DCM–

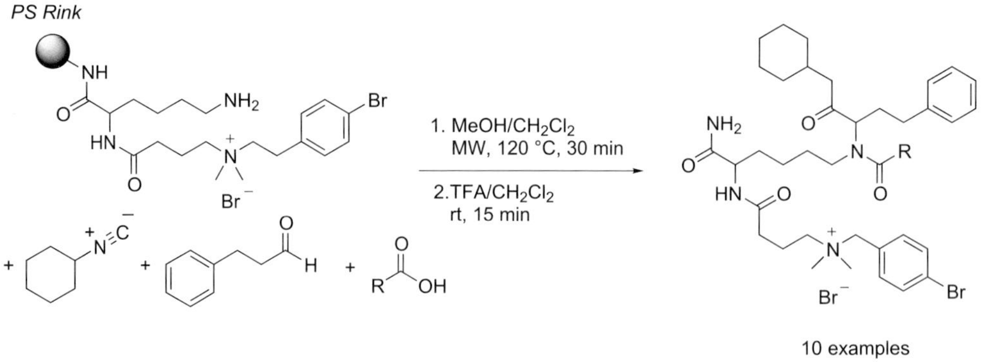

Scheme 16.30. Polymer-bound Ugi-4CC.

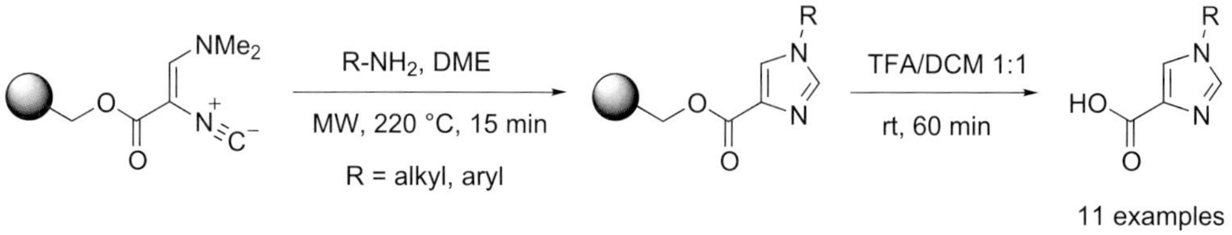

Scheme 16.31. Synthesis of 1-substituted 4-imidazolecarboxylates.

MeOH at 120 °C for 30 min (Scheme 16.30). The reactivity of each acid was measured after cleavage by single ESI+ MS analysis. Detailed results from the reactivity screening can be found elsewhere [53].

Microwave-assisted heterocycle synthesis has also been used to prepare a small set of pharmaceutically important imidazole derivatives [54]. The procedure utilized an initially prepared polymer-bound 3-N,N-(dimethylamino) isocyanoacrylate. The best results in imidazole synthesis were obtained by microwave-assisted reaction of an eightfold excess of the polymer-supported isonitrile suspended in 1,2-dimethoxyethane (DME) with the corresponding amines (Scheme 16.31). Cleavage with 50% trifluoroacetic acid in dichloromethane afforded the desired heterocyclic scaffolds in moderate yields.

Another valuable multicomponent reaction is the Gewald synthesis leading to 2-acyl-3-aminothiophenes; these are of current interest because they are commercially used as dyes or conducting polymers and have much pharmaceutical potential.

In this context, microwave-assisted Gewald synthesis on commercially available cyanoacetic acid Wang resin as solid support has been investigated [55]. By application of microwave irradiation the overall two-step reaction procedure, including the acylation of the initially formed 2-aminothiophenes, could be performed in less than 1 h (Scheme 16.32). This solid-phase "one-pot" two-step microwave-

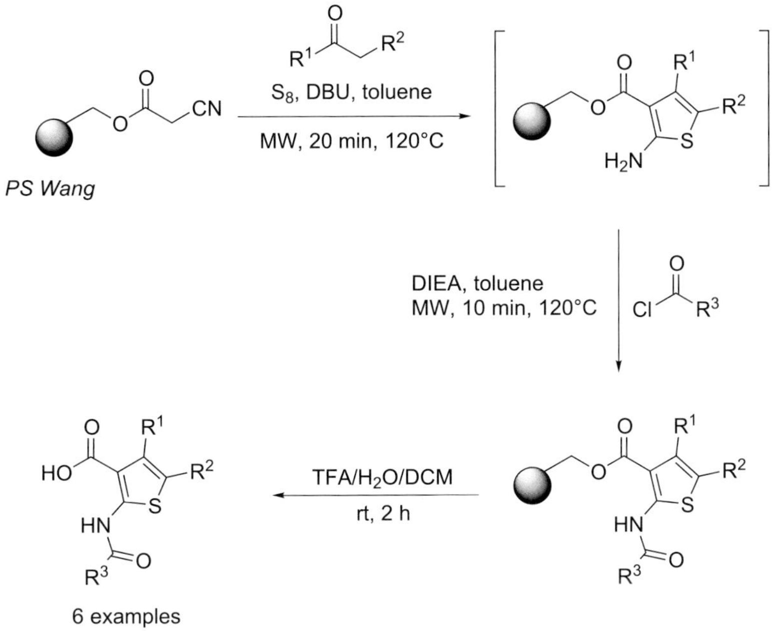

Scheme 16.32. One-pot Gewald synthesis on polymeric support.

promoted process is an efficient route to 2-acylaminothiophenes in high yields and, usually, good purity.

16.1.8
Microwave-assisted Condensation Reactions

As discussed in Section 16.1.4, polymer-bound acetoacetates can be used as precursors for the solid-phase synthesis of enones [30]. For these Knoevenagel condensations, the crucial step is to initiate enolization of the CH acidic component. For the microwave-assisted enolization procedure, however, piperidinium acetate was found to be the catalyst of choice, if temperatures were kept below 130 °C. A 21-member library, of polymer-bound enones was generated (Scheme 16.33) by using a multi-vessel rotor system for parallel microwave-assisted synthesis. Reaction

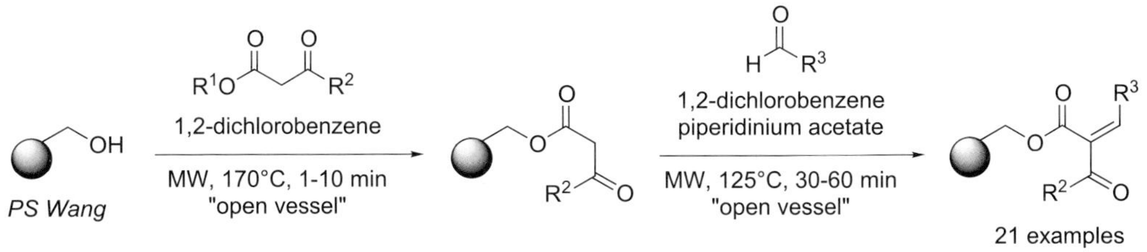

Scheme 16.33. Parallel synthesis of polymer-bound enones.

times could be reduced from 1–2 days to 30–60 min by using parallel microwave-promoted synthesis in open vessels, without effecting the purity of the resin-bound products.

Microwave-assisted Knoevenagel reactions have also been used for preparation of resin-bound nitroalkenes [56]. A variety of resin-bound nitroalkenes has been generated by use of resin-bound nitroacetic acid, which was condensed with a variety of aldehydes under microwave conditions. To demonstrate the potential of these resin-bound products in combinatorial applications, the readily prepared nitroalkenes were subsequently employed in Diels–Alder reactions with 2,3-dimethylbutadiene [56].

In a different approach, microwave-mediated oxazole synthesis utilizing β-ketoesters bound to a novel polymeric resin has been described [57]. The desired polymer support was prepared by transesterification reactions between *tert*-butyl β-ketoesters and hydroxybutyl functionalized JandaJel resin, and subsequent standard diazo transfer. The resulting α-diazo β-ketoesters have been used for synthesis of an array of oxazoles (Scheme 16.34). Because of the thermal sensitivity of the Burgess' reagent used, the temperature was kept rather low, but irradiation for 15 min at 100 °C furnished satisfactory results [57]. Cleavage from the solid support was achieved by diversity-introducing amidation; this led to the corresponding oxazole amides in reasonable yields.

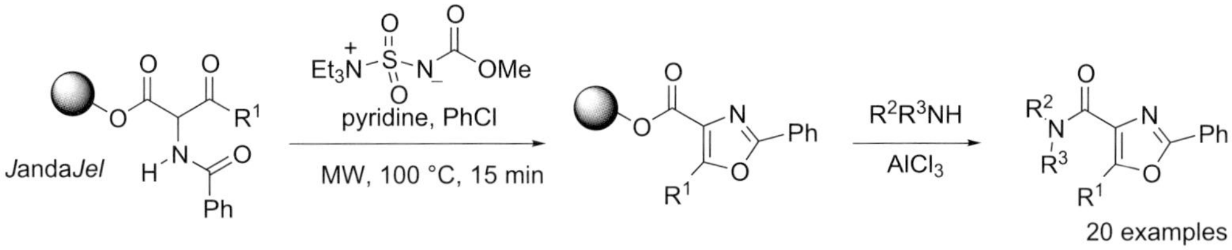

Scheme 16.34. Oxazole synthesis on functionalized JandaJel.

16.1.9
Rearrangements

Microwave-assisted rearrangement reactions on solid-phases have rarely been discussed in the literature – only two examples describing Claisen rearrangements have been reported [58, 59]. In the first study, Merrifield resin-bound O-allylsalicylic esters were rapidly rearranged to the corresponding ortho allylsalicylic esters [58]. Microwave heating in open vessels in a domestic microwave cavity was used (Scheme 16.35). Acid-mediated cleavage of these resin-bound ester products afforded the corresponding ortho allylsalicylic acids.

A more recent report discussed a microwave-assisted tetronate synthesis entailing domino addition/Wittig olefinations of polymer-bound α-hydroxy esters with the cumulated phosphorus ylide $Ph_3P=C=C=O$ [59]. The desired immobilized α-hydroxy esters can be obtained by microwave-mediated ring opening of the corresponding glycidyl esters by OH-, NH-, or SH-terminal polystyrenes of the Merri-

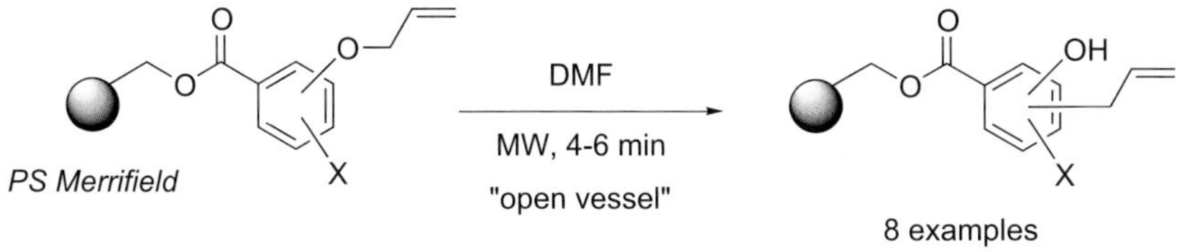

Scheme 16.35. Claisen rearrangement on solid phases.

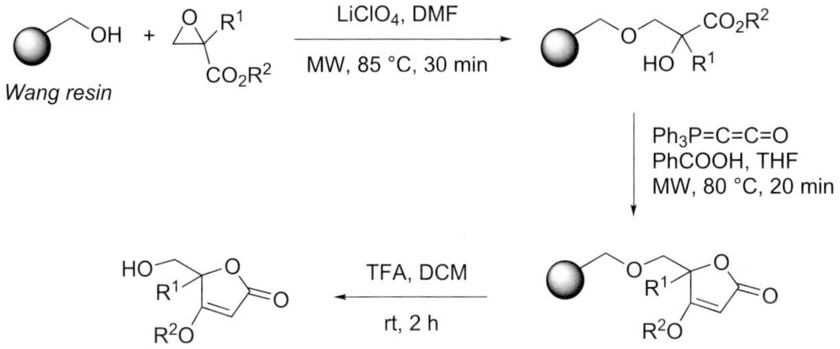

Scheme 16.36. Tetronate synthesis by domino addition and Wittig olefination.

field or Wang type (Scheme 16.36). The subsequent tandem Wittig reaction was performed in tetrahydrofuran, with catalytic amounts of benzoic acid, under the action of microwave irradiation. The formation of the polymer-bound tetronates was complete within 20 min irradiation at 80 °C [59]. Quantitative cleavage of the tetronates was achieved by treatment with 50% trifluoroacetic acid in dichloromethane at room temperature.

16.1.10
Cleavage Reactions

In addition to the aforementioned microwave-assisted reactions on solid supports, several publications describe microwave-assisted resin cleavage. A variety of cleavage procedures have been investigated, depending on the nature of the linker used. In this context it was shown that several resin-bound carboxylic acids (Scheme 16.1) were cleaved from traditionally non-acid-sensitive Merrifield resin by using 50% trifluoroacetic acid in dichloromethane (Scheme 16.37). Microwave irradiation enables these cleavages to be conducted at elevated pressure and/or temperature in sealed Teflon vessels [23]. Evaporation of the filtrate to dryness furnished the recovered benzoic acid in quantitative yield and excellent purity.

In another application of microwave-assisted resin cleavage, N-benzoylated alanine attached to 4-sulfamylbutyryl resin was cleaved (after activation of the linker with bromoacetonitrile by using Kenner's safety catch principle) with a variety of

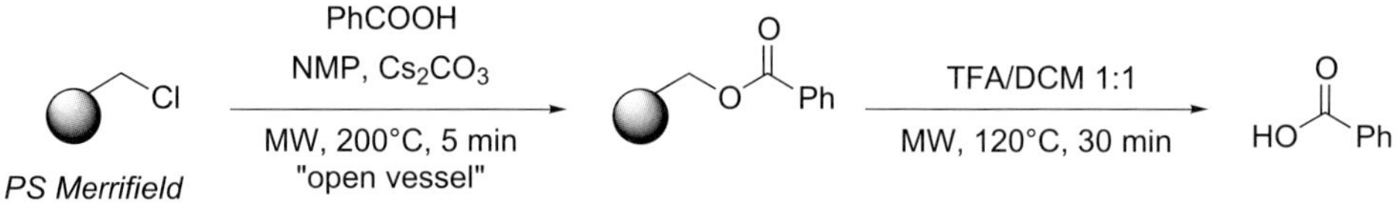

Scheme 16.37. Acidic cleavage from polystyrene resin under elevated pressure.

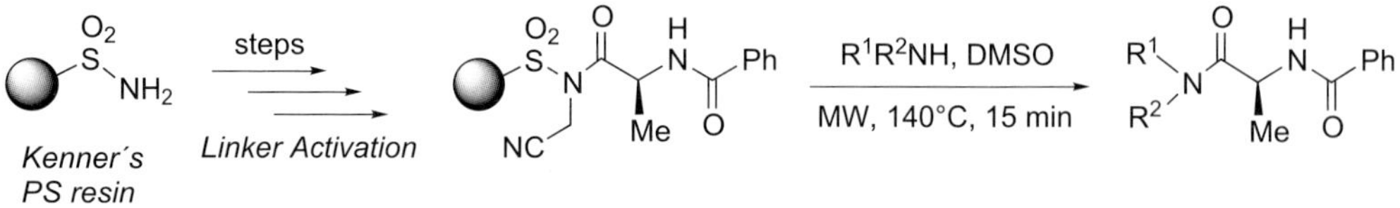

Scheme 16.38. Solid-supported amide synthesis employing the "safety catch" principle.

amines (Scheme 16.38) [17, 18]. Cleavage rates in dimethyl sulfoxide were investigated for N,N-diisopropylamine and aniline under different reaction conditions using both microwaves (domestic oven) and conventional heating (oil bath). The results showed that when experiments were run at the same temperature (80 °C) microwave heating did not accelerate reaction rates compared with conventional heating. When microwave heating was used, however, even cleavage with normally unreactive aniline could be accomplished within 15 min at approximately 140 °C.

The microwave approach has been used for parallel synthesis of an 880-member library in 96 well plates; ten different amino acids, each bearing a different acyl group, were coupled to the 4-sulfambutyryl resin and 88 different amines were used in the cleavage step.

In closely related work, similar solid-phase chemistry was used to prepare biaryl urea compound libraries via microwave-assisted Suzuki couplings followed by cleavage from the resin with amines [18]. The procedure enabled generation of large biaryl urea compound libraries by use of a simple domestic microwave oven.

A more recent study investigated microwave-assisted parallel synthesis of di and trisubstituted ureas from carbamates bound on modified Marshall resin [60]. Modification of the resin was achieved by treatment with p-nitrophenyl chloroformate and N-methylmorpholine (NMM) in dichloromethane at low temperatures. The resulting resin was further modified by attachment of a variety of amines to obtain a set of polymer-bound carbamates (Scheme 16.39).

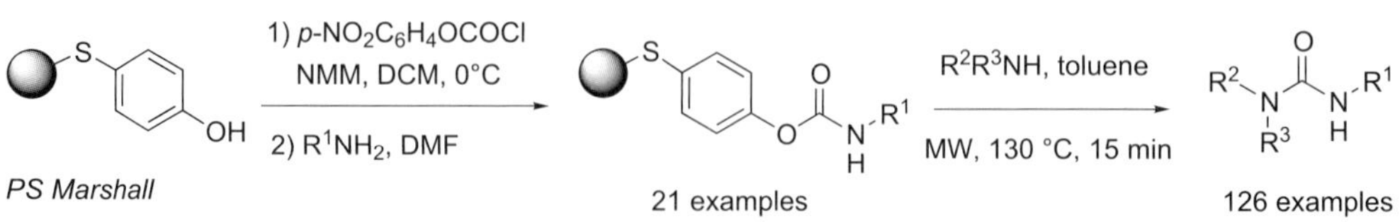

Scheme 16.39. Parallel synthesis of substituted ureas from thiophenoxy carbamate resins.

The immobilized carbamates were transferred to a sealable 96-well Weflon plate and mixed with different primary or secondary amines dissolved in anhydrous toluene. The plate was sealed and irradiated in a multimode microwave instrument, generating a ramp which reached 130 °C in 45 min. This temperature was held for an additional 15 min. After cooling, the resins were filtered and the filtrates were evaporated to furnish the desired substituted ureas in good purity and reasonable yields.

A comprehensive study has investigated multidirectional cyclative cleavage transformations leading to bicyclic dihydropyrimidinones [61]. This approach required synthesis of 4-chloroacetoacetate resin as the key starting material; this was prepared by microwave-assisted acetoacetylation of commercial available hydroxymethyl polystyrene resin under open-vessel conditions. This resin precursor was subsequently treated with urea and a variety of aldehydes in a Biginelli-type multi-component reaction, leading to the corresponding resin-bound dihydropyrimidinones (Scheme 16.40). The desired furo[3,4-d]pyrimidine-2,5-dione scaffold was obtained by a novel procedure for cyclative release under the action of microwave irradiation in sealed vials at 150 °C for 10 min.

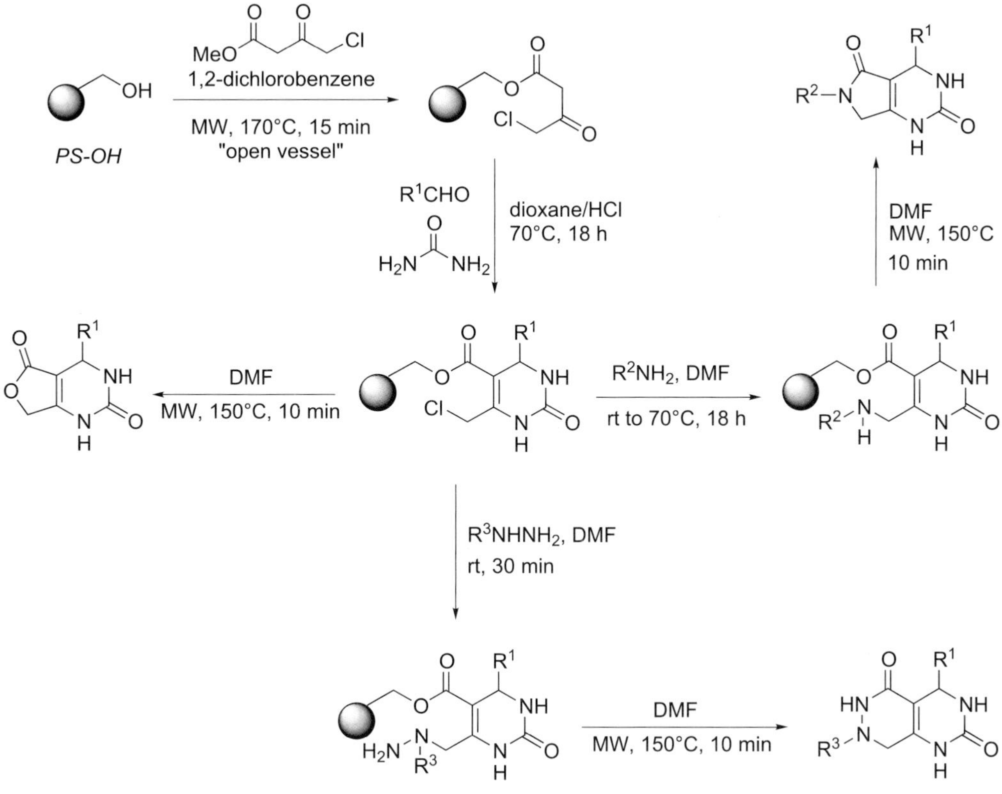

Scheme 16.40. Preparation of bicyclic dihydropyrimidinones via cyclative cleavage.

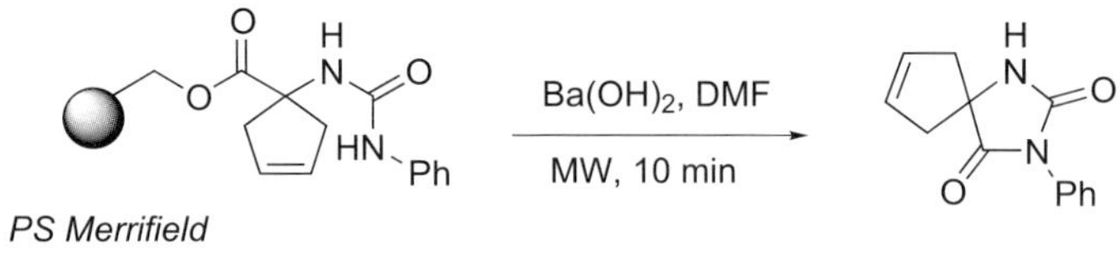

Scheme 16.41. Intramolecular carbanilide cyclization on a solid support.

Alternatively, pyrrolo[3,4-*d*]pyrimidine-2,5-diones and pyrimido[4,5-*d*]pyridazine-2,5-diones have been synthesized by using the same pyrimidine resin precursor, which was first treated with a representative set of primary amines or hydrazines, respectively, to substitute the chlorine. Subsequent cyclative cleavage was performed as already described, leading to the corresponding bicyclic scaffolds in high purity but moderate yield.

In an earlier report, microwave-mediated intramolecular carbanilide cyclization to hydantoins was described [62]. For the solid-phase approach, conventional *i*PrOCH$_2$-functionalized polystyrene Merrifield resin was used. After attachment of the corresponding substrate, the resin was treated with a solution of barium(II) hydroxide in DMF within an appropriate sealed microwave vial. The vial was heated in the microwave cavity for 5 × 2 min cycles with the reaction mixture being left to cool to room temperature between irradiation cycles (Scheme 16.41). The procedure led to comparatively modest isolated yields of hydantoins.

A more recent report discussed the traceless solid-phase preparation of phthalimides by cyclative cleavage from conventional Wang resin [63]. To find the optimum conditions for the cyclative cleavage step, *ortho*-phthalic acid was chosen as the model compound and was attached to the polystyrene resin via a Mitsunobu procedure. Use of *N*,*N*-dimethylformamide as a solvent for the cyclative cleavage furnished the desired compounds in high yield and excellent purity (Scheme 16.42). The type of amine had an effect on the outcome of the reaction, however. Aromatic amines could not be included in the study because auto-induced ring-closure occurred during conversion of the polymer-bound phthalic acid.

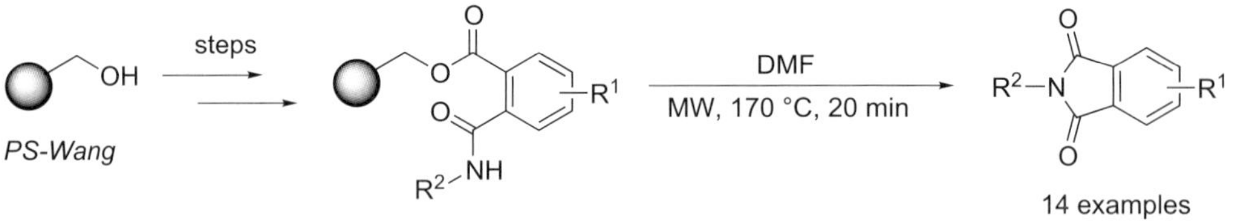

Scheme 16.42. Polymer-supported phthalimide synthesis.

16.1.11
Miscellaneous

Several other types of reaction on solid supports have also been investigated using microwave heating. In early work the addition of resin bound amines to isocya-

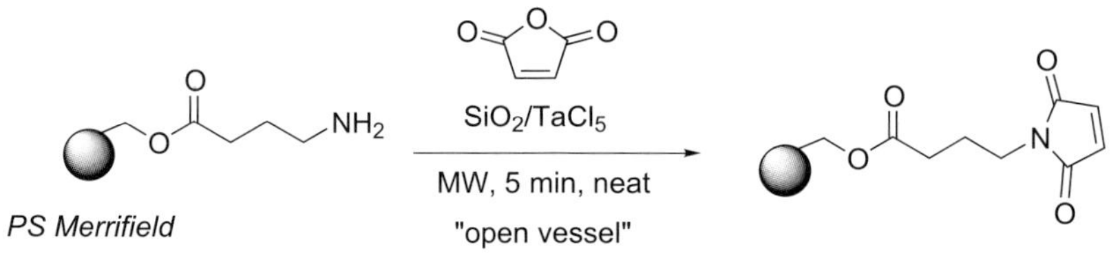

Scheme 16.43. Solvent-free preparation of polymer-bound 4-(1,3-2,3-dihydro-1H-2-isoindolyl)butanoic acid.

nates was monitored by use of on-bead FTIR measurements to investigate differences between reaction progress under microwave and thermal conditions [64].

A different approach toward solid-supported synthesis with microwave heating has enabled synthesis of N-alkyl imides on solid-phases under solvent-free conditions [65]. Tantalum(V) chloride-doped silica gel was used as a Lewis acid catalyst (Scheme 16.43). Surprisingly, dry and unswollen polystyrene resin is involved in this rather unusual method. The reaction was performed in a domestic microwave oven with 1-min irradiation cycles and thorough agitation after each step. Within 5–7 min the reaction was complete and the corresponding N-alkyl imide product was obtained in good yield after subsequent cleavage from the polystyrene resin–silica gel mixture by trifluoroacetic acid–dichloromethane. In addition, use of two resin-bound amines and three different anhydrides in this solid-phase procedure enabled synthesis of a set of six cyclic imides in good yields.

Use of phosphoranes as polymer-bound acylation equivalents has also been reported [66]. Initial alkylation of the polymer-supported triphenylphosphine reagent was achieved with bromoacetonitrile under microwave irradiation conditions (Scheme 16.44). Simple treatment with triethylamine transformed the polymer-bound phosphonium salt into the corresponding stable phosphorane, which could be efficiently coupled with a variety of protected amino acids. After Fmoc deprotection and subsequent acylation the resulting acyl cyano phosphoranes could be released from the polymer support by ozonolysis at −78 °C. The released highly activated electrophiles can be converted in situ with appropriate nucleophiles [66].

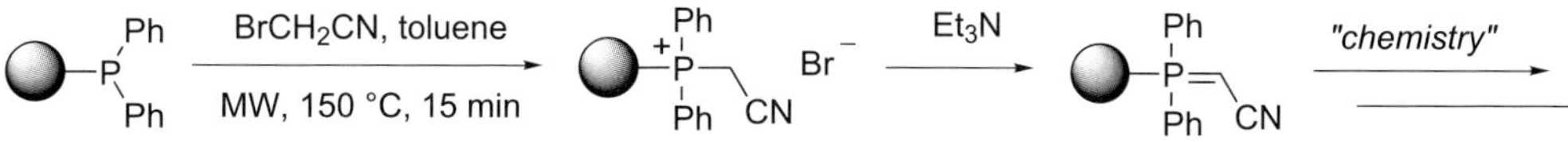

Scheme 16.44. Phosphoranes as polymer-bound acylanion equivalents.

In a more recent study, microwave irradiation was used in the accelerated solid-phase synthesis of aryl triflates [67] (Scheme 16.45). Reducing the reaction times from 3–8 h under conventional heating conditions to only 6 min under the action of microwaves has made this procedure more amenable to high-throughput synthesis. The use of the commercially available chlorotrityl linker as a solid support

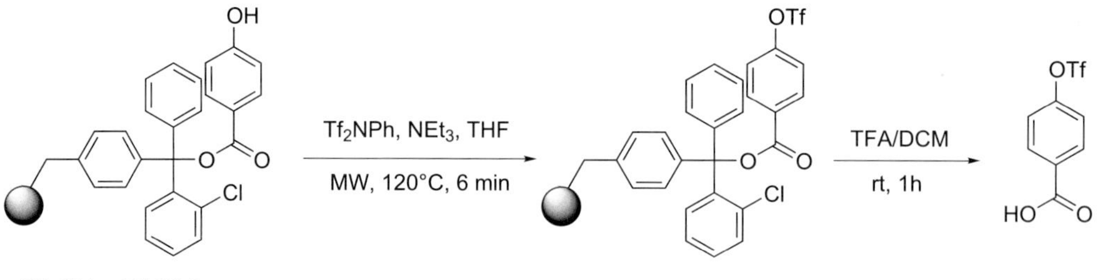

Scheme 16.45. Solid-phase triflating procedure.

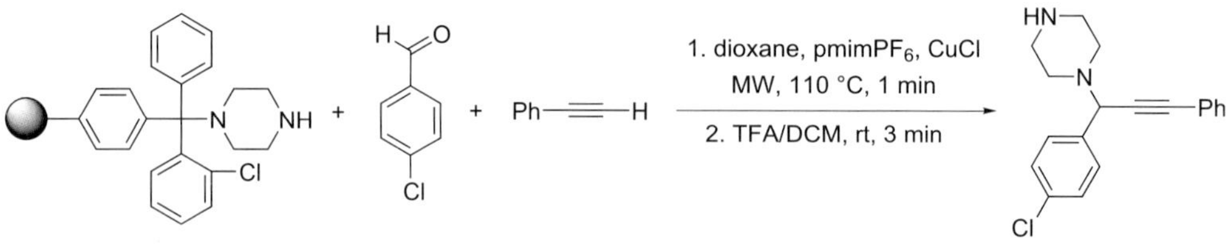

Scheme 16.46. One-pot Mannich reaction using a polymer-supported piperazine.

enables use of mild cleavage conditions to obtain the desired aryl triflates in good yields.

Another interesting study led to a solid-phase procedure for a one-pot Mannich reaction employing the above mentioned chlorotrityl linker [68]. In this approach p-chlorobenzaldehyde and phenylacetylene were condensed with readily prepared immobilized piperazines (Scheme 16.46).

The mixture was heated to 110 °C within 2 min and kept at this reaction temperature for another minute. After cooling to room temperature the product was rapidly released from the polymer support by use of trifluoroacetic acid (TFA) in dichloromethane, furnishing the corresponding bis-TFA salt in moderate yield.

The feasibility of applying an already established microwave-assisted iodination procedure to a polymer-supported substrate was discussed in 2003 [69]. A pyrimidinone attached to a conventional Merrifield polystyrene resin was suspended in N,N-dimethylformamide, treated with N-iodosuccinimide (NIS), and subjected to microwave irradiation for 3 min. Treatment of the polymer-bound intermediate with Oxone resulted in release of the desired 5-iodouracil in almost quantitative yield.

A study of the microwave-assisted solid-phase Diels–Alder cycloaddition reaction of 2(1H)-pyrazinones with dienophiles has also been reported [70]. After fragmentation of the resin-bound primary cyclo adduct formed by Diels–Alder reaction of the 2(1H)-pyrazinone with an acetylenic dienophile, separation of the resulting pyridines from the pyridinone by-products was achieved by applying a traceless-

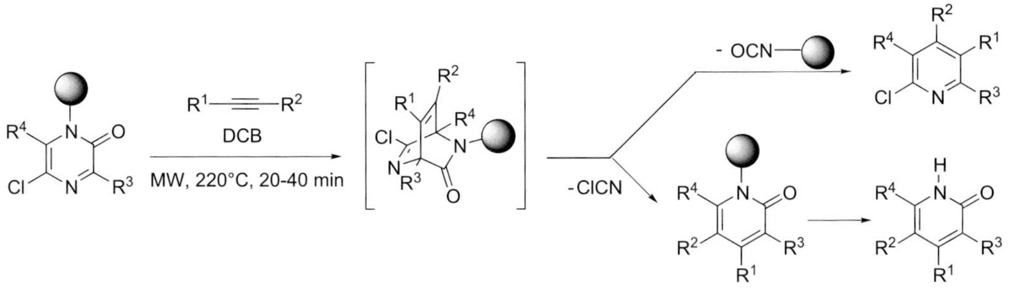

Scheme 16.47. General reaction sequence for 2(1H)-pyrazinone Diels–Alder cycloadditions with acetylenic dienophiles.

linking concept in which the pyridinones remained on the solid support with concomitant release of the pyridine products into solution (Scheme 16.47).

For this approach a novel a tailor-made and readily available linker derived from inexpensive syringaldehyde was developed [70]. The novel linker was produced by cesium carbonate-activated coupling of commercially available syringaldehyde to the Merrifield resin under microwave heating conditions. The aldehyde moiety was subsequently reduced at room temperature within 12 h and the benzylic position was finally brominated by treatment with a large excess of thionyl bromide leading to the desired polymeric support (Scheme 16.48).

In addition, an appropriate strategy for cleavage from the novel syringaldehyde resin was developed by means of a new solution-phase model study of intramolecular Diels–Alder reactions [70]. By using the novel syringaldehyde resin, smooth release from the support could be performed by microwave heating of a suspension of the resin-bound pyridinones in trifluoroacetic acid–dichloromethane at 120 °C for only 10 min.

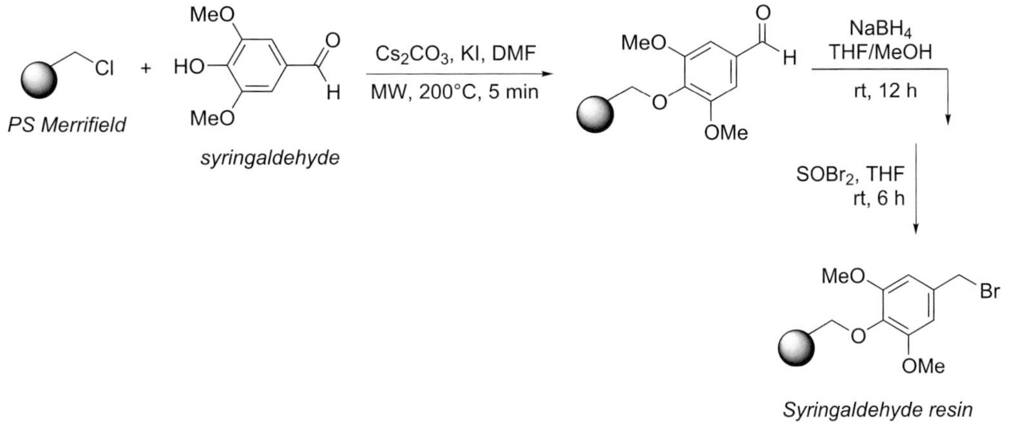

Scheme 16.48. Preparation of brominated syringaldehyde resin.

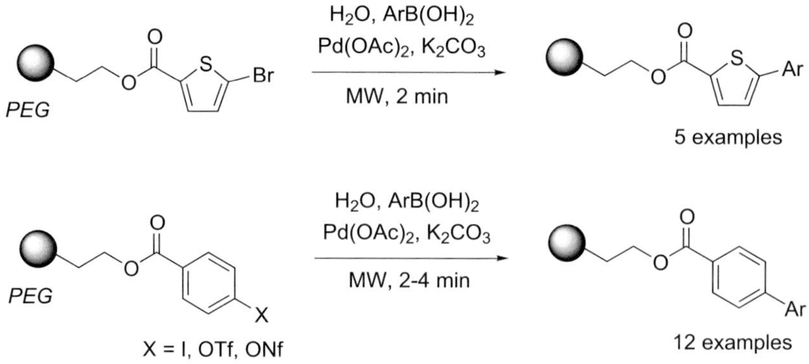

Scheme 16.49. PEG-supported Suzuki couplings.

16.2
Soluble Polymer-supported Synthesis

Chemistry on soluble polymer matrices has recently emerged as a viable alternative to solid-phase organic synthesis (SPOS) involving insoluble cross-linked polymer supports. Separation of the functionalized matrix is achieved by solvent or heat precipitation, membrane filtration, or size-exclusion chromatography. Suitable soluble polymers for liquid phase synthesis should be crystalline at room temperature, with functional groups on terminal ends or side chains, but must not be not cross-linked; they are therefore soluble in several organic solvents.

Among the first examples of so-called liquid-phase synthesis were aqueous Suzuki reactions employing poly(ethylene glycol) (PEG)-bound aryl halides and sulfonates in palladium-catalyzed cross couplings [71]. It was shown that no additional phase-transfer catalyst (PTC) was needed when the PEG-bound electrophiles were coupled with aryl boronic acids in water under microwave irradiation conditions, in sealed vessels, in a domestic microwave oven (Scheme 16.49). Work-up involved precipitation of the polymer-bound biaryl from a suitable organic solvent with ether.

Another palladium-catalyzed coupling reaction performed successfully on soluble polymers is the Sonogashira coupling. In a more recent report it was reported that PEG 4000 acts simultaneously as polymeric support, solvent, and phase transfer catalyst (PTC) in both steps, coupling and hydrolysis [72]. Poly(ethylene glycol) (PEG)-bound 4-iodobenzoic acid was readily reacted with several terminal alkynes under rapid microwave conditions (Scheme 16.50). Cleavage of the coupling products from the PEG support was achieved efficiently by simple saponification with brief microwave irradiation in an open beaker in a domestic microwave oven.

In a related study, a Schiff base-protected glycine was reacted with a variety of electrophiles under the action of microwave irradiation [73]. No extra solvent was necessary to perform these reactions because PEG played the role of polymeric support, solvent, and PTC. The best results were obtained by using cesium carbo-

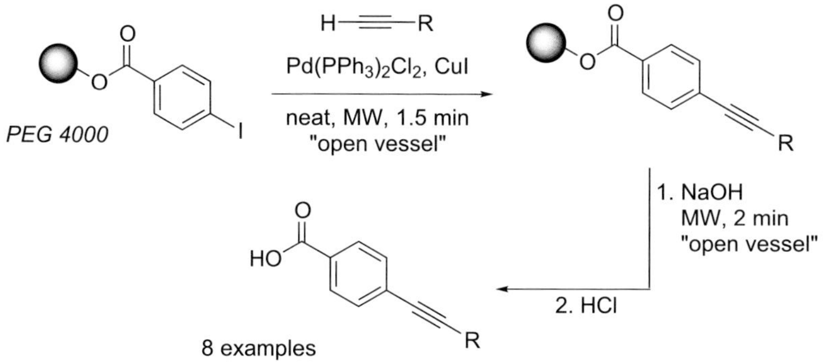

Scheme 16.50. Sonogashira couplings on a PEG support.

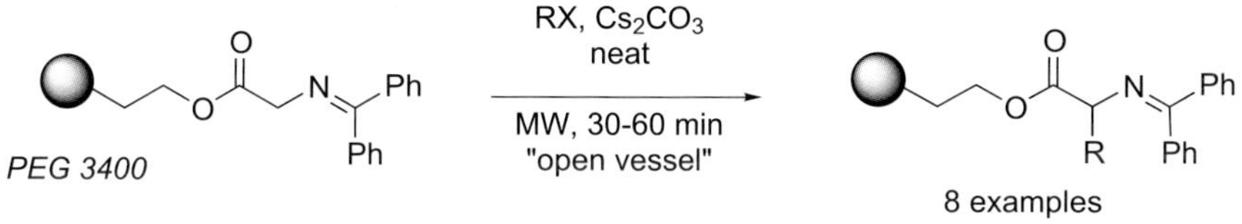

Scheme 16.51. Microwave-assisted alkylations on PEG.

nate as inorganic base (Scheme 16.51). After alkylation the corresponding amino esters were released from the polymer support by transesterification with methanol in the presence of triethylamine.

A method for microwave-assisted transesterifications has been published [74]. The microwave-mediated derivatization of poly(styrene–co-allyl alcohol) was investigated as key step in the polymer-assisted synthesis of heterocycles. The procedure was applied to several β-ketoesters and multigram quantities of products were obtained when neat mixtures of the reagents in open vessels were exposed to microwave irradiation in a domestic microwave oven. The soluble supports obtained were used for preparation of a variety of bicyclic heterocycles, for example pyrazolopyridinediones or coumarins.

Another procedure for preparation of valuable heterocyclic scaffolds involves the Biginelli condensation on a PEG Support [75, 76]. Polymer-bound acetoacetate was prepared by reacting commercially available PEG 4000 with 2,2,6-trimethyl-4H-1,3-dioxin-4-one in toluene under reflux (Scheme 16.52). The microwave-assisted cyclocondensation was performed with nonvolatile polyphosphoric acid (PPA) as a catalyst in a domestic microwave oven [76]. During microwave heating the PEG-bound substrate melted, ensuring a homogeneous reaction mixture. After the reaction diethyl ether was added to precipitate the polymer bound products. The desired compounds were released by treatment with sodium methoxide in methanol at room temperature. All dihydropyrimidines were obtained in high yield; purification was achieved by recrystallization from ethanol.

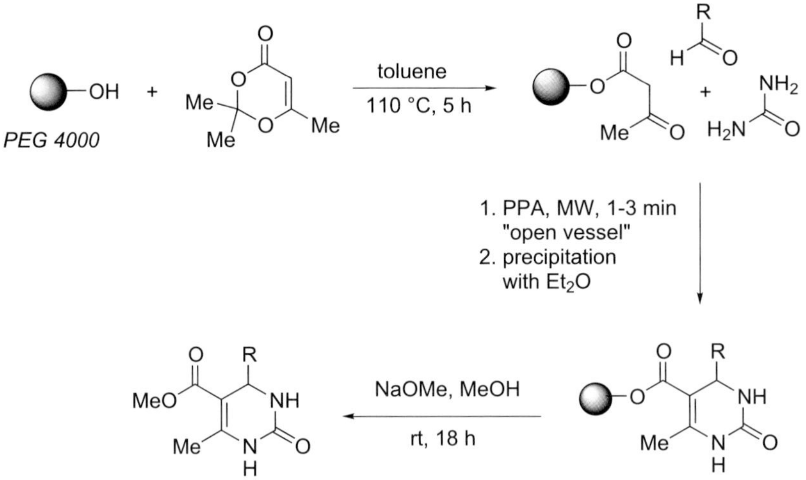

Scheme 16.52. One-pot Biginelli cyclocondensation on a PEG support.

A related support frequently used for liquid phase synthesis is methoxypoly(ethylene glycol) (MeO-PEG) and a general procedure for microwave-assisted synthesis of organic molecules on MeO-PEG has been reported [77]. The use of MeO-PEG under microwave conditions in open vessels simplified the process of polymer-supported synthesis (Scheme 16.53), because the polymer-bound

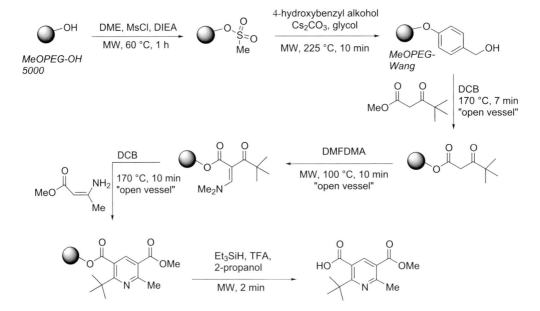

Scheme 16.53. Liquid-phase pyridine synthesis using MeO-PEG.

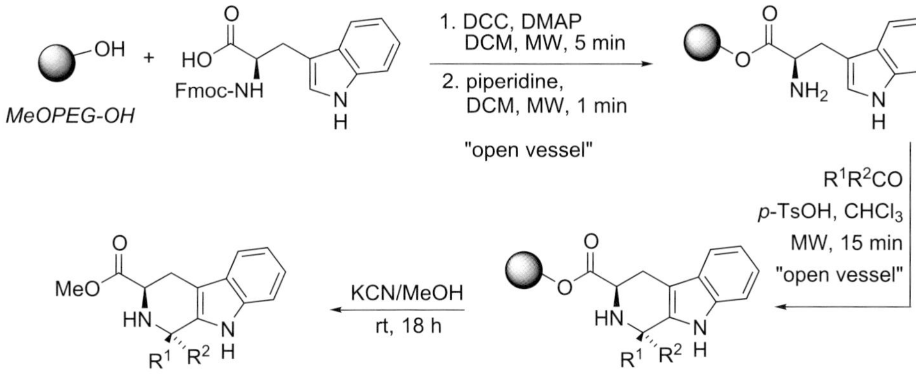

Scheme 16.54. Liquid-phase preparation of 1,2,3,4-tetrahydro-β-carbolines.

products precipitate on cooling after removal from the microwave oven. In addition, cleavage could be performed rapidly under microwave irradiation conditions by using a mixture of trifluoroacetic acid (TFA) and 2-propanol to furnish the desired nicotinic acid as the corresponding pyridinium trifluoroacetate.

Earlier, a versatile procedure for liquid phase synthesis of 1,2,3,4-tetrahydro-β-carbolines was reported [78]. After successful esterification of MeO-PEG-OH with Fmoc-protected tryptophan, one-pot cyclocondensation with a variety of ketones and aldehydes was performed under the action of microwave irradiation (Scheme 16.54). The desired products were released from the soluble support in good yields and high purity.

Generation of the desired polymer-bound tryptophan was achieved rapidly under microwave-irradiation conditions by using a classical esterification procedure, followed by Fmoc-deprotection (Scheme 16.54). Cyclocondensations with a variety of carbonyl compounds were performed with catalytic amounts of p-toluenesulfonic acid (p-TsOH) within 15 min under the action of microwave irradiation, resulting in quantitative conversion. Finally, treatment of the polymer-bound heterocycles with 1 mol% potassium cyanide (KCN) in methanol furnished the desired target structures in high yields.

In a related study microwave-assisted liquid-phase synthesis of quinoxalin-2-ones was achieved [79]. 4-Fluoro-3-nitrobenzoic acid was coupled efficiently with a commercially available poly(ethylene glycol) (PEG 6000). A variety of primary amines were then condensed with the readily prepared immobilized *ortho*-fluoronitrobenzene by nucleophilic ipso-fluoro displacement. Reduction of the resulting ortho nitroanilines furnished the corresponding ortho phenylenediamines, which were subsequently reacted with chloroacetyl chloride to form polymer-bound 1,2,3,4-tetrahydroquinoxalin-2-ones (Scheme 16.55). Every step was performed within a few minutes under the action of microwave irradiation in a dedicated single-mode reactor, with the PEG-bound intermediates precipitating from ethanol and subsequently being isolated by filtration to remove them from the by-products in the ethanolic phase. Finally, cleavage of the target molecules was achieved within

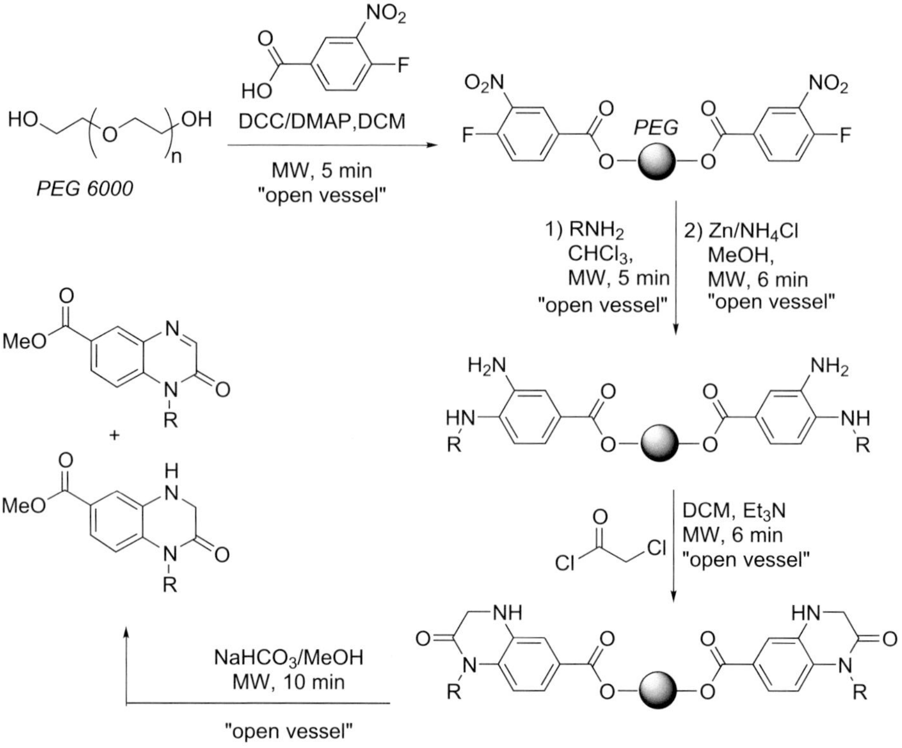

Scheme 16.55. Liquid-phase generation of quinoxalin-2-ones.

10 min under the action of microwave irradiation in a methanolic solution of sodium bicarbonate (NaHCO$_3$).

By using the same aryl-fluoride linker on conventional MeO-PEG polymer, the same authors also achieved microwave-accelerated liquid-phase synthesis of benzimidazoles (Scheme 16.56) [80].

Following the strategy described above with ipso-fluoro displacement and subsequent reduction, the resulting ortho phenylenediamines were treated with several aromatic isothiocyanates to form the corresponding PEG-bound benzimidazoles. Cleavage of the desired compounds was readily achieved by microwave-mediated transesterification with lithium bromide and 1,8-diazabicyclo[5.4.0]undec-7-ene (DBU). MeOPEG-OH was separated by precipitation and filtration to obtain the crude products in high yields and good purity [80].

A closely related study led to microwave-assisted liquid-phase synthesis of chiral quinoxalines [81]. A variety of L-α-amino acid methyl ester hydrochlorides were coupled to MeO-PEG-bound *ortho*-fluoronitrobenzene by the already described ipso-fluoro displacement. Reduction under the action of microwave irradiation resulted in spontaneous synchronous intramolecular cyclization to the corresponding 1,2,3,4-tetrahydroquinoxalin-2-ones.

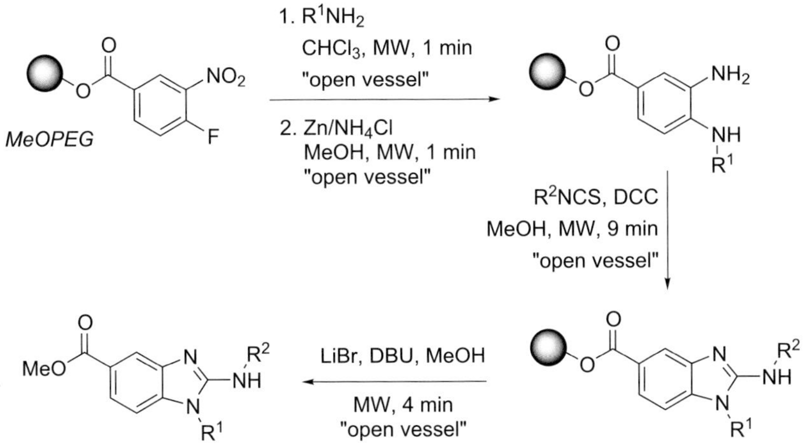

Scheme 16.56. Liquid-phase synthesis of benzimidazoles.

A variation of this method has been used to generate bis-benzimidazoles [82, 83]. The versatile immobilized ortho-phenylenediamine template was prepared as described above in several microwave-mediated steps. Additional N-acylation was performed by utilizing the initially used 4-fluoro-3-nitrobenzoic acid at room temperature exclusively at the primary aromatic amine moiety. A variety of amines have been used to introduce diversity by nucleophilic aromatic substitution. Cyclization to the polymer-bound benzimidazole was achieved by heating in a mixture of trifluoroacetic acid and chloroform, under reflux, for several hours. Individual steps at ambient temperature for selective reduction, cyclization with several aldehydes, and final detachment from the polymer support were necessary to obtain the desired bis-benzimidazoles. A set of thirteen examples was prepared in high yields and good purity [82].

Cyclization to the desired head-to-tail-linked bis-benzimidazoles can also be performed by reaction of aryl or alkyl isothiocyanates with N,N'-dicyclohexylcarbodiimide (DCC) [83]. In a closely related and more recent study by the same group, mercury chloride was used as a catalyst to perform cyclization to the benzimidazoles [84]. Another application of the bis-hydroxylated polymer support PEG 6000, microwave-accelerated liquid-phase synthesis of thiohydantoins, has been reported [85, 86].

The same authors also chose 3-chloropropionyl chloride as the immobilized building block to achieve ring expansion; this led to the generation of a 14-member library of the corresponding thioxotetrahydropyrimidinones [86, 87]. The initially prepared polymer-bound chloropropionyl ester was efficiently transformed into the corresponding diamines by transamination with several primary amines. These diamine intermediates could also be obtained by treatment of the pure polymeric support with acryloyl chloride and subsequent addition of the corresponding amines (Scheme 16.57).

In a procedure similar to the thiohydantoin synthesis, the PEG-bound diamines

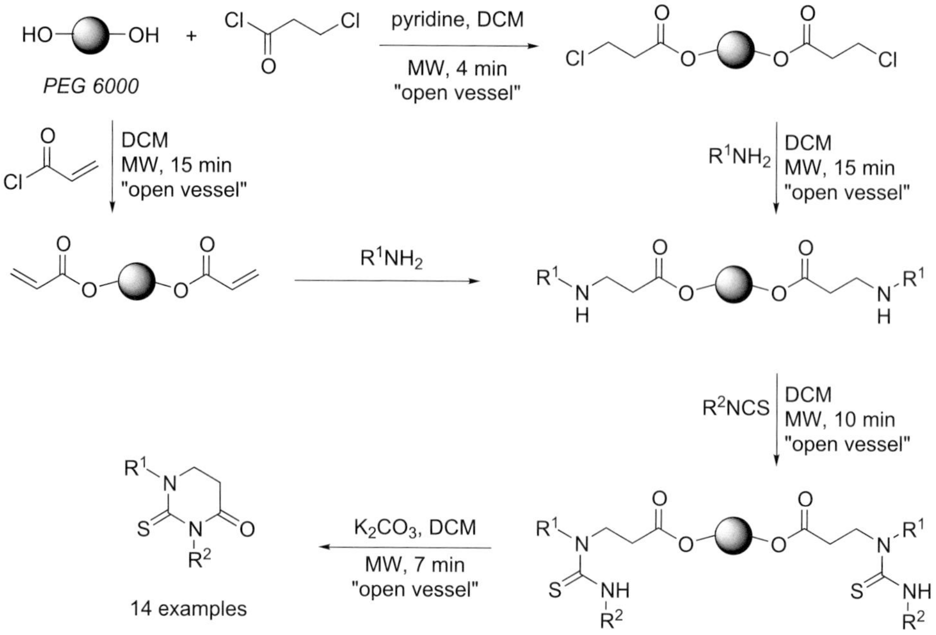

Scheme 16.57. Liquid-phase preparation of 1,3-disubstituted thioxo tetrahydropyrimidinones.

were treated with a variety of alkyl and aryl isothiocyanates and, after traceless cyclative cleavage, the desired thioxotetrahydropyrimidinones were obtained in excellent yields [86].

Finally, another related study achieved the synthesis of hydantoins by using acryloyl chloride to prepare a suitable polymer support under reflux conditions in a dedicated microwave instrument [88]. The soluble polymer support was dissolved in dichloromethane and treated with chloroacetyl chloride for 10 min under the action of microwave irradiation. Subsequent nucleophilic substitution using a variety of primary amines was carried out in N,N-dimethylformamide as solvent. The resulting PEG-bound amines were reacted with aryl or alkyl isothiocyanates in dichloromethane to furnish the polymer-bound urea derivatives after microwave irradiation for 5 min (Scheme 16.58). Finally, traceless release of the desired compounds by cyclative cleavage was achieved under mild basic conditions with microwave irradiation for 5 min. The 1,3-disubstituted hydantoins were obtained in varying yields but high purity.

16.3
Fluorous-phase Organic Synthesis

Fluorous-phase organic synthesis is a novel separation and purification technique for organic synthesis and process development. This technique involves the use of

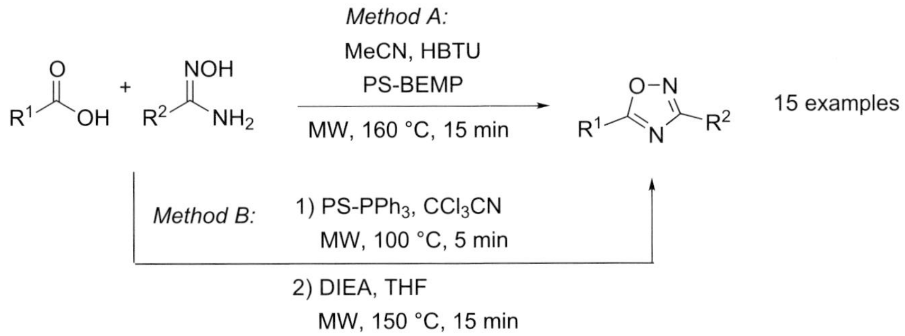

Scheme 16.71. Generation of 1,2,4-oxadiazoles by use of polymer-supported reagents.

compounds are difficult to handle, a microwave procedure was developed in which the acid chlorides are generated *in situ* [109]. For this purpose a combination of polymer-bound triphenyl phosphine and trichloroacetonitrile was most efficient in this two-step one-pot reaction under the action of microwave irradiation (Scheme 16.71). In the first step, the acid chlorides were generated *in situ* almost quantitatively within 5 min at 100 °C. DIEA and the amidoxime were then added and additional heating for 15 min at 150 °C furnished the desired heterocycles. The PS-PPh$_3$ resin did not interfere with the second step, because all 1,2,4-oxadiazoles were obtained in excellent yield and high purity.

Microwave-assisted thionation of amides using a polymer-supported thionating reagent has also been studied [110]. The polymer-supported amino thiophosphate serves as a convenient substitute for its homogeneous analog in microwave-induced rapid conversion of amides to thioamides. Under microwave conditions the reaction is complete within 15 min as opposed to conventional reflux in toluene for 30 h (Scheme 16.72). Use of the ionic liquid 1-ethyl-3-methylimidazolium hexafluorophosphate (emimPF$_6$) to dope nonpolar solvents such as toluene and assist with heating under microwave irradiation conditions was described for the first time (Chapter 7).

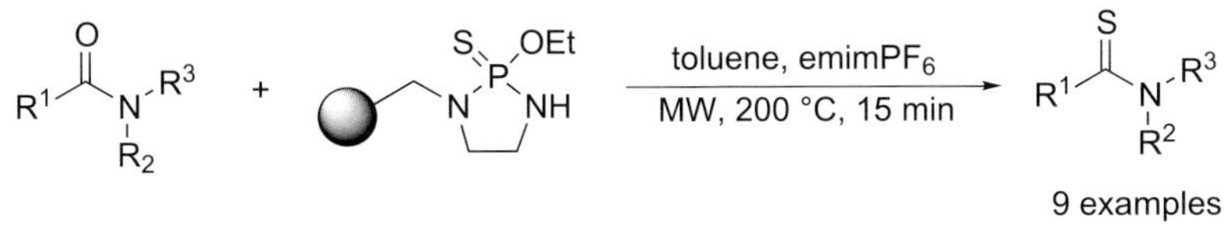

Scheme 16.72. Thionation of amides by use of polymer-bound aminothiophosphate.

A microwave-induced one-pot synthesis of olefins by Wittig olefination on polymer-supported triphenylphosphine has been reported [111]. Preparation of the Wittig reagent with *in situ* formation of the corresponding ylide was achieved in high yield within minutes, in contrast with several days by conventional methods. A variety of aldehydes and organic halides were reacted in the presence of the

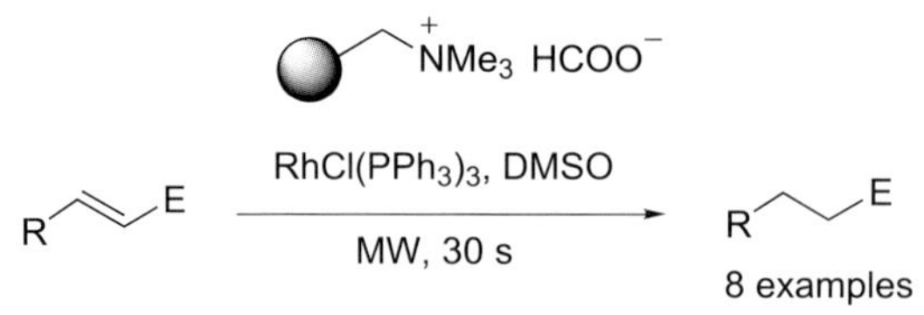

$$\text{R}\diagup\!\!\!\diagdown\text{E} \xrightarrow[\text{MW, 30 s}]{\substack{\bullet\!-\!\text{NMe}_3^+ \text{ HCOO}^- \\ \text{RhCl(PPh}_3)_3,\text{ DMSO}}} \text{R}\diagup\!\!\!\diagdown\text{E}$$

8 examples

Scheme 16.73. Transfer hydrogenations using a resin-bound formate.

solid-supported phosphine, yielding the corresponding olefins in excellent purity within 5 min of microwave heating at 150 °C; potassium carbonate was used as the base and methanol as the solvent.

An interesting report has described a very rapid solid-phase-transfer hydrogenation utilizing a polymer-supported hydrogen donor (ammonium formate derivative) for reduction of electron-deficient alkenes in the presence of Wilkinson's catalyst, $RhCl(PPh_3)_3$, under the action of microwave heating [112]. The formate used, immobilized on Amberlite IRA-938, was mixed with the corresponding substrate in a minimum amount of dimethyl sulfoxide and the reaction mixture was irradiated in a sealed vessel for 30 s only (Scheme 16.73). After cooling, the mixture was diluted with dichloromethane, washed with water, and dried. Evaporation of the solvent furnished the successfully hydrogenated compounds in high yields.

In addition, application of a recyclable polymer-supported brominechloride resin for regio and chemoselective bromomethoxylation of a variety of substituted alkenes under microwave conditions has been investigated [113]. The immobilized brominating reagent used (perbromide resin) was generated from a commercially available chloride-exchange resin by simple bromination at room temperature. Brief microwave irradiation (Scheme 16.74) was performed, in methanol under reflux, in a modified domestic oven equipped with a mounted reflux condenser. The polymer-supported bromine resin could be recycled and regenerated by successive washing with methanol, acetonitrile, and chloroform, then passage through the resin of bromine in tetrachloromethane.

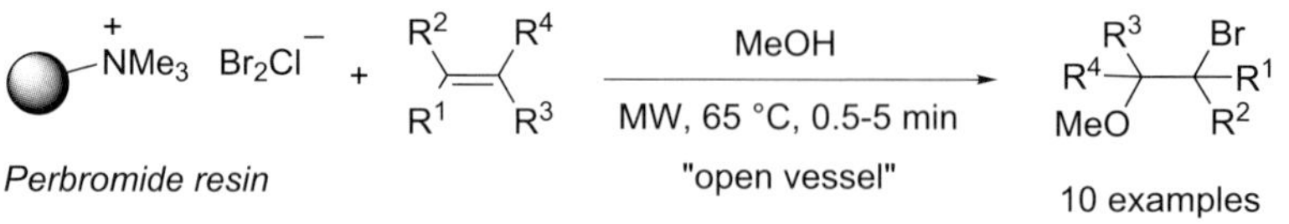

●—NMe_3^+ Br_2Cl^- + (alkene with R^1, R^2, R^3, R^4) $\xrightarrow[\text{MW, 65 °C, 0.5-5 min}]{\text{MeOH}}$ product with R^1, R^2, R^3, R^4, Br, MeO

Perbromide resin   "open vessel"   10 examples

Scheme 16.74. Bromomethoxylations utilizing polymer-supported brominechloride resin.

The conversion of isothiocyanates to isonitriles on polymer-supported [1,3,2]oxaphospholidine under microwave conditions has been intensively studied [114]. Commercial Merrifield resin was treated with aminoethanol and the resulting precursor was condensed with bis(diethylamino)phosphine in anhydrous toluene to generate the desired polymer-supported reagent. When using the novel solid-

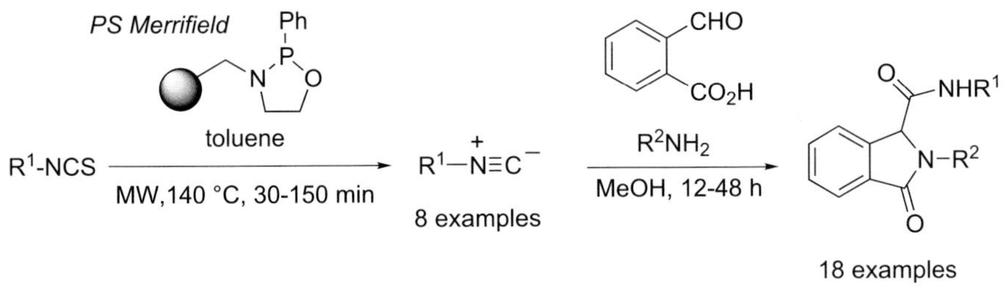

Scheme 16.75. Conversion of isothiocyanates to isonitriles on polymer-supported [1,3,2]oxazaphospholidine.

supported reagent conversion of the isothiocyanates could be afforded in parallel fashion under microwave irradiation conditions, and the corresponding isonitriles were isolated with high purity (Scheme 16.75). To demonstrate feasibility for library generation, the isocyanides produced were subjected to an Ugi three-component condensation with a variety of primary amines and carboxybenzaldehyde. The resulting 2-isoindolinone derivatives were obtained in high to excellent yields.

In related work, solid-phase-mediated synthesis of isonitriles from formamides was achieved by using polystyrene-bound sulfonyl chloride as a suitable supported reagent [115]. The sulfonyl chloride resin could be quantitatively regenerated by treatment of the sulfonic acid resin formed with phosphorus pentachloride (PCl_5) in N,N-dimethylformamide at room temperature.

In a more recent study, polymer-bound borohydride was used for reductive amination of tetrameric isoquinolines [116]. These tetrameric isoquinolines, serving as lead compounds in research to find antibacterial distamicyn A analogues, have been prepared from the corresponding isoquinoline, imidazole, and pyrrole building blocks by standard amide bond formation reactions. The final derivatization by reductive amination was efficiently accelerated by microwave irradiation in the presence of Merrifield resin-supported cyanoborohydride (Scheme 16.76).

Effective O-alkylation of carboxylic acids using a polymer-supported O-methylisourea reagent has also been demonstrated [117]. Microwave heating enabled complete esterification within 15–20 min, requiring only 2 equiv. of the polymer-bound O-methylisourea. Use of the polymer-supported reagent also simplified purification, because the pure methyl esters were obtained by evaporation after straightforward filtration. A polystyrene-bound carbodiimide has also been used for convenient and rapid amide synthesis [118]. An equimolar mixture of 1-methylindole-3-carboxylate, the corresponding amine, and 1-hydroxybenzotriazole (HOBt) in 1-methyl-2-pyrrolidinone (NMP) was admixed with 2 equiv. polymer-bound carbodiimide and irradiated for 5 min at 100 °C (Scheme 16.77). After cooling, the mixture was diluted with methanol and subjected to solid-phase extraction with silica carbonate. Evaporation of the filtrate furnished the desired compounds in excellent purity.

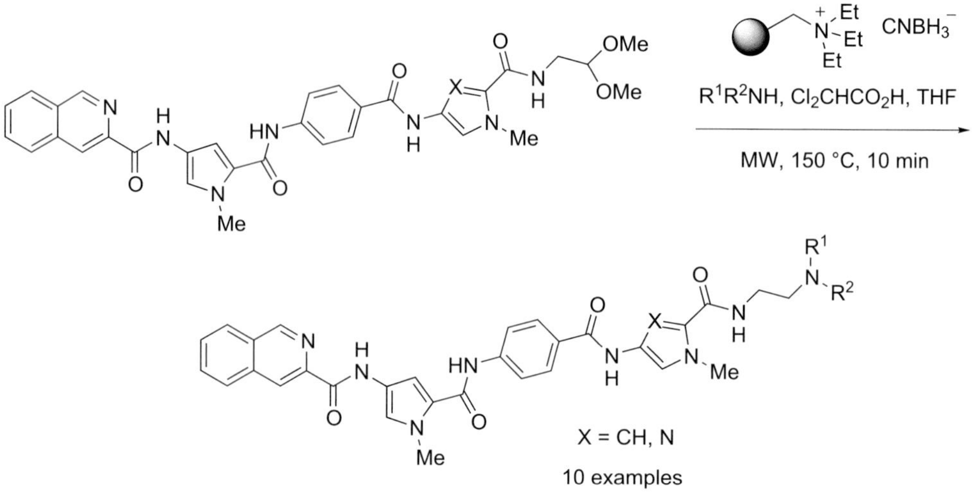

Scheme 16.76. Reductive amination of tetrameric isoquinolines.

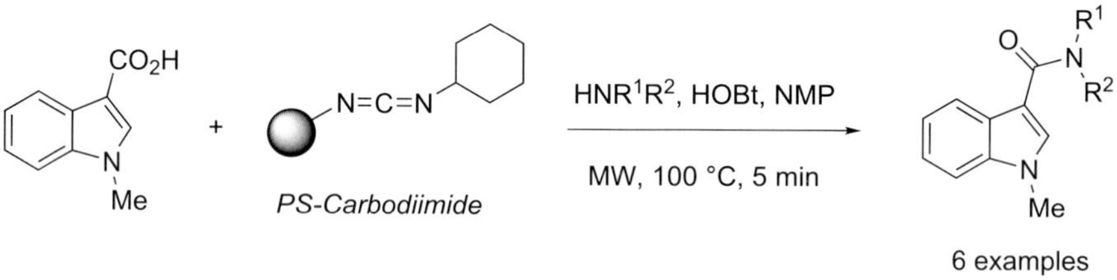

Scheme 16.77. Amide synthesis using polymer-bound carbodiimide.

The preparation and application of solid-supported cyclohexane-1,3-dione as a so-called "capture and release"-reagent for amide synthesis, and a novel scavenger resin, have been reported in detail [119]. A three-step synthesis of polymer-bound cyclohexane-1,3-dione (CHD resin) using inexpensive and readily available starting materials is described. The key step in this reaction is the microwave-assisted complete hydrolysis of the 3-methoxycyclohexen-1-one resin to the desired CHD resin. This novel resin-bound CHD derivative has been used for the preparation of an amide library under the action of microwave irradiation. An enol ester is formed by reaction of the starting resin-bound CHD with an acyl or aroyl chloride. Treatment of this with amines leads to the corresponding amide, regenerating the CHD (Scheme 16.78). This demonstrates the feasibility of use of the CHD resin as a "capture and release" reagent for the synthesis of amides. The "resin capture–release" methodology [120] aids in the removal of impurities and facilitates product purification.

Another example of the "capture and release" strategy, use of polystyrene-bound piperazine for generation of a pyrimidine library, was recently reported [121]. In a

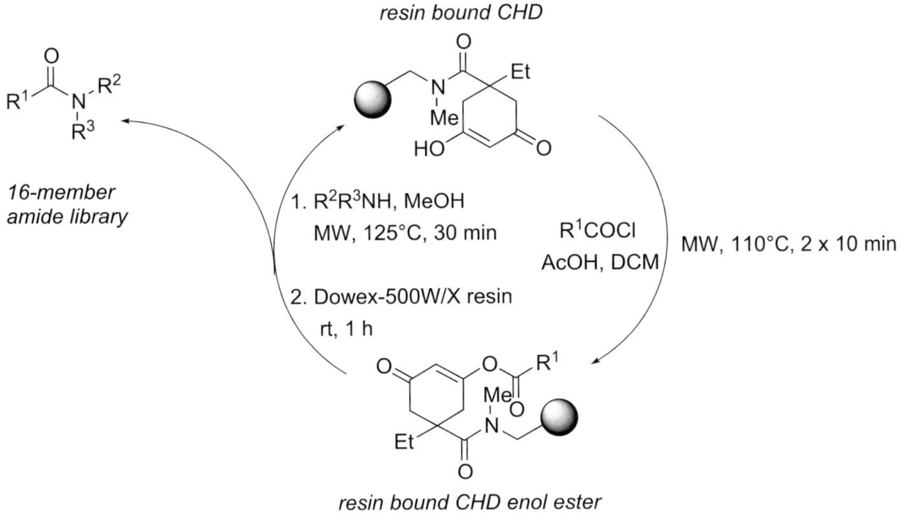

Scheme 16.78. Generation of an amide-library by use of resin capture–release methodology.

novel synthetic route to these valuable heterocycles, polymer-bound enaminones were condensed with guanidines (Scheme 16.79). The enaminones used were readily prepared by condensation of N-formylimidazol diacetal and β-keto compounds with the polymer-bound piperazine in the presence of camphorsulfonic acid (CSA). The mixture was heated in an open flask in a dedicated single-mode microwave reactor for 30 min to achieve full conversion of the substrates. After cooling, the functionalized resin was collected by filtration and subjected to the releasing cyclization by treating the polymer-bound enaminone with the corresponding guanidine. The mixture was irradiated in a sealed flask for 10 min at 130 °C to

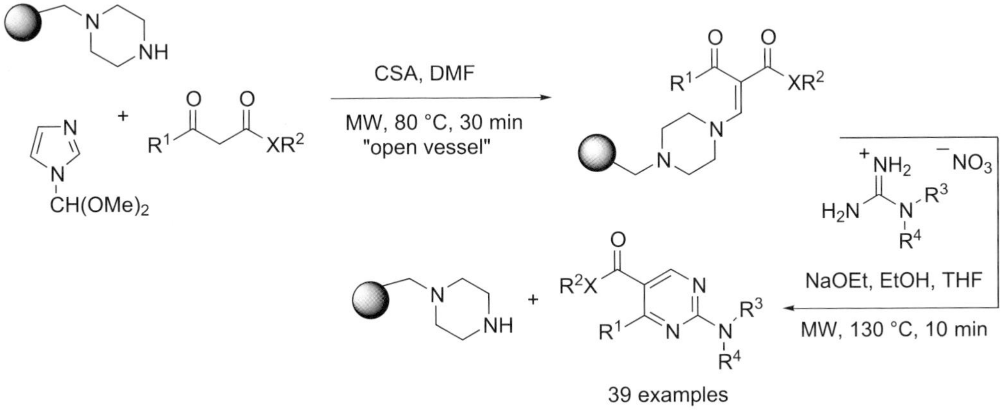

Scheme 16.79. Generation of a pyrimidine library by the capture and release strategy.

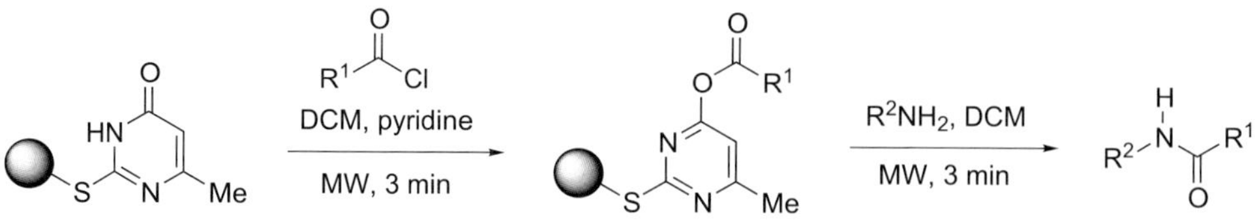

Scheme 16.80. Amide synthesis by use of a re-usable polymer-supported acylation reagent.

furnish the desired 2,4,5-trisubstituted pyrimidines. After the reaction, the piperazine resin could be collected by filtration and reused after several washings with ethanol. After aqueous work-up and evaporation the products were isolated from the filtrate in high yields and with excellent purity.

A recyclable solid supported reagent has been developed for acylations and was easily applied in a parallel synthesis of the corresponding amides starting from amines (Scheme 16.80) [122]. The solid-supported reagent could be removed by simple filtration after the reaction and the product isolated by evaporation. The supported reagent could be re-used at least twice before any decrease in the reaction yield.

In a related and more recent study a variety of solid-supported acylating agents were synthesized and used for microwave-mediated transformation of amines, alcohols, phenols, and thiophenols [123]. In a microwave-mediated procedure, Merrifield resin was first modified by attaching 1,4-butanediol to introduce a spacer unit. Bromination and subsequent reaction with commercially available 6-methyl-2-thiouracil then treatment with corresponding acyl chloride afforded the desired polymer-bound pyrimidines (Scheme 16.81). The acylating ability of this supported reagent has been proven by reaction with benzylamine.

Similarly, a solid-supported imide has been used as an acylating reagent under microwave conditions [124]. The starting imide was immobilized on aminomethyl polystyrene and benzoyl chloride was used to prepare the corresponding acylating reagent (Scheme 16.82). The resin bound acylating agent works efficiently for primary amines and piperazines and was shown to be recyclable after washing with N,N-dimethylformamide and reactivation under microwave conditions.

The first use of a polymer-supported Mukaiyama reagent for microwave-mediated synthesis of amides was presented in 2004 [125]. To prove its effectiveness, even in difficult coupling reactions, it was used in the microwave-accelerated synthesis of an amide from sterically hindered pivalic acid (Scheme 16.83). The mixture was subjected to microwave irradiation at 100 °C for 10 min and the desired product was obtained in 80% yield.

In related work a modified polymer-bound Mukaiyama reagent has been used to achieve microwave-mediated solution-phase synthesis of esters and lactones [126]. After optimization of the method by esterification of phenoxyacetic acid with benzyl alcohol, a small library of twenty esters was produced by use of the optimized procedure (Scheme 16.84). Overall reaction time was set to achieve full conversion of the starting material in cycles of 2 min irradiation and 1 min hold time. The

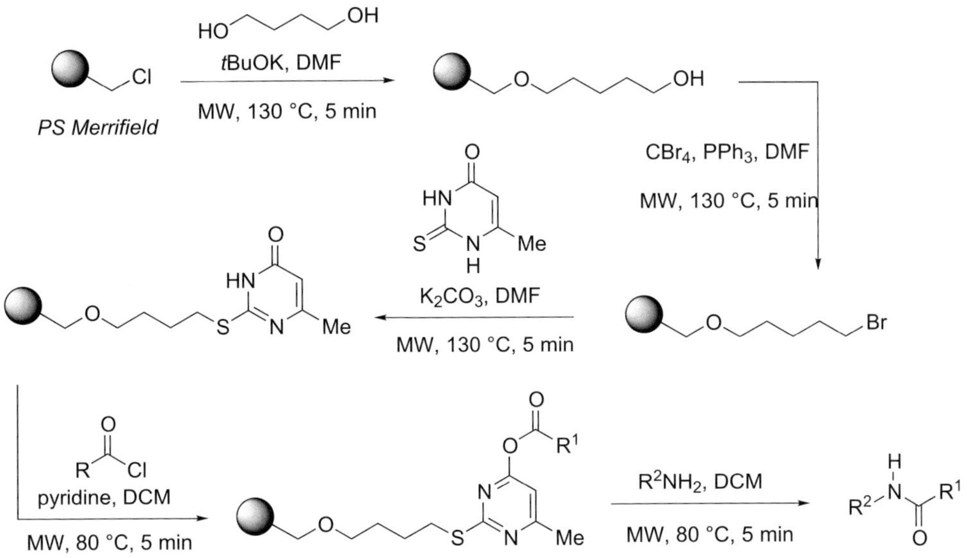

Scheme 16.81. Synthesis of a novel solid-supported reagent and its use for acylation.

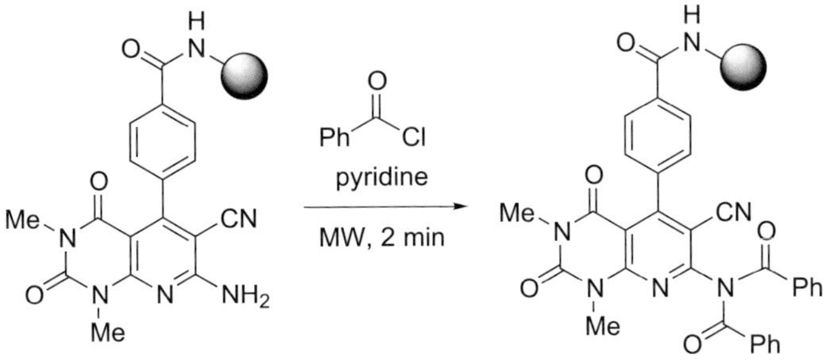

Scheme 16.82. A polymer-supported imide-based acylating agent.

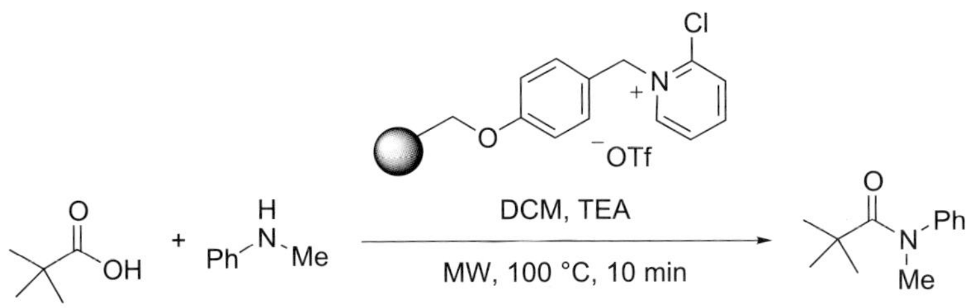

Scheme 16.83. Use of a polymer-supported Mukaiyama reagent for amide synthesis.

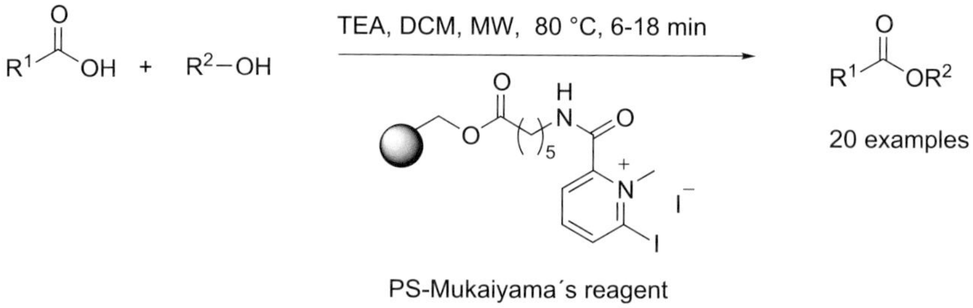

Scheme 16.84. Solution-phase esterification by use of a polymer-supported Mukaiyama reagent.

products were obtained in very good yields and high purity after simple resin filtration and solvent evaporation.

As reported in several publications, microwave heating has been successfully used to accelerate several intermediate slow reactions in the total synthesis of the natural product (+)-plicamine and of several of related spirocyclic templates [127–129]. Microwave heating in combination with polymer-bound reagents, catalysts, and scavengers was beneficial in enabling rapid access to several key intermediates in the synthesis of the target molecule. The penultimate intermediate in the synthesis of (+)-plicamine was isolated in high yields after purification, which was improved by microwave-assisted scavenging (see also Section 16.7).

16.5
Polymer-supported Catalysts

Catalysts immobilized on polymeric supports have an important additional advantage over conventional homogeneous catalysts – after the reaction the spent catalyst can be removed by simple filtration. The catalyst system can often be regenerated and recycled several times without significant loss of activity.

Selective protection of dihydropyrimidinones has been used in a novel approach to improve the synthesis of valuable Biginelli compounds with different substitution patterns [130]. This procedure has been successfully catalyzed by polymer-supported N,N-dimethylaminopyridine (PS-DMAP) (Scheme 16.85). This micro-

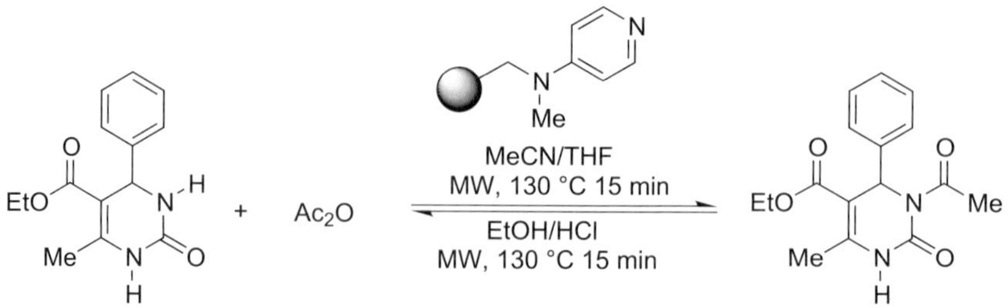

Scheme 16.85. N3-Acylations of dihydropyrimidines utilizing polymer-bound DMAP.

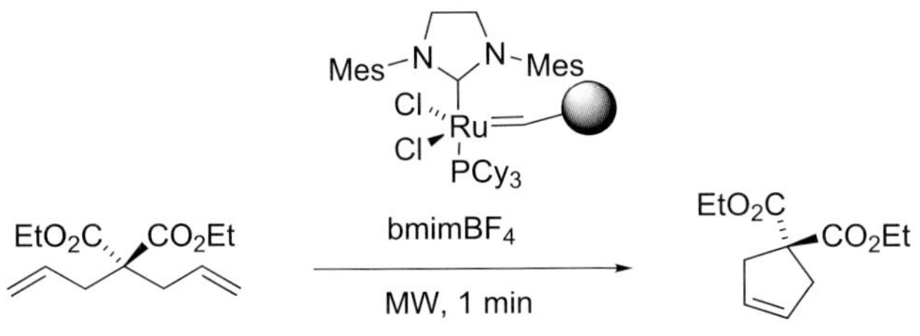

Scheme 16.86. Ring closing metathesis utilizing an immobilized Grubbs catalyst.

wave-induced selective N3-acylation has been successfully applied to diversely substituted dihydropyrimidinone scaffolds within 10–20 min of microwave heating. Purification of the final products was improved by microwave-induced scavenging techniques (see also Section 16.6).

In a different study, the corresponding PS-DMAP was used in a rapid one-pot microwave-induced base-catalyzed reaction of N-aryl and N-alkyl amino acids (or esters) with thioisocyanates for synthesis of a library of thiohydantoins [131]. Although use of PS-DMAP as the base resulted in somewhat lower yields than use of triethylamine, the reaction mixture was cleaner and purification easier.

In this context, a rapid microwave-accelerated ring-closing metathesis (RCM) reaction of diethyl diallylmalonate in the ionic liquid 1-butyl-3-methylimidazolium tetrafluoroborate (bmimBF$_4$) has been reported; both homogeneous and immobilized Grubbs catalysts were used in a sealed pressure tube in a domestic microwave oven (Scheme 16.86) [132]. The polymer bound Grubbs catalyst resulted in significantly lower conversion than the homogeneous analog. Similarly, notable acceleration of ring closing metathesis has also been achieved by use of polymer-supported second-generation Grubbs catalysts under the action of microwave heating, with the advantages of simple purification procedures [133].

Because asymmetric catalysis enables access to several synthetically important compounds, microwave-assisted molybdenum-catalyzed allylic allylation has been studied using both free and polymer-supported bis-pyridylamide ligands [134]. The microwave-assisted catalytic reaction of 3-phenylprop-2-enyl methyl carbonate with dimethyl malonate in the presence of N,O-bis(trimethylsilyl)acetamide (BSA) and a polymer-bound bis-pyridyl ligand was rather slow, however.

The microwave-induced reaction with a polymer-supported bis-pyridyl ligand (Scheme 16.87) was also slow, presumably because the insoluble resin-bound ligand makes the catalyst heterogeneous. After microwave irradiation at 160 °C for 30 min, however, the reaction was complete and the branch-to-linear ratio of the product was 35:1, an enantiomeric excess of 97%. The polymer-supported ligand has obvious advantages over its unsupported analogue because it could be reused at least seven times without any loss of activity [134].

This polymer-supported bis-pyridyl ligand (Scheme 16.87) has also been used in microwave-assisted asymmetric allylic alkylation [134], a key step in the enantioselective synthesis of (R)-baclofen.

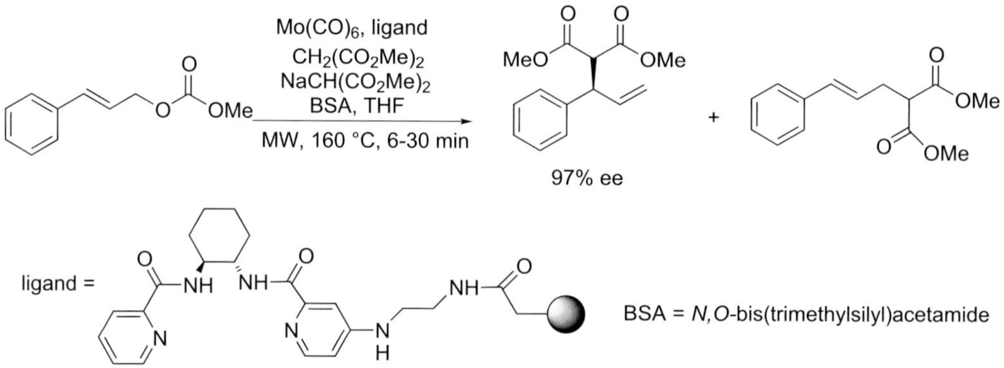

Scheme 16.87. Molybdenum catalyzed allylic allylation using a polymer-bound bis-pyridylamide ligand.

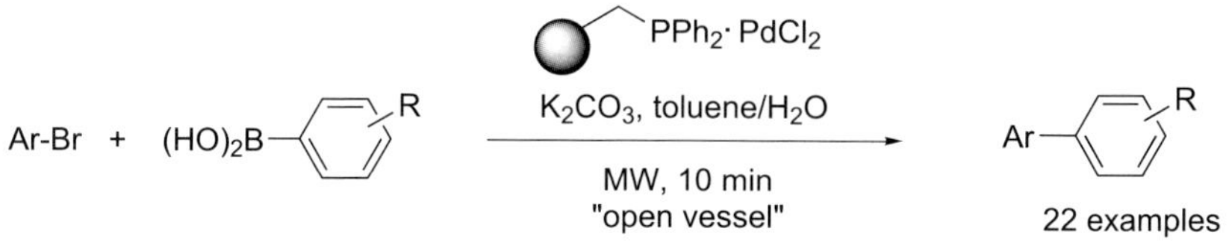

Scheme 16.88. Suzuki coupling using a polymer-supported palladium catalyst.

Microwave-assisted Suzuki coupling using a reusable polymer-supported palladium complex has been achieved in a more recent study [135]. The reaction mixture was treated with the polystyrene-bound palladium catalyst and irradiated in an open flask for 10 min in a domestic microwave oven (Scheme 16.88). After cooling, the mixture was filtered and the catalyst extracted with toluene and dried. The recycled polymer-bound catalyst can be reused five times without loss of efficiency.

In a related study, polymer-supported triphenyl phosphine was used in palladium-catalyzed cyanations [136]. Commercially available resin-bound triphenylphosphine was mixed with palladium(II) acetate in N,N-dimethylformamide to generate the heterogeneous catalytic system under a nitrogen atmosphere. The reagents were then added to the activated catalyst and the mixture was irradiated at 140 °C for 30–50 min (Scheme 16.89). Finally, the resin was removed by filtration and evaporation of the solvent furnished the desired benzonitriles in high yields and excellent purity.

An oligo(ethylene glycol)-bound palladium(II) complex for microwave-mediated Heck coupling of a variety of aryl halides was recently reported [137]. This novel catalyst was prepared in a multistep procedure from poly(ethylene glycol) monomethyl ether via its methanesulfonyl intermediate. The Heck reactions were performed in N,N-dimethylacetamide (DMA), with the catalyst being added as a solution in DMA immediately before subjecting the vessel to microwave irradiation. Heating at 150 °C furnished the desired cinnamate products in moderate to good

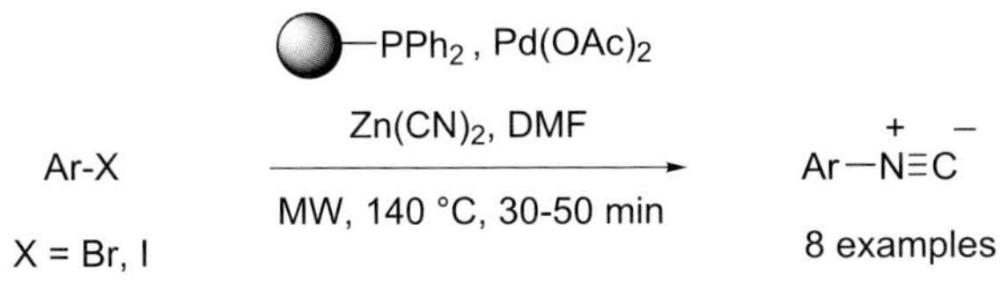

Scheme 16.89. Palladium-catalyzed cyanations utilizing polymer-bound triphenyl phosphine.

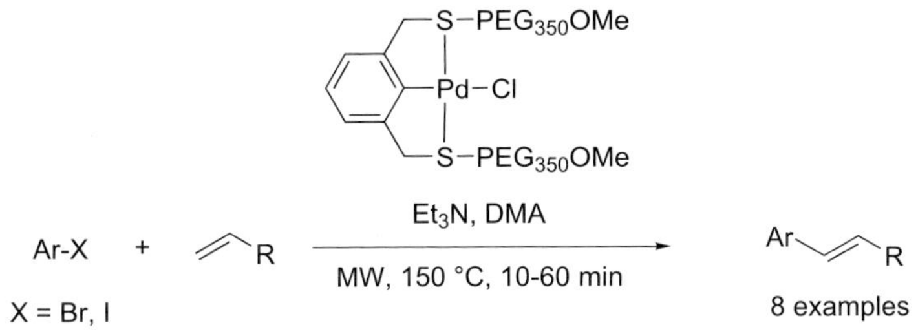

Scheme 16.90. Heck couplings using an oligo(ethylene glycol)-bound SCS-palladium(II) complex as catalyst.

yields (Scheme 16.90). The catalyst can be recycled by performing the reactions in a 10% aqueous DMA–heptane mixture, with slight loss of efficiency for up to four cycles [137].

An application involving a polystyrene-immobilized aluminum(III) chloride for a ketone–ketone rearrangement was published in 2002 [138].

16.6
Polymer-supported Scavengers

Isolation of a clean and homogeneous product is an integral part of any synthesis, and microwave heating has also been instrumental in improving many scavenging and purification techniques utilizing functionalized polymers.

The microwave-induced N3-acylation of the dihydropyrimidine (DHPM) Biginelli scaffold using different anhydrides has been discussed in a comprehensive report [130] (Scheme 16.85). The process included purification of the reaction mixture by means of a microwave-assisted scavenging technique. Several scavenging reagents with different loadings of amino functionalities have been used to sequester excess benzoic anhydride (Bz_2O) from the reaction mixture [130, 139, 140].

A parallel solution-phase asymmetric synthesis of α-branched amines based on stereoselective addition of organomagnesium reagents to enantiomerically pure *tert*-butanesulfinyl imines has been reported [141]. Microwave heating was used in two of the steps for synthesis of asymmetric amines, both for imine formation and for resin capture (Scheme 16.91). After the Grignard addition step, microwave

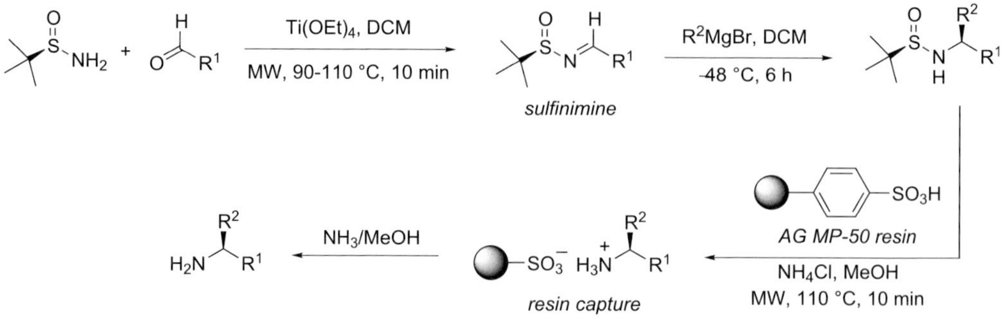

Scheme 16.91. Asymmetric synthesis of α-branched amines.

irradiation was used for acidic alcoholysis of the sulfinimide in the presence of macroporous sulfonic acid resin (AG MP-50). Microwave-assisted resin capture of amines enabled the preparation of a range of analytically pure α-phenylethylamine and diphenylmethylamine derivatives in good overall yields and with high enantiomeric purity.

More recently, use of a high-loading polystyrene Wang aldehyde resin for the scavenging excess amines in the preparation of piperazinium derivatives was reported [142]. In the initial step, chloroacetyl chloride was reacted with excess primary amine at 0 °C for 30 min. After filtration and evaporation the residue was dissolved in dioxane, suspended with Wang aldehyde resin and irradiated in a domestic microwave oven for 20–40 min, with 4-min intervals. The resin was isolated by filtration and the resulting chloroacetamide was treated with the corresponding primary diamines (Scheme 16.92). After the reaction the aldehyde resin was again suspended in the solution and subjected to the microwave-assisted scavenging. After five cycles of irradiation for 4 min, the resin was isolated by filtration and the filtrate was evaporated to obtain the corresponding glycinamides. Further modification and cyclization steps furnished the desired heterocyclic scaffolds [142].

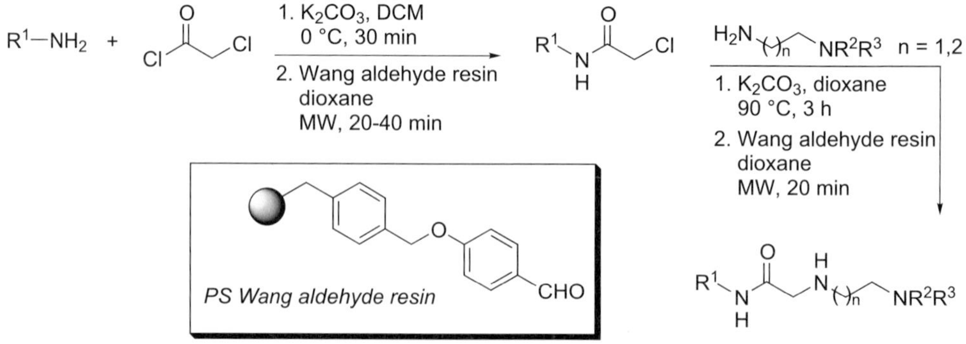

Scheme 16.92. Efficient amine-scavenging using a polystyrene aldehyde scavenger.

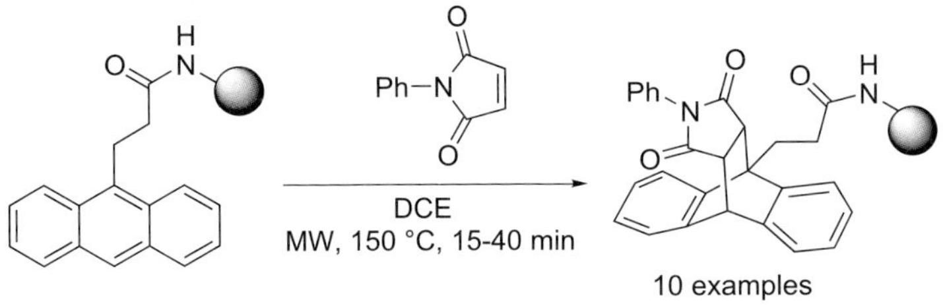

Scheme 16.93. Preparation and application of a novel polystyrene-bound anthracene as a dienophile scavenger.

Finally, the use of polymer-bound anthracene as an effective dienophile scavenger has been described [143]. The scavenging agent can easily be prepared by treatment of the commercially available corresponding Meldrum's acid derivative with aminomethyl polystyrene resin. The reactivity of this novel polymer-bound scavenger was examined for a series of ten different dienophiles, for example N-phenylmaleimide (Scheme 16.93). To demonstrate the effectiveness of this scavenger, it was successfully used in the synthesis of eight different flavonoid Diels–Alder cycloadducts. The two-step microwave-mediated procedure furnished the desired compounds in high yields and excellent purity [143].

16.7 Conclusion

The combination of modern microwave reactor technology and combinatorial chemistry applications is a logical consequence of the increased speed and effectiveness of microwave dielectric heating. Microwave heating can be rapidly adapted to a parallel or automatic sequential processing format. In particular the latter technique enables rapid testing of new ideas and high-speed optimization of reaction conditions. The fact that a "yes or no answer" for a particular chemical transformation can often be obtained within 5 to 10 min (as opposed to several hours in a conventional procedure) has contributed significantly to the acceptance of microwave chemistry in both industry and academia. The recently reported incorporation of real-time, in-situ monitoring of microwave-assisted reactions by Raman spectroscopy will enable a further increase in the efficiency and speed of microwave chemistry [144]. Although this technology is heavily used in pharmaceutical and agrochemical research laboratories already, a further increase in the use of microwave-assisted combinatorial chemistry applications both in industry and in academic laboratories can be expected. This will depend also on the availability of modern instrumentation, for either parallel or sequential processing mode [15]. It is clear that microwave-assisted combinatorial and high-throughput synthesis has a great potential and will probably become a standard tool in most high-throughput synthesis laboratories in a few years.

References

1 K. C. Nicolaou, R. Hanko, W. Hartwig (Eds.), *Handbook of Combinatorial Chemistry*, Wiley–VCH, Weinheim, 2002.

2 F. Z. Dörwald, *Organic Synthesis on Solid Phase*, Wiley–VCH, Weinheim, 2002.

3 R. B. Merrifield, *J. Am. Chem. Soc.* 1963, 85, 2149–2154.

4 Á. Furka, F. Sebestyén, M. Asgedom, G. Dibó, Highlights of Modern Biochemistry In: Proceedings of the 14th International Congress of Biochemistry, Prague, CZ, 1988, 13, p. 14, VSP, Utrecht. Á. Furka, F. Sebestyén, M. Asgedom, G. Dibó, *Int. J. Peptide Prot. Res.* 1991, 37, 487–493.

5 C. O. Kappe, A. Stadler, *Microwaves in Organic and Medicinal Chemistry*, Wiley–VCH, Weinheim, 2005.

6 B. Hayes, *Microwave Synthesis. Chemistry at the Speed of Light*, CEM Publishing, Matthews, NC, 2002. *Microwave-Assisted Organic Synthesis* (Eds.: P. Lidström, J. P. Tierney), Blackwell Publishing, Oxford, 2005.

7 A. Lew, P. O. Krutzik, M. E. Hart, A. R. Chamberlin, *J. Comb. Chem.* 2002, 4, 95–105.

8 C. O. Kappe, *Curr. Opin. Chem. Biol.* 2002, 6, 314–320.

9 P. Lidström, J. Westman, A. Lewis, *Comb. Chem. High Throughput Screen.* 2002, 5, 441–458.

10 M. Larhed, A. Hallberg, *Drug Discovery Today* 2001, 6, 406–416.

11 Anton Paar GmbH, Anton-Paar Strasse 20, A-8054 Graz; phone: (internat.) +43-316-2570; fax: (internat.) +43-316-257257; http://www.anton-paar.com

12 Biotage AB, Hamnesplanaden 5, SH-753 19 Uppsala, Sweden; phone: (internat.) +46-18 4899000; fax: (internat.) +46-18-4899100; http://www.biotage.com

13 CEM Corporation, P.O. Box 200, Matthews, NC 28106, USA; phone: (internat.) +1-800/726-3331; fax: (internat.) +1-704/821-7015; http://www.cemsynthesis.com

14 Milestone Inc., 160B Shelton Road, Monroe, CT 06468, USA; phone: (internat.) +1-203/261-6175; fax: (internat.) +1-203/261-6592; http://www.milestonesci.com

15 C. O. Kappe, A. Stadler, in *Microwaves in Organic and Medicinal Chemistry*, Wiley–VCH, Weinheim, 2005, Chapter 3, p. 29–55.

16 A. P. Combs, S. Saubern, M. Rafalski, P. Y. S. Lam, *Tetrahedron Lett.* 1999, 40, 1623–1626.

17 B. M. Glass, A. P. Combs, Case Study 4–6: Rapid Parallel Synthesis Utilizing Microwave Irradiation in *High-Throughput Synthesis. Principles and Practices*, I. Sucholeiki, (Ed.), Chapter 4.6, pp 123–128, Marcel Dekker, Inc., New York, 2001.

18 B. M. Glass, A. P. Combs, Rapid Parallel Synthesis Utilizing Microwave Irradiation: Article E0027, "*Fifth International Electronic Conference on Synthetic Organic Chemistry*", C. O. Kappe, P. Merino, A. Marzinzik, H. Wennemers, T. Wirth, J. J. Vanden Eynde, S.-K. Lin (eds.) CD-ROM edition, ISBN 3-906980-06-5, MDPI, Basel, Switzerland, 2001.

19 I. C. Cotterill, I. Y. Usyatinsky, J. M. Arnold, D. S. Clark, J. S. Dordick, P. C. Michels, Y. L. Khmelnitsky, *Tetrahedron Lett.* 1998, 39, 1117–1120.

20 C. M. Coleman, J. M. D. MacElroy, J. F. Gallagher, D. F. O'Shea, *J. Comb. Chem.* 2002, 4, 87–93.

21 M. Erdélyi, A. Gogoll, *J. Org. Chem.* 2003, 68, 6431–6434.

22 H.-M. Yu, S.-T. Chen, K.-T. Wang, *J. Org. Chem.* 1992, 57, 4781–4784.

23 A. Stadler, C. O. Kappe, *Eur. J. Org. Chem.* 2001, 919–925.

24 A. Stadler, C. O. Kappe, *Tetrahedron* 2001, 57, 3915–3920.

25 H. Yang, Y. Peng, G. Song, X. Qian, *Tetrahedron Lett.* 2001, 42, 9043–9046.

26 V. A. Yaylayan, M. Siu, J. M. R. Bélanger, J. R. J. Paré, *Tetrahedron Lett.* 2002, 43, 9023–9025.

27 M. Siu, V. A. Yaylayan, J. M. R.

Bélanger, J. R. J. Paré, *Tetrahedron Lett.* **2005**, *46*, 3737–3739.

28 M. Siu, V. A. Yaylayan, J. M. R. Bélanger, J. R. J. Paré, *Tetrahedron Lett.* **2005**, *46*, 5543–5545.

29 J. Westman, R. Lundin, *Synthesis* **2003**, *7*, 1025–1030.

30 G. A. Strohmeier, C. O. Kappe, *J. Comb. Chem.* **2002**, *4*, 154–161.

31 N. Galaffu, M. Bradley, *Tetrahedron Lett.* **2005**, *46*, 859–861.

32 C. Blackburn, *Tetrahedron Lett.* **2005**, *46*, 1405–1409.

33 L.-H. Du, S.-J. Zhang, Y.-G. Wang, *Tetrahedron Lett.* **2005**, *46*, 3399–3402.

34 M. Larhed, G. Lindeberg, A. Hallberg, *Tetrahedron Lett.* **1996**, *37*, 8219–8222.

35 N.-F. K. Kaiser, U. Bremberg, M. Larhed, C. Moberg, A. Hallberg, *Angew. Chem. Int. Ed.* **2000**, *39*, 3596–3598.

36 M. Alterman, A. Hallberg, *J. Org. Chem.* **2000**, *65*, 7984–7989.

37 J. Wannberg, M. Larhed, *J. Org. Chem.* **2003**, *68*, 5750–5753.

38 A. P. Combs, B. M. Glass, S. A. Jackson, *Methods Enzymol.* **2003**, *369*, 223–231.

39 N. Gachkova, J. Cassel, S. Leue, N. Kann, *J. Comb. Chem.* **2005**, *7*, 449–457.

40 N. Kaval, W. Dehaen, E. Van der Eycken, *J. Comb. Chem.* **2005**, *7*, 90–95.

41 P. Walla, C. O. Kappe, *Chem. Commun.* **2004**, 564–565.

42 K. Weigand, S. Pekа, *Mol. Diversity* **2003**, *7*, 181–184.

43 W.-M. Dai, D.-S. Guo, L.-P. Sun, X.-H. Huang, *Org. Lett.* **2003**, 2919–2922.

44 D. Scharn, H. Wenschuh, U. Reineke, J. Schneider-Mergener, L. Germeroth, *J. Comb. Chem.* **2000**, *2*, 361–369.

45 M. D. Bowman, R. C. Jeske, H. E. Blackwell, *Org. Lett.* **2004**, *6*, 2019–2022.

46 P. Grünefeld, C. Richert, *J. Org. Chem.* **2004**, *69*, 7543–7555.

47 T. Ernst, C. Reichert, *Synlett* **2005**, 411–413.

48 E. A. Yates, M. O. Jones, C. E. Clarke, A. K. Powell, S. R. Johnson, A. Porch, P. P. Edwards, J. E. Turnbull, *J. Mater. Chem.* **2003**, *13*, 2061–2063.

49 L. De Luca, G. Giacomelli, A. Porcheddu, M. Salaris, M. Taddei, *J. Comb. Chem.* **2003**, *5*, 465–471.

50 R. E. Austin, J. F. Okonya, D. R. S. Bond, F. Al-Obeidi, *Tetrahedron Lett.* **2002**, *43*, 6169–6171.

51 R. E. Austin, C. Waldraff, F. Al-Obeidi, *Tetrahedron Lett.* **2005**, *46*, 2873–2875.

52 A. M. L. Hoel, J. Nielsen, *Tetrahedron Lett.* **1999**, *40*, 3941–3944.

53 C. Portal, D. Launay, A. Merritt, M. Bradley, *J. Comb. Chem.* **2005**, *7*, 554–560.

54 B. Henkel, *Tetrahedron Lett.* **2004**, *45*, 2219–2221.

55 A. P. Frutos Hoener, B. Henkel, J.-C. Gauvin, *Synlett* **2003**, 63–66.

56 G. J. Kuster, H. W. Scheeren, *Tetrahedron Lett.* **2000**, *41*, 515–519.

57 B. Clapham, S.-H. Lee, G. Koch, J. Zimmermann, K. D. Janda, *Tetrahedron Lett.* **2002**, *43*, 5407–5410.

58 H. M. S. Kumar, S. Anjaneyulu, B. V. S. Reddy, *Synlett* **2000**, 1129–1130.

59 R. Schobert, C. Jagusch, *Tetrahedron Lett.* **2003**, *44*, 6449–6451.

60 B. Martinez-Teipel, R. C. Green, R. E. Dolle, *QSAR Comb. Sci.* **2004**, *23*, 854–858.

61 R. Pérez, T. Beryozkina, O. I. Zbruyev, W. Haas, C. O. Kappe, *J. Comb. Chem.* **2002**, *4*, 501–510.

62 Y.-D. Gong, H.-Y. Sohn, M. J. Kurth, *J. Org. Chem.* **1998**, *63*, 4854–4856.

63 B. Martin, H. Sekijic, C. Chassaing, *Org. Lett.* **2003**, *5*, 1851–1853.

64 A.-M. Yu, Z.-P. Zhang, H.-Z. Yang, C.-X. Zhang, Z. Liu, *Synth. Commun.* **1999**, *29*, 1595–1599.

65 S. Chandrasekhar, M. B. Padmaja, A. Raza, *Synlett* **1999**, 1597–1599.

66 S. Weik, J. Rademann, *Angew. Chem. Int. Ed.* **2003**, *42*, 2491–2494.

67 A. Bengtsson, A. Hallberg, M. Larhed, *Org. Lett.* **2002**, *4*, 1231–1233.

68 N. E. Leadbeater, H. M. Torenius,

H. Tye, *Mol. Diversity* **2003**, *7*, 135–144.

69 L. Paolini, E. Petricci, F. Corelli, M. Botta, *Synlett* **2003**, 1039–1042.

70 N. Kaval, J. Van der Eycken, J. Caroen, W. Dehaen, G. A. Strohmeier, C. O. Kappe, E. Van der Eycken, *J. Comb. Chem.* **2003**, *5*, 560–568.

71 C. G. Blettner, W. A. König, W. Stenzel, T. Schotten, *J. Org. Chem.* **1999**, *64*, 3885–3890.

72 M. Xia, Y.-G. Wang, *J. Chem. Res. (S)* **2002**, 173–175.

73 B. Sauvagnat, F. Lamaty, R. Lazaro, J. Martinez, *Tetrahedron Lett.* **2000**, *41*, 6371–6375.

74 J. J. Vanden Eynde, D. Rutot, *Tetrahedron* **1999**, *55*, 2687–2694.

75 M. Xia, Y.-G. Wang, *Tetrahedron Lett.* **2002**, *43*, 7703–7705.

76 M. Xia, Y.-G. Wang, *Synthesis* **2003**, 262–266.

77 A. Porcheddu, G. F. Ruda, A. Sega, M. Taddei, *Eur. J. Org. Chem.* **2003**, 907–912.

78 C.-Y. Wu, C.-M. Sun, *Synlett* **2002**, 1709–1711.

79 W.-J. Chang, W.-B. Yeh, C.-M. Sun, *Synlett* **2003**, 1688–1692.

80 P. M. Bendale, C.-M. Sun, *J. Comb. Chem.* **2002**, *4*, 359–361.

81 C.-L. Tung, C.-M. Sun, *Tetrahedron Lett.* **2004**, *45*, 1159–1162.

82 M.-J. Lin, C.-M. Sun, *Synlett* **2004**, 663–666.

83 W.-B. Yeh, M.-J. Lin, C.-M. Sun, *Comb. Chem. High Through. Screen.* **2004**, *7*, 251–255.

84 Y.-S. Su, M.-J. Lin, M.-C. Sun, *Tetrahedron Lett.* **2005**, *46*, 177–180.

85 M.-J. Lin, C.-M. Sun, *Tetrahedron Lett.* **2003**, *44*, 8739–8742.

86 W.-B. Yeh, M.-J. Lin, M.-J. Lee, C.-M. Sun, *Mol. Diversity* **2003**, *7*, 185–197.

87 W.-B. Yeh, C.-M. Sun, *J. Comb. Chem.* **2004**, *6*, 279–282.

88 M.-J. Lee, C.-M. Sun, *Tetrahedron Lett.* **2004**, *45*, 437–440.

89 D. P. Curran, *Angew. Chem. Int. Ed.*, **1998**, *37*, 1175–1196. J. A. Gladysz, D. P. Curran, I. Horváth, (Eds). *Handbook of Fluorous Chemistry*, Wiley–VCH, Weinheim, **2005**.

90 M. Larhed, M. Hoshino, S. Hadida, D. P. Curran, A. Hallberg, *J. Org. Chem.* **1997**, *62*, 5583–5587.

91 K. Olofsson, S.-Y. Kim, M. Larhed, D. P. Curran, A. Hallberg, *J. Org. Chem.* **1999**, *64*, 4539–4541.

92 W. Zhang, Y. Lu, C. H.-T. Chen, *Mol. Diversity* **2003**, *7*, 199–202.

93 T. Nagashima, W. Zhang, *J. Comb. Chem.* **2004**, *6*, 942–949.

94 W. Zhang, C. H.-T. Chen, Y. Lu, T. Nagashima, *Org. Lett.* **2004**, *6*, 1473–1476.

95 W. Zhang, T. Nagashima, Y. Lu, C. H.-T. Chen, *Tetrahedron Lett.* **2004**, *45*, 4611–4613.

96 A.-L. Villard, B. Warrington, M. Ladlow, *J. Comb. Chem.* **2004**, *6*, 611–622.

97 W. Zhang, P. Tempest, *Tetrahedron Lett.* **2004**, *45*, 6757–6760.

98 Y. Lu, W. Zhang, *QSAR Comb. Sci.* **2004**, *23*, 827–835.

99 K. S. A. Vallin, Q. Zhang, M. Larhed, D. P. Curran, A. Hallberg, *J. Org. Chem.* **2003**, *68*, 6639–6645.

100 W. Zhang, Y. Lu, C. H.-T. Chen, *Mol. Diversity* **2003**, *7*, 199–202.

101 M. A. Herrero, J. Wannberg, M. Larhed, *Synlett* **2004**, 2335–2338.

102 S. Werner, D. P. Curran, *Org. Lett.* **2003**, *5*, 3293–3296.

103 W. Zhang, C. H.-T. Chen, *Tetrahedron Lett.* **2005**, *46*, 1807–1810.

104 S. V. Ley, I. R. Baxendale, R. N. Bream, P. S. Jackson, A. G. Leach, D. A. Longbottom, M. Nesi, J. S. Scott, R. Ian Storer, S. J. Taylor, *J. Chem. Soc., Perkin Trans. 1* **2000**, 3815–4195. S. V. Ley, I. R. Baxendale, *Nature Rev. Drug Disc.* **2002**, *1*, 573–586.

105 A. Kirschning, H. Monenschein, R. Wittenberg, *Angew. Chem. Int. Ed.* **2001**, *40*, 650–679. C. C. Tzschucke, C. Markert, W. Bannwarth, S. Roller, A. Hebel, R. Haag, *Angew. Chem. Int. Ed.* **2002**, *41*, 3964–4000.

106 C. Brain, J. M. Paul, Y. Loong, P. J. Oakley, *Tetrahedron Lett.* **1999**, *40*, 3275–3278.

107 C. T. Brain, S. A. Brunton, *Synlett* **2001**, 382–384.

108 I. A. Baxendale, S. V. Ley, M. Martinelli, *Tetrahedron* **2005**, *61*, 5323–5349.

109 Y. Wang, R. L. Miller, D. R. Sauer, S. W. Djuric, *Org. Lett.* **2005**, *46*, 925–928.
110 S. V. Ley, A. G. Leach, R. I. Storer, *J. Chem. Soc., Perkin Trans. 1* **2001**, 358–361.
111 J. Westman, *Org. Lett.* **2001**, *3*, 3745–3747.
112 B. Desai, T. N. Danks, *Tetrahedron Lett.* **2001**, *42*, 5963–5965.
113 G. Gopalkrishnan, V. Kasinath, N. D. Pradeep Singh, V. P. Santhana Krishnan, K. Anand Soloman, S. S. Rajan, *Molecules* **2002**, *7*, 412–419.
114 S. V. Ley, S. J. Taylor, *Bioorg. Med. Chem. Lett.* **2002**, *12*, 1813–1816.
115 D. Launay, S. Booth, I. Clemens, A. Merritt, M. Bradley, *Tetrahedron Lett.* **2002**, *43*, 7201–7203.
116 W. Hu, R. W. Bürli, J. A. Kaizerman, K. W. Johnson, M. I. Gross, M. Iwamoto, P. Jones, D. Lofland, S. Difuntorum, H. Chen, B. Bozdogan, P. C. Appelbaum, H. E. Moser, *J. Med. Chem.* **2004**, *47*, 4352–4355.
117 S. Crosignani, P. D. White, B. Linclau, *Org. Lett.* **2002**, *4*, 2961–2963; S. Crosignani, D. Launay, B. Linclau, M. Bradley, *Mol. Diversity* **2003**, *7*, 203–210. S. Crosignani, P. D. White, B. Linclau, *J. Org. Chem.* **2004**, *69*, 5897–5905.
118 D. R. Sauer, D. Kalvin, K. M. Phelan, *Org. Lett.* **2003**, *5*, 4721–4724.
119 C. E. Humphrey, M. A. M. Easson, J. P. Tierney, N. J. Turner, *Org. Lett.* **2003**, *5*, 849–852.
120 A. Kischning, H. Monenschein, R. Wittenberg, *Chem. Eur. J.* **2000**, *6*, 4445–4450.
121 A. Porcheddu, G. Giacomelli, L. De Luca, A. M. Ruda, *J. Comb. Chem.* **2004**, 105–111.
122 E. Petricci, M. Botta, F. Corelli, C. Mugnaini, *Tetrahedron Lett.* **2002**, *43*, 6507–6509.
123 E. Petricci, C. Mugnaini, M. Radi, F. Corelli, M. Botta, *J. Org. Chem.* **2004**, *69*, 7880–7887.
124 R. B. Nicewonger, L. Ditto, D. Kerr, L. Varady, *Bioorg. Med. Chem. Lett.* **2002**, *12*, 1799–1802.
125 S. Crosignani, J. Gonzalez, D. Swinnen, *Org. Lett.* **2004**, *6*, 4579–4582.
126 D. Donati, C. Morelli, M. Taddei, *Tetrahedron Lett.* **2005**, *46*, 2817–2819.
127 I. R. Baxendale, S. V. Ley, C. Piutti, *Angew. Chem. Int. Ed.* **2002**, *41*, 2194–2197.
128 I. R. Baxendale, S. V. Ley, M. Nessi, C. Piutti, *Tetrahedron* **2002**, *58*, 6285–6304.
129 I. R. Baxendale, A.-L. Lee, S. V. Ley, in *Microwave-Assisted Organic Synthesis* (Eds.: P. Lidström, J. P. Tierney), Blackwell Publishing, Oxford, **2005** (chapter 6).
130 D. Dallinger, N. Yu. Gorobets, C. O. Kappe, *Mol. Diversity* **2003**, *7*, 229–245.
131 L. Öhberg, J. Westman, *Synlett* **2001**, 1893–1896.
132 K. G. Mayo, E. H. Nearhoof, J. J. Kiddle, *Org. Lett.* **2002**, *4*, 1567–1570.
133 M. G. Organ, S. Mayer, F. Lepifre, B. N'Zemba, J. Khatri, *Mol. Diversity* **2003**, *7*, 211–227.
134 O. Belda, S. Lundgren, C. Moberg, *Org. Lett.* **2003**, *5*, 2275–2278.
135 L. Bai, Y. Zhang, J. X. Wang, *QSAR Comb. Sci.* **2004**, *23*, 875–882.
136 R. R. Srivastava, S. E. Collibee, *Tetrahedron Lett.* **2004**, 8895–8897.
137 D. E. Bergbreiter, S. Furyk, *Green Chem.* **2004**, *6*, 280–285.
138 G. Gopalakrishnan, V. Kasinath, N. D. Pradeep Singh, *Org. Lett.* **2002**, *4*, 781–782.
139 D. Dallinger, N. Yu. Gorobets, C. O. Kappe, *Org. Lett.* **2003**, *5*, 1205–1208.
140 D. Dallinger, A. Stadler, C. O. Kappe, *Pure Appl. Chem.* **2004**, *76*, 1017–1024.
141 T. Mukade, D. R. Dragoli, J. A. Elman, *J. Comb. Chem.* **2003**, *5*, 590–596.
142 I. Masip, C. Ferrándiz-Huertas, C. García-Martínez, J. A. Ferragut, A. Ferrer-Montiel, A. Messeguer, *J. Comb. Chem.* **2004**, *6*, 135–141.
143 X. Lei, J. A. Porco, *Org. Lett.* **2004**, *6*, 795–798.
144 D. E. Pivonka, J. R. Empfield, *Appl. Spect.* **2004**, *58*, 41–46. For on-line UV monitoring, see: G. S. Getvoldsen, N. Elander, S. A. Stone-Elander, *Chem. Eur. J.* **2002**, *8*, 2255–2260.

17
Multicomponent Reactions Under Microwave Irradiation Conditions

Tijmen de Boer, Alessia Amore, and Romano V.A. Orru

17.1
Introduction

17.1.1
General

Since the synthesis of urea in 1828 by Friedrich Wöhler, synthetic sophistication has been increasing, with many impressive achievements over the past century [1]. Ongoing development of novel synthetic concepts and methods has continued to provide routes to the construction of many complex and challenging synthetic targets.

Despite its scientific merits and its profound effect on the progress of organic chemistry, it has become clear that much of the current synthetic methodology does not meet the conditions required for future purposes. Increasingly severe economic and environmental constraints are forcing the synthetic community to think about novel procedures and synthetic concepts to optimize efficiency. Progress in the efficiency of synthesis can be measured by criteria such as number of steps, overall yield, selectivity, flexibility, cost, scale, resource requirements, waste, development time, and execution time [2]. Ideally, the target molecule must be prepared from readily available starting materials in one simple, safe, environmentally acceptable and resource-effective operation proceeding quickly, selectively, and in quantitative yield [2].

Inspired by the mode of action of Nature, numerous groups have reported the multi-step, single operation construction of complex molecules, in which several bonds are formed in one sequence without isolating the intermediates [3–9]. Such processes, which are commonly referred to as tandem reactions, enable ecologically and economically favorable production of a wide range of organic compounds and will be the subject of discussion of the next section.

17.1.2
Tandem Reactions and Multicomponent Reactions

Generalizing the definition given by Denmark, tandem reactions are best described as one-pot sequences of two or more reactions [6]. Covering a wide range of transformations, the broad class of tandem reactions is further classified according to the order in which the events occur. As suggested by Denmark, distinction is made between three categories, each of which being briefly discussed below.

- *Tandem domino reactions* [10] are processes in which every reaction step brings about the structural change required for the following reaction step. The reaction proceeds without modification of the conditions or addition of supplementary reagents.
- In contrast with tandem domino reactions, *Tandem consecutive reactions* [11] require modification of the reaction conditions or addition of supplementary reagents after completion of the first reaction step to facilitate propagation of the sequence.
- *Tandem sequential reactions* [12] require addition of a supplementary reactant after completion of the first reaction step to enable propagation of the sequence.

Tandem domino and tandem consecutive reactions do not necessarily involve more than one reactant and can be unimolecular, as is exemplified by such processes as polyolefin cyclization [13]. Tandem sequential reactions cannot be unimolecular processes and involve two reactants at least. Both, tandem sequential and higher-order tandem domino reactions are of foremost interest in this text, because they enable the concept of multicomponent reactions (MCRs) to be introduced [4, 7–9]. Depending on the reaction conditions, MCRs can be regarded as a subclass of either tandem sequential (Scheme 17.1a) or higher order tandem domino reactions (Scheme 17.1b). In the former, formation of intermediate AB is followed by addition of a third component C to the reaction mixture to enable formation of reaction product ABC. In the latter, components A, B, and C are simply

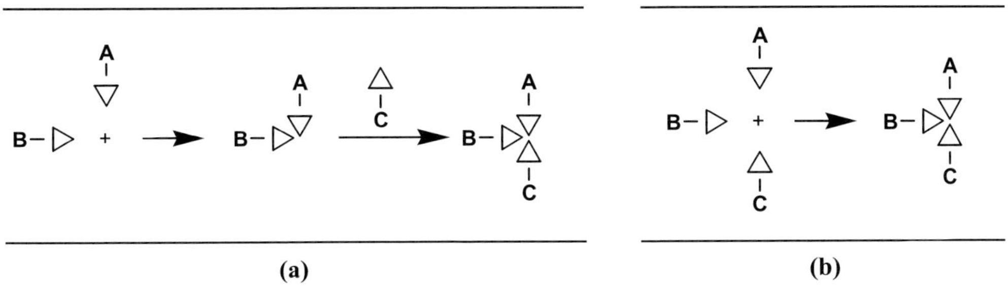

Scheme 17.1

mixed together to afford reaction product ABC in a single step without the need for addition of supplementary reagents or reactants to the reaction mixture.

As defined by Ugi, MCRs are one-pot processes in which at least three easily accessible starting materials react to give a single reaction product, which incorporates essentially all of the atoms of the starting materials [7a]. These reactions have the additional advantage of being highly flexible. Depending on the choice of starting materials, it is possible to generate a series of related compounds that differ only in side-chain functionality. MCRs distinguish themselves from other classes of tandem reaction by their combined efficiency and flexibility and are said to be of high exploratory power [8].

The exploratory power is best represented by a two-dimensional plane in which every point P is associated with a vector $\mathbf{r}$ (r, θ) [8]. The ability to generate molecular complexity is associated with the length r of vectors $\mathbf{r}$, while the tolerance with respect to successful starting materials is associated with the allowed range $\Delta\theta$ of angles θ. The exploratory power (E) is then defined as the scalar product of the allowed range of angles θ and the square of the length r of vectors $\mathbf{r}$. Maximization of exploratory power will be achieved by processes that combine versatility and complexity in a single step. Often MCRs meet these requirements – at least three commercially or easily available starting materials are coupled together by formation of several new bonds to afford complex reaction products in a single step. The concept of exploratory power in multicomponent chemistry is well exemplified by the mild synthesis of highly substituted pyrroles from commercially available primary amines, aldehydes, and α,β-unsaturated nitroalkenes (Scheme 17.2) [14]. In addition to being of high exploratory power, MCRs are selective, convergent, and atom efficient processes; they are therefore of interest in green chemistry.

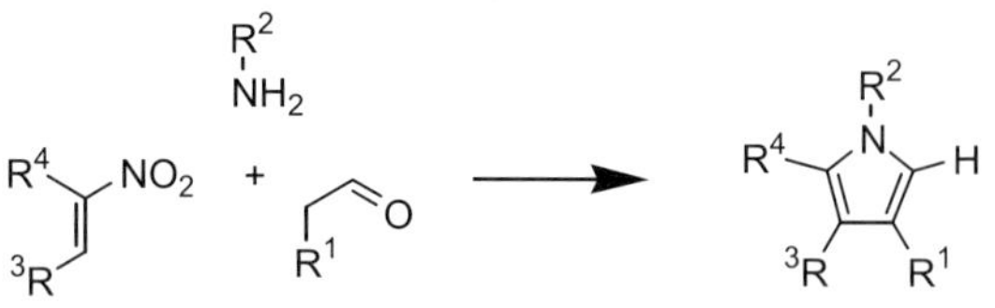

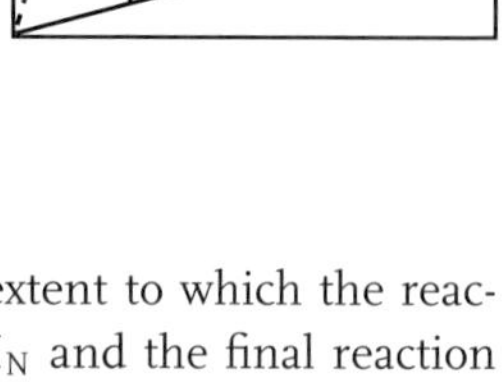

Scheme 17.2

Clearly, much of the success of MCRs depends on the extent to which the reactions leading to intermediate reaction products $I_1, I_2, \ldots, I_N$ and the final reaction product P are reversible (Scheme 17.3).

Distinction can be made between three categories of MCR [7]. Type I MCRs are sequences of reversible reactions whereas type II MCRs are sequences of reversible reactions that are terminated by an irreversible reaction step. Finally, type III MCRs are sequences of irreversible elementary reactions.

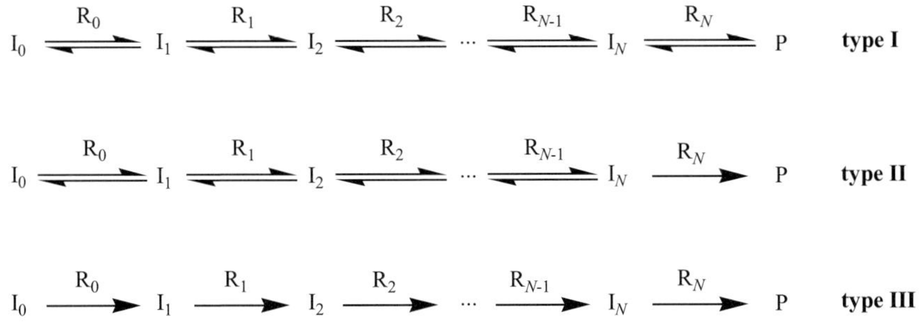

$$I_0 \rightleftharpoons I_1 \xrightarrow{R_0} I_1 \rightleftharpoons I_2 \xrightarrow{R_1} \cdots \rightleftharpoons I_N \xrightarrow{R_{N-1}} P \quad \text{type I}$$

$$I_0 \rightleftharpoons I_1 \xrightarrow{R_0} I_1 \rightleftharpoons I_2 \xrightarrow{R_1} \cdots \rightleftharpoons I_N \xrightarrow{R_N} P \quad \text{type II}$$

$$I_0 \xrightarrow{R_0} I_1 \xrightarrow{R_1} I_2 \xrightarrow{R_2} \cdots \xrightarrow{R_{N-1}} I_N \xrightarrow{R_N} P \quad \text{type III}$$

Scheme 17.3

Type I MCRs are only rarely successful, because the yield of final product P strongly depends on favorable thermodynamics. However, type II and III MCRs are of foremost interest in preparative organic chemistry because thermodynamics at least favors the final reaction step and high yields can be achieved. Most successful multicomponent reactions are type II reactions. Irreversibility of the last reaction step does not exclude formation of side products; because all preceding reactions are equilibrium reactions and more than one reactive species may be present at the same time. Usually, however, possible side reactions are also reversible and the irreversibly formed reaction product P predominates [15].

It is worth mentioning that the ever-growing interest in the development of novel multicomponent procedures was and is closely related to the needs of combinatorial chemistry (Refs [4, 7–9] and Chapter 16 in this book). Combinatorial compound libraries still play an important role in drug discovery and are commonly composed of 10^2 to 10^5 of closely related compounds. Their high exploratory power and simple experimental set-up make MCRs convenient tools for easy and rapid access to such large libraries of organic compounds.

17.1.3
Microwaves and Multicomponent Reactions

17.1.3.1 General Background
Increasing the reaction temperature usually increases the rate of an organic reaction. Thermal instability of the starting materials often limits the temperature increase and forces reactions to be carried out at moderate or low temperatures, however, thus occasionally impeding time-efficient conversion. In the conventional way of heating a reaction mixture (for example by using an oil bath), the reactants are slowly activated by an external heat source. Heat is driven into the substance, passing first through the walls of the vessel to reach the solvent and reactants. This is a slow and inefficient method for transferring energy into the system in comparison with microwave (MW) heating. Such undesired compromises can often be avoided by the use of microwaves instead of conventional heating to accelerate organic reactions [16, 23]. MW couple directly with the molecules of the entire reaction

mixture, leading to a rapid rise in temperature not limited by the conductivity of the vessel [17]. MW irradiation involves selective absorption of MW energy by polar molecules and enables a wide range of reactions to be performed in short times and with high conversions and selectivity [19c].

The use of MW in organic synthesis, including in MCRs, has attracted substantial attention in recent years [18, 19]. The initial slow acceptance of MW-assisted organic synthesis (MAOS) in the late 1980s was, owing to lack of understanding of the basics of MW dielectric heating, because of the lack of power control and reproducibility [19]. Most of these early pioneering experiments were performed in domestic, sometimes modified, kitchen MW ovens.

The heating effect utilized in MAOS is based on MW dielectric heating, which depends on the ability of a specific material (solvent or reagent) to absorb MW energy and convert it into heat (Refs [20, 21] and Chapter 1 in this book).

Two different kinds of MW reactor are currently emerging – multimode and monomode (also referred to as single mode) (Chapters 1 and 2 in this book).

The development of monomode reactors, which focus the electromagnetic waves in an accurately dimensioned wave guide, enables homogeneous distribution of the electric field leading to increased efficiency and reliability. So, the current trend in MAOS is to move away from the multimode MW ovens and use the more dedicated monomode instruments, which have only become available in the last few years [22, 23].

17.1.3.2 Use of Solvent-free MW Techniques in MCR Chemistry

In the early days of MAOS, solvent-free techniques were very popular. In these procedures, the reactants are either mixed directly without use of a solvent or the reactants are preabsorbed on to a solid support before irradiation with MW. The solvent free approach enabled safe use of domestic household microwave ovens and standard open-vessel technology (Refs [16, 19, 23] and Chapters 4 and 8 in this book).

MW-accelerated reactions under solvent-free conditions were shown to be highly efficient and environmentally benign processes, which were successfully applied to the multicomponent synthesis of a variety of heterocyclic compounds (Refs [9, 16, 19] and Chapter 10 in this book). When such conditions are combined with a suitable catalyst, even more efficient processes are conceivable. In fact, this concept has been realized by irradiating mixtures of reagents that have been absorbed on the surface of an insoluble inorganic support [9, 16, 19]. The inorganic supports used are usually inexpensive, commercially available, and easily manageable bulk chemicals. Examples include silica, alumina, Celite, graphite, and mesoporous inorganic solids, for example montmorillonite K-10 clay. Both silica and montmorillonite K-10 clay are believed to owe their catalytic activity to the presence of numerous acidic sites and hence are prone to activate polar functionality toward nucleophilic attack. Both materials have been successfully used in the MW-accelerated multicomponent synthesis of several heterocyclic compounds [9, 14, 24–26]. Experimental procedures are reported to be extremely simple and involve initial absorption of the reagents on the solid support, subsequent irradiation in a domestic MW oven, and final elution of the reaction mixture to give desired reaction products in satisfactory yields after strikingly short reaction times. The solid supports were usually recovered and could be re-used without significant loss of activity.

It should, however, be emphasized that solvent-free methods possibly suffer from technical difficulties relating to nonuniform heating, mixing, and precise determination of the reaction temperature. Nevertheless, these drawbacks are circumvented when operating with efficient mechanical stirring in open vessels in a monomode microwave reactor [23].

MW-assisted MCRs can also be conducted in standard organic solvents under both open and sealed vessel conditions. If solvents are heated in an open vessel, the boiling point of the solvent typically limits the reaction temperature that can be reached. The recent availability of modern monomode MW reactors [23] with on-line monitoring of both temperature and pressure has meant that MAOS in sealed vessels are increasingly commonly employed and this will be the method of choice for performing MW-assisted MCRs in the future, essentially if coupled with solvent-free procedures (cf. Chapters 4 and 8).

The remainder of this chapter is organized according to the class of scaffold accessed via an MCR. When possible, the type of MW equipment (monomode or multimode) is mentioned, as also are the general conditions under which the reactions were performed. For more precise information, the reader is referred to the original literature. In view of the current explosive developments in this area, the overview presented here cannot be exhaustive but is rather meant to provide the reader a useful introduction to this exciting field of chemistry.

17.2
Nitrogen-containing Heterocycles

17.2.1
Dihydropyridines

In 1882, Hantzsch achieved the synthesis of symmetrically substituted dihydropyridines (DHPs) by reacting ammonia, aldehydes, and two equivalents of β-ketoesters [27]. Since then, interest in these types of compound has grown, because of their pharmacological activity [28]. The Hantzsch reaction has successfully been used for synthesis of a wide range of DHPs and is still a popular tool for the construction of members of this class of heterocycles [29]. The classical multicomponent synthesis may require extended reaction times and yields can be low if sterically hindered aldehydes are used [30].

Different research groups have shown that this class of compound can be efficiently synthesized, with high yields and short reaction times, by using MW irradiation [31]. For example, Loupy et al. [31a] obtained unsymmetrical DHPs (**1** and **2**, Scheme 17.4) by reaction (in a one-pot procedure) of an aldehyde, a 1,3 diketone (or a β-ketoester), and methyl β-aminocrotonate. By using a solid inorganic support under solvent-free conditions in a domestic MW oven the desired products

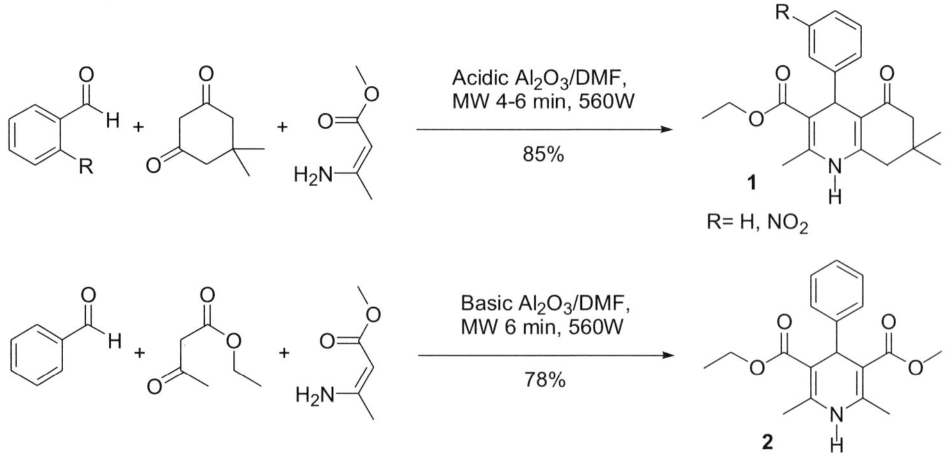

Scheme 17.4

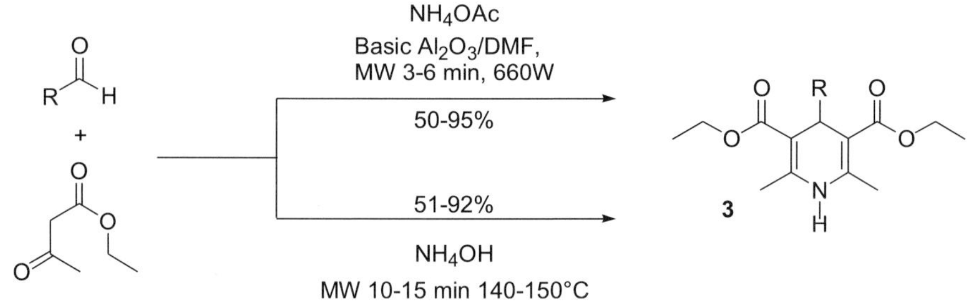

Scheme 17.5

were obtained with shorter reaction times and higher yields than by conventional heating.

The same group successfully obtained symmetrical DHPs (**3**, Scheme 17.5) by using a similar reaction but exchanging the methyl β-aminocrotonate for ammonium acetate [31b]. Later, Ohnberg et al. [31c] performed a similar reaction in a monomode closed-vessel MW synthesizer. In their approach, an aldehyde, a β-keto ester, and aqueous ammonia were reacted in a one-pot condensation to form a small library of DHPs **3** (Scheme 17.5). The reaction mixture was irradiated with MW for 10–15 min at 140–150 °C and the desired DHPs **3** were obtained in 51–92% yield. Under conventional heating conditions the reaction mixture was heated under reflux for 12 h with yields ranging from 15–72% [30].

Fused DHPs have remarkable pharmacological efficiency [32]. Although several methods for synthesis of these compounds are known [33, 34], all fail to provide bis(benzopyrano) fused DHPs **4** (Scheme 17.6). Kidwai et al. reported a MW-

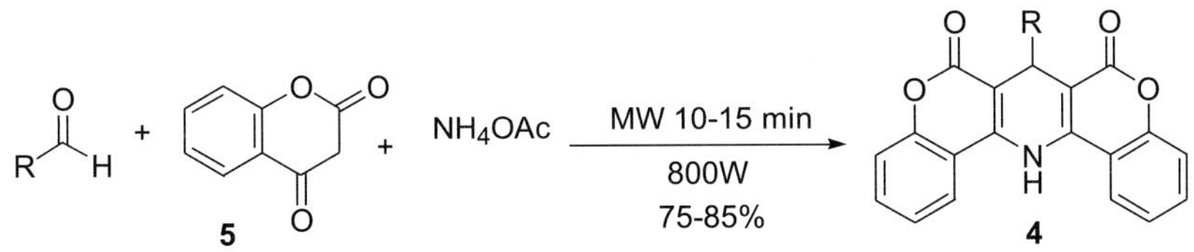

Scheme 17.6

promoted, solvent-free variation of the Hantzsch MCR [35]. In his approach, 4-hydroxy-2H-1-benzopyran-2-one **5**, an aromatic or heteroaromatic aldehyde and NH$_4$OAc were absorbed on two different solid supports (acidic alumina and silica gel) and were irradiated in a domestic MW oven at 800 W (Scheme 17.6). The best results were obtained by using acidic alumina and product **4** was obtained in 75–85% yield in 10–15 min; application of silica gel as support afforded **4** in 60–70% yield and reaction times were slightly longer (15–21 min).

Another way of forming more complex, nonsymmetrical DHPs was introduced by Bagley et al. [36]. The authors developed a one-pot three-component cyclocondensation reaction for preparation of 5-deaza-5,8 dihydropterin **6** (Scheme 17.7) which proceeded with total control of regiochemistry [37]. The three-component reaction was originally conducted under thermal conditions and was catalyzed by zinc(II) bromide but required long reaction times and quite high temperatures to obtain a good yield of the cyclized product **6**.

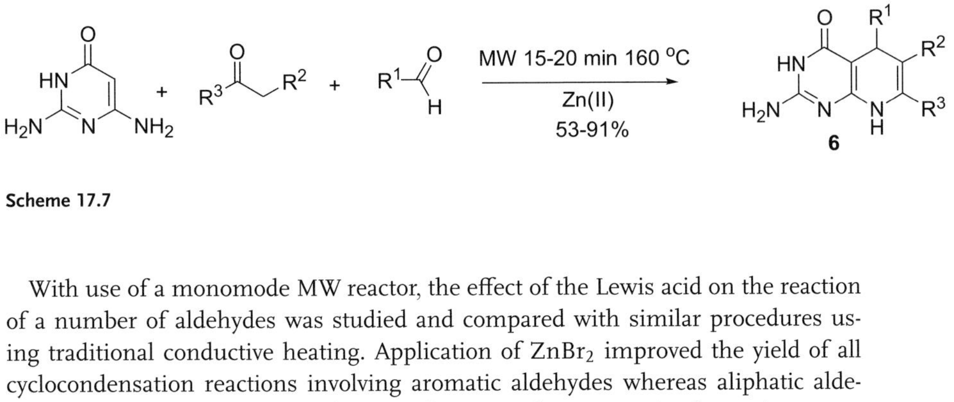

Scheme 17.7

With use of a monomode MW reactor, the effect of the Lewis acid on the reaction of a number of aldehydes was studied and compared with similar procedures using traditional conductive heating. Application of ZnBr$_2$ improved the yield of all cyclocondensation reactions involving aromatic aldehydes whereas aliphatic aldehydes performed best in the absence of ZnBr$_2$. The MW-assisted reactions were almost always superior to previous conventional procedures and gave good to excellent yields (53–91%) of deazadihydropterin product **6**.

Another interesting MW-assisted MCR for construction of the DHP ring as part of a more complex bicyclic scaffold was reported by Quiroga et al. They realized the MW-assisted synthesis of both pyrazolopyridines **7** and pyrido[2,3-d]pyrimidines **8** under solvent-free conditions (Scheme 17.8) [38].

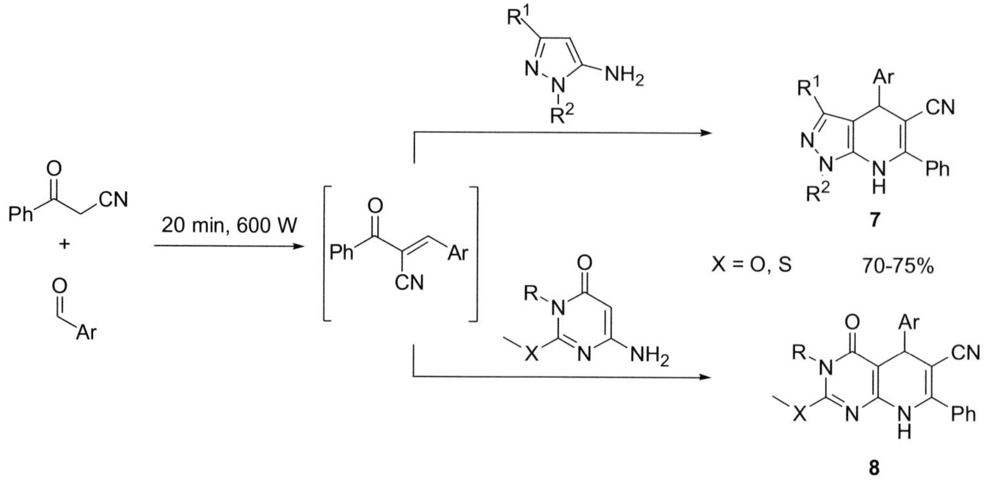

Scheme 17.8

MW irradiation of a mixture of equimolar amounts of starting materials for 20 min at 600 W provided clean reaction products **7** and **8** in yields ranging from 70–75%. In comparison, prolonged heating of equimolar amounts of starting materials in ethanol, under reflux, gave the desired reaction products in poor yields ranging from 21–25%. When the irradiation time was shortened to 12 min, hydrated intermediates (Scheme 17.9, **13**) were detected. The reaction is believed to proceed via Knoevenagel condensation of **9** with aromatic aldehydes. The resulting **10** is then attacked at the β-position by the unsubstituted endocyclic carbon atom of **11** to give Michael adduct **12**. Cyclization followed by final loss of water yields pyrido[2,3-d]pyrimidines **8**.

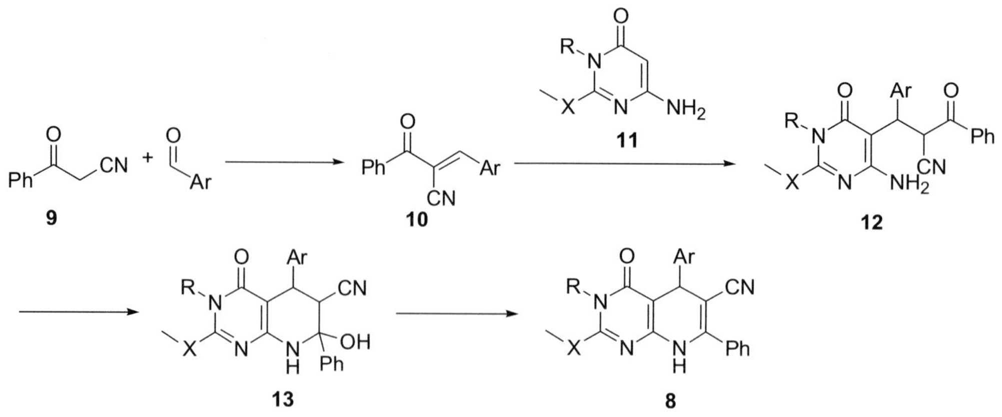

Scheme 17.9

17.2 Nitrogen-containing Heterocycles

Although not shown here, the mechanism for the formation of pyrazolopyridines **7** is similar [38].

N-substituted 1,4-dihydropyridines **14** without substituents on the 2 and 6 positions have broad pharmaceutical activity. For example, these species have been reported to be novel potential inhibitors of HIV-1 protease and are of interest for their anticancer activity [39, 40]. Compounds **14** can be prepared, in yields of 62–94%, by one-pot condensation of an aromatic aldehyde, 2 equiv. ethyl propiolate, and an amine, under solvent-free conditions, in a domestic MW oven (Scheme 17.10) [41].

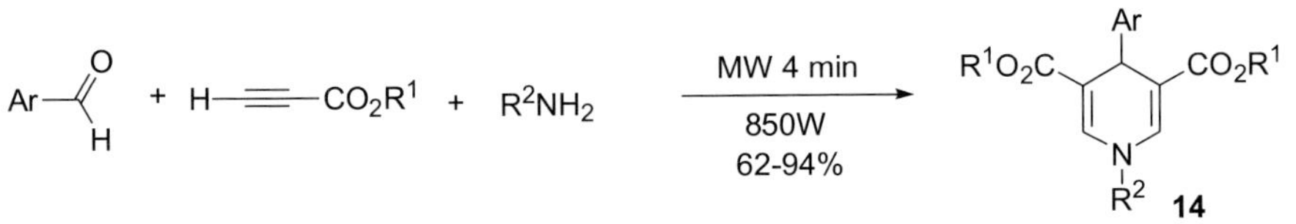

Scheme 17.10

The reaction was performed on four different solid supports (silica gel, acidic alumina, montmorillonite K-10, and zeolite HY). The best results were obtained by use of silica gel.

17.2.2
Pyridones

Another group of compounds which can be synthesized by a MCR under MW conditions are pyridones. Pyridone derivatives **15** have found many applications in the preparation of azo-dyes, especially as disperse dyes for polymeric materials [42]. It is possible to prepare these compounds by one-pot three-component condensation of a β-cyanoester, a β-keto ester, and a primary amine (Scheme 17.11) [43].

The reaction was conducted under solvent-free conditions in a domestic MW oven using four different kinds of solid support (silica gel, acidic alumina, montmorillonite K-10, and zeolite HY). The best results (77–94%) were obtained with silica gel and an optimized irradiation time of 2 min. In the absence of a solid support, under neat conditions, the products were obtained in low yields and the results were not reproducible.

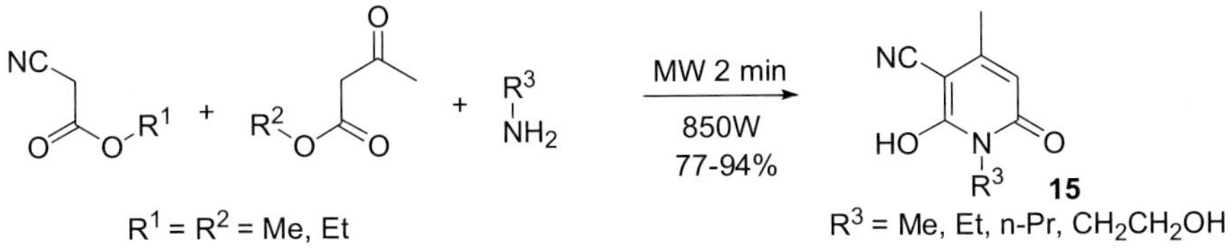

Scheme 17.11

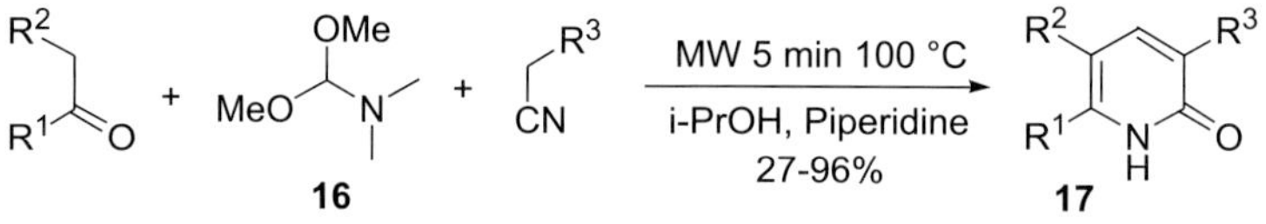

Scheme 17.12

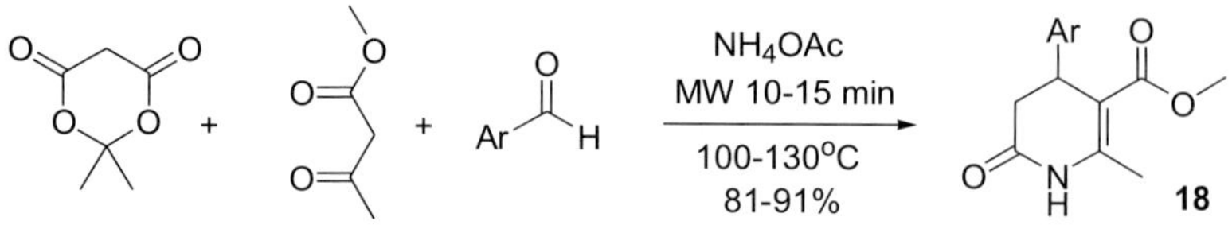

Scheme 17.13

In another MW-assisted synthesis of pyridones, a ketone, dimethylformamide dimethylacetal **16**, and a methylene active nitrile are combined (Scheme 17.12) [44].

The reaction was performed using i-PrOH as solvent in the presence of a catalytic amount of piperidine. The mixture was irradiated in a monomode MW reactor for 5 min at 100 °C to furnish pyridones of type **17** in 27–96% yield.

Substituted dihydropyridones **18** were successfully obtained under the action of MW irradiation in the absence of solvent (Scheme 17.13). The one pot reaction of Meldrum's acid, an arylaldehyde, a β-ketoester, and ammonia provided the desired products in very high yields (>80%, compared with <60% with conventional heating) [45].

17.2.3
Pyridines

MAOS can also be used in a multicomponent synthesis of pyridines (**19**). These compounds can be obtained on a multigram scale by a reaction of an aldehyde, a β-keto ester, and NH_4NO_3 (Scheme 17.14) [29].

The reaction mixture was irradiated in a domestic MW oven for 5 min on a surface of bentonite clay. Not only traditional β-keto esters, but also, for example, cyclic 1,3-diketones can participate in this MCR, thus enabling a four-component Hantzsch-type synthesis of unsymmetrically substituted 1,4-pyridines **20** (Scheme 17.15) [29, 37].

A more complex example of a multicomponent pyridine ring synthesis is the formation of quinoline derivatives, which are very important because of their wide

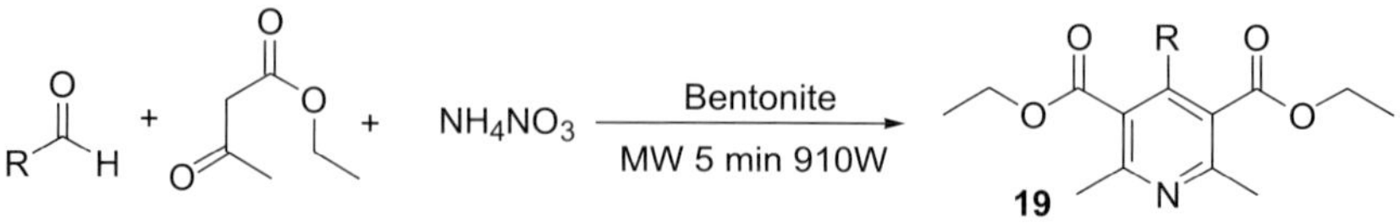

Scheme 17.14

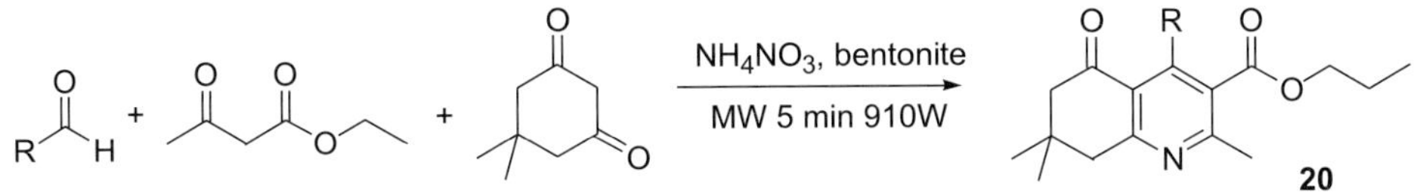

Scheme 17.15

occurrence in natural products and drugs [46]. Because it is known that replacement of hydrogen with fluorine can confer bioactivity on organic molecules [47], a variety of methods have been reported for synthesis of fluorine-containing quinoline derivatives; multiple reaction steps resulting in low yields are, however, hampering convenient access to these scaffolds [48, 49]. Zhang et al. reported a MW-assisted three-component reaction of pentafluorobenzaldehyde, monosubstituted anilines, and alkynes furnishing substituted quinolines **21** (Scheme 17.16) [50].

The reaction occurs on the surface of montmorillonite clay impregnated with catalytic amount of CuBr under solvent-free conditions. By irradiation for 3.5 min in a domestic MW oven the desired **21** are synthesized in 70–92% yield. The same reaction took around 4 h under traditional thermal conditions (80 °C, oil bath) to afford the expected quinoline derivatives **21** in only moderate yield (max. 64%).

Pyrido[2,3-*d*]pyrimidines are annelated uracils which have received substantial attention in recent years because of their biological activity [51, 52]. Several reports have appeared in the literature describing the preparation of these molecules starting from uracils and build-up of the pyridine moiety. These methods usually require long reaction times and complex synthetic pathways, however [53, 54]. Devi et al. developed a one-pot three-component cyclocondensation reaction under solvent-free MW-assisted conditions (Scheme 17.17) [55]. In this procedure, *N,N*-dimethyl-6-aminouracil **22** reacts with equimolar amounts of triethylorthofor-

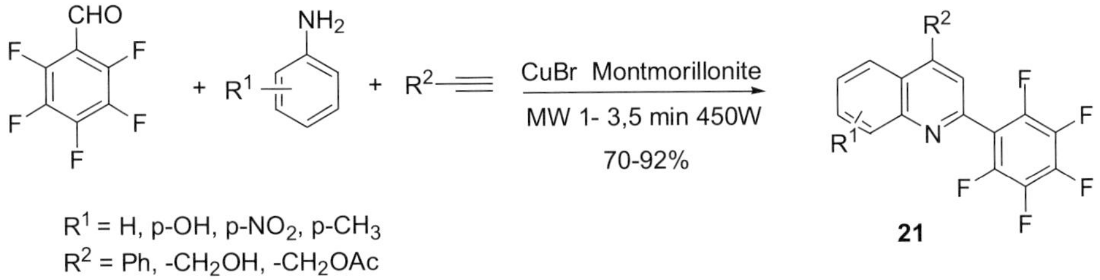

R^1 = H, p-OH, p-NO_2, p-CH_3
R^2 = Ph, -CH_2OH, -CH_2OAc

Scheme 17.16

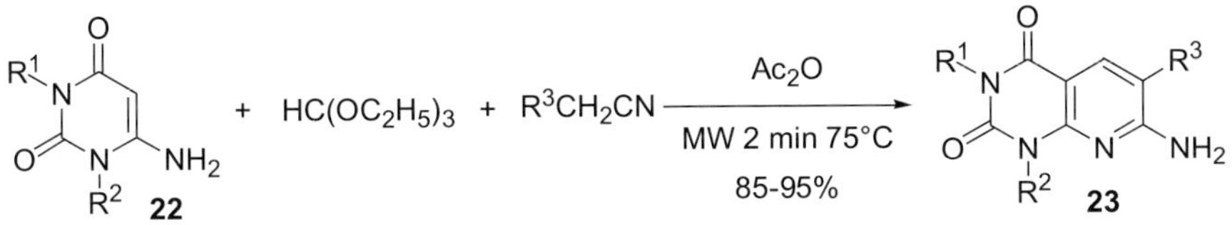

Scheme 17.17

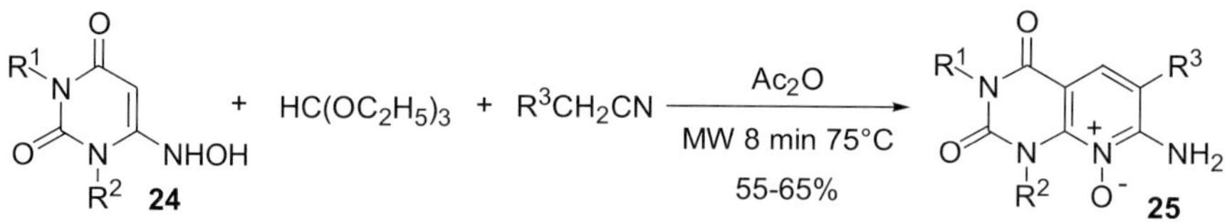

Scheme 17.18

mate and a nitrile in the presence of 3 equiv. acetic anhydride to give pyrido[2,3-d]pyrimidines **23** in high yield (85–95%).

The reaction was conducted in a monomode MW reactor with an irradiation time of 2 min at 75 °C. The reaction was also performed under conventional thermal conditions, which resulted in reduced yields (60–70%) and prolonged reaction times (1 h).

Using the same MW conditions, 6-hydroxyaminouracils **24** were used as input to afford pyrido[2,3-d]pyrimidine N-oxides **25** in good yields (55–65%, Scheme 17.18) [55]. For comparison purposes, this reaction was also performed under conventional thermal conditions, resulting in longer reaction times (2 h) and slightly lower yields (40–50%).

17.2.4
Dihydropyrimidinones

Dihydropyrimidinones (DHPMs) are well-known calcium antagonistic agents and are most commonly prepared by acid-catalyzed cyclocondensation of urea, β-ketoesters, and aromatic aldehydes. Ever since its discovery in 1893, this so-called Biginelli reaction (Scheme 17.19) has been subject of numerous optimization studies and many modified procedures have resulted (*vide infra*). The Biginelli reaction and its modifications are still popular tools for the preparation of DHPMs **26** and have been extensively reviewed by Kappe et al. [56].

Despite its extreme simplicity, the original Biginelli three-component reaction suffers from long reaction times and low to moderate yields of the desired DHPMs **26**. This is particularly true when substituted aliphatic/aromatic aldehydes or thioureas are used. To overcome this problem, several improved reaction procedures for the synthesis of **26** have been developed either by modification of the classical

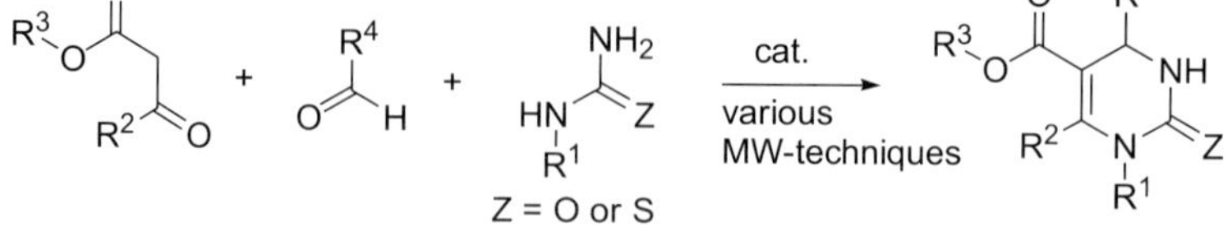

Scheme 17.19

one-pot Biginelli approach itself [57] or by the development of novel, but more complex multistep strategies [58].

Significant rate and yield enhancements have been reported for Biginelli reactions performed under MW irradiation conditions. Most of these procedures are solvent-free and were performed in a domestic MW oven and require various catalysts such as polyphosphate ester (PPE) [59], Cu(II) salts [60], iodine–alumina [61], montmorillonite KSF clay [62, 63], or different chlorine-containing catalytic systems [64].

More advanced MW exploitation involves use of a monomode MW reactor equipped with a robotics interface that can be used for automated sequential library synthesis. The Biginelli reaction performed using this procedure employs a diverse set of starting compounds to prepare a 48-compound library of 26 [65]. AcOH–EtOH was used as solvent and a catalytic amount of Yb(OTf)$_3$ was added. When the unattended automation capabilities of the MW synthesizer are used a library of this size can be synthesized in 12 h.

The Biginelli condensation can also be performed on soluble polymers (Scheme 17.20). This procedure couples the advantages of homogeneous solution chemistry with those of solid-phase chemistry (use of excess reagents and easy isolation and purification of the products) [59].

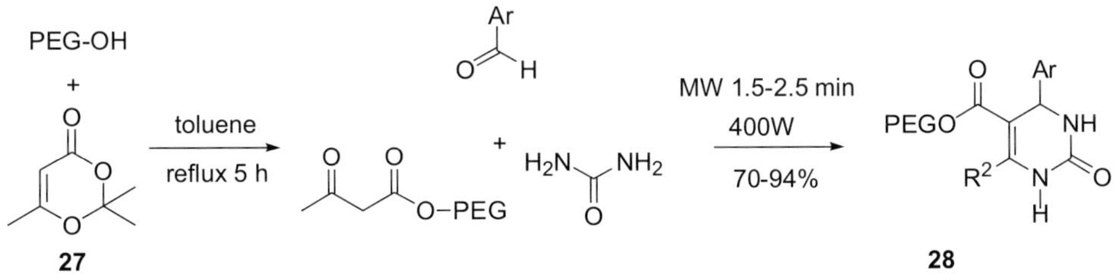

Scheme 17.20

Poly(ethylene glycol) (PEG) is linked to acetoacetate by reaction of PEG with 2,2,6-trimethyl-4H-1,1-dioxin-4-one 27 in anhydrous toluene under reflux for 5 h. The DHPMs synthesis was achieved by mixing the PEG-linked acetoacetate, urea, and the corresponding aldehyde in the ratio 1:2:2 with a catalytic amount of non-oxidant polyphosphoric acid (PPA) [59]. This mixture was irradiated in a domestic MW oven for 1.5 to 2.5 min to obtain 28 in 70–94% yield.

Unlike the classical Biginelli reaction, which only tolerates open-chain β-keto esters, cyclic β-diketones can also participate in a related MCR [66]. This procedure requires Meldrum's acid 29a (X = O and Z = CMe$_2$) or barbituric acid derivatives 29b (X = NH or NMe and Z = CO), urea, and aldehydes (Scheme 17.21). The mixture was irradiated in a domestic MW oven for 4 min under solvent-free conditions. The reaction requires a Brönstedt acid as catalyst to give, selectively, a family of novel heterobicyclic compounds 30 in good yields (70–83%).

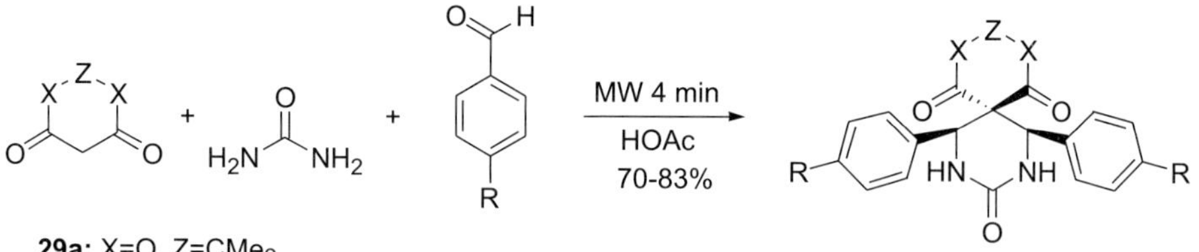

29a: X=O, Z=CMe$_2$
29b: X= NH or NMe, Z=CO

Scheme 17.21

17.2.5
Imidazoles

Compounds that bear an imidazole ring system can have many pharmacological properties and play important roles in biochemical processes [67]. Balalaie et al. synthesized these species under MW conditions by three-component condensation of benzil, benzaldehyde derivatives, and 2 equiv NH$_4$OAc (Scheme 17.22) [68]. The reaction was catalyzed by zeolite HY or silica gel under solvent-free conditions. After MW irradiation for 6 min in a domestic oven at 850 W the desired triarylimidazoles **31** were obtained in good yields (80–94%) by use of zeolite HY; slightly lower yields (54–89%) were obtained for the reactions on silica gel.

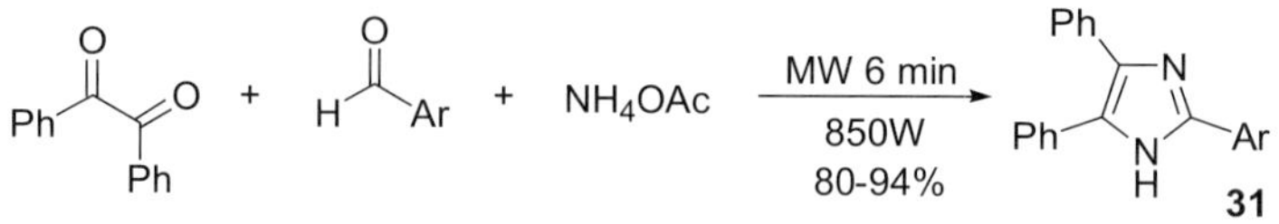

Scheme 17.22

Tetrasubstituted imidazoles **32** can also be synthesized by a three-component condensation [55]. These reactions are usually limited and cannot be performed under neutral conditions [70]. With use of MW irradiation, however, the reaction becomes possible. The procedure requires benzonitrile derivatives, benzil, and primary amines on the surface of silica gel to obtain the tetrasubstituted imidazoles **32** in moderate to good yields (58–92%, Scheme 17.23).

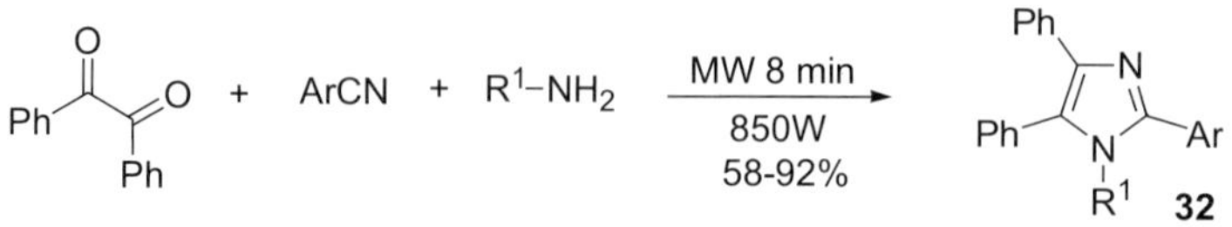

Scheme 17.23

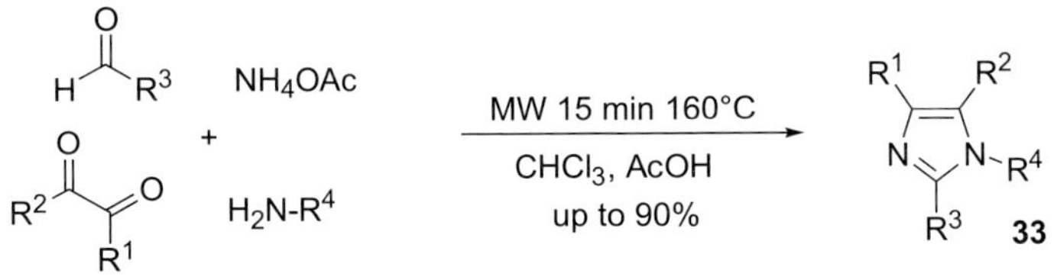

Scheme 17.24

By applying a different procedure, Orru et al. developed a MW accelerated MCR for synthesis of a small 48-membered library of mono, di, tri, and tetrasubstituted imidazoles **33** (Scheme 17.24) [71].

In their variation of Radziszewski's four-component reaction, an aldehyde, NH$_4$OAc, a diketone, and an amine were combined to produce imidazoles **33**. The reaction was performed in an automated monomode MW reactor at 160 °C using a mixture of chloroform and acetic acid as solvent. Product yields were up to 90%, depending on the nature of the different R groups.

Biologically active fused imidazo[1,2-a]pyridines **34** and imidazo[1,2-a]pyrazines **35** were shown to be accessible by reaction of 2-aminopyridine (X = CH) or 2-aminopyrazine (X = N), respectively, with aldehydes and isocyanides in the presence of 5 mol% Sc(OTf)$_3$ (Scheme 17.25) [72].

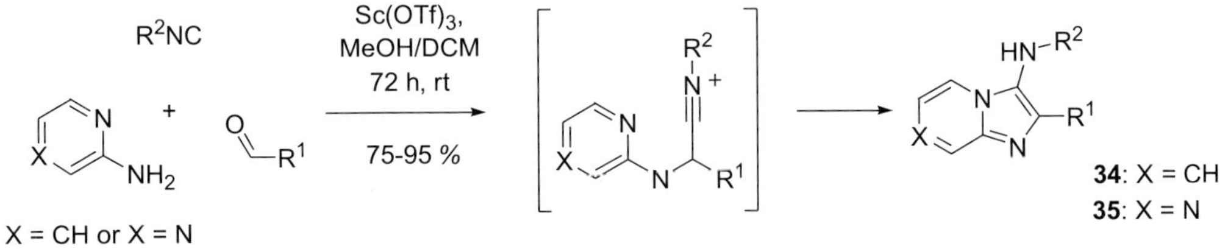

Scheme 17.25

Although flexible, the procedures suffered from poor time efficiency and reaction products were normally obtained only after continuous stirring for 72 h. With the objective of better time efficiency, Varma et al. developed an identical, although MW-accelerated, three-component synthesis catalyzed by montmorillonite K-10 instead of Sc(OTf)$_3$ [24]. Their solvent-free one-pot procedure involves irradiation of a mixture of aldehydes and either 2-aminopyridine or 2-aminopyrazine in the presence of montmorillonite K-10 to generate the corresponding iminium ion. Subsequent addition of isocyanides and further irradiation at reduced power gave the desired imidazo[1,2-a]pyridines **34** and imidazo[1,2-a]pyrazines **35** in good yields (Scheme 17.26). Time-efficiency was improved one-thousandfold compared with previously reported results under conventional heating conditions and reactions were usually complete within minutes. A variety of aldehydes and isocyanides

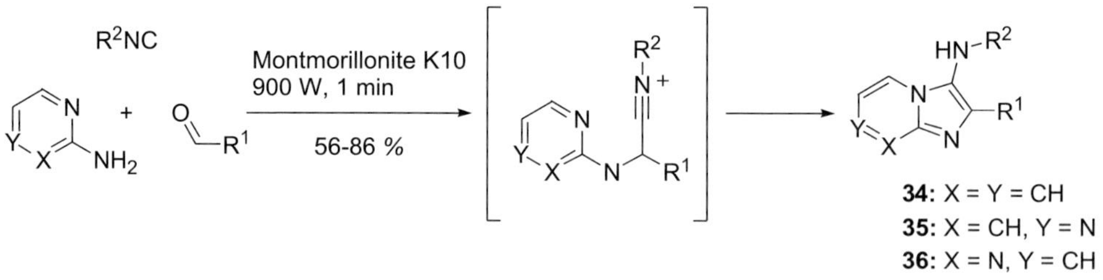

Scheme 17.26

were compatible with this reaction, which enabled generation of a library of **34** and **35**. The preparative scope of the reaction was further extended to the synthesis of a variety of imidazo[1,2-a]pyrimidines **36** by using aldehydes, isocyanides, and 2-aminopyrimidine as starting materials (Scheme 17.26).

MW irradiation can also accelerate the synthesis of fused 3-aminoimidazoles **37** in a similar reaction to that described above but now in MeOH as solvent. Libraries of **37** were produced by Tye et al. (ten compounds) [73] and Zhang et al. (sixty compounds) [74] via a three-component reaction of heterocyclic amidines with isocyanides and aldehydes using a catalytic amount of Sc(OTf)$_3$ (Scheme 17.27).

By use of a monomode MW reactor and MeOH as solvent reaction times could be reduced to 10 min (at 160 °C) compared to the original procedures by Blackburn et al. [72]. The more reactive, electron rich, amidines in combination with benzyl-isocyanide gave high yields (65–93%). Less reactive amidines gave reduced conversions to product and some side products were observed.

Finally, a MW assisted Ugi–de-Boc cyclization sequence can be used for the synthesis of benzimidazole cores **38** (Scheme 17.28) [75].

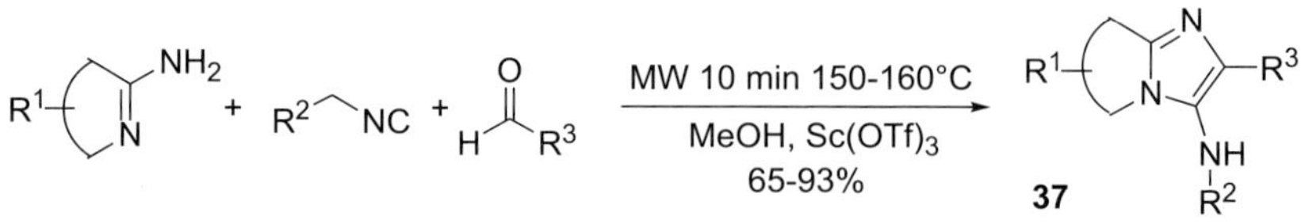

Scheme 17.27

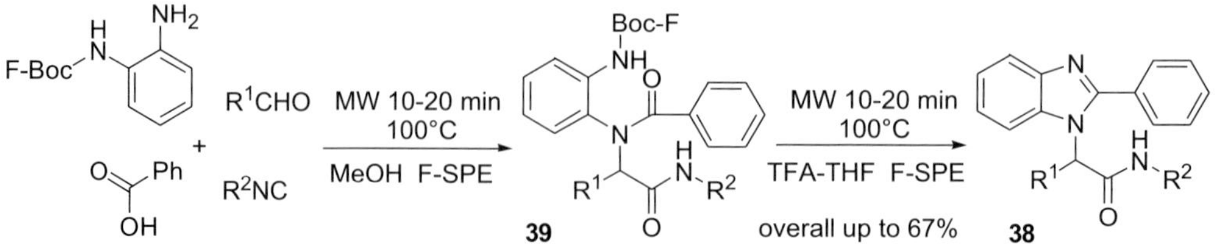

Scheme 17.28

A mixture of an F-Boc-protected diamine with a slight excess of benzoic acid, aldehyde, and isonitrile in MeOH was subjected to MW irradiation for 10–20 min at 100 °C to give the linear Ugi product **39**. Subsequent deprotection of the amine group (de-Boc) induced cyclization and the final product **38** was formed after 10–20 min, again under MW irradiation, now in TFA–THF (1:1). The reaction mixture was purified by fluorous solid-phase extraction (F-SPE) and afforded **38** in up to 67% yield.

17.2.6
Pyrroles

Pyrroles are the core unit of a wide variety of natural products [76]. Although many methods are available for the synthesis of these species, most are multi-step procedures resulting in low yields [77, 78]. However, Hantzsch made another important contribution to the progress of multicomponent chemistry. In his procedure pyrroles were successfully prepared from primary amines, β-ketoesters, and α-halogenated β-ketoesters [79]. Only a few other one-step procedures have been reported for pyrroles but, because of to long reaction times and insufficient scope of substitution at the ring, these are not very satisfactory [80, 81].

An interesting variant is the multicomponent synthesis of pyrroles from carbonyl compounds, primary amines, and nitroalkanes first described by Ishii et al. [82]. When the reaction is performed under the conditions specified by Ishii et al. [82], reaction rates were slow and reaction products were usually obtained in moderate yields. In an attempt to improve the efficiency of the reaction, Ranu et al. [83] developed an identical, but MW-assisted, synthesis on the surface of either silica or alumina. A wide range of structurally different α,β-unsaturated aldehydes and ketones were successfully coupled under mild conditions with primary amines and nitroalkanes to produce the corresponding pyrroles **40** in yields that were usually better than those reported previously (Scheme 17.29). Compared with the procedure described by Ishii et al. [82], time efficiency had been improved substantially and reactions generally proceeded to completion within 10 min. The reaction developed by Ranu et al. was performed in a domestic MW oven with an irradiation time of 5–10 min at 120 W and the desired pyrroles **40** were generated in moderate to good yield (60–72%) [83].

An alternative synthetic route to (bicyclic) pyrroles **41** involves the coupling of a cyclic ketone with an amine and an α,β-unsaturated nitroalkene on the surface of alumina (Scheme 17.30) [14]. After MW irradiation for 13–15 min the product **41** was obtained in 71–86% yield. In this synthesis, substitution on the α-position of

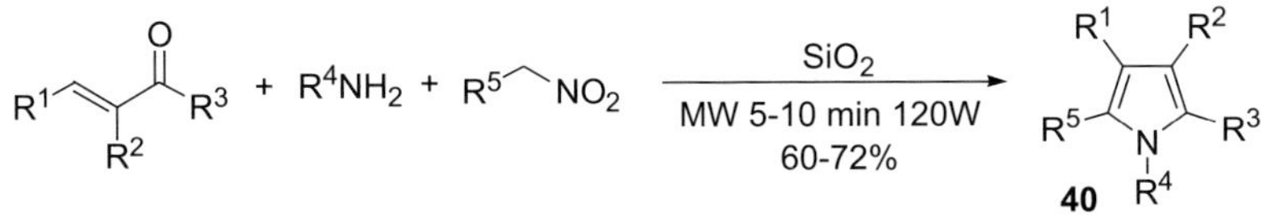

Scheme 17.29

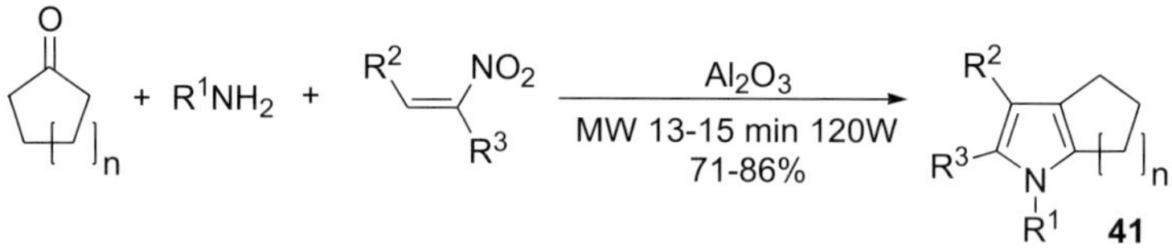

Scheme 17.30

the nitroalkene input is essential, or the reaction takes a different course. Open-chain ketones can also be used in this reaction.

Yet another procedure used to prepare fused pyrroles is three-component condensation of an acyl bromide, pyridine, and an acetylenic compound, catalyzed by basic alumina, to give the corresponding indolizines **42** in 87–94% yield (Scheme 17.31) [84]. The mixture was irradiated for 8 min in a monomode MW reactor with a temperature limit of 250 °C.

It is believed that the N-alkylpyridinium salt, generated *in situ* from condensation of phenacyl bromide and pyridine, is converted into the 1,3-dipole species **43** under basic conditions. Subsequent cycloaddition to ethyl propiolate results in an unstable intermediate, which instantly facilitates aromatization to afford the fused pyrroles **42a** (Scheme 17.32).

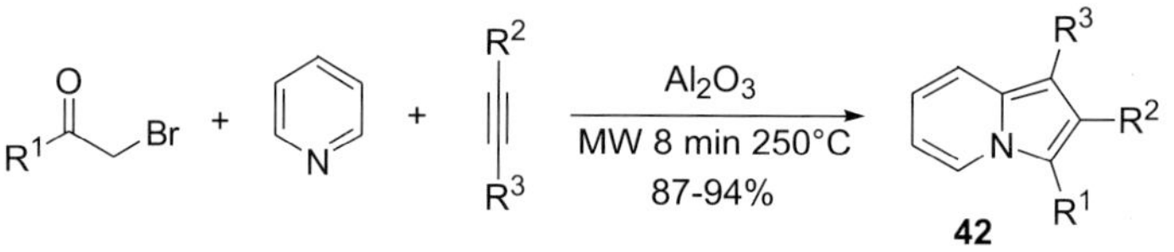

Scheme 17.31

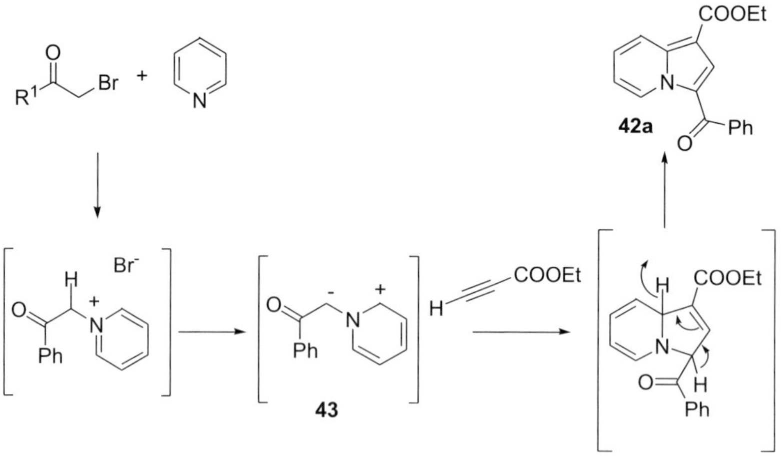

Scheme 17.32

17.2.7
Other N-containing Heterocycles

Yadav et al. reported a three-component coupling reaction under MW conditions for construction of a triazine system containing a thiazole core [85]. In this reaction, a thiazole Schiff's base, an aldehyde, and NH_4OAc react to yield thiazolo-s-triazine C-nucleosides **44** (Scheme 17.33).

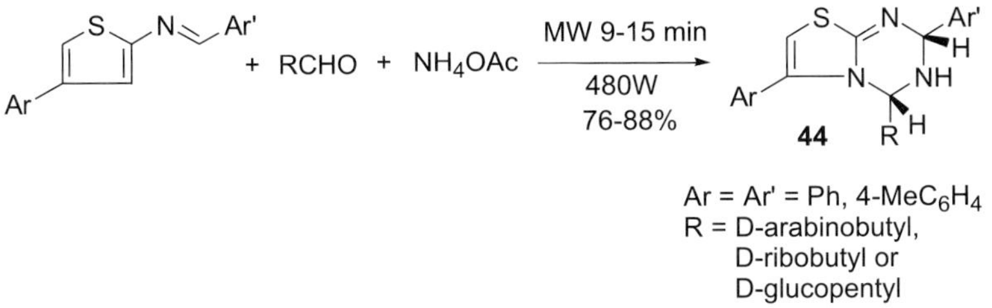

Ar = Ar' = Ph, 4-MeC$_6$H$_4$
R = D-arabinobutyl, D-ribobutyl or D-glucopentyl

Scheme 17.33

This reaction was performed in a domestic MW oven for 9–15 min at a power level of 480 W. The yields were good (76–88%) and triazines **44** were obtained with high diastereoselectivity in favor of the cis isomers (>96:4). This ratio is much higher than that from a comparable experiment in which the reaction was performed under traditional heating conditions. With the same reaction times and temperature (90 °C, measured directly after irradiation) product yields were significantly lower (19–31%) and diastereoselectivity (>56:44) was much lower.

The 1,2,3-triazoles have interesting biological properties [86, 87]. 1,4-Disubstituted triazoles **45** can be synthesized in a one-pot three-component MW-assisted reaction from corresponding alkyl halides, sodium azide, and alkynes (Scheme 17.34) [88].

The alkyl azide is generated *in situ* from the corresponding alkyl halides and sodium azide, whereupon it is captured by copper(I) acetylide forming the desired triazole **45**. The reaction is performed under the action of MW irradiation for 10 min at 125 °C. Although most compounds readily tolerated this temperature, occasionally it resulted in reduced yields. To circumvent these problems these reaction mixtures were irradiated for 15 min at 75 °C. The reaction was performed in a 1:1 mixture of *t*-BuOH and water and the Cu(I) catalyst was prepared *in situ* by

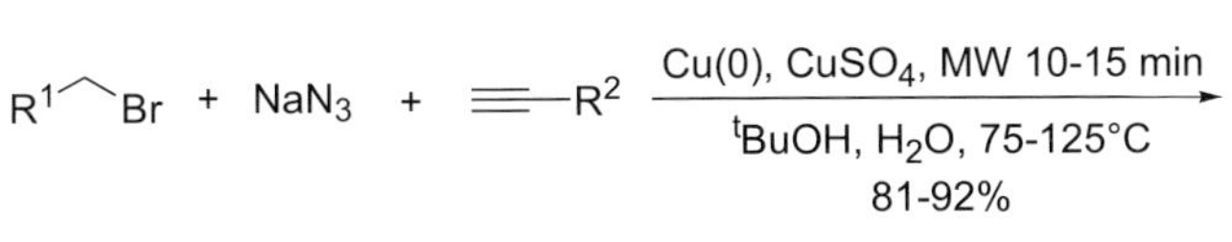

Scheme 17.34

comproportionation of Cu(0) and Cu(II). The desired triazoles (**45**) could then be isolated in 81–92% yield.

Another MW-assisted application of the Ugi–de-Boc cyclization sequence (*vide supra*) is the construction of the interesting N-containing heterocyclic core of a quinoxalinone **48**. When the reaction was performed at room temperature good yields were obtained but it took 36–48 h to complete [89]. Tempest et al. modified the procedure to reduce reaction times and to simplify the purification of the Ugi condensation products [75]. They did this by irradiating a mixture of F-Boc protected diamine with a slight excess of phenylglyoxylic acid **46**, aldehyde, and isonitrile in MeOH in a monomode MW reactor for 10–20 min at 100 °C, initially obtaining intermediate **47** (Scheme 17.35).

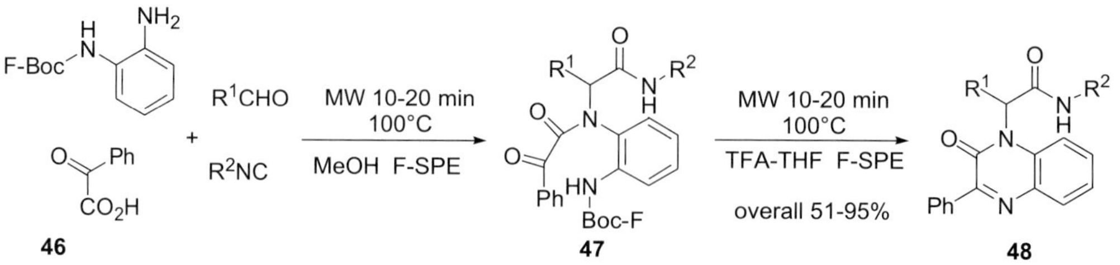

Scheme 17.35

The final quinoxalinone product **48** was generated by de-Boc cyclization of **48** in TFA–THF (1:1) under the same MW conditions. The product **48** was easily separated from the reaction mixture by fluorous solid-phase extraction (F-SPE) in 51–95% yield.

In the next two MCRs, two heterocyclic rings are constructed in the same reaction step. The first example is a rather complicated synthesis of pyrido[2,3-d]pyrimidine scaffolds **49** via a one-pot MW-assisted condensation of a α,β-unsaturated ester, malononitrile or cyanoacetate, and an amidine (Scheme 17.36) [90].

By use of sealed-vessel monomode MW technology full conversions were achieved within 10 min at temperatures of 100–140 °C. Solvent studies showed that the strongly MW-absorbing MeOH resulted in the best yields of the desired products. Initially the reaction was performed without the presence of a base. Although the amidine building block itself is a base, a catalytic amount of a stronger base (5% NaOMe) was necessary to obtain the desired product in high yield.

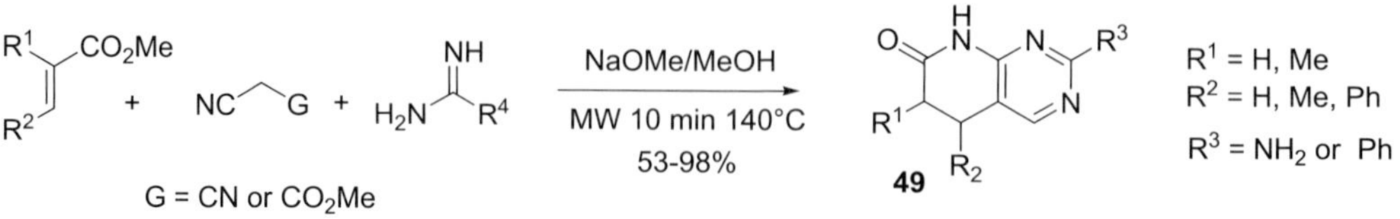

Scheme 17.36

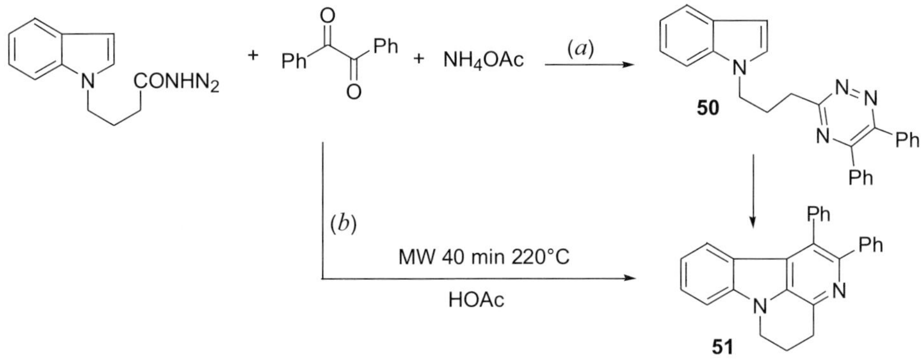

Scheme 17.37

The rather elegant construction of canthines is a second example in which two heterocyclic rings are constructed in a single step by MW-assisted MCR technology (Scheme 17.37). Canthines are a tetracyclic subclass of β-carboline alkaloids with a wide range of pharmacological activity [91]. Several synthetic approaches to the canthine skeleton have been reported [92, 93]. Snyder reported a elegant strategy in which first a three-component reaction was performed to form the triazine **50** (Scheme 17.37, reaction path a) [94].

Intermediate **50** was subsequently heated under reflux in triisopropylbenzene (232 °C) for 1.5 to 20 h to provide the basic canthine skeleton **51**. Recently, Lindsley et al. reported a rapid MW-mediated procedure for synthesis of **51** [95]. This reaction, performed in a monomode MW reactor at 180 °C, required a reaction time of only 5 min. Even more interesting, treatment of the acryl hydrazide-tethered indole input, with benzil in the presence of 10 equiv. NH_4OAc delivered not only the expected triazine **50** but also, directly, the 1,2-diphenyl canthine derivative **51** (Scheme 17.37, reaction path b). The products were formed in a 9:1 ratio of **50** and **51**, respectively. In the one-pot reaction, the indole underwent a three-component condensation to generate **50** followed by an intramolecular inverse-electron-demand Diels–Alder reaction and subsequent chelotropic expulsion of N_2 to generate the 1,2-diphenyl canthine **51**.

It was possible to obtain the desired canthine **51** exclusively, simply by increasing the MW reaction time from 5 to 60 min at 180 °C. The ratio of **50** to **51** could be increased from 9:1 to 1:2. Further increasing the temperature to 220 °C, which is 100 °C above the boiling point of HOAc, for 40 min improved the selectivity to 1:19 in favor of canthine **51**.

17.3
Oxygen-containing Heterocycles

In previous sections, MW-accelerated MCRs were shown to be efficiently catalyzed by several insoluble inorganic supports. Similar results were obtained when mild

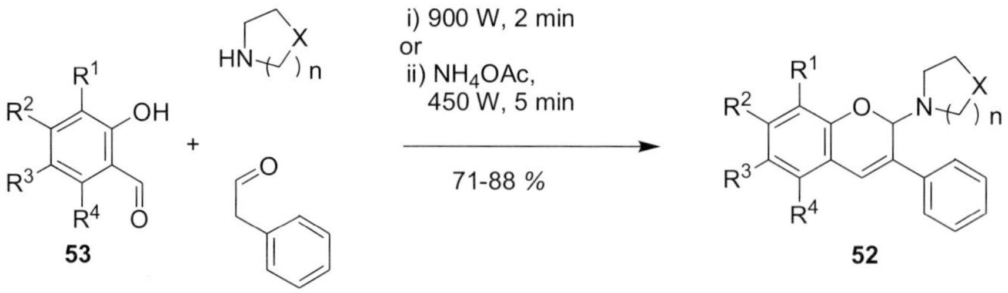

Scheme 17.38

Brönstedt acids were used to catalyze the MW-accelerated multicomponent synthesis of substituted isoflav-3-enes **52** (Scheme 17.38) [96].

Irradiating a mixture of phenyl acetaldehyde and cyclic secondary amines led to the formation of the corresponding enamines, which, on addition of salicylaldehydes, **53**, and a catalytic amount of NH_4OAc, gave isoflav-3-enes **52** in good yields ranging from 71% to 88%. The reaction could be performed in the absence of a solvent and reaction times were strikingly short. Thus an efficient and environmentally benign procedure had been developed for the synthesis of various substituted isoflav-3-enes **52**.

Another important class of O-containing heterocycles with a wide range of biological activity are 4H-benzo[b]pyrans [97]. Conventional synthesis of these species requires the use of organic solvents, for example DMF–acetic acid, that complicate the work up procedure, resulting in poor yields of the products [98]. Kaupp et al. recently reported a novel method for the synthesis of these heterocycles which uses the reactants in the solid or molten state [99]. This two-step reaction was performed at very high temperature and required long reaction times. Devi et al. adjusted this procedure to develop a simple and highly efficient method for synthesis of **54** via a three-component cyclocondensation reaction under MW conditions [100]. In this solvent-free procedure, equimolar amounts of benzaldehyde, alkyl nitrile, and dimedone were mixed with a catalytic amount of NaBr (Scheme 17.39). The reaction mixture was irradiated in a monomode MW reactor for 10–15 min at 70–85 °C to furnish the 4H-benzo[b]pyrans **54** in good to excellent yields (63–95%).

Another multicomponent procedure used to construct oxygen containing heterocycles under MW conditions is the synthesis of pyrano[2,3-d]pyrimidine **56** [55].

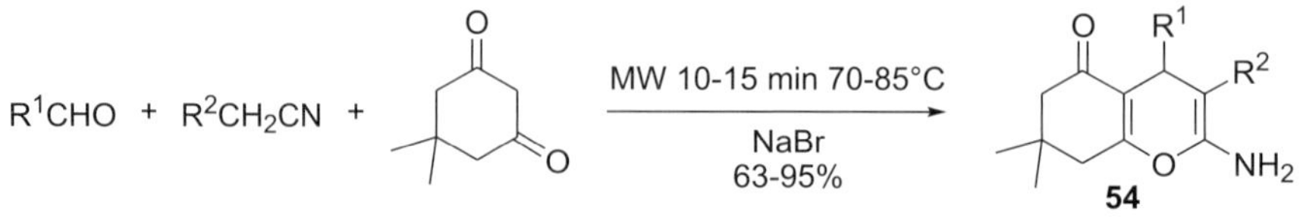

Scheme 17.39

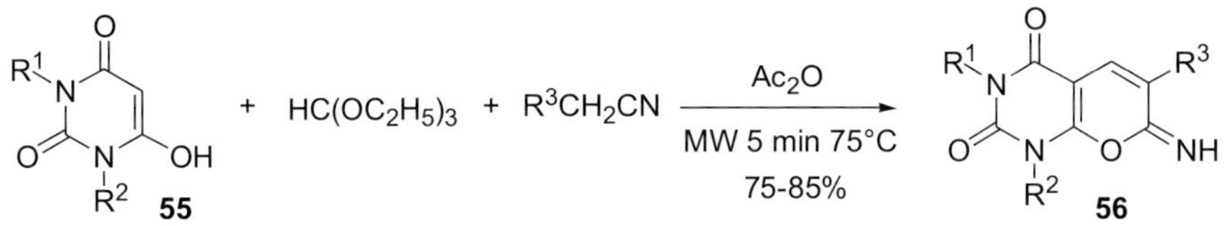

17.3 Oxygen-containing Heterocycles

Scheme 17.40

The reported MCR uses N,N-dimethylbarbituric acid **55** with equimolar amounts of triethylorthoformate, malononitrile and 3 equiv. acetic anhydride (Scheme 17.40). This mixture was irradiated for 5 min at 75 °C to furnish the desired pyrano[2,3-d]pyrimidines **56** in good yield (75–85%) [55].

Heterocyclic systems produced by annulation of a pyran ring to the biologically versatile thiazole nucleus can lead to an attractive scaffold for libraries of drug-like compounds. Yadav et al. reported a synthesis of these species by three-component coupling of glycine, acetic anhydride, and a 5-arylidenerhodanine **57** (Scheme 17.41) [101].

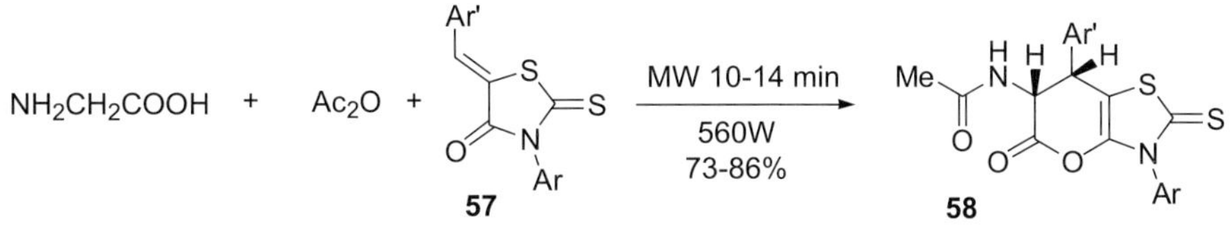

Scheme 17.41

The reaction was performed in an unmodified domestic MW oven with an irradiation time of 10–14 min at 560 W and furnished good yields (73–86%) of the final product **58**. Diastereoselectivity strongly favored the syn isomers (>96:4).

For comparison purposes, the reaction was also performed using a thermostatted oil bath using the same reaction times and temperature; significantly lower yields were obtained (35–43%) and diastereomeric ratio was lower (>57:43).

The reaction can be rationalized in terms of Michael addition of azlactone **59**, generated *in situ*, to thialolone **57**, to afford the corresponding intermediates **60** and **61** (Scheme 17.42). These intermediates then smoothly undergo ring-formation to yield the final product **58**.

It has been argued that the improved diastereoselectivity in this reaction can be attributed to a difference between the polarity of the activated syn complex **60** and the corresponding anti complex. This syn intermediate is more dipolar than the corresponding anti form. Because MW radiation favors reactions occurring via more dipolar activated complexes (Ref. [19c] and Chapter 4 in this book) the syn isomer of **58** is favored.

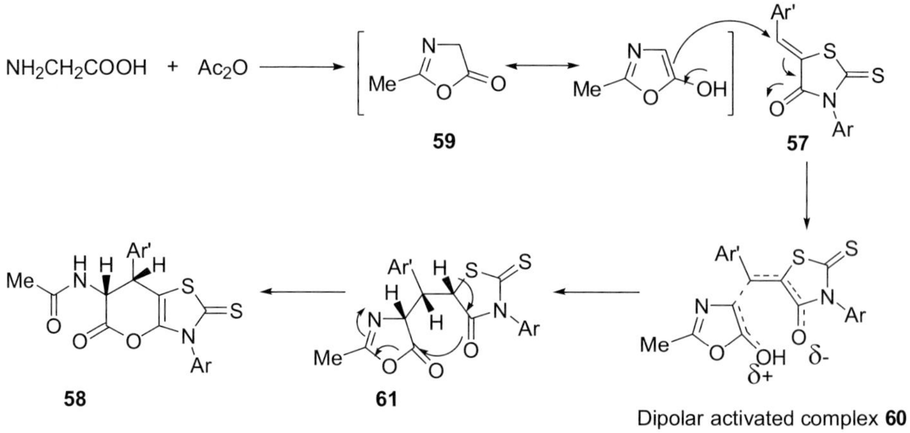

Scheme 17.42

17.4
Other Ring Systems

2,6-Dicyanoanilines **62** have been seen as a basis for artificial photosynthetic systems or molecular electronic devices [102, 103]. These compounds can be prepared from propanedinitrile and α,β-unsaturated ketones but yields are very poor (5–20%) [104]. It is possible to improve the utility of this process by using a practical multicomponent variant under MW conditions [105]. In this procedure the α,β-unsaturated ketones are generated *in situ* from the corresponding aldehydes and ketones, after which they are captured by 2 equiv. propanedinitrile, forming the corresponding polysubstituted 2,6-dicyanoanilines **62** (Scheme 17.43). The irradiation time was 2 min and the product was isolated in an improved 50–63% yield [105].

To improve these yields further, poly(ethylene glycol) (PEG) was used as support. Thus, PEG-bound aldehydes (R^1 = PEG-connection) were mixed with ketones and 3 equiv. propanedinitrile, and NH_4OAc was used as base. The three-component reaction was performed under MW irradiation for 2 min resulting in PEG bound 2,6-dicyanoanilines. These were cleaved by NaOMe–MeOH to afford free polysubstituted 2,6-dicyanoanilines **62** in good yields (65–82%) [105].

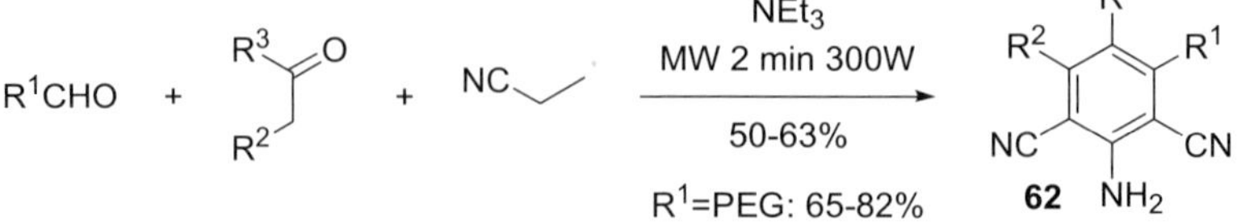

Scheme 17.43

17.4 Other Ring Systems

Thiazolidones are another class of heterocycles that attract much attention because of their wide ranging biological activity [106]. They are usually synthesized by three-component condensation of a primary amine, an aldehyde, and mercaptoacetic acid with removal, by azeotropic distillation, of the water formed [107]. The reaction is believed to proceed via imine formation then attack of sulfur on the imine carbon. Finally, an intramolecular cyclization with concomitant elimination of water occurs, generating the desired product. The general applicability of the reaction is limited, however, because it requires prolonged heating with continuous removal of water. To circumvent these difficulties and to speed up the synthesis, Miller et al. developed a microwave-accelerated three-component reaction for the synthesis of 4-thiazolidinones **63** [108]. In this one-pot procedure, a primary amine, an aldehyde, and mercaptoacetic acid were condensed in ethanol under MW conditions for 30 min at 120 °C (Scheme 17.44). The desired 4-thiazolidinones **63** were obtained in 55–91% yield.

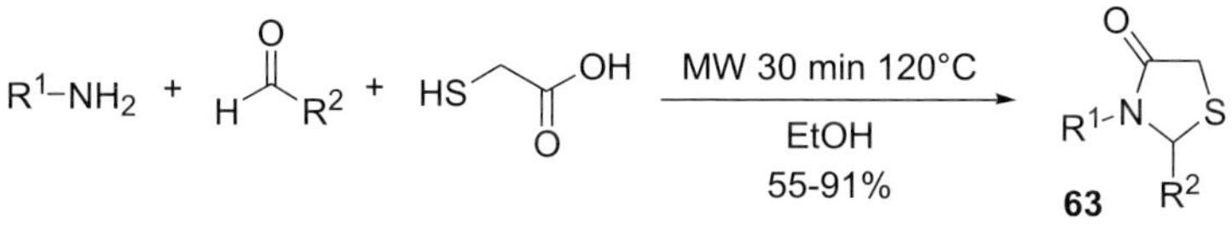

Scheme 17.44

Benzothiazepines are a class of compounds with interesting pharmacological properties. Loupy et al. [109] developed an efficient one pot synthesis of diversely substituted benzothiazepines. Microwave-assisted reaction of isatin and different acetophenones adsorbed on basic alumina leads to the intermediate **64**; this can be further reacted with 5-substituted aminobenzenethiols to furnish the desired benzothiazepines **65** in high yields (Scheme 17.45). This procedure proved to be more efficient than the conventional thermal method, which requires a two-step synthesis involving harsh conditions, long reaction times and gives only moderate yields (max 65%).

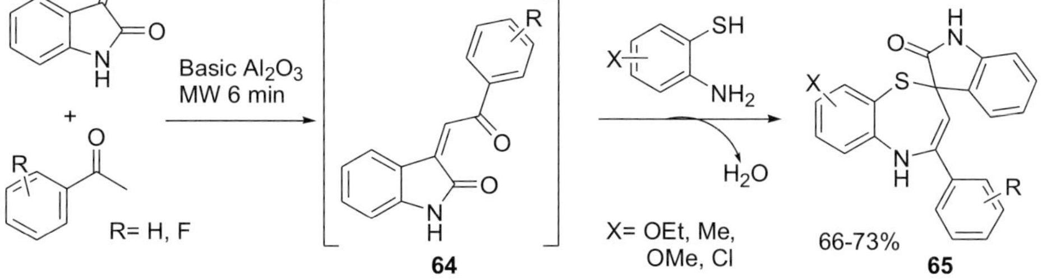

Scheme 17.45

17.5
Linear Structures

Solid-phase Ugi four-component reactions under MW conditions have been used to obtain bisamides **66** [110, 111]. Best results were obtained when the amine input was immobilized, and combined with an aldehyde or ketone, a carboxylic acid, and an isocyanide (Scheme 17.46) [111].

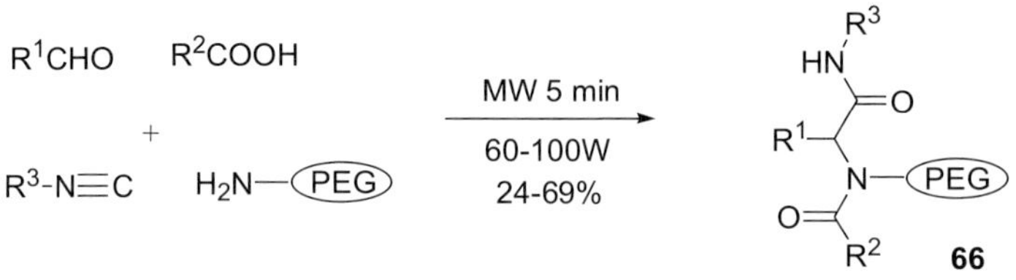

R^1CHO R^2COOH

$R^3-N\equiv C$ H_2N-(PEG)

MW 5 min
60–100W
24–69%

→ bisamide **66** with R^1, R^2, R^3, HN, N–PEG

Scheme 17.46

Although Ugi reactions can proceed rapidly, reaction times from one to several days are not uncommon. MW conditions could speed up these reactions. Thus, after attachment of poly(ethylene glycol) (PEG)-grafted polystyrene to an amino-functionalized building block, the Ugi reaction has been performed in a domestic MW oven, using a mixture of DCM and MeOH (2:1), with a total irradiation time of 5 min. After cleavage from the resin by treatment with TFA:DCM (19:1) for 1 h the desired bisamides **66** were obtained in up to 96% yield. This reduced reaction times by at least a factor of ten compared with conventional reaction conditions.

MW irradiation has also been used in three-component coupling of an aldehyde, an amine, and an alkyne (Scheme 17.47) to generate propargylamines **67** [112, 113]. These compounds can be used to prepare many biologically active nitrogen compounds [114]. Several methods for construction of **67** were known, all required expensive Au or Ag catalysts and the reactions proceeded slowly in water [115, 116]. Tu et al. performed this reaction using a three-component H_2O system in the presence of 15 mol% CuI [112]. The reaction mixture, in a sealed tube, was irradiated in a domestic MW oven for 30 min to give moderate to excellent yields of the desired propargylamines **67** (41–93%).

Thioamides **68** are essential building blocks in the preparation of several biolog-

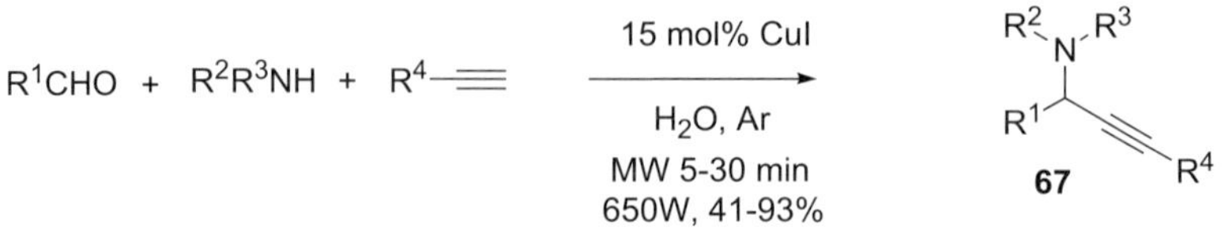

$R^1CHO + R^2R^3NH + R^4-\equiv$

15 mol% CuI
H_2O, Ar
MW 5–30 min
650W, 41–93%

→ **67** with R^1, R^2, R^3, R^4

Scheme 17.47

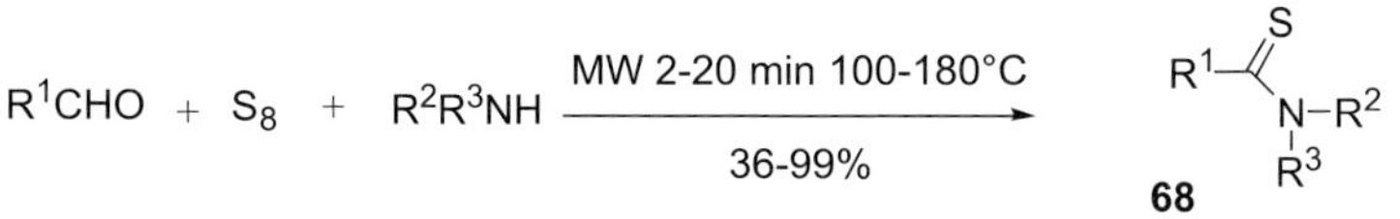

$R^1CHO + S_8 + R^2R^3NH \xrightarrow[36-99\%]{\text{MW 2-20 min 100-180°C}}$ **68** $R^1\underset{R^3}{\overset{S}{\diagdown}}N-R^2$

Scheme 17.48

ically relevant heterocyclic scaffolds [117]. One way of forming these species is by three-component coupling of an aldehyde, elemental sulfur, and an amine (Scheme 17.48). Kindler first reported this one-pot procedure in 1923 [118]. The original method is of limited application only, because of the high reaction temperatures and long reaction times usually required [119]. Another disadvantage is the use of volatile amines or aldehydes, which cannot be used without autoclave technology. For these reasons, MW-accelerated conditions in sealed vessels are beneficial to this reaction [120]. When the same mixture was irradiated for 2–20 min at 100–180 °C in a monomode MW cavity the desired thioamides **68** were obtained in a 36–99% yield.

α-Aminophosphonates are an important class of biologically active compounds and their synthesis has received an increasing attention, because of their structural analogy to α-amino acids [121]. They can usually be prepared by addition of phosphorus nucleophiles to imines in the presence of a Lewis acid. This reaction is not possible in a one-pot procedure with a carbonyl compound, an amine, and diethyl phosphate (Scheme 17.49) [122], however, because both the amine reactant and the water formed can decompose or deactivate the Lewis acid [123]. Some one-pot procedures using lanthanide triflates [124] and indium trichloride [125] catalysts have been reported, but not only are these catalysts expensive, quite long reaction times (10–20 h) are required to obtain the desired products in good yields.

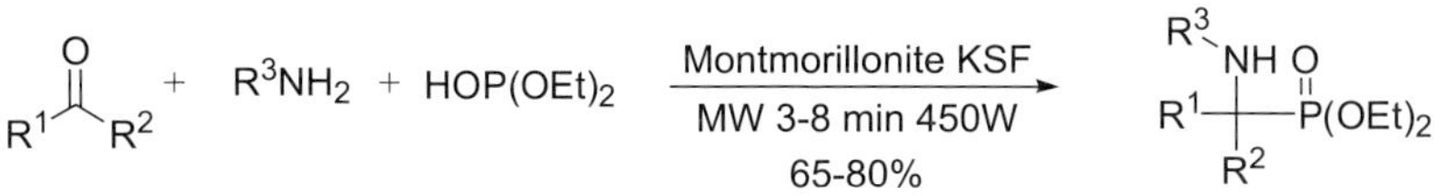

$R^1\underset{R^2}{\overset{O}{\diagdown}} + R^3NH_2 + HOP(OEt)_2 \xrightarrow[\substack{\text{MW 3-8 min 450W} \\ 65-80\%}]{\text{Montmorillonite KSF}}$ **69** $R^1\underset{R^2}{\overset{R^3\diagdown NH}{\diagdown}}\overset{O}{-}P(OEt)_2$

Scheme 17.49

The desired α-aminophosphonates **69** can be synthesized by using an MCR approach under MW-accelerated conditions, however [126]. MW technology enables the use of the inexpensive and reusable montmorillonite KSF clay catalyst. Aromatic and aliphatic aldehydes both afforded excellent yields of products (80–92%) in short reaction times (3–5 min). Ketones gave the corresponding phosphonates **69**, also in good yields (65–80%) after an irradiation time of 6–8 min.

17.6
Conclusions and Outlook

In this chapter a range of different MCRs performed under MW conditions have been described. The reactants in these MCRs are often dipoles, because they usually contain heteroatoms, for example oxygen or nitrogen. In addition, many MCRs proceed via relatively polar intermediates. As a consequence, these reactions may benefit substantially from MW irradiation in terms of time-efficiency, yield, and purity of products (Ref. [19c] and Chapter 4).

Many of the examples reported here make use of solvent-free conditions. The solvent-free approach allows safe use of domestic household MW ovens and standard open-vessel technology (Chapter 8). It should, however, be emphasized that these methods usually suffer from technical difficulties relating to nonuniform heating and mixing, and precise determination of the reaction temperature. The current trend in MAOS is to move away from these multimode MW ovens and to use the more dedicated monomode instruments, which have only become available in the last few years. Reproducibility and some control over the temperature are the main advantages. These monomode MW reactors, including easy-to-use automated equipment for performing many reactions sequentially, are now available in many synthetic laboratories. We can therefore look forward to many more exciting examples of the application of MW technology to MCR chemistry.

References

1 NICOLAOU, K.C.; VOURLOUMIS, D.; WINSSINGER, N.; BARAN, P.S. *Angew. Chem.* **2000**, *112*, 44–122.
2 WENDER, P.A.; HANDY, S.T.; WRIGHT, D.L. *Chem. Ind.* **1997**, 765–766.
3 HALL, N. *Science* **1994**, *266*, 32–34.
4 UGI, I.; DÖMLING, A.; HÖRL, W. *Endeavour* **1994**, *18*, 115–122.
5 TIETZE, L.F. *Chem. Rev.* **1996**, *96*, 115–136.
6 DENMARK, S.E.; THORARENSEN, A. *Chem. Rev.* **1996**, *96*, 137–165.
7 For excellent reviews on isocyanide-based MCRs: (a) DÖMLING, A.; UGI I. *Angew. Chem. Int. Ed.* **2000**, *39*, 3169–3210; (b) ZHU, J.P. *Eur. J. Org. Chem.* **2003**, 1133–1144; (c) BANFI, L.; RIVA, R. *Org. React.* **2005**, *65*, 1–140.
8 BIENAYMÉ, H.; HULME, CH.; ODDON, G.; SCHMITT, PH. *Chem. Eur. J.* **2000**, *6*, 3321–3329.
9 ORRU, R.V.A.; DE GREEF, M. *Synthesis* **2003**, 1471–1499.
10 For example see: BRAVO, P.A.; CARRERO, M.C.P.; GALAN, E.R.; BLAZQUEZ, J.A.S. *Heterocycles* **2000**, *1*, 81–92.
11 For example see: NEUSCHUTZ, K.; VELKER, J.; NEIER, R. *Synthesis* **1998**, *3*, 227–255.
12 For example see: ORGAN, M.G.; WINKLE, D.D.; HUFFMAN, J. *J. Org. Chem.* **1997**, *62*, 5254–5266.
13 For example, JOHNSON, W.S.; GRAVESTOCK, M.B.; MCCARRY, B.E. *J. Am. Chem. Soc.* **1971**, *93*, 4332–4333.
14 RANU, B.C.; HAJRA, A. *Tetrahedron* **2001**, *57*, 4767–4773.
15 WEBER, L.; ILLGEN, K.; ALMSTETTER, M. *Synlett* **1999**, 366–374.
16 VARMA, R.S. *Green Chemistry* **1999**, *1*, 43–55.
17 HAYES, B.L. *Aldrichim. Acta* **2004**, *37*, 66–77.
18 ADAM, D. *Nature* **2003**, *421*, 571–572.

19 (a) KAPPE, C.O. *Angew. Chem. Int. Ed.* **2004**, *43*, 6250–6284. (b) LIDSTRÖM, P.; TIERNEY, J.; WATHEY, B.; WETSMAN, J. *Tetrahedron* **2001**, *57*, 9225–9283. (c) PERREUX, L.; LOUPY, A. *Tetrahedron* **2001**, *57*, 9199–9225.
20 STASS, D.V.; WOODWARD, J.R.; TIMMEL, C.R.; HORE, P.J.; MCLAUGHLAN, K.A. *Chem. Phys. Lett.* **2000**, *329*, 15–22.
21 GABRIEL, C.; GABRIEL, S.; GRANT, E.H.; HALSTEAD, B.S.J.; MINGOS, D.M.P. *Chem. Soc. Rev.* **1998**, *27*, 213–223.
22 CADDICK, S. *Tetrahedron* **1995**, *51*, 10403–10432.
23 LOUPY, A.; PETIT, A.; HAMELIN, J.; TEXIER-BOULLET, F.; JACQUAULT, P.; MATHÉ, D. *Synthesis* **1998**, 1213–1234 and references therein.
24 VARMA, R.S.; KUMAR, D. *Tetrahedron Lett.* **1999**, *40*, 7665–7669.
25 BALALAIE, S.; ARABANIAN, A. *Green Chemistry* **2000**, *2*, 274–276.
26 RANU, B.C.; HAJRA, A.; JANA, U. *Synlett* **2000**, 75–76.
27 HANTZSCH, A. *Justus Liebigs Ann. Chem.* **1882**, *215*, 1.
28 GOLDMANN, S.; STOLTEFUSS, J. *Angew. Chem. Int. Ed.* **1991**, *30*, 1559–1578.
29 COTTERILL, I.C.; USYATINSKY, A.Y.; ARNOLD, J.M.; CLARK, D.S.; DORDICK, J.S.; MICHELS, P.C.; KHMELNITSKY, Y.L. *Tetrahedron Lett.* **1998**, *39*, 1117–1120.
30 LOEV, B.; GOODMAN, M.M.; SNADER, K.M.; TEDESCHI, R.; MACKO, E. *J. Med. Chem.* **1974**, *17*, 956–965.
31 (a) SUAREZ, M.; LOUPY, A.; PÉREZ, E.; MORAN, L.; GERONA, G.; MORALES, A.; AUTIÉ, M. *Heterocyclic. Commun.* **1996**, *2*, 275–280. (b) SUAREZ, M.; LOUPY, A.; SALFRAN, E.; MORAN, L.; ROLANDO, E. *Heterocycles* **1999**, *51*, 21–27. (c) OHNBERG, L.; WESTMAN, J. *Synlett* **2001**, 1296–1298.
32 SUNKEL, C.E.; DECASAJUANA, M.F.; SANTOS, L.; GOMEZ, M.M.; VILLAROYA, M.; GONZALEZ-MORALES, M.A.; PRIEGO, J.G.; ORTEGA, M.P. *J. Med. Chem.* **1990**, *33*, 3205–3210.
33 For example see: ANTAKI, H. *J. Chem. Soc.* **1963**, 4877.
34 For example see: STANKEVICH, E.I.; VANAG, G.Y. *Zhurnal Obshchei Khimii* **1962**, *32*, 1146–1151.
35 KIDWAI, M.; RASTOGI, S.; MOHAN, R. *Bull. Kor. Chem. Soc.* **2004**, *25*, 119–121.
36 BAGLEY, M.C.; SINGH, N. *Synlett* **2002**, 1718–1720.
37 SIMON, C.; CONSTANTIEUX, T.; RODRIGUEZ, J. *Eur. J. Org. Chem.* **2004**, 4957–4980.
38 QUIROGA, J.; CISNEROS, C.; INSUASTY, B.; ABONIA, R.; NOGUERAS, M.; SANCHEZ, A. *Tetrahedron Lett.* **2001**, *42*, 5625–5627.
39 HILGEROTH, A.; BAUMEISTER, U.; HEINEMANN, F.W. *Eur. J. Org. Chem.* **2000**, 245–249.
40 HILGEROTH, A.; WIESE, M.; BILLICH, A. *J. Med. Chem.* **1999**, *42*, 4729–4732.
41 BALALAIE, S.; KOWSARI, E. *Monatsh. Chem.* **2001**, *132*, 1551–1555.
42 For example see: TOWNS, A.D. *Dyes and Pigments* **1999**, *42*, 3–28.
43 BALALAIE, S.; KOWSARI, E.; HASHTROUDI, M.S. *Monatsh. Chem.* **2003**, *134*, 453–456.
44 GOROBETS, N.Y.; YOUSEFI, B.H.; BELAJ, F.; KAPPE, C.O. *Tetrahedron* **2004**, *60*, 8633–8644.
45 RODRIGUEZ, H.; SUAREZ, M.; PEREZ, R.; PETIT, A.; LOUPY, A. *Tetrahedron Lett.* **2003**, *44*, 3709–3712.
46 HUMA, H.Z.S.; HALDER, R.; KALRA, S.S.; DAS, J.; IQBAL, J. *Tetrahedron Lett.* **2002**, *43*, 6485–6488.
47 WANG, Y.; ZHU, S. *Synthesis* **2002**, 1813–1818.
48 GERUS, I.I.; GORBUNOVA, M.G.; KUKHAR, V.P. *J. Fluor. Chem.* **1994**, *69*, 195–198.
49 BILLAH, M.; BUCKLEY, G.M.; COOPER, N.; DYKE, H.J.; EGAN, R.; GANGULY, A.; GOWERS, L.; HAUGHAN, A.F.; KENDALL, H.J.; LOWE, C.; MINNICOZZI, M.; MONTANA, J.G.; OXFORD, J.; PEAKE, J.C.; PICKEN, C.L.; PIWINSKI, J.J.; NAYLOR, R.; SABIN, V.; SHIH, N.-Y.; WARNECK, J.B.H. *Bioorg. Med. Chem. Lett.* **2002**, *12*, 1617–1619.
50 ZHANG, J.M.; YANG, W.; SONG, L.P.; CAI, M.; ZHU, S.Z. *Tetrahedron Lett.* **2004**, *45*, 5771–5773.
51 ANDERSON, G.L.; SHIM, J.L.; BROOM, A.D. *J. Org. Chem.* **1977**, *42*, 993–996.

52 Bennett, L.R.; Blankley, C.J.; Fleming, R.W.; Smith, R.D.; Tessman, D.K. *J. Med. Chem.* **1981**, *24*, 382–389.

53 Hirota, K.; Kuki, H.; Maki, Y. *Heterocycles* **1994**, *37*, 563–570.

54 Srivastava, P.; Saxena, A.S.; Ram, V.J. *Synthesis* **2000**, 541–544.

55 Devi, I.; Bhuyan, P.J. *Synlett* **2004**, 283–286.

56 (a) Kappe, C.O. *Tetrahedron* **1993**, *49*, 6937–6963 and references therein. (b) Kappe, C.O. *Acc. Chem. Res.* **2000**, *33*, 879–888. (c) Kappe, C.O. *Eur. J. Med. Chem.* **2000**, *35*, 1043–1052.

57 For example see: Bigi, F.; Carloni, S.; Frullanti, B.; Maggi, R.; Sartori, G. *Tetrahedron Lett.* **1999**, *40*, 3465–3468.

58 For example see: Shutalev, A.D.; Kishko, E.A.; Sivova, N.V.; Kuznetsov, A.Y. *Molecules* **1998**, *3*, 100–106.

59 Kappe, C.O.; Kumar, D.; Varma, R.S. *Synthesis* **1999**, 1799–1803.

60 Gohain, M.; Prajapati, D.; Sandhu, J.S. *Synlett* **2004**, 235–238.

61 Saxena, I.; Borah, D.C.; Sarma, J.C. *Tetrahedron Lett.* **2005**, *46*, 1159–1160.

62 Krstenansky, J.L.; Khmelnitsky, Y. *Bioorg. Med. Chem.* **1999**, *7*, 2157–2162.

63 Mitra, A.K.; Banerjee, K. *Synlett* **2003**, 1509–1511.

64 Choudhary, V.R.; Tillu, V.H.; Narkhede, V.S.; Borate, H.B.; Wakharkar, R.D. *Catal. Commun.* **2003**, *4*, 449–453.

65 Stadler, A.; Kappe, C.O. *J. Comb. Chem.* **2001**, *3*, 624–630.

66 Shaabani, A.; Bazgir, A. *Tetrahedron Lett.* **2004**, *45*, 2575–2577.

67 For example see: Lombardi, J.G.; Wiseman, E.H. *J. Med. Chem.* **1974**, *17*, 1182–1188.

68 Balalaie, S.; Arabanian, A.; Hashtroudi, M.S. *Monatsh. Chem.* **2000**, *131*, 945–948.

69 Balalaie, S.; Hashemi, M.M.; Akhbari, M. *Tetrahedron Lett.* **2003**, *44*, 1709–1711.

70 Alper, H.; Amaratunga, S. *J. Org. Chem.* **1982**, *47*, 3593–3595.

71 Gelens, E.; de Kanter, F.J.J.; Schmitz, R.F.; Sliedregt, L.A.J.M.; Van Steen, B.J.; Kruse, C.G.; Leurs, R.; Groen, M.B.; Orru, R.V.A. *Mol. Div.* **2006**, *10*, 17–22.

72 (a) Blackburn, C.; Guan, B.; Fleming, P.; Shiosaki, K.; Tsai, S. *Tetrahedron Lett.* **1998**, *39*, 3635–3638. (b) Blackburn, C. *Tetrahedron Lett.* **1998**, *39*, 5469–5472.

73 Ireland, S.M.; Tye, H.; Whittaker, M. *Tetrahedron Lett.* **2003**, *44*, 4369–4371.

74 Lu, Y.M.; Zhang, W. *QSAR Comb. Sc.* **2004**, *23*, 827–835.

75 Zhang, W.; Tempest, P. *Tetrahedron Lett.* **2004**, *45*, 6757–6760.

76 For example see: Rudi, A.; Goldberg, I.; Stein, Z.; Frolow, F.; Benayahu, Y.; Schleyer, M.; Kashman, Y. *J. Org. Chem.* **1994**, *59*, 999–1003.

77 For example see: Periasamy, M.; Srinivas, G.; Bharathi, P. *J. Org. Chem.* **1999**, *64*, 4204–4205.

78 For example see: Berree, F.; Marchand, E.; Morel, G. *Tetrahedron Lett.* **1992**, *33*, 6155–6158.

79 Hantzsch, A. *Ber. Dtsch. Chem. Ges.* **1890**, *23*, 1474.

80 Shiraishi, H.; Nishitani, T.; Sakaguchi, S.; Ishii, Y. *J. Org. Chem.* **1998**, *63*, 6234–6238.

81 Danks, T.N. *Tetrahedron Lett.* **1999**, *40*, 3957–3960.

82 Nishitani, T.; Shiraishi, H.; Sakaguchi, S.; Ishii, Y. *Tetrahedron Lett.* **2000**, *41*, 3389.

83 Ranu, B.C.; Hajra, A.; Jana, U. *Synlett* **2000**, 75–76.

84 Bora, U.; Saikia, A.; Boruah, R.C. *Org. Lett.* **2003**, *5*, 435–438.

85 Yadav, L.D.S.; Kapoor, R. *Tetrahedron Lett.* **2003**, *44*, 8951–8954.

86 Buckle, D.R.; Rockell, C.J.M.; Smith, H.; Spicer, B.A. *J. Med. Chem.* **1986**, *29*, 2262–2267.

87 Genin, M.J.; Allwine, D.A.; Anderson, D.J.; Barbachyn, M.R.; Emmert, D.E.; Garmon, S.A.; Graber, D.R.; Grega, K.C.; Hester, J.B.; Hutchinson, D.K.; Morris, J.; Reischer, R.J.; Ford, C.W.; Zurenko, G.E.; Hamel, J.C.; Schaadt, R.D.; Stapert, D.; Yagi, B.H. *J. Med. Chem.* **2000**, *43*, 953–970.

88 Appukkuttan, P.; Dehaen, W.; Fokin, V.V.; van der Eycken, E. *Org. Lett.* **2004**, *6*, 4223–4225.
89 Nixey, T.; Tempest, P.; Hulme, C. *Tetrahedron Lett.* **2002**, *43*, 1637–1639.
90 Mont, N.; Teixido, J.; Borrell, J.I.; Kappe, C.O. *Tetrahedron Lett.* **2003**, *44*, 5385–5387.
91 Ouyang, Y.; Koike, K.; Ohmoto, T. *Phytochemistry* **1994**, *36*, 1543–1546.
92 Hagen, T.J.; Narayanan, K.; Names, J.; Cook, J.M. *J. Org. Chem.* **1989**, *54*, 2170–2178.
93 Markgraf, J.H.; Finkelstein, M.; Cort, J.R. *Tetrahedron* **1996**, *52*, 461–470.
94 Benson, S.C.; Li, J.H.; Snyder, J.K. *J. Org. Chem.* **1992**, *57*, 5285–5287.
95 Lindsley, C.W.; Wisnoski, D.D.; Wang, Y.; Leister, W.H.; Zhao, Z.J. *Tetrahedron Lett.* **2003**, *44*, 4495–4498.
96 Varma, R.S.; Dahiya, R. *J. Org. Chem.* **1998**, *63*, 8038–8041.
97 Green, G.R.; Evans, J.M.; Vong, A.K. in: *Comprehensive Heterocyclic Chemistry-II*; Eds., Katritzky, A.R.; Rees, C.W.; Scriven, E.F.V. **1995**, *5*, 469.
98 Wang, X.S.; Shi, D.O.; Tu, S.J.; Yao, C.S. *Synth. Commun.* **2003**, *33*, 119–126.
99 Kaupp, G.; Naimi-Jamal, M.R.; Schmeyers, J. *Tetrahedron* **2003**, *59*, 3753–3760.
100 Devi, I.; Bhuyan, P.J. *Tetrahedron Lett.* **2004**, *45*, 8625–8627.
101 Yadav, L.D.S.; Singh, A. *Synthesis* **2003**, 2395–2399.
102 Kurreck, H.; Huber, M. *Angew. Chem. Int. Ed.* **1995**, *34*, 849–866.
103 Metzger, R.M.; Panetta, C.A. *New J. Chem.* **1991**, *15*, 209–221.
104 Victory, P.; Borrell, J.I.; Vidalferran, A.; Seoane, C.; Soto, J.I. *Tetrahedron Lett.* **1991**, *32*, 5375–5378.
105 Cui, S.L.; Lin, X.F.; Wang, Y.G. *J. Org. Chem.* **2005**, *70*, 2866–2869.
106 Singh, S.P.; Parmar, S.S.; Raman, K.; Stenberg, V.I. *Chem. Rev.* **1981**, *81*, 175–203 and references cited therein.
107 Holmes, C.P.; Chinn, J.P.; Look, G.C.; Gordon, E.M.; Gallop, M.A. *J. Org. Chem.* **1995**, *60*, 7328–7333.
108 Gududuru, V.; Nguyen, V.; Dalton, J.T.; Miller, D.D. *Synlett* **2004**, 2357–2358.
109 Dandia, A.; Sati, M.; Arya, K.; Sharma, R.; Loupy, A. *Chem. Pharm. Bull.* **2003**, *51*, 1137–1141.
110 Hoel, A.M.L.; Nielsen, J. *Tetrahedron Lett.* **1999**, *40*, 3941–3944.
111 Short, K.M.; Ching, B.W.; Mjalli, A.M.M. *Tetrahedron* **1997**, *53*, 6653–6679.
112 Shi, L.; Tu, Y.O.; Wang, M.; Zhang, F.M.; Fan, C.A. *Org. Lett.* **2004**, *6*, 1001–1003.
113 Ju, Y.H.; Li, C.J.; Varma, R.S. *QSAR Comb. Sc.* **2004**, *23*, 891–894.
114 For example see: Hattori, K.; Miyata, M.; Yamamoto, H. *J. Am. Chem. Soc.* **1993**, *115*, 1151–1152.
115 Wei, C.M.; Li, C.J. *J. Am. Chem. Soc.* **2003**, *125*, 9584–9585.
116 Wei, C.M.; Li, Z.G.; Li, C.J. *Org. Lett.* **2003**, *5*, 4473–4475.
117 Hurd, R.N.; Delamater, G. *Chem. Rev.* **1961**, *61*, 45–86.
118 Kindler, K. *Liebigs Ann. Chem.* **1923**, *431*, 187–230.
119 Brown, E.V. *Synthesis* **1975**, 358–375.
120 Zbruyev, O.I.; Stiasni, N.; Kappe, C.O. *J. Comb. Chem.* **2003**, *5*, 145–148.
121 Fields, S.C. *Tetrahedron* **1999**, *55*, 12237–12272.
122 Laschat, S.; Kunz, H. *Synthesis* **1992**, 90–95.
123 Zon, J. *Pol. J. Chem.* **1981**, *55*, 643–646.
124 Manabe, K.; Kobayashi, S. *Chem. Commun.* **2000**, 669–670.
125 Ranu, B.C.; Hajra, A.; Jana, U. *Org. Lett.* **1999**, *1*, 1141–1143.
126 Yadav, J.S.; Reddy, B.V.S.; Madan, C. *Synlett* **2001**, 1131–1133.

18
Microwave-enhanced Radiochemistry

John R. Jones and Shui-Yu Lu

18.1
Introduction

The synthesis, analysis and applications of labeled compounds constitute an area in which basic and applied research go hand-in-hand. Over the last quarter of a century the field has seen considerable expansion as reflected in the emergence of a specific journal (*Journal of Labelled Compounds and Radiopharmaceuticals*) and the publication of the proceedings of international symposiums held at 3-yearly intervals. The formation of the International Isotope Society is also an indication of the increasing importance of isotopes and isotopically labeled compounds.

Within the pharmaceutical industry labeled compounds are used for a variety of purposes, e.g. screening new targets, for binding experiments, for identification of metabolites, in absorption, distribution and excretion studies and for quantifying concentrations in target organs [1]. Understanding reaction mechanisms are greatly assisted by the availability of suitably labeled compounds.

As most of the compounds used in the above areas are organic, it follows that most of the isotopes used are those of hydrogen and carbon with oxygen, nitrogen and the halogens being used in a minority of cases [2, 3]. Whilst radioisotopes are preferred in terms of sensitivity the problems associated with the use of radioactivity (separate laboratory facilities required as well as trained personnel) and the production of radioactive waste (costs of storage and disposal) means that increasing use is being made of compounds labeled with stable isotopes, especially as some of the analytical methods, notably nuclear magnetic resonance (NMR) spectroscopy and mass spectrometry are becoming more sensitive and versatile [4–7]. Indeed more and more pharmaceutical companies are developing policies that require drug candidates to be labeled (separately) with both stable and radioactive isotopes. Consequently, there is the added benefit to those interested in the preparation of radio-labeled compounds that the information obtained can be used for preparing the corresponding compounds labeled with a stable isotope. Indeed, for those in academic centers it is customary practice to label the target compound, first with a stable isotope, and then with the radioactive isotope.

In labeling a compound with a stable isotope and with a radioactive isotope there

---

Microwaves in Organic Synthesis, Second edition. Edited by A. Loupy
Copyright © 2006 WILEY-VCH Verlag GmbH & Co. KGaA, Weinheim
ISBN: 3-527-31452-0

are several differences. The former tends to be done on the mg-g scale whereas the radiolabeled compound is usually prepared on the microgram to milligram scale. Purification of the former can be achieved by recrystallization or distillation, as well as by one or more chromatographic methods, whereas radiolabeled compounds are invariably purified by radio-chromatographic methods.

Within the pharmaceutical industry, there is pressure to produce more drug candidates in a shorter time (greater efficiency). This trend makes it necessary to devise new and more efficient methods for preparing labeled compounds. There is also a growing need for a drug candidate to be labeled with several isotopes i.e. multi-deuterated or multi-tritiated rather than mono-deuterated or mono-tritiated. Finally, and as mentioned already, there is a need at the end of particular study, to convert the radioactive waste to a form which can then be re-used, even though the specific activity may not be as high as in the original study. The movement towards a more environmentally friendly chemical industry, with faster, more selective and efficient synthesis, with greatly reduced levels of waste, both radioactive and otherwise, is gathering momentum and it is one in which microwaves are destined to play an important role.

Whilst the work that we focus on in the first part of this chapter concerns the preparation of tritium- and inevitably deuterium-labeled compounds, examples are given where the benefits can also be applied to the carbon (^{11}C, ^{13}C and ^{14}C)-labeled area [8]. Also discussed is the use of microwaves in the synthesis of radio-pharmaceuticals labeled with positron emitters, such as carbon-11 ($t_{1/2} = 20.4$ min) and fluorine-18 ($t_{1/2} = 109.7$ min). The short half lives of these radioisotopes, together with the requirements for high radiochemical yield (RCY), radiochemical purity (RCP) and specific activity (SA) can benefit from the advantages that microwaves provide [8, 9].

18.1.1
Methods for Incorporating Tritium into Organic Compounds

The standard work of Evans [2], as well as a survey of the papers produced in the *Journal of Labelled Compounds and Radiopharmaceuticals* over the last 20 years, show that the main tritiation routes are as given in Table 18.1. One can immediately see, that unlike most ^{14}C-labeling routes, they consist of one-step and frequently involve a catalyst, which can be either homogeneous or heterogeneous. One should therefore be able to exploit the tremendous developments that have been made in catalysis in recent years to benefit tritiation procedures. Chirally catalyzed hydrogenation reactions (Knowles and Noyori were awarded the Nobel price for chemistry in 2001 for their work in this area, sharing it with Sharpless for his work on the equivalent oxidation reactions) immediately come to mind. Already optically active compounds such as tritiated L-alanine, L-tyrosine, and L-dopa *etc* have been prepared in this way.

The development of phase transfer catalysis, of supercritical fluids, of ionic liquids and of course, new reagents, should also have considerable potential in the labeling area. Furthermore, there is the possibility of combining these ap-

Tab. 18.1. Main tritiation procedures.

	Reaction	Example
1	Hydrogen isotope exchange	
	(a) Base-catalyzed	$C_6H_5COCH_3 + OH^- \xrightarrow{HTO} C_6H_5COCH_2T$
	(b) Acid-catalyzed	toluene $\xrightarrow[HTO]{H_3^+O}$ toluene-T
	(c) Metal-catalyzed	naphthalene $\xrightarrow[HTO]{PtO_2}$ naphthalene-T
2	Hydrogenation	
	(a) Homogeneous	$PhCH=CHCOOH \xrightarrow[catalyst]{T_2} PhCHTCHTCOOH$
	(b) Heterogeneous	
3	Aromatic dehalogenation	4-X-toluene $\xrightarrow[catalyst]{T_2}$ 4-T-toluene, X = Br, Cl, I
4	Methylation	benzimidazole-NH $\xrightarrow[CT_3I]{NaH}$ benzimidazole-NCT$_3$
5	Borohydride reduction	$PhCHO \xrightarrow{NaBT_4} PhCHTOH$

proaches with energy-enhanced conditions – in this way marked improvements can be expected.

18.1.2
Problems and Possible Solutions

The tritiation procedures given in Table 18.1 all have serious limitations or disadvantages. Thus, for all three hydrogen isotope exchange reactions, HTO is used as the donor. For health and safety reasons 50 Ci mL^{-1} (1 Ci = 37 GBq), corresponding to close on 2% isotopic abundance, is the highest specific activity that we have used and this inevitably limits the maximum specific activity of the products that can be obtained *via* these routes.

The success of the base-catalyzed hydrogen isotope exchange reaction very much depends on the acidity of the carbon acid – the weaker it is (higher pK_a) the stronger the base required to abstract the proton to form the reactive carbanion [10]. Within the pharmaceutical industry there is a reluctance to use tritium labeled compounds produced *via* this route – this is because of the dangers of "back exchange". If the compound to be labeled contains several acidic sites then the label will no longer be incorporated at one site. This may or may not be a disadvantage depending on what use is to be made of the labeled compound.

Although many pharmaceutical compounds are unable to withstand harsh acidic conditions a surprisingly large number of compounds have been labeled *via* this route. Werstiuk [11] for example has reduced the acid concentration but increased the temperature, one effect more than compensating for the other – however, the time required is frequently very long, extending into days. Ion exchange resins, both acid and base forms, can be used to overcome separation problems [12, 13].

Since the pioneering work of Garnett and Long [14, 15], much progress has been made in increasing the selectivity of one-step metal-catalyzed hydrogen isotope exchange reactions. $RhCl_3$, and iridium(I) catalysts of the type $[Ir(COD)(L)_2]PF_6$ (COD = *cis,cis*-1,5-cyclooctadiene) have been successfully used by, amongst others, Heys, Hesk, Lockley, Salter and their coworkers [16–19]. In some of these studies, HTO has been replaced by T_2 as donor so that compounds of very high specific activity can be obtained. Myasoedov and colleagues [20, 21] have also made extensive use of high temperature solid-state catalytic isotope exchange (HSCIE) for the tritiation of a wide range of organic compounds at high specific activity.

Up until recently, hydrogenation reactions with T_2 were performed on glass gas lines but this is now frowned upon by the environmental and health and safety inspectorate. Fortunately, there are two commercial instruments available, one manufactured in Switzerland and the other in the USA, which are entirely metallic and use an uranium 'getter' for storing T_2 gas; gentle heating allows a pre-determined volume of gas to be transferred to the reaction vessel and on completion of the reaction any excess can be returned to a secondary bed for storage and re-use. T_2 gas is relatively inexpensive and available at 100% isotopic incorporation (specific activity of 56 Ci mmol^{-1}). The main disadvantage now is that it is sparingly soluble in many organic solvents with the result that the catalyzed reactions, under both homogeneous and heterogeneous conditions, are frequently very slow.

Aromatic dehalogenation suffers from the disadvantage that only 50% of the tritium is incorporated, the rest appearing as waste. This situation is even more marked for borotritide reductions but the problem can be overcome by using some of the new tritide reagents that have recently become available as a result of the synthesis of carrier-free lithium tritide (Scheme 18.1) [22]. Their reactivity can be fine-tuned through the elements (e.g. B, Al, Sn) to which the tritium is attached and by the electronic and steric nature of the substituents at the central atom.

Tritiated methyl iodide has the advantage that three tritium atoms can be incorporated in one step so that compounds with a specific activity close to 80 Ci mmol^{-1} can be prepared. CT_3I is available from commercial sources and being a low boiling liquid needs very careful handling. It is stable for short periods, consequently there is a need for new methylating agents that offer greater flexibility.

n-BuLi + T$_2$ $\xrightarrow{\text{TMEDA}}$ LiT + n-BuT
hexane

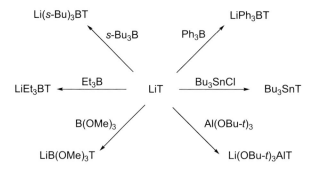

Scheme 18.1. Preparation of tritide reagents from LiT.

18.1.3
Use of Microwaves

The possibility of accelerating chemical reactions through the application of a source of energy, be it in the form of external radiation or *via* an electric or magnetic field, has a long history. The first radiation induced method of labeling was reported by Wilzbach [23] in 1957 and was subsequently described as the Wilzbach gas exposure method. Typically 0.5–4.0 g of substrate was exposed to 7–14 Ci of T$_2$ gas for between 3 and 10 days, leading to 20–600 mCi incorporation with specific activities of the product being in the 1–100 mCi g^{-1} region. This work was done long before the development of ^{3}H NMR spectroscopy [24] so that the pattern of labeling could not be ascertained. Furthermore purification of the products by techniques such as high performance liquid chromatography (HPLC) was not possible. The hopes of the author that 'the availability of T$_2$ gas at low cost and the high levels of radioactivity attainable, even in materials of complex structure, combine to make exposure to tritium gas an attractive method for the preparation of tritium labeled compounds' were not therefore fully realized. In the course of time, more attractive tritiation methods were developed. Nevertheless, the work stimulated a great deal of interest with the objective of minimizing radiation damage and increasing the specific activity of the labeled substrate. Many improved versions of the Wilzbach method were reported, of which the following are best known:

1. recoil labeling (which does, in fact, predate the Wilzbach method);
2. electrical discharge method,
3. low pressure method,
4. ion beam method,

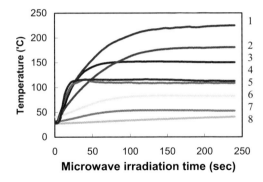

Solvent	b.p (°C)	ε_s	Tanδ
1 DMSO	189	47	0.825
2 DMF	153	36.7	0.161
3 BuOH	118	n/a	n/a
4 H$_2$O	100	80.4	0.123
5 EtOH	78	24.6	0.941
6 CHCl$_3$	61	4.8	0.091
7 CH$_2$Cl$_2$	40	9.1	0.042
8 1,4-dioxane	101	n/a	n/a

Fig. 18.1. Microwave heating profile of organic solvents using a Prolabo Synthewave S402 microwave instrument (5 cm^3 at 300 W).

5. microwave discharge activation (MDA) method,
6. adsorbed tritium method.

Although chemists very frequently refer to two papers [25, 26] published in 1986 as the start of the 'microwave-enhanced era' as an efficient new procedure for organic synthesis, it was Ghanem and Westermark [27], way back in 1960, who first suggested that input of microwave power into a system containing tritium could accelerate the tritiation kinetics. A number of antibiotics, local anaesthetics and amino acids were amongst the early compounds labeled. Wolf et al. [28–30] improved the technique by circulating the tritium gas within the labeling system and successfully tritiated a number of tripeptides, amongst other compounds. Essentially the same system was used by Peng et al. [31, 32] to tritiate a number of steroids. The apparatus was relatively simple to construct and the MDA method was seen to be superior to the other radiation-induced methods of labeling, requiring a short reaction time and using a small amount of tritium gas. However, the chemistry within this kind of microwave plasma is complex with various tritium species being formed e.g. T^+, T_3^+, T and T^-. Consequently, there are several reaction mechanisms taking place concurrently so that the overall tritiation is characterized by low selectivity; extensive purification is also necessary. Being a radiation-induced method the work can not be applied to deuteriation studies, in sharp contrast to the more recently developed microwave-enhanced methods. An excellent account of radiation-induced methods of labeling has been given by Peng [33].

The exponential growth in microwave-enhanced synthetic organic chemistry (see Fig. 1 in Ref. 8 and 34 and now in Chapter 4 of this book) which Gedye and Giguere and their colleagues initiated in 1986 may come to represent one of the most significant events in the development of chemistry over the last half-century. Whilst microwave dielectric heating has been widely used in the food processing area for many years chemists have been slow in recognising the potential, partly because the necessary theory is usually taught as part of a physics degree course and partly because of the lack of suitable instrumentation for making quantitative, as distinct from qualitative, measurements. This second problem will soon not exist as several powerful computer-controlled instruments have come to market within the last few years.

A detailed account of the theory behind the interaction of microwaves with matter is beyond the scope of this Chapter but there are several excellent and recent reviews available [34–38 and other chapters in this book].

18.1.4
Instrumentation

Most of the early work was carried out using the multi-mode or commercial domestic microwave ovens. At Surrey we first used a Matsui M169BT unit (750 W), and later on a Prolabo Synthewave S402 focused microwave reactor (300 W) was acquired. In our earlier studies, the reagents were contained in a thick walled glass tube sealed at the top but experience showed that a small conical flask (20 cm^3) fitted with a septum stopper was well suited for most of the reactions. It was placed in a beaker containing vermiculite and slowly rotated during the course of the irradiation. To avoid undue pressure build-up the vessel can also be left un-stoppered. For parallel reactions, the Radley's RDT 24 place PTFE carousel reaction station together with Pyrex glass tubes (25 cm^3, o.d. = 15 mm) were placed on the turntable of the microwave oven. The reaction vessels for the Synthelabo S402 are quartz tubes of different sizes. For smaller scale reaction specially designed Pyrex tubes (o.d. = 9 mm) can be inserted inside the standard quartz tube (o.d. = 25 mm).

More and more researchers are using the mono-mode instruments and the manufacturers are bringing the second generation of microwave reactors (e.g. The Discover focused microwave synthesis systems by CEM, the 521 Accelerated Microwave Heater by Resonance Instruments, the Initiator by Biotage, and the MicroSYNTH Labstation by Milestone) to the market; these are more effective and user friendly. New standard glassware, capable of withstanding high temperature and pressure is also being developed and marketed.

Investigations of the mechanisms and kinetics of microwave-enhanced reactions is at an early stage but the development [39] of a microwave reactor/ultraviolet/visible hybrid spectrometer will be a valuable tool. It has already been used to study the formation of benzimidazole from the reaction of 1,2-diaminobenzene and formic acid.

Real-time *in situ* Raman analysis of microwave-assisted organic reactions has also been reported [40, 41]. Raman spectroscopy provides a combination of high selectivity along with the ability to conduct analysis directly through the wall of glass reaction vessels and in monitoring the Knoevenegel condensation reaction of salicylaldehyde and benzyl acetoacetate to form 3-acetylcoumarin, observations were made inside the microwave compartment of the synthesizer.

18.2
Microwave-enhanced Tritiation Reactions

18.2.1
Hydrogen Isotope Exchange

This is one of the most versatile reactions known as it can be catalyzed by acids, bases and metals as well as induced by photochemical and other means. It is also amongst the simplest of organic reactions so that studies in this area have greatly improved our understanding of reaction mechanisms. It is also one of the best methods for introducing deuterium/tritium into organic compounds. The knowledge gained from basic research studies can therefore be put to good use within a fairly short time interval.

Koves [42–44] was the first to show that acid-catalyzed aromatic hydrogen isotope exchange could benefit from the use of microwaves and shortly afterwards we showed that both heterogeneous and homogeneous metal-catalyzed reactions could also be greatly accelerated [45, 46]. It was during these studies that we recognized that further benefits could accrue by converting the organic compound (usually neutral) to an ionic form e.g. by protonating the $-NH_2$ group. These preliminary studies were then extended to take in a large number of nitrogen-containing heterocyclic compounds – here protonation on an $-NH_2$ group or the ring-based nitrogen was possible. This work was prompted by the earlier findings of Werstiuk employing a high temperature, dilute acid, approach to the deuteration of many organic compounds. Extensive labeling could be obtained provided the heating times were long (typically 12–50 h). Rarely can compounds of pharmaceutical interest withstand such demanding conditions. Furthermore, the development of new, combinatorial chemistry, requires that labeling reactions should be rapid so that high sample throughput can be achieved.

The results for mono-, di- and fused ring substituted pyridines showed, as for the earlier study on the microwave-enhanced deuteration of o-toluidine, that by first forming the hydrochloride salt extensive labeling could be achieved within 20 min. Furthermore the reaction products were easily isolated and of high purity, a frequently noticed feature of these reactions. The reaction mechanism is represented in Scheme 18.2, the deuterium being inserted in the *ortho* and *para* positions. The time taken to reach a pre-determined temperature depends greatly on the polarity of the solvent used and as the results in Fig. 18.1 show this can be

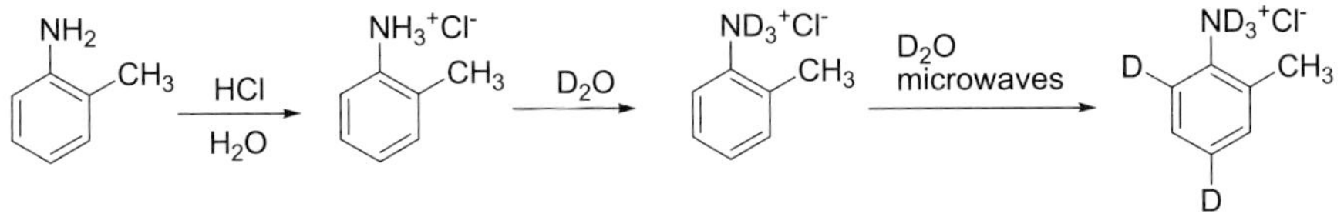

Scheme 18.2. Deuteration of o-toluidine hydrochloride under microwave conditions.

Tab. 18.2. Pattern of labeling for a model carbohydrate under the action of ultrasound.

Position	THF	DMTHF	THP	1,3-Dioxane	1,4-Dioxane	DME
C_2	27	28	20	25	<1	15
C_3	73	84	59	60	62	<1
C_4	89	93	59	76	75	29

THF = tetrahydrofuran, DMTHF = 2,5-dimethyltetrahydrofuran,
THP = tetrahydropyran, DME = 1,2-dimethoxyethane

altered either by changing the solvent or adding a co-solvent. For labeling purposes a D_2O (or HTO)–DMSO mixture has great attraction as the heating time can now be reduced to a matter of 1–2 min.

The choice of solvent can also be beneficial in another respect. This possibility was highlighted by the findings of Cioffi on the Raney Nickel catalyzed hydrogen-deuterium exchange of a model carbohydrate (1-O-methyl-β-D-galactopyranoside), but under ultrasonic irradiation (Table 18.2) [47]. Extensive deuteriation at the C-4 position occurred for a series of etheral solvents, the C-3 position was deuterated by seven solvent systems and the C-2 position deuterated less extensively, also by seven solvent systems. For 1,4-dioxane-D_2O no labeling at the C-2 position occurred and for 1,2-dimethoxyethane-D_2O no C-3 labeling was observed.

We chose the microwave-enhanced Raney Nickel catalyzed hydrogen isotope exchange of indole and N-methylindole as our substrates and D_2O, CD_3COCD_3, CD_3OD and $CDCl_3$ as the solvents. The thermal reaction had already been the subject of a recent study [48]. The microwave-enhanced method was some 500-fold faster than the corresponding thermal reaction (at 40 °C). Furthermore the pattern of labeling (Scheme 18.3) varied with the choice of solvent. Thus in the case of indole itself the more polar solvents such as D_2O and CD_3OD give rise to general labeling whilst $CDCl_3$, for example, give very regiospecific labeling. There is at this stage no indication that the pattern of labeling in the microwave-enhanced reactions is different in any way from that observed for the thermal reactions. However inherent in both sets of results are mechanistic features worthy of more detailed investigation.

The benefits of using ionic compounds in microwave-enhanced reactions (see Chapter 7 in this book) led us to explore the possibility of using ionic solvents i.e. ionic liquids, as donors for both deuterium and tritium. Whilst D_2O is now relatively inexpensive and available at high isotopic enrichment, tritiated water is usually employed, for safety reasons, at low isotopic incorporation (we typically use

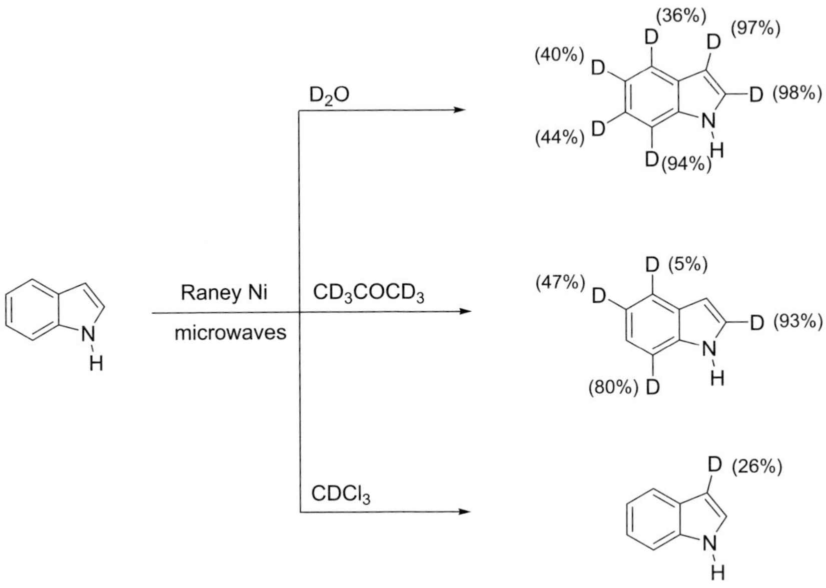

Scheme 18.3. Pattern of deuteration at different sites for indole using a variety of solvents and microwave irradiation.

HTO at 5 or 50 Ci mL^{-1} specific activity corresponding to 0.2–2% isotopic incorporation). This is a serious limitation when there is a need to provide compounds at high specific activity.

Ionic liquids [49–52] are liquids containing only ions, and are fluid at or close to, room temperature; those with higher melting points are called molten (or fused) salts. They are non-volatile and many are both air- and moisture-stable as well as being good solvents for a wide range of both inorganic and organic molecules; they frequently permit unusual combinations of reagents to be brought into the same phase.

Many ionic liquids are based on N,N′-dialkylimidazolium cations (BMI) which form salts that exist as liquids at, or below, room temperature. Their properties are also influenced by the nature of the anion e.g. BF_4^-, PF_6^-. The C-2(H) in imidazole is fairly labile but the C-4(H) and the C-5(H) are less so. Under microwave-enhanced conditions, it is therefore possible to introduce three deuterium atoms (Scheme 18.4). As hydrogen isotope exchange is a reversible reaction, this means that the three deuterium atoms can be readily exchanged under microwave irradiation. For storage purpose, it might be best to back-exchange the C-2(D) so that the 4,5-[^{2}H$_2$] isotopomer can be safely stored as the solid without any dangers of deuterium loss. The recently reported microwave-assisted preparation of a series of ambient temperature ionic liquids may well accelerate their use in this area [53]. We have already carried out some deuteration studies with the 2,4,5-[^{2}H$_3$]-BMI-BF$_4$ ionic liquid and found that for reasonably volatile organic compounds these

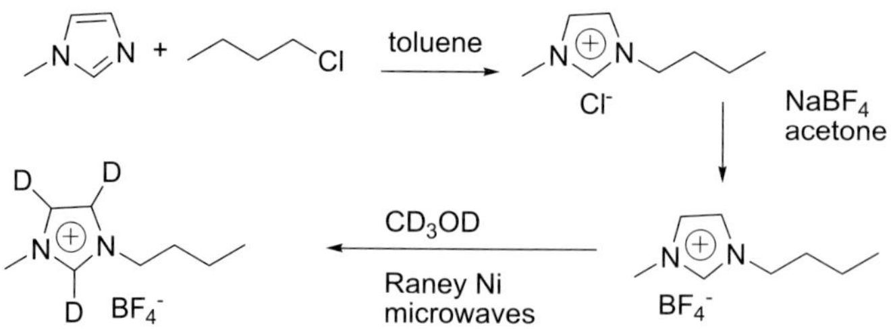

Scheme 18.4. Preparation of 2,4,5-[^{2}H$_3$]-BMI-BF$_4$ ionic liquid.

can be separated off by simple evaporation leaving the deuterated ionic liquid available for repeated use i.e. until all the deuterium has been replaced by hydrogen [54]. Extension of the investigations to the tritium area will require the analogue to be prepared using Raney Nickel and T$_2$ gas under thermal conditions.

Ease of separation of tritiated products from a reaction medium is an important feature in the choice of labeling procedure. Sometime ago we used polymer-supported acid and base catalysts [12, 13] to good effect and with the current interest in Green Chemistry one can expect to see more studies where the rate accelerations observed under microwave-enhanced conditions are combined with the use of solid catalysts such as Nafion, or zeolites.

Several successful examples of microwave-enhanced hydrogen-deuterium exchange have appeared in the literature and these can easily be adapted to the corresponding hydrogen-tritium exchange reaction. Irradiation of estrone in CF$_3$COOD for a short time (2 × 4 min) gave [2, 4, 16, 16-^{2}H$_4$]estrone (Scheme 18.5) in 95% yield, far better than what was achieved by either sonication or prolonged refluxing [55].

Fodor-Csorba et al. [56] adopted a very similar procedure, but now on a preparative scale (i.e. a few grams) to rapidly deuterate a number of ketones that are useful intermediate in the synthesis of different classes of liquid crystals where the deuterated group is directly linked to the mesogenic core. 2-Octanone was one of the successful examples (Scheme 18.6). With an irradiation time of 15 min, five deuterium atoms at 96% isotopic incorporation were incorporated. When thermal conditions are used refluxing for 12 h is typical.

Selective C-2(H)deuteriation of a number of heterocyclic compounds (imidazole,

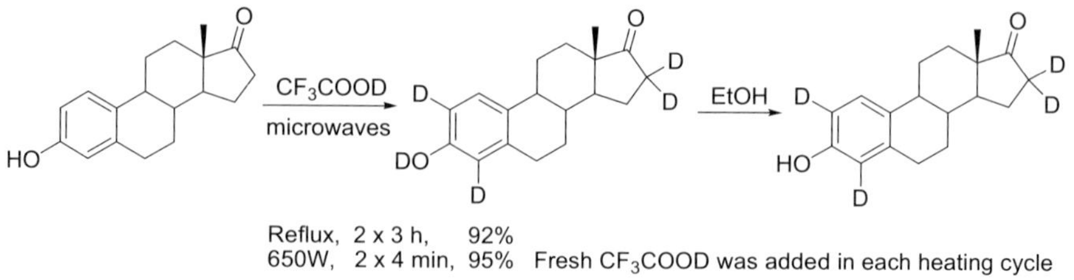

Reflux, 2 x 3 h, 92%
650W, 2 x 4 min, 95% Fresh CF$_3$COOD was added in each heating cycle

Scheme 18.5. Expedient microwave deuteration of estrone in CF$_3$COOD.

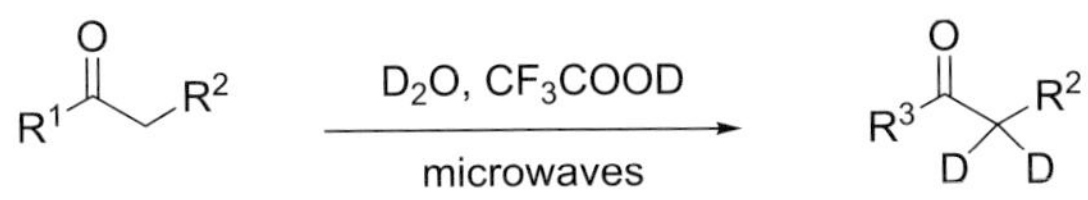

R¹−C(=O)−CH(R²) →[D₂O, CF₃COOD, microwaves] R³−C(=O)−C(D)(D)−R²

R¹	R²	R³	
CH₃	(CH₂)₄CH₃	CD₃	90W, 15-20 min
CH₃	CH₂CH₃	CD₃	reflux, 8-9 hrs
CH₃	CH(CH₃)₂	CD₃	
CH₃	4'-(4-n-heptylbiphenyl)	CD₃	
4'-biphenyl	(CH₂)₄CH₃	biphenyl	

Scheme 18.6. Microwave-assisted deuterium exchange reactions for preparation of various ketones.

thiazole etc.) as well as the corresponding C-8(H)deuteriation of purines is known to occur in D_2O at elevated temperatures over the course of 12–24 h but on microwave irradiation the reaction is greatly accelerated with reaction times as low as 5 min sometimes sufficient. Having used a number of model compounds Masjedizadeh et al. [57] were able to extend their investigations and successfully deuterate bleomycin A_2, an anti-tumour antibiotic. Irradiation at 165 °C for 6 min was sufficient. It is not strictly true to say that these reactions are uncatalyzed as the reaction usually involves the protonated form of the substrate and OD^- (from the D_2O).

Cioffi et al., who previously has studied the Raney Nickel catalyzed hydrogen-deuterium exchange of a model carbohydrate (1-O-methyl-β-D-galactopyranoside) under ultrasonic irradiation, have extended their investigations into the microwave area [58]. Using a D_2O-tetrahydrofuran (THF) solvent mixture the above compound could rapidly (<10 min) be deuterated with 86% ^2H incorporation at one site and an even better incorporation (91%) for sucrose; other sites were labeled less extensively.

Metal catalyzed hydrogen isotope exchange has been extensively used for many years to tritiate a whole range of organic compounds and once ^3H NMR spectroscopy was developed it became an increasingly attractive method. Of the many catalysts used pre-reduced PtO_2 has found most appeal; in most cases elevated temperatures (>100 °C) and long reaction times (typically 12–48 h) are employed. A good example concerns the compound, SCH388714, which was satisfactorily tritiated after heating at 100–140 °C for 48 h. The corresponding microwave-enhanced procedure could be accomplished within 5 min with equally good tritium corporation (Scheme 18.7). The short reaction time also ensures a better radiochemical purity [59].

The homogeneous tristriphenylphosphine ruthenium (II) catalyst [60, 61] effects nearly specific hydrogen-tritium exchange in primary alkanols at the α-methylene group provided the reaction time is kept short (∼0.5 h). Prolonged heating resulted in the β-methylene positions also becoming tritiated. In a recent deuterium labeling study [62] the authors were able, once again, to highlight the time reductions

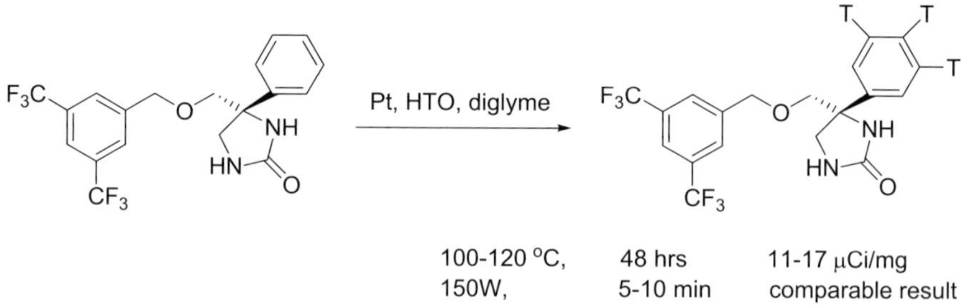

Scheme 18.7.

Scheme 18.7. Tritium labeling of SCH388714 under thermal and microwave conditions.

Conditions: Pt, HTO, diglyme

Thermal: 100–120 °C, 48 hrs, 11–17 µCi/mg
Microwave: 150W, 5–10 min, comparable result

that can be achieved by microwave irradiation. Furthermore, optically active primary alcohols could be deuterated without any racemization. Using a similar procedure regioselective deuteriation of a number of primary and secondary amines was also achieved.

18.2.2
Hydrogenation

Neither tritium or deuterium gas, with zero dipole moments, can be expected to interact with microwave radiation. Their low solubilities are seen as a further disadvantage. Our thoughts therefore turned towards an alternative procedure, of using solid tritium donors and the one that has found most favor with us is formate, usually as the potassium, sodium or ammonium salt. Catalytic hydrogen transfer of this kind is remarkably efficient as the results for α-methylcinnamic acid show [63]. The thermal reaction, when performed at a temperature of 50 °C, takes over 2 h to come to equilibrium whereas the microwave-enhanced reaction is complete within 5 min. A further advantage is that more sterically hindered alkenes such as α-phenylcinnamic acid which are reduced with extreme difficulty when using H_2 gas and Wilkinson's catalyst are easily reduced under microwave-enhanced conditions.

Formate can only donate one hydrogen, the other presumably coming from traces of water present in the solvent or on the surface of the reaction vessel. With the need to increase the specific activity of the product our thoughts turned to the preparation of a di-formate salt and as a first example we synthesised the di-formic acid salt of tetramethylethylenediamine (TMEDA).

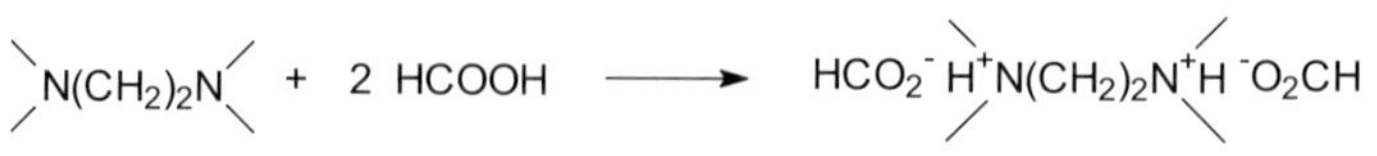

$$\text{N(CH}_2\text{)}_2\text{N} + 2\ \text{HCOOH} \longrightarrow \text{HCO}_2^-\ \text{H}^+\text{N(CH}_2\text{)}_2\text{N}^+\text{H}\ ^-\text{O}_2\text{CH}$$

The tritiated version could be prepared from tritiated formic acid which we had prepared at high specific activity (2.5 Ci mmol^{-1}) by a metal-catalyzed hydrogen-tritium exchange procedure using T_2 gas. The material can be stored either as a solid or as a solution; if the latter any release of tritium by back exchange can be easily monitored by ^{3}H NMR spectroscopy. In our experience very little exchange occurs over several weeks of storage [64].

The use of tritiated formates has further benefits. Firstly the exact amount necessary for 100% hydrogenation can be added so the problems associated with the use of excess T_2 gas are avoided and virtually no radioactive waste is produced. Secondly, the pattern of labeling can be easily varied, as illustrated for the case of cinnamic acid (using deuterium rather than tritium). When formate is used in D_2O there are three possible combinations: $H_2O + DCO_2^-$, $D_2O + HCO_2^-$, and $D_2O + DCO_2^-$ (or the DCO_2D salt of TMEDA on its own). The ^{2}H NMR spectra show that three isotopomers can be prepared – $C_6H_5CHDCH_2COOH$, $C_6H_5CH_2CHDCOOH$ and $C_6H_5CHDCHDCOOH$.

As of now no details of the synthesis of optically active tritiated compounds produced under microwave-enhanced conditions have been published. Another area of considerable interest would be the study of solvent effects on the hydrogenation of aromatic compounds using noble-metal catalysts as considerable data on the thermal reactions is available [65]. Comparison between the microwave and thermal results could then provide useful information on the role of the solvent, not readily available by other means.

Derdau [66] has given an excellent example of the potential of catalytic hydrogen transfer in the ^{2}H/^{3}H labeling area. Piperidines, piperazines and tetrahydroisoquinolines are widely used as building blocks in the pharmaceutical drug design area. Catalytic hydrogenation of pyridines, pyrazines and isoquinolines in an especially direct approach to these compounds but in most cases harsh experimental conditions – high pressures and temperatures – are necessary. On the other hand for the catalytic hydrogen transfer – Pd/C is the catalyst and ammonium formate the in situ hydrogen source. When using $ND_4^+DCO_2^-$ and various substrates good deuterium incorporation was achieved over the course of 12–18 h but in the one case where microwave irradiation was used the desired product was achieved after 20 min at 80 °C, in 86% yield and a comparable deuteriation level.

18.2.3
Aromatic Dehalogenation

As far as preparing tritiated (and deuterated) compounds are concerned aromatic dehalogenation is second only to hydrogenation in importance. Furthermore it suffers from the same disadvantages e.g. slow rates which are caused in part by the poor solubility of the T_2 gas in many organic solvents. The situation is however worse as only 50% of the tritium is introduced into the desired product. Once again we modified the classical dehalogenation by replacing D_2/T_2 gas with labeled formates [67]. There were some previous examples [68–70] of the use of formates in dehalogenation reactions but none of these describe the utility of these agents for labeling organic compounds under the influence of microwave irradiation.

The N-4-picolyl-4-halogenobenzamide system (Scheme 18.8) was chosen as the basic substrate structure because

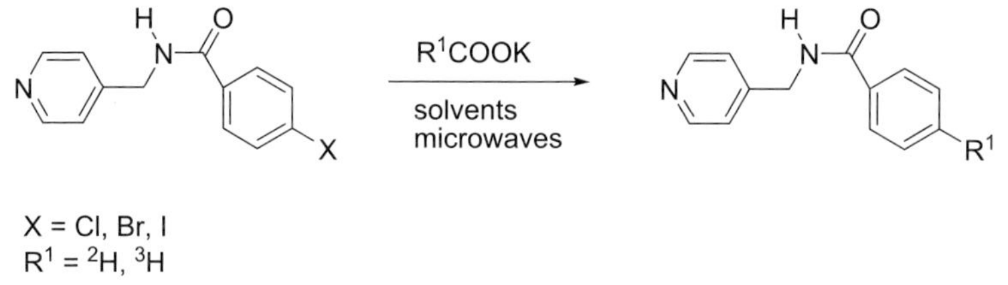

18 Microwave-enhanced Radiochemistry

Scheme 18.8. Dehalogenation of N-4-picolyl-4-halogenobenzamide compounds.

X = Cl, Br, I
R^1 = ^{2}H, ^{3}H

1. the Cl, Br, and I derivatives were all easily synthesized;
2. satisfactory purification procedures were available;
3. they yield strong pseudomolecular ions in both positive and negative-ion HPLC–MS; and
4. they have simple NMR spectra.

Under the experimental conditions dehalogenation proved to be extremely rapid and was complete within 1 min. This contrasts with the 90–270 min at 100 °C required for thermal debromination of 2-bromonaphthalene. No dehalogenation takes place in the absence of the formate donor and when the deuterium is located in the co-solvent rather than the donor (i.e. HCOOK + D$_2$O) hardly any deuterium incorporation takes place. Another interesting observation was that the % deuterium incorporation was always lower when protic solvents, for example alcohols, rather than aprotic solvents, for example dimethyl sulfoxide (DMSO), were used. These are interesting findings which are valuable for proposed tritiation studies.

18.2.4
Borohydride Reductions

One of the most attractive features of borohydride reductions is that under microwave-enhanced conditions they can be performed in the solid state, and rapidly. We were attracted by the work of Loupy et al. [71], and in particular Varma and Saini [72, 73] who have shown that irradiation of a number of aldehydes and ketones in a microwave oven in the presence of alumina doped NaBH$_4$ for short periods of time, led to rapid reduction (0.5–2 min) in good yields (62–93%). In our study [74], seven aldehydes and four ketones were reduced (Table 18.3). Again reduction was complete within 1 min, the products were of high purity (>95%) and of high isotopic incorporation (95%, same as the NaBD$_4$), and the reactions completely selective.

So far this remains the only microwave-enhanced borohydride deuteriation study. Corresponding tritiation studies can be anticipated in due course, especially with the wider range of tritiated reducing agents, referred to previously, becoming available. Significant reductions in radioactive waste can be anticipated.

Many borohydrides are highly unstable and have to be used as freshly prepared ethereal solutions. However there are instances where the polymer-supported ver-

Tab. 18.3. Borodeuteride reduction of carbonyl compounds under microwave conditions (750 W, 1 min).

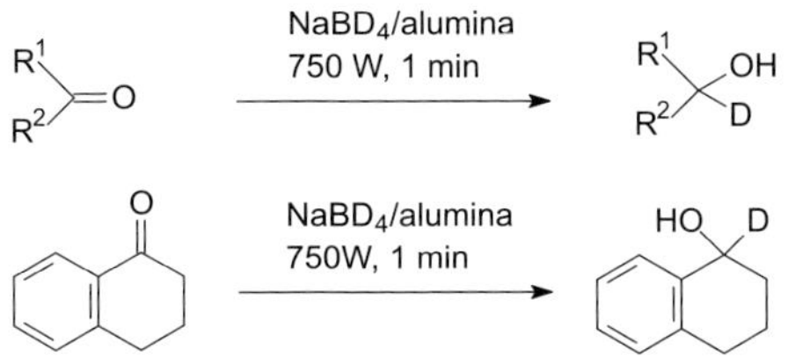

$$R^1R^2C=O \xrightarrow{\text{NaBD}_4/\text{alumina}, 750\text{ W}, 1\text{ min}} R^1R^2C(OH)(D)$$

$$\text{1-tetralone} \xrightarrow{\text{NaBD}_4/\text{alumina}, 750\text{W}, 1\text{ min}} \text{1-tetralol (HO, D)}$$

Aldehydes $R^2 = H$

R^1	Ph	4-NO$_2$C$_6$H$_4$	trans-PhCH=CH	2,4,6-(OMe)$_3$C$_6$H$_2$	1-naphthyl	2-naphthyl	PhCH$_2$
Yield (%)	37	37	77	83	85	89	68

Ketones

R^1	Ph	4-NO$_2$C$_6$H$_4$	3-ClC$_6$H$_4$	5-methyltetralone	
R^2	Me	Me	Me	–	
Yield (%)	82	68	86	87	

sions are more stable e.g. an Amberlyst anion exchange resin supported borohydride and cyanoborohydride [75], polyvinylpyridine supported zinc borohydride [76] and the corresponding zirconium borohydride [77]. Such compounds, in their labeled forms, should turn out to be very useful.

18.2.5
Methylation Reactions

Methyl iodide is the most widely used methylating agent and is the favored route to CT$_3$- and CD$_3$-containing compounds. As it can be rapidly synthesized (from CH$_4$) the [^{11}C]-isotopomer also finds wide application (see Section 18.4.1) in the rapidly expanding positron emission tomography (PET) area [78]. Nevertheless it is generally agreed that [^{11}C]methyl triflate is a better methylating agent – it is more reactive and less volatile so that one does not require cooling for trapping or heating for reaction to take place.

We realized that using a low boiling liquid in a microwave environment, even on a small scale, did not constitute 'best practice' and as for hydrogenations our thoughts turned to using formates in a modified Eschweiler-Clarke reaction [79–81] and successfully methylated a number of primary and secondary amines under microwave conditions (Scheme 18.9) [82].

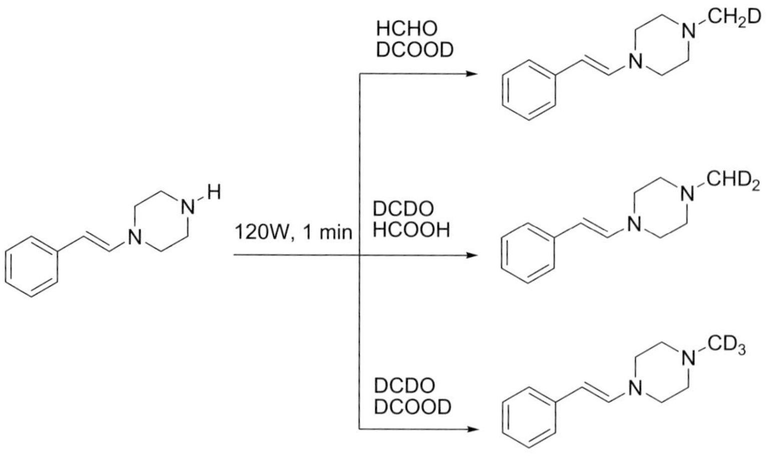

Scheme 18.9. Microwave-enhanced methylation using Eschweiler–Clarke reaction.

The methylation of secondary amines works better than for primary amines because there is no competition between the formation of mono- or di-methylated products. The best results for the microwave-enhanced conditions were obtained when the molar ratios of substrate to formaldehyde to formic acid were 1:1:1, so that the amount of radioactive waste produced is minimal. The reaction can be carried out in neat form if the substrate is reasonably miscible with formic acid–formaldehyde or in DMSO solution if not. Again the reaction is rapid – it is complete within 2 min at 120 W microwave irradiation compared to longer than 4 h under reflux. The reaction mechanism and source of label is ascertained by alternatively labeling the formaldehyde and formic acid with deuterium. The results indicate that formaldehyde contributes two deuterium atoms and the carbon, whilst formic acid contributes one deuterium atom; there is no exchange between the formaldehyde and formic acid.

18.2.6
Aromatic Decarboxylation

As previous examples have shown, the development of microwave-enhanced labeling technology means more than accelerating reactions – it provides alternative opportunities. It follows therefore that some previously used methods now become much more attractive and this is the case for certain aromatic decarboxylations which can now be used for tritiations as well as in the treatment of tritiated waste. In previous studies [83] of the reaction the overriding feature was the harsh experimental conditions required.

2-Unsubstituted indoles, widely used intermediates in organic chemistry, are commonly synthesized through decarboxylation of the parent acid [84]. This is

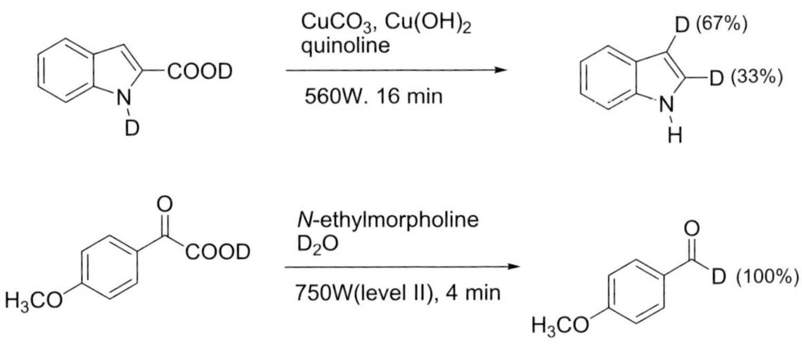

Scheme 18.10. Examples of microwave-enhanced decarboxylation.

achieved by prolonged heating in the presence of Cu (metal/salt) as catalyst and a basic solvent such as quinoline. In our studies [85] prior washing of the acid with CH_3OD to exchange the carboxyl proton with deuterium, followed by brief microwave activation, is sufficient to achieve decarboxylation/deuteriation in ~100% yield (Scheme 18.10). Other studies employing a commercial reactor and thick wall glass tubes capable of withstanding high pressure show that the decarboxylation proceeds in the absence of the environmentally undesirable copper catalysts and that quinoline can be replaced with water [86–88].

For a number of benzylformic acids we used N-ethylmorpholine as catalyst and D_2O as solvent/donor. Once again decarboxylation/deuteriation occurs very rapidly and is complete within 4 min. The range of compounds that can be labeled in this manner has been further widened by the recent observation [89–91] that tributylphosphine and other trivalent phosphorus compounds (R_3P, R = Bu, Ph, Me_2N, OEt) catalyze the decarboxylation of α-iminoacid. By using deuterated/tritiated acetic acid as a D^+/T^+ donor several labeled imines have been prepared. Under microwave-enhanced conditions the reactions would be expected to be much faster which again would be very useful as tritiated imines are frequently used to label β-lactams and other biologically interesting compounds such as α-aminophosphate [92]. It is also worth mentioning that the corresponding phosphites ($R_2HP=O$, R = OEt, OMe) are cheap, non-toxic hydrogen-atom donors and attractive alternatives to organic tin hydrides [93], and have been identified as effective radical reducing agents for organic halides, thioesters and isocyanides. The tritiated version of these reagents thus provide new opportunities, especially when coupled to microwave irradiation.

The potential of the microwave-enhanced decarboxylation route in the radioactive waste area is immediately apparent – washing the tritium waste with a protic solvent leads to exchange of the labile tritium. The solvent can then be used with one of the carboxylic acids mentioned above and after the microwave-enhanced decarboxylation the waste is now in the form of a solid (greatly reduced volume) which may have some further use.

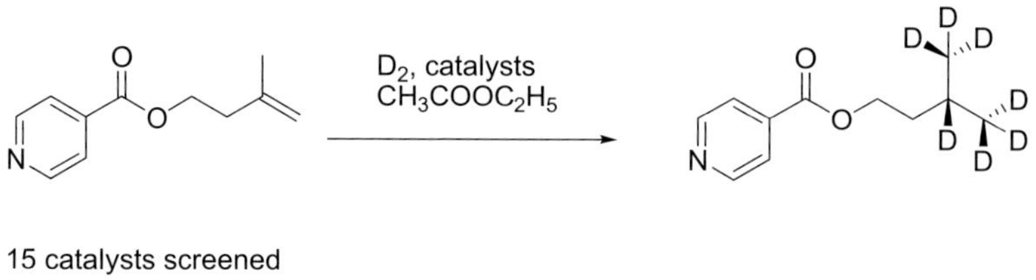

15 catalysts screened

Scheme 18.11. Parallel screening and ranking of catalysts for the reduction of isobutenyl groups.

18.2.7
The Development of Parallel Procedures

Chemistry as a subject has developed through the synthesis of individual compounds in a number of distinct steps. Recently it has benefited from the introduction of combinatorial/parallel chemistry techniques as well as microwave-enhanced technology but so far these studies have not been combined [94]. Lockley and co-workers [95–98] have shown very nicely how parallel chemistry techniques can be used for the rapid screening and ranking of catalysts using the hydrogenation of 3-methyl-3-butenyl-isonicotinate as the model reaction (Scheme 18.11).

The authors constructed an 80-well hydrogenator. Fifteen catalysts were screened and the isotopic incorporation assessed by LC–MS. The regiospecificity was determined by VAST Direct Injection NMR [99] in conjunction with SPADEZ [100], a multi-spectrum analysis tool – this is capable of displaying and quantifying up to 96 spectra. So far it has not been possible to vary the solvents as, under the experimental conditions, the more volatile distil over into the less volatile solvents.

A similar study [97, 98] was performed for *ortho*-directing hydrogen isotope exchange reactions of substituted aromatics. The initial screening showed catalytic activity to reside exclusively with the Group VIII metals, especially the salts and complexes of Ru, Rh and Ir. Iridium based catalysts are superior to those previously used – they are more active, operate at lower temperatures and are applicable to a wider range of substrates. Eventually CODIrAcac (acac = acetylacetone or 2,4-pentanedione) was identified as the catalyst displaying the best activity; further optimization of activity was achieved by varying the ligand structure.

In our own preliminary studies [101] on parallel procedures under microwave-enhanced conditions, we have used the Radley's RDT 24 place PTFE carousel reaction station on the turntable of the Matsui M 169BT microwave oven. In this way, we have studied the catalytic activity of $RhCl_3$ and $Pd(OAc)_2$ towards the reduction or dehalogenation of 4-bromocinnamic acid and structurally similar compounds. A nine-reaction matrix was used under microwave-enhanced conditions as illustrated in Scheme 18.12 – greatly reduced reaction times and easy optimization of reaction conditions are immediate benefits. As robotics come to play an increasingly important role in chemistry, one can immediately see more sophisticated labeling experiments being undertaken.

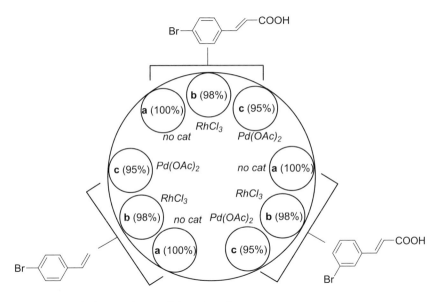

Scheme 18.12. A nine-reaction parallel matrix under microwave-enhanced conditions, and schematic representation of the carousel arrangement.

18.2.8
Combined Methodology

In the search for labeled compounds at higher specific activity/isotope incorporation, we were aware of the fact that the most widely used methods (see Table 18.1) are invariably performed separately. Consequently we embarked on a study to see whether, by choosing an appropriate compound, we could not combine e.g. hydrogenation and dehalogenation, or methylation and dehalogenation, or isotope

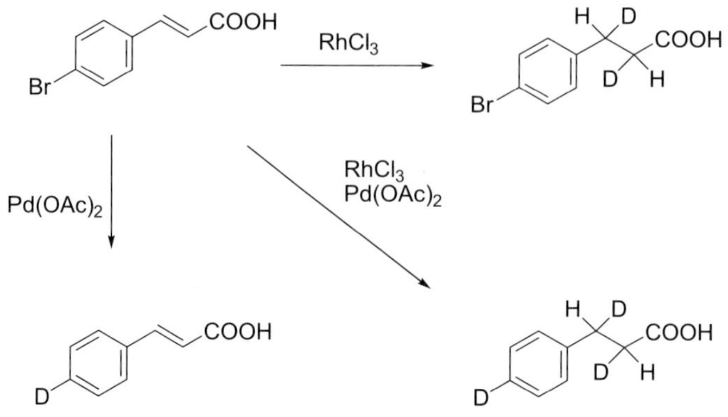

Conditions for all reactions: catalysts, DCOOK/D$_2$O, DMSO, level (II) of 750W, 60s

Scheme 18.13. Example of multiple deuterium labeling using combined methodology.

exchange/hydrogenation/dehalogenation. Whilst the investigations are at an early stage, there are sufficient successful examples to indicate that in future more attention will be paid to using combined methodologies to improve isotopic incorporation.

As an example, p-bromocinnamic acid undergoes debromination in the presence of HCOOK/H$_2$O/DMSO using Pd(OAc)$_2$ as catalyst. The reaction is complete in 60–90 s. On the other hand, the same compound undergoes hydrogenation under the same conditions using RhCl$_3$ as catalyst; again reaction is complete within 60 s. The combined debromination/hydrogenation of 4-bromocinnamic acid was complete within 1 min when the reaction was performed under microwave-enhanced conditions (Scheme 18.13) [102]. Therefore one-pot M+3 deuterium labeling in a multi-functional molecule is achieved using a microwave-enhanced combined hydrogenation/aromatic dehalogenation procedure. Isotopic purity in all three positions was >95%.

The screening of the catalytic activity in these reactions is made possible by the readily available library of various heterogeneous and homogeneous transition metal catalysts. The use of microwaves ensures that two reactions, each requiring different times to reach equilibrium under thermal conditions, can now be completed within a very short interval.

According to Blackwell [103] the application of microwave irradiation to expedite solid-phase reactions could be the tool that allows combinatorial chemistry to deliver on its promise – providing rapid access to large collections of diverse small molecules. Several different approaches to microwave-assisted solid-phase reactions and library synthesis are now available. These include the use of solid-supported reagents, multi-component coupling reactions, solvent-free parallel library synthesis, and spatially addressable library synthesis on planar solid support.

Solid-phase reactions are heterogeneous and often take up to 10 times as long as their homogeneous counterparts to come to equilibrium. Reagent diffusion into the polymer matrix is the main reason for the difference and this becomes very slow for large macro-beads, hence the need for more forcing conditions such as high temperature and prolonged times which themselves encourage by-product formations. This therefore represents as ideal opportunity to exploit the advantages of using microwaves to accelerate reactions. Furthermore, the possibility of using thin layer chromatography – where the plates can serve both as support on which to perform the synthesis and as a medium for separation – under microwave irradiation [104] opens up exciting possibilities, not least, in the labeling area.

18.3
Microwave-enhanced Detritiation Reactions

In the preparation of tritium-labeled compounds there are four stages:

- purchase of the tritiated reagent,
- radiochemical synthesis,
- storage of reagent/product, followed by
- removal of radioactive waste.

The first of these can be expensive as also can the last, consequently there is increasing interest in developing procedures which are more efficient than hitherto so that less radioactive waste is produced in the first place. Some radioactive waste can be disposed of *via* the drains but less so to the atmosphere, consequently there is also increasing interest in the possibility of converting the waste to a suitable reagent which can then be used in subsequent syntheses even though it is likely that the specific activity will be somewhat reduced. This philosophy can clearly be seen in the operation of both the Trisorber and Tritech tritium gas units which have replaced the old glass gas lines. In both instruments, the tritium gas is stored on an uranium 'getter' which on warming, releases a pre-determined amount of T_2 into the reaction vessel. On completion of the reaction, any surplus gas is returned to a secondary bed where it can be stored prior to use in a subsequent reaction.

Of all the methods used to triitate organic compounds hydrogen isotope exchange stands on its own by virtue of the fact that it is the only truly reversible reaction. Consequently the benefits that have emerged from the study of microwave-enhanced hydrogen-tritium exchange should be immediately transferable to tritium-hydrogen exchange. By performing the reaction in a good microwave solvent such as water, a tritiated compound or mixture of compounds could be 'decontaminated' and the HTO formed used in further tritiation studies, albeit at a lower level of specific activity. If the specific activity requirements are not too demanding the whole tritiation/detritiation cycle could be repeated several times, thereby making much better use of the tritiated water.

The first example [105] that we encountered was of an oil that had been exposed to a harsh tritium rich environment for a considerable time and had a level of radioactivity in the 2.1–2.2 KBq mg^{-1} range. An inactive oil, as represented by its ^{1}H

Tab. 18.4. Microwave-enhanced and thermal detritiation of [^{3}H]-oils.

$$\text{CH}_2\text{T}\text{-C(T)-...-C(T)-...} \xrightarrow[\text{Microwaves}]{\text{Raney Nickel, H}_2\text{O}} \text{CH}_2\text{H}\text{-C(H)-...-C(H)-...} + \text{HTO}$$

[^{3}H]-Oil	Original activity (KBq mg^{-1})	Thermal detritiation		Microwave detritiation[a]	
		% Detritiation	Time (h)	% Detritiation	Time (min)
1	140	62[b]	48	60	5 × 2
2	1128	89[c]	48	87	5 × 2
3	2.1	60[c]	66	43	5 × 2
4	2.2	70[c]	66	71	5 × 2

[a] microwave power set at level I (25%) of 750 W
[b] at 120 °C
[c] at 180 °C

NMR spectrum, had an identical composition and was used as a model compound. Based on previous work on shale oils [106] and engine basestocks [107] we knew that Raney Nickel could catalyze the hydrogen-tritium exchange reactions although the thermal reactions required high temperature (>180 °C) and long reaction times (48–66 h). However under microwave-enhanced conditions 5 × 2 min pulses was sufficient to exchange 60–90% of the tritium in the oils; repeating the procedure led to further detritiation (Table 18.4). Although these studies were performed on a small scale there is no reason why, as for other microwave-enhanced reactions, the scale of the operation can not be increased, or, alternatively a flow system designed.

18.4
Microwave-enhanced PET Radiochemistry

Positron emission tomography (PET) employs radiotracers and radioligands containing positron emitters (e.g. ^{11}C, ^{13}N, ^{15}O and ^{18}F) in organic molecules, to image and obtain information on physiological, biochemical and pharmacological functions at the molecular level *in vivo* in animal and human subjects [78, 108, 109]. Carbon-11 ($t_{1/2}$ = 20.4 min) and fluorine-18 (as a hydrogen substitute, $t_{1/2}$ = 109.7 min) have become the two most widely used radioisotopes for PET radiopharmaceuticals. Their short half lives have clear benefits for human subjects undergoing imaging studies with very low radiation exposure doses, but pose enormous challenges to chemists who race against time to produce radioligands of sufficient quantity, purity and specific activity. In order to achieve a high radiochemical yield and high specific activity the labeling procedures must be rapid (usually complete within 2–3 half-lives), simple to perform, the products easy to purify, with the final dose sterile and pyrogen free. Automation provides further benefits whilst the need to use reactants on a micro-scale requires a good appreciation of both the reaction stoichiometry and the factors that influence the rates of chemical reactions.

Because of these requirements, and in particular, the need to perform the reactions rapidly, it is not surprising that the application of microwaves was explored in the PET radiopharmaceutical synthesis area at an early stage [110]. However, the subsequent development did not match the rapid expansion in synthetic organic chemistry area in large part due to more stringent radiation safety considerations. With the development of new instrumentation and understanding of reaction behavior under microwave conditions more PET radioligands and PET radiopharmaceutical research are now carried out under microwave enhanced conditions. Two important reviews [8, 9] are available for an up-to-date picture whilst here we discuss some noteworthy, and more recent examples. The benefits of microwaves to PET radiochemistry are also highlighted in other general review articles [111–114].

18.4.1
^{11}C-labeled Compounds

The most widely used synthetic routes follow on closely from those adopted for ^{14}C syntheses – a small number of key precursors such as $^{11}CO_2$, $^{11}CH_3I$, $^{11}CN^-$, and $H^{11}CHO$ are sufficient to label a wide range of compounds.

One step O-, N- and S-alkylations can be achieved using $^{11}CH_3X$ (X = I, OTf) [115–119]. Through the use of microwaves the time required to synthesize [N-methyl-^{11}C]flumazenil can be reduced from 5 to 0.5 min, accompanied by a 20% improvement in radiochemical yield. Further improvements can be achieved by changing the solvent, in this case from acetone to DMF (Scheme 18.14) [115].

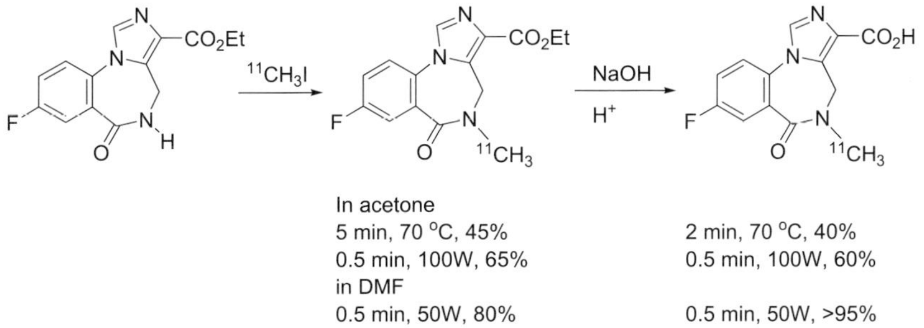

Scheme 18.14. Synthesis of [N-methyl-^{11}C]flumazenil acid.

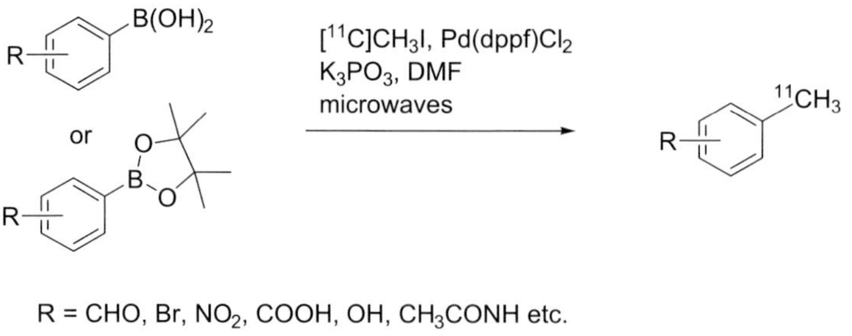

Scheme 18.15. Synthesis of [^{11}C]toluene derivatives via Suzuki coupling with [^{11}C]CH$_3$I.

Even more striking, the total synthesis time was reduced by nearly one half life for the preparation of the β_1 adrenoceptor ligand [N-methyl-^{11}C]CGP20712A [116]. RCYs obtained for 20 min at 110 °C by thermal heating vs. 90 s at 600 W under microwave condition are comparable. Microwaves had no obvious influence on regioselectivity of the various N- and O-methylations, but the addition of a salt, such as NaI, could apparently change the regioselectivity considerably under both microwave irradiation and thermal heating conditions. More difficult reactions, such as O- and S-alkylation with ethyl iodide or propyl iodide that did not give sufficient radiolabeled product are now possible with microwave enhancement [118, 119].

The synthesis of [^{14}C]-labeled esters using a C-alkylation of diethyl malonate under microwave-enhanced solid-liquid phase transfer catalysis (PTC) conditions and a subsequent microwave-enhanced decarboalkoxylation (Krapcho reaction) [167] indicates that an alternative approach to the preparation of [^{11}C]-labeled fatty acids/esters [168] with less harsh conditions may be possible.

The introduction of a methyl group to aromatic rings is attractive but more difficult. The classical methodology for the synthesis of [methyl-^{11}C]toluene derivatives is through Stille coupling of an aryl trialkyltin precursor with [^{11}C]methyl iodide. The application of microwave heating led to a significant increase in the radiochemical yields using Suzuki coupling of the corresponding boron containing precursors and [^{11}C]methyl iodide (Scheme 18.15) [120]. A large number of other functional groups are tolerated. This provides an alternative strategy which is more robust and avoids potentially toxic tin-containing by-products or contaminants.

[^{11}C]Cyanide is a secondary labeling agent, produced from ^{11}CO$_2$ via [^{11}C]methane. The [^{11}C]nitrile is usually an intermediate that may subsequently be converted to other functionalities, such as acids, amines, amides etc. [121, 122]. The nucleophilic displacement of the bromide group could be accelerated with microwaves, with RCY of 60% obtained using 60 W for 30 s, matching the RCY obtained with heating at 90 °C for 7 min. The slower cyano-dechlorination could be enhanced by manipulating the polarity of the salt that is used in the reaction mixture, with KCl and KBr showing the best effects [121].

Additionally, as a result of microwave-enhanced accelerations, a wide range of amines, acids, esters and alcohols can be prepared from [^{11}C]cyanide (Scheme 18.16) [123] and used for the formation of more complex structures [124–126].

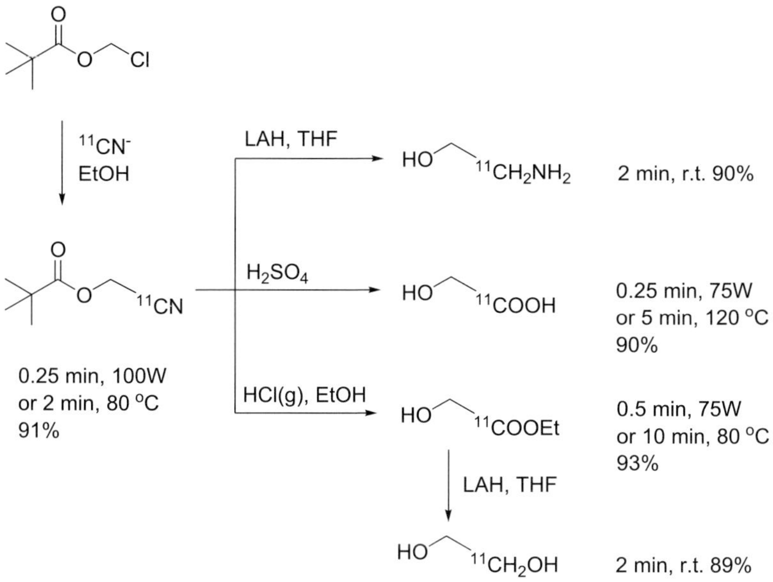

Scheme 18.16. Synthesis of various ^{11}C-labeled amines, acids, esters, and alcohols.

Again, reactions with less reactive substrates can be achieved by increasing the polarity of the reaction medium through the addition of various salts.

[^{11}C]formaldehyde has been widely used for reductive methylation reactions but because of the marked fluctuation in the reported yields, as well as impurities formed in its preparation from ^{11}CO$_2$, the tendency has been to use ^{11}CH$_3$I for direct methylation. However, the recent development [127, 128] of a low temperature no-carrier-added (NCA) method for preparing H^{11}CHO, coupled to the microwave-enhanced Eschweiler–Clarke reaction [79–82], has led to a resurgence of interest in the use of H^{11}CHO. In their study Roeda and Crouzel showed that the LAH reduction of ^{11}CO$_2$ produced 34% H^{11}CHO plus 59% H^{11}COOH. Our experience [82] in using a mixture of DCHO and DCOOH for the N-methylation of both primary and secondary amines suggests that the corresponding [^{11}C]mixture would be useful for the ^{11}C-labeling of amines under microwave-enhanced conditions.

The application of microwaves also opens up new opportunities in carbon-11 labeling. No-carrier-added aromatic and aliphatic [*carbonyl*-^{11}C]amides were traditionally prepared through reacting cyclotron produced [^{11}C]carbon dioxide with organometallic reagents, such as Grignard, to obtain the [^{11}C]carboxymagnesium halides or [^{11}C]carboxylic acid, followed by their conversion into a more reactive species, such as [^{11}C]acid chloride [129]. Now they can be rapidly synthesized in one pot in useful radiochemical yields (20–65%, decay-corrected) by directly coupling amines with NCA [^{11}C]carboxymagnesium halides generated *in situ* from Grignard reagents and [^{11}C]carbon dioxide (Scheme 18.17). The scope of the reaction was widened considerably by the application of microwaves – without micro-

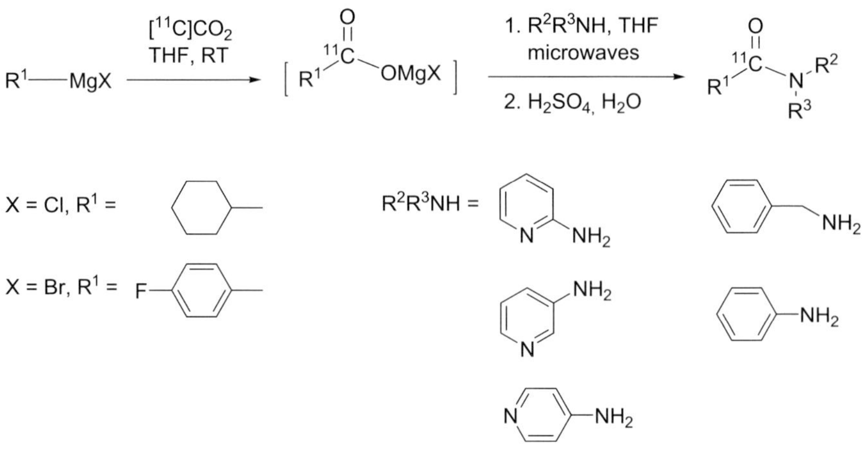

Scheme 18.17. Synthesis of [carbonyl-^{11}C]amides by direct reaction of in situ generated [^{11}C]carboxymagnesium halide with amines.

wave enhancement success was limited only to the most reactive aliphatic Grignard reagents and amines [130, 131].

18.4.2
^{18}F-labeled Compounds

Substitution of fluorine for hydrogen in an organic compound causes little steric perturbation. However, as a result of its greater electron negativity and high C-F bond energy, the chemical and biological properties will be different. Two strategies, namely electrophilic fluorination and nucleophilic fluorination, are used to prepare [^{18}F]-labeled compounds. Electrophilic fluorinations are fast, but suffer from low specific activities resulting from the use of a labeling agent derived from ^{18}F$_2$. They are therefore used less frequently. Nucleophilic fluorinations, either aromatic or aliphatic, rely on no-carrier-added [^{18}F]fluoride as the primary labeling precursor. The most widely used nucleophilic fluorination reaction involves [^{18}F]fluoride ion displacing a leaving group, such as –NO$_2$, –X (halogen), or –N$^+$R$_3$ on an aromatic ring. The frequent need for an activating group within the same aromatic structure, the use of polar solvents and high reaction temperature gives ample scope for microwave enhancement.

Radiofluorination with nitro-precursors is by far the most convenient and reliable choice [132–142]. Three examples of [^{18}F]radioligand production using microwaves are given in Scheme 18.18, demonstrating the improvement in short reaction time and better RCYs that can be achieved. The reduced amount of precursors in microwave-enhanced reactions, which not only saves precious starting materials, but also reduces the burden at the purification stage, is an added advantage.

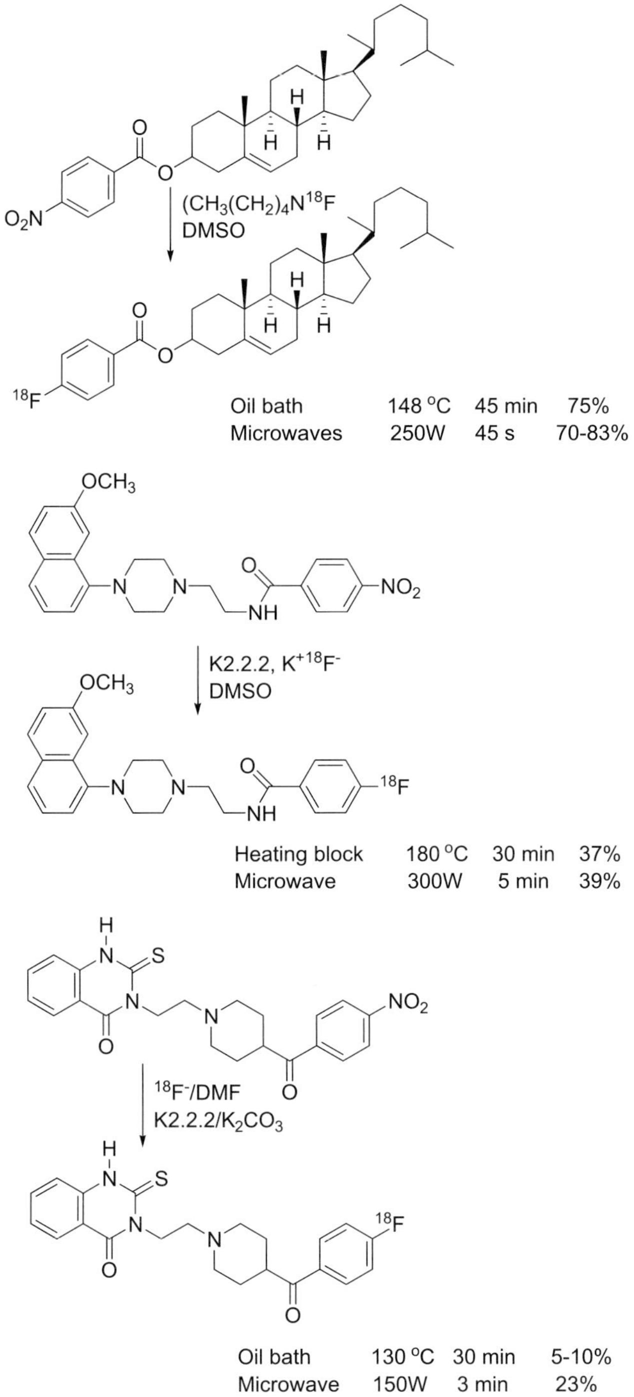

18.4 Microwave-enhanced PET Radiochemistry

Scheme 18.18. Examples of aromatic nucleophilic substitutions used in the synthesis of ^{18}F-labeled radiopharmaceuticals.

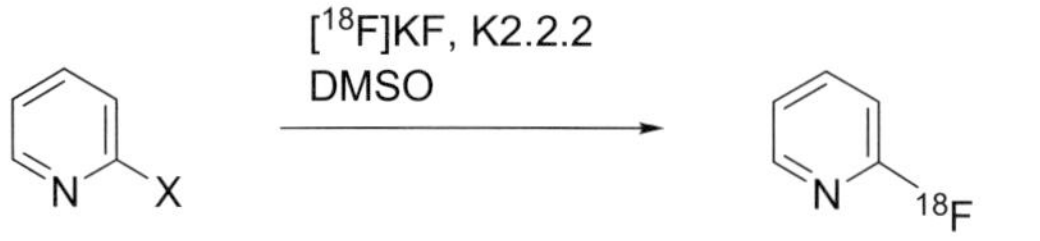

Scheme 18.19

[pyridine-X] → [pyridine-^{18}F]   [^{18}F]KF, K2.2.2, DMSO

	X = Cl	Br	I	NO$_2$	N$^+$(CH$_3$)$_3$
50W, 4 min	26%	68%	8%	76%	96%
100W, 2 min	22%	71%	14%	88%	90%

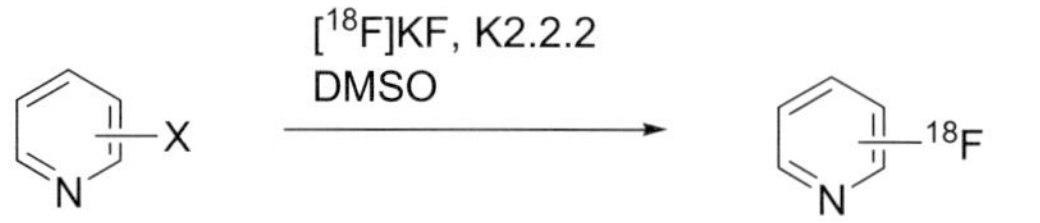

[pyridine with X at different positions] → [^{18}F-pyridine]   [^{18}F]KF, K2.2.2, DMSO

	X = 2-NO$_2$	3-NO$_2$	4-NO$_2$
145 °C, 10 min	66%	1%	60%
100W, 2 min	94%	2%	72%

Scheme 18.19. Substituent and leaving-group regiochemistry effects in the microwave-enhanced nucleophilic fluorination of several pyridine derivatives.

In a series of comparative studies, Dollé et al. [140–146] examined the nucleophilic aromatic substitution of a number of aromatic rings. The [^{18}F]fluorination on a pyridine ring, as summarized in Scheme 18.19, shows that under microwave-enhanced conditions the 2-NO$_2$ and 2-$^+$NMe$_3$ groups led to excellent fluorine incorporation whilst the 2-iodo compound was virtually unreactive. Under thermal conditions, no fluorination was observed for the 2-chloro and 2-bromo compounds. In a separate study, Banks and Hwang [147] again observed the beneficial effects in nitro and trimethylammonium substitution by microwaves. The trimethylammonium triflate salt was also successfully applied in the synthesis of [^{18}F]-labeled COX-1 and COX-2 inhibitors [148]. However, care must be taken to optimize the reaction condition, otherwise a side reaction producing [^{18}F]fluoromethane occurs [148, 149]. This could be potentially useful for a novel microwave-enhanced procedure of producing [^{18}F]fluoromethane [149], with approaching 70% RCY achieved at optimum conditions.

In aliphatic nucleophilic substitution, tosylate proves to be a good leaving group [150–152]. Although in the preparation of [^{18}F]fluoroethyl tosylate, from ethylene ditosylate in acetonitrile, the reaction was relatively easy and did not require microwave enhancement, in the case of [^{18}F]FHBG (Scheme 18.20), when the solvent was acetonitrile or DMSO and thermal heating used, the RCYs were very poor (4–5%) and the results were not reproducible. When DMSO was used with microwaves, RCYs increased up to 30–40%.

The synthesis of 2-deoxy-2-^{18}F-fluoro-D-glucose (2-[^{18}F]FDG) [153–155] is another example where the application of microwaves can reduce the synthesis time and increase the radiochemical yields. An initial optimization of the standard nucleophilic fluorination with mannose triflate under microwave conditions cut short

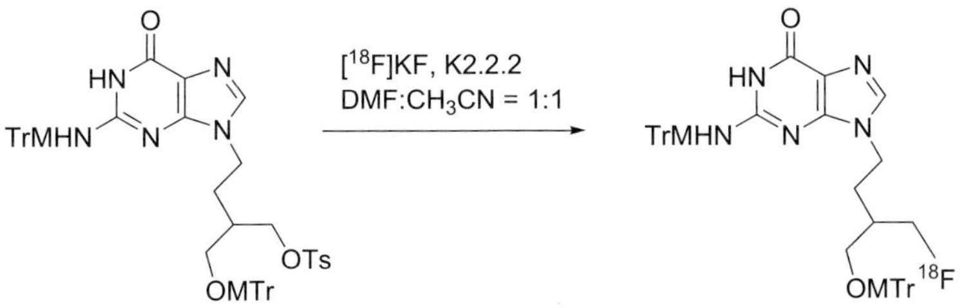

Scheme 18.20. Nucleophilic aliphatic radiofluorination with tosylate as leaving group.

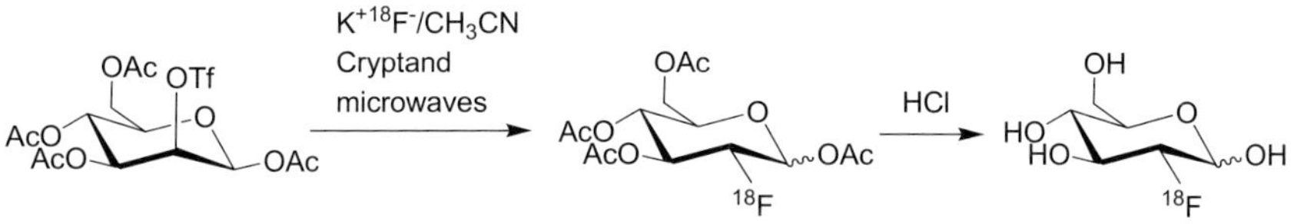

Scheme 18.21. Synthesis of 2-[^{18}F]FDG under microwave conditions.

	RCY(%)	Synthesis time (min)
Thermal module	47	64
Microwave/K2.2.2	68	51
Microwave/TBAHCO$_3$	76	31

the synthesis time from 64 min to 51 min, whilst increasing the decay-corrected RCYs from 47% to 68%. When K2.2.2 was replaced by tetrabutylammonium bicarbonate (TBAHCO$_3$), the synthesis time further decreased to 31 min and RCY increased to 76% (Scheme 18.21).

The use of microwaves, together with the involvement of a suitable protection group, can also be beneficial in reducing isotopic dilution as shown in Scheme 18.22 [156]. Formation of the oxazoline reduced undesirable ^{18}F–^{19}F exchange and the desired product was rapidly (15 s) isolated in good yield (45–55%) with a specific activity (1.1 GBq μmol^{-1}) that was twenty times higher than that obtained when using the direct debromination route. In a study of the carrier-added reaction, ^{19}F-NMR showed the presence of HF$_2^-$ and other unidentified fluorine-containing impurities in the F$^-$/K$_2$CO$_3$/K2.2.2-cryptand residue when conventional heating was used to prepare dry [^{18}F]fluoride ion. It was also demonstrated that microwave heating eliminated unidentified fluorine-containing impurities from the [^{18}F]fluoride residue and from the reaction mixture between F$^-$ and mannose triflate.

The reduction of reaction time and improvement in RCYs by microwave enhancement are of great advantages in the synthesis of radiopharmaceuticals involving multiple steps. In certain cases, the synthesis of the secondary labeling agent was accelerated so that the time saved could be diverted to the next step, or in other

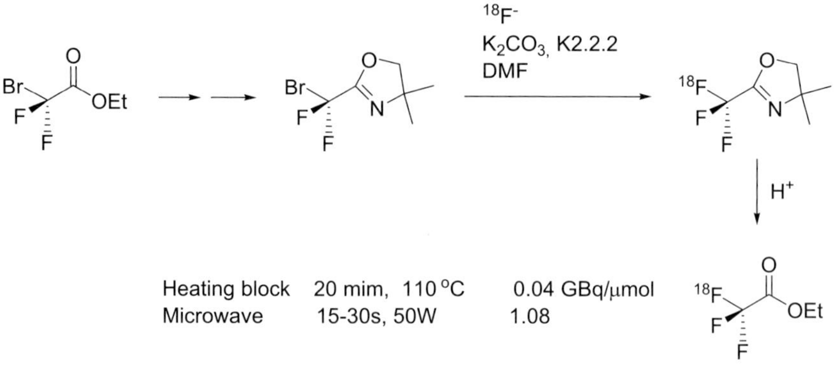

Scheme 18.22. Protection–deprotection strategy in a microwave-enhanced fluoro–debromination reaction.

cases the first step was performed as usual but the subsequent reaction was enhanced. It provides the radiochemist with more choice of precursors and secondary labeling agents so that more rapid, selective and functional group-tolerating reactions can be used [157–160].

The synthesis of 4-[^{18}F]fluoroiodobenzene and its application in Sonogashira cross-coupling reactions exemplifies the first strategy [157]. 4,4'-Diiododiaryliodonium salts were used as precursors for the synthesis of 4-[^{18}F]fluoroiodobenzene, enabling convenient access to 4-[^{18}F]fluoroiodobenzene in 13–70% yield (Scheme 18.23). High reaction temperatures (140 °C or higher) were essential for an efficient thermal decomposition of diaryliodonium salts in a highly polar aprotic solvent in the presence of [^{18}F]fluoride ion to form 4-[^{18}F]fluoroiodobenzene. Microwave activation did not further improve the radiochemical yield of 4-[^{18}F]fluoroiodobenzene. Because the rapid access to high temperature in a short time under microwave conditions, the reaction time was reduced from 40 to 5 min which is of great advantage in performing the subsequent reaction.

Reactions of no-carrier-added [^{18}F]β-fluoroethyl tosylate with amine, phenol or carboxylic acid to form the corresponding [^{18}F]N-(β-fluoroethyl)amine, [^{18}F]β-fluoroethyl ether or [^{18}F]β-fluoroethyl ester, were found to be rapid and efficient under microwave-enhanced conditions (Scheme 18.24) [159]. The preparation of [^{18}F]β-fluoroethyl tosylate did not require microwaves. The subsequent O- and N-alkylation allow reactants to be heated rapidly to 150 °C in a low boiling point solvent, such as acetonitrile, and avoid the need to use high boiling point solvents, such as DMSO and DMF, to promote reaction. The microwave-enhanced reactions gave about 20% greater radiochemical yields than thermal reactions performed at similar temperatures and over similar reaction times.

The synthesis of 2-(4-[^{18}F]fluorophenyl)benzimidazole represents another example of the latter strategy [160]. Cyclocondensation of an aromatic diamine with an aliphatic acid using microwave heating proceeds rapidly; the same reaction using aromatic and sterically hindered alkyl acids requires more vigorous conditions (e.g.

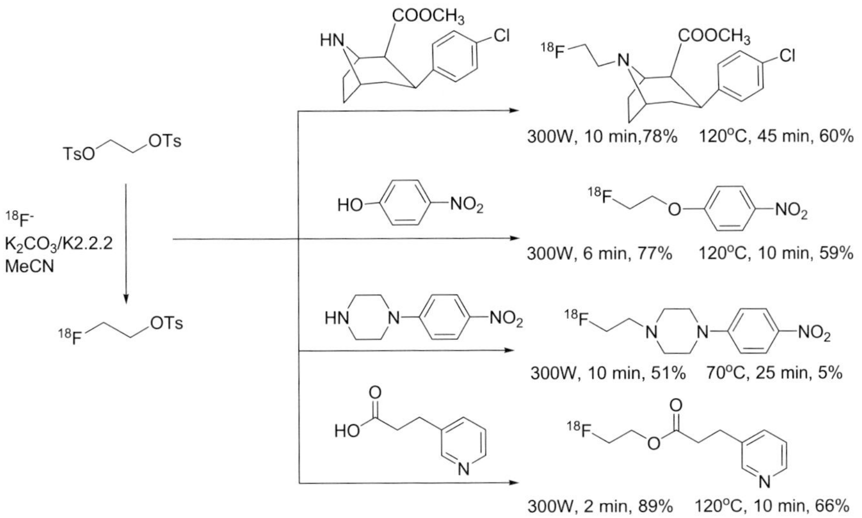

Scheme 18.23. Preparation of NCA [^{18}F]N-β-fluoroethylamine, [^{18}F]β-fluoroethyl ether, and [^{18}F]β-fluoroethyl ester, using [^{18}F]β-fluoroethyl tosylate as labeling agent, in a microwave-enhanced procedure.

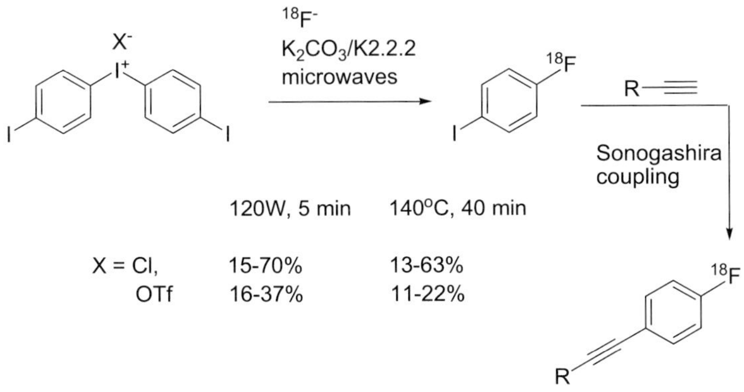

Scheme 18.24. Radiosynthesis of 4-[^{18}F]fluoroiodobenzene and its subsequent application in Sonogashira cross-coupling.

more heat and pressure and longer reaction time). The cyclocondensation of 1,2-diaminobenzene with radiolabeled [4-^{18}F]fluorobenzoic acid in neat methanesulphonic and polyphosphoric acids under microwave heating led rapidly to the cyclised phenylbenzimidazole (Scheme 18.25). This methodology makes the phenylbenzimidazole building block readily available for the construction of other PET radiopharmaceuticals.

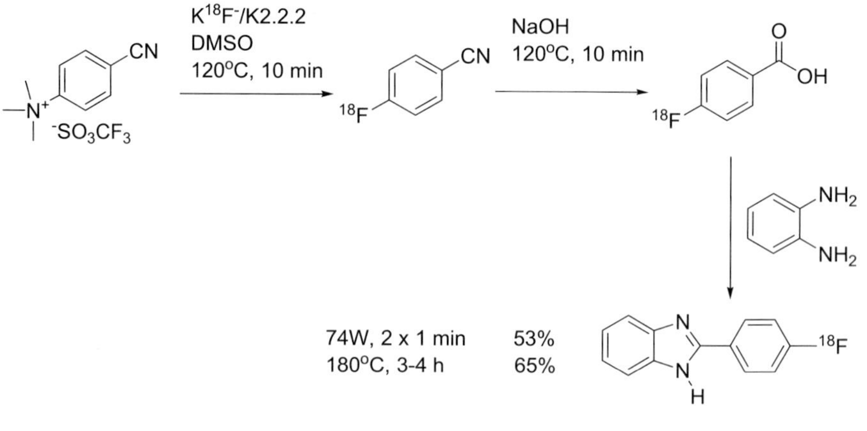

Scheme 18.25. Synthesis of 4-[^{18}F]fluorobenzoic acid and its microwave-enhanced cyclocondensation with 1,2-diaminobenzene.

18.4.3
Compounds Labeled with Other Positron Emitters

^{76}Br ($t_{1/2}$ = 16.2 h) is used as an alternative to ^{18}F for labeling octreotide with PET applications in the diagnosis of somatostatin-positive tumors. N-Succinimidyl-4-[^{76}Br]bromobenzoate and N-succinimidyl 5-[^{76}Br]bromo-3-pyridine-carboxylate are both used for incorporating ^{76}Br into octreotide through conjugation labeling [161]. The use of dichloromethane as the solvent and microwave heating resulted in 70–90% labeling of ε-Boc-octreotide in 30 min whereas the corresponding thermal heating only resulted in less than 20% RCY (Scheme 18.26). The use of

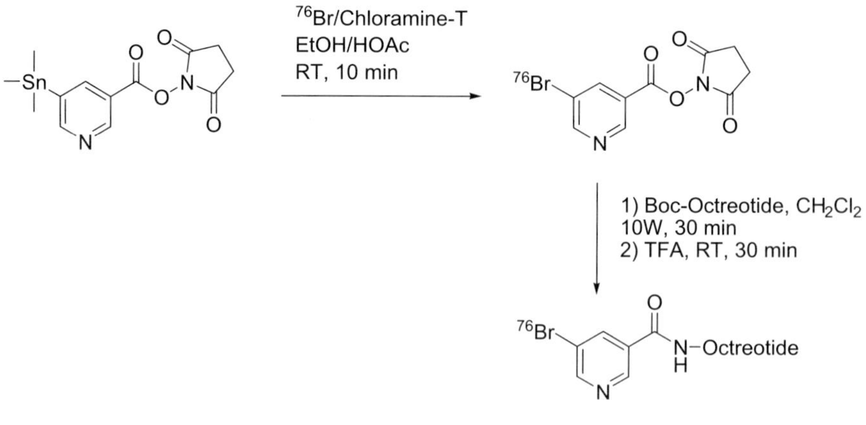

Scheme 18.26. Synthesis of N-succinimidyl 5-[^{76}Br]bromo-3-pyridine carboxylate and microwave-enhanced conjugation with ε-Boc-octreotide.

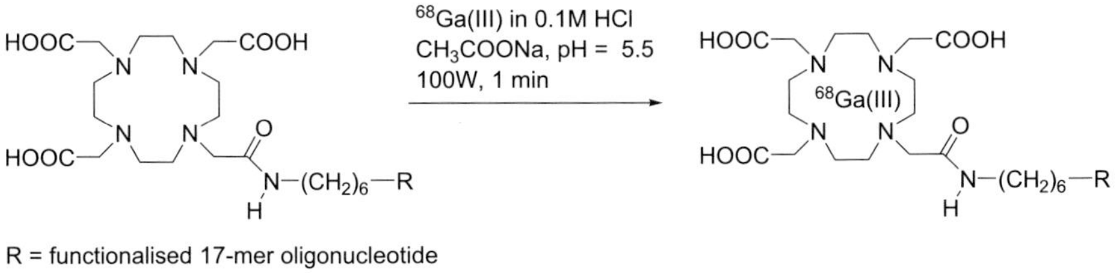

Scheme 18.27. Microwave-enhanced ^{68}Ga labeling of DOTA-oligonucleotides.

dichloromethane, though a poor microwave solvent, gave a convenient reaction medium for the subsequent Boc-protection group removal and simpler HPLC separation procedure.

Oligonucleotides can be labeled with organic positron-emitting radioisotopes, such ^{11}C, ^{18}F and ^{76}Br. They are also labeled with metal positron-emitting radioisotopes, such as ^{68}Ga, ^{99m}Tc etc. The oligonucleotides were conjugated to the bifunctional chelator, 1,4,7,10-tetraazacyclododecane-1,4,7,10-tetraacetic acid (DOTA) through the hexylamine linker, and then labeled with ^{68}Ga($t_{1/2} = 68$ min) using microwave activation (Scheme 18.27) [162, 163]. The application of microwaves shortened the synthesis time considerably (from 20 min to 1 min) and improved the radiochemical yield (from 19% to 30–52%) compared to thermal heating. Owing to the shortened reaction time, the amount of radioactive material and the product specific activity was increased by 21%. Furthermore, microwave activation

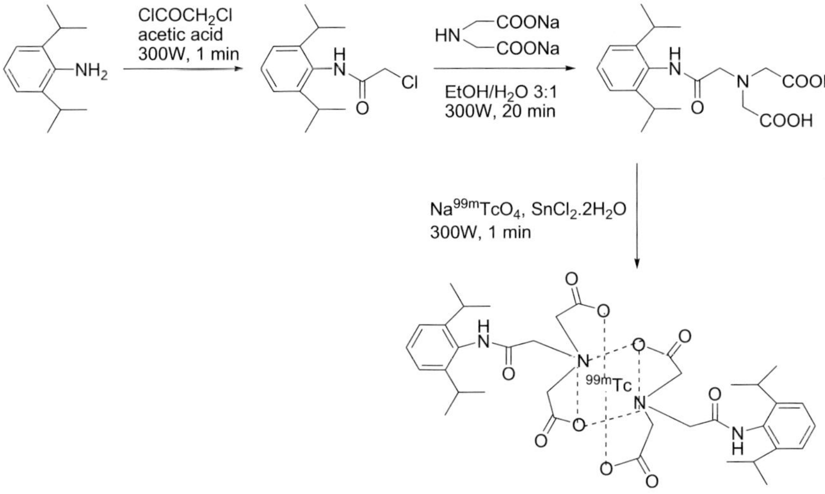

Scheme 18.28. Synthesis of 2,6-diisopropylacetanilidoiminodiacetic acid and ^{99m}Tc labeling under microwave irradiation conditions.

not only reduces the chemical reaction time, but also reduces side reactions, increases the yield, and improves reproducibility as illustrated by the two-fold increase in analytical RCY compared to previous results with conventional heating.

^{99m}Tc labeled compounds are another category that benefit from microwaves [164–166]. For example, the synthesis and ^{99m}Tc labeling of 2,6-diisopropylacetanilidoiminodiacetic acid could be performed on a large scale under microwave irradiation in a very short time (Scheme 18.28) [166]. The synthesis time for 2,6-diisopropylacetanilidoiminodiacetic acid was reduced from 48 h to 20 min, and the radiolabeling synthesis time was reduced from 60 min to 1 min, respectively. However, caution should be exercised to avoid over-heating by microwaves, as suggested by Hung et al. [165], so that the maximum RCY and radiochemical purity (RCP) could be achieved.

18.5
Conclusion

From an early stage in the development of modern chemistry it has been customary practice to simplify complex problems by the selective use of isotopes. For example, in the area of catalysis, relatively simple reactions such as hydrogen isotope exchange or hydrogenation reactions, have been investigated in order to delineate some of the finer details of reaction mechanisms. This information has then been applied so as to optimize the procedures that have been developed for labeling organic compounds. The last 50 years has seen the emergence of several, now large companies, specializing in these areas, as well as many smaller versions.

Now, with the emergence of new, microwave-enhanced technology, the process is destined to be repeated. In this case, faster, more selective and efficient procedures will emerge, where levels of radioactive waste will be greatly reduced and a more favorable, environmentally friendlier image of the chemical industry achieved. The products so produced can then go on to be used, in the words of Gordon Dean, sometime Chairman of the US Atomic Energy Commission, "to treat the sick, to learn more about disease, to improve manufacturing processes, to increase the productivity of crops and livestock, and to help man to understand the basic processes of his body, the living things about him, and the physical world in which he exists."

Acknowledgments

The tritium work at Surrey has been generously funded over many years by EPSRC (previously SERC and, originally, SRC), the EU, NATO, and the chemical industry. Shui-Yu Lu's research is supported by the Intramural Research Program of the National Institutes of Health (National Institute of Mental Health).

References

1. D. Dalvie, *Curr. Pharm. Design* **2000**, 6, 1009.
2. E. A. Evans, *Tritium and Its Compounds* (2nd Ed), Butterworths, London, **1974**.
3. J. A. Elvidge, J. R. Jones (eds), *Isotopes: Essential Chemistry and Applications I*, The Chemical Society, London, **1980**. J. R. Jones (ed.), *Isotopes: Essential Chemistry and Applications II*, The Royal Society of Chemistry, London, **1988**.
4. F. J. Winkler, K. Kuhnl, R. Medina, R. Schwarzkaske, H. L. Schmidt, *Iso. Environ. Health Studies* **1995**, 31, 161.
5. T. J. Simpson, *Biosyn. Top. Curr. Chem.* **1998**, 195, 1.
6. L. Y. Lian, D. A. Middleton, *Prog. Nucl. Mag. Res. Spec* **2001**, 39, 171.
7. K. W. Turteltaub, J. S. Vogel, *Curr. Pharm. Design* **2000**, 6, 991.
8. N. Elander, J. R. Jones, S. Y. Lu, S. Stone-Elander, *Chem. Soc. Rev.* **2000**, 29, 239.
9. S. Stone-Elander, N. Elander, *J. Labelled Compd. Radiopharm.* **2002**, 45, 715.
10. J. R. Jones, *The Ionisation of Carbon Acids*, Academic Press, London, **1973**.
11. N. H. Werstiuk, in *Isotopes in the physical and biological sciences* Vol 1A, E. Buncel, J. R. Jones (Eds), Elsevier, Amsterdam, **1987**, p124.
12. J. R. Brewer, J. R. Jones, K. W. M. Lawrie, D. Saunders, A. Simmonds, *J. Labelled Compd. Radiopharm.* **1994**, 34, 391.
13. J. R. Brewer, J. R. Jones, K. W. M. Lawrie, D. Saunders, A. Simmonds, *J. Labelled Compd. Radiopharm.* **1994**, 34, 787.
14. J. L. Garnett, *Catal. Rev.* **1971**, 5, 229.
15. J. L. Garnett, M. A. Long, in *Isotopes in the physical and biological sciences* Vol 1A, E. Buncel, J. R. Jones (Eds.), Elsevier, Amsterdam, **1987**, p86.
16. R. Salter, I. Bosser, *J. Labelled Compd. Radiopharm.* **2003**, 46, 489.
17. D. Hesk, J. R. Jones, W. J. S. Lockley, D. J. Wilkinson, *J. Labelled Compd. Radiopharm.* **1990**, 28, 1309.
18. D. Hesk, P. R. Das, B. Evans, *J. Labelled Compd. Radiopharm.* **1995**, 36, 497.
19. J. R. Heys, *J. Chem. Soc., Chem. Commun.* **1992**, 680.
20. Y. A. Zolotarev, Y. A. Borisov, N. F. Myasoedov, *J. Phys. Chem. A.* **1999**, 103, 4861.
21. Y. A. Zolotarev, E. V. Laskatelev, V. S. Kozik, E. M. Dorokhova, S. G. Rosenberg, Y. A. Borisov, N. F. Myasoedov, *Russ. Chem. Bull.* **1997**, 46, 1726.
22. D. K. Jaiswal, H. Andres, H. Morimoto, P. G. Williams, C. Than, S. J. Seligman, *J. Org. Chem.* **1996**, 61, 9625.
23. K. E. Wilzbach, *J. Am. Chem. Soc.* **1957**, 79, 1013.
24. E. A. Evans, D. C. Warrell, J. A. Elvidge, J. R. Jones, *Handbook of Tritium NMR Spectroscopy and Applications*, J. Wiley and Sons, Chichester, **1985**.
25. R. Gedye, F. Smith, K. Westaway, H. Ali, L. Baldisera, L. Laberge, J. Rousell, *Tetrahedron Lett.* **1986**, 27, 279.
26. R. J. Giguere, T. L. Bray, S. M. Duncan, G. Majetich, *Tetrahedron Lett.* **1986**, 27, 4945.
27. N. A. Ghanem, T. Westermark, *J. Am. Chem. Soc.* **1960**, 82, 4432.
28. W. C. Hembree, R. L. E. Ehrenkaufer, S. Lieberman, A. P. Wolf, *J. Bio. Chem.* **1973**, 248, 5532.
29. R. L. E. Ehrenkaufer, W. C. Hembree, S. Lieberman, A. P. Wolf, *J. Am. Chem. Soc.* **1977**, 99, 5005.
30. R. L. E. Ehrenkaufer, A. P. Wolf, W. C. Hembree, S. Lieberman, *J. Labelled Compd. Radiopharm.* **1977**, 13, 395.
31. R. Hua, C. T. Peng, *J. Labelled Compd. Radiopharm.* **1987**, 24, 295.
32. G. Z. Tang, C. T. Peng, *J. Labelled Compd. Radiopharm.* **1988**, 25, 585.

33 C. T. Peng, *J. Labelled Compd. Radiopharm.* **1993**, 33, 419.
34 L. Perreux, A. Loupy, *Tetrahedron* **2001**, 57, 9199.
35 A. C. Metaxas, R. J. Meredith, *Industrial Microwave Heating*, Peter Peregrinus Ltd., London, **1988**.
36 A. Miklavc, *Chem Phys Chem.* **2001**, 8/9, 552.
37 C. Gabriel, S. Gabriel, E. H. Grant, B. S. J. Halstead, D. M. P. Mingos, *Chem. Soc. Rev.* **1998**, 27, 213.
38 C. O. Kappe, *Angew. Chem. Int. Ed.* **2004**, 43, 6250.
39 G. S. Getvoldsen, N. Elander, S. Stone-Elander, *Chem. Eur. J.* **2002**, 8, 2255.
40 I. Lewis, *Spectroscopy Eur.* **2005**, 17, 34.
41 D. E. Pivonka, J. R. Empfield, *Appl. Spectroscopy* **2004**, 58, 41.
42 G. J. Koves, *J. Labelled Compd. Radiopharm.* **1994**, 34, 255.
43 G. J. Koves, *J. Labelled Compd. Radiopharm.* **1991**, 29, 15.
44 G. J. Koves, *J. Labelled Compd. Radiopharm.* **1987**, 24, 455.
45 J. M. Barthez, A. V. Filikov, L. B. Frederiksen, M. L. Huguet, J. R. Jones, S. Y. Lu, *Can. J. Chem.* **1998**, 76, 726.
46 S. Anto, G. S. Getvoldsen, J. R. Harding, J. R. Jones, S. Y. Lu, J. C. Rusell, *J. Chem. Soc., Perkin Trans. 2* **2000**, 2208.
47 E. A. Cioffi, *Tetrahedron Lett.* **1996**, 37, 6231.
48 W. M. Yau, K. Gawrisch, *J. Labelled Compd. Radiopharm.* **1999**, 42, 709.
49 K. R. Seddon, *Kinetics and Catalysis* **1996**, 37, 693.
50 J. D. Holbrey, K. R. Seddon, *Clean Products and Processes* **1999**, 1, 223.
51 T. Welton, *Chem. Rev.* **1999**, 99, 2071.
52 N. E. Leadbeater, H. M. Torenius, H. Tye, *Comb. Chem. High Throughput Scr.* **2004**, 7, 511.
53 R. S. Varma, V. V. Namboodiri, *J. Chem. Soc., Chem. Commun.* **2001**, 643.
54 R. N. Garman, G. S. Getvoldsen, J. R. Jones, K. W. M. Lawrie, S. Y. Lu, P. Marsden, J. C. Russell, K. R. Seddon, in *Synthesis and Applications of Isotopically Labelled Compounds* Vol 7, U. Pleiss and R. Voges (eds.), Wiley, Chichester, **2001**, p97.
55 P. S. Kiuru, K. Wähälä, *Tetrahedron Lett.* **2002**, 43, 3411.
56 K. Fordor-Csorba, G. Galli, S. Holly, E. Gács-Baitz, *Tetrahedron Lett.* **2002**, 43, 3789.
57 S. A. de Keczer, T. S. Lane, M. R. Masjedizadeh, *J. Labelled Compd. Radiopharm.* **2004**, 47, 733.
58 E. A. Cioffi, R. H. Bell, B. Le, *Tetrahedron: Asymmetry* **2005**, 16, 471.
59 D. Hesk, M. Calvert, P. McNamara, in *Synthesis and Applications of Isotopically Labelled Compounds* Vol 8, D. C. Dean, C. N. Filer and K. E. McCarthy (eds.), Wiley, Chichester, **2004**, p51.
60 S. L. Regen, *J. Org. Chem.* **1974**, 39, 260.
61 J. M. A. Al-Rawi, J. A. Elvidge, J. R. Jones, R. B. Mane, M. Saieed, *J. Chem. Res. (S)* **1980**, 298.
62 M. Takahashi, K. Oshima, S. Matsubara, *Chem. Lett.* **2005**, 34, 192.
63 M. H. Al-Qahtani, N. Cleator, T. N. Danks, R. N. Garman, J. R. Jones, S. Stefaniak, A. D. Morgan, A. J. Simmonds, *J. Chem. Res. (S)* **1998**, 400.
64 R. N. Garman, PhD Thesis, University of Surrey **1998**.
65 H. Tagaki, T. Isoda, K. Kusakabe, S. Morooka, *Energy & Fuels* **1999**, 13, 1191.
66 V. Derdau, *Tetrahedron Lett.* **2004**, 45, 8889.
67 J. R. Jones, W. J. S. Lockley, S. Y. Lu, S. P. Thompson, *Tetrahedron Lett.* **2001**, 42, 331.
68 S. Rajagopal, A. F. Spatola, *J. Org. Chem.* **1995**, 60, 1347.
69 B. K. Banik, K. Barakat, D. R. Wagle, M. S. Manhas, A. K. Bose, *J. Org. Chem.* **1999**, 64, 5746.
70 H. Wiener, J. Blum, Y. Sasson, *J. Org. Chem.* **1991**, 56, 6145.
71 A. Loupy, A. Petit, J. Hamelin, F. Texier-Boullet, P. Jacquault, D. Mathé, *Synthesis* **1998**, 1213.
72 R. S. Varma, R. K. Saini, *Tetrahedron Lett.* **1997**, 33, 4337.

73. R. S. Varma, *Green Chem.* **1999**, 43.
74. W. Th. Erb, J. R. Jones, S. Y. Lu, *J. Chem. Res. (S)* **1999**, 728.
75. D. C. Sherrington, P. Hodge, *Syntheses and Separations Using Functional Polymers*, Wiley, New York, **1988**.
76. H. Firouzabadi, B. Tamami, N. Goudrazian, *Synth. Commun.* **1991**, 21, 2275.
77. B. Tamami, N. Goudrazian, *J. Chem. Soc., Chem. Commun.* **1994**, 1079.
78. G. Stöcklin, V. W. Pike (Eds), *Radiophamaceuticals for Positron Emission Tomography, Methodological Aspects*, Kluwer Academic Publishers, Dordrecht, **1993**.
79. W. Tarpey, H. Hauptmann, B. M. Tolbert, H. Rapoport, *J. Am. Chem. Soc.* **1950**, 72, 5126.
80. S. H. Pine, B. L. Sanchez, *J. Org. Chem.* **1971**, 36, 829.
81. S. Torchy, D. Barbry, *J. Chem. Res. (S)* **2001**, 292.
82. J. R. Jones, J. R. Harding, S. Y. Lu, R. Wood, *Tetrahedron Lett.* **2002**, 43, 9487.
83. T. Cohen, R. A. Schambach, *J. Am. Chem. Soc.* **1970**, 92, 3189.
84. G. B. Jones, B. J. Chapman, *J. Org. Chem.* **1993**, 58, 5558.
85. L. B. Frederiksen, T. H. Grobosch, J. R. Jones, S. Y. Lu, C. C. Zhao, *J. Chem. Res. (S)* **2000**, 42.
86. J. An, L. Bagnell, T. Cablewski, C. R. Strauss, R. W. Trainor, *J. Org. Chem.* **1997**, 62, 2505.
87. C. R. Strauss, R. W. Trainor, *Aust. J. Chem.* **1998**, 51, 703.
88. C. R. Strauss, *Aust. J. Chem.* **1999**, 52, 83.
89. D. H. Barton, F. Taran, *Tetrahedron Lett.* **1998**, 39, 4777.
90. D. H. R. Barton, E. Doris, F. Taran, *J. Labelled Compd. Radiopharm.* **1998**, 41, 871.
91. F. Taran, E. Doris, J. P. Noel, *J. Labelled Compd. Radiopharm.* **2000**, 43, 287.
92. D. H. R. Barton, D. O. Jang, J. C. Jaszberenyi, *J. Org. Chem.* **1993**, 58, 6838.
93. M. Saljoughian, *Synthesis* **2002**, 13, 1781.
94. R. H. Crabtree, *J. Chem. Soc., Chem. Commun* **1999**, 1611.
95. L. P. Kingston, W. J. S. Lockley, A. N. Mather, S. P. Thompson, D. J. Wilkinson, in *Synthesis and Applications of Isotopically Labelled Compounds* Vol 7, U. Pleiss, R. Voges (eds.), Wiley, Chichester, **2001**, p139.
96. B. McAuley, M. J. Hickey, L. P. Kingston, J. R. Jones, W. J. S. Lockley, A. N. Mather, E. Spink, S. P. Thompson, D. J. Wilkinson, *J. Labelled Compd. Radiopharm.* **2003**, 46, 1191.
97. L. P. Kingston, W. J. S. Lockley, A. N. Mather, E. Spink, S. P. Thompson, D. J. Wilkinson, *Tetrahedron Lett.* **2000**, 41, 2705.
98. L. P. Kingston, W. J. S. Lockley, A. N. Mather, E. Spink, S. P. Thompson, D. J. Wilkinson, in *Synthesis and Applications of Isotopically Labelled Compounds* Vol 7, U. Pleiss, R. Voges (eds.), Wiley, Chichester, **2001**, p79.
99. P. A. Keifer, S. H. Smallcombe, E. H. Williams, K. E. Solomon, G. Mendez, J. L. Belletire, C. D. Moore, *J. Comb. Chem.* **2000**, 2, 151.
100. M. A. Bernstein, J. M. Dixon, D. L. Hardy, R. J. Lewis, in Abstract of the 41st Experimental Nuclear Magnetic Resonance Conference, Asilomar, USA, April 9–14, **2000**.
101. M. R. Chappelle, J. R. Harding, B. B. Kent, J. R. Jones, S. Y. Lu, A. D. Morgan, *J. Labelled Compd. Radiopharm.* **2003**, 46, 567.
102. M. R. Chappelle, B. B. Kent, J. R. Jones, S. Y. Lu, A. D. Morgan, *Tetrahedron Lett.* **2002**, 43, 5117.
103. H. E. Blackwell, *Org. Biomol. Chem.* **2003**, 1, 1251.
104. L. Williams, *Chem. Commun.* **2000**, 435.
105. J. R. Jones, P. B. Langham, S. Y. Lu, *Green Chem.* **2002**, 4, 464.
106. D. S. Farrier, J. R. Jones, J. P. Bloxsidge, J. A. Elvidge, M. Saieed, *J. Labelled. Compd. Radiopharm.* **1982**, 19, 213.
107. L. Carroll, J. R. Jones, P. R. Shore, *J. Labelled. Compd. Radiopharm.* **1986**, 24, 763.

108 J. S. Fowler, A. P. Wolf, *The Synthesis of Carbon-11, Fluorine-18 and Nitrogen-13 Labelled Radiotracers for Biomedical Applications*, Technical Information Center, US Department of Energy, **1982**.

109 M. J. Welch, C. S. Redvanly (eds), *Handbook of Radiopharmaceuticals – Radiochemistry and Applications*, Wiley, Chichester, **2003**.

110 D. R. Hwang, S. M. Moerlein, L. Lang, M. J. Welch, *J. Chem. Soc., Chem. Commun.* **1987**, 1799.

111 M. C. Lasne, C. Perrio, J. Rouden, L. Barre, D. Roeda, F. Dollé, C. Crouzel, *Top. Curr. Chem.* **2002**, 222, 201.

112 R. Bolton, *J. Labelled Compd. Radiopharm.* **2002**, 45, 485.

113 A. Corsaro, U. Chiacchio, V. Pistara, G. Romeo, *Curr. Org. Chem.* **2004**, 8, 511.

114 A. de La Hoz, A. Diaz-Ortiz, A. Moreno, *Curr. Org. Chem.* **2004**, 8, 903.

115 S. Stone-Elander, N. Elander, J. O. Thorell, G. Solås, J. Svennebrink, *J. Labelled Compd. Radiopharm.* **1994**, 34, 949.

116 P. H. Elsinga, A. Van Waarde, G. M. Visser, W. Vaalburg, *Nucl. Med. Bio.* **1994**, 21, 211.

117 J. O. Thorell, M. H. Hedberg, A. M. Johansson, U. Hacksell, S. Stone-Elander, L. Eriksson, M. Ingvar, *J. Labelled Compd. Radiopharm.* **1995**, 37, 314.

118 J. O. Thorell, S. Stone-Elander, M. Ingvar, *J. Labelled Compd. Radiopharm.* **1994**, 35, 496.

119 J. Zhang, C. S. Dence, T. J. McCarthy, M. J. Welch, *J. Labelled Compd. Radiopharm.* **1995**, 37, 240.

120 E. D. Hostetler, G. E. Terry, H. D. Burns, *J. Labelled Compd. Radiopharm.* **2005**, 48, 629.

121 J. O. Thorell, S. Stone-Elander, N. Elander, *J. Labelled Compd. Radiopharm.* **1992**, 31, 207.

122 C. Giron, G. Luurtsema, M. G. Vos, P. H. Elsinga, G. M. Visser, W. Vaalburg, *J. Labelled Compd. Radiopharm.* **1995**, 37, 752.

123 J. O. Thorell, S. Stone-Elander, N. Elander, *J. Labelled Compd. Radiopharm.* **1994**, 34, 383.

124 J. O. Thorell, S. Stone-Elander, N. Elander, *J. Labelled Compd. Radiopharm.* **1993**, 33, 995.

125 J. O. Thorell, S. Stone-Elander, M. Ingvar, L. Eriksson, *J. Labelled Compd. Radiopharm.* **1995**, 36, 251.

126 J. O. Thorell, S. Stone-Elander, T. Duelfer, S. X. Cai, L. Jones, H. Pfefferkorn, G. Ciszewska, *J. Labelled Compd. Radiopharm.* **1998**, 41, 345.

127 N. W. Nader, S. K. Zeisler, A. Theobald, F. Oberdorfer, *Appl. Radiat. Isot.* **1998**, 49, 1599.

128 D. Roeda, C. Crouzel, *Appl. Radiat. Isot.* **2001**, 54, 935.

129 S. Y. Lu, J. Hong, V. W. Pike, *J. Labelled Compd. Radiopharm.* **2003**, 46, 1249.

130 C. Aubert, C. Huard-Perrio, M. C. Lasne, *J. Chem. Soc., Perkin I* **1997**, 2837.

131 C. Huard-Perrio, C. Aubert, M. C. Lasne, *J. Chem. Soc., Perkin I* **2000**, 311.

132 D. R. Hwang, S. M. Moerlein, M. J. Welch, *J. Nucl. Med.* **1989**, 26, 1757.

133 S. D. Jonson, M. J. Welch, *Nucl. Med. Bio.* **1999**, 26, 131.

134 T. L. Collier, M. M. Goodman, G. W. Kabalka, C. P. D. Longford, *J. Nucl. Med.* **1992**, 33, 1025.

135 D. Le Bars, C. Lemaire, N. Ginovart, A. Plenevaux, J. Aerts, C. Brihaye, W. Hassoun, V. Leviel, P. Mekhsian, D. Weissmann, J. F. Pujol, A. Luxen, D. Comar, *Nucl. Med. Bio.* **1998**, 25, 343.

136 C. Lemaire, R. Cantineau, M. Guillaume, A. Plenevaux, L. Christiaens, *J. Nucl. Med.* **1991**, 32, 2266.

137 P. Tan, R. M. Baldwin, R. Soufer, P. K. Garg, D. S. Charney, R. B. Innis, *Appl. Radiat. Isot.* **1999**, 50, 923.

138 M. Vandecapelle, F. Dumont, F. De Vos, K. Strijckmans, D. Leysen, K. Audenaert, R. A. Dierckx, G. Slegers, *J. Labelled Compd. Radiopharm.* **2004**, 47, 531.

139 S. Y. Lu, J. Hong, J. L. Musachio, F. T. Chin, E. S. Vermeulen, H. V.

Wikström, V. W. Pike, *J. Labelled Compd. Radiopharm.* **2005**, 48, 971.

140 M. Karramkam, F. Hinnen, M. Berrehouma, C. Hlavacek, F. Vautrey, C. Halldin, J. A. McCarron, V. W. Pike, F. Dollé, *Bioorg. Med. Chem.* **2003**, 11, 2769.

141 L. Dolci, F. Dollé, H. Valette, F. Vaufrey, C. Fuseau, M. Bottlaender, C. Crouzel, *Bioorg. Med. Chem.* **1999**, 7, 467.

142 F. Dollé, H. Valette, M. Bottlaender, F. Hinnen, F. Vaufrey, I. Guenther, C. Crouzel, *J. Labelled Compd. Radiopharm.* **1998**, 41, 451.

143 L. Dolci, F. Dollé, S. Jubeau, F. Vaufrey, C. Crouzel, *J. Labelled Compd. Radiopharm.* **1999**, 42, 975.

144 M. Karramkam, F. Hinnen, Y. Bramoullé, S. Jubeau, F. Dollé, *J. Labelled Compd. Radiopharm.* **2002**, 45, 1103.

145 M. Karramkam, F. Dollé, H. Valette, L. Besret, Y. Bramoullé, F. Hinnen, F. Vaufrey, C. Franklin, S. Bourg, C. Coulon, M. Ottaviani, M. Delafroge, C. Loc'h, M. Bottlaender, C. Crouzel, *Bioorg. Med. Chem.* **2002**, 10, 2611.

146 M. Karramkam, F. Hinnen, F. Vautrey, F. Dollé, *J. Labelled Compd. Radiopharm.* **2003**, 46, 979.

147 W. R. Banks, D. R. Hwang, *Appl. Radiat. Isot.* **1994**, 45, 599.

148 T. J. McCarthy, A. U. Sheriff, M. J. Graneto, J. J. Talley, M. J. Welch, *J. Nucl. Med.* **2002**, 43, 117.

149 W. R. Banks, M. R. Satter, D. R. Hwang, *Appl. Radiat. Isot.* **1994**, 45, 69.

150 D. E. Ponde, C. S. Dence, D. P. Schuster, M. J. Welch, *Nucl. Med. Biol.* **2004**, 31, 133.

151 T. J. McCarthy, C. S. Dence, M. J. Welch, *Appl. Radiat. Isot.* **1993**, 44, 1129.

152 C. Gilissen, G. Bormans, T. De Groot, A. Verbruggen, *J. Labelled Compd. Radiopharm.* **1998**, 41, 491.

153 R. Chirakal, L. Girard, G. Firnau, E. S. Garnett, G. Rodrigues, B. McCarry, *J. Labelled Compd. Radiopharm.* **1992**, 32, 123.

154 R. Chirakal, B. McCarry, M. Lonergan, G. Firnau, S. Garnett, *Appl. Radiat. Isot.* **1995**, 46, 149.

155 M. D. Taylor, A. D. Roberts, R. J. Nickles, *Nucl. Med. Biol.* **1996**, 23, 605.

156 P. S. Johnström, S. Stone-Elander, *Appl. Radiat. Isot.* **1996**, 47, 401.

157 F. R. Wüst, T. Kniess, *J. Labelled Compd. Radiopharm.* **2003**, 46, 699.

158 J. T. Patt, M. Patt, *J. Labelled Compd. Radiopharm.* **2002**, 45, 1229.

159 S. Y. Lu, F. T. Chin, J. A. McCarron, V. W. Pike, *J. Labelled Compd. Radiopharm.* **2004**, 47, 289.

160 G. Getvoldsen, A. Fredriksson, N. Elander, S. Stone-Elander, *J. Labelled Compd. Radiopharm.* **2004**, 47, 139.

161 U. Yngve, T. S. Khan, M. Bergström, B. Långström, *J. Labelled Compd. Radiopharm.* **2001**, 44, 561.

162 I. Velikyan, G. J. Beyer, B. Långström, *Bioconjugate. Chem.* **2004**, 15, 554.

163 I. Velikyan, G. Lendvai, M. Välilä, A. Roivainen, U. Yngve, M. Bergström, B. Långström, *J. Labelled Compd. Radiopharm.* **2004**, 47, 89.

164 J. C. Hung, M. E. Wilson, M. L. Brown, R. J. Gibbons, *J. Nucl. Med.* **1991**, 32, 2162.

165 J. C. Hung, S. Chowdhury, M. G. Redfern, D. W. Mahoney, *Eur. J. Nucl. Med.* **1997**, 24, 655.

166 S. H. Park, H. J. Gwon, J. S. Park, K. B. Park, *QSAR Comb. Sci.* **2004**, 23, 868.

167 L. Hernandez, E. Casanova, A. Loupy, A. Petit, *J. Labelled Compd Radiopharm.* **2003**, 46, 151.

168 K. Ogawa, M. Sasaki, T. Nozaki, *Appl. Radiat. Isot.* **1997**, 48, 623.

19
Microwaves in Photochemistry

Petr Klán and Vladimír Církva

19.1
Introduction

Chemistry under extreme or nonclassical conditions is currently a dynamically developing issue in applied research and industry. Alternatives to conventional synthetic or waste treatment procedures might increase production efficiency or save the environment by reducing the use or generation of hazardous substances in chemical production.

Microwave (MW) energy is a nonclassical energy source, with ultrasound, high pressure, mechanical activation, or plasma discharge. Since first reports of the use of MW heating to accelerate organic chemical transformations [1, 2], numerous articles have been published on the subject of microwave-assisted synthesis and related topics – microwave chemistry has certainly became an important field of modern organic chemistry [3–14]. Microwave activation increases the efficiency of many chemical processes and can simultaneously reduce formation of the byproducts obtained from conventionally heated reactions. Chemical processes performed under the action of microwave radiation are believed to be affected in part by superheating, hot-spot formation, polarization, and spin alignment [6, 7, 12]. The existence of a specific nonthermal microwave effect in homogeneous reactions has been a matter of controversy in recent years [10, 13, 15–18].

Microwave heating has already been used in combination with other unconventional activation processes. Such combinations might have a synergic effect on reaction efficiencies or, at least, enhance them by summing the individual effects. Application of MW radiation to ultrasound-assisted chemical processes has recently been described by some authors [19–21]. Mechanical activation has also been successfully combined with MW heating to increase the chemical yields of several reactions [22]. There have also been attempts to affect *photochemical* reactions by use of other sources of nonclassical activation, for example ultrasound [23, 24].

Combined chemical activation by use of two different types of electromagnetic radiation, microwave and ultraviolet–visible, is covered by the discipline described in this chapter. The energy of MW radiation is substantially lower than that of UV

Microwaves in Organic Synthesis, Second edition. Edited by A. Loupy
Copyright © 2006 WILEY-VCH Verlag GmbH & Co. KGaA, Weinheim
ISBN: 3-527-31452-0

radiation, certainly not sufficient to disrupt the bonds of common organic molecules. We therefore assume that, essentially, photoinitiation is responsible for a chemical change and MW radiation subsequently affects the course of the subsequent reaction. The objective of microwave-assisted photochemistry is frequently, but not necessarily, connected with the *electrodeless discharge lamp* (EDL) which generates UV radiation when placed in the MW field.

This chapter gives a complete picture of our current knowledge of microwave-assisted photochemistry and contains recent and updated information not included in the preceding edition [25]. It provides the necessary theoretical background and some details about synthetic and other applications, the technique itself, and safety precautions. Although microwave-assisted photochemistry is a newly developing discipline of chemistry, recent advances suggest it has a promising future.

19.2
Ultraviolet Discharge in Electrodeless Lamps

The electrodeless discharge lamp (EDL) [26] consists of a glass tube ("envelope") filled with an inert gas and an excitable substance and sealed under a lower pressure of a noble gas. A low frequency electromagnetic field (radiofrequency or MW, 300–3000 MHz) can trigger gas discharge causing emission of electromagnetic radiation. This phenomenon has been studied for many years [27] and was already well understood in the nineteen-sixties [28]. The term "electrodeless" means the lamps lack electrodes within the envelope. Meggers [28] developed the first EDL using the mercury isotope ^{198}Hg in 1942 (Fig. 19.1); its earliest application was in absorption spectroscopy [29]. EDL are usually characterized by higher emission intensity than cathode lamps, lower contamination, because of the absence of the electrodes [30], and a longer lifetime [31]. The lamps have been used as light sources in a variety of applications, and in atomic spectrometers [32].

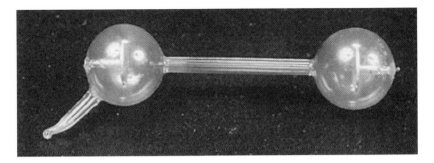

Fig. 19.1. The electrodeless mercury lamp made by William F. Meggers. With permission from the National Institute of Standards and Technology, Technology Administration, US Department of Commerce.

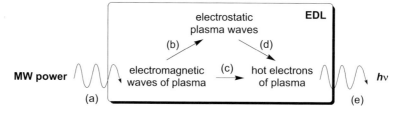

Fig. 19.2. Block diagram illustrating operation of the EDL: (a) energy flows from a MW source into the plasma chamber; (b) collisional or collisionless transformation; (c) normal or nonlinear wave absorption; (d) collisional or collisionless dumping; (e) collisional excitation of atoms and ions followed by emission. Adapted from Ref. [33].

19.2.1
Theoretical Aspects of the Discharge in EDL

The theory of EDL operation, as it is currently understood, is shown in Figs. 19.2 [33] and 19.3 (an example of a mercury EDL, or Hg EDL). Free electrons in the fill (i.e. electrons that have become separated from the environment because of the ambient energy) accelerate as a result of the energy of the electromagnetic (EM) field. They collide with the gas atoms and ionize them to release more electrons. Repetition of this causes the number of electrons to increase significantly over a short period of time, an effect known as an "avalanche". The electrons are gener-

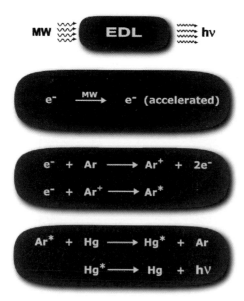

Fig. 19.3. The principle of operation of the mercury EDL and the emission of energy as UV–visible radiation.

ated by processes including *collisional* or *collisionless* transformation of EM waves, and *normal* or *nonlinear wave absorption* [30]. The energetic electrons collide with the heavy-atom particles present in the plasma, exciting them from the ground state to higher energy levels. The excitation energy is then released as EM radiation with spectral characteristics which depend on the composition of the envelope. The excited molecular or atomic species in the plasma can emit photons over very broad portion of the EM spectrum, ranging from X-rays to IR [34].

19.2.2
The Fundamentals of EDL Construction and Performance

The EDL system is modular and consists of two basic parts, a gas-filled bulb and a power supply with waveguides or external electrodes. A typical EDL is made of a scaled (usually quartz) tube envelope, which contains an inert gas (for example a noble gas) and an excitable substance (e.g., Hg, Cd, Na, Ga, In, Tl, Sc, S, Se, or Te) [35]. The envelope material must be impermeable to gases, an electrical insulator, and chemically resistant to the filling compounds at the temperature of operation.

Historically four basic methods have been used to excite discharges without electrodes [36–40]. In the first method, known as *capacitive coupling*, the electric field lines of the applied EM signal (usually 915 MHz) originate from one external electrode, pass through the gas-filled bulb containing the discharge, and terminate at a second external (coaxial) electrode. This discharge is similar to arc discharge in an electrode lamp, but needs a higher current. The second method of exciting EDL, with MW power (typically 2450 MHz), is to place the bulb in the path of radiation from a directional antenna. The *microwave discharge* is excited by both electric and magnetic components of the EM field. Because free propagation of the MW power occurs, however, emission is often inherently inefficient. This method is used for excitation of EDL inside a microwave oven. The third method is called the *traveling wave discharge* – a gap between the external electrodes provides the electric field that launches a surface wave discharge. The fourth method uses *inductive coupling* of the EDL, and the system can be compared with an electrical transformer. An alternating current in the coil causes a changing magnetic field inducing the electric field that drives a current into the plasma. The operating frequency is limited to approximately 50 kHz [41].

The construction of microwave-excited EDL is relatively straightforward but there are several operating conditions in their preparation which must be considered to produce an intense light source. The desired characteristics and requirements for EDL are high intensity, high stability, long lifetime, and, to a lesser extent, low cost and high versatility. In practice, it is very difficult to meet all these characteristics simultaneously.

The performance of EDL depends strongly on many preparation and operating conditions [35]:

- *The inert gas.* The arc chamber contains a buffer noble gas (usually Kr, Xe, or Ar) which is inert to the extent that it does not adversely affect the lamp operation.

Helium has higher thermal conductivity than other noble gases and, therefore, higher thermal conduction loss is observed [42]. The inert gas easily ionizes at low pressure but its transition to the thermal arc is slower and the lamp requires a longer warm-up time. Ionization is more difficult at higher pressures and requires a higher input power to establish the discharge. In general, the recommended pressure of the filling gas is between 0.266 and 2.66 kPa (2–20 Torr) at the operating temperature, which is usually much higher than that of a conventional electrode lamp. Use of argon was regarded as the best compromise between high EDL radiance and long lifetime. Air and nitrogen cannot be used because of their quenching properties in microwave plasmas, similar to water vapor.

- *Choice of fill material* initiating the discharge is very important. Together with a standard mercury fill it is often desirable to incorporate an additive in the fill material with a low ionization potential and sufficient vapor pressure (Cd, S, Se, Zn) [43, 44]. One category of low-ionization-potential materials is the group of alkali metals or their halides (LiI, NaI) but other elements, for example Al, Ga, In, or Tl [45, 46] or Be, Mg, Ca, Sr, La, Pr, or Nd [27, 42, 47], can be used. Other metal-containing compounds have been used to prepare EDL, including amalgams of Cd, Cu, Ag, and Zn. Multi-element EDL have been prepared using combinations of elements (e.g. Li–Na–K, As–Sb, Co–Ni, Cr–Mn, Bi–Hg–Se–Te, Cd–Zn, Ga–In, Se–Te) [48]. The spectral output from each individual element is very sensitive to temperature [49]. It has been found that no interelement interferences occur in the lamp.

- *Temperature of the lamp.* Operation at a high power or high temperatures can increase emission intensity but, at the same time, reduce the lamp lifetime and lead to broadening of the atomic line profile, because of self-absorption and self-reversal effects. It has been found that the optimum operating temperature for mercury filling is 42 °C (for the 254 nm line) [35]. The output is reduced when the temperature is beyond the optimum.

- *The dimensions and properties of the lamp envelope* are based on the discovery that the volume of Hg is critical for effective UV operation [50]. Higher Hg pressures result in the need to use higher microwave power levels. To focus the MW field efficiently into the EDL, a special Cd low-pressure lamp with a metal antenna (a molybdenum foil) was developed for experiments in MW-absorbing liquids [51]. The envelope material must be impermeable to gases, an electrical insulator, and chemically resistant to the filling compounds at the temperature of operation. High quality quartz is the most widely used lamp envelope material but early manufacturers of EDL used glass, Vycor, or Pyrex [52].

- *The nature and characteristics of the EM energy-coupling device.* For coupling of the MW energy to EDL, cavities (e.g. Broida-type or Evenson-type) and antennas (Raytheon) have been used. Optimum conditions for a lamp operation in one type of a MW cavity will by no means be optimum for operation in a different cavity, however. The results obtained in one MW oven will not, therefore, necessarily be the same as those from other tested cavities.

- *The frequency and intensity* of EM energy is determined by the type of a device.

Microwave energy is widely used for excitation of EDL because it is usually more efficient than radiofrequency energy for generation of intense light. Microwave radiation for excitation of gas discharges is usually generated by use of a fixed-frequency (2.45 GHz) magnetron oscillator.

19.2.3
EDL Manufacture and Performance Testing

Although general procedures of EDL manufacture are available in the literature [52–57], many minor details critical for proper lamp function are often omitted. The investigator who wants to make an EDL is thus faced with a very large amount of information dispersed in the literature and finds it very difficult to reproduce these procedures to develop EDL with the properties desired. An experimental vacuum system for EDL (Hg, HgI_2, Cd, I_2, KI, P, Se, S) manufacture has recently been designed by Církva and coworkers (Fig. 19.4) [58]. The technique is very simple and enables the preparation of EDL in a conventional chemistry laboratory. Examples of EDL are shown in Fig. 19.5. EDL performance is tested to prepare the lamps for spectral measurements [58]. A typical experimental system for such testing comprises a round-bottomed flask, placed in a MW oven, containing n-heptane and equipped with fiber-optic temperature measurement, a spectral probe, and a Dimroth condenser (Fig. 19.6).

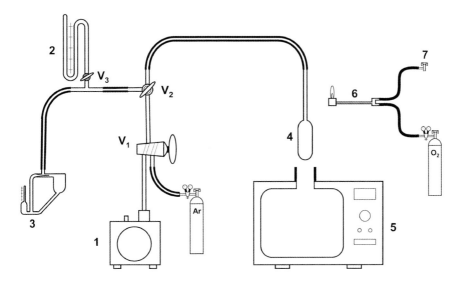

Fig. 19.4. A vacuum system for manufacture of EDL. 1, rotary vacuum pump; 2, mercury manometer; 3, tilting-type McLeod pressure gauge; 4, EDL blank; 5, modified microwave oven; 6, glass-working burner; 7, natural gas; V_1–V_3 are stopcocks. Adapted from Ref. [58].

19 Microwaves in Photochemistry

Fig. 19.5. EDL for photochemical applications.

Fig. 19.6. Testing EDL performance in a Milestone MicroSYNTH Labstation.

19.2.4
Spectral Characteristics of EDL

The spectral characteristics of EDL are of general interest in microwave-assisted photochemistry experiments. The right choice of EDL envelope and fill material can be very useful in planning an efficient course of the photochemical process without the need to filter out the undesirable part of the UV radiation by use of other tools, for example glass or solution filters or monochromators [59, 60].

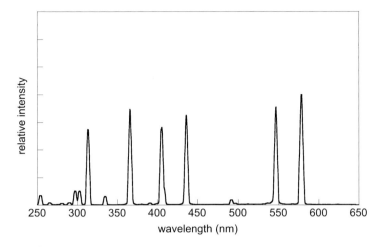

Fig. 19.7. The emission spectrum of an Hg EDL in n-decane (quartz envelope; argon atmosphere). With permission from Elsevier Science [60].

Whereas atomic fills usually furnish line emission spectra, molecular fills give continuous emission bands [61]. The total emission output of the most common lamp – the mercury EDL (Hg EDL) – in the region 200–600 nm is approximately the same as that of the electrode lamp with the same power input [62]. The distribution of the radiation is, however, markedly different, as a result of much higher Hg pressure and the greater number of atoms present in the plasma. EDL emit over three times as much UV and over a half as much IR as a conventional lamp [63]. It has been noted that EDL and electrode lamps provide different spectra when the fill contains a rare-earth material but similar spectra when a non-rare-earth fills are used [64]. Addition of material had very substantial effects on the spectral distributions of EDL [62].

Müller, Klán, and Církva have reported the emission characteristics of a variety of EDL containing different fill materials (for example Hg, HgI_2, Cd, I_2, KI, P, Se, or S) in the region 250–650 nm [60]. Whereas distinct line emission peaks were obtained for the mercury (Fig. 19.7), cadmium, and phosphorus (Fig. 19.8) fills, the iodine, selenium, and sulfur-containing EDL (Fig. 19.9) emitted continuous bands. Sulfur-containing EDL have been proposed for assisting phototransformations of environmental interest, because the emission flux is comparable with that of solar radiation. In addition, the EDL spectra could easily be modified by the choosing a suitable EDL envelope glass material, temperature, MW output power, or solvent, according to the needs of a photochemical experiment [59]. The relative intensities of the individual emission peaks in Hg EDL were found to be very dependent on temperature (35–174 °C); the short-wavelength bands (254 nm) were suppressed with increasing temperature. The emission spectra of quartz and Pyrex Hg EDL

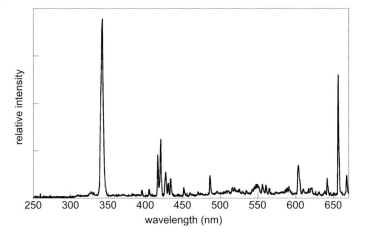

Fig. 19.8. The emission spectrum of a P EDL in n-decane (Pyrex envelope; argon atmosphere). With permission from Elsevier Science [60].

are compared in Fig. 19.10 (Pyrex absorbs most of the UV radiation below 290 nm). Most lamps emitted less efficiently below 280 nm than a standard Hg lamp. Table 19.1 summarizes characteristics, reported in the literature, of EDL filled with a variety of compounds.

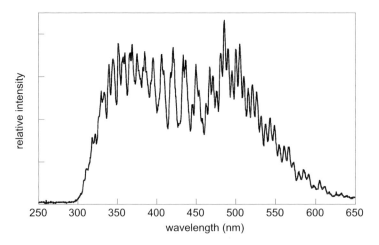

Fig. 19.9. The emission spectrum of an S EDL (quartz envelope; argon atmosphere). With permission from Elsevier Science [60].

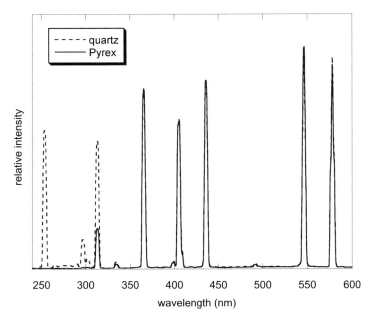

Fig. 19.10. The emission spectrum of a quartz and Pyrex Hg EDL in boiling *n*-hexane. With permission from Elsevier Science [59].

19.3
Photochemical Reactor and Microwaves

The photochemical reactor used for microwave-assisted experiments is an essential tool for experimental work. Such equipment enables simultaneous irradiation of the sample with both MW and UV–visible radiation. The idea of using an electrodeless lamp, in which the discharge is powered by the MW field, for photochemistry was born half a century ago [53, 62]. The lamp was originally proposed as a source of UV radiation only, without considering the effects of microwaves on photochemical reactions. The first applications of EDL were connected with the construction of a high-intensity source of UV radiation for atomic fluorescence flame spectrometry [88–90].

Gunning, Pertel, and their coworkers reported the photochemical separation of mercury isotopes [92–95] in a flow reactor which consisted of a microwave-operated discharge lamp [52, 96] cooled by a flowing film of water. A filter cell and a circulation system, to prevent heating of the filter solution and the cell, were placed concentrically and coaxially with the lamp. A similar reactor, for small-scale laboratory photolysis of organic compounds in the solution or gas phase, has been proposed by Den Besten and Tracy [91]. In this arrangement the EDL was placed in a reaction solution and was operated by means of an external microwave field from a radio or microwave-frequency transmitter (Fig. 19.11). The quantum output of the lamp was controlled by changing the output of the trans-

Tab. 19.1. Filling compounds and wavelengths of EDL emission.

Filling material (filling gas)	Excited species	Main emission bands, λ [nm]	Refs
Hg (Ar)	Hg	185, 254, 297, 313, 365, 405, 436, 546, 577, 579	35, 36, 50, 51, 60, 64–66
Cd (Ar)	Cd	229, 327, 347, 361, 468, 480, 509, 644	35, 51, 60, 67
SnI_2 (Ar)	SnI_2	400–850, 610	68, 69
$SnBr_2$ (Ar)	$SnBr_2$	400–850	70
BiI_3 (Ar)	BiI_3	300–750	71
$FeCl_2$ (Ar)	Fe	248, 272, 358, 372–376	35
Zn (Ar)	Zn	214, 330, 468	35, 51, 72
CuCl (Ar)	Cu	325, 327	35
NaI (Xe, Kr)	Na	589	73, 74
Mg, H_2 (Ar)	MgH	518, 521, 480–560	75
$AlBr_3$ (Ne)	AlBr	278	76
$AlCl_3$ (Ne)	AlCl	261	77, 78
Ga, GaI_3 (Ar)	Ga	403, 417, 380–450	65, 72
InI_3 (Ar)	In	410, 451	72
TlI (Ar)	Tl	277, 352, 378, 535	35, 72
P (Ar)	P	325, 327, 343	60
PCl_4 (Kr)	P_2	380	79
S (Ar)	S	320–850, 525	45, 60, 80–82
Se (Ar, Xe)	Se	370–850, 545	60, 81–84
Te (Xe)	Te	390–850, 565	81, 82, 84
Ar (Ar)	Ar_2	126, 107–165, 812	34, 85
Ar, Cl_2 (Ar)	ArCl	175	34, 85
Xe, Cl_2 (Xe)	XeCl	308	34, 85
B_2O_3, S (Kr)	B_2S_3	812	86
I_2 (Ar)	I_2	342	60
I_2, HgI_2 (Ar)	I	183, 206	87

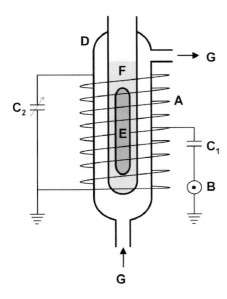

Fig. 19.11. Apparatus for electrodeless UV irradiation. A, antenna; B, transmitter; C_1, capacitor; C_2, variable capacitor; D, jacketed flask; E, EDL; F, reaction mixture; G, circulating coolant. Adapted from Ref. [91].

mitter or by using a dilute ionic solution circulating through the cooling jacket. For maximum lamp output a weakly conducting solution has been proposed. Placing EDL in the solution was quite advantageous, because the full quantum output was used. The authors recommended keeping the sample temperature lower, because EDL produce a substantial amount of heat.

The use of a domestic microwave oven appeared in a patent [97], according to which gaseous reactants were irradiated with microwave and UV–visible radiation to produce desired photoproducts (the EDL was positioned inside the MW cavity, although outside the reaction vessel). Several similar reactors have been proposed for UV sterilization [98–100] or for treatment of waste water containing organic pollutants [101–103].

Církva and Hájek have proposed a simple application of a domestic microwave oven for microwave-assisted photochemistry experiments [105]. In this arrangement the EDL (the MW-powered lamp for this application was specified as a microwave lamp or MWL) was placed in a reaction vessel located in the cavity of an oven. The MW field generated a UV discharge inside the lamp that resulted in simultaneous UV and MW irradiation of the liquid sample. This arrangement provided the unique possibility of studying photochemical reactions under extreme thermal conditions [106].

Klán, Literák, and coworkers published a series of papers that described the scope and limitations of this reactor [104, 107–109]. In a typical design (Figs. 19.12 and 19.13), four holes were drilled into the walls of a domestic oven – one

872 | *19 Microwaves in Photochemistry*

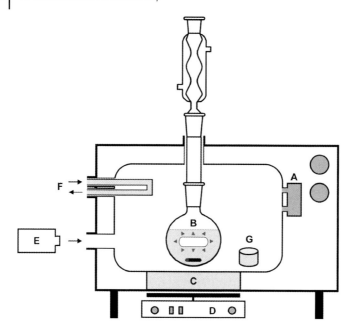

Fig. 19.12. A modified MW oven for microwave-assisted photochemistry experiments. A, magnetron; B, reaction mixture with EDL and magnetic stir bar; C, aluminum plate; D, magnetic stirrer; E, infrared pyrometer; F, circulating water in a glass tube, G, dummy load inside the oven cavity. With permission from Elsevier Science [104].

Fig. 19.13. Photochemistry in a microwave oven (the EDL floats on the liquid surface).

Tab. 19.2. Advantages and disadvantages of EDL applications in photochemistry. Adapted from Ref. [108].

Advantages

Simultaneous UV and MW irradiation of the sample
Possibility of performing photochemistry at high temperatures
Good photochemical efficiency – the EDL is "inside" the sample
Simplicity of the experimental arrangement and a low cost of the EDL
Easy method of EDL preparation in the laboratory
Use of a commercially available microwave oven
"Wireless" EDL operation
Choice of the EDL material might modify its spectral output

Disadvantages

Technical difficulties of performing experiments at temperatures below the solvent b.p.
Greater safety precautions
EDL overheating causes lamp emission failure
Polar solvents absorb MW radiation, thus reducing the UV output efficiency of the EDL

for a condenser tube in the oven top, another in the side for an IR pyrometer, and two ports for a glass tube with circulating water. Part of the oven bottom was replaced with an aluminum plate to enable magnetic stirring. The opening for the IR pyrometer could also serve for an external (additional) source of UV radiation. The vessel was connected to a very efficient water-cooled condenser by means of a long glass tube. The circulating cool water or different amounts of a MW-absorbing solid material (dummy load – basic Al_2O_3, molecular sieve, etc.) were used when a small quantity of a nonabsorbing or poorly absorbing sample was used. This material removed excess microwave power and prevented the magnetron from being destroyed by overheating. The EDL had always to be placed in a position in which the solvent cooled it efficiently, because lamp overheating might cause failure of lamp emission. Intense IR output from the lamp triggered immediate boiling of all solvents including nonpolar (MW-transparent) liquids [107, 108]. Polar solvents, on the other hand, absorbed most of MW radiation, resulting in reduced UV output efficiency. Table 19.2 depicts the most important advantages and disadvantages of EDL applications.

Chemat and his coworkers [110] have proposed an innovative combined MW–UV reactor (Fig. 19.14) based on a commercially available MW reactor, the Synthewave 402 (Prolabo) [8]. This is a monomode microwave oven cavity operating at 2.45 GHz designed for both solvent and dry-media reactions. A sample in the quartz reaction vessel could be mechanically stirred and its temperature was monitored by means of an IR pyrometer. The reaction systems were irradiated by means of an external source of UV radiation (a 240-W medium-pressure mercury lamp).

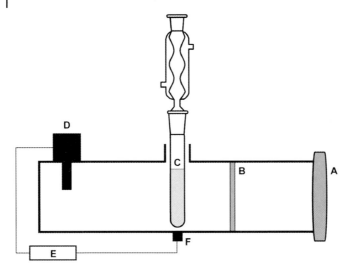

Fig. 19.14. Reactor for microwave-assisted photochemistry based on the Synthewave 402 (Prolabo). A, medium-pressure Hg lamp; B, window opaque to MW radiation; C, reaction mixture; D, magnetron; E, regulator; F, IR sensor. Adapted from Ref. [110].

Similar photochemical applications in a Synthewave reactor using either an external or an internal UV source have been reported by Louerat and Loupy [111].

A microwave-assisted, high-temperature, and high-pressure UV digestion reactor has been developed by Florian and Knapp [51] for analytical purposes. The apparatus contained an immersed electrodeless discharge lamp operating as a result of the MW field in the oven cavity (Fig. 19.15). An antenna, fixed to the top of the EDL enhanced the EDL excitation efficiency. Another interesting MW–UV reactor has been designed by Howard and his coworkers [112]. A beaker-shaped electrodeless discharge lamp, placed in a modified domestic MW oven has been used for mineralization of organophosphate compounds. The samples in quartz tubes were positioned in a carousel inside an open UV beaker; they were thus efficiently photolyzed from the whole surface of the beaker.

Microwave irradiation has also been used for heterogeneous photocatalytic oxidation of solutions of a variety of organic compounds in aqueous TiO_2 dispersions [113, 114] or ethylene in the gas phase using a TiO_2–ZrO_2 mixed oxide [115]. The microwave photocatalytic reactors consisted either of an external UV source irradiating the sample placed inside the MW cavity [114–117], in a manner similar to that shown in Fig. 19.14, or of EDL in the cavity and powered by microwaves [115, 118]. Horikoshi, Hidaka, and Serpone have proposed a flow-through quartz photoreactor for photocatalytic remediation of aqueous solutions (Fig. 19.16) under the action of irradiation from a 1.5-kW microwave generator [119–121]. The incident and reflected MW radiation was determined and the circulating dispersion

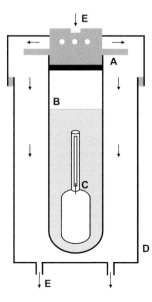

Fig. 19.15. Simplified schematic diagram of a high-pressure digestion vessel with EDL. A, plug and seal; B, quartz pressure reaction vessel with a sample solution; C, EDL with an antenna; D, PEEK vessel jacket with a screw cap; E, air flow. Adapted from Ref. [51].

was tempered in a cooling device. Either conventional or electrodeless mercury lamps were employed as sources of UV radiation [121]. A similar microwave-stimulated flow-through UV reactor was designed for disinfection of drinking, waste, and feed waters [122].

Církva and coworkers have studied a flow MW photoreactor containing a glass

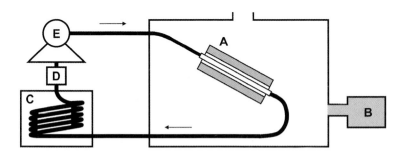

Fig. 19.16. Experimental setup of a flow-through quartz photoreactor used for photocatalytic decomposition in aqueous TiO$_2$ dispersions using cylindrical electrodeless mercury discharge lamps. A, cylindrical EDL with a quartz pipe with the sample solution in a microwave cavity; B, a microwave generator; C, cooling circulator device; D, thermometer; E, peristaltic pump. Adapted from Ref. [119].

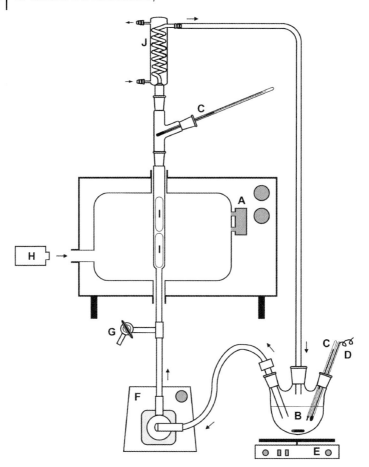

Fig. 19.17. A flow microwave photoreactor. A, microwave oven with magnetron; B, reaction mixture with magnetic stir bar; C, thermometer; D, pH meter with glass electrode; E, magnetic stirrer; F, PTFE membrane pump; G, outlet; H, spectrometer with a fiber-optic probe; I, glass tube with quartz Hg EDL; J, condenser [123].

tube with quartz Hg EDL (254 nm emission) inside a microwave oven, a PTFE membrane pump, a container flask, a cooling condenser, thermometers, and a pH meter with a glass electrode (Fig. 19.17) [123]. Photohydrolysis of chloroacetic acid to hydroxyacetic acid and hydrogen chloride was chosen as model reaction; the course of the reaction was followed by monitoring the change of pH of the solution. The conversion was optimized as a result of a trade-off between the thermal dependence of the quantum yield (which increased with increasing temperature) and the thermal dependence of a relative intensity of a short-wavelength band (which increased with decreasing temperature).

Microwave-enhanced chemistry introduces unique *safety considerations* not encountered by the chemist in other fields of chemistry [124]. Careful planning of all experiments is strongly advised, especially when the results are uncertain, because control of the reaction temperature might be complicated by rapid heat-transfer. It is, furthermore, well known that electronically excited singlet oxygen, capable of causing serious physiological damage, is generated by microwave discharge through an oxygen stream [125]. The combined effect of MW and UV irradiation could increase the singlet oxygen concentration in the MW cavity, particularly in the presence of a photosensitizer.

19.4 Interactions of Ultraviolet and Microwave Radiation with Matter

Although microwave chemistry has already received widespread attention from the chemical community, considerably less information is available about the effect of microwave radiation on photochemical reactions. Photochemistry is the study of the interaction of ultraviolet or visible radiation ($E = 600$–170 kJ mol^{-1} at $\lambda = 200$–700 nm) with matter. The excess energy of electronically excited states significantly alters species reactivity – it corresponds, approximately, to typical reaction activation energies helping the molecules overcome activation barriers. The microwave region of the electromagnetic spectrum, on the other hand, lies between infrared radiation and radio frequencies. Its energy ($E = 1$–100 J mol^{-1} at $v = 1$–100 GHz) is approximately 3–6 orders of magnitude lower than that of UV radiation (a typical MW kitchen oven operates at 2.45 GHz). Microwave heating is not identical with classical external heating, at least at the molecular level. Molecules with a permanent (or induced) dipole respond to an electromagnetic field by rotating, which results in friction with neighboring molecules (thus generating heat). Additional (secondary) effects of microwaves include ionic conduction (ionic migration in the presence of an electric field) or spin alignment.

Simultaneous UV–visible and MW irradiation of molecules, which does not necessarily cause any chemical change, might affect the course of a reaction by a variety of mechanisms at *each* step of the transformation. From many possibilities, let us present a simplified model describing two main distinct pathways (Fig. 19.18). The first route, more probable, is a photochemical reaction starting with a ground state molecule **M**, which is electronically excited to **M***, transformed into an intermediate (or a transition state) **I**, and, finally, a product **P**. Virtually every step may

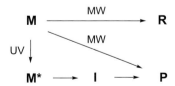

Fig. 19.18. Simplified model of nonsynergistic effects of UV and MW radiation on a chemical reaction.

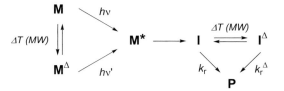

Fig. 19.19. A simplified model of the synergistic effect of UV and MW radiation on a chemical reaction, where Δ denotes "hot" molecules, and k_r and k_r^Δ are the rate constants of the processes leading eventually to the same product **P**.

be complicated by a parallel microwave-assisted reaction enabling a different chemical history. There is a theoretical possibility that MW radiation affects the electronically excited molecule **M*** or a short-lived transition state. In such circumstances the lifetime of the species should be long enough to interact with this low-frequency radiation. The second pathway becomes important when MW initiate a "dark" chemical reaction (essentially through polar mechanisms), competitive with or exclusive to a photochemical pathway, yielding a different (**R**) or the same (**P**) product. Figure 19.18 depicts a model in which MW and UV effects are easily distinguishable – it is assumed there is no synergistic effect during a single step of the transformation.

Let us, on the other hand, assume that the efficiency of a photoreaction is altered by microwave induction. In an example shown in Fig. 19.19 microwave *heating* alters the excitation energy of the starting molecule. Both types of electromagnetic radiation simultaneously affect a single chemical step in which the ground-state molecules **M** and **M**$^\Delta$ (a MW-heated molecule) are being excited. If, furthermore, the intermediates **I** and **I**$^\Delta$ react with different rate constants, the total observed rate constant of the reaction, k_{obs}, is proportional to the sum $k_{obs} \approx (\chi k_r + \chi^\Delta k_r^\Delta)$, where χ and χ^Δ represent the populations of **I** and **I**$^\Delta$.

19.5
Photochemical Reactions in the Microwave Field

19.5.1
Thermal Effects

Baghurst and Mingos have hypothesized that superheating of polar solvents at atmospheric pressure, so that average temperatures are higher than the corresponding boiling points, is a result of microwave dissipation over the whole volume of a liquid [126]. With the absence of nucleation points necessary for boiling, heat loss occurs at the liquid–reactor wall or at liquid–air interfaces. Many reaction efficiency enhancements reported in the literature have been explained as the effect

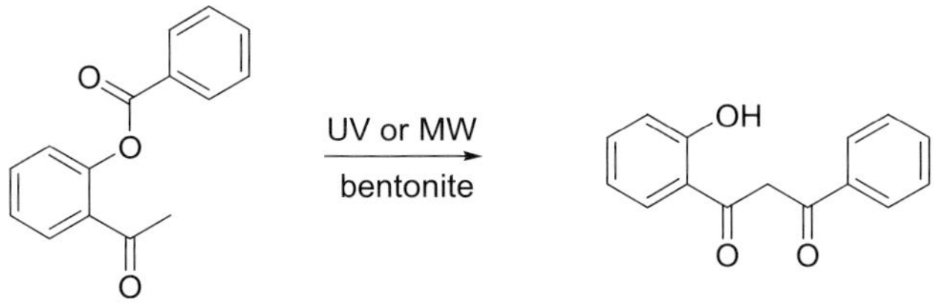

19.5 Photochemical Reactions in the Microwave Field | 879

Scheme 19.1

of superheating when the reactions were essentially performed in sealed vessels without any stirring [127–131]; this effect is also expected in microwave-assisted photochemistry experiments in condensed media. Gedye and Wei [16], for example, have observed enhancements of the rate of several different thermal reactions by factors of 1.05 to 1.44 in experiments accomplished in a domestic-type MW oven but not in a variable-frequency microwave reactor. The enhancement was interpreted as a consequence of solvent superheating or hot-spot formation rather than nonthermal effects. Stadler and Kappe reported similar results in an interesting study of the MW-mediated Biginelli reaction [15].

Chemat et al. [110] reported the UV and MW-induced rearrangement of 2-benzoyloxyacetophenone, in the presence of bentonite, into 1-(o-hydroxyphenyl)-3-phenylpropane-1,3-dione in methanol at atmospheric pressure (Scheme 19.1). The reaction, performed in the reactor shown in Fig. 19.14, was subject to a substantial activation effect under simultaneous UV and MW irradiation; this corresponded at least to the sum of the individual effects (Fig. 19.20). The rearrangement was not studied in further detail, however. Such competitive processes can be described by the diagram in Fig. 19.18, because the product obtained from both types of activation was the same.

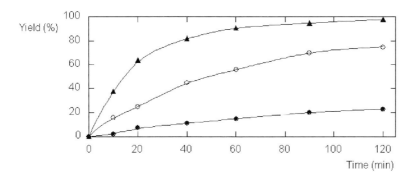

Fig. 19.20. Reaction yields in the rearrangement of 2-benzoyloxyacetophenone induced by microwave heating (●), ultraviolet irradiation (○), or simultaneous UV and MW irradiation (▲). Adapted from Ref. [110].

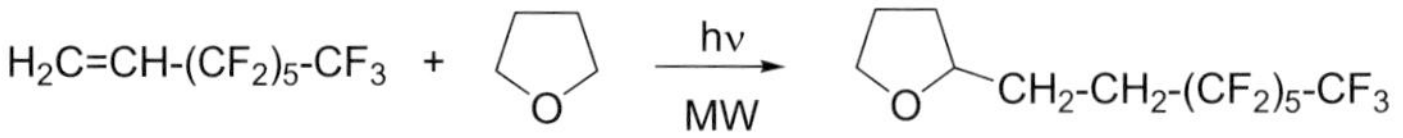

H₂C=CH-(CF₂)₅-CF₃ + [THF] $\xrightarrow{h\nu, MW}$ [THF]-CH₂-CH₂-(CF₂)₅-CF₃

Scheme 19.2

Církva and Hájek have studied the photochemically or microwave-induced addition of tetrahydrofuran to perfluorohexylethene (Scheme 19.2) [105]. Whereas the thermal reaction was too slow, photochemical activation was very efficient, with no apparent thermal effects of MW radiation. Combined UV and MW radiation (Fig. 19.12) has principally been used to initiate EDL operation in the reaction mixture. Another illustration of the MW–UV-assisted reaction has been demonstrated by Nüchter et al. [22] on dehydrodimerization reactions of some hydrocarbons.

Klán et al. [109] successfully evaluated MW superheating effects in polar solvents by use of a temperature-dependent photochemical reaction. It is known that quantum efficiencies of the Norrish type II reaction [132] of *p*-substituted valerophenones depend on the presence of a weak base, because of specific hydrogen bonding to the biradical OH group (**BR**; Scheme 19.3) [106, 107]. The efficiency of this reaction was linearly dependent on temperature over a broad temperature range and the system served as a photochemical thermometer at the molecular level, even for the MW-heated mixtures. The magnitude of the photochemical change in the MW field suggested the presence of a superheating effect (4–11 °C) for three aliphatic alcohols and acetonitrile as reaction solvents. The results were in a perfect agreement with measurements by use of a fiber-optic thermometer.

Klán et al. recently studied temperature-sensitive, regioselective photochemical nucleophilic aromatic substitution of 4-nitroanisole by the hydroxide anion in

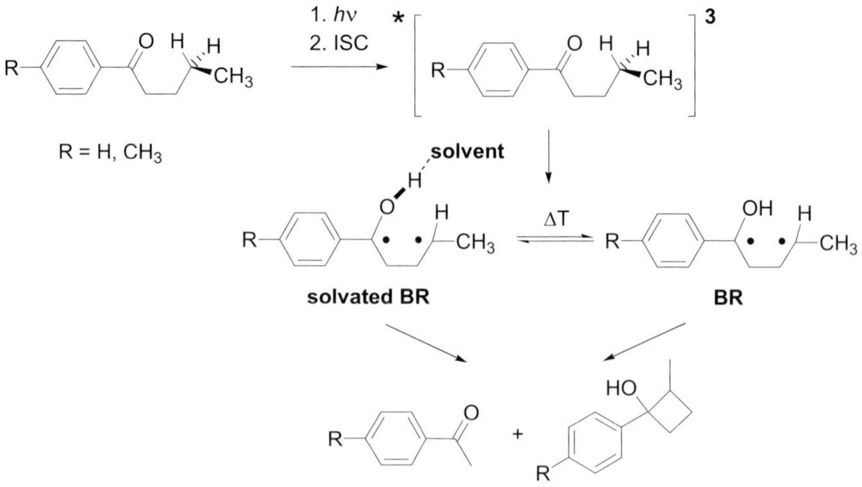

Scheme 19.3

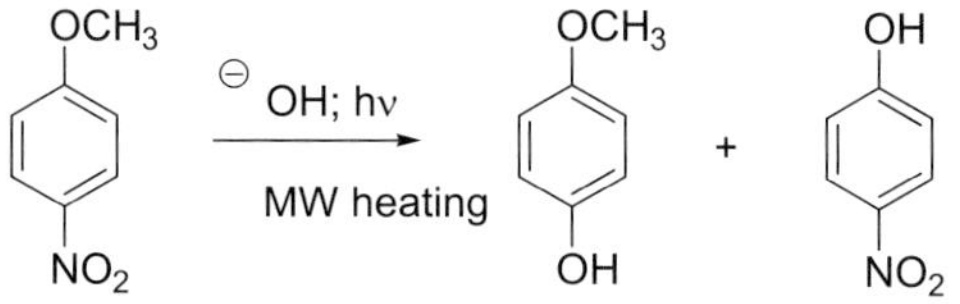

Scheme 19.4

homogeneous solutions by use of microwave heating and an EDL (Fig. 19.12) [133]. The quantum yield for formation of one product (4-methoxyphenol) was found to be independent of temperature, in contrast with that for formation of 4-nitrophenol, suggesting the occurrence of a temperature-dependent process after partitioning between replacement of the nitro and methoxy groups (Scheme 19.4). The technique of microwave-assisted photochemistry was proposed in this paper as an efficient and practical tool for organic synthesis. Subsequent investigation of the release of a photoremovable protecting group, the 2,5-dimethylphenacyl chromophore, from the carboxyl moiety, enhanced by MW heating, showed that quantum yields for ester degradation in polar solvents increased by a factor of three when the temperature was increased from 20 to 50 °C [134]. Distribution of the products and reaction conversions of several different photochemical systems, irradiated by use of a conventional UV source and by an EDL in a MW–UV reactor (Fig. 19.12), were compared to elucidate the advantages and disadvantages of this technique [108]. Some reactions, e.g. photolysis of phenacyl benzoate in the presence of triethylamine or photoreduction of acetophenone by 2-propanol, were moderately enhanced by MW heating. Two temperature-sensitive model photochemical reactions, the Norrish Type II reaction and photochemical nucleophilic aromatic substitution of 4-nitroanisole by the hydroxide ion, have recently been studied in high-temperature water (100–200 °C) in a pressurized vessel under microwave heating [135]. The observed chemoselectivity and the ability to increase the solubility of hydrophobic organic compounds in this solvent were found to be promising results for environment-friendly (photo)chemical applications.

Církva et al. have recently investigated an effect of UV and combined MW–UV irradiation on the transformation of 2-tert-butylphenol (2TBP) in the presence and the absence of sensitizers with different values of singlet and triplet energy, and in the presence of solvents with different polarity [136]. UV or combined UV–MW irradiation of the starting molecules furnished three isomers – 3,3′-di-tert-butylbiphenyl-2,2′-diol (ortho–ortho), 3,3′-di-tert-butylbiphenyl-2,4′-diol (ortho–para), and 3,3′-di-tert-butylbiphenyl-4,4′-diol (para–para) (Scheme 19.5). Their concentration ratios depended on the nature of the solvents and sensitizers used. No significant specific effects of the combined MW–UV radiation on the distribution of the products from 2TBP photolysis were observed.

MW–UV irradiation of 4-tert-butylphenol (4TBP) has also been investigated [137]. Photolysis of this compound furnished 4′,5-di-tert-butyl-2-hydroxydiphenyl ether (ortho–O) and 5,5′-di-tert-butylbiphenyl-2,2′-diol (ortho–ortho), and 2TBP as

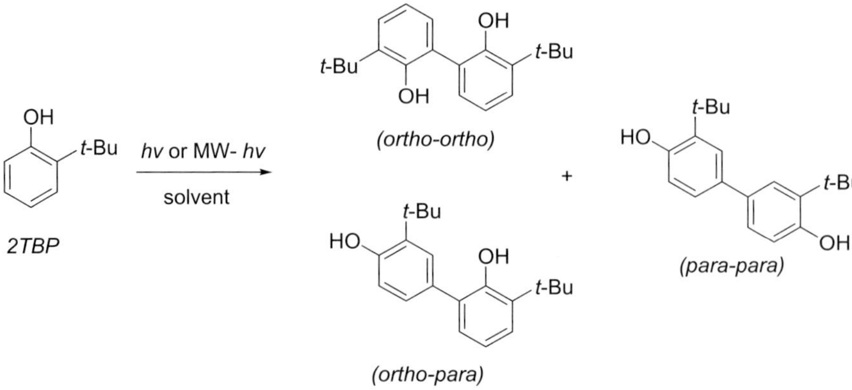

Scheme 19.5

an isomerization by-product (Scheme 19.6). Again, no specific MW effect was observed.

Kunz et al. have studied the hydrogen peroxide-assisted photochemical degradation of EDTA with a microwave-activated ultraviolet source (model UV LAB, UMEX) [138]. The effect of pH and H_2O_2–EDTA molar concentration ratio on the efficiency of degradation was evaluated. Han et al. investigated enhanced oxidative degradation of aqueous phenol in a UV–H_2O_2 system under the action of microwave irradiation [139]. The experimental results, based on the kinetic study, showed that MW irradiation can enhance both phenol conversion and the TOC removal efficiency by up to 50% or above.

Louerat and Loupy studied some photochemical reactions (e.g. stilbene isomerization) in homogenous solutions and on solid supports such as alumina [111] and the Norrish type II photoreaction of alkyl aryl ketones on alumina or silica gel surfaces has been investigated by Klan et al. [140]. On the basis of on these results, a model in which the short-lived biradical intermediate interacts with the surface, in addition to a polar effect on the excited triplet of ketone, has been proposed (Scheme 19.7). Both acidic and basic sites are present on the amphoteric alumina surface; while the acidic OH groups coordinate to the carbonyl oxygen, the basic groups (O^-) are involved in hydrogen bonding with the OH group of the biradical

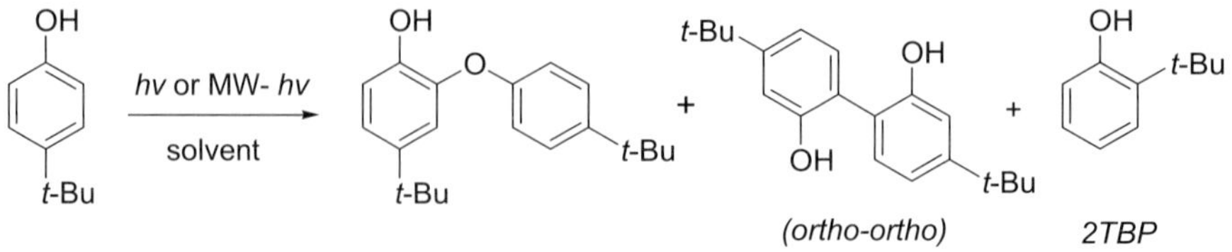

Scheme 19.6

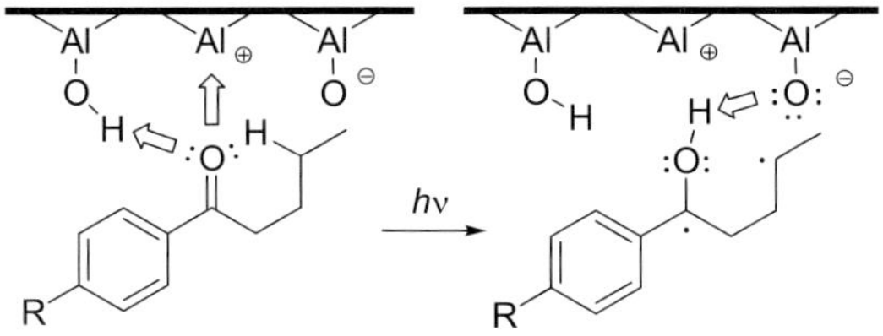

Scheme 19.7

intermediate. The change in the regioselectivity of the reaction as a result of microwave heating was explained in terms of the weakening of such interactions.

19.5.2
Microwaves and Catalyzed Photoreactions

Advanced oxidation processes, for example photocatalysis, have emerged as potentially powerful methods for transforming organic pollutants in aqueous solution into nontoxic substances. Such remediation usually relies on generation of reactive free radicals, especially hydroxyl radicals ($^{\bullet}$OH) [141]. These radicals react rapidly and usually nonselectively with most organic compounds, either by addition to a double bond or by abstraction of a hydrogen atom, resulting in a series of oxidative degradation reactions ultimately leading to mineralization products, for example CO_2 and H_2O [142].

In the past four years, Horikoshi, Hidaka, and Serpone have published results from a series of studies on environmental remediation of a variety of aqueous solutions by simultaneous MW–UV irradiation in the presence of a TiO_2 catalyst. They have clearly shown that this integrated illumination technique is superior to simple photocatalysis in the degradation of a variety of organic compounds, e.g. rhodamine-B dye [114, 119, 120, 143], bisphenol-A [117], 2,4-dichlorophenoxyacetic acid [118, 121], and carboxylic acids, aldehydes, and phenols [116]. Microwave radiation has occasionally been used to power an EDL inside the photochemical reactor [119, 120]. The authors suggested that, in addition to thermal effects, nonthermal effects [117, 121, 143] might govern microwave-assisted reactions in the presence of TiO_2, because combined UV and MW radiation generated more hydroxyl radicals (proved by use of electron spin resonance spectroscopy to detect the radicals) [144]. A solution of 5,5-dimethyl-1-pyrrolidine-N-oxide (DMPO; spin trap) containing TiO_2 was subjected to photolysis and/or thermolysis and the number of DMPO–$^{\bullet}$OH spin adducts was determined. MW irradiation yielded a small quantity of OH radicals; substantially more were produced by UV irradiation. Combined UV–MW treatment at the same sample temperature generated even more radicals, by a factor of approximately 2. It was suggested this increase was a result of nonthermal interactions between the MW field and the surface of the catalyst. The nature of such interactions, however, remains to be elucidated. It was, moreover, suggested that hydrophilic–hydrophobic changes on the TiO_2 surface as a

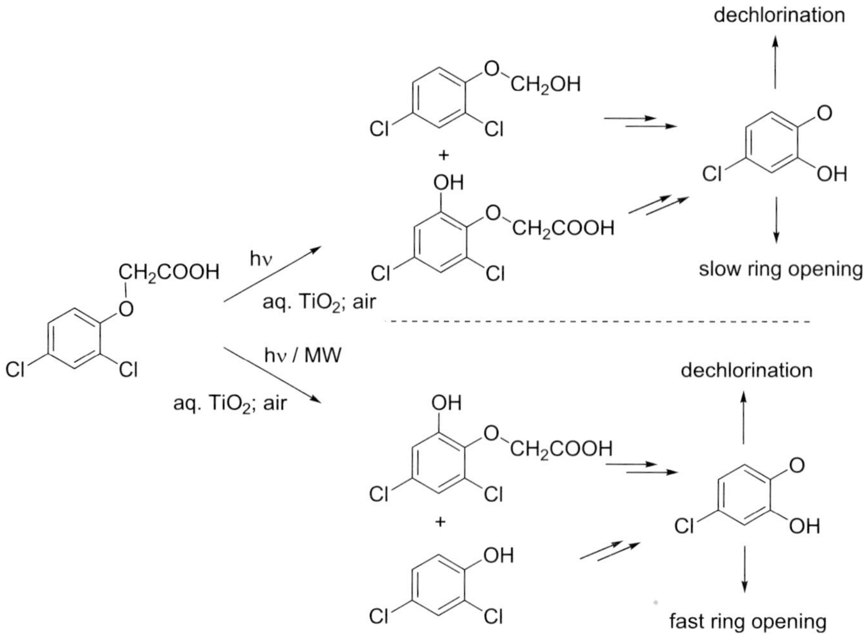

Scheme 19.8

result of MW radiation led to changes in the population of the surface hydroxyls [115]. When rhodamine-B, for example, was subjected to photocatalytic destruction in the absence of MW radiation, it was suggested the two oxygen atoms of the carboxylate moiety of the dye were interacting with the positively charged TiO_2 surface [143]. MW-assisted photolysis, however, increased the hydrophobic nature of TiO_2, as a result of the MW irradiation, and adsorption might be facilitated by the aromatic rings, eventually causing formation of different degradation intermediates than in the absence of MW. In a different project, the effect of dissolved oxygen and MW irradiation on photocatalytic destruction of 2,4-dichlorophenoxy-acetic acid was investigated [121]. The grater efficiency of MW-assisted degradation was ascribed to a nonthermal effect on ring-opening of the aromatic moiety via oxidative reaction (Scheme 19.8).

Heterogeneous catalytic degradation of humic acid in aqueous titanium dioxide suspension under MW–UV conditions was studied by Chemat et al. [110]. Enhancement in this application was reported as substantial – i.e. greater than simple addition of both effects. Zheng et al. [145–147] have recently reported microwave-assisted heterogeneous photocatalytic oxidation of ethylene using porous TiO_2 and $SO_4{}^{2-}$–TiO_2 catalysts. Significant enhancement of photocatalytic activity was attributed to the polarization effect of the high-defect catalysts in the MW field. These studies also included modeling of the photodegradation of organic pollutants in the microwave field. TiO_2–ZrO_2 mixed-oxide catalyst, prepared by sol–gel processing, has also been used in photocatalytic oxidation of ethylene by Anderson and coworkers [115]. The adsorption experiments demonstrated that MW irradiation removed water from the surface of the catalyst better than when heat was supplied by conductive (conventional) heating. Ai et al. [148] have investigated microwave-assisted photocatalytic degradation of 4-chlorophenol by use of an electrodeless discharge UV system. The effects of pH, irradiation intensity, aeration, and amount of H_2O_2 both on direct photolysis and on TiO_2 photocatalysis were evaluated. It was found that the process proceeds through the same mechanism as in the absence of the catalyst. Zhang and Wang have recently reviewed the effects and mechanisms of microwave-assisted photocatalysis [149].

19.5.3
Intersystem Crossing in Radical Recombination Reactions in the Microwave Field – Nonthermal Microwave Effects

Radical pairs and biradicals are extremely common intermediates in many organic photochemical (and some thermal) reactions. A singlet state intermediate is formed from the singlet excited state in reactions that conserve spin angular momentum whereas the triplet intermediate is obtained via the triplet excited state. Radical pairs in solution coherently fluctuate between singlet and triplet electronic states [150, 151] and the recombination reactions are often controlled by electron–nuclear hyperfine interactions (HFI) on a nanosecond time-scale [152, 153]. Only pairs of neutral radicals with singlet multiplicity recombine. A triplet pair intersystem crosses into the singlet pair or the radicals escape the solvent cage and react independently at a later stage (Fig. 19.21) [154]. The increasing efficiency of triplet-to-singlet interconversion ("mixing" of states) leads to a more rapid recombination reaction and *vice versa*. It is now well established that a *static* magnetic field can affect intersystem crossing in biradicals (magnetic field effect, MFE) and the effect has been successfully interpreted in terms of the radical pair mechanism [155, 156]. This concept enabled explanation of nuclear and electronic spin polarization phenomena during chemical reactions, e.g. chemically induced dynamic nuclear polarization (CIDNP) or reaction yield-detected magnetic resonance (RYDMAR).

An external magnetic field stronger than the hyperfine couplings inhibits (because of Zeeman splitting) singlet–triplet interconversions by isolating the triplets T_{+1} and T_{-1} from the singlet (S); these can, therefore, mix only with T_0 (Fig. 19.21a, b). For the *triplet-born radical pair*, the magnetic field reduces the probability of radical recombination. The microwave field, which is in resonance with the energy gaps between the triplet levels (T_{+1} or T_{-1}) and T_0, transfers the excess population from the T_{+1} or T_{-1} states back to a mixed state. Application of a strong magnetic field to the *singlet-born radical pair* leads to an increase in the probability of recombination that can, however, also be controlled by microwave irradiation [156].

This microwave-induced spin dynamics can be regarded as an archetypal *nonthermal* MW effect. Because the radical pair is usually created via a photochemical

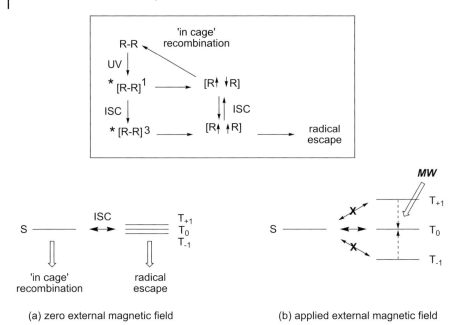

Fig. 19.21. Schematic illustration of magnetic field and MW effects in radical-pair chemistry.

pathway, the topic should certainly be included in this chapter. The literature offers many examples that span photobiology, photochemistry, and photophysics. Wasielewski et al. [157], for instance, showed that the duration of photosynthetic charge separation could be controlled by use of microwave radiation. It was, moreover, possible to observe the dynamics of radical-pair processes involving primary bacterial photochemistry [158]. Okazaki et al. [159] reported the possibility of controlling chemical reactions by inducing the ESR transition of the intermediate radical pair in the photoreduction of anthraquinone micellar solution under the action of an external magnetic field and simultaneous MW irradiation. A similar study with a bifunctional molecule was reported by Mukai et al. [160]. Research in this field is very well covered by several reviews and books [155, 156, 161, 162]. Weak static magnetic fields, smaller than an average hyperfine coupling, also affect radical pair recombination yields [163, 164]. This effect is opposite to the effect of a strong field [165, 166]. The effect of a magnetic field on singlet oxygen production in a biochemical system was reported recently [167].

Until recently, little attention has been devoted to the effects of *time-dependent* magnetic fields (created by electromagnetic waves) in the absence of a strong magnetic field [168]. Hore and coworkers [169–171] recently described this effect, denoted the *oscillating magnetic field effect* (OMFE), on the fluorescence of an exciplex formed in the photochemical reaction of anthracene with 1,3-dicyanobenzene over the frequency range 1–80 MHz. Another study of the electron–hole recombination of radical ion pairs (pyrene anion and dimethylaniline cation) in solution has been reported [172]. Triplet–singlet interconversions as a result of HFI are relatively efficient in a zero applied magnetic field (to be more precise, in the Earth's field of ∼50 µT). All the states are almost degenerate, assuming separation of the radicals is such that their electronic exchange interaction is negligible [155]. Jackson and his coworkers [173] suggested that the resonance energy of the oscillating field should be tuned to the HFI in one of the radicals. With a typical value of HFI in the radicals of 0.1–3.0 mT, the oscillating magnetic field effect, enhancing the conversion of the singlet state to the triplet (as was observed for weak static fields), is expected in the *radiofrequency* region (3–80 MHz) [170]. Canfield et al. calculated the effects and proved them experimentally on the radical pairs involved in coenzyme B_{12}-dependent enzyme systems [174–176]. Other theoretical studies have appeared in recent years [172, 177, 178]. Whether electromagnetic fields affect animal and human physiology is still open question. It has, for instance, been suggested that radiofrequency fields might disorient birds [179]. Detailed experimental studies of OMFE in the microwave region have not yet been performed. Hoff and Cornelissen [180] have reasoned that triplet-state kinetics could be affected by a pulse of resonant microwaves rather than by equilibrium methods in the zero static field.

According to the OMFE model a weak oscillating magnetic field (the magnetic interactions are much smaller than the thermal energy of the molecule [177]) has no effect on equilibrium constants or activation energies; it can, however, exercise immense *kinetic* control over the reaction of the radicals [177]. The simplified kinetic scheme in Fig. 19.22 shows the excitation of a starting material R–R′ into the singlet state, which intersystem crosses to the triplet (k_{isc}) and is followed by cleavage (k_{cl}) to the triplet radical pair. The oscillating magnetic field affects state mixing of the radical pair (k_{TS} and k_{ST}). The probability that the triplet radical pair

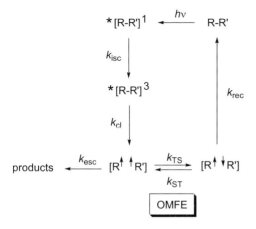

Fig. 19.22. The oscillating magnetic field effect (OMFE) in the triplet-state radical-pair reaction.

will form the products is given by the efficiency of radical escape from the solvent cage (k_{esc}) and of triplet-to-singlet intersystem crossing (k_{TS}). The recombination reaction is very rapid when the tight radical pair reaches the singlet state.

19.6
Applications

19.6.1
Analytical Applications

In addition to analytical applications in which microwaves serve as a power source for the electrodeless discharge lamps (Section 19.2), the first successful use of combined MW–UV irradiation for efficient degradation of a variety of samples before a subsequent analytical application has been reported. Florian and Knapp [51] have proposed a novel MW–UV, high-temperature–high-pressure digestion procedure for decomposition of interfering dissolved organic carbon as a part of trace element analysis of industrial and municipal wastewater or other liquid samples. Very efficient and rapid mineralization was obtained in an original reactor (Fig. 19.15) because of the very high temperature (250–280 °C). The high temperature also enabled dissolution of solid organic matrices by use of dilute mineral acids. A Cd low-pressure electrodeless discharge microwave lamp, strongly emitting at $\lambda = 228$ nm, guaranteed even more efficient degradation than standard mercury UV lamps. The pressurized sealed vessel did not require a separate cooling device to prevent sample evaporation. Efficient decomposition of organophosphate compounds, with the aim of the colorimetric phosphate determination, has been achieved by Howard et al. in a novel beaker-shaped electrodeless MW–UV lamp [112]. Although no details of the organophosphate decomposition mechanism have been presented, the authors suggested two possible pathways. In addition to direct photodegradation, much of the decomposition resulted from photochemical generation of hydroxyl and oxygen radicals from dissolved O_2 in the samples. The concentration of OH radicals could be enhanced by addition of hydrogen peroxide. In addition, Sodre et al. have proposed a new procedure for digestion of natural waters, based on a microwave-activated photochemical reactor, in their speciation studies of copper–humic substances [181].

19.6.2
Environmental Applications

Photodegradation [182] and microwave thermolysis [183] of pollutants, toxic agents or pathogens in waste water, often in combination with a solid catalyst (e.g., TiO_2), are two important methods for their removal. Results from environmentally relevant studies of the combined use of MW and UV [97, 101–103] have already appeared in the scientific literature and the topic is also covered by several patents. Photochemical oxidation is a process in which a strong oxidizing reagent (ozone or hydrogen peroxide) is added to water in a UV-ionizing reactor, resulting in the generation of highly reactive hydroxyl radicals ($^{\cdot}$OH). The first-generation techniques used commercial EDL (high pressure Hg–Xe lamps) immersed in the water tanks. The lamps deteriorated rapidly, however, leading to poor production of hydroxyl radicals. The second-generation technique incorporated manual cleaning mechanisms and use of a polymer coating (PTFE) on the quartz sleeve, additional oxidizers (ozone), and catalytic additives (TiO_2) to enhance the rate of an OH radical production [184]. A novel UV-oxidation system used a highly efficient EDL combined with a simple coaxial flow-through reactor [103]. In this reactor, a liquid containing contaminants (MTBE, 2-propanol, or phenol) was pumped from the bottom and flowed vertically upward through the reactor vessel against gravity. The mercury UV source was mounted above the reactor vessel and the radiation was directed downward through the vessel. An H_2O_2 solution was injected into the liquid being treated and thoroughly mixed by means of an in-line mixer, just before the mixture entered the reactor vessel. It was found by Lipsky et al. that photooxidation of humic acids causes changes in their absorption and luminescence properties that might be of a great importance in environmental photophysics and photochemistry [185]. Aqueous aerated alkaline solutions of the acids were irradiated with a mercury EDL in a flow system and analyzed by means of fluorescence, absorption, and chemiluminescence techniques. Campanella et al. reported minor but positive enhancement of the efficiency of photocatalytic degradation of o and p-chlorophenol in aqueous solutions by microwave heating [113]. The success of these model chemical systems suggested extension to other environmentally interesting compounds, e.g. sodium dodecylbenzenesulfonate or organophosphate pesticides. It has been suggested that microwave-assisted photodegradation of pollutants may be of great interest in the future. Several other research groups, for example those of Chemat [110], Zheng [145], Ai [148], and Hidaka and Serpone [116, 118, 121] have demonstrated improvement of degradative efficiency when microwave radiation is coupled with photocatalytic degradation of pollutants in aqueous solutions, as already described in Section 19.5.2. Spherical or cylindrical EDL have been used to remediate fluids, directly or by excitation of photocatalyst surfaces, which may be located on the lamps themselves or on structures which permeable to the fluids [186].

Noncatalytic remediation of aqueous solutions of a variety of aromatic compounds by microwave-assisted photolysis in the presence of hydrogen peroxide has recently been studied by Klan and Vavrik [187]. The combined degradation effect of UV and MW radiation was always larger than the sum of the isolated effects. It was concluded that this overall increase in efficiency is essentially because of thermal enhancement of subsequent oxidation reactions of the primary photoreaction intermediates. Optimization revealed that this effect is particularly significant for samples containing low concentrations of H_2O_2, although a large excess of H_2O_2 was essential for complete destruction in most experiments. The degradation profiles of the techniques used in the destruction of 4-chlorophenol are com-

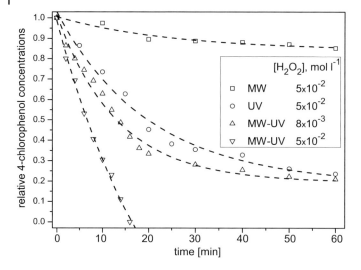

Fig. 19.23. Degradation of 4-chlorophenol ($c = 10^{-3}$ mol L^{-1}) in the presence of H_2O_2 (the concentrations are shown in the inset) in water under given conditions: microwave heating (MW), photolysis at 20 °C (UV), and MW-assisted photolysis (MW–UV). With permission from Elsevier Science [187].

pared in Figure 19.23, which also suggests possible optimum degradation conditions. The results from this work showed that simultaneous MW–UV–H_2O_2 remediation could be an attractive alternative to conventional oxidation or photocatalytic degradation methods for environmental remediation of polluted wastewaters.

Sterilization techniques for intermittent or continuous destruction of pathogens in solid films or in organic and biological fluids, without significantly affecting the properties or physiological characteristics of the medium, are based on the biocidal synergism of UV and MW irradiation. UV radiation induces chemical modification of DNA in bacteria (usually dimerization of thymine). The first apparatus involved a commercial UV-emitting lamp with a separate power source inside the chamber of a MW oven and was used for simple sterilization of biological fluids [188]. An apparatus using a mercury EDL for surface sterilization or disinfection of objects such as bottles, nipples, contact lenses, or food has been proposed by LeVay [98] and Okuda [189]. An apparatus for continuous sterilization of bottle corks and textiles has also been described [190–192]. The sterilization performance of a microwave-powered commercial UV lamp designed to generate active oxygen species for destruction of microorganisms has also been reported [193].

In addition, ozone treatment can be used in combination with UV exposure to sanitize or disinfect a variety of substances [99, 100, 194–197]. Another application of EDL (containing Hg, Cd–Ar, or Kr) for disinfection of aqueous solutions has recently been reported by Michael [198].

19.6.3
Other Applications

Simultaneous use of UV and MW irradiation has found widespread use in industry. The techniques are based on the conventional UV lamps or MW-powered electrodeless lamps [33].

Photolithography is a technique used for manufacture of semiconductor devices (e.g. transistors or integrated circuits). In the process the image of an optical mask is copied, by use of UV radiation, on to a semiconductor wafer coated with a UV-sensitive photoresist. The main goal is to reduce the size of the components and to increase their densities. Application of shorter wavelengths (190–260 nm) results in greater depth of focus, i.e. sharper printing. The first EDL applied were made from a material known as commercial water-containing natural quartz [199]. It was found that transmission of the envelope at vacuum UV wavelengths falls off sharply with time. The lamps developed later from water-free quartz [66] were much more transparent. Excimer lamps used for photoetching and microstructuring of the polymer surface have been developed for applications in standard MW ovens [85].

A photochemical apparatus for generating superoxide radicals ($O_2^{\cdot-}$) in an oxygen-saturated aqueous sodium formate solution by means of an EDL has been described [200]. An interesting method for initiating and promoting chemical processes by irradiation of starting gaseous materials in the EM field under a lower pressure has been proposed by Lautenschläger [97]. EDL (containing GaI_3, InI_3, or AlI_3) with a "blue" output are now often used for dental purposes or for curing polymers. High-power microwave lamps (H and D bulb, fusion UV curing system) have been used for polymerization of maleimide derivatives [201]. The very small size of the lamps makes them particularly useful for supplying light to an optical fiber or light pipe [202]. Another example of microwave photochemical treatment of solutions at different wavelengths has been described by Moruzzi, who used MW for promotion of photochemical reactions [203].

19.7
Concluding Remarks

Understanding, on the molecular scale, of processes relevant to microwave-assisted photochemistry has not yet reached the maturity of other topics in chemistry. Such a challenge is somewhat ambitious, because of several difficulties. Although some obstacles have been overcome, study of the effects of microwaves on photochemical reactions requires a special approach. Microwave-assisted photochemistry involves highly reactive, electronically excited molecules which are exposed to a different kind of reactivity-enhancing stimulation. Microwave heating strongly interferes with possible nonthermal effects that cannot be easily separated in mechanistic studies. One solution seems to be investigation of the spin dynamics of photochemically generated radical pairs. Many photochemical reactions could be affected by a MW treatment if they pass through polar intermediates, e.g. ions or ion-radicals. Application of EDL simplifies the technical procedure, especially in organic synthesis, environmental chemistry, and analysis.

In this review we have discussed how the concept of microwave-assisted photochemistry has become important in chemistry. Although still at the beginning, detailed analysis of past and current literature confirms explicitly the usefulness of this method of chemical activation. The technique is already established in industry and we hope it will also find its way into conventional chemical laboratories.

Acknowledgments

We would like to thank Milan Hájek, Jaromír Literák, and André Loupy for their participation in our research projects, and for fruitful discussions. We also acknowledge the Czech Ministry of Education, Youth and Sport (MSM0021622413) and the Grant Agency of the Czech Republic (203/05/0641 and 104/06/0992) for financial support. We are grateful to Milestone, Inc. (Italy) for a technical support.

References

1 R. Gedye, F. Smith, K. Westaway, H. Ali, L. Baldisera, L. Laberge, J. Rousell, Tetrahedron Lett. **1986**, 27, 279–282.
2 R. J. Giguere, T. L. Bray, S. M. Duncan, G. Majetich, Tetrahedron Lett. **1986**, 27, 4945–4948.
3 A. Fini, A. Breccia, Pure Appl. Chem. **1999**, 71, 573–579.
4 S. A. Galema, Chem. Soc. Rev. **1997**, 26, 233–238.
5 D. M. P. Mingos, A. G. Whittaker, Microwave dielectric heating effects in chemical synthesis, in: Chemistry under Extreme or Non-Classical Conditions, R. van Eldik, C. D. Hubbard, (Eds.), Wiley, New York, **1997**, p. 479–514.
6 R. A. Abramovitch, Org. Prep. Proced. Int. **1991**, 23, 685–711.
7 D. M. P. Mingos, D. R. Baghurst, Chem. Soc. Rev. **1991**, 20, 1–47.
8 A. Loupy, A. Petit, J. Hamelin, F. Texier-Boullet, P. Jacquault, D. Mathe, Synthesis **1998**, 1213–1234.
9 C. R. Strauss, R. W. Trainor, Aust. J. Chem. **1995**, 48, 1665–1692.
10 F. Langa, P. de la Cruz, A. de la Hoz, A. Diaz-Ortiz, E. Diez-Barra, Contemp. Org. Synth. **1997**, 4, 373–386.
11 C. Gabriel, S. Gabriel, E. H. Grant, B. S. J. Halstead, D. M. P. Mingos, Chem. Soc. Rev. **1998**, 27, 213–223.
12 P. Lidström, J. Tierney, B. Wathey, J. Westman, Tetrahedron **2001**, 57, 9225–9283.
13 A. de la Hoz, A. Diaz-Ortiz, A. Moreno, Chem. Soc. Rev. **2005**, 34, 164–178.
14 C. O. Kappe, Angew. Chem.-Int. Edit. **2004**, 43, 6250–6284.
15 A. Stadler, C. O. Kappe, J. Chem. Soc. – Perkin Trans. 2 **2000**, 1363–1368.
16 R. N. Gedye, J. B. Wei, Can. J. Chem. **1998**, 76, 525–532.
17 R. Laurent, A. Laporterie, J. Dubac, J. Berlan, S. Lefeuvre, M. Audhuy, J. Org. Chem. **1992**, 57, 7099–7102.
18 L. Perreux, A. Loupy, Tetrahedron **2001**, 57, 9199–9223.
19 M. Maeda, H. Amemiya, New J. Chem. **1995**, 19, 1023–1028.
20 F. Chemat, M. Poux, J. L. DiMartino, J. Berlan, J. Microw. Power Electromagn. Energy **1996**, 31, 19–22.

21 G. Cravotto, M. Beggiato, A. Penoni, G. Palmisano, S. Tollari, J. M. Leveque, W. Bonrath, *Tetrahedron Lett.* **2005**, *46*, 2267–2271.

22 M. Nüchter, B. Ondruschka, A. Jungnickel, U. Müller, *J. Phys. Org. Chem.* **2000**, *13*, 579–586.

23 A. Gaplovsky, J. Donovalova, S. Toma, R. Kubinec, *Ultrason. Sonochem.* **1997**, *4*, 109–115.

24 A. Gaplovsky, J. Donovalova, S. Toma, R. Kubinec, *J. Photochem. Photobiol. A – Chem.* **1998**, *115*, 13–19.

25 P. Klan, V. Cirkva, *Microwave Photochemistry*, in: *Microwaves in Organic Synthesis*, A. Loupy, (Ed.), Wiley–VCH, Weinheim, **2002**, p. 463–486.

26 F. C. Fehsenfeld, K. M. Evenson, H. P. Broida, *Rev. Sci. Instr.* **1965**, *36*, 294–298.

27 A. J. Shea, A. E. Feuersanger, W. M. Keeffe, C. W. Struck, *Eur. Pat. Appl.* (**1994**) EP 0603014.

28 S. C. Brown, *Introduction to Electrical Discharges in Gases*, John Wiley & Sons, Inc, **1966**.

29 P. G. Wilkinson, Y. Tanaka, *J. Opt. Soc. Am.* **1955**, *45*, 344–349.

30 R. M. Hruda, H. E. De Haven, *U.S. Pat. Appl.* (**1974**) US 3826950.

31 R. Chandler, O. Popov, E. K. Shapiro, J. Maya, *U.S. Pat. Appl.* (**2001**) US 2001000941.

32 A. Ganeev, Z. Gavare, V. I. Khutorshikov, S. V. Khutorshikov, G. Revalde, A. Skudra, G. M. Smirnova, N. R. Stankov, *Spectrochim. Acta B* **2003**, *58*, 879–889.

33 D. M. Spero, B. J. Eastlund, M. G. Ury, *U.S. Pat. Appl.* (**1975**) US 3872349.

34 M. R. Wertheimer, A. C. Fozza, A. Hollander, *Nucl. Instrum. Methods Phys. Res. Sect. B – Beam Interact. Mater. Atoms* **1999**, *151*, 65–75.

35 R. F. Browner, J. D. Winefordner, T. H. Glenn, *U.S. Pat. Appl.* (**1974**) US 3786308.

36 P. O. Haugsjaa, W. F. Nelson, R. J. Regan, W. H. McNeil, *U.S. Pat. Appl.* (**1976**) US 3942058.

37 J. M. Proud, D. H. Baird, *U.S. Pat. Appl.* (**1981**) US 4266167.

38 D. O. Wharmby, *Iee Proceedings-a-Science Measurement and Technology* **1993**, *140*, 465–473.

39 W. E. Garber, A. Gitsevich, D. A. Kirkpatrick, G. K. Bass, J. Proctor, J. F. Copsey, W. G. Grimm, V. H. Kwong, I. Levin, S. A. Ola, R. J. Roy, J. E. Simpson, P. E. Steiner, P. Tsai, B. P. Turner, L. E. Dymond, D. A. MacLennan, *PCT Int. Appl.* (**2001**) WO 0103161.

40 D. A. MacLennan, B. P. Turner, G. K. Bass, D. A. Kirkpatrick, J. E. Simpson, W. C. Trimble, M. G. Ury, *PCT Int. Appl.* (**2001**) WO 0103476.

41 J. M. Anderson, *U.S. Pat. Appl.* (**1977**) US 4017764.

42 J. T. Dakin, T. Berry, M. E. Duffy, T. D. Russell, *U.S. Pat. Appl.* (**1994**) US 5363015.

43 J. T. Dolan, M. G. Ury, B. P. Turner, J. F. Waymouth, C. H. Wood, *PCT Int. Appl.* (**1993**) WO 9321655.

44 J. T. Dolan, B. P. Turner, M. G. Ury, C. H. Wood, *U.S. Pat. Appl.* (**1997**) US 5682080.

45 J. T. Dolan, M. G. Ury, C. H. Wood, *US Pat. Appl.* (**1995**) US 5404076.

46 P. D. Johnson, J. T. Dakin, J. M. Anderson, *U.S. Pat. Appl.* (**1989**) US 4810938.

47 T. D. Russell, T. Berry, J. T. Dakin, M. E. Duffy, *Eur. Pat. Appl.* (**1993**) EP 0542467.

48 G. B. Marshall, T. S. West, *Anal. Chim. Acta* **1970**, *51*, 179–190.

49 B. M. Patel, R. F. Browner, J. D. Winefordner, *Anal. Chem.* **1972**, *44*, 2272–2277.

50 M. G. Ury, C. H. Wood, *U.S. Pat. Appl.* (**1989**) US 4859906.

51 D. Florian, G. Knapp, *Anal. Chem.* **2001**, *73*, 1515–1520.

52 W. S. Gleason, R. Pertel, *Rev. Sci. Instr.* **1971**, *42*, 1638–1643.

53 M. Zelikoff, P. H. Wyckoff, L. M. Aschenbrand, R. S. Loomis, *J. Opt. Soc. Am.* **1952**, *42*, 818–819.

54 J. P. S. Haarsma, G. J. De Jong, J. Agterdenbos, *Spectrochim. Acta B* **1974**, *29*, 1–18.

55 J. Sneddon, R. F. Browner, P. N. Keliher, J. D. Winefordner, D. J.

Butcher, R. G. Michel, *Prog. Anal. Spectrosc.* **1989**, *12*, 369–402.
56 F. R. Pothoven, T. A. Pothoven, *U.S. Pat. Appl.* (**2003**) US 6666739.
57 M. D. Seltzer, R. G. Michel, *Anal. Chem.* **1983**, *55*, 1817–1819.
58 V. Církva, L. Vlková, S. Relich, M. Hájek, *J. Photochem. Photobiol. A – Chem.* **2006**, *179*, 229–233.
59 P. Müller, P. Klan, V. Cirkva, *J. Photochem. Photobiol. A – Chem.* **2003**, *158*, 1–5.
60 P. Müller, P. Klan, V. Cirkva, *J. Photochem. Photobiol. A – Chem.* **2005**, *171*, 51–57.
61 J.-S. Choi, H.-J. Kang, H.-S. Jeon, Y.-S. Jeon, H.-S. Kim, Y. Brodsky, V. Vdovin, S. Golubev, N. Kovalev, Y. Kalyazin, Y. Korotchkov, A. Perminov, *Eur. Pat. Appl.* (**2001**) EP 1119021.
62 R. Phillips, *Sources and Applications of Ultraviolet Radiation*, Academic Press, London, **1983**.
63 D. M. Spero, J. C. Matthews, *J. Radiat. Curing* **1979**, *6*, 6–10.
64 J. M. Kramer, W. H. McNeil, P. O. Haugsjaa, *U.S. Pat. Appl.* (**1980**) US 4206387.
65 K. Yoshizawa, H. Kodama, Y. Minowa, H. Komura, H. Ito, *U.S. Pat. Appl.* (**1985**) US 4498029.
66 P. Mueller, M. G. Ury, C. H. Wood, *U.S. Pat. Appl.* (**1985**) US 4501993.
67 T. Ono, S. Murayama, *Appl. Optics* **1990**, *29*, 3934–3937.
68 H. S. Kim, J. S. Choi, H. J. Kang, Y. S. Jeon, H. S. Jeon, *Eur. Pat. Appl.* (**2001**) EP 1093152.
69 H.-S. Kim, J.-S. Choi, H.-J. Kang, Y.-S. Jeon, H.-S. Jeon, *U.S. Pat. Appl.* (**2003**) US 6633111.
70 Y.-S. Jeon, J.-S. Choi, H.-S. Kim, H.-S. Jeon, J.-Y. Lee, B.-J. Park, *U.S. Pat. Appl.* (**2005**) US 6946795.
71 J.-S. Choi, H.-J. Kang, H.-S. Jeon, Y.-S. Jeon, H.-S. Kim, Y. Brodsky, V. Vdovin, S. Golubev, N. Kovalev, Y. Kalyazin, Y. Korotchkov, A. Perminov, *U.S. Pat. Appl.* (**2004**) US 6734630.
72 A. Hochi, S. Horii, M. Takeda, T. Matsuoka, *Eur. Pat. Appl.* (**1997**) EP 0762476.
73 J. T. Dakin, *U.S. Pat. Appl.* (**1988**) US 4783615.
74 H. L. Witting, *U.S. Pat. Appl.* (**1989**) US 4890042.
75 S. Ukegawa, C. Gallagher, *U.S. Pat. Appl.* (**2000**) US 6121730.
76 W. P. Lapatovich, G. R. Gibbs, J. M. Proud, *U.S. Pat. Appl.* (**1987**) US 4647821.
77 W. P. Lapatovich, G. R. Gibbs, J. M. Proud, *U.S. Pat. Appl.* (**1984**) US 4480213.
78 W. P. Lapatovich, G. R. Gibbs, *U.S. Pat. Appl.* (**1985**) US 4492898.
79 W. P. Lapatovich, S. J. Butler, J. R. Bochinski, *Eur. Pat. Appl.* (**1997**) EP 0788141.
80 D. A. Kirkpatrick, J. T. Dolan, D. A. MacLennan, B. P. Turner, J. E. Simpson, *PCT Int. Appl.* (**2000**) WO 0070651.
81 M. Kamarehi, L. Levine, M. G. Ury, B. P. Turner, *US Pat. Appl.* (**1998**) US 5831386.
82 D. A. Kirkpatrick, D. A. MacLennan, T. Petrova, V. D. Roberts, B. P. Turner, *PCT Int. Appl.* (**2002**) WO 02082501.
83 J. T. Dolan, M. G. Ury, J. F. Waymouth, C. H. Wood, *PCT Int. Appl.* (**1992**) WO 9208240.
84 B. P. Turner, *U.S. Pat. Appl.* (**1997**) US 5661365.
85 U. Kogelschatz, H. Esrom, J. Y. Zhang, I. W. Boyd, *Appl. Surf. Sci.* **2000**, *168*, 29–36.
86 W. P. Lapatovich, S. J. Butler, J. R. Bochinski, *Eur. Pat. Appl.* (**1997**) EP 0788140.
87 J. M. Proud, S. G. Johnson, *U.S. Pat. Appl.* (**1984**) US 4427921.
88 B. J. Russell, J. P. Shelton, A. Wals, *Spectrochim. Acta* **1957**, *8*, 317–328.
89 J. Mansfield, J. M., J. Bratzel, M. P., H. O. Norgodon, D. O. Knapp, K. E. Zacha, J. D. Winefordner, *Spectrochim. Acta* **1968**, *23B*, 389–402.
90 K. E. Zacha, J. Bratzel, M. P., J. D. Winefordner, J. Mansfield, J. M., *Anal. Chem.* **1968**, *40*, 1733–1736.
91 I. E. Den Besten, J. W. Tracy, *J. Chem. Educ.* **1973**, *50*, 303.
92 K. R. Osborn, C. C. McDonald, H. E.

Gunning, *J. Chem. Phys.* **1957**, *26*, 124–133.
93 R. Pertel, H. E. Gunning, *J. Chem. Phys.* **1957**, *26*, 219–219.
94 R. Pertel, H. E. Gunning, *Can. J. Chem.* **1959**, *37*, 35–42.
95 N. L. Ruland, R. Pertel, *J. Am. Chem. Soc.* **1965**, *87*, 4213–4214.
96 H. Okabe, *J. Opt. Soc. Am.* **1963**, *54*, 478–481.
97 W. Lautenschläger, *Eur. Pat. Appl.* (**1991**) EP 0429814.
98 T. C. LeVay, *U.S. Pat. Appl.* (**1992**) US 5166528.
99 T. C. LeVay, J. A. Rummel, *PCT Int. Appl.* (**1996**) WO 9640298.
100 P. Hirsch, *PCT Int. Appl.* (**1989**) WO 8909068.
101 J. M. Rummler, *PCT Int. Appl.* (**1994**) WO 9425402.
102 T. Kurata, *Jpn. Pat. Appl.* (**2000**) JP 2000035215.
103 S. P. Oster, *PCT Int. Appl.* (**2000**) WO 0055096.
104 P. Klan, M. Hajek, V. Cirkva, *J. Photochem. Photobiol. A – Chem.* **2001**, *140*, 185–189.
105 V. Cirkva, M. Hajek, *J. Photochem. Photobiol. A – Chem.* **1999**, *123*, 21–23.
106 P. Klan, J. Literak, *Collect. Czech. Chem. Commun.* **1999**, *64*, 2007–2018.
107 P. Klan, J. Literak, M. Hajek, *J. Photochem. Photobiol. A – Chem.* **1999**, *128*, 145–149.
108 J. Literak, P. Klan, *J. Photochem. Photobiol. A – Chem.* **2000**, *137*, 29–35.
109 P. Klan, J. Literak, S. Relich, *J. Photochem. Photobiol. A – Chem.* **2001**, *143*, 49–57.
110 S. Chemat, A. Aouabed, P. V. Bartels, D. C. Esveld, F. Chemat, *J. Microw. Power Electromagn. Energy* **1999**, *34*, 55–60.
111 F. Louerat, A. Loupy, personal communication.
112 A. G. Howard, L. Labonne, E. Rousay, *Analyst* **2001**, *126*, 141–143.
113 L. Campanella, R. Cresti, M. P. Sammartino, V. G., in SPIE Conference on Environmental Monitoring and Remediation Technologies, Boston, 1998, p. 105–113.
114 S. Horikoshi, H. Hidaka, N. Serpone, *Environ. Sci. Technol.* **2002**, *36*, 1357–1366.
115 S. Kataoka, D. T. Tompkins, W. A. Zeltner, M. A. Anderson, *J. Photochem. Photobiol. A – Chem.* **2002**, *148*, 323–330.
116 S. Horikoshi, F. Hojo, H. Hidaka, N. Serpone, *Environ. Sci. Technol.* **2004**, *38*, 2198–2208.
117 S. Horikoski, A. Tokunaga, H. Hidaka, N. Serpone, *J. Photochem. Photobiol. A – Chem.* **2004**, *162*, 33–40.
118 S. Horikoshi, H. Hidaka, N. Serpone, *J. Photochem. Photobiol. A – Chem.* **2004**, *161*, 221–225.
119 S. Horikoshi, H. Hidaka, N. Serpone, *Environ. Sci. Technol.* **2002**, *36*, 5229–5237.
120 S. Horikoshi, H. Hidaka, N. Serpone, *J. Photochem. Photobiol. A – Chem.* **2002**, *153*, 185–189.
121 S. Horikoshi, H. Hidaka, N. Serpone, *J. Photochem. Photobiol. A – Chem.* **2003**, *159*, 289–300.
122 H. Bergmann, T. Iourtchouk, K. Schöps, K. Bouzek, *Chem. Eng. J.* **2002**, *85*, 111–117.
123 V. Cirkva, S. Relich, M. Hájek, *J. Photochem. Photobiol. A – Chem.*, in print.
124 H. M. "Skip" Kingston, P. J. Walter, W. G. Engelhart, P. J. Parsons, Chapter 16: Laboratory Microwave Safety, in: *Microwave-Enhanced Chemistry: Fundamentals, Sample Preparation, and Applications*, H. M. "Skip" Kingston, S. J. Haswell, (Eds.), ACS Professional Reference Book, Washington, **1997**, p. 697–745.
125 H. H. Wasserman, R. W. Murray, *Singlet Oxygen*, Academic Press, New York, **1979**.
126 D. R. Baghurst, D. M. P. Mingos, *J. Chem. Soc. – Chem. Commun.* **1992**, 674–677.
127 D. A. Lewis, J. D. Summers, T. C. Ward, J. E. McGrath, *J. Polym. Sci. Pol. Chem.* **1992**, *30*, 1647–1653.
128 D. Stuerga, M. Lallemant, *J. Microw. Power Electromagn. Energy* **1994**, *29*, 20–30.
129 D. Stuerga, P. Gaillard, *Tetrahedron* **1996**, *52*, 5505–5510.

130 A. G. Whittaker, D. M. P. Mingos, *J. Microw. Power Electromagn. Energy* **1994**, *29*, 195–219.

131 V. Sridar, *Curr. Sci.* **1998**, *74*, 446–450.

132 P. J. Wagner, P. Klan, Chapter 52: Norrish Type II Photoelimination of Ketones: Cleavage of 1,4-Biradicals Formed by γ-Hydrogen Abstraction, in: *CRC Handbook of Organic Photochemistry and Photobiology*, W. M. Horspool, F. Lenci, (Eds.), CRC Press, Boca Raton, **2003**, p. 1–31.

133 P. Klan, R. Ruzicka, D. Heger, J. Literak, P. Kulhanek, A. Loupy, *Photochem. Photobiol. Sci.* **2002**, *1*, 1012–1016.

134 J. Literak, S. Relich, P. Kulhanek, P. Klan, *Mol. Divers.* **2003**, *7*, 265–271.

135 P. Muller, A. Loupy, P. Klan, *J. Photochem. Photobiol. A – Chem.* **2005**, *172*, 146–150.

136 V. Cirkva, J. Kurfurstova, J. Karban, M. Hajek, *J. Photochem. Photobiol. A – Chem.* **2004**, *168*, 197–204.

137 V. Cirkva, J. Kurfurstova, J. Karban, M. Hajek, *J. Photochem. Photobiol. A – Chem.* **2005**, *174*, 38–44.

138 A. Kunz, P. Peralta-Zamora, N. Duran, *Adv. Environ. Res.* **2002**, *7*, 197–202.

139 D. H. Han, S. Y. Cha, H. Y. Yang, *Water Res.* **2004**, *38*, 2782–2790.

140 J. Literak, P. Klan, D. Heger, A. Loupy, *J. Photochem. Photobiol. A – Chem.* **2003**, *154*, 155–159.

141 O. Legrini, E. Oliveros, A. M. Braun, *Chem. Rev.* **1993**, *93*, 671–698.

142 O. Legrini, E. Oliveros, A. M. Braun, *Photochemical processes for water treatment*, McGraw–Hill, New York, **1994**.

143 S. Horikoshi, A. Saitou, H. Hidaka, N. Serpone, *Environ. Sci. Technol.* **2003**, *37*, 5813–5822.

144 S. Horikoshi, H. Hidaka, N. Serpone, *Chem. Phys. Lett.* **2003**, *376*, 475–480.

145 Y. Zheng, D. Z. Li, X. Z. Fu, *Chinese J. Catal.* **2001**, *22*, 165–167.

146 D. Z. Li, Y. Zheng, X. Z. Fu, *Acta Phys. – Chim. Sin.* **2002**, *18*, 332–335.

147 D. Z. Li, Y. Zheng, X. Z. Fu, *Chem. J. Chin. Univ. – Chin.* **2002**, *23*, 2351–2356.

148 Z. H. Ai, P. Yang, X. H. Lu, *Fresenius Environ. Bull.* **2004**, *13*, 550–554.

149 X. W. Zhang, Y. Z. Wang, *Prog. Chem.* **2005**, *17*, 91–95.

150 J. R. Fox, G. S. Hammond, *J. Am. Chem. Soc.* **1964**, *86*, 4031–4035.

151 P. D. Barlett, N. A. Porter, *J. Am. Chem. Soc.* **1968**, *90*, 5317–5318.

152 H.-J. Werner, K. Schulten, A. Weller, *Biochim. Biophys. Acta* **1978**, *502*, 255–268.

153 R. Haberkorn, M. E. Michel-Beyerle, *Biophys. J.* **1979**, *26*, 489–498.

154 N. J. Turro, G. C. Weed, *J. Am. Chem. Soc.* **1983**, *105*, 1861–1868.

155 A. L. Buchachenko, E. L. Frankevich, *Chemical Generation and Reception of radio and Microwaves*, VCH Publishers, Inc., New York, **1994**.

156 U. E. Steiner, T. Ulrich, *Chem. Rev.* **1989**, *89*, 51–147.

157 M. R. Wasielewski, C. H. Bock, M. K. Bowman, J. R. Norris, *Nature* **1983**, *303*, 520–522.

158 J. R. Norris, M. K. Bowman, D. E. Budil, J. Tang, C. A. Wraight, G. L. Closs, *Proceedings of the National Academy of Sciences of the United States of America – Biological Sciences* **1982**, *79*, 5532–5536.

159 M. Okazaki, Y. Konishi, K. Toriyama, *Chem. Lett.* **1994**, *23*, 737–740.

160 M. Mukai, Y. Fujiwara, Y. Tanimoto, Y. Konishi, M. Okazaki, *Z. Phys. Chem. – Int. J. Res. Phys. Chem. Chem. Phys.* **1993**, *180*, 223–233.

161 K. M. Salikhov, Y. N. Molin, R. Z. Sagdeev, A. L. Buchachenko, *Spin Polarization and Magnetic Field Effects in Radical Reactions*, Elsevier, Amsterdam, **1984**.

162 G. Maret, J. Kiepenheuer, N. Boccara, *Biophysical Effects of Steady Fields*, in: *Springer Proceedings of Physics, Vol. 11.*, Springer, Berlin, **1986**.

163 C. R. Timmel, U. Till, B.

Brocklehurst, K. A. McLauchlan, P. J. Hore, *Mol. Phys.* **1998**, *95*, 71–89.
164 U. Till, C. R. Timmel, B. Brocklehurst, P. J. Hore, *Chem. Phys. Lett.* **1998**, *298*, 7–14.
165 B. Brocklehurst, *J. Chem. Soc. Faraday Trans. 2* **1976**, *72*, 1869–1884.
166 B. Brocklehurst, K. A. McLauchlan, *Int. J. Radiat. Biol.* **1996**, *69*, 3–24.
167 Y. Liu, R. Edge, K. Henbest, C. R. Timmel, P. J. Hore, P. Gast, *Chem. Commun.* **2005**, 174–176.
168 C. R. Timmel, K. B. Henbest, *Philos. Trans. R. Soc. Lond. Ser. A – Math. Phys. Eng. Sci.* **2004**, *362*, 2573–2589.
169 J. R. Woodward, R. J. Jackson, C. R. Timmel, P. J. Hore, K. A. McLauchlan, *Chem. Phys. Lett.* **1997**, *272*, 376–382.
170 D. V. Stass, J. R. Woodward, C. R. Timmel, P. J. Hore, K. A. McLauchlan, *Chem. Phys. Lett.* **2000**, *329*, 15–22.
171 C. R. Timmel, J. R. Woodward, P. J. Hore, K. A. McLauchlan, D. V. Stass, *Meas. Sci. Technol.* **2001**, *12*, 635–643.
172 J. R. Woodward, C. R. Timmel, K. A. McLauchlan, P. J. Hore, *Phys. Rev. Lett.* **2001**, *87*, Art. No. 077602.
173 R. J. Jackson, K. A. McLauchlan, J. R. Woodward, *Chem. Phys. Lett.* **1995**, *236*, 395–401.
174 J. M. Canfield, R. L. Belford, P. G. Debrunner, *Mol. Phys.* **1996**, *89*, 889–930.
175 J. M. Canfield, R. L. Belford, P. G. Debrunner, K. Schulten, *Chem. Phys.* **1995**, *195*, 59–69.
176 J. M. Canfield, R. L. Belford, P. G. Debrunner, K. J. Schulten, *Chem. Phys.* **1994**, *182*, 1–18.
177 C. R. Timmel, P. J. Hore, *Chem. Phys. Lett.* **1996**, *257*, 401–408.
178 C. Eichwald, J. Walleczek, *J. Chem. Phys.* **1997**, *107*, 4943–4950.
179 T. Ritz, S. Adem, K. Schulten, *Biophysical Journal* **2000**, *78*, 707–718.
180 A. J. Hoff, B. Cornelissen, *Mol. Phys.* **1982**, *45*, 413–425.
181 F. F. Sodre, P. G. Peralta-Zamora, M. T. Grassi, *Quim. Nova* **2004**, *27*, 695–700.

182 J. P. Scott, D. F. Ollis, *Environ. Prog.* **1995**, *14*, 88–103.
183 G. Windgasse, L. Dauerman, *J. Microw. Power Electromagn. Energy* **1992**, *27*, 23–32.
184 J. Downey, W. F., *U.S. Pat. Appl.* (**1995**) US 5439595.
185 M. Lipski, J. Slawinski, D. Zych, *J. Fluoresc.* **1999**, *9*, 133–138.
186 T. N. Obee, S. O. Hay, J. J. Sangiovanni, J. B. Hertzberg, *PCT Int. Appl.* (**2003**) WO 03094982.
187 P. Klan, M. Vavrik, *J. Photochem. Photobiol. A – Chem.* **2006**, *177*, 24–33.
188 R. M. G. Boucher, *U.S. Pat. Appl.* (**1975**) US 3926556.
189 S. Okuda, K. Atsumi, *PCT Int. Appl.* (**2003**) WO 03033036.
190 R. A. R. Little, D. Briggs, *Eur. Pat. Appl.* (**1997**) EP 0772226.
191 J. Lucas, J. L. Moruzzi, *PCT Int. Appl.* (**2000**) WO 0032244.
192 H. Linn, *DE Pat. Appl.* (**2001**) DE 10008487.
193 S. Iwaguchi, K. Matsumura, Y. Tokuoka, S. Wakui, N. Kawashima, *Colloid Surf. B – Biointerfaces* **2002**, *25*, 299–304.
194 V. A. Danilychev, *U.S. Pat. Appl.* (**1997**) US 5666640.
195 V. A. Danilychev, *U.S. Pat. Appl.* (**1999**) US 5931557.
196 J. D. Michael, *U.S. Pat. Appl.* (**2001**) US 6171452.
197 B.-U. Hur, Y.-B. Park, *U.S. Pat. Appl.* (**2004**) US 6762414.
198 J. D. Michael, *U.S. Pat. Appl.* (**2000**) US 6162406.
199 J. C. Matthews, M. G. Ury, C. H. Wood, M. Greenblatt, *U.S. Pat. Appl.* (**1985**) US 4532427.
200 R. A. Holroyd, B. H. J. Bielski, *U.S. Pat. Appl.* (**1980**) US 4199419.
201 H. Andersson, U. W. Gedde, A. Hult, *Macromolecules* **1996**, *29*, 1649–1654.
202 A. B. Buninger, W. P. Lapatovich, F. L. Palmer, J. M. Browne, N. H. Chen, *Eur. Pat. Appl.* (**1999**) EP 0962959.
203 J. L. Moruzzi, *PCT Int. Appl.* (**2001**) WO 0109924.

20
Microwave-enhanced Solid-phase Peptide Synthesis

Jonathan M. Collins and Michael J. Collins

20.1
Introduction

With the involvement of peptides and proteins in virtually every biological process, understanding and controlling their expression and function is of major importance. With more than 20 000 papers published annually, peptide research has grown dramatically in recent decades [1]. Peptides are involved in many biochemical processes including cell–cell communication, metabolism, immune response, and reproduction. Peptides also act as hormones and neurotransmitters in receptor-mediated signal transduction. As the enormous role of peptides in many physiological and biochemical processes has become more understood, so has interest in their value as potential drug candidates.

Compared with small-molecule drugs peptides have the advantage of higher potency and specificity with fewer toxicological problems. Small molecules have the advantage of oral availability, lower price, and ability to cross cell membranes. Improvements in peptide drug-delivery systems, for example PEGylation and pulmonary delivery, are increasing the attractiveness of peptide drugs, however [2, 3].

Obtaining peptides from natural sources can often be very difficult. In tissue samples the desired peptides are often at very low concentrations requiring highly sensitive assay methods. The availability and storage of natural tissue samples can also limit availability. Although recombinant genetics has been the major production tool for synthesis of proteins, it is a time-consuming and laborious task laden with difficulties.

Chemical synthesis of proteins enables site-specific control of backbone and side-chain modifications with customizability not available via recombinant strategies. Chemical synthesis also enables protein sequences to be synthesized not only rapidly, but also free from DNA impurities or endotoxins that may be present during recombinant synthesis. This is especially important for *in-vivo* applications requiring high dosages and diagnostic imaging, for which a product free from endotoxins is required. The combination of different chemical synthesis strategies has been successful in producing small proteins containing up to 200 amino acids.

Although chemical synthesis of peptides and proteins has developed substantially in recent decades, it does require use of toxic reagents, large amounts of expensive reagents, and often suffers from incomplete reactions that significantly reduce final product purity. Chemical synthesis can be performed in solution, on a solid phase, or by a combination of both. Solution-phase synthesis has been the method of choice for large-scale preparation, but developments in solid phase tools have enabled its use also.

Although microwave synthesis has become widespread for enhancing chemical reactions in organic synthesis, only recently has it been adopted in peptide synthesis. This is primarily attributed to concern about side reactions and, until recently, lack of availability of the proper tools. Several well-known side reactions have been extensively documented in solid phase peptide synthesis (SPPS) and there has been concern that microwaves, while speeding up reaction rates, may also accelerate racemization, aspartimide formation, and other potential problems. Several publications have reported use of domestic and single-mode microwave devices and enhancements in coupling reactions [40, 41]. In 2003, the complete SPPS process including both deprotection and coupling steps was performed in a single-mode microwave reactor for generation of the $^{65-74}$ACP peptide [42]. The results showed deprotection was complete in 2 min and coupling in 3 min. This enabled complete synthesis of a 10mer peptide in 4.5 h. Not only does the application of energy used for SPPS enable more rapid synthesis, it can also result in higher product purity. Because the need for peptides is growing quickly, it is critically important to enable faster delivery of high purity peptides to the market.

20.2
Solid-phase Peptide Synthesis

Classical solution-phase peptide synthesis suffers from the time consuming processes of purification and characterization at each step. The development of SPPS overcame many of the problems of classical solution-phase synthesis and has become the standard method for research-scale synthesis of peptides containing up to 50 amino acids [4]. This method uses a solid resin bead for assembly of a peptide chain from C to N terminus (Scheme 20.1). The carboxy group of the C-terminal amino acid is attached to a linker group, thereby attaching it to the resin. A temporary protecting group on the α-amino group is first removed. The second amino acid, with N-terminus temporary protection, is then added to the reaction in excess with its carboxy group activated to generate an activated ester capable of amide bond formation. Excess reagents are then removed from the reaction, usually by filtration, and the resin washed repeatedly with solvent. The N-terminus protecting group of the dipeptide attached to the resin is then removed and the cycle continues until the desired sequence is made. The last step involves deprotection of the final N-terminus protecting group and removal of the peptide from the resin. This step also removes any side-chain-protecting groups used during the

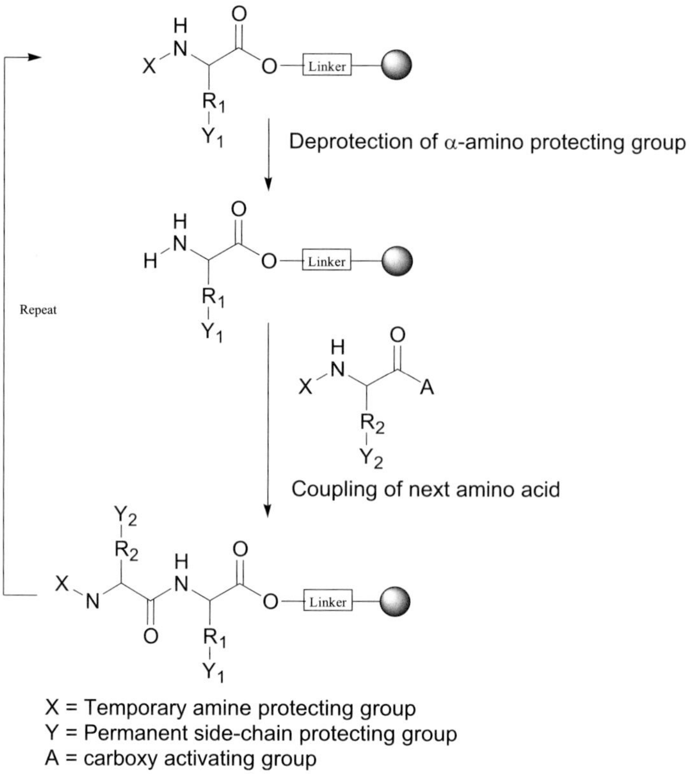

Scheme 20.1. Solid-phase peptide synthesis cycle.

X = Temporary amine protecting group
Y = Permanent side-chain protecting group
A = carboxy activating group

stepwise synthesis, yielding an unprotected peptide in solution. The two main approaches of SPPS used are the Fmoc and Boc methods. The Boc method is the older of the two and has largely been replaced by the Fmoc method.

20.2.1
Boc Chemistry

The original technique developed by Merrifield used a scheme of graduated acidolysis for selective removal of temporary and permanent protecting groups. This method uses the *tert*-butoxycarbonyl (Boc) group for temporary protection of the α-amino group (Fig. 20.1). The Boc group is typically removed with trifluoroacetic acid (TFA), or TFA in dichloromethane (DCM). TFA removal of the Boc group generates a protonated amine, which must be neutralized before coupling of the next amino acid [5]. The graduated acidolysis method employed in Boc synthesis requires a stronger acid to remove both the side-chain-protecting groups and to break the peptide–resin bond. This typically requires the use of hydrofluoric acid (HF). The high toxicity and risks associated with handling HF require the use of special apparatus.

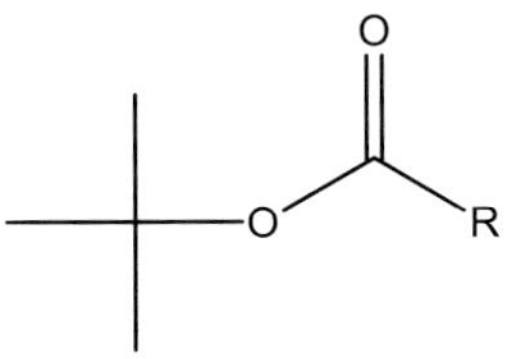

Fig. 20.1. The *tert*-butoxycarbonyl (Boc) group.

Although use of Boc has enabled successful syntheses, its application has been limited, because of the use of HF – a major factor in the current preference for Fmoc chemistry [6]. Little known work has been performed with Boc chemistry using microwave energy. For this reason, this chapter focuses on the application of microwave energy for Fmoc SPPS.

20.2.2
Fmoc Chemistry

The Fmoc approach for SPPS uses an orthogonal protecting group strategy that incorporates the base-labile N-Fmoc group to protect the α-amino group, while incorporating acid-labile side-chain-protecting groups and forming the peptide–resin bond (Fig. 20.2). The advantage of this approach is that it enables use of milder acidic conditions than those used in the Boc method. Typically, TFA is used for removal of side-chain-protecting groups and breaking the peptide–resin bond.

Although the Fmoc method does not require use of HF, it does suffer from inherent difficulties during stepwise synthesis. These can occur because of intermolecular aggregation, β-sheet formation, steric hindrance from protecting groups, and premature termination of the sequence. As a result, standard amino acid cycle times range from 30 min to 2 h and incomplete deprotection and coupling reactions are still common.

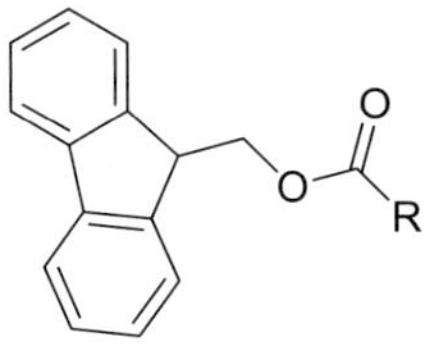

Fig. 20.2. The 9-fluorenylmethoxycarbonyl (Fmoc) group.

20.2.2.1 Activation
Carboxylic acids and amines will react to give an ammonium salt (acid–base equilibrium) instead of a carboxamide. Activation of the carboxylic acid to form an ester species enables peptide bond formation to occur (Fig. 20.3).

$$R-\underset{\underset{}{\overset{\overset{O}{\|}}{}}}{C}-OH + H_2N-R \; \rightleftharpoons \; R-\underset{\underset{}{\overset{\overset{O}{\|}}{}}}{C}-O^{\ominus} + H_3\overset{\oplus}{N}-R$$

Fig. 20.3. Generation of ammonium salt from a carboxylic acid and free amine.

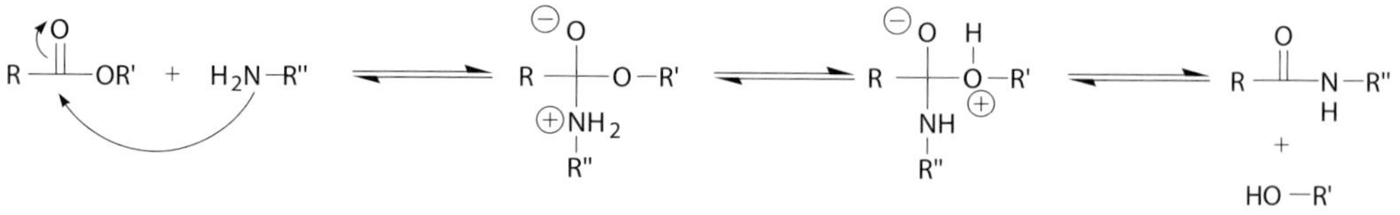

Fig. 20.4. Carboxamide formation from reaction of an ester with an amine.

This reduces the electron density on the C=O group, which favors nucleophilic attack by the amine group (Fig. 20.4). The improved nucleofugicity of the leaving group also increases the reaction rate.

Many different methods of activation have been developed for SPPS. Original activation methods involved use of carbodiimides with an equivalent amount of HOBt for ester formation. The most popular is 1,3-diisopropylcarbodiimide (DIC) because the urea it generates during activation is soluble and allows its use to be easily automated. Microwave has been used to accelerate SPPS couplings with DIC/HOBt [6a]. The most recent developments have led to use of aminium and phosphonium activation methods for *in-situ* generation of active esters as the preferred coupling method in SPPS. These methods use phosphonium and aminium salts based on either 1-hydroxybenzotriazole (HOBt) or 1-hydroxy-7-azabenzotriazole (HOAt) to convert Fmoc-amino acids into OXt esters in the presence of a tertiary base (Fig. 20.5). Both phosphonium and aminium activation strategies based on HOAt have superior coupling efficiency and reduced rates of racemization in SPPS [7–9].

The phosphonium method was originally developed from derivatives of tris-dimethylaminophosphonium salts for the activation of carboxylic acids [10–13]. The first developed phosphonium salt-type reagents were μ-oxo-bis-[tris(dimethylamino)phosphonium]-bis-tetrafluoroborate (Bates Reagent) and benzotriazol-1-yl-N-oxy-tris(dimethylamino)phosphonium hexafluorophosphate (BOP). Although

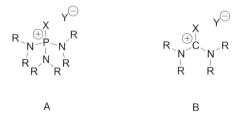

Fig. 20.5. Structures of phosphonium (A) and aminium (B) salts.

BOP has proven to be an efficient reagent in SPPS, its use results in generation of carcinogenic hexamethylphosphoric triamide (HMPA) during the coupling reaction. More recently, PyBOP was developed, which replaces the dimethylamino groups of BOP with pyrrolidine substituents. PyBOP, with its HOAt analogue (PyAOP) have the advantage that they do not form HMPA during the coupling reaction.

Analogues of the phosphonium salts with replacement of the phosphonium atom by a carbon atom have been developed. Originally these were labeled as uronium salts because it was assumed they were analogous to phosphonium salts. X-ray crystallographic studies have shown however, that these reagents crystallize as aminium salts (guanidinium N-oxides) [14, 15].

The aminium salts most widely used in SPPS are N-[(1H-benzotriazol-1-yl)-(dimethylamino)methylene]-N-methylmethanaminium hexafluorophosphate N-oxide (HBTU) and N-[(dimethylamino)-1H-1,2,3-triazolo[4,5-b]pyridino-1-ylmethylene]-N-methylmethanaminium hexafluorophosphate N-oxide (HATU). Replacement of the hexafluorophosphate counter-ion for tetrafluoroborate gives the corresponding TBTU and TATU species that are also used in SPPS. Although tetrafluoroborate is more soluble in DMF than hexafluorophosphate, the counter-ion does not seem to affect the reaction rate in SPPS [16].

1H-Benzotriazolium-1-[bis(dimethylamino)methylene]-5-chlorohexafluorophosphate-(1-),3-oxide (HCTU), based on a Cl-substituted hydroxybenzotriazole was developed recently. The coupling efficiency of HCTU is reported to be superior to that of HBTU and PyBOP, because OCt is a better leaving group. HCTU is also a nonirritant and a nonexplosive reagent; this makes it advantageous over HOBt and HOAt-based coupling species.

The phosphonium and aminium activation methods require use of a base to form the anion of the carboxy component and begin activation. Typically, tertiary amines such as diisopropylethylamine (DIEA), or N-methylmorpholine (NMM) are used. The amines are also hindered to minimize risk of racemization.

20.2.2.2 Aggregation

Whereas some peptide sequences could be synthesized relatively easily, it was noticed early on that other sequences were much more difficult. During chain assembly sudden decreases in reaction rates were observed. Occasionally, repeated or prolonged reaction time resulted in no improvement in chain assembly. Optimum reaction conditions require a fully solvated peptide–polymer matrix that enables efficient reagent penetration. It has been observed during the synthesis of a difficult peptide that the reaction matrix becomes partially inaccessible, typically from the 6th to 12th residue of the chain. This is thought to occur because of formation of secondary structures of aggregates that result in poor solvation of the peptide–polymer matrix [18–20]. As a peptide chain is built stepwise on a resin bead it can form aggregates either with itself or with neighboring chains. This involves inter or intramolecular hydrogen bonding of the peptide backbone that leads to β-sheet formation [21–23].

20 Microwave-enhanced Solid-phase Peptide Synthesis

Tab. 20.1. Phosphonium and aminium activators.

Name	Acronym	Structure
Benzotriazol-1-yl-*N*-oxy-tris(dimethylamino)phosphonium hexafluorophosphate	BOP	
Benzotriazol-1-yl-*N*-oxy-tris(pyrrolidino)phosphonium hexafluorophosphate	PyBOP	
(7-Azabenzotriazol-1-yloxy)-tris(pyrrolidino)phosphonium hexafluorophosphate	PyAOP	
N-[(1*H*-benzotriazol-1-yl)(dimethylamino)methylene]-*N*-methylmethanaminium hexafluorophosphate *N*-oxide	HBTU	
N-[(dimethylamino)-1*H*-1,2,3-triazolo[4,5-*b*]pyridino-1-ylmethylene]-*N*-methylmethanaminium hexafluorophosphate *N*-oxide	HATU	
N-[(1*H*-benzotriazol-1-yl)(dimethylamino)methylene]-*N*-methylmethanaminium tetrafluoroborate *N*-oxide	TBTU	

Tab. 20.1. (continued)

Name	Acronym	Structure
N-[(dimethylamino)-1H-1,2,3-triazolo[4,5-b]pyridino-1-ylmethylene]-N-methylmethanaminium tetrafluoroborate N-oxide	TATU	
1H-benzotriazolium-1-[bis(dimethylamino)methylene]-5-chloro, hexafluorophosphate (1-),3-oxide	HCTU	

20.2.3
Microwave Synthesis

The standard microwave frequency used for synthesis is 2450 MHz. At this frequency, molecular rotation occurs as molecular dipoles or ions try to align with the alternating electric field of the microwave by processes called dipole rotation or ionic conduction [24, 25]. On the basis of the Arrhenius equation, ($k = Ae^{-Ea/RT}$) the reaction rate constant depends on two factors, the frequency of collisions between molecules that have the correct geometry for a reaction to occur, A, and the fraction of those molecules that have the minimum energy required to overcome the activation energy barrier, $e^{-Ea/RT}$.

Although there is some speculation that microwaves can reduce activation energy by dipolar polarization, this has yet to be proven. Microwave energy will affect the temperature of the system, however. In the Arrhenius equation, T measures the average bulk temperature of all components of the system. It is known that for a given temperature the molecules in the system are at a range of temperatures as shown in the Boltzmann equation, $F(E) = 1/Ae^{(E/kT)}$. Because not all components absorb microwave energy equally, the Boltzmann profile of temperatures will tend to flatten to reflect a higher range in molecular temperatures at a given bulk temperature. At a given bulk temperature with microwave energy, a greater percentage of molecules can be at an energy level above the activation energy required for the reaction and increases in reaction rates are observed. The rapid and selective energy transfer that results from microwave irradiation leads to higher instantaneous heating of the reactants and a broader distribution of molecular energies. This is believed to be the primary reason for the rate enhancements observed.

The N-terminus amine group and the backbone in peptides are polar, which makes them constantly try to align with the alternating electric field of the microwave. During peptide synthesis this can help break up the chain aggregation which results from intra and interchain association and enable easier access to the growing end of the chain. One must also consider the solvent used for SPPS. Both DMF and NMP are medium absorbers of microwave energy whereas DCM is a lower absorber. The dielectric loss for NMP, DMF, and DCM is 8.855, 6.070, and 0.382, respectively. Conventional heating relies on migration of energy from outside the vessel, which is a slow and nonspecific transfer. With microwaves, energy transfer occurs in 10^{-9} s with each cycle of electromagnetic energy. The kinetic molecular relaxation from this energy is approximately 10^{-5} s. Thus a large amount of energy can be applied directly to the molecules involved in the reaction in a very efficient manner.

Some laboratories have reported that use of simultaneous cooling with microwave irradiation leads to improvements in reaction rates compared with microwaves alone [26–28]. This strategy enables greater amounts of microwave energy to be delivered to a sample while keeping the bulk of the solution at a lower temperature.

20.2.4
Tools for Microwave SPPS

During the SPPS cycle two distinct reactions are required for each amino acid addition. Complete reaction requires both proper addition of reagents and ensuring that the resin is completely submerged in solution. Also, during the coupling reaction excess of aminium-based activators compared with Fmoc amino acid can cause capping of the peptide chain. Thus, accurate delivery of several reagents is required throughout the synthesis.

Automated peptide synthesizers have been in use for several decades and have provided much needed assistance for manual synthesis of long peptides. In 2003, CEM Corporation introduced the first automated microwave peptide synthesizer, Odyssey. This system was updated shortly afterwards to the current version, Liberty (Fig. 20.6).

The Liberty system is a sequential peptide synthesizer capable of complete automated synthesis, including cleavage, of up to twelve different peptides. The system uses the Discover single-mode microwave reactor that has been widely used in organic synthesis.

The Liberty features a standard 30-mL Teflon glass fritted reaction vessel for 0.1–1.0 mmol synthesis. A 100-mL reaction vessel is also available for multi-mmol scales. The reaction vessel features a spray head for delivery of all reagents and a fiber-optic temperature probe for monitoring microwave power delivery. The system uses up to 25 stock solutions for amino acids and seven reagent ports that perform the functions: main wash, secondary wash, deprotection, capping, activator, activator base, and cleavage. The system uses nitrogen pressure for transfer of all reagents and to provide an inert environment during synthesis. Nitrogen bubbling is also used for mixing during deprotection, coupling, and cleavage reactions. The

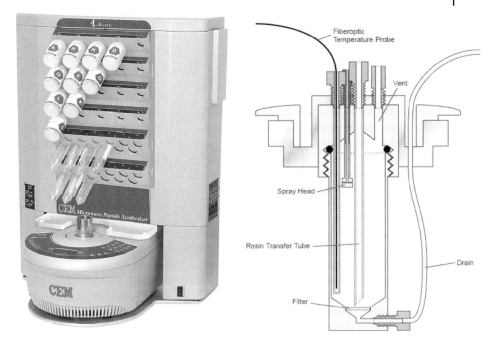

Fig. 20.6. The CEM Liberty automated peptide synthesizer, shown with reaction vessel.

system uses a metered sample loop for precise delivery of all amino acid, activator, and cleavage solutions. The Liberty is controlled by an external computer, which enables complete control of each step in every cycle.

CEM also offers a manual peptide synthesizer, Discover SPS (Fig. 20.7). This system is also based on the Discover microwave reactor. It uses a 25 mL polypropylene SPE cartridge that is held in place in the reactor. Fiber-optic temperature monitoring is included in the system and enables control of microwave power delivery. A vacuum manifold station is provided that enables rapid washing and filtration of reagents during the cycle. The system can be upgraded to the automated Liberty.

20.2.5
Microwave Enhanced *N*-Fmoc Deprotection

The *N*-Fmoc protecting group is labile to organic bases and is removed by base-catalyzed elimination (Scheme 20.2). Deprotection is most efficient with unhindered secondary amines, but is also susceptible to primary and tertiary amines. Typically, a 20% piperidine in DMF solution is used to form a dibenzofulvene (DBF) intermediate that is immediately trapped by the secondary amine to form an inert adduct.

For difficult peptides, incomplete Fmoc deprotection can be a problem and use of the stronger tertiary base, 1,8-diazabicyclo[5.4.0]undec-7-ene (DBU) has been shown to increase reaction efficiency. Typically, small amounts of piperidine are

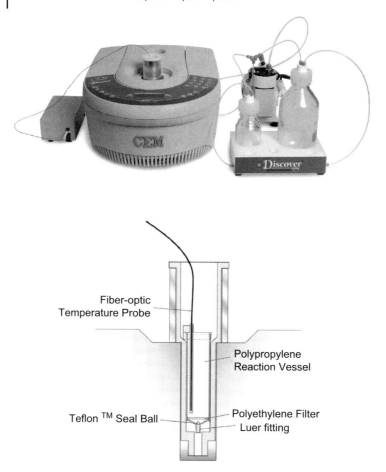

Fig. 20.7. CEM Discover SPS manual peptide synthesizer shown with reaction vessel.

added to DBU to scavenge the free DBF because this is not achieved with DBU alone. Free DBF can potentially react with the resulting resin-bound terminal N^α-amine preventing acylation of the next amino acid.

The deprotection reaction is often performed in two stages to prevent DBF alkylation of the resin-bound terminal N^α-amine. The first stage is typically shorter than the second and serves to remove a significant amount of DBU from the reaction vessel before a longer deprotection reaction is employed with fresh reagent.

20.2.5.1 Aspartimide Formation

Base-catalyzed aspartimide formation, which has been described in detail in the literature, can be a serious problem during chain assembly in peptide synthesis [29]. This side reaction involves attack by the nitrogen atom attached to the α-carboxy group of either aspartic acid or asparagine on the side chain ester or amide group,

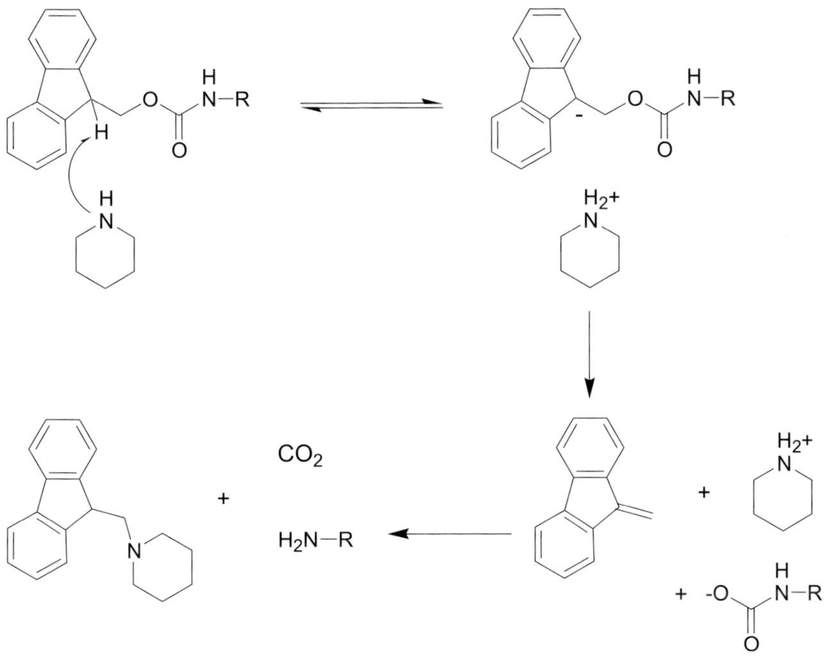

Scheme 20.2. Fmoc removal by piperidine.

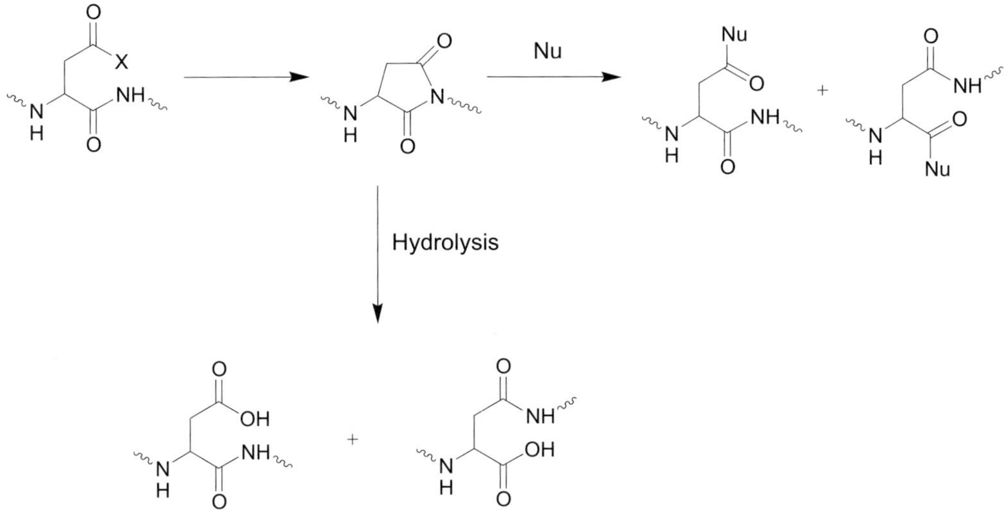

Scheme 20.3. Aspartimide formation of Asp(X) sequences.

Tab. 20.2. Effect of deprotection reagent on aspartimide formation of VKDGYI.

Deprotection reagent	Amount aspartimide (%)
20% Piperidine in DMF	10.90
20% Piperidine with 0.1 M HOBt in DMF	5.55
Hexamethyleneimine–N-ethylpyrrolidine–HOBt in NMP–DMSO	1.49
20% Piperidine in DMF (with Hmb backbone protection)	Not detected

respectively (Scheme 20.3). Nucleophilic attack then causes subsequent ring opening, which gives rise to a mixture of α-aspartyl and β-aspartyl peptides.

Aspartimide formation has been shown to occur in sequences containing the "Asp-X" moiety, where X = Gly, Asn, Ser, or Thr. Each subsequent deprotection cycle after the "Asp-X" sequence further increases aspartimide formation. This can be a serious problem in longer peptides with multiple Asp residues.

The process occurs naturally in biological systems with proteins containing aspartic acids. Inclusion of β-tert-butyl ester protection is thought to reduce this because of its bulkiness, although this side reaction is well documented in routine peptide synthesis even with side-chain-protection of aspartic acid. Incorporation of 0.1 M HOBt in the deprotection solution has been shown to reduce aspartimide formation [30, 31]. Often, however, significant amounts of aspartimide are still formed. The hexapeptide "VKDGYI" has been shown to produce substantial amounts of aspartimide-related products during SPPS [32]. This peptide was synthesized manually in a single-mode microwave reactor with three 30-s, 100-W, cycles with ice bath cooling between each irradiation cycle [33]. Maximum temperature was measured to be approximately 40 °C. Significant aspartimide formation was detected when 20% solution of piperidine in DMF was used. This was reduced by alteration of the deprotection solution, but only completely eliminated by use of Hmb dipeptide insertion of DG (Table 20.2).

The use of piperazine in place of piperidine results in much lower levels of aspartimide formation [34] (Fig. 20.8). Piperidine is a precursor in the synthesis of phenylcyclidine (angel dust) and is regulated by the Drug Enforcement Agency. Piperazine is not a controlled substance and so is more accessible to laboratories than piperidine. Piperazine is also an oral medication used to treat roundworm infection and is less odorous and toxic than piperidine [35]. Piperazine, with a pK_a of 9.8 compared with 11.1 for piperidine, is however, a slower deprotection reagent. In conventional synthesis of hydrophobic sequences, use of piperazine can lead to more incomplete Fmoc removal. DBU is often used as a stronger deprotection agent than piperidine, but can cause large amounts of aspartimide formation.

Microwave energy can substantially accelerate Fmoc deprotection with piperazine. Complete Fmoc removal can be accomplished with piperazine in 3 min. This enables efficient deprotection with a very desirable reagent. For example, aspartimide formation was minimized on a 20mer peptide with a "Gly–Asp" C-

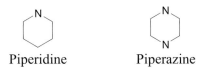

Fig. 20.8. The structures of piperidine and piperazine.

terminal sequence by use of microwave SPPS [36] (Table 20.3). Fmoc removal was performed with a range of deprotection chemistry. The deprotection reaction was performed in two steps, with each containing a fresh deprotection solution. The first step was a 30-s microwave method with constant power giving a maximum temperature of approximately 40 °C. The second step used a 3-min method with the maximum temperature reaching 80 °C.

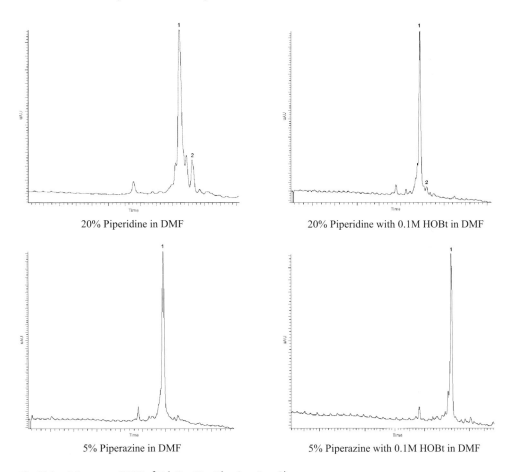

Fig. 20.9. Microwave SPPS of Val–Tyr–Trp–Thr–Ser–Pro–Phe–Met–Lys–Leu–Ile–His–Glu–Gln–Cys–Asn–Arg–Ala–Asp–Gly–NH$_2$: 1, product; 2, α and β-piperidides.

Tab. 20.3. Amount (%) of D-Asp measured by GC–MS after hydrolysis of Val–Tyr–Trp–Thr–Ser–Pro–Phe–Met–Lys–Leu–Ile–His–Glu–Gln–Cys–Asn–Arg–Ala–Asp–Gly–NH$_2$ with 6 M DCl in D$_2$O.

Deprotection reagent	Amount of aspartimide (%)	Amount of D-Asp (%)
20% Piperidine in DMF	31.50	9.60
20% Piperidine with 0.1 M HOBt in DMF	9.10	3.83
5% Piperidine with 0.1 M HOBt in DMF	3.15	1.18

Deprotection with piperazine led to an overall reduction of side-products resulting from aspartimide formation, and piperazine led to less aspartimide formation than piperidine. With both reagents, addition of 0.1 M HOBt reduced aspartimide formation further. When piperidine-containing deprotection solutions were used, α and β-piperidides resulting from base catalyzed imide ring opening were detected by LC–MS analysis (Fig. 20.9). The corresponding products were not detected when piperazine-containing deprotection solutions were used. Racemization of the aspartic acid, which can occur by hydrolysis of the imide ring in solution, was also substantially less when piperazine was used in place of piperidine, as shown in Table 20.3.

20.2.6
Microwave-enhanced Coupling

The coupling reaction has been a major focal point throughout the history of SPPS. It is well known for its difficulties in aggregating sequences and has led to the development of exotic activators, solvent mixtures, lengthy reaction times, and use of double couplings. In addition, many laboratories employ a capping step after each coupling reaction. This involves acetylation of the peptide chain, typically by use of acetic anhydride, effectively preventing the chain with uncoupled amino acid from continuing to grow.

There have been several reports of use of elevated temperature to accelerate the coupling reaction [37–39]. Although results were indicative of increased rates of acylation, there was concern about the potential increase in racemization. Microwave energy was used to accelerate the coupling reaction of the $^{65-74}$ACP peptide, enabling completion in 6 min in a domestic microwave oven [40]. This resulted in better coupling efficiency than in a conventional automated peptide synthesizer using reaction monitoring of the deprotection reaction.

In 2002, a single mode microwave synthesizer was also shown to accelerate the coupling reaction of several 3mer peptides [41]. Temperatures of up to 120 °C were used for the coupling reaction. In 2003, microwave energy was applied to the complete Fmoc SPPS cycle, including both deprotection and coupling steps [42]. The

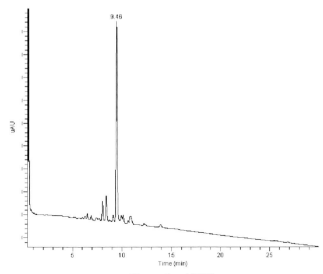
Conventional SPPS

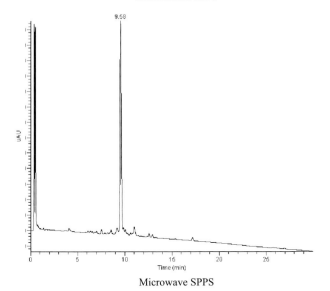
Microwave SPPS

Fig. 20.10. Microwave SPPS of [65–74]ACP Sequence Val–Gln–Ala–Ala Ile–Asp–Tyr–Ile–Asn–Gly [42].

[65–74]ACP peptide was assembled by using a 1-min deprotection step and a 2-min coupling step. Without microwave energy, multiple deletions were observed, as shown in Fig. 20.10.

Galanin is a unique neuroendocrine peptide found in the brain and gut. It is involved in a variety of physiological processes including cognition, memory, neuro-

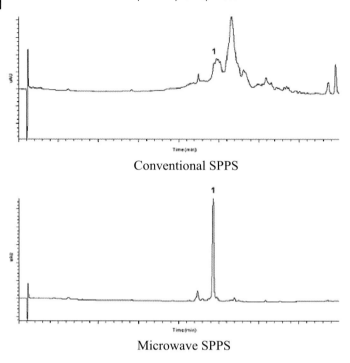

Fig. 20.11. SPPS of Gly–Trp–Thr–Leu–Asn–Ser–Ala–Gly–Tyr–Leu–Leu–Gly–Pro–Gln–Gln–Phe–Phe–Gly–Leu–Met–NH$_2$; 1, product. Synthetic conditions: 20% piperidine in DMF = deprotection; HBTU–HOBt–DIEA 0.9:1:2, ×10 excess = coupling; reagent K = cleavage.

transmitter secretion, and feeding behavior. This peptide was synthesized on a Liberty system with a 3-min deprotection reaction and a 4-min coupling reaction, both with and without microwave energy [43]. The maximum temperature reached during both the deprotection and coupling reactions was 75 °C in the microwave experiment. With microwave energy the product was obtained in 84.8% purity whereas without use of microwaves the product was less than 10% pure, as shown in Fig. 20.11.

20.2.6.1 Racemization

The properties of proteins and peptides are critically dependent on the configuration of their chiral centers. Alteration of a single chiral center can have a drastic effect on biological activity. With the exception of glycine, all 20 standard amino acids contain a chiral center at the α-carbon atom. Isoleucine and threonine also have a chiral center in their side-chains. In SPPS, racemization has been extensively documented during the coupling reaction. It is generally thought to occur via formation of an oxazolone intermediate [44].

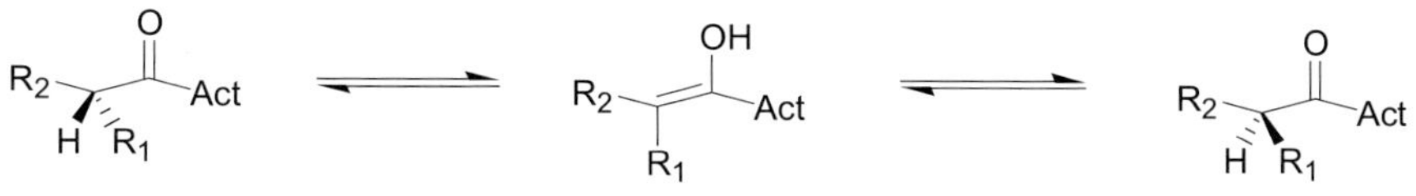

Scheme 20.4. Enolization leading to racemization.

Enolization Direct enolization can occur during the coupling reaction. During coupling, conversion of the incoming amino acid to an activated ester increases the acidity of the α-carbon (Scheme 20.4). This can tend to favor enolization, which leads to rearrangement about the α-carbon. In SPPS, sterically hindered tertiary amines are used in an effort to minimize base-catalyzed removal of the α-carbon proton.

Oxazolone formation Enantiomerization of the incoming amino acid residue is possible by formation of an oxazolone ring (Scheme 20.5). This ring can then re-open leading to racemization about the α-carbon. Enantiomerization by oxazolone formation occurs more frequently in fragment condensation when connecting peptide fragments together. Some N-urethane protected amino acids, particularly histidine and cysteine, have been shown to be susceptible to this reaction, however [45–48].

20.2.6.2 Quantifying Racemization During SPPS

Whenever a new condition is introduced to the coupling reaction (activator, solvent, base, resin, protecting group, etc.), it is important to study the effect it may have on racemization of the incoming amino acid. Although enantiomerization about a single chiral center can have a major impact on a protein or peptide's activity, it can be very difficult to quantify. In smaller peptides, racemization can often be quantified by efficient HPLC separation. When a peptide increases in length, separation of centers of racemization can become an impossible task. Although chiral columns are an improvement on standard C_{18} columns, this is often not sufficient.

An accurate and sensitive method for detection of racemization during peptide

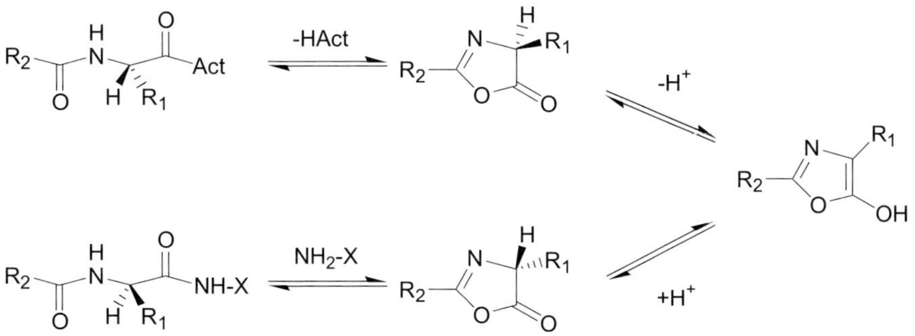

Scheme 20.5. Oxazolone formation.

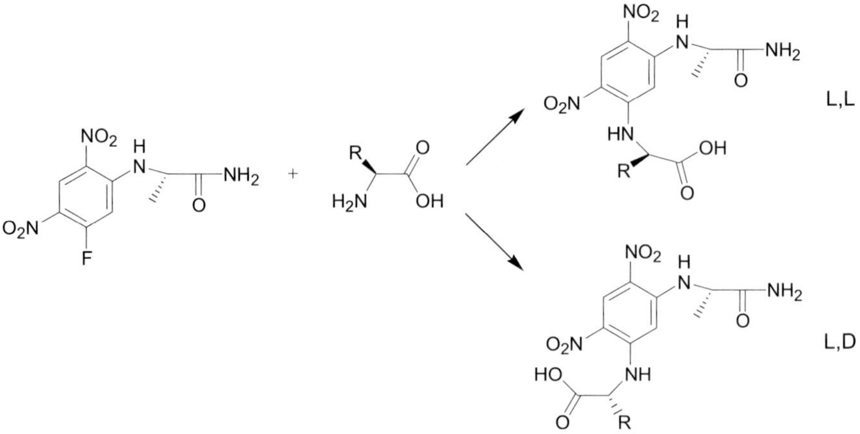

Scheme 20.6. Derivatization with Marfey's reagent.

synthesis involves hydrolysis of the peptide into free amino acids followed by derivatization and analysis by gas chromatography. Derivatization is usually performed with Marfey's reagent or 1,fluoro-2-4-dinitrophenyl-5-L-alanine amide (FDAA). Marfey's reagent reacts by nucleophilic substitution, through its aromatic fluorine of the amine group on the amino acid. Marfey's reagent also contains a chiral center, in the L configuration, in its alanine group.

Reaction of Marfey's reagent with a racemic mixture of amino acids results in a mixture of L–L and L–D diastereomers. In the L–L diastereomer an intramolecular hydrogen bond is formed between the carboxyl and carboxyamide groups; this does not occur (or does to a much lesser extent) in the L–D diastereomer. This feature enables the derivatives to be separated by RP HPLC.

One problem with this approach is that hydrolysis of the peptide can contribute to racemization. Frank et al. developed an approach that involves hydrolysis of the peptide in 6 M DCl in D_2O, which leads to deuterium exchange on racemization [49, 50]. The amount of racemization resulting from hydrolysis can then be measured by mass spectrometric determination and subtracted to give a measure of racemization from synthesis of the peptide.

Racemization studies have recently been performed on two model peptides with serine and aspartic acid residues [51]. In this study, the microwave coupling reaction was performed using six cycles of 40 W for 25 s. Between cycles the reaction vessel was externally cooled to maintain the maximum temperature below 30 and 50 °C in different experiments, respectively. Racemization analysis was performed by separation of the L and D enantiomers by HPLC. Analysis showed that after use of microwave energy in the coupling reaction, racemization was less than or comparable with that after conventional coupling (Table 20.4).

Racemization studies were performed during microwave synthesis of the peptide Ala–Cys–Asp–Glu–Phe–Gly–His–Ile–Lys–Leu–Met–Asn–Pro–Gln–Arg–Ser–Thr–Val–Trp–Tyr–NH$_2$ under a variety of coupling conditions using an Odys-

Tab. 20.4. Amount of racemization (%) found by HPLC analysis of H–Gly–Xxx–Phe–NH$_2$.

Sequence	Method	Amount D (%)	Amount L (%)
H–Gly–L-Ser–Phe–NH$_2$	Conventional (RT)	0.73	99.27
	MW-SPPS $T_{max} = 30\,°C$	0.09	99.91
H–Gly–D-Ser–Phe–NH$_2$	Conventional (RT)	98.00	2.00
	MW-SPPS $T_{max} = 30\,°C$	98.44	1.56
	MW-SPPS $T_{max} = 50\,°C$	98.44	1.56
H–Gly–L-Asp–Phe–NH$_2$	Conventional (RT)	0.63	99.37
	MW-SPPS $T_{max} = 30\,°C$	0.25	99.75
H–Gly–D-Asp–Phe–NH$_2$	Conventional (RT)	99.75	0.25
	MW-SPPS $T_{max} = 30\,°C$	99.63	0.37
	MW-SPPS $T_{max} = 50\,°C$	99.68	0.32

sey system [52]. Racemization analysis was performed by C.A.T. GmbH via GC–MS analysis after hydrolysis in 6 M DCl in D$_2$O. In this study substantial racemization occurred on the aspartic acid and cysteine residues during most activation schemes. Optimum results were observed with the PyBOP–DIEA and HATU–DIEA activation schemes.

More recently, a detailed study was undertaken with a 20mer peptide that contained each of the natural 20 amino acids, but with a selectively placed C-terminal Asp–Gly segment to encourage maximum aspartimide formation [36]. The peptide was synthesized on a Liberty system, on the 0.1 mmol scale, using a Rink MBHA amide resin with fivefold excess of reagents for coupling. Microwave irradiation was used during the deprotection in two stages, a 30-s method repeated with a new solution for 3 min to give a maximum temperature of 80 °C. A 5-min coupling reaction was performed with the maximum temperature also reaching 80 °C. Racemization analysis was performed by GC–MS after hydrolysis in 6 M DCl in D$_2$O. Results indicated that the sequence was susceptible to racemization at the histidine and cysteine residues, and at aspartic acid as a by-product of aspartimide formation with standard piperidine deprotection. Addition of HOBt to the coupling solution did not reduce racemization for all the amino acids tested. Racemization of aspartic acid was controlled by reduction in aspartimide formation as a result of addition of 0.1 M HOBt and substitution of piperidine with piperazine during deprotection. Results are summarized in Table 20.5.

20.2.6.3 Racemization of Histidine and Cysteine

Histidine and cysteine are both very susceptible to racemization during routine SPPS. If not optimized, microwave SPPS has been shown capable of causing substantial racemization during the coupling of both histidine and cysteine [36]. Cysteine racemization has been attributed to α-carbon proton abstraction by

Tab. 20.5. Amount of racemization (%) measured by GC–MS after hydrolysis, with 6 M DCl in D_2O, of Val–Tyr–Trp–Thr–Ser–Pro–Phe–Met–Lys–Leu–Ile–His–Glu–Gln–Cys–Asn–Cys–Arg–Ala–Asp–Gly–NH_2.

Amino Acid	Synthetic conditions							
	Conventional[a]	Microwave						
	20% Piperidine–DMF	20% Piperidine–DMF (80 °C)	20% Piperidine w/ 0.1 M HOBt in DMF (80 °C)	20% Piperidine w/ 0.1 M HOBt in DMF (80 °C)	5% Piperazine w/ 0.1 M HOBt in DMF (80 °C)	5% Piperazine w/ 0.1 M HOBt in DMF (80 °C)	5% Piperazine w/ 0.1 M HOBt in DMF (80 °C)	5% Piperazine w/ 0.1 M HOBt in DMF (80 °C)
	HBTU–DIEA in DMF	HBTU–DIEA in DMF (80 °C)	HBTU–HOBt–DIEA in DMF (80 °C)	PyBOP–DIEA in DMF (80 °C)	HBTU–DIEA in DMF (60 °C)	HBTU–HOBt in DMF (80 °C) (Cys, His[b])	HBTU–TMP in DMF (80 °C)	HBTU–HOBt in DMF (80 °C) (Cys, His[c])
D-Asp	1.19	9.60	3.50	3.83	0.92	1.62	0.91	Not determined
D-Ala	0.21	0.50	0.34	0.18	0.22	0.63	0.12	Not determined
D-Arg	0.18	0.17	0.21	0.18	0.10	0.20	<0.10	Not determined
D-Cys	1.09	<4.48	<7.00	<6.95	<2.71	<3.16	<1.03	<1.14
D-Glu	1.46	<1.14	<1.45	<1.04	<0.98	<1.51	<1.26	Not determined
D-His	0.65	9.40	8.40	12.30	3.96	1.59	8.49	0.67
D-Ile	<0.10	<0.10	<0.10	<0.10	<0.10	<0.10	<0.10	Not determined
L-allo Ile	<0.10	<0.10	<0.10	<0.10	<0.10	<0.10	<0.10	Not determined
D-allo Ile	<0.10	<0.10	<0.10	<0.10	<0.10	<0.10	<0.10	Not determined
D-Leu	0.17	0.14	0.22	0.21	0.13	0.24	0.11	Not determined
D-Lys	0.10	0.11	0.15	0.16	<0.10	0.14	<0.10	Not determined
D-Met	0.48	0.37	0.81	0.44	0.28	0.49	0.39	Not determined
D-Phe	0.28	0.20	0.85	0.35	0.16	0.41	0.14	Not determined
D-Pro	<0.10	<0.10	<0.10	<0.10	<0.10	<0.10	0.13	Not determined
D-Ser	0.46	0.54	0.99	1.02	0.25	1.83	0.31	Not determined

D-Thr	<0.10	<0.10	<0.10	<0.10	<0.10	Not determined
L-allo Thr	<0.10	<0.10	<0.10	<0.19	0.13	Not determined
D-allo Thr	<0.10	<0.10	<0.10	<0.10	0.12	Not determined
D-Trp	0.19	0.20	0.57	0.30	0.80	0.17
D-Tyr	0.42	0.35	0.72	0.43	0.88	0.64
D-Val	<0.10	<0.10	<0.10	<0.10	<0.10	<0.10

[a] Conventional synthesis was performed with deprotection = 5 and 15 min, coupling = 30 min
[b] Microwave methods were modified for Cys and His coupling by using no power for 2 min then 4 min at 50 °C maximum temperature
[c] The Cys and His coupling cycles were performed conventionally [36]

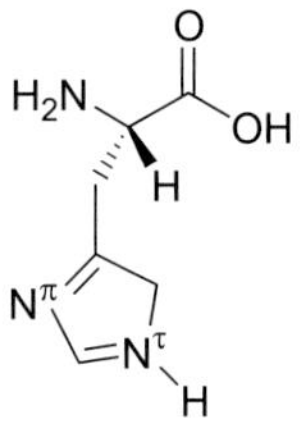

Fig. 20.12. The structure of histidine.

the tertiary amine in the coupling reaction. This has led to the use of very bulky hindered tertiary amines, which limits their ability to reach the α-carbon proton. Substitution of DIEA with the more hindered base collidine, also known as TMP, led to substantial reduction of cysteine racemization. Use of TMP, however, did lead to lower coupling efficiency, as was shown by substantial valine deletion during synthesis of the 20mer peptide. Reduction of coupling temperature from 80 °C to 50 °C was also successful in substantially reducing cysteine racemization.

Histidine is susceptible to racemization through its own side chain (Fig. 20.12). Thus, alteration in the base used during coupling has no real effect on histidine racemization. Typical side-chain-protection for histidine involves protecting the τ-nitrogen, because it is more accessible [53]. The π-nitrogen is actually closer to the α-carbon proton, however, and responsible for nucleophilic attack leading to α-carbon rearrangement. Reducing the coupling temperature to 50 °C for histidine substantially reduced racemization. Derivatives of histidine are available with side-chain-protection on the π-nitrogen; although this eliminates concern about histidine racemization, they are substantially more expensive and not an attractive option for routine synthesis.

Although coupling temperatures up to 80 °C can be used during microwave SPPS, both cysteine and histidine require special attention. Reducing the coupling temperature to 50 °C had a positive effect for both amino acids. When both histidine and cysteine have been coupled, no further increase in racemization is observed during subsequent chain synthesis steps with microwave energy. This enables coupling of both amino acids to be performed at room temperature with no increased susceptibility to racemization at elevated temperatures of up to 80 °C during other amino acid coupling and deprotection steps.

20.2.6.4 γ-Lactam Formation from Arginine

During coupling, the nucleophilic side-chain of arginine is susceptible to γ-lactam formation [54] (Scheme 20.7). This irreversible reaction will effectively render the activated arginine derivative inactive during the coupling reaction.

During normal synthesis this may not be apparent as a significant problem, because the coupling reaction will occur before substantial γ-lactam formation occurs. In a difficult coupling reaction, however, γ-lactam formation may become more favorable than coupling. In these circumstances, extending the reaction time will not promote completion of the coupling. This was shown recently with the mi-

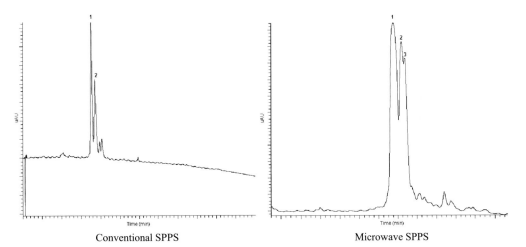

Scheme 20.7. γ-Lactam formation.

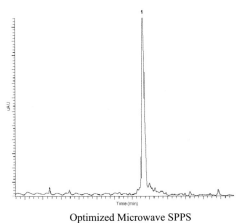

Fig. 20.13. SPPS of Gly–Val–Arg–Gly–Asp–Lys–Gly–Asn–Pro–Gly–Trp–Pro–Gly–Ala–Pro–Tyr–NH$_2$: 1, product; 2, del[Arg]; 3, del[Arg] w/ dehydration. Synthetic conditions: 20% piperidine in DMF = deprotection; HBTU–HOBt–DIEA 0.9:1:2, ×10 excess = coupling; reagent K = cleavage [36].

crowave synthesis of the 1992 ABRF peptide sequence – almost 50% deletion of arginine was observed in product analysis [55]. The use of a pseudo-proline derivative was beneficial in enabling access to the N-terminus of the peptide chain and thus increasing coupling of arginine before lactam formation could occur. Complete elimination of any arginine deletion in the sequence was made possible by use of an all PAL–PEG resin from ChemMatrix [36]. This resin enabled improved solvation of the peptide-resin matrix and enabled the coupling reaction to reach completion before complete lactam formation. Results from synthesis of the 1992 ABRF peptide are shown in Fig. 20.13.

20.2.7
Longer Peptides

Because longer peptides enable more coverage of a protein and better ability to study its interactions, the ability to synthesize longer chains is highly desirable. Conventional Fmoc SPPS has enabled routine assembly of peptide sequences of up to 30 to 40 amino acids when aggregation is not an important issue.

The 31mer C-peptide fragment naturally cleaved from proinsulin has been synthesized on a Liberty system in 0.1 mmol quantities with a crude purity of 75.0%, as shown in Fig. 20.14 [43]. When the synthesis was repeated, but with ten additional amino acids, including two arginine residues added to the chain, the crude purity dropped to 48.2%, as shown in Fig. 20.15, primarily because of incomplete coupling of arginine. As shown previously, when lactam formation occurs, activated arginine residues can no longer couple.

Microwave synthesis of the 30mer peptide Val–Tyr–Trp–Thr–Ser–Pro–Phe–Met–Lys–Leu–Ile–His–Glu–Gln–Cys–Asn–Arg–Ala–Asp–Gly–Val–Gln–Ala–Ala–Ile–Asp–Tyr–Ile–Asn–Gly with crude purity of 80.0% has also been reported.

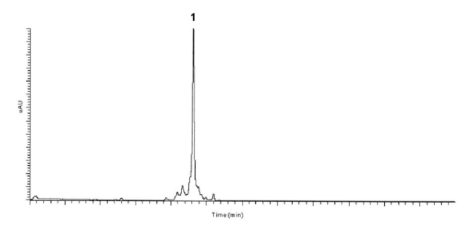

Fig. 20.14. Microwave-SPPS of Glu–Ala–Glu–Asp–Leu–Gln–Val–Gly–Gln–Val–Glu–Leu–Gly–Gly–Gly–Pro–Gly–Ala–Gly–Ser–Leu–Gln–Pro–Leu–Ala–Leu–Glu–Gly–Ser–Leu–Gly–NH$_2$; 1 = Product. Synthetic conditions: 20% piperidine in DMF = deprotection, PyBOP–HOBt–DIEA 0.9:1:2, ×6 excess = coupling, reagent K = cleavage [43].

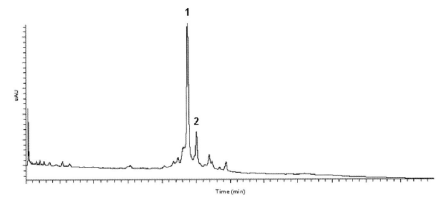

Fig. 20.15. Microwave-SPPS of Gly–Phe–Phe–Tyr–Thr–Pro–Lys–Thr–Arg–Arg–Glu–Ala–Glu–Asp–Leu–Gln–Val–Gly–Gln–Val–Glu–Leu–Gly–Gly–Gly–Pro–Gly–Ala–Gly–Ser–Leu–Gln–Pro–Leu–Ala–Leu–Glu–Gly–Ser–Leu–Gly–NH$_2$; 1 = product; 2 = del[Arg]. Synthetic conditions: 20% piperidine in DMF = deprotection; PyBOP–HOBt–DIEA 0.9:1:2, ×6 excess = coupling, reagent K = cleavage.

The synthesis was performed at the 0.1 mmol level using 5% piperazine in DMF as the deprotection reagent and HBTU activation with a fivefold excess of reagents and was complete in 12.5 h [36].

Erythropoietin (EPO) is a naturally occurring protein responsible for red blood cell production. A 20mer peptide mimic of this natural protein has been synthesized in 8 h by use of microwave SPPS. Purity was higher than that obtained in 5 days by conventionally synthesis [33] and the biological activity was reported to be equal to that of the conventionally synthesized material, as shown in Fig. 20.16.

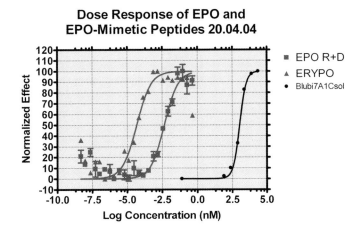

Fig. 20.16. Biological activity of Gly–Gly–Thr–Tyr–Ser–Cys–His–Phe–Gly–Pro–Leu–Thr–Trp–Val–Cys–Lys–Pro–Gln–Gly–Gly [33].

Microwave SPPS has been successfully used to synthesize the well-known difficult $^{1-42}\beta$-amyloid fragment [36]. In addition to being difficult to synthesize this sequence also can be challenging to analyze [56–59]. When microwave energy was used the crude purity was nearly twice that from conventional synthesis, as shown in Fig. 20.17.

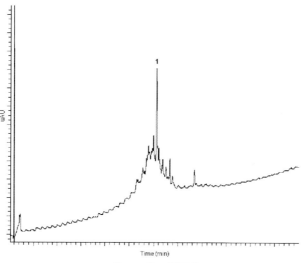

Conventional SPPS

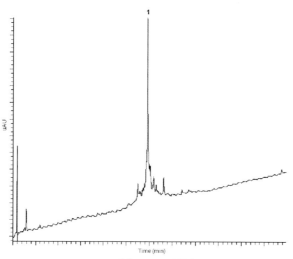

Microwave SPPS

Fig. 20.17. Microwave-SPPS of Asp–Ala–Glu–Phe–Arg–His–Asp–Ser–Gly–Tyr–Glu–Val–His–His–Gln–Lys–Leu–Val–Phe–Phe–Ala–Glu–Asp–Val–Gly–Ser–Asn–Lys–Gly–Ala–Ile–Ile–Gly–Leu–Met–Val–Gly–Gly–Val–Val–Ile–Ala; 1 = product. Synthetic conditions: 5% piperazine with 0.1 M HOBt in DMF = deprotection; HBTU–HOBt–DIEA 0.9:1:2; ×5 excess [36].

20.3
Conclusion

Microwave energy has been used to successfully enhance Fmoc solid-phase peptide synthesis. Because successful synthesis of a peptide necessitates almost 100% completion of two reactions per cycle, microwave energy is an efficient way of driving these reactions to completion. Common problems in SPPS, typically from aggregation, necessitate much time and reagent cost to drive the reaction to completion. Microwave energy is a rapid and efficient alternative that is useful for providing efficient energy to accelerate the deprotection and coupling steps. The key to using microwave energy for SPPS is proper control and feedback of the microwave power. Temperature monitoring is available for control of microwave power output and should be used during both deprotection and coupling. Common side-reactions (aspartimide formation and racemization) are observed during microwave SPPS, but can be controlled by optimizing the methods.

The deprotection reaction is prone to aspartimide formation that can arise from "Asp–Xxx" sequences. Synthesis conditions must be controlled for these sequences, to prevent excess aspartimide formation. This can be easily accomplished by using HOBt, as used conventionally, is equally powerful for limiting aspartimide formation during microwave SPPS. A key advantage of microwaves are their ability to accelerate Fmoc deprotection with reagents, for example piperazine, that are more desirable to use and advantageous for prevention of aspartimide formation, but not powerful enough to be used conventionally.

During the coupling reaction, racemization is a concern with histidine and cysteine in both conventional and microwave SPPS. Increases in racemization resulting from excess microwave energy can, however, be overcome by reducing the temperature below 50 °C for histidine and by use of a more hindered base, for example collidine, for cysteine [36]. Alternatively, derivatives of histidine that protect its π-nitrogen are available that enable racemization-free coupling, but are substantially more expensive and not desirable for routine synthesis. Coupling of both histidine and cysteine can also be performed without microwave energy, and the rest of the peptide can synthesized using microwaves because this will have no increased effect on racemization.

It is obvious that microwave energy must be controlled during SPPS or the potential for some side reactions is increased. This is, however, easy to achieve with commercially available instrumentation, both manual and automated. With the reduced cycle times and ability to drive difficult reactions to completion, microwave energy is a major tool in SPPS. Whereas typical conventional cycles times range from 60 to 180 min, microwave energy enables routine cycle times to be reduced to 15 to 30 min. For a 20mer peptide, therefore, use of microwave energy enables synthesis time to be reduced by approximately 15 to 30 h. Further optimization will enable preparation of hydrophobic and longer peptides, currently a major challenge for researchers, of even greater purity.

20.4
Future Trends

Microwave SPPS is enabling technology currently involved in much active research. As new hardware is installed throughout the world, use and knowledge of microwave SPPS is increasing daily. For example, microwave energy was recently used to substantially increase the purity of several long β-peptides, which suffer from low synthetic efficiency when conventional techniques are used [60].

As researchers have a demand for longer synthetic peptide sequences, convergent SPPS (CSPPS) methods, for example fragment condensation and chemical ligation, have developed. Fragment condensation of fully protected peptide segments has the advantage of purifying each fragment with the potential to achieve higher purities at each segment step. The development of trityl based resins and handles enable cleavage in less than 1% TFA or mixtures of acetic acid and trifluoroethanol (TFE) in DCM that generate fully protected peptide fragments. Fully protected peptides are poorly soluble in water and most organic solvents, however. Use of a special solvent system or segment structural modification is most commonly used to increase solubility. In either of these, limited solubility of protected fragments is a serious issue and can lead to reduced coupling efficiency. Microwave energy has the potential to enhance these reactions by overcoming some of the difficulties associated with solubility and large fragment size.

Solubility problems in fragment condensation led to the development of chemical ligation methods, which enable coupling of unprotected segments with greater solubility in most solvents. This method is based on non-amide bond formation, which enables protecting groups to be eliminated. Although the resulting peptide backbone has nonnatural structures at ligation points, it is still predominately natural and has often been shown to have complete biological activity. A variation of chemical ligation, known as thiol capture ligation, has been developed to enable native backbone formation, but requires an N-terminal Cys residue on one of the segments [61]. Microwave energy will have to be optimized for application during chemical ligation to ensure that temperature increases are kept to a minimum, to prevent side-reactions with unprotected side-chain groups.

If microwaves can sufficiently accelerate coupling of protected fragments, this would usefully eliminate many of the extra synthetic steps required when using unprotected fragments. Microwave energy also has the potential to increase the length of peptide fragments that can be used at each step; this would result in more selectivity at C-terminal coupling sites. Development of optimized microwave methods should enable increased coupling efficiency of all the different CSPPS approaches.

With future opportunities for microwave energy in the peptide synthesis field, more advanced tools will be needed. Applications requiring peptide libraries may lead to the development of a parallel system capable of 96-well plate formats. With currently available hardware, however, preparing the plate may limit overall peptide production because increasing reaction speed may not substantially increase throughput. In this format, however, microwaves do result in increases in overall product purity, which may be a substantial benefit.

Phosphopeptides are an important research tool for investigating the effects of phosphorylation on the structure of peptides and proteins and the corresponding regulatory process of protein kinases. The development of Fmoc phosphoamino acids derivatives enables direct incorporation of these amino acids during SPPS instead of difficult post-synthetic phosphorylation [62]. In SPPS, however, incorporation of phosphoamino acids of Ser, Thr, and Tyr require extremely difficult couplings that also make subsequent couplings difficult. This is area in which microwave energy has the benefit of overcoming the steric hindrance associated with these bulky derivatives.

Proteomics has led to a great need for large numbers of protein-binding molecules. Peptidomimetic compounds that are similar to native proteins but are immune to proteases are an intriguing class of compounds with potential target-binding applications. One class of peptidomimetic is the peptoids, which are similar to peptides except that each amino acid R group is shifted from the α-carbon to the neighboring nitrogen atom [63]. Peptoid monomers are available that enable the solid-phase principle with deprotection and coupling to be used. Typically, conventional coupling of peptoid monomers takes 2.5 to 3 h. Microwave energy has been used to accelerate peptoid coupling to less than 1 min, giving results comparable with those from conventional coupling [64]. Use of this method enables a nine-residue peptoid to be synthesized in 3 h compared with 20 to 32 h with conventional methods.

As researchers obtain peptides of interest in research quantities they will ultimately need to develop strategies for synthesis of greater amounts. Whereas current instrumentation enables synthesis of multi-mmole amounts, development of larger-scale systems will be important. Larger microwave cavities are available and will need to be adapted to enable proper manipulation for SPPS. Scale-up of microwave synthesis, not just in peptide synthesis, but in drug discovery also has generated massive interest in recent years; this should lead to the development of new tools for applying microwave energy on a larger scale.

Abbreviations

Boc	tert-Butoxycarbonyl
BOP	Benzotriazol-1-yl-N-oxy-tris(dimethylamino)phosphonium hexafluorophosphate
DBU	1,8-Diazabicyclo[5.4.0]undec-7-ene
DBF	Dibenzofulvene
DCM	Dichloromethane
DIC	N,N'-diisopropylcarbodiimide
DIEA	Diisopropylethylamine
DMF	N,N-Dimethylformamide
FDAA	1-Fluoro-2,4-dinitrophenyl-5-L-alanine amide
Fmoc	9-Fluorenylmethoxycarbonyl

HATU	N-[(Dimethylamino)-1H-1,2,3-triazolo[4,5-b]pyridino-1-ylmethylene]-N-methylmethanaminium hexafluorophosphate N-oxide
HBTU	N-[(1H-Benzotriazol-1-yl)(dimethylamino)methylene]-N-methylmethanaminium hexafluorophosphate N-oxide
HCTU	1H-Benzotriazolium-1-[bis(dimethylamino)methylene]5-chloro hexafluorophosphate (1-),3-oxide
HF	Hydrofluoric acid
Hmb	N-2-Hydroxy-4-methoxybenzyl
HMPA	Hexamethylphosphoric triamide
HOAt	7-Azabenzotriazole
HOBt	Hydroxybenzotriazole
NMM	N-Methylmorpholine
NMP	N-Methylpyrrolidinone
HPLC	High-performance liquid chromatography
PAL	5-[4-(Aminomethyl)-3,5-dimethoxyphenoxy]pentanoic acid
PEG	Poly(ethylene glycol)
PyBOP	Benzotriazol-1-yl-N-oxy-tris(pyrrolidino)phosphonium hexafluorophosphate
PyAOP	(7-Azabenzotriazol-1-yloxy)-tris(pyrrolidino)phosphonium hexafluorophosphate
SPPS	Solid-phase peptide synthesis
TATU	N-[(Dimethylamino)-1H-1,2,3-triazolo[4,5-b]pyridino-1-ylmethylene]-N-methylmethanaminium tetrafluoroborate N-oxide
TBTU	N-[(1H-Benzotriazol-1-yl)(dimethylamino)methylene]-N-methylmethanaminium tetrafluoroborate N-oxide
TFA	Trifluoroacetic acid

References

1 SEWALD, N.; JAKUBKE, H.; Peptides: Chemistry and Biology, 2002, Wiley–VCH.
2 DELGADO, C.; FRANCIS, G.E.; FISHER, D.; Crit. Rev. Ther. Drug Carrier Syst., 9, 249, 1992.
3 PETTIT, D.K.; BONNERT, T.P.; EISENMAN, J.; SRINIVASAN, S.; PAXTON, R.; BEERS, C.; LYNCH, D.; MILLER, B.; YOST, J.; GRABSTEIN, K.H.; GOMBOTZ, W.R.; J. Biol. Chem., 272, 2312, 1997.
4 MERRIFIELD, R.B.; J. Am. Chem. Soc., 85, 2149, 1963.
5 ALEWOOD, P.; ALEWOOD, D.; MIRANDA, L.; LOVE, S.; MEUTERMANS, W.; WILSON, D.; Methods Enzymol., 289, 14–29, 1997.
6 SMITH, A.J.; YOUNG, J.D.; CARR, S.A.; MARSHAK, D.R.; WILLIAMS, L.C.; WILLIAMS, K.R.; Techniques in Protein Chemistry III, 1992, Academic Press, Orlando.
6a X-FARA, M.; DIAZ-MOCHON, J.; BRADLEY, M.; Tetrahedron Lett., 47, 1011–1014, 2006.
7 CARPINO, L.A.; EL-FAHAN, A.; MINOR, C.A.; ALBERICIO, F.; J. Chem. Soc. Chem. Commun., 201–203, 1994.
8 CARPINO, L.A.; EL-FAHAN, A.; ALBERICIO, F.; Tetrahedron Lett., 35, 2279–2282, 1994.
9 XU, Y.P.; MILLER, M.J.; J. Org. Chem., 63, 4314–4322, 1998.
10 GAWNE, G.; KENNER, G.W.; SHEPPARD, R.C.; J. Am. Chem. Soc., 91, 5670–5671, 1969.

11 Castro, B.; Dormoy, J.R.; *Bull. Soc. Chim. Fr.*, 3034–3036, **1971**.
12 Barstov, L.E.; Hruby, V.J.; *J. Org. Chem.*, 36, 1305–1306, **1971**.
13 Yamada, S.; Takeuchi, Y.; *Tetrahedron Lett.*, 3595–3598, **1971**.
14 Abdelmoty, I.; Albericio, F.; Carpino, L.A.; Foxman, B.M.; Kates, S.A.; *Lett Peptide Sci.* 1, 57–67, **1994**.
15 Henklein, P.; Costisella, B.; Wray, V.; Domke, T.; Carpino, L.A.; El-Faham, A.; Kates, S.A.; Abdelmoty, A.; Foxman, B.M.; In: Ramage, R.; Epton, R.; eds. Peptides 1996. Proceedings of the 24th European Peptide Symposium. Kingswinford, England: Mayflower Scientific, **1998**, pp 465–466.
16 Bailén, M.A.; Chinchilla, R.; Dodsworth, D.; Nájera, C.; Soriano, J.M.; Yus, M.; Peptides **1998**, Proceedings of the 25th European Peptide Symposium, Bajusz, S.; Hudecz, F.; eds. Akadémiai Kiadó, Budapest, **1999**, pp 172–173.
17 Marder, O.; Shvo, Y.; Albericio, F.; *HCTU and TCTU: New Coupling Reagents: Development and Industrial Applications* Poster Presentation Gordon Conference February **2002**.
18 Kent, S.B.H.; Peptides: Structure and Function, Proc. Am. Pept. Symp. 9th; Deber, C.M., Hruby, V.J., Kopple, K.D., Eds.; Pierce Chemicals: Rockford, **1985**, 407–414.
19 Kent, S.B.H.; *Annu. Rev. Biochem.*, 57, 959, **1988**.
20 Bedford, J.; Hyde, C.; Johnson, T.; Jun, W.; Owen, D.; Quibell, M.; Sheppard, R.C.; *Int. J. Peptide Protein Res.*, 40, 300, **1992**.
21 Pillai, V.; Mutter, M.; *Acc. Chem. Res.*, 14, 122, **1981**.
22 Narita, M.; Chen, J.Y.; Sato, H.; Lim, Y.; *Bull. Chem. Soc. Jpn.*, 58, 2494, **1985**.
23 Mutter, M.; Altmann, K.H.; Bellof, D.; Floersheimer, A.; Herbert, J.; Huber, B.; Klein, B.; Strauch, L.; Vorherr, T.; Gremlich, H.U.; Peptides: Structure and Function, Proc. Am. Pept. Symp. 9th; Deber, C.M.; Hruby, V.J.; Kopple, K.D., Eds.; Pierce Chemicals: Rockford, **1985**, 397–405.
24 Loupy, A.; Microwaves in Organic Synthesis; Wiley–VCH, Weinheim, **2002**.
25 Hayes, B.L.; Microwave Synthesis: Chemistry at the Speed of Light; CEM Publishing: Matthews, NC, **2002**.
26 Katritzky, A.; ARKIVOC; (xv) 47–64, **2003**.
27 Chen, J.J.; Deshpande, S.V.; *Tetrahedron Lett.*, 44, 8873–8876, **2003**.
28 Turner, N.J.; *Org. Lett.*, 5, 849–852, **2003**.
29 Nicolas, E.; Pedroso, E.; Giralt, E.; *Tetrahedron Lett.*, 30, 497–500, **1989**.
30 Dölling, R.; Beyermann, M.L.; Jaenel, J.; Kernchen, F.; Krause, E.; Franke, P.; Brudel, M.; Biernet, M.; *J. Chem. Soc. Chem. Commun.*, 853–854, **1994**.
31 Lauer, J.L.; Fields, C.G.; Fields, G.B.; *Lett. Peptide Sci.*, 1, 197–205, **1994**.
32 Mergler, M.; Dick, F.; Sax, B.; Weiler, P.; Vorherr, T.; *Pept. Sci.*, 9, **2003**.
33 Rybka, A.; "Microwave Assisted Peptide Synthesis" CEM International Microwave Conference, **2005**, Orlando, FL.
34 Tregear, G.; Macris, M.; Mathieu, M.N.; Wade, J.D.; *Pept. Sci.*, 7, 107, **2000**.
35 Swanson, L.; Stone, W.; Wade, A.; *J. Am. Vet. Med. Assoc.*, 130, 252–254, **1957**.
36 Palasek, S.; Cox, Z.; Collins, J.; (In Press).
37 Krchnak, V.; Vagner, J.; *J. Pept. Res.*, 3, 182, **1990**.
38 Lloyd, D.H.; Petrie, G.M.; Noble, R.L.; Tam, J.P.; Peptides: Proc. Am. Pept. Symp. 11th; Rivier, J.E., Marshall, G.R., Eds.; Escom: Leiden, Netherlands, **1990**, 909–910.
39 Rabinovich, A.K.; Rivier, J.E.; *Am. Biotechnol. Lab.*, 12, 48, **1994**.
40 Yu, H.M.; Chen, S.T.; Wang, K.T.; *J. Org. Chem.*, 57, 4781, **1992**.
41 Erdélyi, M.; Gogoll, A.; *Synthesis*, 1592–1596, **2002**.
42 Steorts, R.C.; Collins, M.J.; Collins, J.M.; Novel Method for Solid-phase peptide synthesis Using Microwave

Energy; Poster Presented at the 18th American Peptide Symposium, Boston, MA, **2003**.
43 KING, J.E.; FERGUSON, J.D.; COX, Z.J.; HEGEMAN, J.T.; GREENE, G.R.; LAMBERT, J.J.; KING, E.E.; COLLINS, J.M.; Microwave Enhanced Solid-phase peptide synthesis; Poster Presented at the 28th European Peptide Symposium, Prague, Czech Republic, **2004**.
44 BENOITON, L.N.; *Int. J. Peptide Protein Res.*, 44, 399, **1994**.
45 KAISER, T.; NICHOLSON, G.J.; KOHLBAU, H.J.; VOELTER, W.; *Tetrahedron Lett.*, 37, 1187–1190, **1996**.
46 HAN, Y.; ALBERICIO, F.; BARANY, G.; *J. Org. Chem.*, 62, 4307–4312, **1997**.
47 ANGELL, Y.M.; HAN, Y.; ALBERICIO, F.; BARANY, G.; Minimization of cysteine racemization during stepwise solid-phase peptide synthesis. In *Peptides: Frontiers of Peptide Science Proceedings of the 15th American Peptide Symposium;* TAM, J.P.; KAUMAYA, P.T.P., eds.; Kluwer, Dordrecht, The Netherlands, **1999**, 339–340.
48 HAN, Y.; ALBERICIO, F.; BARANY, G.; *J. Org. Chem.* 62, 4307–4312, **1997**.
49 MANNING, M.; COY, E.; SAWYER, W.H.; *Biochemistry*, 9, 3925, **1970**.
50 KEMP, D.S.; *The Peptides: Analysis, Synthesis, Biology*, Volume 1, E. GROSS, J. MEIENHOEFER, Eds.; Academic Press, New York, **1979**, 315.
51 CARENBAUER, A.L.; CECIL, M.R.; CZERWINSKI, A.; DARLAK, K.; DARLAK, M.; LONG, D.W.; VALENZUELA, F.; BARANY, G.; Microwave Assisted Solid Phase Synthesis on CLEAR Supports; Presented at the 19th American Peptide Symposium, San Diego, C.A. **2005**.

52 ARAL, J.; TAGARI, P.; HOLDER, R.; LONG, J.; DIAMOND, S.; MIRANDA, L.; Effects of Coupling Reagent, Base, and Solvent Choice on Microwave-Assisted Solid-phase peptide synthesis; presented at 3rd International Microwave Conference, Orlando, FL, **2005**.
53 HARDING, S.J.; JONES, J.H.; SABIROV, A.N.; SAMUKOV, V.V.; *J. Pept. Sci.*, 5, 368, **1999**.
54 CEZARI, M.H.; JULIANO, L.; *Pept. Res.*, 9, 88–91, **1996**.
55 WHITE, P.; COLLINS, J.; COX, Z.; Comparative study of conventional and microwave assisted synthesis; presented at the 19th American Peptide Symposium, San Diego, C.A., **2005**.
56 HYDE, C.; JOHNSON, T.; OWEN, D.; QUIBELL, M.; SHEPPARD, R.C.; *Int. J. Peptide Protein Res.*, 43, 431, **1994**.
57 HAMDAN, M.; MASIN, B.; ROVATTI, L.; *Rapid Commun. Mass Spec.*, 10, 1739, **1996**.
58 VYAS, S.B.; DUFFY, L.K.; *Protein Peptide Lett.*, 4, 99, **1997**.
59 TICKLER, A.; CLIPPINGDALE, A.; WADE, J.D.; *Protein Peptide Lett.*, 11, 377–384, **2004**.
60 MURRAY, J.K.; GELLMAN, S.H.; *Org. Lett.*, 7, 1517, **2005**.
61 CLIPPINGDALE, A.B.; BARROW, C.J.; WADE, J.D.; *Pept. Sci.*, 6, 225, **2000**.
62 VORRHERR, T.; BANNWARTH, W.; *Bioorg. Med. Chem. Lett.*, 5, 2661, **1995**.
63 FIGLIOZZI, G.M.; GOLDSMITH, R.; NG, S.C.; BANVILLE, S.C.; ZUCKERMANN, R.N.; *Methods Enzymol.*, 267, 437, **1996**.
64 OLIVOS, H.J.; ALLURI, P.G.; REDDY, M.M.; SALONY, D.; KODADEK, T.; *Org. Lett.*, 4, 4057–4059, **2002**.

21
Application of Microwave Irradiation in Fullerene and Carbon Nanotube Chemistry

Fernando Langa and Pilar de la Cruz

21.1
Fullerenes Under the Action of Microwave Irradiation

21.1.1
Introduction

Since Krätschmer and Huffmann discovered a procedure for preparation of bulk quantities of C_{60} [1] in the early nineties, the physical properties of fullerenes have been intensively investigated. Members of this new family of compounds are insoluble or sparingly soluble in most solvents, however [2], and, consequently, C_{60} is difficult to handle. Chemical functionalization enables the preparation of soluble C_{60} derivatives while maintaining the electronic properties of fullerenes. For this reason, the chemistry and derivatization of fullerene has continued to attract attention aimed at preparing derivatives with interesting physical properties and biological activity [3]. New materials based on C_{60} derivatives have shown promise in materials science [4], solar energy conversion [5], nonlinear optical behavior [6], superconductivity [7], ferromagnetism [8], medicinal chemistry [9], and HIV-1 protease inhibition [10].

Among suitable procedures for functionalization of fullerenes (Scheme 21.1) [11], cycloadditions are a powerful tool because C_{60} behaves as an electron-deficient olefin with a relatively low-lying LUMO (−3.14 eV). As a consequence, one of the main ways of functionalizing fullerenes involves cycloaddition reactions [12] in which C_{60} (**1**) is a reactive 2π component. In this context, [1+2], [2+2], [3+2], and [4+2] cycloadditions have all been performed (Scheme 21.2) and the conditions for cycloadduct formation strongly depend on the gap between the controlling orbitals. For this reason it is frequently necessary to use conditions involving several hours (or days) under reflux in high-boiling solvents. It therefore seemed interesting to investigate the potential of microwave irradiation in the preparation of fullerene derivatives when this type of reaction was involved. It should, however, be noted that it is necessary to use a solvent in these reactions, because of some of the characteristics of [60]fullerene (**1**), e.g. the absence of a dipole moment and the need to work on a very small scale.

Microwaves in Organic Synthesis, Second edition. Edited by A. Loupy
Copyright © 2006 WILEY-VCH Verlag GmbH & Co. KGaA, Weinheim
ISBN: 3-527-31452-0

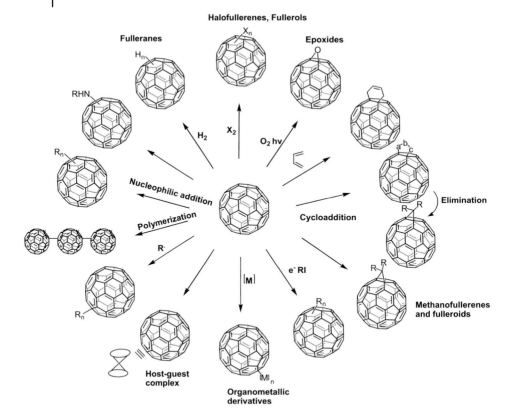

Scheme 21.1

Cycloaddition reactions are readily performed under microwave irradiation and problems concerning the reversibility and decomposition of reagents and/or products can be overcome by use of this methodology.

In this context, most fullerene derivatives have been obtained by Diels–Alder [13] or 1,3-dipolar cycloaddition [14] reactions.

21.1.2
Functionalization of Fullerenes

21.1.2.1 Diels–Alder Cycloaddition Reactions
[4+2] Cycloaddition reactions selectively afford adducts on 6,6-ring junctions [12] and the products occasionally undergo a facile retro-Diels–Alder reaction as a consequence of the low thermodynamic stability of the adduct. Very stable Diels–Alder cycloadducts have, however, been prepared by using different substituted o-quinodimethanes; the success of this reaction is probably because of stabilization by aromatization of the resulting adducts [15].

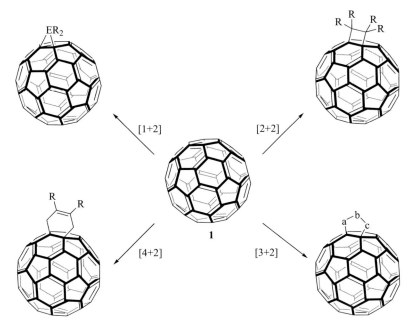

Scheme 21.2

Langa et al. [16] described the microwave-induced Diels–Alder reaction of *o*-quinodimethane, generated *in situ* from 4,5-benzo-3,6-dihydro-1,2-oxathiin-2-oxide (**2**) (sultine) [17], which led to cycloadduct **3** (Scheme 21.3). This reaction was the first application of microwave irradiation in the preparation of a functionalized C_{60} [16].

Because the reaction occurred in solution and, in these early days, in an effort to avoid explosion hazards [18], a modified domestic microwave oven was used. By use of microwave irradiation higher yields of the desired products were obtained more quickly (39% yield after irradiation for 20 min) than under thermal condi-

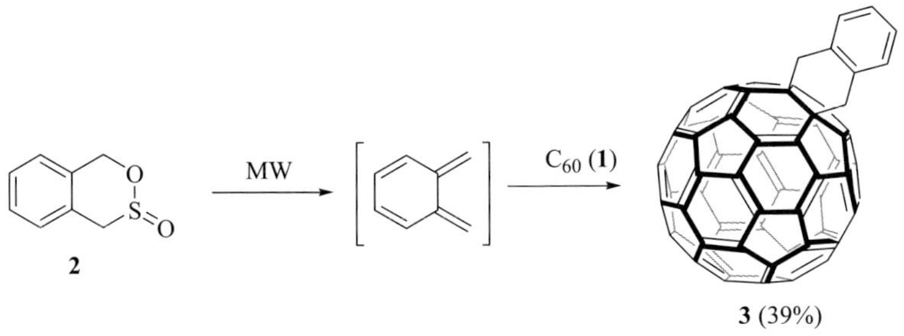

Scheme 21.3

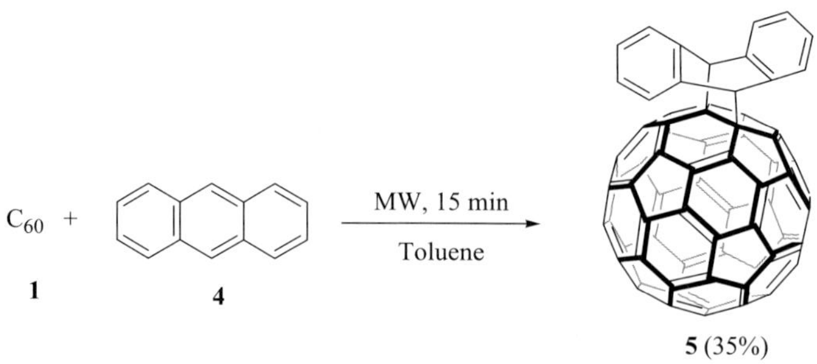

$$C_{60} + \text{anthracene} \xrightarrow[\text{Toluene}]{\text{MW, 15 min}} 5\ (35\%)$$

1 **4**

Scheme 21.4

tions. Longer irradiation times led to a decrease in the yield of **3**, because of increased bis adduct formation.

The scope of the microwave technique in the preparation of fullerene derivatives was assessed by investigating the well known Diels–Alder reaction of C_{60} with anthracene (**4**) [19], which has been reported to proceed under thermal conditions (13% reflux, toluene, 3 days [19a]; 25%, reflux, benzene, 12 h [19b]) (Scheme 21.4). In addition to **5**, multiply-substituted adducts were formed; these underwent cycloreversion to the starting materials.

Application of microwave irradiation in conjunction with use of toluene as solvent gave cycloadduct **5** in 35% yield after 15 min at 800 W [20], an improvement on the yields obtained by conventional heating. This higher yield is probably because of reduced reversion of the cycloaddition in the shorter period of time needed for the irradiated reaction. It is remarkable that under microwave conditions the formation of bis adducts was not observed in these reactions.

Novel pyrazine-containing cycloadducts **9a–c** were synthesized by Diels–Alder reactions between [60]fullerene and the corresponding 2,3-bis(bromomethyl)-pyrazine derivatives **8a–c** (Scheme 21.5) [21]; the 2,3-pyrazinoquinodimethanes were trapped as the Diels–Alder adducts by reaction with [60]fullerene in o-dichlorobenzene (ODCB) under reflux with classical heating and under microwave irradiation conditions in a focused microwave reactor. Once again, use of microwave irradiation led to greater yields than classical heating for **9a** and **9b** (up to 4.5 times) and the reaction times were significantly reduced from 24 h to 0.5 h. With **9c** large amounts of polyadducts were detected when microwave irradiation was used; this led to a decrease in the yield.

Interestingly, in the syntheses of thiophene cycloadducts **11** (Scheme 21.6) use of microwave irradiation led to lower yields [21] than conventional heating (23% compared by means of 43%), although reaction times were substantially reduced [22].

This type of [60]fullerene derivative (**13a–e**), which contains a thiophene ring, was, nevertheless, synthesized by Chung and coworkers [23] using thienosultines (**12a–e**) as precursors for the corresponding *ortho*-quinodimethanes (Scheme 21.7). The reaction was markedly accelerated by microwave irradiation and gave yields

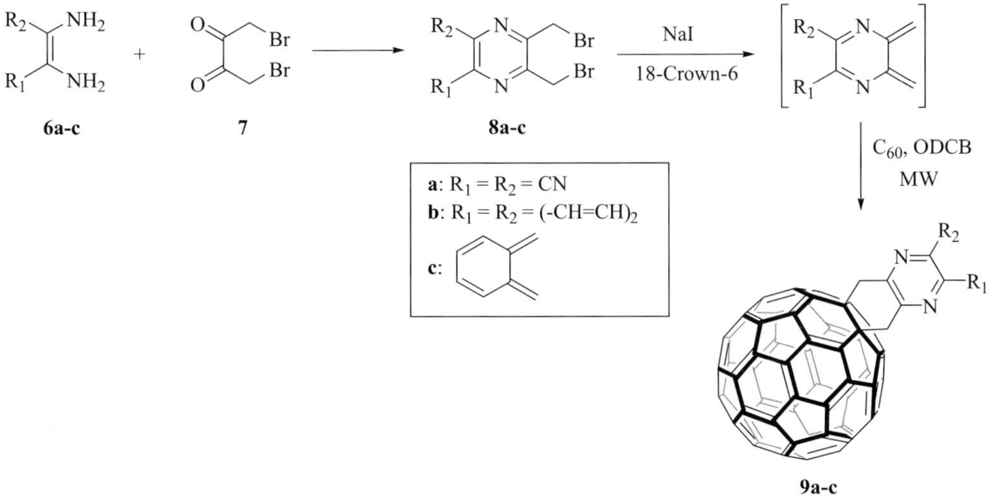

Scheme 21.5

comparable with those of the thermal reaction. The ratio of monoadduct (**13a–e**) to bis adduct also increased when microwave radiation was used.

Hetero-Diels–Alder Reactions The hetero-Diels–Alder reaction is one of the most important methods for the synthesis of heterocyclic compounds. Although it is a potentially powerful synthetic tool it has, however, been used relatively seldom.

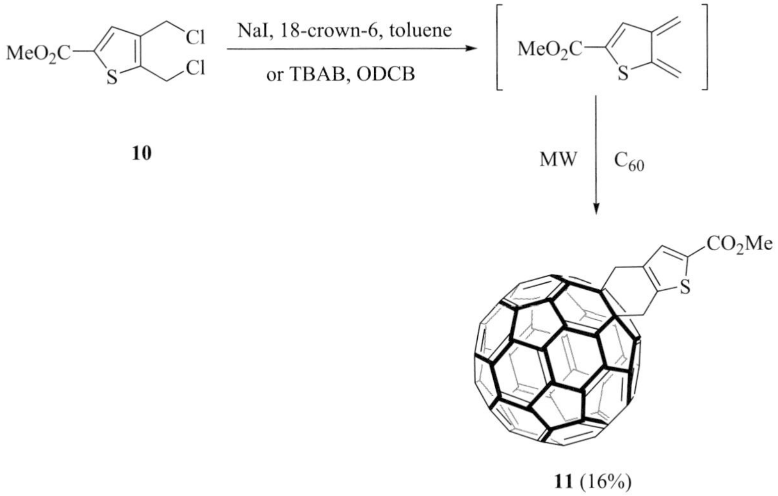

Scheme 21.6

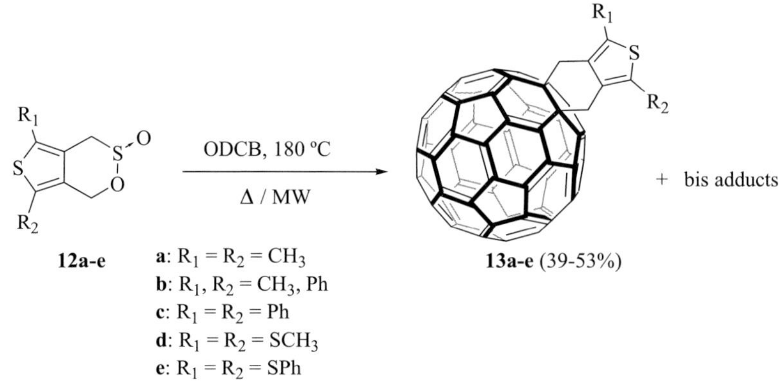

Scheme 21.7

Microwave irradiation has been used to improve reactions involving heterodienophiles and heterodienes of low reactivity.

Reaction of C_{60} with o-quinonemethide, prepared from o-hydroxybenzyl alcohol (**14**) (Scheme 21.8), was performed in a modified domestic oven at 800 W and gave **15** in 27% yield after only 4 min [20]. Although Eguchi et al. [24] reported slightly better yield (31%) when the same reaction was conducted by thermolysis in a sealed vessel, the microwave approach to this adduct is a simple procedure which avoids the risks associated with high-pressure conditions.

21.1.2.2 1,3-Dipolar Cycloaddition Reactions

Addition of a 1,3-dipole to an alkene to give a five-membered ring is a classical organic reaction. Indeed, 1,3-dipolar cycloaddition reactions are useful for formation of carbon–carbon bonds and for preparation of heterocyclic compounds.

The transition state of the concerted 1,3-dipolar cycloaddition is determined by the frontier molecular orbitals of the substrates. The low value of the LUMO level of C_{60} (-3.14 eV) suggests that interaction between the HOMO of the dipole and

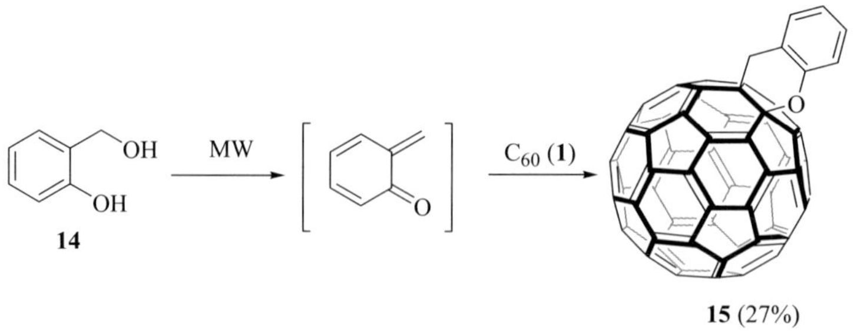

Scheme 21.8

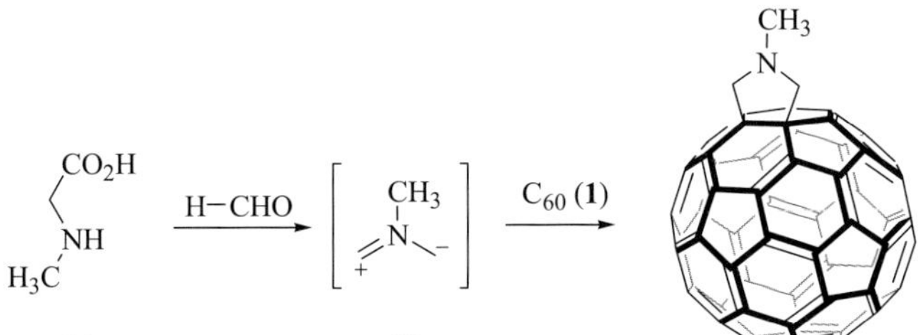

Scheme 21.9

the LUMO of [60]fullerene (**1**) should be the most favored approach and, consequently, those dipoles with higher HOMO levels should react more easily with C_{60}.

In this way, diazomethanes [25], azides [26, 27], azomethine ylides [28, 29], and nitrile oxides [30, 31] have been used to prepare fullerene derivatives in attempts to increase the solubility of C_{60} in common solvents, to reduce aggregation (to make materials that are easier to handle), or to prepare tailor-made novel fullerene derivatives for specific applications in medicinal or material chemistry.

Azomethine ylides A general method for functionalization of C_{60} (**1**) involves 1,3-dipolar cycloaddition of azomethine ylides. This process was first described by Prato and leads to the formation of pyrrolidino[60]fullerenes; this method has been widely used [32].

Azomethine ylides can be generated *in situ* from a wide variety of readily available starting materials. For example, *N*-methylfulleropyrrolidine **18** can be obtained by cycloaddition to C_{60} (**1**) of the azomethine ylide **17** which results from decarboxylation of *N*-methylglycine in the presence of paraformaldehyde in toluene under reflux (Scheme 21.9).

The simplicity of this procedure and the good yields achieved explain the popularity of this approach in fullerene chemistry. Fulleropyrrolidines have been extensively used in different areas of chemistry. For example, water-soluble or protein-modified fulleropyrrolidines have been prepared for medical applications [33]. In 1993, C_{60} was proposed as an inhibitor of HIV protease [24] and, more recently, fulleropyrrolidine derivatives bearing two ammonium groups on the surface of the fullerene sphere were shown to have substantial anti-HIV activity [33a, c].

Microwave-induced 1,3-dipolar cycloadditions involving azomethine ylides have been reported in the literature. The first example was reported by Langa et al. [20] and involved the synthesis of several fulleropyrrolidines (**20a–c**). These authors observed that microwave irradiation again compares favorably with thermal heating and, in this way, **20a** was prepared in 37% yield by use of a focused microwave reactor. In addition, **20b** and **20c**, which had not been previously been reported, were prepared in 30% and 15% yield, respectively (Scheme 21.10).

This procedure was recently used to prepare other pyrrolino[60]fullerene systems – 2,5-disubstituted C_{60}-pyrrolidine derivatives **25** and **26** – using microwave irradi-

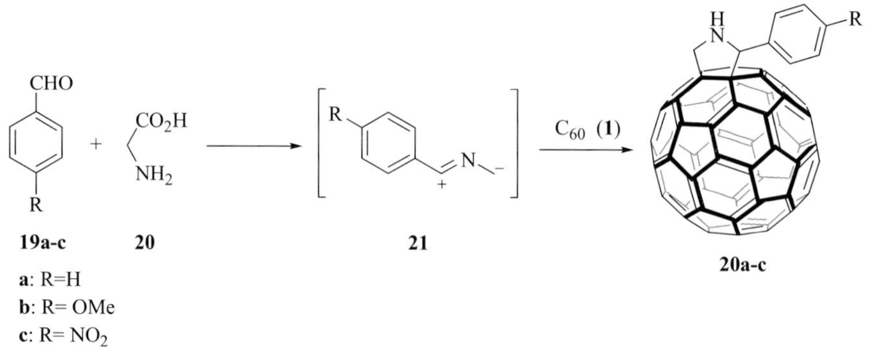

938 *21 Application of Microwave Irradiation in Fullerene and Carbon Nanotube Chemistry*

Scheme 21.10

19a–c
a: R=H
b: R= OMe
c: R= NO$_2$

ation (Scheme 21.11) [34]. The authors described the synthesis of these structures by classical heating and compared the results with those obtained by microwave irradiation. Reduction of the reaction time was the main advantage of the microwave approach, because yields were equivalent in both. Indeed, even the ratios of cis (**25**) to trans (**26**) isomers, as determined by ^{1}H NMR spectroscopy, were found to be similar.

The design of covalently linked donor–fullerene systems capable of undergoing photoinduced electron-transfer processes has been widely studied as a result of the remarkable photophysical [35] and electronic [36] properties of fullerenes. Porphyrins, phthalocyanines, tetrathiafulvalenes, carotenes, and ferrocene [37] have been covalently attached to the fullerene sphere, usually as pyrrolidine[60]fullerene derivatives by 1,3-dipolar cycloaddition reactions.

Langa et al. [38] described the first synthesis of D–A dyads (**30a–c**) based on C$_{60}$ and used ruthenocene as the electron-donor fragment (Scheme 21.12). Synthesis of these pyrrolidine[60]fullerene systems was achieved by 1,3-dipolar cycloaddition of ruthenocenecarboxaldehyde (**28**), N-methylglycine, and C$_{60}$ in toluene under the action of microwave irradiation in a focused microwave oven. The adducts **30a–c** were obtained in moderate yields (26–31%).

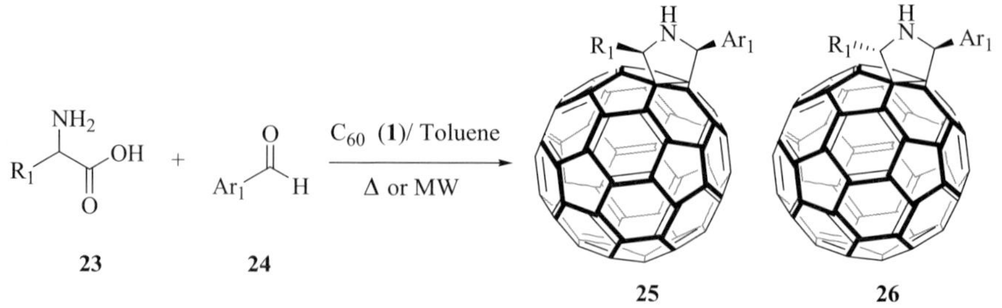

Scheme 21.11

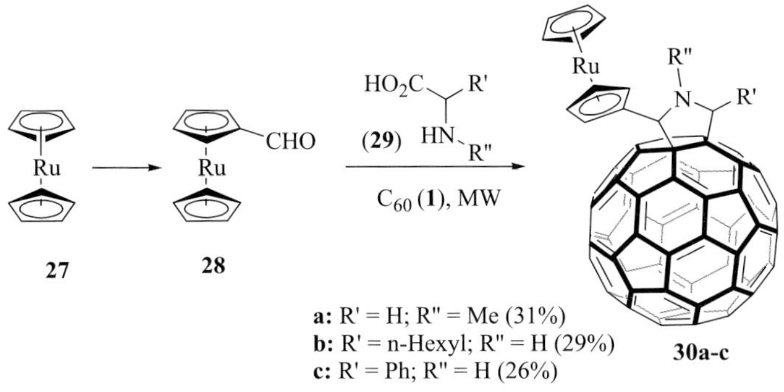

Scheme 21.12

Zeng and coworkers [39] described the synthesis, under microwave irradiation, of several donor–acceptor systems based on amino-pyrrolidino[60]fullerene in which the C_{60} (electron acceptor) and amino group (electron donor) were covalently bonded with a short linker (Scheme 21.13).

As indicated above, the yields of fullerene[60]-ethylcarbazole and the fullerene[60]-triphenylamines were not particularly high, but were reasonable in fullerene chemistry because of the need to avoid bis and/or tris addition. Photophysical studies on these compounds (**32a–c**) showed that efficient photoinduced electron transfer occurs in these amino-pyrrolidino[60]fullerene derivatives.

As already indicated, the unique structure and reactivity of fullerenes C_{60} and C_{70} have attracted considerable attention in a variety of research [40]. Although

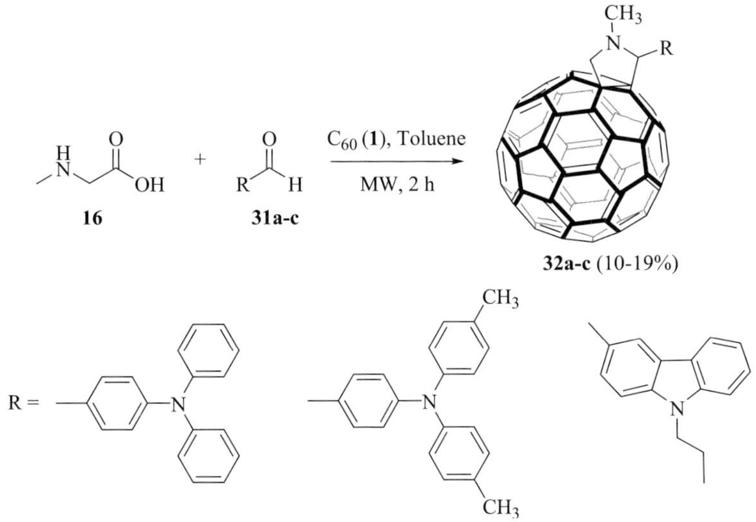

Scheme 21.13

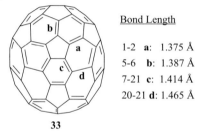

Bond Length		
1-2	**a**:	1.375 Å
5-6	**b**:	1.387 Å
7-21	**c**:	1.414 Å
20-21	**d**:	1.465 Å

33

Fig. 21.1. The [6,6] bond lengths of C_{70} fullerene.

the chemistry of C_{60} has been extensively studied [15], the derivatization and study of C_{70} is still relatively unexplored, because of its low abundance and high cost. Some common trends in reactivity can, however, be identified from the limited data available. The lower symmetry of C_{70} (D_{5h} symmetry) gives rise to a larger number of isomers than C_{60} – whereas C_{60} contains one type of [6,6] bond, C_{70} contains four different [6,6] bonds (Fig. 21.1). Similarly to C_{60}, cycloaddition reactions with C_{70} occur exclusively on [6,6] bonds (Fig. 21.1), with the 1–2 and 5–6 bonds [41] being most reactive, in this order [42]. This is supported by theoretical calculations which show that the product in the 1–2 position is the most stable, followed by the 5–6 isomer [43]. Some reports indicate that C_{60} and C_{70} differ in their reactivity and it has been found that hydroboration [44], addition of hydroxide [45], or 1,3-dipolar cycloaddition with azomethine ylides [46] or nitrile oxides [47] proceed more slowly with C_{70} than with C_{60}.

In this context Langa et al. evaluated the potential of microwave irradiation to modify the regioselectivity in the formation of cycloadducts with [70]fullerene. They described the cycloaddition of *N*-methylazomethine ylides to C_{70} (**33**) to give three regioisomers (**34a–c**) by attack at the 1–2, 5–6 and 7–21 bonds, respectively (Scheme 21.14) [48]. Under conventional heating conditions the 7–21 isomer (**34c**) was formed in small amounts only and the 1–2 isomer (**34a**) was found to predominate. Use of microwave irradiation and *o*-dichlorobenzene (ODCB), a polar solvent that absorbs microwaves efficiently, led to significant changes (Fig. 21.2). In contrast with classical conditions, **34c** was not formed under the action of microwave irradiation, irrespective of irradiation power, and isomer **34b** was found to predominate at higher power (Scheme 21.14).

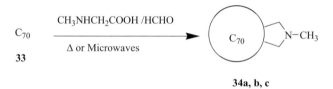

34a, b, c

Scheme 21.14

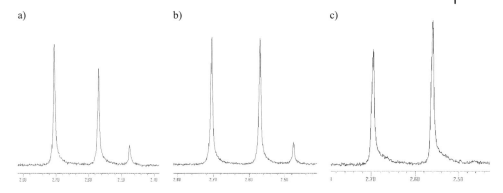

Fig. 21.2. ^{1}H NMR spectra (methyl groups) of the adducts **34a–c** (from left to right): (a) classical heating in toluene; (b) classical heating in ODCB; (c) microwave irradiation at 180 W in ODCB.

A computational study [48] of the mode of cycloaddition showed that the reaction is stepwise, with the first step consisting of nucleophilic attack on the azomethine ylide. The most negative charge on the fullerene moiety in the transition states leading to **34a** and **34b** is located on the carbon adjacent to the carbon–carbon bond being formed. In the transition state that affords **34c**, however, the negative charge is delocalized throughout the C_{70} subunit. The relative ratio of isomers **34a–c** is related to the greatest hardness, and formation of the product via the harder transition state should be favored under microwave irradiation conditions. It is worthy of note that purely thermal arguments predict the predominance of **34c** under microwave irradiation conditions, in contrast with the result found experimentally.

Nitrile oxides Nitrile oxides have been used in conjunction with microwaves in fullerene chemistry. For example, the 3′-(N-phenylpyrazolyl)isoxazolino[60]-fullerene dyad **38a** was prepared in 22% yield from the corresponding nitrile oxide (Scheme 21.15) [49]. Longer reaction times afforded larger amounts of bis adducts. The same reactions under thermal conditions produced markedly lower yields (14–17%). A significant accelerating effect (10 min compared with 24 h) was observed on using microwave irradiation.

The scope of the reaction has been demonstrated by preparation of a series of isoxazolino[60]fullerenes (**38a–h**) (Fig. 21.3) [50]. The nitrile oxides were prepared from the corresponding oximes (**36**) by reaction with NBS or NCS and subsequent treatment of the mixture with triethylamine under the action of microwave irradiation.

Nitrile imines Hydrazones undergo a thermal 1,2-hydrogen shift to afford azomethine imine intermediates which in turn can undergo 1,3-dipolar cycloaddition [51]; bipyrazoles were obtained from pyrazolylhydrazones under the action of microwave irradiation. The synthesis involved [3+2] cycloaddition with alkenes after

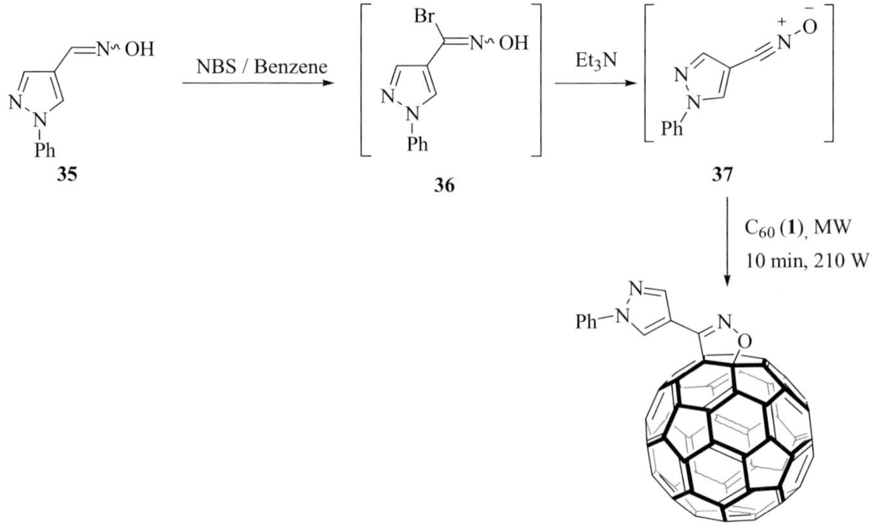

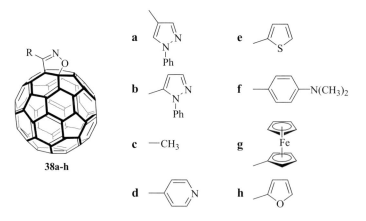

Scheme 21.15

Fig. 21.3. Prepared of isoxazolino[60]fullerenes.

efficient hydrazone–azomethine imine tautomerization (Scheme 21.16). It should be noted that this reaction does not proceed under classical heating conditions [52].

This procedure has been used to react different pyrazolylhydrazones (readily prepared from aldehydes in almost quantitative yields) with C_{60} under microwave irradiation conditions to form the corresponding fullerene cycloadducts [53]. The yields obtained were lower than 6%, however.

An alternative route to the pyrazolino[60]fullerenes involves use of nitrile imines as dipoles – these are prepared in a one-pot procedure starting from hydrazones.

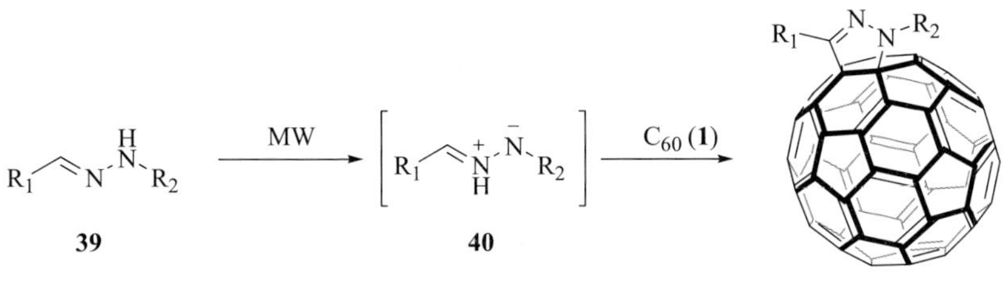

Scheme 21.16

Formation of several pyrazolyl-pyrazolino[60]fullerene adducts (**45a–c**) from nitrile imines by use of this procedure has been described (Scheme 21.17) [53, 54]. The nitrile imines are generated *in situ* from the corresponding hydrazone **42a–c** and NBS in the presence of Et$_3$N and then reacted with C$_{60}$ (**1**) under the action of microwave irradiation, using a focused microwave oven. The route is simpler than the previously described method, which involved cycloaddition of C$_{60}$ to nitrile imines prepared from the corresponding *N*-chlorobenzylidene derivatives [55].

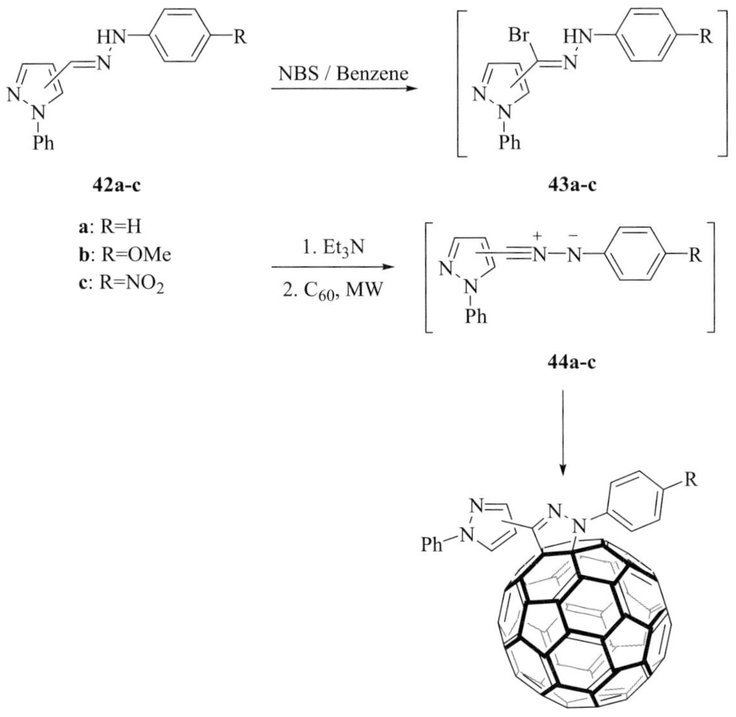

Scheme 21.17

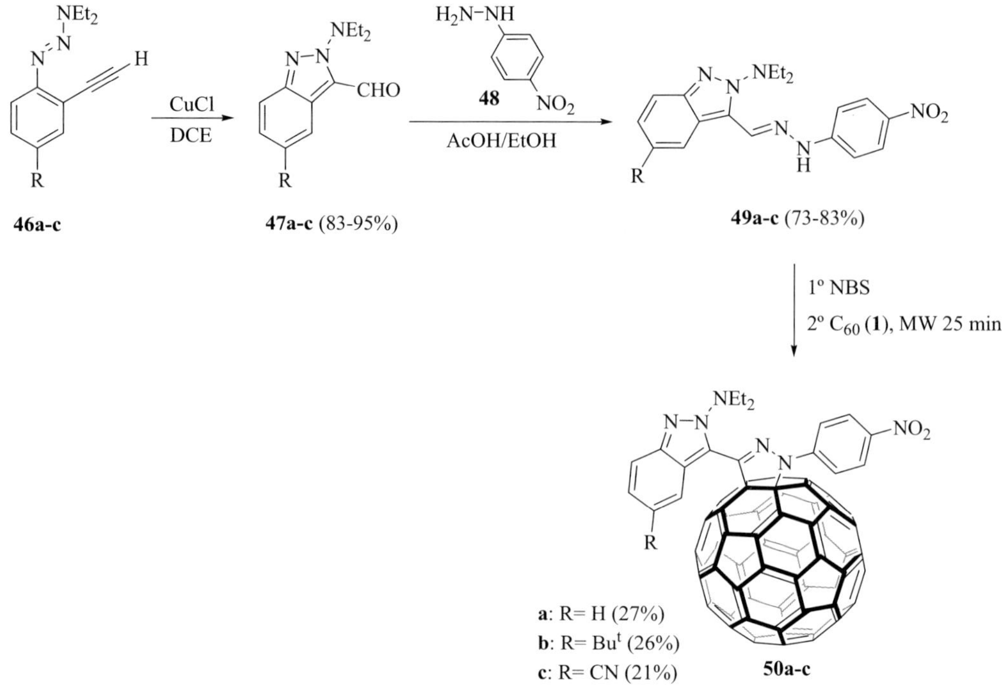

Scheme 21.18

The synthetic approach used to prepare isoindazolylpyrazolino[60]fullerenes **50a–c** (Scheme 21.18), in 21–27% isolated yield [56], relies on 1,3-dipolar cycloaddition of the isoindazolyl nitrile imines under the action of microwave irradiation. The nitrile imines can be generated *in situ* from the corresponding isoindazole hydrazones (**49a–c**), which in turn are synthesized from the respective aldehydes [57] (**47a–c**).

Langa et al. used this method to obtain fullerodendrimers (Scheme 21.19) [58]. Dendrimers with oligoarylenes absorb light efficiently and can act as light-harvesting antennas, enabling photoinduced electron-transfer (PET) processes when attached to a suitable acceptor, e.g. [60]fullerene.

The fullerodendrimers **52a,b** were prepared in 31–34% yield by the procedure outlined in Scheme 21.19, with microwave irradiation used as the energy source.

Oligophenylenevinylenes (OPV) play a prominent role in science and, owing to their optical luminescence and electronic properties, have attracted much attention from researchers in many fields [59]. The electronic properties of oligophenylenevinylenes (OPV) make them versatile photo and/or electro-active components for preparation of photochemical molecular devices [60], particularly when this fragment is bonded to the C_{60} sphere [61]. Indeed, an oligophenylenevinylene moiety (OPV) has been attached to C_{60} using a pyrazoline ring as a linker (Scheme 21.20) [62].

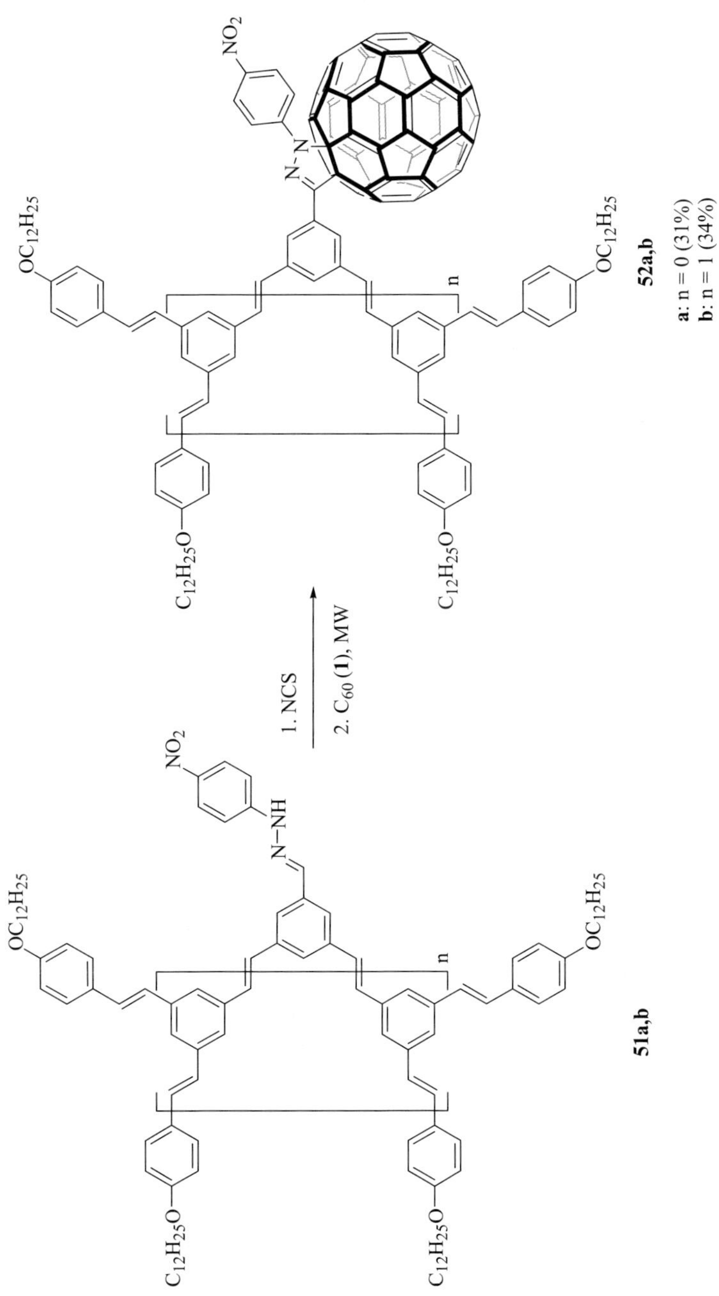

Scheme 21.19

51a,b → (1. NCS; 2. C$_{60}$ (**1**), MW) → **52a,b**

a: n = 0 (31%)
b: n = 1 (34%)

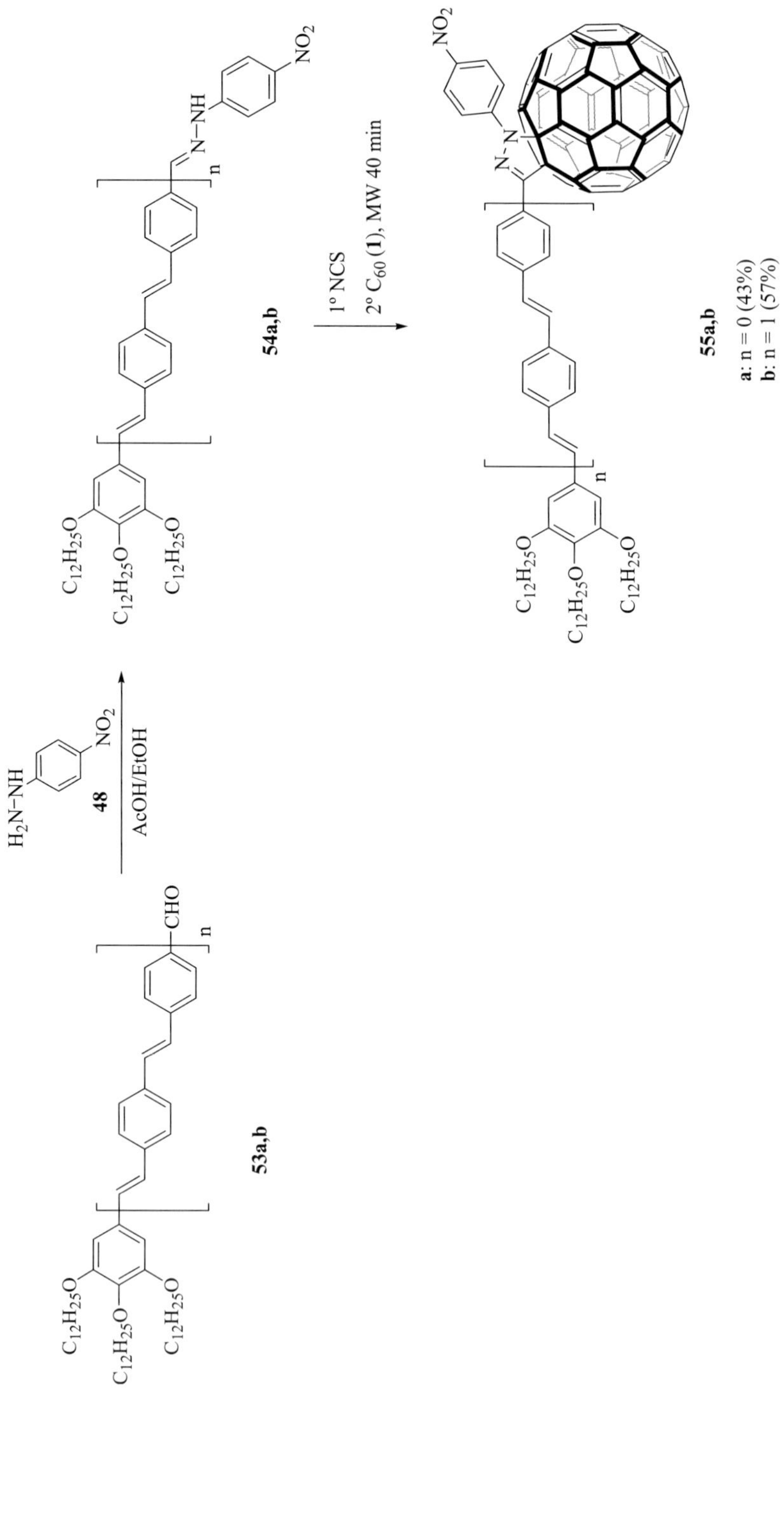

Scheme 21.20

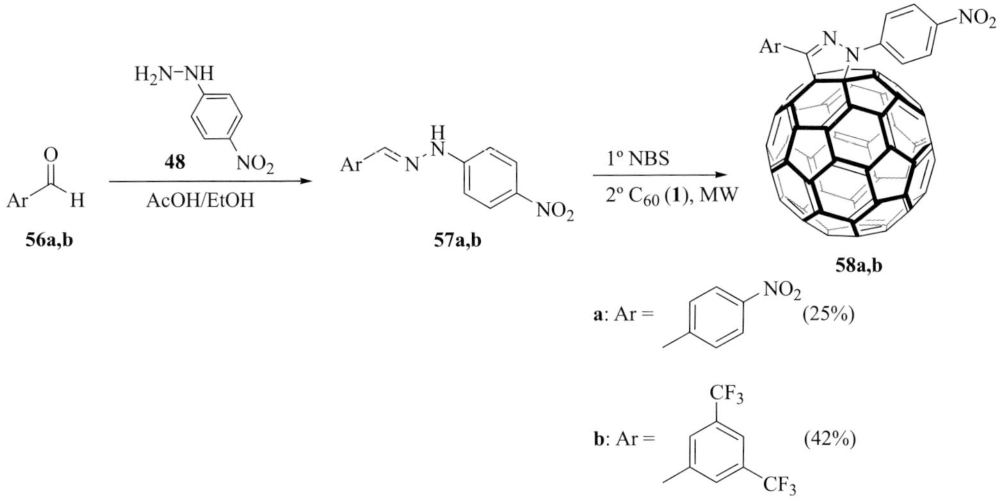

Scheme 21.21

The hydrazones **54a,b** were obtained from aldehydes **53a,b** by standard methods and were reacted with NCS to afford the corresponding nitrile imine intermediate, which was reacted *in situ* with C_{60} under the action of microwave irradiation (210 W, 40 min) in a focused microwave oven. The corresponding pyrazolino[60]-fullerenes **55a,b** were obtained in good yields (43–57%).

One of the most important goals in fullerene chemistry is to synthesize molecules with better electron-acceptor ability than pristine C_{60} for applications in the preparation of photovoltaic cells. This is not an easy task, because saturation of one of the double bonds of the C_{60} cage occurs when it is functionalized and this shifts the first reduction potential of the C_{60} cage to more positive values by approximately 150 mV.

Langa, who has reported several C_{60} derivatives with intramolecular PET processes, recently described the synthesis under microwave irradiation conditions of two pyrazolino [60]fullerene (**58a,b**) bearing electron-withdrawing substituents. The resulting fullerene derivatives were highly soluble in several polar and nonpolar solvents, and electrochemical studies revealed improved electron-acceptor properties compared with unmodified C_{60} [63].

The synthesis of compounds **58a,b** was achieved according to the procedure outlined in Scheme 21.21, in two steps from the corresponding aldehydes (**56a,b**).

Photophysical studies on a nanosecond time scale provided clear evidence of intermolecular PET from TTF, an excellent electron donor, to the pyrazolino[60]-fullerene systems (**58a,b**).

21.1.2.3 Other Reactions

The synthesis of *NH*-pyrrolidino[60]fullerenes under the action of microwave irradiation has been described in Section 21.1.2.2. These compounds provide access to

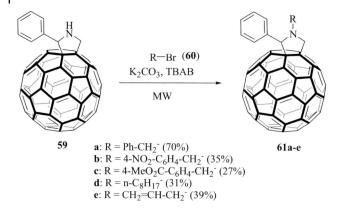

a: R = Ph-CH$_2^-$ (70%)
b: R = 4-NO$_2$-C$_6$H$_4$-CH$_2^-$ (35%)
c: R = 4-MeO$_2$C-C$_6$H$_4$-CH$_2^-$ (27%)
d: R = n-C$_8$H$_{17}^-$ (31%)
e: R = CH$_2$=CH-CH$_2^-$ (39%)

Scheme 21.22

derivatives that are further functionalized at the nitrogen atom. This system is, nevertheless, of low reactivity, because pyrrolidino[60]fullerenes are less basic than the corresponding pyrrolidine by six orders of magnitude [64].

Synthesis of a series of N-alkylpyrrolidino[60]fullerenes (**61a–e**) was, nevertheless, achieved by use of a combination of microwave irradiation and PTC in the absence of solvent (Scheme 21.22) [65]. Solvent-free PTC has been successfully applied to a wide range of organic reactions and the technique is very useful when combined with microwave irradiation, especially for anionic reactions that proceed slowly, give low yields, or require harsh conditions.

This method enabled the synthesis of several N-alkylpyrrolidino[60]fullerenes (**61a–e**) by treatment of 2′-phenylpyrrolidino[3′,4′:1,2][60]fullerene (**59**) with the corresponding alkyl or benzyl bromide and potassium carbonate using TBAB as the phase-transfer catalyst (Scheme 21.22).

[60]Fullerene has important properties as a nonlinear material, particularly as an optical limiting material with applications in device fabrication. The poor solubility of this material and the processing difficulties associated with it severely limit its direct application, however. One possibility of overcoming these disadvantages involves binding polymers on to C$_{60}$ to obtain fullerene-based polymeric materials with peculiar physical and/or chemical properties and good processability.

An application of microwave irradiation in the synthesis of fullerene-functionalized polycarbonates has recently been reported. The process is based on direct reaction of C$_{60}$ (**1**) and polycarbonate (**62**) in the presence of AIBN under the action of microwave irradiation [66]. The resulting C$_{60}$-PC (**63**) material can be dissolved in common organic solvents and has optical limiting properties similar to those of C$_{60}$ itself. This material can be used for optical applications.

[60]Fullerene (**1**) has also been used as an initiator of the polymerization of vinyl monomers. The [60]fullerene-initiated polymerization of N-vinylcarbazole (**64**) under microwave irradiation conditions has been reported [67]. The main advan-

tages of this technique are the remarkable reduction in reaction time and the substantial improvement in the yield of poly(vinylcarbazole) (**65**) in comparison with results obtained when polymerization is conducted in a water bath at 70 °C.

Other derivatives of [60]fullerene that have been the subject of substantial attention are the alkali-metal fullerides – mainly since the discovery of superconductivity in K_3C_{60} (**66**). Synthesis of these systems usually involves combination of stoichiometric quantities of alkali metal and C_{60}. Solid-state synthesis required long reaction times, however – typically several days. Use of a microwave-induced argon plasma led to reaction times in the order of seconds [68].

21.2
Microwave Irradiation in Carbon Nanotube Chemistry

Carbon nanotubes (CNs) were discovered by Iijima in 1991 [69], since when they have been regarded as materials of exceptional interest. This high level of interest is not only because of their remarkable electronic properties, their extremely desirable mechanical properties (strength and flexibility), or their good physical and chemical properties [70], but also because of to their potential applications (hydrogen storage, chemical sensors, nanoelectronic devices, components for high performance composites) [71].

Carbon nanotubes are shells of sp^2-hybridized carbon atoms that form a hexagonal network within a tubular motif closed at the ends by hemispherical endcaps. Manufactured as single-walled (SWNTs) or multi-walled (MWNTs) systems with different diameters and lengths, each type of CN has different properties.

21.2.1
Synthesis and Purification

Several techniques are currently used to produce carbon nanotubes – arc-discharge [72], pulse laser vaporization [73], chemical vapor deposition [74] and the HIPCO (high-pressure carbon monoxide) process [75]. CNs can also be synthesized by heating – by use of microwave irradiation – a catalyst loaded on different supports (carbon black, silica powder, or an organic polymer) [76]. In this procedure, cobalt or some of its salts (sulfide, naphthenate), which act as a catalyst, are deposited on the substrates placed in a quartz reactor. This is then introduced into a microwave reactor through which a mixture of acetylene, hydrogen, and hydrogen sulfide is flowing. In this way fibrous nanocarbons that are either multi-walled CNs or graphitic nanofibers, carbon nanoparticles, or amorphous carbon are formed. The CN yields and morphologies depend on the reaction conditions.

CNs of exceptionally high purity are required for optimum performance in applications but the synthetic products usually contain impurities such as amorphous carbon, carbon nanoparticles, and some metal catalyst. Different methods have been described for the purification of nanotubes [77]; most involve oxidation by use of mineral acids and/or gas phase oxidation to remove catalytic metal particles and amorphous carbon. Microwave energy also has been used to purify MWNTs. Ko and coworkers [78] described the purification of MWNTs by treatment with nitric acid in a closed vessel. The mixture was irradiated at 160 °C for 30 min inside a commercial microwave oven modified to enable control of the temperature and pressure of the process. The amount of metal impurities was reduced substantially in a short time. Microwave energy was also applied by Chen in a microwave digestion system [79]. In the same way, microwave irradiation can be used as a tool to purify SWNTs by heating in a domestic multimode microwave reactor [80] or by use of a microwave digestion unit [81].

21.2.2
Functionalization of Carbon Nanotubes

Despite much interest in CNs, manipulation and processing of these materials has been limited by their lack of solubility in most common solvents. Many applications of CNs (mainly SWNTs) require chemical modification of the materials to make them soluble and more amenable to manipulation. Understanding the chemistry of SWNTs is critical for rational modification of their properties, and several different procedures for chemical derivatization of CNs have been described in the last four years. These methods have been developed in an effort to understand the chemical derivatization and to control the properties of these systems. There is substantial interest in studying the photophysical properties of single-walled carbon nanotube (SWNT) derivatives obtained by covalent [82] and noncovalent [83] functionalization, with the overall objective of obtaining materials with new properties [84]. Functionalization of SWNTs by covalent bonding can be achieved by two different approaches – the bonds can be formed either at the tube opening or on the lateral walls.

Initial attempts to functionalize these compounds were limited to oxidation reactions [85] that resulted in shortened nanotubes with carboxylic acid groups on the open edges. Haddon and coworkers first reported use of these acid groups to attach long alkyl chains by means of amide linkages [86]. Later, Sun and coworkers showed that esterification can also be used to functionalize SWNTs [87]. These procedures afforded solubilized SWNTs, which enabled characterization and further solution-based investigations.

In general, functionalization reactions of SWNTs are very slow and take several days to proceed. In this respect, microwave irradiation seems to be a potentially powerful tool to functionalize SWNTs but only a few such reactions have been described to date. One example of the application of microwaves was described by Della Negra et al. [88]. Soluble single-walled carbon nanotubes were synthesized by grafting poly(ethylene glycol) (PEG) chains on to SWNTs. Use of microwave irradiation enhanced reaction rates in comparison with similar syntheses using conventional heating. An amidation reaction has also been performed, in two steps, under microwave irradiation conditions (Scheme 21.23) [89]. Amide-SWNT derivative **68** was synthesized by reaction of 2,6-dinitroaniline and the carboxylic acid-

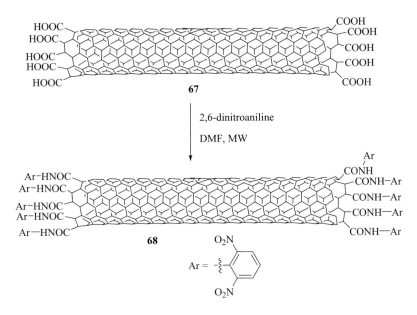

Scheme 21.23

grafted SWNT **67** (described above) using DMF as the solvent. The reaction was performed for 15–20 min under the action of microwave irradiation. The reaction time is between 3 and 5 days when conventional heating is used.

These two examples are based on functionalization of SWNTs at the tube opening but microwave irradiation has also been applied to the synthesis of SWNTs functionalized at the lateral walls.

As described above, cycloaddition reactions are among the most powerful for functionalization of fullerenes, and microwave irradiation has proven a very useful technique in this kind of chemistry. The reactivity of the walls of SWNTs should be similar to that of fullerenes – albeit significantly slower. It is therefore of interest to apply microwaves for lateral functionalization of SWNTs. The first such example of Diels–Alder addition to the lateral wall of a SWNT has been described [90], and involved use of microwaves as a source of energy (Scheme 21.24). Ester functionalized SWNT **69** [91] was reacted with o-quinodimethane **71**, generated in situ from 4,5-benzo-1,2-oxathiin-2-oxide (**70**), under the action of microwave irradiation for 45 min. The reaction was also performed under classical heating conditions for 72 h in ODCB under reflux but under these conditions the ^{1}H NMR spectrum was indicative of very low conversion.

1,3-Dipolar cycloaddition reactions of azomethine ylides are probably the most widely used reactions for functionalization of [60]fullerene. This reaction has also been used to obtain SWNT derivatives and, occasionally, microwave irradiation has been used [89]. Pyrrolidino-SWNT **76** was synthesized by reaction of pristine

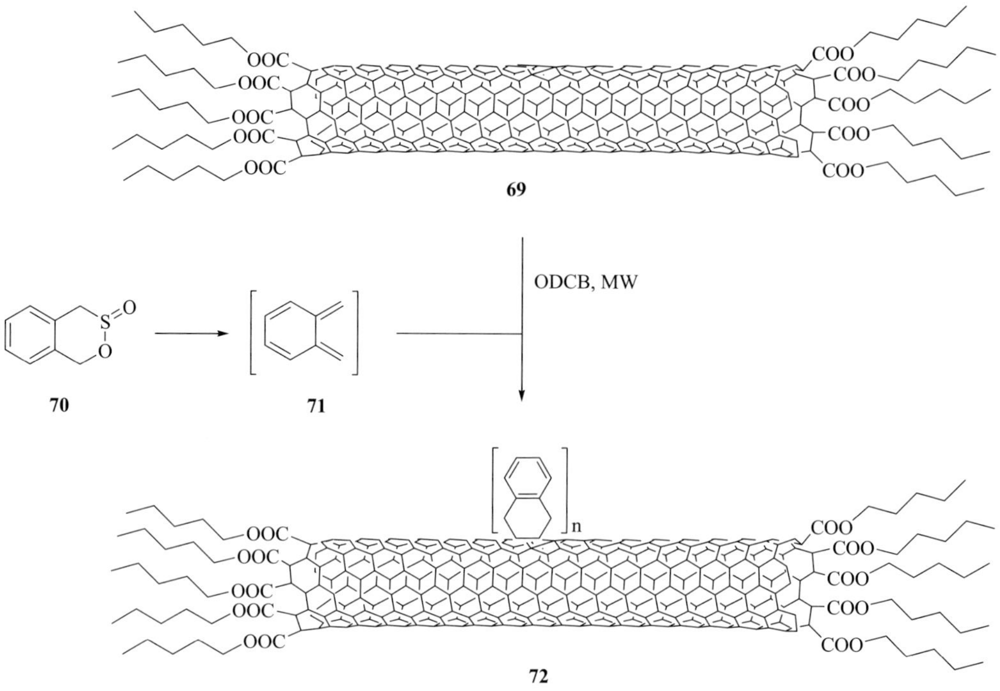

Scheme 21.24

SWNT (**73**) with salicylaldehyde (**74**) and methionine (**75**) in DMF as a solvent (Scheme 21.25). The mixture was irradiated under microwave irradiation for about 15 min. The reaction time is significantly shorter than that required for conventional methods (approx. 5 days).

As described above, 1,3-dipolar cycloaddition reactions of nitrile oxides have also been used for preparation of fullerene derivatives. Theoretical calculations performed for this type of reaction with SWNTs predict that the 1,3-dipolar cycloaddition reaction of nitrile oxides to the sidewall of the nanotubes is possible but not as

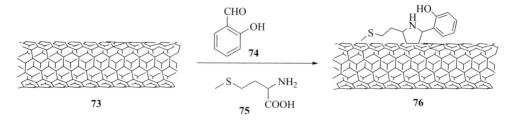

Scheme 21.25

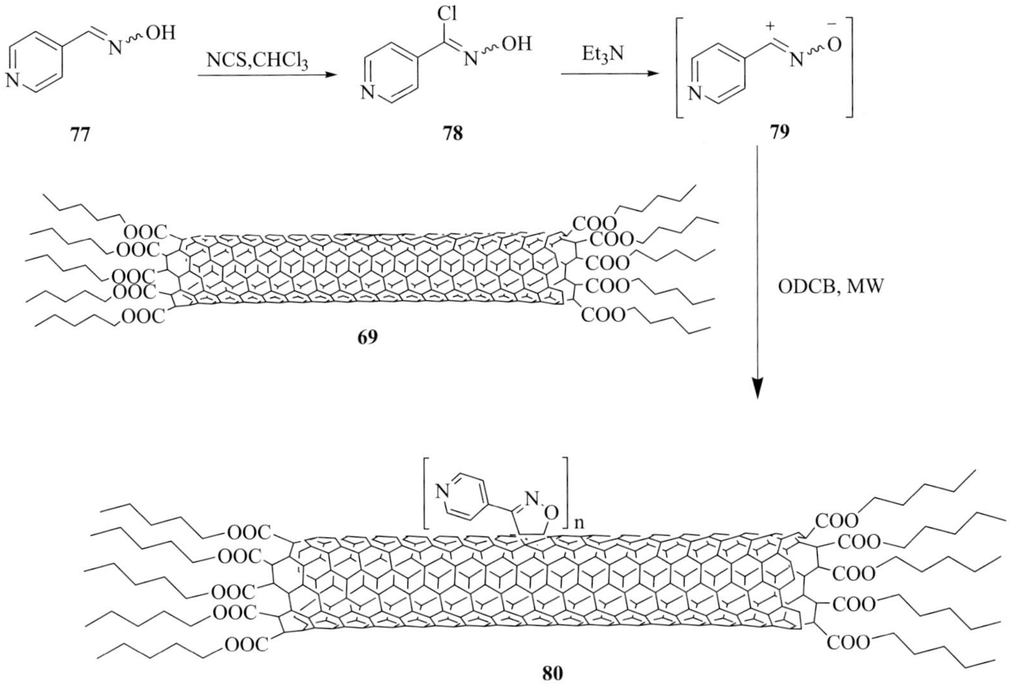

Scheme 21.26

favorable as that involving azomethine ylides [92]. Langa and coworkers verified this [93] by synthesizing the isoxazolino-SWNT derivative **80** (Scheme 21.26). The route involved the 1,3-dipolar cycloaddition of nitrile oxide **79**, generated *in situ* from 4-pyridylcarboxaldehyde oxime (**77**) by treatment with N-chlorosuccinimide (NCS) and triethylamine, and the ester-SWNT **69**. The success of this approach has provided a new means of preparation of SWNT derivatives.

21.3 Conclusions

We have shown that use of microwaves enables efficient fullerene and carbon nanotube chemistry, both of which are of much interest in materials science. Microwaves have been used for preparation of both forms of carbon and for functionalization reactions performed with the objective of solubilizing or modifying the properties of these materials. The yields obtained by use of this technique are higher than those obtained by classical heating and reaction times are significantly shorter.

References

1 W. Krätschmer, L. D. Lamb, K. Fotstiropoulos, D. R. Huffmann, *Nature* **1990**, *347*, 354–358.

2 R. S. Ruoff, D. S. Tse, R. Malhotra, D. C. Lorents, *J. Phys. Chem.* **1993**, *97*, 3379–3383.

3 T. Braun, A. P. Schubert, R. N. Kostoff, *Chem. Rev.* **2000**, *100*, 23–38.

4 (a) M. Prato, *J. Mater. Chem.* **1997**, *7*, 1097–1109; (b) F. Diederich, M. Gómez-López, *Chem. Soc. Rev.* **1999**, *28*, 263–277; (c) J. F. Nierengarten, *Chem. Eur. J.* **2000**, *6*, 3667–3670.

5 Y. P. Sun, J. E. Riggs, Z. Guo, H. W. Rollins, *Optical and Electronic Properties of Fullerenes and Fullerene-Based Materials*. (Eds.: J. Shinar, Z. V. Vardeny, Z. H. Kafafi), Marcel Dekker, Inc., **2000**, chapter 3.

6 L. W. Tutt, A. Kost, *Nature* **1992**, *356*, 2225–2226.

7 K. Tanigaki, *Handbook of Organic Conductive Molecules and Polymers* (Ed.: H. S. Nalwa), John Wiley & Sons, **1997**, Vol. 1, chapter 6.

8 P. M. Allemand, K. C. Khemani, A. Koch, F. Wudl, K. Holczer, S. Donovan, G. Gruner, J. D. Thompson, *Science* **1991**, *253*, 3012–303.

9 (a) N. Tagmatarchis, H. Shinohara, *Mini Reviews in Medicinal Chemistry* **2001**, *1*, 339–348; (b) T. D. Ros, G. Spalluto, M. Prato, *Croatica Chemica Acta* **2001**, *74*, 743–755.

10 S. H. Friedman, D. L. Decamp, R. P. Sijbesma, G. Srdanov, F. Wudl, G. L. Kenyon, *J. Am. Chem. Soc.* **1993**, *115*, 6506–6509.

11 (a) N. Martín, L. Sánchez, B. Illescas, I. Pérez, *Chem. Rev.* **1998**, *98*, 2527–2547; (b) M. Prato, M. Maggini, *Acc. Chem. Res.* **1998**, *31*, 519–526; (c) F. Diederich, M. Gómez-López, *Chem. Soc. Rev.* **1999**, *28*, 263–277; (d) H. Imahori, Y. Sakata, *Eur. J. Org. Chem.* **1999**, 2445–2457; (e) H. Imahori, K. Tamaki, H. Yamada, K. Yamada, Y. Sakata, Y. Nishimura, I. Yamazaki, M. Fujitsuka, O. Ito, *Carbon* **2000**, *38*, 1599–1605; (f) D. I. Schuster, *Carbon* **2000**, *38*, 1607–1614; (g) J. L. Segura, N. Martín, *Chem. Soc. Rev.* **2000**, *29*, 13–25; (h) D. M. Guldi, *Chem. Soc. Rev.* **2002**, *31*, 22–36; (i) J.-F. Nierengarten, N. Armaroli, G. Accorsi, Y. Rio, J.-F. Eckert, *Chem. Eur. J.* **2003**, *9*, 36–41.

12 (a) A. Hirsch, in *The Chemistry of the Fullerenes*, Thieme, 1994; (b) A. Hirsch, *Synthesis* **1995**, 895–913; (c) A. Hirsch, M. Brettreich, in *Fullerenes: Chemistry and Reactions*, Wiley-VCH, **2005**.

13 Developments in Fullerene Science, vol. 4. *Fullerenes: From Synthesis to Optoelectronic Properties*. Eds. Guldi, D. M., Martin, N., Springer, **2003**.

14 N. Tagmatarchis, M. Prato, *Synlett* **2003**, *6*, 768–779.

15 (a) Y. Rubin, S. Khan, D. I. Freedberg, C. Yeretzian, *J. Am. Chem. Soc.* **1993**, *115*, 344–345; (b) P. Belik, A. Gügel, J. Spickerman, K. Müllen, *Angew. Chem. Int. Ed. Engl.* **1993**, *32*, 78–80.

16 B. Illescas, N. Martín, C. Seoane, P. de la Cruz, F. Langa, F. Wudl, *Tetrahedron Lett.* **1995**, *36*, 8307–8310.

17 (a) J. L. Segura, N. Martín, *Chem. Rev.* **1999**, *99*, 3199–3246; (b) W. F. Jarvis, M. D. Hoey, A. L. Finocchio, D. C. Dittmer, *J. Org. Chem.* **1991**, *56*, 1947–1948.

18 H. M. Kingston, P. J. Walter, W. G. Engelhart, P. J. Parsons, in *Microwave-Enhanced Chemistry*, H. M. (Skip) Kingston, S. J. Haswell, Eds., American Chemical Society, **1997**, chapter 16.

19 (a) J. A. Schlueter, J. M. Seaman, S. Taha, H. Cohen, K. R. Lykke, H. H. Wang, J. M. Williams, *J. Chem. Soc., Chem. Commun.* **1993**, 972–974; (b) M. Tsuda, T. Ishida, T. Nogami, S. Kurono, M. Ohashi, *J. Chem. Soc., Chem. Commun.* **1993**, 1296–1298; (c) K. Komatsu, Y. Murata, N. Sugita, K. Takeuchi, T. S. M. Wan, *Tetrahedron Lett.* **1993**, *34*, 8473–8476.

20 P. de la Cruz, A. de la Hoz, F.

Langa, B. Illescas, N. Martín, *Tetrahedron* **1997**, *53*, 2599–2608.

21 U. M. Fernández-Paniagua, B. Illescas, N. Martín, C. Seoane, P. de la Cruz, A. de la Hoz, F. Langa, *J. Org. Chem.* **1997**, *62*, 3705–3710.

22 U. M. Fernández-Paniagua, B. Illescas, N. Martín, C. Seoane, *J. Chem. Soc., Perkin Trans. 1* **1996**, 1077–1079.

23 C.-C. Chi, F. Pai, W.-S. Chung, *Tetrahedron* **2004**, *60*, 10869–10876.

24 M. Ohno, T. Azuma, S. Eguchi, *Chem. Lett.* **1993**, 1833–1834.

25 F. Wudl, A. Hirsch, K. C. Khemani, T. Suzuki, P.-M. Allemand, A. Koch, H. Eckert, G. Srdanov, H. M. Webb, in *Fullerenes: Synthesis, Properties, and Chemistry*, Hammond, G. S. and Kuck, V. J., Eds., ACS Symposium Series **1992**, *481*, 161–175.

26 M. Cases, M. Durán, J. Mestres, N. Martín, M. Sola, *J. Org. Chem.* **2001**, *66*, 433–442.

27 A. Hirsch, B. Nuber, *Acc. Chem. Res.* **1999**, *32*, 795–804.

28 (a) M. Maggini, E. Menna, in *Fullerenes: From Synthesis to Optoelectronic Properties*; D. M. Guldi, N. Martin, Eds., **2002**, 1–50; (b) F. Novello, M. Prato, T. da Ros, M. da Amici, A. Bianco, C. Toniolo, M. Maggini, *Chem. Commun.* **1996**, *8*, 903–904; (c) A. Bianco, M. Maggini, G. Scorrano, C. Toniolo, G. Marconi, C. Villani, M. Prato, *J. Am. Chem. Soc.* **1996**, *118*, 4072–4080.

29 (a) X. Tan, D. I. Schuster, S. R. Wilson, *Tetrahedron Lett.* **1998**, *39*, 4187–4190; (b) B. Illescas, J. Rifé, R. M. Ortuño, N. Martín, *J. Org. Chem.* **2000**, *65*, 6246–6248; (c) T. da Ros, M. Prato, V. Luchini, *J. Org. Chem.* **2000**, *65*, 4289–4297.

30 (a) M. S. Meier, M. Poplawska, *J. Org. Chem.* **1993**, *58*, 4524–4525; (b) M. S. Meier, M. Poplawska, *Tetrahedron* **1996**, *52*, 5043–5052; (c) T. da Ros, M. Prato, F. Novello, M. Maggini, M. de Amici, D. de Micheli, *Chem. Commun.* **1997**, *1*, 59–60.

31 (a) H. Irngartinger, A. Weber, T. Escher, P. W. Fettel, F. Gassner, *Eur. J. Org. Chem.* **1999**, *9*, 2087–2092; (b) H. Irngartinger, A. Weber, T. Escher, *Eur. J. Org. Chem.* **2000**, *8*, 1647–1651; (c) H. Irngartinger, P. W. Fettel, T. Escher, P. Tinnefeld, S. Nord, M. Sauer, *Eur. J. Org. Chem.* **2000**, *3*, 455–465.

32 M. Prato, M. Maggini, *Acc. Chem. Res.* **1998**, *31*, 519–526.

33 (a) F. Wudl, S. H. Friedman, D. L. Decamp, R. P. Sijbesma, G. Srdanov, G. L. Kenyon, *J. Am. Chem. Soc.* **1993**, *115*, 6506–6509; (b) T. da Ros, M. Prato, *Chem. Commun.* **1999**, *8*, 663–670; (c) G. L. Marcorin, T. da Ros, S. Castellano, G. Stefancich, I. Bonin, S. Miertus, M. Prato, *Org. Lett.* **2000**, *2*, 3955–3958; (d) K. Kordatos, T. da Ros, S. Bosi, E. Vázquez, M. Bergamin, C. Cusan, F. Pellarini, V. Tomberli, B. Baiti, D. Pantarotto, V. Georgakilas, G. Spalluto, M. Prato, *J. Org. Chem.* **2001**, *66*, 4915–4920.

34 S. Wang, J.-M. Zhang, L.-P. Song, H. Jiang, S.-Z. Zhu, *J. Fluorine Chem.* **2005**, *126*, 349–353.

35 (a) D. M. Guldi, *Chem. Commun.* **2000**, *5*, 321–328; (b) D. Gust, T. A. Moore, A. L. Moore, *Acc. Chem. Res.* **2001**, *34*, 40–48.

36 L. Echegoyen, L. E. Echegoyen, *Acc. Chem. Res.* **1998**, *31*, 593–601.

37 (a) M. Prato, *Top. Curr. Chem.* **1999**, *199*, 173–187; (b) H. Imahori, Y. Sakata, *Eur. J. Org. Chem.* **1999**, 2445–2457; (c) J. F. Nierengarten, *Top. Curr. Chem.* **2003**, *228*, 87–110.

38 J. J. Oviedo, P. de la Cruz, J. Garín, J. Orduna, F. Langa, *Tetrahedron Lett.* **2005**, *46*, 4781–4784.

39 H.-P. Zeng, T. Wang, A. S. D. Sandanayaka, Y. Araki, O. Ito, *J. Phys. Chem. A* **2005**, *109*, 4713–4720.

40 (a) T. Da Ros, M. Prato, *Chem. Commun.* **1999**, 663–669; (b) D. M. Guldi, *Chem. Commun.* **2000**, 321–327; (c) J.-F. Nierengarten, *New J. Chem.* **2004**, *28*, 1177–1191.

41 Numbered according to: E. W. Godly, R. Taylor, *Pure & Appl. Chem.* **1997**, *69*, 1441–1434.

42 C. Thilgen, A. Herrmann, F. Diederich, *Angew. Chem. Int. Ed. Engl.* **1997**, *36*, 2268–2280.

43 J. Mestres, M. Durán, M. Solà, *J. Phys. Chem.* **1996**, *100*, 7449–7454.

44 C. C. Henderson, C. M. Rohlfing, K. T. Gillen, P. A. Cahill, *Science* **1994**, *264*, 397–399.

45 A. Naim, P. B. Shevlin, *Tetrahedron Lett.* **1992**, *33*, 7097–100.

46 S. R. Wilson, Q. Lu, *J. Org. Chem.* **1995**, *60*, 6496–6498.

47 M. S. Meier, M. Poplawska, A. L. Compton, J. P. Shaw, J. P. Selegue, T. F. Guard, *J. Am. Chem. Soc.* **1994**, *116*, 7044–7048.

48 F. Langa, P. de la Cruz, A. de la Hoz, E. Espíldora, F. P. Cossío, B. Lecea, *J. Org. Chem.* **2000**, *65*, 2499–2507.

49 P. de la Cruz, E. Espíldora, J. J. García, A. de la Hoz, F. Langa, N. Martín, L. Sánchez, *Tetrahedron Lett.* **1999**, *40*, 4889–4892.

50 F. Langa, P. de la Cruz, E. Espíldora, A. González-Cortés, A. de la Hoz, V. López-Arza, *J. Org. Chem.* **2000**, *65*, 8675–8684.

51 (a) R. Grigg, J. Kemp, N. Thompson, *Tetrahedron Lett.* **1978**, *19*, 2827–2830; (b) R. Grigg, *Chem. Soc. Rev.* **1987**, *16*, 89–121.

52 A. Arrieta, J. R. Carrillo, F. P. Cossío, A. Díaz-Ortiz, M. J. Gómez-Escalonilla, A. de la Hoz, F. Langa, A. Moreno, *Tetrahedron* **1998**, *54*, 13167–13180.

53 P. de la Cruz, A. Díaz-Ortiz, J. J. García, M. J. Gómez-Escalonilla, A. de la Hoz, F. Langa, *Tetrahedron Lett.* **1999**, *40*, 1587–1590.

54 F. Langa, P. de la Cruz, E. Espíldora, A. de la Hoz, J. L. Bourdelande, L. Sánchez, N. Martín, *J. Org. Chem.* **2001**, *66*, 5033–5041.

55 Y. Matsubara, H. Tada, S. Nagase, Z. Yoshida, *J. Org. Chem.* **1995**, *60*, 5372–5373.

56 J. L. Delgado, P. de la Cruz, V. López-Arza, F. Langa, D. B. Kimball, M. M. Haley, *J. Org. Chem.* **2004**, *64*, 2661–2666.

57 (a) D. B. Kimball, A. G. Hayes, M. M. Haley, *Org. Lett.* **2000**, *2*, 3825–3827; (b) D. B. Kimball, R. Herges, M. M. Haley, *J. Am. Chem. Soc.* **2002**, *124*, 1572–1573; (c) D. B. Kimball, T. J. R. Weakley, M. M. Haley, *J. Org. Chem.* **2002**, *67*, 6395–6405; (d) D. B. Kimball, T. J. R. Weakley, R. Herges, M. M. Haley, *J. Am. Chem. Soc.* **2002**, *124*, 13463–13473.

58 F. Langa, M. J. Gómez-Escalonilla, E. Díez-Barra, J. C. García-Martínez, A. de la Hoz, J. Rodríguez-López, A. González-Cortés, V. López-Arza, *Tetrahedron Lett.* **2001**, *42*, 3435–3438.

59 (a) R. E. Martin, F. Diederich, *Angew. Chem. Int. Ed.* **1999**, *38*, 1350–1377; (b) K. Müllen, G. Wegner, *Electronic Materials: The Oligomer Approach*, Wiley–VCH, Weinheim, **1998**; (c) G. P. Bartholomew, G. C. Bazan, *Acc. Chem. Res.* **2001**, *34*, 30–39.

60 (a) J. F. Eckert, J. F. Nicoud, J. F. Nierengarten, S. G. Liu, L. Echegoyen, F. Barigelletti, N. Armaroli, L. Ouali, V. Krasnikov, G. Hadziioannou, *J. Am. Chem. Soc.* **2000**, *122*, 7467–7479; (b) E. H. A. Beckers, P. A. Van Hal, A. Schenning, A. El-ghayoury, L. Sanchez, J. C. Hummelen, E. W. Meijer, R. A. J. Janssen, *Chem. Commun.* **2002**, *23*, 2888–2889; (c) N. Armaroli, J. F. Eckert, J. F. Nierengarten, *Chem. Commun.* **2000**, *21*, 2105–2106.

61 (a) J. M. Tour, *Acc. Chem. Res.* **2000**, *33*, 791–804; (b) S. Welter, K. Brunner, J. W. Hofstraat, L. De Cola, *Nature* **2003**, *421*, 54–57.

62 N. Armaroli, G. Accorsi, J.-P. Gisselbrecht, M. Gross, V. Krasnikov, D. Tsamouras, G. Hadziioannou, M. J. Gómez-Escalonilla, F. Langa, J.-F. Eckert, J.-F. Nierengarten, *J. Mater. Chem.* **2002**, *12*, 2077–2087.

63 J. L. Delgado, P. de la Cruz, V. López-Arza, F. Langa, Z. Gan, Y. Araki, O. Ito, *Bull. Soc. Jpn.* **2005**, *78*, 1500–1507.

64 A. Bagno, S. Claeson, M. Maggini, M. L. Martini, M. Prato, G.

Scorrano, *Chem. Eur. J.* **2002**, *8*, 1015–1023.

65 P. de la Cruz, A. de la Hoz, L. M. Font, F. Langa, M. C. Pérez-Rodríguez, *Tetrahedron Lett.* **1998**, *39*, 6053–6056.

66 R. Tong, H. Wu, B. Li, R. Zhu, G. You, S. Qian, *Physica B* **2005**, *366*, 192–199.

67 Y. Chen, J. Wang, D. Zhang, R. Cai, H. Yu, C. Su, Z.-E. Huang, *Polymer* **2000**, *41*, 7877–7880.

68 R. E. Douthwaite, M. L. H. Green, M. J. Rosseinsky, *Chem. Mater.* **1996**, *8*, 394–400.

69 S. Iijima, *Nature* **1991**, *354*, 56–58.

70 (a) W. A. D. Heer, A. Chatelain, D. Ugarte, *Science* **1995**, *270*, 1179–1180; (b) Q. H. Wang, A. A. Setlur, J. M. Lauerhaas, J. Y. Dai, E. W. Seelig, R. P. H. Chang, *Appl. Phys. Lett.* **1998**, *72*, 2912–2913; (c) J.-M. Bonard, J.-P. Salvetat, T. Stockli, L. Forro, A. Chatelain, *Appl. Phys. Lett.* **1998**, *73*, 918–920.

71 (a) S. M. Lee, Y. H. Lee, *Appl. Phys. Lett.* **2000**, *76*, 2877–2879; (b) P. G. Collins, K. Bradley, M. Ishigami, A. Zettl, *Science* **2000**, *287*, 1801–1804; (c) P. G. Collins, A. Zettl, H. Bando, A. Thess, R. E. Smalley, *Science* **1997**, *278*, 100–103.

72 C. Journet, W. K. Maser, P. Bernier, A. Loiseau, M. Lamy de la Chapelle, S. Lefrant, P. Deniard, R. Lee, J. E. Fischer, *Nature* **1997**, *388*, 756–758.

73 J. Liu, A. G. Rinzler, H. Dai, J. H. Hafner, R. K. Bradley, P. G. Boul, A. H. Lu, T. Iverson, K. Shelimov, C. B. Huffman, F. Rodriguez-Macias, Y. S. Shon, T. R. Lee, D. T. Colbert, R. E. Smalley, *Science* **1998**, *280*, 1253–1256.

74 (a) W. Z. Li, S. S. Xie, L. X. Qian, B. H. Chang, B. S. Zou, W. Y. Zhou, R. A. Zhao, G. Wang, *Science* **1996**, *274*, 1701–1703; (b) S. Fan, M. G. Chapline, N. R. Franklin, T. W. Tombler, A. M. Cassell, H. Dai, *Science* **1996**, *283*, 512–514; (c) K. Hernadi, A. Fonseca, J. B. Nagy, D. Bernaerts, A. Riga, A. Lucas, *Synth. Met.* **1996**, *77*, 31–43; (d) Z. P. Huang, J. W. Xu, Z. F. Ren, J. H. Wang, M. P. Siegal, P. N. Provecio, *Appl. Phys. Lett.* **1998**, *73*, 3845–3847; (e) L. C. Qin, D. Zhou, A. R. Krauss, D. M. Gruen, *Appl. Phys. Lett.* **1998**, *72*, 3437–3439.

75 P. Nikolaev, M. J. Bronikowski, R. K. Bradley, F. Rohmund, D. T. Colbert, K. A. Smith, R. E. Smalley, *Chem. Phys. Lett.* **1999**, *313*, 91–97.

76 (a) E. H. Hong, K.-H. Lee, S.-H. Oh, C.-G. Park, *Adv. Funct. Mater.* **2003**, *13*, 961–966; (b) E. H. Hong, K.-H. Lee, S.-H. Oh, C.-G. Park, *Adv. Mater.* **2002**, *14*, 676–679.

77 (a) T. W. Ebbesen, P. M. Ajayan, H. Hiura, K. Tanigaki, *Nature* **1994**, *367*, 519; (b) H. Hiura, T. W. Ebbesen, K. Tanigaki, *Adv. Mater.* **1995**, *7*, 275–276; (c) E. Dujardin, T. W. Ebbesen, A. Krishnan, M. J. Treacy, *Adv. Mater.* **1998**, *10*, 611–613.

78 (a) F.-H. Ko, C.-Y. Lee, C.-J. Ko, T.-C. Chu, *Carbon* **2005**, *43*, 727–733; (b) C.-J. Ko, C.-Y. Lee, F.-H. Ko, H.-L. Chen, T.-C. Chu, *Microelectron. Eng.* **2004**, *73/74*, 570–577.

79 C.-M. Chen, M. Chen, F.-C. Leu, S.-Y. Hsu, S.-C. Wang, S.-C. Shi, C.-Fu Chen, *Diam. Relat. Mater.* **2004**, *13*, 1182–1186.

80 (a) A. R. Harutyunyan, B. K. Pradhan, J. Chang, G. Chen, P. C. Eklund, *J. Phys. Chem. B* **2002**, *106*, 8671–8675; (b) E. Vázquez, V. Georgakilas, M. Prato, *Chem. Commun.* **2002**, 2308–2309.

81 M. T. Martínez, M. A. Callejas, A. M. Benito, W. K. Maser, M. Cochet, J. M. Andrés, J. Schreiber, O. Chauvet, J. L. G. Fierro, *Chem. Commun.* **2002**, 1000–1001.

82 (a) V. Georgakilas, K. Kordatos, M. Prato, D. M. Guldi, M. Holzinger, A. Hirsch, *J. Am. Chem. Soc.* **2002**, *124*, 760–761; (b) C. A. Dyke, J. M. Tour, *J. Am. Chem. Soc.* **2003**, *125*, 1156–1157; (c) M. Holzinger, J. Abraham, P. Whelan, R. Graupner, L. Ley, F. Hennrich, M. Kappes, A. Hirsch, *J. Am. Chem. Soc.* **2003**, *125*, 8566–8580; (d) H. Murakami, T. Nomura, N. Nakashima, *Chem. Phys. Lett.* **2003**, *378*, 481–485; (e) D. M.

Guldi, M. Marcaccio, D. Paolucci, F. Paolucci, N. Tagmatarchis, D. Tasis, E. Vazquez, M. Prato, *Angew. Chem., Int. Ed.* **2003**, *42*, 4206–4209.

83 (a) A. Star, D. W. Steuerman, J. R. Heath, J. F. Stoddart, *Angew. Chem., Int. Ed.* **2002**, *41*, 2508–2512; (b) T. Fukushima, A. Kosaka, Y. Ishimura, T. Yamamoto, T. Takigawa, N. Ishii, T. Aida, *Science* **2003**, *300*, 2072–2074; (c) J. Sun, L. Gao, M. Iwasa, *Chem. Commun.* **2004**, 832–833.

84 (a) A. Fujiwara, Y. Matsuoka, H. Suematsu, N. Ogawa, K. Miyano, H. Kataura, Y. Maniwa, S. Suzuki, Y. Achiba, *Jpn. J. Appl. Phys.* **2001**, *40*, L1229–L1231; (b) M. Freitag, Y. Martin, J. A. Misewich, R. Martel, Ph. Avouris, *Nano Lett.* **2003**, *3*, 1067–1071; (c) L. Cao, H. Chen, M. Wang, J. Sun, X. Zhang, F. Kong, *J. Phys. Chem. B* **2002**, *106*, 8971–8975.

85 K. C. Hwang, *J. Chem. Soc., Chem. Commun.* **1995**, 173–174.

86 J. Chen, A. M. Rao, S. Lyuksyutov, M. E. Itkis, M. A. Hamon, H. Hu, R. W. Cohn, P. C. Eklund, D. T. Colbert, R. E. Smalley, R. C. Haddon, *J. Phys. Chem. B* **2001**, *105*, 2525–2528.

87 Y. Sun, K. Fu, Y. Lin, W. Huang, *Acc. Chem. Res.* **2002**, *35*, 1096–1104.

88 F. Della Negra, M. Meneghetti, E. Menna, *Fullerene Sci. Techn.* **2003**, *11*, 25–34.

89 Y. Wang, Z. Iqbal, S. Mitra, *Carbon* **2005**, *43*, 1015–1020.

90 J. L. Delgado, P. de la Cruz, F. Langa, A. Urbina, J. Casado, J. T. López Navarrete, *Chem. Commun.* **2004**, 1734–1735.

91 M. Alvaro, P. Atienzar, P. De la Cruz, J. L. Delgado, H. García, F. Langa, *Chem. Phys. Lett.* **2004**, *386*, 342–345.

92 X. Lu, F. Tian, N. Wang, Q. Zhang, *J. Am. Chem. Soc.* **2003**, *125*, 10459–10464.

93 M. Alvaro, P. Atienzar, P. de la Cruz, J. L. Delgado, V. Troiani, H. Garcia, F. Langa, A. Palkar, L. Echegoyen, *J. Am. Chem. Soc.* **2006**, *128*, 6626–6635.

22
Microwave-assisted Extraction of Essential Oils

Farid Chemat and Marie-Elisabeth Lucchesi

22.1
Introduction

Herbs and spices are invaluable resources, useful in daily life as food additives, flavors, fragrances, pharmaceuticals, colors, or directly in medicine. This use of plants has a long history all over the world, and over the centuries humanity has developed better methods for extraction of essential oils from such materials.

Essential oils and aromas are complex mixtures of volatile substances usually present at low concentrations. Before such substances can be used or analyzed, they must be extracted from the plant matrix. Different methods can be used for this purpose, e.g. hydrodistillation, steam distillation, cold pressing, and simultaneous distillation–extraction. The molecules extracted are well known to be thermally sensitive and vulnerable to chemical changes, however [1–4]. Losses of some volatile compounds, low extraction efficiency, degradation of unsaturated or ester compounds by thermal or hydrolytic effects, and toxic solvent residue in the extract may be encountered as a result of use of these extraction methods. These shortcomings have led to the consideration of the use of new "green" techniques in essential-oil extraction, which typically use less solvent and energy; examples include microwave extraction [5], supercritical fluid extraction [6], the headspace method [7], ultrasound extraction [8], the controlled pressure drop process [9], and subcritical water extraction [10]. Extraction under extreme or nonclassical conditions is currently a dynamically developing subject in applied research and industry. Alternatives to conventional extraction procedures may increase production efficiency and contribute to environmental preservation by reducing the use of solvents and fossil energy and the generation of hazardous substances.

Microwave energy is well known to have a significant effect on the rate of a variety of processes in the chemical and food industry. Much attention has been devoted to the application of microwave dielectric heating for extraction of natural products in processes that typically needed hours or days to reach completion with conventional methods. By use of microwaves, extractions can now be completed in seconds or minutes with high reproducibility, reducing the consumption of solvent, simplifying manipulation and work-up, giving a final product of higher

Microwaves in Organic Synthesis, Second edition. Edited by A. Loupy
Copyright © 2006 WILEY-VCH Verlag GmbH & Co. KGaA, Weinheim
ISBN: 3-527-31452-0

purity, eliminating post-treatment of waste water, and consuming only a fraction of the energy normally needed for a conventional extraction method such as steam distillation or solvent extraction. Several classes of compound, for example essential oils, aromas, pigments, antioxidants, and other organic compounds have been extracted efficiently from a variety of matrices (mainly animal tissues, food, and plant materials) [5].

Microwave extraction is a research topic which affects several fields of modern chemistry. All the reported applications have shown that microwave-assisted extraction is an alternative to conventional techniques for such matrices. The main benefits are decreases in extraction times and the amounts of solvents used. The advantages of using microwave energy, which is a noncontact heat source, for extraction of essential oils from plant materials, include: more effective heating, faster energy transfer, reduced thermal gradients, selective heating, reduced equipment size, faster response to process heating control, faster start-up, increased production, and elimination of process steps [11]. Extraction processes performed under the action of microwave radiation are believed to be affected in part by polarization, volumetric, and selective heating [12, 13].

This chapter presents a complete picture of current knowledge on microwave extraction of essential oils. It provides the necessary theoretical background and some details about essential oils and their extraction, the technique, and safety precautions. We will also discuss some of the factors which make the combination of extraction and microwaves one of the most promising topics of research in modern chemistry.

22.2
Essential Oils: Composition, Properties, and Applications

Essential oils have probably been used since the discovery of fire. Egyptians and Phoenicians, Jews, Arabs, Indians, Chinese, Greeks, Romans, and even Mayas and Aztecs all possessed a fragrance culture of great refinement. The Egyptian art of perfumery was pre-eminent in the civilized antique world, hence the use of aromatic fumigation in religious ceremonies and in mummification. The advent of Christianity and the fall of the Roman Empire caused the art and science of perfumery to move into the Arabic world, where it reached an unequalled level of refinement. During the middle ages, the Crusaders introduced the art of alchemy to Europe [14]. The process used by alchemists was the distillation technique using the alambic to produce *spirit* or *Quinta essentia*, precisely what we know today as essential oils [15].

Essential oils are also known as volatile oils in contrast with fatty vegetable, animal, and mineral oils. Thus a drop of essential oil on a piece of cloth or paper disappears within a few minutes or few days at most, depending on the temperature; this is not so for fatty oils.

Essential oils are highly concentrated aromatic oily liquids obtained from a variety of spices and aromatic plant materials. Numerous publications have presented data on the composition of a variety of essential oils, which can contain more than hundred individual components. Major components can constitute up to 85% of the essential oil, whereas other components are present in trace amounts only [16]. Essential oils can be classified into two main groups:

1. hydrocarbons consisting of terpenes, for example monoterpenes, sesquiterpenes, and diterpenes; and
2. oxygenated compounds, for example esters, aldehydes, ketones, alcohols, phenols, oxides, acids, and lactones. Occasionally nitrogen and sulfur compounds are also present.

Terpene is the generic name of a group of natural products, structurally based on isoprene (2-methylbutadiene) units having the molecular formula $(C_5H_8)_n$. The term may also refer to oxygen derivatives of these compounds; these are known as the terpenoids. They are normally classified into groups based on the number of isoprene units from which they are biogenetically derived. Monoterpenes contain two isoprene units. These are widely distributed in nature, particularly in essential oils. They are important in perfumery and in the flavor industry. Sesquiterpenes contain three isoprene units. They are found in many living systems but particularly in higher plants. Diterpenes contain twenty carbon atoms in their basic skeletons, made from four isoprene units. They occur in almost all plant families and belong to more than 20 major structural types [17]. Table 22.1 contains some of the most common naturally occurring essential oil constituents grouped according to their molecular formulae.

Essential oils are obtained from a variety of aromatic plant material including flowers, buds, seeds, leaves, twigs, bark, herbs, wood, fruits and roots. These aroma compounds are formed by plants as byproducts or, indeed, as final metabolic products and stored in particular organs of the plant, for example:

- thyme, sage, and rosemary (Lamiaceae family) – in glandular cells, hairs and scales;
- cinnamon, laurel, and cassia (Lauraceae family) – in essential oil and resin cells;
- caraway, anis, and coriander (Umbelliferae family): in essential oil canals occurring as intercellular spaces in plant tissue;
- lemon, orange, and bergamot (Rutaceae family) – in lysigenous secretory reservoirs formed inside the plant.

The yield and quality of essential oil from a plant depends mainly on cultivar (chemotype and genetic variability), environment (fertilization, climatic conditions, and crop protection), and physiological stage (plant development stage) [18].

Approximately 3000 essential oils are known; of these approximately 300 are commercially important and widely used in the perfume, cosmetics, pharmaceutical, agricultural, and food industries. It has long been recognized that some have antimicrobial, antioxidant, antiviral, antimycosic, antitoxigenic, antiparasitic, and

Tab. 22.1. Most common naturally occurring essential oil constituents.

	Molecular formula	Boiling point (°C)	Solubility in water (g L^{-1})
Monoterpene hydrocarbons			
Limonene	$C_{10}H_{16}$	175.4	$<10^{-3}$
Pinene	$C_{10}H_{16}$	157.9	$<10^{-3}$
Sabinene	$C_{10}H_{16}$	164	$<10^{-3}$
Myrcene	$C_{10}H_{16}$	167	$<10^{-3}$
γ-Terpinene	$C_{10}H_{16}$	183	$<10^{-3}$
para-Cymene	$C_{10}H_{16}$	173.9	$<10^{-3}$
Sesquiterpene hydrocarbons			
β-Caryophyllene	$C_{15}H_{24}$	268.4	$<10^{-3}$
α-Santalene	$C_{15}H_{24}$	247.6	$<10^{-3}$
α-Zingeberene	$C_{15}H_{24}$	270.7	$<10^{-3}$
β-Curcumene	$C_{15}H_{24}$	266	$<10^{-3}$
Derivative of diterpene hydrocarbon			
Phytol	$C_{20}H_{40}O$	335.5	$<10^{-3}$
Oxygenated derivatives			
Alcohols			
Geraniol	$C_{10}H_{17}OH$	229.5	0.67
Linalool	$C_{10}H_{17}OH$	198.5	0.67
Aldehydes			
Citral	$C_{10}H_{16}O$	210.9	2.61
Cuminic aldehyde	$C_{10}H_{12}O$	254.6	0.26
Ketones			
Camphor	$C_{10}H_{16}O$	207.4	0.92
Carvone	$C_{10}H_{14}O$	230.5	1.60
Phenols			
Thymol	$C_{10}H_{14}O$	233	0.85
Eugenol	$C_{10}H_{12}O_2$	255	2.52
Acetates			
Neryl acetate	$C_{12}H_{20}O_2$	247.5	0.71
Linalyl acetate	$C_{12}H_{20}O_2$	220	0.57
Oxides			
1,8-Cineol	$C_{10}H_{18}O$	174.8	5.8×10^{-3}
Linalool	$C_{10}H_{18}O$	198.5	0.67

insecticidal properties. These characteristics are possibly related to their function in the plants. For example, monoterpenes and sesquiterpenes serve as antiherbivore agents that have substantial insect toxicity while having negligible toxicity to mammals [19].

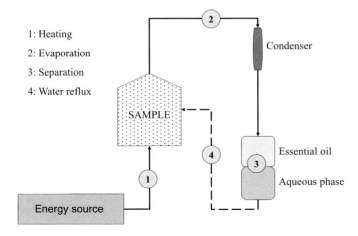

Fig. 22.1. Schematic representation of conventional recovery of essential oils.

22.3
Essential Oils: Conventional Recovery Methods

According to ISO and AFNOR standards, essential oils are defined as products obtained from raw plant material which must be isolated by physical means only. The physical methods used are distillation (steam, steam–water and water), expression (also known as cold pressing for citrus peel oils), or dry distillation of natural materials. After distillation, the essential oil is physically separated from the aqueous phase (Fig. 22.1) [20–22].

The traditional way of isolating volatile compounds as essential oils from plant material is distillation. During distillation, fragrant plants exposed to boiling water or steam release their essential oils by evaporation. Recovery of the essential oil is facilitated by distillation of two immiscible liquids, i.e., water and essential oil, on the basis of the principle that, at the boiling temperature, the combined vapor pressures equals the ambient pressure. Thus the essential oil ingredients, for which boiling points normally range from 200 to 300 °C, are evaporated at a temperature close to that of water. The essential oil-laden steam rises and enters narrow tubing that is cooled by an outside source. As steam and essential oil vapors are condensed, both are collected and separated in a vessel traditionally called the "Florentine flask" [14]. The essential oil, being lighter than water, floats at the top while the water settled to the bottom and can be easily separated. The amount of essential oil produced depends on four main criteria: distillation time, temperature, operating pressure, and, most importantly, the type and quality of the plant material. Typically, the yield of essential oils from plants is between 0.005 and 10% [23].

Historically, three types of distillation have been used: water distillation, water–steam distillation, and steam distillation. Water distillation is sometimes referred to as "indirect" steam distillation. In this method, plant material is soaked in water and heated until it boils. The resulting steam from the boiling water carries the volatile oils with it. Cooling and condensation subsequently separate the oil from the water. Apart from its slowness, the disadvantage of this technique is that both materials and scent deteriorate as a result of constant heat exposure. In the water–steam method, the leafy plant material is placed on a grill above the hot water, and the steam passes through the plant material. The leaves must be carefully distributed on the grill to enable even steaming and thorough extraction. "Direct" steam distillation is the most common method used for extraction of essential oils. In this process, no water is placed inside the distillation tank itself. Instead, steam is directed into the tank from an outside source. The essential oils are released from the plant material when the steam bursts the sacs containing the oil molecules. From this stage, the process of condensation and separation is conventional.

In addition to those mentioned above, there are numerous other improved methods of producing natural fragrance materials and essential oils including turbo-distillation, hydro-diffusion, vacuum-distillation, continuous-distillation, cold-expression, dry-distillation, and molecular-distillation [24, 25]. All these conventional extraction techniques have important drawbacks, for example low yields, formation of by-products, and limited stability. For steam distillation and hydrodistillation, the steam is percolated through the flask with plants, from the bottom, and the oil evaporates. The emerging mixture of vaporized water and oil moves through a coil usually cooled with running water, where the steam is condensed. The mixture of condensed water and essential oil is collected and separated by decantation or, rarely, by centrifugation. The elevated temperatures and prolonged extraction time can cause chemical modification of the essential oil components and often a loss of the most volatile molecules. Manufacturers use dry distillation to extract high boiling point oils from wood. In this process, heat is applied, usually as a direct flame, to the vessel containing the plant material. The high temperature releases essential oils by evaporation. The vapor is piped away and condensed to give a mixture of liquid oils. Some of the components in the plant material are degraded (pyrolysis) at the high temperatures used and add a burnt smoky character to the odor of the oil. Cade and birch tar oils are the two major oils produced by this method. When attempts are made to extract essential oils by dry distillation, many organic compounds decompose.

When using the cold pressing technique, citrus essential oil is agitated vigorously with water and gradual diminution in citral and terpene alcohols occurs. Furthermore, during agitation, air is entrained into the liquid, creating conditions favorable for hydrolysis, oxidation, and resinification. Another important aspect of the essential oils industry is enhancement of the quality and stability of the oil. Some essential oils contain more than 80% of monoterpene hydrocarbons, mainly limonene. Deterpenation is performed to remove some of the limonene with other, unstable, terpenes and to concentrate the oxygenated fraction. This process improves the stability of the oil, increases the solubility of the essential oil in food-grade solvents, and reduces storage and transportation costs [26–29].

Tab. 22.2. Twenty years of microwave chemistry and extraction.

1986	Gedye (Canada) and Giguere (USA)	First microwave-assisted organic synthesis	Ganzler (Hungary)	First microwave-assisted extraction
1994	Strauss (Australia)	First continuous microwave reactor	Mengal and Monpon (France)	First patent on vacuum microwave hydrodistillation
2004	Loupy (France) and Varma (USA)	Microwave green chemistry	Chemat and Lucchesi (France)	Solvent-free microwave extraction

22.4
Microwave Extraction Techniques

Use of microwave energy in the chemical laboratory was described for the first time in 1986 simultaneously by Gedye [30] and Giguere [31] in organic synthesis and by Ganzler [32] and Lane [33] for extraction of biological samples for analysis of organic compounds. Since then, numerous laboratories have studied the synthetic and analytical possibilities of microwaves as a nonclassical source of energy. Over 2000 and 500 articles, respectively, have since been published on the subject of microwave synthesis and extraction (Table 22.2) [34, 35].

In the last decade there has been an increasing demand for new extraction techniques, amenable to automation, with shortened extraction times and reduced organic solvent consumption, to prevent pollution and reduce the cost of sample preparation. Driven by these goals, advances in microwave extraction have resulted several techniques such as microwave-assisted solvent extraction (MASE) [32, 36–39], vacuum microwave hydrodistillation (VMHD) [40, 41], microwave hydrodistillation (MWHD) [42, 43], compressed air microwave distillation (CAMD) [44], microwave headspace (MHS) [5], and solvent-free microwave hydrodistillation (SFME) [45, 46]. Table 22.3 summarizes the most common microwave extraction techniques for plant matrices and lists their advantages and drawbacks. Over the years procedures based on microwave extraction have replaced some of the conventional processes and other thermal extraction techniques that have been used for decades in chemical laboratories.

22.4.1
Microwave-assisted Solvent Extraction (MASE)

MASE uses microwave energy to heat a liquid organic solvent in contact with a sample. There are two types of microwave extraction – microwave assisted extraction (MAE), which is performed under controlled pressure and temperature in a

Tab. 22.3. Comparison of microwave-assisted extraction techniques for plant matrix.

	Extraction techniques				
	MAE	VMHD	MWHD	CAMD	SFME
Name	Microwave-assisted extraction	Vacuum microwave hydrodistillation	Microwave hydrodistillation	Compressed air microwave distillation	Solvent-free microwave extraction
Ref.	Paré et al. (1992) [36–39]	Archimex S.A. (1994) [40–41]	Stashenko et al. (2004) [42, 43]	Craveiro et al. (1989) [44]	Chemat et al. (2004) [45–46]
Brief description	Sample is immersed in a microwave-absorbing solvent in a closed vessel and irradiated with microwave energy	Sample is submitted to a microwave heating combined with application of a sequential vacuum	Sample is immersed in water and irradiated with microwave energy at atmospheric pressure	Sample is irradiated with microwaves under an air flow remove the volatile oil	Sample without any solvent is submitted to microwaves at atmospheric pressure
Extraction time	3–30 min		30 min	5 min	10–60 min
Sample size	1–10 g	3–30 kg h^{-1}	100 g	30–40 g	0.1–1 kg
Solvent usage	10–40 mL	No solvent	1 L Water	Air pump	No solvent
Investment	Moderate	High	Low	Low	Low
Advantages	Fast and multiple extraction Low solvent volumes Elevated temperatures	Fast Head note accentuated	No clean-up step No change in volatile oil composition	Fast and easily to realize	Fast, green and economical No clean-up step Volatile oil highly odoriferous

Drawbacks	Extraction solvent must be able to absorb microwaves Clean-up step needed Waiting for the vessels to cool down	Difficult and expensive to implement	Clean-up step needed
Product	Microwave extract	Essential oil	Microwave extract
Type of matrix	Leaves, roots, bulb, wood, seeds	Leaves	Leaves
Examples	Peppermint Yield: 0.296 to 0.474% (EV: 0.3% in 2 h)	Sage 2.55 mL d'HE (HD: 2.77 mL)	Basil (var. American) Yield: 2.1% (EV: 2.1%)
		Lippia alba L. Yield: 0.69% (in 30 min) (HD: 0.7% in 2 h)	
			Essential oil
			Leaves, seeds
			Thyme Same yield, ESSAM: 30 min HD 4 h 30

closed vessel, and focused microwave assisted extraction (FMAE), which is performed with an open vessel under atmospheric pressure in a single-mode microwave oven [47, 48].

MASE was first used in the extraction of several compounds from environmental samples, for example soil or polluted water. Numerous classes of compound, for example pesticides, phenols, dioxins, and organometallic compounds have been extracted efficiently, i.e. rapidly and reproducibly, from a variety of matrices. MASE has also been applied to plant material to extract aroma compounds. The technique was patented in 1990 as a microwave assisted process (MAP) [37]. Typically, plant material is immersed in a microwave-nonabsorbing solvent, for example hexane, and irradiated with microwave energy. When the oil glands of the plant are subjected to severe thermal stress and localized high pressures, as in microwave heating, the pressure build-up within the glands exceeds their capacity for expansion, and causes their rupture more rapidly than in conventional extraction. Volatile oil dissolves in the organic solvent before being separated by liquid–liquid extraction. Aroma compounds from different types of plant have been extracted by MASE from wood, roots, leaves, or seeds. Yields and composition of microwave extracts are always comparable with those obtained by classical solvent extraction, for example Soxhlet extraction, but have been achieved with reduced extraction time [49–51].

22.4.2
Compressed Air Microwave Distillation (CAMD)

One of the first methods using microwave-assisted extraction of essential oil was presented in 1989 by Craveiro [44]. The essential oil of *Lippia sidoides* was extracted using microwave energy and compressed air only. Inspired by classical steam distillation, the CAMD technique used compressed air instead of vapor to extract the volatile oil. Typically, plant material is placed in a reactor inside the microwave cavity and heated. At the same time, a compressor, located outside the cavity, forces compressed air into the reactor. Volatile oil and vapor are then driven to the recovery flask outside the cavity. In 5 min CAMD provides an essential oil which is qualitatively and quantitatively identical with that produced by the conventional hydrodistillation method.

22.4.3
Vacuum Microwave Hydrodistillation (VMHD)

VMHD was elaborated and patented by Archimex in 1994 [40]. This technique is based on selective heating by microwaves combined with sequential application of a vacuum. The plant material is placed in a microwave cavity with water to refresh the dry material. The plant material is then exposed to microwave radiation to release the natural extract. Reducing the pressure to between 100 and 200 mbar enables the evaporation of the azeotropic water–volatile oil mixture from the biological matrix. The procedure is repeated in a stepwise fashion to extract all the volatile oil from the plant. Up to 30 kg h^{-1} material can be treated [40, 41].

According to the patents, VMHD provides yields comparable to those obtained by traditional hydrodistillation but with extraction times only one tenth those required with hydrodistillation. The thermally sensitive crude notes seem to be preserved with VHMD, in contrast to conventional hydrodistillation. VMHD is suggested as an economical and efficient technique to extract high-quality natural products on a large scale [52–54].

22.4.4
Microwave Headspace (MHS)

Microwave-assisted desorption coupled to *in situ* headspace solid-phase micro-extraction (HS–SPME) was first proposed as a possible alternative pretreatment of samples collected from workplace monitoring. Therefore, pretreatment that takes a short time and uses little or no organic solvents has led to the recent development of a new extraction technique. Solid-phase micro-extraction (SPME) coupled with GC analysis has been used successfully to analyze pollutants in environmental matrices. MHS has been developed to achieve one-step, *in situ* headspace sampling of semivolatile organic compounds in aqueous samples, vegetables, and soil [7, 55–58].

22.4.5
Microwave Hydrodistillation (MWHD)

Stashenko [42, 43] used the classical technique of hydrodistillation in association with microwave energy. Part of the conventional equipment, in which the plant material is usually immersed in water, is placed inside the microwave cavity, whereas the cooler and the recovery system for the essential oil are situated outside the microwave cavity. Essential oils are obtained more rapidly but with yields and quality comparable with those obtained by hydrodistillation. Because there is no clean-up step, this technique is faster.

22.4.6
Solvent-free Microwave Hydrodistillation (SFME)

SFME is a recent method of extraction, patented in 2004, with the specific objective of obtaining essential oil from plant material [45, 46]. Based on a relatively simple principle, SFME involves placing the vegetable material in a microwave reactor without addition of solvent or water. SFME is a combination of microwave heating and distillation, and is performed at atmospheric pressure. In terms of quality and quantity, SFME seems to be more competitive and economic than classical methods such as hydro or steam distillation [59, 60].

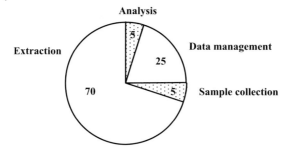

Fig. 22.2. Relative time consumed by different steps in an analytical procedure for essential oils.

22.5
Importance of the Extraction Step

An analytical procedure for essential oils or aromas from plants or spices usually comprises two steps – extraction (e.g. steam distillation, hydrodistillation, or simultaneous distillation–extraction) and analysis (e.g. gas chromatography, GC, or gas chromatography coupled to mass spectrometry, GC–MS). Although the analysis step is complete after only 15 to 30 min, extraction takes at least several hours. It frequently involves prolonged heating and stirring in boiling water. The principal limiting step of sample treatment is extraction of the essential oil from the matrix, which consists in transferring the compounds into boiling water then azeotropic distillation, condensation, and physical separation. The essential oil contains not only the compounds of interest at high concentration or trace levels but also co-extracted compounds (e.g. water, endogenous compounds, and other contaminants) which can interfere with the analysis. It is necessary to purify extracts, to re-concentrate or to dilute them. Analysis is performed by means of a high-resolution separation step combined with a sensitive and selective detector, typically gas chromatography coupled with mass spectrometry (GC–MS). Figure 22.2 shows the duration of each treatment step and extraction procedure. Distillation of the essential oil accounts for 70% of total processing time. It is thus important to control this limiting step, which includes extraction and separation. The choice of technique is the result of a compromise among efficiency and reproducibility of extraction, ease of the procedure, and considerations of cost, time, degree of automation, and safety.

22.6
Solvent-free Microwave Extraction: Concept, Application, and Future

Microwave energy is a key enabling technology in achieving the objective of sustainable (green) chemistry. It has been shown that solvent-free conditions are especially suited to microwave-assisted organic synthesis, because reactions can be run safely under atmospheric pressure in the presence of significant amounts of products [34]. Solvent free microwave extraction has been conceived by following the concepts of solvent-free microwave synthesis. When coupled with microwave radiation, solvent-free techniques have proved to be of special efficiency as clean and economic procedures. Major improvements and simplifications over conventional methods originate from their rapidity, and their enhancement of yield and product purity. SFME extraction extends from the analytical scale to industrial applications for laboratory or commercial purposes. This section will deal with topics of interest to the essential oil analysis laboratory.

22.6.1
Concept and Design

A new method for extracting natural products without added solvent or water by use of microwave energy has been developed and patented in collaboration with Milestone [45, 46]. The solvent-free microwave extraction apparatus is an original combination of microwave heating and distillation at atmospheric pressure. SFME was conceived for laboratory scale applications in the extraction of essential oils from different kinds of aromatic plant. Based on a relatively simple principle, this method involves placing plant material in a microwave reactor, without added solvent or water. Internal heating of *in situ* water within the plant material distends the plant cells and leads to rupture of the glands and oleiferous receptacles. Thus, this process liberates essential oil which is evaporated by the *in situ* water of the plant material. A cooling system outside the microwave oven condenses the distillate continuously. Excess water is returned, by reflux, to the extraction vessel to restore the *in situ* water to the plant material. SFME is neither modified micro-

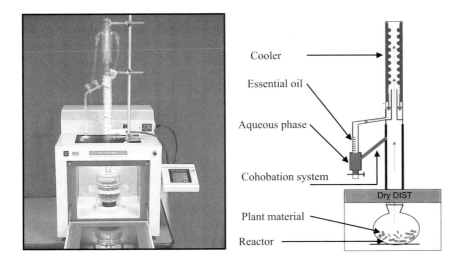

Fig. 22.3. Solvent-free microwave extraction: "Microwave Clevenger".

Tab. 22.4. Composition, yields, and olfactory notes for the major components of essentials oils obtained by SFME and HD.

n°	Compound	AROMATIC HERBS						SPICES						Olfactive Note
		Basil		Crispate mint		Thyme		Ajowan		Cumin		Star anise		
		HD	SFME	HD	SFME	HD	SFME	HD	SFME	HD	SFME	HD	SFME	
	Terpene Hydrocarbons													
1	β-Pinene	1.1		1.4	0.4			4.8	1.5	16.2	5.9			Smell of dry wood, resinous, slightly clinging
2	p-Cymene					11.1	7.5	29.2	21.2	18.4	12.1			Citrus fruit note, reminiscent of lemon and bergamot
3	Limonene			20.2	9.7	0.9	0.6			0.5		11.6	6.6	Fresh, light, perfume of sweet Citrus fruit
4	γ-Terpinene	0.2		0.8	0.2	22.8	17.1	28.6	16.4	22.3	12.9			Herbaceous, smell of Citrus fruit
	Oxygen terpene derivatives													
5	1,8-Cineole	5.8	1.3	0.0	1.5	0.7	0.5				0.7	1.5	1.5	Fresh, light, freshly camphorated
6	Linalol	39.1	25.3	0.4	0.4	4.0	4.6					1.1	1.7	Fresh and floral with a slight note of Citrus fruit

22.6 Solvent-free Microwave Extraction: Concept, Application, and Future | 973

	C1	C2	C3	C4	C5	C6	C7	C8	C9	C10	C11	C12	
7 Carvone			52.3	64.9									Warm-herbaceous, penetrating, spicy
8 Anethole											78.0	81.4	Herbaceous, warm and sweet
9 Thymol			1.9	5.2	40.5	51.0	35.4	60.3					Strong, herbaceous, and warm
10 Eugenol	11.0	43.2	0.2	1.2	0.3	1.5							Powerfull, warm-spicy
Aldehydes													
11 Cumin aldehyde							22.8	37.4					Irritant, pungent, and herbaceous
12 α-Terpinen-7-al									14.4	29.1			
Extraction time (min.)	275	30	220	30	260	30	455	60	495	60	460	60	
Yield of essential oil (%)	0.03	0.03	0.09	0.09	0.16	0.16	3.34	1.41	1.43	0.63	4.16	1.38	

wave assisted extraction (MAE) which uses organic solvents, nor modified hydrodistillation (HD) which uses a large quantity of water. The SFME apparatus is illustrated in Fig. 22.3.

22.6.2
Applications

To investigate the potential of the SFME technique, comparisons have been made with hydrodistillation for extraction of essential oil from spices – ajowan (*Carum ajowan*; *Apiaceae*), cumin (*Cuminum cyminum*; *Umbelliferae*), star anise (*Illicium anisatum*; *Illiciaceae*) – and from fresh aromatic herbs – basil (*Ocimum basilicum*; *Labiaceae*), crispate mint (*Mentha crispa*; *Labiaceae*), thyme (*Thymus vulgaris*; *Labiaceae*) [59, 60]. The yields of essential oil and their chemical composition for the two extraction methods are reported in Table 22.4.

One of the advantages of SFME is rapidity. The extraction temperature is the boiling point of water at atmospheric pressure (100 °C) for both SFME and HD extraction. Figure 22.4 shows the temperature profiles during SFME and HD of essential oil from aromatic herbs. To reach the extraction temperature (100 °C) and thus obtain the first essential oil droplet, it is necessary to heat for only 5 min with SFME compared with 90 min for HD. It is important to note that there is no superheating effect because of the heterogeneity of the medium and the available water used for SFME is provided from the plant matter moisture or the so called *"in situ"* water.

The essential oils of fresh aromatic herbs (90% moisture) extracted by SFME for 30 min were quantitatively (yield) similar to those obtained by conventional hydrodistillation for 4.5 h. For extraction from dry spices, the plant material is soaked in water before extraction to achieve a maximum moisture content of 70%. Yields of essential oils obtained from spices by HD were higher than were obtained by SFME.

Substantially greater amounts of oxygenated compounds and smaller amounts

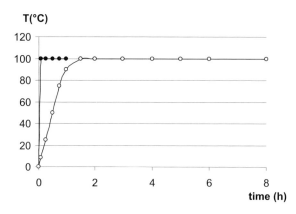

Fig. 22.4. Temperature profiles during HD and SFME (○ HD ● SFME).

of monoterpene hydrocarbons are present in the essential oils of aromatic plants and spices extracted by SFME compared with HD. Monoterpene hydrocarbons are less valuable than oxygenated compounds in terms of their contribution to the fragrance of the essential oil. Conversely, the oxygenated compounds are highly odoriferous and, hence, the most valuable. Linalol and eugenol were the main components of the essential oil extracted from basil but the relative amounts differed for the two extraction methods. Eugenol is the most abundant component of the SFME extract (43.2%) and linalol the second most abundant (25.3%), whereas the HD extract is dominated first by linalol (39.1%) and then by eugenol (11.0%). The essential oil of garden mint isolated by SFME and HD is characterized by the yield of the most important oxygenated compound carvone; yields were, respectively, 64.9% and 52.3%. Limonene, a monoterpene which is the second most abundant compound, is present at 9.7% and 20.2% in SFME and HD extracts, respectively. The essential oil of thyme isolated by SFME and HD contains the same three dominant components: thymol (51.0% and 40.5%, respectively), γ-terpinene (17.1% and 22.8%), and p-cymene (7.5% and 11.1%).

Star anise essential oil isolated by HD and SFME is dominated by *trans*-anethole (oxygenated compound) – 78% and 81.4%, respectively. Limonene, the second most important compound, is only present at 11.6% and 6.6%, respectively, in the oils obtained by HD and SFME. Cumin essential oil isolated by SFME or HD contains the same five dominant compounds but in different quantitative amounts – cumin aldehyde (37.4% and 22.8%, respectively), α-terpin-7-enal (29.1% and 14.4%), γ-terpinene (12.9% and 22.3%), p-cymene (12.1% and 18.4%), and β-pinene (5.9% and 16.2%). Ajowan essential oil isolated by SFME is characterized by a substantial amount of the oxygenated monoterpene thymol (60.3%) whereas the oil extracted by HD is dominated by three compounds in relatively similar amounts – thymol (35.4%), p-cymene (29.2%) and γ-terpinene (28.6%). γ-Terpinene and p-cymene are also present in the essential oil extracted by SFME, but in smaller amounts – 16.4% and 21.2%, respectively.

For fresh aromatic herbs SFME and HD furnish similar yields of essential oils but the proportion of oxygenated compounds is greater for SFME. For soaked spices SFME furnishes lower yields of essential oil than HD but, again, a greater proportion of oxygenated compounds. The greater proportion of oxygenated compounds in the SFME essential oils is probably because of diminution of thermal and hydrolytic effects compared with hydrodistillation, which uses a large quantity of water and is time and energy-consuming. Water is a polar solvent which accelerates many reactions, especially reactions via carbocations as intermediates.

Essential oils obtained by SFME or HD have been evaluated olfactorily by Jean Claude Ellena (Perfurmer from Symrise, France) and Daniel Maurel (Robertet, France). Better reproduction of the natural aroma was more likely for essential oils extracted by SFME than for those extracted by hydrodistillation. Essential oils extracted by SFME and HD have also been analyzed by the AFNOR standard method to determine the usual physical properties characterizing the oils – specific gravity, refractive index, optical rotation, and solubility in 95% ethanol at 20 °C [20, 21]. There was no significant difference between the physical constants of essential oils obtained by SFME and HD.

22.6.3
Cost, Cleanliness, Scale-up, and Safety Considerations

The reduced cost of extraction is clearly advantageous for the proposed SFME method in terms of time and energy. Hydrodistillation required an extraction time of 90 min for heating of 6 L water and 500 g aromatic plants to the extraction temperature, followed by evaporation of the water and essential oil for 180 min. The SFME method required irradiation for 3 min only and evaporation for 27 min of the *in-situ* water and essential oil of the same material. The energy required to perform the extractions is 4.5 kW h for HD, and 0.25 kW h for SFME. The power consumption was determined with a Wattmeter at the microwave generator supply and the electrical heater power supply.

With regard to environmental impact, the calculated quantity of carbon dioxide emitted to the atmosphere is greater for HD (3600 g CO_2 g^{-1} essential oil) than for SFME (200 g CO_2 g^{-1} essential oil). These calculations were based on the assumption that to obtain 1 kW h by combustion of fossil fuel 800 g CO_2 will be emitted to the atmosphere [61].

SFME is proposed as an "environmentally friendly" extraction method suitable for sample preparation before essential oil analysis. SFME is a very clean method which avoids the use of large quantities of water and voluminous extraction vessels, in contrast with HD. SFME could also be used to produce larger quantities of essential oils by using existing large-scale microwave extraction reactors [62]. These microwave reactors are suitable for the extraction of 10, 20, or 100 kg of fresh plant material per batch. These reactors could be easily modified and used for SFME extractions.

The microwave extraction process is simple and can be readily understood in terms of the operating steps to be performed. Application of microwave energy can pose serious hazards in inexperienced hands, however. High levels of safety and attention to detail must be exercised by all persons planning and performing experiments involving microwaves. Personnel must ensure they seek proper information from knowledgeable sources and not attempt to implement this technique unless proper guidance is provided. Only approved equipment and scientifically sound procedures should be used at all the times.

22.7
Solvent-free Microwave Extraction: Specific Effects and Proposed Mechanisms

22.7.1
Effect of Operating Conditions

Moisture content of the plant material, microwave power, and extraction time affect not only the extraction yield of the essential oil but also its composition. Sample moisture content during microwave treatment is critical, because water is an excellent absorber of microwave energy. This strong absorption leads to the temperature increase inside the sample which causes rupture of the essential oil cells by the *in situ* water, followed by evaporation of the water and the essential oils. The microwave input power required is directly related to sample size and weight. The power must be sufficient to achieve the boiling point of the water (100 °C), which fixes the extraction temperature. The power should not be too high, however, or loss of volatile compounds will result. Extraction time is the a factor directly affecting the yield of essential oil from microwave SFME extraction. As time increases, the yield increases almost linearly. The extraction time must be optimized to maximize the yield of the extraction without affecting the quality of the essential oil. The extraction time for SFME must, moreover, be much lower than that for hydrodistillation if it is to be economically and environmentally viable [63].

The efficiency of SFME extraction, in terms of yield and composition of the essential oil, can usually be increased by increasing all three of these factors.

22.7.2
Effect of the Nature of the Matrix

Plant tissue consists of cells surrounded by walls. Some cells exist as glands (external or internal) filled with essential oil. A characteristic of such glands (when external) is that their skin is very thin and can be very easily destroyed. For internal glands the amount of milling of the plant material plays an important role. The plant matrix can be compared with a grain comprising an impermeable core coated with a water boundary layer (Fig. 22.5). Essential oils are extracted in several steps – desorption from the matrix surface or release from internal glands, diffusion through the boundary layer to the boiling water, and separation by azeotropic

1: Desorption

2: Diffusion

3: Solubilisation

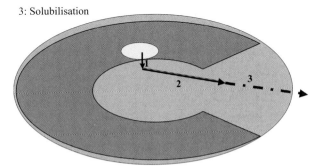

Fig. 22.5. Schematic representation of the individual steps in the process of extraction of essential oils from plant materials.

distillation. Extraction recovery can be limited by one or more of these steps. The interactions involved depend partly on the composition of the matrix.

Essential oils are classically extracted by Clevenger distillation [64]. This method proceeds by iterative distillation and boiling of the aromatic matrix by re-condensed vapors of water, and generally uses large quantities of water and energy. The extraction time can vary between 6 and 24 h.

The different extraction methods (SFME and HD) produce distinguishable physical changes in the plant material. Figures 22.6 and 22.7 enable comparison of scanning electronic micrographs of cardamom (dry seed) and garden mint (fresh material) – the structures of the untreated plant material can be compared with those of the material treated by SFME or HD. After SFME extraction at 100 °C, cells and cell walls have been affected to different degrees. We observed a huge perforation on the external surface of the particle and some starch is dispersed. The husk is clearly damaged. Part of this protective cover was destroyed. In contrast, plant material subjected to hydrodistillation appeared very similar to untreated

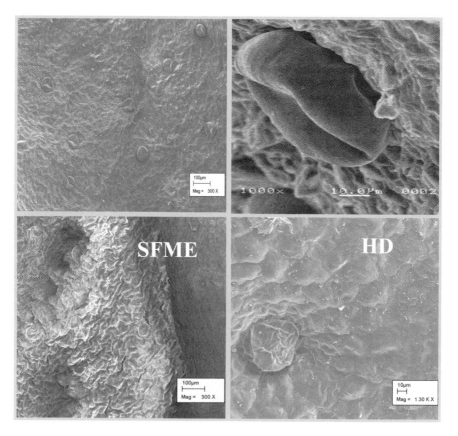

Fig. 22.6. Electron micrographs of garden mint: untreated; after HD extraction; and after SFME extraction.

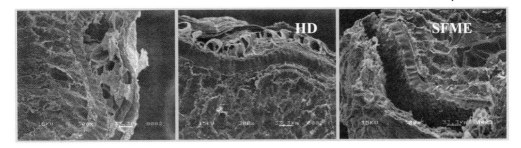

Fig. 22.7. Electron micrographs of cardamom seeds: untreated, after HD extraction, and after SFME extraction.

material. Some parts are still filled, although the albumen is also damaged. The changes observed for SFME extraction were markedly different from those observed by HD, showing clearly that the cells are broken and damaged during SFME. This indicates that the mechanical strain induced by the rapid decompression and the violent vaporization of water have two main effects – the dehydrating effect of vaporization and a subsequent change in the surface tension of the glandular wall, causing it to crumble or rupture more readily.

Similar effects have been reported by Paré and Bélanger [5] and by Chen and Spiro [65, 66] after microwave extraction of rosemary leaves in hexane. When the glands were subjected to more severe thermal stresses and localized high pressures, as in microwave heating, pressure build-up within the glands could have exceeded their capacity for expansion and caused their rupture more rapidly than in conventional extraction.

22.7.3
Effect of Temperature

The relationship between the boiling point of the aroma compounds and the composition (oxygenated and nonoxygenated fraction) of the essential oil extracted by SFME and HD will be examined more closely. Essential oils extracted by SFME are generally dominated by oxygenated compounds. To understand the relationship between boiling point and composition, we will examine the composition of the essential oil of garden mint which contains two major compounds – an oxygenated monoterpene (carvone) and a monoterpene hydrocarbon (limonene). The essential oil of garden mint isolated by SFME and HD is characterized by carvone – 64.9% and 52.3%, respectively. Limonene, a monoterpene, which is the second most abundant compound, is present at 9.7% and 20.2%, respectively, for SFME and HD. The boiling point of carvone, the main component in garden mint essential oil, is 231 °C at atmospheric pressure. In contrast, the boiling point of limonene is 175 °C.

If a solution containing limonene and carvone is distilled, limonene (bp 175 °C) will be isolated before carvone (bp 231 °C). In microwave extraction or distillation of garden mint, the essential oil of which contains primarily these two compounds, oxygen-containing high-boiling components are isolated more rapidly than lower-boiling substances, however.

During the process of distillation, boiling water (100 °C) will penetrate the plant material and remove essential oil by azeotropic distillation. The extraction temperature is equal to the boiling point of water at atmospheric pressure (100 °C), which has been confirmed by experiments using either microwave SFME or conventional HD. Hydrocarbons and oxygenated compounds will distill at the same temperature, 100 °C, irrespective of their own boiling point. Boiling point is not, therefore, the factor determining the yield or composition of essential oils during microwave extraction.

22.7.4
Effect of Solubility

A more direct way of understanding the composition of the essential oils obtained by use of different extraction methods is to consider the solubility of compounds. Solubility reflects the extent to which a substance dissolves in a particular mixture, e.g. an organic solvent or water. Solubilization is the last step of the extraction process after desorption from the matrix surface and diffusion through the solvent boundary layer to the solvent, which is simply water for aqueous distillation of aroma compounds.

Extraction recovery can be limited by one step or several steps. Microwaves cause more damage and destroy the essential oil cells in leaves or seeds more rapidly and effectively than conventional extraction methods, thus the desorption step, which can often be the limiting factor, is of minor importance in SFME. Solubility is rarely the limiting factor in solvent extraction if the solvent is well chosen. In the distillation of essential oils the solvent is always water, however, and aroma compounds can be totally different in structure and chemical characteristics, especially in their solubility.

The importance of solubility in distillation was first mentioned by Von Rechenberg [67] in 1910 with the famous "Theorie der Gewinnung und Trennung der ätherischen Öle durch Destillation" and next emphasized by Koedam [27] in 1982. According to Von Rechenberg: "the compounds vaporize according to their degree of solubility in the distillation water rather than following the order of their boiling point".

In 1910 von Rechenberg gave a very acceptable explanation of the fact that oxygen-containing high-boiling components are isolated more rapidly than lower-boiling substances. In the process of extraction by distillation with water, water at 100 °C will penetrate the plant material. As a result, it becomes necessary for the essential oil components first to dissolve in the aqueous phase and, second, to diffuse through this barrier (hydrodiffusion). Because oxygen-containing compounds dissolve more readily than the more apolar substances, the latter will remain in the plant material for longer. Because of this phenomenon, hydrodistillation is very unattractive technically. It is, furthermore, almost impossible to achieve quantitative extraction of volatile components, for two reasons. The plant material often contains components, mostly triglycerides, with strong affinity for the volatile components. This prevents total separation of the volatiles from the biological material. In addition, the extraction is stopped prematurely at a point when the amount of essential oil extracted per kilogram of steam injected is no longer economically viable.

In his study of the distillation of the volatile seed oil from dill (*Anethum graveolens* L.), Koedam demonstrated the theory of Von Rechenberg by following the relative amounts of limonene (bp 175 °C) and carvone (bp 231 °C) as a function of distillation time. It was observed that carvone distils first, despite its higher boiling point, and then the amount recovered declines with an attendant increase in the amount of limonene (175 °C) recovered. In summary, after distillation for one hour carvone is the major compound, accounting for approximately 90% of the extract, whereas after 16 h the carvone content has decreased to approximately 60% and the limonene content has increased from 10% to 40%.

Similar results have been found for the SFME of seeds and aromatic herbs. After only 30 min for fresh basil, the compounds with the highest boiling point are largely predominant. For HD, in contrast, after distillation for 3 h differences between the concentrations of the compounds are definitely less than for SFME and sometimes the differences are completely reversed. Eugenol is the most abundant component of the SFME extract (43.2%) with linalool second (25.3%) whereas the essential oil obtained by HD is dominated first by linalool (39.1%) and then by eugenol (11.0%). It seems that the phenomenon called "hydrodiffusion" described by Von Rechenberg is more pronounced in microwave extraction by SFME.

22.7.5
Effect of Molecular Polarity

Aromatic plants are usually constituted from cellulose, essential oil, and water. If these three compounds are heated by microwaves at a fixed radiation power and for a set time, the heating rate will be the highest for water, followed by essential oil and cellulose, respectively. One of the interactions of the microwave energy with the matrix is called the dipolar polarization mechanism. A substance can generate heat when irradiated with microwaves if it has a dipole moment, for example that of the water molecule. A dipole is sensitive to external electric fields and will attempt to align itself with the field by rotation.

In the microwave radiation region, the frequency of the changing electromagnetic field is such that the rotating dipoles are unable to realign completely during each cycle. This results in the current and voltage vectors being out of phase, which implies that energy is being dissipated as heat. On a molecular level it can be visualized that the energy is generated from the dipole by molecular friction and collisions, giving rise to dielectric heating. Thus, in the earlier example it becomes clear why cellulose, which lacks the dipole characteristics necessary for microwave dielectric heating, does not heat whereas water or essential oil components, which have large dipole moments, heat readily.

Tab. 22.5. SFME for selective extraction of carvone in the presence of limonene.

	Carvone	Limonene
Structure	(structure)	(structure)
% SFME	64.9	9.7
% HD	52.3	20.2
Boiling point (°C)	231	175
Solubility (g L^{-1})	1.6	0.00042
Dipole moment (D)	24	14
Dielectric constant	2.44	0.75

Essential oils contain a variety of compounds divided into two main groups: hydrocarbons and oxygenated compounds. For garden mint, the essential oil is mainly carvone (oxygenated monoterpene) and limonene (monoterpene hydrocarbon). How does the microwave energy effect differ for these two different aroma compounds? It would be reasonable to believe that the more polar the compound the more readily the microwave irradiation is absorbed, because of the better interaction between the electromagnetic wave and compound, so the oil obtained contains more of the more polar aromatic components. This would seem to correspond well to the observed behavior of carvone (polar compound) and limonene (non polar compound) (Table 22.5).

22.7.6
Proposed Mechanisms

Solvent-free microwave extraction and partitioning of aroma compounds may occur by one of these two proposed extraction mechanisms or by a combination:

- *Mechanism I.* For a sample with high dielectric loss (high water content), efficient extraction can be performed using microwaves only, i.e. without added solvent or water. This is possible because the water inside the sample matrix will be locally heated. Microwaves interact with the free water molecules present in the glands and vascular systems. These systems therefore undergo dramatic expansion, with subsequent rupture of the tissue, enabling flow of the essential oil toward the water layer. This mechanism also depends on the solubility of the essential oil components in water. Solubilization is the limiting step and solubility

becomes the essential property determining the selectivity of microwave SFME extraction.
- *Mechanism II.* Essential oils contain organic compounds that strongly absorb microwave energy. Compounds with high and low dipolar moments could be extracted in different proportions by microwave extraction. Organic compounds with a high dipole moment will interact more vigorously with the microwaves and be extracted more easily than aroma compounds with low dipole moments.

22.8
Conclusions

One of the great success stories of modern chemistry has been the evolution of increasingly more efficient methods for translating knowledge directly into technology and commercial products. Use of microwave technology for extraction of essential oils and natural extracts is a process which has evolved to keep the wheel of development rolling. Microwave extraction makes use of physical and chemical phenomena that are fundamentally different from those which are important in conventional extraction techniques. This novel process can produce essential oil in a concentrated form, free from residual solvents, contaminants, or artifacts. The new systems developed to date indicate that microwave extraction has advantages in terms of yield and selectivity, with better extraction time and essential oil composition, and is also environmentally friendly.

Understanding, on the molecular scale, of processes relevant to microwave extraction has not yet reached the maturity of understanding of other topics in chemistry. Such a challenge is somewhat ambitious and requires a special approach. Microwave extraction interferes with polarization effects that cannot be readily separated from the physical and chemical properties of the extracted molecules. In this chapter we have discussed how the concept of microwave extraction has already become an important issue in the chemistry of natural products. Detailed analysis of past and present literature confirms explicitly the usefulness of this extraction technique. We have hope that this chapter will widen the scope of laboratory and commercial success for the potential applications of microwave technology in essential oil extraction.

References

1 M.D. LUQUE DE CASTRO, M.M. JIMENEZ-CARMONA, V. FERNANDEZ-PEREZ, *Trends Anal. Chem.* **1999**, *18*, 708–716.
2 P. POLLIEN, A. OTT, L.B. FAY, L. MAIGNIAL, A. CHAINTREAU, *Flavour Frag. J.* **1998**, *13*, 413–423.
3 C.M. DIAZ-MAROTO, S.M. PEREZ-COELLO, D.M. CABEZUDO, *J. Chromatogr. A* **2002**, *947*, 23–29.
4 M.M. JIMENEZ-CARMONA, J.L. UBERA, M.D. LUQUE DE CASTRO, *J. Chromatogr. A* **1999**, *855*, 625–632.
5 J.R.J. PARÉ, J.M.R. BÉLANGER, Microwave-assisted Process (MAPTM): principles and applications, Instru-

mental methods in food analysis, J.R.J. PARÉ, J.M.R. BÉLANGER (eds.), Elsevier Sciences BV, Amsterdam, **1997**.
6 E. REVERCHON, *J. of Supercrit. Fluids* **1997**, *10*, 1–37.
7 Y.Y. SHU, S.S. WANG, M. TARDIF, Y. HUANG, *J. Chromatogr. A.* **2003**, *1008*, 1–12.
8 M. VINATORU, M. TOMA, T.J. MASON, *Advances in Sonochemistry* **1999**, *5*, 209–247.
9 S.A. REZZOUG, C. BOUTEKEDJIRET, K. ALLAF, *Journal of Food Engineering* **2005**, *71*, 9–17.
10 M.Z. OZEL, F. GOGUS, A. LEWIS, *Food Chemistry* **2003**, *82*, 381–386.
11 J.L. LUQUE-GARCIA, M.D. LUQUE DE CASTRO, *Talanta* **2004**, *64*, 571–577.
12 C. PONNE, Interaction of electromagnetic energy with vegetable food constituents, Eindhoven, **1996**.
13 A.C. METAXAS, R.J. MEREDITH, Industrial Microwave Heating. Peter Peregrinus Ltd., London, **1983**.
14 E. GUENTHER, The Essential Oils, Van Nostrand Reinhold Company, New York, **1948**.
15 J. FRENCH, The Art of Distillation, Richard Cotes Editions, London, **1651**.
16 Y.R. NAVES, G. MAZUYER, Natural Perfume Materials, Reinhold Publishing Corporation, New York, **1947**.
17 K.D. PERRING, The chemistry of Fragrances, D.H. PYBUS, C.S. SELL (eds.), Royal Society of Chemistry, Cambridge, **1999**.
18 P.B. KAUFMAN, L.J. CSEKE, S. WARBER, J.A. DUKE, H.L. BRIELMANN, Natural Products from Plants, CRC Press LLC, Boca Raton, **1999**.
19 P.J. TEISSEIRE, Chimie des Substances Odorantes, Technique et Documentation Lavoisier, Paris, **1991**.
20 Huiles Essentielles, Monographie relative aux huiles essentielles, Tome 2, Volume 1 et 2, AFNOR, Paris, **2000**.
21 Huiles Essentielles, Echantillonnage et méthodes d'analyse, Tome 1, AFNOR, Paris, **2000**.
22 Pharmacopée Européenne 1 Conseil de l'Europe, Maisonneuve S.A. Editions, Sainte Ruffine, **1996**.
23 Y.R. NAVES, Technologie et Chimie des parfums Naturels, Masson Edit., Paris, **1974**.
24 A. KOEDAM, Capillary Gas Chromatography in Essential Oil Analysis, P. SANDRA, C. BICCHI (eds.), Huethig, New York, **1987**.
25 B. MEYER-WARNOD, *Perfum. Flavor.* **1984**, *9*, 93–103.
26 G. DUGO, A. DI GIACOMO, The genus Citrus, Taylor and Francis Publishing, London, **2002**.
27 N. MARGARIS, A. KOEDAM, D. VOKOU, Aromatic plants: Basic and Applied Aspects. Martinus Nijhoff publishers, The Hague, **1982**.
28 G. DUGO, A. DI GIACOMO, The genus Citrus, Taylor and Francis Publishing, London, **2002**.
29 M.C. MARTINI, M. SEILLER, Actifs et additifs en cosmétologie, Editions Tec and Doc, Paris, **1999**.
30 R.N. GEDYE, F.E. SMITH, K.C. WESTAWAY, H. ALI, L. BALDISERA, L. LABERGE, J. ROUSSEL, Tetrahedron Lett. **1986**, *27*, 279–282.
31 R.J. GIGUERE, T.L. BRAY, S.M. DUNCAN, G. MAJETICH, Tetrahedron Lett. **1986**, *27*, 4945–4948.
32 K. GANZLER, A. SALGO, K. VALKO, *J. Chromatogr.* **1986**, *371*, 299–306.
33 D. LANE, S.W.D. JENKINS, Microwave extraction: a novel sample preparation method for chromatography. Polynuclear Aromatic Hydrocarbons. Chemical Characterisation Carcinogenic 9th International Symposium, **1986**, 437–449.
34 L. PERREUX, A. LOUPY, Tetrahedron, **2001**, *57*, 9199–9223.
35 M. LETTELIER, H. BUDZINSKI, *Analusis* **1999**, *27*, 259–271.
36 J.R.J. PARÉ, Microwave assisted process for extraction and apparatus therefore. *Canadian patent*, CA 2055390, **1992**.
37 J.R.J. PARÉ, M. SIGOUIN, J. LAPOINTE, Extraction de produits naturels assistée par micro-ondes. *European patent*, EP 398798, **1990**.
38 J.R.J. PARÉ, M. SIGOUIN, J. LAPOINTE, Microwave-assisted natural product extraction. *American patent*, US 5 002 784, **1991**.

39 J.R.J. PARÉ, Microwave extraction of volatile oils. *American patent*, US 5 338 557, **1994**.
40 P. MENGAL, B. MOMPON, Procédé et installation d'extraction sans solvant de produits naturels par micro-ondes. *International patent*, WO 94/26853, **1994**.
41 P. MENGAL, B. MOMPON, Procédé et installation d'extraction sans solvant de produits naturels par micro-ondes. *European patent*, EP 698 076 B1, **1996**.
42 E.E. STASHENKO, B.E. JARAMILLO, J.R. MARTINEZ, *J. Chromatogr. A.* **2004**, *1025*, 105–113.
43 E.E. STASHENKO, B.E. JARAMILLO, J.R. MARTINEZ, *J. Chromatogr. A.* **2004**, *1025*, 93–103.
44 A.A. CRAVEIRO, F.J.A. MATOS, J.W. ALENCAR, M.M. PLUMEL, *Flavour and Frag. J.* **1989**, *4*, 43–44.
45 F. CHEMAT, J. SMADJA, M.E. LUCCHESI, Extraction sans solvant assistée par micro-ondes de produits naturels. *European patent*, EP 1 439 218 A1, **2004**.
46 F. CHEMAT, M.E. LUCCHESI, J. SMADJA, Solvent-free microwave extraction of volatile natural substances. *American patent*, US 2004/0187340A1, **2004**.
47 C.S. ESKILSSON, E. BJÖRKLUND, *J. Chromatogr.* **2000**, *902*, 227–250.
48 J.L. LUQUE-GARCIA, M.D. LUQUE DE CASTRO, *Talanta* **2004**, *64*, 571–577.
49 M.J. ALFARO, J.M.R. BÉLANGER, F.C. PADILLA, J.R.J. PARÉ, *Food Research International* **2003**, *36*, 499–504.
50 J. BÉLANGER, L. DEXTRAZE, M.J. ISNARDI, J.C. CHALCHAT, R. Ph. GARRY, G. COLLIN, *Journal of Essential Oil Research* **1997**, *9*, 657–662.
51 F.I. JEAN, G.J. COLLIN, D. LORD, *Parfum. Flavor.* **1992**, *17*, 35–41.
52 P. TOURSEL, *Process* **1997**, *1128*, 38–41.
53 P. MENGAL, D. BEHN, M. BELLIDO GIL, B. MONPON, *Parfums, Cosmétiques, Aromes* **1993**, *114*, 66–67.
54 P. MENGAL, B. MOMPON, Method and plant for solvent-free microwave extraction of natural products. *Canadian patent*, CA 2161127, **1994**.
55 H.-P. LI, G.-C. LI, J.-F. JEN, *J. Chromatogr. A.* **2003**, *1012*, 129–137.
56 A. LATORRE, S. LASCORTE, D. BARCELÒ, M. MONTURY, *J. Chromatogr. A.* **2005**, *1065*, 251–256.
57 W.-H. HO, S.-J. HSIEH, *Anal. Chim. Acta* **2001**, *428*, 111–120.
58 C.-T. YAN, T.-S. SHIH, J.-F. JEN, *Talanta* **2004**, *64*, 650–654.
59 M.E. LUCCHESI, F. CHEMAT, J. SMADJA, *J. Chromatogr. A.* **2004**, *1043*, 323–327.
60 M.E. LUCCHESI, F. CHEMAT, J. SMADJA, *Flavour and Frag. J.* **2004**, *19*, 134–138.
61 J. BERNARD, *Sciences et Vie* **2001**, *214*, 68–73.
62 www.archimex.com
63 M.E. LUCCHESI, F. CHEMAT, J. SMADJA, S. BRADSHAW, W. LOUW, *Journal of Food Engineering* **2005**, under press.
64 J.F. CLEVENGER, *American Perfumer and Essential Oil Review* **1928**, 467–503.
65 S.S. CHEN, M. SPIRO, *Flavour and Frag. J.* **1995**, *10*, 101–112.
66 M. SPIRO, S.S. CHEN, *Flavour and Frag. J.* **1995**, *10*, 259–272.
67 C. VON RECHENBERG, *Theorie der Gewinnung und Trennung der ätherischen öle*. Selbstverlag, Von Schimmel, Berlin, 1910.

Index

a
ab initio calculations   150, 182
absorbers   6
acetalation   583
acetalization   625
acetic acid   803
acetic anhydride   800, 811
acetoacetate   801
acetoacetylation   731, 733, 751
acetonitrite   141
acetophenone   813, 881
acid anhydrides   668
acid-base complexes   36
acidic alcoholysis   782
acidic alumina   395 ff, 795, 797
acidic cleavage   739, 750
acidic supports   339
acrylamide   680
acrylonitrile   654
activated carbon   631
activated complex   628
activation energy   28, 138, 148, 199
activators   904
active site   628
– acidic   628
acyl bromide   806
acyl isothiocyanate   306
acyl sulfonamides   720
3-acylaminoindanones   721
acylation   417, 438, 442, 580 ff, 609, 742, 753, 776 f
– acetylation   580
– anisole   441
– aromatic compounds   438
– benzoylation   581
– dodecanoylation   583
– enzymatic   609
– Friedel-Crafts   196, 438, 625
– naphthalene   441
– of aromatic compounds   417
– peracetylation   580
– pivaloylation   582
– toluene   441
addition of vinylpyrazoles to imine systems   205
adiabatic reaction   63
adsorbents   618
aggregation   903
– peptide sequences   903
agricultural   961
alambic   960
aldehyde   793 ff, 800 ff, 805, 807, 812 ff, 883
aldol condensation   302
aldoximes   547
aliphatic ethers   285
aliquat   283 f, 290, 302 ff, 307 f, 310, 312, 320
aliquat 336   316
alkenes   646
alkyl aryl ketones   882
alkyl dibromides   667
alkyl halides   807
alkyl nitrile   810
N-alkyl phthalimides   295
N-alkyl succinimides   295
N-alkyl acrylamide   656
alkylation   191, 757
– C-alkylation   205
– N-alkylation   205
– allylic alkylation   779
– allylic allylation   779 f
– malonate anion   191
– selective   191
– with dihalogenoalkanes   300
alkylation of
– furfuryl alcohol   300
– methylnaphthalene   628
– phenylacetonitrile   299
– potassium benzoate   190

Index | 987

– pyrazoles   356
N-alkylation of
– 2-halopyridines   178
– 6-amino-2-thiouracil   296
– anilines   626
– saccharin   293
S-alkylation of n-octyl bromide   321
N-alkylations   293
O-alkylations   281, 300
S-alkylations   301, 843
C-alkylations of active methylene groups   297
N-alkylpyridinium salt   806
alkyne   799, 807, 814
Allan-Robinson synthesis   391
alloc protecting group   742
alternative reaction engineering   89
alumina   528, 587 f, 606, 621, 792, 806, 882
alumina catalysts   15
alumina-supported potassium fluoride   300, 310, 365, 378 ff, 397 ff
Amberlite 200C   624
amberlyst-15   622
(AMEBA)-linked polystyrene   744
amidation   748
– of carboxylic acids   162
amide-library   775
amide synthesis   750, 773 ff
amides   294
amidine   804, 808
amine   797, 803, 805, 810, 813 ff
amine scavenging   782
amine spacer   740
aminium salt   902
α-amino acids   298, 815
aminobenzenethiols   813
aminocarbonylations   735 f
aminocoumarins   173
3-aminoimidazoles   804
aminomethyl polystyrene   776
aminomethyl resin   732 f
α-aminophosphonates   815
aminopropenoates   731
aminopropenones   731
2-aminopyrazine   803
2-aminopyridine   803
2-aminopyrimidine   804
2-amino-substituted isoflav-3-enes   392
– *in situ*-generated enamines   392
aminotoluenesulfonamides synthesis   167
aminotriazine   183
ammonia   793 f, 798
ammonium acetate   794, 802 f, 807, 809 f
analytical applications   888
1,6-anhydro-β-D-hexopyranoses   586

aniline   657, 799
anionic β-elimination   189
anisothermal conditions   54
anomeric ratio   122
anthracene   886, 934
anthranilic acid   428 ff
anticancer   797
antimicrobial oxazolidinones   698
apogalanthamine   695
appropriate temperatures   210
aqueous conditions   688 f
aqueous solutions of NaCl   34
arcing   630
Argand diagram   29
aromas   959
aromatic amines   295
aromatic decarboxylation   836
aromatic dehalogenation   823, 833, 840
aromatic esters   283
aromatic ethers   182
aromatic nucleophilic substitution (S_NAr)   156, 181, 309
aromatic plants   960
aromatization of propane   620
2-aroylbenzofurans   394
Arrhenius law   137
N-aryl azacycloalkanes   181
aryl chlorides   687, 691, 704, 710, 712, 719
aryl halide exchange reactions   716
aryl nitriles   707
aryl phosphonates   713
aryl triflates   708, 712, 753
3-aryl-4-hydroxyquinolin-2(1H)-ones   395 f
arylaldehyde   798
arylazetidine   567
aryldiazepinone synthesis   144
5-arylidenerhodanine   811
aryl-nitrogen bond formation   708
aryl-nitrogen coupling   708
N-arylsulfonamide   712
aryl-sulfur bond formation   715
aspartic acid   675
aspartimide formation   908, 925
asymmetric catalysis   779
asymmetric induction   175
asynchronous mechanism   149 f, 156, 207
athermal   52
atom efficient   790
atomic polarization   13
atom-transfer radical polymerization (ATRP)   659 f
attenuation factor   47
automation   965
average moment   20

azaheterocycles 296
azetidinones 607
azides 562, 937
azlactone 811
azo-dyes 797
azolic fungicides 207
azomethine ylides 556, 937, 951

b

Baker-Venkataraman rearrangement 391
Bal-resin 736
barbitone 294
barbituric acid 801
base-catalyzed
– isomerization of allylic aromatic compounds 307
– transesterification 305
base-free ester aminolysis 171
basic alumina 813
basic medium 193
batch reactor 63, 115, 697
Battle of Britian 3
Beckmann rearrangement 339, 370 ff, 380 ff
– aldoximes 381
bed, fluidized 639, 645
bentonite clay 798
benzaldehyde 802, 810
benzaldehyde-supported ionic liquids 343
benzaxazinones 293
benzil 374, 802, 809
benzil diimines 374
benzimidazole 428 ff, 709 f, 761, 804
benzo(b)furans 292
4H-benzo[β]pyrans 810
benzoic acid 661, 805
benzonitrile 802
– oxide 603
2H-benzopyrans (coumarins) 432
– Pechmann reaction 432
benzothiazepines 175, 393 ff, 813
benzothiazinones 293
benzothiazol 431, 715
benzoxazinones 397
benzoylation 196
– of anisole 625
benzoyloxyacetophenone 879
benzylation 339
– 2-pyridone 205
benzylidene cineole derivatives 302
N-benzyl-N-ethylaniline 295
2′-benzyloxyacetophenone 288
bicyclic prolines 768
Biginelli 399 ff, 751, 801
– compounds 778

– condensation 400 ff, 711, 757
– cyclocondensation 758
– dihydropyrimidine synthesis 141
– reaction 119, 800
– scaffold 781
Biltz synthesis 169
biocidal synergism of UV and MW irradiation 890
Bio-Diesel 623
biological fluids 890
biologically active amino ethers 291
biomimetic reaction 125
biotage 75
biradical intermediate 882
biradicals 885
bis(chlorophthalimide)s 668
bis(indolyl)nitroethanes 376
bisamides 814
Bischler-Napieralski 608
bisphenol-A 665, 668, 883
N,O-bis(trimethylsilyl)acetamide (BSA) 585
bmimPF$_6$ 718
Boc (tert-butoxycarbonyl) 368, 900
boiling point 962, 974
Boltzman's law 20
borohydride reductions 834
boron carbide 623
boronic acids 317
^{76}Br 852
Bredereck-type condensation 744
broadening 24
bromination of alkanones 402
– dioxane dibromide 1.1 402
– dioxane dibromide 2.5 402
– silica 402
bromoacetals 309
bromomethoxylation 772
2,3-bis(bromomethyl)-pyrazine 934
p-bromostyrene 660
Brönstedt acid 801, 810
Brownian ion 12
Brownian motion 6, 24
BSA 585
Buchwald-Hartwig amination 739
buflavine 695
Burgess reagent 748, 769
butylphenol 881
4-t-butylphenol 627
butynoate 563

c

C$_{60}$ 931
C$_{70}$ 940
Cadogan synthesis 694

calcium antagonistic agents   800
Cannizzaro reaction   390
canthines   809
capacitor   11
ε-caprolactone   661 f, 679
capture and release strategy   774
carbasugars   593, 599
carbazole   658
2-(9-carbazolylethyl)ethyl methacrylate   658
carbene generation   308
carbene ligands   687 f
carbenes   710
carbohydrates   552, 742
β-carboline alkaloids   809
carbon   416 f
carbon black powder   28
carbon deposition   621
carbon monoxide   704, 720
carbon nanotubes (CNs)   540, 949
carbon-11   821, 842
carbon-carbon bond formation   765
carbon-carbon coupling   737, 764
carbon-nitrogen cross-coupling reaction   739
carbon-sulfur cross-coupling   764
carbon-supported catalysts   632
[carbonyl-^{11}C]amides   845
carbonylation   719
carboxylic acid   814, 883
1,6-bis-(p-carboxyphenoxy)hexane   671
carotenes   938
carvone   973
catalysis   62, 615, 617, 624, 647
– asymmetric   624
– drying   617
– environmental   647
– heterogeneous   615
catalyst   15, 73, 529, 619, 631, 884
catalyst-free   376
catalyst preparation   617
catalytic activity   620
catalytic hydrogen transfer   832 f
catalyzed photoreactions   883
causality principle   16
β-CD resins   731
celite   792
cellobiose   609
cellulose   680, 740
cellulose beads   744
cellulose phosphorylation   681
CEM Cooperation   75
chain termination   288
chalcone   742
– synthesis   740
characteristic time   18

charge space polarization   136
CHD resin   774
chelotropic   809
chemical microwave process   43 f
chemically induced dynamic nuclear
   polarization (CIDNP)   885
ChemMatrix   922
chiral ionic liquids   332
chitosan   679 f
chlorodiazines   311
chloroform   803
o-chlorophenol   889
p-chlorophenol   889
chloroquinoline ring   201
N-chlorosuccinimide   548
chlorotrityl linker   753 f
chocolate   28
cinnamonitrile   554
citrus oil   963
Claisen rearrangement   118, 381, 748 f
– bis(4-allyloxyphenyl)sulfone   381
Claisen-Schmidt condensation   740
Clausius-Mosotti-Lorentz equation   10, 20
Clayan   368 ff, 372ff, 401
claycop   384 ff
claycop-hydrogen peroxide   384
Clayfen   368 ff, 401
clays   15, 527
clean-up   966
cleavage   371 ff
– basic KF-alumina surface   372
– hydrazones   371
– methoxyphenyl methyl (MPM)   372
– montmorillonite K10 clay impregnated with
   ammonium persulfate   371
– phenylhydrazones   371
– semicarbazones   371
– sulfonamides   372
– sulfonates   372
– tetrahydropyranyl (THP) esters   372
cleavage of ethers   443 ff
– acylative   443
click reactions   562
cobalt catalyst   618
coconut reactor   96
coenzyme B$_{12}$ 887
coke formation   621
cold pressing   963
Cole-Cole model   30
Cole-Davidson model   31
collision   14, 19, 24
– inner friction   19
– molecular orientations   19
– resistive couple   19

colloids 41 ff
– charge redistribution 42
– double layer 42
– space charge polarization 43
– surface charge 42
colored sparks 48
combinatorial chemistry 726 f, 791
combined degradation effect of UV and MW radiation 889
combined methodology 839
cominatorial synthesis 400
competitive β-elimination 301
competitive reactions 156
complexes 36
compressed air 968
concept 971
condensation 119, 372, 378 ff
– 1,6-anhydro-β-hexopyranoses 379
– benzoxazinone nucleosides 379
– Bignelli 400 ff, 711, 757
– Claisen-Schmidt 740
– cyclic ethers 379
– Knoevennagel 144, 372 ff, 731, 747, 796
– Ugi-type three-component 119
– quinoline-3,4-dicarbonximide library 378
condensed phases 12, 24, 32
conducting material 27
conduction losses 27
conductive heating 2
conductivity 27, 73
consecutive reactions 63, 789
continous-distillation 964
continuous flow reactor 622 f, 687
continuous microwave reactor 110
controlled living radical polymerization 658
convective motion 44
convective patterns 49
conversion of alkyl alcohols to halides 341
cooking 3 f
– baking 4
– drying process 3
– food applications 4
– freeze-drying 3
– frozen foods 4
– pasta drying 4
– poultry-processing system 4
– sterilization 4
– vacuum drying 4
cooling 635, 638, 640
– microscopic 638
– simultaneous 635, 640
Coolmate system 156
copper catalysts 633

copper sulfate 384
core enabling technology 114
corn starch 680
corroles 397
cosmetic 961
cosmetic ingredients 284
coumarin 374, 477, 536, 628
coupling loops 44
COX-inhibitors 698
cracking 646
cracking catalysts 619
cracking of benzene 633
critical temperature 22
cross coupling 686, 756, 765
– carbon-nitrogen 739
– carbon-sulfur 764
– reaction 734, 738
– Sonogashira 850
– Stille-type 735
crossed-Cannizzaro reaction 390
– paraformaldehyde 390
crown ethers 278
18-crown-6 182, 185, 310, 313
cryptates 278
crystalline polymer 292
crystals lattices 38
cyanide 844
2-cyano benzothiazoles 393
cyanoacetate 808
β-cyanoester 797
cybernetic modeling 44
cyclative cleavage 751 f, 762, 764
cyclic acetals 173
cyclic ketene acetals 309
cyclic β-ketoesters 188, 311
cyclization 424 ff, 744, 752, 760 f, 763, 766, 775
– intramolecular 424
– reactions 199
cycloaddition 148, 206, 418, 423, 524, 602 f, 768, 931
– [3+2] 546
– [4+2] 532
– Alder reaction 423
– Diels-Alder 420
– 1,3-dipolar 148 f, 152, 342, 936, 951 f
– intramolecular 533
– to C70 fullerene 207
– reactions 951
cyclocondensation 199, 757 f, 759
cyclodehydration 769
β-cyclodextrin (CD) 731 f
cyclopropane derivatives 299
cycloreversion 934

d

DABCO 339
DBU 586, 600
– see 1,8-diazabicyclo[5,4,0]undec-7-ene
deacetalation 588
deacylation 366, 587
– debenzoylation 587
– depivaloylation 587
dealkoxycarbonylation 188
– of activated esters 311
dealkylation reaction 638
– simultaneous 638
5-deaza-5,8 dihydropterin 795
Debye's model 15, 18
decarbonization 621
decarboxylation 444 ff
– barium 445
– calcium 445
– magnesium 445
decomplexation 433
– metal complexes 433
decomposition of chloromethane 634
decomposition of hydrogen sulfide 637
decomposition of olefins 633
dehalogenation 838
dehydrating agent 769
dehydration 621
dehydrogenation 633
– of ethylbenzene 620
delta/μ agonists 697
demethylation 185 f, 313
deoximation 369 ff
– silica-supported ammonium persulfate 370
– silica-supported periodate 370
– zeolite-supported thallium(III) nitrate 370
deprotection 364, 366 ff, 586, 742 f
– aldehyde diacetates 366
– allyl esters 203, 367
– t-butyldimethylsilyl (TBDMS) 368
– carboxylic esters 367
– cleavage 366 ff
– deacylation 366, 586
– 9-fluorenylmethoxycarbonyl (Fmoc) 367, 901
– N-tert-butoxycarbonyl groups 368
– deacetalation 586
– desilylation 586
– trimethylsilyl ether 368
deprotection reaction 925
deprotection reagent 910, 912
deprotonations 307
design 971
desilylation 368, 588
– τ-butyldimethylsilyl (TBDMS) 368

desorption 621
dethioacetalization 368
– ketals 368
– thioacetals 368
dethiocarbonylation 371
detritiation 841
deuteriation 389, 830, 832
deuterium 821, 828, 830
deuterium labeling 597
DFMBA (N,N-diethyl-α,α-difluoro-(m-methyl-benzyl)amine) 596
DHP (dihydropyridine) 793 f
3,4-dialkoxybenzaldehydes 303
dialkylation 186
4,4'-diaminodiphenylmethane 664
4,4'-diaminodiphenyl sulfone 664
dianhydrohexitols 186 f, 286, 301
1,8-diazabicyclo[5,4,0]undec-7-ene (DBU) 586
diazoketones 568
diazomethanes 937
2,5-dibromopyrazine 677
1,1-dichloro-2,2-bis(4-hydroxyphenyl) ethylene 666
dichloronorcarane 308
2,4-dichlorophenoxyacetic acid 883 f
2,6-dicyanoanilines 812
1,3-dicyanobenzene 886
dielectric 11, 529
– capacitor 11
– constant 10, 67, 127
– frequency 26
– induction current 11
– loss 16, 45, 67, 127
– loss tangent 67
– permittivity 9, 12, 26
– temperature 26
– vacuum capacitance 11
– water 22
Diels-Alder 144, 420, 524, 604, 809, 951
– [2+2] 524
– [2+2+1] 524
– [3+2] 524
– [4+2] 524
– [6+4]–[4+2] 524
– [6+4] 524
– reaction 206, 355, 748, 755
– reaction of tetrazines 420
– reagents 421
Diels-Alder cycloaddition 129, 148, 150, 157, 418 ff, 524, 754, 768, 932
Diels-Alder cycloadducts 783
– reaction 206, 355, 748, 755
diene 524
– scavenger 768

dienophile 524, 531
– scavenger 783
diethers from dianhydrohexitols 286
diethyl fumarate 552
diethyl phosphate 815
N,N-diethyl-α,α-difluoro-(m-methylbenzyl)amine (DFMBA) 596
diethylacetylenedicarboxylate 541
differnetial scanning calorimetry (DSC) 164
diffusion 621
digestion 109, 888
diglycidyl ether of bisphenol-A 664 f
dihydrobenzothiazepines 394 ff
dihydroisoxazoles 549
dihydropyridines 793 f, 795
1,4-dihydropyridines 321
dihydropyridones 798
3,4-dihydropyrimidin-2(1H)-ones 399
dihydropyrimidin-2-thione 716
dihydropyrimidine 740, 757, 778, 781
dihydropyrimidinones 751, 778, 800
dihydropyrimidones 711
N,N-(diisopropyl)aminoethylpolystyrene 580
diketone 803
β-diketone 801
1,2-diketone 719
1,3-diketone 793, 798
dimedone 810
dimensional resonances 48
– dielectric resonators 48
– spherical shapes it 48
dimethyl carbonate 586
5,5-dimethyl-1-pyrrolidine-N-oxide 883
N,N-dimethylacetamide (DMA) 317
N,N-dimethyl-6-aminouracil 799
dimethylacetylenedicarboxylate 541
2-(dimethylamino)ethyl ester 658
N,N-dimethylbarbituric acid 811
N,N-dimethylformamide diethylacetal (DMFDEA) 731
dimethylformamide dimethylacetal 798
2,5-dimethylphenacyl chromophore 881
2,6-dinitroaniline 950
diphenylbutadiene 544
diphenylhydrazonoyl 560
diphenylnitrilimine 560 f
α,N-diphenylnitrones 552
1,3-dipolar cycloaddition 148f, 152, 342, 936, 951f
– of diphenylnitrilimine 312
dipolar intermediates 154
dipolar moment 14
dipolar polarization 138
dipolar transition states 154, 180, 200
dipolarophiles 528
dipole moment 9, 38, 139, 141, 150, 544
dipole rotation 127
dipole-dipole-type interactions 135, 148
1,3-dipoles 528, 549
discharge 630
discharge in EDL 862
disinfection 890
displacement 744
dissolved organic carbon 888
distillation 960
distortion polarization 17
distribution function 20
diterpene 962
dithiazoles 430 ff
DMA (N,N-dimethylacetamide) 317
DMF (dimethylformamide) 141, 147, 167, 297, 306, 317, 584, 600, 704
DMSO (dimethylsufoxid) 136, 141, 195, 311
DNA synthesis 742
domestic microwave (MW) oven 108, 282, 294, 713, 792–816
domestic oven 68, 135, 285, 304
domino reactions 789
donor-acceptor systems 939
doping agents 526, 533
dry distillation 964
dry media 146, 280, 527
dry media reactions 108
dynamic consequences 24

e

eco-friendly approach 134
EDL (electrodeless discharge lamp) 861 ff, 888 f
– applications 873f
– beaker-shaped 888
– dimensions and properties 864
– emission characteristics 866 f
– envelope 863
– fill material 864
– inert gas 863
– manufacture and performance testing 865
– microwave discharge 863
– nature and characteristics of the EM energy-coupling device 864
– temperature of the lamp 864
– UV digestion reactor 873f
EDTA (ethylen diamine tetra autate) 882
effects on selectivity 156, 204
electric field 71
electric field effects 46
– perturbation 46
– reactor-to-applicator volume ratio 46

electric field-focusing effects 48
electrodeless discharge lamp (EDL) 861
electrodeless MW-UV lamp 888
electrolytes 33
electromagnetic energy 7
electromagnetic spectrum 5
electron spin resonance spectroscopy 883
electron-nuclear hyperfine interactions 885
electron-rich olefins 719
electron-transfer processes 938
electronic polarizability 13
electrophilic aromatic substitution 120, 198
electrophilic fluoridation 846
electrostatic polar effects 135
elemental sulfur 815
α-elimination 308
β-elimination 153, 189, 309
– competitive 301
elimination reaction 402
– hydroxypyrrolidines 402
– pyrrolines 402
Emrys™ 84
enamines 373 ff, 810
β-enamino ketone 375
enantioselective reactions 735
enantioselectivity 624
ene reaction 423 ff
energy 7 f, 73
– Born-Oppenheimer approximation 7
– electronic 7
– Euler angle 8
– excited states 8
– Hamiltonian operators 7
– of electric 7
– one-photon process 8
– partition 7
– rational 7
– vibrational 7
energy barrier 52
energy-consuming 975
energy efficiency 73, 687
energy profiles 48
energy transfer 297
enolization 747
enthalpy of activation 52
enthalpy of formation 139
entropy of activation 52
envirocat reagent 374
environmental remediation 883
environmentally benign media 123
environmentally relevant studies 888 f
environment-friendly 881
enzymatic reactions 608
enzyme-catalyzed 405

– novozym (*candida antarctica*) 405
– *pseudomonas* 405
– *pyrococcus furiosus* 405
– *sulfolobus solfataricus* 405
epichlorhydrin 291, 668
epimerization 699
epoxidation 634
epoxide 606
epoxide ring-opening 300
– epoxyisophorone 299
epoxy resin 663, 665
3,4-epoxycyclohexylcarboxylate 666
equilibrium 63, 620
– nonthermodynamic 620
Erlenmeyer reaction 166
erythropoietin (EPO) 923
Eschweiler-Clarke reaction 835, 845
essential oils 959
ester aminolysis 305
– base free 171
– basic medium 193
ester saponification 303
ester synthesis 285
esterification 141, 437, 729, 733 f, 759, 773, 776 f
– acetic acid 622, 628
– fusel oil 172
– stearic acid 437, 622
esters 375
ester-SWNT 953
ethanol 27
etherification 183
– of heterocyclic compounds 310
Ethos 81, 590
ethoxycarbonylnitrile oxide 550
ethyl acrylate 626
ethyl propiolate 552, 797, 806
ethylene 539
eugenol 307
eugenol-isoeugenol 380
exothermal processes 108
exploratory power 790 f
extraction 959
– concept and design 971
– essential oils 959
– herbs 959
extraction step 970
extraction temperature 974
extraction time 970

f

fatty acids 167
fatty compounds 193
FC acylations 444 ff

FDMP resin 733 f
Ferrier rearrangement 381, 591, 599
ferrites 29
ferrocene 938
field homogeneity 72
field penetration 48
Fischer glycosylation 590
Fischer-Helferich synthesis 121
Fischer indole synthesis 126, 396
five-membered nitrogen heterocycles 296
flash heating 117, 718, 721
flavones 391 f
– o-hydroxydibenzoylmethanes 392
– K10 clay 392
flavour industry 961
fluorene 675
9-fluorenylmethoxycarbonyl (Fmoc) 901
fluorinated dipolarophiles 149
[^{18}F]fluorination 848
fluorinations 311
fluorine-18 821, 842
[^{18}F]fluoroethyl tosylate 848
[^{18}F]β-fluoroethyl tosylate 850
fluorous chemistry 692, 701
fluorous ligands 767
fluorous-phase organic synthesis 762 f
fluorous solid-phase extraction (F-SPE) 764, 767, 805, 808
fluorous tag 763 ff
Fmoc (9-fluorenylmethoxycarbonyl) 367, 901
N-Fmoc deprotection 907
focused microwaves 129
foodstuffs 36
β-formyl enamides 569
free charges 9
free energy 52
free-radical polymerization 654
freezing 15, 35
frequency 6
Friedel-Crafts acylation 196, 438, 444, 625
fructose 605
fullerene 149, 526, 951
fullerodendrimers 944
fulvenes 570
fumigation 960
functionalization 729 ff
functionalization of SWNTs by covalent bonding 950
functionalized acetates 297
functionalized tartramides 165
furanic diethers 286

g
^{68}Gallium 853
Gabriel amine synthesis 294
galactal 592
galacto-oligosaccharides (GOS) 610
galactosamine 591
galactose 581, 589, 591, 595, 605
gas-phase reactions 628
gel surfaces 882
gel-type resin 732
geometrical shape 51
germanium 435
– tetrabromide 435
germanium tetrahalide 434
Gewald synthesis 746 f
Glarum's generalization 31
glucat 592
glucosamine 581, 589, 591
glucose 580, 590 f, 605
glycals 593, 599 f, 603
glycine 811
glycodendrimers 603
glycosyl nitrones 556
glycosylation 122, 589
– enzymatic 609
C-glycosylation 594
N-glycosylation 594
O-glycosylation 589, 591
– Ferrier's rearrangement 591
– Fischer glycosylation 589
Goldberg reaction 711
GOS (galacto-oligosaccharides) 610
Graft polymerization 679
gram 86
graphite 138, 160, 379, 417, 421 ff, 425 f, 437 ff, 530, 792
– acylation 439 ff
– adsorption of organic molecules 449
– Carbon-Carbon bond 417
– catalytic activity 417, 437, 441 f
– catalytic effect 440, 442
– oxidation of propan-2-ol 425
– reagents 421
– retention of reagents 421
– safety measures 417, 447
– sensitizer 418 ff, 437
– synthetic commerical 438
– temperature measurement 448
– thermal reactions 427
– typical procedures 447
green chemistry 64, 134, 145, 210, 310, 328, 492, 790, 965
greener solvents 405
Grignard reagent 704, 845
ground state 139
Grubbs catalyst 204, 779
guar gum 680
guar-g-polyacrylamide 680

h

α-halogenation 805
halogenation 595 ff
– bromination 595
– chlorination 595
– fluorination 596
– iodination 597
halogermanes 434
3-halopyridines 182
Hammond postulate 155
Hantzsch 798, 805
Hantzsch MCR 795
Hantzsch reaction 129, 321, 399
HBTU 770
head note 966
headspace 969
heat captor 622
heating 635
– aids 354
– conversation 9
– effect 67
– rapid 635
– selective 635
– volumetric 635
heating rate 45, 50
– specific heat 45
– thermal 45
– thermal diffusivity 45
Heck reaction 316, 337, 403, 717, 767
– arylation 647
– coupling 129, 780 f
– vinylations 766
Henry reaction 375
hepatitis C virus NS3 protease inhibitor 720
hetero Diels-Alder (IMHDA) 356, 533, 935
heterocycle synthesis 731, 746
S-heterocycle-containing liquid crystalline targets 393
heterocycles 602
heterocyclic 393 ff, 528
– bridgehead thiazoles 393
– chemistry 456
– compounds 391, 693
heterogeneous reaction 64
heterogeneous catalytic degradation 884
heterogeneous kinetics 56
heterogeneous media 146
heterogeneous photocatalytic oxidation 874
– flow-through quartz photoreactor 874
– TiO_2 874
– TiO_2–ZrO_2 874
heteropolyacids 623
Hickman 49
high-frequency 9

high-temperature water 123, 881
high-throughput synthesis 728
history of microwaves 2
HIV-1 protease inhibitors 698 f, 705
Hiyama coupling 318
Hiyama reaction 705
homogeneous 64
homogeneous catalysis 161
hot spot 56, 90, 630 f, 634 ff
– macroscopic 637
– microscopic 637
Hughes-Ingold theory 146
humic acid 884, 889
humic substances 888
hybrid polymer 730
hydantoins 762, 765
hydrazone synthesis 162
hydrazones 376, 942, 947
– azomethine imine tautomerization 942
– carbonyl compounds 376
– hydrazines 376
hydroacylation of 1-alkenes 161
hydrocarbon 646
hydrocarbon oxidation 628, 634
hydrochalcogenation 390
hydrocracking 620, 633
hydrodechlorination 633
– of chlorobenzene 619 f
hydro-distillation 964
hydrodynamic 44
hydrodynamic aspects 49
hydrogen bonding 37
hydrogen isotope exchange 822 f, 838, 841
– base-catalyzed 823
– metal catalyzed 823
hydrogen peroxide 319, 889
hydrogen peroxide-assisted photochemical degradation 882
hydrogen peroxide oxidation 626
hydrogen peroxide urea adduct (UHP) 320
hydrogenation 333, 594, 623 f, 633, 823, 832, 838, 840
– catalytic transfer hydrogenation 594
hydrogenation of alkenes 633
hydrogenation of benzene 619, 628
hydrogenation of nitrobenzene 623
hydrogenation reaction 404
– $H_2/D_2/T_2$ 404
hydrogen-bond 33
hydrogen-isotope exchange 827 f
– acid catalyzed 827
– metal-catalyzed 827 f
hydrogenolysis 624, 632 f
hydrolysis of sucrose 624
hydrostannylation 700, 702

hydroxide ion 27
4-hydroxy-2H-1-benzopyran-2-one 795
6-hydroxyaminouracils 800
o-hydroxybenzyl alcohol 936
hydroxyl radicals 883
hydroxylamine hydrochloride-clay 400
hydroxylation of benzene 627
8-hydroxyquinolines 290
Hyflo Super Gel 590
hyperfine coupling 885 f

i

imaginary parts 9
imidazol[1,2-α-]annulated N-heterocygles 399
– imidazol[1,2-α]pyrazines 399, 803
– imidazol[1,2-α]pyridines 399, 803 f
– imidazol[1,2-a]pyrimidines 399
imidazole 626, 802 f
imidazole synthesis 746
imidazolines 167
imidization 138, 198
imine 159, 373 ff, 567, 813, 815
iminium ion 803
impregnated reagents 146
indium trichloride 815
indole 428 ff, 809
indole formation 739 f
indolizines 806
induced selectivity 54
Industrial Scientific and Medical frequencies 6
inhibitors of HIV-1 protease 797
inorganic solid supports 321
inorganic supports 809
input 44
instrumentation 925
insulating materials 9
interactions with matter 877
– microwave 877
– ultraviolet 877
– visible radiation 877
interfacial reactions 146
interfacial relaxation 40
intermolecular distances 14
intermolecular interactions 36
intersystem crossing 885
intramolecular cyclization 316
intramolecular cycloaddition 533
intramolecular Diels-Alder (IMDA) 533
intramolecular Heck reactions 317
intramolecular Michael additions 202
intramolecular nucleophilic aromatic substitution 201

iodine-alumina 399
ionic dissociation 194
ionic liquid 180, 281, 327, 405 f, 526, 628, 718, 771, 779, 829
– 1-alkyl-3-methylimidazolium tetrachloroindate(III) 406
– catalysts 628
– cyclic carbonates 406
– 3-methylimidazolium tetrachloro-gallate 406
ionic product 124
ion-pair 36, 185
– loose 185
– tight 185
ion-pair dissociation 139
ion-pair exchange 280
ipso-fluoro displacement 759 f
Irori Kan 691
irreversible 63
isatin 813
Isay condensation 605
ISM band 6
isocyanate resins 732
isocyanide 804, 814
isoflav-3-enes 810
isoflavones 695
isoidide 286 f, 667
isomannide 186, 286
isomeric inversion 55
isomerization 379, 633
– of 2-methylpentene 621
– of safrole 142, 308
isonitrile 805, 808
isophorone 300
isopolar transition-state reactions 148
isoprene 655
isosorbide 186 f, 286 f, 301, 667 f
isothermal 63
isotopes 820
isoxazole 530, 549, 744
isoxazoline 603
isoxazolino[60]fullerenes 941
isoxazolino-SWNT derivative 953

j

JandaJel resin 748
jasminaldehyde 302

k

K10 clay 202
Kabachnik-Fields reaction 349
Kenner's safety catch principle 749
ketene acetal 189, 528
ketenes 567

β-ketoester 793 f, 797 f, 800, 805
ketone 798, 805 f, 812, 814 f
Kindler 815
kinetic conditions 52
kinetic control 156, 887
kinetic effects 43
kinetic products 120
kinetics measurements 113
Knoevenagel 400 ff
– condensation 144, 372 ff, 731, 747, 796
– hetero Diels-Alder reactions 303, 343, 478, 748
Krapcho reaction 188, 311
KSF clay 364
Kumada reaction 704

l

labeled compounds 820
^{14}C-labeled esters 191, 298
labeled formates 833
laboratory-scale 721
lactam 294, 536
β-lactam 567, 607
– stereo control 206
γ-lactam formation 920
D,L-lactide 661 f
lactones 536
lactose 590, 610
Langevin s function 14
lanthanide triflates 815
large-scale 704
lattice defects 39 f
– adsorbed phases 40
– color centers 40
– interstitial ions 39
– substitutional 39
laurydone 97
Lawesson's reagent 168, 393 ff
leaving group 179, 180, 198, 205
Leuckart reductive amination 211
Lewis acids 538
library 352
limonene 972
linked charges 9
liquid-liquid PTC 278
liquid phase synthesis 756, 758 f, 760 f
localized high temperatures 138
long-chain esters 283
long-chain halides 281, 290
loose ion pairs 141, 154, 184, 189
lossy dielectric material 48
low-bolling solvents 112
luotonine A 430

m

macromolecules 37 f
– atactic 38
– dipole moment 38
– isotatic 38
– peptide 38
– poly(vinyl acetate)s 37
– poly(vinyl chloride)s 37
– polyacrylates 37
– polyethylene 37
– polypeptides 38
– polystyrene 37
– polytetrafluoroethylene 37
macroscopic description 17
macroscopic theory 16
magnetic field 886
magnetic field effect, (MFE) 885
magnetic losses 29
magnetic susceptibility 9
magnetron 2, 69
magnetron power 69
magtrieve 138
MALDITOF mass spectrometry 287
maleic acid 675
maleic anhydride 544, 657, 664 f, 668
maleimide derivatives 891
malonic alkylation 191
malononitrile 808, 811
maltose 590
Mannich reaction 119, 754
mannose 581, 589 ff
MAOS (microwave-assisted organic synthesis) 792 f, 798, 816
p38 MAP kinase inhibitors 709 f
Marangoni 49
Marfey's reagent 916
Marshall resin 750
mastery 43
Maxwell's equations 4
Maxwell-Wagner effect 40
MCR 790 ff, 797, 801, 803, 809, 811, 815 f
MDAD 530
mechanical stirring 75
melatonin derivatives 702
Meldrum's acid 798, 801
mercaptoacetic acid 813
3-mercaptopropionic acid 656
Merrifield resin 376, 730 f, 737, 752, 755, 772, 776, 900
mesoporous inorganic solids 792
metal-catalyzed 554
metal oxide 629
metal oxide catalyst 632
metallophthalocyanines 170

methacrylamide 657
methacrylic acid 656 ff
2-methallyloxyphenols 289
methane combustion 620
methyl β-aminocrotonate 793 f
methyl acrylate 556
methyl iodide 823, 835, 844
[^{11}C] methyl iodide 843
methyl methacrylate 654 f, 658 ff
methyl triflate 835
methylation 586, 844
methylation reactions 835
methylenation of 3,4-dihydroxybenzahldehyde 289
N-(methylpolystyrene)-4-(methylamino) pyridine (PDS-DMAP) 580, 583 f
Meyer-Schuster acidic rearrangement 380
Michael addition 118, 157, 174, 356, 392, 400 ff, 626, 811
Michael-addition reactions 377 f
– 1,4-diketones 378
– diethyl ethoxymethylenemalonate (EMME) 377
– α,β-unsaturated ketones 377
Michael condensation 315
Michael reaction 293
microwave 5, 362, 788, 791, 906
– household kitchen MW oven 362
– tools 906
microwave 793 f, 796 ff, 801 f, 804 ff, 812, 814 ff, 861
– absorption 128
– apparatus 417
– applicator 57, 71
– batch reactors 111
– cavity 68
– clevenger 971
– dielectric heating 135
– effect 102, 627
– energy 128
– -enhanced coupling 913
– extract 966
– frequency 905
– oven 3
– oven cascade 90
– peptide synthesizer 728
– photochemistry 91
– photon 6
– power 73
– technology 130
– thermolysis 888
– -ultrasound 92
microwave-assisted 524
– extraction 93

– nucleophilic substitution 509
– organic chemistry 109
– organic synthesis (MAOS) 64, 529, 792
– reactions 456
– resin cleavage 749
microwave heating 127
– properties 127
microwave irradiation 363
– solvent-free methods 363
microwave-mediated condensation 464
milestone (MLS) 75, 641
mineral support surface 379 ff
mineral supports 527
mineralization 883
Mitsunobu 752
MnO$_2$ catalyst 631
modification of chemoselectivity and regioselectivity 233, 236, 241, 247, 253, 257
– cycloaddition reactions 247
– electrophilic aromatic substitution 236
– miscellaneous 257
– polymerization 253
– protection and deprotection of alcohols 233
– synthesis and reactivity of heterocyclic compounds 241
modification of stereo and enantioselectivity 264
molecular behavior 29
molecular-distillation 964
molecular reorientation 13
molecular sieves 621
molten salts 328
molybdenum-catalyzed alkylations 735
molybdenum hexacarbonyl 720, 735, 767
monomode MW 308, 525, 792 f, 808, 816
monomode MW cavity 291, 815
monomode MW reactor 134, 178, 287, 304, 795, 798, 800 ff, 806, 808, 810
montmorillonite 424, 432, 584, 588, 592, 622, 627
montmorillonite [67] 626
montmorillonite clay 799
montmorillonite K10 clay 159, 199, 203, 364, 374, 395 ff, 792, 797, 803
montmorillonite KSF clay 175, 179, 815
MORE 592
Mukaiyama reagent 776 ff
multicomponent reaction 398, 745, 751, 766, 788 f, 791
multicomponent synthesis 349
multimode 68, 792 f
multimode microwave ovens 816

n

N-alkylation 180
naphthalene 645
natural gas 629
NCS 548
neat conditions 797
neat reactants 528
neat reagents 146, 528
Negishi couplings 738
Negishi reaction 703 f
new reactors 112
Nicholas reaction 737
nickel catalyst 618, 631 ff
Niementowski 428, 430 ff
Niementowski reaction 160
nitration of styrenes 402
nitric acid 950
nitrile imines 556, 941, 943
nitrile oxide 547, 937, 941, 952
nitrile sulfides 551
nitriles 400, 566, 798, 800
nitroalkanes 805 f
nitroalkenes 373 ff, 549
4-nitroanisole 880
nitrocyclohexanols 176
nitrogen-containing heterocycles 395, 793
– 2′-aminochalcones 395
– 1,2,3,4-tetrahydro-4-quinolones 395
nitrones 149, 602
β-nitrostyrenes 402, 557
NMP 138, 142, 198
^{3}H NMR spectroscopy 824, 833
no-carrier-added 845 f
nonlinear feedback 43
nonmetallic oxidants 385
– alumina impregnated with Iodobenzene diacetate (IBD) 385
– bis(trifluoroacetoxy)iodobenzene (BTI) 385
– Dess-Martin periodinane 385
– iodobenzene diacetate (IBD) 385
nonpolar 526
nonpolar organic solvent 279
nonpolar solvents 143, 210
nonthermal activation 74
nonthermal microwave effects 625, 627, 635, 885
– specific 335
NO_x decomposition 619
nucleation regulator 137
nucleophilic addition to carbonyl compounds 153, 159, 192, 302
nucleophilic aromatic substitution (S_NAr) 181, 761, 880
– mechanism 157
– reactions 158
nucleophilic fluoridation 846
nucleophilic substitution (S_N2 reaction) 153, 176, 198, 731, 744
– tetralkylammonium salb 198
C-nucleoside 605
C-nucleoside synthesis 297

o

octupole moment 11
n-octyl acrylate 660
OH radicals 888 f
olefin metathesis 334
olfactory notes 972
oligomerization of methane 631
oligophenylenevinylenes 944
open vessels 686
optical limiting material 948
organic synthesis 881
organoboron reagents 318
organometallic reaction 403
organophosphate compounds 888
organozinc reagents 703
orienting effect 13
ortho-hydroxychalcones 202
ortholakylation of ketimines 161
ortho-quinodimethanes 934
oscillating magnetic field effect (OMFE) 886
output 44
oxadiazole synthesis 769
1,2,4-oxadiazoles 770 f
1,3,4-oxadiazoles 769 f
oxazole synthesis 748
oxazolidines 552
oxazolidinones 736
oxazoline 662
2-oxazolines 165
oxidation 319, 382, 384 ff, 595, 950
– activated manganese dioxide-silica 382
– arenes 386
– chromium trioxide-wet alumina 382
– manganese dioxide on bentonite clay 382
– claycop-hydrogen peroxide 384
– Copper sulfate 384
– enamines 386
– [hydroxyl(tosyloxy)iodo]benzene (HTIB) 387
– iron(III) chloride 387
– α-ketoesters 387
– iodobenzene diacetate-alumina 386
– of alcohol 382
– of methane 631
– of toluene 633
– oxon-alumina 384

- potassium permanganate-alumina   386
- sodium periodate-silica   385 f
- sulfides   382, 386
- sulfones   385
- sulfoxides   385
oxidation catalyst   620
oxidation with Clayfen   383 ff
oxidative coupling of methane   629
oximes   941
oxindoles   711 f
oxone-alumina   384
ozone treatment   890
ozonolysis   753

p

Paal-Knorr cyclization   200
palladacycle   337, 719
palladium acetate   317
palladium catalyst   623
palladium-catalyzed cyanations   780
- synthesis of diaryl acetylenes   319
palladium complexes   403
palladium compounds   316
palladium on carbon   338
Panasonic   644
parallel procedures   838
parallel reactions   63, 826
parallel synthesis   728, 747, 750, 776
partial oxidation of methane   629
pathogens   888
PCC (pyridinium chlorochromate)   595
Pd (Palladium)   619
Pd(II)   553
PDS-DMAP (N-(methylpolystyrene)-4-(methyl-amino)pyridine   580
Pechmann approach   373
Pechmann reaction   173
PEG (Polyethylenglucol)   319, 680, 689 f, 730, 756, 759, 801, 812, 814
PEG Support   757 f
PEGylated Merrifield resin   732
PEGylation   730 f
PEG 400   317, 321
penetration   47
penetration depth   48, 73, 646
pentafluorobenzaldehyde   799
peptide sequences   922
peptide synthesizer   906 f
- automated   907
- manual   908
perbromide resin   772
perfluorohexylethene   880
perfumery   960
pericyclic reactions   148, 209

PET (positron emission tomography) radiochemistry   842 f
pharmaceutical   961
phase lag   16
phase-transfer (PT)   530
phase-transfer agent   185, 194, 279
phase-transfer catalysis (PTC)   182, 186, 191, 278, 668
phase-transfer catalysis conditions   188 ff, 667
phase-transfer catalyst   194, 756
phenacyl benzoate   881
phenacylation of 1,2,4-triazole   147
phenethylation of pyrazole   176
phenol oxidation   620
phenolic ethers   288
phenolic polyethers   292
phenols   883
phenyl acetaldehyde   810
p-phenylene   676
m-phenylenediamine   664
phenyl thioureas   307
2-phenyl-2-oxazoline   662
phenylglyoxylic acid   808
(R)-N-(1-phenylethyl) methacrylamide   656
(R)-1-phenylethylamine   656 f
N-phenylmaleinimide   657
N-phenylpyrrolidino(60)fullerene   295
phosgenation   339
phosphine alkylations   152
phosphonium salt   178, 902 f
photocatalysis   883
photocatalytic activity   884
photochemical oxidation   889
photochemical reactions   860 ff
photochemical reactor   869 ff
- domestic microwave oven   871
- flow reactor   869
photochemistry   860 ff, 889
photocleavable linkers   733
photoinitiation   861
photolithography   891
photolysis   734
photolytic cleavage   734
photoresist   891
phthalic anhydride   668
phthalimide derivatives   294
phthalimide synthesis   752
phthalocyanines   200, 938
physical constants   975
Pictet-Spengler   608
Pictet-Spengler reactions   350
pinacol-pinacolone rearrangement   380
piperazine   479
piperazine resin   776

Planck's law   135
plasma   620, 630 ff
plasmepsin I and II inhibitors   699 f
Platinium (II) and (IV)   553
platinium (Pt) catalyst   621
platinum dioxide   831
polar groups   38
polar intermediates   164, 816
polar mechanisms   210
polar molecules   9, 525
polar solvents   141
polar transition state   169
polarity enhancement   154
polarity of the solvent   144
polarization   9, 12
pollutants   888
polyacrylonitrile   680
polyamides   671
polyanhydrides   671
poly(alkylene hydrogen phosphonate)s   678
poly(amic acid)   138, 198, 674
poly(aspartic acid)   675
polydispersity index   287
poly(dichlorophenylene oxide)   678
poly(ε-caprolactone)   661
polyesters   287, 667 ff
polyheterocyclic systems   485, 515
poly(ether imide)s   668, 670 ff
poly(ethylene glycol) (PEG)   319, 680, 689, 730, 756, 759, 801, 812, 814
poly(ethylene oxide)   681
polyimides   671 f
polyisoprene   655
polymer chains   288
polymer resins   729
polymer supported catalysts   690, 700, 707, 729
polymerbound enones   731
polymer-bound quinones   733
polymerization   163, 186, 206, 891, 948 f
polymer-supported   830, 834
polymer-supported PTC catalyst   308
polymer-supported reagents (PSR)   769
polymer-supported thionating reagent   771
polymer-supported triphenylphosphine   753, 771
poly(methyl methacrylate)   654
poly(pyrazine-2,5-diyl)   677
poly(vinyl acetate)   654
poly(4-vinylpyridinium-p-toluenesulfonate) (PPTS)   588
poly-N,N-dimethylacrylamide   655
poly-N-[3-(dimentylamino)propyl]acrylamide   655

poly-N-isopropylacrylamide   655
polyphosphoric acid (PPA)   757, 801
polysaccharides   601
polystyrene   654
polythioureas   674
polytropic   63
polyureas   674
porphyrins   938
position of the transition state   155, 284
positively charged reactants   196
positron emission tomography (PET)   842
positron emitters   821, 842
potassium carbonate   296, 316
power density   45
power penetration   48
PPTS (poly(4-vinylpyridinium-p-toluene sulfonate))   588
pre-exponential factor   137
pressure   22
pressure vessels   109
pressurized conditions   716
pressurized microwave reactors   116
– advantages   116
pressurized vessel   881
primary amines   802
primary bacterial photochemistry   886
process intensification   62, 407
processor   44
profiles of temperature   134, 165
prolabo   77, 590
proline   559
propagation constant   47
– phase factor   47
– attenuation factor   47
propanedinitrile   812
propargylamines   814
O-propargylic salicylaldehydes   556
prophyrin   647
protected carbohydrates   305
protection   364 f, 580
– formation of acetals and dioxolanes   364
– N-alkylation reactions   365
– thioacetals   365
proton transfer   36
PS-DIEA (N,N-(diisopropyl)aminoethyl-polystyrene)   580
PS-PEG-HMDI polymer   731
Pt (Platinium) catalyst   621
PTC (phase-transfer catalysis) conditions   188ff, 667
pterins   605
pulsar system   99
purification of nanotubes   949
purine synthesis   745

purines 744 f
– derivatives 296
pyran ring 811
pyranol[2,3-d]pyrimidine 810 f
pyrazine-containing cycloadducts 934
2(1H)pyrazinone 695, 702
pyrazole 177, 190, 201, 608, 744
– alkylation 190
pyrazolines 561
pyrazolino[60]fullerenes 942, 947 f
pyrazolo(3,4-β)quinolines 376
pyrazolo(3,4-χ)pyrazoles 376
pyrazolopyridines 795, 797
pyrazolopyrimidines 694
pyridine 754 f, 798 f, 806
pyridine halides 311
pyridine synthesis 758
pyridinium chlorochromate (PCC) 595
pyridinones 755
pyridinyl pyrimidines 703 f
pyrido fused-ring systems 397
– benz-1,3-oxazine formation 397
– pyridopyridazine 397
– quinoline derivatives 397
pyrido[2,3-d]pyrimidine 795 f, 799 f, 808
pyrido[2,3-d]pyrimidine N-oxides 800
pyridones 797 f
pyridopyrimidinone 694
pyrimidine library 775
pyrimidines 535, 740 ff, 765, 776
– derivatives 296
pyrolysis 436
pyrolysis of methane 631
pyromellitic acid 671
pyromellitic anhydride 670
pyromellitic dianhydride 672, 674
α-pyrones 570
pyrroles 556, 805 f
pyrrolidine 152, 556
pyrrolidine[60]fullerenes 937

q

quadrupole moment 10
qualification 101
quaternary ammonium salt 282
quaternary onium salts 278
quinazoline 301, 417, 428 ff, 450 ff
quinocalinone 808
quinodimethane 540
o-quinodimethanes 932
quinolin-2(1H)ones 695 f
quinoline 798 ff
– derivatives 184

quinolinones 536
quinolones 395
quinoxaline derivatives 400
– montmorillonite K10 clay 400
– quinoxaline derivatives 400

r

racemization 914, 917 f
– amount of 918
– cysteine 917
– histidine 917
racemization studies 916
radar 2
radial distributions 48
radiation 631
– pulsed 631
radical pairs 885
radical recombination 885
radioactive waste 820, 834, 837, 841
radiofluoridation 846
radiofrequency 861
radiofrequency region 887
radioisotopes 820
radiolabeled compounds 404
– deuteration 404
– tritiation 404
radioligands 843
radiopharmaceuticals 821, 842
Radziszewski's four-component reaction 803
Raman spectroscopy 826
ramo system 95
Raney nickel 624, 626
Raytheon Company 2
reaction 616, 629
– catalytic 616
reaction kinetics 63
reaction mechanisms 147
reaction rate 63
reaction yield-detected magnetic resonance (RYDMAR) 885
reacton 144
reactor 624, 634, 792
– continuous flow 624, 634
– stirred tank 624
real 9
rearrangement 208, 379
– ammonium ylides 208
– benzil-benzilic acid 380
– Claisen 118, 381, 748 f
– Fries 380
– pinacol-pinacolone 380
– Stevens 209
recovery methods 963

reduction 391 ff
– alumina-supported hydrazine 391
– aromatic nitro compounds 391
reduction reactions 388
– aluminium alkoxides 388
– borohydride-alumina 388
– carbonyl compounds 388
– regioselective reduction 388
– sodium 388
– trans-cinnamaldehyde 388
reductive amination 389, 773 f
– of carbonyl compounds 211
reductive decyanation of
   alkyldiphenylmethanes 321
reforming 646
reforming catalysts 632
refractive index 10, 18
regioselectivity 147, 940
– photochemical 880
relative permittivity 10
relaxation process 40
relaxation time 17, 20 f, 25
– acid chlorides 21
– alcohol 21
– alcohol ether 21
– aliphatic 21
– alkanes 21
– aromatic amines 21
– aromatic halogens 21
– aromatic ketones 21
– esters 21
– nitriles 21
– solvents 21
reproducibility 102
resin 814
resin capture 782
resin catalyst 622
resonant devices 48
retention of reactants on graphite 449
reversible 63
rhodamine-B 884
– dye 883
ring-closing metathesis 779
– olefin 204
ring-expansion transformation 380
– KSF clay 380
– AgBF$_4$-Al$_2$O$_3$ 380
ring-opening
– of epoxides 177, 297
– polymerization 661 ff
Rink amide polystyrene resin 745
Rink amide resin 739
Robinson annulation 315

Rosemund von Braun reaction 336
Ruhmkorff coil 4
rutaecarpine 430
ruthenocenecarboxaldehyde 938

S
S1000 314
S402 314
saccharide sequencing 598
safety considerations 876
salicylaldehyde 291, 536, 810, 952
saponification of hindered aromatic esters
   192
saponifications of hindered mesitoic esters
   155
saturation effects 14
Sc(OTf)$_3$ 803 f
Scaffold-decoration 738
scale-down 80
scale-up 43, 80
scavenger 778, 781
scavenger resin 774
scavenging reagents 781
Schiff's base 389, 807
Schuster acidic rearrangement 380
sealed tube 74
sealed vessels 562
sebacinic acid 671
selective alkylation 207
– 1,2,4-triazole 207
selective dealkylation 185
– of aromatic ethers 313
selective heating 220, 223, 230, 232, 622,
   638, 640
– catalysts 223
– molecular radiators 230
– solvents 220
– susceptors 232
selective hydrolysis of nitriles to amides 315
selective monoalkylation 118
selectivity 627, 629 ff, 638
semi-batch 64
semiconductor 136
semiconductor devices 891
semi-empirical calculations 181
sensitizer 418, 449 ff
sequential irradiation 331
sequential reactions 789
sequential vaccum 966
sesquiterpene 962
SFME 969
Shikoku Instrumentation 641
short-range interactions 33

silica   792, 805, 882
silica catalyst   623
silica gel   395 ff, 528, 588, 592 f, 608, 621, 753, 795, 797, 802
silicon carbide   621
silver catalyst   634
silyl ketene acetals   312
silylation   585
simultaneous cooling   628, 689
simultaneous UV-visible and MW irradiation   877
single mode   68, 686, 697, 700 f, 719, 792
single-walled carbon nanotube (SWNT)   950
singlet-born radical pair   885
singlet excited state   885
skin depths   47
slower reacting systems   210
SmithSynthesizer   84
S_N2, see nucleophilic substitution
S_NAr, see nucleophilic aromatic substitution
sodium acrylate   680
sodium azide   807
sodium dodecylbenzenesulfonate   889
software technology   129
solid state   390
solid support   279 f, 792, 797
solid surfaces   15
solid tritium donors   832
solid-liquid PTC   279
solid-liquid solvent-free PTC   284
solid-phase chemistry   801
solid-phase organic synthesis   726
solid-phase peptide synthesis (SPPS)   898 f, 911, 914 f
– enolization   915
– microwave SPPS   911
– microwave-enhanced   898
– oxazolone formation   915
– properties of proteins   914
– racemization   915
solid-phase reactions   840
solid-phase synthesis   376, 727
solids   38, 40
solid-state synthesis   949
solubility in water   962
solubilized SWNTs   950
soluble ionic liquid supports   350, 757
solute-solute effects   33
solvent   525 f, 639
– nonpolar   639
solvent-free   64, 376 ff, 492, 527, 586 ff, 592 ff, 608, 706, 793–810, 816, 965
solvent-free conditions   108, 210, 280, 374 ff, 753

solvent-free microwave processes   109
– techniques   792
solvent-free PTC   948
solvent-free reactions   145
solvent-free synthesis   492 ff, 510
– 1,8-cineole derivatives   510
– aziridines   492
– benzoxazines   514
– coumarins   511
– creatinine   500
– cyclic ureas   499
– dioxolans   503
– furans   498
– hydantoin   500
– imidazoles   500
– imidazolones   502
– imides   493
– lactams   493
– N-formylmorpholine   513
– oxazoles   503
– pyrazines   512
– pyrazoles   500
– pyridines   506
– pyrimidines   511
– pyrroles   495
– tetrahydrofuran   497
– thiazoles   505 f
– thiiranes   492
– thiophenes   499
– triazines   514
solvent-free systems   279
Sonogashira coupling   403, 728, 737, 756 f
Sonogashira coupling reaction   706
Sonogashira cross-coupling   850
Sonogashira reaction   318
Sonogashira-Nicholas reaction   737
sorbose   605
soxhlet   968
spatial distribution   71
spatial nonuniformity   74
specific activation   74
specific effects   52, 635
specific hydrogen bonding   880
specific microwave effect   164, 530
specific nonthermal effects   136
– microwave effects   137
Spencer   2
spices   959
spin dynamics   885
spin tray   883
spiro-fused heterocycles   400
spiro(indole-pyridol) thiazines   377
– 3-indolylimine   377
– 2-mercaptonicotinic acid   377

split-and-mix   727
SPME   969
SPOS   726, 742, 744, 756, 769
SPOT-synthetic techniques   740
SPPS (solid-phase peptide synthesis)   899, 926
– Boc   900
– Fmoc   900
stabilization of the transition state   148
stable isotopes   820
standing wave effect   69, 73
static magnetic field   885
static permittivity   17 f
Staudinger   594
steam   963
sterilization techniques   890
– food   890
– surfaces   890
Stevens rearrangement   209
Stilbene Isomerization
Stille coupling   763, 844
Stille reaction   700
Stille-type cross-coupling   735
stirred vessel   63
storage of electromagnetic energy   9
styrene   654
N-substituted amides   294
2-substituted 2-oxazolines   166
substitution reactions   740
sucrose   584, 595, 598 f, 610
sugar labeling   598
sulfonic acid resin   782
sulfonyl chloride resin   773
N-sulfonylimines   160, 373 ff, 375
– sulfonamides   375
sulfoximine   712
sultine   933
superacid   619
superconductivity   949
supercritical conditions   23
supercritical fluids   23
superheating   56, 74, 437, 449, 525, 622, 628, 635 ff
– localized   635 ff
superheating effect   137, 880
superheating of polar solvents   878
– hot-spot formation   879
– nucleation point   878
– superheating   879
superoxide radicals   891
supported reagents   362 f
– heterogeneous reactions   362
– inorganic oxide   362

– mineral-supported reagents   363
– solvent-free methods   363
Suzuki   735, 764
Suzuki coupling   129, 349, 403, 736, 750, 756, 768, 780, 844
– palladium-doped alumina   403
Suzuki-Miyaura reaction   686
Suzuki reaction   317, 349, 625, 765
SWNT (single-walled carbon nanotube)   950
symmetrical anhydride procedure   729
synergistic effect of UV and MW radiation   878
synthesis   165, 169, 172, 178, 203, 456 f, 459, 462, 465 ff, 477, 479, 483 f, 492, 922
– acyl isothiocyanates   306
– alkyl p-toluenesulfinates   172
– aziridines   456
– creatinine   463
– cyclic ureas   463
– coumarins   373
– diaryl-a-tetralones   315
– dibenzyl diselenides   314
– esters   282
– furans   462
– hydantoin   463
– hydantoins   169
– imidazoles   466
– β-lactam   142, 312, 457
– longer peptides   922
– nitriles   336
– oxadiazoles   468
– oxazoles   466
– oxazolidines   466
– oxiranes   457
– pyrazines   483
– pyrazoles   465
– pyridines   469 ff
– pyrido-fused ring systems   203
– pyrimidines   479
– pyrroles   459
– symmetrical disulfides   321
– tetrahydrofurans   462
– tetrahydropyrans   477
– tetrahydropyrimidine   479
– tetrazoles   469
– thiazoles   467 f
– thiazolidines   467
– thiazolidinones   352
– thiohydanthoin   463
– thiohydantoins   169
– thiophenes   462
– triazines   342, 484
synthetic vanillin   307
synthewave   77, 596, 609

synthewave 402 (S402) 282
synthewave 1000 (S1000) 282
synthos 70
syringaldehyde resin 755
systemic approach 43
systemic theory 44

t

tandem reactions 788 ff
TBAB (tetrabutylammonium bromide) 280 ff, 285 f, 294 ff, 316 f, 321, 581 ff, 588, 667, 689, 706
^{99m}Technetium (Tc) 854
temperature 21
temperature control 642
temperature gradients 626, 630, 633, 636, 638 f, 644
temperature gradients rather 625
temperature measurement 71
temperature profile 646
temperature sensor 71
TEMPO 660
TentaGel 745
TentaGel resin 735
TentaGel-support 707
terpenes 961
tert-amino effect 203
tert-butoxycarbonyl (Boc) 900
N-tert-butoxycarbonyl groups 368
tetrabutylammonium bromide, *see* TBAB
tetrahydroquinolones 202
tetrasubstituted imidazoles 802
tetrathiafulvalenes 938
tetrazole 562, 695 f, 707, 736, 749
tetronate synthesis 749
theoretical calculations 207 f
thermal 421
– decomposition 421
thermal conductivity 418
thermal conversion 9
thermal dependency 25, 45
thermal effects 136, 635, 878
thermal feedback 45
thermal fluctuations 56
thermal gradients 57
thermal path effect 54
thermal runaway 45, 69
thermochemical reaction 128
thermodynamic 63
thermodynamic effects 51
thermolysis 163, 421, 425 ff, 436
– esters 425
– urea 436
thermostar system 100

thia-Fries rearrangement 380
– arylsulfonates 380
thialolone 811
1,3-thiazines 400
thiazole 393, 417, 431, 450 ff, 811
thiazolidones 813
thienosultines 541
thiiranes 393, 395 ff
– arylmethyl ketones 393
– epoxides 395
– [hydroxy(tosyloxy)iodo]benzene (HTIB) 393 ff
thioamides 814
thiohydantoin synthesis 761
thionation of carbonyl compounds 168
γ-thionolactones 168
thiophene 676
thiophene cycloadducts 934
thiosemicarbazides 307
thiourea 658, 800
– *N*-aryl 306
– *N*-acyl 306
2-thioxotetrahydropyrimidin-4-(1H)-ones 354
three-component condensation reaction 766
three-phase extraction 763
tiazole 428 ff
tight ion pairs 154, 189, 304
time-dependent 886
titanium dioxide 883 f, 889
– TiO$_2$–ZrO$_2$ 884
Tipson-Cohen elimination 600
p-toluenesulfonic acid (PTSA) 364
toxic agents 888
transamination 761
transesterification 193, 757, 760
– PTC 193
– reactions 748
transfer hydrogenation 333, 772
transformation 729
transformation of 2-t-butylphenol 638
transimination 161
transition-metal catalysis 739
transition metal-free reaction 706
transition state 139, 155
– product-like 155
– reactant-like 155
transition temperatures 292
transparent 527
trehalose 609
triaryl phosphines 714
triarylimidazoles 802
2,4,5-triarylimidazoline derivatives 396
triazine 807, 809
triazine synthesis 741

triazole 807 f
1,2,3-triazoles 562, 807
trichloroacetonitrile 552
trichloroethylene 645
triethylorthoformate 799 f, 811
N,O-bis-(trimethylsilyl)acetamide 585
triplet excited state 885
triplet-born radical pair 885
triplet-to-singlet interconversion 885
– mixing of states 885
tritiated formates 833
tritiation 821
tritium 821, 824 f, 828
tube 2
tubular-flow reactor 63
tungsten carbide 633
turbo-distillation 964

u
Ugi 805, 808, 814
Ugi-de-Boc cyclization 808
Ugi-de-Boc cyclization sequence 804
Ugi four-component condensation 745, 766
Ugi three-component condensation 773
Ullmann coupling 710 f
Ullmann reactions 713
ultrasound 622, 628, 860
ultrasound (US) 189
ultraviolet-visible 860
unimolecular reactions 154, 198
α,β-unsaturated aldehydes 805
α,β-unsaturated carbonyl compound 175
α,β-unsaturated ester 808
α,β-unsaturated ketones 812
unsaturated monosaccharides 599
α,β-unsaturated nitroalkenes 375
– amines 375
– ketones 375
– N-substituted hydroxylamines 375
– oximes 375
– substituted oximes and ketones 375
up-scaling 721
uracils 397
urea 800 f

v
vacuum-distillation 964
vacuum pyrolysis 329
valerophenones 880
validation 101
Vilsmeier reagent 197
vinyl acetate 646, 654
vinylation 767
N-vinylcarbazole 658, 948
vinylic nucleophilic substitutions 184
vinyl monomers 948

w
Wang aldehyde resin 782
Wang resin 729, 731, 746, 752
waste water 888
water 27, 689, 718
water distillation 963
wave-lengths 5
Welfon 751
Wilkinson's catalyst 772
Williamson reaction 285
Wittig olefination 129, 155, 373, 748 f, 771
– phosphonium salts 373
Wittig reagent 771
Wittig strategy 391
Wolff-Kichner reduction 376

x
X-ray crystallography 903
p-xylene 645
o-xylene oxidation 618

y
Yb(OTf)₃ 801

z
Zeeman splitting 885
zeofen 383
zeolite 527, 619, 626 f, 631, 634, 640
zeolite HY 797, 802
zeolite HZ SM5 383
zink powder 196 f